R.
Capilano College

Basic Stats
Jarrell

DATE	ISSUED TO

PRINTED IN CANADA

BASIC STATISTICS

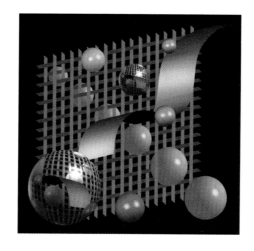

STEPHEN B. JARRELL
Western Carolina University

WCB Wm. C. Brown Publishers
Dubuque, Iowa • Melbourne, Australia • Oxford, England

Book Team

Editor *Paula-Christy Heighton*
Developmental Editor *Jane Parrigin*
Publishing Services Coordinator *Julie Avery Kennedy*

Wm. C. Brown Publishers
A Division of Wm. C. Brown Communications, Inc.

Vice President and General Manager *Beverly Kolz*
Vice President, Publisher *Earl McPeek*
Vice President, Director of Sales and Marketing *Virginia S. Moffat*
National Sales Manager *Douglas J. DiNardo*
Marketing Manager *Julie Joyce Keck*
Advertising Manager *Janelle Keeffer*
Director of Production *Colleen A. Yonda*
Publishing Services Manager *Karen J. Slaght*
Permissions/Records Manager *Connie Allendorf*

Wm. C. Brown Communications, Inc.

President and Chief Executive Officer *G. Franklin Lewis*
Corporate Senior Vice President, President of WCB Manufacturing *Roger Meyer*
Corporate Senior Vice President and Chief Financial Officer *Robert Chesterman*

Cover and interior design by John Rokusek

Copyediting and production by Lachina Publishing Services

Credits: **2**: © Leonard Lessin/Peter Arnold, Inc. **22**: © IFA/Peter Arnold, Inc. **38**: Tables reprinted with permission of the Helen Dwight Reid Educational Foundation. Published by Heldref Publications, 1319 Eighteenth St., N.W., Washington, DC 20036-1802. Copyright © 1949. **74**: © Y. Arthus-Bertrand/Peter Arnold, Inc. **148**: © David J. Cross/Peter Arnold, Inc. **196**: © Richard Hutchings/Photo Researchers, Inc. **246**: © Borrfdon/Explorer/Photo Researchers, Inc. **288**: © Simon Fraser/Science Photo Library **326**: Courtesy East Ohio Gas Company **390**: © Blair Seitz/Photo Researchers, Inc. **451**: © Benelux/Photo Researchers, Inc. **490**: © Bob Daemmrich/The Image Works **576**: © NASA/Science Source **618**: © Shumsky/The Image Works

Copyright © 1994 by Wm. C. Brown Communications, Inc. All rights reserved

A Times Mirror Company

Library of Congress Catalog Card Number: 93–72115

ISBN 0–697–21595–4

No part of this publication may be reproduced, stored in a retrieval system, or transmitted, in any form or by any means, electronic, mechanical, photocopying, recording, or otherwise, without the prior written permission of the publisher.

Printed in the United States of America by Wm. C. Brown Communications, Inc., 2460 Kerper Boulevard, Dubuque, IA 52001

10 9 8 7 6 5 4 3 2 1

To my wife, Marcia

Brief Contents

Chapter 1   What Is Statistics?, 2
Chapter 2   Descriptions of the Distribution of a Data Set, 22
Chapter 3   Measures of Location and Dispersion Within a Data Set, 74
Chapter 4   Introduction to Probability, 148
Chapter 5   Random Variables and the Binomial Distribution, 196
Chapter 6   The Normal Random Variable, 246
Chapter 7   Sampling Distributions, 288
Chapter 8   Statistical Inference: Estimation, 326
Chapter 9   Hypothesis Testing, 390
Chapter 10 Test for the Difference between Two Population Means, 450
Chapter 11 Correlation and Regression Analysis, 490
Chapter 12 Analysis of Variance, 576
Chapter 13 Chi-Square and Nonparametric Tests, 618

Appendixes

A: Rounding Rules, A1
B: Air Quality Data Base, B1
C: State SAT Data Base, C1
D: NCAA Probation Data Base, D1
E: North Carolina Census Data Base, E1
F: Counting: Combinations and Permutations, F1
G: Binomial Distribution Tables, G1
H: Standard Normal Distribution Table, H1
I: Random Number Table, I1
J: Student's *T* Distribution Table, J1
K: State Health Data Set, K1
L: *F* Distribution Tables, L1
M: NCAA Tournament Data Base, M1
N: Chi-Square Distribution Table, N1

Answers, ANS-1

Index, IN-1

Contents

Preface, xv

■ **Chapter 1 What Is Statistics?, 2**

Objectives, 2
Introduction, 2
1.1 Initial Concepts and Definitions, 5
   Section 1.1 Problems, 9
1.2 Types of Variables and Measurement Scales, 10
   Section 1.2 Problems, 13
1.3 Mathematical Operations, 13
   1.3.1 Order of Operations, 13
   1.3.2 Symbols and Notations: Mathematical Shorthand, 14
   Section 1.3 Problems, 17
■ Summary and Review, 18
   Multiple-Choice Problems, 18
   Word and Thought Problems, 20

   Box 1-1: How Nutty Numbers Distort the Facts, 4
   Box 1-2: Not Just Any Number Will Do, 5
   Box 1-3: Inferences and Representative Samples, 9

■ **Chapter 2 Descriptions of the Distribution of a Data Set, 22**

Objectives, 22

Unit I: Tabular Presentation of Frequency Distributions
   Introduction: *River Cleanup*, 22
2.1 Frequency Distributions, 24
   Section 2.1 Problems, 24
   2.1.1 Class Limits, Boundaries, Midpoint, and Size, 25
   2.1.2 Constructing a Frequency Distribution, 26
   2.1.3 The Process of Constructing a Frequency Distribution, 30
   2.1.4 Open-Ended Classes, 34
   Section 2.1 Problems (Continued), 36
2.2 Relative Frequency Distributions, 36
   Section 2.2 Problems, 37
2.3 Cumulative Frequency Distributions, 38
   Section 2.3 Problems, 41
■ Summary and Review, 41
   Multiple-Choice Problems, 41
   Word and Thought Problems, 43

Unit II: Pictorial Presentations of Frequency Distributions
   Introduction: *River Cleanup (continued)*, 47
2.4 Frequency Polygons and Frequency Curves, 47
   Section 2.4 Problems, 48
2.5 Histograms, 49
   Section 2.5 Problems, 51
2.6 Ogives, 51
   Section 2.6 Problems, 53
2.7 Stem-and-Leaf Displays, 54
   Section 2.7 Problems, 56
2.8 Visualizing the Distribution of a Data Set by Using Computers, 57
   Section 2.8 Problems, 59
■ Summary and Review, 60
   Multiple-Choice Problems, 62
   Word and Thought Problems, 66
   Review Problems, 70

   Box 2-1: Born Smart or Became Smart?, 38
   Box 2-2: Describing the Psychological Effects of Disasters with Statistics, 57
   Box 2-3: Florence Nightingale, the Pioneer Statistician, 63

■ **Chapter 3 Measures of Location and Dispersion Within a Data Set, 74**

Objectives, 74

Unit I: Location Measures
   Introduction: *Tennis Survey*, 75
3.1 Unweighted Measures of the Center, 76

3.1.1 Mean, 76
3.1.2 Median, 78
3.1.3 Mode, 82
Section 3.1 Problems, 83
3.2 Comparing Measures and Finding the Typical Value, 84
Section 3.2 Problems, 87
3.3 Weighted Measures of Central Location, 88
3.3.1 Weighted Mean, 88
3.3.2 Estimated Mean from Grouped Data, 91
Section 3.3 Problems, 93
3.4 Other Measures of Location, 94
3.4.1 Quartiles, 94
Section 3.4.1 Problems, 96
3.4.2 Percentiles, 97
Section 3.4.2 Problems, 99
3.5 Location Measures and the Computer, 100
Section 3.5 Problems, 101

■ Summary and Review, 102
Multiple-Choice Problems, 102
Word and Thought Problems, 104

Unit II: Dispersion Measures

Introduction: *Horse Races,* 108
3.6 Dispersion in a Data Set, 109
3.7 Alternative Measures of Dispersion, 109
3.7.1 Range, 109
3.7.2 Interquartile Range, 110
3.7.3 Mean Absolute Deviation, 111
Section 3.7 Problems, 113
3.8 The Population Standard Deviation, 113
3.8.1 Calculation, 113
3.8.2 Interpretation, 115
3.8.3 Applications of the Standard Deviation, 116
Section 3.8 Problems, 121
3.9 The Sample Standard Deviation, 122
Section 3.9 Problems, 124
3.10 Shortcut Formulas, 125
Section 3.10 Problems, 127
3.11 Estimated Standard Deviation from Grouped Data, 128
Section 3.11 Problems, 129
3.12 Descriptive Statistics and the Computer, 130
3.12.1 Standard Deviation, 130
3.12.2 Box-and-Whiskers Diagrams (Optional), 131
Section 3.12 Problems, 133

■ Summary and Review, 134
Multiple-Choice Problems, 136
Word and Thought Problems, 139
Review Problems, 141

Box 3-1: Measuring Risk with the Coefficient of Variation, 117

Box 3-2: Astronomical Measures, 121
Box 3-3: Norwegian Fisheries, 122

■ **Chapter 4** Introduction to Probability, 148

Objectives, 148

Unit I: Basic Probability Concepts

Introduction: *Earthquake Prediction,* 149
4.1 Basic Terminology, 149
4.2 Methods to Determine Probabilities, 151
4.2.1 Relative Frequencies, 152
4.2.2 Equiprobable Outcomes, 155
4.2.3 Subjective Probabilities, 157
Section 4.2 Problems, 158
4.3 Visualizing Probabilities with Venn Diagrams, 158
4.4 Upper and Lower Boundaries on $P(A)$, 159
Section 4.4 Problems, 160
4.5 The Negation or Complement of an Event, 160
Section 4.5 Problems, 161
4.6 Conditional Probabilities, 162
Section 4.6 Problems, 165

■ Summary and Review, 165
Multiple-Choice Problems, 166
Word and Thought Problems, 167

Unit II: Solving Probability Problems for Compound Events

Introduction: *School Bus Route,* 170
4.7 $P(A$ and $B)$, 171
4.7.1 $P(A$ and $B)$ for Any Two Events, 171
Section 4.7.1 Problems, 175
4.7.2 $P(A$ and $B)$ for Independent Events, 176
Section 4.7.2 Problems, 180
4.8 $P(A$ or $B)$, 180
4.8.1 $P(A$ or $B)$ for Any Two Events, 180
Section 4.8.1 Problems, 183
4.8.2 $P(A$ or $B)$ for Mutually Exclusive Events, 183
Section 4.8.2 Problems, 186
4.9 A Strategy for Probability Problems, 186

■ Summary and Review, 187
Multiple-Choice Problems, 188
Word and Thought Problems, 190
Review Problems, 192

Box 4-1: Figuring Odds, 162
Box 4-2: Probabilities of Winning an Educational Scholarship, 164
Box 4-3: Determining the Probability of Pregnancy: Bayes' Formula, 174

Contents

Chapter 5 Random Variables and the Binomial Distribution, 196

Objectives, 196

Unit I: Random Variables and Their Properties

Introduction: *National Health Insurance*, 197

5.1 Random Variables and the Data Set of Experimental Outcomes, 198
Section 5.1 Problems, 202

5.2 Describing the Values of a Random Variable, 202
5.2.1 Probability Distribution of a Random Variable, 202
Section 5.2.1 Problems, 205
5.2.2 The Mean or Expected Value of a Random Variable, 205
Section 5.2.2 Problems, 211
5.2.3 The Standard Deviation of a Random Variable, 211
Section 5.2.3 Problems, 216

Summary and Review, 216
Multiple-Choice Problems, 217
Word and Thought Problems, 219

Unit II: The Binomial Random Variable

Introduction: *Committee Composition*, 222

5.3 The Binomial Formula, 223
5.3.1 Assumptions, 223
5.3.2 Formula, 226
Section 5.3 Problems, 230

5.4 The Binomial Probability Table, 231
Section 5.4 Problems, 232

5.5 Properties of the Binomial Random Variable, 232
5.5.1 Probability Distribution, 232
Section 5.5.1 Problems, 236
5.2.2 Mean, 236
5.5.3 Standard Deviation, 237
Sections 5.5.2 and 5.5.3 Problems, 239

Summary and Review, 239
Multiple-Choice Problems, 239
Word and Thought Problems, 241
Review Problems, 242

Box 5-1: Expected Values, Casinos, and Insurance Companies, 210
Box 5-2: Checking for a Fair Coin, 238

Chapter 6 The Normal Random Variable, 246

Objectives, 246

Unit I: Continuous Random Variables, The Standard Normal

Introduction, 247

6.1 The Probability Distribution of Any Continuous Random Variable, 247
6.2 The Probability Distribution of a Normal Random Variable, 250
6.3 The Standard Normal Random Variable, 252
6.3.1 Finding Probabilities, 253
Section 6.3.1 Problems, 255
6.3.2 Finding Z Values, 256
Section 6.3.2 Problems, 259

Summary and Review, 259
Multiple-Choice Problems, 260
Word and Thought Problems, 261

Unit II: Working with Normal Random Variables

Introduction: *Club Fundraiser,* 263

6.4 The Transformation Formula, 263
Section 6.4 Problems, 266
6.5 Finding Probabilities, 267
Section 6.5 Problems, 271
6.6 Finding X Values, 271
Section 6.6 Problems, 273
6.7 Normal Approximation of a Binomial Probability Distribution, 274
6.7.1 Review of Binomial Random Variable, 275
6.7.2 The Normal Approximation Procedure, 276
6.7.3 Conditions for Good Normal Approximations, 278
Section 6.7 Problems, 280

Summary and Review, 280
Multiple-Choice Problems, 281
Word and Thought Problems, 282
Review Problems, 285

Box 6-1: The Probability of Winning a Football Game, 270
Box 6-2: IQ Scores, 274

Chapter 7 Sampling Distributions, 288

Objectives, 288

Unit I: Introduction to Sampling Distributions

Introduction: *Waste Treatment,* 289

7.1 The Sampling Procedure, 290
7.1.1 Purposes of Sampling, 291
7.1.2 Random Sampling, 291
Section 7.1 Problems, 294
7.2 Sampling Distributions, 294
Responses, 295
7.2.1 Constructing a Sampling Distribution, 296

7.2.2 Solving Probability Problems with Sampling Distributions, 302
Section 7.2 Problems, 303
- Summary and Review, 303
Multiple-Choice Problems, 303
Word and Thought Problems, 305

Unit II: Describing the Statistic, $\overline{X}$, as a Random Variable

Introduction: *Quality Control,* 308
7.3 The Mean of $\overline{X}$, 308
7.4 The Standard Error of the Mean, 309
Sections 7.3 and 7.4 Problems, 311
7.5 Sampling Distribution of $\overline{X}$: The Central Limit Theorem, 312
Section 7.5 Problems, 315
7.6 Sampling Distributions of $\overline{X}$: Normal Population, 315
Section 7.6 Problems, 315
7.7 Sampling Distribution of $\overline{X}$: Finite Populations, 315
Section 7.7 Problems, 317
- Summary and Review, 318
Multiple-Choice Problems, 318
Word and Thought Problems, 320
Review Problems, 322

Box 7-1: Time to Quit, 294
Box 7-2: Leading Questions and Elicited Responses, 295
Box 7-3: Acceptable Survey Methods and Findings, 296
Box 7-4: Are Chemists Girl Crazy?, 297

Chapter 8 Statistical Inference: Estimation, 326

Objectives, 326

Unit I: Confidence Interval Estimates of the Population Mean, μ

Introduction: *Heat Cost,* 327
8.1 Choosing an Estimator for μ, 328
8.2 Confidence Interval for Population Mean, μ: Known Population Standard Deviation, σ, 329
8.2.1 Large Sample from Any Population, 329
Section 8.2.1 Problems, 334
8.2.2 Normal Population, 334
Section 8.2.2 Problems, 335
8.3 Confidence Interval for Population Mean, μ: Unknown Population Standard Deviation, σ, 335
8.3.1 The *T* Probability Distribution, 335
Section 8.3.1 Problems, 338
8.3.2 Normal Population, 338
Section 8.3.2 Problems, 341
8.3.3 Large Samples, 341
Section 8.3.3 Problems, 342

8.4 A Flowchart for Constructing Confidence Intervals for μ, 343
8.5 Confidence Intervals and the Computer, 344
Section 8.5 Problems, 345
- Summary and Review, 346
Multiple-Choice Problems, 349
Word and Thought Problems, 352

Unit II: Errors of Point Estimates of the Population Mean, μ, 354

Introduction: *Heat Cost,* 355
8.6 Sampling Error versus Nonsampling Error, 355
8.7 Components of Sampling-Error Assessment, 357
8.7.1 The Maximum Error, E, 357
Section 8.7.1 Problems, 358
8.7.2 The Approximate Sample Size, n, 358
Section 8.7.2 Problems, 361
8.7.3 Confidence Level Associated with Claimed Error, $100(1 - \alpha)\%$, 361
Section 8.7.3 Problems, 363
- Summary and Review, 364
Multiple-Choice Problems, 364
Word and Thought Problems, 366

Unit III: Estimating p: The Population Proportion of Successes

Introduction: *Sidewalk Ordinance,* 369
8.8 An Estimator of p, 369
8.9 Sampling Distribution of the Sample Success Rate, 370
Section 8.9 Problems, 374
8.10 Confidence Interval for p, 374
Section 8.10 Problems, 376
8.11 Evaluating Point Estimates of p, 376
8.11.1 The Maximum Error, E, 376
Section 8.11.1 Problems, 377
8.11.2 The Appropriate Sample Size, n, 377
Section 8.11.2 Problems, 378
8.11.3 Confidence Associated with Claimed Error, $100(1 - \alpha)\%$, 379
Section 8.11.3 Problems, 380
- Summary and Review, 380
Multiple-Choice Problems, 381
Word and Thought Problems, 383
Review Problems, 385

Box 8-1: Statistics and Nonsmokers' Exposure to Smoke, 347
Box 8-2: Nonsampling Errors, 356
Box 8-3: The Central Limit Theorem and the Sampling Distribution of $\hat{p}$, 373
Box 8-4: Reporting Technical Aspects of Polls, 379

Contents

■ Chapter 9 Hypothesis Testing, 390

Objectives, 390

Unit I: Hypothesis Testing Basics

Introduction: *Cost Estimate,* 391

- 9.1 The Basic Logic behind Hypothesis Testing, 391
  Section 9.1 Problems, 393
- 9.2 Basic Terminology of Hypothesis Testing, 393
  Section 9.2 Problems, 397
- 9.3 The Hypothesis Testing Procedure, 398
  Section 9.3 Problems, 404
- 9.4 The Computer and Hypothesis Testing for Population Means, 404
  Section 9.4 Problems, 405

■ Summary and Review, 406
  Multiple-Choice Problems, 406
  Word and Thought Problems, 409

Unit II: Hypothesis Testing Variations

Introduction: *Absentee Rate,* 412

- 9.5 One-Tailed Tests, 412
  Section 9.5 Problems, 416
- 9.6 Shortcuts and Testing on the Z Axis, 416
  Section 9.6 Problems, 417
- 9.7 Small-Sample Test for μ, 418
  Section 9.7 Problems, 419
- 9.8 Test for Population Proportion, p, 419
  Section 9.8 Problems, 420
- 9.9 Setting Up Hypotheses, 420
  - 9.9.1 Determine the Parameter to Be Tested, 420
  - 9.9.2 Determine the Value Hypothesized for the Parameter, 421
  - 9.9.3 Determine the Appropriate Equality or Inequality Symbol, 421
  Section 9.9 Problems, 422
- 9.10 More about Computers and Hypothesis Testing for Population Means, 424
  Section 9.10 Problems, 425

■ Summary and Review, 426
  Multiple-Choice Problems, 426
  Word and Thought Problems, 428

Unit III: *P*-Values (Optional Unit)

Introduction: *Day-Care Situation,* 431

- 9.11 Test Information from *P*-Values, 432
  - 9.11.1 *P*-Value Definition and Interpretation, 432
  Section 9.11.1 Problems, 435
  - 9.11.2 *P*-Value Calculation, 435
  Section 9.11.2 Problems, 435
  - 9.11.3 Decisions with the *P*-Value, 436
  Section 9.11.3 Problems, 437
- 9.12 Two-Tailed *P*-Values, 437
  Section 9.12 Problems, 438
- 9.13 *P*-Values for the Population Proportion of Successes (Small Samples), 438
  - 9.13.1 One-Tailed Test, 439
  - 9.13.2 Two-Tailed Test, 439
  Section 9.13 Problems, 440
- 9.14 Computers and *P*-Values, 440
  Section 9.14 Problems, 442

■ Summary and Review, 443
  Multiple-Choice Problems, 443
  Word and Thought Problems, 445
  Review Problems, 446

Box 9-1: Maps and the Basic Idea of Hypothesis Testing, 393
Box 9-2: Using Hypothesis Testing to Make a Practical Decision, 398
Box 9-3: Errors and Speed Traps, 399
Box 9-4: Hypothesis Testing to Verify a Theoretical Prediction, 423
Box 9-5: Using a Two-Tailed *P*-Value to Do a One-Tailed Test, 441

■ Chapter 10 Test for the Difference between Two Population Means, 450

Objectives, 450

Unit I: Basics of the Test

Introduction: *Productivity Differences between Pools,* 451

- 10.1 The Hypotheses, 453
  Section 10.1 Problems, 454
- 10.2 Test for Large Independent Random Samples, 454
  - 10.2.1 The Test Statistic, 454
  - 10.2.2 The Sampling Distribution of $\overline{X}_1 - \overline{X}_2$, 455
  Section 10.2.1 and 10.2.2 Problems, 456
  - 10.2.3 The Decision, 457
  Section 10.2.3 Problems, 459

■ Summary and Review, 460
  Multiple-Choice Problems, 461
  Word and Thought Problems, 462

Unit II: Variations of the Test

Introduction: *Viewership Experiment,* 465

- 10.3 Test for Small Independent Samples, 465
  Section 10.3 Problems, 468
- 10.4 Paired Differences (Dependent Samples), 469
  Section 10.4 Problems, 472
- 10.5 The Computer and Tests for Differences in Means, 472
  - 10.5.1 Independent Samples, 473
  - 10.5.2 Paired Differences, 475
  Section 10.5 Problems, 476

- Summary and Review, 479
  Multiple-Choice Problems, 479
  Word and Thought Problems, 482
  Review Problems, 484

  Box 10-1: China: Mass Medical Laboratory, 453
  Box 10-2: Gender Differences among Adolescents from Divorced Homes, 460
  Box 10-3: Hog Socialization, 475

Chapter 11 Correlation and Regression Analysis, 490

Objectives, 490

Unit I: Correlation Analysis

Introduction: *Skills Test and Productivity Experiment*, 491

11.1 Data Sets, 491
  11.1.1 Assembling Data Sets: Time Series or Cross Section, 491
  11.1.2 Displaying Data Sets: Scatter Diagrams, 492
  Section 11.1 Problems, 494

11.2 The Correlation Coefficient, 495
  11.2.1 A Measure of the Direction of Association, 495
  11.2.2 A Measure of Strength and of Direction of Association: Calculation, 498
  11.2.3 Information Relayed by Correlation Coefficient, 499
  Section 11.2 Problems, 501

11.3 Testing to Establish Correlation between Two Variables, 501
  Section 11.3 Problems, 504

- Summary and Review, 504
  Multiple-Choice Problems, 505
  Word and Thought Problems, 509

Unit II: Simple Regression Models: Introduction

Introduction: *Advertising-Sales Relationship*, 513

11.4 Linear Models, 515
  11.4.1 Functions, Predictability, and Causality, 515
  11.4.2 Linear Functions, 516
  Section 11.4 Problems, 518

11.5 Criterion for the Estimated Regression Line, 518

11.6 The Estimators, $\hat{a}$ and $\hat{b}$, 519
  Section 11.6 Problems, 521

11.7 The General Form and Assumptions of a Regression Model, 522
  11.7.1 The Line of Means, 522
  11.7.2 The Error Term and Its Properties, 523
  Section 11.7 Problems, 526

11.8 Regression and the Computer, 527
  Section 11.8 Problems, 531

- Summary and Review, 533
  Multiple-Choice Problems, 533
  Word and Thought Problems, 535

Unit III: Simple Regression Models: Evaluating and Forecasting

Introduction: *Sales-Advertising Relationship*, 541

11.9 Model Evaluation, 542
  11.9.1 R^2: The Coefficient of Determination, 542
  Section 11.9.1 Problems, 545
  11.9.2 Hypothesis Tests for the Regression Coefficient, b, 545
  Section 11.9.2 Problems, 547

11.10 Forecasting the Dependent Variable with a Simple Regression Model, 548
  11.10.1 A Prediction Interval for Y, 548
  Section 11.10.1 Problems, 549
  11.10.2 Precautions for Regression Forecasts, 549

11.11 General Evaluation of the Model, 552

11.12 Regression Analysis and the Computer, 553
  Section 11.12 Problems, 556

11.13 Multiple Regression Models, 559

- Summary and Review, 561
  Multiple-Choice Problems, 562
  Word and Thought Problems, 567
  Review Problems, 570

  Box 11-1: Superstition, Correlation, Causation, and Life Lines, 503
  Box 11-2: Estimating Historical Behavior with Regression, 551
  Box 11-3: Speed Limits and Traffic Fatalities, 561

Chapter 12 Analysis of Variance, 576

Objectives, 576

Unit I: One-Way Analysis of Variance

Introduction: *Training Method Experiment*, 577

12.1 Setting Up the Problem, 578
  12.1.1 The Problem with Paired Tests (Optional), 578
  12.1.2 Controlled Experiments, 578
  Section 12.1.2 Problems, 580
  12.1.3 The Hypotheses, 581
  Section 12.1.3 Problems, 582
  12.1.4 Variations That Affect the Sample Values, 582
  12.1.5 Measuring the Variation and Its Components, 584
  Section 12.1.5 Problems, 589

12.2 The Test Statistic, 589
  12.2.1 SSW and MSW, 589
  12.2.2 SSB and MSB, 590
  12.2.3 The Ratio of Mean Squared Deviations, 591
  Section 12.2 Problems, 592

Contents

12.3 The *F* Distribution, 593
  Section 12.3 Problems, 594
12.4 The Decision for ANOVA, 594
  Section 12.4 Problems, 598
12.5 One-Way ANOVA on the Computer, 598
  Section 12.5 Problems, 602
12.6 Two-Way Analysis of Variance, 603

■ Summary and Review, 606
  Multiple-Choice Problems, 606
  Word and Thought Problems, 609
  Review Problems, 613

  Box 12-1: Child Nutritional Development and Working Mothers, 597
  Box 12-2: Parenting, 602
  Box 12-3: Deadly Statistics, 605

■ **Chapter 13 Chi-Square and Nonparametric Tests, 618**

  Objectives, 618

Unit I: The Chi-Square Test for Independence of Two Variables

  Introduction: *Loan Applicant Evaluation*, 619
13.1 Nonparametric and Parametric Tests, 620
13.2 Setting Up the Problem, 621
  13.2.1 The Hypotheses, 622
  13.2.2 Expectations When H_0 Is True, 622
  Section 13.2 Problems, 624
13.3 The Test, 625
  13.3.1 The Chi-Square Test Statistic, 626
  Section 13.3.1 Problems, 626
  13.3.2 The Chi-Square Distribution, 627
  Section 13.3.2 Problems, 628
  13.3.3 Making the Decision, 629
  Section 13.3.3 Problems, 631
13.4 The Chi-Square Test and the Computer, 632
  Section 13.4 Problems, 632

■ Summary and Review, 634
  Multiple-Choice Problems, 634
  Word and Thought Problems, 637

Unit II: More Nonparametric Tests

  Introduction: *Construction Cost Situation; Productivity Difference Situation; Training Method Situation*, 641
13.5 The Sign Test, 642
  13.5.1 Alternative Test for Population Central Tendency, 642
  13.5.2 The Test Statistic, 642
  Section 13.5.2 Problems, 644
  13.5.3 The Sampling Distribution, 644
  13.5.4 The Decision and Its Interpretation, 645
  Section 13.5.4 Problems, 646

13.6 Wilcoxon-Mann-Whitney Rank Sum Test, 646
  13.6.1 An Alternative Test for the Difference between Two Population Means with Independent Samples, 647
  13.6.2 The Test Statistic, 648
  Section 13.6.2 Problems, 650
  13.6.3 The Sampling Distribution of SR, 651
  13.6.4 The Decision and Its Interpretation, 651
  Section 13.6.4 Problems, 653
13.7 The Kruskal-Wallis Test, 655
  13.7.1 An Alternative Test for One-Way ANOVA, 655
  13.7.2 The Test Statistic, 655
  Section 13.7.2 Problems, 657
  13.7.3 The Sampling Distribution, the Decision, and Its Interpretation, 657
  Section 13.7.3 Problems, 660
13.8 The Computer and Nonparametric Tests, 660
  13.8.1 Computer Output for the Sign Test, 660
  13.8.2 Computer Output for the Wilcoxon-Mann-Whitney Rank Sum Test, 661
  13.8.3 Computer Output for the Kruskal-Wallis Test, 662
  Section 13.8 Problems, 663

■ Summary and Review, 665
  Multiple-Choice Problems, 666
  Word and Thought Problems, 670
  Review Problems, 674

  Box 13-1: Chi-Square Test for Goodness of Fit, 648
  Box 13-2: Gender and Children's Musical Preferences, 654

■ **Appendixes**

A: Rounding Rules, A1
B: Air Quality Data Base, B1
C: State SAT Data Base, C1
D: NCAA Probation Data Base, D1
E: North Carolina Census Data Base, E1
F: Counting: Combinations and Permutations, F1
G: Binomial Distribution Tables, G1
H: Standard Normal Distribution Table, H1
I: Random Number Table, I1
J: Student's *T* Distribution Table, J1
K: State Health Data Set, K1
L: *F* Distribution Tables, L1
M: NCAA Tournament Data Base, M1
N: Chi-Square Distribution Table, N1

■ **Answers, ANS-1**

■ **Index, IN-1**

Preface

Statistics surround us. Some we like (batting averages, average starting salary for new graduates, ratings for our favorite television program or movie). Others we like less (average test scores, incidence of diseases, pollution counts). Some are reliable, some are not. Regardless of how we value these numbers, however, we cannot escape them. Statistical information is everywhere in today's world of accessible electronic technology that easily generates numbers and speedily disseminates them as information. It is important that we understand such reports, even if we do not produce them, in order to ensure proper usage and avoid abuses.

This book offers the student an opportunity to do more than passively read over materials and perfunctorily compute peculiar numbers. It offers active participation with the basic materials that comprise an introductory statistics course and, as a result, it can make this subject and its procedures accessible, interesting, and, yes, even fun. Challenging students and helping them achieve these goals is the purpose of this book.

To achieve these goals, the presentations simplify difficult topics and relate them to relevant professional and everyday situations in order to tempt the reader to make an investment in learning the concepts. Technical aspects, which are important to users and consumers, are appropriately provided. Emphasis on the results, interpreting them or reporting them using words familiar to the situation and/or persons involved rather than technical jargon, climaxes most of the examples and exercises. Once the concepts are learned, those "statistics" and other numbers that we experience in the electronic media and in print will no longer be a mystery. In addition, we discover that we already approach common events and everyday decisions with unquantified and nontechnical, yet statistical, reasoning.

However, this book does not sacrifice difficult topics. Beginning with descriptive statistics and continuing through regression, analysis of variance, and nonparametric statistics, it focuses on statistical inference—how we describe a group when we only obtain information from a subset of the group. The reader thus prepares to apply basic statistical principles in common decision-making situations as well as in more advanced courses.

Multiple examples and a generous supply of colorful, informative diagrams illustrate the topics, procedures, and caveats for the practitioner and for those who must understand the statistical work of others. Most chapters are organized into at least two major units. The compact orientation of these units provides many opportunities for exercise and experience with small logical steps in the process of building a strong basis for more advanced topics in the latter part of the book or in later courses. Although these topics are more complex, simple expositions as well as the emphasis on relevance are preserved without sacrificing important details for the user. The diversity of material allows for variety when choosing and presenting topics for classes with different emphases or goals.

Some high school algebra is required, although mathematical rigor is not stressed. Rather, intuition and personal experience form the basis for most measures and techniques. For example, hypothesis testing is introduced by references to everyday instances of decision making and refutation, such as when we assume a clock keeps accurate time, until we are late several times when we rely on it. With such common situations in mind, you can come to understand statistics as technical renderings of familiar ideas.

Principles of probability are presented in sufficient detail so that we can use the concept and basic rules to understand probability distributions and their characteristics. We need these basic principles in order to comprehend uncertainty in decision making and in everyday situations as well as the role probability plays in statistical inferences. Two distributions, binomial and normal, are presented and employed later in the inferential chapters, where they are complemented with the T, F, and chi-square distributions.

Features

While we're on the subject of statistics, let's discuss a few numbers related to this book. At last count (expositions in the text result from a dynamic process that incorporates changes, additions, deletions, and corrections based on the comments of numerous reviewers, accuracy checkers, editors, colleagues, and students), the average characteristics of the chapters were:

- 165 problems per chapter;
- 25 Question-Answer sequences (situated in the presentations) that allow students to reflect on the issues and contribute to the development of a concept or formula;
- 12 situations that serve for examples with solutions included, many of which appear several times to illustrate different concepts;
- 30 figures that include computer printouts and diagrams; and
- 3 information boxes that supplement the chapter material with interesting statistical applications and concepts.

Now for some elaboration to make sense of all these numbers.

Many modern topics are covered, beginning with stem-and-leaf and box-and-whiskers plots. Because the computer is a primary tool of statistical analysis today, output from the Minitab statistical program is included. This output provides practice in reading printouts in the context of questions and problems that require statistical analysis of available information. P-values are provided by Minitab and many other statistical packages. Because p-values facilitate decision making and provide a more informative measure of the test results, Chapter 9 devotes a separate unit to their development.

The art work summarizes vast amounts of information, then packs it all into colorful, informative illustrations that visually teach as well as reinforce material in the text itself. And there are lots of illustrations. Colors are coordinated in each chapter, with unusual effort devoted to carefully choosing colors so they identify specific concepts and measures discussed in the chapter. Consequently, colors not only attract students to the material but simplify the effort required to comprehend the message within. The diagrams are not subsidiary to the text, but often the primary focus in making a point. Using informative balloons and pointers, the diagrams make it easier to summarize several points at once, notice similarities and differences between situations, and motivate and visualize inquiry into a new point or solution. In addition, there are several figures that illustrate beyond simple curves and labels, using instead characters and images from a problem or situation. All in all, the art work is one of the most salient features of this book. The color key shown on page xix and on the inside front cover shows the pedagogical use of color in the art program.

No less important are the questions interspersed throughout the text to encourage the reader to pause and construct the answer. The answers follow these text questions. Such frequent self-tests provide opportunities not only to check comprehension and intuition but also to develop concepts and techniques. This pedagogy leads the reader to discover results rather than simply having those results presented. Thus, the reader can develop the ability to think critically and speculate about outcomes of different situations.

Many other problem opportunities are provided—over 2,100 total in the text. All major units except the first chapter and the first units of Chapters 2 and 4 contain an initial set of multiple-choice questions covering material from previous chapters and units that is relevant to the new material. These questions furnish a review and an assessment of the student's readiness for the new material. Problem sets follow most sections and relate solely to the topics in that section. Multiple-choice questions and word problems conclude each unit, and a set of review problems, organized randomly from previous material (except in the last four chapters, which are independent of one another), concludes each chapter. These exercises relate to everyday experiences as well as to more formal examples from the social sciences, hard sciences, business, education, and other disciplines. Many are taken from real situations and the sources are cited. Students do not have to wonder if they will ever again employ statistical procedures—the evidence of the opportunities is in front of them.

Exercises contain thought (discussion, conceptual) questions as well as computation problems, and most computations request a verbal statement of the result or meaning of the numerical answer. Such verbal summaries, even when not requested in the problem, benefit the statistics student because a meaningless number is of no use to anyone. The questions and problems present numerous opportunities for the student to master concepts and calculations, to relate the topics to everyday experiences, and to creatively extend the ideas presented in the chapter. These problem sets as well as the text presentation have been revised several times after extensive use in different classes.

To help the student understand the steps required to solve a problem and the meaning of the computations, there are many worked-out examples. Often a single situation is examined several times to illustrate different concepts and the different procedures we can use to analyze the same set of circumstances, so the average number of these examples does not convey the actual number of solutions available to the student.

Answers to the interactive questions within the chapter, to the multiple-choice questions at the beginning and end of each unit, and to the remaining odd problems are provided in the text. However, a separate solutions manual for instructors provides solutions and detailed computation for all exercises except those multiple-choice questions that need no explanation beyond the correct choice. A student solution manual with the same information for odd-numbered problems is also available.

In addition to the two types of solution manuals, other ancillaries are available from the publisher. These include a test bank, testing software, a study guide for students, and a Minitab workbook for students.

The appendixes include six different data bases that cover such diverse areas as North Carolina census data, pollution, sports, and education. These data can be used for classroom demonstrations, class projects, or additional exercises, especially when computers are integrated into the course.

The informative boxes in each chapter offer insight into other situations where statistical analysis enlightens and enlivens debate. Other boxes expand the topics beyond the basics in each chapter.

The numerous examples and exercises, along with sections on objectives and review, provide the typical features of a study guide; consequently, the book itself is a complete package on modern statistical analysis for the student.

Acknowledgments

I would like to briefly thank the following: my students, my colleagues, the staffs, and the administrations at Western Carolina University and Western Kentucky University for their encouragement and support; J. P. Lenney, David Dietz, and Judy Hauck, from Mosby-Yearbook, Inc., for their patient guidance and faith in the project; Earl McPeek, Jane Parrigin, Paula Heighton, Karen Storlie, Julie Kennedy, Susan W. Gilday, Marilyn Sulzer, Julie Schmidt, Carrie Burger, and the many others at Wm. C. Brown who participated in this creation and brought it to fruition; Lachina Publishing Services for their help in discovering and correcting ambiguities, inconsistencies, and other blunders; Keith Stiles, one of my former students at Western Carolina University, for his help in checking for errors in the problem solutions; Professors Giles Maloof (Boise State University) and Jon W. Scott (Montgomery College) for their accuracy checks of the examples and problems with answers during the hectic latter stages of production; and the following professors for their helpful reviews, comments, suggestions, and corrections (though I bow to tradition and accept all responsibility for remaining transgressions and credit for all of the good stuff):

Timothy L. Hardy—University of Northern Iowa
Gerry Hobbs—West Virginia University
Michael R. Karelius—American River College
Jerald T. Ball—Las Positas College
Alice Murillo—Hartnell College
Gaspard T. Rizzuto—University of S.W. Louisiana
Pat Deamer—Skyline College
Douglas H. Frank—Indiana University of Pennsylvania
Sharon E. Navard—Virginia Commonwealth University
Phillip E. Johnson—University of North Carolina, Charlotte
Dee E. Oyler—Utah Valley Community College
Shu-ping Hodgson—Central Michigan University
Alice Kelly—Santa Clara University
Arthur Dull—Diablo Valley College
Mary Teegarden—San Diego Mesa Community College
Eric Lubot—Bergen Community College
Mary Sue Beersman—Northeast Missouri State University
Michael Orkin—California State University, Hayward
Joseph M. Moser—San Diego State University
Thomas Roe—South Dakota State University
Lynne L. Doty—Marist College
Gael T. Mericle—Mankato State University
Suren Shrivastava—San Antonio College
Donald Gray—Iowa Western Community College

I would like to especially acknowledge the effort of Judy Hauck from Mosby. Although not initially involved with the project, she joined and influenced the book in many and major ways. She convinced me to shift information to the illustrations to stimulate learning rather than simply reiterate and fine-tune ideas stated in the text. The impact can hardly go unnoticed as one progresses through the book. She and Sue Gilday made many of the color choices and worked to use colors consistently within each chapter. She encouraged and convinced me to make additional passes through the manuscript to incorporate new presentations after earlier rewrites (ask any author or editor how easy this is). Each time, the manuscript, art work, and I benefitted

Preface

creatively and instructionally (and sometimes begrudgingly). I can hardly thank her enough for her thoroughness and abiding interest.

Many thanks are due Jane Parrigin for her fortitude in turning a completed manuscript into a book. It's not easy to tackle a project with much of the creative input intact and oversee the thankless thousand-and-one tasks and details that turn endless pages of words, lines, X's, arrows, insertions, and deletions into a colorful, informative book.

Thanks to Earl McPeek at Wm. C. Brown for his calm and steady influence during those few periods of frustration, isolation, and panic that dotted the meadows and slopes on our way to the crest.

Also, a special thanks to Professors Tom Stanley and Dick Cantrell. Over the years both have conversed at length with me about statistics and contributed to my intuition, which I hope I have imparted in the following pages. My kids have been very patient and are probably more thankful than I to see the book completed. Finally, a very special thanks to my wife, Marcia, who continued to listen to my extended conversations about statistics and the book, encouraged me, proofed for me, checked solutions, handled permissions, heightened her own interest in statistics (some of which waned before we finished), and survived another effort.

Color Key

- Populations
- Samples
- Values of any variable
- Population mean
- Sample mean
- Specific value in a sample
- Median
- Mode
- Location of a value
- Event outcomes
- Probability
- Expected value of a random variable
- Variance of X
- Z-values
- Standard deviation
- Standard error of the mean
- Confidence intervals
- Degrees of freedom
- Sampling error
- Level of significance
- x-values
- p-values
- Critical values
- Acceptance region
- Correlation coefficient

BASIC STATISTICS

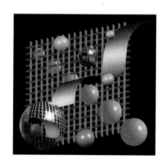

Chapter 1

What Is Statistics?

Objectives

This chapter is an introduction to the basic concepts and definitions of statistics, including the summation symbol Σ. This symbol frequently appears in the formulas and measures you will use. Upon completing this chapter, you should be able to

1. State why learning statistics is useful in the contemporary world.
2. State three reasons why samples are used to infer information about a population.
3. Differentiate among different types of variables and different types of measurement scales.
4. Use the proper order for performing mathematical operations to compute the value of a mathematical expression.
5. Round solutions to problems.
6. Sum a series of terms involving arithmetic operations as indicated with a summation symbol $\left(\sum_{i=1}^{n} X_i\right)$, also expressed as $(\Sigma_{i=1}^{n} X_i)$.
7. Define statistics, descriptive statistics, inferential statistics, population (census or universe), sample, destructive sample, random sample, representative sample, qualitative variable, quantitative variable, nominal scale, ordinal scale, interval scale, and ratio scale.

Introduction

When a large set of numbers is presented to you, you may react with an "ugh" or not react at all. Suppose, however, that the numbers contain useful information and your task is to recover the particular information you want. The numbers might represent your future weekly income. Immediately, you become more interested. (If you do not, consider weekly profits, scores on tests, number of traffic tickets, hours spent with a significant other, or something else that is important to you.)

The large box of data in Figure 1-1 displays a set of 52 weekly income values. As you scan the numbers, questions may arise. Are the numbers reliable? How are they arranged? Are they arranged chronologically, so that you can find in which weeks you will have more to spend? If not, what is the average income you can expect? Are most of

Chapter 1 — What is Statistics?

Figure 1-1
Weekly income data set

52 Weekly Income Values ($)

75	40	55	70
95	100	80	105
250	130	160	110
245	250	250	110
85	260	300	110
75	260	415	245
190	375	625	250
300	300	385	260
280	285	250	345
145	160	260	325
250	150	250	400
315	180	230	315
180	185	225	305

What is the highest weekly income you can expect?

How often will you receive at least 300 dollars?

the values close to this average, so that you can assume the average when you plan your budget, or are the values spread out, so that you must watch the list closely? Which week will you get the most income? What percent of the time will you receive at least $300? Perhaps more important, is there information that will help you predict unlisted values?

Statistical techniques help you obtain such information. *Statistics,* the discipline, is the study of processing information in a set of values to produce a quantitative summary or analysis of them (a *statistic* is a number that summarizes information in a data set). This process includes the collection, organization, and description of a data set as well as the formation of inferences from the set. We might find answers to many questions on our own, but knowing how to use statistical techniques allows us to objectively collect and interpret data in a more orderly, time-saving fashion. Many concepts we will study here are based on common-sense approaches to problems, but these approaches are defined more precisely and are, consequently, more technical, mathematical, and consistent. If we keep the common-sense foundation in mind as we proceed through the book, we can reduce the complexity that sometimes accompanies the precision.

Learning statistical methods can improve the way you use information to make decisions when you face uncertain outcomes. In addition, you will learn to interpret and evaluate data sets and statistical analyses performed by others. The media, government agencies, scientists, educators, doctors, lawyers, academicians, all types of professionals as well as everyday people report and use statistics (numerical values), although the last group may not call the process or logic behind the numbers statistics or use the techniques as precisely and carefully as would a person who has studied the subject. Any of these users may employ the numbers to persuade us to accept a particular point of view or to describe or explain the world around us, such as a batting average used to demonstrate the prowess of a baseball player or the Consumer Confidence Index to show changes in people's attitude about the economy (see Figure 1-2). It is often tempting to believe that a report is more accurate or scientifically respectable when it includes numbers to reinforce a point. In fact, the untrained observer may be misled by the mere appearance of a number, whether the usage is correct or incorrect (see Boxes 1-1 and 1-2).

Figure 1-2
One way of using numbers to describe the world around us

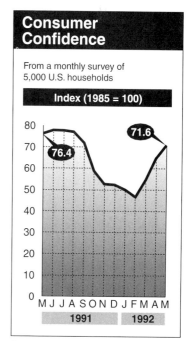

Source: AP/Wide World Photos.

Box 1-1 How Nutty Numbers Distort the Facts

Washington—To the untrained eye, it was just another of the numbing numbers by which journalism calls attention to this or that crisis: "Every year, the World Health Organization estimates, 220,000 people die from pesticide poisoning." To the trained eye of Richard McGuire, New York's commissioner of agriculture and markets, that assertion in an upstate New York newspaper's editorial looked implausible.

It was. Follow McGuire as he follows the slithering number to a lesson about the strange life led by some statistics and the terrible data on which government often makes decisions.

A call from McGuire's office to the upstate editor revealed that he had received the editorial from a California syndicate. A call there revealed that the 220,000 number was from information supporting Sen. Patrick Leahy's (D-Vt.) bill to prohibit U.S. companies from exporting pesticides whose use is banned in America. Leahy was concerned about America importing foods containing residues of chemicals banned here.

Leahy's office directed McGuire to the WHO, which directed him to a WHO report. McGuire wrote to the author in Switzerland, who wrote back to say the figure of 220,000 deaths came from another WHO publication.

The author had warned readers that "reliable data on pesticide poisonings are not available, and the figures given are derived from various estimates." Unfortunately, he said, quoted figures often acquire misplaced momentum because they are shorn of their tentativeness.

Here is what the WHO publication the author relied on actually said: "Of the more than 220,000 intentional or unintentional deaths from acute [pesticide] poisoning, suicides account for approximately 91 percent, occupational exposure for 6 percent and other causes, including food contamination, for 3 percent." Of the 3 percent (itself a guess), we are left to guess what portion involved food contamination.

WHO's basic message was that there were actually 20,000 deaths from unintentional pesticide poisoning in a world population of 5 billion. The numbers floated downstream, from the WHO to the senator's office to the editorial writer's office where this was written:

"Every year, the World Health Organization estimates, 220,000 people die from pesticide poisoning; 25 million fall victim to injury or illness. There are no reliable numbers on how many of these casualties result from exposure to unlicensed chemicals imported from this country.... But there is no question that the American manufacturers who continue to traffic in these poisons are a significant part of the problem."

That is, American traffickers in poisons are unquestionably a significant part of the problem, if there is a significant problem. (U.S. Food and Drug Administration tests on imported foods reveal no significant problem with chemical residues on food imports.)

The use of nutty numbers to advance political agendas may result from cynicism or from confusions born of carelessness. The result can be foolish public policies, feeding on and fed by the journalism of apocalypse.

Twenty years ago, The Public Interest published "The Vitality of Mythical Numbers" by Max Singer, then president of the Hudson Institute. He dissected a then-commonly cited number, that New York City's "100,000-plus" heroin addicts were stealing upward of $5 billion worth of property a year.

The assumptions behind the numbers were: 100,000 addicts were each spending an average of $30 a day on their habits, or $1.1 billion a year (100,000 $\times$ 365 $\times$ $30). Stolen property is fenced for about one-quarter of its value, so addicts must steal upward of $5 billion worth.

Singer was skeptical.

Most stealing by addicts then was by shoplifting and burglary. All retail sales in the city then totaled $15 billion (including cars, carpets, diamonds and other goods not susceptible to shoplifting). All losses from all forms of theft and embezzlement were about 2 percent. Even if shoplifters accounted for half of that (they don't; employees steal much more), and if all shoplifters were addicts (they aren't), the addicts' shoplifting total would be $150 million.

Burglary? Even if one-fifth of the city's 2.5 million households had been burglarized each year (they weren't), and accepting the police estimate that the average loss from a burglary was property worth $200, the burglary total ($200 $\times$ 500,000) was $100 million.

So even with inflating assumptions, the burglary and shoplifting sum was a quarter of a billion dollars' worth of property. That is not chopped liver, but it is one-20th of $5 billion.

Probably the "100,000-plus" number of addicts was inflated. A pertinent question about such numbers is: Whose interests are served by a numerical exaggeration? The answer often is: The people whose funding or political importance varies directly with the perceived severity of a particular problem.

Here, then, is a helpful number: 2. When an advocacy group cites hair-raising numbers about the problem for which they are advocating solutions, or a bureaucracy cites such numbers about the problem its programs address (homelessness, drug abuse, teen-age prostitution, whatever), divide the numbers by 2.

Similarly, when the Office of Management and Budget issues deficit projections, multiply by 3.

Source: George Will, "How Nutty Numbers Distort the Facts." © 1992, Washington Post Writers Group. Reprinted with permission.

Box 1-2 Not Just Any Number Will Do

Suppose a newspaper reports that the average number of people in a region who acquire AIDS from sexual contact in spite of using a condom is 1,500 per year and that among nonusers the average is 800 people. The report concludes that condoms are not very effective in preventing AIDS. However, this conclusion masks the effects of population size and consequently the risks of contracting AIDS. A better statistic would be the ratio of users who contract AIDS to the number of users in the region, that is, the proportion who contract AIDS among the region's condom users. Similarly, we could determine the proportion who contract AIDS among nonusers. A comparison of these two proportions illuminates the different chances of contracting AIDS, depending on the use or nonuse of condoms. Suppose there are 1,000,000 users and 250,000 nonusers in the region whose only contact with other people's body fluids is through sexual activity. The risks of AIDS for condom users are thus $1,500/1,000,000 = 0.0015$ and for nonusers more than twice as great, $800/250,000 = 0.0032$, in spite of the larger absolute number who contract AIDS among users.

This situation offers an opportunity for statistics to provide information that would be difficult to obtain otherwise. We can easily obtain the number who contract AIDS in the region from government agencies or area hospitals and clinics, but the number of condom users and nonusers in the region is not so readily available. A random sample of the region's inhabitants is likely to be the only practical solution for obtaining information about regional AIDS infection counts, condom usage, sexual activity, and opportunities for contact with other people's body fluids. If a survey provides confidentiality and only reports information in summary form, not individually, respondents are more likely to provide accurate responses on such sensitive personal topics. The proportions who contract AIDS for users and for nonusers (among sexually active people with no other body-fluid contacts) in the sample would be estimates of the corresponding ratios for the entire population in the region. Suppose the sample contains 1,024 users with two AIDS-infected people and 311 nonusers with one AIDS-infected person. The corresponding estimated risks would be 0.0020 and 0.0032, respectively.

Thus, statistics can be useful in analyzing specific problems and improving decision making, but we must take care to choose the appropriate statistic. Not just any number will do.

In our computer-dominated world we continually find new technical improvements increasing our ability to collect information and, subsequently, to produce numerical values. It is important for us as informed decision makers to be aware of the benefits as well as the limitations of statistics so that we can recognize correct and incorrect analyses.

1.1 Initial Concepts and Definitions

In many situations the proper application of statistics can improve the chances of saving time or money. We can condense sets of numbers to one measure or a few measures that reveal information about the general nature of the entire set. For instance, the average weekly income from the data in the blue box in Figure 1-3 is $222.98. We call this process of condensing and summarizing values *descriptive statistics*. Most people are familiar with the average or mean of a set of numbers—a descriptive statistic. The average value provides information about all of the values but avoids an investigation of each and every individual observation.

A set of values will be a population or a sample. The entire set of entities (items, persons, firms, countries) under consideration is a *population* (also called a *census* or a *universe*). If we wish to study the wages of the workers in a firm with 100 employees and the set of values contains 100 observations—one for each employee—the set is called a population. The blue box in Figure 1-3 contains an income value for *every* week during the year, an example of a population if we want to know about weekly incomes during that year. We can imagine some populations that do not physically exist. Consider a cooking-utensil manufacturer who wants to measure the temperature at which the utensils melt. To obtain a population of the temperatures would require melting every utensil, leaving the firm with nothing to sell.

Figure 1-3
Taking samples from a population

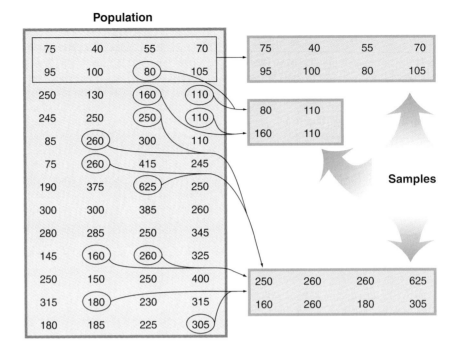

A *sample* is a subset of a population, usually a set of observed data values. If the data set consists of wages for only 10 of the 100 employees, it is a sample of all the employees' wages. Notice that a sample may be the same as the population if we are only interested in the wages of these 10 employees. The green boxes of Figure 1-3 show three of the many possible samples that can be selected from the population of weekly incomes.

Statistics can also save time and money by providing information about a population without our having to observe a value for every item in the population. *Inferential statistics* is the process of gaining information about a population based on information from a sample. We draw inferences or conclusions about the population or we estimate population characteristics from the sample data. Inference is an important function of statistics.

We use samples in statistics because often the time or money necessary to obtain a population is overwhelming. In addition, it may be impossible to obtain a population, such as the melting temperature of the cooking utensils mentioned above. Obviously, it would make more sense for the melting-temperature study to use a sample. We call such sampling *destructive sampling,* because the items tested are no longer useful after we obtain the information. When researchers test new drugs or medical treatments to determine the good and bad effects on humans, a similar situation exists. They use samples, rather than populations, in order to avoid harming a large number of people from unknown side effects. We must use samples, not populations, when we analyze soil, test water and air for pollution, and determine the heat content of a load of coal.

Briefly, whenever (1) high costs, (2) large allotments of time, or (3) destruction of items would be required to obtain a population, we collect a sample instead. We use the information from the sample to infer information about the population.

A well-designed sampling procedure increases the chances of obtaining accurate estimates from a relatively small sample. Random sampling is the basis for most procedures. A sample of *n* observations is a *random sample* when every possible set of *n* items from the population has an equal chance of selection as the sample. *A consequence of random sampling is that every observation in the population will have an equal chance of being included in the chosen sample.* In effect, random choices signify that no population members receive special consideration or priority for selection. Chapter 7 shows more details about random selection.

Random sampling is more likely to produce a *representative sample* (one that approximately duplicates the characteristics of items in a population on a smaller scale) than is a sample containing observations that were the easiest or most convenient to collect or that were systematically chosen in a manner that excludes or underrepresents some population members from the sample.

Consider the top sample in Figure 1-3. It consists of the first eight weekly income values in the population, a convenient selection. If these were January and February values then the values may be unusually low, perhaps because many businesses experience a slow period during these months. This convenient selection process produces a sample of nontypical values. It is not a representative sample. The average or mean income of this sample is $77.5, not so close to the population average, $222.98.

The bottom sample in Figure 1-3 consists of randomly selected observations. This time the mean is $287.50, which is closer to the population average. *Random selection processes do not guarantee a representative sample.* In fact, it is possible that the top sample in the figure could be randomly selected. We will spend a lot of time discussing and quantifying the accuracy of the inferential process. We do not expect inferences to be exact predictions, but often we can compute the likelihood of different errors.

Question: A senator receives 76 duplicate letters (identical except for the signature) and an assortment of 48 other letters, all 124 from her constituents, concerning the location of a waste site near a community in the state. The duplicate letters are copies of a suggested draft from the newsletter of the selected community's Clean Air Committee that opposes the selection. Should the senator accept this set of information as a random sample of her constituents' opinions?

Answer: Probably not. It is likely that most readers of the newsletter were from the selected community, not evenly distributed across the state or constituents. In addition, the newsletter may have motivated respondents to write, and its suggested letter definitely facilitated negative responses. Other constituents around the state who are less threatened may also be less motivated to write. Many may even ignore the issue. The sample is certainly not random; its representation of the population is questionable, for it is likely to be biased against the site.

Question: Is the advertisement in Figure 1-4 likely to produce a random sample of responses? A representative sample? What is the desired population?

Answer: We can speculate about respondents, but the likely characteristics of those who had time for seven visits include unemployed, impoverished residents of the Boston vicinity, who have seen the advertisement. Although none of these characteristics necessarily influences a subject's responses to the allergy medication, the selection procedure is not the same as randomly selecting subjects from the population of healthy individuals. Thus, it does not guarantee every healthy individual an equal chance of participating. Proper screening of applicants is necessary, in the event that unhealthy individuals or drug addicts volunteer. Either characteristic is likely to unduly influence the experimental results.

Figure 1-4
Soliciting a sample

Part a of Figure 1-5 represents parts of the inferential process. The blue oval is the population of times (in minutes) that a mouse requires to complete different runs through a maze. It encloses the values 25, 18, 3, 22, 10, 4, and 6. The population also has a mean or average, 12.6 minutes, which we represent with the Greek letter μ. The small red oval represents a sample of these times, 10, 4, 22, collected by a researcher during an experiment. $\overline{X}$ (pronounced "ex bar") symbolizes the sample mean. In this case it is 12. The thick arrow in Part a of the figure represents the use of $\overline{X}$ as the estimate of μ. Inferential statistics relies on this process of estimating the population characteristics

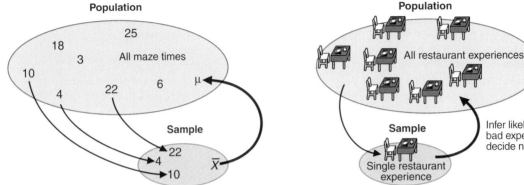

Figure 1-5
The inferential process

from a sample. This process usually provides good estimates, saves time and money, and makes population information available without destroying or harming the entire population in the process.

Although the preceding description makes the inferential process appear very formal, it is a process that many people use frequently (perhaps with less attention to detail or accuracy than that used by the statistician). Have you ever informally polled some of your friends on an issue, such as which professor to select for a course, and then acted on this information? Have you ever had a bad experience with a business, such as poor service at a restaurant, and decided never to return? In both cases you made decisions from sample information: Your friends are a sample of the entire student population for a particular time period, and your single experience in the restaurant is a subset of all the customers' experiences or of all your potential experiences (see Part b of Figure 1-5). Inferential statistics will help you understand and evaluate the validity of such estimates and decisions.

■ **Example 1-1:** In order to determine if a production process meets standards, a carburetor manufacturer decided to test a sample of finished carburetors, using a computer-generated set of random numbers to select 20 finished carburetors from the 1,500 that are produced each hour. In one particular hour, there were 15 defective carburetors out of the 20 that were tested, prompting the manager to halt the production process to check for problems. Describe possible populations for which the 20 selected carburetors form a sample. Which of these populations was the manager considering when inferring that a problem existed?

■ **Solution:** If the sample is the 20 randomly selected carburetors, possible populations include the 1,500 carburetors produced that hour, all the carburetors the company produced that day, all it has ever produced, all it will produce from this time forward (with or without solutions to the problem on the production line), or just the 20 selected carburetors. The 20 carburetors form a *random* sample for only the 1,500 carburetors produced that hour, because these are the only ones that had an equal chance of selection for the sample. Thus, the information is most useful for determining that a problem existed during that hour. The manager may well assume that this sample is representative of all carburetors that will be produced in the future if nothing is done to correct the problem.

In summary, statistics provides us with methods for quantitatively representing and analyzing sets of values. The objective of this text is to acquaint you with some of these methods and techniques, as well as to help you understand statistical concepts in an

Box 1-3 Inferences and Representative Samples

In 1863, the U.S. Commissioner of Agriculture estimated the loss of sheep from attacks by dogs using the following procedure:

> The loss of sheep by dogs may be closely approximated. For a series of years in Ohio, the average of ascertained damages was $11,548 per year, when sheep were very low in price. In 1863, the ascertained loss was $144,658. The Secretary of the New York State Agricultural Society estimates the loss in New York in 1862 at fifty thousand sheep worth $175,000. . . . It is a moderate assumption to take Ohio as a basis for the country, as Ohio had 4,448,229 sheep in 1862, the loyal states 23,000,000 in round numbers, and (as) the average loss of that state was $40,764, the entire loss would be $229,102 in killed: and a similar calculation upon the basis of 25,483 injured in Ohio would show a total of 142,219 maimed.*

The estimates in this report are made by applying the loss ratio for Ohio sheep to the sheep population of the rest of the Union in 1863. The Ohio kill loss rate is $40,764/4,448,229 sheep = 0.00916409 dollars per sheep. Applying this rate to 25 million sheep nationally results in $229,102.40. A similar calculation using the number of maimed instead of kill losses produces 143,219.92 maimed nationally.†

Notice that the commissioner believes that the set of Ohio sheep is a representative sample of the population of national sheep: "It is a moderate assumption to take Ohio as a basis for the country." He then follows with inferences from this sample to the national population. Given the state of statistical knowledge and the practical objectives of the commissioners at the time, this inference may have suited him; that is, he was content with the risks and size of the possible error from his estimation. Perhaps he knew more about the similarities of Ohio sheep-raising conditions and those of the rest of the country at that point in history than we do. Given your knowledge of the desirability of random samples to produce representative samples, how would you have arranged to obtain a random sample of sheep losses? What considerations, such as time and money, would affect your ability to obtain such a sample?

*U.S. Commissioner of Agriculture, *Annual Report*, 1863, p. 452, as reported in Edward Norris Wentworth, *America's Sheep Trails* (Ames: Iowa State College Press, 1948), p. 488.
† It appears that some transcription errors were made either by the author of the original text or the author quoting the text, because 25 million, rather than 23 million, will make the computations correct. Also the number of maimed animals would be 143,219.92, not 142,219.

---

intuitive fashion. Many of the topics to be covered may seem complex, but they are actually based on simple, common-sense ideas. If you keep the simple idea in mind, the more detailed development and analysis can often be kept in perspective. The answers to problems will have more meaning. Also, you will be better equipped to recognize unreasonable answers and thus incorrect applications and solutions (see Box 1-3).

Section 1.1 Problems

1. Fifty percent of 810 adult Iowans preferred General Norman Schwarzkopf, leader of the U.N. forces in the Gulf War of 1991, for a neighbor over 11 other celebrities and television characters. The order of selection of the other 11 was Oprah Winfrey, Michael Jordan, Bob Vila, Jay Leno, Murphy Brown, Martha Stewart, Tom and Roseanne Arnold, Steve Erkel, Dr. Ruth Westheimer, and Madonna. (Source: "Want Tom, Roseanne as Neighbors?" *Asheville Citizen-Times*, July 20, 1992, p. 2A.)
   a. What inference can you make using the information about the general? To what sample and population does your inference refer?
   b. A newspaper report about the poll begins by noting that the Arnolds selected Iowa for another home in addition to Hollywood, but a majority of Iowans preferred someone else for a famous neighbor. What is the sample, inference, and population with regard to the statement about the "majority"?

2. "A federal study said about half the heavy trucks on U.S. highways have brake problems that pose dangers," according to a newspaper report. This conclusion is based on the result of surprise inspections of 1,500 heavy trucks in several states and finding 46.1% with maladjusted brakes and another 10% with serious brake problems. What is the inference, sample, and population? (Source: "Study: Many Large Trucks Have Dangerous Brakes," *Asheville Citizen-Times*, July 20, 1992, p. 6D.)

3. A researcher studied 970 characters from 30 daytime public broadcasting television programs, 30 prime time network programs, and 30 Saturday morning network programs (a total of 90 television programs directed toward families and children). A newspaper summary says that the researcher found that few television characters work and the distribution of the job types among the working characters is different from the actual distribution in the U.S. population. Specifically, the study found that

58% of the characters' roles either do not involve them working or place them in fictional jobs, such as a ghostbuster. Eighty-one percent of the Saturday morning roles fit one of these descriptions. Seventy-five percent of the characters who do work are employed in white-collar positions, whereas the actual percentage is 56% according to the U.S. Bureau of Labor Statistics. (Source: "Study: TV Children Have Poor Role Models," *Asheville Citizen-Times*, July 21, 1992, p. 2A.)

   a. What inference can we draw from the 58% figure? What is the sample and population relevant to this inference? Is there a statement in this report that matches your inference? If so, what is it?
   b. Repeat Part a for the 81% figure.
   c. Repeat Part a for the 75% figure. The 56% figure is different from the 75% figure. Is this relevant? (Hint: see the title of the report.)

4. To expedite a large number of trials in which plaintiffs sought damages for injuries from asbestos, attorneys for the defendants and for the plaintiffs chose a sample of three plaintiffs each from the 8,555 plaintiffs. The six plaintiffs were tried to see if any were likely to receive compensation. The court awarded $11.2 million to the three plaintiffs chosen by the plaintiff's lawyers and nothing to the three selected by the defendants' lawyers. Discuss the apparent sampling methodology that the two sides employed. What, if any, inference can we draw from this sample of six plaintiffs? (Source: "Asbestos Damages Awarded," *Asheville Citizen-Times*, July 24, 1992, p. 8A.)

1.2 Types of Variables and Measurement Scales

A *variable* is a property or characteristic whose value varies among entities in a population or sample, such as hometown, grade point average, or number of years until graduation for a set of students. We use the word *data* to describe the values of a variable, the information we observe and collect to describe and analyze. The data may be numbers or words. Although many of the techniques discussed in this book require numbers, a more useful distinction of variables is the way we express or measure different values of the variable, qualitative versus quantitative. Data values for a *quantitative variable* measure amounts of the property associated with each observation in the population or sample. *Qualitative variable* values categorize or rank different observations.

Qualitative data may be words or numbers. If the pieces of information or values differ by categories, we say the scale of measure is *nominal* (from "name"). For instance, a data set that lists the property stolen by 30 thieves in a large metropolitan area last year, such as televisions, electronic equipment, computer time, telephone services, automobiles, weapons, personal property, money, or some combination of these properties, constitutes nominal data. See the data in the column labeled "Stolen Property" in Table 1-1. There is no information about the amounts of the property or severity of the theft, at least not in the sense of a definite amount. However, we might classify the crimes by using words such as "Class 1 Theft" when the theft involves personal property, Class 2 for business property, Class 3 for theft of business services, and Class 4 for government property (see the column in Table 1-1 labeled "Class"). Again this is a nominal scale that provides some information about the crime but not a specific amount. Notice that these numbers designate the different types of theft but not some level of criminality. We do not know how much worse one crime is than another.

Question: Which other variables in Table 1-1 are examples of nominal measures?

Answer: Gender and Race are nominal values, because we can classify the criminal according to these characteristics, but the terms "amount of gender" or "quantity of race" are nonsensical, because we cannot measure quantities of these characteristics.

We can also express another type of qualitative variable as a number. Values that rank the different observations in the data set according to the amount of the property they possess are *ordinal* measures. The "Severity Rank" column of Table 1-1 lists a judge's subjective ranking of the severity of the thirty crimes. These values order the severity of the individual crimes. They distinguish more severe crimes from less severe ones but do not tell us how much more severe one crime is than another.

Section 1.2   Types of Variables and Measurement Scales

Table 1-1   Theft Data

Crime	Stolen Property	Class	Severity Ranking	Thief's Gender	Thief's Race (white = 0, nonwhite = 1)	Insurance Payments	Market Value
1	Television	1	28	M	0	450	700
2	Money	1	30	M	0	1,250	1,500
3	Electronic equipment	4	7	F	1	10,850	10,850
4	Telephone services	3	11	M	0	645	745
5	Money	2	1	M	0	8,750	9,000
6	Television	1	3	M	0	900	1,000
7	Jewelry	1	14	F	0	2,000	3,000
8	Combination	1	23	F	0	5,500	6,000
9	Money	2	20	M	0	4,200	4,450
10	Combination	2	26	F	0	25,000	26,000
11	Personal property	1	12	M	1	300	1,300
12	Automobile	1	15	M	0	9,000	10,000
13	Automobile	1	17	M	0	15,600	15,650
14	Automobile	2	24	F	0	12,250	12,500
15	Road sign	4	2	M	0	225	475
16	Money	2	22	F	0	23,500	23,750
17	Personal property	1	9	F	1	850	1,100
18	Money	2	8	M	1	9,460	9,510
19	Electronic equipment	1	4	M	1	1,200	1,700
20	Computer time	3	27	M	1	800	1,800
21	Automobile	2	19	M	1	15,000	15,250
22	Automobile	1	6	F	0	18,900	18,950
23	Television	1	18	M	1	700	950
24	Automobile	1	16	M	1	7,500	8,000
25	Money	1	10	M	1	500	1,500
26	Money	2	21	F	0	2,500	2,550
27	Automobile	1	25	M	1	18,000	18,250
28	Combination	1	13	M	1	1,680	1,780
29	Automobile	1	29	M	0	5,920	6,420
30	Automobile	1	5	M	0	7,240	7,740

We can measure the amounts of a property reflected by quantitative variables along interval or ratio scales. The type of scale depends on the meaningfulness of the number produced by a mathematical comparison of different observations. The comparison may be a difference (one value minus another value) or a ratio (one value divided by another). To understand these scales and why ordinal measures lack the level of information contained in quantitative data, consider any of the two severity ranks from the theft data of Table 1-1. For the second and third most severe crimes the difference in ranks is $3 - 2 = 1$, and their ratio is $3/2 = 1.5$. The difference between any two consecutive ranks will always be one, but it may conceal vast differences in the levels of severity of the two crimes. For example, the second may be stealing a relatively inexpensive and noncrucial road sign, the third a television theft, and the fourth a complete cleanout of an electronics store. Similarly, the ratio, 1.5, cannot assure us that the third crime is 1.5 times as severe as the second.

If the difference between two values is meaningful but their ratio is not, the value is measured along an *interval scale*. In this case, we can discuss how much more or how much less of the property one entity possesses over another. The traditional example of an interval scale is temperature values. If we compare the difference in oven heat between 300° Fahrenheit and 325° Fahrenheit, we know the second setting is 25° hotter than the first. Similarly, 450° is 50° hotter than 400°. Moreover, we can perform arithmetic operations on these differences and produce meaningful results, such as the fact that the 50° difference is twice as much as the 25° difference. However, 450° is not 1.5 times as hot as 300°, because, even when the oven is set to 0°, there will be heat in the oven. Only 0° Kelvin (absolute zero) measures a state in which there is no heat. Consequently, a setting of 2° Fahrenheit does not produce twice as much heat as 1° Fahrenheit. The value zero on an interval scale does not reflect the absence of the property.

Question: Insurance company records of their reimbursement of victims over and above the deductible amount (all victims had insurance) provide an alternative measure of the severity of each theft. The column labeled "Insurance Payments" in Table 1-1 reflects these net payments by the insurance companies. Are these values measured on an interval scale if we use them to indicate severity of the crime? Why or why not?

Answer: This is an example of an interval scale, because the extent of a crime with an insurance payment of $1,600 is $1,200 more than one with a payment of $400. However, the actual severity of the first crime may not be four times that of the second ($1,600/$400 = $4), because deductibles are not included in the measures. Suppose both victims have a $100 deductible policy. Then, the first crime is actually 3.4 times as severe as the second ($1,700/$500) if we assume the value of stolen goods and services measures the severity. Additional discrepancies occur when the deductibles differ. Note also that a zero payment does not mean there was no theft.

Question: If we use jail sentences for each criminal to measure severity of the theft, is the resulting scale an interval scale?

Answer: No. A criminal might receive no jail time if it were the first offense, if jails were overcrowded, or if judges decide that a theft must exceed some arbitrary level before sentencing a thief to jail. Thus, a zero sentence does not reflect an absence of theft. The differences between two sentences may reflect how much more severe one theft is than another. However, a more basic problem muddles the severity measurement here, because judges consider hardships and prior records when sentencing convicted thieves. Thus, a difference in two sentences may not reflect the actual difference in severity of the two crimes.

If both the difference and the ratio of two values of a variable are meaningful, the scale is a *ratio scale*. Now we can measure how many times larger or smaller one amount is than another as well as how much more or less it is. This is because the zero now indicates the absence of the property. The actual market values of the stolen goods and services listed in the data set under "Market Value" measure severity along a ratio scale. A theft of $1,000 worth of goods and services is twice as severe as a theft worth $500, and a $2,000 theft is $700 more than a $1,300 theft.

Question: Are a firm's annual profit values for the last 20 years measured on a ratio scale? Explain.

Answer: Yes, because zero means an absence of profit and $5,000 is twice as much profit as $2,500. Also, differences between profit values represent how much more or less one profit is than the other.

The quantitative/qualitative distinction will be sufficient to apply the statistical techniques covered in this book. Most of these techniques accommodate quantitative variables measured on an interval or ratio scale. Chapter 12 specifically addresses techniques for qualitative variables, measured both nominally and ordinally.

Section 1.2 Problems

Classify each of the following variables, first, as qualitative or quantitative and, second, as a nominal, ordinal, interval, or ratio measure.

5. Pollen counts provided as numbers between 1 and 10, where 1 implies there is almost no pollen and 10 that it is rampant, but for which the values do not represent an actual count of grains of pollen.
6. The atomic number of elements, that is, the number of protons (or electrons) in the nucleus of an atom of the element.
7. Month of the year—January, February, and so on.
8. Month of the year, designated by the month value in date expressions, such as the 2 in 2/3/78.
9. Dividend income, measured by taxable dividend income, where taxpayers deduct the first $400 of dividend income and pay taxes on the remainder.
10. Times for swimmers to complete a 50-meter race.

1.3 Mathematical Operations

1.3.1 Order of Operations

Standard procedure for evaluating arithmetic expressions requires performing the following levels of operations in the order listed below. Always work from left to right when performing each level of operation.

1. All exponential and logarithmic operations (such as squaring, taking square roots, and finding the log of a value).
2. Multiplication and division.
3. Addition and subtraction.

■ **Example 1-2:** Compute $25 + 30\sqrt{29} - 8(6)^2 + (5)^3$.

■ **Solution:** The correct order is

1. $25 + 30(5.3851648) - 8(36) + 125$.
2. $25 + 161.55494 - 288 + 125$.
3. 23.55494.

In practice, you may not need to write out each step as we did here.

When a formula or equation contains expressions enclosed in parentheses or brackets, obtain a value for the expression within the innermost set of parentheses, then continue with the next most innermost, until all enclosed expressions are completed before evaluating nonenclosed statements.

■ **Example 1-3:** List the steps for evaluating

$35.6(18.88)^2 - (53.6 - 25) + [100 + (66.543 - 6.543)^2 + \sqrt{16 + 25 + 36}]$.

■ **Solution:**
1. $35.6(18.88)^2 - 28.6 + [100 + (60)^2 + \sqrt{77}]$.
   Notice the implied parentheses enclosing the expression under the square root symbol. In general, $\sqrt{X + Y} \neq \sqrt{X} + \sqrt{Y}$.
2. $35.6(18.88)^2 - 28.6 + [100 + (60)^2 + 8.7749644]$.
3. $35.6(18.88)^2 - 28.6 + [100 + 3,600 + 8.7749644]$.
4. $35.6(18.88)^2 - 28.6 + 3,708.775$.
5. $35.6(356.4544) - 28.6 + 3,708.775$.
6. $12,689.777 - 28.6 + 3,708.775$.
7. $16,369.952$.

Example 1-4: A teacher can grade an essay question in 15 minutes and a computation question in 3 minutes. If there are two essay questions and four computation questions on a test, the total time to grade a test would be $2 \times 15 + 4 \times 3$. Instead of the prescribed order of operations, perform each operation in turn from left to right. Use the result to show that this new rule does not work.

Solution: The grading time will not be the same from both rules as shown below.

New Rule (left-to-right)	Correct Order
1. $30 + 4 \times 3$	1. $30 + 4 \times 3$
2. 34×3	2. $30 + 12$
3. 102 minutes per test	3. 42 minutes per test

We know that 42 minutes is correct because two essays require 30 minutes and four computations require 12 minutes for a total of 42 minutes per test, not 102 minutes. In addition, changing the order of the expression to $4 \times 3 + 2 \times 15$ will not change the answer when we use the correct rule. However, the answer without the order would be 210 minutes. Also $3 \times 4 + 15 \times 2$ and $15 \times 2 + 3 \times 4$ produce further discrepancies when we use the incorrect rule for order, but always 42 with the correct rule.

Discrepancies in solutions may also appear because of different methods of rounding numbers. Whenever we compute results from data that contain values with different degrees of accuracy or the result appears to be more accurate than is necessary or practical, we will need to round values. For instance, we may know the total of our wages to the nearest cent for income tax purposes, but the IRS only requires accuracy to the nearest dollar. To help ensure a common solution to any problem, we establish general rules to use for rounding in Appendix A. Reading this short appendix will help you understand rounding and future answers.

1.3.2 Symbols and Notations: Mathematical Shorthand

Often the investigation of ideas is easier if they are expressed mathematically, avoiding the ambiguity of words and lengthy explanations. Mathematical symbols can speed this process. The following symbols and notations will be used in the remainder of the text.

The letters X, Y, and Z represent variables, characteristics or measures that can vary with each observation. For example, assume that the purity of a certain chemical compound can vary, depending on its source or seller. The compound obtained from a certain source could be 90% pure. If we let X represent purity, we could rewrite the previous sentence as $X = 90$.

A subscript on a variable, such as X_6, is a means of differentiating observations of a particular variable. The sixth observation of variable X is X_6. The general form is X_i, which is the ith observation of variable X. Suppose you obtain the above compound from three different dispensers and determine purity values of 89%, 75%, and 96%. You can record the values as

$$X_1 = 89 \quad X_2 = 75 \quad X_3 = 96.$$

A *constant*, such as k, is a value that does not vary with each observation. For example, the price is the same per container for all three compounds. A constant price of $5 can be expressed as $k = 5$.

Often values of a variable must be summed. If there are lots of values, a lengthy string of mathematical expressions can be avoided by using the symbol for summation Σ (the upper case Greek letter sigma). The sum of the purities is $X_1 + X_2 + X_3$, which can be expressed more compactly as $\sum_{i=1}^{3} X_i$, or (to avoid spacing problems in this book)

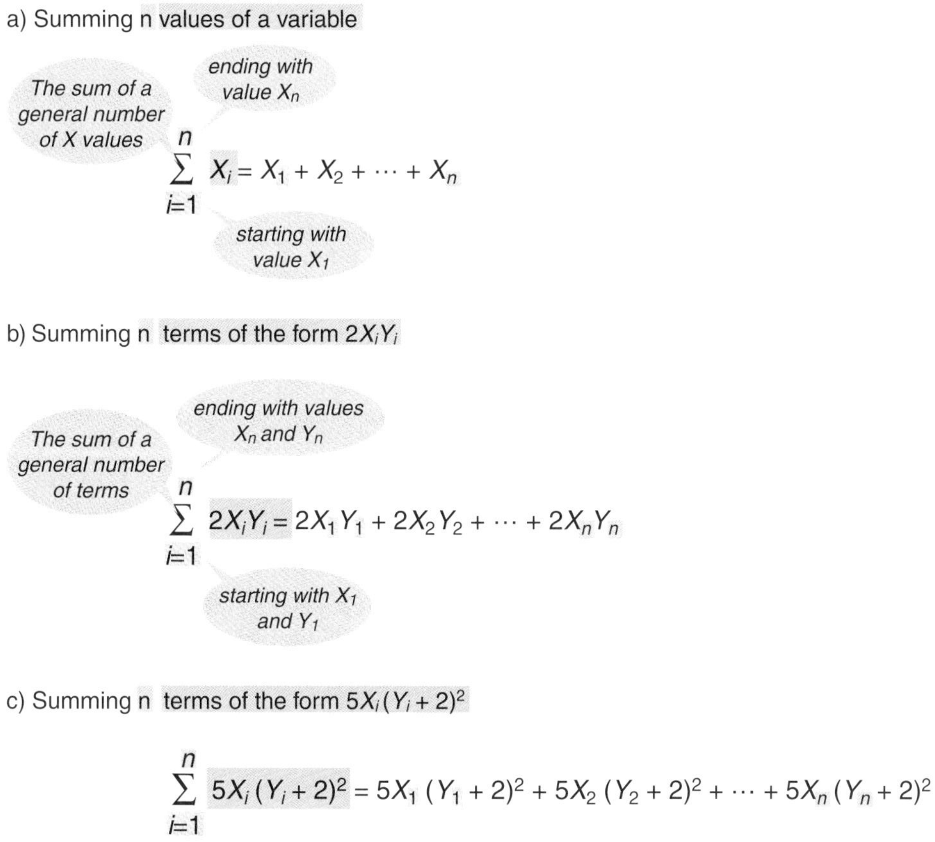

Section 1.3 Mathematical Operations 15

Figure 1-6
Interpreting the summation symbol, Σ

a) Summing n values of a variable

The sum of a general number of X values (ending with value X_n, starting with value X_1):

$$\sum_{i=1}^{n} X_i = X_1 + X_2 + \cdots + X_n$$

b) Summing n terms of the form $2X_iY_i$

The sum of a general number of terms (ending with values X_n and Y_n, starting with X_1 and Y_1):

$$\sum_{i=1}^{n} 2X_iY_i = 2X_1Y_1 + 2X_2Y_2 + \cdots + 2X_nY_n$$

c) Summing n terms of the form $5X_i(Y_i + 2)^2$

$$\sum_{i=1}^{n} 5X_i(Y_i + 2)^2 = 5X_1(Y_1 + 2)^2 + 5X_2(Y_2 + 2)^2 + \cdots + 5X_n(Y_n + 2)^2$$

as $\sum_{i=1}^{3} X_i$, which means the sum of several values of X, beginning with X_1 and ending with X_3. We write the sum of n values of X, starting with the first and continuing to the nth, as $\sum_{i=1}^{n} X_i$ (see Part a of Figure 1-6). The expression $\sum_{i=1}^{n} X_i$ is read "the sum from i equals 1 to n of X_i."

For the chemical compound data, $\sum_{i=1}^{3} X_i = X_1 + X_2 + X_3 = 89 + 75 + 96 = 260$.

■ **Example 1-5:** The manager of an amusement park assembles data on daily paid admissions, the number of people who paid to get in to the park, for five days. X_i = paid admissions on day i. $X_1 = 200$, $X_2 = 300$, $X_3 = 100$, $X_4 = 100$, and $X_5 = 50$. Find the total number of people who paid to enter the park for the five-day period, and express the necessary computation using Σ notation.

■ **Solution:** The total number of people over the five-day period is $\sum_{i=1}^{5} X_i = 200 + 300 + 100 + 100 + 50 = 750$ persons.

We also use the summation sign to sum a series of arithmetic operations that are identical except for the values substituted in the operations, such as $\sum_{i=1}^{n} 2X_iY_i$. In this case we sum the n values that result from multiplying 2 times X_i times Y_i as shown in Part b of Figure 1-6.

The identical arithmetic operations that follow the summation sign must be a *term*, an expression that combines mathematical operations other than addition and subtraction. For us, the usual operations will be multiplication, division, and exponentiation. Some examples include $X_iY_i/15$ and $4.2X_i^2Y_i$. Operations inside parentheses may form a term even when they include addition and subtraction, such as $(X_i + Y_i)$, $-(2X_i - 18)$, and $5X_i(Y_i + 2)^2$. Part c of Figure 1-6 shows how the summation symbol streamlines the actual statement of the work that must be done for the latter term. In general, the summation symbol shortens an otherwise very long statement of the sum of n terms involving identical mathematical operations.

■ **Example 1-5 (continued):** The manager also lists data for Y, admission price, over the five-day period, because the price structure encourages entry on the first few days and takes advantage of increased demand for recreation during the latter days of the period. So X_i = number of people who paid to enter the park and Y_i = admission price (in dollars), both on day i.

Day 1	Day 2	Day 3	Day 4	Day 5
$X_1 = 200$	$X_2 = 300$	$X_3 = 100$	$X_4 = 100$	$X_5 = 50$
$Y_1 = 1$	$Y_2 = 1$	$Y_3 = 2$	$Y_4 = 3$	$Y_5 = 4$

Find the following totals for the five-day period, and express the necessary computations using Σ notation:

1. Total revenue collected from entry fees;
2. Total revenue using a $5 per person daily admission price rather than the present fee structure; and
3. Total number of people who enter the park if 20 people are allowed free entry each day.

Then perform the following calculation, and interpret the result in terms of the amusement park problem:

4. $\sum_{i=1}^{5} X_i + 20$ (the 20 means 20 people).

■ **Solution:**

1. Total revenue collected from entry fees when the entry price, Y_i, varies from day to day, is

$$\sum_{i=1}^{5} X_i Y_i = (200 \times 1) + (300 \times 1) + (100 \times 2) + (100 \times 3) + (50 \times 4) = \$1,200.$$

2. Total revenue over the period, if a $5 entry fee is maintained over the five-day period, would be

$$\sum_{i=1}^{5} 5X_i = 5(200) + 5(300) + 5(100) + 5(100) + 5(50) = \$3,750.$$

3. If 20 individuals are allowed to enter free each day, total number of people is

$$\sum_{i=1}^{5} (X_i + 20) = (200 + 20) + (300 + 20) + (100 + 20) + (100 + 20) + (50 + 20) = 850 \text{ persons.}$$

4. $\sum_{i=1}^{5} X_i + 20 = (200 + 300 + 100 + 100 + 50) + 20 = 770$ people. Note the absence of the parentheses following the Σ symbol, so the 20 is a separate term. Twenty is added once to the sum of the X values, which could denote 20 free admissions over the entire period rather than each day.

When solving Part 1 of Example 1-5, you may have been tempted to total the X values, total the Y values, and multiply the two totals, because the calculations seem to be fewer or easier than finding the five products before summing. This alternative operation would be written as $(\sum_{i=1}^{5} X_i)(\sum_{i=1}^{5} Y_i)$, and the result is $(750)(11) = \$8,250$, obviously not $1,200. In general, the answers will not be the same for $\sum_{i=1}^{n} X_i Y_i$ and $(\sum_{i=1}^{n} X_i)(\sum_{i=1}^{n} Y_i)$. It is important that you read the symbols correctly and perform the correct operation. Note the following rule.

Problems

> In general,
> $$\sum_{i=1}^{n}(X_iY_i) \neq \left(\sum_{i=1}^{n}X_i\right)\left(\sum_{i=1}^{n}Y_i\right),$$
> and
> $$\sum_{i=1}^{n}(X_i)^2 \neq \left(\sum_{i=1}^{n}X_i\right)^2.$$

■ **Example 1-5 (continued):** Calculate and compare $\sum_{i=1}^{5}(Y_i)^2$ and $(\sum_{i=1}^{5}Y_i)^2$. Although we cannot easily interpret the results here, we often perform such computations to derive statistics (such as the standard deviation as we shall see in Chapter 3).

$$Y_1 = 1 \quad Y_2 = 1 \quad Y_3 = 2 \quad Y_4 = 3 \quad Y_5 = 4.$$

■ **Solution:** $\sum_{i=1}^{5}(Y_i)^2 = 1^2 + 1^2 + 2^2 + 3^2 + 4^2 = 31$. This is the sum of the squared Y values. $(\sum_{i=1}^{5}Y_i)^2 = (1 + 1 + 2 + 3 + 4)^2 = 11^2 = 121$. This is the square of the sum of the Y values. As the rule states, these calculations do not result in the same value.

Sometimes summation notation for $\sum_{i=1}^{n}X_iY_i$ is shortened to $\sum_{1}^{n}X_iY_i$, $\sum_i X_iY_i$, $\sum X_iY_i$, or just $\sum XY$. All of these notations represent the same operation: to sum the terms over all values of X_i and Y_i. Usually, by convention, the possible values of i go from 1 to N if the set of values is a population, and they go from 1 to n if the data set is a sample. Sometimes we use letters other than i for subscripts.

Question: Does $(\sum X_i + \sum Y_i)$ make any sense for the amusement park example?
Answer: No. We can certainly sum the values as indicated by $(\sum X_i + \sum Y_i)$, but to obtain the result we added people to dollars, which produces a meaningless result.

We can always substitute numbers in a formula and get a numerical answer. However, it is important to remember to consider the formula, the operations, and the data we use to solve problems to be certain that the answer is reasonable. We should check that the magnitude of the answer seems reasonable as well. For instance, if the calculation for the total attendance in Example 1-5 resulted in 105,832 people, when the five daily entry totals were each less than 500, we should suspect that something is wrong. Similar problems arise when we employ computers to do calculations. The computer will perform operations with sets of numbers as *we* instruct it. It is *our* responsibility to make sure that the data and instructions we supply will not produce implausible results.

Section 1.3 Problems

11. Evaluate $26/13 \cdot 2 + 16$.
12. Evaluate $5^2 - 3(4-2)^3/24$.

Use the following information to solve Problems 13–21.

$X_1 = 82 \quad X_2 = 101 \quad X_3 = 0 \quad X_4 = 10 \quad X_5 = 1.$
$Y_1 = 0 \quad Y_2 = 0.1 \quad Y_3 = 0.2 \quad Y_4 = 0.3 \quad Y_5 = 0.4.$

13. $\sum_{i=1}^{5}Y_i$. **14.** $\sum_{i=1}^{5}X_i$. **15.** $\sum_{i=1}^{5}X_iY_i$.
16. $\sum_{i=1}^{5}2X_iY_i$. **17.** $(\sum_{i=1}^{5}Y_i)(\sum_{i=1}^{5}X_i)$.
18. $\sum_{i=1}^{5}X_i/5$. **19.** $\sum_{i=1}^{5}X_i^2$.
20. $\sum_{i=1}^{5}(X_i - Y_i)^2$. **21.** $\sum_{i=1}^{5}(X_i - 1)^2/5$.

Summary and Review

Statistics is the study of collecting and processing information in a set of values in order to produce summaries and analyses of the data. It is important in our contemporary world, because the widespread use of computers provides opportunities for us to employ statistics to improve our own decision making and because we need to be able to evaluate statistical analyses performed by other people.

The two major divisions of statistics are descriptive statistics and inferential statistics. *Descriptive statistics* deals with condensing and summarizing a set of values. *Inferential statistics* concerns gaining information about characteristics of a population from sample data. Much can be learned about a set of values in a short time without studying each observation. Inferential statistics enables us to save time and money in gathering information about a population, because a randomly selected sample usually contains a good indication of the information being sought. Inferences about populations that would be destroyed if we collected every observation (as in the example of the melting point of cookware) are also possible. Both areas of statistics assist the decision maker in organizing data and information into a useful form for dealing with uncertain situations.

Different statistical techniques employ different types of data. *Qualitative variables* may be measured on a *nominal* or *ordinal* scale. In these cases, we obtain information that classifies the data or orders the observations according to whether they possess more or less of some property being measured. *Quantitative variables* may be measured on an *interval* or *ratio* scale. We interpret the difference in interval-scaled values as how much more or how much less an entity possesses of a property. A value measured on a ratio scale reflects the amount of some property possessed by the entity. We interpret a ratio of ratio-scaled values as how many times larger or smaller one value is than another, which is to say, how many times more or less of a property resides with one entity than with another. The quantitative/qualitative distinction is most important for the techniques discussed in this book.

Although many of the concepts in statistics are based on simple ideas, precision requires the technical expression of these concepts. This book stresses concepts; however, we cannot avoid some mathematical symbols and expressions. Actually, the mathematics allows us to express ideas clearly and concisely. The first mathematical symbol we use is the summation sign, Σ. For example,

$$\sum_{i=1}^{n} X_i = X_1 + X_2 + \cdots + X_n.$$

$\sum_{i=1}^{n} X_i$ says to add up the n observations on the variable X.

Multiple-Choice Problems

22. The primary objective of statistics is
  a. To prove a preconceived notion.
  b. To summarize and analyze the information in a set of values.
  c. To separate samples from populations.
  d. To inspect unusual values in a set of values.
  e. To explain discrepancies in different sets of values.

23. Which of the following is a sample and not a population of a psychologist's clients last year?
  a. A list of all clients.
  b. A list of clients whose appointments were on Saturdays.
  c. A list of clients from two years ago.
  d. An estimate of the total number of clients.
  e. A count of the total number of clients.

24. A businessman notes that for the past few months his competitor has been running a newspaper ad on Mondays. He infers that next Monday's paper will contain another ad. The sample involved in this inference is the contents of
  a. Next Monday's paper.
  b. Last Monday's paper.
  c. All Monday papers.
  d. All past Monday papers.
  e. Monday papers for the last few months.

25. Statistics allows us to
  a. Trade off complete knowledge of a group for reliable estimates about the group and savings in time and money.
  b. Obtain reliable estimates about the lifetimes of all of a firm's products without destroying or using up all of the products in the process.
  c. Make educated guesses without complete information in a decision-making process.
  d. All of the above.

26. The process of condensing and summarizing data to avoid observing each listed observation or number in a set of values is called
  a. Descriptive statistics.
  b. Inferential statistics.
  c. Destructive sampling.
  d. Census taking.
  e. Technical statistics.

27. To learn more about the effect of a new instructional method on sixth-graders in Wyoming, a random sample of students from Mrs. Robinson's sixth-grade class in

Wheatland is selected and subjected to the new method. Which of the following statements does *not* describe a potential problem with making inferences for all Wyoming sixth-graders with the information from the sample?
   a. Mrs. Robinson does not teach all of the sixth-graders in Wyoming. She should have been randomly selected from Wyoming teachers.
   b. The sample is drawn from the correct population, but it is not randomly selected from the population of interest.
   c. The sample is truncated rather than rounded.
   d. The sample may not be representative of all Wyoming sixth-graders.
   e. More time and money would be required to increase the accuracy of inferences for the population of Wyoming sixth-graders.

28. If an archaeologist only digs in one 40-foot by 40-foot section of a 5-acre site in order to describe typical meal preparations of an ancient civilization, we may
   a. Suspect biased conclusions, because different types of households may have occupied different areas of the site.
   b. Question the conclusions, because random spots within the site were not selected.
   c. Find that the archaeologist accepted greater risks of incorrect conclusions to offset the costs of obtaining more information or more accurate information.
   d. Reason that, if digging destroys information, the preservation of other sections for later research necessitates random selection of only one small area.
   e. All of the above.

Use the following data set for Problems 29–38.

A teacher uses one or more quizzes every term. The point value of a quiz is always one of four values. Y_i is the ith potential quiz score, and X_i is the number of quizzes with a potential score of Y_i given during a particular term.

$$X_1 = 0 \quad X_2 = 1 \quad X_3 = 0 \quad X_4 = 1.$$
$$Y_1 = 10 \quad Y_2 = 20 \quad Y_3 = 30 \quad Y_4 = 40.$$

29. $\sum_{i=1}^{4} X_i =$
   a. 0.   b. 1.   c. 2.   d. 3.   e. 4.
30. The solution to Problem 29 represents
   a. The number of quizzes given during the particular term.
   b. The average possible quiz points for the term.
   c. The total number of possible quiz points during the term.
   d. The number of quizzes worth 20 points.
   e. The number of points available on either quiz.
31. Calculate $\sum X_i Y_i$.
   a. 100.   b. 60.   c. 2.   d. 62.   e. 0.
32. The solution to Problem 31 represents
   a. The number of quizzes given during the particular term.
   b. The average possible quiz points for the term.
   c. The total number of possible quiz points during the term.
   d. The number of quizzes worth 20 points.
   e. The number of points available on either quiz.
33. $\sum_{i=1}^{4}(X_i + Y_i) =$
   a. 0.   b. 100.   c. 60.   d. 102.   e. 4.
34. $\sum_{i=1}^{4} 2(X_i)^2 =$
   a. 0.   b. 1.   c. 2.   d. 3.   e. 4.
35. $\sum_{i=1}^{4}(Y_i - X_i) =$
   a. 0.   b. 100.   c. -98.   d. 98.   e. 2.
36. $\sum_{i=1}^{4}(X_i^2) =$
   a. 0.   b. 1.   c. 2.   d. 3.   e. 4.
37. $(\sum_{i=1}^{4} X_i)^2 =$
   a. 0.   b. 1.   c. 2.   d. 3.   e. 4.
38. $\sum_{i=1}^{4}(X_i/3) =$
   a. 0.7.   b. 0.67.   c. 2.0.   d. 0.667.   e. 1.
39. A variable that purports to measure the severity of the thefts from Table 1-1 uses a 0 to signify that no victims were present during the crime, a 1 for a single victim present, and a 2 for more than one present victim. What type of measurement scale is employed here?
   a. Nominal.   d. Ratio.
   b. Ordinal.   e. Empirical.
   c. Interval.
40. Whitewater rafting enthusiasts subjectively rate the difficulty of each major rapid in a stream according to an international rating system that considers different characteristics of the rapid. The less difficulty they have negotiating the rapid, the lower the rating on a 1-to-6 scale. A rafter selects a sample of 50 streams and records the largest rating for any rapid in that stream. What is the type of variable and measurement scale associated with the data set?
   a. Qualitative, ordinal.   d. Quantitative, ratio.
   b. Quantitative, ordinal.   e. Qualitative, interval.
   c. Qualitative, nominal.
41. If the rafter orders the 20 rapids in a stream according to her perception of the least to the most difficult rapid, these ratings would be
   a. Nominal scaled.   d. Ratio scaled.
   b. Ordinal scaled.   e. Empirically scaled.
   c. Interval scaled.
42. Elderly people's legs tend to jerk regularly during sleep. Researchers reached this conclusion by monitoring the sleep of a randomly selected group of people 65 and older. Forty-five percent of the group experienced five or more leg jerks per hour of sleep. What is the most appropriate population for inferences beyond the sample? (Source: "Leg Jerks Common Sleep Ailment," *Asheville Citizen-Times*, February 24, 1992, p. C1.)
   a. The people whose legs jerked.
   b. All adults.
   c. The randomly selected group.
   d. People 65 and older.
   e. The people 65 and older whose legs jerked.

Word and Thought Problems

43. A survey of 44 hospital administrators from large metropolitan hospitals finds that 33 have degrees specifically related to hospital administration. Apparently, to be hired as a hospital administrator, people with a degree in hospital administration have greatly improved chances. What is the sample? What is the population? What is the inference? Do you think it is correct? What additional information would you need to assess this inference?

44. You are embarking on a course in statistics. Do you expect it to be difficult or easy? How did you form your opinion? Was it from friends or acquaintances who have already taken a statistics course? How many? Was there a pattern to your information gathering, or was it random? How accurate are your sources? How reliable are your information-gathering techniques?

45. Suppose the first time you dine at a restaurant you receive badly prepared food and poor service. You decide never to return to the restaurant. What is the sample? The population? The inference? How good is this inference? What could you consider in deciding to enlarge the sample (for example, cost in food, money, and frustration versus an unusually bad time for the restaurant coinciding with your first visit)?

46. What are some advantages of sampling over investigating an entire population?

47. Is your grade in a course a descriptive statistic, an inferential statistic, or both? Discuss.

48. Suppose someone infers from a sample of 50 observations that the average age of a person in Kentucky is 65. What is the sample? What is the population? What is the inference? Does this seem reasonable? What questions might you ask about the process that produced this inference?

Use the following information to solve Problems 49–54.

X_i is the number of repetitions of an experiment performed at temperature Y_i (measured in degrees Celsius), and Z_i is the corresponding change in volume of the mixture (measured in cubic centimeters).

$$X_1 = 0 \quad X_2 = 1 \quad X_3 = 2 \quad X_4 = 3.$$
$$Y_1 = 25 \quad Y_2 = 20 \quad Y_3 = 15 \quad Y_4 = 10.$$
$$Z_1 = -1 \quad Z_2 = 0 \quad Z_3 = 1 \quad Z_4 = 5.$$

49. $\sum_{i=1}^{4} X_i$. Interpret this solution.
50. $\sum_{i=1}^{4} Y_i$.
51. $\sum_{i=1}^{4} X_i Z_i$. Interpret this solution.
52. $\sum_{i=1}^{4} X_i Y_i$.
53. $\left(\sum_{i=1}^{4} Z_i\right)^2$ and $\sum_{i=1}^{4} (Z_i)^2$. What general rule do your solutions demonstrate?
54. $\sum_{i=1}^{4} (Y_i - X_i)$.
55. Police believe that tickets for running red lights are not being prosecuted by the district attorney's office. A check of 14 such tickets written during a recent month reveals that 11 were pretrial diverted for 12 months (the offender does not pay the fine unless there is another traffic violation within 12 months).
    a. What is the sample?
    b. What is the population?
    c. What is the inference?

56. During the 1980 U.S. Census, sampling was used to obtain information about income, education, and housing conditions. One household in six was mailed a long form to answer the basic questions everyone was asked plus extra questions. Many did not return the long forms. What type of household in your opinion tended not to return the form? Discuss the undercounting implications for the census with regard to estimating income, education, and housing conditions.

57. At the end of the day, a child explains to her father that she does not need a bath, because it rained that day and she did not play outside. As additional evidence, she tells her father to look at her leg, which does appear clean. If the leg is a sample, what are the population and the inference the child wishes the father to make? What would be a better sample for the father to use in determining if his daughter is dirty or clean?

58. Discuss the importance of sampling for maintaining affordable quality standards in a production process.

59. Warnings that the substance should first be tried on a small, hidden area of the material to be treated often accompany strong laundry chemicals or wood-staining materials. What are the populations, samples, and inferences in these situations?

60. Identifying a population can be difficult, yet it is important. Consider a national political party holding primaries to select a national candidate. Presumably, primary results reflect the party's preference for a candidate; that is, the population is the set of all party members. Discuss the implications or potential problems for the outcome of the national election in November that could result from assuming that voters in the primary are a sample with characteristics similar to the population, if
    a. Only 25% of registered voters turn out for the primaries.
    b. Only 30% of registered voters belong to this party.
    c. Primary voters must vote in the party primary corresponding to the party affiliation claimed when registering to vote.
    d. Primary voters declare party affiliation on primary day, just before voting.

61. New drugs are usually thought to have beneficial effects in treating certain diseases, but they may also have deleterious side effects. If a sample is to be used to determine how a drug will affect all patients with a particular disease, learning the effect on each patient in the sample may be very costly or even fatal. To reduce this risk, the drug is first tested on animals. What are the population and inference in this setting? Discuss the

Word and Thought Problems

possibility of the animal sample representing the population of diseased humans and the corresponding inferences.

62. When a child finds himself being reprimanded for behavior that he had repeatedly exhibited in the past without parental comment, he remarks that his parent had never said anything before to correct the behavior. When the child misbehaved this most recent time, what was the population, sample, and inference that made him act without hesitating?

63. When are sports statistics for a team or an individual from a season descriptive statistics, and when are they inferential statistics?

64. a. Check Appendix A on rounding before rounding:
  i. 184.65 to the nearest whole number.
  ii. 29,265.83 to the nearest hundredth.
  iii. 4.3005 to the nearest thousandth.
  iv. 14.15 and 14.25 to the nearest tenth.
  b. What place (such as hundreds, ones, hundredths, thousandths) should the last digit in a final answer occupy, if you use the following values in the calculation?
  i. 53.6; 93.3; 10.2; 100.5; 4.6; 28.9; 240.1.
  ii. 20; 1; 53; 95; 26; 3; 120; 422.
  iii. 82.2; 15.66; 74; 93.1294; 2.513; 35.
  iv. 35.22; 91.28; 34.92; 29.153; 902.382; 892.22; 1.52; 7.342.

65. Evaluate the following expressions, and round your answer according to the rounding rules.
  a. $15.6(28.5) + 0.3(0.1) - 18.9(20.5)$.
  b. $31(20.5)^2$.
  c. $20 + 20.5 + 20.55 + 20.555$.
  d. $(5.2 - 8.76)^2 + (8.6 - 8.76)^2 + (12.5 - 8.76)^2$.
  e. $\sqrt{34.2 + 98.44 - 55.298}$.
  f. $\sqrt{7 + 3(5) + 2(9 - 2)}$.

66. a. A data set consists of the values of two variables, local cable channel location and audience rating (percentage of audience watching the program), for 25 television programs. Decide if each variable is qualitative or quantitative and the type of measurement scale associated with each. Explain.
  b. A data set consists of the starting times for different classes listed in military time (rounded to the nearest hour). Is this variable qualitative or quantitative? What is the measurement scale? Explain.
  c. If pro-choice advocates use an individual's age, counted from the day of birth, and pro-life advocates were to count years beginning with conception, for which group is age measured on an interval scale? Ratio scale? Explain.

67. Researchers surveyed 11,000 U.S. high school students from all 50 states, the District of Columbia, Puerto Rico, and the Virgin Islands about their smoking habits. Using the findings that 16% of white students were frequent smokers compared to 2% of blacks, the U.S. Centers for Disease Control concluded that almost no black students were frequent smokers and that there was a large racial difference among high school smokers. (Source: "Study: One in Five High School Seniors Are Frequent Smokers," *Asheville Citizen-Times*, September 13, 1991, p. 6A.)
  a. What two inferences are drawn?
  b. What is the sample for each inference?
  c. What is the population for each inference?

68. The Environmental Protection Agency selected 35 streams in the Blue Ridge Mountains to represent the effect of increased emission levels over the next 50 years. They concluded that 159 streams would be acidic and 340 would experience acidification periodically. What are the population, sample, and inference in this circumstance? (Source: "EPA Study: Blue Ridge Due Acidity," *Asheville Citizen-Times*, August 28, 1989, pp. 1, 6.)

Chapter 2

Descriptions of the Distribution of a Data Set

The process of condensing and summarizing a data set is called descriptive statistics. In this chapter and the next one we study various techniques for describing a data set that allow us to avoid a tedious examination of every observation in a list of values. This chapter specifically addresses ways to report and display how often different values appear in the data set.

Objectives

A frequency distribution organizes the values in a data set and reports the distribution of these values in tabular form. Unit I focuses on understanding the information content of a frequency distribution, while alerting users to possible misinterpretations and misrepresentations of the raw data set. Coverage includes details for constructing various types of frequency distributions. Upon completing the unit, you should be able to

1. Describe a data set with a frequency distribution, a relative frequency distribution, and a cumulative frequency distribution.
2. Explain the advantages and disadvantages of using a frequency distribution to describe data.
3. Define *class limits, class boundaries, class midpoint, class size, range,* and *outlier*.

Unit II shows how to display graphically, rather than to list, the information for a frequency distribution. Such displays speed the formation of a visual impression and aid the later recollection of an image of how the values are arranged. Upon completing the unit, you should be able to

1. Draw and interpret a histogram, frequency polygon, frequency curve, and ogive.
2. Construct a stem-and-leaf display for a data set.
3. Read and interpret computer output for histograms and stem-and-leaf displays.
4. Define *symmetric* and *skewed*.

Unit I: Tabular Presentation of Frequency Distributions

Introduction

River Cleanup An environmental group and an outdoor recreation group conduct regular volunteer cleanup drives to collect trash, garbage, and other items that have been discarded in and along a local mountain river. Then the sponsoring groups separate the

33	20	11	24	14	4	6	13	3	24	22	11	5	23	2	13	14
30	28	21	40	2	29	23	26	4	5	20	4	5	3	22	6	39
6	22	34	14	23	9	5	5	8	22	5	21	28	33	18	5	9
11	14	22	9	11	23	5	22	22	18	11	3	28	33	25	26	33
15	23	4	15	29	7	6	32	22	16	15	14	15	21	38	29	15
13	16	3	5	8	18	4	23	25	17	25	7	25	22	3	29	17
5	17	16	8	21	22	20	8	24	21	3	8	21	5	26	3	23
16	21	23	24	8	11	21	17	13	16	6	37	21	13	23	15	24
5	23	21	33	5	26	32	37	16	9	16	11	36	11	13	12	4
24	29	39	14	9	4	21	24	2	13	1	4	11	37	22	28	24
39	15	16	8	14	15	11	28	14	16	19	12	14	8	2	32	
6	13	19	24	5	37	15	7	12	5	14	4	29	14	12	28	
15	5	4	19	13	13	21	13	18	9	18	5	3	21	37	21	
4	22	14	23	23	4	14	24	24	14	6	34	16	25	3	21	
23	7	15	15	15	5	15	23	7	15	19	21	16	15	14	5	

Table 2-1
Data for Truckloads of Materials from River Cleanups

materials and truck them to recycling centers and dispose of the remainder in proper locations.

A group member kept records of the number of pickup-truck loads collected in the last 250 cleanups as shown in Table 2-1. Using these records, the group can plan ahead and prearrange hauling for the materials they collect during future cleanups. Looking over 250 values in order to estimate the number of trucks or truckloads needed for future cleanups is a confusing prospect and could result in an inaccurate estimate. The group would also like to demonstrate the extent of pollution caused by private individuals to the local governing board and newspaper. How can the sponsors condense the numbers in order to easily and accurately determine the information they need?

One way would be to organize the values in ascending or descending order. Although this makes reading through the values easier, there are still 250 individual values to read in order to develop some mental image of their nature. To reduce the reading further, we could arbitrarily choose cutoff values, such as 5, 10, 15, . . . , 40, and tally the number of values that fall within the classes created by these cutoff values. For instance, there were 56 cleanups that required 10 to 15 truckloads. Table 2-2 displays the more organized data, now showing the distribution of truckloads over the 250 drives. Most required 25 or fewer truckloads; after 25 the number of truckloads tapers off to a maximum possible value of 40. It is certainly easier to use such a tabular summary, called a frequency distribution, than it is to look at the 250 individual numbers.

In the following sections, we will discuss the use and construction of frequency distributions. They assume various forms appropriate for different situations.

Truckloads of Material	Number of Cleanups
1–5	45
6–10	26
11–15	56
16–20	27
21–25	58
26–30	17
31–35	10
36–40	11
	250

Table 2-2
Frequency Distribution for Material Collected on River Cleanups

2.1 Frequency Distributions

Frequency distributions fall within the area of descriptive statistics. Remember that this area covers condensing and summarizing a data set. Suppose you have a large set of numbers to study, such as the truckloads listed in Table 2-1. Looking down a list of hundreds of numbers to discover the information they contain can be a useless effort, especially if time is short. It is difficult to remember specific numbers or number patterns if they are not arranged in any particular order. Studying each individual number is a time-consuming task. Some organization of the data and a condensed version that tells the story contained in the large data set are probably more useful. Frequency distributions do these things.

A *frequency distribution* is a table that arranges a data set into different classes of values and lists the number (frequency) of observations that fall in each class. *Its purpose is to display the arrangement, pattern, or general tendencies within the data set.* If we want to know where most of the observations in a data set lie and where few of them lie, we are looking for a distribution of the frequencies.

■ **Example 2-1:** Notice that any of the river cleanups represented in Table 2-2 collected somewhere between 1 and 40 truckloads of materials. The general pattern appears to be a data set with most observations falling between 1 and 25 truckloads, then frequencies taper off as truckload values approach 40. Specifically, 45 cleanups collected 1 to 5 truckloads, 26 cleanups collected 6 to 10 truckloads, 56 cleanups collected 11 to 15 truckloads, and so on. How many cleanups collected more than 20 truckloads?

■ **Solution:** The fifth through eighth classes listed in Table 2-2 (21–25 through 36–40) include all drives collecting more than 20 truckloads. Using f_i to symbolize the frequency of the ith class, the solution is $\sum_{i=5}^{8} f_i = 58 + 17 + 10 + 11 = 96$. So 96 cleanups collected more than 20 truckloads of materials.

Question: Can you tell from the information in Table 2-2 if any of the original truckload values were exactly 16?

Answer: We know that 27 values fall somewhere between 16 and 20 truckloads, but we do not know the exact magnitude of these values. In the extreme cases, all of them or none of them could be 16.

A frequency distribution is especially useful for condensing the information in a very large data set, such as the income distribution in a certain region. Imagine looking at a list of everyone's income in Dallas or Texas or in the southwestern United States to determine the incomes of most citizens, the incomes that few people make, or the number of incomes below the poverty level. A frequency distribution can provide us with much of the information in the detailed list, and do it in a fraction of the time needed to study each value.

Section 2.1 Problems

1. Use the following frequency distribution of peak ozone levels (in parts per million, ppm) for 63 metropolitan statistical areas (MSA) from the Air Quality Data Base (Appendix B) to answer the following questions.

Peak Ozone Level (ppm)	Number of MSAs
0.09–0.13	44
0.14–0.18	14
0.19–0.23	2
0.24–0.28	2
0.29–0.33	1

   a. How many MSAs experienced a peak level between 0.14 and 0.18 ppm?
   b. How many MSAs experienced a peak level 0.24 ppm or higher?
   c. How many MSAs experienced a peak level of 0.24 ppm?
   d. How many MSAs experienced a peak level of 0.23 ppm or higher?
   e. Did most MSAs experience low or high peak levels?
   f. The smallest half of the 63 levels are between what two values?

2. The following frequency distribution organizes the 63 peak particulate matter (PM10) levels (in micrograms per cubic meter, $\mu g/m^3$) from the Air Quality Data Base (Appendix B). These are particles with aerodynamic diameters smaller than 10 micrometers that can reach extended regions of the respiratory tract. Describe the basic distribution or pattern of the data as depicted by this frequency distribution. Compare this frequency distribution with the one in Problem 1, then decide if their patterns are similar or different.

PM10 ($\mu g/m^3$)	Number of MSAs
20–29	7
30–39	28
40–49	17
50–59	4
60–69	3
70–79	3
80–89	0
90–99	1

3. The following frequency distributions reflect data from the State SAT Data Base (Appendix C). The left frequency distribution of SAT scores is for states where 13% or fewer of the high school seniors take the test, while the one on the right is for states where 52% or more take the test. Draw a conclusion from comparing these diagrams about the difference or similarity of the distributions of scores based on the percentage taking the test.

States Where 13% or Fewer Take SAT		States Where 52% or More Take SAT	
SAT Scores	Number of States	SAT Scores	Number of States
950–974	1	830–844	2
975–999	7	845–859	2
1000–1024	4	860–874	1
1025–1049	3	875–889	5
1050–1074	2	890–904	6
1075–1099	1	905–919	1
	18	920–934	1
			18

2.1.1 Class Limits, Boundaries, Midpoint, and Size

When we construct an actual frequency distribution, it will be useful to refer to certain aspects of classes. Three of these aspects are points along the number line, the class limits, boundaries, and midpoint. Figure 2-1(a) labels these points for the frequency distribution of truckloads in Table 2-2 on a number line. *Class limits* are the endpoints of the values in each class. They confine the set of possible values in the class. The class limits of the second class are 6 and 10 truckloads. The lower limit is 6. The upper limit is 10, as shown in Figure 2-1(b).

Question: What are the class limits of the following first three classes of a frequency distribution: 10.0–10.9, 11.0–11.9, and 12.0–12.9?

Answer: The lower limit of the first class is 10, and the upper limit is 10.9. The limits of the second class are 11.0 and 11.9, and the limits of the third class are 12.0 and 12.9.

Figure 2-1
Number line for the river cleanup frequency distribution

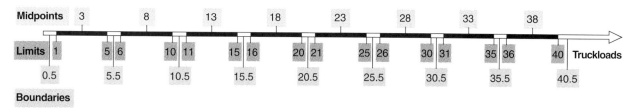

a) Class limits, boundaries, and midpoints

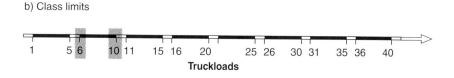

b) Class limits

A *class midpoint* is the middle value in a class, the midpoint of the class limits. We find midpoints of two numbers by summing the two numbers and dividing by 2. The midpoint of the second class in Table 2-2 is $(6 + 10)/2 = 8$ truckloads.

Question: Find the midpoints of the three classes: 10.0–10.9, 11.0–11.9, and 12.0–12.9.

Answer: The midpoints for the first class is 10.45, which is $(10.0 + 10.9)/2$. Following this operation on the other classes produces midpoints of 11.45 and 12.45. Notice the pattern of equal spacing between midpoints that emerges here and in Figure 2-1(a). This occurs when classes are all the same size, a situation we will discuss in Section 2.1.2.

Class boundaries are the midpoints of the space between different classes. The boundaries of the second class in Table 2-2 are 5.5 and 10.5. As the figure shows, once we find one boundary we should know the pattern for the others. In this case, we simply add or subtract 0.5 to limits to form boundaries. This pattern enables us to form the lower boundary of the first class and the upper boundary of the last class, even though there is no adjacent class. Thus, the lower boundary of the first class is 0.5, and the upper boundary of the last class is 40.5.

Question: Find the boundaries of the three classes: 10.0–10.9, 11.0–11.9, and 12.0–12.9.

Answer: The upper boundary of the first class (hence, the lower boundary of the second class) is 10.95, the point halfway between 10.9 and 11.0. Consequently, the lower boundary of the first class is 9.95 and the upper boundary of the second class is 11.95.

Class size is the length of the interval that the class occupies along a number line. We can determine the size of a class by finding the difference in a class's boundaries. The size of the second class in Table 2-2 is $10.5 - 5.5 = 5$ truckloads. The size of the fourth class is $20.5 - 15.5 = 5$ truckloads.

Question: What are the sizes of the three classes: 10.0–10.9, 11.0–11.9, and 12.0–12.9?

Answer: The size of all three classes is 1.0; $(10.95 - 9.95)$, $(11.95 - 10.95)$, and $(12.95 - 11.95)$.

Class size is also the difference between the upper limits, the lower limits, or the midpoints of two consecutive classes. *Notice that it is not the difference of the upper and lower limits of the same class.* For the second class in Table 2-2, using the upper limits, we have $10 - 5 = 5$.

2.1.2 Constructing a Frequency Distribution

To determine the frequency for each class, we find the class in which each observation belongs and keep a tally of the number of observations falling in each class. Figure 2-2 shows the tallying process for the first column of data in Table 2-1 for the classes for the frequency distribution of the truckload values listed in Table 2-2. The color of the box enclosing an observation corresponds to the color of its class and tally mark in the diagram. Although the tallying process is tedious, it is simple compared to deciding on the interval of values that compose each class or how many classes to include in the frequency distribution.

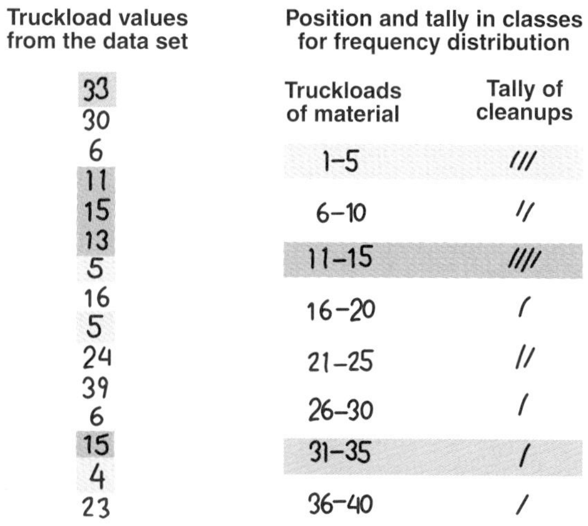

Truckload values from the data set	Position and tally in classes for frequency distribution	
	Truckloads of material	Tally of cleanups
33		
30		
6	1–5	///
11		
15	6–10	//
13		
5	11–15	////
16		
5	16–20	/
24		
39	21–25	//
6	26–30	/
15		
4	31–35	/
23	36–40	/

Figure 2-2
Tallying the frequencies of each class of truckload values by using the first column of truckload data from Table 2.1

■ **Example 2-2:** The quantities of liters of oil sold to 39 customers of a Canadian auto-supply store are:

6	13	6	10	14	2	10	10	6	3	6	14	7
10	2	6	6	10	1	10	3	15	15	6	14	3
8	13	6	6	2	1	2	10	15	10	10	14	8

Determine the frequencies for the potential sets of classes listed below (in liters):
a. 1–2, 3–4, 5–6, 9–10, 11–12, 13–14, 15–16.
b. 1–3, 4–6, 7–9, 10–12, 13–15.
c. 1–5, 6–10, 11–15.

■ **Solution: a.**

Liters	Number of Customers
1–2	6
3–4	3
5–6	9
7–8	3
9–10	9
11–12	0
13–14	6
15–16	3
	39

b.

Liters	Number of Customers
1–3	9
4–6	9
7–9	3
10–12	9
13–15	9
	39

c.

Liters	Number of Customers
1–5	9
6–10	21
11–15	9
	39

As you can see from this example, the same set of data can yield a number of different frequency distributions. There is no unique method or set of rules for constructing a frequency distribution. As we proceed, we will offer suggestions. Keep in mind that the *function of a frequency distribution is to present a condensed version of the data that depicts the distribution or pattern of the original data set.* We must be sensitive to the effects of our construction decisions on the resulting impression and take care not to create a misrepresentation of the true pattern of the values. Similarly, when gleaning information from a frequency distribution constructed by someone else, we need to be aware of the potential for distortion and abuse.

> **Suggestion:** Use a moderate number of classes.

Some statisticians recommend using between 4 and 10 classes, but the situation or distribution of the data may dictate other values. If there is a large number of classes, the intervals can become so small that many classes may have zero frequencies. Eventually the list of classes would approach the original list of values, and nothing would be condensed.

On the other hand, using too few classes obscures the distribution of the values. Remember, the exact value of an observation is lost in the frequency distribution. For example, the nine customers who bought between 1 and 5 liters of oil in the frequency distribution in Part c of Example 2-2 could all be 1-liter customers, 5-liter customers, or equally distributed over the interval 1–5. Too few classes can distort and obscure the distribution of the data.

Question: Upon examining the oil purchase observations from Example 2-2, we find that most of the observations lie at two points, 6 and 10. Compare and summarize the impressions of the distribution of the data that result from each of the frequency distributions in the example.

Answer: Notice that there is an increasing loss of information and a possible distortion of the underlying distribution as the number of classes declines in these frequency distributions.

1. The first frequency distribution with the most classes also most directly reflects the actual distribution of the data set. Notice the dense areas near 6 and 10 and the sparse areas around them.
2. Using five larger classes for the second frequency distribution, rather than eight smaller ones, tends to disguise the original data set. Now the two dense areas still appear, 6 or below and 10 or above; however, the heavy concentration at 6 and 10 has disappeared. The actual number of values below 6 and above 10 declines. Observing this frequency distribution, we might say that the data are almost equally distributed, because all the classes except one have the same frequency.
3. A further reduction in the number of classes with a simultaneous expansion in their size further obscures the reader's image of the original data values. There are no longer two dense areas of values in this version. Instead, most observations appear to lie in the middle of the data set with only a few trailing values on either end of it.

In some cases, a small number of classes may be appropriate. The question to be answered from a data set may dictate the number of classes. For example, a person deciding crate size for liters of oil may find the frequency distribution with three classes in Part c of Example 2-2 most useful. In the river cleanup situation, the groups might include three members who regularly volunteer their pickups, so classes such as 1–3 truckloads, 4–6 truckloads, and so on, may prove helpful in trying to estimate and minimize the number of hauls for each truck during the cleanups. This case requires fourteen classes. If each truck can only make one haul, then the sponsors of the cleanup need to know how often there are 1–3 and 4–40 truckloads. In this case, two classes are sufficient to condense the available information and provide insight for estimating the need for additional trucks.

Section 2.1 Frequency Distributions

In summary, when making a general frequency distribution in an attempt to discover the general nature or pattern of the data, use a moderate number of classes, such as 4–10, to avoid making the process tedious with too many classes or obscuring the distribution with too few classes. A different number of classes may prove useful for answering specific questions about the data set.

> **Suggestion:** Use classes of equal size.

All the classes are the same size in the examples thus far, an advantage for readers, because they can grasp the information more easily. Different class sizes in the same frequency distribution require readers to scrutinize class sizes as well as frequencies. Because equal-size classes are common, readers may expect this and not notice the unequal sizes, getting a distorted impression of the data set. It is important for us as readers of frequency distributions to notice the class delineations as well as the frequencies.

At times, there is some importance attached to the choice of the limits causing the class sizes not to be equal. For example, age groups, such as youth, young adults, adults, the middle-aged, and the elderly, may be arbitrarily defined by 0–20, 21–28, 29–40, 41–59, and 60 and up.

Question: Reconsider the oil customer data with two dense areas of observations at 6 and 10 liters. The frequency distribution with classes, 1–2, 3–4, 5–10, and 11–15 liters is as follows. Describe your impression of the distribution.

Liters	Number of Customers
1–2	2
3–4	1
5–10	7
11–15	3

Answer: At first glance, or to a careless observer, the data seem to be more heavily concentrated in the classes with larger values, although in reality the values are equally spaced on both sides of 7.

In summary, use equal-size classes, unless varying sizes will provide the needed information from the data.

> **Suggestion:** Make the class size a multiple of 5 if this is reasonable.

Once again, using this suggestion will make the information's presentation clearer. Intervals of 10 or 25 are easier to visualize than intervals of 1.3 or 16. When the data values cover only a small interval, then multiples of 0.5, 0.05, 0.005, or even smaller may be appropriate for class sizes. For instance, a class size of 0.05 might be appropriate for a set of 200 values that range between 1.23 and 1.56.

Other class sizes may be appropriate, depending on the purpose of the information display. Suppose a local business considers offering its larger truck for hauling materials for the river cleanups. This truck hauls the equivalent of four pickup loads in a single trip. In this case, a class size of four is appropriate to demonstrate the equivalent number of large truckloads. When we construct a frequency distribution as part of our own search for a description or impression of the data, this suggestion may be less important than when we construct a frequency distribution as part of a formal presentation of the data for someone else to use.

Table 2-3
Student Registration Data for 33 Courses

92	3	29	3	124	111	53	2	116	79	21
10	6	34	89	22	71	85	101	110	89	118
81	51	92	39	72	63	117	51	34	43	99

2.1.3 The Process of Constructing a Frequency Distribution

Now we will construct a frequency distribution, starting with these three guidelines. In the process, we will make a few more suggestions to help produce a generally reliable, easy-to-understand impression of the original data.

■ **Example 2-3:** Table 2-3 presents data for the number of students registered for the 33 courses offered during a summer term. The registrar wants to determine if there are excess courses few students are interested in or classes with an excessive number of students. Either case would precipitate a revision of the current scheduling of summer courses. Because there are only 33 observations, we do not need many classes to condense the data. Suppose we try six classes to get started. Now, use the earlier suggestions to decide class size.

■ **Solution:** These six classes must incorporate all the values in the data set from the smallest to the largest (in this case, between 2 and 124 students). The difference between the largest and smallest value in a data set is called the *range* of the values, in this case $124 - 2 = 122$ students. Six equally spaced intervals would have $122/6 = 20.33$ units in each interval. Often we round the result of this calculation up to the nearest multiple of 5 in order to provide an easy-to-understand set of classes. To use 6 equally spaced intervals that are multiples of 5 and that will include all 33 values between 2 and 124, the class size needs to be 25 students.

■ **Example 2-4:** A set of times for different mice to run a maze contains values between 62.9 and 350.6 seconds. What would be an appropriate size for 10 classes?

■ **Solution:** Range = $350.6 - 62.9 = 287.7$; approximate size = $287.7/10 = 28.77$. A size that is a multiple of 5 and that results in a set of 10 classes containing all of the data is 30.0 seconds.

Now, where do we start the first interval?

> **Suggestion:** Choose a convenient starting point, usually 0 or a multiple of 5 that is equal to or near the smallest value in the data set.

People generally assume that all the values in a class are equally dispersed throughout the class or else concentrated at the class midpoint. If the first class contains only one or a few values, all near its upper limit, it is possible to get a false impression of the data in that class, as well as of the entire distribution. Also, if the smallest value is near the upper limit of the first class, it may be necessary to add an extra class to cover all of the values in the data set. The values in that last class are unlikely to extend beyond the lower portion of the interval. The smallest number of registered students listed in Table 2-3 is 2, so the initial class will begin at 0. The classes will be 0–24, 25–49, and so on. It is not wrong to start at 2, but the classes would be 2–26, 27–51, 52–76, and so on. It is more difficult to visualize the values included in each of these classes and whether the classes are of equal size. Consequently, it is more difficult to gain an impression of the distribution of the data.

Section 2.1 Frequency Distributions 31

Question: The smallest maze time in Example 2-4 is 62.9 seconds, and the suggested class size is 30.0 seconds. What is an appropriate starting point?
Answer: 60.0, because it is a multiple of 5 that is near 62.9.

> **Suggestion:** Make the class limits as accurate as the data.

If the data contain observations given to one decimal place, then the limits should be expressed with one place past the decimal. For the maze times, the limits would be 60.0–89.9, 90.0–119.9, and so on. If you do not follow this suggestion, you will encounter difficulties. Suppose you try limits such as 60–89, 90–119, and so on, for the maze times. There would be no class to contain the observation 89.7 seconds. On the other hand, there is no need to go beyond the accuracy of the data. The last 9 in a class limit of 89.99 would be superfluous for placing observations with only one decimal place, such as 89.7.

Question: Reconsider the registration data listed in Table 2-3. The earlier suggestions for class size and lower limit of the first class are 25 and 0 students, respectively. List the appropriate class limits for the six classes.
Answer: The values have no places past the decimal, so 0–24, 25–49, . . . , 125–149 are appropriate. These values are counts of people, so fractions would be nonsensical anyway.

Notice that the last class will have no values from the data set, because the largest value is 124, leaving this frequency distribution with only five classes, not the six we anticipated. At this point we can continue constructing this frequency distribution with the option of changing the class size, number of classes, or starting point if the resulting frequency distribution seems misrepresentative or inappropriate in some other way. One class holding most of the values would constitute such a situation.

Question: What problem would arise with the registration data if the limits had been 0–25, 25–50, 50–75, and so on?
Answer: A value of 25 or of 50 students would have to be placed in two classes. The result would be double counting of these values. If the overlapping values are placed in only one of the classes, the reader cannot determine which one without more information.

This problem leads to the next suggestion.

> **Suggestion:** Do not overlap class limits.

The following list summarizes the six suggestions (not rules) for constructing a frequency distribution.

> Suggestions for constructing a frequency distribution:
> 1. Use a moderate number of classes (between 4 and 10).
> 2. Use classes of equal size.
> 3. Make the class size a multiple of 5 or another value suitable for answering a specific question about the data set.
> 4. Choose a convenient starting point, usually 0 or a multiple of 5 that is equal to or near the smallest value in the data set.
> 5. Make the class limits as accurate as the data.
> 6. Do not overlap class limits.

The completed frequency distribution for student registrations that incorporates all of the suggestions appears as follows:

Number of Students	Number of Classes
0–24	7
25–49	5
50–74	6
75–99	8
100–124	7
Total	33

Question: Describe the distribution of the data set based on this frequency distribution.

Answer: It appears that course sizes are relatively evenly or uniformly distributed. There are about the same number of values in each class.

To construct a frequency distribution, use the preceding suggestions and follow the flowchart in Figure 2-3.

Figure 2-3
Procedure for constructing a frequency distribution

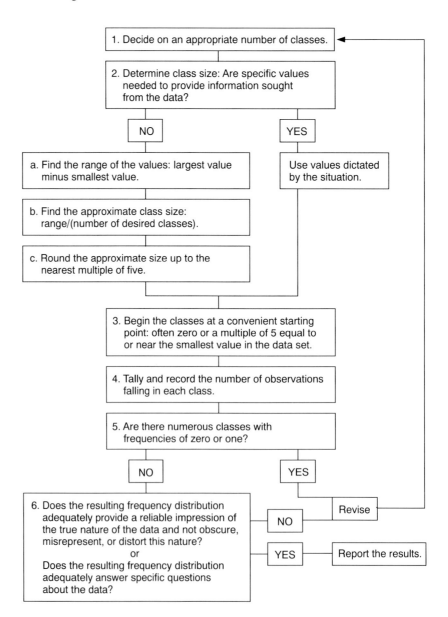

Section 2.1 Frequency Distributions 33

39	44	36	47	42	42	42	47	47	44	44	45	45
45	45	46	51	48	48	50	50	48	48	48	49	46
50	53	53	50	54	51	51	52	52	52	55	59	53
52	57	52	54	51	53	60	61	60	60	56	48	55
57	62	62	62	61	63	55	63	64	64	56	61	51
65	54	65	60	67	64	63	67	65	65	66	64	66
67	64	72	69	70	70	69	70	70	68	72	69	69
71	68	71	72	72	71	70	74	73	74	73	72	71
72	74	75	72	75	74	75	74	74	73	75	73	74
74	75	74	76	76	75	76	75	77	76	77	76	75
77												

Table 2-4

Estimated Life Expectancies for 131 Nations (years)

Source: United Nations Children's Fund, *The State of the World's Children, 1988* (New York: Oxford University Press, 1988), pp. 72–73.

■ **Example 2-5:** The figures in Table 2-4 are life expectancy values (in years) estimated by the U.N. Population Division for 131 nations. Construct a frequency distribution. Aim for 10 classes in order to determine if there is substantial variation in life expectancies or if they are about the same in most countries with only a few extreme cases.

■ **Solution:** Step 1: We have already decided to start with 10 classes.
Step 2: No specific values require special interest in the problem statement, so we proceed rather mechanically to determine the class size.
Step 2a: Range: 77 − 36 = 41 years.
Step 2b: Approximate class size: 41/10 = 4.1 years.
Step 2c: We can try a class size of five years (4.1 rounded up to the nearest multiple of 5). If we only want this frequency distribution to satisfy our own curiosity, we might go ahead and work with 4.1 years, but we would be aware of the otherwise awkward class size.
Step 3: An appropriate starting point might be 35 years, because 35 is a multiple of 5 near 36, the smallest value in the data set. If we begin at 0, the first seven classes of five years will have 0 for frequencies. We will only need nine classes if we start at 35, but after trying this set of classes, we may revise the number of classes, class size, or starting value, if too many observations seem to be grouped in too few classes. However, such an unbalanced grouping might tell us all we wish to know about life expectancies. For instance, if a substantial proportion have life expectancies between 65 and 69 years with few falling in other classes, there would be little reason to use life expectancy to distinguish the difference in health among the different countries in the data set.
Step 4: The frequency distribution shown in Table 2-5 records the result of the tallying stage.
Step 5: There are no classes with frequencies of 0 or 1. Although a few classes have small frequencies, no single class or small set of classes contains most of the observations.

Question: Describe the distribution of the data set based on this frequency distribution.

Answer: Most countries have longer life expectancies, between 60 and 79 years. Very few have life expectancies less than 45 years.

Table 2-5

Life Expectancy (years)	Number of Nations
35–39	2
40–44	6
45–49	16
50–54	21
55–59	8
60–64	18
65–69	15
70–74	29
75–79	16
	131

Step 6: The answer to the last question implies that we can adequately answer the original question concerning the variation in life expectancies among the countries. There does seem to be quite a bit of difference in the values with no accumulation in a small range of ages.

Without computers, we may be satisfied that the procedure outlined in the text and flowchart will produce an unbiased representation of the distribution of the data, rather than rework another frequency distribution to see if the impression changes. With computers, it is easy to experiment with the number of classes, class size, and starting point to determine if the description presented by a frequency distribution is sensitive to these three values, especially if there are empty classes or classes with unusually large frequencies compared to the other classes. If the description does vary considerably when one or more of these three values change, the data themselves may be ordered or tallied by individual values so they can be further scrutinized before we choose an appropriate frequency distribution to represent the data.

2.1.4 Open-Ended Classes

Sometimes data sets contain outliers. An *outlier* (also called an *extreme value*) is a value that is set apart or is extremely different from most of the other values in the data set. If a data set contains one value of 950.6, but all other values fall between 200 and 600, then 950.6 would be an outlier. Similarly, 52.8 would be an outlier for this data set. We will be more specific about what "set apart" and "extremely different" mean after we study more descriptive measures in Chapter 3.

■ **Example 2-6:** The values in Table 2-6 are the number of radio receivers per 1,000 people for 128 nations. Find the outlier(s) in the data.

■ **Solution:** 2,101 is an outlier. It is clearly identifiable among the three four-digit numbers in the data set, because of the distance between it and the other two, 1,016 and 1,274. These two appear to be much closer to the other values in the data set.

Applying the suggestions for forming a frequency distribution for this data set, we obtain the following table. The observation in the last class, 2,101, is obviously an outlier, because there are two empty classes between the class containing 2,101 and the next nonempty class. This evidence is even stronger when we note that 125 of 128 observations fall in 4 of the 9 classes.

Section 2.1 Frequency Distributions 35

91	16	222	245	184	30	43	32	21	26	49
219	34	58	109	132	228	58	110	22	70	14
30	53	40	74	251	89	581	85	21	96	22
90	100	104	644	224	95	206	66	133	184	28
30	256	203	222	175	117	115	78	43	366	221
219	44	321	309	244	130	189	126	100	293	63
391	81	342	160	65	190	139	238	163	131	225
787	175	162	113	422	264	654	424	183	172	936
598	238	141	193	656	332	321	385	274	85	212
222	574	271	327	405	272	470	904	2,101	620	457
596	259	281	430	568	298	1,016	1,274	586	879	863
416	787	828	821	780	988	868				

Table 2-6
Number of Radio Receivers per 1,000 People for 128 Nations

Source: United Nations Children's Fund, *The State of the World's Children, 1988* (New York: Oxford University Press, 1988), pp. 70–71.

Radios per 1,000 People	Number of Nations
0–249	78
250–499	26
500–749	10
750–999	11
1,000–1,249	1
1,250–1,499	1
1,500–1,749	0
1,750–1,999	0
2,000–2,249	1
	128

If a data set contains an outlier and you follow the steps for forming a frequency distribution, then, very likely, there will be several intermediate intervals with relatively small frequencies, usually zero or one, as appeared in the example just completed. Simultaneously, the nonextreme observations will be squeezed into a few classes, obscuring the distribution of these values.

To avoid this problem, we can ignore the outlier (or outliers) while constructing the frequency distribution. Then we can either describe them in a footnote or, more commonly, put them in an *open-ended class*, a class with only one class limit. For example, "1,500 or more radios" labels an open-ended class.

■ **Example 2-6 (continued):** Form a frequency distribution for the data on radios per 1,000 people in Table 2-6 using an open-ended class for the value 2,101. Aim for 10 classes, ignoring the 2,101.

■ **Solution:** Range: $1{,}274 - 14 = 1{,}260$ radios; approximate class size: $1{,}260/10 = 126.0$, so we will try 150 radios. Zero appears to be a good starting point. The result is as follows.

Radios per 1000 People	Number of Nations
0–149	50
150–299	38
300–449	14
450–599	8
600–749	4
750–899	8
900–1,049	4
1,050–1,199	0
1,200–1,349	1
1,350 or more	1
	128

Although a few classes with large frequencies still dominate, we do know more about the distribution of the radio data after reconstructing the table. To learn more about the distribution of the smaller values, it might be worthwhile to consider the observation 1,274 as an outlier (or even more of the larger values) and reconstruct the distribution again. The new open-ended class would have a frequency of 2.

Section 2.1 Problems (Continued)

Use the suggestions in this section to construct a frequency distribution with the specified number of classes for the four variables in Problems 4–7. Then, compare your result with the frequency distribution displayed in the cited problem. Tell whether your new frequency distribution reveals a pattern that is different from or similar to the earlier frequency distribution.

4. Peak ozone levels (in ppm) for the 63 MSAs from the Air Quality Data Base in Appendix B. Nine classes. Compare to Problem 1 (p. 24).
5. Peak particulate matter (PM10) levels (in $\mu g/m^3$) from the Air Quality Data Base in Appendix B. Five classes. Compare to Problem 2 (p. 25).
6. SAT scores for states where 13% or less of the high school seniors take the test, from the State SAT Data Base in Appendix C. Five classes. Compare to Problem 3 (p. 25).
7. SAT scores for states where 52% or more of the high school seniors take the test from the State SAT Data Base in Appendix C. Five classes. Compare to Problem 3 (p. 25).
8. Consider each frequency distribution in Problems 1–7. Are there any that would be better represented with an open-ended class? If so, state which ones and then reconstruct each one with an open-ended class.

2.2 Relative Frequency Distributions

Sometimes the frequencies listed in a frequency distribution may be difficult to interpret. Is a frequency of 160 large or small? In a set of 200 observations, 160 values falling in one class is a significant fraction of the data set. On the other hand, in a set of one million values, 160 is a very small fraction. To determine whether the frequency is relatively large or relatively small, we compare the frequency with the total size of the data set.

We use a special kind of frequency distribution to summarize such comparisons. A *relative frequency distribution* reports the proportion of the total observations that fall in each class, rather than the absolute frequency. This proportion is found by dividing the number of observations in the class by the total number of observations in the data set:

$$\text{Relative class frequency} = \frac{\text{Class frequency}}{\text{Total number of observations in data set}}.$$

Sometimes, the relative proportion is expressed as a percentage by simply multiplying the above proportion by 100.

When we form a proportion from two numbers, we may lose information if we round only one place past the decimal. For instance, a class that contains observations, but whose relative frequency is smaller than 0.05, would be rounded to 0.0, making it appear that the class contains no observations. To avoid this, we will *keep all places past the decimal or round four places past the decimal (whichever produces fewer places past the decimal) when converting two values to a proportion.*

■ **Example 2-1 (continued):** Consider the frequency distribution of truckload values for the river cleanup data displayed in Table 2-2 (p. 23). Transform this frequency distribution to a relative frequency distribution. Show the proportions and percentages.

Problems

■ **Solution:** The following results include the original classes and absolute frequencies. Dividing these frequencies by 250, the total number of observations, produces the relative frequencies expressed as proportions. Multiplying these proportions by 100 results in the percentages in the last column. Ordinarily, only the classes and one expression of the relative frequencies appear in a relative frequency distribution.

Truckloads	Absolute Frequency Number of Cleanups	Relative Frequency Proportion	Relative Frequency Percentages
1–5	45	0.180	18.0
6–10	26	0.104	10.4
11–15	56	0.224	22.4
16–20	27	0.108	10.8
21–25	58	0.232	23.2
26–30	17	0.068	6.8
31–35	10	0.040	4.0
37–40	11	0.044	4.4
	250		

Question: What is the sum of the relative frequency columns in the table above? Explain.

Answer: Because the frequency distribution (relative or absolute) accounts for all, or 100%, of the values in the data set, the sum of the relative frequencies is 1 when expressed as a proportion and 100 when expressed as a percentage. There may be a small deviation from these totals because of rounding. If the sum is very different from 1, you would be wise to recheck the calculations.

Question: What percent of the cleanup drives collected at least 26 truckloads?

Answer: The classes that include cleanups collecting at least 26 truckloads are 26–30, 31–35, and 35–40. The corresponding relative percentages are 6.8%, 4.0%, and 4.4%, which sum to 15.2%.

■ **Example 2-5 (continued):** Construct a relative frequency distribution of the life expectancies from the frequency distribution in Table 2-5 (p. 34), using the same classes as before. Express the frequencies as proportions.

■ **Solution:** The results from dividing class frequencies by the 131 total observations follow.

Life Expectancy (years)	Proportion of Nations
35–39	0.0153
40–44	0.0458
45–49	0.1221
50–54	0.1603
55–59	0.0611
60–64	0.1374
65–69	0.1145
70–74	0.2214
75–79	0.1221

Box 2-1 shows various frequency distributions, including relative frequencies.

Section 2.2 Problems

9. Make a relative frequency distribution from the distribution of peak ozone levels in Problem 1 (p. 24). Use proportions.

10. Make a relative frequency distribution from the distribution of peak PM10 levels in Problem 2 (p. 25). Use percentages.

Box 2-1 Born Smart or Became Smart?

To provide information about the role of heredity and environment on children's intelligence, a research team in Iowa studied a group of children who were adopted before they were six months old. They collected several types of information about this group at different intervals during a ten-year period. To anticipate possible findings from comparing true and foster parent characteristics, the researchers presented summary information before performing more complex statistical analyses. For example, they employed frequency distributions and relative frequency distributions shown in the table to present education levels of each true and each foster parent. Notice that the classes contain only one value rather than an interval of values.

prospective foster parents. In addition, the study excluded most children placed in the homes of relatives or other homes through private sources. The researchers suggested that such children came from better educated parents who were also better off economically and socially. The researchers further concluded that foster parents' education levels tended to be above the average for their age and region, whereas the true parents' levels tended to be below average.

Further analysis demonstrated no evidence of an increase in intelligence when the education level of the adopted parents increased. However, the IQ levels tended to be high and very similar among the children, as shown in the frequency distributions of two sets of IQ measures (see table). This similarity made it difficult to demonstrate that a change in the level of any variable had an effect on intelligence.

Distribution of True and Foster Parent Education

School Attainment (years)	True Fathers No.	True Fathers %	True Mothers No.	True Mothers %	Foster Fathers No.	Foster Fathers %	Foster Mothers No.	Foster Mothers %
20	—	—	—	—	3	3.0	—	—
19	—	—	—	—	—	—	1	1.0
18	—	—	—	—	6	6.0	—	—
17	—	—	—	—	2	2.0	1	1.0
16	2	3.4	—	—	10	10.0	14	14.0
15	1	1.7	—	—	7	7.0	9	9.0
14	2	3.4	2	2.2	2	2.0	10	10.0
13	3	5.0	6	6.5	14	14.0	16	16.0
12	15	25.4	24	26.0	16	16.0	18	18.0
11	7	11.8	9	9.8	3	3.0	5	5.0
10	3	5.0	7	7.6	6	6.0	8	8.0
9	3	5.0	9	9.8	5	5.0	3	3.0
8	13	22.0	23	25.0	23	23.0	9	9.0
7	5	8.5	8	8.7	2	2.0	4	4.0
6	2	3.4	2	2.2	—	—	1	1.0
5	2	3.4	—	—	—	—	—	—
Less than 4	—	—	2	2.2	1	1.0	1	1.0

IQ Score	Number of Children	
	1916 Stanford-Binet	1937 Stanford-Binet
150–154	—	1
145–149	—	1
140–144	1	3
135–139	—	7
130–134	2	12
125–129	7	6
120–124	11	14
115–119	12	13
110–114	12	14
105–109	15	10
100–104	13	9
95–99	10	2
90–94	5	3
85–89	5	—
80–84	2	3
75–79	2	1
70–74	2	1
65–69	1	—

The distributions indicated that foster parents are better educated than true parents. This result did not surprise the researchers for two reasons. First, adoption agency officials employed a very selective process for choosing among

Source: Marie Skodak and Harold M. Skeels, "A Final Follow-Up Study of One Hundred Adopted Children," Journal of Genetic Psychology, Vol. 75, 1949, pp. 85–125.

11. Make a relative frequency distribution from the distribution of SAT scores in Problem 3 (p. 25) for states where 13% or less of the high school seniors take the SAT test. Use proportions.

12. Make a relative frequency distribution from the distribution of SAT scores in Problem 3 (p. 25) for states where 52% or more of the high school seniors take the SAT test. Use percentages.

2.3 Cumulative Frequency Distributions

A *cumulative frequency distribution* reports accumulated frequencies of data values in a class and all preceding (or all following) classes, rather than the individual class frequencies. The result quickly answers questions about how many observations in the

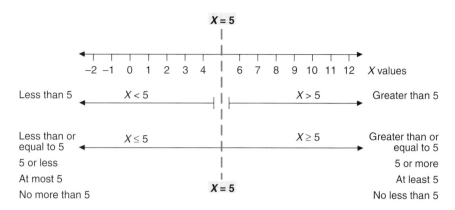

Figure 2-4

Relational expressions: The relation of a set of X values to a specified value (5)

data set are smaller (or larger) than certain values. Figure 2-4 demonstrates phrasing and relational expressions for cumulative distributions.

■ **Example 2-5 (continued):** Using the classes shown in Table 2–5 (p. 34), construct a cumulative "more than" frequency distribution of life expectancies.

■ **Solution:** Figure 2-5 illustrates the construction of the cumulative "more than" distribution from the regular frequency distribution. Two steps are necessary.

Step 1: To construct a "more than" distribution from a frequency distribution, we must use the upper class limits, because we know how many observations are more than this value. We do not know how many observations are more than a lower class limit, because the values in the class can lie anywhere in the interval. For example, the 16 observations in the last class are "more than" 74 years, but we do not know how many are more than 75 years. Starting with raw data avoids this consideration.

Step 2: Next, we calculate the accumulated frequencies, that is, the number of observations more than specified values. To do this for a particular upper class limit, we sum the frequencies from all classes including larger data values as shown in Figure 2-5. Forty-five observations, 16 in the last class and 29 in the second-to-last class, are all "more than" 69 years. All 131 values must be "more than" 34 years, not 35 years. For clarity we add a last class to show that no observations are more than 79 years.

■ **Example 2-1 (continued):** The sponsors of the river cleanup want a table to show how often they collected a certain number or less of truckloads to determine how often the number of trucks available for a given cleanup would have sufficed in the past. Using the classes from the river cleanup frequency distribution shown in Table 2-2 (p. 23), construct the cumulative "or less" distribution they require.

■ **Solution:** The original frequency distribution is shown in Table 2-7 as well as the cumulative distribution. Remember, a number that is "X or less" can be X or any value smaller than X, so we must use the upper class limits as the specified values.

Notice that the difference between two cumulative frequencies is the frequency of observations between the two data values. For instance, the difference between the 154 and 71 in the truckload cumulative frequency distribution, 83, is the number of cleanups that collected more than 10 but not more than 20 truckloads.

It is possible to report a cumulative frequency distribution by using relative frequencies rather than absolute frequencies. The choice of presentation for a particular data set will depend on the purpose of constructing the frequency distribution or the question it needs to answer.

Figure 2-5
Construction of a cumulative "more than" frequency distribution from a regular frequency distribution

Regular Frequency Distribution

Life expectancy classes (years)	Number of nations		Number of nations with life expectancies longer than stated number of years
35-39	2		131
40-44	6		129
45-49	16		123
50-54	21		107
55-59	8		86
60-64	18		78
65-69	15		60
70-74	29		45
75-79	16		16
>79	0		0
TOTAL	131		

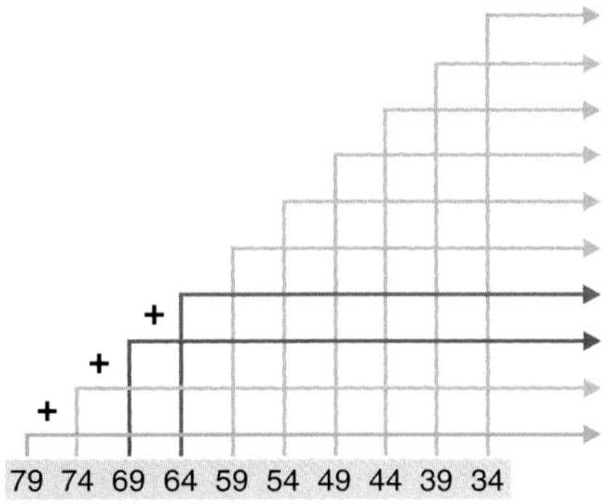

Cumulative Frequency Distribution

Life expectancies (years)	Number of nations with life expectancies longer than stated number of years
34	131
39	129
44	123
49	107
54	86
59	78
64	60
69	45
74	16
79	0

Table 2-7

Original Frequency Distribution		Cumulative Frequency Distribution	
Truckloads	Number of Cleanups	Truckloads	Number of Cleanups That Collect Stated Number of Truckloads "or Less"
1–5	45	5	45
6–10	26	10	71
11–15	56	15	127
16–20	27	20	154
21–25	58	25	212
26–30	17	30	229
31–35	10	35	239
36–40	11	40	250
	250		

Section 2.3 Problems

13. Make a cumulative "more than" frequency distribution from the distribution of peak ozone levels in Problem 1 (p. 24).
14. Make a cumulative "at least" frequency distribution from the distribution of peak PM10 levels in Problem 2 (p. 25).
15. Make a cumulative "less than" frequency distribution from the distribution of SAT scores for states where 13% "or less" of the high school seniors take the SAT test in Problem 3 (p. 25).
16. Make a cumulative relative "or less" frequency distribution from the distribution of SAT scores for states where 52% "or more" of the high school seniors take the SAT test in Problem 12 (p. 38). Use percentages.

Summary and Review

A useful technique in descriptive statistics is a tabular display of the values in a data set. A *frequency distribution* is a table that shows the number of observations from a data set that fall in different classes of values. A *relative frequency distribution* shows the proportion or percentage of the total observations in the data set that occur in each class. A *cumulative frequency distribution* shows the number, proportion, or percentage of the values in the data set that are larger or smaller than certain values.

All three forms of presentation are useful for condensing data so that not every observation has to be read. They also depict the distribution or pattern of the data set. However, the condensing process causes us to lose information, because we no longer know the exact value of each observation. In addition, the same data set may exhibit a different pattern for a different frequency distribution, which can result in a misrepresentation of the true nature of the data set's distribution and, in extreme cases, in abuse of the technique. Constructing a frequency distribution forces us to notice common and uncommon values in the data set. One type of uncommon value is called an *outlier*, a value that is extremely different from the other values in the data set. It is important to be aware of these benefits and complications when we construct or read a frequency distribution.

Multiple-Choice Problems

17. Which of the following is not true about frequency distributions?
    a. The data are organized rather than randomly recorded.
    b. The data are condensed into a unique description.
    c. The distributions show the number of observations contained in different classes of values.
    d. Frequency distributions fall under the category of descriptive statistics.
    e. The distributions can misrepresent the distribution of the entire data set.

For Problems 18–22, consider the following frequency distribution of daily high temperatures for the first 50 days of the year in a mountain community.

Degrees Fahrenheit	Days
27–29	15
30–32	26
33–35	3
36–38	4
39–41	2

18. How many days do the temperatures rise above the freezing point (32°)?
    a. 41.
    b. 35.
    c. 9.
    d. 26.
    e. Cannot be determined from preceding information.
19. How many days do the temperatures fall below the freezing point?
    a. 9.
    b. 15.
    c. 41.
    d. 35.
    e. Cannot be determined from the preceding information.
20. What proportion of the temperatures are between 27 and 29°?
    a. 15.
    b. 0.15.
    c. 0.286.
    d. 30.
    e. 0.30.

21. A relative frequency distribution
  a. Displays the proportion of values that fall in each class.
  b. Highlights the largest value in the data set.
  c. Highlights the smallest value in the data set.
  d. Corrects for a regular frequency distribution that misrepresents the pattern of the data.
  e. Eliminates any problem caused by outliers.
22. A cumulative frequency distribution constructed from this frequency distribution could relate that
  a. Nine values are at least 32°.
  b. 41 values are less than 32°.
  c. Nine values are more than 33°.
  d. Nine values are at least 33°.
  e. 41 values are at most 33°.
23. What is wrong with the following frequency distribution of number of votes for a gubernatorial candidate in different precincts?

Votes	Number of Precincts
000–100	22
100–200	18
200–300	35
300–400	6
400–500	10
500–600	30

  a. There is ambiguity about which class contains certain observations.
  b. There are too few classes.
  c. There are too many classes.
  d. The class sizes are unequal.
  e. The class size is not a multiple of 5.

Use the following frequency distribution of the number of bookkeeping errors in a company's ledger to answer Questions 24–28.

Error (dollars)	Number of Errors
0.01–1.00	1
1.01–2.00	0
2.01–3.00	1
3.01–4.00	2
4.01–5.00	1
5.01–6.00	1
6.01–7.00	2
7.01–8.00	1
8.01–9.00	3
9.01–10.00	2
10.01–11.00	1

24. What is wrong with the frequency distribution?
  a. There is ambiguity about which class contains certain observations.
  b. There are too few classes.
  c. There are too many classes.
  d. The class sizes are unequal.
  e. The class size is not a multiple of 5.

25. What are the class limits of the first class?
  a. 0.00 and 1.995.   **d.** 0.01 and 1.00.
  b. 0.00 and 1.00.   **e.** 0.005 and 1.005.
  c. 0.00 and 0.99.
26. What are the class boundaries of the first class?
  a. 0.00 and 1.995.   **d.** 0.01 and 1.00.
  b. 0.00 and 1.00.   **e.** 0.005 and 1.005.
  c. 0.00 and 0.99.
27. What is the midpoint of the first class?
  a. 0.495.   **d.** 0.5.
  b. 1.005.   **e.** 0.505.
  c. 0.55.
28. What is the size of the first class?
  a. 0.99.   **d.** 1.05.
  b. 1.00.   **e.** 0.55.
  c. 1.01.
29. A data set contains the number of movements of subjects' hands during a monitored period of sleep. The values lie between 36 and 324 motions. A frequency distribution of these values with 12 classes is desired. Using the suggestions in this unit, recommend a class size for a frequency distribution to be used at a medical convention.
  a. 24.   **d.** 2.
  b. 25.   **e.** 10.55.
  c. 12.
30. Suppose you need to make a frequency distribution from some daily sales receipts. In which of the following cases would you most likely make a frequency distribution with 4–10 classes?
  a. The reader wants to know how many days the receipts were above the break-even point, the point where receipts equal costs.
  b. There are 10 observations in the data set.
  c. The observations vary over a wide range.
  d. The sales receipts are identical over the period under consideration.
  e. The reader needs the distribution of receipts according to 12 local tax rate categories that are based on different receipt values.
31. Which of the following statements is true when constructing frequency distributions?
  a. A small range of values can still encompass a large data set.
  b. Use fewer classes to accommodate an outlier.
  c. Suggest more classes as the range of a data set increases.
  d. An outlier does not affect the range of a data set.
  e. An outlier that is a small value is a good starting point for the first class.
32. Which of the following is an outlier?
  a. An annual income of $50,000.
  b. A textbook with 400 pages.
  c. An Olympic-size swimming pool.
  d. A tomato that is 10 inches in diameter.
  e. A standard horsepower rating for an automobile engine.

Word and Thought Problems

33. A change in the pattern depicted by a frequency distribution will *not* likely result when which of the following actions occurs?
   a. Decrease class size.
   b. Increase number of classes.
   c. Extend the accuracy or number of places past the decimal of the class limits.
   d. Decrease the lower limit of the first class.
   e. Separate overlapping class limits.

34. Because the suggestions in this unit are not rules
   a. There is not a unique frequency distribution associated with a particular data set.
   b. Equal-sized classes are necessary to answer any question that might arise about the data set.
   c. Readers must be alert for distortions from a relative frequency distribution that would not occur with a regular or absolute frequency distribution.
   d. Frequency distributions cannot be used to describe data sets with outliers.
   e. Cumulative frequency distributions must be used for data sets with outliers.

Use the following frequency distribution to answer Questions 35 and 36.

Test Scores	Number of Students
90–100	8
80–89	12
70–79	11
60–69	9
Below 60	10

35. This frequency distribution exhibits the following characteristic:
   a. Relative frequencies.   **d.** Open-ended class.
   b. Cumulative frequencies.   **e.** Equal-size classes.
   c. Outlier.

36. Describe the distribution of the data.
   a. Scores are either very good or very bad.
   b. There are no very high scores and no very low scores.
   c. The number of scores is uniformly distributed over the five possible grades.
   d. Almost everyone made the same score.
   e. There are lots of good scores and only a few outliers.

37. Determine the number of students in a freshman biology course who are "more than" 20 years of age but not "more than" 25.

Age (years)	Number of Students Who Are "at Most" Stated Number of Years
20	49
25	78
30	100
35	110
40	110
45	118
50	120
55	124
60	125

   a. 78 students.   **d.** 22 students.
   b. 5 students.   **e.** 127 students.
   c. 29 students.

38. Which suggestions listed in Unit I would you follow to construct a frequency distribution of grades on a math test at your school? Explain why or why not for each suggestion.

39. The following frequency distribution describes the pattern of estimated ages of different fossils found in a certain region. The values are expressed in millions of years.

Fossil Age (millions of years)	Number of Fossils
0.50–0.99	20
1.00–1.49	15
1.50–1.99	25
2.00–2.49	43
2.50–2.99	30
3.00–3.49	96
3.50–3.99	125
4.00–4.49	215
4.50–4.99	198
	767

Word and Thought Problems

   a. What are the class limits of the first class? Boundaries? Midpoint?
   b. How many fossils are less than two million years old?
   c. How many fossils are more than two million years old?
   d. Make a relative frequency distribution from this information. Express the frequencies as proportions.
   e. Make a cumulative "at least" frequency distribution and cumulative relative "at least" frequency distribution from this information.
   f. Can we do a cumulative "more than" distribution starting from this frequency distribution? If yes, state the values in the Fossil Age column. If no, explain why not.

40. What are some advantages and disadvantages of using a frequency distribution to describe a set of data?

41. A large chemistry class is measuring the speed of an enzymatic reaction in the lab. All of the students get their chemicals and materials from the same source; however, measurements differ slightly for a number of reasons, such as variation in instruments, stopwatches, and students' technical acumen. Theoretically, the reaction should occur after 2.67 seconds, but the students with little experience at this point are slow to recognize it. Two students each form a frequency distribution from the resulting data set of reaction times. Do each of the following frequency distributions provide the same perception of the distribution of the values? Explain. If your answer is no, which distribution do you think most aptly describes the original data's distribution? Explain.

Reaction Times (seconds)	Number of Students	Reaction Times (seconds)	Number of Students
2.75–2.99	11	2.00–2.99	11
3.00–3.24	6	3.00–3.99	13
3.25–3.49	2	4.00–4.99	14
3.50–3.74	3	5.00–5.99	7
3.75–3.99	2		
4.00–4.24	5		
4.25–4.49	4		
4.50–4.74	2		
4.75–4.99	3		
5.00–5.24	3		
5.25–5.49	3		
5.50–5.74	1		

42. A count of sentences that begin with prepositional phrases for a set of randomly selected theses from the English Department contains observations between 124 and 298 sentences. If nine classes are desired,
   a. What should the class size be (follow suggestions)?
   b. Suggest a starting point.
   c. What are the class limits of the first class? Boundaries? Midpoint?

43. When hauled aboard a fishing boat, a net is found to contain a 25-pound lobster, which is unusually heavy for a lobster. What term defined in this unit applies to the weight of this lobster as compared with the weights of most lobsters?

44. A psychology experiment includes a record of the IQ scores of the participants. Use the frequency distribution of these values to answer the following questions.

IQ Score	Number of Participants
80–89	2
90–99	12
100–109	36
110–119	112
120–129	40
130–139	9
140–149	1

   a. Describe the distribution of the data set.
   b. How many participants had IQ scores below 100?
   c. What is the greatest possible difference between the highest and lowest IQ score? Can you determine the exact difference between these two values?

45. A first-aid instructor uses the following procedure to see whether students remember not to move a victim before checking for injuries. Each student enters a testing area alone and finds a person, acting as a victim, calling for help. Many students respond to the victim's pleas by helping her stand before they check her for injuries. The exercise dramatizes the point of not moving a victim until she has been examined. The instructor has kept records of the percentage of students in each class who moved the victim without checking for injuries. The results are presented in the following frequency distribution.

Percentage Who Moved the Victim	Number of Classes
1–20	15
21–40	9
41–60	7
61–80	0
81–100	1
	32

   a. Describe the distribution of the data set.
   b. Describe how to construct a frequency distribution that retains more detail about the actual values in the data set.

46. Grades on a test have the following frequency distribution:

Grades	Number of Students
90–100	34
80–89	42
70–79	18
60–69	4
below 60	1

   a. Why do you think the last class is labeled "below 60"? What do we call this kind of class?
   b. Are the other classes all the same size? If so, do you think they are adequate for displaying the grade distribution? If not, is this a reasonable choice for class sizes, even though it does not follow the suggestions given in this unit?
   c. Judging from the distribution, is there reason to request a curve of grades from the instructor? Does the test seem to have been "too hard" or "too easy"? Justify your statements.

47. The values listed below are average 1982 hourly earnings (in dollars) of production workers for the 50 states and the District of Columbia.

Average 1982 Hourly Earnings for Production Workers ($)

7.22	6.95	7.35	7.58	6.61	8.23	8.35
8.60	8.63	10.07	9.79	9.31	11.18	9.37
9.11	10.00	8.46	7.50	7.36	8.43	8.77
8.64	8.78	9.37	7.36	9.40	6.35	6.68
6.75	7.02	8.38	7.16	7.33	6.41	6.69
9.38	8.69	8.60	9.85	8.62	8.53	8.63
7.21	8.73	8.40	8.80	11.23	10.02	9.24
11.74	7.96					

Source: *Statistical Abstract of the U.S. 1984 Census*, 104th ed. (Washington, D.C.: Bureau of the Census, 1983), p. 433.

a. Construct a frequency distribution for the data. Aim for eight classes.
b. Form a relative frequency distribution. Use proportions.
c. The smallest one-third (approximately) of the average hourly earnings is at most how many dollars?
d. There appear to be three outliers. What are they?
e. Construct a frequency distribution for the nonextreme hourly earnings. Use an open-ended class for the outliers. Again, aim for eight closed-ended classes.

48. What do you think will be the consequence of not following each of the suggestions for constructing a frequency distribution discussed in this section?

49. Construct a frequency distribution of radios per 1,000 people with an open-ended class. Use the information from Example 2-6 (p. 34), allowing for 1,274 and 2,101 as outliers and aim for 10 closed-ended classes.

50. Average prices of single bottles of Château Mouton-Rothschild wine of different vintages, which were auctioned during January to November of 1990, are listed. The producer's yearly change of artist for the label is an added investment attraction of this wine. Display these data using a frequency distribution with five classes. Discuss the result and possible ways to improve the display.

Auction Prices of Single Bottles

377	879	53	87	99	63	59
331	77	250	60	75	74	38
451	548	33	75	41	60	76
173	179	150	66	84	152	86

Source: Thomas Matthews, "Labels Key to Pricing Mouton," *The Wine Spectator*, February 28, 1991, p. 15.

51. The following data set lists 68 randomly selected effective annual yields (in percents) of tax-exempt money market mutual funds. (Source: *Wall Street Journal*, Thursday, July 11, 1991, p. C21.)

1.82	2.94	2.97	2.99	3.00	3.01	3.06
3.08	3.11	3.11	3.12	3.14	3.24	3.27
3.30	3.32	3.32	3.33	3.34	3.35	3.37
3.38	3.39	3.40	3.40	3.42	3.43	3.43
3.44	3.46	3.46	3.47	3.49	3.50	3.50
3.53	3.56	3.56	3.59	3.61	3.65	3.66
3.69	3.69	3.71	3.73	3.74	3.75	3.76
3.76	3.76	3.77	3.80	3.82	3.82	3.90
3.91	3.95	3.95	3.97	3.97	3.98	4.01
4.09	4.15	4.18	4.21	4.30		

a. Construct a frequency distribution with nine classes.
b. Convert the frequency distribution in Part a to a relative frequency distribution. Use proportions.
c. Convert the frequency distribution in Part a to a cumulative "or more" distribution.
d. Identify any outliers in the data. If there are outliers, then reconstruct the frequency distribution using an open-ended class or classes.

52. The number of people in an experiment who could recollect traveling between two points when presented with 21 pairs of familiar points are listed. Make a frequency distribution with four classes from these values. (Source: Tommy Gärling, Anders Böök, Erik Lindberg, and Constantino Arce, "Evidence of a Response-Bias Explanation of Noneuclidean Cognitive Maps," *Professional Geographer*, Vol. 43, No. 2, 1991, pp. 143–49.)

12	9	4	10	3	13	9
2	7	18	10	2	10	8
12	6	14	3	1	10	6

53. The daily per capita calorie supply as a percentage of requirements for 125 nations is listed below.
a. Construct a frequency distribution with nine classes and a corresponding cumulative "less than" distribution.
b. Given that the values represent percentages of requirements, could two classes provide an interesting presentation of this information? Explain and describe the two classes. (Source: United Nations Children's Fund, *The State of the World's Children, 1988* (New York: Oxford University Press, 1988), pp. 66–67.)

92	69	85	95	94	85	91	68	87	86	97
79	105	92	109	97	103	87	85	93	93	88
99	78	94	93	99	88	92	79	124	109	93
96	96	118	89	97	94	102	78	100	85	127
84	152	108	109	108	87	84	95	119	99	132
118	105	125	118	95	97	111	88	79	107	117
91	110	101	126	111	129	127	117	117	101	102
118	111	114	95	122	110	98	126	117	103	118
127	134	128	102	126	112	118	124	146	135	126
127	145	143	119	131	140	130	139	143	143	114
133	140	130	129	114	119	142	130	129	106	128
126	114	111	114							

54. The atomic weights for 103 elements are listed below to the nearest integer. Construct a frequency distribution and a relative frequency distribution with percentages. Aim for 14 classes. (Source: "Standard Atomic Weights Abridged to Five Significant Figures," *Pure and Applied Chemistry: Official Journal of the International Union of Pure and Applied Chemistry*, Vol. 63, No. 7, July 1991, pp. 987–88, Table 6.)

1	4	7	9	11	12	14	16	19	20	23
24	27	28	31	32	35	39	40	40	45	48
51	52	55	56	59	59	64	65	70	73	75
79	80	84	85	88	89	91	93	96	99	101
103	106	108	112	115	119	122	128	127	131	133
137	139	140	141	144	147	150	152	157	159	162
165	167	169	173	175	178	181	184	186	190	192
195	197	201	204	207	209	210	210	222	223	226
227	232	231	238	237	239	241	244	249	252	252
257	258	259	262							

55. The milligrams of sodium per three tablespoons of peanut butter for different varieties, types, and brands

are listed. (Source: "The Nuttiest Peanut Butter," *Consumer Reports*, September, 1990, p. 590.)
  a. Form a frequency distribution and cumulative "at least" frequency distribution with seven classes.
  b. Reform the frequency distribution using an open-ended class.

220	15	0	0	168	225	165	240	225	187	225
3	225	15	225	255	225	225	225	15	162	211
0	0	195	165	188	195	255	225	180	208	3
225	225	210	195							

56. Salaries (in hundred-thousand-dollar units) of selected well-known hitters in professional baseball are listed, followed by similar data for pitchers.
  a. Form two relative frequency distributions, one for each data set. Aim for eight classes.
  b. Compare the frequency distributions and describe the differences or similarities that you can discern about the two sets of salary data.

Hitters

11.90	19.00	18.04	17.00	16.00	21.61	16.30	20.00	7.75
12.50	13.50	10.85	21.00	7.60	5.50	5.70	11.00	14.00
12.00	12.25	14.67	18.33	18.33	11.90	13.00	21.20	20.00
15.60	12.75	2.93	8.00	22.00	20.00	22.44	2.12	8.75
20.00	21.00	8.90	3.10	3.50	23.40	14.20	11.00	

Pitchers

4.25	2.00	14.50	2.25	12.25	10.25	5.05	23.00	15.67
6.00	5.88	6.63	8.13	8.88	10.68	24.17	13.75	27.67
15.25	6.50	2.25	5.00	0.75	13.00	1.50	7.35	0.95
11.92	19.89	11.50	8.00	3.25	14.50	9.00	18.00	13.00
13.00	5.50	9.50	15.00	9.25	15.50	11.33	3.65	8.75

Source: Lawrence Hadley and Elizabeth Gustafson, "Major League Baseball Salaries: The Impacts of Arbitration and Free Agency," *Journal of Sport Management*, Vol. 5, 1991, pp. 111–27.

57. The number of weeks on the best-seller list is recorded below for each of 20 books listed as the top 10 best-sellers each for fiction and nonfiction during a July week. (Source: "Best Sellers," compiled by Publishers Weekly, reprinted in *Wall Street Journal*, July 11, 1991, p. A8.)

3	19	10	7	10	2	5	4	11	10
9	32	6	4	4	4	2	3	6	7

  a. What is the term we use to describe the value 32?
  b. Make a frequency distribution with four classes, then use an open-ended class, and reform the frequency distribution. Describe the different impressions, if any, that derive from the two presentations.

Unit II: Pictorial Presentations of Frequency Distributions

Preparation-Check Questions

Use the following frequency distribution of the time until maximum effect from different pain remedies to answer Questions 58–61.

Time until Maximum Effect (minutes)	Number of Remedies
1–15	1
16–30	3
31–45	12
46–60	7
61–75	2

58. Most remedies reach their peak effectiveness in how many minutes?
  a. 1–15.   d. 46–60.
  b. 16–30.   e. 61–75.
  c. 31–45.
59. The number of remedies in the data set is
  a. 25.   c. 74.   e. 50.
  b. 75.   d. 12.
60. What is the relative frequency of the second class?
  a. 0.12.   c. 0.28.   e. 0.08.
  b. 0.04.   d. 0.48.

61. Which of the following can be determined from this information?
  a. The number of remedies peaking in 30 or more minutes.
  b. The number of remedies peaking in 31 or less minutes.
  c. The number of remedies peaking in less than 45 minutes.
  d. The number of remedies peaking in more than 45 minutes.
  e. The number of remedies peaking in at least 15 minutes.
62. The purpose of frequency distributions is
  a. To determine the largest value in the data set.
  b. To display the pattern of the values in the data set.
  c. To determine the range of the data set.
  d. To dispose of outliers.
  e. To reduce the description of the data set to a single value.

Answers 58. c; 59. a; 60. a; 61. d; 62. b.

Introduction

River Cleanup (continued) The environmental and recreation group members want to appeal to the local governing body for funding to help haul the collected materials. The agenda for the next board meeting allocates them only a short time to present their case. Obviously, they must present evidence of need: the large number of truckloads they hauled away and recycled during the past 250 Saturdays. However, they do not want to use their limited time presenting all 250 values. They want a visual presentation that clearly shows the need for hauling assistance on previous drives evidenced in the raw data and the frequency distribution.

Often, the information in a frequency distribution can be communicated more effectively if it is presented in a visual form. However, we must proceed cautiously as we execute these graphic displays, because optical illusions can distort the true information. We will examine four types of displays here: frequency polygons, frequency curves, histograms, and ogives. These visual forms do not usually distort the data's underlying pattern, unless they are derived from a frequency distribution that does so.

2.4 Frequency Polygons and Frequency Curves

A *frequency polygon* is a line graph of a frequency distribution. Figure 2-6 pictures the frequency polygon corresponding to the frequency distribution from the river cleanups in Table 2-2 (p. 23). The frequency of each class is plotted above its midpoint. The polygon (a closed, many-sided figure) acquires its shape from a line connecting these consecutive points that represent class frequencies. Notice that we close the figure with two extra points from fictitious classes with zero frequencies at both ends of the frequency distribution, rather than leave the ends of the line graph suspended in the diagram. Thus, the polygon provides a visual impression of the extent or range of the observations, as well as picturing the values that occur often and those that do not.

The following box details the procedure for graphing a frequency polygon from a frequency distribution.

Figure 2-6
The relation of a frequency polygon to a frequency distribution, using data from the river cleanup case

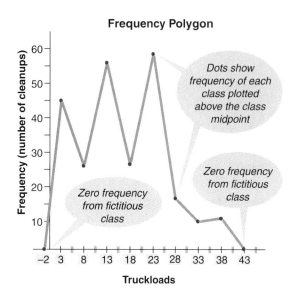

Frequency Distribution

Trucksloads	Number of Cleanups	Class Midpoints
(−4)-0	0	−2
1-5	45	3
6-10	26	8
11-15	56	13
16-20	27	18
21-25	58	23
26-30	17	28
31-35	10	33
36-40	11	38
41-45	0	43

Fictitious classes

Procedure for Graphing a Frequency Polygon

1. Label the vertical axis *frequency* the number of observations in a class. Label the horizontal axis in units corresponding to the actual measurements in the data set.
2. Locate the midpoints of each class on the horizontal axis, including the midpoints of two extra classes of the same size, one at the beginning and one at the end of the existing set of classes.
3. Plot the frequency of each class above its midpoint. The frequency associated with both extra classes is zero.
4. Draw a line to connect the consecutive frequencies.

This procedure applies also when frequencies are replaced with relative frequencies.

■ **Example 2-2 (continued):** Form a frequency polygon for the following frequency distribution from the oil customer example:

Liters	Number of Customers
1–3	9
4–6	9
7–9	3
10–12	9
13–15	9
	39

■ **Solution:** Figure 2-7 displays the result.

This distribution is *symmetric* because the left half and the right half of the diagram are mirror images. If the graph were folded at the middle point (8 liters), the two sides would coincide. We use the term *skewed* to describe a distribution with disproportionately more values in one direction (large or small) than in the opposite direction, such as a data set with an outlier or outliers all lying in the same direction. The diagrams in Figure 2-8 display the two possibilities. Skewness is labeled according to the direction of the tail. These descriptions of distributions occur frequently in real-world data sets, although the data sets are usually neither perfectly symmetric nor perfect replicas of the diagrams in Figure 2-8.

Question: Do the river cleanup data in Figure 2-6 appear to be symmetric or skewed?
Answer: Because the large frequencies correspond to the small truckload values and a small proportion of the truckload values are large, the data are skewed to the right.

The diagrams in Figure 2-8 are *frequency curves* rather than frequency polygons. Think of the curves as frequency polygons that connect a large number of very small classes. The numerous short straight lines appear to be a smooth curve. We often sketch frequency curves to quickly represent and analyze situations visually.

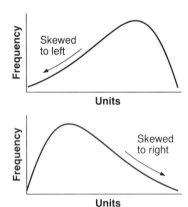

Figure 2-7
The symmetric frequency polygon for Example 2.2 (oil customers)

Figure 2-8
Skewed data sets

Section 2.4 Problems

63. Construct a frequency polygon for the frequency distribution of peak ozone levels in Problem 1 (p. 24). Describe the distribution of the data, using terms from this section, such as skewed and symmetric.

64. Construct a frequency polygon for the frequency distribution of particulate matter (PM10) levels in Problem 2 (p. 25). Describe the distribution of the data.

65. Construct a frequency polygon for the frequency distribution of SAT scores for states where 13% or fewer of high school seniors take the test from Problem 3 (p. 25). Describe the distribution of the data.

66. Construct a frequency polygon for the frequency distribution of SAT scores for states where 52% or more of high school seniors take the test from Problem 3 (p. 25). Describe the distribution of the data, then contrast the diagram and the description with those from Problem 65 for the low-percentage states.

2.5 Histograms

A *histogram* is a bar graph of a frequency distribution. Figure 2-9 displays the relation of a histogram to a frequency distribution, using the river cleanup data from Table 2-2 (p. 23). The height of each bar represents the class's frequency. The base of the bar extends from the class's lower boundary to its upper boundary, although we designate the class's limits below the base so that readers can understand exactly which values are represented by the bar. This alignment of the bases causes the width of each base to equal the class size.

Question: What is the base and height of the bar for the second class in Figure 2-9?
Answer: The base stretches from the boundaries, 5.5 to 10.5 (but is labeled 6-10), and the height is 26.

Notice that the taller bars give the impression of more observations. Actually, the areas of the bars are proportional to the number of observations in the class. For instance, the bar for the third class seems to be about twice the size of the one for the fourth class, because there are 56 versus 27 observations, respectively. This ratio is approximately 2 to 1. Because the bases of all the bars are fixed at five truckloads, which correspond to equal-size classes, the frequency or height determines the area of all the rectangles (area is base times height).

If the classes are not of equal size, base widths will differ and the height of the bar for a different-size class must be adjusted so that all the bar areas remain proportional. If one class is three times the size of all other classes, we should reduce its bar height to one-third of its frequency. If it is four-fifths the standard size, its bar height would be five-fourths of its frequency. This adjustment preserves the relative proportionality of the areas of the different bars. The general formula for this kind of adjustment is

$$\text{Adjusted height} = (1/d) \times \text{Frequency,}$$

Figure 2-9
Histogram for the frequency distribution of riverbank cleanups

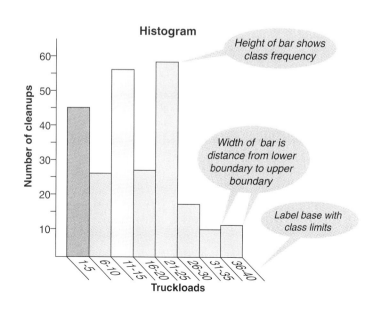

where

$$d = \frac{\text{Different-sized class's size}}{\text{Standard class size}}.$$

■ **Example 2-1 (continued):** The local business with the large truck that hauls the equivalent of four pickup loads volunteers the truck for up to five trips (20 pickup loads) per cleanup. To display this ability, the group combines the first four classes in the frequency distribution shown in Figure 2-9 to obtain:

Truckloads	Number of Cleanups
1–20	154
21–25	58
26–30	17
31–35	10
36–40	11
	250

Construct a histogram of this frequency distribution, and make the necessary adjustments.

■ **Solution:** If we constructed a histogram from these data without making adjustments, the diagram shown in Figure 2-10a results. The visual impression has changed drastically from the histogram in Figure 2-9. The number of smaller cleanups that the sponsors can handle with private trucks appears to dominate the number of large cleanups for which they need funding. The bar areas are no longer proportional to the frequencies. We adjust the height using the general formula, $d = 20/5 = 4$, so the adjusted height of this new class is $(1/4) \times 154 = 38.5$ (the total frequency of the class divided by 4). Now we draw Figure 2-10b. Notice that we print the frequencies at the top of each bar in order to avoid the impression that the frequency of the large bar is smaller than it is.

Figure 2-10

Histograms with unequal class sizes: (a) not adjusted for unequal class sizes; (b) adjusted for unequal class sizes

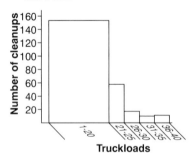

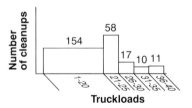

The following box details the procedure for diagramming a frequency distribution with a histogram.

> **Procedure for Constructing a Histogram**
>
> 1. Label the vertical axis frequency or number of observations in a class. Label the horizontal axis in units corresponding to measurements of the original data set.
> 2. Draw the bases of the bars from the class's lower boundary to its upper boundary. Label the bases with the class limits.
> 3. Make the height of each bar equal to the class frequency, unless a class has a size that is d times the size of most of the classes. The height for such a bar must be $(1/d)$ times the frequency, and the frequencies should be labeled for each bar.

■ **Example 2-2 (continued):** Make a histogram for the following frequency distribution of the quantity purchased by oil customers:

Liters	Number of Customers
1–3	9
4–6	9
7–9	3
10–12	9
13–15	9

■ **Solution:** Figure 2-11 displays the result.

Question: Describe the distribution of the data displayed in this histogram.
  Answer: It is symmetric.

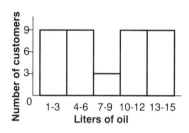

Figure 2-11
Histogram for Example 2.2 (oil customers)

Histograms can picture relative frequency distributions as well. The height of each bar is the class-relative frequency rather than its absolute frequency, so that the only difference between the histograms for a frequency distribution and a corresponding relative frequency distribution is the scale on the vertical axis. This statement holds as well for frequency polygons. Also, a frequency curve results from connecting the bar tops of a histogram with a large number of small-size classes.

Section 2.5 Problems

67. Construct a histogram for the frequency distribution of peak ozone levels in Problem 1 (p. 24).

68. Construct a histogram for the frequency distribution of particulate matter (PM10) levels in Problem 2 (p. 25).

69. Construct a histogram from Problem 3 (p. 25) for the frequency distribution of SAT scores for states where 13% or fewer of high school seniors take the test.

70. Construct a histogram from Problem 3 (p. 25) for the frequency distribution of SAT scores for states where 52% or more of high school seniors take the test.

2.6 Ogives

Histograms, frequency polygons, and frequency curves provide a visual impression of the overall distribution of a data set, but for a quick visual impression of the proportion of a data set that lies on either side of one of a given set of values, an ogive is the appropriate diagram. An *ogive* is the graph of a cumulative frequency distribution. Figure 2-12 shows an ogive that corresponds to the cumulative "more than" frequency distribution for the life expectancies of Figure 2-5 (p. 40).

52 ■ Descriptions of the Distribution of a Data Set Chapter 2

Life expectancy (years)	Number of nations with life expectancies *more than* stated value
34	131
39	129
44	123
49	107
54	86
59	78
64	60
69	45
74	16
79	0

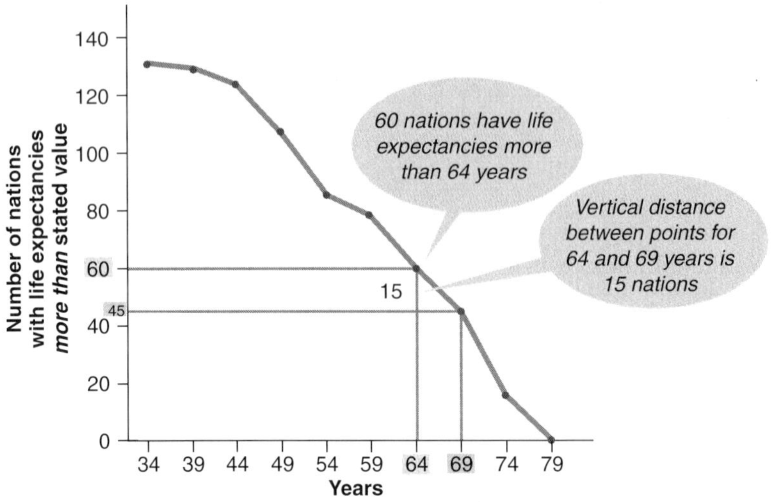

Figure 2-12
Ogive for Example 2.5 (life expectancies)

The horizontal axis measures units of the original data values (years). We label the vertical axis with frequency and a relational statement, such as "number of observations more than...." The height of points indicates accumulated frequencies for the corresponding number on the horizontal axis. For example, 60 nations in Figure 2-12 have life expectancies that are more than 64 years. Finally, we connect these points to form the ogive.

Figure 2-12 shows also that the vertical distance between two points on an ogive is the number of observations between the two values. The 15 nations between the points for "more than 64 years" and "more than 69 years" must be more than 64 and not more than 69 years. These are the 15 nations corresponding to the class 65–69 years in the frequency distribution of the life expectancies listed in Table 2-5 (p. 34).

■ **Example 2-1 (continued):** Construct an ogive corresponding to the cumulative frequency distribution for the river-cleanup data.

Collections, Truckloads	Number of Cleanups That Collect Stated Number of Truckloads "or Less"
5	45
10	71
15	127
20	154
25	212
30	229
35	239
40	250

■ **Solution:** Figure 2-13 shows the result.

Depending on the relational term (Figure 2-4, p. 39, summarizes these expressions), the curve that connects these points will slope upward or downward as one looks left to right on the diagram. Terms that limit a set of values on the upper end, such as "less than" and "at most," will slope upward to the right, because the upper limit increases as we move to the right (see Figure 2-13). More values can be at most 100 than can be at most 5. The opposite slope occurs when the relation limits a set of values on the lower end (see Figure 2-12).

Problems

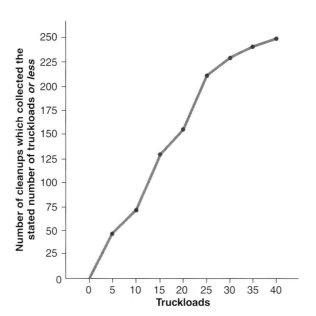

Figure 2-13
Ogive for Example 2.1 (river cleanup data)

Question: Consider the line segments connecting points on an ogive. What is indicated by a steeply sloped line segment? What is indicated by a line segment with a near-level slope?

Answer: A large increase in the cumulative frequency generates a large vertical distance. Consequently, the line segment will be steep. A small increase in the cumulative frequency makes the distance and the slope small. What does it mean if the segment is exactly level?

We must be careful not to interpret the direction of the slope of an ogive as we would a frequency polygon, histogram, or frequency curve. An ogive will always increase or always decrease (or remain level), depending on the relation used for accumulating values. The direction of slope of the other diagrams indicates that larger or smaller data values occur more often, downward when smaller values occur more often and upward when larger values occur more often. These diagrams' slopes may change directions, but the ogive's cannot.

One final note, we can create ogives from cumulative relative frequency distributions as well. As the number of data values distinguished on the horizontal axis becomes very large, the ogive approaches a smooth curve rather than a series of straight line segments.

Section 2.6 Problems

71. Prepare an ogive from the solution for Problem 13 (p. 41), a cumulative "more than" frequency distribution for peak ozone levels. Should a "more than" ogive slope upward to the right, downward to the right, or can we tell before diagramming the data?

72. Prepare an ogive from the solution for Problem 14 (p. 41), a cumulative "at least" frequency distribution for peak PM10 levels. In what direction should an "at least" ogive slope?

73. Prepare an ogive from the solution for Problem 15 (p. 41), a cumulative "less than" frequency distribution of SAT scores for states where 13% or less of the high school seniors take the SAT test. Is there a different slope for "less than" and "at least?"

74. Prepare an ogive from the solution for Problem 16 (p. 41), a cumulative relative "or less" frequency distribution from the distribution of SAT scores for states where 52% or more of the high school seniors take the SAT test.

2.7 Stem-and-Leaf Displays

Another device for revealing the pattern of a data set is a *stem-and-leaf display*. Basically, this device arranges the data in ascending order, but the result is a tabular and visual presentation similar to a frequency distribution and a histogram. The stem-and-leaf display shown here originates from the course registration data listed below.

Table 2-3 (repeated)
Student Registration Data for 33 Courses

92	3	29	3	124	111	53	2	116	79	21
10	6	34	89	22	71	85	101	110	89	118
81	51	92	39	72	63	117	51	34	43	99

Corresponding Stem-and-Leaf Display

Number of Registrants		Frequency
Stem	Leaves	Number of Classes
0	2, 3, 3, 6	4
1	0	1
2	1, 2, 9	3
3	4, 4, 9	3
4	3	1
5	1, 1, 3	3
6	3	1
7	1, 2, 9	3
8	1, 5, 9, 9	4
9	2, 2, 9	3
10	1	1
11	0, 1, 6, 7, 8	5
12	4	1
Total		33

The display separates each observation into two parts, its stem and its leaf. A *stem* is like a class. It is the leading (first) digit or digits of each number. The trailing digits that follow the stem form the *leaf*. For the numbers 21, 22, and 29 in the example, 2 is the stem, and 1, 2, and 9 are the leaves.

Question: List the leaves that correspond to the stems 11 and 0. Then, compose the actual numbers from their stems and leaves.

Answer: The leaves with an "11" stem are 0, 1, 6, 7, and 8, and the leaves that accompany the "0" stem are 2, 3, 3, and 6. We simply write each leaf after the stem to derive the values: 110, 111, 116, 117, 118 and 2, 3, 3, 6.

Question: If the stem for the five values 110, 111, 116, 117, and 118 is 1 (meaning put a 1 in the hundreds position), what are their leaves?

Answer: The leaves would be the trailing values, 10, 11, 16, 17, and 18.

Because the leaves are all the same size, the lengths of the rows that contain the leaves are proportional to the frequency of observations in that class. Thus, visually, we obtain an impression of the distribution of the data just as we do from a histogram. Only this time, the histogram is displayed horizontally, rather than vertically.

Question: Describe the distribution of the data displayed in the stem-and-leaf display for the registration data.

Answer: The data appear to be approximately evenly distributed, because there is no unusually large or small row of leaves (frequency).

Section 2.7 Stem-and-Leaf Displays

Question: How would you recognize the presence of an outlier with a stem-and-leaf display?

Answer: A string of stems with no leaves preceded or followed by a nonempty row signals the presence of an outlier.

To construct an easy-to-read, informative stem-and-leaf display, we begin by observing the data, especially the largest and smallest value, in order to determine the appropriate stems. As with frequency distributions, we do not want a stem-and-leaf display to contain too many or too few stems.

Consider the course registration data again. The values are stated to the nearest ones place, so the stems could include leading digits through the tens position (leaving one-digit leaves) or through the hundreds position (leaving two-digit leaves). If we choose stems stated to the nearest hundreds position, we obtain only two stems, 0 and 1. This disguises the information in the raw data set in the same way that a frequency distribution with only two classes obscures the distribution of the data. Stems stated through the tens position, a better choice, would be those listed in the stem-and-leaf display that we showed.

The final step is to organize the data by aligning the leaf of each value in the row corresponding to its stem. We record a leaf each time a value appears in the data set, so a leaf can be repeated in the same row. They are usually arranged in ascending order, though this is not required.

■ **Example 2-4 (continued):** Table 2-8 lists the maze times (in seconds) for the mice in an experiment. Form a stem-and-leaf display for these data.

■ **Solution:**

Step 1: The largest and smallest values are 62.9 and 350.6, respectively, and each value is expressed to the nearest tenth, so the stem candidates are to the nearest one, ten, or hundred. Using ones would require 289 consecutive stems, a large and awkward set, although there are only 24 values in the data set, none of which repeats the same digits through the ones position (we still list stems with no leaves). We might just as well list the values. Using tens would require 30 stems, still a large and awkward set, many with no leaves. Using four stems at the hundreds position seems the reasonable solution in this situation.

Step 2: The digits past the tens position form each observation's leaf. Locating each leaf in the appropriate row produces

	Maze Times (seconds)	Number of Mice
Stem	**Leaves**	**Frequency**
0	62.9, 72.8, 77.3, 80.2, 85.1, 89.2, 90.0, 95.8	8
1	08.5, 38.7, 59.2, 62.9, 83.2, 99.9	6
2	01.4, 43.3, 44.3, 78.8, 84.6, 87.3	6
3	05.4, 25.0, 42.6, 50.6	4

To provide information more readily about the distribution, we might split each stem to have two rows of leaves and produce the display on the following page. Leaves between 00.0 and 49.9 fall in the first row. The second row lists values between 50.0 and 99.9. Now the skewness of the data is apparent.

Table 2-8
Maze Times for Mice (seconds)

85.1	77.3	72.8	159.2	244.3	89.2	62.9	350.6
199.9	201.4	284.6	325.0	305.4	90.0	342.6	80.2
162.9	243.3	278.8	287.3	108.5	95.8	138.7	183.2

	Maze Times (seconds)	Number of Mice
Stem	Leaves	Frequency
0		0
	62.9, 72.8, 77.3, 80.2, 85.1, 89.2, 90.0, 95.8	8
1	08.5, 38.7,	2
	59.2, 62.9, 83.2, 99.9	4
2	01.4, 43.3, 44.3,	3
	78.8, 84.6, 87.3	3
3	05.4, 25.0, 42.6,	3
	50.6	1

Question: Suppose you construct a frequency distribution from this stem-and-leaf display. The seven resulting classes correspond to the intervals represented by the split stems, and the frequencies correspond to the values in the frequency column. List the class limits and midpoints that would result.

Answer: The digits in the original stems belong in the hundreds position, so the classes would have been 0–99, 100–199, 200–299, and 300–399. Because the stems are split, the necessary number of classes doubles.

Maze Time (seconds)	Midpoint	Number of Mice
50.0–99.9	74.95	8
100.0–149.9	124.95	2
150.0–199.9	174.95	4
200.0–249.9	224.95	3
250.0–299.9	274.95	3
300.0–349.9	324.95	3
350.0–399.9	374.95	1

Question: Compare this frequency distribution with the stem-and-leaf display for the maze times. Which is more informative?

Answer: We derived the frequency distribution from the stem-and-leaf display, so the stem-and-leaf must contain at least as much information. It contains more information, because the actual data values are listed. Although the frequency column is not always included, it is easy to obtain from counting the leaves in each row.

It may be necessary to round values so that the leaves will all be the same size in order to preserve the visual properties of a histogram. Otherwise, the lengths of different rows of leaves will not be proportional to the number of observations represented in the rows.

A stem-and-leaf display is similar to a histogram turned on its side. The length of the string of leaves is like the height of the bar, and relative lengths of these rows gives us a visual impression of the distribution of the data. If we list the frequencies, the stem-and-leaf display synthesizes the frequency distribution and the histogram. It can also be useful in constructing other types of frequency distributions, because it organizes the data. Its chief drawback is that it is cumbersome for very large data sets. One of its chief advantages is that it retains information about actual values while communicating the pattern or distribution of the data. (See Box 2-2.)

Section 2.7 Problems

75. Make a stem-and-leaf display for the peak ozone levels from the Air Quality Data Base (Appendix B).

76. Make a stem-and-leaf display for the peak PM10 levels from the Air Quality Data Base (Appendix B).

77. Make a stem-and-leaf display for the SAT scores in the State SAT Data Base (Appendix C) for states where 13% or fewer of a state's high school seniors took the test.

78. Make a stem-and-leaf display for the SAT scores in the State SAT Data Base (Appendix C) for states where 52% or more of a state's high school seniors took the test.

Box 2-2 Describing the Psychological Effects of Disasters with Statistics

We frequently bandy about many numbers regarding disasters; fatalities, injuries, severity, longevity, damages, and so on. News stories describe qualitative characteristics and impacts as well. Perhaps you know that many victims require psychological help as well as other kinds of aid. Psychologists study and try to document the psychological consequences of such events. Not surprisingly, the results of these studies and estimates of the effects vary.

Important questions about ourselves, our society, our environment, and our universe usually generate multiple research efforts to provide answers. These answers often differ in some degree because of different designs of experiments among social as well as natural scientists. They employ different methods, data, subjects, measures, time periods, and other factors in an attempt to achieve greater understanding of the phenomenon, adding details to our knowledge of complex phenomena and of variations in results from altering some facet of the experiment.

Over time, these research efforts become widespread and the findings numerous. Compiling these findings facilitates and guides further research as well as the public's awareness and understanding of the present state of knowledge about the phenomenon.

Such is the case regarding the question of the psychological impact from a disaster. Two researchers, seeking to update, describe, and analyze previously published quantitative results, collected 31 measures from various experiments. They employ a stem-and-leaf display to present these values.

These statistical measures, whose procedure and interpretation we will discuss in more detail in Chapter 11, measure the psychological effect of a disaster. For now, you only need to understand that the larger this value, the greater the demonstrated psychological effect. A larger positive value indicates that the psychological effect is greatest for victims. If it is negative, the effect is greatest for nonvictims. One of the negative measures occurred when the nondisaster group measured for comparison purposes consisted of rescue workers and persons closely related to victims but not involved in the disaster.

The display follows and shows that most researchers discovered a positive effect on the level of psychological problems from the disaster. The values are skewed to the right with most values concentrated around 0.000. Such skewness can result from what is called the "file-drawer problem," the phenomenon of editors being reluctant to publish weak results or results not confirming accepted theory. The researchers offer evidence that their findings would probably change very little if they were able to investigate all unpublished experiments.

Effect Size Estimates

Stem	Leaves
0.7	49
0.6	—
0.5	16, 33
0.4	61
0.3	28, 38, 96
0.2	08, 43
0.1	10, 41, 52, 77, 83, 88, 89
0.0	00, 00, 00, 00, 03, 04, 29, 54, 60, 69, 73, 89
−0.0	57, 84
−0.1	85

Source: Anthony V. Rubonis and Leonard Bickman, "Psychological Impairment in the Wake of Disaster: The Disaster-Psychopathology Relationship," Psychological Bulletin, Vol. 109, No. 3, 1991, pp. 384–99. Copyright 1991 by the American Psychological Association. Reprinted by permission.

2.8 Visualizing the Distribution of a Data Set by Using Computers

The accessibility of computers makes it easy for organizations as well as individuals to use statistics. Computers can perform complex calculations on very large data sets rapidly. They store, organize, present, and analyze data with many types of tables, displays, and measures. Without computers, complicated operations on large data sets would be impractical.

There are software packages available to perform statistical procedures. They require little or no knowledge of programming, but they require considerably more knowledge of the statistics we want generated. At various points in this book, we will observe and discuss output from Minitab, a commonly used software for learning and performing statistics.

Figure 2-14 displays a histogram and a stem-and-leaf display generated by Minitab from the maze time data set shown in Table 2-8 (p. 55). The histogram output includes

Figure 2-14

Minitab output of maze times for mice

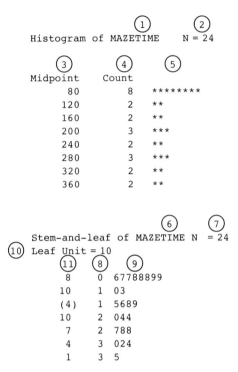

the frequency distribution as well. Although much of the printout is self-explanatory, we need to know some information about Minitab's procedures in order to use the results better.

The circled numbers in the histogram section of the output correspond to the following information:

1. The assigned name for the observations being studied, in this case MAZETIME.
2. The total number of observations in the data set.
3. Midpoint of each class represented by a bar in the histogram. Users can either instruct Minitab about class size and the starting point or use the program's default procedures.
4. Count of observations or frequency associated with each midpoint or class.
5. Bars of the histogram formed horizontally, rather than vertically, with asterisks. In this case, there is one asterisk for each observation. A legend above the histogram details when asterisks symbolize more than one case.

Question: Describe the distribution of the maze times portrayed in this histogram.
Answer: They appear to be skewed to the right.

To form classes from these midpoint displays, recall that the distance between consecutive midpoints (80 to 120, for instance) is the class size. Half of the distance goes on either side of the midpoint to form a class. The class size in Figure 2-14 is 40, so 20 is added and subtracted to each midpoint to form a class. *Minitab resolves the apparent overlap of the class limits by placing any data value on the boundary in the higher-valued class.* Thus, the class limits for this histogram are: 60–99, 100–139, 140–179, 180–219, 220–259, 260–299, 300–339, 340–379.

The circled values in the stem-and-leaf display correspond to the following information:

6. The assigned name for the observations being studied. Again, this is MAZETIME.
7. The total number of observations in the data set.

Problems 59

8. Stems. Notice that Minitab split stems 1, 2, and 3, so that leaves between 0 and 4 fall on the first row for each stem and leaves between 5 and 9 fall on the second row. The user can split the stem by specifying increments for the rows.
9. Leaves. Minitab rounds all values to the same number of positions, so that leaves are always a single digit and not separated by commas.
10. Leaf unit or the position occupied by the leaf digit. In Figure 2-14 the leaf unit is 10. Thus, combining each leaf with its stem forms the value as a number of tens. If we multiply the combination by 10, we obtain the actual value. Consequently, the second row represents the values 100 and 130 seconds. If the leaf unit were tenths, the same stem-leaf combinations (10 and 13) would represent 1.0 and 1.3. Because Minitab only uses one digit for leaves, some trailing digits may be lost, as happens here (see Table 2-8, p. 55). In such a case, the output still provides a quick-and-easy opportunity to get an impression of the data's distribution.
11. *Depth*, a cumulative frequency for observations on the same row and "beyond." To determine the direction of beyond you must first locate the depth value printed in parentheses (the 4 in Figure 2-14). This value reports the number of observations on that specific row and also locates the row containing the middle value of the data set (or middle values when there are two). If there are two middle values that are not both on the same row, parentheses are omitted. Depth values above the middle tell how many observations are in that row and the rows above, and depth values below the middle tell the number of observations on that row and the rows below. The 7 in this column in Figure 2-14 means there are seven observations in the last three rows of the display.

■ **Example 2-6 (continued):** Figure 2-15 shows the histogram and stem-and-leaf display for the number of radios per 1,000 inhabitants (RADIO/C) for different nations. The last stem-and-leaf display in the figure excludes the outlier ($n = 127$). Describe the distribution of the remaining values in this data set.

■ **Solution:** Omitting the outlier removes very little of the skewed appearance of the data. Notice in the histogram that each asterisk represents two observations.

Section 2.8 Problems

79. Figure 2-16 displays a Minitab output of a histogram and stem-and-leaf display of the 63 values for peak carbon monoxide (CO) levels (in ppm) from the Air Quality Data Base (Appendix B). Use this information to answer the following questions:
    a. Describe the general pattern of the data values.
    b. What is the outlier?
    c. All of the leaves in the stem-and-leaf display are zeros. Can this be correct? If yes, explain why. If not, suggest new stems that will produce a more accurate stem-and-leaf display.
80. Figure 2-17 displays a Minitab output of a histogram and stem-and-leaf display of the 63 values for peak lead (PB) levels (in micrograms per cubic meter, UGM) from the Air Quality Data Base (Appendix B). Use this information to answer the following questions:
    a. Describe the general pattern of the data values.
    b. What are the four outliers?
    c. What type of class might be useful for depicting the pattern of this data set?
81. Figure 2-18 displays a Minitab output of a histogram and stem-and-leaf display of the SAT scores for all 50 states (and D.C.) in the State SAT Data Base (Appendix C). Use this information to answer the following questions:
    a. Describe the general distribution of the data set.
    b. Are there outliers? If so, what are they?
    c. The scores for the low-percentage states—those with 13% or fewer high school seniors who take the SAT—are skewed right, whereas the scores for the high-percentage states—those with 52% or more high school seniors who take the SAT—are skewed left. Most values for the first group of states occur in the interval 975–999, whereas most values for the second group occur in the interval 875–904. Relate this information to the shape of the distribution.

Figure 2-15
Minitab output for radio data

```
Histogram of RADIO/C    N = 128
Each * represents 2 obs.

Midpoint    Count
       0       35   ******************
     200       53   ***************************
     400       16   ********
     600       10   *****
     800        8   ****
    1000        4   **
    1200        1   *
    1400        0
    1600        0
    1800        0
    2000        0
    2200        1   *


Stem-and-leaf of RADIO/C    N = 128
Leaf Unit = 100

    62    0   000000000000000000000000000000000001111111111111111111111111111
   (35)   0   22222222222222222222222222333333333
    31    0   4444444555555
    18    0   6666777
    11    0   88888999
     3    1   0
     2    1   2
     1    1
     1    1
     1    1
     1    2   1


Stem-and-leaf of RADTRIM    N = 127
Leaf Unit = 10

    35    0   1122222233333444445556667778888999 9
    62    1   000011112333334666777888899
   (26)   2   001112222222233445556777899
    39    3   022234689
    30    4   0122357
    23    5   678899
    17    6   2455
    13    7   888
    10    8   22667
     5    9   038
     2   10   1
     1   11
     1   12   7
```

Summary and Review

There are several different graphic ways of presenting the distribution of a data set. These display the information in different types of frequency distributions. A *frequency polygon* is a line graph of a frequency distribution. A smoothed frequency polygon is a *frequency curve*. A *histogram* displays a frequency distribution as a bar graph. An *ogive* is a line graph of a cumulative frequency distribution. All give a quick visual impression of the distribution of the data.

A *stem-and-leaf display* gives a visual impression of the data by recording ending digits of each data value (leaf) in a row that corresponds to the beginning digits of the value (stem). Thus, we retain information about the actual values along with the visual impression.

(continued on page 62)

Summary and Review

```
Histogram of CO    N = 63

Midpoint   Count
     4       6    ******
     6      16    ****************
     8      22    **********************
    10       9    *********
    12       6    ******
    14       3    ***
    16       0
    18       1    *


Stem-and-leaf of CO         N  = 63
Leaf Unit = 0.10

      1      3  0
      6      4  00000
     15      5  000000000
     22      6  0000000
    (10)     7  0000000000
     31      8  000000000000
     19      9  000000
     13     10  000
     10     11  000
      7     12  000
      4     13  000
      1     14
      1     15
      1     16
      1     17
      1     18  0
```

Figure 2-16
Minitab output for carbon monoxide levels

```
Histogram of PB    N = 63

Midpoint   Count
    0.0     34    **********************************
    0.2     20    ********************
    0.4      5    *****
    0.6      0
    0.8      0
    1.0      1    *
    1.2      1    *
    1.4      0
    1.6      0
    1.8      1    *
    2.0      0
    2.2      1    *


Stem-and-leaf of PB         N  = 63
Leaf Unit = 0.10

    (50)    0  00000000000000000000000000000000001111111111111111
     13     0  222233
      7     0  444
      4     0
      4     0
      4     0  01
      2     1
      2     1
      2     1  7
      1     1
      1     2
      1     2  2
```

Figure 2-17
Minitab output for lead levels

Figure 2-18
Minitab output for SAT scores

```
Histogram of ALL    N = 51

Midpoint    Count
   840        3    ***
   860        2    **
   880        7    *******
   900        9    *********
   920        5    *****
   940        3    ***
   960        3    ***
   980        3    ***
  1000        8    ********
  1020        2    **
  1040        3    ***
  1060        2    **
  1080        1    *


Stem-and-leaf of ALL      N = 51
Leaf Unit = 10

    1      8   3
    4      8   445
    6      8   67
   16      8   8888889999
   22      9   000001
   (5)     9   22223
   24      9   44
   22      9   6667
   18      9   8899999
   11     10   00011
    6     10   33
    4     10   4
    3     10   66
    1     10   8
```

Computers construct frequency distributions, histograms, stem-and-leaf displays, and other descriptive information easily and quickly. They are especially useful for large data sets and for obtaining a quick visual impression of the data's distribution. It is important to understand the statistics being calculated and constructed in order to recognize and check for problems and to correctly interpret the results. Box 2-3 refers to Florence Nightingale, a historical figure not usually associated with statistics, and to her role in the development of statistical tables and diagrams, in an earlier time when computers were unknown.

Multiple-Choice Problems

82. How would you describe the distribution of the data set pictured below?
   a. Symmetric.
   b. Skewed left.
   c. Skewed right.
   d. Distorted.
   e. Unique.

Dollars

83. What type of visual display is the diagram in Problem 82?

   a. Stem-and-leaf display.
   b. Histogram.
   c. Frequency curve.
   d. Ogive.
   e. Frequency polygon.

Consider the following frequency distribution of daily high temperatures for the first 50 days of the year in a mountain community to answer Questions 84–85.

Degrees Fahrenheit	Days
27–29	15
30–32	26
33–35	3
36–38	4
39–41	2

84. How would you describe the distribution of the data values?

Box 2-3 Florence Nightingale, the Pioneer Statistician

We remember Florence Nightingale mostly for her role in nursing, but she also played a role in the development of statistics and in applying them to solve social problems. Her nursing fame originated in the Crimean War, where she found that many British fatalities occurred in the military hospitals, not in battle, or from noncombat causes. She installed sanitary procedures that drastically reduced mortality rates.

After the war, Nightingale returned to England and proceeded to inform officials and improve medical care and facilities in the military. Military medical administrators tried to prevent investigations and change, but she fought back and overcame them. Statistics was a primary weapon she used to convince government officials that a problem existed and something could be done.

In the process she developed graphical depictions of the situation and influenced the collection of statistics to combat social problems and improve the general welfare of society. The figure below displays the diagram she employed to demonstrate the mortality rates and show the effect of her sanitary regulations on them.

Source: I. Bernard Cohen, "Florence Nightingale," Scientific American, Vol. 250, No. 3, March 1984, pp. 128-33, 136-37. Diagram and caption from same article.

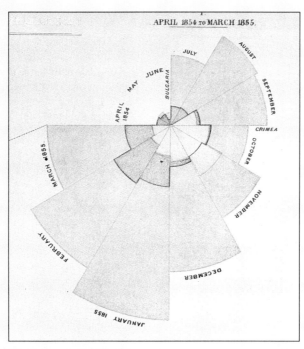

Photo courtesy Tom Pantages

POLAR-AREA DIAGRAM was invented by Florence Nightingale to dramatize the extent of needless deaths in British military hospitals during the Crimean War (1854–56). She called such diagrams "coxcombs." The area of each colored wedge, measured from the center, is proportional to the statistics being represented. Blue wedges represent deaths from "preventable or mitigable zymotic" diseases (contagious diseases such as cholera and typhus), pink wedges deaths from wounds and gray wedges deaths from all other causes. Mortality in the British hospitals peaked in January, 1855, when 2,761 soldiers died of contagious diseases, 83 of wounds and 324 of other causes, for a total of 3,168. Based on the army's average strength of 32,393 in the Crimea that month, Nightingale computed an annual mortality rate of 1,174 per 1,000. The diagram is taken from Nightingale's book *Notes on Matters Affecting the Health, Efficiency and Hospital Administration of the British Army* (1858); half of the diagram, representing the period from April, 1855, to March, 1856, does not appear.

a. Symmetric.   d. Distorted.
b. Skewed left. e. Unique
c. Skewed right.

85. What is the midpoint of the first class?
a. 27.5.   d. 28.5.
b. 27.55. e. 28.495.
c. 28.

86. What is the advantage of a histogram as a display device?
a. Unequal class sizes do not affect the presentation.
b. Data skewed left and data skewed right are visually indistinguishable.
c. Patterns are easily detected, because areas of bars are proportional to class frequencies.
d. Negative values never occur, which reduces the likelihood of computation errors.
e. Outliers are easily distinguishable for any class size.

Use the following stem-and-leaf display of attendance for home basketball games for Questions 87–88.

Home Basketball Game Attendance

Stem	Leaves	Frequency
234	0, 1, 1, 5, 7	5
235	1, 3, 5, 5, 5, 5, 6	7
236	9, 9, 9	3
237		0
238	8	1
239		0
240	9	1
		17

87. Which value in the data set occurs most often?
a. 235.   d. 2,400.
b. 2,355. e. 2,357.
c. 2,369.

88. How would you describe the distribution of the data?
a. Symmetric.   d. Distorted.
b. Skewed left. e. Unique.
c. Skewed right.

89. Which type of display do we use to graph cumulative frequency distributions?
a. Stem-and-leaf display. d. Ogive.
b. Histogram.   e. Frequency polygon.
c. Frequency curve.

90. What does the vertical distance between two consecutive points in an ogive indicate?
a. The number of observations between the two corresponding values on the horizontal axis.
b. The skewness of the data between the two points.
c. The symmetry of the data between the two points.
d. The midpoint of the class.
e. The direction to look for outliers.

Figure 2-19 shows Minitab output for condensing and describing the set of hourly earnings values in Problem 47 of Unit I (p. 44). Use this information to answer Questions 91–93.

91. Without the outliers this data set is approximately
a. Symmetric.   d. Distorted.
b. Skewed left. e. Unique.
c. Skewed right.

92. The number of observations in the class 8.25–8.74 is
a. 51.   d. 21.
b. 14.   e. 12.
c. 16.

93. The values in the stem-and-leaf display are rounded to the nearest
a. Cent.   d. Tens of dollars.
b. Dime.   e. Hundreds of dollars.
c. Dollar.

94. The choice of stems for a stem-and-leaf display is important because
a. The height of the bars will have to be adjusted for unequal stems.
b. Too many or too few stems defeat the purpose of a stem-and-leaf display.
c. There can be no more than 10 stems in the display.
d. Outliers tend to make stems carry more positions rather than fewer and make the leaves inconsequential.
e. The stems must be single digits to accommodate statistical computer packages.

95. The display that preserves most of the information in the original data set is a
a. Stem-and-leaf display. d. Ogive.
b. Histogram.   e. Frequency polygon.
c. Frequency curve.

96. A psychologist wants a diagram to depict a set of 1,000 IQ scores for a public demonstration. She wants to approximate values, such as the 500th and 750th highest score or any other position requested by the audience. Which display will most facilitate the location process?
a. Histogram.   d. Stem-and-leaf.
b. Frequency polygon. e. Frequency curve.
c. Ogive.

97. What is wrong with this statement? "The ogive of fish counts in a public pond is flat, then slopes upward, so the number of fish in the pond must be increasing after a period of stagnant growth."
a. Ogives always exhibit nonincreasing or nondecreasing slopes, no matter what the trend of the fish count is.
b. Ogives cannot show flat segments; they must always fluctuate up and down.
c. Ogives can slope downward but never upward.
d. The slope must decrease during stagnant periods.
e. The conclusion about the distribution of the underlying data requires knowledge of the cumulative frequency distribution as well as its ogive.

98. If a segment of an ogive between 20 and 40 minutes is flat, before sloping downward for values larger than 40, then
a. The largest value in the data set has been exceeded.
b. There are no observations in the data set between 20 and 40 minutes.

Multiple-Choice Problems

```
Histogram of EARNINGS     N = 51

Midpoint    Count
    6.5       5    *****
    7.0       6    ******
    7.5       6    ******
    8.0       2    **
    8.5      14    **************
    9.0       5    *****
    9.5       5    *****
   10.0       5    *****
   10.5       0
   11.0       2    **
   11.5       1    *


Stem-and-leaf of EARNINGS     N = 51

     2     6  34
     7     6  66679
    15     7  01223333
    18     7  559
    24     8  233444
   (12)    8  566666667778
    15     9  1233334
     8     9  78
     6    10  000
     3    10
     3    11  12
     1    11  7
```

Figure 2-19
Minitab output for hourly earnings data

c. The middle value in the data set is between 20 and 40 minutes.
d. The number of observations in the two classes is the same.
e. The relative frequency of the two classes is the same.

99. The frequency of zero for the outcome A in the frequency polygon indicates

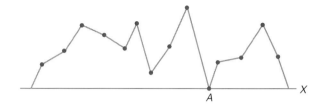

a. An outlier.
b. Skewness.
c. Symmetry.
d. A potential stem with no leaves.
e. An unequal-size bar on the corresponding histogram.

Consider the following histogram of the number of occupied campsites in a national forest campground for 200 nights. Use this information to answer Questions 100–102.

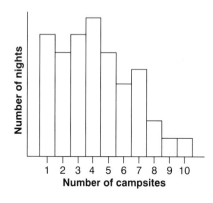

100. What is the most campsites used on any one night?
a. 0.   d. 4.
b. 5.   e. Insufficient information to answer.
c. 10.

101. Which campsite value occurs most often?
a. 0.   d. 4.
b. 5.   e. Insufficient information to answer.
c. 10.

102. Overall, the distribution of occupied campsites is approximately
a. Uniformly distributed.
b. Symmetric.
c. Skewed right.
d. Skewed left.
e. Cyclic, a regular up-and-down or flowing pattern.

Word and Thought Problems

103. Choose one type of diagram presented in Unit II as an answer to each of the following questions and justify your choice.
    a. Which diagram is most informative about the actual data values?
    b. Which diagram is least useful for very large data sets?
    c. Which diagram would you use for a quick sketch of the essence of the data's pattern displayed by a frequency distribution?
104. Suppose you participate in a public forum discussing the hours that students spend studying. Part of the information is a data set of the number of hours that 450 students spent in a monitored study hall during the previous term. Which, if any, of the diagram discussed in this unit would you use in your presentation? Justify your decision.
105. Tell how you can detect an outlier from viewing each of the following diagrams. Then, sketch an example for Parts b and c.
    a. Frequency polygon and frequency curve.
    b. Histogram.
    c. Ogive.
    d. Stem-and-leaf display.
106. Sketch a generalized frequency curve and histogram of data that are
    a. Symmetric.   c. Skewed right.
    b. Skewed left.   d. None of the above.
107. a. Construct a stem-and-leaf display of the following data from a psychology experiment that counted the number of eye blinks during a random spectral display. Describe the distribution of the data set.

    | 57 | 40 | 1 | 76 | 23 | 69 |
    | 8 | 50 | 18 | 22 | 39 | 30 |
    | 23 | 10 | 31 | 2 | 42 | 72 |
    | 24 | 6 | 6 | 53 | 60 | 74 |
    | 24 | 15 | 25 | 54 | 65 | 63 |

    b. Construct the corresponding relative frequency distribution.
    c. Construct a frequency polygon from the relative frequency distribution. Compare the visual result and impression with that achieved by the stem-and-leaf plot.
108. The following frequency distribution, taken from Problem 39 of Unit I, presents estimated ages (in millions of years) of different fossils found in a certain region.
    a. What are the midpoints of the classes?
    b. Sketch the frequency curve corresponding to this frequency distribution. Describe the distribution.
    c. Construct the corresponding frequency polygon.
    d. Construct a frequency polygon from the relative frequency distribution you constructed from this information to answer Problem 39 Part d in Unit I (p. 43). Compare the result and impression produced

Fossil Age (millions of years)	Number of Fossils
0.50–0.99	20
1.00–1.49	15
1.50–1.99	25
2.00–2.49	43
2.50–2.99	30
3.00–3.49	96
3.50–3.99	125
4.00–4.49	215
4.50–4.99	198
	767

   here with those from the frequency polygon of Part c of this problem.
   e. Construct an ogive from the cumulative "at least" frequency distribution you constructed from this information to answer Problem 39 Part e of Unit I.
109. Construct a "more than" ogive corresponding to the following frequency distribution of the percentage of students in a first-aid class who move an injured person too soon (see Problem 45 of Unit I). Describe the effect of the outlier on the ogive.

Percent Who Moved the Victim	Number of Classes
1–20	15
21–40	9
41–60	7
61–80	0
81–100	1
	32

110. The following values are the number of members of the House of Representatives from each state, based on the 1980 census and the 1990 census.

1980 Census

7	6	2	5	11	2	3	6	1	10
1	6	2	7	18	3	34	5	9	8
5	1	22	8	8	2	11	23	27	4
4	19	10	2	5	2	1	2	3	9
45	10	6	8	9	14	21	6	1	1

1990 Census

9	2	6	4	5	7	9	12	13	1
1	2	6	3	9	9	16	6	6	2
2	1	3	1	4	6	19	11	31	2
5	1	30	1	7	20	3	23	2	1
52	3	6	8	5	10	11	21	10	8

   a. Form a stem-and-leaf display for each data set.
   b. Form frequency distributions with classes that correspond to the stems for each data set. What are the midpoints of each class?
   c. Make histograms with bars that correspond to the stems from Part a for each data set.

Word and Thought Problems

d. Draw frequency polygons corresponding to the frequency distributions in Part b.
e. Are these data sets symmetric? Skewed? Different from one another? Explain.

111. Use the following frequency distribution of the 51 hourly earnings values from Problem 47 of Unit I to construct the corresponding frequency polygon and histogram.

Hourly Earnings ($)	Number of States
6.30–6.99	7
7.00–7.69	10
7.70–8.39	4
8.40–9.09	15
9.10–9.79	8
9.80–10.49	4
10.50–11.19	1
11.20–11.89	2
	51

Source: *Statistical Abstract of the U.S. Census, 1984*, 104th ed. (Washington, D.C.: Bureau of the Census, 1983), p. 433.

112. Use the following frequency distribution and cumulative frequency distribution for the daily per capita calorie supply as a percent of requirements from Problem 53 in Unit I to construct the corresponding frequency polygon and ogive, respectively.

Daily per Capita Calorie Supply (% of requirements)	Number of Nations
65–74	2
75–84	7
85–94	24
95–104	22
105–114	21
115–124	16
125–134	22
135–144	8
145–154	3
	125

Percentage of Requirements	Number of Nations with Daily per Capita Calorie Supply "Less than" Stated Value
65	0
75	2
85	9
95	33
105	55
115	76
125	92
135	114
145	122
155	125

Source: United Nations Children's Fund, *The State of the World's Children, 1988* (New York: Oxford University Press, 1988), pp. 66-67.

113. Can the two histograms at the top of the next page originate from the same data set? Explain.

114. Use the following frequency distribution and cumulative frequency distribution for the sodium content in peanut butter from Problem 55 of Unit I to construct the corresponding histogram and ogive, respectively.

Milligrams of Sodium per Three Tablespoons of Peanut Butter	Number of Varieties
0–39	9
40–79	0
80–119	0
120–159	0
160–199	10
200–239	15
240–279	3
	37

Milligrams of Sodium per Three Tablespoons of Peanut Butter	Number of Varieties with at Least Stated Amount of Sodium
0	37
40	28
80	28
120	28
160	28
200	18
240	3
280	0

Source: "The Nuttiest Peanut Butter," *Consumer Reports*, September, 1990, p. 590.

115. Can a symmetric data set contain outliers? Explain.

116. Use the following frequency distribution of the number of weeks on the best-seller list of the 20 best-selling fiction and nonfiction books from Problem 57 in Unit I to construct the corresponding "more than" ogive.

Number of Weeks on Best-Seller List	Number of Books
0–4	8
5–9	6
10–14	4
15–19	1
20 or more	1
	20

Source: "Best Sellers," compiled by *Publishers Weekly*, reprinted in *Wall Street Journal*, July 11, 1991, p. A8.

117. The following data set lists attendance figures (in millions) for the 40 largest fairs in the United States in 1988. (Source: "The Fairest Fairs of All," *U.S. News and World Report*, August 21, 1989, p. 66, using data from International Association of Fairs and Expositions.)

3.30	2.90	1.70	1.60	1.40	1.20	1.20	1.20	1.20	1.10
1.00	1.00	0.98	0.91	0.85	0.83	0.81	0.80	0.79	0.72
0.72	0.72	0.72	0.69	0.69	0.68	0.65	0.64	0.64	0.64
0.63	0.62	0.61	0.60	0.59	0.59	0.58	0.58	0.58	0.57

a. Form a stem-and-leaf display for the data.
b. Based on the result of Part a, use an open-ended

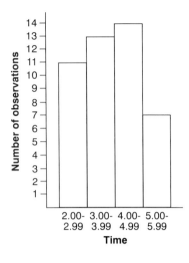
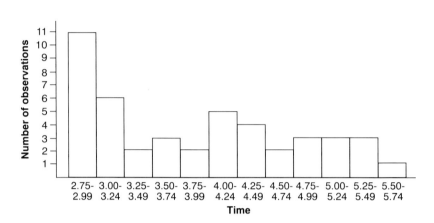

class for the two largest observations in order to construct a frequency distribution that will give a more detailed impression of the distribution of the data set. Aim for five classes.

118. Consider the following frequency polygon of prices for a certain configuration of computer and accessories. Construct the corresponding frequency distribution and relative frequency distribution.

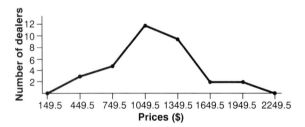

119. Your assistant in the personnel department is supposed to construct a histogram of the distribution of raises. The assistant decides to use Minitab and obtains the result shown in Figure 2-20. Suggest an alternative construction to your assistant that will provide a more accurate impression of the distribution.

120. Suppose you are in charge of a professional bowling tournament. Sketch a frequency curve for the distribution of participants' games scores that you consider an ideal, but realistic, distribution for producing an exciting and interesting tournament. Explain your diagram. (Bowling scores must be between 0 and 300 per game. Higher scores are better.)

121. Figure 2-21 shows Minitab output of a histogram and stem-and-leaf display for 1990 census counts for the 50 states and the District of Columbia. The second stem-and-leaf display is the same as the first, except that the split stems with no observations in the first display are removed. Describe the distribution of the data. Does the second stem-and-leaf display change your description? (Source: "Population Trends and Congressional Apportionment," *1990 Census Profile*, U.S. Department of Commerce, Economics and Statistics Administration, Bureau of the Census, No. 1, March 1991, p. 4.)

122. A summary of studies of the influence of viewing media violence on subsequent viewer aggressive behavior reports the studies' results in the stem-and-leaf display at the bottom of page 69. (The display is split to conserve space.) Each value represents the magnitude of the difference between a group of subjects who view presentations with violence and a control group who do not view the presentation. The values are standardized to avoid variation in differences from using different

Figure 2-20
Minitab output for raise data

```
Histogram of RAISE    N = 42

Midpoint    Count
   1.0        41    *******************************************
   1.5         0
   2.0         0
   2.5         0
   3.0         0
   3.5         0
   4.0         0
   4.5         0
   5.0         0
   5.5         1    *
```

Word and Thought Problems

Figure 2-21
Minitab output of census data for the 50 states and the District of Columbia

```
Histogram of COUNT    N = 51

Midpoint    Count
       0      18   ******************
 4000000      20   ********************
 8000000       6   ******
12000000       4   ****
16000000       2   **
20000000       0
24000000       0
28000000       1   *


Stem-and-leaf of COUNT      N = 51
Leaf Unit = 1000000

    18    0   000000001111111111
   (11)   0   22222333333
    22    0   444444455
    13    0   66667
     8    0   9
     7    1   011
     4    1   2
     3    1
     3    1   67
     1    1
     1    2
     1    2
     1    2
     1    2
     1    2   9


Stem-and-leaf of COUNT      N = 51
Leaf Unit = 1000000

    18    0   000000001111111111
   (11)   0   22222333333
    22    0   444444455
    13    0   66667
     8    0   9
     7    1   011
     4    1   2
     3    1   67
     1    2   9
```

units of measure. Positive values indicate more aggressiveness from those who view violent presentations, and negative values indicate control groups are more aggressive.

a. How many observations are in the data set? List them.

c. The authors describe the distribution as "relatively flat and not skewed in either direction," (p. 375). Do you agree or disagree? Explain.

c. What do your answers to Parts a and b imply about the "file-drawer problem" discussed in Box 2-2?

d. Form a frequency distribution with four classes from the original values using the suggestions in Unit I. Does the result alter your responses in Parts b and c?

Stem	Leaves	Stem	Leaves
1.0	4	0.2	9
0.9		0.1	
0.8	3	0.0	
0.7	8	−0.0	7
0.6	3, 5	−0.1	7
0.5	6	−0.2	
0.4		−0.3	4
0.3	2, 5		

Source: Wendy Wood, Frank Y. Wong, and J. Gregory Chachere, "Effects of Media Violence on Viewers' Aggression in Unconstrained Social Interaction," *Psychological Bulletin*, Vol. 109, No. 3, 1991, pp. 371–83.

Review Problems

123. The following is a relative frequency distribution of incomes:

Class ($)	Relative Frequency (%)
Below $5,000	16.5
5,000–10,000	20.4
10,000–15,000	17.9
15,000–20,000	15.6
20,000–25,000	11.5
25,000–50,000	16.0
50,000 or more	2.1
	100.0

  a. Critique the construction of this relative frequency distribution.
  b. Are the data symmetric, skewed, or neither?

124. Suppose you have to make a short presentation to an audience. Your data are approximately 2,700 utility bills. Discuss how you would communicate the information in the 2,700 values by using techniques we have discussed in this chapter.

125. What is wrong with the following histogram of teenage hourly wages?

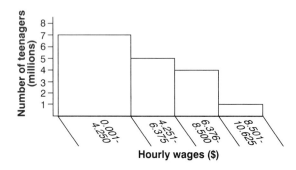

126. Use the frequency polygon of monthly unemployment rates in a county to answer the following questions.

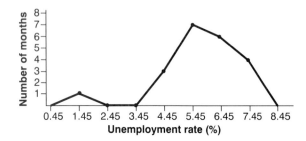

  a. What are the values on the horizontal axis called?
  b. What are the class limits of classes corresponding to the points labeled on the horizontal axis?
  c. What are the frequencies for each of the classes you found in Part b?
  d. What do we call the single value graphed above 1.45? Offer some explanation for this value's appearance in the data set.
  e. Describe the distribution of the data.
  f. Derive the corresponding "or larger" cumulative frequency distribution, and construct the matching ogive.

127. A runner has been recording her mileage per week over the last five years. Use the following frequency distribution to answer Parts a–e.

Mileage/Week	Frequency
0–9	47
10–19	72
20–29	96
30–39	27
40–49	18
	260

  a. How often has the runner run at least 20 miles per week?
  b. How often has the runner run at least 25 miles per week?
  c. If the 260 values are arranged in ascending order, in which class would the middle two values (130th and 131st) be located?
  d. Construct a histogram and frequency polygon for these data.
  e. Sketch a frequency curve for the data. Are the data better described as somewhat symmetric or somewhat skewed?

128. Must you alter a frequency distribution if you learn the original data set is a sample rather than a population? Why or why not? If it is a random sample, will the frequency distribution change?

129. If f_i is the frequency of the ith class and there are K classes, what does $\Sigma_{i=1}^{K} f_i$ represent?

130. Describe how to recognize an outlier in each of the following: frequency distribution, relative frequency distribution, cumulative frequency distribution, frequency polygon, histogram, ogive, and stem-and-leaf display.

131. The following values are the number of games won by teams in a basketball league.

52	27	33	27	32	46	18	34	34
39	43	39	47	50	30	46	47	48
20	28	40	41	16	17	38	24	47
55								

  a. Form a stem-and-leaf display.
  b. Form a frequency distribution from the stem-and-leaf display.
  c. Draw the corresponding histogram.
  d. Draw the corresponding frequency polygon.
  e. Sketch a frequency curve from Part d.

Review Problems

132. Suppose that the following "at least" ogives represent daily fat content of meals for randomly selected people from three different countries. Each volunteered to record daily menus. Use this information to answer the following questions.

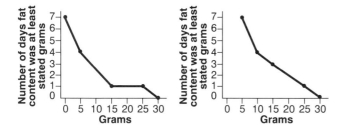

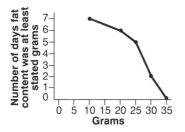

a. How many days did each person consume between 25 and 29 grams of fat?
b. Identify outliers in any diagram. If there are none, describe how you can tell.

133. If you collect a random sample of 100 values between 1 and 1,000, describe the likely frequency distribution, frequency polygon, histogram, and ogive that would result.

134. Define descriptive statistics. Explain why the tabular and visual displays discussed in this chapter are part of descriptive statistics. What are the advantages and disadvantages of using frequency distributions?

135. a. Use the following set of counts of African-Americans in the 50 states and the District of Columbia according to the 1990 census to construct a frequency distribution with six classes. Use the suggestions from this chapter. (Source: "Race and Hispanic Origin," *1990 Census Profile*, U.S. Department of Commerce, Economics and Statistics Administration, Bureau of the Census, No. 2, June, 1991, p. 4.)

5,000	1,037,000	1,090,000	39,000	274,000
2,859,000	1,292,000	245,000	1,155,000	432,000
1,694,000	4,000	3,000	95,000	48,000
548,000	1,190,000	400,000	57,000	143,000
112,000	1,040,000	1,747,000	1,163,000	56,000
1,456,000	1,021,000	915,000	1,760,000	263,000
778,000	2,022,000	2,000	374,000	1,299,000
234,000	30,000	111,000	3,000	4,000
133,000	46,000	2,209,000	12,000	79,000
150,000	2,000	300,000	22,000	27,000
7,000				

b. Repeat Part a for the following count of Hispanics. (Source: "Race and Hispanic Origin," *1990 Census Profile*, U.S. Department of Commerce, Economics and Statistics Administration, Bureau of the Census, No. 2, June, 1991, p. 5.)

7,000	11,000	4,000	288,000	46,000
213,000	2,214,000	740,000	232,000	140,000
99,000	904,000	202,000	93,000	54,000
33,000	62,000	5,000	5,000	37,000
94,000	16,000	125,000	33,000	160,000
8,000	77,000	31,000	109,000	1,574,000
22,000	33,000	25,000	16,000	20,000
93,000	86,000	4,340,000	12,000	53,000
26,000	424,000	579,000	688,000	85,000
124,000	215,000	113,000	7,688,000	18,000
81,000				

c. Compare the frequency distribution from Parts a and b.

136. Construct a frequency distribution with 10 classes using the variable, WINPCT, from the NCAA Probation Data Base located in Appendix D and the suggestions from this chapter. Construct the corresponding histogram.

137. Construct a relative frequency distribution from the following set of movie lengths (in minutes). Aim for five classes. Construct the corresponding cumulative "at least" relative frequency distribution and ogive from the frequency distribution.

109	99	85	105	109	109	100	74	90	90	110
110	133	89	105	97	84	82	107	140	90	101
101	76	90	83	67	75	91	68	89	81	93
105	92	118	120	90	106	90	120	93	105	120
123	73	96	120	90	82	82	98	99	92	85
109	120	95	120	89	95	100	105	106	105	82
96	106	91	110	77	120	120	101	91	93	109
135	90	59	120	75	95	120	76	62	90	86
96	105	102	110	145	99	90	105	101	111	120
107	180	95	93	96	94	170	110	120	108	91
64	55	91	105	94	100	90	152	66	87	70
120	67	102	134	132	98	90	88	88	84	103
102	95	122	81	120	153	110	90	62	135	70
95	95	89	105	117	114	75	135	94	104	102
95	114	88	84	82	89	101	108	105	73	89
132	71	90	139	94	104	88	96	90	97	101
100	100	145	105	102	100	90	153	128	93	87
90	120	105								

138. Use the following list of the daily number of diapers soiled by a baby to construct a stem-and-leaf display. Then describe the data set.

8	4	4	8	7	4	9	7
7	5	12	8	7	5	8	7
8	4	7	5	12	6	8	6
6	6	6	11	10	6	9	7

Figure 2-22
Minitab output for movie lengths of Problem 137

```
Histogram of MOVIE     N = 190

Midpoint   Count
   60        5     *****
   70       10     **********
   80       17     *****************
   90       48     ************************************************
  100       40     ****************************************
  110       34     **********************************
  120       19     *******************
  130        5     *****
  140        5     *****
  150        5     *****
  160        0
  170        1     *
  180        1     *


Stem-and-leaf of MOVIE    N = 190
Leaf Unit = 1.0

    2      5  59
    9      6  2246778
   21      7  001334555667
   47      8  11222223444556778888999999
  (53)     9  00000000000000000111112233333444455555555666667788999
   90     10  000000111111222223445555555555555666778899999
   45     11  00000014478
   34     12  0000000000000000238
   16     13  22345559
    8     14  055
    5     15  233
    2     16
    2     17  0
    1     18  0
```

139. Figure 2-22 displays Minitab output for the movie data in Problem 137. Contrast these results with your frequency distribution. Does it change your impression of the data? Explain.

140. Figure 2-23 is the Minitab output for monthly sales of a local furniture store. Describe the distribution.

141. Figure 2-24 presents Minitab output for a library's monthly expenditures on periodicals. Describe the distribution.

142. Use the 68 randomly selected effective annual yields of tax-exempt money market mutual funds from Problem 51 of Unit I to answer the following questions. (Source: *Wall Street Journal,* Thursday, July 11, 1991, p. C21.)

```
1.82  2.94  2.97  2.99  3.00  3.01  3.06  3.08  3.11  3.11  3.12
3.14  3.24  3.27  3.30  3.32  3.32  3.33  3.34  3.35  3.37  3.38
3.39  3.40  3.40  3.42  3.43  3.43  3.44  3.46  3.46  3.47  3.49
3.50  3.50  3.53  3.56  3.56  3.59  3.61  3.65  3.66  3.69  3.69
3.71  3.73  3.74  3.75  3.76  3.76  3.76  3.77  3.80  3.82  3.82
3.90  3.91  3.95  3.95  3.97  3.97  3.98  4.01  4.09  4.15  4.18
4.21  4.30
```

a. Construct a stem-and-leaf display.
b. Construct the frequency polygon and histogram that correspond to the stem-and-leaf display.
c. Describe the distribution of this data set based on your answers to Parts a and b.

Review Problems

```
Histogram of REVENUE    N = 60

Midpoint    Count
     0       17    *****************
 50000       18    ******************
100000        5    *****
150000        4    ****
200000        4    ****
250000        3    ***
300000        5    *****
350000        2    **
400000        2    **


Stem-and-leaf of REVENUE    N = 60
Leaf Unit = 10000

    26    0   00000000000111112223333444
   (11)   0   55555566679
    23    1   011334
    17    1   589
    14    2   0134
    10    2   7888
     6    3   122
     3    3   7
     2    4   11
```

Figure 2-23
Minitab output for furniture store sales

```
Histogram of LIBRARY    N = 84

Midpoint    Count
  7000        2    **
  8000        4    ****
  9000        5    *****
 10000        7    *******
 11000       12    ************
 12000       13    *************
 13000       14    **************
 14000        8    ********
 15000        7    *******
 16000        5    *****
 17000        5    *****
 18000        1    *
 19000        1    *


Stem-and-leaf of LIBRARY    N = 84
Leaf Unit = 100

     1     6   8
     5     7   0577
     8     8   059
    13     9   24459
    23    10   0012477899
    36    11   0113333667999
   (12)   12   000122455899
    36    13   011122334599
    24    14   022346688
    15    15   1225899
     8    16   3569
     4    17   13
     2    18   3
     1    19   3
```

Figure 2-24
Minitab output for library's periodical expenditures

Chapter 3

Measures of Location and Dispersion Within a Data Set

In this chapter we further condense the information presented in a frequency distribution of the values. The condensed information defines specific locations within the data set, such as the center and the arrangement of the other values about this location. These are measures that in most cases summarize the information about the data set in a single value.

Objectives

Unit I details different ways to describe a set of numbers by locating various positions, especially the center, of the data set. Upon completing the unit, you should be able to

1. Locate the center of a data set using the mean, median, mode(s), and weighted mean and check for a reasonable result.
2. Distinguish situations in which the mean may not be the best measure of central tendency.
3. Locate other positions using a percentile and quartile.
4. Interpret computer output for location measures.
5. Associate the symbols μ, $\overline{X}$, N, n, L, $\overline{X}_W$, $\overline{X}_G$, M_i, and f_i with the corresponding statistical term.

Because the location measures consolidate, and thus obscure, most of the original data values, in Unit II we will pursue a measure to recover some of this lost information, again in condensed form. After completing the unit, you should be able to

1. Compute and interpret the range, interquartile range, mean absolute deviation, variance, and standard deviation for both a population and a sample, then use the values to rank the dispersion of different data sets.
2. Calculate and interpret the standard or Z scores for a data set, then use them to identify outliers.
3. Determine the likely location of values according to the Empirical Rule and according to Chebyshev's inequality.
4. Recognize five ways that we apply the standard deviation to provide information about data sets.
5. Interpret dispersion information from Minitab computer output, including a box-and-whisker diagram.
6. Understand the concepts of deviation and standardization.
7. Associate the symbols σ, S, σ_G, and S_G with the corresponding statistical term.

Unit I: Location Measures

Preparation-Check Questions

1. Which of the following is a sample, and not a population, of all car owners?
   a. Ford truck owners.
   b. Those who lease their automobiles.
   c. Those who drive their parents' car.
   d. Imported-car owners.
   e. All automobile owners.

2. Use the following information to find $\sum_{i=1}^{5} X_i Y_i$:

   $X_1 = 10; X_2 = 20; X_3 = 30; X_4 = 20; X_5 = 10$
   $Y_1 = 9; Y_2 = 6; Y_3 = 3; Y_4 = 5; Y_5 = 1$

   a. 90.   d. 2,160.
   b. 410.   e. 5.
   c. 24.

3. Figure 3-1 displays a frequency polygon for commuting hours per week for students at a rural university. Which value occurs most often?
   a. 0.   d. 3.   g. 6.   j. 60.
   b. 1.   e. 4.   h. 20. k. 80.
   c. 2.   f. 5.   i. 40.

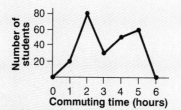

Figure 3-1
Frequency polygon for the commuter data

4. An outlier in a data set is one that
   a. Occurs more often than all the other values.
   b. Is extremely different from the other values.
   c. Is smaller than all the other values.
   d. Is larger than all the other values.
   e. Falls in the class of a frequency distribution with the most observations.

Answers: 1. d; 2. b; 3. c; 4. b.

Introduction

Tennis Survey Suppose you own a tennis club that provides indoor courts, lockers, showers, a reservation system, and other services. You decide to review the advertising policy of your club, because you believe you can target your advertising toward groups with the highest potential for becoming customers. To start the process, you have been asking your 2,000 current customers to complete a questionnaire. Soon you have 1,500 observations for each question about age, income, number of subscriptions to tennis magazines, number of subscriptions to sports magazines, and number of hours per week spent watching TV. This is a lot of information, but you want to summarize it to describe your typical customer.

As noted, 1,500 responses to each question comprise a large set of information, too much for anyone to absorb. One way to condense this information is to make a frequency distribution of the responses to each question. This procedure will depict the distribution of each set of data but is subject to some distortion, depending on the choice of class size and the starting point. Because you are interested in the typical customer, you might prefer a measure that focuses on the center of all the responses to a given question. You might want a single value, rather than a table, to summarize the responses to each question.

Statisticians have developed several measures of this typical or central value in a data set. These measures are called *averages*. Contrary to what you may have learned in grade school, there are several ways to find an average.

3.1 Unweighted Measures of the Center

Frequently we need a measure that will condense an entire set of observations of a variable, such as all of the responses to a specific survey question, into a single value. The information we wish to retain in this value is the data set's center or typical value, also known as its average. No such number can represent so much information and simultaneously differentiate the many distributions that produce an identical value as the center. Unfortunately, information is lost in the averaging process, but the result succinctly pinpoints a central value, thus avoiding an investigation of each individual item.

There are several ways to describe the center of a data set and a different quantitative measure for each description. In this section we describe the center with three alternative concepts; the mean, the point that balances distances of all of the values from the center that it locates, the median, the middle value, and the mode, the value or values that occur most often in the data set. These measures assume that all of the data values are equally important. Later in this unit, we will examine the weighted mean that accounts for the importance of different observations by weighting them differently.

3.1.1 Mean

The *mean* is the measure that balances the distances of the values from the center. If the mean test score in your class is 80, then we know that if you scored better than 80, say 85, there was another test score or set of test scores, such as 77 and 78, worse than 80 by an equal amount, in this case worse by 5 points. Any distance between an observation and the mean is offset by the distances of another point or combination of points from the mean on the other side.

The mean is probably the "average" that you are accustomed to using. When you calculate an average grade, you generally sum your test scores and divide the total by the number of tests. Statisticians call this average a mean. The term *average* refers to a measure that condenses a data set to a single value to describe the typical value, but the terms *average* and *mean* are used synonymously throughout this book.

In general, the *mean* of a set of X values is the sum of the X values divided by the number of X values. Symbolically, the mean for a data set containing N observations of variable X is

$$\text{Mean} = \frac{\sum_{i=1}^{N} X_i}{N}.$$

■ **Example 3-1:** The nine oldest siblings in a neighborhood report the following number of hours spent baby-sitting their younger brothers and sisters during a given week: 2.5, 1, 8.25, 2, 4.75, 2, 5.5, 2, and 12. Find the mean number of hours.

■ **Solution:** The mean is

$$\frac{2.5 + 1 + 8.25 + 2 + 4.75 + 2 + 5.5 + 2 + 12}{9} = \frac{40}{9} = 4.4 \text{ hours.}$$

Notice that the mean does not have to be one of the values in the data set. Thus, a typical baby-sitting time for these older siblings is 4.4 hours.

Question: Can an observer, who simply looks at the nine values and then claims the mean must be about 15 hours, be correct? Explain.

Section 3.1 Unweighted Measures of the Center **77**

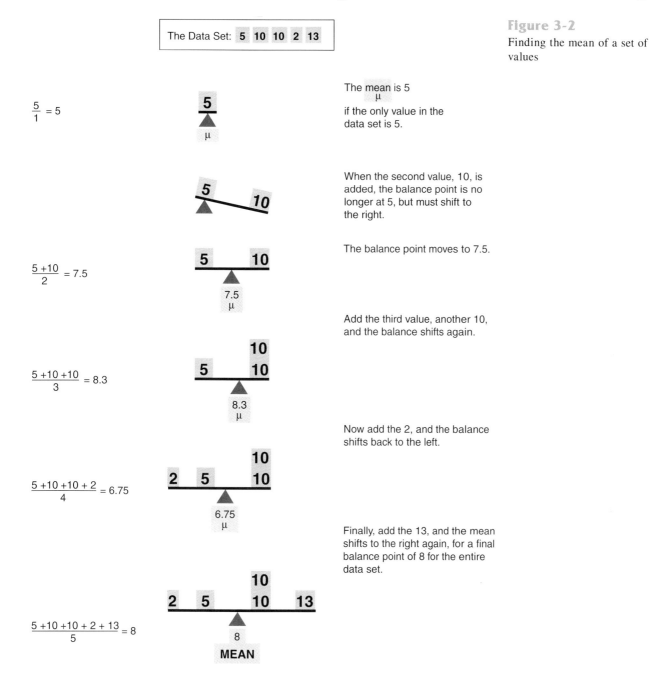

Figure 3-2
Finding the mean of a set of values

Answer: No. The mean indicates the center of the data set, but 15 hours is larger than the largest value, 12 hours.

To better understand the relationship among the data represented by the mean, consider a seesaw. The mean is the center of gravity or the point where the fulcrum is positioned to balance the values, each weighted equally, on the seesaw (or number line). Figure 3-2 demonstrates this concept for the data set containing 5, 10, 10, 2, and 13 with mean 8 by accumulating and balancing (averaging) the values in successive order.

Notice that the sum of the distances of the values from the mean 8 is the same for points below 8 and for those above 8.

	Points below 8		Points above 8
X_i	Distance from 8	X_i	Distance from 8
2	6	10	2
5	3	10	2
	9	13	5
			9

Total of distances → 9

Thus, the mean balances the observations in a data set to locate the center. This characteristic will affect our choice of a measure of dispersion in the next unit.

At this point, we must develop some statistical notation that will facilitate the writing and reading as we continue:

N = number of observations in a population; n = number of observations in a sample; μ = mean of a population (pronounced "mu"); $\overline{X}$ = mean of a sample (pronounced "ex bar").

> The formula for a population mean is
> $$\mu = \frac{\sum_{i=1}^{N} X_i}{N}.$$
> The formula for a sample mean is
> $$\overline{X} = \frac{\sum_{i=1}^{n} X_i}{n}.$$

The procedure for determining the mean is the same, whether the data set is a sample or a population, but their symbolic expressions are slightly different. The different symbols are the generally accepted ones for these concepts and will be useful shorthand for distinguishing population and sample means in later material. The same is true for N and n.

■ **Example 3-2:** Summarize the weights of solids (measured in milligrams per liter) found in a *sample* of 39 different types and brands of bottled water as listed in Table 3-1. Interpret the result.

■ **Solution:** The sum of the 39 values is 6,318, so the sample mean, $\overline{X}$, is 6,318/39 = 162 milligrams per liter. The typical or central value is 162 milligrams per liter.

3.1.2 Median

The next measure we will consider is the *median*. Remember, we are searching for a single value to represent the center of a data set; in this case, the middle value of a data set whose values are arranged in ascending or descending order. Unlike the mean, it

Table 3-1
Weights of Solids Found in Bottled-Water Sample

34	200	300	75	130	160	160	71	360
100	140	73	340	27	530	120	78	8
960	0	14	4	71	67	2	1	260
1	5	18	280	270	140	470	360	36
290	67	96						

Source: "How Good Is Bottled Water?" *Consumer's Research,* June, 1991, pp. 14–15.

Section 3.1 Unweighted Measures of the Center

does not attempt to balance distances among the values. It is a useful measure when the mean's balancing act results in a nontypical value as shown below and in Section 3.2.

■ **Example 3-3:** The grades on a test are 76, 95, 76, 70, 88, 77, and 78. Your score is the 78. Find the median score on the test.

■ **Solution:** In this case, if we make a list, the middle value, 77, is apparent.

$$70 \quad 76 \quad 76 \quad \underset{\text{Median}}{77} \quad 78 \quad 88 \quad 95$$

Three scores are smaller than 77 (70, 76, 76), and three are larger (78, 88, 95).

Because your 78 is above the median, you scored in the top half. A different picture emerges if we locate the middle score with the mean, which is (76 + 95 + 76 + 70 + 88 + 77 + 78)/7 = 80. Now, your 78 appears to be in the bottom half of the class. This apparent discrepancy occurs because the mean shifts upward and away from the scores in the 70s to balance the large values 88 and 95. Observing the position of the points and the distances between them on a number line helps us distinguish the different tasks accomplished by the mean and median.

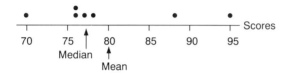

The set of grades contains an odd number of observations. If there were an even number, then we need one more step to find the value in the middle.

■ **Example 3-3 (continued):** Suppose there are eight grades rather than seven. Someone scored 100, so the values arranged in ascending order are: 70, 76, 76, 77, 78, 88, 95, 100. There are two middle values in this situation.

Question: Do you have any ideas for picking a median in this case?

Answer: The median is defined to be the midpoint or mean of the two middle values when there is an even number of observations. In this case, we have

$$\frac{77 + 78}{2} = 77.5.$$

Four values are smaller than 77.5, and four are larger.

Thus, the median divides the data set into two subsets with an equal number of observations; one with values smaller than the median and the other with values larger than the median. (Either subset may contain values equal to the median, when the median value repeats in the data set.) Each subset contains exactly or approximately half of the observations, depending on whether the total number of observations is even or odd, respectively.

■ **Example 3-4:** Find the median change in stock indices over the last year's value from the following list of percentage changes of six different indices (Source: "Markets Diary," *Wall Street Journal*, July 31, 1991, p. C1):

3.82   9.26   8.58   13.70   11.58   −23.08

■ **Solution:** Organized in ascending order, the list of values becomes:

$$-23.08 \quad 3.82 \quad 8.58 \quad 9.26 \quad 11.58 \quad 13.70$$

Because there are an even number of observations, we use the middle values after arranging the data in ascending order, 8.58 and 9.26, to determine the median percentage change:

$$\frac{8.58 + 9.26}{2} = 8.920.$$

Question: Do outliers affect the median?
Answer: No, because extreme values do not determine the median, only the middle or middle two values.

A useful formula for finding the location of the median is to take half of the sum of the number of values in the data set plus one. Symbolically, the location is $L = (n + 1)/2$, where n is the number of observations in the data set (N if the data set is a population). If there is an odd number of observations, L will be an integer pointing to the position of the center value of the data set. There are seven test scores in Example 3-3, so $L = (7 + 1)/2 = 4$. This means that the median is the fourth value in the data set after it has been arranged in ascending order as shown in Figure 3-3a. If n is even, as in the set of eight grades, L will not be an integer. Instead, it will be a number ending with a ".5" which signals that there are two middle values. For the eight test values, $L = (8 + 1)/2 = 4.5$. Thus, the value of the median is the average of the fourth and fifth observations, in the ordered set as shown in Figure 3-3b.

$$L = \frac{n + 1}{2} = \text{Location of the median value among the } n \text{ observations.}$$

Notice two things about the formula for L. First, L helps us *locate* the median; it is not the value of the median. Second, if we know the median's location, we only need to arrange half of the data in ascending order to find the actual median.

■ **Example 3-2 (continued):** Find the median weight of solids in bottled water from the list in Table 3-1.

■ **Solution:** Step 1: Find the middle L. $L = (39 + 1)/2 = 20$, so the median will be in the 20th position.
Step 2: Arranging the 20 smallest values produces

| 0 | 1 | 1 | 2 | 4 | 5 | 8 | 14 | 18 | 27 | 34 | 36 |
| 67 | 67 | 71 | 71 | 73 | 75 | 78 | 96 | | | | |

Thus, the median is 96 milligrams per liter.

■ **Example 3-5:** Find the median monthly ale production (in gallons) produced in a medieval household between October, 1412 and September, 1413 from the following list (Source: Ernest Rubin, "Statistical Exploration of a Medieval Household Book," *The American Statistician*, Vol. 26, No. 5, December 1972, pp. 37–39):

| 560 | 560 | 448 | 560 | 448 | 560 | 560 | 448 | 560 | 448 |
| 1,008 | 560 | | | | | | | | |

Section 3.1 Unweighted Measures of the Center 81

a) **The Data Set: A sample of 7 test grades**

70, 76, 76, 77, 78, 88, 95

An ODD number of data

1	2	3	4	5	6	7

There are $n = 7$ possible locations for the seven test grades' values.

The Location of the median value is at

$$L = \frac{n+1}{2} = \frac{7+1}{2} = 4$$

So the median value is the number in Location 4

70	76	76	77			
1	2	3	**4**			

b) **The Data Set: A sample of 8 test grades**

70, 76, 76, 77, 78, 88, 95, 100

An EVEN number of data

There are $n = 8$ possible locations for the eight test grades' values.

1	2	3	4	5	6	7	8

The Location of the median value is at

$$L = \frac{n+1}{2} = \frac{8+1}{2} = 4.5$$

The value of the median, then, is

70	76	76	77	78			
1	2	3	4	5	6	7	8

$$\frac{77 + 78}{2} = 77.5$$

Figure 3-3
Finding the location and value of the median in a data set

■ **Solution:** $L = (12 + 1)/2 = 6.5$. When we arrange the seven smallest values in ascending order, the list becomes

The sixth and seventh values are identical, 560 gallons, which is the median.

The following box lists the steps for finding the median.

Procedure for Finding the Median

1. Find the position of the median using $L = (n + 1)/2$.
2. If L is an integer, arrange the smallest L values in ascending order to determine the median, the Lth value.
3. If L is not an integer, round it *down* to an integer, and arrange the smallest $L + 1$ values in ascending order to determine the median, the average of the Lth and $(L + 1)$st values.

3.1.3 Mode

The third measure of the center of a data set is called the *mode*. This measure indicates the value (or values) that occur most often within the data set. Often, we think of an average as a value that is not only an actual value in the data set but also one that we would most likely observe when randomly inspecting the data set. If we want a typical value to satisfy these characteristics, the mode is the appropriate measure.

■ **Example 3-6:** Suppose an investor is building a motel and must decide how many people to accommodate in each room. Knowing the number of people per room requested most often by travelers would be helpful. This measure is called the *modal party size*. Suppose a national survey of motels and other accommodations reveals the following:

Party Size (persons per room)	Number of Requests
1	1,583
2	2,967
3	2,500
4	2,486
5	1,029
6	474
	11,039

Find the modal party size.

■ **Solution:** The modal party size is two persons, which occurs 2,967 times, more times than any of the other party-size values. Therefore, the builder might put a double bed or two singles in most of the rooms.

■ **Example 3-5 (continued):** Find the mode of the ale production data for the medieval household.

| 560 | 560 | 448 | 560 | 448 | 560 | 560 | 448 | 560 | 448 |
| 1,008 | 560 |

■ **Solution:** The value 560 gallons is the mode because it occurs seven times, while 448 gallons occurs four times, and 1,008 gallons occurs only once. The household produced 560 gallons of ale during most months.

■ **Example 3-2 (continued):** How would you describe the mode for the data on solids in bottled water in Table 3-1?

■ **Solution:** Although most values occur only once, several occur twice: 1, 67, 71, 140, 160, and 360. Because they all occur the same number of times, which is more than any of the other values, we call all of them modes.

Thus, *there can be more than one mode*. When there are two, the data set is bimodal. If there are three, it is trimodal. If there are more than three, it is *multimodal. It is also possible for a data set not to have any mode.*

■ **Example 3-4 (continued):** Find the modal change in stock indices over the last year from the earlier list of percentage changes.

| 3.82 | 9.26 | 8.58 | 13.70 | 11.58 | −23.08 |

Problems

Solution: There is no mode. Each observation occurs the same number of times (in this case once each).

The following box summarizes the steps to locate the mode or modes.

Procedure for Locating the Mode or Modes

1. Note whether any values in the data set occur more than once.
2. Count the number of times such observations occur.
3. The value or values that occur most often are the modes.
4. If no value occurs more often than any other, there is no mode.

If we are choosing a value at random from a data set or using the set of values to predict future values, then the mode is the value that is most likely to occur. It is possible that several values are equally likely or more likely to occur than the other values. It is also possible that there is no value that is more likely to occur than any of the others.

Section 3.1 Problems

5. The Natural Resources Defense Council compiled the following set of numbers of 1991 beach closings from 14 coastal states (California, Connecticut, Delaware, Florida, Hawaii, Louisiana, Maine, Maryland, Massachusetts, New Hampshire, New Jersey, New York, Rhode Island, and Virginia): 745, 293, 11, 299, 106, 1, 47, 24, 59, 1, 108, 314, 0, and 2, respectively. Find the mean, median, and mode(s) of the closing values. (Source: "Dirty Beaches: Contamination Closed 2,000 Beaches in 1991," *Asheville Citizen-Times*, July 24, 1992, p. 3A.)

6. The market share of the top 12 software firms are 1.8%, 2.1%, 2.4%, 2.7%, 3.0%, 3.3%, 4.0%, 4.2%, 7.5%, 8.9%, 12.0%, 25.4%. Find the mean, median, and mode(s) of the market shares. Do these measures indicate that a typical firm from these 12 controls a sizable portion (say 10% or more) of the software industry? (Source: William M. Bulkeley, "Software Industry Loses Start-Up Zest as Big Firms Increase Their Dominance," *Wall Street Journal*, August 27, 1991, pp. B1, B5.)

7. The Sentencing Project, a group that analyzes criminal justice matters, compiled the following list of incarceration rates (citizens in jail/total population count) for the listed countries from available data. The values are expressed as percentages. Find the mode(s), mean, and median rate.

Country	Incarceration Rate (%)
United States	0.455
South Africa	0.311
Venezuela	0.177
Hungary	0.117
Canada	0.111
China	0.111
Australia	0.079
Portugal	0.077
Czechoslovakia	0.072
Denmark	0.071
Albania	0.055
Netherlands	0.046
Republic of Ireland	0.044
Sweden	0.044
Japan	0.042
India	0.034

Source: "Study Depicts U.S. as World's Biggest Jailer," *Asheville Citizen-Times*, February 11, 1992, p. 6D.

8. Find the mean, median, and mode(s) of the following enrollment figures (rounded to the nearest thousand) for the 16 campuses of the University of North Carolina. (Source: "UNC's Overall 1991 Enrollment Reaches an All-Time High," *Board of Governors Quarterly*, Fall 1991, p. 6.)

11,000	17,000	2,000	4,000	7,000	5,000
1,000	27,000	3,000	3,000	24,000	15,000
12,000	8,000	6,000	3,000		

9. Find the mean, median, and mode(s) of the number of participants (in millions) in different airline's frequent flyer programs (Source: Asra Q. Nomani and Bridget O'Brian, "Frequent Fliers Scurry as Clouds Gather," *Wall Street Journal*, August 13, 1991, p. B1.)

14.2	13.3	10.1	9.1	8.6	7.8	6.2	1.7
1.2	0.7	0.6					

Figure 3-4

Frequency curve for antenna spacing

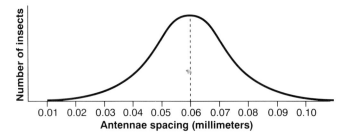

3.2 Comparing Measures and Finding the Typical Value

In the course of introducing the three measures of central tendency, we calculated all three for the data on solids in bottled water in Table 3-1. The mean is 162 milligrams per liter, the median is 96 milligrams per liter, and the modes are 1, 67, 71, 140, 160, and 360 milligrams per liter. Which one is correct? Which one is best? What do we know about this data set from the three pieces of information? If we know the similarities and differences of the three measures, it will be easier to select a proper one. This knowledge will also aid in interpreting a measure someone else reports.

A frequency curve is useful for comparing the behavior of the measures for different types of distributions. Figure 3-4 pictures different measurements of the space between an insect's antennae (in millimeters). The important thing to notice here is the symmetric distribution and the number of modes.

Question: Where is the mode in Figure 3-4?

Answer: The mode is at 0.06 millimeters, because it corresponds to the highest point on the frequency curve. Because height corresponds to frequency of occurrence, this observation occurs more often than any of the others.

Question: Where is the mean in Figure 3-4?

Answer: The mean is the same as the mode (0.06 millimeters here) when the data are symmetric. There are an equal number of observations above and below 0.06, and they are equally spaced, so that they balance one another and keep the mean at the center value. For instance, there are just as many 0.05s as there are 0.07s (0.05 and 0.07 are both 0.01 millimeter from 0.06).

Question: Where is the median in Figure 3-4?

Answer: The median is also 0.06 millimeters. Half of the observations must be smaller than 0.06 and half larger than 0.06, because of the symmetry.

Choosing a measure to represent the center of a symmetric data set depends on the number of modes. *When the data are symmetric and have a single mode (unimodal), the mean, median, and mode are identical, as shown in Figure 3-5a. Choose the measure that will be easiest to find.* If the data are symmetric with more than one mode, the three measures are not the same. If we report the modes, along with the median or mean, they will help identify the shape of the distribution and its center or centers. See Figure 3-5b.

As the symmetric or near-symmetric shape disappears, the measures tend to diverge. In such cases, reporting more than one measure may be helpful for identifying general tendencies in the data set.

For instance, if the data set contains an outlier, a value that is very different in magnitude from the typical value or most of the observations in a data set, the mean is pulled toward the outlier to balance the distances on either of its sides. Suppose the data set is skewed to the left or right. This refers to a data set dominated by values of a certain relative size (can be either the larger values or the smaller values from the

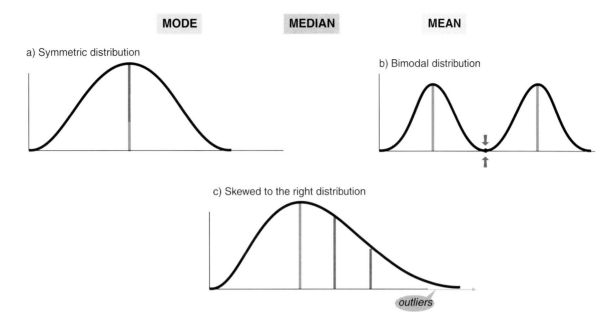

Figure 3-5
Measures of central tendency for different data distributions

data set) with very few values of the opposite size. The few values may contain one or more outliers. In any case, the outlier or the smaller set of relative values pull the mean in their direction, away from the location of the mass of the values in the data set. Although an outlier will not affect the median, a skewed data set will tend to pull it away from the location of the mass of values, but not as far away as the mean. In such cases, the mode still locates the position where the values are massed.

Question: The value 100 is an outlier in the following data set: 10, 10, 11, 12, 100. Most of the observations are clustered within an interval of relatively smaller values (10–12). Find the mean, median, and mode of this data set.

Answer: Mean: 28.6, median: 11, mode: 10. As predicted, the mean moves away from the cluster of small values toward the outlier, 100.

A frequency curve of data that is skewed to the right appears in Figure 3-5c. The mode is the value corresponding to the highest point of the curve. The outliers in the tail cause the mean to move away from most of the values in the data set toward the extreme values. Finally, the median will generally be between the mode and mean—"generally" because most data sets are not so smoothly distributed as pictured here. Thus, odd arrangements of these measures do occur.

In general, the median and mode more accurately describe a typical value for a skewed data set. When data are skewed, the median is frequently chosen to measure the typical value. For example, incomes are typically skewed to the right because in many groups a few people make extremely large incomes, whereas most people have incomes clustered at lower levels. So, we often find median income reported rather than mean income. Consider the effect on mean income when a factory owner's income is included in the set of factory employees' incomes or the set of incomes of a small factory town. Another example in which medians often are reported is when nationwide new (or used) home prices are being condensed. Here again, the number of extremely expensive homes is few compared to the bulk of home prices.

We mentioned at the outset of this section that the three measures differ for the data on solids in bottled water from Table 3-1 (mean: 162, median: 96, modes: 1, 67, 71, 140, 160, 360), so we should expect a nonsymmetric distribution. We might predict skewness to some extent in the direction in which the mean diverges from the other

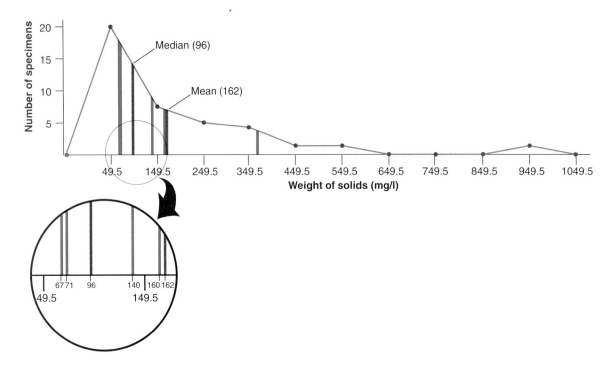

Figure 3-6
Frequency polygon for the weight of solids in bottled water

measures. In this case it is larger than the median, so we predict skewness to the right. There are six modes that vary from relatively small to relatively large values in the data set, although three of them are smaller than the median and five are smaller than the mean (though one, 160, is very close to the mean, 162). The frequency polygon for the data, which appears in Figure 3-6, agrees with this skewness assessment.

Choosing which measure of central tendency to use often depends on the objective of the measurement. If the objective is to determine *the exact value or values that occur most often*, the mode is the obvious choice. This is the value that is more likely to occur than any other, when we randomly observe a value from the data set. A shoe salesperson planning inventory is interested in modal sizes (size 9, for example). However, the mode does not account for outliers when the data are skewed, and remember, there may be no mode. If the modal home price is presented, it might well be the price of a very popular prefabricated model that has been mass advertised and mass purchased across the nation. This measure essentially ignores information on prices of other designs and construction forms.

In the remaining chapters, most of the discussion of central tendency employs the mean. The reasons for this are:

1. The mean is probably the familiar measure for most people.
2. The mean takes the magnitude of each individual observation into account, whereas the median and mode do not. Of course, accounting for each magnitude causes problems when there are outliers.
3. Statistical theory has evolved around the mean more than around the other two measures.

■ **Example 3-6 (continued):** The mode of the motel party sizes is 2 people per room. The mean is 3.0 persons per room (Section 3.3 will show an easy method for finding this value when the information is provided in a table). Find the median party size and describe the typical value.

Persons per Room	Number of Requests
1	1,583
2	2,967
3	2,500
4	2,486
5	1,029
6	474
	11,039

■ **Solution:** The median is 3 persons because the 5,520th observation [(11,039 + 1)/2] would be halfway through the 11,039 observations when arranged in ascending order. There are 4,550 observations smaller than 3 (1,583 requests from parties of a single person, and 2,967 requests from parties of 2 people), and 2,500 requests from parties of 3 people, so the 5,520th value must be in the last category of 3 people per room. The ogive in Figure 3-7 illustrates this location.

The frequency polygon in Figure 3-8 displays the near symmetry of the data in the motel example. There are fewer large-party sizes, but the large values, 5 and 6 persons per room, are not extremely different in magnitude from the other values, 1 through 4 persons per room. The mean and median may better portray a typical value here, though the motel designer needs the mode in order to accommodate as many groups as possible in the given space. Knowledge of the median and mean, which occur almost as often as the mode here, would probably be useful information to include in deciding alternative accommodation sizes.

Section 3.2 Problems

10. Check the mean, median, and modes of the data sets in Problems 5–9 from Section 3.1. In which problems do the measures indicate that the data are approximately symmetric? skewed? If skewed, in what direction?

11. People often compare average or mean SAT scores by state.
   a. Consider the histogram of state scores in Figure 2-17 (p. 61). Does it appear that the mean is a good representation of the typical value? Explain.
   b. Now consider Problem 3 of Chapter 2 (p. 25) that displays the frequency distribution for states where 13% or fewer of high school seniors took the SAT and the frequency distribution where 52% or more took the test. Will the typical scores of these two groups be more alike if we use the mean or median to compare typical scores? What if we use the mode? Suggest which measure will best represent the typical value in each of the two data sets.

Figure 3-7 Locating the median from an ogive of the motel data

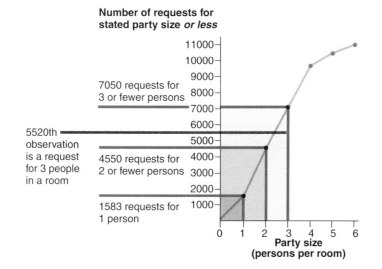

Figure 3-8

Frequency polygon for the motel example

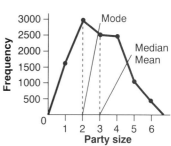

12. Suggest which measure of central tendency will best represent the typical peak level of ozone and of PM10, based on the frequency distributions in Problems 1 and 2 of Chapter 2 (pp. 24–25). Explain your answers.
13. A newspaper reporter calculates the median pay raise nationwide (expressed in percentage terms) to condense a national sample of last year's pay raises. Suggest a rationale for the use of this measure that is probably less familiar to most of the newspaper's readers.

3.3 Weighted Measures of Central Location

3.3.1 Weighted Mean

At the end of the last section in Example 3-6, we stated that the mean of the set of motel party sizes is 3.0 persons per room. A simple or unweighted mean of these six values, repeated in the table below, is $(1 + 2 + 3 + 4 + 5 + 6)/6 = 3.5$ persons per room. However, to obtain the mean of all 11,039 requests for rooms, we must account for the frequency of request for each party size or its importance. There are 1,583 ones, 2,967 twos, and so forth, so the mean is

$$\frac{1{,}583(1) + 2{,}967(2) + 2{,}500(3) + 2{,}486(4) + 1{,}029(5) + 474(6)}{11{,}039} = 3.0 \text{ persons per room.}$$

The smaller party-size values, which occur more often, receive more weight and pull the mean of all 11,039 values away from the simple mean of 3.5 persons per room.

Persons per Room	Number of Requests
1	1,583
2	2,967
3	2,500
4	2,486
5	1,029
6	474
	11,039

This calculation exemplifies another measure of central location, the weighted average or, more accurately, the *weighted mean*. As the name implies, we calculate a mean, but now we count each observation differently. We weight each according to its importance, such as its frequency of occurrence for the six party sizes in the motel example. When demonstrating the mean using the seesaw, each observation "weighed" the same. Now some observations "weigh" more than others, and the center of gravity or mean will shift toward the "heavier" observations.

Figure 3-9 demonstrates the formula for calculating a weighted mean, using the motel values once again. Notice that we multiply each observation, X_i, by its weight or measure of importance before summing. Then, we divide the sum of the weighted X values by the sum of the weights.

The formula for the weighted mean is

$$\overline{X}_W = \frac{\sum_{i=1}^{n} X_i W_i}{\sum_{i=1}^{n} W_i},$$

where W_i = weight of the ith observation.

Section 3.3   Weighted Measures of Central Location

$$\overline{X}_w = \frac{\text{weighted sum of } X \text{ values}}{\text{sum of weights}}$$

Original data values to be averaged

$$\overline{X}_w = \frac{\sum_{i=1}^{n} X_i W_i}{\sum_{i=1}^{n} W_i} = \frac{1(1583) + 2(2967) + 3(2500) + 4(2486) + 5(1029) + 6(474)}{1583 + 2967 + 2500 + 2486 + 1029 + 474}$$

Weight or importance of individual values of X

Figure 3-9
Interpreting the formula for the weighted sample mean for the motel example

■ **Example 3-7:** Five economic forecasters have published their predictions for the unemployment rate for the next year. The projections are listed along with the percent of each forecaster's previous predictions that fell within 0.5 percentage points of the actual unemployment rate. The additional information about the past forecasting performance weights the attention we devote to each forecast in determining for ourselves or business what the likely future rate will be. We give greater weights to more accurate past performance.

Often, combining forecasts produces better forecasts over time than following one individual's forecast. Combine these forecasts by finding the weighted mean of the projected rates to produce a single overall prediction.

Forecast Unemployment Rate (%)	Forecaster's Predictions within 0.5% of True Rate (%)
7.0	58
6.5	30
6.8	47
6.5	38
7.1	42

■ **Solution:** The values we wish to average are the forecasts, so these are the X values in the formulas. The W values are the percent of correct predictions for each forecaster. An easy way to follow the calculation and substitution into the formula is to make the chart that follows. The product of each X and W combination forms a new column that is summed to produce the numerator of the weighted mean formula. Similarly, the sum of the W column is the denominator.

X_i	W_i	$X_i W_i$
7.0	58	406.0
6.5	30	195.0
6.8	47	319.6
6.5	38	247.0
7.1	42	298.2
	215	1,465.8

The weighted mean is $1,465.8/215 = 6.82\%$. The simple mean of the five predictions is $(7.0 + 6.5 + 6.8 + 6.5 + 7.1)/5 = 6.78\%$, but in the weighted mean, the heavier weights for the larger predictions pull the mean up from 6.78 to 6.82%.

■ **Example 3-8:** Suppose that a professor's evaluation of your performance in a course depends on the mean of six different test scores. However, the second and third tests count twice as much as the first. The fourth counts half as much, the fifth counts one-fourth as much, and the sixth is three times as much. Find the weighted mean if the six consecutive scores are 50, 62, 71, 43, 55, and 95.

■ **Solution:** We want the mean test score, so these are the X values. The weights are the emphasis determined by the professor. If we let $W_1 = 1$, then we can easily express the other weights as multiples of W_1. We could use any value for W_1 and apply the appropriate factors, but using the value 1 will be the simplest.

Test Score (X_i)	Emphasis (W_i)	XW
50	1.00	50.00
62	2.00	124.00
71	2.00	142.00
43	0.50	21.50
55	0.25	13.75
95	3.00	285.00
	8.75	636.25

The weighted mean is $(636.25)/8.75 = 72.7$ points. Notice that all of the scores except the last one fall below the mean. The heavier weight makes the 95 an especially influential outlier.

■ **Example 3-9:** Suppose concert tickets sell for $5 in the balcony, $10 in the rear of the concert hall, and $15 in the front. There are 2,200 seats in the balcony, 1,750 in the rear, and 2,100 in the front.

Price ($)	Quantity Available
5	2,200
10	1,750
15	2,100
	6,050

We want to find the mean price of *all* tickets offered to the public or a weighted mean price.

Question: Is price the X_i or W_i in the formula?

Answer: Price is the X_i, because we want to find a mean price, not the mean quantity available.

■ **Solution:**

X_i	W_i	$X_i W_i$
5	2,200	11,000
10	1,750	17,500
15	2,100	31,500
	6,050	60,000

$\overline{X}_W = 60,000/6,050 = \$9.9$.

Question: If we reverse the X_i and W_i in this situation and check the answer to see if it is reasonable, we obtain

$$\frac{\$5(2,200) + \$10(1,750) + \$15(2,100)}{\$5 + \$10 + \$15} = \frac{60,000}{30} = 2,000.0.$$

2,000.0 cannot be a mean of $5, $10, and $15. What is it?

Answer: It is a weighted mean of the number of seats, the weights for the mean price problem.

Question: What is the mean of the ticket prices a potential buyer can choose or the mean price without accounting for quantity available?

Answer: This answer is the simple mean of the prices offered when you are ready to purchase a ticket.

$$\frac{\$5 + \$10 + \$15}{3} = \$10.$$

The Consumer Price Index, calculated by the Bureau of Labor Statistics and widely reported in the media each month, is an example of a weighted measure of price changes. The change of each item is weighted according to a predetermined measure of the importance of the item in a typical nonfarm family budget. A sample of detailed household budget data provides information about the choice of items whose prices will be included in the calculation, and estimates of the percent of this typical budget allocated to the different items are used as weights.

3.3.2 Estimated Mean from Grouped Data

We can use the weighted mean concept to estimate the mean of a data set from a frequency distribution of its values. Consider the following frequency distribution of the first year of operation (ORIGIN) for the schools listed in the NCAA Probation Data Base to estimate the mean starting date. (See Appendix D.)

Starting Date	Number of Schools
1780–1799	5
1800–1819	5
1820–1839	6
1840–1859	16
1860–1879	26
1880–1899	17
1900–1919	6
1920–1939	3
1940–1959	1
	85

Because the frequency distribution obscures each school's actual starting date, we must find a substitute value for each school in order to estimate the mean.

Question: From the frequency distribution we know that five schools started between 1780 and 1799. Suggest and justify a single value for a suitable starting date for the group of five schools.

Answer: All five dates can be different or the same. They may be uniformly distributed, skewed, symmetric, or irregularly distributed in the interval. Usually, we assume that all of the points fall at the class midpoint, M_i (the mean of the class limits); that is, we treat the class midpoint as the X_i. In this case, the midpoint is the year $(1780 + 1799)/2 = 1789.5$. We expect starting dates later than the midpoint to offset those before this date when the points are dispersed in the interval. In general, we expect the midpoint to offset unknown imbalances better than any other point in the class.

The following chart lists the resulting class midpoints, M_i for the nine classes (also called groups). To find the mean of the 85 school starting values, we first need their sum. In a process similar to summing the 11,039 motel party sizes, we multiply each of the nine midpoints by the corresponding class frequency, f_i, to include each observation that falls in each class, and then we sum the resulting products, $M_i f_i$. Thus, the five schools with an estimated starting date of 1789.5 account for 8,947.5 in the sum, and 158,487.5 is the estimated sum of all 85 starting dates.

Starting Date	M_i Midpoint	f_i Number of Schools	$M_i f_i$ Products
1780–1799	1,789.5	5	8,947.5
1800–1819	1,809.5	5	9,047.5
1820–1839	1,829.5	6	10,977.0
1840–1859	1,849.5	16	29,592.0
1860–1879	1,869.5	26	48,607.0
1880–1899	1,889.5	17	32,121.5
1900–1919	1,909.5	6	11,457.0
1920–1939	1,929.5	3	5,788.5
1940–1959	1,949.5	1	1,949.5
		85	158,487.5

The weighted mean estimate of the 85 starting dates is $(158{,}487.5)/85 = 1{,}864.6$. The actual mean is 1,864.8. The close approximation here is fortunate, but not to be expected in all cases.

From this example, we find that we treat each class midpoint as the X_i in the weighted mean formula. The weight, W_i, is the frequency of the corresponding class, f_i, so the midpoint of a class with a large frequency counts more in the averaging process than one with a small frequency. ΣW_i is the same as the sum of the frequencies, Σf_i, which must total to the number of observations in the data set (N for a population and n for a sample). Figure 3-10 illustrates the components of the formula that represents this process.

Figure 3-10
Interpreting the formula for the mean of grouped data as a special case of a weighted mean

Graphic Summary

Class midpoints which approximate the X_i values to be averaged

$$\bar{X}_G = \frac{\sum_{i=1}^{9} X_i W_i}{\sum_{i=1}^{9} W_i} = \frac{\sum_{i=1}^{9} M_i f_i}{\sum_{i=1}^{9} f_i}$$

Class frequencies which account for the importance of the data values

$$= \frac{1789.5\,(5) + 1809.5\,(5) + 1829.5\,(6) + 1849.5\,(16) + 1869.5\,(26) + \cdots + 1949.5\,(1)}{5 + 5 + 6 + 16 + 26 + \cdots + 1}$$

Estimated sum of original values

n, the number of values in the data set

$$= \frac{158487.5}{85}$$

> The formula for finding the mean from grouped data is
>
> $$\overline{X}_G = \frac{\sum_{i=1}^{k} M_i f_i}{\sum_{i=1}^{k} f_i} = \frac{\sum_{i=1}^{k} M_i f_i}{n},$$
>
> where k is the number of classes in the frequency distribution.

This procedure works with relative class frequencies as well.

■ **Example 3-10:** Estimate the mean amount people say they would contribute for a party from the following relative frequency distribution.

Contribution	Percentage Willing to Contribute
40.01–50.00	10
30.01–40.00	15
20.01–30.00	45
10.01–20.00	25
0.01–10.00	5

■ **Solution:** The following table charts the class midpoints. We substitute the relative frequencies just as we would absolute frequencies.

Contribution ($)	Midpoint	Proportion Willing to Contribute	$M_i f_i$
40.01–50.00	45.005	0.05	2.25025
30.01–40.00	35.005	0.15	5.25075
20.01–30.00	25.005	0.45	11.25225
10.01–20.00	15.005	0.30	4.50150
0.01–10.00	5.005	0.05	0.25025
		1.00	23.50500

The estimated mean is 23.50500/1.00 = $23.505, a reasonable value, because 75% of the values are between $10.01 and $30.00 with more observations in the $20.01–$30.00 category.

Notice that when we use relative frequencies, the denominator will always sum to 1 or 100% when we use percentages rather than proportions (except for small discrepancies caused by rounding).

Section 3.3 Problems

14. The following table lists the rejection rates of five professional journals (the higher the rate, the more articles the journal refuses to publish) along with an academic departments' number of publications in each of the journals. Use the rejection rates to calculate a weighted mean of the number of publications per journal.

Journal	Number of Publications	Rejection Rate (%)
1	10	80
2	23	10
3	18	20
4	5	90
5	14	50

15. Approximate the mean peak ozone level from the frequency distribution of these levels in Problem 1, Chapter 2 (p. 24).

16. Approximate the mean peak particulate matter (PM10) level from the frequency distribution of these levels in Problem 2, Chapter 2 (p. 25).

17. Approximate the mean SAT score for low-percentage and for high-percentage states from the frequency distributions in Problem 3, Chapter 2 (p. 25).

3.4 Other Measures of Location

Sometimes it is useful to locate positions other than the center of a data set. Following the procedure to obtain the median, which divides the data into two equal-size groups, we can find quartiles that divide the data into four equal groups and percentiles that divide the data into 100 equal parts. Figure 3-11 demonstrates these relationships and the similarities in the measures using the 85 actual university starting dates from the NCAA Probation Data Base (see Appendix D). The figure also shows that the formulas for the positions of these measures are also similar to the formula for the median's position. First we determine the fraction of the total set of values that are smaller than the desired measure, such as three-fourths for the third quartile or 0.3 for the 30th percentile. Then, we multiply this fraction by the sum of the number of observations in the data set plus 1. For instance, we use $(1/2)(n + 1)$ for locating the median, $(3/4)(n + 1)$ for the third quartile, and $(30/100)(n + 1)$ for the 30th percentile.

3.4.1 Quartiles

A *quartile* divides the data into quarters. Figure 3-11 shows that 1848 is the first quartile of the starting-date values. So, approximately one-fourth or 25% of the dates are before 1848 (the 21 earlier dates out of all 85 dates), and approximately 75% are later (the 64 later dates out of all 85 dates). Similarly, the third quartile is the year 1885.5, so approximately 75% of the dates are before 1885.5 and 25% are later. Of course, the second quartile is the same as the median.

Question: If 922 is the third quartile of a data set, what do we know about the other values?
Answer: 75% are smaller than 922 and 25% are larger.

Question: A committee that chooses the recipient of a certain scholarship automatically considers students with grade point averages (GPAs) in the top 25% of their class. The first, second, and third quartiles of the entire class's GPAs are 2.03, 2.88, and 3.33, respectively. Which students will the committee consider if their GPAs are: 2.25, 2.88, 3.51, 1.99, 3.25, 3.98?
Answer: They will consider the students with GPAs of 3.51 and 3.98. Their averages are above 3.33, the third quartile, so they are in the top 25% of the GPAs.

We can adapt the formula for locating the median to locate quartiles. One-fourth of the values are smaller than the first quartile, so the location will be $(1/4)(n + 1)$. For the 85 values in the starting date data, we obtain $(1/4)(85 + 1) = 21.5$. Because the result is not an integer, we use the midpoint of the 21st and 22nd values, after the data are arranged in ascending order. In this case, 1848 is midway between 1847 and 1849. The location formula for the third quartile would be $(3/4)(n + 1)$; however, we can arrange the largest 25% rather than the smallest 75% of the values. Consequently, we can retain the formula $(1/4)(n + 1)$ for the third quartile but arrange the values in descending rather than ascending order. If you renumber the observations starting with 1 at the bottom, you will find that the 21st and 22nd values are the years 1886 and 1885, respectively, with a midpoint of 1885.5. The following box summarizes the steps for locating these quartiles.

Procedure for Finding the First and Third Quartiles

1. Determine the position, L, using $L = (1/4)(n + 1)$.

(continued on page 96)

Figure 3-11 Relationships and similarities among the median, quartiles, and percentiles; university starting dates (ORIGIN) from the NCAA Probation Data Base

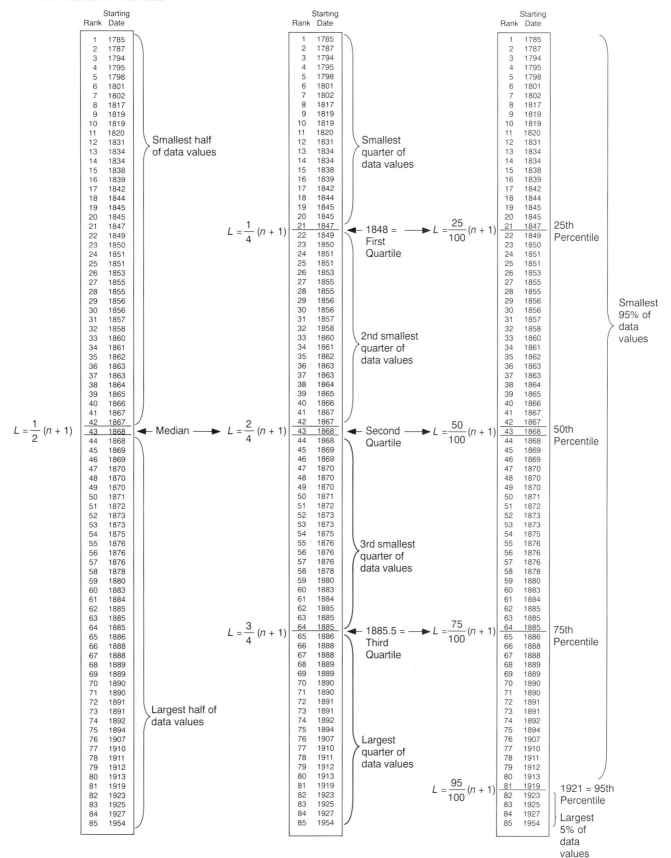

2. If L is an integer, arrange the smallest L values in ascending order for the first quartile and the largest L values in descending order for the third quartile. The Lth value is the quartile in question.

3. If L is not an integer, round it *down* to an integer, and arrange the smallest $L + 1$ values in ascending order for the first quartile and the largest $L + 1$ values in descending order for the third quartile. The midpoint of the Lth and $(L + 1)$st values is the quartile in question.

Question: If a data set has 1500 observations, how many values must be arranged in ascending or descending order to locate the first and third quartile?

Answer: $(1/4)(1501) = 375.25$, so 376 observations must be arranged. The quartiles are the midpoints of the 375th and 376th values for values arranged in ascending order for the first quartile and descending order for the third quartile. These points are also the two largest values in the set of the smallest 376 values and the two smallest values in the set of the largest 376 values, respectively.

Question: How many values need to be arranged in ascending or descending order to locate the second quartile?

Answer: We need to arrange half (or one more than half) of the values, because the second quartile is the median.

■ **Example 3-11:** Find the first, second, and third quartiles of the number of previous math courses taken by a random sample of students in a statistics class: 5, 2, 8, 3, 3, 5, 1, 2, 3, 9, 4, 6, 11, 13, 4, 3, 7, 5, 6, 3, 4, and 5.

■ **Solution:** To find the first and third quartiles we proceed as follows:

Step 1: There are 22 values, one-fourth of $(22 + 1)$ yields 5.75. We will need the midpoints of the fifth and sixth values.
Step 2: We must arrange six values in ascending order and six in descending order.
Smallest six values: 1, 2, 2, 3, 3, 3.
Largest six values: 13, 11, 9, 8, 7, 6.
Step 3: The first quartile is the midpoint of the two largest values in the ordered set of the smallest six values. The result is three courses. The third quartile is the midpoint of the two smallest values in the ordered set of the six largest values or $(6 + 7)/2 = 6.5$ courses.

To find the second quartile, we use the rules for locating the median.

Step 1: $L = (22 + 1)/2 = 11.5$.
Step 2: The first 12 values need to be arranged in ascending order. Supplementing the smallest six, we obtain: 1, 2, 2, 3, 3, 3, 3, 3, 4, 4, 4, 5.
Step 3: The second quartile, which is also the median, is $(4 + 5)/2 = 4.5$.

Section 3.4.1 Problems

18. Find the first and third quartiles of the peak carbon monoxide (CO) levels from the Air Quality Database listed in Appendix B.

19. Find the first and third quartiles of the county population values in the North Carolina Census Data Base (see Appendix E).

20. The following frequency distribution presents the last 200 closing prices of a stock. Find the first and third quartile of the closing prices.

Closing Price	Number of Days
15.000	18
15.125	32
15.250	53
15.375	44
15.500	27
16.125	16
16.375	10

21. Find the first and third quartile of the following set of 300 SAT scores.

453	454	458	458	461	468	470	473	478	479	481	483
486	488	493	497	497	504	504	507	509	517	519	520
521	530	533	534	537	540	542	545	551	554	566	573
576	579	582	595	598	599	604	605	614	622	643	653
657	662	666	667	672	676	684	685	691	693	695	699
703	704	705	714	719	726	728	730	731	732	736	737
747	752	758	763	766	768	769	772	775	777	783	787
789	797	798	804	804	805	809	812	813	815	817	817
832	836	841	842	843	845	859	864	865	866	868	869
870	876	882	883	887	887	890	898	906	907	922	924
926	927	932	933	936	937	940	946	953	953	967	968
973	975	981	982	983	984	994	997	1,000	1,007	1,008	1,010
1,011	1,011	1,017	1,018	1,021	1,022	1,022	1,026	1,027	1,028	1,028	1,029
1,034	1,047	1,050	1,063	1,063	1,063	1,065	1,081	1,085	1,087	1,088	1,089
1,094	1,100	1,105	1,106	1,115	1,118	1,123	1,132	1,132	1,137	1,141	1,143
1,145	1,147	1,148	1,161	1,163	1,165	1,167	1,170	1,172	1,173	1,174	1,175
1,177	1,193	1,197	1,198	1,203	1,204	1,212	1,214	1,214	1,221	1,222	1,224
1,228	1,228	1,230	1,233	1,236	1,236	1,239	1,244	1,245	1,245	1,245	1,248
1,248	1,250	1,251	1,259	1,260	1,260	1,263	1,263	1,268	1,270	1,271	1,273
1,274	1,280	1,288	1,289	1,292	1,296	1,297	1,304	1,306	1,307	1,310	1,313
1,316	1,316	1,330	1,333	1,334	1,337	1,345	1,347	1,347	1,361	1,374	1,389
1,396	1,396	1,399	1,405	1,412	1,416	1,419	1,422	1,434	1,435	1,440	1,445
1,445	1,454	1,457	1,470	1,472	1,474	1,474	1,481	1,482	1,483	1,487	1,488
1,493	1,497	1,525	1,525	1,530	1,541	1,542	1,542	1,546	1,554	1,554	1,561
1,562	1,564	1,565	1,565	1,569	1,572	1,574	1,577	1,579	1,581	1,583	1,590

3.4.2 Percentiles

If you score 72 on a test, you might wonder what percent of the scores are poorer or what percent are better than yours. A percentile would answer your question.

A *percentile* divides the data set into 100 parts. It tells us where in 100 possible places a value, such as your test score, falls. Suppose the teacher says that approximately 46% of the scores are lower than your 72. Then, 72 is the 46th percentile of the scores. This means that approximately 54% of the scores are higher than your 72.

Question: If the 80th percentile of a set of annual state highway expenditures is $89 million, what can we say about the other values?

Answer: Eighty percent of the expenditures are smaller than $89 million, and 20% are larger.

Again we adapt the formula for locating the median to locate percentiles. Ninety-five percent, or 95/100, of the values must precede the 95th percentile, so the location will be (95/100)(n + 1). For the 85 values in the starting-date data, we obtain (95/100)(85 + 1) = 81.7. Again, the result is not an integer, so we use the midpoint of the 81st and 82nd values after we arrange the data in ascending order. For the dates, the 95th percentile is 1921, the midpoint of 1919 and 1923, the 81st and 82nd values, respectively, as shown in Figure 3-11. Notice that locating this percentile required arranging almost all of the 85 dates to find the 95th percentile! If we recall that approximately 5% of the values are larger than the 95th percentile, we can reduce our task to arranging fewer values in descending order. $(5/100)(85 + 1) = 4.3$, so we can arrange the largest five values in descending order and find the midpoint of the fourth and fifth values. Observe in Figure 3-11 that 1921 is indeed the midpoint of the fourth and fifth largest values.

The next box summarizes the location procedure for a percentile.

Procedure for Finding a Particular Percentile of a Data Set

For the 50th percentile or a smaller percentile:

1. Determine the position, L, the percent of $n + 1$ that corresponds to the percentile (for instance, use 30% of $n + 1$ for the 30th percentile).
2. If L is an integer, arrange the L smallest values in ascending order. The Lth value is the percentile in question.
3. If L is not an integer, round it *down* to an integer, and arrange the $L + 1$ smallest values in ascending order.

The midpoint of the Lth and $(L + 1)$st values is the percentile in question. To locate a percentile larger than the 50th percentile: use the percent of the data values that will be as large or larger than the percentile (for instance, to find the 80th percentile, replace 80 with 20%) in the steps above. Arrange the values in descending rather than ascending order, and locate the Lth or $(L + 1)$st value as necessary.

Question: To find the 31st percentile of a set of 863 observations, how many observations and which ones (largest or smallest) must we arrange in ascending order?

Answer: 268, because $(31/100)(863 + 1) = 267.84$. The midpoint of the 267th and 268th smallest values is the 31st percentile. We arrange the data in ascending order, because the 31st percentile is below the 50th.

Question: If we wanted the 80th percentile, how many and which values must we use?

Answer: Because the 80th percentile is larger than the 50th, we use $(100 - 80)\% = 20\%$ of $(863 + 1) = 172.8$. We need to arrange the largest 173 values in descending order. The midpoint of the two smallest values in this ordered set of 173 values is the 80th percentile.

■ **Example 3-12:** Find the 87th percentile of starting dates for the 85 schools in the NCAA Probation Data Base (Appendix D). The values are reproduced in ascending order.

1785	1787	1794	1795	1798	1801	1802	1817	1819	1819	1820
1831	1834	1834	1838	1839	1842	1844	1845	1845	1847	1849
1850	1851	1851	1853	1855	1855	1856	1856	1857	1858	1860
1861	1862	1863	1863	1864	1865	1866	1867	1867	1868	1868
1869	1869	1870	1870	1870	1871	1872	1873	1873	1875	1876
1876	1876	1878	1880	1883	1884	1885	1885	1885	1886	1888
1888	1889	1889	1890	1890	1891	1891	1892	1894	1907	1910
1911	1912	1913	1919	1923	1925	1927	1954			

■ **Solution:** Step 1: The 87th percentile is larger than the 50th, so we use $13\% = (100 - 87)\%$ of 86, which is 11.2, as the location, L. We will need to arrange 12 values in order.

Step 2: Also, because 87 is greater than 50, we need to arrange the 12 largest values in descending order. These values are:

1954   1927   1925   1923   1919   1913   1912
1911   1910   1907   1894   1892.

Step 3: The 87th percentile is the midpoint of 1892 and 1894 or (1892 + 1894)/2 = 1893. Approximately 87% of the starting dates are before 1893 and approximately 13% afterward.

■ **Example 3-12 (continued):** What percentile is the date 1831?

■ **Solution:** Eleven of the 85 values or approximately 13% are smaller than 1831: 1785, 1787, 1794, 1795, 1798, 1801, 1802, 1817, 1819, 1819, and 1820. So 1831 is the 13th percentile.

Question: What is another statistical term for the 50th percentile?
  Answer: The median, because approximately 50% of the observations are smaller than the median and 50% are larger.

Sometimes, students find the percentile concept confusing. For instance, it may seem that the 95th percentile should be 95. Notice that the value 95 makes no sense for the starting-date data. As there is no year 95 in the data set, this result is nonsensical. Other mistakes include finding 95% of the largest value. For the starting dates, this would be 0.95(1954) = 1856.3, a much earlier date in the data set, not near the end. Finally, some would add 95% of the distance between the smallest and largest value to the smallest value. The distance between the years 1785 and 1954 is 169 years. Ninety-five percent of 169 years is 160.55 years. If we add 160.55 years to 1785, the result is 1945.55, much closer to the actual 95th percentile, but incorrect. For a value, X, to be the 95th percentile, approximately 95% of the observations in the data set must be smaller than X and approximately 5% of the data must be larger than X.

Percentiles increment a data set into 100 groups, then locate values that describe these positions. They must assume values similar to those in the data set, just as the magnitudes of the mean, median, and mode must fall between the largest and smallest values in a data set.

Section 3.4.2 Problems

22. Find the 90th and 95th percentiles of the peak carbon monoxide (CO) levels from the Air Quality Data Base listed in Appendix B. Sacramento's value is 13. What percentile is this?

23. Find the 40th and 45th percentiles of the county population values in the North Carolina Census Data Base (see Appendix E). What is the percentile for Jackson County's value 26,846?

24. Find the 33rd and 67th percentiles of the following stock's closing prices.

Closing Price ($)	Number of Days
15.000	18
15.125	32
15.250	53
15.375	44
15.500	27
16.125	16
16.375	10

25. Find the 5th and 87th percentiles of the set of 300 SAT scores from Problem 21. What percentile is the score 1,273?

Figure 3-12

Minitab output for starting dates of schools in the NCAA Probation Data Base

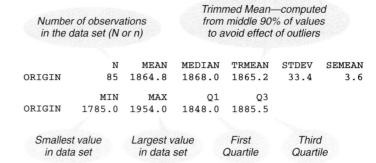

```
                N     MEAN   MEDIAN   TRMEAN   STDEV   SEMEAN
ORIGIN         85   1864.8   1868.0   1865.2    33.4      3.6

              MIN      MAX       Q1       Q3
ORIGIN     1785.0   1954.0   1848.0   1885.5
```

Smallest value in data set   Largest value in data set   First Quartile   Third Quartile

3.5 Location Measures and the Computer

Minitab computes and displays most of the measures we have discussed in this unit. Figure 3-12 shows the output for a description of the starting date of the schools from the NCAA Probation Data Base in Appendix D and Example 3-12.

Question: According to Figure 3-12, what is the largest and smallest possible values for the mean in this data set?

Answer: The minimum value is 1785, and the maximum value is 1954, so the mean must lie somewhere between these two values.

Question: The mean is less than the median and less than the trimmed mean. What does this suggest about the location of outliers if there are any?

Answer: Because the mean is smaller than both other measures that are supposed to be less influenced by outliers, the results suggest that the outlier or outliers are small values. There is no strong suggestion of an outlier, because all three values are very similar in the midst of a data set spanning 1785 through 1954.

The computer makes it easy for us to experiment and observe the effect on descriptive measures when we transform every value in a data set. For instance, suppose we add 100 years to each value in the starting-date data and call this XPLUS. Similarly, XMINUS is the starting data minus 100 years, XTIMES is the product of 100 and the starting date, and XDIVIDE is the quotient of the starting date and 100. Figure 3-13 shows the descriptive output by the computer for each of these variables along with the original descriptive output for starting dates.

Question: Describe the effects of these changes on the mean, median, and quartiles.

Answer: *Each transformation affects the mean in the same way that it affects each value in the data set.* For instance, the mean of the new starting dates after adding

Figure 3-13

Minitab output for transformations of starting dates of schools in the NCAA Probation Data Base

```
                N      MEAN    MEDIAN    TRMEAN    STDEV   SEMEAN
ORIGIN         85    1864.8    1868.0    1865.2     33.4      3.6
XPLUS          85    1964.8    1968.0    1965.2     33.4      3.6
XMINUS         85    1764.8    1768.0    1765.2     33.4      3.6
XTIMES         85    186482    186800    186519     3339       362
XDIVIDE        85    18.648    18.680    18.652    0.334     0.036


                  MIN       MAX        Q1        Q3
ORIGIN         1785.0    1954.0    1848.0    1885.5
XPLUS          1885.0    2054.0    1948.0    1985.5
XMINUS         1685.0    1854.0    1748.0    1785.5
XTIMES         178500    195400    194800    188550
XDIVIDE        17.850    19.540    18.480    18.855
```

100 years, XPLUS, is 1964.8, exactly 100 years more than the original mean. The mean of XDIVIDE is 1864.8/100 = 186.48. *Similar changes occur to the original median and quartiles.* Note that Minitab interpolates between the values in the L and $L + 1$ positions to find the quartile rather than approximating quartiles as the midpoints of the two values.

We employ the descriptive output to learn quickly and easily about the values and information contained in a data set, especially a large one. It is also useful for checking for incorrect data values that result from inaccurately recording the original data (responses or experimental result), transcription errors when recopying at different stages of data collection and presentation, and data-entry errors at the computer. Often we can detect mistakes by checking for unreasonable values for the location measures, such as a mean very different from the value we expect.

Section 3.5 Problems

26. Figure 3-14 displays the Minitab output that describes the variable, peak carbon monoxide (CO) levels from the Air Quality Data Base (see Appendix B). What are the mean and median? What are the first and third quartiles? Study the largest and smallest values in the data set along with the other information to locate any possible outliers.

27. Figure 3-15 displays the Minitab output that describes the county census counts from the North Carolina Census Data Base (see Appendix E). What clues in the figure indicate that this data set is skewed? In which direction?

28. Figure 3-16 shows the Minitab output of descriptive statistics of the closing stock prices from Problem 24. What are the median, mean, and quartiles? Is this data set "roughly" symmetric?

29. Figure 3-17 shows the descriptive statistics for the 300 SAT values from Problem 25. The mean and the trimmed mean are approximately the same value. What does this indicate about the distribution of the values?

	N	MEAN	MEDIAN	TRMEAN	STDEV	SEMEAN
CO	63	7.714	7.000	7.561	2.802	0.353

	MIN	MAX	Q1	Q3
CO	3.000	18.000	6.000	9.000

Figure 3-14
Minitab output for carbon monoxide levels

	N	MEAN	MEDIAN	TRMEAN	STDEV	SEMEAN
TOTAL	100	66286	39443	53072	82057	8206

	MIN	MAX	Q1	Q3
TOTAL	3856	511433	21151	81536

Figure 3-15
Minitab output for North Carolina Census county counts

	N	MEAN	MEDIAN	TRMEAN	STDEV	SEMEAN
CLPRICE	200	15.395	15.250	15.363	0.352	0.025

	MIN	MAX	Q1	Q3
CLPRICE	15.000	16.375	15.156	15.500

Figure 3-16
Minitab output for closing stock prices (Problem 24)

	N	MEAN	MEDIAN	TRMEAN	STDEV	SEMEAN
SATSCORE	300	1019.8	1022.0	1019.7	324.8	18.8

	MIN	MAX	Q1	Q3
SATSCORE	453.0	1590.0	759.2	1269.5

Figure 3-17
Minitab output for 300 SAT values (Problem 25)

Summary and Review

We employ four basic measures to locate the center of a data set. All result in a single-valued expression of the typical value except the mode. A data set can have no mode or many modes. The information we obtain from each is as follows.

1. The mean is the value that balances the data points' distances from the center that it locates.
2. The median is the middle value after the values have been arranged in ascending or descending order.
3. The mode is the value (or values) that occurs more frequently than all of the other values.
4. The weighted mean is a mean that takes into account the varying importance (weight) of values.

It is also possible to estimate the mean of a data set when the only available information is a frequency distribution of the data, by using a procedure based on the weighted mean.

The measure of choice in a particular case depends on the information desired and the distribution of the data.

The mean, median, and mode are identical if the data are symmetrical and have a single mode. They diverge as the data become more irregularly dispersed. The mean is adversely affected by outliers, because it is pulled away from the mass of values in the direction of the extreme point. However, it is the common measure of the center and a major focus of the remainder of this book.

Other useful measures of location (not necessarily the center) are the percentile and quartile. Both provide information about the proportion of the observations that are smaller and larger than a particular value.

The computer provides quick and easy access to descriptive information about data sets, especially large ones. With the output of several location measures we can learn about the probable distribution of the data and check for errors by searching for unreasonable values for these measures.

Multiple-Choice Problems

30. Which of the following is *not* a measure of the central value in a data set?
    a. Mode.
    b. Frequency distribution.
    c. Mean.
    d. Weighted mean.
    e. Median.

31. If we wanted a measure of a typical value in a data set that takes magnitudes and importance of individual observations into account, we would choose the
    a. Mode.
    b. Frequency distribution.
    c. Mean.
    d. Weighted mean.
    e. Median.

32. If we wanted a measure of a typical value in a data set that is unaffected by outliers, can be found for any data set, and is a single value, we would choose the
    a. Mode.
    b. Frequency distribution.
    c. Mean.
    d. Weighted mean.
    e. Median.

33. Find the mean of the following sample of distances of stars from the earth (in light-years): 18.2, 56.9, 24.6, 13.5.
    a. $\bar{X} = 28.30$ light-years.
    b. $\bar{X} = 43.40$ light-years.
    c. $\mu = 28.30$ light-years.
    d. $\mu = 43.40$ light-years.
    e. No mean.

34. Find the median of the data set in Problem 33.
    a. 24.60 light-years.
    b. 21.40 light-years.
    c. 28.30 light-years.
    d. 43.40 light-years.
    e. No median.

35. Find the mode of the data set in Problem 33.
    a. 28.30 light-years.
    b. 21.40 light-years.
    c. 43.40 light-years.
    d. 20.00 light-years.
    e. No mode.

36. Which of the labeled points in Figure 3-18 is the mode?
    a. A.
    b. B.
    c. C.
    d. D.
    e. E.

37. Consider the following scores of course projects.

Grade	Frequency	Grade	Frequency
A	10	D	3
B	12	F	1
C	15		

What is the modal grade?
    a. A.
    b. B.
    c. C.
    d. D.
    e. F.

Figure 3-18
Frequency curve for Problem 36

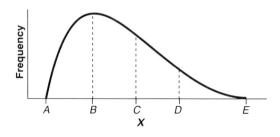

Multiple-Choice Problems

38. Consider the data in Problem 37 and find the median grade.
  a. A.   d. D.
  b. B.   e. F.
  c. C.

39. A manufacturer buys a certain part from several suppliers. Use the following information to find the weighted mean of price changes the manufacturer will experience for the parts that will be purchased and the simple mean of the price changes of the suppliers.

Supplier	Price Change ($)	Number of Parts to Be Purchased
A	+2.0	10
B	−0.5	20
C	+2.0	30
D	+1.0	40
E	0.0	50

  a. +$0.73, +$0.90.   d. +24.44, +0.90.
  b. +$22.0, +$0.90.   e. +24.44, +0.73.
  c. +$0.73, −$0.73.

40. Several construction firms repair and pave roads for a city. The cost per mile for different firms and the number of miles contracted are as follows.

Price per Mile ($)	Miles Contracted
250,000	0.75
200,000	2.05
225,000	0.93
175,000	3.61
180,000	2.98

If a weighted mean of the price per mile paid by the city is calculated to be $1.92, what would be your conclusion?
  a. A mean should have been calculated, rather than a weighted mean.
  b. The bimodal calculation is more representative.
  c. The mode is the best average here, because the price per mile is skewed to the left.
  d. Price per mile has been used as a weight, rather than as a value to be averaged.
  e. The heaviest weight should be given to $250,000, because it is the highest price.

A city's fire department collected a sample of 36 months of total answers to fire alarms. The ten smallest values in the data set are 42, 806, 97, 102, 105, 100, 100, 520, 50, and 600. Use this information to answer Questions 41–44.

41. The 15th percentile is
  a. 100.0 answers.   d. 101.0 answers.
  b. 98.5 answers.   e. 102.0 answers.
  c. 806.0 answers.

42. If the 75th percentile is 938 answers, then
  a. 25% of the values are between 42 and 938 answers.
  b. 75% of the values are smaller than 938 answers.
  c. The 100th value must be 1,000 answers.
  d. The 938th value is the midpoint of 75 and 76 answers.
  e. The location is 75 answers.

43. Using the information in Problem 42, determine the third quartile.
  a. 75 answers.   d. 62 answers.
  b. 25 answers.   e. 703.5 answers.
  c. 938 answers.

44. The first quartile is
  a. 806 answers.   d. 101 answers.
  b. 98.5 answers. e. 703 answers.
  c. 600 answers.

45. What is another name for the 50th percentile of any set of values?
  a. Mean.   d. 50%.
  b. Median. e. Standard deviation.
  c. Mode.

46. Approximate the mean number of music faculty recital performances during the year from the following frequency distribution.

Number of Individual Performances	Number of Faculty
0–2	8
3–5	6
6–8	10
9–11	2
12–14	1
	27

  a. 5.0.   d. 7.0.
  b. 135.0. e. 5.4.
  c. 27.0.

Use the Minitab output of the interval (in months) between vaccination and onset of an illness for 17 vaccine failures as shown in Figure 3-19 to answer Questions 47–50. (Source: David P. Greenberg et al., "Protective Efficacy of *Haemophilus influenzae* Type b Polysaccharide and Conjugate Vaccines in Children 18 Months of Age and Older," *Journal of the American Medical Association*, Vol. 265, No. 8, February 27, 1991, pp. 987–92.)

47. Approximate the mean by using the information in the histogram output.
  a. 17.5 months.   d. 8.2 months.
  b. 15.3 months.   e. 1 month.
  c. 2.1 months.

48. If we describe the distribution as slightly skewed to the right based on the histogram, what relationship of the location measures supports this description?
  a. The third quartile is larger than the first quartile.
  b. The mean is outside the range of values in the data set.
  c. The mean is larger than the median.
  d. The median is between the quartiles.
  e. The second quartile is not given, and the mean is between the first and third quartile.

Figure 3-19
Minitab output for vaccine failures

```
                N      MEAN    MEDIAN    TRMEAN    STDEV   SEMEAN
INTERVAL       17     15.59    15.00     15.33      8.10     1.96

               MIN      MAX       Q1        Q3
INTERVAL      2.00    33.00     9.50     20.50


         Histogram of INTERVAL   N = 17
   Midpoint   Count
       0        1   *
       5        2   **
      10        3   ***
      15        6   ******
      20        2   **
      25        1   *
      30        1   *
      35        1   *
```

49. The second quartile is
  a. 17.00.   d. 15.33.
  b. 15.00.   e. The same as the mode.
  c. 15.59.

50. The middle 50% of the intervals are between which two values?
  a. 15.00 and 15.59.   d. 14.41 and 15.59.
  b. 14.67 and 15.33.   e. 9.50 and 20.50.
  c. 2.00 and 33.00.

Word and Thought Problems

51. What problem can occur from using the mean to locate the center of a data set?

52. Relate each of the following symbols to a concept from this unit: μ, $\overline{X}$, N, n, $\overline{X}_W$, $\overline{X}_G$, M_i, and f_i.

53. Find the mean, median, and mode(s) of the following data sets.
  a. 62, 21, 83, 95, 21, 102, 21.
  b. 0, 0, 1, 2, 10, 9, 9, 8.
  c. 5, 5, 5, 5.
  d. 3.0, −1.2, 8.6, 5.3, −8.0, −8.6.

54. Differentiate among the four measures of central location discussed in this unit by the type of information each conveys. What information do percentiles and quartiles convey?

55. Find the mean, median, and mode(s) of the following U.S. murder rates for the period 1960–1990. Is the data set symmetric? How can you tell? (Source: "A Grisly Record: U.S. Murder Toll Rising to New Level," Associated Press Report in *Asheville Citizen-Times*, August 5, 1991, pp. 1A, 3A.)

5.1	4.8	4.6	4.6	4.9	5.1	5.6	6.2
6.9	7.3	7.9	8.6	9.0	9.4	9.8	9.6
8.8	8.8	9.0	9.7	10.2	9.8	9.1	8.3
7.9	7.9	8.6	8.3	8.4	8.7	9.4	

56. Find the mean, median, and mode(s) of the number of times the teams in the NCAA Probation Data Base (Appendix D) have been on regional television. Based on these measures, predict the distribution of the values, then form a stem-and-leaf display to check your prediction. Does your answer remain the same or change after studying the stem-and-leaf plot?

57. Find the problem with the following statement made by a school superintendent who wishes to stress the educational plight of the district. "About half of our students score below the median national ACT score," he claims. Correct the statement so that it reflects a district with a problem.

58. Locate the mean, median, and mode(s) of annual personal consumption expenditures (in billions of dollars) between 1980 and 1988 for the following categories (Source: U.S. Bureau of the Census, *Statistical Abstract of the United States: 1990*, 110th edition. Washington, D.C.: Bureau of the Census, 1990, U.S. Bureau of Economic Analysis, *The National Income and Product Accounts of the United States, 1929–1982*; and *Survey of Current Business*, July issues):
  a. Total recreation expenditures

  | 115.0 | 128.6 | 138.3 | 152.1 | 168.3 | 185.7 |
  | 201.2 | 224.5 | 246.8 | | | |

  b. Books and maps

  | 5.6 | 6.2 | 6.6 | 7.2 | 7.8 | 8.1 | 8.6 |
  | 9.4 | 9.8 | | | | | |

  c. Radio and television receivers, records, and musical instruments

  | 19.9 | 22.0 | 24.5 | 28.2 | 31.5 | 37.0 |
  | 38.8 | 42.5 | 48.8 | | | |

Word and Thought Problems

    d. Radio and television repair
        2.6   2.7   2.8   2.8   2.8   3.2   3.3
        3.7   3.9
    e. Admissions to motion picture theaters
        2.7   2.9   3.3   3.6   3.9   3.6   3.9
        4.2   4.2
    f. Spectator sports
        2.0   2.0   2.3   2.6   2.9   2.9   2.9
        3.0   3.2
    g. Parimutuel net receipts
        2.1   2.2   2.2   2.3   2.6   2.6   2.6
        2.7   2.8

59. Find a student's mean grade in a course if the scores are 74, 83, 82, 61, 97, 71, and 53, if each test counts 10% of the grade, except for the fifth and sixth scores, which count 30% and 20%, respectively. Do you think the student would prefer a simple mean rather than the weighted mean? Explain.

60. Twelve racing experts have estimated the finishing position of a particular horse in a race. The past accuracy of each expert is listed below, along with the corresponding forecast for the horse. Use this information to find an average estimate of the horse's finishing position that incorporates the information on the relative accuracy of the experts.

Finishing Position	Expert's Previous Accuracy
2	0.80
1	0.72
5	0.53
1	0.48
5	0.72
4	0.66
3	0.82
3	0.60
3	0.55
2	0.61
1	0.60
1	0.58

61. a. Approximate the mean age of children with *Haemophilus influenzae* Type b during an experimental study period by using the following frequency distribution.

Age (months)	Number of Cases
0–17	250
18–23	30
24–35	31
36–59	27
	338

Source: David P. Greenberg et al., "Protective Efficacy of *Haemophilus influenzae* Type b Polysaccharide and Conjugate Vaccines in Children 18 Months of Age and Older," *Journal of the American Medical Association*, Vol. 265, No. 8, February 27, 1991, pp. 987-92.

    b. This frequency distribution appears to disregard suggestions from Chapter 2. Which ones are violated? Can you think of a reason why the violations are reasonable in this case?

62. a. Approximate the mean weekly allowance for first-graders from the following frequency distribution.

Allowance ($)	Number of First-Graders
0.00–0.99	56
1.00–1.99	75
2.00–2.99	66
3.00–3.99	52
4.00–4.99	52
5.00–5.99	72
6.00–6.99	12
7.00–7.99	6
8.00–8.99	2
9.00–9.99	1
10.00–10.99	25
	419

    b. Interpret the result of switching allowance values and number of first-graders in the weighted mean procedure. Perform this calculation to demonstrate your answer.

63. Suppose there are three firms that employ a particular type of worker. Use the following information to determine the weighted mean of wage rates for this type of worker.

Firm	Number of Employees	Wage Rate ($)
A	400	15.00
B	356	14.50
C	92	12.00

64. Why is a weighted mean of price increases, where the weights are the percentage of a typical family's budget spent on the items under consideration, a better measure to use in finding the average price increase for a month (or measuring inflation with the Consumer Price Index) than an unweighted mean of price increases?

65. Explain how an unweighted mean is a special case of a weighted mean.

66. According to *The Wall Street Journal* (April 17, 1984), each year Alexander Grant and Co. of Chicago calculates a rating of states' business climates. Each state's rating is a weighted average of 22 such quantifiable factors as energy costs, taxes, wage rates, days lost through strikes, growth in government spending and revenue, and debt level. The weights of each factor are determined annually by 32 state manufacturers' associations. They choose and weigh the factors that they consider to be most important in selecting a factory site. Discuss the problems that might arise in making comparisons of the data from year to year. Suggest some solutions that minimize the problems of making comparisons among different annual data.

67. A metropolitan hospital has records for the last 100 quarters on the number of births in the hospital. The 26 largest values are

400	360	400	500	550	380	220
100	95	460	500	500	280	300
950	500	200	250	170	420	480
310	250	150	100	270		

a. What is the 85th percentile?
b. What is the third quartile?

68. a. What is the 43rd percentile of the set consisting of consecutive integers from 1 to 100?
b. What are the first, second, and third quartiles of this data set?

69. A four-point grading system awards four quality points per credit hour for an A in a course. For example, an A in a three-hour course equals 12 quality points. Similarly, the following points are awarded per credit hour: three for a B, two for a C, one for a D, and none for an F. The total number of quality points for each student is then divided by the number of credit hours taken. A weighted mean is the result.
a. What are the X_i and W_i in this calculation?
b. What are the limits on the quality-point values? On the credit-hour values?
c. Suppose a student's record consists of 40 credit hours of A work, 45 credit hours of B work, 40 credit hours of C work, 3 credit hours of D work, and 3 credit hours of F work. Find the student's quality-point average.
d. A three-point grading system awards three quality points per credit hour for an A, two for a B, one for a C, and none for a D or F. Recompute the quality-point average by using this system for the student in Part c.
e. Under the four-point system, a student is considered a B student if his or her quality-point average is between 2.5 and 3.5. Under the three-point system, a B student has a point average that is between 1.5 and 2.5. Which system is better for the academic high achiever? Academic low achiever? "Average" student? Explain your answers.

70. a. In diving competitions, several judges observe a dive and each provides a numerical rating of the dive. The high and low values are excluded in calculating the *total score*. Why?
b. If the scorers use the remaining values for a dive to calculate the average rating, which measure, the mean, median, or mode, will be most affected by the exclusion of the two end scores?
c. A diver's *total score* on each dive is multiplied by the dive's degree of difficulty, a predetermined numerical measure of the skill required to perform a dive (a higher value implies that more skill is required). Officials determine the rank of each entrant from the sum of these products for all rounds of diving. Would the rankings change if officials calculated weighted means of a diver's scores by using the degree of difficulty as the weight rather than simply totaling the weighted scores of each dive? Would this ranking method produce the same winner as the method that uses the total score?
d. Suppose scores are weights and degree of difficulty is the X_i in a weighted-mean method used to rank divers. Would use of this method cause divers to choose dives with higher or lower degrees of difficulty?
e. Based on your answers to Parts c and d, can you explain why total scores, rather than means, are used to determine diver rankings?

71. The median measure of tensile strength of an experimental alloy is 152.7 pounds per square inch. The first and third quartiles are 145.3 and 175.6 pounds per square inch. Describe the likely distribution of this set of measurements.

72. Check the Minitab output of location measures in Figure 3-20 with your own calculations of the same measures from the following original data for the 25 airlines that flew revenue passengers the largest number of kilometers (in billions) in 1990. (Source: Airline reports and U.S. Transportation Department data compiled by Air Transport World, in Asra Q. Nomanij and Laurie McGinley, "Airlines of the World Scramble for Routes in Industry Shakeout," *Wall Street Journal*, July 23, 1991, pp. A1, A8.)

244	124	123	95	84	66	63	57	56
55	51	42	37	33	31	28	27	27
27	24	23	23	22	20	19		

73. The Minitab output for 134 death rates from different automobiles for the 1984–1988 models, shown in Figure 3-21, has been altered. Find the location measures that are unreasonable. The stem-and-leaf display is shown to help your detective work. (Source: "Status Report," The Insurance Institute for Highway Safety, April 13, 1991, as reported in "Death Rates in Popular Cars," *Consumers' Research*, July 1991, pp. 23–25.)

74. A psychologist administers a battery of three tests to determine the IQ of an individual. In a random survey of psychologists, 80% rated test 1 as a good measure of IQ, 65% rated test 2 as good, and 50% rated test 3 as good. The psychologist administering the tests uses these ratings and the test data to determine an average IQ to report for the individual, because she believes each test provides varying amounts of different information about IQ. The tested individual scores 118, 110, and 125 on the respective tests. What value would the psychologist report for IQ?

Figure 3-20
Minitab output for airline data

	N	MEAN	MEDIAN	TRMEAN	STDEV	SEMEAN
TRAFFIC	25	56.04	37.00	49.48	49.57	9.91

	MIN	MAX	Q1	Q3
TRAFFIC	19.00	244.00	25.50	64.50

```
               N      MEAN   MEDIAN   TRMEAN   STDEV   SEMEAN
RATE          134    0.2955  1.7500   1.8275   0.8756  0.0756

              MIN     MAX      Q1       Q3
RATE        0.5000  4.7000   1.3000   4.7000


Stem-and-leaf of RATE    N = 134
Leaf Unit = 0.10

    14    0  56677788889999
    45    1  000001111112222223333344444444444
   (36)   1  555555555555666667777788888888999999
    53    2  000000112222222222333334444
    27    2  5555566777889
    15    3  00123
    10    3  556
     7    4  00333
     2    4  57
```

Figure 3-21
Altered Minitab output for death rates in car accidents

75. If we survey television stations across the nation to learn about the levels of employment and advertising, why might we report medians rather than means?

76. Determine the mean, median, and mode(s) of 1989 state legislative appropriations (in thousands of dollars) for state arts agencies from the following list of values. (Source: U.S. Bureau of the Census, *Statistical Abstract of the United States: 1990,* 110th edition. [Washington, D.C.: Bureau of the Census, 1990], from National Assembly of State Arts Agencies, Washington, D.C., unpublished data.)

1,476	1,308	3,248	825	5,971	4,913
22,760	10,024	3,119	457	1,695	2,117
6,747	1,073	19,539	726	710	2,670
338	3,771	1,545	786	339	2,368
12,426	893	55,962	1,431	3,506	1,756
1,021	206	7,509	728	3,177	269
5,005	12,753	3,310	1,783	14,604	20,838
1,970	622	496	465	214	1,440
1,603	1,881				

Unit II: Dispersion Measures

Preparation-Check Questions

77. Suppose you want to study utility bills for May 1983 for people in Louisville, Kentucky. Which of the following is the population?
    a. The set of all May 1983 Louisville utility bills.
    b. A subset of all May 1983 Louisville utility bills.
    c. The set of all May 1983 Kentucky utility bills.
    d. The set of all May 1983 Louisville gas bills.
    e. The set of May 1983 utility bills for randomly selected household in Louisville.

78. What is the mean and range of the following set of blooms per stem value for a hybrid rose: 8, 7, 2, 1, 7?
    a. 7, 7.   d. 7.5, 1.
    b. 5, 7.   e. No mean, range is 7.5.
    c. 4.5, 1.

79. The mean is a measure of
    a. How the observations in a data set are spread out or dispersed.
    b. Central location or a typical value in a set of data.
    c. Variation in a set of data relative to another set of data.
    d. How the data are skewed.
    e. The number of peaks in a data set.

80. If the 3rd quartile in a set of grades is 83, then
    a. 83% of the grades are smaller than 83.
    b. 17% of the grades are smaller than 83.
    c. 25% of the grades are smaller than 83.
    d. 75% of the grades are smaller than 83.
    e. 83% of the grades are smaller than 75.

Answers: 77. a; 78. b; 79. b; 80. d.

Introduction

Horse Races Suppose you are at Churchill Downs deciding which horse to bet on in the Kentucky Derby. You have narrowed your choices to three horses who seem to be identical in all respects:

1. The jockeys have identical records and weights.
2. The horses have run and won the same number of races.
3. Their mean or average time for running the distance is the same, two minutes and nine seconds, $\mu = 2.15$ minutes. We use μ, because we are interested in all of the times they have run this distance.

Tossing a coin may seem the only option at this point to choose among the horses, but only if it is post time. If not, there may yet be some useful information available to you. If you have actual times for these horses in races of the same distance, you could examine them to see which horse most consistently runs its average time.

Figure 3-22 depicts times along the number line for the three race horses.

Question: Which horse appears to be more consistent in running a time near the mean?

Answer: Horse B is more consistent in running the mean time than is Horse A. However, most of Horse C's times are tightly clustered about a time slightly faster than the mean time. This time shifts toward a larger mean time, because of the *outlier*, the one much longer run time.

Question: How would you use the information to choose a horse, if you favor a horse that consistently runs a fast time? How would your response differ, if you favor a risky horse that is sometimes quite fast and sometimes quite slow?

Answer: If you favor consistency about the fastest time, Horse C looks good, but only if we can explain the one peculiar race time. Perhaps further investigation might reveal a sloppy track or other circumstances that make this time incomparable with the other races. If you prefer risky bets, especially those with a higher expected payoff, you would choose Horse B, which, along with Horse A, has run the fastest time, but B has run fewer slow races than A.

Although the mean of all three data sets in Figure 3-22 is 2.15 minutes, the data sets clearly differ in the way they are *dispersed*, that is, the way they are spread out about the mean. Horse A's times are more dispersed than B's, because many of B's times are close to the mean time and none are farther from the mean. All but one of Horse C's times are tightly clustered about the mean, but the one is an outlier. It is not always so easy to determine which set of numbers is more dispersed when comparing two or

Figure 3-22
Distribution of race times for three horses

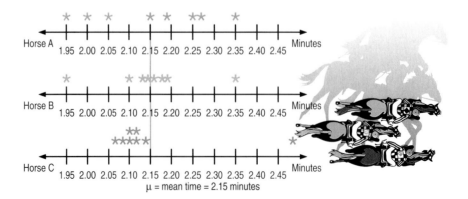

more sets. Statisticians have developed a measure that summarizes the spread of the data about the mean. It is the standard deviation, the major topic of this unit.

Often further research into specific outliers provides information about conditions that alter the outcome of a phenomenon we are studying. For example, if the outlier for Horse C occurred because of a sloppy track, we would be alert for the track conditions when choosing among the three horses, and we might study the other horses' records on muddy fields. In this unit we establish more precise standards for identifying outliers by explicitly defining "extremely different," using the new dispersion measure.

3.6 Dispersion in a Data Set

Means, medians, and modes are single-value measures of central location. In this chapter, we develop another value that adds to the information obtained from the mean. It is a *single value for measuring the spread or dispersion of the data points*—especially as they relate to the mean.

The mean race time for each horse in Figure 3-22 is 2.15 minutes. This measure condenses all three sets of times to the same value, so that we cannot distinguish the differences in the data sets. Times for Horses B and C tend to be more compact with a few exceptions, while those of Horse A are more spread out. We need a measure to differentiate among the data sets, even though the means are identical.

In addition, we use this measure when we want to know about the dispersion of a single data set, just as we use the mean to tell us about its center. For instance, using the mean to represent a widely dispersed data set may be misleading, if we think of most values being similar in magnitude to the mean. If the mean lifetime associated with an automobile is advertised as 110,000 miles, a surprised customer may find a wide variety of lifetimes represented by the single value. Often a report that includes the mean but lacks a measure of dispersion dissatisfies us or at least makes us wary.

We will explore several alternative measures for the dispersion of values in a data set. The behavior of each and the problems we encounter with each lead us finally to the usual measure of dispersion called the standard deviation. This evolution should give you a more intuitive understanding of what the standard deviation measures as well as how to compute it.

3.7 Alternative Measures of Dispersion

3.7.1 Range

The first alternative measure we will consider is the *range*, the difference between the largest and smallest values in the data set. Actually, we employed the range when we determined class sizes for a frequency distribution.

■ **Example 3-13:** Compute the range for each of the three populations of race times for Horses A, B, and C.

Horse A times: 1.95, 2.00, 2.05, 2.15, 2.19, 2.25, 2.26, 2.35.
Horse B times: 1.95, 2.10, 2.13, 2.14, 2.16, 2.18, 2.19, 2.35.
Horse C times: 2.08, 2.09, 2.10, 2.10, 2.11, 2.11, 2.13, 2.48.

■ **Solution:** The range for all three populations is 0.40 minutes; however, the calculations involve different values. Both calculations for Horses A and B are 2.35 − 1.95 = 0.40 minutes. For Horse C the computation is 2.48 − 2.08 = 0.40 minutes.

The range is inadequate to indicate a difference in spread for these three populations of race times, which we know are very dissimilar.

There is another problem with using the range as a measure of dispersion: It is especially sensitive to outliers. Look at Horse C. Most of the observations in this population cluster more tightly than those in the first two populations, yet the three ranges are identical. If the largest value for Horse C were 4.00 minutes, the range would be 2.92 minutes, much larger than the other ranges in spite of the fact that seven out of eight values lie within 0.05 minutes (3 seconds) of one another. This contradiction occurs because range only measures the distance between two end values in a data set and ignores the middle observations.

The range is useful, as we have seen, in developing frequency distributions and graphs. However, we need a measure that is less influenced by outliers and that accounts for more of the values in the data set and their distribution.

3.7.2 Interquartile Range

One alternative to the range is the *interquartile range*. This is the difference between the third and first quartiles. We know that the first and third quartiles of the university starting dates from the NCAA Probation Data Base (Appendix D) are the years 1848 and 1885.5, respectively. Thus, the interquartile range would be 1885.5 − 1848 or 37.5 years.

Recall that the first quartile is a value, such that one-fourth of the observations are smaller than it and three-fourths are larger, while three-fourths are smaller than the third quartile and one-fourth are larger. Consequently, the interquartile range is actually the range of the middle half of the values in the data set.

■ **Example 3-13 (continued):** Find the interquartile range for each of the three horses' times.

> Horse A times: 1.95, 2.00, 2.05, 2.15, 2.19, 2.25, 2.26, 2.35.
> Horse B times: 1.95, 2.10, 2.13, 2.14, 2.16, 2.18, 2.19, 2.35.
> Horse C times: 2.08, 2.09, 2.10, 2.10, 2.11, 2.11, 2.13, 2.48.

■ **Solution:** Because each data set has eight observations, the first quartile is the midpoint of the second and third smallest values, and the third quartile is the midpoint of the sixth and seventh smallest values (the values are already listed in ascending order). The quartiles are 2.025 and 2.255 for Horse A, 2.115 and 2.185 for Horse B, and 2.095 and 2.120 minutes for Horse C. The corresponding interquartile ranges are 0.230, 0.070, and 0.025 minutes.

The different dispersions are apparent when we compare the interquartile ranges of the original populations. The interquartile range is unaffected by the outlier for Horse C, and the clustering of points for this horse relative to the other two is obvious from its interquartile range. In fact, an outlier like 4.00 minutes would have no impact on the interquartile range. In addition, the interquartile range picks up the relative compactness of Horse B's times compared to Horse A's.

The range and interquartile range are useful measures of dispersion. We can determine their values relatively easily to obtain a quick idea of the interval spanned by the data. In addition, using the interquartile range eliminates the outlier effect associated with the range as a measure of dispersion. However, in the process of computing the range or interquartile range, we ignore information from all but two or four of the observations. It is generally preferable to use a measure that is not subject to a large influence by so

Section 3.7 Alternative Measures of Dispersion 111

few of the values in the data set. We favor a measure that takes each observation into account, not just the end points or the points on either side of the quartiles.

3.7.3 Mean Absolute Deviation

To understand the other alternative measures of dispersion, we must define a deviation. A *deviation* is the difference between a value in the data set and the mean, $X_i - \mu$. It locates X_i relative to the mean by giving us the distance between the two points and the direction to proceed from the mean to reach the point, toward smaller or toward larger values. The deviations for Horse A are

X_i (minutes)	$X_i - \mu$ (minutes; $\mu = 2.15$ minutes)
1.95	−0.20
2.00	−0.15
2.05	−0.10
2.15	0.00
2.19	0.04
2.25	0.10
2.26	0.11
2.35	0.20

The deviation for 1.95 is −0.20 minutes, which indicates that this observation is 0.20 minutes from the mean. The minus sign indicates that the observation is below or smaller than the mean:

```
           −0.20 minutes
      ←──────────────────
                          → Minutes
         1.95          2.15
                       Mean
```

There is a deviation for every observation. Just as we employ a mean to avoid studying numerous values in a data set, we need to summarize the set of deviation values with a single value. We also want a measure that accounts for each observation—in this case, each observation is a deviation. The logical choice is to find the mean of all of the deviations. But summing the column of deviations in the list for Horse A produces a total of zero, $\Sigma(X_i - 2.15) = 0$. Consequently, the mean of the deviations is $0/8 = 0$.

■ **Example 3-13 (continued):** Find the deviations and calculate the mean for each of the other horses.

■ **Solution:**

	Horse B		Horse C	
	Times	Deviations	Times	Deviations
	1.95	−0.20	2.08	−0.07
	2.10	−0.05	2.09	−0.06
	2.13	−0.02	2.10	−0.05
	2.14	−0.01	2.10	−0.05
	2.16	0.01	2.11	−0.04
	2.18	0.03	2.11	−0.04
	2.19	0.04	2.13	−0.02
	2.35	0.20	2.48	0.33
$\Sigma(X_i - 2.15)$		0.00		0.00

Again, the means of the deviations are zero. It happens every time: *the sum of the deviations from the mean for any data set will be zero*, because the mean balances the positive and negative distances. This measure will not distinguish one population from any other. It even gives a false impression of the distribution of the deviations.

Question: When we say that the typical deviation is zero, what does this imply about the set of values?

Answer: Technically, we know that this zero balances the deviations, no matter how large or small each one may be, but we often think of the mean as a typical value. So, a mean deviation of zero may leave an impression that there is no (or zero) dispersion in the set of numbers. This conclusion incorrectly describes the horse data sets and many other ones.

Although this measure is impractical for distinguishing the dispersion of different data sets, finding the mean of the deviations is appealing, because it is a direct application of a familiar measure. The other measures we introduce modify this basic procedure to avoid the zero-sum problem.

The first of these measures of dispersion avoids the zero-sum problem of the mean deviation by ignoring the sign (+ or −) associated with the deviation (called its absolute value) and concentrating on the distance of each from the mean. It is the mean of the absolute values of the deviations, called the *mean absolute deviation* (often referred to as the MAD):

$$\text{Mean absolute deviation} = \frac{\sum_{i=1}^{N} |X_i - \mu|}{N}.$$

■ **Example 3-13 (continued):** Find the MAD for the three horses' times.

■ **Solution:**

Horse A		Horse B		Horse C	
$X_i - \mu$	$X_i - \mu$	$X_i - \mu$	$X_i - \mu$	$X_i - \mu$	$X_i - \mu$
−0.20	0.20	−0.20	0.20	−0.07	0.07
−0.15	0.15	−0.05	0.05	−0.06	0.06
−0.10	0.10	−0.02	0.02	−0.05	0.05
0.00	0.00	−0.01	0.01	−0.05	0.05
0.04	0.04	0.01	0.01	−0.04	0.04
0.10	0.10	0.03	0.03	−0.04	0.04
0.11	0.11	0.04	0.04	−0.02	0.02
0.20	0.20	0.20	0.20	0.33	0.33
0.00	0.90	0.00	0.56	0.00	0.66

Dividing the sum of the absolute value of the deviations by 8 produces 0.112, 0.07, and 0.082 as the respective MADs.

This alternative measure does display a difference between the three populations and is a valuable alternative measure. Notice that, because this measure accounts for each deviation, Horse C's outlier pulls the MAD away from most of the deviations that lie between 0.02 and 0.07. In spite of this problem, the MAD is intuitively appealing, because it does account for each observation and it directly averages the deviations with a familiar process. However, it is not used as widely as the measure to be explained next, because it is more difficult to manipulate absolute values algebraically. Meanwhile, most statistical theory and application is based on the next measure, the standard deviation.

Section 3.7 Problems

Fill the empty cells in each row of the following table.

Problem	Data from Problem	μ	Range	Quartile 1st	Quartile 3rd	Interquartile Range	Sum of Deviations	Mean Absolute Deviation
81	5	143.6						
82	6	6.44						
83	7	0.1154						
84	8	9,200						
85	9	6.68						

3.8 The Population Standard Deviation

3.8.1 Calculation

Another way to eliminate the effect of the minus signs associated with the deviations is to square each value. This procedure is the basis for the new measure, the standard deviation. For Horse A, the squared values are

X_i	$X_i - \mu$	$(X_i - \mu)^2$
1.95	−0.20	0.0400
2.00	−0.15	0.0225
2.05	−0.10	0.0100
2.15	0.00	0.0000
2.19	0.04	0.0016
2.25	0.10	0.0100
2.26	0.11	0.0121
2.35	0.20	0.0400
		0.1362

Next, we find the mean of the squared values: $0.1362/8 = 0.017025$ (we do not round yet, because this is an intermediate step for the standard deviation). This value is called the *population variance*, and its symbol is σ^2 (the lowercase Greek letter sigma, squared).

> The population variance is
> $$\sigma^2 = \frac{\sum_{i=1}^{N}(X_i - \mu)^2}{N}.$$

The following box details the steps to obtain the population variance.

Procedure for Obtaining the Population Variance

1. Calculate the population mean, μ.
2. Find the deviation, $X_i - \mu$, for each observation, X_i.

(continued on the next page)

3. Square each of the deviations from Step 2.
4. Sum the squared deviations from Step 3.
5. Divide the sum from Step 4 by the number of observations in the population, N, to produce the population variance.

Note that using a table or series of columns to complete Steps 2 and 3 facilitates hand computations.

■ **Example 3-13 (continued):** Find the variance of each population of the other two horses' times.

■ **Solution:** Step 1: We already know $\mu = 2.15$ minutes for both horses from earlier calculations.

Step 2: Next we find the deviations shown in the second columns below for each horse.

Horse B			Horse C		
X_i	$X_i - \mu$	$(X_i - \mu)^2$	X_i	$X_i - \mu$	$(X_i - \mu)^2$
1.95	−0.20	0.0400	2.08	−0.07	0.0049
2.10	−0.05	0.0025	2.09	−0.06	0.0036
2.13	−0.02	0.0004	2.10	−0.05	0.0025
2.14	−0.01	0.0001	2.10	−0.05	0.0025
2.16	0.01	0.0001	2.11	−0.04	0.0016
2.18	0.03	0.0009	2.11	−0.04	0.0016
2.19	0.04	0.0016	2.13	−0.02	0.0004
2.35	0.20	0.0400	2.48	0.33	0.1089
		$\Sigma(X_i - \mu)^2 = 0.0856$			$\Sigma(X_i - \mu)^2 = 0.1260$

Step 3: The third columns display the square of each deviation.
Step 4: We sum the squared deviations, shown at the bottom of the third column of each table.
Step 5: Finally we average the squared deviations by dividing the sum from Step 4 by the number of observations, in this case, 8, to obtain $\sigma^2 = 0.0856/8 = 0.01070$ for Horse B and $\sigma^2 = 0.1260/8 = 0.01575$ for Horse C.

The population variance is the mean or average of the squared deviations. It is quite useful in statistics, but our major concern at this point is the population standard deviation, represented by σ, which is the square root of the variance.

> The population standard deviation is
> $$\sigma = \sqrt{\sigma^2} = \sqrt{\frac{\sum_{1}^{N}(X_i - \mu)^2}{N}}.$$

The standard deviation is a type of "average" of the deviations, in the sense that it provides a typical or central value for the set of deviations, although we do not compute it in the traditional manner introduced in the discussion of averages and means. Notice that the variance deals in square units. Consider the steps in the computations for each horse. The original values are times given in minutes, so the mean is 2.15 minutes. Hence, the deviation is in minutes also. For example, 1.95 minutes − 2.15 minutes = −0.20 minutes. Squaring deviations results in squared units, such as $(-0.20 \text{ minutes})^2 = 0.0400$ square minutes. Thus, we express the variance, the mean

Section 3.8   The Population Standard Deviation

of the squared deviations, in square units, 0.017025 square minutes for Horse A. To return to the original units (minutes) and to obtain a *standard* deviation (because we are trying to find an "average" deviation rather than an "average" squared deviation), we take the square root of the variance.

To obtain the population standard deviation, follow Steps 1 through 5 (see box on pages 113–114) for calculating the population variance, then the final step is:
6. Take the square root of the variance, the result of Step 5.

Thus, the standard deviation for Horse A is
$$\sigma = \sqrt{0.017025} \text{ square minutes} = 0.130 \text{ minute.}$$

■ **Example 3-13 (continued):** Find the standard deviation for the other horses.

■ **Solution:** For Horse B:
$$\sigma = \sqrt{0.01070} \text{ square minutes} = 0.103 \text{ minute.}$$

For Horse C:
$$\sigma = \sqrt{0.01575} \text{ square minutes} = 0.125 \text{ minute.}$$

■ **Example 3-14:** Find the standard deviation for the following population of test scores: 20, 90, 100, 50, 80.

■ **Solution:** The following table keeps track of these steps.

X_i	$X_i - 68$	$(X_i - 68)^2$
20	−48	2,304
90	22	484
100	32	1,024
50	−18	324
80	12	144
340		4,280

Step 1: The mean of the values is $340/5 = 68$.
Step 2: We find the deviations by subtracting 68 from each value. The second column of the table records the results.
Step 3: Column three lists each of the squared deviations.
Step 4: The mean of the squared values is $4,280/5 = 856$, the population variance.
Step 5: The standard deviation is the square root of the variance, $\sigma = \sqrt{856} = 29.3$ points.

3.8.2 Interpretation

A standard deviation is a single value that measures the dispersion of data about the mean. A larger standard deviation indicates a more dispersed set of data points.

Consider again the populations of race times as shown in Figure 3-22 (p. 108). These graphs remind us that Horse A's values appear to be more dispersed than the others, resulting in a larger standard deviation. Meanwhile, most of C's values are tightly clustered, but the outlier will expand the average deviation. B's values tend to cluster less than the nonoutliers of C's, but there is no outlier in B. Consequently, the relative magnitudes of the standard deviations, 0.130, 0.103, and 0.125 for A, B, and C, respectively, do reflect the rank of the degree of dispersion in each of the data sets, given that this measure takes each value, even outliers, into account.

It is more difficult to interpret the actual magnitudes of these values. In an important sense, each value is an "average deviation" of the set of deviations it represents.

Although the averaging process is not as simple as the traditional calculation, the new "average" has an interpretation similar to the traditional mean, because it is a measure of the central value for all of the deviation values.

Question: Referring to Horse A's race times in Figure 3-22 (p. 108), would a standard deviation of 0.50 minute be reasonable for Horse A?

Answer: No, because the largest deviation is 0.20 minute. The other deviations are smaller than 0.20, and the "average" of values between 0.00 and 0.20 cannot be 0.50.

In a similar vein, 0.005 minute is too small to be the standard deviation of Horse B's times, because the smallest deviation is 0.01 minute.

Question: Suppose a set of values falls in the interval 500–800. Is it reasonable to suspect that the mean could be 650 and that the standard deviation is 200?

Answer: The mean of 650 is reasonable, because it is between 500 and 800. The standard deviation could not be 200, because the largest possible deviation is 800 − 650 = 150. The 200 is too large to be the "average" of the deviations.

In addition, both the standard deviation and the mean entail similar problems with outliers. In fact, outliers especially influence the standard deviation, because in a sense we weight each deviation by its own size when we square the deviation. Thus, not only is the deviation for an outlier larger, but also we multiply it by the same large value in the process of finding the standard deviation.

Nevertheless, in statistical analysis the standard deviation is the primary measure of dispersion of values in a data set. More specifically, its magnitude represents the typical or "average" distance between the mean and a value in the data set.

3.8.3 Applications of the Standard Deviation

Accuracy Gauge Remembering that the standard deviation is the typical distance of a data point from the mean will help when we employ standard deviations in different situations. Generally, when the standard deviation is large, we expect descriptions and inferences about data sets to be more difficult and less accurate. The mean of a data set with a small standard deviation provides a better impression of the typical value than does the mean of one with a large standard deviation. Thus, the standard deviation is a gauge of the precision of descriptive and inferential analyses.

Data Set Comparisons We can use standard deviations for comparing dispersion among different data sets as we demonstrated in Example 3-13 with the horse data. In cases where the means differ, we often include the standard deviation in a different comparison device called the coefficient of variation (see Box 3-1).

Empirical Rule Standard deviations are also useful when the distribution of the values fit a normal or "bell-shaped" curve as depicted in Figure 3-23. Many variables do follow such a pattern. Examples include psychological and biological characteristics of large populations, such as heights and intelligence test scores. We will soon learn to work with this curve and with variables whose distributions fit this configuration, but for now we will simply demonstrate the curve's relation to the mean and standard deviation, called the *Empirical Rule*. The curve is symmetric, so the mean is the center value. About 68% of the population values will be within one standard deviation of the mean, about 95% within two standard deviations, and almost 100% within three standard deviations (see Figure 3-23). If the mean height of basketball players across the nation were 6 feet with a standard deviation of two inches, then about 68% of all the players' heights would lie between 5'10" and 6'2", about 95% between 5'8" and 6'4", and about 100% between 5'6" and 6'6". Knowing the mean and standard deviation of a variable,

Box 3-1 Measuring Risk with the Coefficient of Variation

Some stock prices are more volatile than others, meaning that the price tends to fluctuate up and down by larger amounts and more often, increasing the likelihood of making a sizable profit and also the risk of sustaining a larger loss. A stock with no or very little change in price offers little opportunity for profit or loss. By comparing information on previous prices of each stock, investors can assess the risk associated with each one and then choose according to the investor's preference for gambling.

An obvious measure for assessing volatility is the standard deviation, because it measures the dispersion of a data set about its mean. But if you compare the standard deviation alone for each of the stocks being considered, you will miss an important element of the profit-and-loss picture.

Consider that a large-price stock's deviations vary more in absolute magnitude than a small-price stock. Stock A with an average price of $1 will likely experience variation in the cents range rather than in dollars. Suppose we use $0.50. On the other hand, finding a standard deviation of several dollars, say $2, for Stock B, if it has an average price of $50, should not surprise us.

If we like taking risks in hopes of making more money and choose B, we would be mistaken. By contrasting the typical return on each dollar we invest in a stock rather than the return on a share of the stock, we would buy A rather than B. The typical return or typical loss per share is measured by the standard deviation of the price, the "average" price change. For A, the risk for each dollar invested is 0.50 (50% when expressed as a percentage), because the average price is $1. For B, the risk is $2 when you purchase a share for the average price of $50, so the risk per dollar would be $2/$50 or 0.04 or 4%. Clearly, A affords the greater risk per dollar spent on the stock. (Remember, you could buy 50 shares of A, so the standard deviation of A multiplied by 50, in this case 0.5(50) = 25.0, represents a greater risk than the value of 2 associated with B.)

This ratio of the standard deviation of a data set to its mean is called the *coefficient of variation*. The result is a proportion or a percentage when the result is multiplied by 100. Economists and financiers use this measure to standardize data for comparisons. The results for the two stocks, 50% and 4%, are readily comparable with the relatively assured return from interest for $1 placed in a savings account.

Another benefit of the coefficient of variation for comparison purposes is that measurement units do not matter. Standard deviations of 10 yen and $10 are not necessarily equally risky. However, the mean price of the stock expressed in yen, say 25 yen, produces a coefficient of variation without units, 10 yen/25 yen = 0.4 or 40% that can be compared with the coefficient of variation for a stock whose price is expressed in dollars. If the coefficients of variation for two stocks are the same, we judge the risks to be equal.

Try visualizing the information represented in the coefficient of variation by considering the following problem. To keep customers and inspectors satisfied, manufacturers strive for consistency of operation when filling the contents of food containers. The standard deviations of two production lines for filling catsup bottles are both 1 ounce. One line fills 20-ounce bottles and the other 40-ounce bottles. Which line presents the bigger problem for consistency of fills?

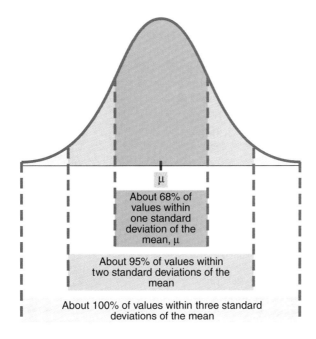

Figure 3-23
Standard deviations and the normal curve

Figure 3-24
The empirical rule applied to the university starting dates

and that it follows this pattern, allows us to extract information about the values without examining the entire set of values individually.

Figure 3-24 demonstrates this rule with a real data set, the university starting dates. Part a shows that the data is mound shaped and approximately symmetric, although not

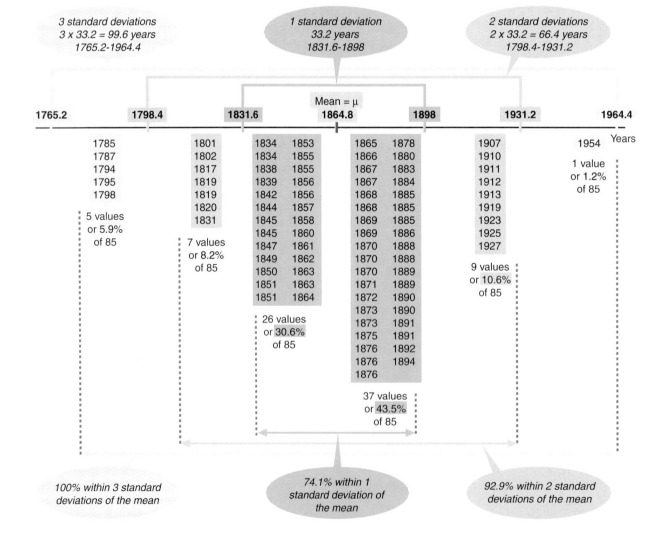

Section 3.8 — The Population Standard Deviation

perfectly bell shaped. Consequently, we should expect the empirical rule to be approximately correct as Part b shows.

Chebyshev's Inequality Not all data sets conform to the normal curve, but we do know something about the relationship between the standard deviation and the proportion of observations that lie within certain distances of the mean for any data set. Chebyshev's inequality (named for the man who first proved it) states that *the proportion of observations falling within K standard deviations of the mean, μ, must be at least $1 - (1/K^2)$, if K is any positive value greater than 1*. The "at least" part of the rule means that the proportion we calculate will be the minimum proportion falling in the prescribed interval. It is possible that more than this proportion will fall in the interval.

If we want to know the minimum proportion of observations in a data set falling within two standard deviations of the mean, we use $K = 2$ and determine that the minimum proportion is $1 - (1/2)^2 = 3/4$ or 75%. Similarly, at least 8/9 or 89% fall within three standard deviations ($K = 3$), as shown in Figure 3-25 for a set of 134 death rates (deaths per 10,000 registered passenger cars) from different automobiles for the 1984–1988 models. The stem-and-leaf display in Figure 3-21 (page 107) shows the skewness of the data. If we consider these data to be a population, the mean death rate is 1.896 and the standard deviation is 0.872. In this case, 94.8% of the values lie within two standard deviations, which is more than 75%. Similarly, 99.3% lie within three standard deviations, which is more than 89%.

Figure 3-25
Chebyshev's inequality: Using the standard deviation and mean to obtain information from data set (death rates in car accidents)

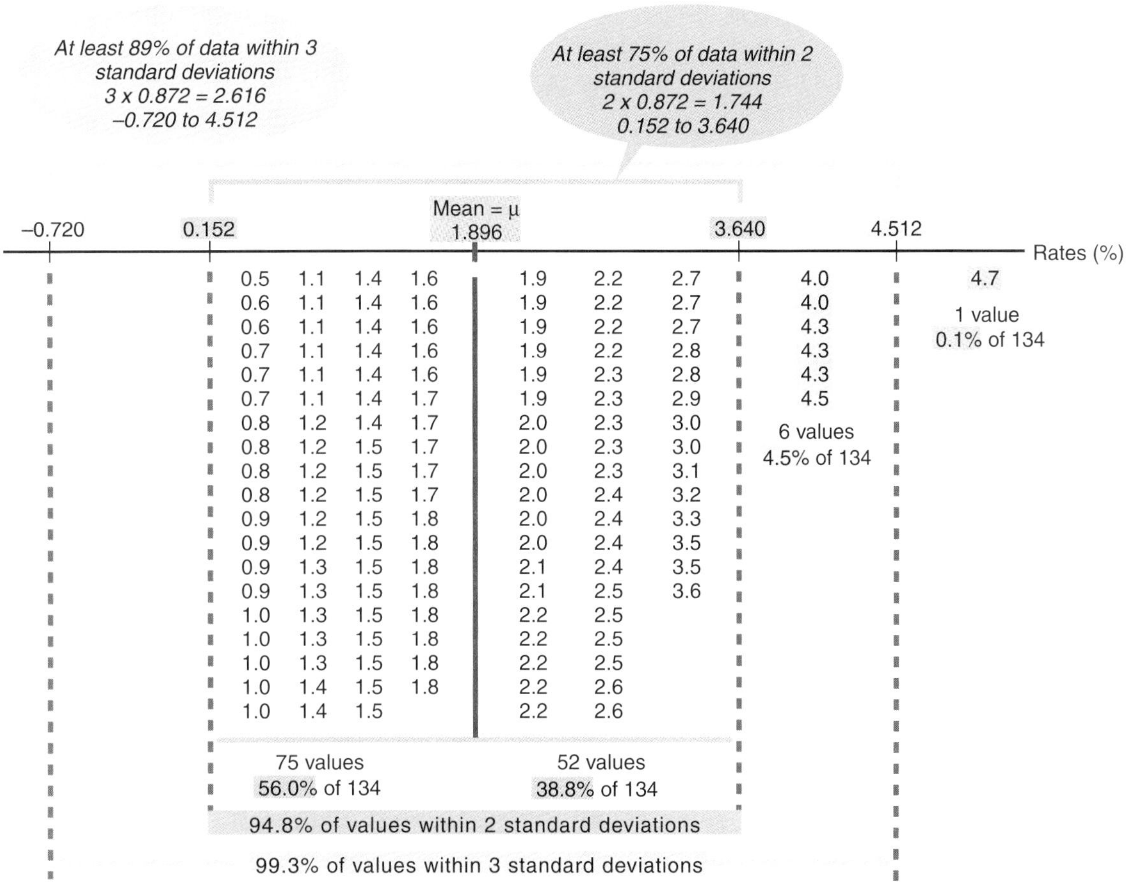

At least 89% of data within 3 standard deviations
3 × 0.872 = 2.616
−0.720 to 4.512

At least 75% of data within 2 standard deviations
2 × 0.872 = 1.744
0.152 to 3.640

Mean = μ
1.896

−0.720   0.152   1.896   3.640   4.512   Rates (%)

0.5	1.1	1.4	1.6	1.9	2.2	2.7	4.0	4.7
0.6	1.1	1.4	1.6	1.9	2.2	2.7	4.0	
0.6	1.1	1.4	1.6	1.9	2.2	2.7	4.3	
0.7	1.1	1.4	1.6	1.9	2.2	2.8	4.3	
0.7	1.1	1.4	1.6	1.9	2.3	2.8	4.3	
0.7	1.1	1.4	1.7	1.9	2.3	2.9	4.5	
0.8	1.2	1.4	1.7	2.0	2.3	3.0		
0.8	1.2	1.5	1.7	2.0	2.3	3.0		
0.8	1.2	1.5	1.7	2.0	2.3	3.1		
0.8	1.2	1.5	1.7	2.0	2.4	3.2		
0.9	1.2	1.5	1.8	2.0	2.4	3.3		
0.9	1.2	1.5	1.8	2.0	2.4	3.5		
0.9	1.3	1.5	1.8	2.1	2.4	3.5		
0.9	1.3	1.5	1.8	2.1	2.5	3.6		
1.0	1.3	1.5	1.8	2.2	2.5			
1.0	1.3	1.5	1.8	2.2	2.5			
1.0	1.3	1.5	1.8	2.2	2.5			
1.0	1.4	1.5	1.8	2.2	2.6			
1.0	1.4	1.5		2.2	2.6			

75 values
56.0% of 134

52 values
38.8% of 134

6 values
4.5% of 134

1 value
0.1% of 134

94.8% of values within 2 standard deviations
99.3% of values within 3 standard deviations

■ **Example 3-15:** Determine the minimum proportion of the values that must lie within two-and-a-half standard deviations of the mean for the automobile data in Figure 3-25. Then, verify your answer by finding the actual proportion that lie within these limits.

■ **Solution:** The minimum proportion within two-and-a-half standard deviations ($K = 2.5$) is $1 - (1/2.5)^2 = 1 - (1/6.25) = 0.84$, or 84%. The distance spanned by 2.5 standard deviations would be $2.5(0.872) = 2.18$ deaths, so values within 2.5 standard deviations of 1.896 must lie within the interval -0.284 to 4.076. Notice that only five values (3.7% of the 134 values) lie outside this interval: 4.3, 4.3, 4.3, 4.5, and 4.7. Thus, 96.3% of the values lie within the interval, which is greater than the minimum 84% predicted by Chebyshev's inequality.

Chebyshev's rule is more general than the Empirical Rule, because it works for any data set, whereas the Empirical Rule only pertains to bell-shaped data distributions. Thus, 95% of bell-shaped distributed data lie within two standard deviations of the mean, and the 95% exceeds Chebyshev's minimum requirement of 75% within two standard deviations.

Outlier Detection Knowing the proportions described by the Empirical Rule and Chebyshev's inequality helps us establish guidelines for locating outliers. Because both require that a large proportion of values fall within 2.5 standard deviations of the mean, *we will call any value more than 2.5 standard deviations from the mean an outlier*. The chances of any value exceeding this bound is 16% and only 1% if we know the data set fits the bell-shaped curve.

To determine the number of standard deviations between a point and the mean, we convert the distance between the value and the mean, its deviation, to standard deviations. Dividing the deviation by the standard deviation accomplishes this transformation, called *standardizing* a value. The result is called a *standard score* or *Z score*.

To determine the Z score for a value expressed in original units use

$$Z = \frac{\text{Deviation}}{\text{Standard deviation}} = \frac{X_i - \mu}{\sigma}.$$

Because the standard deviation, σ, relates the typical or average distance of a point in the data set from the mean, the formula specifies the relative distance of a specific point from the mean compared to the typical point's distance. When this relative distance is large, we suspect that the point is an outlier.

■ **Example 3-13 (continued):** Find the Z scores for the race time values for all three horses, and identify any outliers using the new definition.

■ **Solution:**

Horse A $Z_i = \frac{(X_i - 2.15)}{0.130}$		Horse B $Z_i = \frac{(X_i - 2.15)}{0.103}$		Horse C $Z_i = \frac{(X_i - 2.15)}{0.125}$	
X_i	Z_i	X_i	Z_i	X_i	Z_i
1.95	−1.54	1.95	−1.94	2.08	−0.56
2.00	−1.15	2.10	−0.49	2.09	−0.48
2.05	−0.77	2.13	−0.19	2.10	−0.40
2.15	0.00	2.14	−0.10	2.10	−0.40
2.19	0.31	2.16	0.10	2.11	−0.32
2.25	0.77	2.18	0.29	2.11	−0.32
2.26	0.85	2.19	0.39	2.13	−0.16
2.35	1.54	2.35	1.94	2.48	2.64

Box 3-2 Astronomical Measures

Scientists repeat experiments because measures of results usually vary with each repetition. Reasons for the variation include measuring-instrument accuracy, human error, and unexpected or uncontrolled circumstances, such as using an impure compound. Frequently, scientists combine the various values for a given measure and report only a single representative value, such as the mean.

Astronomers use this procedure when measuring the brightness of distant objects in space. Powerful telescopes electronically repeat the brightness measures of a single object at time intervals specified by the astronomer. This process easily produces a large set of measures of what is in actuality a single value.

Astronomers do not automatically average these measurements. They carefully inspect the data set for outliers produced by such events as a bird or airplane flying in the path of the light being measured and distorting the image or the information for measuring the distance. They discard obvious mismeasures, often five or 10 standard deviations from the mean. The reported measure includes only credible values. In addition, astronomers report the standard deviation to assess the error involved in the measurement process.

Source: Conversation with Dr. Paul Heckert, Astronomer, Western Carolina University, Department of Chemistry and Physics, July 4, 1991.

The only outlier more than 2.5 standard deviations from the mean (the new definition) is the time of 2.48 minutes for Horse C.

This definition of an outlier provides a conservative method of identification, meaning that some outliers may not be identified as such. This occurs because we employ the outlier when calculating the mean and standard deviation before searching for outliers. Remember that the outlier pulls the mean in its direction, which decreases its deviation. In addition, the standard deviation weights a large deviation more heavily, so that outliers enlarge the magnitude of the standard deviation. The smaller deviation decreases the numerator of the standardizing formula. The larger standard deviation increases the denominator, (σ). Together, these changes tend to decrease the standardized value or distance from the mean. Thus, if we identify a value as an outlier, it has passed a strenuous identification test. Box 3-2 shows how astronomers use this concept as well as other measures we have discussed.

The standard deviation is important for (1) informing us of the "average" distance of a point in the data set from the mean; (2) ranking data sets by their level of dispersion; (3) describing the likely location of values according to the Empirical Rule; (4) describing the likely location of values by applying Chebyshev's inequality; and finally (5) identifying outliers. Box 3-3 identifies an instance where a farmer uses this information to make a business decision.

We will subsequently develop new topics involving the standard deviation. Experience calculating and interpreting the standard deviation at this point should give you a good feel for the information contributed by this measure and confidence as we proceed to more complex issues.

Section 3.8 Problems

86. **a.** Find the standard deviation of the beach-closing values from Problem 5 (p. 83). The mean is 143.6 closings.
   b. Interpret the result.
   c. Verify that at least 75% of the values are within two standard deviations of the mean.

87. **a.** Compute the standard deviation of the software firms' market shares in Problem 6 (p. 83). The mean is 6.44%.
   b. Interpret this value.
   c. Verify that at least 89% of the values are within three standard deviations of the mean.

88. **a.** Calculate the standard deviation of the incarceration rates in Problem 7 (p. 83). The mean proportion is 0.1154.
   b. Interpret the answer.
   c. Calculate the Z score for the largest and smallest values in the data set. How many standard deviations are they from the mean? Are they outliers?

Box 3-3   Norwegian Fisheries

A Norwegian statistician studied the fisheries industry to help the farmers determine the optimal time to harvest their fish. Different weight categories command different prices in the market, with heavier fish worth more on a per-unit weight basis.

After studying the fish farms, the statistician learned that the coefficient of variation, the ratio of the standard deviation to the mean (see Box 3-1), was about the same in each weight class. Consequently, the farmers, who knew nothing about measuring or guessing the dispersion of the weights of the current crop but who are able to guess the typical size of fish in their hatcheries, could provide enough information for the statistician to estimate the standard deviation of the weights.

He also found that the distribution of weights in most fisheries was approximately bell shaped. Even when the farmer selectively kills some fish, the distribution returns to a bell-shaped distribution within a few weeks.

Thus, armed with estimates of the mean and standard deviation along with the bell-shaped distribution, the statistician could estimate the number of fish in each weight class and the current market value of the fish versus the additional revenue and cost that the farmer would incur by postponing the sale of the fish to increase their size. (In order to compare costs, an estimate of the relationship between weight gain and amount of feeding is necessary as well.)

Source: Conversation with Jostein Lillestøl, Norwegian School of Economics and Business Administration, 5035 Bergen, Norway, June 24, 1992, and Jostein Lillestøl, "On the Problem of Optimal Timing of Slaughtering in Fish Farming," Modeling, Identification and Control, 1986, Vol. 7, No. 4, pp. 199-207.

89. a. Determine the standard deviation of the University of North Carolina campus enrollments from Problem 8 (p. 83). The mean is 9,200 students. (Hint: Express each value in thousands of students to simplify computations. For example, 9,200 students would be 9.2 thousand students.)
   b. Interpret the solution.
   c. Are there outliers in this data set?

90. a. Find the standard deviation of the number of participants (in millions) in different airlines' frequent flyer programs from Problem 9 (p. 83). The mean is 6.68 million participants.
   b. Interpret the result.
   c. Find the Z scores for the maximum and minimum values in the data set. Interpret these scores. Is either value an outlier?

3.9 The Sample Standard Deviation

Similar to the message of the population standard deviation, the sample standard deviation relays information about the "average" distance of a point in the sample from the sample mean. Usually we calculate the sample standard deviation when we need an estimate of the population standard deviation.

If we measure the spread of sample values with the same formula we use for the population standard deviation, we are likely to underestimate the spread of the population. Figure 3-26 shows 30 random samples of five race times selected from the population of eight times of Horse A. The standard deviation of all eight values is 0.130 minute. The "σ Formula" column shows the standard deviation estimates that result from calculating the sample standard deviation by a computation identical to that used for the population standard deviation. The mean of these values is only 0.116 compared to the actual value of 0.130 minute. Statistical theory verifies that the average of such sample standard deviations will be smaller than the standard deviation of the population from which the sample originated. We could guess this intuitively. Remember, the sample usually does not contain all of the population values, and it is more likely to contain typical values than outliers. Thus, the dispersion of the sample values is likely to be less than that of the population values.

For this reason, when the data set is a sample rather than a population, the variance and standard deviation formulas change slightly to increase the results: The denominators are $n - 1$ rather than n. (N is the number of elements in a population, and n is the

Section 3.9 — The Sample Standard Deviation

Five values in each sample

$$\sqrt{\frac{\Sigma(X_i - \bar{X})^2}{n}} \qquad \sqrt{\frac{\Sigma(X_i - \bar{X})^2}{n-1}}$$

30 samples:

Sample	Time 1	Time 2	Time 3	Time 4	Time 5	σ Formula	S Formula
1	2.00	1.95	2.26	2.15	2.25	0.127	0.142
2	2.25	2.19	2.19	2.35	2.00	0.114	0.128
3	1.95	2.05	2.00	1.95	2.00	0.037	0.042
4	1.95	2.00	2.00	2.25	2.15	0.112	0.125
5	2.19	2.35	2.35	2.25	2.35	0.112	0.125
6	1.95	2.00	2.00	2.00	2.25	0.107	0.119
7	2.05	1.95	1.25	1.95	2.35	0.162	0.182
8	2.15	2.19	2.19	2.19	2.35	0.070	0.078
9	2.35	2.26	2.25	2.00	2.05	0.134	0.150
10	2.19	2.25	2.00	2.25	1.95	0.128	0.143
11	1.95	2.26	2.00	2.19	2.26	0.132	0.147
12	2.05	2.19	2.26	2.25	2.19	0.075	0.084
13	2.05	2.15	2.19	1.95	1.95	0.099	0.111
14	2.05	2.19	2.00	2.26	2.35	0.130	0.145
15	2.19	2.25	2.19	1.95	2.26	0.113	0.126
16	2.25	2.35	2.25	2.05	2.35	0.110	0.122
17	2.19	2.35	2.19	1.95	2.35	0.147	0.164
18	2.15	2.26	2.19	2.26	2.05	0.078	0.088
19	2.26	2.19	2.25	2.26	2.00	0.099	0.111
20	1.95	2.26	2.25	2.25	1.95	0.149	0.166
21	2.05	1.95	2.19	2.25	2.35	0.142	0.159
22	2.19	1.95	2.25	1.95	2.35	0.162	0.181
23	2.15	2.15	2.05	1.95	1.95	0.089	0.100
24	2.26	2.00	2.25	2.26	2.05	0.115	0.128
25	2.15	2.00	2.15	2.35	2.26	0.118	0.132
26	2.25	2.26	2.05	1.95	2.19	0.121	0.135
27	2.26	2.25	2.19	2.26	2.05	0.080	0.090
28	2.26	2.00	2.26	2.15	1.95	0.129	0.144
29	2.19	2.26	2.35	2.00	1.95	0.152	0.170
30	2.05	1.95	2.26	2.05	2.35	0.149	0.166
						0.116	0.130

Mean of σ values computed from 30 samples using n=5 as the denominator

Mean of S values computed from 30 samples using n−1=4 as the denominator

number of elements in a sample.) The sample standard deviations calculated with this formula are larger, as shown in the last column of Figure 3-26. The figure also demonstrates that their mean is larger and closer to the true value. (They happen to be the true value, 0.130, for this set of 30 samples, but this would not always be true. The mean of the standard deviations from *all possible samples of five observations* would be 0.130.)

Figure 3-26
Using $n - 1$ to calculate a sample standard deviation to avoid underestimating a population standard deviation

Because $n - 1$ is not very different from n, especially when n is large, these modified formulas do not alter the interpretation of the results as a measure of the average squared deviation and "average" deviation, respectively. Thus, we determine the *sample variance*, S^2, and *sample standard deviation*, S, from the following formulas.

> To find the sample variance and sample standard deviation use
> $$S^2 = \frac{\sum_{1}^{n}(X_i - \overline{X})^2}{n - 1} \quad \text{and} \quad S = \sqrt{\frac{\sum_{1}^{n}(X_i - \overline{X})^2}{n - 1}}.$$

■ **Example 3-16:** Find the standard deviation of the following sample of acceptance speech lengths (in minutes) at an awards ceremony:

2.2   1.8   5.6   0.7   4.0   1.2   4.0   3.2   0.3   2.0

■ **Solution:**

X_i	$X_i - \overline{X}$	$(X_i - \overline{X})^2$
2.2	−0.3	0.09
1.8	−0.7	0.49
5.6	3.1	9.61
0.7	−1.8	3.24
4.0	1.5	2.25
1.2	−1.3	1.69
4.0	1.5	2.25
3.2	0.7	0.49
0.3	−2.2	4.84
2.0	−0.5	0.25
25.0		25.20

$$\overline{X} = \frac{25.10}{10} = 2.5 \text{ minutes} \qquad S = \sqrt{\frac{25.20}{10 - 1}} = \sqrt{2.80} = 1.67 \text{ minutes.}$$

Question: What is the sample variance for the speech lengths?
Answer: 2.80 is the sample variance, the value under the radical in the last solution, or the square of the standard deviation.

■ **Example 3-13 (continued):** If the three sets of horses' race times were samples rather than populations, how would the standard deviation calculation change and would the results be larger or smaller.

■ **Solution:** The computations are the same, except we divide the sum of squared deviations by $n - 1 = 8 - 1 = 7$ before taking the square root. The resulting standard deviations are larger than when we divide by 8. The population standard deviations are 0.130, 0.103, and 0.125. The corresponding sample results are 0.139, 0.111, and 0.134.

Section 3.9 Problems

91. Assume the beach closing counts in Problem 5 (p. 83) form a sample, and find the standard deviation. Check that the result is larger than the population standard deviation from Problem 86.

92. Assume the market share values from Problem 6 (p. 83) form a sample, and find the standard deviation. What is the sample variance?

93. Assume the incarceration rates from Problem 7 (p. 83) are a sample, and find the variance and standard deviation. Interpret both values.

94. Assume the campus enrollments from Problem 8 (p. 83) are a sample, and calculate the standard deviation. Interpret the result.

95. Assume the frequent flyer participant counts from Problem 9 (p. 83) form a sample, and find the standard deviation. Interpret the result.

3.10 Shortcut Formulas

We can develop alternative formulas for σ and S algebraically from the definitional formulas. Recall that Σ alone is the shorthand symbol that indicates we perform the operation for *all* values in the data set and sum the results.

Formulas

$$\sigma = \sqrt{\frac{N(\Sigma X_i^2) - (\Sigma X_i)^2}{N^2}} \quad \text{or} \quad \sigma = \frac{1}{N}\sqrt{N(\Sigma X_i^2) - (\Sigma X_i)^2};$$

$$S = \sqrt{\frac{n(\Sigma X_i^2) - (\Sigma X_i)^2}{n(n-1)}}.$$

These formulas are shortcuts, because we can skip the intermediate step of finding deviations. We usually have the sum of the observations (ΣX_i) already from finding the mean. We simply square the observations and then sum the squares (ΣX_i^2). Remember from Chapter 1 that the two expressions, (ΣX_i^2) and $(\Sigma X_i)^2$, usually have different values, so do not confuse the operations.

■ **Example 3-17:** Suppose that a potential investor is interested in the closing stock prices for two aerospace firms only over the specific ten-day period when major national media stories appeared that concerned government research and development programs. Compare the data sets for the investor using the means and standard deviations of the stock prices (in dollars rounded to the nearest cent) listed below. Compute the standard deviations using the definitional and shortcut formulas.

Firm A		Firm B	
78.25	84.50	43.62	48.12
77.88	85.75	43.88	48.38
79.75	84.75	44.00	48.62
79.50	85.25	44.75	48.50
78.75	84.75	46.38	48.50

■ **Solution:** The data sets are populations, because the investor only cares about prices during this ten-day period. The computations follow.

Firm A

X_i (\$)	$X_i - \mu$	$(X_i - \mu)^2$	X_i^2
78.25	−3.663	13.417569	6,123.0625
77.88	−4.033	16.265089	6,065.2944
79.75	−2.163	4.678569	6,360.0625
79.50	−2.413	5.822569	6,320.2500
78.75	−3.163	10.004569	6,201.5625
84.50	2.587	6.692569	7,140.2500
85.75	3.837	14.722569	7,353.0625
84.75	2.837	8.048569	7,182.5625
85.25	3.337	11.135569	7,267.5625
84.75	2.837	8.048569	7,182.5625
819.13		98.836210	67,196.2319

$\mu = 819.13/10 = 81.913$.

Using the definitional formula for firm A, we obtain

$$\sigma = \sqrt{\frac{98.836210}{10}} = \sqrt{9.8836210} = 3.144.$$

Using the shortcut formula for firm A, we obtain the same result. (Differences may occur because of rounding.)

$$\sigma = \sqrt{\frac{10(67196.2319) - (819.13)^2}{10^2}} = \sqrt{\frac{988.3621}{100}} = \sqrt{9.883621} = 3.144.$$

Firm B

X_i (\$)	$X_i - \mu$	$(X_i - \mu)^2$	X_i^2
43.62	−2.855	8.151025	1,902.7044
43.88	−2.595	6.734025	1,925.4544
44.00	−2.475	6.125625	1,936.0000
44.75	−1.725	2.975625	2,002.5625
46.38	−0.095	0.009025	2,151.1044
48.12	1.645	2.706025	2,315.5344
48.38	1.905	3.629025	2,340.6244
48.62	2.145	4.601025	2,363.9044
48.50	2.025	4.100625	2,352.2500
48.50	2.025	4.100625	2,352.2500
464.75		43.132650	21,642.3889

$\mu = 464.75/10 = 46.475$.

The definitional formula results in

$$\sigma = \sqrt{\frac{43.132650}{10}} = \sqrt{4.3132650} = 2.077,$$

and the shortcut formula follows suit:

$$\sigma = \sqrt{\frac{10(21,642.3889) - (464.75)^2}{10^2}} = \sqrt{\frac{431.3265}{100}} = \sqrt{4.313265} = 2.077.$$

Firm A's stock prices tend to be higher as evidenced by the means, \$81.913 for A and \$46.475 for B. However, there is greater variation about the mean in the stock prices of A with a standard deviation of \$3.144, whereas B's standard deviation is \$2.077.

Question: Suppose that the high and low prices for firm A over the last year are 85.75 and 77.88, respectively. The corresponding values for firm B are 48.62 and 43.62. Are the results for σ consistent with these values (use μ from the 10 observations above)?

Answer: Yes, they are consistent, because σ is smaller than the largest deviation. The potential maximum deviations for A are $85.75 - 81.913 = 3.837$ and $77.88 - 81.913 = -4.033$, so the largest possible deviation is 4.033 (we do not care about the sign). Because 3.144, the standard deviation that we calculated, is smaller than 4.033, it is a reasonable value for the standard deviation. Similarly for B, the possible largest deviations are 2.145 and 2.855, so 2.855 is the largest possible deviation, and it is larger than the value we calculated, 2.077.

Question: Suppose that you consider purchasing 100 shares of one of the company's stock. Do you stand the best chance of making (or losing) the most money with the stock of firm A or firm B?

Answer: You have the best chance with firm A, because the standard deviation is larger. There is more variation, so presumably you can buy low and wait for a large price increase for each of the 100 shares.

■ **Example 3-18:** A patient's temperature has been taken and charted once an hour for the last 19 hours. The following values are a sample of the patient's temperature

Problems

during the time in the hospital. Use this information to estimate the population standard deviation, σ, by finding S from the sample. Use the definitional and shortcut formulas.

103.7	102.0	104.1	104.6	101.0	101.0	99.7	103.2	102.0
102.0	101.2	100.0	102.0	102.5	102.2	100.8	100.0	103.5
102.5								

■ **Solution:**

X_i (Degrees)	$X_i - \overline{X}$	$(X_i - \overline{X})^2$	X_i^2
103.7	1.7	2.89	10,753.69
102.0	0.0	0.00	10,404.00
104.1	2.1	4.41	10,836.81
104.6	2.6	6.76	10,941.16
101.0	−1.0	1.00	10,201.00
101.0	−1.0	1.00	10,201.00
99.7	−2.3	5.29	9,940.09
103.2	1.2	1.44	10,650.24
102.0	0.0	0.00	10,404.00
102.0	0.0	0.00	10,404.00
101.2	−0.8	0.64	10,241.44
100.0	−2.0	4.00	10,000.00
102.0	0.0	0.00	10,404.00
102.5	0.5	0.25	10,506.25
102.2	0.2	0.04	10,444.84
100.8	−1.2	1.44	10,160.64
100.0	−2.0	4.00	10,000.00
103.5	1.5	2.25	10,712.25
102.5	0.5	0.25	10,506.25
1,938.0		35.66	197,711.66

$\overline{X} = (1,938.0)/19 = 102.00$.

Definitional formula

$$S = \sqrt{\frac{35.66}{19 - 1}} = 1.41.$$

Shortcut formula

$$S = \sqrt{\frac{19(197,711.66) - (1,938.0)^2}{19(19 - 1)}} = 1.41.$$

So we would estimate the standard deviation of the patient's temperature to be $1.41°$ during the hospital stay.

Question: If a nurse wanted to pick a typical temperature during this 19-hour period, would it matter which temperature she picked, given the apparently small S value of $1.41°$?

Answer: 1.41 may appear to be a small number, but this much variation in body temperature may be significant. There is a lot of variation during the period, so one particular temperature is not very representative or likely to be informative for purposes of diagnosis and treatment. Information on temperature fluctuation, the standard deviation or range, and the boundaries ($99.7°$ to $104.6°$), may be more informative.

Section 3.10 Problems

96. Rework the standard deviation of the beach closing counts from Problem 5 with the shortcut formula. Assume the data is a sample. Compare your answer with that of Problem 91.

97. Redo the standard deviation of the market shares from Problem 6 with the shortcut formula. This time assume the data is a population. Compare the answer with that of Problem 87.

98. Compare the shortcut calculation of the *sample* standard deviation of the incarceration rates from Problem 7 with the definitional formula used in Problem 93.
99. Assume the campus enrollments in Problem 8 are a population, and find the standard deviation with the shortcut formula. Compare the result with that of Problem 89. Recall that expressing the values in thousands simplifies the computations.
100. Use the shortcut formula to find the standard deviation of the frequent flyer participant counts from Problem 9, assuming they form a sample. Does the shortcut formula produce the same result as Problem 95?

3.11 Estimated Standard Deviation from Grouped Data

We can estimate the standard deviation of a data set when the only available information is a frequency distribution of the values. We did this for the mean in Section 3.3.2. Again, we assume that all of the observations in each class lie at the midpoint of the class, M_i. We treat M_i like an observation, X_i, in the regular formulas, but we account for multiple observations of the same value by multiplying each squared deviation associated with M_i by its class frequency, f_i. We incorporate these changes in the formulas shown in the following box and elaborate in Figure 3-27.

To approximate the standard deviation from a frequency distribution, use

$$\sigma_G = \sqrt{\frac{\sum_{i=1}^{k}(M_i - \mu)^2 f_i}{\sum_{i=1}^{k} f_i}}; \quad \text{and} \quad S_G = \sqrt{\frac{\sum_{i=1}^{k}(M_i - \overline{X})^2 f_i}{\left(\sum_{i=1}^{k} f_i\right) - 1}},$$

where k is the number of classes.

The expression beneath the radical estimates the variance. Remember, it is the average squared deviation, and notice that, in this case, it is a weighted average of the squared deviations, where f_i is the weight. Also, note that $\sum_{i=1}^{n} f_i$ is N for the population formula and n for the sample formula.

Figure 3-27
Formulas for the calculation of the standard deviation from grouped data

■ **Example 3-19:** Estimate the mean and standard deviation of the number of lawyers per state in 1985 from the following frequency distribution. Assume this is a sample of state values from a population of values covering several years.

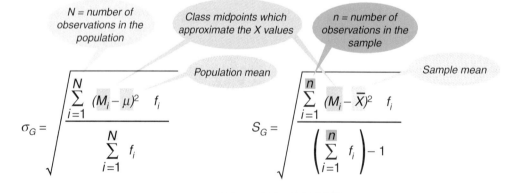

Class frequencies: the weights for averaging squared deviations. Note: Sum of frequencies equals number of observations in the data set

Number of Lawyers	Number of States
0–9999	30
10,000–19,999	10
20,000–29,999	5
30,000–39,999	3
40,000–49,999	0
50,000–59,999	0
60,000–69,999	0
70,000–79,999	1
80,000–89,999	1
	50

Source: American Bar Foundation, Chicago, Ill., *The Lawyer Statistical Report: A Statistical Profile of the U.S. Legal Profession in the 1980s*, 1985 and *Supplement to the Lawyer Statistical Report: The U.S. Legal Profession in 1985*, 1986, listed in U.S. Bureau of the Census, *Statistical Abstract of the United States: 1990*, 110th edition. (Washington, D.C.: U.S. Bureau of the Census, 1990), p. 182.

■ **Solution:** Because the number of lawyers is large, we can manage the problem better if we divide these values by 1,000 and talk about thousands of lawyers rather than just lawyers. Then we simply multiply our answers by 1,000 to get back to lawyers. The relevant midpoints and frequencies along with the calculations are

Number of Lawyers M_i	Number of States f_i	$M_i f_i$	$M_i - \mu$	$(M_i - \mu)^2$	$(M_i - \mu)^2 f_i$
4.9995	30	149.9850	−8.8	−7.44	2,323.20
14.9995	10	149.9950	1.2	1.44	14.40
24.9995	5	124.9975	11.2	125.44	627.20
34.9995	3	104.9985	21.2	449.44	1,348.32
74.9995	1	74.9995	61.2	3,745.44	3,745.44
84.9995	1	84.9995	71.2	5,069.44	5,069.44
	50	689.9750			13,128.00

$\overline{X} = 689.9750/50 = 13.7995$ thousand lawyers or 13,799.5 lawyers

$$S_G = \sqrt{\frac{13,128}{50 - 1}} = \sqrt{267.91837} = 16.3682,$$

or 16,368.2 lawyers.

Section 3.11 Problems

101. Approximate the standard deviation of the sample of 200 closing prices of a stock shown in the following frequency distribution.

Closing Price ($)	Number of Days
15.000	18
15.125	32
15.250	53
15.375	44
15.500	27
16.125	16
16.375	10

102. Use the frequency distribution below for the fat contents in milligrams in single servings of all snack foods produced by a certain company to produce an approximate of the standard deviation of these values for the company. The company is only interested in its current product line.

Fat Contents (milligrams)	Number of Products
0–4	3
5–9	14
10–14	10
15–19	12
20–24	28
25–29	9

103. Use the following frequency distribution of the population of the number of people playing baseball in countries around the world (for countries with 1,000 or more nonprofessional participants) to find the standard deviation for these countries. The value over 4.0 is 20.0 from the United States.

Number of Players per Country (millions)	Number of Countries
0.001–1.000	40
1.001–2.000	4
2.001–3.000	1
3.001–4.000	1
Over 4.000	1
	47

Source: "Batters Up!" *World Monitor*, April 1992, p. 11.

104. Use the following frequency distribution of monthly unemployment rates to find the standard deviation for this sample of all such rates.

Monthly Unemployment Rate (%)	Number of Months
3.0–3.9	2
4.0–4.9	113
5.0–5.9	181
6.0–6.9	159
7.0–7.9	85
8.0–8.9	36
9.0–9.9	4
	580

3.12 Descriptive Statistics and the Computer

3.12.1 Standard Deviation

The Minitab output of descriptive statistics includes the sample standard deviation. Its position is boxed in Figure 3-28, Minitab output for the patient temperature data from Example 3-18 and several transformations of these data. Notice that the standard deviation of the actual temperature values (in the first row of the output) replicates our earlier finding, 1.41°.

The transformations of the original temperatures include: adding 100, XPLUS; subtracting 100, XMINUS; multiplying by 100, XTIMES; and dividing by 100, XDIVIDE.

Question: Describe the effect of each transformation on the standard deviation.

Answer: Adding and subtracting have no effect on the standard deviation (the 1.408 for XMINUS rounds to 1.41), because all we have done is shift every value including the mean, in the same direction by the same amount along the number line. We have not changed the spread or dispersion of the values about the new mean. Multiplying and dividing the original values by some factor transform the standard deviation by the same amount. Think of the multiplication and division operations as stretching or shrinking the distances between the points and the new mean in the same way that the transformation expands or contracts the original values. The effects of the transformations that we have demonstrated hold true for all data sets.

Figure 3-28
Minitab output for patient temperature data (Example 3-18)

	N	MEAN	MEDIAN	TRMEAN	STDEV	SEMEAN
DEGREES	19	102.00	102.00	101.98	1.41	0.32
XPLUS	19	202.00	202.00	201.98	1.41	0.32
XMINUS	19	2.00	2.000	1.982	1.408	0.323
XTIMES	19	10200.0	10200.0	10198.2	140.8	32.3
XDIVIDE	19	1.0200	1.0200	1.0198	0.0141	0.0032

	MIN	MAX	Q1	Q3
DEGREES	99.70	104.60	101.00	103.20
XPLUS	199.70	204.60	201.00	203.20
XMINUS	−0.300	4.600	1.000	3.200
XTIMES	9970.0	10460.0	10100.0	10320.0
XDIVIDE	0.9970	1.0460	1.0100	1.0320

	N	MEAN	MEDIAN	TRMEAN	STDEV	SEMEAN
NURSING	100	xxxx	319.5	394.5	517.4	51.7
COLLEGE	100	713	0	382	xxxxx	176

	MIN	MAX	Q1	Q3
NURSING	0.0	3049.0	173.7	606.2
COLLEGE	0	8481	0	593

Figure 3-29

Minitab output for 1990 Census count of people in nursing homes and college dormitories

Question: Figure 3-29 displays the Minitab output for the 1990 count of people in nursing homes and college dormitories in North Carolina. The mean value for nursing homes and the standard deviation of the college dormitory counts are concealed. Use the available information to determine the range of plausible values for the missing values.

Answer: The mean nursing home count must fall between the smallest possible value, 0 people, and the largest possible value, 3,049 people. The actual mean is 470.1 people. To determine the missing standard deviation, we must first note that the difference between the minimum college dormitory count and the mean is $0 - 713 = -713$ people. For the maximum college dormitory value, this difference is $8,481 - 713 = 7,768$. Thus, the approximate largest possible deviation is 7,768. ("Approximate" because the sample standard deviation formula uses $n - 1$ in the denominator, which may inflate the "average" deviation, but only slightly for a large data set.) The smallest possible deviation is zero, because there may be an observation of 713 people. Thus, the possible values for an average of these deviations are contained in the interval, 0–7,768 people. The actual value for the 100 values, assuming they compose a sample of the counts for North Carolina counties in any given year, is 1,765 people.

3.12.2 Box-and-Whiskers Diagrams (Optional)

Another method that we use to detect outliers is to analyze a diagram based on several location and dispersion measures, the median, quartiles, and interquartile range, called a *box-and-whiskers diagram* (also called a boxplot). Figure 3-30 displays Minitab's box-and-whiskers diagram for the university starting dates from the NCAA Probation Data Base (Appendix D). It also shows the actual dates and a histogram to illustrate the relation of this diagram to a genuine data set.

This diagram forms a box that encloses values between the two years, 1849 and 1885, called the upper and lower *hinges* (approximately the first and third quartiles, so the length of the box is approximately the interquartile range). The actual quartiles are 1848 and 1885.5. A plus sign locates the median, 1868, in this box of the middle half of the observations.

The dashed lines are the whiskers. They extend from each hinge to the most extreme values in the data that are still within 1.5 interquartile ranges of the nearest hinge. Asterisks (*) and the letter O expose potential outliers, with the O symbols representing the most likely outlier candidates. In Figure 3-30, the earliest asterisk is for 1785, the earliest starting date in the data set, and the latest asterisk locates 1954, the latest starting date.

Question: The median appears to be close to the middle of the box that encloses the middle half of the data in Figure 3-30. In addition, both whiskers extend about 40 or 50 years, 1795 to 1849 and 1885 to 1927. Finally, the distance from each hinge to the most extreme value on the same side of the box is about 85 years, 1785 to 1868 and 1868 to 1954. How would you describe this data set, symmetric, skewed, or unclassifiable?

Answer: The data must be approximately symmetric if the median is approximately halfway between the earliest and latest date and halfway between the hinges. The gap between the end of the dashed line and rightmost asterisk raises suspicions of an outlier and skewness, and the proximity of the asterisks on the left to the dashed

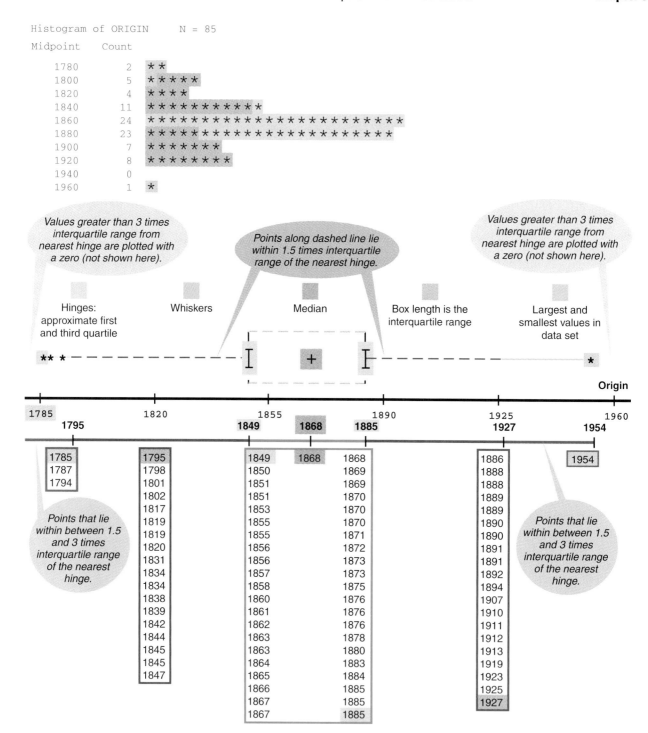

Figure 3-30
Box-and-whiskers diagram for university starting dates

line is less suspicious, as we shall discover shortly. Notice the histogram that shows the approximate symmetry and potential outlier.

Question: Figure 3-31 shows a box-and-whiskers plot for the data on solids in the bottled water from Example 3-2 (p. 78). Do you think these data are symmetric, skewed, or unclassifiable? Explain your answer.

Answer: The most obvious clue of skewness is the length of the right whisker compared to the short left whisker. There is also a potential outlier on the right. Finally,

Problems

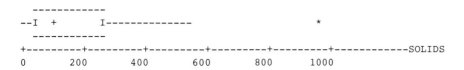

Figure 3-31
Box-and-whiskers diagram for the bottled-water data

the median is closer to the left side of the box, which adds to the indication that the smallest half of the data is compact compared to the larger half.

The procedure for finding the hinges approximates the quartile procedure. Consequently, the hinges will approximate the first and third quartile, when observations cluster near these two values. If not, we often interpret the results as approximate quartiles, upper boundaries of the lower one-fourth and lower three-fourths of the data, because the procedures are similar. We rarely trouble ourselves with all of the computations and ordering necessary to construct this diagram. Rather, we depend on the computer. A large distance between the median (+) and a hinge may signal a gap between the quartile and hinge, but it may also signal data that are very dispersed in this interval. As with all summary statistics, we must be alert to possible misrepresentations of the underlying data set.

In summary, dashes extend to the smallest and largest observations that fall outside the box, but not more than one and one half times the distance of the interquartile range (actually, the upper hinge minus the lower hinge) beyond the hinges. Asterisks represent those observations that are between one and one half and three times this distance beyond the hinges. Finally, the letter O represents values more than three times this distance, the most likely outliers. (No letter O appears in Figure 3-31.)

Question: Figure 3-32 shows the box-and-whiskers diagram for the twelve medieval household ale production values from Example 3-5 (p. 80). The median falls on the upper hinge without a dashed line in either direction. In fact, there is no left whisker. What do these clues tell us about the data set?

Answer: The lack of a dashed portion of a whisker indicates that the upper and lower hinge values repeat a large proportion of the time in the data set. This is especially true for the upper hinge that is also the median and that, together with the single value outside the box, means that all of the values in the upper half of the data set, except for one of them, must be the same number. Altogether, the clues indicate a data set with few different values, in this case 448 gallons occurs 4 times, 560 gallons occurs 7 times, and 1,008 gallons occurs once.

Question: Would you describe the data set shown in Figure 3-32 as symmetric, skewed, or unclassifiable?

Answer: The data set is skewed, because of the outlier on the right.

Section 3.12 Problems

105. Figure 3-33 shows the descriptive statistics for the baseball participant data from Problem 103. The computer used the data expressed in thousands, rather than millions, as was done in Problem 103.
  a. Compare the standard deviation in the figure with the value from the solution of Problem 103, 2.8432 million players (or 2,843.2 thousand players), and give two reasons for the difference in the values.
  b. Describe the distribution of the data from the box-and-whiskers diagram.

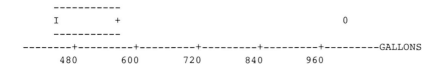

Figure 3-32
Box-and-whiskers diagram for medieval ale production data

Figure 3-33
Minitab output for baseball participants data in Problem 103

(a) Descriptive Statistics

```
                N      MEAN   MEDIAN   TRMEAN   STDEV   SEMEAN
PLAYERS        47       744        8      255    2964      432


              MIN       MAX       Q1       Q3
PLAYERS         1     20000        3      400
```

(b) Box-and-Whiskers

106. Determine if the minimum and maximum values from the descriptive statistics in Figures 3-14 through 3-17 (p. 101) and Figure 3-33 for the baseball participation data are outliers.

107. Figure 3-34 shows the box-and-whiskers diagrams for the data sets from Problems 26–29. Do the diagrams verify the outliers you found for these data sets in Problem 106? Which data sets appear to be skewed?

Summary and Review

The standard deviation is a single value that summarizes the spread or dispersion of a set of numbers about their mean. The use of the mean or of another measure of central location as the only value to describe different data sets can disguise the differences among them, as shown in Figure 3-35. Both data sets in the figure have the same mean, but X is more dispersed than Y, so the standard deviation of X is larger than the standard deviation of Y.

Several devices for speeding standard deviation calculations include shortcut formulas, formulas that employ frequency distribution values, and computers. The formulas all derive from the definition of the standard deviation based on the idea of an "average" deviation. Each observation in a data set is a certain distance from the mean, $X_i - \mu$, called a deviation. The standard deviation averages these deviations in a nontraditional manner, because it is the square root of the variance, the mean of the squared deviations. Thus, it is an "average" of the unsquared deviations.

The formula for the sample standard deviation differs from that for a population. Its value is inflated by dividing by $n - 1$ rather than by n. This alteration improves the properties of S as an estimator of σ, because the population is likely to be more dispersed than a subset of its values.

Standard deviations have several applications at this point. They are useful for envisioning the average "deviation," comparing the relative dispersion of different data sets, describing the likely location of values according to the Empirical Rule or Chebyshev's inequality, and determining which observations, if any, in a data set are outliers by computation of their Z scores. A Z score measures the distance of any observation from the mean in standard deviations rather than the original units. Other useful measures of dispersion include the range, interquartile range, and mean absolute deviation (MAD). In addition, the box-and-whiskers diagram assists in picturing the distribution of values in a data set and in detecting outliers. However, much statistical analysis focuses on the standard deviation, and we will find other uses for it as we proceed.

The formulas introduced in this chapter and their purposes are presented at the top of page 136.

Summary and Review

Figure 3-34 Box-and-whiskers plots for Problem 107

(a) Peak Levels of Carbon Monoxide (CO) from Air Quality Database

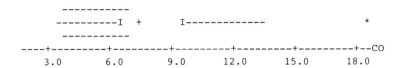

(b) North Carolina Country Census Counts from North Carolina Database

(c) Stock Closing Prices

(d) SAT Scores

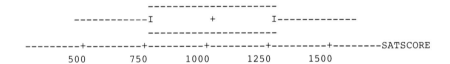

Figure 3-35 Histograms of X and Y

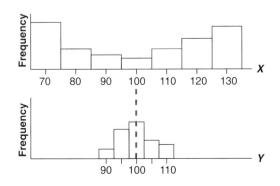

1. Mean

$$\mu = \frac{\sum X_i}{N} \quad \text{and} \quad \overline{X} = \frac{\sum X_i}{n}.$$

2. Location

Median: $\frac{n+1}{2}$;   First Quartile: $\frac{n+1}{4}$;

Third Quartile: $\frac{3}{4}(n+1)$; rth Percentile: $\left(\frac{r}{100}\right)(n+1)$.

3. Weighted Mean

$$\overline{X}_W = \frac{\sum X_i W_i}{\sum W_i}.$$

4. Mean from Grouped Data

$$\overline{X}_G = \frac{\sum_{i=1}^{k} X_i f_i}{\sum_{i=1}^{k} f_i} \quad \text{or} \quad \frac{\sum_{i=1}^{k} X_i f_i}{n},$$

where k is the number of classes.

5. Range

Largest value − Smallest value.

6. Interquartile Range

Third quartile − First quartile.

7. Mean Absolute Deviation

$$\text{MAD} = \frac{\sum |X_i - \mu|}{n}.$$

8. Standard Deviation
   a. Population

   Definition: $\sigma = \sqrt{\dfrac{\sum (X_i - \mu)^2}{N}}$;

   Shortcut: $\sigma = \dfrac{1}{N}\sqrt{N\left(\sum X_i^2\right) - \left(\sum X_i\right)^2}$;

   Grouped: $\sigma_G = \sqrt{\dfrac{\sum_{i=1}^{k}(M_i - \mu)^2 f_i}{N}}$,

   where k is the number of classes.

   b. Sample

   Definition: $S = \sqrt{\dfrac{\sum (X_i - \overline{X})^2}{n-1}}$;

   Shortcut: $S = \sqrt{\dfrac{n\left(\sum X_i^2\right) - \left(\sum X_i\right)^2}{n(n-1)}}$;

   Grouped: $S_G = \sqrt{\dfrac{\sum_{i=1}^{k}(M_i - \overline{X})^2 f_i}{n-1}}$,

   where $k =$ is the number of classes.

9. Z Score or Standardized Score

$$Z = \frac{X_i - \mu}{\sigma} \quad \text{or} \quad Z = \frac{X_i - \overline{X}}{S}.$$

Multiple-Choice Problems

108. The standard deviation is a single value that measures
   a. The general significance of a set of data.
   b. The spread or dispersion of the data points about the mean.
   c. The middle value in a data set.
   d. The difference between the highest and lowest values in a data set.
   e. The number of data points that differ from the mean.

109. The population standard deviation is an "average" of the
   a. X_i.
   b. μ.
   c. $\overline{X}$.
   d. $X_i - \mu$.
   e. $(X_i - \mu)^2$

110. A correct formula for the population standard deviation is

   a. $\dfrac{\sum_{i=1}^{n} X_i}{n}$

   b. Highest X_i − lowest X_i.

   c. $\dfrac{\sum_{i=1}^{N}(X_i - \mu)}{N}$.

   d. $\dfrac{\sum_{i=1}^{N}|X_i - \mu|}{N}$.

   e. $\sqrt{\dfrac{\sum_{1}^{N}(X_i - \mu)^2}{N}}$.

Questions 111–115 are based on the following *population* of the number of students in seven sections of music appreciation offered during a fall term: 26, 52, 26, 10, 16, 10, 14.

111. What is the range of the data set?
   a. 10 students.   c. 42 students.   e. 21 students.
   b. 52 students.   d. 26 students.

Multiple-Choice Problems

112. What is the interquartile range?
  a. 16 students.   d. 0 students.
  b. 42 students.   e. 26 students.
  c. 10 students.

113. Which of the following values would be the numerator for the standard deviation calculation? (Ignore the square root for now, and use the definitional formula.)
  a. 1,320 square-students.   d. 0 square-students.
  b. 154 square-students.   e. 22 square-students.
  c. 76 square-students.

114. The denominator in the calculation would be
  a. 22.   d. 7.
  b. 5.   e. 8.
  c. 6.

115. If the set of data were a sample rather than a population, the denominator in the calculation would be
  a. 22.   d. 7.
  b. 5.   e. 8.
  c. 6.

116. Find the standard deviation of the following population of percentage discounts offered during various promotions: 10, 20, 30. Use the definitional formula.

  a. $\sqrt{200/3} = 8.2\%$.   d. $\sqrt{200/2} = 10.0\%$.
  b. $\sqrt{200} = 14.1\%$.   e. $\sqrt{20/3} = 2.6\%$.
  c. $\sqrt{10} = 3.2\%$.

117. Find the standard deviation of the following sample of current scouts' longevity with the organization (in years): 3, 8, 11, 14.
  a. $\sqrt{66} = 8.1$.   d. $\sqrt{14} = 3.7$.
  b. $\sqrt{33} = 5.7$.   e. $\sqrt{7} = 2.6$.
  c. $\sqrt{22} = 4.7$.

118. Consider the histograms in Figure 3-36. Which data set will have the smallest standard deviation? (Assume that each data set has the same number of observations.)
  a. A.   d. D.
  b. B.   e. E.
  c. C.

119. Which of the following sets of heights is likely to have the smallest standard deviation?
  a. All people.
  b. All males.
  c. All U.S. males.
  d. All U.S. males over 30 years old.
  e. All U.S. males under 30 years old.

Figure 3-36
Histograms for Problem 118

Data set A
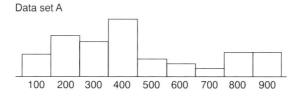

Data set B
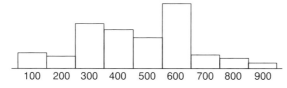

Data set C
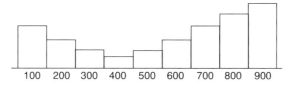

Data set D
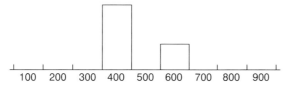

Data set E
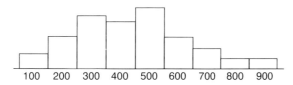

120. Which of the following sets of income values is likely to have the smallest standard deviation?
  a. Corporation presidents.
  b. Restaurant owners.
  c. Plumbers on a union job.
  d. Life insurance salespeople.
  e. Real estate brokers.

121. Suppose you are about to purchase a car and have narrowed your choices to two models. All things are about equal (price, options, even the average annual maintenance costs). You find an owner survey in an auto magazine that indicates the standard deviation of maintenance costs is smaller for one of the models. Which model is preferable, if you like to reduce the risks in a purchasing decision?
  a. They are equally acceptable, because standard deviations are not useful for comparisons of data sets.
  b. The one with the smaller standard deviation is preferable, because the smaller value implies that the mean is a more reliable representation of maintenance costs.
  c. The one with the smaller standard deviation is preferable, because the lower value ensures lower maintenance costs.
  d. The one with the larger standard deviation is preferable, because the large value implies a good chance of having low maintenance costs.
  e. The one with the larger standard deviation is preferable, because the larger value implies a smaller amount of variation in the data.

122. Which of the following combinations of μ and σ is plausible for a population that contains *different* values between 1 and 10?
  a. $\mu = 4, \sigma = 3$.   d. $\mu = 0, \sigma = 12$.
  b. $\mu = 5, \sigma = 0$.   e. $\mu = 5, \sigma = 10$.
  c. $\mu = 12, \sigma = 1$.

Figure 3-37 presents computer-generated descriptive statistics of invoices for Salespeople A, B, and C. Use this information to answer Questions 123–125.

123. Which salesperson's invoices are skewed?
  a. A.   c. C.   e. All three.
  b. B.   d. None of the three.

124. Which salesperson made the largest number (not amount) of sales?
  a. A.   c. C.
  b. B.   d. All made the same number of sales.

125. Which salesperson's invoices all have the same value?
  a. A.   d. None of the three.
  b. B.   e. All three.
  c. C.

126. Which of the following descriptive statistics will tell us that the samples of hourly rates (in dollars) for a certified public accountant between April 15 and May 15 (50, 60, 80, 90) and (50, 55, 70, 85, 90) are different?
  a. Mean.   d. Range.
  b. Median.   e. Standard deviation.
  c. Mode.   f. None of the above.

127. We can use the standard deviation to
  a. Visualize the "average" deviation in a data set.
  b. Compare the dispersion of different data sets.
  c. Describe the likely location of values with the Empirical Rule, if appropriate, or Chebyshev's inequality.
  d. Identify outliers.
  e. All of the above.

128. Approximate the standard deviation of music faculty recital performances during the year from the following frequency distribution of the number of individual performances by the set of all music faculty.

Number of Individual Performances	Number of Faculty
0–2	8
3–5	6
6–8	10
9–11	2
12–14	1
	27

  a. 10.7.   d. 3.3.
  b. 7.   e. 14.
  c. 9.

129. Which of the following Z scores indicates an outlier?
  a. 1.76.   d. −1.99.
  b. −0.83.   e. 0.45.
  c. 2.65.

Figure 3-37
Minitab output for three salespersons' invoices

	N	MEAN	MEDIAN	TRMEAN	STDEV	SEMEAN
A	16	858.60	698.06	809.15	396.05	99.01
B	18	742.26	741.32	742.08	40.30	9.50
C	6	750.12	750.12	750.12	0.00	0.00

	MIN	MAX	Q1	Q3
A	584.44	1824.98	666.52	727.47
B	685.19	802.25	701.02	780.62
C	750.12	750.12	750.12	750.12

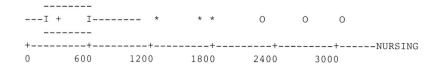

Figure 3-38
Minitab box-and-whiskers diagram for North Carolina Census count of people in nursing homes

130. The box-and-whiskers diagram in Figure 3-38 is based on the North Carolina Census count of nursing home inhabitants. Based on the probable population distribution indicated by the diagram, which of the following is appropriate for finding the proportion of observations within a prescribed number of standard deviations from the mean?
   a. Interquartile range.
   b. Range.
   c. Empirical Rule.
   d. Chebyshev's inequality.
   e. Standard or Z score.

Word and Thought Problems

131. Calculate the variance and standard deviation for the following data sets. Assume that sets a and c are populations and sets b and d are samples. The mean is provided.
    a. 62, 21, 83, 95, 21, 102, 21; $\mu = 57.9$.
    b. 0, 0, 1, 2, 10, 9, 9, 8; $\overline{X} = 4.9$.
    c. 5, 5, 5, 5; $\mu = 5$.
    d. 3.0, −1.2, 8.6, 5.3, −8.0, −8.6; $\overline{X} = -0.15$.

132. a. What information about a data set is contained in the variance? Standard deviation? Range? Interquartile range? MAD?
    b. Which of these measures are affected by outliers?
    c. Can a data set have more than one standard deviation?

133. What might a standard deviation of zero indicate for a set of lengths of bird beaks?

134. What is the difference between σ and S?

135. The following sets of numbers represent a compilation of opinions about 37 movies from a sample of media reviews. Rank the data sets from most to least dispersed using the range, interquartile range, MAD, variance, and standard deviation. (Source: "What the Critics Say about Movies," *Consumer's Research*, July 1991, p. 39.)

Number of Positive Opinions Expressed for Each Movie

0	4	1	0	3	1	3	0	0	0	1
2	2	2	9	2	3	1	0	6	8	0
1	1	5	0	2	4	0	0	1	2	9
3	9	2	4							

Number of Dubious Opinions Expressed for Each Movie

0	1	17	3	1	3	4	1	1	1
1	12	4	3	6	12	4	3	2	5
10	12	12	8	15	0	0	3	4	3
8	2	8	7	11	16	5			

Number of Negative Opinions Expressed for Each Movie

2	0	3	2	0	1	0	9	4	10
4	5	0	12	3	2	0	1	3	0
3	8	5	10	2	3	1	3	3	0
13	0	4	0	3	1	2			

Word and Thought Problems

136. Find the standard deviation of the following list of the number of citations of twenty academic journals in three major sociology journals over a three-year period. Assume the data is a sample of citations over a longer time span. Interpret the result. (Source: Michael Patrick Allen, "The Quality of Journals in Sociology Reconsidered: Objective Measures of Journal Influence," *Footnotes*, Vol. 18, 1990, pp. 4–5.)

6,065	17,358	7,613	5,443	12,074	1,972
15,685	5,126	5,046	27,234	12,895	3,595
1,701	1,554	18,969	2,643	5,388	3,471
2,836	1,211				

137. The journals from Problem 136 published the following number of articles over the sample three-year span. (Source: Michael Patrick Allen, "The Quality of Journals in Sociology Reconsidered: Objective Measures of Journal Influence," *Footnotes*, Vol 18, 1990, pp. 4–5.)

72	516	140	156	188	106	445	85
273	820	201	143	63	82	194	121
170	153	86	59				

   a. Calculate and interpret the standard deviation of this data set.
   b. Add the value 2,000 to the data set, and recalculate the standard deviation of these 21 values. Compare the mean and standard deviation with those from Part a.
   c. Increase each of the 20 observations in the original sample by 2,000 (for example, 72 becomes 2,072). What happens to the mean and standard deviation?
   d. Repeat Parts a, b, and c for the range and interquartile range.

138. Compare the means and standard deviations of the typical family's median income before and after direct federal taxes over the period 1980–1990 listed below. Which data set is most dispersed? (Source: Various sources reported in "U.S. Family's Real Income Drops in 1990," *Consumer's Research*, January 1991, p. 23)

		Before Taxes			
29,627	32,224	34,515	36,106	38,713	40,593
42,492	44,536	46,658	49,090	51,421	

		After Taxes			
23,761	25,695	27,752	29,387	31,369	32,944
34,296	36,061	37,536	39,381	41,130	

139. Two different brands of watches have the same mean lifetime; however, the standard deviations of the two lifetimes differ. Which one has a better chance of lasting longer?

140. a. Two production techniques have the same mean output. However, the standard deviation of technique A is larger than that of technique B. Which technique is more reliable for producing the mean output?
    b. Suppose that the mean output of technique B is larger than the mean output of technique A. Assume that the relationship of their standard deviations is the same as in Part a. If being consistent about the mean output is more important than the amount of output, which technique would be preferable?

141. Frequently, controversy over the scores in Olympic diving competitions occurs because the judges are citizens of the participating divers' countries. Suppose that we record the scores given by each judge throughout the competition. The mean scores given vary from judge to judge; however, the standard deviation for each judge's scores is the same. If judges are indeed biased, would you expect the size of the standard deviation of a judge's scores to be large or small relative to the size of the mean of the judge's scores? Explain your answer. Then, explain why each of the following situations supports or does not support the contention that the judges are biased.
    a. A large dispersion of the means for scores given by each judge for all the dives of a single diver.
    b. A small dispersion of all of the judges' scores for a single dive by the same diver.
    c. A large dispersion of the overall total scores among the divers.

142. On some surveys, respondents use a numbering system to represent verbal responses, such as 1 = poor, 2 = fair, 3 = average, 4 = good, 5 = excellent. If the mean response to a certain question is 4.12, what information does the standard deviation of the responses to the question provide for evaluating the overall response to this question?

143. A certain cold-relief medicine has provided Connie with relief for the following number of hours after taking the prescribed dosage: 4.5, 3.75, 3.8, 3.5, 4, and 3.45. Find the standard deviation of relief times, assuming that the data set is a population. Interpret your answer.

144. Calculate the standard deviation of sentences for burglary (in years) for a sample of eight inmates' terms: 2, 2, 1, 5, 3, 10, 2, 7. Interpret your answer.

145. Check the estimation of the mean and standard deviation of the number of lawyers per state in Example 3-19 in Section 3.11, using the actual data listed below. (Source: American Bar Foundation, Chicago, Ill., *The Lawyer Statistical Report: A Statistical Profile of the U.S. Legal Profession in the 1980s*, 1985, and *Supplement to the Lawyer Statistical Report: The U.S. Legal Profession in 1985*, 1986, listed in U.S. Bureau of the Census, *Statistical Abstract of the United States: 1990*, 110th edition. [Washington, D.C.: Bureau of the Census, 1990], p. 182.)

6,679	11,174	2,787	5,555	22,154	2,006
3,164	8,302	1,230	13,390	1,749	11,397
2,031	7,017	20,445	4,289	72,575	7,385
8,782	11,507	7,535	1,597	37,873	10,569
11,343	2,276	9,265	27,798	39,028	2,835
3,741	30,444	9,499	2,528	4,270	2,139
1,290	2,527	3,446	10,315	82,001	13,652
6,182	12,866	12,439	23,301	27,191	5,021
1,372	1,116				

146. Describe the effect on the distance between data values when each of the following transformations occurs to the original values of a data set: addition of a constant, subtraction of a constant, multiplication by a constant greater than 1, division by a constant greater than 1. Describe the expected effect on the mean, range, interquartile range, and standard deviation of each of the transformations, given your answer about the effect on the distance between points.

147. Approximate the standard deviation and mean from the following frequency distribution of the estimated ages of a *sample* of specimens (in millions of years) collected on a three-year geology project. Interpret your answer.

Estimated Specimen Age (millions of years)	Number of Specimens
0–4	60
5–9	100
10–14	250
15–19	30
	440

148. Approximate the population mean and standard deviation of apartment-complex monthly rents from the following frequency distribution of all rents within five blocks of the center of campus. Interpret your answers.

Rent ($)	Number of Apartment Units
100–199	52
200–299	88
300–399	93
400–499	156
500–599	130
600–699	42
700–799	25
	586

Review Problems

149. Find the mean and standard deviation of the following sample of the number of annual cases on the U.S. Supreme Court docket. Can you think of an explanation for an outlier in this data set? Then, compute the Z score of each observation, and identify outliers, if any. Interpret the Z score for the first value. (Source: Office of the Clerk, Supreme Court of the United States, in U.S. Bureau of the Census, *Statistical Abstract of the United States 1990*, 110th edition. [Washington, D.C.: Bureau of the Census, 1990], p. 183.)

4,212	4,761	5,144	5,079	5,100	5,006
5,158	5,123	5,268	5,657		

150. Suppose that voter turnouts in gubernatorial elections have a mean of 450,000 voters and a standard deviation of 75,000 voters.
  a. If the turnouts conform to the normal curve, then, between which two values will the middle 95% of the turnouts be?
  b. Which of the following turnouts are outliers?

583,465	450,000	605,002	300,000
640,000	260,000		

  c. If we do not know that the turnouts are normally distributed, use Chebyshev's inequality to find the minimum proportion of the values that are between 300,000 voters and 600,000 voters.

151. Scores on a test taken by several thousand students have a mean of 73.1 and a standard deviation of 10.8.
  a. Which of the following values are outliers?

99	90	80	70	60	50	40	30
20	75	65	45	35	25.		

  b. Use Chebyshev's inequality to find the interval that contains at least the middle 75% of the scores.
  c. If the scores fit the normal curve, what proportion of them fall between 62.3 and 83.9?

152. Figure 3-39 displays Minitab output describing a random sample of cents-off coupons from the Sunday newspapers for a month. The display also includes a list of the original values and corresponding Z scores, generated by Minitab commands and descriptive output for the Z scores. Use this information to answer the following questions.
  a. What is the range and interquartile range of the coupon values?
  b. List the outliers, if any.
  c. Use Chebyshev's inequality to find the interval that contains at least the middle 75% of the coupon values.
  d. What is the mean and standard deviation for the set of Z scores?

153. The Kentucky Derby is a race for three-year-old horses, most of which have racing and winning experience. The weight of the jockey is supplemented if necessary, so that each horse carries the same amount of weight, 126 pounds, which helps ensure that the jockey's skill and the horse's ability, rather than other factors, determine the outcome of the race. Data on the following variables were recorded for each of 13 horses participating in the 1985 Derby.

AGE: horse's age.
WEIGHT: Weight carried by horse during race.
STARTS: Number of races by horse in 1985 prior to Derby.
WINS: Number of wins by horse in 1985 prior to Derby.
FINISH: Position of horse at finish (1 = first, 2 = second, and so on).
POST: Position of horse relative to inside rail at beginning of race (1 = closest; 2 = next closest, and so on).
TIME: Amount of time for horse to finish race.

Each of these variables is described in the Minitab outputs of Figure 3-40. Someone has incorrectly typed the values when entering them in the computer, which results in unreasonable descriptive statistics. Use the computer output to determine which variables contain mistaken entries. Tell why the value of the descriptive statistic is unreasonable.

154. The box-and-whiskers plot in Figure 3-41 is based on the North Carolina Census count of college dormitory inhabitants.
  a. Which value appears to occur most often?
  b. Describe the distribution of the data.
  c. Why is there no left whisker?

155. Describe the distribution of a data set with a box-and-whiskers plot as described.
  a. Small box and long whiskers.
  b. Small box and short whiskers.
  c. Wide box and long whiskers.
  d. Wide box and short whiskers.
  e. Single-value box and no whiskers.
  f. Single-value box and long whiskers.
  g. All box and no whiskers.

156. Study the Minitab output in Figure 3-42 to determine if the set of counts of skiers (in millions) for the top 10 ski countries is skewed. Are there outliers? (Source: Marj Charlier, "Troubled U.S. Ski Resorts Hope to Cure Ills with an Infusion of Foreign Tourists," *Wall Street Journal*, November 25, 1991, p. B1.)

157. a. Find the mean, median, mode(s), range, interquartile range, and standard deviation of the percentage of income taken for taxes from the following percentages for each of the 50 states in 1991. Consider the data sets to be samples drawn from a period of several years.

Figure 3-39
Minitab output for cents-off coupons

	N	MEAN	MEDIAN	TRMEAN	STDEV	SEMEAN
COUPON	80	44.22	45.00	43.65	13.13	1.47
ZSCORE	80	0.000	0.059	-0.043	1.000	0.112

	MIN	MAX	Q1	Q3
COUPON	10.00	100.00	35.00	50.00
ZSCORE	-26.06	4.248	-0.702	0.440

ROW	COUPON	ZSCORE	ROW	COUPON	ZSCORE	ROW	COUPON	ZSCORE
1	35	-0.70221	28	50	0.44021	55	45	0.05941
2	50	0.44021	29	50	0.44021	56	40	-0.32140
3	45	0.05941	30	75	2.34425	57	40	-0.32140
4	30	-1.08302	31	60	1.20183	58	40	-0.32140
5	45	0.05941	32	50	0.44021	59	50	0.44021
6	50	0.44021	33	50	0.44021	60	30	-1.08302
7	50	0.44021	34	50	0.44021	61	30	-1.08302
8	50	0.44021	35	50	0.44021	62	25	-1.46382
9	10	-2.60625	36	40	-0.32140	63	25	-1.46382
10	35	-0.70221	37	35	-0.70221	64	35	-0.70221
11	25	-1.46382	38	45	0.05941	65	40	-0.32140
12	45	0.05941	39	50	0.44021	66	45	0.05941
13	25	-1.46382	40	50	0.44021	67	50	0.44021
14	50	0.44021	41	45	0.05941	68	50	0.44021
15	50	0.44021	42	38	-0.47372	69	45	0.05941
16	50	0.44021	43	60	1.20183	70	60	1.20183
17	35	-0.70221	44	50	0.44021	71	50	0.44021
18	100	4.24829	45	50	0.44021	72	55	0.82102
19	75	2.34425	46	25	-1.46382	73	50	0.44021
20	25	-1.46382	47	25	-1.46382	74	50	0.44021
21	25	-1.46382	48	30	-1.08302	75	55	0.82102
22	35	-0.70221	49	25	-1.46382	76	25	-1.46382
23	40	-0.32140	50	40	-0.32140	77	45	0.05941
24	50	0.44021	51	50	0.44021	78	60	1.20183
25	50	0.44021	52	50	0.44021	79	45	0.44021
26	40	-0.32140	53	50	0.44021	80	45	0.05941
27	40	-032,10	54	50	0.44021			

Figure 3-40
Minitab output for Derby data

	N	MEAN	MEDIAN	TRMEAN	STDEV	SEMEAN
AGE	13	3.000	3.000	3.000	1.354	0.376
WEIGHT	13	126.00	126.00	126.00	0.00	0.00
STARTS	13	4.692	4.000	4.636	1.377	0.382
WINS	13	5.692	5.000	5.636	1.377	0.382
FINISH	13	7.0000	7.0000	7.0000	0.0000	0.0000
POST	13	7.00	7.00	7.00	3.89	1.08
TIME	13	2.0000	2.0000	2.0000	0.0000	0.0000

	MIN	MAX	Q1	Q3
AGE	1.000	5.000	2.000	4.000
WEIGHT	126.00	126.00	126.00	126.00
STARTS	3.000	7.000	3.500	6.000
WINS	4.000	8.000	4.500	7.000
FINISH	7.0000	7.0000	7.0000	7.0000
POST	1.00	13.00	3.50	10.50
TIME	2.0000	7.0000	2.0000	2.0000

Review Problems

```
         -----
   +   I-----**       0            000          00        0  00
         -----
   +---------+---------+---------+---------+---------+------COLLEGE
   0       1600      3200      4800      6400      8000
```

Figure 3-41
Minitab box-and-whiskers plot for North Carolina Census count of people in college dormitories

```
                  N      MEAN    MEDIAN    TRMEAN     STDEV    SEMEAN
   SKIERS        10      5.31      3.85      4.84      3.81      1.20

                MIN       MAX        Q1        Q3
   SKIERS      2.00     12.39      2.80      7.13
```

Figure 3-42
Minitab output for ski data

```
                  ---------------
          -----I      +       I                             *   *
                  ---------------
      ----+---------+---------+---------+---------+---------+--SKIERS
         2.0       4.0       6.0       8.0      10.0      12.0
```

(Source: Tax Foundation methodology and published and unpublished data from U.S. Department of Commerce, Bureau of Economic Analysis; and Government Finance Division, Bureau of the Census, in "Taxes as a Percentage of Income by State," *Consumer's Research*, June 1991, p. 18.)

Total

31.6	36.0	34.8	31.8	34.8	33.4	37.5
38.7	33.9	33.6	37.3	32.4	35.3	33.5
34.7	34.2	32.6	33.9	35.0	35.8	35.0
36.3	36.8	31.2	32.5	34.1	34.7	35.0
32.6	36.7	34.9	40.0	33.8	34.6	34.6
33.5	35.5	34.7	36.1	33.0	32.2	32.5
34.1	33.2	35.6	33.3	35.6	33.0	36.0
34.5						

Federal

21.2	19.6	21.5	21.2	22.8	21.7	25.9
26.3	23.3	22.0	22.3	20.7	23.9	22.0
21.8	22.2	21.0	21.3	21.2	23.2	23.2
23.2	22.8	19.6	22.2	21.4	22.2	24.4
23.5	25.0	21.3	23.4	21.9	22.1	22.8
21.6	22.5	23.2	24.0	20.9	21.1	22.3
22.7	20.6	22.7	22.2	23.2	20.7	22.4
21.0						

State/Local

10.4	16.4	13.3	10.6	12.0	11.7	11.6
12.5	10.6	11.6	15.0	11.7	11.4	11.5
13.0	11.9	11.6	12.6	13.7	12.7	11.9
13.1	14.0	11.6	10.3	12.7	12.5	10.6
9.1	11.7	13.5	16.6	11.8	12.4	11.8
11.9	13.0	11.5	12.1	12.1	11.1	10.2
11.4	12.6	12.9	11.1	12.4	12.3	13.6
13.5						

b. Find the 70th percentile of the total values.

158. If the populations of all cities and towns in the United States are to be condensed to a single value in order to represent the typical city or town population, which measure would you choose? Why?

159. Which measure of the typical income of a group is likely to make the group appear better off, the mean or the median? Which measure of the center of individual grades, mean or median, makes a student with a few high grades and many low grades appear to be a better student (have a higher overall grade)? Which measure makes the student with many high scores and a few low scores have a higher overall score?

160. Which of the following measures do you expect to change if we drop the largest and smallest score (or scores if either value occurs more than once) from a set of grades: mean, median, mode, range, interquartile range, or standard deviation. Explain your choices.

161. The following sample is a list of the annual number of crimes (in thousands) known to police for the years 1979–1988. (Source: U.S. Federal Bureau of Investigation, *Crime in the United States*, in U.S. Bureau of the Census, *Statistical Abstract of the United States: 1990*, 110th edition. [Washington, D.C.: Bureau of the Census, 1990], p. 170.)

12,250	13,408	13,424	12,974	12,109
11,882	12,431	13,212	13,509	13,923

a. Find the mean and standard deviation of these values.
b. Calculate the Z scores for each value, and locate outliers, if any.

162. The following sample is a list of the annual number of automobile thefts (in thousands) known to police for the years 1979–1988.

1,113	1,132	1,088	1,062	1,008	1,032
1,103	1,224	1,289	1,433		

(Source: U.S. Federal Bureau of Investigation, *Crime in the United States*, in U.S. Bureau of the Census, *Statistical Abstract of the United States, 1990*, 110th edition. [Washington, D.C.: Bureau of the Census, 1990], p. 170.)

    a. Find the mean and standard deviation of these values.

    b. Calculate the Z scores for each value, and locate outliers, if any.

163. Suppose your construction company loses $1,000 on about one-tenth of its jobs, breaks even on about one-quarter of its jobs, and makes $1,000 on the rest. What is the mean amount of earnings of your company?

164. A certain recipe calls for three kinds of flour in different quantities.

Type of Flour	Price/lb.	Pounds in Recipe
Self-rising	$0.69	2
Plain	$0.65	1
Whole wheat	$0.79	3

A cook claims that the mean price of a pound of flour used in the recipe is $1.11. Is this claim correct? Explain your answer.

165. a. Find the mean, median, mode, range, and standard deviation of the following population of import values (in millions of dollars): 53.6, 82.1, 10.0, 28.9, 53.6. Interpret the results.

    b. Suppose we add the import value $42.5 million to the data set. Now, calculate the measures for the observations: 53.6, 82.1, 10.0, 28.9, 53.6, and 42.5. Which measures change with the additional observation?

166. A government agency is looking for a performance index for different centers involved in a program to keep teenagers from dropping out of school. Rather than just record the mean number of successes per district in a center's region, weights equal to the number of children living in households with income below the government-established poverty level in each district are to be applied in the averaging process to account for the variation in the difficulty to attain success from county to county. Find the performance score for the center with the following performance record.

District Successes	Children in Households below Poverty Level
5	5,828
14	3,336
21	6,941
35	9,823
9	5,775

167. A home economist reports that the mean expenditure on living-room furnishings is $1,875.82 with a standard deviation of $462.55 and a range of $398.22. What is wrong with this statement?

168. People in temperate climates often assume that extreme temperatures usually occur in the summer and winter. If daily high temperatures are used to calculate standard deviations for temperatures during each of the four seasons, for which season(s) would you expect the largest standard deviation(s)?

169. Two express-mail services have the following means and standard deviations for delivery time:

Service	Mean	Standard Deviation
A	8.0 hours	1.0 hour
B	8.5 hours	0.75 hour

    a. Which service is more "consistent" in its delivery time? Explain your answer.

    b. If you are in a hurry, which service is more likely to deliver the letter sooner? Explain your answer.

170. Why might we use the median, rather than the mean, to locate the center of a set of apartment rents from a large metropolitan area?

171. Find the mean, median, mode, and standard deviation of the number of meals served each month in a medieval household to members of the household, workers, and guests between October 1412 and September 1413 from the following list. Interpret the mean and standard deviation results. (Source: Ernest Rubin, "Statistical Exploration of a Medieval Household Book," *The American Statistician*, Vol. 26, No. 5, December 1972, pp. 37–39.)

1,431	1,346	1,297	1,505	1,165	1,137
1,202	1,341	1,260	1,325	2,220	1,274

172. Two cuts of meat have the same mean amount of cooked lean meat per pound of meat. The standard deviation of cooked lean meat per pound is greater for cut A than for cut B. If the prices per pound of the two cuts are the same, which is the better buy for a family on a tight budget, who must make this choice every week?

173. Approximate the sample mean and sample standard deviation from the following frequency distribution of total income taken as taxes from the list of percentages in Problem 157. Compare the approximations with the actual results from Problem 157.

Total Income Taken as Taxes (%)	Number of States
31.0–31.9	3
32.0–32.9	6
33.0–33.9	11
34.0–34.9	12
35.0–35.9	8
36.0–36.9	6
37.0–37.9	2
38.0–38.9	1
39.0–39.9	0
40.0–40.9	1
	50

174. Approximate the mean and standard deviation from the following frequency distribution of a sample of

Review Problems

	N	MEAN	MEDIAN	TRMEAN	STDEV	SEMEAN
SITES	50	24.18	13.50	20.59	25.84	3.65

	MIN	MAX	Q1	Q3
SITES	1.00	109.00	9.00	30.00

Figure 3-43
Minitab output for hazardous waste sites data

```
STEM-AND-LEAF OF SITES    N = 50
LEAF UNIT = 1.0

    13      0   12333566778899
   (17)     1   00011111222345667
    20      2   00123459
    12      3   358
     9      4   025
     6      5   1
     5      6
     5      7   9
     4      8   3
     3      9   17
     1     10   9
```

ROW	SITES	ZSITES	ROW	SITES	ZSITES
1	12	-0.47136	26	10	-0.54876
2	6	-0.70356	27	6	-0.70356
3	11	-0.51006	28	1	-0.89706
4	11	-0.51006	29	16	-0.31656
5	91	2.58591	30	109	3.28251
6	16	-0.31656	31	10	-0.54876
7	15	-0.35526	32	83	2.27632
8	20	-0.16176	33	22	-0.08437
9	51	1.03793	34	2	-0.85836
10	13	-0.43266	35	33	0.34133
11	7	-0.66486	36	12	-0.47136
12	9	-0.58746	37	8	-0.62616
13	38	0.53483	38	97	2.81811
14	35	0.41873	39	11	-0.51006
15	21	-0.12307	40	23	-0.04567
16	11	-0.51006	41	3	-0.81966
17	17	-0.27786	42	14	-0.39396
18	11	-0.51006	43	29	0.18653
19	9	-0.58746	44	12	0.18653
20	10	-0.54876	45	8	-0.62616
21	25	0.01373	46	20	-0.16176
22	79	2.12152	47	45	0.80573
23	42	0.68963	48	5	-0.74226
24	3	-0.81966	49	40	0.61223
25	24	-0.00697	50	3	-0.81966

cholesterol counts in a medical laboratory during one month.

Cholesterol Level (mg/dL)	Number of Patients
100–149	24
150–199	183
200–249	95
250–299	67
300–349	55
350–399	12
	436

175. Figure 3-43 shows Minitab descriptive output, a stem-and-leaf display, and Z scores for data for the 50 states on the number of hazardous waste sites on the national priority list for 1989. Use this information to answer the following questions. (Source: U.S. Environmental Protection Agency, press release, October, 1989, reported in U.S. Bureau of the Census, *Statistical Abstract of the United States: 1990*, 110th edition. [Washington, D.C.: Bureau of the Census, 1990], p. 205.)

Figure 3-44

Minitab output for highest points in the 50 states and the District of Columbia

	N	MEAN	MEDIAN	TRMEAN	STDEV	SEMEAN
HIGHEST	51	1845	1340	1749	1555	218

	MIN	MAX	Q1	Q3
HIGHEST	105	6198	595	3428

a. Describe the distribution of the data set. Do the locations of the mean and median support your description?
b. Identify any outliers in the data set.
c. What is the interquartile range?
d. What is the 25th percentile? The 75th?

176. Consider the Minitab output in Figure 3-44 of the highest point in the 50 states and the District of Columbia (in meters) as you answer the following questions.
    a. What is the highest point in the United States and the District of Columbia?
    b. What is the lowest high point in the United States and District of Columbia?
    c. What is the range and interquartile range of the high points? Do these values lead you to believe that the distribution of high points may be skewed?
    d. Consider the first quartile and minimum value versus the third quartile and maximum value. Does this comparison lead you to believe there may be outliers?
    e. Compute the Z scores for the minimum and maximum values and for the first and third quartiles. What is your opinion concerning outliers now? Explain.

177. Sketch a box-and-whiskers diagram to depict each of the following data distributions.
    a. Symmetric.
    b. Skewed.
    c. Very dispersed.
    d. Slightly dispersed.

Chapter 4

Introduction to Probability

This chapter sets the stage for later work, when we will make generalizations, or inferences, about a population based on sample information. Without complete information we will use the available information to make our best guess, a statistical inference. Probability is our means of measuring the degree of uncertainty associated with this process. But uncertainty looms in many other situations, making probability a useful tool of analysis in general.

Objectives

Unit I introduces probability and related concepts. Upon completing the unit, you should be able to

1. Recognize that probability is a measure of uncertainty with alternative methods of determining basic probability values. Specific values relate the proportion of times an event occurs during a large number of repetitions of the same experimental conditions.
2. State the lower and upper limits of a numerical probability value and recognize an incorrect solution that violates these limits.
3. Find the probability that an event will not occur.
4. Recognize conditional probability as a method of incorporating additional information when determining the probability of some event and as a means of contracting the set of possible outcomes of an experiment before determining the probability.
5. Become familiar with the concepts of statistical experiment, trial, outcome, sample space, event, probability, success, equiprobable outcomes, mutually exclusive, complement, and conditional probability.
6. Recognize the symbols $P(A)$, $P(A')$, and $P(A|B)$.

Unit II introduces compound events and means for calculating their probabilities. It incorporates the basic concepts developed in Unit I into the solution process to expand the set of solvable problem situations and to simplify the process or formulas. Upon completing the unit, you should be able to

1. Recognize the meaning of compound events connected with "and" and with "or."
2. Understand the rules for calculating probabilities and use them to solve problems.
3. Check your solutions for reasonable results.
4. Become familiar with the term *independent events*.
5. Recognize the symbol statements $P(A \text{ or } B)$ and $P(A \text{ and } B)$.

Unit I: Basic Probability Concepts

Preparation-Check Questions

1. Use the following frequency distribution to determine the relative frequency of patients who stay in therapy less than one year.

Length of Psychotherapy (years)	Number of Patients
0.00–0.99	55
1.00–1.99	48
2.00–2.99	29
3.00–3.99	12
4.00 or more	4
	148

   a. 0.3716.
   b. 0.6284.
   c. 55.
   d. 93.
   e. 55%.

2. Find the mean and standard deviation of the following sample of the number of homework problems worked correctly by a college algebra class: 8, 10, 10, 5, 9, 8.
   a. 8.3 and 2.5 problems.
   b. 8.3 and 1.7 problems.
   c. 8 and 5 problems.
   d. 8.3 and 5.0 problems.
   e. 8.3 and 1.9 problems.

   Answers: 1. a; 2. e.

Introduction

Earthquake Prediction   An earthquake analyst predicted a 50% chance of an earthquake along the New Madrid Fault within two days of December 3, 1990. The earthquake did not occur. Was the analyst wrong?

People must often guess about the outcome of an event in order to make a decision. For instance, a doctor's or psychologist's prognosis is an estimate of the likely outcome of a set of symptoms and corresponding treatments. Teachers choose a specific quantity of material to present or a method of presentation based on the probable class response and comprehension. Scientists estimate the appropriate time to launch space vehicles and have them complete their missions successfully. Businesspeople consider their chances of selling different quantities of an item when they are deciding how much of the item to order. Politicians study polls to guess their likelihood of winning and to plan strategies for victory. Investors guess about their future income from various possible investment opportunities. When choosing your major course of study or deciding to take a statistics course, you may have considered your likelihood of obtaining a job in which you will use statistics in the future. In these cases and in many others, people do not always literally calculate probabilities, but the potential is often there for determining a measure of the chances that an event will occur.

Using the statistical rules of probability, we can estimate the uncertainty associated with an event or judge the accuracy associated with a statistical estimate, such as a weather forecast. We use probability to determine the reliability of the inferences we reach about populations based on limited (that is, sample) information. Thus, probability is a very useful and important concept in decision making and in statistics.

4.1 Basic Terminology

Probability is a measure of the uncertainty associated with the outcomes of a particular activity. The activity can be observing if a four appears when a die is thrown, if it rains tomorrow, if a company's sales reach $5 million this year, or if a student from an impoverished region will attend college. The observation procedure may also include controlling a set of circumstances to observe the outcome in a prescribed situation, such as controlling the temperature and humidity in a chemical experiment. We call the process of observing outcomes a *statistical experiment* (often shortened to *experiment*). *Trials* are repetitions of an experiment.

The outcomes we observe when we conduct a trial of an experiment fall into different categories depending on the question we wish to answer. If we toss a die, we may observe a 4, an even value, or a value smaller than 3. We reserve the term *outcome* to mean the most simple, distinct observations that we can make. "Simple" means that the outcomes are elementary, that is, they cannot be broken down further. The outcomes from tossing a die are the numbers on its six sides, 1, 2, 3, 4, 5, and 6.

"Distinct" means that the outcomes are different and distinguishable from one another. If the die shows a 2, we know it cannot be any of the other five values. We say that the outcomes are *mutually exclusive,* because when one outcome occurs, it automatically precludes the occurrence of the other outcomes on the same trial. One and only one outcome can result from a single trial. For example, attending college and not attending college are mutually exclusive and represent the outcomes for the experiment of observing a student's decision regarding college attendance the year after high school graduation.

We call the set of all possible outcomes the *sample space*. Figure 4-1 illustrates the sample space and outcomes for the die tosses. In this case, the sample space contains the six values on the sides:

$$\text{Sample space} = \{1, 2, 3, 4, 5, 6\}.$$

Sometimes, the sample space is very large, and it may even contain an infinite set of outcomes, such as the amount of rainfall a region experiences tomorrow.

Figure 4-1
Basic probability terms and the toss of a die

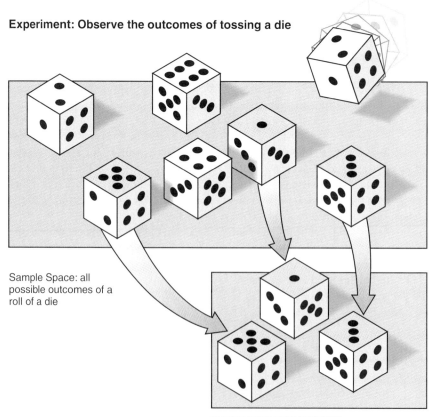

Question: A questionnaire asked marketing managers to check one of five responses to rate how ethical certain behaviors by employees were in general, rather than personally. For instance, one behavior was "ordering the most expensive item on the menu when the company is paying for it." The possible responses were "definitely wrong," "somewhat wrong," "neither wrong nor right," "somewhat right," and "definitely right." Are the possible responses mutually exclusive? Describe the sample space for the responses to this question. (Source: Judith A. Wiles, Charles R. Wiles, and Peter J. Gordon, "Ethical Attitudes of Marketing Managers: A Comparison of Small Businesses and *Fortune 500* Companies," *Business*, Vol. 5, No. 2, Fall/Winter 1989, pp. 34–40.)

Answer: Assuming each respondent checked only one response for the behavior, checking any one response, such as "somewhat wrong," means that none of the other four responses were checked, and the responses are mutually exclusive. The sample space consists of the five possible responses:

Sample space = {"definitely wrong," "somewhat wrong," "neither wrong nor right," "somewhat right," "definitely right"}.

Each possible response is an outcome.

An *event* is a set of outcomes from the sample space, a subset of the sample space. For instance, observing an odd number when tossing a die includes the outcomes 1, 3, and 5, so an odd number is an event for this experiment (see Figure 4-1 again). Observing a number between 1 and 6 is an event that encompasses the entire sample space. Observing a rainfall between 1.5 and 2.0 inches tomorrow is an event.

Question: Is responding that a behavior is "wrong" on the marketing manager questionnaire an event? If so, state the outcomes that comprise the event.

Answer: "Wrong" is an event, because it includes the two outcomes "definitely wrong" and "somewhat wrong."

Question: Are the events "wrong" and "right" mutually exclusive?

Answer: Yes, because a "wrong" response precludes a "right" response. The latter includes the outcomes "somewhat right," "definitely right," both of which do not overlap with the "wrong" outcomes.

Because the future is uncertain, we need some method to gauge the likelihood that an event will occur. We also need to gauge the likelihood that statistical estimates are "good," that is, close to the value being estimated. The gauge in both cases is probability. The terms from this section focus on the experimental situation to which a probability value applies.

4.2 Methods to Determine Probabilities

We can use several methods to assign a value to the probability of an event. The basic information we seek from such a value, no matter the method, is *the proportion of times the event occurs during a large number of trials of the experiment*. For example, the statement, "the probability of obtaining a head when a coin is tossed is 1/2" means that when a coin is flipped repeatedly, a head will appear about one-half of the times.

There is no set amount of times in the statement. Theoretically, we refer to infinite repetitions or repetitions over the long run, but in practice, we must fall short of infinite trials. Often, as we shall see, we expediently use available information to estimate or approximate probabilities.

Although we base the probability value on infinite trials of an experiment, once we obtain the value, we use it as a guide to predict the outcome of a single trial. It is often the best information we possess about the likely result in this instance or the best summary of previous results in similar situations. It may guide us in decisions requiring risks. If we know that the probability of a plane crashing is 0.00001, we interpret this as approximately one in 100,000 flights results in a crash, but it indicates that we can proceed with our next flight with little concern for our survival.

Notice also that probabilities are not the proportion of all time. Any experiment occurs in a specific context or set of circumstances. When we say that the probability of a forest fire is 20%, we mean that 20% of the times when we check if the forest is burning, the answer will be affirmative. Without more information about the experimental circumstances, we might assume that at any given moment, there is a 20% chance of fire. Other information that circumscribes the experimental situation improves our interpretation and use of the probability. Such information describes the context in which we can expect to observe forest fires 20% of the time. Perhaps the experiment (checking if the forest is burning) only occurred during windy, long-lived droughts, when more firefighters are on alert in these conditions than during normal conditions. Strictly speaking, the probability only refers to this situation, although we may apply an available value to similar circumstances as an approximation. For example, the probability estimated in Colorado might be applied in Wyoming or New Mexico.

Question: Today an economic forecaster says that there is a 30% chance that interest rates are going to decline. Relate this statement to the concept of probability.

Answer: The probability of a decline in interest rates is 0.3 (we commonly express probabilities as proportions); that is, 30% of the times when current conditions recur, interest rates will drop.

Before proceeding, we note that we abbreviate probability operations and events with symbols. Capital letters usually denote events. So, H may stand for a head when a coin is tossed. R may symbolize a rise in profits. $P(A)$ means "the probability that event A will occur." If H is tossing a head, then $P(H) = 1/2$ reads that the probability of tossing a head is $1/2$.

4.2.1 Relative Frequencies

One method of determining the probability of an event is to find the long-run relative frequency of the event's occurrence. We begin by letting n be the number of trials of an experiment. For instance, if the experiment is checking a 1-acre cornfield to determine if the yield is more than 10 bushels and the experimenter checks 30 different acres for yields, then $n = 30$.

Next, let X be the number of times trial outcomes fit the event in question. We call such an outcome a *success* (even when the event is not something we consider a positive occurrence, such as patients who experience side effects from a new drug). For the agricultural example, the event is a yield of more than 10 bushels per acre. Successes include the outcomes of 11 bushels per acre, 12 bushels per acre, and so forth. Suppose the crop yield is more than 10 bushels per acre on six of the fields. Then $X = 6$. See Figure 4-2.

The *relative frequency* of the occurrence of an event is the proportion of the n trials whose outcomes fit the event, or

$$\text{Relative frequency of an event} = \frac{\text{Number of successes}}{\text{Number of trials}} = \frac{X}{n}.$$

Figure 4-2 shows that the relative frequency for the agricultural example would be $6/30 = 1/5 = 0.20$. This tells us that the yield is more than 10 bushels per acre on one-fifth of the 30 fields. Without conducting more trials and as a practical expedient, we might use this relative frequency as an estimate of the probability of producing more

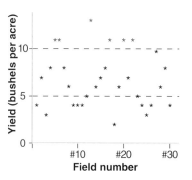

Figure 4-2

Behavior of the yield per acre on 30 different fields

The Experiment:
Is the yield of a one acre field <u>more than 10 bushels of corn</u>?

n = 30 trials: We observe the harvest of 30 cornfields

X = 6 successes: 6 fields produce more than 10 bushels

X/n = 6/30 = Relative frequency of **event**
= Relative frequency of "**more than 10 bushels**"

than 10 bushels per acre, but the actual probability involves many more trials, as we shall see.

Question: A political candidate's campaign workers approach 50 potential contributors. Thirty of them make a donation. If the event is donating, what is the relative frequency of a donation?

Answer: There are 50 trials of requesting donations and 30 successes, receiving a donation, so $n = 50$, $X = 30$, and the relative frequency is $X/n = 30/50 = 3/5$, or 0.60.

Return to the agricultural example. Consider the fact that what happens on the 30 acres might not be a good indication of what would happen on another set of 30 acres, on 100 acres, or on 100,000 acres. To obtain a better indication of the relative frequency of harvesting more than 10 bushels, we incorporate more information into the measure by using a very large number of trials. In addition, we must be careful, because the magnitude of the relative frequency of an event from a small number of trials is very sensitive to the outcome of any single trial. In the agricultural example, if X were 5 rather than 6, the relative frequency would change by about 0.03, from 0.20 to 0.17. If the relative frequency of our event for 100,000 acres is 0.20 ($X = 20,000$) and X decreases by 1, the relative frequency becomes 0.19999, a change so small, 0.0001, that it would not affect the relative frequency at the second decimal place.

As the number of trials, n, grows very large, if the relative frequency of an event becomes very insensitive to changes in the outcome of a single trial (that is, it stabilizes about a certain value), we call that value the *probability of the event*. A formal statement of this approach follows.

Long-Run Relative Frequency Method of Determining the Probability of an Event

The relative frequency of an event is the number of successes, X, divided by the number of trials of the experiment, n. The probability of an event is the value that its relative frequency, X/n, approaches and where it finally stabilizes as the number of trials approaches a very large number (n becomes infinite). In essence, the *probability of the event* is its long-run relative frequency.

Figure 4-3
The long-run relative frequency of tossing a head

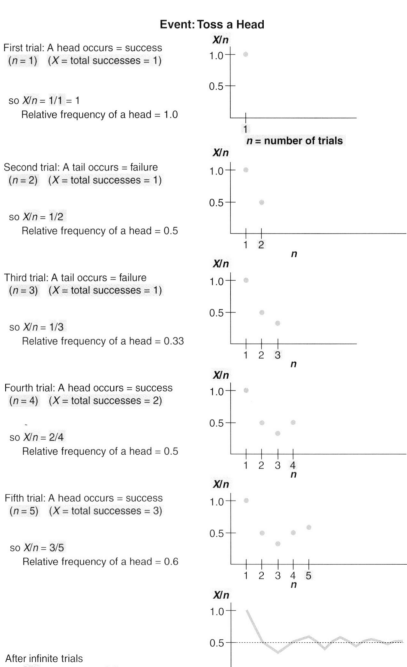

First trial: A head occurs = success
($n = 1$) (X = total successes = 1)

so $X/n = 1/1 = 1$
Relative frequency of a head = 1.0

Second trial: A tail occurs = failure
($n = 2$) (X = total successes = 1)

so $X/n = 1/2$
Relative frequency of a head = 0.5

Third trial: A tail occurs = failure
($n = 3$) (X = total successes = 1)

so $X/n = 1/3$
Relative frequency of a head = 0.33

Fourth trial: A head occurs = success
($n = 4$) (X = total successes = 2)

so $X/n = 2/4$
Relative frequency of a head = 0.5

Fifth trial: A head occurs = success
($n = 5$) (X = total successes = 3)

so $X/n = 3/5$
Relative frequency of a head = 0.6

After infinite trials
 X/n converges on 0.5

The different parts of Figure 4-3 demonstrate this approach to finding the probability of obtaining a head when tossing a fair coin. Technically, we would say that the limit of X/n, as n approaches infinity, is the probability of the event. In this case, the limit is 0.5, so the probability of tossing a head is 0.5, or $P(H) = 0.5$.

The long-run relative frequency approach provides a convenient and intuitive way of interpreting probabilities developed from any of the three methods. Again, for practical situations we employ a finite number of trials that is much smaller than infinity to estimate the probability of an event. This is often the best available information that we possess. The next method we discuss uses another type of information when it is available to determine the probability of an event. We can still construe this value as the proportion of times the event would occur during a long-run set of trials.

4.2.2 Equiprobable Outcomes

As we continue to explore more complicated events and to determine their probability, we will find the equiprobable outcomes approach useful. It is also valuable for finding probabilities of simple events or situations as well. When each outcome in the sample space has the same chance of occurring, we call such outcomes *equiprobable outcomes.*

Consider finding the probability of obtaining an odd number on a single toss of a die. Each side contains a unique number of dots, so there are six outcomes in the sample space. If the die is fair, each side has an equal chance of being on top after the toss, so there are six equiprobable outcomes to the experiment. Three ($X = 3$) of the six outcomes are odd (1, 3, 5). So there are three chances out of six of tossing an odd number or $P(\text{Odd}) = 3/6 = 1/2$.

The following box states the equiprobable outcome method more formally.

> **The Equiprobable Outcomes Method of Determining the Probability of an Event**
> If there are n equiprobable outcomes to an experiment, X of which comprise an event, then the probability of the event is X/n.

This approach not only gives insight into what probability tells us—the proportion of all possible outcomes of an experiment that fall in a particular category or event—but it is also useful for formulating actual values for probabilities, as in the following examples.

■ **Example 4-1:** There are five equally fast and easy routes to a hospital from the current location of an ambulance as shown in Figure 4-4. If the driver chooses randomly, what is the probability route 3 is selected? What is the sample space and event in this problem?

■ **Solution:** The sample space is the set of five routes. The event, choosing road 3, consists of the single outcome, labeled road 3. Thus, there are five equiprobable choices, $n = 5$, and only one success, $X = 1$, so $P(\text{road 3 chosen}) = 1/5$.

■ **Example 4-2:** What is the probability of obtaining a heart in a single random draw from a standard deck of cards?

■ **Solution:** Figure 4-5 displays the standard deck as well as the solution to this problem. There are 52 cards or outcomes, and each card is equally likely to be selected when choosing randomly ($n = 52$). There are 13 cards that are hearts ($X = 13$). Thus, the probability of drawing a heart, $P(H)$, is $13/52 = 1/4$.

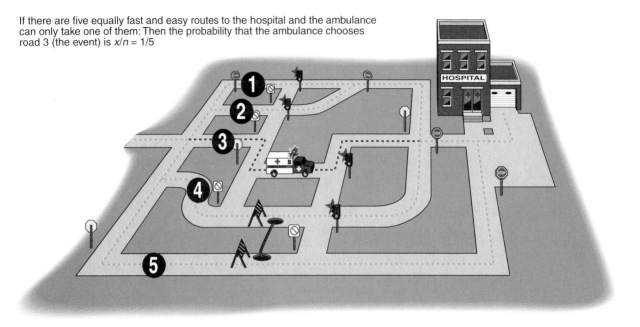

If there are five equally fast and easy routes to the hospital and the ambulance can only take one of them: Then the probability that the ambulance chooses road 3 (the event) is $x/n = 1/5$

Figure 4-4
The equiprobable outcomes approach

■ **Example 4-3:** A bin of 350 parts contains 10 defectives. If an employee grabs a part at random from the bin, what is the probability it will be defective?

■ **Solution:** $X = 10$, $n = 350$, $P(\text{defective}) = X/n = 10/350 = 1/35$.

■ **Example 4-4:** According to Environmental Information Ltd., there are 20 commercial hazardous waste dumps distributed among 15 states as follows: California has three; Illinois, Louisiana, and Texas each have two; Alabama, Idaho, Indiana, Michigan, Nevada, New York, Ohio, Oklahoma, Oregon, South Carolina, and Utah each have one. Use this information to determine the probability that a missing load of waste is deposited in:

a. A dump in California.
b. A dump in a state with *only one* additional dump.
c. A dump in a state with at least one other dump.
d. A dump in a state bordering the Pacific Ocean.
e. A dump in New England.

(Source: Jeff Bailey, "Toxic Waste Sparks War between States," *Wall Street Journal*, August 16, 1991, p. B1.)

■ **Solution:** Each of the 20 different dumps are equiprobable sites for the missing load, because there is no information that indicates that one dump is a more likely depository for the missing load than any other. Thus, $n = 20$ for all five problems. The number of successes, X, varies.

a. There are three dumps in California, so $P(\text{California}) = 3/20 = 0.15$.
b. There are three states with two dumps; that is, there are six dumps with only one other dump in the same state, so $P(\text{only one other dump in the same state}) = 6/20 = 0.30$.
c. There are nine dumps in the four states with at least two dumps. So each dump has at least one other dump in the same state, $P(\text{at least one other dump in the same state}) = 9/20 = 0.45$.

Section 4.2 Methods to Determine Probabilities 157

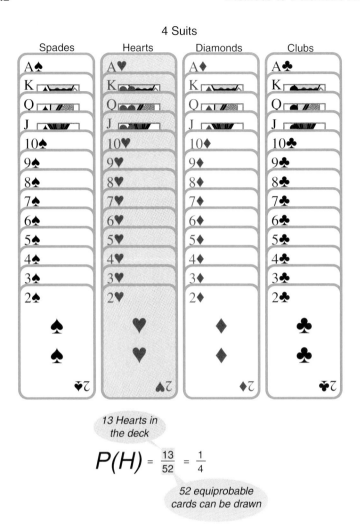

Figure 4-5
The standard deck of 52 cards and the probability of drawing a heart in one random draw

d. Two states, California and Oregon, border the Pacific Ocean. There are four dumps in these states, so P(dump in state bordering the Pacific Ocean) = 4/20 = 0.20.

e. There are no dumps in New England, so P(dump in New England) = 0/20 = 0.

This approach is useful when we can determine all of the possible outcomes and that the outcomes are equiprobable—especially in games of chance. Appendix F discusses procedures for determining the number of possible outcomes when this set is large in some situations. If the outcomes are not equiprobable, we must turn to other approaches, relative frequencies, or the next method, subjective probabilities.

4.2.3 Subjective Probabilities

The final approach to determining the probability of an event is a subjective estimate of the likelihood that an event will occur. There are often no data on previous trials and no way to determine that the outcomes are equiprobable to produce a more objective probability estimate. In other cases, the user may consider the values produced by these methods faulty or inapplicable. Instead, the user produces a probability estimate based on his or her recollected experience or intuition. A person ordering a new product to place in a store may decide that the probability that 100 or more of the items will sell is 0.7, the probability that 80 or more will sell is 0.75, the probability that 50 or more will sell is 0.9, and so on. These probability estimates represent the extent of the individuals's belief that the event would occur or the belief that the event would occur the stated proportion of the time if infinite trials were conducted.

In summary, all three approaches for formulating, calculating, and analyzing probability values in situations involving uncertainty are used, although we will employ the relative frequency and equiprobable outcomes methods in this book. Keep in mind that we can interpret the probability value generated by any of the approaches as *the proportion of trials that result in a specific event*. We expect that after infinite trials we would find that the relative frequency of the event approaches the value generated by any of the methods.

Section 4.2 Problems

3. The probability of throwing a 7 with a pair of dice is 1/6. Interpret this statement.
4. What is the probability of drawing a 7 in a single draw from a standard deck of cards?
5. A survey question about household income listed the following possible responses: (a) not more than $5,000; (b) greater than $5,000 but not more than $10,000; (c) greater than $10,000 but not more than $20,000; and (d) greater than $20,000. Suppose that we randomly select a survey respondent's answer to this question. What is the sample space? Can we use the equiprobable outcomes method to find the probability that the response will be greater than $10,000? Explain your answer. If not, what method(s) would we use to determine the probability of this event?
6. A slate of student council candidates includes six females and four males. If positions on the ballot are randomly assigned, what is the probability the first name on the ballot belongs to a male? to a female?
7. A biologist randomly selects mice from a cage that contains 10 black mice, 5 white ones, and 12 gray ones. What is the probability that the first mouse drawn will be gray?
8. The table below classifies 500 teachers in a local school system according to major subject area and years of experience. If one teacher is randomly selected, what is the probability the teacher is:
   a. An English teacher?
   b. A veteran with 3–5 years' experience?
   c. A veteran English teacher with 3–5 years' experience?

Experience (years)	Subject Area						Total
	English	Math	History	Science	Social Studies	Physical Education	
1	5	8	3	14	0	4	34
2	15	22	1	28	21	14	101
3–5	20	17	12	30	25	8	112
More than 5	83	71	39	26	32	14	265
Total	123	118	55	98	78	40	512

4.3 Visualizing Probabilities with Venn Diagrams

A useful tool for visualizing probability analysis is a Venn diagram. Figure 4-6a illustrates $P(A)$ and $P(B)$. The area of the large box encloses the sample space, the set of all possible outcomes of an experiment. The orange circle A represents the outcomes that form the event A, so by the equiprobable outcomes approach, $P(A)$ would be the ratio of the area of A to the area of the box, like X/n. Because we are looking at the proportion, it makes no difference if we let the area of the box be 1, 100% of all possible outcomes. The area of circle A (divided by 1) is $P(A)$. Similarly, the area inside circle B is $P(B)$. Alternatively, if the large box represents the outcomes of all or 100% of the trials

Section 4.4 Upper and Lower Boundaries on *P(A)* 159

a) *P(A)* and *P(B)*

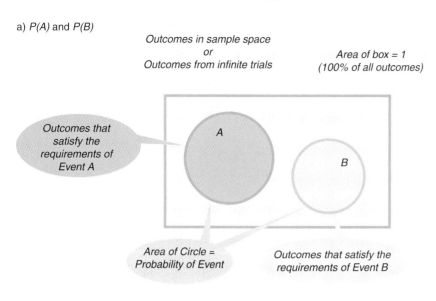

Figure 4-6
Illustrating probabilities with Venn diagrams

b) Corn harvests and Venn diagrams

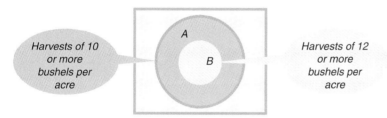

of an experiment, then the area of circle A represents the proportion of the trial outcomes that satisfy the requirements of event A.

Question: Which event pictured in Figure 4-6a is more likely to occur, *A* or *B*? If circle A encloses circle B, as shown in Figure 4-6b, what do we know about the occurrence of *A* and *B*?

Answer: *A* is more likely to occur, because circle A is larger than circle B in both parts of Figure 4-6. If circle B is inside circle A, then whenever *B* occurs, *A* must necessarily occur. There are times when *A* occurs and *B* does not. For instance, if *A* is harvests of 10 or more bushels per acre and *B* is harvests of 12 or more bushels per acre, then when *B* occurs, *A* must occur as well. However, if the output is 11 bushels per acre, *A* occurs and *B* does not. Thus, *P(A)* must exceed *P(B)*.

4.4 Upper and Lower Boundaries on *P(A)*

We have seen several examples of probability values now, and they were all fractions. Now, we need to establish boundaries on acceptable or reasonable values for probabilities.

Question: What is the smallest possible value for the probability that event A will occur, *P(A)*?

Answer: The least number of times that *A* can occur in *n* trials is zero, which means it does not occur. For example, the fewest number of coin tosses that can be heads is zero. So the proportion of trials that result in heads would be zero. Thus, $P(A)$ cannot be a negative number, $P(A) \geq 0$.

Question: What is the largest possible value of $P(A)$?
Answer: If A occurs every time we perform an experiment, then, X and n are identical values and X/n is 1. Then, the probability is 1. So $P(A)$ cannot exceed 1, $P(A) \leq 1$.

These answers lead us to the following conclusion.

> The lower and upper boundaries on values of $P(A)$ are 0 and 1; $0 \leq P(A) \leq 1$.

This result may seem rather obvious, but keep it in mind when working problems. If a calculated probability value is negative or greater than 1, then there is an error.

Section 4.4 Problems

Check any of the following values that is a legitimate probability.
- **9.** 0.5.
- **10.** 0.28.
- **11.** 1.72.
- **12.** 1/5.
- **13.** 1.
- **14.** −1.
- **15.** 3/7.
- **16.** −0.22.
- **17.** 12/7.
- **18.** 0.
- **19.** 3/4.
- **20.** 0.66.

4.5 The Negation or Complement of an Event

The symbol for an event not occurring is a prime beside the symbol that represents the event. A' means "A does not occur." So $P(A')$ means "the probability that A does not occur." Figure 4-7a shows that when A does not occur, then some outcome other than those composing event A occurs. This set of non-A outcomes completes the set of trial outcomes when combined with A outcomes, so we call it the *complement of A*.

If $P(A)$ represents the proportion of trials in which A results, then $P(A')$ represents the rest of the trials, when A does not result. Thus, $P(A) + P(A') = 1$, which means that

Figure 4-7
The complement of event A

a) Outcomes in the complement of A

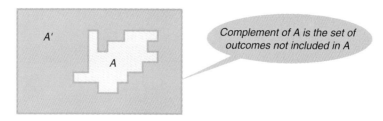

b) A and A' account for all experimental trial outcomes

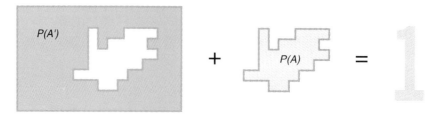

we have accounted for 100% of the trials as Figure 4-7b demonstrates. If we solve this equation for $P(A')$ by subtracting $P(A)$ from both sides, the following rule results.

> **Rule 1**
> To determine the probability that A does not occur, A', use
> $$P(A') = 1 - P(A).$$

Example 4-5: If we randomly select one card from a standard deck, what is the probability it is *not* a face card? (Assume aces are not face cards.)

Solution: Let F = face card. There are 12 face cards (4 jacks, 4 queens, and 4 kings), so $P(F) = 12/52$. Then,

$$P(F') = 1 - P(F) = 1 - 12/52 = 40/52 = 10/13.$$

Alternatively, and more directly in this example, we can use the equiprobable outcomes method to find $P(F')$. There are 40 cards that are not face cards (4 twos, 4 threes, ..., 4 tens, 4 aces), so $X = 40$. There are 52 cards in the deck that are the outcomes to this experiment, so $n = 52$. Hence, $P(F') = X/n = 40/52 = 10/13$.

Example 4-6: An employment agency places 90% of its registrants. What is the probability that a registrant will not obtain a job?

Solution: Let J = placed in a job. Then, $P(J) = 0.9$. Thus,

$$P(J') = 1 - 0.9 = 0.1.$$

Box 4-1 discusses odds, an alternative presentation of the probability of an event that relates the probability of occurrence to the probability of nonoccurrence.

Section 4.5 Problems

21. What is the probability of not drawing a club in a single draw from a deck?

22. What is the probability of not throwing an even number in a single roll of a die?

23. If approximately half of new babies are male, what is the probability a newborn will not be male?

24. If a teacher is randomly selected from the 512 teachers charted below, what is the probability the teacher's specialty is not history?

25. If we select a U.S. senator at random, what is the probability the senator is not from California?

Experience (years)	Subject Area						Total
	English	Math	History	Science	Social Studies	Physical Education	
1	5	8	3	14	0	4	34
2	15	22	1	28	21	14	101
3–5	20	17	12	30	25	8	112
More than 5	83	71	39	26	32	14	265
Total	123	118	55	98	78	40	512

Box 4-1 Figuring Odds

The term "odds" often arises in probability contexts as a substitute form of probability. Actually, odds reflect probabilities, but they are not probabilities.

For example, the odds that event X does not occur represent the net return or payback from a dollar bet that X does occur. Suppose you bet $1 that X occurs. Five-to-one odds means that you will receive $6 if X occurs (you win the bet), which is a net winnings of $5, because you pay $1 to engage in the bet. If X does not occur, you lose the $1.

If the odds (or the bet) are "fair," then you and your opponent would not gain or lose any money after a large number of (infinite) repetitions of the bet, if you bet the same way each time. Because you win some of the time and lose some of the time, the amount you win and lose must be equal to ensure fairness. If these statements are untrue, then the bet is not "fair," and the relationship between odds statements and probability statements will not hold.

With the bet, if X occurs one-sixth of the time, $P(X) = 1/6$, you will win the $5 one-sixth of the times you bet. Consequently, X does not occur five-sixths of the time, $P(X') = 5/6$, and you lose your $1 five-sixths of the time. Notice that $\$5(1/6)$ is the same as $\$1(5/6)$ or $5(\tfrac{1}{6}) = 1(\tfrac{5}{6})$. If we solve this equation for the odds $\$5/\$1$, we obtain

$$\frac{5}{1} = \frac{(5/6)}{(1/6)} = \frac{P(\text{loss})}{P(\text{win})} = \frac{P(X')}{P(X)}.$$

In general, odds that X does not occur $= P(X')/P(X)$. The odds that X does occur would be the reciprocal of this ratio. Because the odds are a ratio of probabilities, they also tell us the relative likelihood of experimental outcomes. If the odds that X does not occur are 10-to-1, then the nonoccurrence of X is 10 times more likely than the occurrence of X.

If we represent the odds that X does not occur with W-to-1 (3-to-2 is really 1.5-to-1), we can figure the probabilities associated with occurrences of X from the odds ratio. The following expressions derive from algebraic manipulation of the relationship between odds and probabilities.

$$P(X) = \frac{1}{W+1} \qquad P(X') = \frac{W}{W+1}.$$

For instance, if the odds are 5-to-1 that X does not occur, then $P(X) = 1/(5+1) = 1/6$ and $P(X') = 5/(5+1) = 5/6$.

4.6 Conditional Probabilities

Another useful probability concept is conditional probability. A *conditional probability* is the probability an event A occurs *if* (*conditional on* or *given that*) an event B occurs. Knowing event B occurs (has occurred or will occur) provides additional information that may affect the probability of A that we would obtain in the absence of the extra details. The chances that a firm's sales will increase during a month might ordinarily be 60% (or 0.6). If the firm finds that its competitor is in the midst of an advertising blitz, the chances the firm's sales will increase might drop to a value such as 40%. This example demonstrates that conditional probability simply formalizes our common behavior and reasoning. As we receive and synthesize information, we continually update our perception of what is happening and likely to happen in the world about us.

Question: If asked to guess the sex of an anonymous student, you might assume that the probability of the student's being female is 0.5 and toss a coin to decide your response. If, however, someone provided the additional information that the student is a nursing student, what adjustment, if any, would you make to this probability?

Answer: Experience in hospitals and doctors' offices suggests that many, but certainly not all, nursing students are female, so we would probably adjust the probability of the student being female upward. The information would tip our guess in favor of female.

The symbol for conditional probability is $P(A|B)$. The vertical line between A and B is not a fraction or division symbol. It stands for the words *if*, *conditional on*, *given that*, or *after*. The information that follows the vertical line narrows the set of trials or the sample space that we must consider to learn the probability that A occurs. If we already know that a card randomly selected from a deck is red, we eliminate the 26 black cards

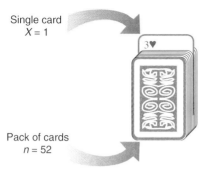
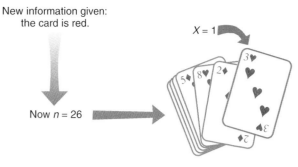

from consideration. Figure 4-8 shows the remaining 26 possible outcomes (red cards) that form a new sample space after we take the new information into account. This is the relevant sample space to consider when we find the probability of some other event related to the card, such as it being a 3 of hearts. There is only one 3 of hearts in this set of 26 cards, so $P(3 \text{ of hearts} \mid \text{red card}) = 1/26$.

Figure 4-8
Conditional probabilities

Question: What is the probability we will throw a 4 in a single roll of a die?
Answer: $n = 6$, $X = 1$, and $P(4) = 1/6$.

Question: If you roll a die and learn that the result is an even value, what is the probability you have thrown a 4?
Answer: Because you know that the result is even, there are only three possibilities (2, 4, or 6). Thus, $P(4 \mid \text{even}) = 1/3$.

■ **Example 4-7:** There are 10 applicants for a scholarship, including Philip. The scholarship committee must narrow the choice to three finalists.
a. With no information about who the finalists are, what value would Philip assess for his chances of getting the scholarship?
b. What is the probability that Philip will get the scholarship if he learns he is one of the three finalists?
c. What is the probability that Philip will *not* get the scholarship if he learns that he is one of the three finalists?

■ **Solution:** Let S = obtains the scholarship and F = is one of three finalists.
a. If Philip has no information about the finalists, then $n = 10$ and $P(S) = 1/10$.
b. If he is one of the finalists, then $n = 3$ and $P(S \mid F) = 1/3$.
c. From Part b and Rule 1, we find $P(S' \mid F) = 1 - 1/3 = 2/3$.

The given part in the last example, knowing that Philip is a finalist, diminishes the sample space. If the problem only stated that there were three finalists and Philip was among them, then we would work Parts b and c as if they were unconditional rather than conditional probabilities and apply Rule 1 without the conditional probability symbols. Just remember that conditional probabilities are like regular probabilities, they just let us know that the value incorporates extra information. (Box 4-2 presents actual conditional probabilities related to scholarship competitions.)

■ **Example 4-8:** The table below records the number of students who were randomly assigned to three different classes, where each class used a different teaching

Box 4-2 Probabilities of Winning an Educational Scholarship

Knowing the likelihood of winning a contest often affects our decision to enter, scholarship competitions included. In one case, 26% of the entrants in a competition, sponsored by a consortium of large related organizations for its employees, received scholarships. There are not many contests where the chances of financial reward are quite so high.

Do these chances mean you should automatically apply? If it were a random drawing, perhaps. However, the group listed major criteria for awards, such as need, academic achievement, high school activities, and teacher recommendations. Although these criteria encouraged some and discouraged others, they did not provide entrants precise information about probabilities of winning an award that are associated with different combinations of credentials.

The following table lists conditional probabilities of winning the scholarship, estimated by using data from the pool of entrants with a 26% success rate. The values, estimated by using more advanced statistical techniques, provide information to entrants as well as those who provide scholarship about the kinds of entrants and behaviors that are rewarded. For instance, the probability of winning is 0.32 for a male with little need of financial assistance who scored 600 on the verbal section of the SAT and 650 on the math. Notice that probabilities are greater for females, for higher SAT scores, and, generally, for higher need. Now entrants have more precise estimates of the likelihood of winning and can decide if the costs of entry, such as filling out forms, studying to improve SAT scores, or joining a club, are likely to be worth the effort.

Some implications of these values are not intuitive or supportive of the criteria initially advertised by the consortium. The authors of the study suggest that higher values for females may reflect a better academic performance or greater allocation of time to academic activities, including preparing for the SAT. Meanwhile, males respond to larger average wage rates and allocate their time to income-generating activities. Such a time allocation may account for the lower probabilities for high-need males than for medium-need males in several cases. The high-need males may feel more pressure to pursue income-earning activities versus academic activities. The probabilities do suggest that more time be devoted to academic activities in order to succeed in obtaining one of these scholarships.

The diverse set of probabilities affects the incentive to join the competition as well as the exertion to win. Certain behaviors are encouraged, and others are not.

Source: A.G. Holtmann and Todd L. Idson, "Winning an Educational Scholarship," American Economist, Vol. 35, No. 2, Spring 1991, pp. 30–39.

Probability Estimates of Winning a Scholarship

SAT Score		Need (Female)			Need (Male)		
Verbal	Math	Low	Medium	High	Low	Medium	High
400	450	0.000	0.014	0.078	0.000	0.008	0.010
450	500	0.000	0.066	0.206	0.000	0.043	0.040
500	550	0.000	0.211	0.411	0.000	0.154	0.125
550	600	0.007	0.459	0.644	0.000	0.376	0.289
600	650	0.274	0.726	0.833	0.032	0.650	0.515
650	700	0.895	0.904	0.941	0.500	0.862	0.737
700	750	0.999	0.978	0.984	0.968	0.963	0.891
750	800	1.000	0.997	0.997	1.000	0.994	0.966

method, and the number who made different grades in the classes. Use this information to find $P(\text{grade} = B \mid \text{method-1 class})$ and $P(\text{grade} \neq C \mid \text{method-3 class})$.

Grade	Teaching Method			Total
	1	2	3	
A	10	8	12	30
B	14	14	6	34
C	16	21	10	47
D	3	3	2	8
F	1	0	1	2
Total	44	46	31	121

Solution: There are 44 students studying under method 1 ($n = 44$) and 14 students in this group score a B ($X = 14$), so $P(\text{grade} = B \mid \text{method-1 class}) = 14/44$. There are 31 students in the method-3 class ($n = 31$), 10 of which have a C ($X = 10$), so $P(\text{grade} = C \mid \text{method-3 class}) = 10/31$. Consequently, $P(\text{grade} \neq C \mid \text{method-3 class}) = 1 - 10/31 = 21/31$.

Question: Express the statement, "68% of spring vacations are spent in Florida" as a conditional probability.

Answer: Let S = have a spring-vacation choice to make and F = choose Florida as vacation spot. Then, we realize the statement addresses spring vacations, not all vacations, so the set of trial outcomes is diminished to spring vacations. Then, we know that 68% of spring-vacation choices are decided in favor of Florida. Thus, $P(F \mid S) = 0.68$.

Conditional probabilities are useful for reevaluating the probability of an event, such as $P(A)$, when we acquire more related information, B, to obtain $P(A \mid B)$. The probability a student scores well on a test will probably change if we know how much the student studies. Conditional probabilities also figure in computations of more complex events as we shall see in the next unit.

Section 4.6 Problems

26. Find $P(\text{king} \mid \text{face card})$ for a single draw from a standard deck. (Assume aces are not face cards.)
27. Give an everyday situation as an example where the additional information would cause a conditional probability of an event to increase from the unconditional probability. Give a situation where it would decrease.

28. $P(\text{experience} = 2 \mid \text{subject} = \text{science})$.
29. $P(\text{subject} = \text{science} \mid \text{experience} = 2)$.
30. $P(\text{experience} \geq 3 \mid \text{subject} = \text{physical education})$.
31. $P(\text{subject} = \text{math} \mid \text{experience} \geq 3)$.
32. $P(\text{subject} \neq \text{math} \mid \text{experience} \geq 3)$.

Use the following table, which classifies 512 teachers according to subject area and experience, to find the probabilities listed in 28–32.

Experience (years)	Subject Area						Total
	English	Math	History	Science	Social Studies	Physical Education	
1	5	8	3	14	0	4	34
2	15	22	1	28	21	14	101
3–5	20	17	12	30	25	8	112
More than 5	83	71	39	26	32	14	265
Total	123	118	55	98	78	40	512

Summary and Review

A sample space is the set of all possible outcomes of a statistical experiment. An event is a set of outcomes with pertinent common characteristics sometimes labeled A.

We use probabilities to gauge uncertainty associated with different events. The probability of an event is a measure of the proportion of experimental trials that result in the event. It can also be viewed as the proportion of the set of equiprobable outcomes of an experiment that comprise the event. Sometimes probabilities are subjective judgments about the likelihood that the event occurs. These approaches allow us to formulate probabilities for events that assist in decision making and evaluating statistical inferences. Venn diagrams illustrate events and their probabilities.

Numerical probability values are proportions that must fall within the interval of values bounded by 0 and 1, $0 \leq P(A) \leq 1$. An event, A, cannot occur less than 0% or more than 100% of the time.

The negation or complement of an event A is a new event, A', that includes all the outcomes not included in A. To find the probability of A' we use $P(A') = 1 - P(A)$.

A conditional probability value, $P(A \mid B)$, is the proportion of time an event A occurs given that some other event B occurs. The latter event may have some effect on the occurrence of the first event and, consequently, on the probability of the first event. We use this concept to evaluate the probability of an event when we acquire additional information about a related event. In the next unit, we use conditional probabilities to define and determine independence of events.

Multiple-Choice Problems

33. Interpret the following statement: In one roll of a pair of dice, the probability of throwing a total of 7 is 1/6.
    a. A 7 will occur on every sixth roll.
    b. A 7 occurs on about 1/6 of a very large number of rolls.
    c. A 7 must occur when the dice are rolled.
    d. The dice are rolled 1/6 of an infinite number of times.
    e. A 1 on one die and a 6 on the other die occur on 1/6 of a very large number of rolls.

34. If an earthquake analyst forecasts a 50–50 chance of a major earthquake along the San Andreas Fault that does not occur, then
    a. The forecaster is a quack.
    b. The 50–50 chance or $P(\text{earthquake}) = 0.50$ should have been another value.
    c. 50% of earthquakes are unpredictable.
    d. 50% of earthquakes are predictable.
    e. To say the analyst is wrong we need to know the proportion of times earthquakes occur in similar circumstances.

35. Which of the following is a legitimate value for a probability?
    a. 1.63.   d. 0.25.
    b. −0.05.   e. −2.08.
    c. 5.27.

36. Which method to determine a probability value requires no experimentation or experience with the event and no detailed information about the outcomes?
    a. Trial-and-error method.
    b. Equiprobable outcomes method.
    c. Subjective method.
    d. Long-run method.
    e. Relative frequency method.

37. Which of the following is the sample space for one roll of a pair of dice? The first number inside the parentheses represents the value rolled on the first die, and the second value represents the second die's roll.
    a. {1, 2, 3, 4, 5, 6}.
    b. {(1, 1), (2, 2), (3, 3), (4, 4), (5, 5), (6, 6)}.
    c. {(1, 2), (1, 3), (1, 4), (1, 5), (1, 6),
        (2, 3), (2, 4), (2, 5), (2, 6),
        (3, 4), (3, 5), (3, 6),
        (4, 5), (4, 6),
        (5, 6)}.
    d. {(1, 1), (1, 2), (1, 3), (1, 4), (1, 5), (1, 6),
        (2, 2), (2, 3), (2, 4), (2, 5), (2, 6),
        (3, 3), (3, 4), (3, 5), (3, 6),
        (4, 4), (4, 5), (4, 6),
        (5, 5), (5, 6),
        (6, 6)}.
    e. {(1, 1), (1, 2), (1, 3), (1, 4), (1, 5), (1, 6),
        (2, 1), (2, 2), (2, 3), (2, 4), (2, 5), (2, 6),
        (3, 1), (3, 2), (3, 3), (3, 4), (3, 5), (3, 6),
        (4, 1), (4, 2), (4, 3), (4, 4), (4, 5), (4, 6),
        (5, 1), (5, 2), (5, 3), (5, 4), (5, 5), (5, 6),
        (6, 1), (6, 2), (6, 3), (6, 4), (6, 5), (6, 6)}.

38. $P(A \mid B)$ stands for the probability that
    a. A occurs conditional on B occurring.
    b. A occurs, and B occurs.
    c. A and B both do not occur.
    d. A occurs, and B does not.
    e. B occurs, and A does not.

39. Suppose you are throwing a pair of dice (one is white, and one is red). Let T = total dots showing from the dice. Let R = number of dots showing on the red die. Find $P(T = 8 \mid R = 2)$.
    a. 5/36.   d. 1/36.
    b. 1/2.   e. 4/52.
    c. 1/6.

40. The probability of drawing an ace in a single random draw from a standard deck is
    a. 4/52.   d. 39/52.
    b. 48/52.   e. 13/52.
    c. 1/52.

41. Estimating proportions from a graph produced the following: Of those aged 25 or less, 17% owned homes in 1989, 36% of those aged 25 to 29, and 54% of those aged 30 to 34. (Source: National Association of Realtors in *Consumer's Research,* July 1991, p. 32.) From this information, we can say
    a. Something is wrong, because more than 100% of people aged 34 or less own homes.
    b. The probability of a person 34 or less owning a home is 1.07.
    c. $P(\text{own home} \mid \text{person aged 30–34}) = 0.54$.
    d. $P(\text{person aged 30–34} \mid \text{own home}) = 0.54$.
    e. $P(\text{own home}) = 0.36$.

Word and Thought Problems

42. $P(A')$ stands for
   a. The proportion of trials on which A occurs.
   b. The proportion of trials on which A does not occur.
   c. The probability that A occurs.
   d. The probability that A occurs given the occurrence of another event.
   e. The probability that A does not occur given the occurrence of another event.

43. There are 35 scores on a test: 5 As, 9 Bs, 13 Cs, 4 Ds, and 4 Fs. Without more information about a particular student in this class, what is the probability this student passes with "at least" a C?
   a. 13/35.   d. 8/35.
   b. 27/35.   e. 13/31.
   c. 27/31.

44. The probability of selecting an even-number card in a single draw from a standard deck is
   a. 2/52.   d. 10/52.
   b. 20/52.   e. 40/52.
   c. 5/52.

Use the following analysis of data from the Defense Department and other branches of the military service regarding the Persian Gulf War to find the probability a randomly selected file for one of these soldiers belongs to one who is described in problems 45–48.

Cause of Death	Number of Deaths
Deaths related directly to battles	
Killed in unspecified type	
of contact with enemy	16
Explosion	21
Scud missile attack on barracks	28
Friendly fire	35
Aircraft shot down	48
Death from nonaction activities	
Unexploded ordnance	12
Natural causes	17
Accidental shooting, explosion,	
suicide, trauma	24
Drowning	24
Unknown or under investigation	32
Vehicle accident	58
Aircraft crashes	70
	385

Source: Knight-Ridder analysis of casualty data from Defense Department, Army, Air Force, Marines, and Navy by Judy Treible, *Knight-Ridder Tribune News*, in "Accidental Fatals Greatly Outnumber Gulf Action Deaths," *Asheville Citizen-Times*, August 17, 1991, p. 8A.

45. Died from natural causes.
   a. 0.3065.   d. 0.3844.
   b. 0.0857.   e. 0.0909.
   c. 0.0442.

46. Died in an aircraft-related incident.
   a. 0.3065.   d. 0.3844.
   b. 0.0857.   e. 0.0909.
   c. 0.0442.

47. Died during action with the enemy.
   a. 0.3065.   d. 0.3844.
   b. 0.0857.   e. 0.0909.
   c. 0.0442.

48. Did *not* drown.
   a. 0.0623.   d. 0.9377.
   b. 0.9558.   e. 0.8746.
   c. 0.0442.

49. The probability of *not* drawing an ace on a single random draw from a standard deck of cards is
   a. 50/52.   d. 4/52.
   b. 48/52.   e. 13/52.
   c. 39/52.

The proportion of Utah's population that is 5 to 11 years old is 15% and 12 to 17 years old is 11%. The comparable U.S. proportions are 10% and 8%. Use this information to find the probabilities listed in Questions 50–52. (Source: "West Is Youngest Area, in More Ways Than One," *Wall Street Journal*, July 19, 1991, p. B1.)

50. P(a randomly selected American is *not* 5 to 11 years old)
   a. 0.10.   d. 0.85.
   b. 0.82.   e. 0.73.
   c. 0.90.

51. P(being 5 to 11 years old | live in Utah)
   a. 0.15.   d. 0.08.
   b. 0.11.   e. 0.6667.
   c. 0.10.

52. P(live in Utah | 5 to 11 years old)
   a. 0.15.   d. 0.6667.
   b. 0.85.   e. Cannot tell from the information given.
   c. 0.1765.

Word and Thought Problems

53. The probability of obtaining a head with a coin is 0.5. If you flip the coin 10 times and count the number of heads, what totals do you expect *not* to happen? Explain your answer.

54. A fast-food restaurant plans to engage in a promotion game that provides each customer a game card with a spot to rub off and determine the prize, if any. There are 2 million cards, 5 that hide $1,000 awards, 10 that hide

$500, 2,000 that hide $5, 25,000 that hide free hamburgers, 25,000 that hide free french fries, and 50,000 that hide free soft drinks; the rest award no prizes. Let W represent the outcome of a game card, and answer the following questions.
   a. What is the sample space?
   b. Find $P(W = \$500)$ for a randomly distributed card.
   c. Find $P(W \geq \$500)$ for a randomly distributed card.
   d. Find $P(W$ is not a food prize) for a randomly distributed card.
   e. Find $P(W = \$500 \mid$ customer wins money) for a randomly distributed card.
55. We might describe the relative frequency method of determining a probability value as empirical or practical and the equiprobable outcomes method as theoretical. Why? State the long-run relative frequency interpretation that we can use for a probability value developed from either approach.
56. Use the following table that classifies experimental subjects by gender (G) and body weight (W) to find the following probabilities for a randomly selected subject.

	Body Weight			
Gender	Underweight	Appropriate Weight	Overweight	Total
Female	15	56	27	98
Male	23	77	41	141
Total	38	133	68	239

   a. $P(W = $ overweight$)$.
   b. $P(G = $ female$)$.
   c. $P(G = $ female $\mid W = $ overweight$)$.
   d. $P(W = $ overweight $\mid G = $ female$)$.
   e. $P(W \neq $ appropriate weight$)$.

57. Suppose you draw two cards from a standard deck, with replacement after each card. $A =$ card drawn first, and $B =$ second card. Complete the three cases in the following table.

A	B	$P(B \mid A)$
Black 4	Diamond	
4 of any suit	4 of any suit	
Club 4	Diamond 4	

58. Redo Problem 57, assuming you do not replace the first card.

A	B	$P(B \mid A)$
Black 4	Diamond	
4 of any suit	4 of any suit	
Club 4	Diamond 4	

59. Suppose you draw one card from a standard deck where A and B are events that could occur on this single draw. Complete the following table.

A	B	$P(B \mid A)$
4 of any suit	Diamond	
Club 4	Diamond 4	
Red	Heart	

60. Extract a section from each of the Venn diagrams in Figure 4-9 to depict the new sample space when finding $P(A \mid B)$. Shade the area that corresponds to $P(A \mid B)$.

Figure 4-9
Original Venn diagrams for Problem 60

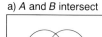
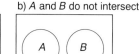

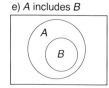

Word and Thought Problems

61. The members of a fire department know that there will be a drill some time during their next 24-hour shift (from midnight one night until midnight the next night), but they do not know the exact hour when it will occur. What is the probability that the hour for the drill will be between 6 and 7 A.M., if
  a. It is standard time throughout the shift?
  b. Daylight saving time begins during this shift?
  c. Daylight saving time ends during this shift?
  d. The drill must occur either before 11 A.M. or after 6 P.M. on a standard-time day?

62. Elizabeth has three duties in her job: scheduling personnel, orienting new customers, and listening to customer complaints. If she divides her time equally among the three duties, find
  a. The probability that she will be scheduling personnel at any particular time.
  b. The probability that she will be dealing with customers at any particular time.
  c. The probability that she will *not* be dealing with customers at any particular time.

63. A historical association has narrowed its choices for sites for a Mississippi River Museum to Minneapolis, St. Louis, Memphis, Lake Village (Arkansas), and New Orleans. The final selection will be a random draw. What is the probability that the chosen site will be
  a. Memphis.
  b. A city with two words in its name.
  c. St. Louis or New Orleans.
  d. Lake Village, if the chosen city has two words in its name.

  Suppose the first city drawn will be the site of the museum, but the second city drawn will host the annual convention for the next three years. Find the probability that Minneapolis will host the convention, if
  e. The museum site can also host the convention, that is, it is returned to the group of convention-site possibilities before the second draw.
  f. The museum-site city is not allowed to compete for the convention.

64. If there are 10,000 registered voters in your district and potential jurors are selected randomly from voter lists, what is the probability that you will *not* be selected for jury duty this year, if you can be selected only once during the year.

65. A sample of six soil tests pH readings are 7, 5, 5, 6, 8, and 4. Study the values and consider the possible values of $\overline{X}$ for this sample to find the following probabilities (you do not need to calculate $\overline{X}$).
  a. $P(\overline{X} < 4)$.
  b. $P(\overline{X} \geq 8)$.
  c. $P(4 \leq \overline{X} \leq 8)$.
  d. $P(4 < \overline{X} < 8)$.

66. The number of children under age 18 involved in divorces varies as follows based on information from 32 states in 1988 (rounding causes the percentages to sum to 101%).

Number of Children Involved	Divorces (%)
0	47
1	26
2	20
3	6
4 or more	2
	101

Source: "More Divorces Involve Fewer Children Apiece," *Wall Street Journal*, July 19, 1991, p. B1.

  a. Find P(at least one child involved in a divorce).
  b. Find P(more than one child involved in a divorce).
  c. Find P(no children involved in a divorce).

67. The following information is from the same source as Problem 66.
  P(children involved | white couple divorcing) = 0.52.
  P(children involved | black couple divorcing) = 0.50.
  P(children involved | other-race couple divorcing) = 0.48.
  Does children involvement appear to *depend* on the race of the divorcing couple? Explain.

68. Explain the limits for numerical probability values.

69. Suppose we want to determine the probability of a recession beginning next month. We continue to go back as far as economic records allow (say 2,400 months) and calculate X/n each time we add a month. However, the value X/n continues to vacillate and never settles to one value. Is there a probability of X/n? What would you suggest for the probability for the next month: the last value of X/n or some other measure? Explain.

70. Create a game based on one random draw from a single deck of cards, so that the probability of winning is 16/52, losing is 32/52, and a tie is 4/52.

71. The 10 states with the largest 1990 American Indian, Eskimo, or Aleut (AIEA) population are listed below along with the count (in thousands) and also the state population count (also in thousands). There are 1,959,000 total AIEA in the United States in a total 1990 population of 248,710,000. Use this information to answer the following questions.

State	AIEA Count	Population Count
Oklahoma	252	3,146
California	242	29,760
Arizona	204	3,665
New Mexico	134	1,515
Alaska	86	550
Washington	81	4,867
North Carolina	80	6,629
Texas	66	16,987
New York	63	17,990
Michigan	56	9,295

Source: "Race and Hispanic Origin," *1990 Census Profile*, U.S. Department of Commerce, Economics and Statistics Administration, Bureau of the Census, No. 2, June 1991, pp. 4, 6.

  a. P(AIEA) if we select randomly from U.S. citizens
  b. P(AIEA | from California).
  c. P(from California | AIEA).

d. P(from California).
e. P(from California | from one of top 10 AIEA states).
f. P(AIEA from one of top 10 AIEA states) if we randomly select from U. S. citizens
g. P(AIEA from east of Mississippi River | from one of top 10 AIEA states).

72. Bike commuters make up about 1.67% of Americans according to a poll conducted for *Bicycling* magazine. If you choose an American at random, what is the probability that they do not commute by bicycle? (Source: Nelson Pena, "Power in Numbers," *Bicycling*, April 1991, pp. 44, 46.)

Unit II: Solving Probability Problems for Compound Events

Preparation-Check Questions

73. $P(A)$ symbolizes
   a. The multiplication of P and A.
   b. The chance of A occurring given that another event occurs.
   c. The proportion of trials that result in A occurring over the long run.
   d. The relative frequency of A occurring over a short period of time.
   e. The experimental outcomes that together form event A.

74. Which of the following is *not* a permissible probability value?
   a. −1.   d. 42/586.
   b. 1.   e. 0.0424.
   c. 0.5.

75. The probability of drawing a face card (king, queen, or jack) in a single random draw from a standard deck is
   a. 3/12.
   b. 3/52.
   c. 32/52.
   d. 12/52.
   e. 12/26.

76. $P(A|B)$ symbolizes
   a. The probability that A and B occur simultaneously.
   b. The probability that A occurs and B does not occur.
   c. The probability that B occurs and A does not.
   d. The probability that B occurs given that A occurs.
   e. The proportion of trials that result in A among the long-run trials that result in B.

Answers: 73. c; 74. a; 75. d; 76. e.

Introduction

School Bus Route Suppose a school bus has a route that passes through a four-way stop. After numerous trips through this intersection, the driver determines the following information about waiting times.

Waiting Time	How Often
Less than 1 minute	50%
1–3 minutes	30%
Over 3 and up to 5 minutes	15%
Over 5 minutes	5%

Also, a major road project is under construction on the route. The traffic interference from the construction varies due to weather, breakdowns, work breaks, progress, and apparent bursts of energy by the workers. The driver determines the following information about waiting times due to the construction:

Waiting Time	How Often
5 minutes or less	75%
Over 5 minutes	25%

Section 4.7 $P(A$ and $B)$ **171**

This route seems to involve a lot of waiting. Has the time come to find an alternative route or reschedule times for pickup and delivery of the children? Perhaps.

This unit concerns measuring the likelihood of combinations of events. Using rules for calculating probability, we can find the probability of waiting more than five minutes at both obstacles on a single trip. We can calculate the probability of waiting more than five minutes at one obstacle or the other. After considering these calculations, the driver may well decide that over a large number of trips, the chances of a long wait are too high, leading him to search for an alternate route or a new schedule. Similar calculations of the probability of delays along any new route would provide a basis for comparing and selecting routes.

There are several rules that we employ to calculate unknown probabilities from known probabilities. We will discuss and demonstrate them by using examples.

4.7 $P(A$ and $B)$

The symbols A and B each represent a single event. A *compound event* is a combination of single events. The word *and* connects single events to form the first compound event we will discuss. We will find the $P(A$ and $B)$. Then, we will discuss compound events formed with the word "or," to find the $P(A$ or $B)$.

In order for the new compound event, "A and B" to occur, both events must occur. Suppose A is Jackie runs fast and B is Jackie throws well. Then for "A and B" to be true, Jackie must do both, run fast and throw well. To draw a heart *and* a 7 on a single draw requires the 7 of hearts. There is no card that is an ace *and* a king, so this compound event can never occur. Satisfying one event is not enough. This use of the word "and" is straightforward. The formality here will help you understand how the probability rules that we develop later will work, even though right now we may seem to be overworking a simple idea.

Question: Sketch a Venn diagram that shows two events A and B, where the compound event "A and B" contains outcomes. Shade these outcomes. Repeat this exercise, but for a diagram when "A and B" contains no outcomes.

Answer: Figure 4-10a and b displays the two diagrams. At least one point must be common to circles A and B for "A and B" to occur, so circles A and B must intersect (Part a of the figure). The area of this intersection represents $P(A$ and $B)$. When there are no common points, then no outcome satisfies events A and B, so $P(A$ and $B) = 0$ (Part b).

4.7.1 $P(A$ and $B)$ for Any Two Events

General Formula If we know that females compose half of the students in a university of 10,000 students and that, in addition, one-fourth of the female students are juniors, then we can determine the number of female juniors at the university. We simply say that 5,000 students, half of the 10,000, must be females. Then, 1,250, one-fourth of the 5,000 females or one-fourth of half of the university students must be juniors. As a mathematical expression, we have:

Number of students who are female and juniors = (1/4) (1/2) (10,000)
= (1/4) (5,000) = 1250.

We can express this finding as a relative frequency or probability of randomly selecting a female junior (a female and a junior or "F and J") by the ratio 1,250/10,000 = 1/8.

Figure 4-10
Venn diagrams for the compound event "A and B"

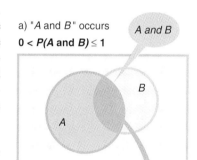

a) "A and B" occurs
$0 < P(A$ and $B) \leq 1$

The intersection area represents the relative frequency or probability of the event "A and B" in the set of all trial outcomes

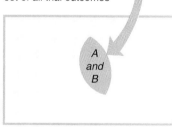

b) "A and B" never occurs
$P(A$ and $B) = 0$

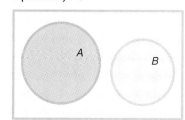

Thus, 1/8 of the university students are female juniors or, alternatively, 1/4 of 1/2 of the university students are female juniors, because $1/8 = (1/2)(1/4)$. Symbolically, we have:

$$P(\text{female and junior}) = P(F \text{ and } J) = (1/8) = (1/2)(1/4).$$

But 1/2 is the relative frequency of a female among the university students or $P(F) = 1/2$. Likewise, 1/4 is the relative frequency of a junior among the female students or $P(J|F) = 1/4$. Together, these imply the following formula for finding the probability of randomly selecting a female junior.

$$P(F \text{ and } J) = (1/2)(1/4) = P(F)P(J|F).$$

The Venn diagrams in Figure 4-11 illustrate this formula.

The following box summarizes this rule using the general symbols A and B rather than F and J, which are specific for this example. The second form, $P(A \text{ and } B) = P(B)P(A|B)$, works as well. For instance, if we knew the proportion of the university students who are juniors and the proportion of juniors who are female, we could compute the proportion who are female juniors once again. The answer would be the same, but the magnitudes that we multiply would probably be different.

> **Rule 2:** To find the probability of A and B both occurring for any events A and B, use
>
> $$P(A \text{ and } B) = P(A)P(B|A) = P(B)P(A|B).$$

Figure 4-11
Relation of $P(J \text{ and } F)$ with other probabilities

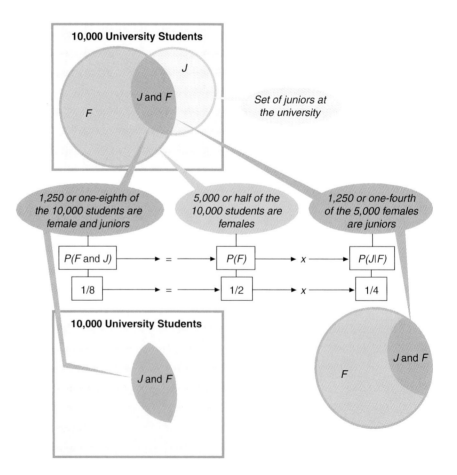

All three of these terms are equal, so we may use any two at one time to calculate probabilities, such as $P(A \text{ and } B) = P(A)P(B|A)$, or $P(A)P(B|A) = P(B)P(A|B)$.

■ **Example 4-9:** Use Rule 2 to find the probability of drawing two aces in two draws from a standard deck of cards, when the first card is not replaced before drawing the second.

■ **Solution:** Let $A1$ be an ace on the first draw and $A2$ be an ace on the second draw. Then, we want the probability that both happen or $P(A1 \text{ and } A2)$. We know that $P(A1) = 4/52$ and $P(A2|A1) = 3/51$, because on the second draw only 3 aces remain in the deck, and there are only 51 cards left. Thus, we have

$$\begin{aligned} P(A1 \text{ and } A2) &= P(A1)P(A2|A1) \\ &= (4/52)(3/51) \\ &= 12/2652 = 1/221. \end{aligned}$$

■ **Example 4-10:** Jim uses his special bread recipe about 60% of the times he bakes bread. The dough for this recipe fails to rise about 5% of the time, because of a special yeast mixture (the secret to the recipe). Jim is considering making a loaf of bread. What is the probability that he will make the special recipe *and* that it will fail to rise?

■ **Solution:** Let S be that Jim uses the special recipe and F be that the bread fails to rise. "S and F" means that both happen, so we want $P(S \text{ and } F)$. We know that $P(S) = 0.6$ and $P(F|S) = 0.05$. Therefore, $P(S \text{ and } F) = P(S)P(F|S) = (0.6) \times (0.05) = 0.03$. So about 3% of Jim's long-run bread baking episodes will be unsatisfactory special recipes.

■ **Example 4-11:** Virginia does her homework 90% of the time. Seventy-five percent of the time that she does the homework, she does it correctly. On a randomly selected assignment, what is the probability that she did the work and got the correct answers?

■ **Solution:** Let W be did homework and C be got correct answers. We want the probability that both occur or $P(W \text{ and } C)$. We know that $P(W) = 0.9$ and $P(C|W) = 0.75$. Then, $P(W \text{ and } C) = P(W)P(C|W) = (0.9)(0.75) = 0.675$. So about 67.5% of her homework assignments will be done and done correctly.

Question: Suppose a friend worked the last problem and told you that the solution is 0.95, rather than 0.675. How would you recognize this as a wrong answer without working the problem?

Answer: Virginia only does her work 90% of the time, so the percent of time she does the work and gets it correct definitely cannot exceed 90% (0.9), so 0.95 must be incorrect. (See Figure 4-12.)

■ **Example 4-12:** A study reports variation in mortality rates for the patients of 12 surgeons who perform bypass surgery in five Philadelphia teaching hospitals as follows. (The mortality rate is the death rate adjusted for patient age and seriousness.)

Surgeon	Adjusted Mortality Rate (%)	Surgeon	Adjusted Mortality Rate (%)
1	4.9	7	6.8
2	4.9	8	5.9
3	4.4	9	7.4
4	5.1	10	9.3
5	6.4	11	9.9
6	5.4	12	9.1

Source: Study in *Journal of the American Medical Association*, August 14, 1991, as reported by Ron Winslow, "Heart Bypass Death Rates Vary Widely," *Wall Street Journal*, August 14, 1991, p. B1.

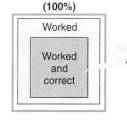

Figure 4-12
Demonstration of $P(W \text{ and } C) \leq P(W)$

Box 4-3 Determining the Probability of Pregnancy: Bayes' Formula

After listening to a woman describe her condition (or symptoms), the obstetrician judges whether or not she is pregnant. Although the doctor may not state the probability, the uncertainty is there, because other health conditions exhibit symptoms common to pregnancy. The next step is a test to obtain more information. The doctor uses the results to revise the initial probability estimate of pregnancy. It is hoped that the final estimate is large or small enough to confirm or reject the diagnosis.

In general, doctors make an initial educated judgment of the probability a person is afflicted with a certain malady. They base these probabilities on experience, the prevalence of the suspected condition, and other information about the disease that may be causing the patient's symptoms. Next, doctors use test results to revise this initial estimate, to confirm or suspend the diagnosis.

Tests are not foolproof. They vary in their ability to confirm or deny the presence of a pregnancy. For example, a brand's experiment might suggest that the sensitivity of the test (medical jargon for the probability of a positive result for a pregnant woman) is 0.992 for the urine assay and 0.995 for the serum assay. Concentrating on the urine segment of the test, we say $P(POS|PR) = 0.992$, where POS = positive test result and PR = pregnant. In other words, the test will indicate pregnant for about 99.2% of pregnant women. Sources of positive results for a nonpregnant woman include body conditions that produce the substance detected by the pregnancy test. Similarly, the specificity of the urine test, $P(NEG|NPR)$, might be 98.0%, where NEG = negative test result and NPR = not pregnant. The test indicates not pregnant for 98.0% of nonpregnant women. Despite its lack of certainty, this information is valuable for revising the doctor's initial estimate with a probability rule called Bayes' formula.

Often, we update or revise an initial probability estimate when new information becomes available by using an expression called Bayes' formula. A simple version is Rule 2'. Another simple version follows from substituting $P(B)P(A|B)$ for $P(A \text{ and } B)$ in the numerator of 2'.

Another Version of Bayes' Formula for Updating Probability Estimates:

$$P(A|B) = \frac{P(A)P(B|A)}{P(B)}.$$

Let's see how this works with the woman and doctor who suspect pregnancy. Doctors probably correctly identify at least 95% of pregnant women after listening to the patient, so we will use $P(PR) = 0.95$. Other diseases are more difficult to diagnose. Sometimes the prevalence—the incidence of the disease among the relevant population—is the initial probability.

Suppose the woman tests positive for pregnancy. If we use Bayes' formula to revise the doctor's estimate of pregnancy, $P(PR) = 0.95$, we obtain the following expression. Incidentally, $P(PR|POS)$ is called the positive predictive value of the test, medically speaking.

$$P(PR|POS) = \frac{P(PR)P(POS|PR)}{P(POS)}.$$

We know the sensitivity is $0.992 = P(POS|PR)$. We need to know the probability of obtaining a positive test result for any woman, $P(POS)$.

Test results are positive for 99.2% of pregnant women, and 95% of the women taking the test (the ones with symptoms related to pregnancy) are pregnant. Thus, 99.2% of these 95% or 94.24% [(0.992)(0.95)(100)] of all tested women test positive and are pregnant. Test results are also positive for 2.0% of nonpregnant women, because they are negative for the other 98.0% who are not pregnant (see specificity above). Five percent of the women taking the test are not pregnant, so 2.0% of 5% is 0.10% of all tested women test positive and are not pregnant. Together, these two cases account for all of the positive results that occur or 94.34% [94.24% + 0.10%] of the tested women, $P(POS) = 0.9434$.

Now we can solve for the probability this specific woman is pregnant.

$$P(PR|POS) = \frac{P(PR)P(POS|PR)}{P(POS)} = \frac{(0.95)(0.992)}{(0.9434)}$$
$$= 0.9989.$$

In general, 99.89% of the women suspected of pregnancy who test positive are pregnant. Thus, the pregnancy is confirmed.

By the way, the chances for the woman being pregnant if the test result is negative is 0.1343. The initial estimate of 0.95 takes a nose dive, and the doctor will consider alternative diagnoses.

Source: Conversation with Dr. Dan Southern, Clinical Laboratory Sciences Program, Western Carolina University, Cullowhee, NC, August 28 and 30, 1991.

Question: If S is the surgeon and M is patient dies, express the adjusted mortality rates with probability symbols.

Answer: The adjusted mortality rates are the probabilities that a patient dies for (given) a particular surgeon or $P(M|S)$.

Problems

Change the mortality rates to survival rates, $P(M'|S)$, then use these values to determine the probability for each surgeon of an emergency patient being randomly assigned to the surgeon and surviving.

■ **Solution:** First, finding the conditional survival rate, $P(M'|S)$, is like finding the $P(M')$ when we know $P(M)$. We know $P(M|S)$, so the survival rate for a particular surgeon, S, is $P(M'|S) = 1 - P(M|S)$. The values for surgeons 1 through 13 are 95.1%, 95.1%, 95.6%, 94.9%, 93.6%, 94.6%, 93.2%, 94.1%, 92.6%, 90.7%, 90.1%, and 90.9%, respectively. The probability of being randomly assigned to a particular surgeon and surviving is $P(S$ and $M')$. By Rule 2, we can find this value using $P(S)P(M'|S)$. $P(S) = 1/12$ because of the random assignment. Consequently, the probability for the first surgeon is $P(M'$ and $S) = (1/12)(0.951) = 0.0792$. The values for the remaining surgeons are 0.0792, 0.0797, 0.0791, 0.0780, 0.0788, 0.0777, 0.0784, 0.0772, 0.0756, 0.0751, 0.0758, respectively.

In the last problem and in general, *we will round probabilities four places past the decimal*. This avoids losing information when working with very small values.

Simplified Version of Bayes' Formula If the question is $P(A|B)$ or $P(B|A)$, then Rule 2 or a variation of it can be used. The known information can be substituted in the rule to determine the required probability, such as $P(A|B)$, or Rule 2 can be manipulated algebraically to produce a new rule, 2′. This rule is a simplified version of an important statistical proposition called Bayes' formula. See Box 4-3 for more information.

Rule 2′: Simple Version of Bayes' Formula
To find $P(A|B)$ or $P(B|A)$, use

$$P(A|B) = \frac{P(A \text{ and } B)}{P(B)} \quad \text{and} \quad P(B|A) = \frac{P(A \text{ and } B)}{P(A)}.$$

■ **Example 4-13:** The temperature and humidity must be controlled for 60% of a laboratory's tests, although 75% of the tests require only temperature control. What is the likelihood of needing humidity control if a test demands temperature control?

■ **Solution:** Let T be requires temperature control and H be requires humidity control. Then we want to find $P(H|T)$. We know that $P(H$ and $T) = 0.60$ and $P(T) = 0.75$, so using Rule 2′ we obtain $P(H|T) = P(H$ and $T)/P(T) = 0.60/0.75 = 0.80$. Thus, 80% of the tests that require temperature control also require humidity control.

In conclusion, Rule 2 works for finding $P(A$ and $B)$ for any two events. We can simplify the work if we have a special set of events, independent ones, as we shall see in the next section.

Section 4.7.1 Problems

77. Eighty percent of the graduates of a government training program obtain jobs. Fifty-five percent of the students graduate. What is the probability an entering trainee will graduate and get a job?

78. Find the probability of drawing two 7s in a row without replacement from a standard deck.

79. Biologists estimate that 20% of a certain species of dogs have rabies. This species makes up 15% of the local dog population. If a dog bites a child and the child cannot describe anything about the dog, what is the likelihood the dog belongs to this species and is infected with rabies?

80. One student's mother sends chocolate chip cookies 90% of the time when she prepares cookies for class parties. She prepares cookies for 25% of the parties. If a parent randomly chooses a party to attend for his child, what is the probability he will find this mother's chocolate chip cookies served?

81. The following box shows the distribution of the 10,000 students in the example in this section. Use this information to find $P(J)$ and $P(F \mid J)$. Then, check if the product of these values is indeed $P(F \text{ and } J) = 1/8$.

Class Standing	Gender		Total
	Male	Female	
Freshman	1,750	1,550	3,300
Sophomore	1,250	1,200	2,450
Junior	1,100	1,250	2,350
Senior	900	1,000	1,900
Total	5,000	5,000	10,000

82. Rain falls about 65% of the time during the tourist season, May–October, at a vacation spot. If you randomly choose a month to vacation in this spot, what is the probability you choose a month during the tourist season and it will rain?

4.7.2 *P(A and B)* for Independent Events

We can simplify the formula for finding $P(A \text{ and } B)$, if we know that A and B are independent events.

Independent Events Suppose we want to know if scoring an A on a test and the amount of time we study for the test are related. Of course, most of us know from experience or we are willing to assume that the more we study, the better we expect to score when we take a test. We believe that the test score *depends* on the study time that precedes the test. Suppose we obtain the information that the probability of scoring an A on a test is ordinarily 0.2, but it becomes 0.4 for a student who studies everyday.

Question: Do these probabilities confirm our intuition?
Answer: The additional information that a student studies everyday affects the chances of scoring an A on the test. The probability of an A increases from 0.2 to 0.4.

If the probability of scoring an A were the same value, whether a student studied everyday or not, we would interpret this finding to mean that studying everyday would have no consequence on making an A. In this circumstance, where $P(A)$ and $P(A \mid B)$ are equal, we say the events are *independent*. The following box formalizes this definition.

> A and B are independent, if $P(A \mid B) = P(A)$.
> If $P(A \mid B) \neq P(A)$, then A and B are dependent.

■ **Example 4-14:** Demonstrate the formal definition of independence when the two events are A, draw a king on the first random draw from a standard deck, and B, draw a king on the following draw if the first card is replaced before the second is drawn. Then, assume the first card is not replaced.

■ **Solution:** Using the equiprobable outcomes method, we know that the probability of drawing a king is ordinarily 4/52. If the first card is replaced, knowing that we get a king on the second (even knowing ahead of time) does not affect the likelihood of a king on the first draw.

$P(\text{king on first draw}) = P(A) = 4/52.$

$P(\text{king on first} \mid \text{king on second}) = P(A \mid B) = 4/52.$

Similarly,

$P(\text{king on second}) = P(B) = 4/52.$

$P(\text{king on second} \mid \text{king drawn on first and replaced}) = P(B \mid A) = 4/52.$

Thus, the events are independent when we replace the first card, because $P(A \mid B) = P(A)$, or $P(B \mid A) = P(B)$.

Not replacing the first card affects the likelihood of a king on the second draw.

$P(\text{king on second} \mid \text{king on first}) = P(B \mid A) = 3/51,$

because there are only three kings and a total of 51 cards remaining in the deck for the second draw. This probability is different from the usual probability of a king

$P(\text{king on second when replace earlier drawn cards}) = P(B) = 4/52.$

This inequality $P(B \mid A) \neq P(B)$ is sufficient evidence for us to conclude that A and B are dependent. We can reverse the position of A and B in the definition, $P(B \mid A)$ will be $P(B)$ when the events are independent. Besides, we can choose either event to be A or to be B in any problem.

Question: According to a recent study, 57% of teenagers between 15 and 17 years old experience some level of behavioral problems when we do not account for their living arrangements, $P(\text{behavioral problem}) = 0.57$. The behavior problems include hyperactivity, anxiety, dependency, and headstrong behavior among others. The probability of problems conditional on the teenager's living arrangement (the number and types of parents in the household) is reflected in the following list. According to this information, are living arrangements and behavioral problems of teenagers between 15 and 17 years old independent or dependent? (Source: National Center for Health Statistics, as reported by Diana Crispell of *American Demographics* magazine in "Children's Troubles Tied to Living Arrangements," *Wall Street Journal,* August 21, 1991, p. B1.)

$P(\text{behavioral problem} \mid \text{mother and father}) = 0.52.$

$P(\text{behavioral problem} \mid \text{never-married mother and no father}) = 0.49.$

$P(\text{behavioral problem} \mid \text{mother and stepfather}) = 0.76.$

$P(\text{behavioral problem} \mid \text{formerly married mother and no father}) = 0.69.$

Answer: They are dependent, because the conditional probabilities of behavioral problems are all different from the unconditional probability, 0.57.

We often interpret the definition of independence less formally. *If A and B are independent, the occurrence of one does not affect the occurrence of the other.* Hence, knowing that B occurs is useless information for predicting whether A will occur. Knowing you ate your breakfast is useless for predicting whether you will go to the movies, so $P(\text{go to movies} \mid \text{ate breakfast}) = P(\text{go to movies})$ or $P(A \mid B) = P(A)$. The result of a coin flip, be it heads or tails, has no effect on the result of another flip. Thus, the outcomes of the flips are independent. Ten tosses of a fair die are independent, because the result of any one toss does not help us predict the result of any of the other nine tosses.

On the other hand, *if A and B are dependent, then knowing that one event occurs affects the chances that the other occurs.* Not replacing a card affects the possible cards that we can draw next from the deck. Possessing a car and attending a movie are dependent events for many people.

Some situations lack the necessary probabilities for us to be sure that A and B are independent, so we must consider the context of the events and determine if there is reason to assume that one event will affect the other, at least in the time span under consideration. If we have to stretch our sense of reasonableness to make the events related, they are probably independent for the situation at hand. Similarly, extreme rationalization for independence may well indicate that events are dependent.

Question: If A is concert attendance and B is the price of the concert tickets, for a randomly selected concert, are A and B independent?
Answer: A and B are dependent, because as the price increases, we curtail our concert attendance, and as the price decreases, we escalate our attendance.

Question: If researchers choose two children K and J, randomly from a list of all children born in a hospital in the last two years in order to measure their IQs, are the two IQ scores independent?
Answer: They are independent. The random selection ensures that whoever is selected first will not influence our second choice.

Questions about the independence of events are important in situations when we want to understand how one of the events affects the other. If you grasp the notion of independence and dependence among events intuitively, it will aid your problem solving and make the technical definitions more meaningful. In addition, we must often make a reasoned judgement about the independence of events before we can choose appropriate rules and proceed to a solution.

The Special Rule for $P(A \text{ and } B)$ The next rule concerns $P(A \text{ and } B)$ when A and B are independent events. If A and B are independent, we can substitute $P(A)$ for $P(A|B)$ and $P(B)$ for $P(B|A)$ in Rule 2 to obtain Rule 2″.

> **Rule 2″:** To find the probability of the event "A and B" when A and B are *independent*, use
>
> $$P(A \text{ and } B) = P(A)P(B).$$

■ **Example 4-15:** What is the probability of getting tails on two consecutive tosses of a coin?

■ **Solution:** Let $T1$ = tails on first toss and $T2$ = tails on second toss. We want both to occur or $P(T1 \text{ and } T2)$. Because the results of coin tosses are independent events, we can use Rule 2″, $P(T1 \text{ and } T2) = P(T1)P(T2)$. We know that $P(T1) = 0.5$ and $P(T2) = 0.5$, so $P(T1 \text{ and } T2) = (0.5)(0.5) = 0.25$.

■ **Example 4-16:** The utility company's computer is down about 5% of the time, and the clinic's is down about 3% of the time. If you go to these places to pay bills, what is the probability you will encounter inactive computers in both locations?

■ **Solution:** Let U be utility computer down and C be clinic computer down. We are concerned with both happening or $P(U \text{ and } C)$. If both places use separate computer systems, we can reasonably assume that the events are independent. Therefore, $P(U \text{ and } C) = P(U)P(C) = (0.05)(0.03) = 0.0015$.

■ **Example 4-9 (continued):** Is Rule 2 or Rule 2″ applicable to finding the probability we draw two aces in two consecutive draws from a standard deck of cards, when we replace the first card before drawing the second?

■ **Solution:** Both are applicable, because Rule 2 works for dependent and independent events. It is more general than Rule 2″, which requires independent events. The events in this problem are independent, because the first draw has no effect on the second draw after replacement. Let $A1$ be an ace on the first draw and $A2$ an ace on the second. We express the probability of drawing both aces as $P(A1 \text{ and } A2)$. Using Rule 2, $P(A1) = 4/52$, and $P(A2|A1) = P(A2) = 4/52$, we obtain $P(A1 \text{ and } A2) = (4/52)(4/52) = 16/2{,}704 = 1/169$.

Question: When we solved this example before without replacing the first-drawn card, the answer was $1/221$. Compare the two answers. Do you think that their relative magnitudes seem reasonable?

Answer: Yes, because the probability of getting the two aces is higher with replacement, $1/169 > 1/221$. We have a better chance of getting an ace on the second draw with replacement ($4/52$ versus $3/51$), so the probability of getting aces on both draws should be higher.

■ **Example 4-17:** About 1.67% of Americans are bike commuters, according to a poll conducted for *Bicycling* magazine. Hallmark Cards reports that 9% of Americans were born in August. If you choose an American at random, what is the probability that this person commutes by bicycle and celebrates a birthday in August? (Sources: Nelson Pena, "Power in Numbers," *Bicycling*, April 1991, pp. 44, 46; "August Hot Birthday Month," *Asheville Citizen-Times*, August 13, 1991, p. C1.)

■ **Solution:** Let A be commutes by bicycle and B be born in August. Again, we want the probability that both conditions occur, or $P(A \text{ and } B)$. It seems reasonable that A and B are independent, so we use Rule 2″. $P(A \text{ and } B) = P(A)P(B) = (0.0167)(0.09) = 0.0015$. Less than 1% of Americans fit into both categories at once.

■ **Example 4-18:** Using the bus driver's tables repeated from the School Bus Route situation, find the probability that the bus waits more than five minutes at both obstructions.

Waiting Time at Stop Sign	How Often
Less than 1 minute	50%
1–3 minutes	30%
Over 3 and up to 5 minutes	15%
Over 5 minutes	5%

Waiting Time at Construction	How Often
5 minutes or less	75%
Over 5 minutes	25%

■ **Solution:** Let WS be waiting more than five minutes at the stop sign and WC be waiting more than five minutes at the construction site. We want the probability of waiting more than five minutes at both locations or $P(WS \text{ and } WC)$. If these waiting times are independent, $P(WS \text{ and } WC) = P(WS)P(WC) = (0.05)(0.25) = 0.0125$. However, both sites are on the same route, so if backups at one location hinder progress at the other, the events would be dependent. If we administered bus routes for the system, we might want to check for dependency before making an important decision with this probability value. Because there is no information to this effect in the description of the situation, we accept the solution with caution.

Rule 2″ can easily be generalized to handle more than two events.

> **Rule 2‴:** If $A, B, \ldots, Z$ are independent events, then $P(A \text{ and } B \text{ and } \cdots \text{ and } Z) = P(A)P(B) \cdots P(Z)$.

■ **Example 4-19:** Find the probability of tossing five tails in five flips of a coin.

■ **Solution:** $P(T1$ and $T2$ and $T3$ and $T4$ and $T5)$
$= P(T1)P(T2)P(T3)P(T4)P(T5)$
$= (0.5)(0.5)(0.5)(0.5)(0.5) = 0.03125.$

In short, after determining that we want $P(A$ and $B)$, the next step is to check A and B for independence. If they are, use Rule 2″; otherwise, use Rule 2. We will soon find that $P(A$ and $B)$ is not always the objective, it may only be a step in calculating the probability of other compound events.

Section 4.7.2 Problems

Label the sets of events in Problems 83–86 that we can reasonably assume to be independent. Otherwise, label them dependent.

83. Your body weight and your food consumption.
84. The profits of a firm and its advertising expenditures.
85. The sex of a newly born child and the outside temperature on the birth date.
86. Selecting a series of cards from a standard deck without replacing between draws.
87. $P(R) = 0.6$ and $P(R|T) = 0.8$, where R is a "reaction when two chemicals combine" and T is the "temperature of the combination is above 150° Fahrenheit." Are the events R and T independent?
88. Find the probability of drawing two successive hearts with replacement from a standard deck.
89. The probability any customer at the express lane will have twice the allowable number of items is 0.08. What is the probability the next three customers will wish to purchase this many items?
90. John estimates the chances of getting the summer job he wants to be 0.6 and the chances of winning an expense-paid trip over the summer in a sweepstakes to be 0.015. What is the probability both events happen?
91. If the following box shows a different distribution of the 10,000 students in the example in an earlier section, are the events Sophomore and Female independent? Find $P(S$ and $F)$, using one of the versions of Rule 2. Then, use the equiprobable outcomes method to verify your solution.

Class Standing	Gender		Total
	Male	Female	
Freshman	1,550	1,550	3,100
Sophomore	1,200	1,200	2,400
Junior	1,250	1,250	2,500
Senior	1,000	1,000	2,000
Total	5,000	5,000	10,000

4.8 $P(A$ or $B)$

When the event is "A or B," then one of three possibilities must occur.

1. A occurs and B does not.
2. B occurs and A does not.
3. A and B both occur.

This "or" is called an *inclusive or*, because Possibility 3 is allowed. Thus, "John plays football or baseball" will be a true statement if John plays either sport or both. $P(A$ or $B)$ symbolizes the proportion of time that outcomes in all three of these categories occur.

4.8.1 $P(A$ or $B)$ for Any Two Events

Suppose we want to find the probability that the House of Representatives (H) or the Senate (S) will pass crime-control legislation during this session. Figure 4-13 displays data from 45 sessions and illustrates how the four different possibilities in the table appear in a Venn diagram. If we compare Figure 4-13 with the definition of *or*, the pale

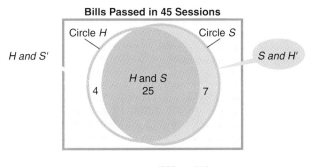

Figure 4-13
Relation of $P(H \text{ or } S)$ to the definition of inclusive "or"

orange region of circle H represents the relative frequency of sessions that the House passes crime-control legislation and the Senate does not (H and S'), 4/45 (possibility 1 in the definition). The medium orange region of circle S contains the sessions when the Senate passed such legislation and the House did not (S and H'), 7/45 (outcomes satisfying possibility 2). The darkest orange intersection region represents the proportion of sessions where both passed crime-control legislation (H and S), 25/45 (possibility 3). These three regions sum to the proportion of time the House *or* the Senate passed crime control legislation, $P(H \text{ or } S)$, 36/45.

Often, the available information arrives in forms other than these three pieces of information or other than a table of values, such as the one on crime-legislation activity. We might obtain the proportions of sessions during which the House passed legislation (circle H), during which the Senate did so (circle S), and during which they both did so (the intersection of circles H and S). We can use these pieces of information to find the total area for "H or S," as shown in the Venn diagrams in Figure 4-14. If we try summing $P(H)$ and $P(S)$ [the total areas of both circles] to find $P(H \text{ or } S)$, we discover that the intersection region, $P(H \text{ and } S)$, would be included twice. The answer would be too large (sometimes, it would be greater than 1). To avoid this double counting, $P(H \text{ and } S)$ must be subtracted once, and we arrive at the next rule.

> **Rule 3:** To determine the probability that A or B occurs for any two events, use
>
> $$P(A \text{ or } B) = P(A) + P(B) - P(A \text{ and } B).$$

■ **Example 4-20:** Find the probability of drawing a heart or an ace in a single draw from a standard deck of cards.

■ **Solution:** Let H be a heart and A an ace. The word "or" appears in the problem and makes our objective, $P(H \text{ or } A)$, obvious. We know that $P(H) = 13/52$, $P(A) =$

Figure 4-14
Demonstration of rule for "*H* or *S*"

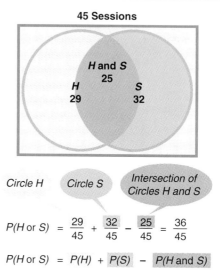

$4/52$, and $P(H \text{ and } A) = 1/52$, because only the ace of hearts will satisfy "*H* and *A*". Thus, we obtain

$$P(H \text{ or } A) = P(H) + P(A) - P(H \text{ and } A)$$
$$= 13/52 + 4/52 - 1/52$$
$$= 16/52.$$

Notice in this solution that $P(H) = 13/52$ includes the ace of hearts. The same is true of $P(A) = 4/52$. So, we subtract $1/52$ to avoid counting the ace of hearts twice.

■ **Example 4-21:** A construction firm has a project in a locality where heavy rain falls 40% of the time, gusting winds occur 20% of the time, and both occur together 5% of the time. Work is delayed when heavy rain or gusting winds occur. Find the probability that work on this project will be delayed.

■ **Solution:** Let R be heavy rain and W be gusting winds. R or W delay the project, so we want $P(R \text{ or } W)$. From the statement of the problem, we know that $P(R) = 0.40$, $P(W) = 0.20$, and $P(R \text{ and } W) = 0.05$. Using Rule 3, we have

$$P(R \text{ or } W) = P(R) + P(W) - P(R \text{ and } W)$$
$$= 0.4 + 0.2 - 0.05 = 0.55.$$

That is, there is a 55% chance of delay for the project.

■ **Example 4-22:** John watches college football on television four months out of the year, September through December. He watches college basketball during five months, November through March. If we randomly select a month and ask John what he is watching, what is the probability that he will respond either "college football" or "college basketball"?

Solution: Let F be football and B, basketball. We want to know $P(F \text{ or } B)$. We already know that $P(F) = 4/12$ and $P(B) = 5/12$. We need $P(F \text{ and } B)$ to use rule 3. Because he watches both sports in November and December, $P(F \text{ and } B) = 2/12$. Thus,

$$P(F \text{ or } B) = P(F) + P(B) - P(F \text{ and } B)$$
$$= 4/12 + 5/12 - 2/12 = 7/12.$$

Question: Suppose a friend tells you that the answer to the last example is 3/12. How would you recognize this wrong answer without working the problem?

Answer: John watches football 4/12 of the time. One possibility of the event F or B is to watch football only, so the answer must be at least 4/12, not smaller. Refer to summing the areas in Figure 4-13.

Rule 3 works for finding $P(A \text{ or } B)$ for any two events. In the next section, we will see that the rule can be simpler, but *only if* A and B are mutually exclusive.

Section 4.8.1 Problems

92. Find the probability of drawing a face card (king, queen, or jack) or a black card in a single draw from a deck of cards.

93. Use the table in Figure 4-13 to find $P(H' \text{ or } S')$.

The following table classifies 10,000 university students. Let G be gender and C, class standing. Consider upperclass to be junior or senior standing. Use this information to work Problems 94–96.

Class Standing	Gender		Total
	Male	Female	
Freshman	1,550	1,550	3,100
Sophomore	1,200	1,200	2,400
Junior	1,250	1,250	2,500
Senior	1,000	1,000	2,000
Total	5,000	5,000	10,000

94. $P(G = \text{female or } C = \text{junior})$.

95. $P(G = \text{male or } C = \text{sophomore})$.

96. $P(G = \text{female or } C = \text{upperclass})$.

97. An artist uses female subjects in 60% of her paintings and males in 75%. Both sexes appear in 50% of these paintings. If an art exhibitor randomly chooses one of her paintings for a show, what is the probability a male or female subject will appear in the painting?

98. Forty percent of a species of cat have blue eyes. Thirty percent of the same species are spotted. Five percent possess both characteristics. If you randomly select a subject from this species, what is the probability it will exhibit one characteristic or the other?

4.8.2 $P(A \text{ or } B)$ for Mutually Exclusive Events

Mutually Exclusive Events We have already encountered this term when we defined experimental outcomes as distinct. If A and B are *mutually exclusive events,* then when one happens the other cannot happen on the same trial of an experiment. Obtaining a head on a single toss of a coin precludes a tail result.

A problem may not always state the trial situation or time period exactly, but we can usually understand from the context. For instance, for an individual, being 5 feet tall and being 6 feet tall are not mutually exclusive over a lifetime or period of years, but usually the context would pinpoint a given time, and the individual cannot be both at that time.

Figure 4-15

Venn diagrams for mutually exclusive events

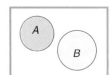

Question: Are the events A, finding a defendant guilty, and B, finding the defendant not guilty of a crime, mutually exclusive?

Answer: Yes. The jury cannot decide on both A and B for a single crime.

Question: If M is being married and K is having kids, are M and K mutually exclusive?

Answer: No. Both events are possible simultaneously.

Figure 4-15 illustrates mutually exclusive events with a Venn diagram.

Question: Each cell in the following table reports the number of trials with the stated combination of outcomes, such as neither A nor B occurred, in 35 trials. What value should fill the blank cell, if A and B are mutually exclusive?

	A	A'
B		72
B'	43	35

Answer: There should be a 0 in the empty cell, because A and B cannot both occur if they are mutually exclusive.

The Special Rule for P(A or B) Figure 4-15 pictures mutually exclusive events A and B. Now circle A represents $P(A)$ or the outcomes satisfying event A but not B (because A occurring means B cannot). This is possibility 1 in the definition of P(A or B). Circle B represents $P(B)$, or the outcomes in event B and not in A (possibility 2). The combination of these two proportions $[P(A) + P(B)]$ results in the proportion of time that possibility 1 or 2 occurs.

Question: What is P(A and B) or possibility 3 of the definition in this situation?

Answer: Because A and B are mutually exclusive, the circles do not intersect. Possibility 3 is impossible. There are no trial outcomes that satisfy A and B simultaneously. Thus, P(A and B) must be 0.

These conclusions form the basis of Rule 3'.

> **Rule 3':** To determine the probability that the event "A or B" occurs, when A and B are mutually exclusive, use
>
> $$P(A \text{ or } B) = P(A) + P(B).$$

■ **Example 4-23:** Find the probability that a single random draw from a standard deck produces a heart or a diamond.

■ **Solution:** Let H be a heart and D, a diamond. We want P(H or D). Because this is an "or" problem, we next want to know if H and D are mutually exclusive. They are because we cannot draw a heart and a diamond in a single draw, so we can use the simpler "or" rule, Rule 3'. We know $P(H) = P(D) = 1/4$, so $P(H \text{ or } D) = P(H) + P(D) = 1/4 + 1/4 = 1/2$.

Question: Is Rule 3' applicable to finding the probability of drawing a heart or a jack on a single draw?

Answer: No, the events are not mutually exclusive, because of the jack of hearts.

■ **Example 4-24:** In the Philippines 6.0% of central government expenditures were allocated to health in 1985, 20.1% to education, and 11.9% to defense. What is the probability a hypothetical voucher for a randomly selected dollar spent that year pertains to health or education? (Source: United Nations Children's Fund, *The State of the World's Children* [New York: Oxford University Press, 1988], p. 75.)

■ **Solution:** If H is health and E is education, then we want to find $P(H$ or $E)$. Presumably, these categories are separate budget items and, thus, mutually exclusive. Then $P(H$ or $E) = P(H) + P(E) = 0.060 + 0.201 = 0.261$. So 26.1% of the expenditures went for health or education.

■ **Example 4-18 (continued):** What is the probability that the school bus will wait more than three minutes at the four-way stop? Use the following information collected by the bus driver.

Waiting Time	How Often
Less than 1 minute	50%
1–3 minutes	30%
Over 3 and up to 5 minutes	15%
Over 5 minutes	5%

■ **Solution:** A wait of more than three minutes means either that the schoolbus waits three to five minutes (call this $W1$) or that it waits for more than five minutes ($W2$). These events are mutually exclusive, because a wait of more than five minutes means that the schoolbus did not wait three to five minutes. Thus,

$$P(\text{wait more than three minutes}) = P(W1 \text{ or } W2)$$
$$= P(W1) + P(W2)$$
$$= 0.15 + 0.05 = 0.20.$$

Rule 3' can easily be generalized to handle more than two events.

> **Rule 3″:** If $A, B, \ldots, Z$ are mutually exclusive events, then $P(A$ or B or $\cdots$ or $Z) = P(A) + P(B) + \cdots + P(Z)$.

■ **Example 4-18 (continued):** Find the probability that the schoolbus waits at least one minute.

■ **Solution:** Let $W3$ = wait one to three minutes. Then we seek $P(W1$ or $W2$ or $W3) = P(W1) + P(W2) + P(W3)$, because the times are mutually exclusive. Using Rule 3″ and the values from the table of waiting times,

$$P(W1 \text{ or } W2 \text{ or } W3) = P(W1) + P(W2) + P(W3)$$
$$= 0.15 + 0.05 + 0.30 = 0.50.$$

Question: Is Rule 3 or Rule 3' applicable to finding the probability of drawing a 7 or an 8 in a single draw from a standard deck of cards?

Answer: Both are applicable, because Rule 3 works whether the outcomes are mutually exclusive or not. $P(7$ and $8)$ will be 0, because we cannot draw one card that is a 7 and an 8. If we use Rule 3, we find

$$P(7 \text{ or } 8) = P(7) + P(8) - P(7 \text{ and } 8)$$
$$= 4/52 + 4/52 - 0 = 8/52.$$

From this solution, we can see that Rule 3 is more general than Rule 3', because it handles events that are mutually exclusive and those that are not. If you must jump to a formula for an *or* problem, be sure it is the general one. Better yet, pause after you have decided that it is an *or* problem, and decide if the events are mutually exclusive. If they are, use Rule 3'; otherwise, use Rule 3.

Section 4.8.2 Problems

Label the pairs of events in Problems 99–103 as mutually exclusive or not mutually exclusive.

99. Living in Kentucky and attending the Kentucky Derby.
100. Working one job in the morning and another job at night.
101. Using gasoline and diesel fuel to power your automobile.
102. Typing a manuscript and making mistakes.
103. Drawing a heart and drawing a jack on a single draw from a standard deck of cards.
104. Are A and B mutually exclusive, if $P(A \text{ and } B) = 0.5$? Explain.
105. Find the probability a randomly selected kindergarten teacher will be involved in instructing or disciplining the students, if ordinarily these teachers spend 75% of their class time in instruction, 10% in recess, 12% in administrative duties, and 3% in disciplining students.
106. A student drifting as a summer vagabond eventually runs out of money and decides to apply for a randomly chosen temporary job from the want ads. On the day of the decision, there are 15 grocery jobs, 24 clerical jobs, 16 lawn-mowing jobs, and 33 fast-food jobs. What is the probability the student will apply for a grocery or fast-food job?
107. If 10% of the scores on a test are As, 20% Bs, 50% Cs, 15% Ds, and the rest Fs, what is the probability a randomly selected student will have scored an A or a B?
108. If one-third of the available TV shows at 2 P.M. one day are movies, one-sixth news programs, one-third soap operas, one-twelfth kids shows, and one-twelfth old sitcoms, what is the probability a randomly selected channel will be showing a movie or an old sitcom or a news program?

4.9 A Strategy for Probability Problems

With several rules available for solving many types of probability problems, deriving solutions may seem difficult. The flowchart in Figure 4-16 presents a useful strategy to attack and solve many of these problems, especially the word problems. The dashed lines represent alternative formulas that will work as well.

■ **Example 4-25:** Forty percent of the students in a class received an A on their first paper, and 30% did so on the second. Seventy percent of the ones who received an A on the first paper also did so on the second. What is the probability that a randomly selected student scored an A on one paper or the other?

Figure 4-16
Flowchart for probability solutions

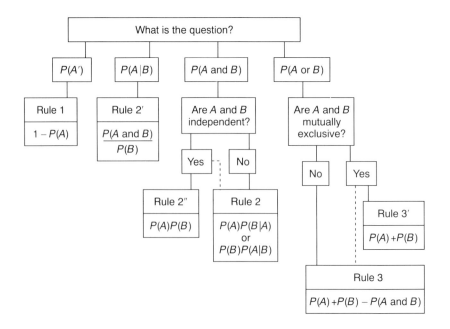

Solution: Step 1: The word "or" appears in the problem, guiding us to the question $P(A1 \text{ or } A2)$, where $A1$ is an A on the first paper and $A2$ is an A on the second paper.

Step 2: Because a student could score an A on both papers, the events are not mutually exclusive, so the solution requires Rule 3:

$$P(A1 \text{ or } A2) = P(A1) + P(A2) - P(A1 \text{ and } A2).$$

Step 3: We need $P(A1)$, $P(A2)$, and $P(A1 \text{ and } A2)$ to finish the solution. The problem information provides us with $P(A1) = 0.40$ and $P(A2) = 0.30$. However, $P(A1 \text{ and } A2)$ is not given, so we must perform a side problem to determine this value.

Sidestep 1: We know the problem is an "and" problem, so we need to determine if $A1$ and $A2$ are independent. It seems reasonable that making an A on both papers might well be related or dependent. We also find in the problem that $P(A2 \mid A1) = 0.7$, which is not equal to $P(A2) = 0.30$, so the events are dependent. Thus, we must employ Rule 2.

$$P(A1 \text{ and } A2) = P(A1)P(A2 \mid A1) = P(A2)P(A1 \mid A2).$$

Sidestep 2: As you can see, we have two formulas to choose from to determine $P(A1 \text{ and } A2)$, but we already know $P(A1)$ and $P(A2 \mid A1)$, so we will use the expression involving these values.

$$P(A1 \text{ and } A2) = 0.40(0.70) = 0.28.$$

Step 4: Now we know the three values that we need to substitute in the original formula, $P(A1) = 0.40$, $P(A2) = 0.30$, and $P(A1 \text{ and } A2) = 0.28$, so

$$P(A1 \text{ or } A2) = 0.40 + 0.30 - 0.28 = 0.42.$$

So 42% of the students made an A on one paper or the other.

Summary and Review

Sets of events can be mutually exclusive (occurrence of one precludes the occurrence of others), independent (occurrence of one has no effect on the occurrence of others), or neither. We use these relationships between and among events and probabilities for single events to simplify computations for compound events, such as "A and B," "A or B," and "A given B," according to the basic rules for probability calculation:

1. $P(A') = 1 - P(A)$.
2. $P(A \text{ and } B) = P(A)P(B \mid A) = P(B)P(A \mid B)$.
   2'. Simplified version of Bayes' formula:

   $P(A \mid B) = P(A \text{ and } B)/P(B)$ and
   $P(B \mid A) = P(A \text{ and } B)/P(A)$.

   2". If A and B are independent, $P(A \text{ and } B) = P(A)P(B)$.
   2'''. If $A, B, \ldots, Z$ are independent, then
   $P(A \text{ and } B \text{ and } \ldots \text{ and } Z) = P(A)P(B)\ldots P(Z)$.

3. $P(A \text{ or } B) = P(A) + P(B) - P(A \text{ and } B)$.
   3'. If A and B are mutually exclusive,
   $P(A \text{ or } B) = P(A) + P(B)$.
   3". If $A, B, \ldots, Z$ are mutually exclusive events, then

   $P(A \text{ or } B \text{ or } \cdots \text{ or } Z) = P(A) + P(B) + \cdots + P(Z)$.

A good strategy for solving probability problems starts with determining, then expressing symbolically, the question asked in a problem. Often, determining whether events are mutually exclusive or independent is important next, in order to select the appropriate rule to use for *or* and *and* problems, respectively. Having found the correct rule, we will know what kinds and the form of information that we need in order to solve the problem.

Multiple-Choice Problems

109. The "*A* or *B*" in *P*(*A* or *B*) does not include which of the following?
  a. Neither *A* nor *B* occurs.
  b. *A* occurs, and *B* does not.
  c. *B* occurs, and *A* does not.
  d. *A* and *B* both occur.

110. If the composer of a symphony and the conductor of its first performance are independent selections, then
  a. Knowing the composer is no help in predicting who the first conductor of the piece will be.
  b. The composer cannot conduct the first performance of the piece.
  c. Knowing the first conductor is instrumental in guessing who the composer is.
  d. $P(\text{conducting piece } A) \neq P(\text{composing piece } A)$.
  e. $P(\text{conducting piece } A) \neq P(\text{composing piece } B)$.

111. Which of the following pairs of events lists mutually exclusive events?
  a. Participating in a psychology experiment and being a social worker.
  b. Scoring an A and scoring an F on a test.
  c. Doing a chemistry experiment and involving the laws of physics to theorize and predict the outcome.
  d. Passing a hearing exam and wearing a hearing aid.
  e. Painting a picture and sculpting a work for the same art show.

112. The probability of drawing a diamond or a face card in a single random draw from a standard deck of cards is
  a. $13/52 + 12/52 - 3/52$.
  b. $13/52 + 12/52$.
  c. $13/52 \times 12/52$.
  d. $13/52 \times 12/51$.
  e. $13/52 \times 11/51$.

113. The probability of drawing a red card in a single random draw from a standard deck of cards is smaller than 1/4. Criticize this statement.
  a. The statement is true as it stands.
  b. The probability is equal to $1/4 = 1/2 \times 1/2$.
  c. Because the probability of drawing a heart is 1/4, and a diamond is also red, the probability of drawing a red card is greater than 1/4.
  d. The probability of drawing a red card is 1 minus the probability of drawing a black card, so the answer must be smaller than 1/2.
  e. Because red and black are mutually exclusive, the probability is 1.

114. A thunderstorm occurs 20% of the days of the year in a certain locality. Fifty percent of the days when there is a thunderstorm, there is damage from the lightning. What proportion of the days of the year is there a thunderstorm and lightning damage?
  a. 0.5.   **d.** 0.1.
  b. 0.7.   **e.** 0.2.
  c. 0.6.

115. Suppose 50% of new-car buyers purchase an AM-FM radio for their car, 30% purchase an AM radio only, and 20% want no radio. Find the probability that the next new-car buyer will want some type of radio in the car.
  a. 0.8.   **d.** 0.3.
  b. 0.2.   **e.** 0.7.
  c. 0.5.

116. During the last 100 quarters, your rival in the wholesale industry has employed three basic strategies (only one strategy at a time; there was no strategy mix during a quarter). In 35 of the 100 quarters, prices changed. In 20 of the 100 quarters, promotion expenditures changed. In 45 of the 100 quarters, nothing changed. Estimate the probability that your rival will employ an active ("do-something") strategy this quarter.
  a. 80/100.   **d.** 45/100.
  b. 55/100.   **e.** 20/100.
  c. 35/100.

117. The probability of drawing three hearts in three consecutive draws from a standard deck of cards with replacement after each draw is
  a. 1/4.
  b. 1/16.
  c. 1/64.
  d. 1/2.
  e. 3/4.

118. Suppose you are drawing two cards from a standard deck without replacing them. If you want to find the probability of drawing a heart first (*H*1) and a diamond second (*D*2), criticize the following statement:

$$P(H1 \text{ and } D2) > 1/4.$$

  a. There is nothing wrong with this statement.
  b. The answer is equal to 1/4, because $1/2 + 1/2 = 1/4$.
  c. The answer is $1/4 + 1/4 = 1/2$, because the outcomes are mutually exclusive.
  d. The answer is $1/4 + 1/4 - 1/16 = 7/16$, because the outcomes are not mutually exclusive.
  e. Because we expect a heart to occur 1/4 of the time on the first draw, and a diamond to follow only 1/4 of these first draws, $P(H1 \text{ and } D2)$ must be smaller than 1/4.

119. According to a survey for *Bicycling* magazine, 1.67% of Americans commute by bicycle. Another survey for Hilton Hotels indicates that the goal for the 1990s for 72% of Americans is to save money. Do we have sufficient information to find the probability that a randomly selected American will be a bike commuter and plan to save money during the decade? (Sources: Nelson Pena, "Power in Numbers," *Bicycling*, April 1991, pp. 44, 46; Hilton Hotels time values survey, as reported in Carol Hymowitz, "Trading Fat Paychecks for Free Time," *Wall Street Journal*, August 5, 1991, p. B1.)
  a. Yes, because all we need is the probability of each event, *P*(*A*) and *P*(*B*), whenever we find *P*(*A* and *B*).
  b. Yes, because riding a bicycle and saving money are not mutually exclusive.

Multiple-Choice Problems

  c. No, because we need to multiply the probability of saving money given that you ride a bicycle by the probability you ride a bicycle given that you are a saver.
  d. No, because saving money and riding a bicycle can reasonably be considered dependent events.
  e. Cannot say for sure, because we need to know the probability of the negation of each event in order to solve it.

120. The birthdays of 35.4% of Americans occur during July–September, 7.7% in February, and 7.8% in April. Can 50% of Americans have their birthday in the set of unmentioned months? (Source: "August Hot Birthday Month," *Asheville Citizen-Times*, August 13, 1991, p. C1.)
  a. No, because $0.077 \leq P$(birthday in another month) ≤ 0.354.
  b. Yes, because the birth months are not mutually exclusive.
  c. Yes, because the birth months are independent.
  d. No, because P(birthday month mentioned) + P(birthday month unmentioned) would exceed 1.
  e. No, because this probability is conditional on whether it is a leap year.

121. Suppose 55% of the students in the first statistics course passed and continued into the second course. Ninety-five percent of the students in the first course passed. What is the probability a student went on with the second course given that he passed the first course?
  a. 0.5789.   d. 1.5000.
  b. 1.7273.   e. 0.9211.
  c. 0.4000.

Suppose some of the books for 90% of the classes during one term could be purchased used. At least one had to be purchased new in 15% of the classes. A mix of required new books and possible used books was necessary in 25% of the classes. Some classes did not require book purchases. Use this information to answer Questions 122 and 123.

122. What is the probability a student in a randomly selected class must have purchased a new or a used book?
  a. 1.1000.   d. 1.3000.
  b. 0.8000.   e. 0.5000.
  c. 0.9150.

123. What is the probability that a student in a randomly selected class was not required to purchase a text?
  a. 0.2000.   d. 0.1000.
  b. 0.9150.   e. 0.7500.
  c. 0.0850.

Use the mortality rates of the Philadelphia teaching hospital surgeons to find the probabilities requested in Problems 124–126.

Surgeon	Adjusted Mortality Rate (%)
1	4.9
2	4.9
3	4.4
4	5.1
5	6.4
6	5.4
7	6.8
8	5.9
9	7.4
10	9.3
11	9.9
12	9.1

Source: Study in *Journal of the American Medical Association*, August 14, 1991 as reported by Ron Winslow, "Heart Bypass Death Rates Vary Widely," *Wall Street Journal*, August 14, 1991, p. B1.

124. Find the probability that a random assignment of surgeons will yield the first surgeon and the patient will die.
  a. Approximately 0.   c. 0.0041.   e. 0.0490.
  b. Approximately 1.   d. 0.0833.

125. If patients are randomly assigned to these twelve surgeons, find the probability that the next three patients will go to surgeon 1 through surgeon 3 in consecutive order.
  a. 0.0006.   c. 0.2742.   e. Approximately 0.
  b. 0.0142.   d. 0.0008.

126. If patients are randomly assigned to these twelve surgeons, find the probability that the next patient will go to surgeon 1 or 2 or 3.
  a. 0.0833.   d. 0.0142.
  b. 0.0006.   e. Approximately 0.
  c. 0.2500.

127. If 60% of Americans do not know about environmental symbols used for recognizing recyclable products and containers, as a survey reports, what is the probability that five randomly selected Americans will all be unaware of the symbols? (Source: American Opinion Research poll for industry clients reported in "Environmental Terms Catch on Very Slowly," *Wall Street Journal*, July 11, 1991, p. B1.)
  a. 0.6000.
  b. 0.0778.
  c. 0.0102.
  d. 0.4000.
  e. Cannot be computed because the answer will exceed 1.

128. Of 1,358 respondents to a question on the status of medical education in the United States, 121 were deans. Of these deans, 67.8% thought "medical student education today has many good attributes but needs *fundamental changes*." What is the probability of randomly selecting a survey respondent who is a dean and who gave the stated response? (Source: Joel C. Cantor, et al., "Medical Educators' Views on Medical Education Reform," *Journal of the American Medical Association*, Vol. 265, No. 8, February 27, 1991, p. 1003.)
  a. 0.0891.   d. 0.0673.
  b. 0.7609.   e. 0.8204.
  c. 0.0604.

129. If A and B are mutually exclusive, then
  a. $P(A) = P(B)$.
  b. The occurrence of A means B will not occur.
  c. $P(A) \neq P(B)$.
  d. $P(A \mid B) = P(A)$.
  e. The occurrence of A does not affect the occurrence of B.

The proportion of Utah's population that is 5 to 11 years old is 15% and 12 to 17 years old is 11%. The comparable U.S. proportions are 10% and 8%. Use this information to answer Questions 130–131. (Source: "West Is Youngest Area, in More Ways than One," *Wall Street Journal*, July 19, 1991, p. B1.)

130. Being 5 to 11 years old and living in Utah are
  a. Dependent events.
  b. Independent events.
  c. Mutually exclusive events.
  d. Conditional events.
  e. Equiprobable events.

131. Being 5 to 11 years old and being 12 to 17 years old are
  a. Dependent events.
  b. Independent events.
  c. Mutually exclusive events.
  d. Conditional events.
  e. Equiprobable events.

Word and Thought Problems

132. What is $P(A \mid B)$ if A and B are independent? What is $P(A \text{ and } B)$ if A and B are mutually exclusive?

133. State the meaning of the following symbols and the rules that are applicable to the solution of each: $P(A)$, $P(A')$, $P(A \mid B)$, $P(A \text{ and } B)$, and $P(A \text{ or } B)$.

134. Decide if A and B are independent, then find $P(A \text{ and } B)$ for each of the three following cases. Use the same three cases to decide if A and B are mutually exclusive, then find $P(A \text{ or } B)$.
  a. $P(A) = 0.2$; $P(B) = 0.6$; $P(A \mid B) = 0.1$
  b. $P(A) = 0.2$; $P(B) = 0.6$; $P(A \mid B) = 0.2$
  c. $P(A) = 0.2$; $P(B) = 0.6$; $P(A \mid B) = 0.0$

135. Suppose you draw two cards from a standard deck, with replacement and then without replacement. A is the card drawn first and B is the second card. Complete table I below for replacement and then without replacement.

136. Suppose you draw one card from a standard deck where A and B are events that could occur on this single draw. Complete table II below.

137. Do you expect D, drunk drivers on the road, and B, being in an automobile accident, to be independent or dependent events? Explain.

138. Tell whether each of the following pairs of events is mutually exclusive:
  a. You spend tonight in Bowling Green; you spend tonight in Seattle.
  b. You sell a customer a hammer and sell the same customer nails.
  c. A customer spends a total of $50 on groceries today, and the same customer spends a total of $30 on groceries today.
  d. A firm locates its new plant in Nashville, and it locates its new plant in Atlanta.
  e. It rains in Los Angeles today, and Congress is in session today.
  f. You toss a die twice, which results in a 2 on the first toss and a 4 on the second toss.
  g. You toss a die once, which results in a 2 and a 4.
  h. You pick two cards from a standard deck, with replacement after each card, and obtain
    i. A red 6 and a red 7.
    ii. The queen of hearts and the queen of diamonds.
    iii. The king of clubs both times.
    iv. A red card both times.

Table I

A	B	Mutually Exclusive	Independent	$P(A \text{ and } B)$
Black 4	Diamond	Yes or no	Yes or no	
4 of any suit	4 of any suit	Yes or no	Yes or no	
Club 4	Diamond 4	Yes or no	Yes or no	

Table II

A	B	Mutually Exclusive	Independent	$P(A \text{ and } B)$	$P(A \text{ or } B)$
4	Diamond	Yes or no	Yes or no		
Club 4	Diamond 4	Yes or no	Yes or no		
Red	Heart	Yes or no	Yes or no		

Word and Thought Problems

  i. Repeat part h, assuming no replacement.
  j. Impressions of a particular brand of soft drink are obtained from randomly selected shoppers in a mall. One shopper prefers the brand, and another shopper does not.
  k. A bill passes in a Senate subcommittee, and the same bill passes on the Senate floor.

139. Is it reasonable to assume independence for each event pair from Problem 138 (consider only Parts a, b, e, f, h, i, j, and k)?

140. In 1982 0.27% of the surface area of the United States was forestland in Montana and 4.86% of the surface area of the United States was in Montana. Find the percentage of Montana's surface area that was forestland. (Source: U.S. Department of Agriculture, Soil Conservation Service, and Iowa State University, Statistical Laboratory; Statistical Bulletin No. 756, *Basic Statistics—1982 National Resources Inventory*, September 1987, as reported in U.S. Bureau of the Census, *Statistical Abstract of the United States: 1990*, 110th edition. [Washington, D.C.: Bureau of the Census, 1990], p. 198.)

141. Roy has purchased win tickets on two horses in the same race: Mule Runner and Zebra Zipper. He figures the probability that Mule Runner will win is 0.03, the probability that Zebra Zipper will win is 0.22, and, given the electronic technology and cameras, no chance for a tie.
  a. Are Mule Runner winning and Zebra Zipper winning independent?
  b. Are Mule Runner winning and Zebra Zipper winning mutually exclusive?
  c. What is the probability that Mule Runner and Zebra Zipper will win?
  d. What is the probability that Mule Runner or Zebra Zipper will win?

142. Five percent of the local citizens are members of the Audubon Society, 6.3% are members of the Sierra Club, and 9.8% are members of the Humane Society. No one holds a dual membership. What is the probability that a letter requesting donations for help in protecting endangered species will be received by a local member of one of these groups?

143. Suppose a salesperson's income increases about 10 months out of 12. The salesperson increases investments about 20% of the time that income increases. Find the probability that next month, income and investments will increase.

144. A baseball player gets a hit 36.3% of the time. Of the hits 12.5% are home runs. What is the probability that the player will hit a home run the next time he comes up to bat?

145. A 100-member band has 25 members who play brass and 15 who play strings. Twelve of the players in each of these groups play both string and brass instruments. A band member is randomly chosen to speak to a music-appreciation class. Find the probability that the member plays brass or strings?

146. A baseball pitcher allows a hit to five-ninths of the batters he faces, walks one-ninth, and strikes out the rest. What is the probability that a batter will get on base when he next faces this pitcher?

147. A newspaper editor sends two photographers to cover a very important news event. One obtains a suitable front-page picture 72% of the time. The other obtains such a picture 78% of the time. What is the probability that one or the other will obtain a suitable photograph of the occasion? What assumptions did you make in obtaining your answer?

148. If 75% of adults watch television as their primary evening activity, and 25% of these viewers do something else simultaneously, what is the probability that a randomly selected adult will be involved in two activities, one of which is watching television?

149. Suppose that you place a classified advertisement in a newspaper for two days to sell a car. There are usually about 80 car advertisements per day in the paper, and page position is randomly assigned each day. What is the probability that your advertisement will be first on the page both days?

150. The following relative frequency distribution displays the number of times that business firms change the prices of their major products during a typical year based on a preliminary sample of 72 firms during 1990. Use the information to answer the following questions.

Number of Price Changes Annually	Percent of Firms
More than 12.00	10.1
4.01–12.00	4.3
2.01–4.00	10.1
1.01–2.00	20.3
1.00	37.7
Less than 1.00	17.4
	99.9

Source: Alan S. Blinder, "Why Are Prices Sticky? Preliminary Results from an Interview Study," *American Economic Review*, Vol. 81, No. 2, May 1991, p. 93, with permission.

  a. P(prices change more than twice a year).
  b. P(prices change four or fewer times per year).
  c. P(prices change only once a year).

151. Mr. Joiner purchases components from Company E and cabinets from Company C. Ninety-eight percent of the electronic components are workable, and 90% of the cabinets are workable. When Joiner is ready to assemble a stereo, what is the probability that
  a. Both the components and cabinet will work.
  b. Both the components and cabinet will not work.
  c. The components or the cabinet will be defective.

152. Thirty-three of 1,358 respondents to a survey about medical education were deans who responded, "on the whole, medical student education is sound and requires only *minor changes*." Of the respondents, 8.91% were deans. What is the probability a randomly selected dean will have given the "minor change" response? (Source: Joel C. Cantor et al., "Medical Educators' Views on

Medical Education Reform," *Journal of the American Medical Association*, Vol. 265, No. 8, February 27, 1991, p. 1003.)

153. Ms. Cunningham, a district attorney, figures that she wins 75% of the cases for which she can get an indictment. Mr. Gumsneaker, chief of detectives, claims that for 95% of the cases in which his detectives make an arrest, there is sufficient evidence for an indictment. What are the chances of a person, arrested by the local detectives, being indicted and convicted? Are indictment and conviction mutually exclusive? Are they independent?

154. a. Sixty-eight percent of the population of Peru was urban in 1985, and 73% of the urban population had access to drinking water between 1983 and 1986. Use these values to estimate the likelihood of a Peruvian citizen being an urban dweller with access to water.
   b. Use the information in Part a to find the probability a randomly selected Peruvian citizen will be rural and have access to water if 18% of rural citizens have access to water.
   c. Should your answers to Parts a and b sum to 1? Explain.
   d. Forty-seven percent of the Egyptian population was urbanized in 1985. Also, 41.36% was both urbanized in 1985 and had access to health services between 1980 and 1986. Use these values to estimate the likelihood of an urban dweller having access to health services.
   e. Use the information in Part d and the fact that 33.92% of the Egyptian population had access to health services and lived in rural areas to determine the likelihood of a rural dweller having access to health services. (Source: United Nations Children's Fund, *The State of the World's Children* [New York: Oxford University Press, 1988], pp. 68, 72.)

155. Eighty percent of a firm's customers reside in the same city. Forty percent of its customers pay cash. Thirty-five percent of the local customers pay cash.
   a. What percentage of the cash customers are local?
   b. What percentage of the cash customers live elsewhere?
   c. What percentage of the customers are local or pay cash?
   d. What percentage of the customers live elsewhere or pay cash?

156. The following table classifies songs played on a radio station by time of day (D) and type of music (M). Assume you turn the radio on at a random time of day and determine the type of music playing. Use this information to answer the following questions.
   a. Are D = Morning and M = Jazz independent? Explain.
   b. Find $P(D$ = morning and M = jazz).
   c. Are D = morning and M = jazz mutually exclusive? Explain.
   d. Are D = morning and D = early evening mutually exclusive?
   e. Find $P(M$ = classical).
   f. Find $P(M$ = classical or D = afternoon).
   g. Find $P(M$ = classical or M = country).

Music Type	Time of Day			
	Morning	Afternoon	Early Evening	Late Evening
Rock and Roll	42	52	27	44
Jazz	0	12	82	39
Classical	27	36	8	17
Rap	7	15	5	36
Country	28	22	12	35

Review Problems

157. Interpret the information that the probability of rain tomorrow is 60%.
158. According to the Bureau of Justice Statistics, 13% of convicted offenders in local jails during 1989 admitted committing their crimes for money to purchase illegal drugs. The percentage is 33.33% for robbers and burglars. Of violent offenders, 25%, and almost 33.33% of property offenders, admitted using drugs when they committed their crimes. Use this information to determine the following. (Source: "Drug Abuse, Crime Linked," *Asheville Citizen-Times,* August 26, 1991, p. 6B.)
   a. If type of crime and the objective being drug money are independent. Explain.
   b. If use of drugs during a crime and type of crime are independent. Explain.
   c. If type of crime and the objective being drug money are mutually exclusive. Explain.

d. If use of drugs during a crime and type of crime are mutually exclusive. Explain.

159. A construction firm does both small and large projects. To learn more about its customers and for help in advertising, the firm's management randomly selects 200 present and former customers from its files and mails them a survey. Out of the 200 customers, 150 return the survey with usable responses. The results related to project size and primary method of contractor selection are as follows:

Selection Method	Project Size Large	Small
Word of mouth	20	28
Newspaper	5	38
Directory	12	32
Salesperson	15	0

Use this information to answer the following questions.
- **a.** A potential customer has left a message to call. What is the probability that this customer will want a small project done?
- **b.** The company is considering changing its marketing strategy. Its objective is to attract more customers, large and small. When customers choose a firm to contact, which selection method has the highest probability of being employed?
- **c.** What is the probability that a customer was not contacted by a salesperson first?
- **d.** What does $P(\text{newspapers}|\text{small})$ equal?
- **e.** What does $P(\text{newspaper and small})$ equal?
- **f.** What does $P(\text{newspaper or salesperson})$ equal?
- **g.** What does $P(\text{newspaper or small})$ equal?
- **h.** What does $P(\text{salesperson or small})$ equal?

160. According to the League of Women Voters' Education Fund, 18- to 24-year-olds turned out to vote as follows. Find the indicated probabilities.

Election Year	18–24 Turnout (%)
1972	49.6
1976	42.2
1980	39.9
1984	40.8
1988	36.2

Source: "18 Vote Hits New Low," *Asheville Citizen-Times*, August 26, 1991, p. 6B.

- **a.** $P(\text{turnout} > 40\%)$.
- **b.** $P(\text{turnout} \leq 40\%)$.
- **c.** $P(\text{median turnout} \leq 35\%)$.
- **d.** $P(\text{mode of turnouts} > 50\%)$.

161. Ms. Joiner has a customer in her store 75% of the time. Twenty percent of the time when there are customers, she can still work on her bookkeeping. Find the probability that she will have a customer and simultaneously accomplish some bookkeeping in the next hour.

162. Elizabeth has three duties in her job: scheduling personnel, orienting new customers, and listening to customer complaints. If she divides her time equally among the three duties, find
- **a.** The probability that she will be orienting new customers or scheduling personnel at any particular time.
- **b.** The probability that she will be orienting new customers and scheduling personnel at any particular time.

163. About 5% of cars produced in a certain model year had diesel engines. If 25% of a used-car dealer's inventory consists of automobiles from this model year, and you randomly select one of the cars to test drive, what is the probability that you will pick a car from this model year and that it will have a diesel engine?

164. In a particular location, 85% of the time that it rains 1 inch, the reservoir level rises by 1.5 inches. The reservoir is 18 inches below its seasonal level at the end of April. Data collected over the last 100 years suggests that, in 57 of those years, there has been at least 12 inches of rain in May. Figure the probability that sufficient rain will fall and the reservoir will reach its regular level by the end of May.

165. Two traffic lights are supposedly synchronized, so that a driver who makes the first light will make the second light 80% of the time. A driver within two blocks of the first light when it turns green will be able to pass through the first intersection 70% of the time. Both of these statements assume the driver is moving between 25 and 35 miles per hour. If you are such a driver and you are two blocks from the first light and it turns green, what is the probability that you will pass through both lights without stopping?

166. John watches the local late-night talk show 50% of the time, and Doris watches it 60% of the time. If John and Doris meet for the first time in a bar on Friday evening, what is the probability that
- **a.** They will both be able to discuss their impressions of Thursday's show.
- **b.** John or Doris will be able to discuss Thursday's show.

167. There are five clerks in a government office. A single line of patrons forms, and the person at the front of the line is served by the next available clerk. Because services and, hence, time per customer vary, the clerk who will service the last patron in a long line is essentially chosen at random. Ms. Monroe is especially interested in Clerk Cassidy and finds excuses to visit this government office frequently. What is the probability that she will be served by Clerk Cassidy in both of two trips on a given day?

168. A machine produces tables with five legs, but not always perfectly. In a lot of 1,000 tables, the following combinations of defects have been observed:
50 tables have defect A, only four legs.
25 tables have defect B, hole in table top.
10 tables have defects A and B (these tables are included in the groups of 50 and 25 above).

If a table is selected at random from this lot of 1,000, find
   a. $P(A \text{ and } B)$.
   b. $P(A \text{ or } B)$.
   c. $P(A|B)$.
   d. $P(A' \text{ and } B)$.
   e. Are A and B independent?
   f. Are A and B mutually exclusive?

169. A multiple-choice question has five possible responses, A, B, C, D, and E, only one of which is correct. If you randomly select a response, find the probability that you choose A or C.

170. According to the survey for *Bicycling* magazine, 3% of 1,254 adults who represent a cross-section of Americans with household incomes of less than $15,000 commute by bicycle. Bike commuters make up 1.6% of those with incomes above $35,000; 1.67% of Americans are bike commuters, and 20% would be bicycle commuters if better storage and shower facilities were available. (Sources: Nelson Pena, "Power in Numbers," *Bicycling*, April 1991, pp. 44, 46.)
   a. If 15% of Americans fit in the low-income category, find the probability of randomly selecting a bike commuter who is also in the low-income group for another survey.
   b. If 47% of Americans fit the high-income category, find the probability of randomly selecting a bike commuter who is also in the high-income group for another survey.
   c. Are bicycle commuting and availability of facilities independent? Explain.
   d. Are bicycle commuting and availability of facilities mutually exclusive? Explain.

171. Twelve percent of unidentified flying object (UFO) sightings are eventually determined to be meteors, 25% are burned-out satellites returning to Earth's atmosphere, and 18% are high-altitude airplanes seen on clear nights. What is the probability that a single UFO sighting will not be explicable by any of the above explanations?

172. Over the period 1978–1989, the Chinese economy experienced the following inflation rates. Increases in state food prices, matched with offsetting urban-wage increases precipitated three inflationary jumps in 1980, 1985, and 1988, followed by government checking of the price rises. Also, during the 1979–1986 reform period, food prices comprised a large segment, 40%, of urban household expenditures, so food price behavior explains 78% of all price increases during this period. Afterward, during the 1986–1989 period, their movement explains only 28% of all price increases as the pace of price rises of all goods picked up speed. During the third quarter of 1988, the buildup of inflation combined with an announcement of approaching price reforms caused panic buying and hoarding. During July and August the inflation rate rose above a 50% annual rate.

Year	Inflation Rate
1978	0.7
1979	1.9
1980	7.5
1981	2.5
1982	2.0
1983	2.0
1984	2.7
1985	11.9
1986	7.0
1987	8.8
1988	20.7
1989	16.3

Source: Data and explanations from Barry Naughton, "Why Has Economic Reform Led to Inflation?" *American Economic Review*, Vol. 81, No. 2, May 1991, pp. 207–11.

   a. Use Z scores to demonstrate the extent of the three jumps in the inflation rate compared to the mean rate. Are these jumps outliers? Does the text of this question help you explain the Z score results?
   b. Find the probability of experiencing an inflation rate greater than 10% during this period.

173. Snakes tend to lie in the warm road about 85% of the time when the temperature exceeds 75°. The temperatures surpass this barrier about 92% of the time in a certain locale during the summer. What is the probability of encountering a snake on the road in the summer in this locale?

174. A driveway sealer meets the following qualities for 78% of all purchases: It covers 400 square feet as advertised, and the solid materials in the sealer suspend well throughout the liquid. The solid materials suspend well for 85% of all purchasers. Find the probability the sealer will cover the 400 square feet, if the solids suspend well in the liquid. What is the probability you will not cover 400 square feet, if you have no solid suspension difficulties?

175. Jason finds the bands at a certain nightclub very good on about 60% of the evenings he is there. Twenty percent of the time, they are just good. Fifteen percent of the time, they are satisfactory. The remainder of the time, he wishes they would stop and turn on the juke box. If he goes to the nightclub on a random night without knowing anything about the entertainment, what is the probability he will not regret entering? What is the probability he will regret entering?

176. JP is always providing Steve with a long list of worthwhile activities that each require an extensive proportion of a week to perform and there is no overlap between tasks, so that accomplishing one would mean there is less to do on another. There is no way they can all be accomplished in a week, yet JP expects them to

be done before the week is done. Use probability concepts to explain the problem with JP's behavior.

177. We mentioned Chebyshev's inequality in Chapter 3. This rule states that the proportion of observations, X, in a data set that fall within K standard deviations of the mean must be at least $1 - (1/K)^2$ for K greater than 1. We can state this, using probability, as

$$P(\mu - K\sigma < X < \mu + K\sigma) \geq 1 - (1/K)^2$$

for a population and

$$P(\overline{X} - KS < X < \overline{X} + KS) \geq 1 - (1/K)^2$$

for a sample.

If the mean of a population of IQ scores is 100 and the standard deviation is 10, find the minimum probability of events described in Parts a and b.

a. Randomly selecting a person within 1.5, 2, and 3 standard deviations of the mean, μ.
b. A randomly selected person's IQ being between 85 and 115; between 87.5 and 112.5; and between 60 and 140.

If the data set is a sample of IQ scores, find the minimum probability of a randomly selected person's IQ being

c. between 75 and 125; between 77.5 and 122.5.
d. within three standard deviations of the mean; within four standard deviations.

178. What are the three methods of determining a probability value? Once we obtain a value, how do we usually interpret it?

179. The 10 states with the largest 1990 American Indian, Eskimo, or Aleut (AIEA) population are listed below along with the count (in thousands) and also the state population count (also in thousands). There are 1,959,000 total AIEA in the United States in a total 1990 population of 248,710,000. Use this information to answer the following questions.

State	AIEA Count	Population Count
Oklahoma	252	3,146
California	242	29,760
Arizona	204	3,665
New Mexico	134	1,515
Alaska	86	550
Washington	81	4,867
North Carolina	80	6,629
Texas	66	16,987
New York	63	17,990
Michigan	56	9,295

Source: "Race and Hispanic Origin," *1990 Census Profile*, U.S. Department of Commerce, Economics and Statistics Administration, Bureau of the Census, No. 2, June 1991, pp. 4, 6.

a. Are being from California and being part of AIEA independent?
b. Should we expect independence for being from the top 10 AIEA states (or a single state in this group) and being part of AIEA? Explain.
c. Being from Wisconsin and being part of AIEA are very close to being independent. $P(\text{AIEA} \mid \text{from Wisconsin}) = 0.0079$ and $P(\text{AIEA}) = 0.0080$. If you are aware that a person is from Wisconsin, does this help you determine if they are part of AIEA? Explain.

Chapter 5

Random Variables and the Binomial Distribution

This chapter expands the probability concepts introduced in Chapter 4, by subjecting the outcomes of a statistical experiment to the same types of descriptive statistics that we explored in Chapters 2 and 3, the relative frequency distribution, mean, and standard deviation. An application of these techniques, the development and description of a special and very useful random variable, the binomial, follows. It is the basis for inferences about the proportion of successes in a population, such as the proportion of eligible voters in favor of a candidate.

Objectives

Using the association between data sets and experimental outcomes, Unit I develops the concept of a random variable and its properties, including probability distributions and expected-value operations. Upon completing the unit, you should be able to

1. Solve probability problems using information from a probability distribution of a random variable.
2. Use expected-value operations to generate the mean and standard deviation of a random variable.
3. Associate the information in a random variable's probability distribution, expected value or mean, and standard deviation with their counterparts, the descriptive measures of an ordinary data set or population. Interpret these measures in terms useful for decision making.
4. Become familiar with the terms *random variable, probability distribution, discrete random variable, continuous random variable, expected value of a random variable,* and *fair game.*
5. Recognize the symbols X, x, $P(X = x)$, $P(x)$, and $E(X)$.

Unit II introduces a specific discrete random variable, the binomial, and applies the principles of probability and random variables to develop and demonstrate its properties. Upon completing the unit, you should be able to

1. Identify a binomial random variable and situations where it is applicable for problem solving.
2. Employ the binomial formula and the binomial table to determine a probability, then interpret the result and relate it to a decision or resolution of a problem.
3. Use shortcut formulas to compute the mean and standard deviation of a binomial random variable.
4. Graph a binomial random variable and its probability distribution.
5. Become familiar with the terms *Bernoulli trial, binomial experiment, binomial random variable, binomial probability distribution, success,* and *failure.*
6. Recognize the symbols p, X, and n.

Unit I: Random Variables and Their Properties

Preparation-Check Questions

1. The probability of an event is *not*
   a. A gauge of the uncertainty associated with the occurrence of a specified event.
   b. The proportion of long-run experimental trials whose outcome satisfies the requirements or description of the event.
   c. The total number of equiprobable outcomes in the sample space.
   d. The long-run relative frequency of occurrence of the event.
   e. The ratio of the number of outcomes of an experiment (that are also elements of the event subset) to the total number of outcomes, if the outcomes are equiprobable.

A group of physicists measured the noise level, L, at the center of campus every night at 10 P.M. for a year with the following results. Use this information to answer Questions 2–4.

Noise Level (decibels)	Number of Nights
50	96
60	121
70	72
80	21
90	28
100	15
110	12
	365

2. If they randomly choose a night to determine the noise level, find $P(L \leq 80)$.
   a. 0.0575.   d. 0.1507.
   b. 0.8493.   e. 0.2082.
   c. 0.7918.

3. The mean of the L values is
   a. 66.1 decibels.   d. 1 decibel.
   b. 31,400 decibels.   e. 8 decibels.
   c. 80 decibels.   f. 32.1 decibels.

4. The standard deviation of the L values is (using 66 as the mean and assuming the data set is a population)
   a. 90,900.0 decibels.   d. 60.0 decibels.
   b. 249.0 decibels.   e. 596.0 decibels.
   c. 15.8 decibels.   f. 24.4 decibels.

5. The standard deviation measures the "average" of deviations of outcomes about the mean by
   a. Summing the deviations and dividing by the number of deviations.
   b. Summing the absolute value of the deviations and dividing by the number of deviations.
   c. Subtracting the smallest observation from the largest observation.
   d. Taking the square root after dividing the range by the number of observations.
   e. Taking the square root of the mean of the squared deviations.

6. A variable is
   a. A trial or repetition of a statistical experiment with only one observable outcome.
   b. A value that we can predict precisely.
   c. A statistical experiment with a single outcome.
   d. An unknown that changes value.
   e. A measure of the dispersion of a data set or the standard deviation.

Answers: 1. c; 2. b; 3. a; 4. c; 5. e; 6. d.

Introduction

National Health Insurance   Although many people would support access to medical care for everyone, they do not agree on how to accomplish this goal, especially the financing. One question on a nationwide survey of 1,004 registered voters about national health insurance asked respondents about extra amounts they would be willing to pay as a monthly tax for national health insurance. The table below records their responses and the relative frequency of each response.

Monthly Payments	Percent of All Responses
100	31
90	2
80	3
70	3
60	3
50	17
40	3
30	2
20	5
10	6
Won't pay	20
Not sure	5
	100

Source: Peter D. Hart Research Associates, as reported in Judy Treible, of Knight-Ridder Tribune, "Survey: People Want National Insurance," *Asheville Citizen-Times*, May 17, 1992, p. 5A.

If we infer that a poll of all registered American voters would produce the same results, then apparently 31% would be willing to pay $100 per month and almost 60% (59% to be exact) would pay $600 or more per year ($50 or more per month). A more useful guide for legislators or regulators in forming an acceptable or least objectionable policy might be the mean and standard deviation of the amounts. Similar information on individual medical expenses or current insurance payments would help policy formulations. Using information like that shown on the table and incorporating the mean and standard deviation of the possible outcomes to plan or make decisions are the subjects of this unit.

5.1 Random Variables and the Data Set of Experimental Outcomes

A variable is a property or characteristic whose value varies among a set of entities, such as a student's grades on different tests, the percentage of the popular vote received by the winning presidential candidates since American presidential elections started, or the number of moles damaging a lawn. A *random variable* is a variable whose values correspond to outcomes of trials of a statistical experiment. The value of the random variable can vary with each repetition, because of the uncertainty of the outcome.

If the experiment is the number showing after rolling a die, then the result can vary with each roll. The possible values may be the numbers 1, 2, 3, 4, 5, and 6. If the experiment is to determine whether the number that shows after the roll is even or odd, we might assign the value 0 to all even numbers and 1 to the odd numbers. We can assign the same value to more than one outcome, as we did for the odd-even experiment, to form an event. But we cannot assign more than one value to the same outcome for an experiment.

Question: Are the events represented by different values of the random variable mutually exclusive?

Answer: Yes, because we can assign only one value to an outcome for an experiment. If an event corresponds to the random variable value 3, an event that corresponds to a different value, say 5, must consist of different outcomes in the sample space.

Unlike a nonrandom variable, the values of a random variable cannot be predicted precisely or controlled. In any instance, we must understand its experimental nature to judge whether a variable is random or nonrandom. For instance, a nonrandom variable might be the number of days of vacation individual employees receive as the personnel manager develops a report of vacation days earned. The amounts might be any positive value from 0 to 365 days. The company's policies and employees' work records dictate this amount. If the vacation report is a prediction of the amount each employee will earn during the next five years, based on relative frequencies of employees' previous work records, then the variable is random.

Question: If a variable is the amount of rainfall at the local weather station, describe an experiment that would make this a random variable. Then, describe a situation where the variable is nonrandom.

Answer: First, rainfall can vary with the day chosen for the measurement. Possible values start with 0 inch and increase indefinitely, although large values are certainly unusual, if not impossible. If a meteorologist uses scientific theories and previous experience with similar weather conditions to predict tomorrow's rainfall, the prediction would be a random variable value, because the outcome is uncertain. These predictions have some probability of occurring, but they are not definite. If an analyst records rainfall amounts from an earlier period, say yesterday, these values are nonrandom.

Question: If the analyst chooses a random day from the past to determine the rainfall, is the rainfall value a random or nonrandom variable?

Answer: Random, because the outcome depends on the chance selection of the day to examine.

Sometimes, experimental outcomes are not numerical values. If the random variable is the suit of a card selected in a single draw from a standard deck, the results include spades, hearts, diamonds, and clubs. Figure 5-1 shows that we can designate arbitrary values to represent these results, in this case, the integers 1 through 4, for the respective

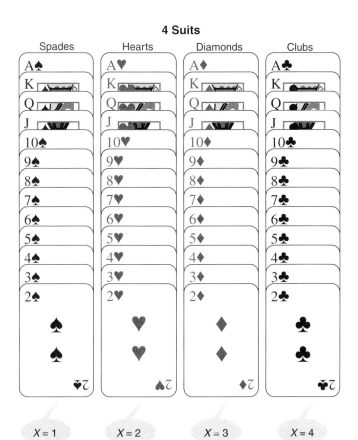

Figure 5-1

The random variable X represents the suit of the card selected in a single random draw from a standard deck of 52 cards

suits. Notice that we assign the same value to all 13 possible outcomes that fit the event, the 2 applies to all 13 heart cards. In most cases, we assign meaningful values to the various outcomes. For instance, the responses to the survey question about monthly payments for national health insurance include "won't pay" and "not sure." An obvious choice for "won't pay" is $0. Associating a meaningful number with "not sure" is more problematical. Any unused value, such as −100 or 1,523 would signal this response but not be useful for later computations and descriptions of the results. We might eliminate these responses as uninformative or not useful, decide to allocate these responses equally among the other 11 responses, or we might assume that people who are unsure would probably not want to pay or would pay the average amount from the other responses. These random variable values constitute mathematical expressions rather than verbal expressions of the result of a statistical experiment. Hence, we can perform mathematical operations on these values and produce values that we can interpret and associate with the experiment itself.

We classify random variables into one of two types, discrete or continuous. The difference depends on whether or not we can count the individual values the random variable can assume.

The random variables representing the roll of a die and the monthly payments fit into the category of *discrete random variables*. In both cases we can count the different possible values of the random variable. We can think of making a list of the outcomes and successively numbering them beginning with one. A random variable with infinitely many outcomes can still be discrete, if we can theoretically list or count the outcomes. The random variable equal to the number of cars passing a toll booth is an example. There could be zero car, one car, two cars, and so on. There is no limit to the number of outcomes, but theoretically we can still count them.

On the other hand, there are infinite values of a *continuous random variable*, but we have no way of listing or counting the individual values. Generally, we measure these values, such as length, weight, volume, temperature, and time, with some type of instrument, we do not count them. A thermometer, for example, may be graduated in degrees, so we read the temperature to the nearest degree, but, actually, any other point along the thermometer is a possible temperature value as well. An infinite number of values lie along the thermometer, and we cannot count them all. Figure 5-2 illustrates infinite values that we cannot count.

A good way to illustrate the difference between the two types of random variables is Figure 5-3. We purchase oil for our automobiles in discrete units, by the quart. On a number line, the possible values of this random variable fall only at certain points with gaps between. The values filling the gaps are not possible outcomes, we cannot purchase 3.3 quarts of oil. Gasoline purchases are continuous instead of discrete, because we can buy any amount that does not overfill the tank. The quantity we purchase can be any value in an interval on the number line. There are no gaps.

Figure 5-2

Infinite outcomes on a scale (they cannot be counted)

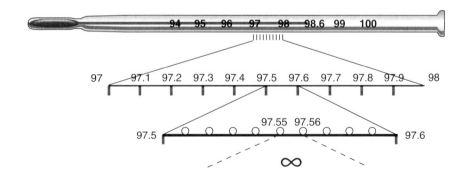

Discrete (countable)
We buy oil for cars in quart containers.

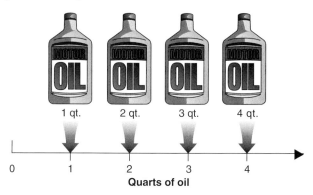

Continuous (measurable)
We can buy any amount of gasoline, including any fraction of a gallon.

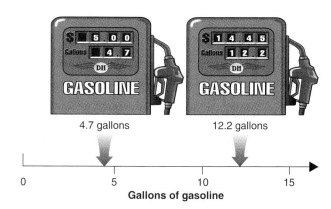

Figure 5-3

Discrete and continuous random variables

Question: Consider random variables that represent the progress of an athletic team during a game. In football, we determine progress in yards gained, X, and in baseball, progress is measured in bases attained, Y. Classify each random variable, X and Y, as discrete or continuous. Explain your classification.

Answer: X, football progress, is a continuous random variable, because the team can realize any fraction of a yard, even though commentators and statisticians usually report progress in whole number amounts. The football field is like a measuring instrument, graduated in yards, but any point along the field is a possible position for the lie of the football and the point to begin *measuring* further progress. Y, baseball progress, is discrete, because there are only four possible bases or points for determining progress. Baseball players do not get credit for running part of the way to the next base. We *count* their progress in bases.

Whether a random variable is discrete or continuous will determine the mathematical operations we use to extract details and summary measures about its values. In the remainder of this chapter, we will use discrete random variables to develop and demonstrate the measurement of different characteristics of random variables. These measures convey the same information about the values of a continuous random variable, but their calculation requires more advanced mathematics not covered in this book.

Section 5.1 Problems

7. Consider the number of questions on a test. When is this a random variable for the teacher constructing the test? When is it a random variable for the students taking the test? Is it discrete or continuous? Explain.
8. Is the number of questions on a test that a randomly selected student will answer correctly a random variable? Explain.
9. Consider a producer deciding the number of 30-second commercials allotted during a television production. Is this variable random if the program is a movie? Is it random if the program is a live event, such as a football game?
10. Is the random variable in Problem 9 discrete or continuous? If the values of the random variable are total time available for commercials rather than number of commercials, is it discrete or continuous?
11. One exercise in a television broadcasting class is to stage and tape volunteer performances, such as those of a singer, band, actor, or magician. To assist in scheduling studio time, the professor has kept time sheets from previous exercises in the laboratory studio over many previous terms. According to these records, the amount of time required for this exercise varies, as displayed in the following relative frequency distribution:

Studio Time (minutes)	Proportion of Total Projects
30	0.15
40	0.20
50	0.22
60	0.27
70	0.08
80	0.06
90	0.02

a. If the experiment is determining the studio time required, then studio time is a random variable. Suggest values for this random variable.
b. Suppose that the random variable is the type of performance being taped rather than the time spent on the project. Possible performers include singers, bands, actors, and others. Suggest a number scheme to associate with each of these types of performers.

5.2 Describing the Values of a Random Variable

Each time an experiment occurs, the random variable assumes one of its possible values, depending on the outcome of the trial. The set of values produced by many repetitions of an experiment form a data set. If the value of the random variable, D, is the number showing from rolling a die, then, after recording the values from a large number of rolls, we would have a large data set consisting of the values 1, 2, 3, 4, 5, and 6. Just as we avoid studying each individual observation in a large data set with descriptive statistics, we shall condense the information in this infinite set of outcomes by using various measures and operations analogous to descriptive statistics, such as the relative frequency distribution, mean, and standard deviation.

When we discuss the general result of an experiment, uppercase letters signify random variables. Lowercase forms of the same letter symbolize specific outcomes or values of the random variable. If X represents the suit of the selected card, then x is a particular value that indicates to us the chosen suit. If we assign the value 2 to hearts, then $X = 2$ is the shorthand way of expressing that the randomly selected card is a heart. We will want to know the probabilities of certain events, $P(X = x)$, such as $P(X = 2)$ in the card example. When there are no misunderstandings about the meaning of the symbols, we will substitute $P(x)$ for $P(X = x)$, such as $P(2)$ in the card example. This simplifies notation and some formulas.

5.2.1 Probability Distribution of a Random Variable

Sometimes when we study situations or make decisions, we need to know the probability associated with different values of a random variable. We may need to know all of the alternative values and the probability of each one. We call this information the probability distribution of the random variable. More formally, a *probability distribution* is a representation (table, formula, or graph) of all possible values of a random variable and the probability associated with each value.

Suppose the random variable D represents outcomes of the experiment, observing the number showing after rolling a die. If we assign the number shown after the roll as the value of D, we obtain the following probability distribution of D. Remember D is the general term for the random variable, d is a specific value of D.

d = Specific Value of Toss	$P(d)$
1	1/6
2	1/6
3	1/6
4	1/6
5	1/6
6	1/6
	1

A random variable's probability distribution summarizes the values produced from many trials of an experiment. It condenses this data set to a form similar to a frequency distribution, where the frequencies are long-run relative frequencies, that is, probabilities associated with the different values of the random variable. In the case of D, the values obtained by tossing a single die, we expect each of the six sides to occur on about one-sixth of the trials or rolls. Hence, $P(1) = P(2) = P(3) = P(4) = P(5) = P(6) = 1/6$.

■ **Example 5-1:** Use the probability distribution for D, to find $P(D > 4)$ and $P(D \geq 4)$.

■ **Solution:** Because the values of D are mutually exclusive, we find

$$P(D > 4) = P(D = 5 \text{ or } D = 6) = 1/6 + 1/6 = 2/6 = 1/3.$$
$$P(D \geq 4) = P(D = 4 \text{ or } D = 5 \text{ or } D = 6) = 1/6 + 1/6 + 1/6 = 1/2.$$

■ **Example 5-2:** If we omit the unsure responses to the national health insurance survey question and use the value $0 for those who are unwilling to make a monthly payment, we can form the random variable T, the amount of monthly tax or payment for national health insurance that the respondent is willing to pay. To find the probabilities, we form new relative frequencies, based on 954 definite responses (1,004 responses − 50 unsure responses). The probabilities apply to the experiment of observing the response of a randomly selected respondent from the 954 definite responders or to registered voters in general if we accept the survey results as an approximate representation of the behavior of this group. Use the following result to find $P(T = 70)$, $P(T > 70)$, and $P(30 < T \leq 70)$.

t	$P(t)$
100	0.3260
90	0.0210
80	0.0314
70	0.0314
60	0.0314
50	0.1792
40	0.0314
30	0.0210
20	0.0524
10	0.0629
0	0.2107
	0.9988

Solution: $P(T = 70) = 0.0314$
$P(T > 70) = P(T = 80 \text{ or } T = 90 \text{ or } T = 100)$
$= 0.0314 + 0.0210 + 0.3260 = 0.3784.$
$P(30 < T \leq 70) = P(T = 40 \text{ or } T = 50 \text{ or } T = 60 \text{ or } T = 70)$
$= 0.0314 + 0.1792 + 0.0314 + 0.0314 = 0.2734.$

Example 5-3: The following probability distribution of the intensity of earthquakes, X, in California and western Nevada is compiled from intensities measured between 1800 and 1980. The intensity values correspond to the Modified Mercalli Intensity Scale of 1931, where an intensity value of I indicates that the earthquake escapes notice by almost everyone, whereas the largest value, XII, indicates total damage. The intensities in California and western Nevada range between IV ("During the day felt indoors by many, outdoors by few. At night some awakened. Dishes, windows, and doors disturbed; walls make creaking sound. Sensation like heavy truck striking building. Standing motorcars rocked noticeably.") to XI ("Few, if any [masonry], structures remain standing. Bridges destroyed. Broad fissures in ground. Underground pipelines completely out of service. Earth slumps and land slips in soft ground. Rails bent greatly."). When intervals of intensities are reported rather than a single intensity, we assign the outcome the highest value, so the following information is a "worst-case scenario." Finally, we assign the Arabic number equivalent to the Roman numeral intensity as the values of X. The probability distribution is the last two columns presented in the following table. The first column is just details for the reader. Assume that the historical relative frequencies over the 180-year time span apply to the experiment of observing the intensity of an earthquake in this region. Use this information to express the probability of the following events symbolically, and then find the answer.

a. The next earthquake's intensity will be a VII.
b. The next earthquake's intensity will be a VII or less.
c. The next earthquake's intensity will be less than a VII.
d. A building constructed to withstand a VI earthquake will survive the next earthquake intact.

Modified Mercalli Intensity Scale	x = intensity	$P(x)$
IV	4	0.0057
V	5	0.1276
VI	6	0.5048
VII	7	0.2324
VIII	8	0.0705
IX	9	0.0324
X	10	0.0171
XI	11	0.0095
		1.0000

Source: Jerry L. Coffman, Carl A. von Hake, and Carl W. Stover, eds., *Earthquake History of the United States,* Publication 41-1, rev. ed. (through 1970), reprt. 1982 with suppl. (1971–1980), U.S. Department of Commerce, National Oceanic and Atmospheric Administration, and U.S. Department of the Interior, Geological Survey, Boulder, Colorado, 1982, pp. 137–49, 35a–36a. Descriptions and quotations concerning the Mercalli scale are taken from Harry O. Wood, and Frank Neumann, *Bulletin of the Seismological Society of America,* Vol. 21, No. 4, December 1931, as reported in the above detailed source, pp. 3–4.

Solution: a. Symbolically, we express the probability that the next earthquake's intensity is VII as $P(7)$. From the chart, this probability is 0.2324.

b. The probability the next intensity is VII or less is $P(X \leq 7) = P(X = 4 \text{ or } X = 5 \text{ or } X = 6 \text{ or } X = 7) = 0.0057 + 0.1276 + 0.5048 + 0.2324 = 0.8705$.

c. The probability the next earthquake's intensity is less than VII is $P(X < 7) = P(X = 4 \text{ or } X = 5 \text{ or } X = 6) = 0.6381$.

d. If the building withstands a VI, it should withstand any lower intensity earthquake, so the problem is $P(X \leq 6) = 0.6381$ from Part c.

Question: Why do the probabilities in a probability distribution sum to 1 (or approximately 1 when rounding occurs)?

Answer: They sum to 1 because the values of X account for all possible outcomes; the total of the probabilities for X must reflect 100% of the trials or outcomes.

Now that we can condense the values of a random variable that result from infinite repetitions of the experiment to a probability distribution, we proceed to further summarize the values with a mean and standard deviation.

Section 5.2.1 Problems

12. Form the probability distribution for the random variable X, number of heads in two tosses of a coin.

13. Find the missing value in the following probability distribution of X, the number of jurors on a six-member panel who vote for acquittal in a hung jury. A criminal justice analyst compiled the values from 50 years of trial data.

x	P(x)
1	0.12
2	0.18
3	0.42
4	
5	0.08

14. Form the probability distribution for the random variable R, the total count from tossing a pair of dice. Then find $P(10)$, $P(R > 10)$, and $P(R \text{ is even})$.

15. One exercise in a television broadcasting class is to stage and tape volunteer performers, such as a singer, band, actor, or magician. To assist in scheduling studio time, the professor has kept time sheets from previous exercises in the laboratory studio over many previous terms. According to these records, the amount of time required for this exercise, S, varies, as displayed in the following probability distribution.

s	P(s)
30	0.15
40	0.20
50	0.22
60	0.27
70	0.08
80	0.06
90	0.02

It appears that half-hour slots accommodate only 15% of the projects, while 84% of the students will finish in an hour or less. Verify and express these two probability statements about values of S with symbols rather than words.

16. Use the following probability distribution of V, the number of guests who volunteer to clean up after a party, to find $P(2)$ and $P(V > 4)$. What do we know about the probability of V values greater than 6?

v	P(v)
0	0.33
1	0.28
2	0.14
3	0.08
4	0.08
5	0.05
6	0.04

5.2.2 The Mean or Expected Value of a Random Variable

The primary single-value measure of the center or typical value of a data set is the mean, μ. Often, we need this same information about values of a random variable, X. As we saw in the preceding section, the probability distribution of X is like a relative frequency distribution based on many trials of the experiment that generate the values of X. Because it accounts for each and every time that we perform the experiment, the mean condenses the *population* of X values, so we represent it with the symbol μ. This is not a simple mean of the possible values of X that we list when presenting a probability distribution, but a mean of the many X values that occur after many trials of the experiment. We must account for the fact that some values occur more often than others.

Suppose a company constructs the following probability distribution for the random variable representing its daily profits from data over a long span of years.

Profit (Dollars)	P(Profit)
1,000	0.1667
3,000	0.5000
6,000	0.3333

If we list the individual values as we observe profit a large number of days, $1,000 should show up about 0.1667 (one-sixth) of the days; $3,000, half of the days; and $6,000, one-third of the days. Suppose we observe profit N days, where N is a very large number.

Question: If N is 300,000 days of profit values, about how many days should $1,000 occur among these observations?

Answer: $1,000 should occur about one-sixth of the N days or $0.1667N$ days, in this case, $300,000(0.1667) = 50,010$ days of $1,000.

Thus, to find the mean of the N profit values, we must sum the $0.1667N$ observations of $1,000, the $0.5000N$ days of $3,000, and the $0.3333N$ days of $6,000. Then, to find the mean, we would divide this sum by N, the total number of days ($0.1667N + 0.5000N + 0.3333N = N$). We can write this procedure as

$$\mu = \frac{\$1,000(0.1667N) + \$3,000(0.5000N) + \$6,000(0.3333N)}{N}.$$

If we factor N from the numerator, we obtain

$$\mu = \frac{N[\$1,000(0.1667) + \$3,000(0.5000) + \$6,000(0.3333)]}{N}.$$

The N in the numerator cancels the N in the denominator, leaving

$$\mu = \$1,000(0.1667) + \$3,000(0.5000) + \$6,000(0.3333) = \$3,666.50.$$

This equation is the sum of the results of multiplying each value of the random variable by its probability, or, substituting symbols for the values, we obtain

$$\mu = \sum_{1}^{3} (\text{profit})[P(\text{profit})].$$

(Subscripts are omitted to simplify expressions, but summation operations must be completed for all observations in a data set in the remainder of this text.)

We use the same operation to find the mean for any discrete random variable. Suppose there are k possible values of X:

$$\mu = \sum_{1}^{k} xP(x).$$

Figure 5-4 illustrates this procedure with the profit example.

This equation is the formula for finding the *mean, average,* or, more commonly, the *expected value of a random variable* (the terms are used interchangeably in this text). $E(X)$ is the shorthand symbol for the operation $\sum_{1}^{k} xP(x)$, so we can say $\mu = E(X)$. The average daily profit from a large number of observations of the profit value is $3,666.50 (see Figure 5-4). We expect a $1,000 profit to occur about one-sixth of the days we compute profit, $3,000 about half of the days, and $6,000 about one-third of the days,

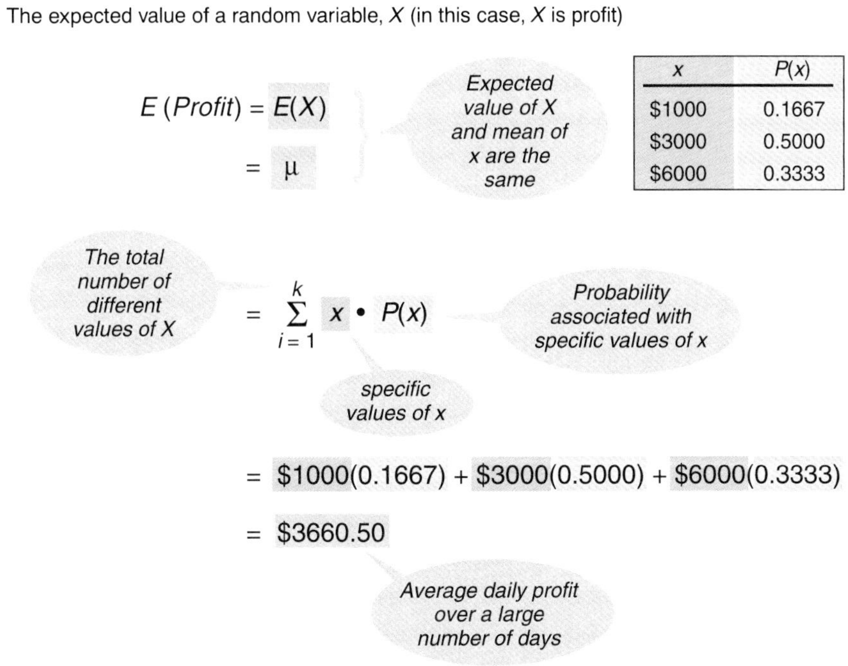

Section 5.2   Describing the Values of a Random Variable   **207**

The expected value of a random variable, X (in this case, X is profit)

$$E(\text{Profit}) = E(X) = \mu = \sum_{i=1}^{k} x \cdot P(x)$$

x	P(x)
$1000	0.1667
$3000	0.5000
$6000	0.3333

- Expected value of X and mean of x are the same
- The total number of different values of X
- Probability associated with specific values of x
- specific values of x

$$= \$1000(0.1667) + \$3000(0.5000) + \$6000(0.3333)$$

$$= \$3660.50$$

Average daily profit over a large number of days

Figure 5-4
The relation between the expected value of a random variable and its probability distribution

but the average we expect is $3,666.50. Notice that the expected value, like an average, does not have to be one of the possible X values, but it must be representative of these values; that is, between the largest and smallest X values. Also, it is not a simple mean of the listed values, $3,333 = ($1,000 + $3,000 + $6,000)/3 in this case.

Remembering the formula for finding the expected value of X, E(X), is easy if we remember that it is the mean X value, μ, as well. Because we have the probability distribution, not an actual data set, we use the weighted mean (see Chapter 3) to account for the importance or frequency of occurrence of each X value. The weighted-mean formula is $(\Sigma xw/\Sigma w)$. The weights w are $P(x)$. The denominator is the sum of the probabilities in the probability distribution, which we know must sum to 1, so $\mu = E(X) = \Sigma xP(x)$. Again, we arrive at the same formula for E(X), with the same interpretation of the result, the mean of the values of X, although we used the weighted-mean formula to take advantage of the condensed data set presented in the probability distribution. Associating a probability distribution with a relative frequency distribution and using the formula for finding the mean of grouped data (Chapter 3) would also duplicate the results. The following box summarizes these findings.

To find the expected value (average or mean) of a random variable, X, with k possible values, use

$$E(X) = \mu = \sum_{1}^{k} xP(x)$$

■ **Example 5-1 (continued):** Find and interpret the expected value of D, the number of dots obtained when tossing a die. Remember that the probability distribution is

d	P(d)	d	P(d)
1	1/6	4	1/6
2	1/6	5	1/6
3	1/6	6	1/6

■ **Solution:** $E(D) = \mu$
$$= 1(1/6) + 2(1/6) + 3(1/6) + 4(1/6) + 5(1/6) + 6(1/6)$$
$$= 3.5 \text{ dots.}$$

Thus, the average number of dots shown from infinite tosses of a single die is 3.5 dots.

■ **Example 5-2 (continued):** Find and interpret the expected value of T, the monthly amount a respondent is willing to pay for national health insurance.

t	$P(t)$
100	0.3260
90	0.0210
80	0.0314
70	0.0314
60	0.0314
50	0.1792
40	0.0314
30	0.0210
20	0.0524
10	0.0629

■ **Solution:** $E(X) = 100(0.3260) + 90(0.0210) + 80(0.0314) + 70(0.0314)$
$$+ 60(0.0314) + 50(0.1792) + 40(0.0314) + 30(0.0210)$$
$$+ 20(0.0524) + 10(0.0629)$$
$$= 53.6070.$$

Thus, we expect the average monthly amount that a very large number of registered voters are willing to pay to be \$53.6.

■ **Example 5-3 (continued):** Find and interpret the expected value of X, the intensity of earthquakes.

x	$P(x)$
4	0.0057
5	0.1276
6	0.5048
7	0.2324
8	0.0705
9	0.0324
10	0.0171
11	0.0095

■ **Solution:** $E(X) = 4(0.0057) + 5(0.1276) + 6(0.5048) + 7(0.2324)$
$$+ 8(0.0705) + 9(0.0324) + 10(0.0171) + 11(0.0095)$$
$$= 6.4.$$

Thus, the average intensity on the modified Mercalli intensity scale is 6.4, or between a VI and VII.

An average outcome is a useful concept for making decisions. The following examples illustrate this.

■ **Example 5-4:** Suppose you have just entered Pot Luck Casino. At the first table, the following game commences: A deck is shuffled, and you draw one card. If the card is an ace, you win \$10. There is no charge for playing. Would you want to play if your objective is to pocket more money when you leave than when you arrived? Explain.

Solution: If you like a no-risk game, this game is great, because you risk none of your money. If you continue to play, you will draw an ace and receive $10 on about 4/52 of the deals. The other 48/52 of the deals will pass with no exchange of funds. You cannot lose your money. Let X be player winnings, expressed in dollars.

x	$P(x)$
10	4/52
0	48/52

$E(X) = \$10(4/52) + \$0(48/52) = \$0.77$. If you play this game a very long time (infinite number of times), the average you would win is about 77 cents per deal.

Question: What is the dealer's expected value for this game?

Answer: Obviously, the dealer can only lose money, because you never have to pay. If you average a 77-cent win per deal, then the dealer must average a 77-cent loss per deal. Using expected value formulations, with Y being dealer winnings in dollars, we obtain

y	$P(y)$
−10	4/52
0	48/52

$E(Y) = (-\$10)(4/52) + \$0(48/52) = -\$0.77$.

■ **Example 5-5:** Needless to say, the first game is not a profitable game for the casino. Suppose now the casino alters the game by charging you $1 to play the game each deal. What is the expected value for you now? Would you still play?

■ **Solution:** Subtracting $1 from the previous outcomes gives us the possible winnings for a player. The $1 must be paid and thus subtracted from the result, lose or win. If we let $Z = X - 1$, we obtain

z	$P(z)$
$(10 - 1) = 9$	4/52
$(0 - 1) = -1$	48/52

$E(Z) = 9(4/52) + (-1)(48/52) = -0.23$.

Notice that subtracting $1 from the expected or average amount won from the game when there is no charge to play, $0.77, gives us the −$0.23 also.

You average losing 23 cents each time you play, if you play long enough. Note that if you win on the first try and quit, the expected value is not the average outcome.

Question: A *fair game* is one with an expected value of 0 for all players. Why does this seem a proper definition?

Answer: On average, everyone breaks even—no winners and no losers.

Question: Suppose you can choose between two games. Their expected values are $0.03 and $0.04, respectively. Which game should you choose to win the most money?

Answer: Choose the game with an expected value of $0.04, because on average you will win more per game than when the expected value is $0.03.

For more information about expected values and games of chance, risk reading Box 5-1.

A possible strategy for nongame decision making is to do what we have just done: Compare expected values of different alternatives, and choose the appropriate alternative for achieving our objective. Sometimes, we want the largest possible value, for instance, when X is profits. Other times we prefer the minimum value, say when X is costs.

Box 5-1 Expected Values, Casinos, and Insurance Companies

If every player went home a winner, how would a casino continue to fund the games? Someone has to lose, and generally the situation is set up so that everyone loses a little each time they play, on average. The expected value is negative for players and positive for the house. The casino can survive with a very small expected value in its favor, if players participate a large number of times. A single gambler would find playing a large number of (infinite) repetitions of a game difficult, but the house can come close, because it plays every time any player engages the dealer, table, or machine. If the overall expected value for games is 0.02 in favor of a casino, then the house expects to gross 2% of the total money that is bet. If you have ever been to a casino, you know that this 2% of gross can be a tremendous amount of money.

Some poker machines actually advise the participant of the likely consequences, such as informative labels claiming a 92% payback; that is, you can expect to lose eight cents out of every dollar you bet. For those who like to take risks, there is some chance of going home ahead and letting the casino reach its average with other luckless losers. This works for those strong-willed souls who quit while they are ahead.

The same principles hold for insurance companies. Suppose you insure against the occurrence of some event, say a fire, during the period covered by the insurance premium paid to the company. Then, the expected payout to you from a fire must be less than the expected premium you forfeit because the event does not occur, Damages $\times$ P(fire) $<$ Premium $\times$ P(no fire), in order for the insurance company to make a profit.

■ **Example 5-6:** Suppose a local restaurant must repaint dining rooms frequently. The owner purchases paint from two stores that offer the same quality of paint, but usually not at the same price. To save time and energy spent comparing costs for each paint purchase (especially when running short and wishing to finish promptly), the owner wants to determine which store tends to charge less for their paint, on average. Then she will always buy at that store when short of time. Using receipts from prior years and advertisements from old issues of the local paper, she derives the following probability distribution for wall paint prices per gallon for the stores, A and B:

a	$P(a)$	b	$P(b)$
9.00	0.25	8.00	0.05
10.00	0.25	10.00	0.35
15.00	0.50	12.00	0.20
		14.00	0.40

Which store should she choose to minimize expected cost?

■ **Solution:** $E(A) = \$9(0.25) + \$10(0.25) + \$15(0.50) = \$12.25$.
$E(B) = 8(0.05) + 10(0.35) + 12(0.20) + 14(0.40) = 11.90$.

She should choose B, which has the smaller expected value, that is, the lower average price.

If we have a one-time decision to make, expected values are useful for incorporating all of the available information to assess different alternatives, just as a mean takes each observation into account. Remember, though, that an expected value is an average over the long run, making this decision-making method better suited for recurring decision situations with the same alternatives and similar probabilities. If the restaurant owner purchases paint often, especially without checking prices each time, then she should choose the store with the lowest expected price. This strategy minimizes the purchase price of paint over the long run.

In summary, the expected value is the average outcome from repeating the experiment a large number of times. It is useful in inferential statistics and in decision making. Consequently, we will encounter it again as we progress in this book.

Section 5.2.2 Problems

17. Find and interpret the expected value of the following random variable, S, the studio time (to the nearest minute) required for an exercise in a television broadcasting class where students stage and tape volunteer performances.

s	P(s)	s	P(s)
30	0.15	70	0.08
40	0.20	80	0.06
50	0.22	90	0.02
60	0.27		

18. Find and interpret the expected value of V, the number of guests who volunteer to clean up after a party.

v	P(v)	v	P(v)
0	0.33	4	0.08
1	0.28	5	0.05
2	0.14	6	0.04
3	0.08		

19. The random variable X is the result of rolling a pair of dice. Find and interpret its expected value, using its probability distribution which follows.

x	P(x)	x	P(x)
2	1/36	8	5/36
3	2/36	9	4/36
4	3/36	10	3/36
5	4/36	11	2/36
6	5/36	12	1/36
7	6/36		

20. X represents the number of jurors on a six-member panel who vote for acquittal in a hung jury. Find and interpret the expected value of X.

x	P(x)
1	0.12
2	0.18
3	0.42
4	0.20
5	0.08

21. D symbolizes the number of days for the post office to deliver a letter to a private household. Find and interpret the expected value of D.

d	P(d)
1	0.15
2	0.19
3	0.24
4	0.20
5	0.09
6	0.05
7	0.03
10	0.02
14	0.01
21	0.01
30	0.01

5.2.3 The Standard Deviation of a Random Variable

Just as we describe the dispersion of values in an ordinary data set with a standard deviation, we can measure the dispersion of the set of random variable values from the outcomes of all possible trials with the standard deviation. Because we do consider all possible trials, the set of values that result is a population, and we use the symbol σ to represent this standard deviation. The standard deviation measures the "average" distance of data from the mean or the distance we expect to lie between a typical point and the mean.

The probability distribution of the daily profit random variable from the last section is repeated below.

Profit	P(profit)
1,000	0.1667
3,000	0.5000
6,000	0.3333

The standard deviation measures the dispersion of the outcomes of a large number of repetitions, say, N of them. One-sixth of these days (0.1667N) should result in $1,000, 0.5000$N$ days result in $3,000, and 0.3333$N$ days in $6,000. Using the formula from Chapter 3 for finding the standard deviation of a population of values,

$$\sigma = \sqrt{\frac{\Sigma(X-\mu)^2}{N}},$$ we begin by finding the deviation for each of the N outcomes. A deviation is the difference between the outcome and the mean, which we determined to be $3,666.50 from the expected-value operation in the last section. Actually, we only need to find three deviations, one for each of the three profit values ($1,000, $3,000, and $6,000), because we know the rest of the N observations duplicate one of these three values. Then, we square the three deviations. If we wrote out the squared deviation for each outcome, we would have $0.1667N$ of the squared deviations $(1,000 - 3,666.50)^2$, $0.5000N$ of the $(3,000 - 3,666.50)^2$, and $0.3333N$ of the $(6,000 - 3,666.50)^2$. Rather than write all the N terms, we can just write

$$\sigma = \sqrt{\frac{(0.1667N)(1,000 - 3,666.50)^2 + (0.5000N)(3,000 - 3,666.50)^2 + (0.3333N)(6,000 - 3,666.50)^2}{N}}.$$

Factoring N from the numerator and canceling it with the N in the denominator, we obtain

$$\sigma = \sqrt{0.1667(1,000 - 3,666.50)^2 + 0.5000(3,000 - 3,666.50)^2 + 0.3333(6,000 - 3,666.50)^2}$$
$$\sigma = \sqrt{3222277.75} = 1,795.07.$$

Thus, the "average" distance from the mean of profit values of many repeated trials is $1,795.07.

We can translate this process into a general formula that will save us some work, compared to the calculation using all N outcomes. To create the formula, any individual value of the random variable (such as $1,000) becomes x, the probability (such as 0.1667) becomes $P(x)$, and the mean (3667.50) becomes μ. The general formula, then, for the standard deviation of a discrete random variable with k different values becomes

$$\sigma = \sqrt{\sum_{1}^{k}(x-\mu)^2 P(x)}.$$

Figure 5-5 illustrates the formula with the profit random variable.

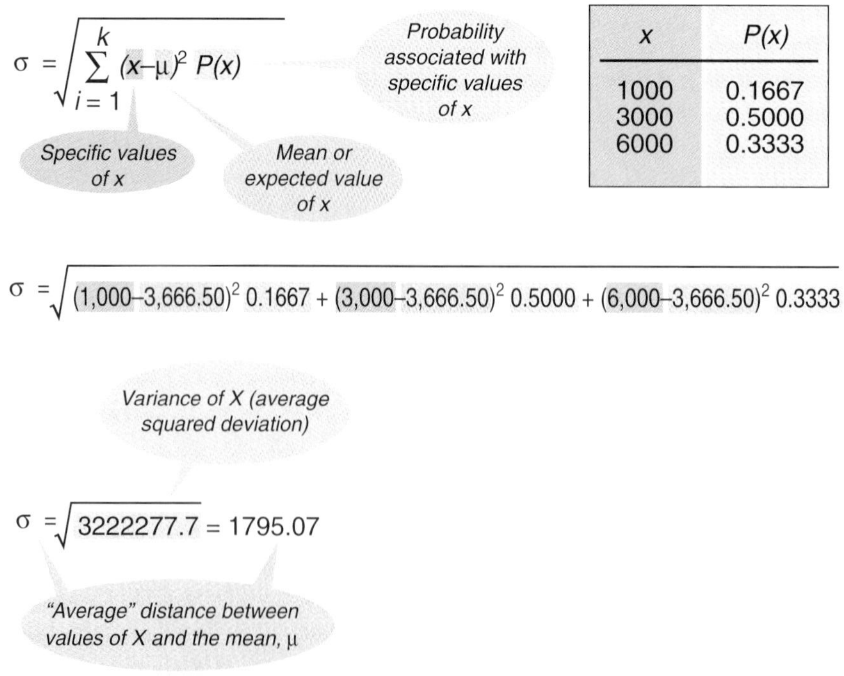

Figure 5-5
Relationship between the standard deviation of a random variable and its probability distribution

$$\sigma = \sqrt{\sum_{i=1}^{k}(x-\mu)^2 P(x)}$$

Specific values of x — Mean or expected value of x — Probability associated with specific values of x

x	P(x)
1000	0.1667
3000	0.5000
6000	0.3333

$$\sigma = \sqrt{(1,000 - 3,666.50)^2\, 0.1667 + (3,000 - 3,666.50)^2\, 0.5000 + (6,000 - 3,666.50)^2\, 0.3333}$$

Variance of X (average squared deviation)

$$\sigma = \sqrt{3222277.7} = 1795.07$$

"Average" distance between values of X and the mean, μ

Section 5.2   Describing the Values of a Random Variable

Notice that the operation under the radical is an expected-value operation. We find the expected value of the squared deviations, the result when we multiply each possible squared deviation by its probability and then sum the products. This procedure copies $E(X)$, where we multiply each X value by its probability and then sum the products. Because finding an expected value is the same as finding an average, we have the average squared deviation, which we call the variance, σ^2:

$$\sigma^2 = E(X - \mu)^2 = \sum_{1}^{k} (x - \mu)^2 P(x).$$

Thus, the standard deviation of a random variable is the square root of the expected value of the squared deviations, or

$$\sigma = \sqrt{E(X - \mu)^2} = \sqrt{\sum_{1}^{k} (x - \mu)^2 P(x)}.$$

The following box lists the necessary steps to complete the formula and to obtain the standard deviation of a random variable. Note the similarity with finding the standard deviation from grouped data in Chapter 3.

Procedure for Obtaining the Standard Deviation of Random Variable X with k Different Values

1. Find the mean of X, which is $\mu = E(X)$.
2. Find the deviation for each value of X, by subtracting μ from each specific x, which is $(x - \mu)$.
3. Square each deviation from step 2 to produce $(x - \mu)^2$.
4. Multiply each squared deviation from Step 3 by the probability of the corresponding x value, $P(x)$, which yields $(x - \mu)^2 P(x)$.
5. Sum all of the products from Step 4 (to produce the average of the squared deviations, which is the expected value of the squared deviation, $(X - \mu)^2$). This is the variance of the random variable, $\sigma^2 = \Sigma(x - \mu)^2 P(x)$.
6. Take the square root of the sum from Step 5 to arrive at σ, the standard deviation of the random variable X.

■ **Example 5-7:** Find the standard deviation of the random variable, Profit, using the formula.

Profit	P(profit)
1,000	0.1667
3,000	0.5000
6,000	0.3333

■ **Solution:** Step 1: We already know that the mean, μ, is $E(X) = \$3,666.50$ from our earlier work.

Step 2: The third column in the following table displays the deviations for each possible value of profit.

(1)	(2)	(3)	(4)	(5)
Profit	P(profit)	(Profit − 3666.50)	(Profit − 3666.50)²	(Profit − 3666.50)² × P(profit)
1,000	0.1667	−2,666.50	7,110,222.25	1,185,274.049
3,000	0.5000	−666.50	444,222.25	222,111.125
6,000	0.3333	2,333.50	5,445,222.25	1,814,892.576
				3,222,277.7

Step 3: The fourth column shows the squared deviations.

Step 4: The fifth column shows the product of the squared deviation and the probability for the particular value of the random variable associated with each row of the table.

Step 5: The value below the fifth column is the sum of the products from step 4. This is the variance of the random variable, 3,222,277.75 square dollars.

Step 6: The square root of 3,222,222.75 square dollars is $1,795.07, the standard deviation of the profit random variable.

The table is a simple method for completing the steps of the σ formula, but we can substitute values in the formula in order to achieve the same result, as follows:

$$\sigma = \sqrt{(1{,}000 - 3{,}666.50)^2(0.1667) + (3{,}000 - 3{,}666.50)^2(0.5000) + (6{,}000 - 3{,}666.50)^2(0.3333)}$$
$$= \sqrt{3{,}222{,}277.75} = 1{,}795.07.$$

■ **Example 5-6 (continued):** Find and interpret the standard deviation of A, paint prices at store A. $E(A) = \$12.25$.

a	P(a)
9.00	0.25
10.00	0.25
15.00	0.50

■ **Solution:** The following table shows the first part of the calculation.

a	P(a)	a − μ	(a − μ)²	(a − μ)²P(a)
9.00	0.25	−3.25	10.5625	2.640625
10.00	0.25	−2.25	5.0625	1.265625
15.00	0.50	2.75	7.5625	3.781250
				7.687500

$$\sigma^2 = 7.687500 \quad \text{and} \quad \sigma = \sqrt{7.687500} = \$2.77.$$

The "average" distance between a paint price at store A and the mean price at store A is $2.77.

We can derive a shortcut formula for the standard deviation from the definitional formula. After several steps of algebra, we obtain

$$\sigma = \sqrt{E(X^2) - \mu^2}.$$

$E(X^2)$ represents the average squared value of X. Mathematically, it is

$$E(X^2) = \sum_{1}^{k} x^2 P(x)$$

■ **Example 5-7 (continued):** Use the shortcut formula to verify that the standard deviation of profit random variable is $1,795.07. The mean is $3,666.50.

Profit	P(profit)
1,000	0.1667
3,000	0.5000
6,000	0.3333

■ **Solution:** $E(X^2) = (1{,}000)^2(0.1667) + (3{,}000)^2(0.5000) + (6{,}000)^2(0.3333)$
$$= 16{,}665{,}500$$

Section 5.2 Describing the Values of a Random Variable   215

$$\sigma = \sqrt{E(X^2) - \mu^2}$$
$$= \sqrt{16{,}665{,}500 - (3{,}666.50)^2}$$
$$= \sqrt{3{,}222{,}277.75}$$
$$= 1{,}795.07.$$

Other than differences because of rounding, we get the same result with the definitional and shortcut formulas.

■ **Example 5-1 (continued):** Now, try the shortcut formula on the die-rolling random variable, D. Remember that $E(D) = 3.5$ dots. Interpret the answer.

d	$P(d)$
1	1/6
2	1/6
3	1/6
4	1/6
5	1/6
6	1/6

■ **Solution:** $E(D^2) = 1^2(1/6) + 2^2(1/6) + 3^2(1/6) + 4^2(1/6) + 5^2(1/6) + 6^2(1/6)$
$$= 15.17.$$
$$\sigma = \sqrt{15.17 - (3.5)^2} = \sqrt{2.92} = 1.71 \text{ dots}.$$

The "average" deviation of the values rolled is 1.71 dots.

■ **Example 5-6 (continued):** Find the standard deviation of B, paint prices in store B, using the shortcut formula and $\mu = \$11.90$. Interpret the result.

■ **Solution:**

b	$P(b)$	b^2	$b^2P(b)$
8.00	0.05	64.00	3.2
10.00	0.35	100.00	35.0
12.00	0.20	144.00	28.8
14.00	0.40	196.00	78.4
			145.4

$$\sigma^2 = 145.4 - (11.9)^2 = 3.79 \quad \text{and} \quad \sigma = \sqrt{3.79} = \$1.95$$

The "average" deviation between the mean and individual paint prices at store B is $1.95.

Question: The standard deviations for prices in stores A and B are $2.77 and $1.95, respectively. The restaurant owner chose to purchase paint at store B, because it had the lowest expected price. Do the standard deviation results reinforce that decision or alert the owner to proceed more cautiously?

Answer: There is more price variation in store A. So, store B's prices vary less (are more consistent) about the mean price. Prices in store A fluctuate more than prices in store B. On any given trip, there is a chance the price in A may be lower than the price in B. In fact, it may be much lower, because the standard or "average" deviation is $2.77 for store A and $1.95 for store B, although the difference in the mean prices is only $0.35 ($12.25 − $11.90). If the owner chooses one store to buy from everytime, then past pricing behavior indicates a lower average price over the long run at store B. However, the standard deviations indicate that price checking at purchase time might produce added savings.

Remember that σ is a measure of the spread or dispersion of the values about the mean, where the values are the theoretical results of repeating an experiment an infinite number of times. This measure will be important as we proceed through other topics in statistics.

Section 5.2.3 Problems

22. Find and interpret the standard deviation of the following random variable, S, the studio time required for an exercise in a television broadcasting class where students stage and tape volunteer performances.

s	P(s)	s	P(s)
30	0.15	70	0.08
40	0.20	80	0.06
50	0.22	90	0.02
60	0.27		

23. Find and interpret the standard deviation of V, the number of guests who volunteer to clean up after a party.

v	P(v)	v	P(v)
0	0.33	4	0.08
1	0.28	5	0.05
2	0.14	6	0.04
3	0.08		

24. The random variable X is the result of rolling a pair of dice. Find and interpret its standard deviation, using the following probability distribution.

x	P(x)	x	P(x)
2	1/36	8	5/36
3	2/36	9	4/36
4	3/36	10	3/36
5	4/36	11	2/36
6	5/36	12	1/36
7	6/36		

25. X represents the number of jurors on a six-member panel who vote for acquittal in a hung jury. Find and interpret the standard deviation of X.

x	P(x)
1	0.12
2	0.18
3	0.42
4	0.20
5	0.08

26. D symbolizes the number of days for the post office to deliver a letter to a private household. Find and interpret the expected value of D.

d	P(d)
1	0.15
2	0.19
3	0.24
4	0.20
5	0.09
6	0.05
7	0.03
10	0.02
14	0.01
21	0.01
30	0.01

Summary and Review

A random variable is a variable whose values correspond to the outcomes of trials of an experiment. The values can vary with each repetition, because of the uncertainty of the outcome. In essence, we assign a number to the outcomes of a statistical experiment. If we can count the possible values of the random variable (or represent them with only certain points on the number line, with gaps between the points), it is a discrete random variable. We cannot count the possible values of a continuous random variable. Often, we determine these values with measuring instruments, such as a scale, and the values can fall anywhere on the scale (or in intervals on a number line).

This unit regards values of a random variable as a set that contains an infinite number of data values, specifically as a population, the set of outcomes for every performance of the experiment. The probability distribution, expected value, and standard deviation describe the set of values, just as the familiar relative frequency distribution, mean, and standard deviation do for any other population of numerical values. These important properties are similar for both discrete and continuous random variables.

The probability distribution of a random variable is a representation of all the possible values of the random variable and the probability associated with each value. It provides information on the proportion of a large number of trials that result in each value.

The mean of a random variable, such as X, is more often called its expected value, $E(X)$. This value is the average outcome from repeating the experiment an infinite number of times. The standard deviation is an "average" deviation of the values of the random variable over infinite repetitions of the experiment. Expected values and standard deviations are useful for incorporating all available information about alternatives for comparison and decision making.

Eventually, we will estimate population characteristics based on sample data, which includes selecting the sample, an experiment with uncertain outcomes. The random variable concept and the descriptions that condense and summarize the set of possible outcomes are important for gauging the credibility of these estimates.

Multiple-Choice Problems

27. A random variable is a
  a. Random sample of quantitative values.
  b. Standard deviation of a random occurrence.
  c. Way to measure the variability of a probability distribution.
  d. Set of unknown values where the particular value to be used is chosen randomly.
  e. Variable whose values represent different outcomes of experimental trials.

28. The probability distribution of a random variable is
  a. The sum of its values multiplied by their associated probabilities.
  b. The product of the values and their associated probability sums.
  c. A representation of all its possible values and the probabilities associated with each value.
  d. A subset of the population of the observed phenomenon and its associated probabilities.
  e. A frequency distribution of all observed probabilities.

29. Someone claims that the following table is a probability distribution for a random variable, X, representing articles of clothing needing to be cleaned. Why is the claim incorrect?

Clothing	x	$P(x)$
Shoes	−1	0.05
Shirt	0	0.15
Pants	1	0.10
Hat	25	0.02

  a. The values of X must be positive.
  b. Not all possible values of X have been listed.
  c. The number of values is larger than the number of repetitions of the experiment.
  d. The values must be intervals on a number line.
  e. The probabilities must all be equal.

30. Consider the following probability distribution of the random variable of turnaround time, T, at a computer center:

t	$P(t)$
5	0.001
6	0.002
10	0.008
30	0.011
60	0.058
90	0.263
120	0.315
180	0.342

From this information, we can generalize and say
  a. The center returns most computer runs in 90 or more minutes.
  b. The probability of getting a run back in 30 minutes is twice the chances for 60 minutes.
  c. The probability of getting a run back in 60 minutes is twice the chances for 30 minutes.
  d. A short turnaround time occurs most frequently at this computer center.
  e. That most problems submitted to this computer center are simple and small, and the computer is very efficient and fast.

31. Which of the following is a continuous random variable?
  a. Potential heights of a child.
  b. Points scored in a football game.
  c. Number of congressional representatives from a state.
  d. Quantity of raffle tickets a group sells.
  e. Number of employees of a company.

X is the number of music lessons taken before a young starting student can read and play several notes unassisted. The probability distribution of X is shown below. Use this information to work Problems 32 and 33.

x	$P(x)$
5	0.5
6	0.3
9	0.2

32. Find the expected value of X.
  a. 6.7 lessons.   d. 1.0 lesson.
  b. 7.5 lessons.   e. No expected value for this problem.
  c. 6.1 lessons.

33. Find the standard deviation, σ, of X. Remember that $\mu = 6.1$ lessons.
  a. 2.3 lessons.   d. 1.4 lessons.
  b. 4.0 lessons.   e. 1.5 lessons.
  c. 4.1 lessons.

34. The expected value for the quantity of televisions sold annually by a company is 240. This statement means that
  a. The company sells 240 every year.
  b. The average annual company sales are 240, over the long run.
  c. 240 is the quantity that has the highest probability of being sold this year.
  d. If the company sells 220 this year, it must sell 260 next year.
  e. There is a 50–50 chance of selling 240 units this year.

35. A card game consists of a single draw from a standard deck. If the card is a 10, you win $52. If it is a face card (assume aces are not face cards), you win $26. Otherwise, you must pay $13. What is the expected value of this game?
  a. $91.   d. $1.
  b. $30.33. e. $65.
  c. $19.

Your sole competitor's bidding distribution for jobs of type Y is:

y = Bid (dollars)	$P(y)$
4.00	0.1
4.50	0.3
5.00	0.6

Another competitor is the only other bidder for jobs of type Z. This competitor's bidding distribution is:

z = Bid (dollars)	P(z)
4.50	0.5
5.00	0.5

Your bidding strategy is to undercut your estimate of a competitor's bid by a constant small amount, say a few cents. The resources for both jobs are different and not interchangeable but cost the same. Use this information to work Problems 36 and 37.

36. If you have time and resources to finance only one job type, which should be your specialty to make the most money over time?
 a. Job Y.   c. Neither.
 b. Job Z.   d. Does not matter which you choose.

37. Calculate the standard deviations to determine whether income fluctuates more with type Y or type Z jobs.
 a. Y jobs fluctuate more, because the standard deviation of $0.112 is larger than $0.0625 for Z.
 b. Y jobs fluctuate more, because the standard deviation of $0.335 is larger than $0.250 for Z.
 c. Y jobs fluctuate more, because the standard deviation of $0.417 is larger than $0.250 for Z.
 d. Y jobs produce more consistent income, because the standard deviation of $0.35 is smaller than $0.50.
 e. We do not need to calculate the standard deviations, because the range of possible values indicates that the standard deviation of Y will be larger than that for Z.

38. If you pay a nickel for a single draw from a standard deck and a chance to win money if you draw a club, what must you be paid when you draw a club to make the game fair?
 a. 20 cents.   d. 35 cents.
 b. 15 cents.   e. Nothing.
 c. 30 cents.

39. A probability distribution is similar to which descriptive statistic for a regular data set?
 a. Randomly distri-   h. Long-run relative
    buted set of values.   frequency.
 b. Mean.   i. Interquartile range.
 c. Quartile.   j. Percentile.
 d. Frequency distribution.   k. Range.
 e. Mode.   l. Mean absolute deviation.
 f. Standard deviation.   m. Independence.
 g. Median.   n. Average deviation.

40. The expected value of a random variable corresponds to which descriptive statistic for a regular data set?
 a. Randomly distri-   h. Long-run relative
    buted set of values.   frequency.
 b. Mean.   i. Interquartile range.
 c. Quartile.   j. Percentile.
 d. Frequency distribution.   k. Range.
 e. Mode.   l. Mean absolute deviation.
 f. Standard deviation.   m. Independence.
 g. Median.   n. Average deviation.

The following tables are probability distributions for the random variables average faculty salary in 1990–1991 for selected fields (rounded and expressed in thousands of dollars) for public (*PUB*) and private (*PR*) institutions, respectively. Use this information to work Problems 41–48.

pub	P(pub)	pr	P(pr)
35	0.019	28	0.019
36	0.038	32	0.019
37	0.170	33	0.077
38	0.094	34	0.038
39	0.113	35	0.096
40	0.075	36	0.115
41	0.094	37	0.058
42	0.132	38	0.096
43	0.075	39	0.058
44	0.038	40	0.077
45	0.075	41	0.096
46	0.019	42	0.038
47	0.019	44	0.058
48	0.019	45	0.038
49	0.019	47	0.058
		48	0.019
		49	0.019
		55	0.019

Source: "Fact File: Average Faculty Salaries by Rank in Selected Fields, 1990–91," *The Chronicle of Higher Education*, May 22, 1991, p. A14.

41. What is the probability that the average salary in a discipline at a private school will exceed $40,000?
 a. 0.578.   d. 0.655.
 b. 0.422.   e. 0.500.
 c. 0.345.

42. What is the probability that the average salary in a discipline at a public school will exceed $40,000?
 a. 0.509.   d. 0.566.
 b. 0.490.   e. 0.625.
 c. 0.434.   f. 0.688.

43. Find $E(PUB)$.
 a. $42,000.   d. $27,055.
 b. $28,000.   e. $41,000.
 c. $40,600.

44. Find $E(PR)$.
 a. $40,167.   c. $39,500.   e. $39,000.
 b. $22,315.   d. $21,676.

45. Use the answers to Problems 43 and 44 to choose the type of school, public or private, at which to interview to maximize expected income for a professor making a career decision.
 a. Interview private schools, because $39,000 < $40,600.
 b. Interview public schools, because $40,600 > $39,000.
 c. Interview half of one and half of the other, because the result is larger than either type by itself.
 d. There is no difference between the two types of schools.
 e. Interview private schools, because the largest possible value is from this type of school.

Word and Thought Problems

46. Find the standard deviation of the public school salaries.
  a. $14,000.   d. $3,300.
  b. $933.3.   e. $3,216.5.
  c. $4,000.

47. Find the standard deviation of the private school salaries.
  a. $27,000.   d. $5,166.7.
  b. $5,000.   e. $5,500.
  c. $1,500.

48. What do the standard deviations indicate about the earlier decision to interview public schools?
  a. The difference in expected values is small compared to the standard deviations of both random variables, so the professor should investigate further, perhaps discipline-specific information.
  b. The public school salaries are more consistent about their mean than the private school salaries, assuring a higher salary in the public school sector.
  c. The public school salaries fluctuate more than the private school salaries, making their salaries even lower than expected.
  d. The standard deviations do not provide additional information that is helpful, because the expected value incorporates all of the given information.
  e. Comparing the standard deviations is a better choice of criteria, because the expected value formulation is best suited to decisions that must be repeated, which is not likely for a professor making a career change.

Word and Thought Problems

49. What is a random variable? What is a probability distribution? What is the sum of the probabilities of all possible values of a random variable?

50. Cholesterol test results report an integer value to indicate the amount of cholesterol in the blood system. When considering possible readings from a large group, such as those served by a county health system, is the reading a random variable? If so, is it discrete or continuous?

51. To estimate the size of the wild boar herd in a mountainous region, the ranger randomly selects sites for positioning talliers, people who seclude themselves and count the boars for a specified amount of time. The total from all of the different sites estimates the herd size. Is the set of site totals a sample or a population? Is the set of possible site count values a random variable? If so, is it discrete or continuous? Is this process descriptive or inferential statistics?

52. What is the formula for finding the expected value of X, $E(X)$? What information does $E(X)$ relate?

53. Suppose the probability distribution for R, July rainfall (in inches) at a spot in eastern Montana, is as follows:

r	$P(r)$
0.0	0.40
0.5	0.30
1.0	0.15
1.5	0.10
2.0	0.05

What is the expected rainfall for this spot in July? Is there much variation about the average? Explain.

54. Tuesday sales, S, in a department store have the following probability distribution:

s	$P(s)$
10,000	0.30
15,000	0.25
18,000	0.25
19,000	0.12
20,000	0.08

What is the mean of this random variable? What is σ? Interpret the numerical values for the store manager.

55. A change in interest rates will affect your business operations. A forecaster suggests the probability is 0.02 that rates will drop by 5%, 0.18 that rates will drop by 2%, 0.5 that rates will not change, and 0.3 that rates will increase by 1%.
  a. What is the expected change in interest rates?
  b. If interest rates are 12% now, find the expected level of interest rates.

56. A game is played in the following manner: A player selects one card from a standard deck and receives $1 if the card is a heart; otherwise, the player pays the dealer 40 cents. What are the expected winnings for the player? Interpret this number by using "long-run average" terminology. What is the expected value for the dealer? Is the game fair?

57. Another game is played as follows: The player throws a fair die once. The outcome, X, is used to calculate Y:

$$Y = \begin{cases} \dfrac{X}{6-X} & \text{if } X \leq 3. \\ \dfrac{X}{6+X} & \text{if } X > 3. \end{cases}$$

Note: Y is a fraction of a dollar. If $X \leq 3$ on the toss, the player receives Y. If $X > 3$, the dealer receives Y. What is the expected value of this game for the player? Would you rather be the player of this game or the one described in Problem 56?

58. The following probability distribution for X, the intensity of an earthquake in the central region of the United States, stems from records from 1811 through 1980. The X values correspond to ratings on the Modified Mercalli Intensity Scale (described in Example 5-3). Determine and interpret the average intensity and the standard deviation of the intensities.

x	P(x)
4	0.0038
5	0.5556
6	0.2989
7	0.1149
8	0.0153
12	0.0095

Source: Jerry L. Coffman, Carl A. von Hake, and Carl W. Stover, eds., *Earthquake History of the United States*, Publication 41-1, rev. ed. (through 1970), reprt. 1982 with suppl. (1971–1980), U.S. Department of Commerce, National Oceanic and Atmospheric Administration, and U.S. Department of the Interior, Geological Survey, Boulder, Colorado, 1982, pp. 37–43, 11a.

59. A wheel of values in a quiz game consists of the following question values: −$100, $0, $100, $500, $1000. The number of times each value appears on the wheel are 3, 4, 5, 6, and 1, respectively. The wheel is spun to determine the question value. What is the expected question value? What is the standard deviation of the question values? (Round probabilities to the nearest hundredth.) Can players expect to spin consistently close to the expected value?

60. The winning proportions of teams in the NCAA Probation Database in Appendix D (rounded to the nearest tenth and expressed as a percent) produce the following probability distribution for the random variable X, representing these winning percentages.

x	P(x)
30	0.0588
40	0.1765
50	0.3529
60	0.2471
70	0.1294
80	0.0353

a. If you choose a team randomly for an opponent, what is the probability the teams record is better than 50% wins?
b. What is $E(X)$? Interpret the result.
c. What is σ for X? Interpret the result.
d. Can any of the X values be considered outliers? Explain.

61. An obstetrician's records indicate that 42% of his problem-free baby deliveries occur one week earlier than their due dates, 45% on time, 8% half a week late, and 5% a week late. What is the expected (or average) adjustment to the doctor's prediction necessary to obtain a better prediction of the due date? (Use a minus sign for early.)

62. Roger has been purchasing the same brand of spark plugs for a long time. His experience has been that one-fourth of them must be replaced after 3 months, one-third after 6 months, one-third after 9 months, and the rest after 12 months. State the probability distribution for time before replacement, then find the mean time before replacement. Find the standard deviation of time before replacement.

63. A manufacturer can purchase a certain part from two suppliers. Supplier A delivers two weeks early 5% of the time, one week early 23% of the time, on time 50% of the time, a week late 17% of the time, and 1.5 weeks late 5% of the time. Supplier B delivers 3 weeks early 8% of the time, 1.5 weeks early 10% of the time, 0.5 week early 40% of the time, on time 40% of the time, and 0.5 week late 2% of the time. Use expected values to make the following decisions. If it is important to have the part as soon as possible, which supplier should the manufacturer choose? If early and late arrivals cause bottlenecks in production, so that on-time delivery is best, which supplier should be chosen?

64. Random variables D and R represent the percent of the voting-age population that voted in the 1988 Democratic and Republican presidential primaries, respectively, for states holding primaries. The following tables display their probability distributions. Consider these results applicable to future elections, and answer the following questions.

d(%)	P(d)	r(%)	P(r)
1	0.028	2	0.056
6	0.028	4	0.111
7	0.028	5	0.083
8	0.028	6	0.056
10	0.028	7	0.056
11	0.028	8	0.139
12	0.083	9	0.083
13	0.056	10	0.194
14	0.167	11	0.056
15	0.083	12	0.028
16	0.111	13	0.028
17	0.083	15	0.028
18	0.056	17	0.028
19	0.056	18	0.028
20	0.028	19	0.028
21	0.028		
24	0.028		
28	0.028		
29	0.028		

Source: Committee for the Study of the American Electorate, "Non-Voter Study, '88–'89," Washington, D.C., as reported in U.S. Bureau of the Census, *Statistical Abstract of the United States: 1990*, 110th ed. (Washington, D.C.: 1990), p. 266. Turnout percentages are rounded to the nearest percent.

a. Which party has the larger expected state turnout for the presidential primary?
b. Which party's values fluctuate more?
c. Several Democrats and only a few Republicans ran in 1988. Given this information, do the relative magnitudes of expected values and standard deviations seem reasonable for the two random variables?

65. A reusable package of freezable material, for use in picnic coolers, is supposed to remain frozen for 17 hours after it is placed in a cooler. A camping association survey included a question about experience with the

Word and Thought Problems 221

product. The results are: 18% said it stayed frozen for about 12 hours, 20% said 15 hours, 55% said 18 hours, and 7% said 24 hours.
   a. What is the expected frozen time, based on this information?
   b. Another brand of the product is also the subject of a survey question. The results are that 38% said the product stayed frozen for 15 hours, 60% said 18 hours, and 2% said 24 hours. Your objective is to keep the product frozen for 17 hours. Which is the better brand if you use expected frozen time to make your decision?

66. Soft drinks are frequently on sale to attract customers to grocery stores. Suppose store A sells Miller's favorite brand for $1.98, 4% of the time; $1.78, 50% of the time; $1.49, 38% of the time; and their regular price, $2.29, the rest of the time. Store B sells the same brand at their regular price, $2.10, 75% of the time, but the other 25% of the time, the price is $1.29. If Miller make up his mind to buy his soft drinks at only one store when making spur-of-the-moment purchases, with no time for comparison shopping, where will he spend less money over the long run?

67. Management is comparing two machines to determine which one to purchase for the production line. An output close to 450 units is most important for avoiding bottlenecks in the process, which occur when there are too few or too many units at this point on the line. Use the following information where A is the output of machine A and B is the output from machine B and expected values to justify your choice of a machine.

a	P(a)	b	P(b)
400	0.40	400	0.1
440	0.25	420	0.1
480	0.25	440	0.3
500	0.10	480	0.2
		500	0.3

68. A lottery is conducted for the men's NCAA basketball tournament. The 64 team names are written on slips of paper, folded so no name shows, and then placed in a fishbowl. For $1, a person gets to choose a slip from the bowl. That person receives the $64 pool, if the team on that slip of paper eventually wins the tournament.
   a. What is the expected value of this lottery?
   b. What is the probability that the same person will win the lottery two consecutive years?
   c. If the players each agree to pay an extra dollar to the winner the second year, if the same person won the previous year, what is the expected value of the lottery to a player over the two-year period? Show the probability distribution that you use in the solution.

69. A machine produces tables with five legs, but not always perfectly. In a lot of 1,000 tables, the following combinations of defects have been observed:
   50 tables have defect A, only four legs.
   25 tables have defect B, hole in table top.
   10 tables have defects A and B (these are included in the groups of 50 and 25).
   Define the random variable, X, in the following way:
   $X = 10$, if A occurs and B does not.
   $= 50$, if B occurs and A does not.
   $= 30$, if both A and B occur.
   $= 100$, if neither A nor B occurs.
   a. Make a table showing the probability distribution associated with X.
   b. Find $P(X \geq 25)$.
   c. Find $P(X \geq 25$ or $X < 10)$.
   d. What is $E(X)$?
   e. What is the mean of X?
   f. What is the standard deviation of X?

Unit II: The Binomial Random Variable

Preparation-Check Questions

70. If A and B are independent events, then we know that A
   a. And B cannot both occur the same proportion of times in an experiment.
   b. Cannot occur simultaneously with B.
   c. Does not affect the chance that B will occur.
   d. Is the negation of B.
   e. Has the same probability of occurring as B.

71. Which of the following pairs of outcomes are mutually exclusive?
   a. Obtaining a 4 and a 6 on one toss of a die.
   b. Obtaining a red card and a jack on one draw from a standard deck.
   c. Residing in New York City and New York State.
   d. Advertising a product on television and radio.
   e. Costs of $1,000 and revenue of $500 for one month.

72. Fill in the blanks associated with each line of the following statement. If $A, B, C, D, \ldots, Z$ are _____ events,
   then $P(A\_\_B\_\_C\_\_D\_\_\cdots\_\_Z) = P(A)\_\_P(B)\_\_P(C)\_\_P(D)\_\_\cdots\_\_P(Z)$.

(continued on next page)

	Line-1 blank	Line-2 blanks	Line-3 blanks
a.	Independent	and	×
b.	Independent	and	+
c.	Independent	or	×
d.	Independent	or	+
e.	Mutually exclusive	and	×
f.	Mutually exclusive	and	+
g.	Mutually exclusive	or	×
h.	Mutually exclusive	or	+

i. Both choices a and h will correctly fill the blanks.
j. Both choices b and g will correctly fill the blanks.
k. Both choices c and f will correctly fill the blanks.
i. Both choices d and e will correctly fill the blanks.
j. None of the choices a through h will correctly fill the blanks.

73. What is the probability that when you toss a fair die three times, you will get a 2 on the first toss, a 5 on the second toss, and anything but a 5 on the last toss?
    a. $(1/6)(1/6)(1/6)$.
    b. $(1/6) + (1/6) + (1/6)$.
    c. $(1/6)(1/6)(5/6)$.
    d. $(1/6) + (1/6) + (5/6)$.
    e. $(1/6) + (1/6) + (5/6) - (1/6)(1/6)(5/6)$.

74. A representation of all possible values of a random variable and their associated probabilities is called a(n)
    a. Expected value.
    b. Range.
    c. Frequency distribution.
    d. Probability distribution.
    e. Standard deviation.

Use the following information to answer Questions 75–77. The random variable, X, represents the number of calls per day by the truant officer in a local school.

x	$P(x)$
1	0.06
2	0.12
10	0.82

75. A reasonable value for the mean of X would be
    a. Between 0 and 10.
    b. Between 1 and 10.
    c. Between 0 and 1.
    d. Between 0.1 and 0.8.
    e. Between 0 and 8.5.

76. $\mu = E(x) = 8.50$, so a reasonable value for the standard deviation of X would be
    a. Any positive value.
    b. Between -10 and $+10$.
    c. Greater than 0 but less than 7.5.
    d. Greater than 0 but less than 1.5.
    e. Greater than 1.5 but less than 7.5.

77. Find the standard deviation of X.
    a. 8.5.   c. 9.   e. 3.2.
    b. 7.5.   d. 80.9.

Answers: 70. c; 71. a; 72. i; 73. c; 74. d; 75. b; 76. e; 77. e.

Introduction

Committee Composition Mr. Shippey, school superintendent over a district with four high schools, wants to form a committee to study school policies and practices that prolong racial confrontations. He wants a small committee (one student from each high school) that will promote informal and frank discussion.

To avoid any sign of favoritism, he wants to select the four students randomly, but he also wants minorities represented on the committee. Of course, one way to ensure minority representation is to decide the minority/majority composition of the committee, such as two from each group, and choose the students randomly from each group. However, minorities make up only 20% of the student body in each high school, so a committee with two members of each group is not exactly representative. Because Mr. Shippey wants to give every student an equal chance of selection for the committee, he still wants to select randomly from each high school.

Before selecting randomly he wants to know the likelihood of different committee compositions if he employs this method. Hence, he needs to know the probability of selecting 0 minorities out of 4 students, 1 out of 4, 2 out of 4, 3 out of 4, and, finally, 4 out of 4.

If finding these five probabilities does not seem like enough work, consider just one of the problems, say the probability that 2 out of the 4 on the committee will be minorities.

There are a number of possible selection sequences that would result in two minorities (six in fact). They could be the first two, the last two, the first and third, or some other combination. Each sequence represents a new probability problem. The following table shows the number of sequences or arrangements for each number of minority students on the committee:

Number of Minorities on the Committee	Number of Arrangements
0	1
1	4
2	6
3	4
4	1
	16

There are a grand total of 16 problems involved in finding the probabilities for 0, 1, 2, 3, and 4 minorities on the committee. The good news is that there is an easier way to do this. The binomial formula condenses the arrangement problems so that we simply find the probability of a specified number of committee members being minority students.

5.3 The Binomial Formula

5.3.1 Assumptions

Many statistical problems analyze situations similar to the committee-composition situation. If a method for solving this problem applies to other like problems, then we can save time by duplicating the general method rather than reworking each problem from scratch. The subject of this unit, the binomial formula, is a simple way to solve problems like the committee-composition problem. For this formula to apply to a situation, certain conditions must be true.

In the committee-selection situation two experiments occur. The first experiment is to determine the minority status of a student selected for the committee. We call such an experiment a *Bernoulli trial,* an experiment with only two possible outcomes, which we designate as *success* and *failure.* A success is simply the particular outcome on which we choose to focus. Calling it a success seems ironic at times. For instance, treating a patient with a medication may cure or kill the patient. If we focus on the number who die, then a success means a patient dies. In the committee-selection problem, a success is to select a minority student, because the superintendent is interested in the number of minority students on the committee. The probability associated with a success is p, so the probability of a failure must be $1 - p$. The probability of a success in the committee-selection problem, selecting a minority, is 0.20, because 20% of each high school's students are minorities. The probability of failure, selecting a nonminority student, must be $0.8 = 1 - 0.2$. Figure 5-6 depicts the general situation of a Bernoulli experiment and compares it with the specific committee selection situation.

If there are more than two possible outcomes in a situation, we must be able to group them into two outcomes to form a Bernoulli trial. For example, if we designated students as African-American, Hispanic, Asian, and Caucasian, we can combine the first three groups into the event "minority" versus the other event "nonminority."

The second experiment is to count the number of successes that occur in a series of n independent and identical Bernoulli trials. Identical trials have a common value for p, and the same definition of a success. We will call such an experiment a *binomial experiment.*

Figure 5-6
Bernoulli trials

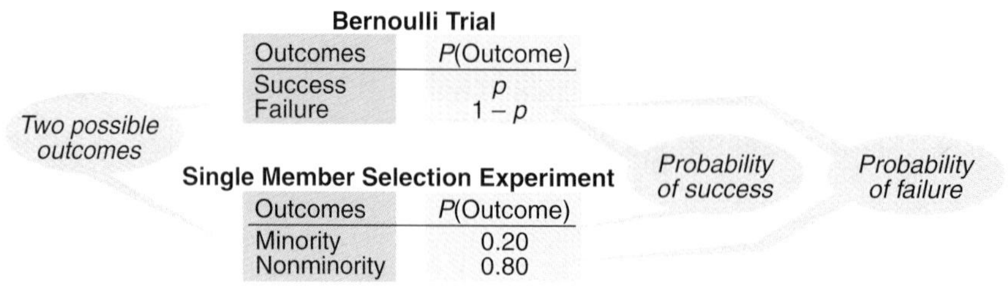

Bernoulli Trial

Outcomes	P(Outcome)
Success	p
Failure	1 − p

Two possible outcomes

Single Member Selection Experiment

Outcomes	P(Outcome)
Minority	0.20
Nonminority	0.80

Probability of success

Probability of failure

■ **Example 5-8:** Show that randomly selecting four committee members and counting the number of minority students selected is a binomial experiment.

■ **Solution:** 1. We have already demonstrated that each selection is a Bernoulli trial. We repeat the trial four times, $n = 4$.
2. The selections are independent trials, because the outcome of one trial has no effect on the outcome of any of the other trials. Random selection ensures this.
3. They are identical, because we conduct the exact same experiment each time. We do not change the experiment on any trial. For instance, we do not check to find if the selection is Hispanic or not Hispanic instead of minority or nonminority at one high school. Also p is 0.20 in each school, because minorities make up 20% of each high school's students.

Figure 5-7
The relation between Bernoulli and binomial experiments

Bernoulli Experiment:
Determine success or failure on single trial

p = probability of success on single trial

Outcomes	P(Outcomes)
Minority	0.2
Nonminority	0.8

Binomial Experiment:
Count number of successes in n independent, identical Bernoulli trials with a common value for p and the same definition for "success"

Number of successes

Probability of x successes in n trials

x	P(x)
0	0.4096
1	0.4096
2	0.1536
3	0.0256
4	0.0016

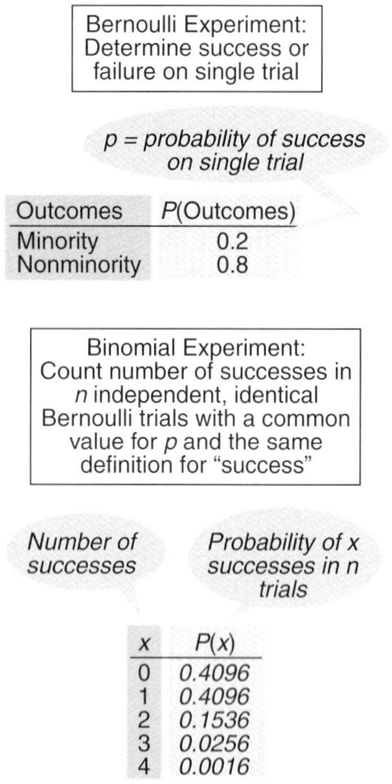

We let X represent the number of successes in the n trials. For instance, $x = 0$ corresponds to no minority students being selected for the committee, and $X = 3$ means that the superintendent selects three minority students out of the four selections. $P(x)$ is the proportion of long-run trials of the binomial experiment that result in x successes or the proportion of binomial trials that result in x successes in the n Bernoulli trials. This probability associated with a particular value of X is usually stated as the *"probability of x successes in n trials."* Figure 5-7 illustrates the connection between the Bernoulli and binomial experiments. We will demonstrate the probability computations shortly. For now note from the Bernoulli experiment that about 20% of individual selections (over a large number of trials) would be minorities. From the probability distribution of X, the result of the binomial experiment, we know that about 15.36% of the time we would obtain a four-member committee with two minority students.

Question: Compare the values of p and $P(X = 0)$.
Answer: p remains 0.2, because the trials are identical, 20% are minorities in each of the four schools. $P(X = 0) = 0.4096$, about 41% of randomly selected, four-member committees would have no minority student members.

If the values of a random variable, X, represent the outcomes of a binomial experiment, then X is called a *binomial random variable*.

Question: Is a binomial random variable discrete or continuous?
Answer: It is discrete, because we can count the possible values of X. There are $n + 1$ possible values, the integers that begin at 0 and end at n. For the committee-selection problem, there are five possible values, 0, 1, 2, 3, and 4.

In short, in order for the random variable X to be binomial (or to use the binomial formula, presently) a situation must meet the assumptions detailed in the following box.

Section 5.3   The Binomial Formula   **225**

> To find the probability of X successes by using the binomial formula, the following assumptions must hold:
> 1. There are n trials of the experiment.
> 2. There are only two possible outcomes of each trial, success or failure.
> 3. Outcomes of each trial are independent.
> 4. The probability of success, p, is the same for each trial.

Figure 5-8 illustrates these four conditions for the committee-selection situation.

■ **Example 5-9:** Before we develop and use the formula, practice determining if a situation fits these assumptions. Answer yes if it does and no if it does not after each description. If it does, determine the values of n, X, and p.
a. What is the probability of 90 heads out of 200 tosses of a coin?
b. Customers can choose from five colors. What is the probability that eight of the next ten customers will choose red and two will choose green?
c. What is the probability it will rain on 15 of the next 20 days?

■ **Solution:** a. Yes. $n = 200$, $X = 90$, and $p = P(\text{head on one toss}) = 1/2$.
           b. No. There are more than two possible outcomes (color) for each trial (a customer chooses a color). We can make this binomial by asking for the probability that eight of the next ten choose red (and two do not choose red) if the probability of choosing red remains the same for each customer and is independent of other customers' choices.
           c. No, because the weather on one day is not always independent of the weather on previous days.

Figure 5-8
The four assumptions for binomial problems:
$P(x)$ = probability of x successes in n trials

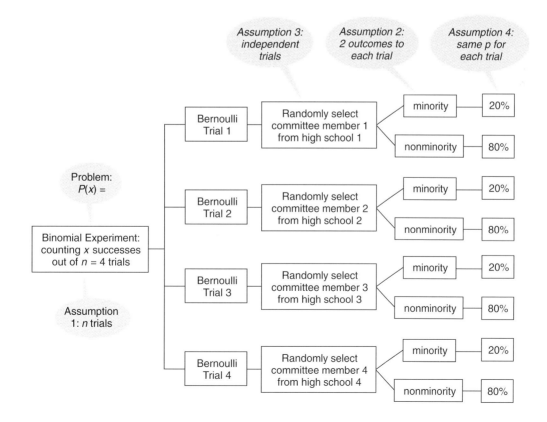

Frequently, events are dependent at least theoretically, because of the interdependence of most behavioral and physical phenomena, such as the rain or customers choosing colors. We must use our judgement and whatever evidence is available in such cases to determine if we can safely assume that the influence is very small or nonexistent for practical purposes, that is, the outcome of one trial should not affect the probability of a success on any other trial.

Reminder: a note of caution: There are two probabilities involved in binomial situations. There is p, the probability of success on one trial (such as the probability of a head on one toss of a coin). Then, there is the probability of X successes in n trials (the probability of 90 heads in 200 tosses). Sometimes, dual probabilities in the same problem confound the issue, so take a moment to study these differences and distinguish them when working a problem.

5.3.2 Formula

The solution to the specific committee-selection problem generalizes to the binomial formula. To illustrate this, we will find the probability of the event that three members of the four-member committee are minority students. Being a minority student is a success, being a nonminority student is a failure, so we want to find the probability that $X = 3$, $P(3)$. For the Bernoulli trial we observe the minority status of a randomly selected student from a high school where 20% of the students are members of a minority group, $p = 0.2$. We conduct four of these trials, so $n = 4$. The minority status of the selected students is independent; because they are randomly selected, one selection has no impact on the minority status of the other selections.

The left half of Figure 5-9 shows that there are only four possible ways in which the event three successes in four trials can occur. S represents a success, being a minority student, and F is a failure, being a nonminority student. To ensure that the sequence occurs as shown, three minority students out of four, we must connect the symbols in each sequence with the word *and*. Then, we connect each of the four sequences with the word *or*, because any one of them will satisfy the event, three minority students out of four. The extended statement on the left is a large-size compound event compared to our usual ones, such as "A and B." Think how the statement would appear if there had been a committee of 20 or 200! Nevertheless, we want the probability of this compound event:

$$P[(S \text{ and } S \text{ and } S \text{ and } F) \text{ or } (S \text{ and } S \text{ and } F \text{ and } S) \text{ or }$$
$$(S \text{ and } F \text{ and } S \text{ and } S) \text{ or } (F \text{ and } S \text{ and } S \text{ and } S)].$$

Figure 5-9

The four ways to obtain three successes in four trials

S = success = being a minority student
F = failure = being a nonminority student

	High schools	Probability of each sequence
	1 2 3 4	1 2 3 4
Sequence 1	S and S and S and F	0.2 x 0.2 x 0.2 x 0.8 = 0.0064
	or	*mutually exclusive*
Sequence 2	S and S and F and S	+ 0.2 x 0.2 x 0.8 x 0.2 = 0.0064
	or	
Sequence 3	S and F and S and S	+ 0.2 x 0.8 x 0.2 x 0.2 = 0.0064
	or	
Sequence 4	F and S and S and S	+ 0.8 x 0.2 x 0.2 x 0.2 = 0.0064
	independent	
		0.0256

Probability of 3 successes in 4 trials = 4 x 0.0064

Section 5.3 The Binomial Formula

Question: To begin solving this problem, compare the compound events enclosed in parentheses, say, (S and S and S and F) versus (S and F and S and S). Are these two outcomes mutually exclusive?

Answer: Yes, because if students from high schools 1, 2, and 3 are minorities and the other is not, it cannot be true that students from high schools 1, 3, and 4 are minorities and the one from school 2 is not.

All four of the sequences represent different mutually exclusive events. Thus, applying rule 2''' from Chapter 4, we replace the *or*s with plus signs and find the sum of the probabilities of each sequence:

$$P[(S \text{ and } S \text{ and } S \text{ and } F) + (S \text{ and } S \text{ and } F \text{ and } S)$$
$$+ (S \text{ and } F \text{ and } S \text{ and } S) + (F \text{ and } S \text{ and } S \text{ and } S)],$$

as shown on the right side of Figure 5-9. There are four separate probability problems now.

Question: What is the next step in solving any one of the four problems with events connected with *and*?

Answer: The next step for an *and* probability problem is to determine if the results of each trial are independent. We said earlier that they are, because each selection is a random draw.

Because trial results are independent and are connected with the word *and*, we apply rule 3'' from Chapter 3 to restate the probability of the first sequence as $P(S) \times P(S) \times P(S) \times P(F)$. We know that the probability of a success on a single trial, $P(S)$, is $p = 0.2$, because 20% of the students in each school are minorities.

Question: What is $P(F)$?

Answer: F is the same as the negation of S, S', so by rule 1 from Chapter 3, we obtain $P(F) = 1 - P(S) = 1 - p = 1 - 0.2 = 0.8$.

Substituting the 0.2 for $P(S)$ and 0.8 for $P(F)$ for the first sequence yields:

$$P(S) \times P(S) \times P(S) \times P(F) = 0.2 \times 0.2 \times 0.2 \times 0.8 = 0.0064,$$

which is the restatement of the first sequence in Figure 5-9 as shown in the first line of the right side of the diagram. Continuing the restatement process for the other three sequences produces the right side of the figure.

When we complete the solution by summing the probabilities for the four sequences, we obtain 0.0256. Thus, the probability that three of the four students will be minorities is 0.0256. Notice some other details about the problem.

Question: What is the probability that any one of the four sequences will occur?

Answer: This probability is 0.0064. The particular sequence does not matter, because we always find the probability by multiplying the same values, and there are three 0.2 factors and one factor of 0.8 in each of these four sequences. Because the order of multiplication makes no difference in the product, the product of these four factors is $(0.2)^3(0.8)^1$ for each sequence.

In the general case of X successes in n trials, we would have n factors to multiply for any one sequence. Remember that $p = P(S)$ for one trial and $1 - p$ is $P(F)$. There would be X of the p factors for the successes.

Question: How many failures must there be?

Answer: $n - X$. If there are $n = 50$ trials and $X = 20$ successes, then there must be $n - X = 50 - 20 = 30$ failures.

Thus, we obtain the product of X of the p factors and $n - X$ of the $1 - p$ factors for the probability of one sequence, or

$$\underbrace{(p)(p)\cdots(p)}_{X \text{ successes}} \underbrace{(1-p)(1-p)\cdots(1-p)}_{n - X \text{ failures}} = p^X(1-p)^{n-X}.$$

In the committee-selection situation there are three of the 0.2 factors and one (from $n - X = 4 - 3$) of the 0.8 factors, so the probability for any one sequence is $(0.2)^3(0.8)^1$.

Thus, we only need to compute the probability for one sequence in order to know the result for each of the other sequences. Notice now from Figure 5-9, that there are four possible sequences of three successes in four trials and $4(0.0064) = 0.0256$, the solution to $P(2)$. In general, if we determine the number of sequences of X success in n trials, then we can find the probability of X successes in n trials by multiplying the probability for any one sequence by the number of sequences, $4(0.0064) = 0.0256$ in the committee example.

The easy way to determine the number of sequences of X successes in n trials is to compute

$$\frac{n!}{X!(n-X)!}$$

where $n!$ means to multiply n by all the positive integers smaller than n. For example,

$$5! = 5 \times 4 \times 3 \times 2 \times 1 = 120.$$
$$2! = 2 \times 1 = 2.$$

In general, $n! = n \times (n-1) \times (n-2) \times \cdots \times 3 \times 2 \times 1$ and $0!$ is defined to be 1. Only the use of this computational formula for the number of sequences of X successes in n trials is explained here. For more explanation of its development and logic, see Appendix F. Calculators with a factorial key (!) or a key to compute the number of sequences by using the formula (sometimes labelled $_nC_x$) accelerate this calculation.

Question: What is 4!?
Answer: $4 \times 3 \times 2 \times 1 = 24$.

In the committee example we consider three successes in four trials, so $n = 4$ and $X = 3$. Consequently, the number of sequences of three successes in four trials is

$$\frac{4!}{3!(4-3)!} = \frac{4!}{3!(1!)} = \frac{4 \times 3 \times 2 \times 1}{(3 \times 2 \times 1)(1)} = \frac{4 \times \cancel{3} \times \cancel{2} \times \cancel{1}}{\cancel{3} \times \cancel{2} \times \cancel{1}} = 4.$$

These are the four sequences we labored with earlier. Imagine the effort this factorial process can save when n becomes much larger.

Question: How many ways can there be two successes in five trials?
Answer: $n = 5, X = 2$. So

$$\frac{5!}{2!(5-2)!} = \frac{5!}{2!(3!)} = \frac{5 \times 4 \times 3 \times 2 \times 1}{(2 \times 1)(3 \times 2 \times 1)} = \frac{5 \times \overset{2}{\cancel{4}} \times \cancel{3} \times \cancel{2}}{\cancel{2} \times \cancel{3} \times \cancel{2}} = 10.$$

They would be:

```
S  S  F  F  F        F  S  F  S  F
S  F  S  F  F        F  S  F  F  S
S  F  F  S  F        F  F  S  S  F
S  F  F  F  S        F  F  S  F  S
F  S  S  F  F        F  F  F  S  S
```

Section 5.3   The Binomial Formula   ■ **229**

Question: How many ways can there be five successes in five trials?
Answer: $\dfrac{5!}{5!(5-5)!} = \dfrac{5!}{5!\,0!} = \dfrac{5!}{5!} = \dfrac{\cancel{5!}}{\cancel{5!}} = 1.$

Remember that $0! = 1$. This one sequence is $S\,S\,S\,S\,S$.

The binomial formula to compute the probability of x successes in n trials results from multiplying the probability for one sequence by the number of possible sequences. In the general case, we have

$$P(x) = \begin{bmatrix} \text{number of sequences of} \\ x \text{ successes in } n \text{ trials} \end{bmatrix} \times \begin{bmatrix} \text{probability of } x \text{ successes} \\ \text{in } n \text{ trials for one} \\ \text{particular sequence} \end{bmatrix}$$

Thus, the *binomial formula* is as follows.

> Binomial formula to find the probability of x successes in n independent, identical Bernoulli trials where the probability of success on each trial is p:
>
> $$P(x) = \left[\dfrac{n!}{x!\,(n-x)!}\right] p^x (1-p)^{(n-x)}.$$

For the committee example, $n = 4$, $X = 3$, $p = 0.2$, so

$$P(2) = \dfrac{4!}{3!\,(4-3)!}\,(0.2)^3 (1-0.2)^{4-3}$$
$$= 4(0.2)^3 (0.8)^1$$
$$= 0.0256.$$

Again, a calculator is very helpful when X becomes large. The "y^x" key allows us to find quickly the Xth power of y.

■ **Example 5-10:** What is the probability that one out of the four students will be a minority? All four of the four?

■ **Solution:** For one out of four, we have $n = 4$, $X = 1$, $p = 0.2$

$$P(1) = \dfrac{4!}{1!\,(4-1)!}\,(0.2)^1 (1-0.2)^{4-1}$$
$$= 4(0.2)^1 (0.8)^3$$
$$= 0.4096.$$

For four out of four, we have $n = 4$, $X = 4$, and $p = 0.2$

$$P(4) = \dfrac{4!}{4!\,0!}\,(0.2)^4 (0.8)^0$$

Remember that $p^0 = 1$. Any non-zero value raised to the zero power is 1. Also $0! = 1$.

$$= 1 \times (0.2)^{4 \times 1}$$
$$= (0.2)^4$$
$$= 0.0016.$$

Question: Distinguish between $p = 0.2$ and $P(4) = 0.0016$.
Answer: $p = 0.2$ is the probability that any one student selection will be a minority. (Try solving with $n = 1$, $X = 1$, and $p = 0.2$.) $P(4) = 0.0016$ is the probability that all four of the four selections are minorities.

Example 5-11: Use the binomial formula to find the probability of getting two heads in three tosses of a fair coin.

Solution: $n = 3$, $X = 2$, $p = 0.5$.

$$P(2) = \frac{3!}{2!(3-2)!}(0.5)^2(1-0.5)^{3-2}$$
$$= 3(0.25)(0.5)$$
$$= 0.375.$$

Example 5-12: According to a survey conducted by the Roper Organization, 39% of all wine drinkers are Democrats, and another 47% are Republicans. What is the probability that an inquisitive bartender will find that five of 20 customers, randomly selected from those served wine in a large nightclub over several weekends, will be Democrats? Justify the use of the binomial formula. (Source: Marvin R. Shanken, and Thomas Matthews, "What America Thinks about Wine," *The Wine Spectator*, February 28, 1991, p. 25.)

Solution: To justify the formula, we must demonstrate that the situation satisfies the four conditions, n independent, identical Bernoulli trials.

1. We want to find the probability of five successes out of 20 trials, so $n = 20$ and $X = 5$.
2. There are two outcomes on each trial. There are more political choices for any wine drinker, because the two parties account for only 86% of wine drinkers' political affiliations. However, the question only concerns Democrats and non-Democrats.
3. Outcomes from quizzing the wine drinkers are independent, because the bartender randomly selects interviewees from a large set of wine drinkers.
4. The probability any wine drinker will be a Democrat is $p = 0.39$ according to the poll.

Substituting $n = 20$, $X = 5$, and $p = 0.39$ in the formula, we obtain

$$P(5) = \frac{20!}{5!(20-5)!}(0.39)^5(1-0.39)^{20-5}$$
$$= 15,504 \times 0.0090224 \times 0.0006024$$
$$= 0.0843.$$

So only 8.43% of the times when the bartender quizzes 20 wine drinkers will five out of the 20 wine drinkers be Democrats.

Section 5.3 Problems

78. Discuss which of the following are situations in which the binomial formula could be used.
   a. The probability that the gross national product (GNP) will rise in three out of four quarters in a given year.
   b. The probability that you get a head the next time you toss a coin.
   c. The probability that five items on a production line out of 25 randomly selected items will be defective.
   d. The probability that three out of four randomly selected dentists recommended chewing gum.

79. Find the number of possible sequences for the following situations.
   a. 4 successes in 6 trials.
   b. 11 successes in 12 trials.
   c. No successes in 15 trials.
   d. 15 successes in 15 trials.
   e. 3 successes in 20 trials.

80. Identify n, X, and p, then solve the following problems. The probability of:
   a. Obtaining 2 hearts in 12 draws with replacement.
   b. Obtaining 4 reds in 7 draws with replacement.
   c. Obtaining 5 tens in 8 draws with replacement.

81. Find the probability that none of the four students in the committee-selection problem from this section will be minorities. Remember that 20% of the students in each high school are minority students.

82. Find the probability a football team won two of a set of three randomly selected games, if the team won 60% of its games.

83. Suppose 10% of a company's seeds never germinate. If a packet contains 100 seeds, find the probability that ten never germinate.

84. An automobile service station performs only two operations, oil changes and brake repairs, in a building with seven bays. If 65% of its customers need oil changes, find the probability the station will be doing seven oil changes at a randomly selected time when all of the bays are full.

85. A Republican Senator wants to interview 20 randomly selected voters from his state. Thirty-five percent of these voters are registered Republicans. Find the probability that only 5 of those selected will be registered as Republicans.

86. A random sample of 17 American university students will be selected for a study of the effect of a nontraditional summer program. If 9% of college students are in California schools, find the probability that two of the 17 will be from a California school.

87. Suppose Ms. Williams arranges for five students who are performing poorly to form a special study group. However, each of these students comes prepared with the proper calculators only 75% of the time. Ms. Williams wants to figure the most likely number of calculators she will need to supply in these sessions. Find the probability that 0 out of 5 students, 1 out of 5, 2 out of 5, ..., and 5 out of 5 will come prepared with the proper calculator, and determine which is the most likely outcome.

5.4 The Binomial Probability Table

There are alternative methods for determining a binomial probability that do not directly use the formula. One such method is to find the desired probability in a binomial probability table such as the one in Appendix G.

This table is useful for cases in which n is one of the values 2 through 20 and p is among the set of values provided in this table. To find the probability of four successes in seven trials when p is 0.05, first find the page with $p = 0.05$ in the top row of the table and $n = 7$ in the leftmost column of the table. Then find $X = 4$ in the next column. Finally, locate the intersection of the $p = 0.05$ column with the $n = 7$ and $X = 4$ row. The value at this intersection, 0.0002, is the probability of four successes in seven trials when the probability of success for one trial is 0.05.

■ **Example 5-12 (continued):** The Roper survey indicates that 40% of non-wine drinkers are Democrats and 27% are Republicans. Find the probability that the snoopy bartender finds five Democrats in a set of 20 customers randomly selected from non-wine drinkers.

■ **Solution:** We satisfied ourselves that a similar situation (for wine drinkers) is binomial earlier. The parameters now are $n = 20$, $X = 5$, and $p = 0.40$. Using the table, we learn that $P(5) = 0.0746$.

You may have noticed that the p values across the top row of the table are all 0.5 or less. We still use the table for p values that are larger than 0.5, because *a binomial problem for X successes in n trials can be stated as the probability of (n − X) failures in n trials*. If p is the probability of success for one trial, then $(1 − p)$ is the probability of failure for one trial. Essentially, we interchange the events we name success and failure.

The last problem found the probability of five Democrats among 20 non-wine drinkers, when 40% of all non-wine drinkers are Democrats, $P(5)$ when $n = 20$ and $p = 0.40$. We obtain the same answer if we find the probability of 15 non-Democrats among 20 non-wine drinkers, when the probability of being a non-Democrat is $0.60 = 1 − 0.40$, $P(15)$ when $n = 20$ and $p = 0.60$. This complementary relationship allows us to use the same table for problems when $p > 0.50$. In such cases, the X values are in the third column from the left, and the p values are along the second row at the top of the table. Using these values to solve the last problem, $P(15)$ when $n = 20$ and $p = 0.60$ is 0.0746, the same answer as before.

Example 5-12 (continued): Seventy-five percent of wine drinkers are beyond 29 years of age, according to the survey. Find the probability that 8 of 15 randomly selected wine drinkers will fall in the 29-and-older age group.

Solution: $n = 15$, $X = 8$, and $p = 0.75$. Using the table, we find $P(8) = 0.0393$.

Using the table seems much easier than using the binomial formula, so you may wonder why we bother to learn the formula. The problem is that the table does not cover all cases. The table in the appendix takes several pages to cover n values of 2 through 20 with p values between 0.01 and 0.10, in increments of 0.01, and between 0.10 and 0.50 in increments of 0.05 (along with the complementary p values). Imagine the size of a table that covered every case! So we must either use the binomial formula or find a more detailed table to solve such problems as $P(25)$ when $n = 50$ and $p = 0.42$. Also, calculators relieve some of the computational stress when we use the formula. Another alternative would be to approximate the answer, using a normal random variable (see Chapter 6), especially for probabilities of numerous values of X or when the number of trials is large. But this does not produce the exact answer that we require in some cases. We have only two methods for determining exact binomial probabilities, the formula or a table.

Section 5.4 Problems

88. Verify your solution to Problems 80–87 (except Problem 83) from the preceding section, using the binomial table.

89. The genetic configuration of a set of parents is such that a recessive trait has a 25% chance of occurring in the offspring. What is the probability that three offspring out of nine will be born with the recessive trait?

90. A stream has been stocked, so that approximately 25% of its fish are bass and the rest catfish. Find the probability that a string of 20 fish will contain 10 catfish.

91. Six lanes for swimming laps at a local pool are each in use 80% of the time that the pool is open. Find the probability that exactly one will *not* be in use when Sanford arrives to swim laps. Find the probability that all will be in use when he arrives.

92. Studies indicate that 10% of stream water in a region is polluted. Find the probability that 12 water samples from 16 randomly selected streams in the region will show pollution.

93. On the day a poll is conducted, 40% of the registered voters in a district plan to vote for Ms. Rainey. Find the probability that four out of 10 randomly selected registered voters from the district will intend to vote for Ms. Rainey.

94. Bill's truck starts right away about 95% of the time in the morning, but with no apparent predictability from one morning to the next. Find the probability it will start on 10 of the next 15 mornings.

95. A multiple-choice test consists of 20 questions each with five choices. If Helen randomly selects answers, what is the probability she will answer 10 correctly?

5.5 Properties of the Binomial Random Variable

The number of successes, X, from a set of n independent, identical Bernoulli trials is a binomial random variable. Now, we will explore the properties of this random variable, as we did in Unit I for discrete random variables in general.

5.5.1 Probability Distribution

Table Representation We can use the binomial formula or the binomial table to calculate the probability of each value of X in a set of n trials. A table that lists each of these values and their probabilities forms a *binomial probability distribution*.

Question: List the possible values of X, the number of minority students in a four-member committee from the committee-selection situation (Example 5-10).

Answer: 0, 1, 2, 3, and 4 students.

Properties of the Binomial Random Variable

■ **Example 5-10 (continued):** We can make a table for the probability distribution of the number of minority students on the committee if we copy probability values corresponding to the five different X values from the binomial table where $n = 4$ and $p = 0.2$.

x	$P(x)$
0	0.4096
1	0.4096
2	0.1536
3	0.0256
4	0.0016

The original question in the committee-selection situation was the likelihood of different numbers of minority students that would result from random selection rather than constraining the committee composition. Based on the information, is it likely that there will be at least one minority student on the committee?

■ **Solution:** P(at least one minority student) $= P(X \geq 1)$
$$= P(X = 1 \text{ or } 2 \text{ or } 3 \text{ or } 4)$$
$$= P(1) + P(2) + P(3) + P(4)$$
$$= 0.4096 + 0.1536 + 0.0256 + 0.0016$$
$$= 0.5904.$$

Alternatively, we can use the fact that probability distributions include all possible values of X, so their probabilities must sum to 1. The only value that is not greater than or equal to 1 is the value 0. Thus, the sum of the probabilities for the X values greater than or equal to 1 is the same as subtracting the excluded probability, $P(0)$, from 1, or $P(X \geq 1) = 1 - P(0) = 1 - 0.4096 = 0.5094$.

This manipulation says that to find the probability of at least one success, you can use 1 minus the probability of no successes. It is an important time-saver, so the rule is boxed below.

Simplification Rule for Binomial Probability Problems
To find the probability of one or more successes in n trials, use 1 minus the probability of no successes. Mathematically, we say
$$P(\text{at least } 1) = 1 - P(\text{none})$$
$$P(X \geq 1) = 1 - P(0).$$

■ **Example 5-13:** Suppose a production process has a 5% rate of defective products, which the firm accepts as normal operating procedure. Every hour, ten items are randomly selected from the line and tested. If one or more items are defective, the production line is checked for malfunctions. Find the probability of the sample resulting in a malfunction check when the process is working acceptably. Justify the use of the binomial random variable.

■ **Solution:** We want the probability of one or more successes in 10 trials, the number of items checked for quality. Each item is either defective or not defective. Random selection provides independent outcomes, and if the line is working acceptably, the probability of a defective is 0.05 for any item. Thus, the binomial random variable is appropriate for this problem. Continuing, $n = 10$, $p = 0.05$, and $X \geq 1$ ("one or more"), which means we can simplify the solution to finding $1 - P(0) = 1 - 0.5987 = 0.4013$.

Question: Is the firm's rule likely to trigger malfunction checks a reasonable proportion of the time?

Answer: The value 0.4013 means that more than 40% of the samples will trigger a check, even though the defective rate is the acceptable value of 0.05—which is like a check about two out of every five hours. This rule obviously produces an excessive amount of checks, especially if the checks are time-consuming.

■ **Example 5-13 (continued):** To eliminate excessive checks, the firm decides to use samples of 12 items and check the production line when two or more are defective. Find the probability such a sample will trigger a check when the defective rate is 0.05.

■ **Solution:** If we refer to the sum of the probabilities in the probability distribution again and proceed one step further with the simplification rule, we can say $P(X \geq 2) = 1 - [P(0) + P(1)] = 1 - (0.5404 + 0.3413) = 1 - 0.8817 = 0.1183$, or unnecessary checks will occur about 12% of the time.

■ **Example 5-14:** Toxicologists found blood, urine, or tissue ethanol concentrations of more than 40 mg/dL from ingestion (rather than postmortem putrefaction) in 2.9% of 377 Federal Aviation Administration aviation fatalities (pilots usually, but sometimes flight crew members and other related personnel). Assume this value will be representative for other airline fatalities and represent the probability distribution of the number of such findings in a random sample of ten cases. Use 3% to approximate 2.9%. Then find the probability of at least one such finding in the sample. (Source: J.J. Kuhlman, Jr., B. Levine, M. L. Smith, and J. R. Hordinsky, "Toxicological Findings in Federal Aviation Administration General Aviation Accidents," *Journal of Forensic Sciences*, Vol. 36, No. 4, July 1991, pp. 1121–28.)

■ **Solution:** Using the binomial table with $n = 10$ and $p = 0.03$ (rounding 0.029 for an approximate answer using the table), we obtain

x = Number of High Ethanol Concentrations from Ingestion	$P(x)$
0	0.7374
1	0.2281
2	0.0317
3	0.0026
4	0.0001
5	0.0000
6	0.0000
7	0.0000
8	0.0000
9	0.0000
10	0.0000

The probability of at least one finding of high ethanol concentration from ingestion in 10 cases is $1 - P(0) = 1 - 0.7374 = 0.2626$.

Graphical Representation Another way of presenting the probability distribution of a random variable is with a graph. Constructing a histogram or bar graph of the probability distribution of a binomial random variable will be helpful later when we approximate binomial probabilities. In addition, it will also facilitate the presentation of finding probabilities for continuous random variables.

We label the vertical axis of this histogram *probability*, and the labels on the horizontal axis represent the positive integers between 0 and n that correspond to the possible values of X. The height of each bar is the probability of the corresponding X value.

Section 5.5 Properties of the Binomial Random Variable 235

We conveniently choose the base for each X value to span the interval $X - 0.5$ to $X + 0.5$. Thus, the base for the bar for $X = 1$ will go from 0.5 to 1.5. The base for $X = 5$ extends from 4.5 to 5.5, and for $X = 0$, from -0.5 to $+0.5$.

■ **Example 5-10 (continued):** Draw the histogram for the probability distribution of the random variable that represents the number of minority students on the four-member committee.

x	$P(x)$
0	0.4096
1	0.4096
2	0.1536
3	0.0256
4	0.0016

■ **Solution:** Consider $X = 3$, for three minority students on the committee. The bar's height is 0.0256, and its base width goes from 2.5 to 3.5. See Figure 5-10 where the graph is complete for all X values.

This graphing procedure reveals an important property of the area of the bars.

Question: What is the magnitude of the base of each bar?
Answer: The magnitude is 1. The boundaries have been chosen so that the base, $X - 0.5$ to $X + 0.5$, is always one unit in length.

Thus, the area of each bar is $P(X)$, the result of multiplying the width of the base, which is 1, and the height, which is $P(X)$. Hence, the area of each bar is the probability of its corresponding X value. See Figure 5-10. This relationship will be important when we develop the probability distribution of continuous random variables and when we use a continuous random variable to approximate these areas to determine binomial probabilities.

Question: What is the sum of the areas of all the bars?
Answer: The sum is 1, because we are summing the probabilities of all possible X values, which means we account for 100% of the binomial experiments when there are four trials.

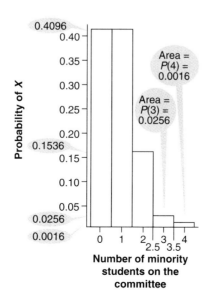

Figure 5-10

Histogram of a binomial random variable ($n = 4$ and $p = 0.2$)

Section 5.5.1 Problems

These situations are taken from Problems 89–95 of Section 5.4, but the questions are different.

96. The genetic configuration of a set of parents is such that a recessive trait has a 25% chance of occurring in the offspring. What is the probability that at least one offspring out of nine will be born with the recessive trait? What is the most likely number of children who will be born with the recessive trait? Sketch a histogram of the probability distribution associated with this binomial random variable, and shade the bars corresponding to the probability of at least one out of nine.

97. A stream has been stocked so that approximately 25% of its fish are bass and the rest catfish. Find the probability that a string of 20 fish will contain at least one catfish. What number of catfish is most likely? Least likely?

98. Six lanes for swimming laps at a local pool are each in use 80% of the time that the pool is open. Find the probability that at least one will *not* be in use when Sanford arrives to swim laps. Find the probability that at least one will be in use when he arrives. Sketch two histograms of the binomial probability distributions associated with this problem. Shade the probabilities that correspond to your two answers to this problem.

99. Studies indicate that 10% of stream water in a region is polluted. Find the probability that at least two water samples from 16 randomly selected streams in the region will show pollution. What number of streams out of the 16 are most likely to show pollution? Sketch the histogram associated with this binomial random variable, and shade the bars corresponding to the probability of at least two out of 16.

100. On the day a poll is conducted, 40% of the registered voters in a district plan to vote for Ms. Rainey. Find the probability that at least four out of 10 randomly selected registered voters from the district will intend to vote for Ms. Rainey.

101. Bill's truck starts right away about 95% of the time in the morning, but with no apparent predictability from one morning to the next. Find the probability it will start on at most 14 of the next 15 mornings. Sketch the histogram for this binomial random variable, and shade the bars corresponding to the solution.

102. A multiple-choice test consists of 20 questions each with five choices. If Helen randomly selects answers, what is the probability she will answer at least 10 correctly?

5.5.2 Mean

We can find the mean value of a random variable by finding $E(X)$.

■ **Example 5-10 (continued):** Find $E(X)$ for $X =$ number of minority students on the four-member committee, $E(X) = \Sigma_1^k x P(x)$.

x	$P(x)$
0	0.4096
1	0.4096
2	0.1536
3	0.0256
4	0.0016

■ **Solution:** $E(X) = 0(0.4096) + 1(0.4096) + 2(0.1536) + 3(0.0256) + 4(0.0016)$
 $= 0.8$ student.

Question: Interpret the 0.8.
 Answer: If a large number of four-member committees are randomly selected from the four high schools with the same minority student ratios, 0.2, then the average number of minority students on the committees is 0.8 students.

Question: If 100 high schools have 20% minority ratios and a student is randomly selected from each school, how many minority students should we expect to select?
 Answer: Twenty percent of 100 students, or $(0.2)(100) = 20$ minority students.

In the case of the four-member committee, we would expect $(0.2)4 = 0.8$ minority students, which is the exact value we obtained when we computed $E(X)$ above. This procedure, multiplying the number of trials, n, by the probability of success on a single trial, p, is a shortcut method to find the mean, μ, of a binomial random variable, rather than use the $E(X)$ formula.

> **Shortcut Formula for Finding the Mean (Expected Value) of a Binomial Random Variable**
>
> $$\mu = E(X) = np.$$

■ **Example 5-14 (continued):** Find the mean of the ethanol test random variable: The number of findings of high ingested ethanol concentrations in a random sample of 10 airline fatalities when the probability of such findings for any one fatality is about 3%.

■ **Solution:** $n = 10$ and $p = 0.03$, so $\mu = 0.3$ findings of high concentrations. Thus, the average number of such findings is 0.3, when the toxicologists investigate random samples of 10 cases many times.

■ **Example 5-13 (continued):** Find the mean number of defectives in the sample of the malfunction-check example, when the sample contains twelve items and the operation is functioning as usual with a 5% defective rate.

■ **Solution:** $\mu = 12(0.05) = 0.6$ defectives.

Remember that this formula is a useful shortcut only for a binomial random variable—not for every discrete random variable.

5.5.3 Standard Deviation

■ **Example 5-10 (continued):** Find the standard deviation of X, the number of minority students on the four-member committee. Remember that the mean is 0.8 minority students, the product of $n = 4$ and $p = 0.2$.

x	$P(x)$
0	0.4096
1	0.4096
2	0.1536
3	0.0256
4	0.0016

■ **Solution:** Using the following table, we can proceed through the definitional formula from Unit I, $\sigma = \sqrt{\sum_1^k (x - \mu)^2 P(x)}$.

x	$x - \mu$	$(x - \mu)^2$	$(x - \mu)^2 P(x)$
0	-0.8	0.64	0.262144
1	0.2	0.04	0.016384
2	1.2	1.44	0.221184
3	2.2	4.84	0.123904
4	3.2	10.24	0.016384
			0.640000

$\sigma = \sqrt{0.640000} = 0.8$ minority students (it is a coincidence that the mean and standard deviation are the same value). Thus, the "average" distance between the values of X and the mean is 0.8 students.

There is also a shortcut formula for this calculation.

> The standard deviation for the binomial random variable is
> $$\sigma = \sqrt{np(1-p)}.$$

We can verify that the formula works from the preceding problem, where $n = 4$, $p = 0.2$, and the standard deviation, σ, is 0.8.

$$\sigma = \sqrt{4(0.2)(1 - 0.2)} = \sqrt{0.64} = 0.8 \text{ minority student.}$$

■ **Example 5-13 (continued):** Find the standard deviation for the malfunction-check random variable with a random sample of 12 items and a defective rate of 5%.

■ **Solution:** $\quad \sigma = \sqrt{12(0.05)(0.95)} = \sqrt{0.57} = 0.7550 \text{ defectives.}$

■ **Example 5-14 (continued):** Find the standard deviation of the number of ethanol concentration findings in a sample of 10 cases, when the likelihood of such a finding is 3% for any single case.

■ **Solution:** $\quad \sigma = \sqrt{10(0.03)(0.97)} = \sqrt{0.291} = 0.5394 \text{ findings.}$

These values measure the spread or dispersion of the values of X about the mean. They are "average" deviations. Box 5-2 illustrates an application of the descriptive concepts, mean and standard deviation, for the binomial random variable, as well as its probability distribution.

Box 5-2 Checking for a Fair Coin

A fair coin should land on heads or tails approximately the same number of times, if tossed a large number of times. Using the binomial random variable, its expected value and standard deviation, and Chebyshev's inequality, we can devise several ways to test or check that a particular coin is fair.

Using the binomial table, we find that the probability of tossing 5 *or fewer* heads in 20 tosses is 0.0207. The same value holds for the chances of tossing 15 *or more* heads in 20 tosses. There is very little possibility (the probability is 0.0414) of tossing 5 *or fewer* or 15 *or more* in a single experiment consisting of 20 tosses. Thus, we could say that if the result is 5 *or fewer* heads or 15 *or more* heads, then we believe the coin to be unfair. Of course, even a fair coin can land on heads 20 times in a row, so we risk making a wrong decision using this criterion. Nevertheless, when an unlikely outcome occurs, it indicates that something is perhaps amiss. We may want to check the coin further (or at least swap coins to make a decision or play a game).

Suppose we proceed further and base our decision on the result of an experiment with many trials, each consisting of counting the number of heads in 20 tosses of the coin. We know the number of heads in this situation is a binomial random variable with mean 10 heads and standard deviation, 2.24 heads. Furthermore, Chebyshev's inequality helps us predict the likely and unlikely occurrences from repeated trials. This rule states that the proportion of observations, X, in a data set that fall within K standard deviations of the mean must be at least $1 - (1/K)^2$ for K greater than 1, or with probability symbols

$$P(\mu - K\sigma < X < \mu + K\sigma) \leq 1 - (1/K)^2.$$

If K is 2, then at least three-fourths of the X values should be within two standard deviations of the mean. In our case, each X value is the number of heads in 20 tosses, the mean outcome from a large number of such trials is 10 heads, and two standard deviations is 4.48 heads. The latter two statements are based on the coin being fair. Thus, if the coin is fair, at least three-fourths of the 20 toss trials should result in between 6 and 14 heads, inclusive. If we do the experiment 100 times and find that 25 or more trials result in fewer than 6 heads or more than 14 heads, then we should suspect that the coin is not behaving like a fair coin. Our original assumption, that the coin is fair, appears to be a false assumption.

Sections 5.5.2 and 5.5.3 Problems

103. Find the mean and standard deviation of a binomial random variable with the following specifications.
    a. $n = 50$ and $p = 0.72$.
    b. $n = 200$ and $p = 0.03$.
    c. $n = 285$ and $p = 0.5$.
104. Find and interpret the mean and standard deviation of a binomial random variable that represents the number of female raffle winners in a group of 25 participants, each with five tickets or a 0.005 chance of winning.
105. Find and interpret the mean and standard deviation of X, the number of hearts in five draws from a deck with replacement.
106. Find and interpret the mean and standard deviation of X, a count of absent students in a class of 30 students, if students tend to miss 3% of the classes.

Summary and Review

The values of a binomial random variable, X, represent the number of successes that result from n independent, identical Bernoulli trials (trials of an experiment with a common definition of a "success" and the same value for p, the probability of success on a single trial). We use the binomial formula to find the probability that specific values of X occur. A binomial probability table is also useful for solving problems for common p values when n is small. Together, the possible X values and their associated probabilities form the binomial probability distribution.

We can use shortcut formulas to find the mean and standard deviation of the binomial random variable and to avoid the long expected value calculations. When we construct a histogram of the probability distribution of X, the area of each bar is equivalent to the probability of its corresponding X value.

The new formulas introduced in this chapter are:

1. Expected value (mean) of a random variable with k values
$$E(X) = \mu = \sum_{1}^{k} xP(x).$$

2. Standard deviation of a random variable with k values
$$\sigma = \sqrt{\sum_{1}^{k} (x - \mu)^2 P(x)}.$$

3. Binomial probability
$$P(x) = \left[\frac{n!}{x!(n-x)!}\right] p^x (1-p)^{(n-x)}.$$

4. Mean of binomial random variable $\mu = np$.
5. Standard deviation of binomial random variable
$$\sigma = \sqrt{n(p)(1-p)}.$$

Multiple-Choice Problems

107. Which of the following is a binomial situation?
    a. Find the probability of being dealt a flush, five cards all in the same suit, in a poker game.
    b. Find the probability that three of the next 10 customers at a restaurant order a hamburger and seven order prime rib.
    c. Find the probability that 20 people out of 100 randomly selected people will refuse to respond to survey questions.
    d. Find the probability that the first five numbers selected in a bingo game will all fall under B.
    e. Find the probability that a presidential aspirant will win 15 out of 20 primaries in a given election year.
108. How many ways are there to obtain eight successes in 11 trials?
    a. 165.   d. 11.
    b. 88.   e. 6.
    c. 8.
109. Find the probability of getting exactly three sixes out of four rolls of a die, in the following exact order: 6, 6, 6, not 6.
    a. 5/1,296.   d. 5/16.
    b. 1/6.   e. 20/1,296.
    c. 1/1,296.
110. Find the probability of getting three sixes out of four tosses of a die, when we do not dictate the order of occurrence.
    a. 5/1,296.   c. 1/1,296.   e. 20/1,296.
    b. 1/6.   d. 5/16.
111. A firm has a 4% defect rate. If 100 items are randomly selected, find the probability that *at most one* out of the 100 is defective.
    a. $(0.96)^{100}$.
    b. $(0.04)^{100}$.
    c. $(0.96)^{100} + 100(0.04)(0.96)^{99}$.
    d. $(0.04)^{100} + 100(0.04)(0.96)^{99}$.
    e. $(0.96)^{100} + 100(0.96)(0.04)^{99}$.
112. Find the probability of *at least one* defective for the information in Problem 111.
    a. 0.07.   d. 0.09.
    b. 0.91.   e. 0.98.
    c. 0.02.

113. The probability of getting one 6 in five tosses of a die is about 0.4. The *p* for this problem is 1/6. Which of the following statements concerning this problem is true?
    a. 1/6 is the $P(X = 1)$, and 0.4 is the probability of success on one trial.
    b. The probability of getting one 6 in five tosses is 1/6.
    c. The probability of getting one 6 in five tosses is 1/5.
    d. *p* is the probability of getting one 6 in one toss.
    e. The problem does not meet the binomial assumptions.

114. A random variable represents the number of people planning a Florida vacation in a sample of 400 people, when 20% of people in this population plan such a trip. Find μ and σ, respectively, for this binomial random variable.
    a. 400, 0.2.   d. 80, 8.
    b. 400, 0.8.   e. 80, 320.
    c. 320, 8.

115. When doing arithmetic calculations, you make random errors 1% of the time. If you must repeatedly perform sets of 150 calculations, what is the mean number of errors you expect to make on each set?
    a. 149.
    b. 0.01.
    c. 1.5.
    d. Cannot be figured, because this is a discrete random variable.
    e. Cannot be figured, because this is a continuous random variable.

116. If we know the area of the bar over the outcome 25 in the graph of a binomial distribution, we know
    a. The expected value of the distribution.
    b. The probability of the outcome 25.
    c. The number of successes in 25 trials.
    d. The mean of the distribution.
    e. The standard deviation of the distribution.

117. Five hundred items are randomly selected from a production line with a 4% defect rate and checked to determine if they are defective. If you want to know the probability that 200 are defective, what are *n*, *X*, and *p*, respectively?
    a. 500; 300; 0.96.
    b. 500; 200; 0.4.
    c. 500; 200; 0.04.
    d. 300; 200; 0.5.
    e. 500; 200; 0.5.

118. The toxicologist from Example 5-14 also found that about 3% of 374 cases tested positive for abused drugs. Use this information to find the probability that a random sample of eight cases will contain two positive cases. (Source: J. J. Kuhlman, Jr., B. Levine, M. L. Smith, and J. R. Hordinsky, "Toxicological Findings in Federal Aviation Administration General Aviation Accidents," *Journal of Forensic Sciences*, Vol. 36, No. 4, July 1991, pp. 1121–28.)
    a. 0.0210.
    b. 0.2213.
    c. 0.2224.
    d. 0.0014.
    e. 0.7787.

119. Which of the following is *definitely not* a binomial probability problem?
    a. The probability that 15 students in a class of 450 will pass a memorization test if the typical student memorizes 75% of the information to be tested.
    b. The probability that five Democrats and two independents are elected, if 50% of the registered voters are Democrats and 10% are independents.
    c. The probability one egg hatches in a nest of a dozen eggs, if ordinarily 90% of the eggs of this species hatch.
    d. The probability that 50 criminals, from those paroled in Colorado last year, will become repeat offenders, if ordinarily 30% of all paroled offenders repeat the offense.
    e. The probability that 758,000 people in a regional television audience will see a commercial, if Nielsen estimates indicate that roughly 25% of the audience will be watching the station during the commercial.

120. Find the probability that two of five students randomly selected to participate in a psychological experiment are altruistic, if the class of 10 students contains three altruistic students.
    a. 0.4.
    b. 0.3.
    c. 0.3087.
    d. 0.2592.
    e. The problem does not meet the binomial assumptions.

About 20% of households spent between $750 and $999 on automobile insurance in 1990. Use this information to solve Problems 121–123. (Source: "What Consumers Pay for Auto Insurance," *Household Insurance Expenditures 1990*, Alliance of American Insurers, as reported in *Consumer's Research*, May 1991, p. 31.)

121. Find the probability that at least one household in 20 households randomly selected for a survey will respond that they did *not* spend in this interval.
    a. 0.9885.   d. 0.0000.
    b. 0.9424.   e. 1.0000.
    c. 0.0576.

122. What is the mean number of households in random samples of 20 households that would report *not* spending in this interval?
    a. 0.05 households.   d. 80.0 households.
    b. 4.0 households.   e. 16.0 households.
    c. 8.0 households.

123. What is the standard deviation of the number of households in random samples of 20 households that would report *not* spending in this interval?
    a. 3.2 households.   d. 4.5 households.
    b. 0.4 households.   e. 16.0 households.
    c. 1.8 households.

Word and Thought Problems

Use the following information to solve Problems 124–126.

Number of Nobel Prize Laureates in Physics, Chemistry, and Physiology/Medicine, 1901–1988

Country	Total	Physics	Chemistry
United States	150	52	33
United Kingdom	69	21	24
West Germany[1]	56	16	28
France	22	8	7
Soviet Union	10	7	1
Japan	4	3	1
Others	82	27	18
	393	134	112

[1] Includes Germany prior to 1946.

Source: U.S. National Science Foundation, as reported in U.S. Bureau of the Census, *Statistical Abstract of the United States: 1990*, 110th ed. (Washington, D.C.: Bureau of the Census, 1990), p. 591.

124. Find the probability that a random sample of four winners during this period will contain no winners from the United Kingdom. Round p one place past the decimal to avoid changing the p value with different selections.

 a. 0.4096. d. 0.0016.
 b. 0.2500. e. 0.1756.
 c. 0.0000.

125. Find the probability that a random sample of four winners during this period will contain two winners from the United States and two from France. Round p one place past the decimal to avoid changing the p value with different selections.
 a. 0.5000.
 b. 0.0855.
 c. 0.3456.
 d. 0.0486.
 e. Cannot use the binomial formula to answer.

126. Find the probability that a random sample of three Physics winners during this period will contain at least one winner from an unlisted country. Round p one place past the decimal to avoid changing the p value with different selections.
 a. 0.3840.
 b. 0.4880.
 c. 0.1040.
 d. 0.3333.
 e. Cannot use the binomial formula to answer.

Word and Thought Problems

127. What are the requirements that define a binomial experiment?

128. Which of the following are situations in which the binomial formula could be used. Explain why not when the answer is no.
 a. The probability that 3 out of 10 tosses of a die will be a 6.
 b. The probability that 3 out of 10 tosses of a die will be a 6, and the other 7 tosses will be a 2.
 c. The probability that 50 out of the next 100 customers will want to see a menu.
 d. The probability that 30 out of 50 students in a classroom have blue eyes.

129. a. If the rate of defective products for a firm's production process is 4%, find the probability that one out of five randomly selected items will be defective.
 b. Find the probability that less than two will be defective.
 c. Find the probability that at least one will be defective.

130. Identify n, x, and p, then solve the following problems. The probability of
 a. Obtaining two fives in 12 tosses of a die.
 b. Obtaining at least two fives in 12 tosses of a die.
 c. Obtaining at most two fives in 12 tosses of a die.
 d. Obtaining two odd values in 12 tosses of a die.
 e. Obtaining two odd values in 20 tosses of a die.

131. In how many ways can you obtain 17 successes in 20 trials of an experiment?

132. If 60% of the American people favor the job performed by the president, what is the probability that 15 people in a random sample of 20 Americans indicate they favor the job being done?

133. Suppose 10% of all people are left-handed. If there are no left-handed people in a sample of 80 people being tested for hand-eye coordination, is there reason to believe that the sample is not representative of all people? Should the handedness of people in any sample be determined as a check for representativeness of the sample?

134. The high-school dropout rate in 1988 was 28.9% for the entire United States, 37.7% in New York, and 19.8% in Kansas. Use this information to find the following probabilities. (Source: U.S. Department of Education, as reported in *The Chronicle of Higher Education Almanac*, August 28, 1991, p. 4.)
 a. Find the probability that a random sample of 13 high-school-age youngsters contains no dropouts.
 b. Find the probability that a random sample of 13 high-school-age youngsters from New York contains no dropouts.
 c. Find the probability that a random sample of 13 high-school-age youngsters from Kansas contains at least one dropout.

135. A disk jockey is supposed to play songs from the 1960s approximately 8% of the time. A listener who tunes in one afternoon at the beach counts three such songs out of 88 songs played that afternoon. How likely is this result, if the disk jockey is really playing 1960s songs as frequently as policy requires?

136. 20% of a firm's customers are overdue on their accounts. Find the probability that all of the next four customers have overdue accounts.
137. Ninety-eight percent of a store's customers write good checks. Find the probability that at least one of the next 10 checks will be a *bad* check.
138. Find the probability that two heads will appear in 20 tosses of a fair coin.
139. If 70% of the people in a large suburb are in favor of a new shopping center locating nearby, what is the probability that 35 out of 40 randomly selected people will be in favor of the center?
140. Mail processing of film is usually cheaper than using local processors, but problems with developing are not as speedily resolved. Suppose problems arise about 3% of the time. What is the probability that there will be five problems in the next 100 rolls that Albertine processes?
141. Items are randomly selected from a production line, tested, and then labeled "workable," "repairable," or "nonrepairable." If 100 items are selected, and you want to know the probability that 40 are workable and 60 repairable, is the situation binomial (can you use the binomial formula to solve the problem)? Explain your answer.
142. A psychologist has data on 15 members of a group of local homeless individuals. Presumably, the 15 observations were randomly selected; however, the psychologist discovers that 12 of the 15 are teenagers, though teenagers comprise only 25% of the local homeless. What is the likelihood of the resulting sample composition? Should the psychologist question the randomness of the selection process?
143. A decision process that buyers sometimes employ in deciding to accept a shipment of items is to randomly select and then inspect a sample of four items from a shipment of the product. If none of the four items is defective, the shipment is accepted, but if one or more items are defective, the shipment is rejected. Irene checks boxes of garbage bags using such a procedure. What is the probability that she accepts a shipment with 3% defectives? 8% defectives? 12% defectives? Use these results to determine if the procedure leads Irene to accept poor-quality shipments frequently.
144. A psychologist misdiagnoses a patient's problem about 5% of the time. What is the probability that at least one of the psychologist's 62 present patients is misdiagnosed?
145. Five customers (who do not know one another) each tend to pay 30 days or more after billing, 20% of the time. Find the probability distribution for the number of customers out of the five who will pay late after the next bill. What are the mean and standard deviation?
146. Make a histogram of the probability distribution in Problems 136 and 145.
147. Gordon's hay baler randomly misbales about 1% of the time. What is the probability that a field yielding 150 bales will yield two imperfect bales?
148. A doctor has six appointments in the next two hours. Each patient complains about the same symptoms related to a virus that generally shows up in 50% of patients with these symptoms. The proper medication is in short supply, and this doctor has none. However, a nearby associate has six doses and is willing to share the number of doses the doctor is most likely to need for these six patients. If the patients do not know or associate with one another, what is the proper number of doses to borrow?

Review Problems

149. Compare a probability distribution with a relative frequency distribution. How do we find the mean and standard deviation of a random variable? How do their interpretations differ from those for the same measures of a regular data set?
150. To learn more about the social interactions of students, a group of sociologists keep records of students' memberships in organizations. Basing their research on students' responses to a questionnaire about their activities during the previous semester, which they voluntarily complete during registration, the sociologists compile the following table. Find the expected number of organizations joined per student, and measure the dispersion of the responses. What is the term from this chapter for the number of organizations? What is the term from this chapter for the table?

Student Organization Count (SOC)	$P(SOC = soc)$
0	0.48
1	0.32
2	0.16
3	0.02
5	0.01
8	0.01

151. If a state institutes a lottery and plans to use the proceeds to finance state services, what do you know about the expected value of the lottery for the player? What do you know about the expected value for the state?
152. Maximum hurricane wind speed experienced along the Atlantic coast is random variable W (measured in miles per hour, mph). Use the following probability distribution for W to find the $E(W)$ and σ. Interpret the results. Are there outliers? Explain.

w	P(w)
100	0.56
125	0.32
150	0.12

153. We obtain the following random variable and probability distribution by combining two groups of data: previous students' taping projects in a broadcasting class, the usual time required to tape each type of performance, and past observations of the proportion of time the students chose a particular type of performance.

Type of Performance	x = Usual Time Required	P(x)
Magician	10	0.05
Solo actor	15	0.11
Singer	20	0.24
Juggler	30	0.02
Singing group	35	0.20
Chamber music group	40	0.10
Band	45	0.10
Gymnast	60	0.17
One-act play	75	0.01

    a. Find $P(X = 75)$.
    b. Find $P(X \geq 45)$.
    c. Find and interpret $E(X)$.
    d. Find the standard deviation for X, using $E(X) = 34$.

154. To estimate the amount of time needed to catch a ton of food fish during season, a group of professional fishers randomly select sites about a specified location in the ocean to fish. At the specific agreed-upon time the group members individually cast their nets until they land a ton of usable fish. The group plans to average the set of times to estimate the average time required to catch the ton. Is the set of times a sample or population of the times required to catch the ton of fish? Is the required fishing time a random variable? If so, is it discrete or continuous? Is this process descriptive or inferential statistics?

155.

x	P(x)
50	0.05
150	0.03
200	0.45
450	0.02
550	0.05
800	0.25
1,100	0.15

    a. The information given here about X is called a _____.
    b. Suppose you are told that $E(X) = 1,500$. Explain why this value is reasonable or unreasonable.
    c. Suppose you are told that $\sigma = 2,500$. Explain why this value is reasonable or unreasonable.

156. Consider the following sample of diamond exports (in millions of pula) from Botswana between 1976 and 1983. (Source: Bank of Botswana, *Annual Reports: 1977, 1979, 1981, 1982, and 1983* (Gaborone, various years), as reported in Robert L. Curry, Jr., "Botswana's Macroeconomic Management of Its Mineral Based Growth," *The American Journal of Economics and Sociology*, Vol. 46, No. 4, October 1987, p. 479.)

37   47   76   184   237   136   246   462

    a. What is $P(\overline{X} < 37)$?
    b. What is $P(\overline{X} > 500)$?
    c. What is $P(\overline{X} < 37 \text{ or } \overline{X} > 500)$?
    d. Find the sample mean and standard deviation.
    e. Find the Z values for each observation, and determine if there are outliers.

157. From his journal entries, Lewis finds that out of 82 trips to a particular park, 25 have resulted in sighting one moose, 19 resulted in two moose, 10 resulted in three moose, and the rest resulted in no moose sightings. If these findings are representative of sightings on a trip to this park, what is the expected number of sightings on a trip to this park?

158. Matt figures that 10% of the time he can shovel snow from five driveways in a day, 30% of the time he can do four, 40% of the time he can do two, and 20% of the time, only one. What is the mean number of driveways he will shovel? Find the standard deviation of shoveled driveways.

159. Two nearby schools' football teams have the following records for total points per game (*TP*) scored by both teams and margins of victory per game (*MV* = school score minus opponent score) for the last few years.

School A				School B			
tp	P(tp)	mv	P(mv)	tp	P(tp)	mv	P(mv)
0	0.1	−10	0.05	14	0.2	−14	0.2
10	0.3	−7	0.15	17	0.4	−7	0.3
14	0.4	0	0.25	21	0.3	3	0.3
21	0.1	3	0.35	35	0.1	7	0.1
27	0.1	7	0.20			21	0.1

If you are planning to buy a season ticket for only one team's games, which school should you choose, if your objective is:
    a. To see a lot of offense or scoring.
    b. To see a lot of close games.
    c. To watch a winning team.

160. Forty percent of the participants in a soccer league are nine years old, 30% are eight, 20% are seven, and 10% are six. Find the mean and standard deviation of ages.

161. Eighteen percent of patients with a certain problem recover 15 days after the onset, 23% in 22 days, 29% in 29 days, 21% in 36 days, and 9% in 43 days. What is the expected recovery time?

162. Find the binomial probability distribution for the number of successful canine lobotomies in a random sample of four such surgeries by a state's veterinarians, if canine lobotomies succeed about 60% of the time. What are the mean and standard deviation?

163. A drug is supposed to be effective for 90% of the patients who use it. Find the probability of successful drug therapy for at least 18 of 20 patients using the drug.

164. Consider the following information, where $P(c)$ represents the probability a domestic customer complains about baggage handling by the stated airline.

Airline	$P(c)$
Eastern	0.0103
Delta	0.0084
USAir	0.0080
Northwest	0.0079
America West	0.0076
Continental	0.0073
United	0.0072
American	0.0072
TWA	0.0070
Alaska	0.0065
Pan Am	0.0049
Southwest	0.0041

Source: Department of Transportation Data for March 1990, as reported in "Lost Baggage Complaints," *USA Today*, May 29, 1990, p. 5B.

   a. Is this table a probability distribution of a random variable? If so, is it a binomial random variable?
   b. Find the probability that none of six randomly selected Eastern customers complain.
   c. Find the probability that nine of 10 randomly selected Eastern customers do *not* complain.

165. A backpacker plans to take rations for himself and two extra people, because he usually finds ill-prepared hikers in the backcountry about 70% of the time. How many people can he randomly meet on the trail in total, yet have approximately a 65% chance of two or fewer of the total being ill-prepared?

166. Toxicologists found ethanol concentrations above 40 mg/dL for 7.4% of 377 Federal Aviation Administration airline fatalities in fiscal year 1989. Twenty-two percent of these high-concentration cases were probably from ingestion rather than postmortem putrefaction. If we randomly select one of the cases, what is the probability it will be a high-concentration case from ingestion? (Source: J. J. Kuhlman, Jr., B. Levine, M. L. Smith, and J. R. Hordinsky, "Toxicological Findings in Federal Aviation Administration General Aviation Accidents," *Journal of Forensic Sciences*, Vol. 36, No. 4, July 1991, pp. 1121–28.)

167. If the average person uses a bathroom sink about 2% of the time, find the probability that a family of six will need more than two sinks in their home at a randomly chosen time.

168. You are making arrangements for a trip during a school holiday to a vacation area. You plan to invite five friends whom you have met in different areas of your state on previous trips to join you at a rented cottage. A large, nonrefundable deposit must be paid early, because of the crowds expected during the holidays. You decide to reserve a cottage that is the proper size for yourself and the number of friends most likely to accept your invitation. Past experience suggests that 85% of the time these friends accept such offers. Find the most likely number of acceptances. Draw a histogram of the random variable involved in this problem. Find the mean on the histogram, and calculate the standard deviation.

169. Sixty percent of cellular phone calls attempted in a 24-hour period are completed in Manhattan and only 28% if the attempt falls between 4 and 6 P.M. Find the probability distribution for the number of completed calls in nine random attempts. Then compute its mean and standard deviation. Repeat this exercise for three random attempts between 4 and 6 P.M. (Source: "Business Bulletin," *Wall Street Journal*, August 22, 1991, p. A1.)

170. Trainers at a guide dog school find that the necessary time for preliminary training before a dog joins its new master varies with different dogs. Records over many years and many dogs indicate the following training times (weeks), D. Weaning the new couple, master and dog, also requires various time spans (weeks), W, as indicated in the following.

d	$P(d)$	w	$P(w)$
14	0.05	3	0.02
15	0.10	4	0.95
16	0.55	5	0.03
17	0.25		
18	0.05		

   a. Use this information to form a new random variable, T, the total training time from receiving a new dog until the pair departs from the training school. Assume that once a dog is trained, the length of that training time does not affect the time required to merge with a new master. The required times for the two training modes are independent. The following table lists the smallest three possible values of T to get you started.

t	$P(t)$
17	0.0010
18	0.0495
19	0.1075

   b. Find and interpret the expected values of all three random variables. Compare the results to determine if there is a relationship among the expected values based on the generation of T from D and W.
   c. Repeat Part b, using the standard deviation, σ, rather than the expected value.

Chapter 6

The Normal Random Variable

This chapter explores the properties and graphs of one particular continuous random variable, the normal random variable. More commonly we refer to this variable or to situations from which it derives as fitting a "bell-shaped curve." This random variable fits or closely approximates many situations and variable behaviors, making it one of the most important random variables in statistics and many other disciplines.

Objectives

Unit I introduces normal random variables, in particular, the standard normal as the first and very important example of a specific continuous random variable. Upon completing this discussion, you should be able to

1. Understand what information is provided by an entry in the standard normal table.
2. Draw a diagram that depicts the given and table information as well as the solution to a problem or situation with outcomes that are standard normal.
3. Find the probability that a value in an interval of values occurs, using the standard normal table.
4. Find the boundaries of an interval of values when given the probability that values in the interval occur, using the standard normal table.
5. Differentiate a normal random variable from a standard normal random variable.

Unit II utilizes the transformation formula introduced in Chapter 3 for converting observations from a data set to Z values. The observations we discuss here are values of a normal random variable, not necessarily standard normal, that represent experimental outcomes that fit a normal probability distribution. Once we transform any normal random variable to standard normal, we can employ the standard normal table to find relevant information and resolve problems. Upon completing this unit, you should be able to

1. Find the probability that a value in an interval of values of any normal random variable occurs.
2. Find the boundaries of an interval of values of any normal random variable when given the probability that a value in the interval occurs.
3. Distinguish X and Z on a diagram that represents a problem situation described by a normal random variable.
4. Use the standard normal transformation to approximate a binomial probability.
5. Describe the characteristics of a binomial random variable that affect the closeness of the approximation to the true solution, and state a condition for obtaining reasonable approximations.
6. Understand the function of the continuity correction factor.

Unit I: Continuous Random Variables, The Standard Normal

Preparation-Check Questions

1. A representation of all possible values of a random variable and their associated probabilities is called a(n)
   a. Expected value.
   b. Binomial situation.
   c. Frequency distribution.
   d. Probability distribution.
   e. Standard deviation.
2. The mean, μ, and standard deviation, σ, of a random variable tell us
   a. The average value and spread of the values about the mean.
   b. The median value and average deviation of the values.
   c. The expected value and the range of the values.
   d. Which value occurs most and how often it occurs.
   e. The largest value and the location of other values relative to this value.
3. The area of the bar for X on the histogram of a binomial is
   a. 1.
   b. 0.
   c. $P(x)$.
   d. $P[(x - 0.5) < X < (x + 0.5)]$.
   e. $P(X \leq x)$.
4. If your score, 82, is the seventieth percentile of a set of test scores, what can you say about the other scores?
   a. About 70% of the scores are below 70.
   b. About 70% of the scores are below 82.
   c. About 82% of the scores are below 70.
   d. 70 is the average test score.
   e. 82 is the average test score.

Answers: 1. d; 2. a; 3. c; 4. b.

Introduction

Many measurable situations produce a normal or bell-shaped distribution of values, like the one shown in Figure 6-1. In the classroom, we sometimes expect grades to fit this pattern. Most students would score in the middle range, with fewer students making very low or very high scores. Of course, this pattern may not occur in a small class, but we expect a very large set of observations to approximate this pattern. Intelligence, for instance, as measured by IQ test scores, would approach this pattern as we accumulate more and more scores. Many other variables, such as peoples' heights and weights, follow such a pattern.

Sometimes, we need specific information about the probabilities or proportions of normal values that fall within certain intervals. We might want to know what proportion of students score 85 or above on a national test. What is the probability of passing a course? What should be the cutoff value for an A, if the top 10% of the scores will be graded A?

We can answer such questions once we position the curve associated with a particular situation on the number line. Unit I shows how to use a table of values to obtain answers from a particular curve that corresponds to the standard normal random variable. It lies at one position on the number line. The normal curves in Unit II lie in any position. However, we can transform all of them to the single position occupied by the standard normal and use the table once again to solve problems.

Figure 6-1
A bell-shaped distribution of test grades

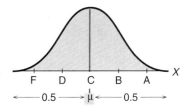

6.1 The Probability Distribution of Any Continuous Random Variable

Before we focus entirely on the normal random variable, we will show how to find probabilities associated with values of any continuous random variable. To understand how to do this, we extend the procedure of determining probabilities from a graph of a discrete random variable to finding the probabilities on the graph of a continuous random variable.

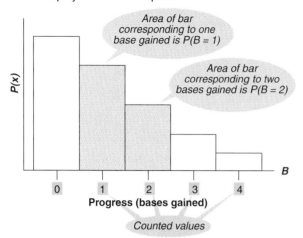
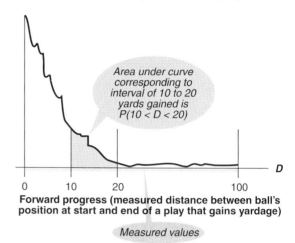

Figure 6-2

Comparison of discrete and continuous random variables

Figure 6-2 exemplifies the difference between a discrete and a continuous random variable that we discussed in the last chapter. B, the progress of a baseball player on the next play of a game (the batter's response to a pitch) is a discrete random variable, because we can count its different values even when there are infinite possible values. In this case, B's values are 0 base, 1 base, 2 bases, 3 bases, and 4 bases. We represent the values as points on a number line with gaps between the points.

D, the *forward* progress of a football player on the next play of a game (the distance between the position of the ball at the start and end of a play that gains yardage and zero if the play loses yardage) is a continuous random variable, because we cannot count its infinite possible values. *Forward* progress can be any value between (but not including) 0 and 100 yards. Values of a continuous random variable usually represent magnitudes on a measuring instrument, perhaps a tape measure in this instance. We can only represent these values as an interval (or intervals) on a number line.

To determine the probability associated with a particular value of a discrete random variable from a bar graph, we find the area of the bar corresponding to that value as shown for 1 base and for 2 bases in Figure 6-2. The area is the width of the bar times the height of the bar. The graph here is similar to the one for a binomial random variable in Chapter 5, where the base is one unit and the height is $P(x)$, although B is not a binomial random variable (because getting another base is dependent on reaching the preceding base).

Question: Describe what mathematical operation we would perform with the areas of different bars to obtain $P(1 \leq B \leq 3)$.

Answer: B can be 1 or 2 or 3 to satisfy the event $1 \leq B \leq 3$, so we would sum the areas of the three bars.

Question: What is the sum of the areas of all of the bars?

Answer: Because the diagram is a pictorial representation of the probability distribution of the random variable, B, it accounts for all five possible values. Hence, the areas (probabilities) must sum to 1.

The bars for values of a continuous random variable collapse into lines, as shown in Figure 6-3. There are no gaps between values and, hence, no space to form a base for a bar. Any attempt to do so would immediately overlap other values. The top points of these bars or lines form a smooth curve like the one in Figure 6-3d, which we use to

Section 6.1 The Probability Distribution of Any Continuous Random Variable ■ **249**

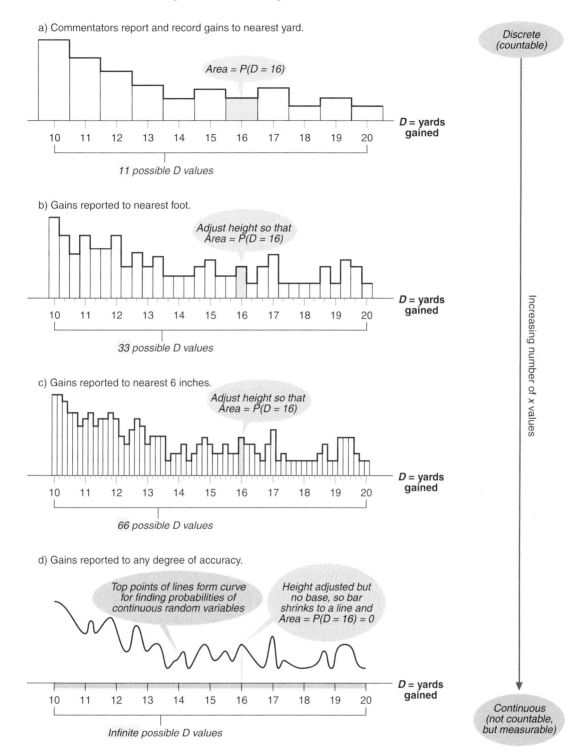

find probabilities for continuous random variables. Different continuous random variables have different probability distributions that will produce different curves.

Look again at Figure 6-2, where we assemble the lines for all of the values in the interval between 10 and 20 yards to obtain the shaded geometric figure with an area that represents the probability of the event gaining between 10 and 20 yards on the

Figure 6-3
Probability distribution of a continuous random variable: yardage gained values between 10 and 20 yards

football play, $P(10 < D < 20)$. The area of such geometric figures (often irregularly shaped), formed from the area under the curve that corresponds to a particular event (interval of values), represents the probability of the event. We will use tables of areas for specific random variables in this text rather than use more advanced mathematics to determine the equation and areas.

Question: What is $P(0 < D < 100)$?
Answer: It must be 1, because the total area under the curve must be 100% or 1 to account for all the possible values of the random variable, in this case 0 to 100.

In summary, we use a graph to depict the probability distribution of a continuous random variable. We can only phrase questions about intervals of values, $P(a < X < b)$, not about an individual value, $P(a)$, because there is no bar with area for the outcome a, only a vertical line. To measure the probability associated with an interval, we find the area under the curve between a and b. Because no probability exists for an individual value, the probability will not change if an interval includes or excludes the boundary points. For instance, the probability that D falls between 10 and 20 in Figure 6-2 is the same whether or not we include the 10 or the 20, $P(10 \leq D \leq 20) = P(10 < D < 20)$.

To find probabilities for a continuous random variable, X, from the curve that represents the probability distribution of X, we use:

$P(a < X < b) = P(a \leq X < b) = P(a < X \leq b) = P(a \leq X \leq b) =$
Area under the curve that corresponds to an interval of values bounded by a and b.

Finally, continuous random variables have a mean, μ, and a standard deviation, σ. Their calculation involves mathematics beyond the scope of this book, but their interpretations are the same as for discrete random variables. The mean is the average outcome or X value from infinite repetitions of the experiment and the standard deviation measures the "average" distance of the X values from the mean.

6.2 The Probability Distribution of a Normal Random Variable

The preceding information describes all continuous random variables. One particular continuous random variable, very important in statistics and many other areas, is the *normal random variable*. Its probability distribution is called the normal distribution or normal curve. It is sometimes known as the bell curve.

Several characteristics of this distribution facilitate the mathematics of problem solving. These include:

1. The curve is symmetric about the mean, μ. Remember, X values lie on the number line, as does μ, the center of the values. Areas under the curve are probabilities. Half of the total area lies under the curve on either side of the mean.

Question: What are the limits on any area that corresponds to an interval of X values?
Answer: Because areas under the curve represent probabilities, their magnitudes must be constrained, as any probability is, to range between 0 and 1.

2. X can be any value between negative and positive infinity.

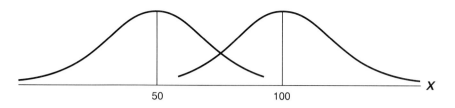

Figure 6-4
Normal curves with different means

3. The tails of the curves approach the number line that represents X values but never actually touch it. Consequently, the total area under the curve approaches 1. Symmetry dictates that the area on either side of the curve is 0.5. Almost all of the total area is within three standard deviations of the mean.
4. The mean, μ, determines the position of the curve on the number line or axis.

Question: Draw two normal curves along a number line; one with a mean of 50 and the other with a mean of 100.

Answer: When the mean changes, the curve shifts along the number line as shown in Figure 6-4.

5. The standard deviation, σ, determines the compactness of the values about the mean, illustrated by the shape of the curve about μ.

Question: Draw two normal curves with the same mean but different standard deviations. Which one has the smaller standard deviation?

Answer: The smaller standard deviation means the X values cluster around μ; that is, the probability an X value will be near μ is greater. The major portion of the total area under the curve in the center near μ reflects this fact. See Figure 6-5.

Figure 6-5
Normal curves with different standard deviations

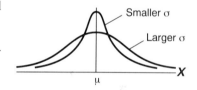

We will use these five properties when we work with probabilities of normally distributed X values.

■ **Example 6-1:** The mean height of a sample of 6,682 men between the ages of approximately 17.6 and 26.8 is 68.707 inches with a standard deviation of 2.602 inches. The mean of a similar sample of 1,330 women between the ages of approximately 17.6 and 28.5 is 64.154 inches with a standard deviation of 2.567 inches. The samples consist of U.S. Army soldiers, but the results compare well with other large sample measures of height, so they appear to represent healthy, young people in the United States. Heights generally follow a normal distribution. Use this information to sketch a curve for males and one for females along a single axis measuring heights. Then, determine the probability that a male in this age group is taller than 68.707 inches and, finally, that a female is shorter than 64.154 inches. Indicate these heights on the diagram and shade the area corresponding to these probabilities. (Source: E. Giles and P. H. Vallandigham, "Height Estimation from Foot and Shoeprint Length," *Journal of Forensic Sciences*, Vol. 36, No. 4, July 1991, pp. 1134–51.)

■ **Solution:** Figure 6-6 displays the diagrams with the curve for female heights located below the curve for male heights, because the female mean is smaller, and with the female curve more compact about its mean, because the female standard deviation is slightly smaller (the figure exaggerates this difference for the sake of comparison). The probabilities, $P(X_{male} > 68.707)$ and $P(X_{female} < 64.154)$, are both 0.5, because the boundary specified in each interval of values is the mean for each group. These areas are shaded in the figure.

Figure 6-6

Comparison of U.S. male and female heights for young, healthy individuals

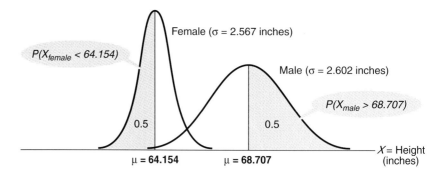

6.3 The Standard Normal Random Variable

Every time μ or σ changes, there is a new normal random variable with a curve positioned or shaped differently. To handle all of these occasions we must first be able to handle one special normal random variable, called the *standard normal*, depicted in Figure 6-7.

> The standard normal random variable is a normal random variable with a mean of 0 and a standard deviation of 1. The symbol for standard normal random variable values is Z throughout this book.

■ **Example 6-2:** Consider Elizabeth's driving speed. She tries to obey the speed limit very closely. In fact, her speed rating, Z, the difference between her actual speed and the speed limit (actual − limit) is standard normally distributed. Thus, half the time she drives below the speed limit (Z is negative) and half the time above (Z is positive). Her average speed rating is zero miles per hour (mph) over the limit, which means that, on average, she drives at the speed limit. A standard deviation of 1 mph implies that her deviations from the speed limit are typically very small. Suppose we want to know the proportion of times her speed rating is between 0 and 1.35 mph. Sketch a standard normal curve and locate this interval of Z values. Then, show where $P(0 < Z < 1.35)$ falls on the diagram.

■ **Solution:** The green interval along the number line in Figure 6-8 denotes speed ratings between 0 and 1.35. The medium orange area under the curve depicts the proportion of her driving time during which she speeds between 0 and 1.35 mph over the limit, $P(0 < Z < 1.35)$.

Figure 6-7

The standard normal random variable

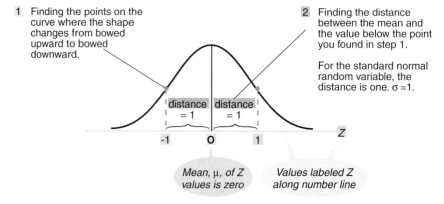

Section 6.3 The Standard Normal Random Variable ■ **253**

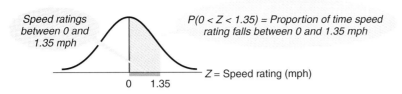

Figure 6-8
Diagram for Elizabeth's speed ratings

6.3.1 Finding Probabilities

To determine the area under a curve, we use a standard normal probability table. Figure 6-9 displays a segment of this table, and Appendix H contains the complete table.

We use the values along the leftmost column under Z and in the top row following Z to locate specific values along the Z axis. The Z column gives the units and tenth places of the value, and the row gives the hundredth places. The table in Appendix H spans Z values between 0 and 3.09. Remember that most of the area is within three standard deviations of the mean (where $\mu = 0$ and $\sigma = 1$ for the standard normal), so we usually do not need a larger table. *Entries inside the table represent areas under the standard normal curve between 0 and the corresponding Z value, z,* such as the medium orange area in Figure 6-8.

To find the probability that a standard normal value falls between 0 and z, that is, $P(0 < Z < z)$, proceed as follows:
1. Locate the units and tenth portions of z along the leftmost column of the table.
2. Locate the hundredth portions of z along the top row.
3. The intersection of the row from Step 1 and the column from Step 2 locates the probability.

■ **Example 6-2 (continued):** Find the proportion of Elizabeth's driving time during which she rides between 0 and 1.35 mph over the speed limit. Then, find the probability she will be driving between 0 and 0.01 mph over the limit.

Hundredths place of Z value

Z	.00	.01	...	.05	...	.09
.0	.0000	.0040	...	.0199	...	.0359
.1	.0398	.0438	...	.0596	...	.0753
⋮	⋮	⋮	...	⋮	...	⋮
1.3	.4032	.4049	...	.4115	...	.4177
⋮	⋮	⋮	...	⋮	...	⋮
2.9	.4981	.4982	...	.4984	...	.4986
3.0	.4987	.4987	...	.4989	...	.4990

Units and tenths places of Z value

Probability values

Figure 6-9
Selected portions of the standard normal table

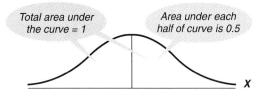
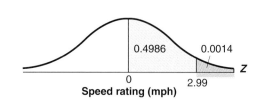

Figure 6-10

Diagrams for solving the speed-rating problem

■ **Solution:** We look down the Z column until we find 1.3. The "5" in the hundredths place is in the top row listed as 0.05. The intersection of the 1.3 row and the 0.05 column contains the value 0.4115. This value is the area we seek, $P(0 < Z < 1.35) = 0.4115$. Thus, about 41% of the time, Elizabeth drives between 0 and 1.35 mph above the speed limit.

The intersection of the 0.0 row and the 0.01 column locates the answer to the second part of the problem, 0.0040. $P(0 < Z < 0.01) = 0.0040$. We could sketch a diagram like Figure 6-8, except 0.01 would replace 1.35 along the Z value axis. The orange area would now be 0.0040.

Often, we will want to find probabilities of Z values that fall in intervals that do not begin or end at zero. The table does not provide this information directly. To determine these probabilities, *we combine, by addition and subtraction, areas that we find in the table with areas that we already know from general information and from the symmetry of the distribution.* Figure 6-10a reviews the areas we already know. Sketches of the normal curve will help organize and visualize the information in the problem and its solution.

■ **Example 6-2 (continued):** Draw a diagram and shade the area that corresponds to Elizabeth's driving more than 2.99 mph over the limit, $P(Z > 2.99)$. Then find this probability using information from the table and properties of the standard normal distribution.

■ **Solution:** Figure 6-10b displays the Z values as a green interval along the number line, and the probability is the dark orange area in the right-hand tail. Looking up $Z = 2.99$ in the table gives us an area value of 0.4986, but this is the lighter medium orange area under the curve between 0 and 2.99, not the area between 2.99 and positive infinity. However, we do know that the entire area from 0 to the right under the curve is 0.5. If the light orange area of this half of the diagram is 0.4986, then the orange area must be $0.5 - 0.4986 = 0.0014$ or $P(Z > 2.99) = 0.0014$.

Figure 6-11

Z values equidistant from 0

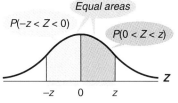

a) Areas that correspond to Z values equidistant from 0

b) Equal-sized intervals that are not equidistant from 0 do not have an equal probability of occurring

Note that all of the Z values in the table are positive. However, because the shape of the curve is the same on both sides of 0, the areas under the curve will be equal for Z value intervals that span the same distance from the mean (0) on either side. Figure 6-11 reinforces this statement and also shows that it does not apply to equal-sized intervals in any position. Symbolically, we can write $P(0 < Z < z) = P(-z < Z < 0)$. We can use this information to find probabilities associated with negative Z values.

■ **Example 6-2 (continued):** Find the probability that Elizabeth drives between 1.35 mph below the speed limit and the actual speed limit, $P(-1.35 < Z < 0)$. Then, find the proportion of driving time during which she stays within 3 mph of the speed limit.

Solution: Because $P(-1.35 < Z < 0) = P(0 < Z < 1.35)$, we only need to find the area that corresponds to $Z = 1.35$ in the table, which we did earlier. The area and the proportion of her driving time spent between 1.35 below the limit and exactly at the limit is 0.4115.

The proportion of driving time during which she stays within 3 mph of the speed limit is the area corresponding to values between -3 and $+3$, or $P(-3 < Z < 3)$. We can find the area under the curve between 0 and 3 (which is also the area between -3 and 0) in the table. These areas are each 0.4987. Summing these areas produces the desired area as shown in Figure 6-12, or $P(-3 < Z < 3) = P(-3 < Z < 0) + P(0 < Z < 3) = 0.4987 + 0.4987 = 0.9974$.

Because $\sigma = 1$ for the standard normal distribution, $P(-3 < Z < 3)$ is the same as saying *the proportion of values that are within three standard deviations of the mean.* Note that 0.9974, or almost all of the values, fall in this region.

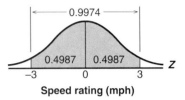

Figure 6-12
$P(-3 < Z < 3) = 0.9974 =$ proportion of time during which Elizabeth drives within 3 mph of the speed limit

■ **Example 6-2 (continued):** Find the probability that Elizabeth drives 1.3 to 2.9 mph below the speed limit.

■ **Solution:** Figure 6-13 shows that the area corresponding to the answer is between -2.9 and -1.3. Looking up $Z = -1.30$, we find an area of 0.4032, the area under the curve between -1.3 and 0. Similarly, we find the area 0.4981 corresponding to $Z = -2.90$. The position of this area overlaps the area between -1.3 and 0, so we have an area of 0.4981, of which 0.4032 is not shaded, and the part we want to know is shaded. The shaded portion must be the difference in the two values or 0.0949. Thus, 9.49% of the time Elizabeth drives between 2.9 and 1.3 mph below the speed limit.

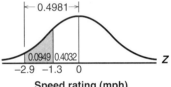

Figure 6-13
$P(-2.9 < Z < -1.30) = 0.0949 =$ proportion of time during which Elizabeth drives between 1.3 and 2.9 mph below the speed limit

One final note, *we will round Z values to two places past the decimal to use the table to find probabilities,* as we proceed. Recall that when a value is exactly halfway between two values, we round so that the rounded value ends in an even digit.

Section 6.3.1 Problems

5. Find $P(0 < Z < 1.48)$.
6. Find $P(0 < Z \leq 0.05)$.
7. Find $P(-2.84 \leq Z \leq 0)$.
8. Find $P(Z \geq 1.01)$.
9. Find $P(Z > -0.88)$.
10. Find $P(2.44 < Z < 2.91)$.
11. Find $P(0.80 < Z \leq 0.85)$.
12. Find $P(Z \geq -2.55)$.
13. Find $P(-0.36 < Z < 2.65)$.
14. Find $P(Z < -1.29)$.
15. Find $P(Z < 1.29)$.
16. $P(1.22 < Z < 2.90)$.
17. Find $P(-0.23 \leq Z \leq -0.03)$.
18. Find $P(-1.28 < Z < 1.28)$.
19. Trent employs a meter that measures optimum reception as 0 for locating radio stations on an FM receiver. The distance between the actual dial setting and the optimum setting is negative when the dial is too far to the left of the station and positive when it is too far to the right. Suppose that the distance from optimum on the first stop when first tuning in a station is standard normally distributed. Find

   a. The probability that when Trent first stops turning the dial to find a particular station, the tuning is off by a distance between -0.75 and 0.58.
   b. The probability the distance is between -2 and 2 on the meter.
   c. The probability the distance is more than 0.99 to the right on the meter.
   d. The probability the distance is more than 0.99 to the left on the meter.

20. Suppose the balance of payments of a country, reported in billions of dollars, is standard normally distributed. Positive values represent surpluses, and negative values represent deficits.

   a. What percentage of the time does the country have a *surplus* of $1.75 billion or more?
   b. What percentage of the time is there a *deficit* between $1.5 and $2.08 billion?
   c. What proportion of the time is the balance between a *deficit* of $0.88 billion and a *surplus* of $0.10 billion?

Figure 6-14

Finding Z value given a probability for an interval of Z values, bounded by z

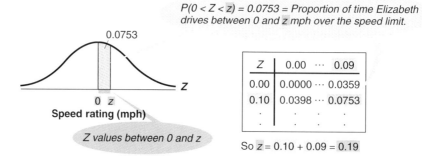

6.3.2 Finding Z Values

So far, we have used the table to find the probabilities that correspond to given intervals of values on the Z axis. Sometimes, though, situations exist where we know the probabilities and must locate the corresponding values along the axis.

■ **Example 6-2 (continued):** Find the specific value of Elizabeth's speed rating, z, such that 7.53% of the time, she drives between 0 and z mph over the limit, that is, $P(0 < Z < z) = 0.0753$. Remember her speed rating, the deviation between the speed she travels and the speed limit, is a standard normal random variable.

■ **Solution:** This time, we know the area under the standard normal curve between 0 and z is 0.0753, so we look within the table until we find an area entry of 0.0753. Moving back along the row containing 0.0753 and up the column containing this entry, we find the Z value corresponding to an area of 0.0753. This value is $Z = 0.19$ mph as shown in Figure 6-14.

This problem is simple, because the given area corresponds exactly to the location of areas given in the table, namely, areas between 0 and a positive Z value. When this is not the case, there are three steps to finding the Z value that corresponds to some area. Determining if z is positive or negative is the first step in locating z.

Question: Suppose the problem is to find z so that $P(Z > z) = 0.485$. Is z a positive or negative Z value?

Answer: z is positive. For this type of problem we can inspect two conceivable diagrams, one where z is positive and the other with a negative z. We shade the area that corresponds to all Z values larger than z in both diagrams, because the problem concerns Z values *greater than* z. Then, we examine each area to determine which one can be 0.485 and which one cannot. In Figure 6-15a, the shaded area is greater than 0.5, because it includes more than half of the area under the curve. Thus, the shaded area cannot be 0.485. This means that z cannot be negative. Figure 6-15b is the correct diagram for this problem.

Figure 6-15

Diagrams to decide if z is positive or negative

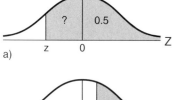
a)

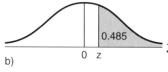
b)

The second step is to locate an area on the diagram that corresponds to the areas provided in the standard normal table. That is, find the area under the curve between 0 and z.

Question: Search the table for the area 0.485, and determine if the corresponding z (with the positive sign attached) is the answer to the problem, $P(Z > z) = 0.485$.

Answer: The table only contains areas that correspond to values between 0 and z. If we look up 0.485 as an area, the corresponding z is 2.17 (see the normal table in Appendix H and Figure 6-16a). 0.485 is the proportion of time the Z value is between 0 and 2.17, not the proportion of time the Z value is greater than 2.17.

a) Wrong area for $P(Z > z) = 0.485$ b) Correct area for $P(Z > z) = 0.485$

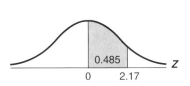
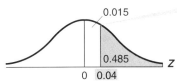

The moral is to avoid automatically looking up any area or probability given in a problem. The area must correspond to the information in the table. Because the area in the tail of the curve is 0.485, the area under the curve between 0 and z must be the difference between 0.5 and 0.485, or 0.015, as shown in Figure 6-16b.

Finally, we work from this area back to the corresponding magnitude of z in the table. We attach the sign determined earlier to this magnitude to obtain the answer.

Question: Using 0.015 as the area under the curve between 0 and z, locate z in the standard normal table.

Answer: The closest area in the table is 0.0160, which corresponds to a Z value of 0.04 as shown in Figure 6-16b.

Attaching the positive sign determined earlier, we obtain $z = 0.04$. Figure 6-16b shows that $P(Z > 0.04) = 0.485$.

These three steps are all that is necessary to find the boundary or boundaries of an interval corresponding to a given area or probability.

Figure 6-16
Finding the area between 0 and the z value

Procedure for Determining the Boundaries of an Interval of Standard Normal Values That Correspond to a Given Area or Probability

1. Determine if z is positive or negative.
   a. Construct two diagrams and locate a positive z on one diagram and a negative z on the other, along with any other given information about the Z values in the situation.
   b. Shade the relevant area from the given information, an interval with z as one of the boundaries.
   c. Determine the valid diagram by eliminating the diagram with a shaded area from Step b that contradicts the properties of the normal distribution, such as a shaded area that is supposed to exceed 0.5, but it covers less than the entire area on one side of the mean.
2. Determine the area under the curve between the mean, 0, and z. Be careful not to use the area mentioned in a problem automatically.
3. Search for the area determined in Step 2 among the area entries of the standard normal table, and determine the magnitude of the z corresponding to this area. Attach the sign determined in Step 1 to this magnitude for the solution for z that satisfies the given information.

Note: Solutions to boundary problems in this book employ the closest probability value in the table when the exact value is not an entry, unless it lies exactly halfway between two Z values, in which case we will use the midpoint of the two values. For instance, an area of 0.45 is exactly midway between the Z values 1.64 and 1.65 in the table, so we will use 1.645.

■ **Example 6-2 (continued):** Find the speed rating, z, if 75% of Elizabeth's speed rating will be larger than z, $P(Z > z) = 0.75$.

Figure 6-17
Finding Z so that $P(Z > z) = 0.75$

a) Positive z causes inconsistency

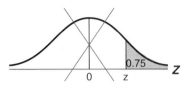

b) Negative z makes $P(Z > z) = 0.75$ true

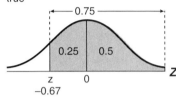

Figure 6-18
Finding the smallest loss, z, in the largest 38% of all losing amounts, $P(Z < z) = 0.38$

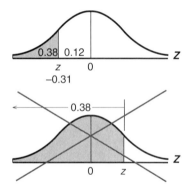

Figure 6-19
Lowest 95% of the balances or Z values, the 95th percentile of the bank balances

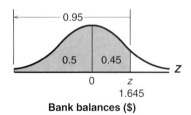

■ **Solution:** Step 1: A comparison of diagrams with positive and negative z values dictates that z must be negative in order for the area under the curve to the right of z to be 0.75, as shown in Figure 6-17. Remember, the area under each side of the curve is 0.5.

Step 2: The area under the curve between z and 0 is 0.25 (the difference between the shaded area 0.75 and the known area to the right of the mean, 0.5). Again, refer to Figure 6-17.

Step 3: The closest entry in the table to 0.25 is 0.2486, which corresponds to $z = 0.67$, but z must be negative, so $z = -0.67$ mph. Seventy-five percent of the time, Elizabeth's speed rating is -0.67 mph or more (which means that during the fastest 75% of her driving time, the least she travels below the legal speed limit is 0.67 mph).

■ **Example 6-3:** Suppose a player's financial outcomes per night, over several years in a friendly penny-ante poker game, fit a standard normal distribution. (Negative values represent the amount lost after an evening's play.) Find the smallest amount lost, z, in the largest 38% of losing amounts.

■ **Solution:** The largest losing amounts lie on the extreme left side of the Z axis as the largest negative Z values. The largest 38% of the losses define an interval on this end of the axis with a shaded area of 0.38 above it. The smallest loss in this interval is the value that bounds the interval on the right, so we want to find z to satisfy $P(Z < z) = 0.38$. Now, the question is whether z is positive or negative. Figure 6-18 reveals that a valid diagram results when z is negative. The area between 0 and z is the difference between 0.5 and 0.38 or 0.12. The area in the table closest to 0.12 is 0.1217, which corresponds to $z = 0.31$. Affixing the negative sign produces an amount of $-\$0.31$ or -31 cents. The largest 38% of losing amounts are more than this amount.

This Z value, $-\$0.31$, is the 38th *percentile* of the poker outcome values, because 38% of the amounts are smaller than $-\$0.31$. This may seem confusing at first, because we equate more negative values with larger losses. However, if we just consider the Z values on a number line when finding percentiles, negative values are smaller than positive values and the more negative a value, the smaller it is. For instance, -2 is smaller than -1.

■ **Example 6-4:** The balance in a financial wizard's checking account follows a standard normal distribution. (Think about it, would you want such a checking account?) Find the 95th percentile of these balances.

■ **Solution:** The 95th percentile, z, must fall on the right side of the curve, so that the area corresponding to balances smaller than z can be 0.95, as shown in Figure 6-19. The area between 0 and z must be the difference between 0.95 and 0.5 or 0.45. The area 0.45 is exactly midway between 1.64 and 1.65 in the table, so we use 1.645. Because the sign is plus, the 95th percentile of the balances is $1.645 (if you do not like three places past the decimal for dollar values, you could round this to $1.64 so that the last digit is even).

■ **Example 6-3 (continued):** Find the limits, $+z$ and $-z$, of the middle 95% of poker outcome values.

■ **Solution:** Because the normal curve is symmetric, the two values, z and $-z$, must be equidistant from the mean, 0, in order for the interval to contain the middle 95% of

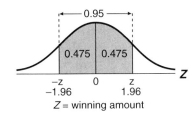

Figure 6-20
Finding z so that $P(-z < Z < z) = 0.95$, the boundaries of the middle 95% of winning amounts

the winning amounts, or $P(-z < Z < z) = 0.95$. Thus, the area corresponding to the interval $-z$ to 0 is identical to the area corresponding to the interval 0 to z. If the total area is 0.95, then half of this total, or 0.475, must lie above each interval, as shown in Figure 6-20. The z that corresponds to this area is 1.96. Thus, $P(-1.96 < Z < 1.96) = 0.95$, or the middle 95% of the poker amounts lie between $-\$1.96$ and $\$1.96$; that is, between losing and winning $1.96.

Section 6.3.2 Problems

In Problems 21–28, find z, so that

21. $P(0 < Z \leq z) = 0.37$. **22.** $P(-z \leq Z \leq 0) = 0.44$.
23. $P(Z < z) = 0.16$. **24.** $P(Z > z) = 0.86$.
25. $P(Z > z) = 0.25$. **26.** $P(-z \leq Z \leq z) = 0.95$.
27. $P(Z \leq z) = 0.55$. **28.** $P(-z < Z < z) = 0.80$.

29. Trent employs a meter that measures optimum reception as zero for locating radio stations on an FM receiver. The distance between the actual dial setting and the optimum setting is negative when the dial is too far to the left of the station and positive when it is too far to the right. Suppose that the distance from optimum on the first stop when first tuning in a station is standard normally distributed. Find

a. The 65th percentile of the meter readings.
b. The 10th percentile of the meter readings.
c. The boundaries of the interval containing the middle 98% of the distances from first stops.

30. Suppose the balance of payments of a country, reported in billions of dollars, is standard normally distributed. Positive values represent surpluses, and negative values represent deficits.
a. Find the smallest surplus in the largest 40% of balances.
b. Find the smallest deficit in the largest 35% of deficits.
c. What is the forty-third percentile of the balances?

Summary and Review

A curve depicts the probability distribution of the values of a continuous random variable. We measure values of the random variable along the horizontal axis, while probabilities are the areas under the curve that correspond to intervals on the value axis. Continuous random variables have a mean, μ, and standard deviation, σ, that we interpret like the μ and σ for discrete random variables.

The normal probability distribution is one of the most important distributions in statistics. Many phenomena exhibit patterns similar to it. The properties of this distribution and its curve are:

1. The curve is symmetric about the mean, μ.
2. Any value between negative and positive infinity is possible, and most values occur within three standard deviations of μ.
3. The curve approaches the Z axis on either side, forming a bell shape, so that the total area under the curve approaches 1, with an area of 0.5 on either side of the mean.
4. The mean, μ, positions the curve on the number line.
5. The standard deviation, σ, determines how compact the curve is about μ.

The standard normal random variable is one of the infinite possible normal random variables, but it is the basic one to which we return to answer questions about normal random variables that are not standard normal. The mean of standard normal values is 0, and their standard deviation is 1. The standard normal table and knowledge of the properties of the normal curve enable us to compute probabilities or locate their associated Z values easily. An example of the latter is to find a percentile.

Multiple-Choice Problems

31. We depict continuous random variables with a
  a. Frequency distribution.
  b. Table.
  c. Graph.
  d. Histogram.
  e. Chart.

32. To find probabilities of a continuous random variable's values, we must ascertain
  a. n, x, and p.
  b. The area in a bar of a histogram.
  c. The area under the curve corresponding to the interval of values.
  d. The number of outcomes that are successes and all the possible outcomes.
  e. If the problem is $P(a < X < b)$ or $P(a \leq X \leq b)$.

33. Values on the horizontal axis of the graph of a continuous random variable represent
  a. Probabilities.
  b. The discrete points that are possible values.
  c. The height of the curve representing the distribution.
  d. Frequencies.
  e. Experimental outcomes.

34. The mean of a normal random variable
  a. Positions the normal curve on the axis or number line.
  b. Measures the average distance between possible values of the random variable.
  c. Determines the shape of the normal probability curve.
  d. Differs from the median because of symmetry.
  e. Is one if the random variable is standard normal.

35. The characteristic that distinguishes a standard normal random variable from other normal random variables is
  a. The fact that the standard normal has a standard deviation and the others do not.
  b. The fact that the specific mean and standard deviation of the standard normal random variable are 0 and 1, respectively.
  c. Almost all values of the standard normal random variable fall within three standard deviation of the mean, whereas the others require a greater distance.
  d. The standard normal is continuous and the other normal random variables are discrete.
  e. The symmetry about the mean of the standard normal.

36. Which of the curves in Figure 6-21 has the smallest standard deviation, σ?
  a. A.   **d.** D.
  b. B.   **e.** E.
  c. C.

37. Find $P(0 < Z < 1.86)$.
  a. 0.4686.   **d.** 0.0314.
  b. 0.4641.   **e.** 0.4800.
  c. 0.3888.

38. Find $P(Z < -1.5)$.
  a. 0.9332.   **d.** 0.5000.
  b. 0.0668.   **e.** 0.4332.
  c. 0.5668.

39. Find z so that $P(Z > z) = 0.1$.
  a. 0.25.   **d.** 1.28.
  b. -0.25.   **e.** 0.0398.
  c. -1.28.

An economic forecaster's errors from predicting sales (forecast value $-$ actual value) fit a standard normal distribution. The forecast and actual values are expressed in billions, so errors are expressed to nine decimal places, and thus almost any value along the number line is a possible error value. Use this information to answer Questions 40–45.

40. What proportion of the errors are between $-\$2$ and $+\$2$?
  a. 0.0228.   **d.** 0.9772.
  b. 0.4772.   **e.** 0.9544.
  c. 0.0456.

41. What proportion of the errors are between $1 and $2?
  a. 0.4772.   **d.** 0.1359.
  b. 0.3413.   **e.** 0.0228.
  c. 0.8185.

42. Find the error value, z, if 42.65% of the errors are negative but larger than z (closer to 0).
  a. $-\$1.45$.   **d.** $\$0.1664$.
  b. $\$1.45$.   **e.** $-\$0.19$.
  c. $-\$0.1664$.

43. Find the boundaries of an interval containing the middle 98% of error values. ($\pm z$ means the positive z and the negative z.)
  a. $\pm\$0.02$.   **d.** $\pm\$0.1879$.
  b. $\pm\$2.33$.   **e.** $\pm\$0.05$.
  c. $\pm\$0.3365$.

Figure 6-21

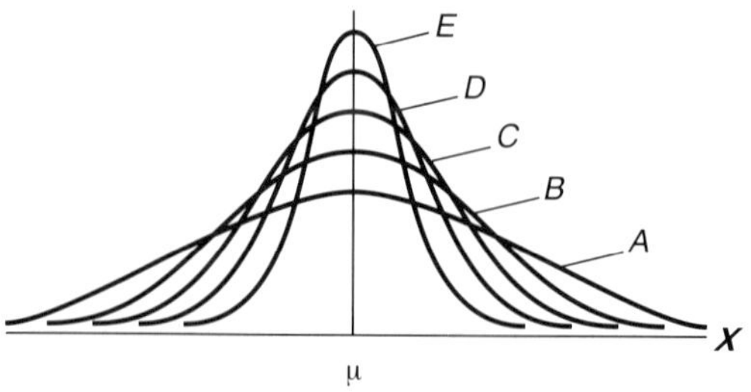

44. $0.44 is the _____ percentile of the error values.
  a. 44th.   d. 83rd.
  b. 17th.   e. 33rd.
  c. 67th.

45. Find the 91st percentile of the error values.
  a. $91.00.   c. $1.96.   e. $1.34.
  b. $0.91.   d. −$1.34.

46. Which of the following is *not* characteristic of normally distributed phenomena?
  a. Values usually occur in the center of possible values.
  b. Over half of the values occur exactly at the mean.
  c. Most values occur within three standard deviations of the mean.
  d. The mean, median, and mode are the same.
  e. The values that are least likely to occur are very small and very large.

47. Find the approximate 75th percentile of a standard normal random variable.
  a. 0.75.   c. 0.67.   e. 1.
  b. 0.25.   d. 0.2764.

48. What proportion of standard normal values fall within one standard deviation of the mean?
  a. 0.3413.   d. 1/3.
  b. 0.6826.   e. 0.3174.
  c. 1.

49. Which of the following tends to be skewed rather than normally distributed?
  a. Body weights.   d. Income.
  b. Body temperatures.   e. Cholesterol values.
  c. IQ scores.

50. The mean hospital occupancy rate for the 50 states and District of Columbia in 1987 was 67.4%, and the standard deviation was about 7.6 percentage points. If these values fit a normal probability distribution, what proportion of the states would have an occupancy rate smaller than 67.4%? (Source: American Hospital Association, *Hospital Statistics*, 1988, as reported in U.S. Bureau of the Census, *Statistical Abstract of the United States: 1990*, 110th ed. [Washington, D.C.: Bureau of the Census, 1990], p. 106.)
  a. 0.5000.   d. 0.0076.
  b. 0.6740.   e. 0.2500.
  c. 0.3260.

51. Find $P(0 \leq Z \leq 2.86)$.
52. Find $P(-1.23 < Z < 0)$.
53. Find $P(Z > 0.57)$.
54. Find $P(0.32 \leq Z \leq 1.66)$.
55. Find $P(-1.96 < Z < 1.96)$.
56. Find $P(Z < 0.86)$.
57. Find $P(Z \leq -2.40)$.
58. Find $P(-3 < Z \leq -1.11)$.
59. Find $P(Z < -2.5 \text{ and } Z > -3)$.
60. Find $P(Z \leq -2.53 \text{ or } Z > 2.53)$.
61. Find $P(Z < 2.53 \text{ and } Z > 2.53)$.
62. Find $P(Z \leq 0.56 \text{ and } Z < -2)$.

In Problems 63–67, find z, so that

63. $P(Z > z) = 0.32$.
64. $P(0 \leq Z \leq z) = 0.45$.
65. $P(Z > z) = 0.64$.
66. $P(-1.5 < Z < z) = 0.16$.
67. $P(-z < Z < z) = 0.99$.

68. Sketch each of the following pairs of normal curves on a single horizontal axis.
  a. The means differ, but the standard deviations do not.
  b. The means differ, and the standard deviations differ.
  c. The means are identical, but the standard deviations differ.

69. What proportion of standard normal values falls within two standard deviations of the mean? 2.5? 2.75?

70. Three levels of distinguished graduates from academic institutions, in order of decreasing distinction, are summa cum laude, magna cum laude, and cum laude. Assuming that the requirements are more restricted (the range of possible grade-point averages contracts) for higher levels of distinction and that the grade point averages within different distinctions are normally distributed, sketch three normal probability curves on a single horizontal axis that reflect this information.

71. Suppose a student's arithmetic errors fit a standard normal distribution. Find
  a. The proportion of errors that are larger than 1.
  b. The proportion of errors that are smaller than 2.83.
  c. The proportion of errors that are larger than 1 in absolute value.
  d. The 28th percentile of the error values.
  e. The two values that define the middle 80% of the errors.

72. The mean change in IQ scores (latest minus older score) between the approximate ages of 2.17 years and 7.00 years for a group of adopted children is −1.8 points, and the standard deviation is 14.6 points. Use this information and assume the changes are normally distributed to answer the following questions. (Source: Marie Skodak and Harold M. Skeels, "A Final Follow-Up Study of One Hundred Adopted Children," *The Journal of Genetic Psychology*, Vol. 75, 1949, pp. 95, 101.)
  a. Sketch the curve depicting this random variable. Shade and determine the probability that a child's score dropped by more than 1.8 points.
  b. Is the probability a child's scored increased equal to, more than, or less than 0.5? Demonstrate your answer by using a sketch of the distribution.

73. The net price change (new minus old price) of a certain stock is standard normal.
   a. What percentage of the time does the price fall more than $0.85?
   b. What percentage of the time is the net change (+ or −) less than $0.50?
   c. What is the 92nd percentile of changes?
74. Suppose arrival time for babies (birth date minus due date measured in weeks) is a standard normal random variable (minus is early and plus is late). Use this information to answer the following questions:
   a. What percentage of arrival times are earlier than 2.5 weeks before the due date?
   b. What percentage of babies arrive within one week of the due date?
   c. The middle 95% of arrival times are between which two values?
   d. The latest 20% of arrival times are at least how many weeks late?
75. The difference between the time at which you remove cookies from the oven for desired moistness and the time listed in the recipe is standard normally distributed. For example, −1 means remove the cookies one minute early. Use this information to find
   a. The proportion of time that you remove the cookies at most two minutes later than the recipe time.
   b. The proportion of time that you remove the cookies between one minute early and three minutes late.
   c. The proportion of time that you remove the cookies within one minute of the recipe time.

Unit II: Working with Normal Random Variables

Preparation-Check Questions

76. If a certain set of values is standard normal, then
   a. Only values between −1 and +1 are possible.
   b. Values smaller than −3 and greater than +3 are most likely to occur.
   c. A binomial situation is being described.
   d. The mean value is 0, and the standard deviation of the values is 1.
   e. The mean value is 1, and the standard deviation of the values is infinite.
77. Find $P(-3.02 < Z < -2)$.
   a. 0.4987.   d. 0.0215.
   b. 0.4772.   e. 0.0013.
   c. 0.9759.
78. Find z so that $P(Z < z) = 0.6255$.
   a. 0.32.   d. −1.15.
   b. −0.32. e. 0.23.
   c. 1.15.
79. To find the Z score for an X value in a population, we compute $Z = (X - \mu)/\sigma$. $Z = 2$ indicates that
   a. The population mean is 2.
   b. The population standard deviation is 2.
   c. The population standard deviation is twice the magnitude of the mean.
   d. The average deviation is 2.
   e. The X value is 2 standard deviations above the mean.
80. Which of the following situations fits the binomial assumptions? Find the probability of
   a. Three fives and three twos in six tosses of a die.
   b. A spade on the first and a red card on the second of two draws from a standard deck with replacement.
   c. A spade on the first and a red card on the second of two draws from a standard deck without replacement.
   d. Three red cards out of 10 draws from a standard deck with replacement.
   e. Three red cards out of 10 draws from a standard deck without replacement.
81. Pick the true statement about the diagram in Figure 6-22.
   a. The area of the bar above $2 = P(2)$.
   b. The length along the X axis from 0 to 5 is 1 unit.
   c. The base of the bar above the 2 goes from 1.75 to 2.75.
   d. $X = 2$ has the highest probability of occurring.
   e. $X = 2$ has the lowest probability of occurring.

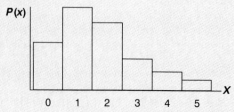

Figure 6-22

(continued on next page)

82. Specify n, X, and p, respectively, for finding the probability of three red cards out of 10 draws from a standard deck with replacement.
   a. 3, 10, 0.3.   d. 10, 3, 0.3.
   b. 10, 3, 0.5.   e. 10, 7, 0.12.
   c. 3, 10, 0.5.
83. Suppose you are concerned with the binomial random variable for 450 trials with a probability of success on a single trial equal to 1/3. What are the mean, μ, and the standard deviation, σ, of this random variable?
   a. 450, 1/3.   d. 150, 10.
   b. 450, 2/3.   e. 300, 100.
   c. 150, 1/3.

Answers: 76. d; 77. d; 78. a; 79. e; 80. d; 81. a; 82. b; 83. d.

Introduction

Club Fundraiser A school club plans a fundraiser, selling plastic party streamer material with a special design featuring the school's mascot and logo. Fans and other customers will be able to purchase any desired length from large rolls of the material. A local gift and party store owner suggests that the lengths customers purchase are approximately normally distributed. She also provides an estimate of the mean and standard deviation of these sizes. The club wants to take advantage of this information and precut some streamers in order to fill orders speedily from an appropriate inventory of sizes that satisfy as many customers as possible and avoid excess stock.

Using our knowledge of the standard normal and the information collected by the store owner, we can easily determine the appropriate inventory, once the normal distribution is transformed into a standard normal one. This transformation is useful for solving many types of problems, including approximating probabilities for other random variables, such as the binomial, as we shall see in this unit.

6.4 The Transformation Formula

The number of different possible normal random variables is infinite. Their distributions all have the same basic symmetrical bell shape, but every change in the mean, μ, or standard deviation, σ, produces a different normal random variable and alters the position on the number line or the compactness of the curve. We use the standard normal table to find probabilities when $\mu = 0$ and $\sigma = 1$. There is no possible way to produce tables for each of the infinite possible normal random variables. Instead, we convert any normal random variable to a standard normal and use the one table to investigate situations that involve this normal random variable.

Converting normally distributed values to standard normally distributed values is similar to changing inches to feet or degrees Fahrenheit to degrees Celsius. If we know the conversion factors (like 12 inches to 1 foot), some simple arithmetic allows us to convert from one measure to the other. In the present case, we convert X, any normal value expressed in X units, to a standard normal value expressed in Z units. To convert X to Z, we use the Z score transformation that was introduced in Chapter 3,

$$Z = (X - \mu)/\sigma.$$

Recall that *a Z value indicates the distance of a value from the mean measured in standard deviations.*

■ **Example 6-5:** Suppose the checking account balances of a financial sorcerer are normally distributed, with a mean, μ, of $100 and a standard deviation, σ, of $10. Label the mean and the points that are one and two standard deviations above and below the mean.

Figure 6-23

X values one and two standard deviations from μ

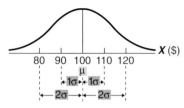

Solution: Begin at the mean ($\mu = \$100$). Moving to the right one standard deviation, \$10, locates the point \$110. One more σ puts us at \$120. The corresponding points one and two standard deviations below the mean are \$90 and \$80 (refer to Figure 6-23).

Alternatively, we can label these points by using the form \$100 plus or minus some multiple of $\sigma = \$10$, beginning at the mean:

$$\$100 = \$100 + 0(\$10).$$

and moving one and two standard deviations to the right:

$$\$110 = \$100 + 1(\$10)$$
$$\$120 = \$100 + 2(\$10).$$

The points one and two standard deviations below the mean become:

$$\$90 = \$100 - 1(\$10)$$
$$\$80 = \$100 - 2(\$10).$$

Actually we can express the value at any point on the X axis in this format. For example, \$115 is 1.5 standard deviations from μ, so

$$\$115 = \$100 + 1.5(\$10).$$

Question: Express \$106 and \$94 in this form.
Answer: \$106 is \$6 from the mean which is 0.6 (\$6/\$10) of the standard deviation, \$10, so $\$106 = \$100 + 0.6(\$10)$. Similarly, $\$94 = \$100 - 0.6(\$10)$.

Next, we draw two new axes parallel to the X axis, one to express X values in this new format and one below it with the points labeled with just the coefficients of \$10 from the second axis as shown in Figure 6-24.

We label each point on each of the axes with an alternative expression for X. There is a one-to-one correspondence between X and the points immediately below it on the other axes, although we label them differently. Remember, Z values measure distances of values from the mean in standard deviations, so one Z equals one standard deviation (in this case \$10). Thus, if we use the bottom axis, the mean of the values is 0, and the

Figure 6-24

Axes for alternative expressions of X values

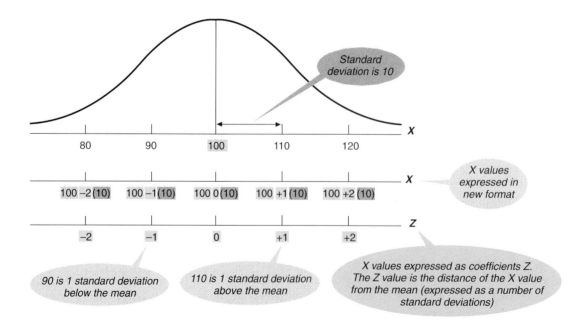

Section 6.4 The Transformation Formula

standard deviation is 1. We have a standard normal random variable, and this axis is the Z axis we used before to solve problems in conjunction with the standard normal table.

To organize this information into the transformation formula, we return to the second axis and substitute symbols. On the second axis, the coefficients of $10 are -2, -1, 0, 1, and 2. We express points on the second axis (which is still an X axis) as

$$X = \$100 + Z(\$10).$$

Question: What are the symbols for the values represented by $100 and $10 in this example?

Answer: $\$100 = \mu$ and $\$10 = \sigma$.

Substituting these symbols in the expression for X, we obtain

$$X = \mu + Z\sigma.$$

This formula allows us to convert a Z value to an X value.

> Transformation formula for converting a standard normal value, Z, to the original units of a normal random variable, X:
>
> $$X = \mu + Z\sigma.$$

In words, the Z values tells us how many standard deviations to move from the mean and the direction in which to move, so multiply Z times σ to find the distance from the mean in the X measurement units. Then, add the result to or subtract it from the mean according to the sign of the Z value to obtain the corresponding X value.

Suppose we know that $Z = 2$ for some bank balance value, X. We can determine this X by first finding $2\sigma = 2(\$10) = \$20$, the distance of X from μ in the original measurement units. Then, we add this amount to the mean, $100, because the Z is positive to obtain $X = \$120$.

Question: Convert $Z = -1.43$ to the corresponding bank balance in dollars, X.

Answer: X is 1.43 standard deviations below the mean. 1.43 standard deviations is the same as $\$14.30 = 1.43(\$10)$. We subtract this amount from the mean $100, because X is below the mean, to obtain $\$85.70 = \$100 - 1.43(\$10)$.

Question: X is normally distributed with mean 63 and standard deviation 0.5. Find the X value that is 1 standard deviation above the mean and the X value that is 2.22 standard deviations below the mean.

Answer: If X is 1 standard deviation above the mean, it must be 0.5 units (σ) more than the mean, 63, or $X = 63.5$. Similarly, if X is 2.22 standard deviations below the mean, it must be $2.22\sigma = 2.22(0.5) = 1.11$ below 63 or $X = 61.89$.

Often we know the X value, but need to convert it to a Z so that we can use the standard normal table. To do this we solve the transformation equation for Z. We begin by subtracting μ from both sides:

$$X - \mu = \cancel{\mu} + Z\sigma - \cancel{\mu}$$
$$X - \mu = Z\sigma.$$

Divide both sides by σ:

$$\frac{X - \mu}{\sigma} = \frac{Z\cancel{\sigma}}{\cancel{\sigma}} = Z.$$

This is the formula that we use to transform any normal value to a standard normal value. Sometimes we call this procedure *standardizing* a value. It is a very important formula in statistics.

Transformation Formula for Converting a Normal Value to a Standard Normal Value

$$Z = \frac{X - \mu}{\sigma}.$$

Remember, changing X to Z with this formula reveals how far an X value is from the mean in terms of standard deviations. In this transformation formula, the numerator, $X - \mu$, is the distance of an X value from the mean, measured in the original X units. The sign indicates whether the value is above or below the mean. To convert this magnitude to standard deviations, we divide by the standard deviation, which is also expressed in X units, just as we would convert 36 inches to feet by dividing 36 inches by one foot, expressed as 12 inches, to obtain 3 feet. The result in our transformation formula is a distance converted to standard deviations. Thus, when $\mu = 100$ and $\sigma = 10$, $X = \$115$ becomes

$$Z = \frac{\$115 - \$100}{\$10} = 1.5,$$

and $115 is 1.5 standard deviations above the mean, $100.

Question: Convert $94 to a Z.

Answer:
$$Z = \frac{\$94 - \$100}{\$10} = -0.6.$$

$94 is six-tenths of a standard deviation below μ.

Question: X is normally distributed with mean 63 and standard deviation 0.5. Convert $X = 64.25$ to a Z.

Answer:
$$Z = \frac{64.25 - 63}{0.5} = 2.5.$$

64.25 is 2.5 standard deviations above the mean.

Section 6.4 Problems

Fill in the blank cell in each row of the following table using the other information in each row. Sketch a normal curve with the given mean and standard deviation. Then, locate x and z on the diagram.

Problem	μ	σ	x	z
84.	100	10	74	
85.	252	53	300	
86.	10	0.72	11	
87.	53.61	1.01	55.02	
88.	100	10		1.13
89.	25	0.1		5.5
90.	345.62	23.44		−2.32

91. X is a normal random variable with a mean of 27.3 and a standard deviation of 0.04. How many standard deviations from the mean does 28 lie? 27? Sketch a normal curve and position this information on it.

92. X is a normal random variable with a mean of 1,440 and a standard deviation of 50. If x is 0.8 standard deviations above the mean, what is x? What is x, if it lies 1.005 standard deviations below the mean? 1.96 standard deviations below the mean? Sketch a normal curve, and position this information on it.

93. Suppose that the result of a single play (gain or loss) by a football team is normally distributed with mean 3.22 yards and a standard deviation of 5.84 yards. Sketch an appropriate normal curve, and locate the values involved in each question.

   a. On one particular play, the team gained 8.1 yards. How many standard deviations from the mean is this measurement?

   b. On another play, the team lost 8.1 yards. How many standard deviations from the mean is this measurement?

   c. A statistician figures that the result of another play is half a standard deviation above the mean. How many yards resulted from the play?

6.5 Finding Probabilities

Once we convert X to a Z value, we can use the standard normal table to solve probability problems.

■ **Example 6-5 (continued):** We want to find the proportion of the sorcerer's balances that exceed $115. Remember, the mean balance is $100, and the standard deviation is $10. Sketch a normal curve, and locate X values that satisfy $X > \$115$, then shade the area that corresponds to this probability.

■ **Solution:** $X > \$115$ corresponds to the green interval on the X axis in Figure 6-25. To find $P(X > 115)$, we must find the area under the curve above X values that are 115 or larger.

Next, we draw a Z axis under the X axis to see how the problem appears in Z terms. Remember that $Z = (115 - 100)/10 = 1.5$. Now the problem becomes $P(Z > 1.5)$, which Figure 6-25 shows is 0.0668. The result is $P(X > 115) = P(Z > 1.5) = 0.0668$.

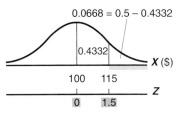

Figure 6-25
Converting a normal probability problem from X values to Z values

Follow these steps to find a normal probability when the values are given, such as $P(x_1 \le X \le x_2)$, where x_1 and x_2 are specified values.
1. Locate the interval of given X values, and shade the area to be found.
2. Transform any measurable boundaries of the X interval to Z values.
3. Determine the areas under the curve between 0 and the transformed boundaries.
4. Use the information from Step 3 and properties of the normal curve to determine the shaded area from Step 1.

■ **Example 6-5 (continued):** Find the probability the sorcerer's bank balance is between $90 and $115 at a randomly selected time. $P(\$90 < X < \$115)$.

■ **Solution:** Step 1: The area under the curve between the values $90 and $115 on the X axis represents $P(\$90 < X < \$115)$ (see Figure 6-26).
Step 2: Transforming $90 to a Z results in $Z = (\$90 - \$100)/\$10 = -1$. The earlier conversion of $115 to 1.5 does not change, because the mean and standard deviation have not changed. Thus, the problem becomes $P(-1 < Z < 1.5)$.
Step 3: The Z table contains the area between -1 and 0 and the area between 0 and 1.5. These are 0.3413 and 0.4332 as shown in Figure 6-26.
Step 4: Finally, we can see from Figure 6-26 that the sum of these two areas produces the area under the curve between $X = \$90$ and $X = \$115$; $0.3413 + 0.4332 = 0.7745$.

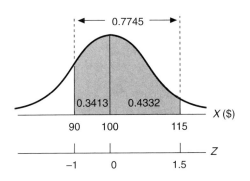

Figure 6-26
Solution diagram for $P(\$90 < X < \$115)$

■ **Example 6-6:** In 1978, the last year of the full Current Population Survey, the estimated mean number of hours usually worked during a week was 41.888 hours for those who usually worked at least 20 hours per week. The standard deviation was 6.291 hours. The sample was 7,594 white males aged 18–64 (who usually worked at least 20 hours per week). Some economic researchers are currently using this information to study the effect of the Fair Labor Standards Act passed in the 1930s that initiated federal minimum wage and overtime regulations. (Source: Stephen J. Trejo, "The Effects of Overtime Pay Regulation on Worker Compensation," *American Economic Review*, Vol. 81, no. 4, September 1991, pp. 719, 723, 725, 726.) Assume these descriptive statistics for the number of hours usually worked are valid for the set of all males described earlier and that the values are normally distributed to

a. Find the proportion of the group that work between 30 and 40 hours per week (hpw).
b. Find the proportion who work more than 40 hpw.
c. Check the validity of the phrase "who usually work at least 20 hours per week" by determining the proportion who work at least 20 hpw.

■ **Solution:** Figure 6-27a shows the given intervals, as well as pertinent areas for Parts a and b. The shaded areas are the ones we seek.

For Part a we need to convert 30 hpw and 40 hpw to Z values to use the Z table. The first, 30 hpw, becomes $(30 - 41.888)/6.291 = -1.89$, and 40 hpw transforms to $(40 - 41.888)/6.291 = -0.30$. The proportion who worked between 30 and 40 hours per week is the difference $P(30 < X < 41.888) - P(40 < X < 41.888) = P(-1.89 < Z < 0) - P(-0.30 < Z < 0) = 0.4706 - 0.1179 = 0.3527$.

Figure 6-27
Hours worked per week

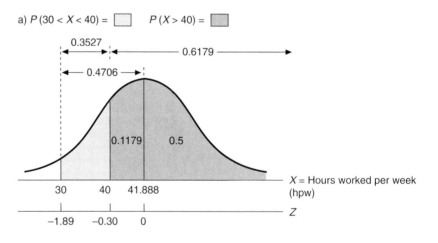

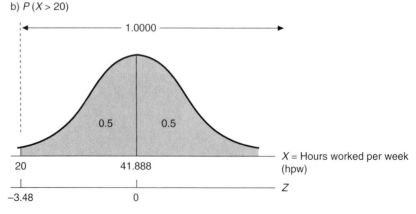

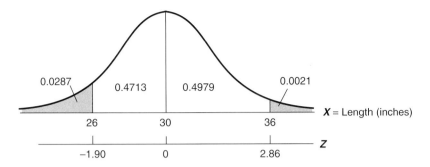

Figure 6-28
Pants-inventory problems

For Part b the proportion who work more than 40 hours per week is $P(40 < X < 41.888) + P(X > 41.888) = 0.1179 + 0.5000 = 0.6179$.

For Part c, Figure 6-27b indicates the interval and area in question. Twenty hpw converts to $(20 - 41.888)/6.291 = -3.48$, a Z value indicating that 20 hpw is more than 3 standard deviations from the mean. There is almost no chance of obtaining a respondent who works 20 hpw or less. Because this value surpasses the Z values in the table, we assume that approximately 0.5, the total area on this side of the curve, lies between -3.48 and 0. Thus, approximately 100% of the respondents must work 20 or more hours per week (as stipulated).

■ **Example 6-7:** If the lengths of men's pants are normally distributed with mean 30 inches and standard deviation 2.1 inches, what proportion of male customers will want pants longer than 36 inches or shorter than 26 inches?

■ **Solution:** Figure 6-28 shows the intervals of X values smaller than 26 and larger than 36 along with the shaded areas to be determined. Converting 26 and 36 to Z values, we obtain $(26 - 30)/2.1 = -1.90$ and $(36 - 30)/2.1 = 2.86$. Continuing,

$$P(X < 26 \text{ or } X > 36)$$
$$= P(X < 26) + P(X > 36) \text{ (mutually exclusive events)}$$
$$= P(Z < -1.90) + P(Z > 2.86)$$
$$= (0.5 - 0.4713) + (0.5 - 0.4979)$$
$$= 0.0287 + 0.0021$$
$$= 0.0308.$$

■ **Example 6-7 (continued):** The store can order a certain style of pants in even lengths only. Assume customers purchase pants that are too long and alter them to the desired length, when they cannot find an exact fit. The manager plans to order 100 pairs between the lengths of 28 inches and 36 inches to accommodate men desiring lengths 36 inches or shorter with minimal alterations. This should accommodate almost everyone, because 36 is approximately three standard deviations above the mean. How many pairs of length 28, 30, 32, 34, and 36 should the manager order?

■ **Solution:** First, we find the proportion of customers that will purchase each size: $P(X \leq 28)$ for anyone desiring length 28 inches or less (assuming even much smaller men will buy this size and shorten it), $P(28 < X \leq 30)$ for length 30, $P(30 < X \leq 32)$ for length 32, $P(32 < X \leq 34)$ for length 34, and $P(34 < X \leq 36)$ for length 36. Figure 6-29 depicts these areas and intervals along with the transformed lengths and table information used to solve for the inventory proportions. The proportions for each size are 0.1711, 0.3289, 0.3289, 0.1424, and 0.0266, respectively. The product of these values and 100 results in the following inventory levels; that is, the number of pairs of

Figure 6-29
Inventory for 100 pairs of pants

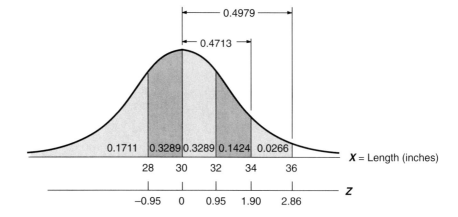

each size to purchase, 17.11, 32.89, 32.89, 14.24, and 2.66, respectively. Practically speaking, the manager would translate these values into orders of 17, 33, 33, 14, and 3, respectively.

In Box 6-1, we see that the normal random variable proves useful for predicting the winner in football games.

Box 6-1 The Probability of Winning a Football Game

The margin of victory, *MV*, in a professional football game is the difference between the favorite's score and the underdog's score. Based on 672 professional games played during the 1981, 1983, and 1984 seasons (an extended player strike occurred in 1982), Stern found *MV* to be approximately normally distributed with a mean of *S*, the point spread, and a standard deviation of 13.861 points. A histogram of the data set appears very close to normal. By using more advanced statistical techniques, Stern substantiates the normal fit, then he employs 1985 and 1986 data to further support his conclusions.

Thus, if we want to know the probability that a team favored by one point wins ($MV > 0$), then we would find the shaded area in the figure or $P(Z > -0.07) = 0.5279$. Using this process, we can determine the $P(MV > 0)$ when the point spread is 3, 5, 7, or 9 to be 0.5871, 0.6406, 0.6950, and 0.7422, respectively. The actual proportion of games won by teams favored by 1, 3, 5, 7, and 9 points are 0.571, 0.582, 0.615, 0.750, and 0.650, respectively, very close to the estimated values.

Stern also points out that if we use these actual proportions as estimates of the difference in quality of the two teams, the result is less intuitive than the estimates based on the normal distribution. The normal estimates increase as the point spread increases (both signs of the difference in quality of the two teams), but the *actual* proportions do not (the actual proportion of the nine-point spread decreases from the proportion for the seven-point spread, from 0.750 down to 0.650).

Source: Hal Stern, "On the Probability of Winning a Football Game," *American Statistician*, Vol. 45, No. 3, August 1991, pp. 179–83.

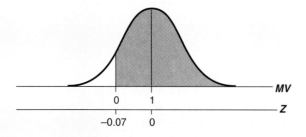

Section 6.5 Problems

94. X is a normal random variable with a mean of 500 and a standard deviation of 50. Use this information to find the probabilities of the following events.
  a. $X < 600$.
  b. $500 \leq X < 525$.
  c. $X \geq 450.5$.
  d. $475 \leq X \leq 575$.
  e. $400 < X < 463.2$.
  f. X is within 2.98 standard deviations of the mean.

95. Suppose that the result of a single play (gain or loss) by a football team is normally distributed with mean 3.22 yards and a standard deviation of 5.84 yards. Use this information to find the probabilities of the following events on the next play.
  a. The team gains 10 or more yards.
  b. The team gains yardage.
  c. The team loses yardage.
  d. The team gains between 5 and 10.5 yards
  e. The team loses 10 or more yards.
  f. The result is between a 5-yard loss and a 5-yard gain.

96. The club fundraiser situation describes the efforts of a school club to sell plastic party streamer material in individual customer-designated lengths. A local gift and party store owner suggests that the lengths customers purchase are approximately normally distributed with a mean of 18 inches and a standard deviation of 2 inches. The club wants to take advantage of this information and precut some streamers in order to fill orders speedily from an appropriate inventory of sizes that satisfy as many customers as possible and avoid excess stock. What proportions of the precut streamers should satisfy the following lengths?
  a. Between 15.55 and 20.52 inches.
  b. More than 20.12 inches.
  c. Less than 17.98 inches.
  d. Between 20 and 23 inches.
  e. Between 15 and 17 inches.
  f. Within 1.22 standard deviations of the mean.

6.6 Finding X Values

Now we consider questions about the values that correspond to given probabilities. Suppose the store manager from Example 6-7 wants to know what lengths of sports pants to order to satisfy the middle 95% of customer needs, using the mean of 30 inches and standard deviation of 2.1 inches.

First, we sketch the situation. We must proceed equal distances on either side of the mean to form the interval containing the middle 95% of the customers. Call the endpoints of this interval x_1 and x_2. We can state the problem as $P(x_1 < X < x_2) = 0.95$. Part a of Figure 6-30 depicts this situation.

Question: What is the area under the curve between 30 and x_2? What is the z value that corresponds to this area?

Answer: Because the curve is symmetrical, half of the area is on each side of the mean. Because the manager seeks the middle 95%, we obtain the area on one side or half of 95%; $0.5(0.95) = 0.475$. Searching the Z table for this area, we obtain the corresponding value, expressed in Z units, of 1.96. Similarly, the z that corresponds to x_1 is -1.96. See Figure 6-30b.

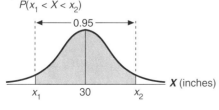

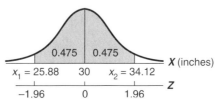

Figure 6-30

Boundaries of the middle 95% of pant sizes

Question: At this point, we know that the values x_1 and x_2 are -1.96 and 1.96 standard deviations from the mean, respectively. Use this information along with the mean, 30 inches, and standard deviation, 2.1 inches, to determine the values of x_1 and x_2.

Answer: 1.96 standard deviations is $1.96\sigma = 1.96(2.1) = 4.116$. We subtract 4.116 from the mean, 30, to obtain $x_1 = 25.884$ and add 4.116 to 30 to obtain $x_2 = 34.116$, which we express as 25.88 and 34.12 inches. (Alternatively, we can substitute the values for z, μ, and σ in the formula that transforms Z to X.)

Thus the middle 95% of customers desire lengths between 25.88 and 34.12 inches. This answer assumes that we can buy any length, not just discrete integer lengths, such as 26, 27, 31, and so forth. There are times when this assumption is not valid and corrections must be made (see the next section). *We assume in this book that any normally distributed value on a measurement scale is possible—not just certain points.*

The procedure to locate random variable values from given areas is the same whether the random variable is normal or standard normal, except for one additional step. We must convert Z values back to the original units of measure for the X values. Notice the similarities and the additional step in the following box that organizes these steps.

To find a specific value, x, of a normally distributed random variable that corresponds to a given area or probability:

1. Determine if x is to the right or left of the mean.
   a. Construct two diagrams, and locate an x larger than the mean on one diagram and smaller than the mean on the other, along with any other given information about the X values in the situation.
   b. Shade the relevant area from the given information, an interval of X values with x as one of the boundaries.
   c. Determine the valid diagram, by eliminating the diagram with a shaded area from Step b that contradicts the properties of the normal distribution, such as a shaded area that is supposed to exceed 0.5 but covers less than the entire area on one side of the mean.
2. Determine the area under the curve between the mean and x. Be careful not to automatically use the area mentioned in a problem.
3. Search for the area determined in Step 2 among the area entries of the standard normal table and determine the magnitude of the z corresponding to this area. Attach the sign determined in Step 1 to this magnitude for the z value that relates the distance (in standard deviations) and direction of x from the mean, μ.
4. Transform z to x using $x = \mu + z\sigma$, or link the distance $z\sigma$ to the mean μ with the sign of z.

■ **Example 6-8:** Winter snow depths at a Montana ski resort are normally distributed with a mean of 42 inches and a standard deviation of 14 inches. The resort wishes to be open for skiing 80% of the winter, and those open days should be when the snow is deepest. What is the minimum depth of the snow when the resort should open, to be open only 80% of the winter?

■ **Solution:** Step 1: Sketch two curves, one with an x smaller than the mean and one with an x larger than the mean. The deepest 80% encompasses all values to the right of x. The shaded area corresponding to an x on the right of the mean would create an impossible situation, because the area to the right of x must be 0.8, which exceeds 0.5, the total area to the right of the mean (see

Figure 6-31a). The appropriate x is on the left of the mean, because the shaded area of 0.8 is larger than 0.5, as shown in Figure 6-31b. Thus, x must be some value smaller than the mean, 42 inches.

Step 2: The area between x and the mean is 0.30, the difference between the 0.80 and the area to the right of the mean, 0.50.

Step 3: The area corresponding to a z of 0.84 is closest to 0.30 in the Z table. z is negative, because it is smaller than the mean, so $Z = -0.84$.

Step 4: Transforming -0.84 to X units requires subtracting 0.84 standard deviations (11.76 inches = 0.84 × 14 inches) from the mean (42 inches). The result is $X = 30.24$ inches. If the depth is 30.24 inches or more, the resort should open. Usually, this will mean that the resort is open 80% of the winter.

Question: What percentile is 30.24 inches?

Answer: Because 20% of the depths are smaller than 30.24 inches, it is the 20th percentile.

■ **Example 6-8 (continued):** If the resort opens when the depth is 24 inches or more, what percent of the time will it be open?

■ **Solution:** This problem asks for a probability, not an X value. Applying the procedures from the last section, $P(X > 24) = P(Z > -1.29) = 0.4015 + 0.5 = 0.9015$ or 90.15% (see Figure 6-32).

Now we have the basic information and tools for working with normal random variables. These include the Z table and procedures for finding probabilities for given intervals of X values or for finding boundaries of an interval of X values associated with a given probability. We can apply these basics in any situation where the X values occur in a normally distributed fashion. (See Box 6-2 for a practical application with IQ scores.) In the next section, we will apply the principles in a situation where the outcomes are binomially distributed, but a normal random variable approximates the behavior.

Figure 6-31
Diagram for deepest 80% of snowfall

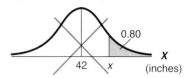

a) Contradictory diagram

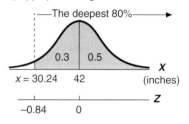

b) Appropriate diagram

Figure 6-32
Proportion of time snow is 24 inches deep or more

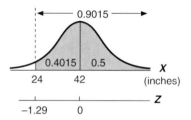

Section 6.6 Problems

97. X is a normal random variable with a mean of 500 and a standard deviation of 50. Use this information to find the x value (or values) that satisfy the following statements.
   a. $P(X < x) = 0.58$.
   b. $P(500 \leq X < x) = 0.2324$.
   c. $P(X \geq x) = 0.448$.
   d. $P(X \leq x) = 0.3333$.
   e. $P(x_1 < X < x_2) = 0.95$.
   f. x is 2.98 standard deviations above the mean.

98. The store manager from Example 6-7 (p. 269) believes that many of the taller men will purchase the pants. To be prepared to serve the tallest 20%, what lengths should the store stock?

99. Suppose that the result of a single play (gain or loss) by a football team is normally distributed with mean 3.22 yards and a standard deviation of 5.84 yards. Use this information to find the X value (or values) that satisfy the following statements.
   a. The 76th percentile of results.
   b. 34.2% of the results are between 3.22 and x.
   c. The longest 28% of results all equal or exceed x.
   d. The smallest 28% of results all equal or exceed x.
   e. The middle 80% of the results are between x_1 and x_2.
   f. x is three standard deviations below the mean.

100. Sales of party streamer material in individual customer-designated lengths are approximately normally distributed with a mean of 18 inches and a standard deviation of 2 inches. Use this information to find the X value (or values) that satisfy the following statements.
   a. The shortest 63% are shorter than x.
   b. The 26th percentile.
   c. The middle 90% of the lengths are between x_1 and x_2.
   d. x is the shortest length in the longest 13%.
   e. x is 2.564 inches below the mean.

Box 6-2 IQ Scores

Psychologists, educators, and other practitioners assess individuals' abilities and capabilities with IQ scores. But what information does a particular score provide about the individual's performance?

The answer varies with the particular test, because different IQ tests define intelligence differently or measure different aspect of intelligence. However, testers report IQ scores in a common form to facilitate comparisons among different test scores.

How do they formulate comparable scores? One way to report results in common units is to convert to Z scores, (score $- \mu)/\sigma$. When μ and σ represent the mean and standard deviation for the population of tested individuals under consideration—often just the general population—the resulting Z locates a score relative to the mean in standard deviation units. Because not many people know about standard deviations or Z scores, learning that a child's score is one or one standard deviation above the mean is often meaningless and a score of negative one sounds terrible. However, practitioners now can easily compare different IQ measures by expressing them in Z values. A Z of 2 on one test and 0 on another readily compares each performance to the mean of the test, indicating that for the characteristics measured by the first test the individual is two standard deviations above the mean and average on the other test.

An adjustment that produces values more commonly accepted and "understood" by individuals is to convert the Z score to a *standard score*. Standard scores all have the same mean and standard deviation, just like Z values all have a mean of 0 and standard deviation of 1. Common values for the mean and standard deviation of IQ standard scores are 100 and 15 or 16, respectively. Using a standard deviation of 15, if we know an individual is two standard deviations above the mean, then the person's standard scores would be $100 + 2(15) = 130$. A Z score of -1 is $100 - 1(15) = 85$ as a standard score. The values are easily compared. Now no one has a negative intelligence score connoted by a negative Z value.

Some practitioners and government agencies use the normal probability distribution to assess an individual's relative performance based on IQ scores for decision-making purposes. Often, they convert scores to percentiles from the standard normal table or special tables that list the standard scores and corresponding percentiles, easily created from the normal table. The figure shows a portion of a page from an evaluation manual used by speech pathologists in North Carolina as one component in an overall assessment of individual strengths and weaknesses to plan therapies and treatments for individual patients and clients.

Source: M. Ray Loree, *Psychology of Education* (New York: The Ronald Press, 1965), p. 562, and personal conversations with Ms. Beth Baxley, Speech and Hearing Clinic, Western Carolina University, September 17, 1991, and Dr. Robert Pittman, Administrative Curriculum and Instruction Department, Western Carolina University, September 22–23, 1991. Illustration from Test Service Notebook No. 148. Reprinted with permission of The Psychological Corporation.

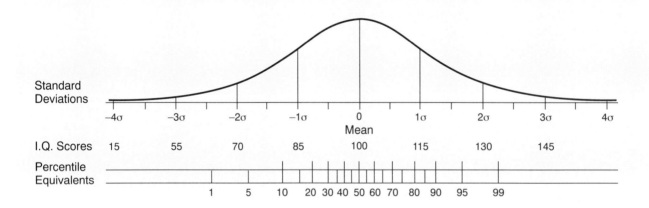

6.7 Normal Approximation of a Binomial Probability Distribution

■ **Example 6-9:** While Mr. Abernathy, a carpenter, is doing some work on a construction site, he finds quite a few bent, stripped, or otherwise defective screws among the shipment from the building-supply store. Realizing that some defectives slip

Section 6.7 Normal Approximation of a Binomial Probability Distribution

by undetected in any manufacturing process and wishing to avoid the time and expense of returning to the building-supply store, he allows for a 10% rate of defectives, based on his prior experience with the product. Perplexed by the apparent large number of defectives in this shipment, he decides to select a sample of 100 screws randomly and count the number of defectives to determine if the number is unreasonably high for a 10% defective rate. If he finds 12 or more defectives among the sample of 100 screws, he plans to return the shipment. What is the probability that he will return the shipment when the defective rate is really 10%?

■ **Preliminary Solution:** If the quality of each screw is independent of the quality for any other screw in the sample (which is likely to be the case, because he randomly selects them), then the problem is to find the binomial probability of obtaining 12 or more defectives in 100 checks for defectiveness (trials), when the chance of a defective result on any specific trial is 0.10. There are two possible ways to solve this problem:

1. P(he returns the shipment given that $p = 0.10$)
   $= P(X = 12 \text{ or } X = 13 \text{ or } \cdots \text{ or } X = 100)$,
   where X = the number of defective screws.
2. $1 - P$(he keeps the shipment given that $p = 0.10$)
   $= 1 - P(X = 0 \text{ or } X = 1 \text{ or } \cdots \text{ or } X = 11)$.

Either case involves quite a bit of calculation if we use the binomial formula. Tables do not always provide probabilities for the desired number of trials (the one in the appendix does not include 100). If an approximation rather than the exact probability will suffice, there is a shortcut. We can approximate binomial probabilities quickly and easily with a normal curve. This section describes such a procedure.

6.7.1 Review of Binomial Random Variable

To approximate a binomial probability with a normal distribution, we need first to recall facts about the binomial random variable. A quick review follows.

To find the probability of X successes when using the binomial formula, the following assumptions must hold:

1. The binomial experiment consists of n trials of an experiment.
2. There are only two possible outcomes of each trial, success or failure.
3. Outcomes of each trial are independent.
4. The probability of success, p, is the same for each trial.
   The formula for finding the *exact* binomial probability is

$$P(X = x) = \frac{n!}{x!(n-x)!} p^x (1-p)^{n-x}.$$

■ **Example 6-10:** Use this formula or the binomial table to construct the binomial probability distribution for the number of heads in three tosses of a coin.

■ **Solution:** This probability distribution consists of all possible values of X = number of heads in three tosses and the probability for each X value. X can be 0, 1, 2, or 3. To find the probabilities using the formula, we must solve problems for each of the four possible X values. For instance, if we want to find the probability of two heads in the three tosses, we substitute $n = 3$, $X = 2$, and $p = 0.5$ in the formula to obtain

$$\frac{3!}{2!\,1!}(0.5)^2(1-0.5)^{3-2} = 0.375.$$

Alternatively, we can find these probabilities in the binomial table and then form the following binomial probability distribution:

x	$P(x)$
0	0.125
1	0.375
2	0.375
3	0.125

The shortcut formula for the mean of a binomial random variable is $\mu = np$, which is $3(0.5) = 1.5$ heads in Example 6-10. The shortcut formula for the standard deviation is

$$\sigma = \sqrt{np(1-p)} = \sqrt{3(0.5)(1-0.5)} = \sqrt{0.75} = 0.866 \text{ heads.}$$

Because the binomial random variable is discrete, its graph is a bar graph, where 0.5 is added to and subtracted from each X value to obtain its base. For example, the base for the bar for $X = 1$ spans the interval 0.5 to 1.5. For $X = 0$, the base is -0.5 to 0.5. Consequently, the base for each bar is 1 and the height is the probability that the X value associated with the bar occurs, resulting in an area of each bar equal to the probability of the corresponding X value. Figure 6-33 displays the complete bar graph.

Figure 6-33
Bar graph of number of heads (X) in three tosses

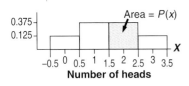

6.7.2 The Normal Approximation Procedure

Figure 6-34
Approximating the bar area with the area under a normal curve

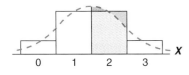

Now compare Figure 6-33 with Figure 6-34 where we superimpose a normal curve on the histogram. In particular, consider the bar that corresponds to an X value of 2 and the area under the curve that corresponds to the base of the bar for 2. The green region of the bar's area is excluded from the area under the curve, and the purple region under the curve is not part of the bar. If these two areas are approximately equal, then the area under the curve will approximate the area of the bar, which is the probability of obtaining two heads.

Question: To solve a normal probability problem, what two things must we know about the normal random variable?
Answer: We need μ and σ to convert to $Z = (X - \mu)/\sigma$.

Remember that μ and σ for the binomial random variable for the number of heads in three tosses are $\mu = np = 1.5$ and $\sigma = \sqrt{np(1-p)} = 0.866$. Using this information, and Figure 6-35, we proceed to approximate $P(2)$, the area of the bar, with the area under the curve that corresponds to the base of the bar, 1.5 to 2.5.

$P(2)$ (exact binomial probability)
$= P(1.5 < X < 2.5)$ (normal approximation)
$= P(0 < Z < 1.15)$ (converting 1.5 and 2.5 to Z values for a standard normal problem)
$= 0.3749$ (from Z table)

Note that the exact binomial probability of two heads in three tosses is 0.375. The approximation is 0.3749. In this case, the approximation missed by 0.0001, which would be a small, probably inconsequential, difference in many instances. However, the approximation is not always this close, as we shall soon see.

Figure 6-35
Approximate probability of two heads in three tosses

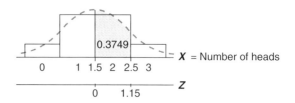

Section 6.7 Normal Approximation of a Binomial Probability Distribution **277**

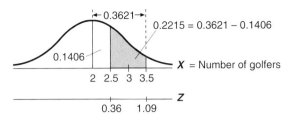

Figure 6-36
Approximation of P(3)

■ **Example 6-11:** If 5% of golfers tend to par (get the golf ball in the hole with the designated number of shots called par) on hole number 6, approximate the probability that three of 40 randomly chosen golfers will do so.

■ **Solution:** Each golfer's outcome is a trial, so $n = 40$. Three golfers scoring par on the hole is three successes, so $X = 3$, and the probability of a success for each golfer is $p = 0.05$. Thus, the mean and standard deviation for the curve are $\mu = 40(0.05) = 2$ golfers and $\sigma = \sqrt{40(0.05)(0.95)} = 1.378$ golfers. We approximate this probability with $P(2.5 < X < 3.5)$ or the area under the curve between 2.5 and 3.5, as shown in Figure 6-36. At this point, we proceed as we would for any normal probability problem. The Z values that correspond to 2.5 and 3.5 are $(2.5 - 2)/1.378 = 0.36$ and $(3.5 - 2)/1.378 = 1.09$. We locate areas 0.1406 for $Z = 0.36$ and 0.3621 for $Z = 1.09$ from the Z table (shown in Figure 6-36) and find their difference to produce the approximate answer 0.2215. The approximate probability of three golfers out of 40 scoring par on this hole is 0.2215.

The approximation is 0.2215. The exact probability from the binomial formula is 0.1851. Here the difference is 0.0364, which still is not necessarily large—depending on the use to be made of the calculated probability.

To approximate these probabilities, use the area under the curve that corresponds to the base of the bar. The base for 3 in the last problem is 2.5 to 3.5 or 3 ± 0.5 (which means, take 3 and subtract 0.5 to get one value, 2.5, and then add 0.5 to 3 to get the other value, 3.5). To find the base for any X value, we add and subtract 0.5. We call the ± 0.5 the *continuity correction factor*. We make this adjustment because we approximate a *discrete* probability distribution (the binomial) with a *continuous* probability distribution (the normal)—thus, continuity correction.

This approximation is especially useful if a problem requires numerous calculations with the binomial formula. An instance would be finding the probability that three or more of the 40 golfers par in the preceding example. This problem involves finding binomial probabilities for several X values, $3, 4, \ldots, 40$.

■ **Example 6-9 (continued):** Approximate the probability that Mr. Abernathy finds 12 or more defective screws in 100 randomly selected screws, if the defective rate is 10%.

■ **Solution:** Figure 6-37a displays the bar graph for this distribution and highlights the bars for the X values that are 12 or more, that is, $12, 13, \ldots, 100$. The sum of the areas in these bars equals $P(X \geq 12)$. This is the sum of 89 separate bars or binomial problems, $P(12) + P(13) + \cdots + P(100)$.

Question: To approximate the $P(X \geq 12)$, what interval on the number line will cover the values?

Answer: The interval includes the bases for each of the bars, beginning with the base for 12, 11.5 to 12.5, then picks up the base for 13, 12.5 to 13.5. This pattern continues until the last base is connected to the others, 99.5 to 100.5. The result is one interval of connected bases that stretches from 11.5 to 100.5, so we use the normal curve to find $P(11.5 < X < 100.5)$. (See Figure 6-37.)

Figure 6-37

Approximating the probability for an interval of binomial values

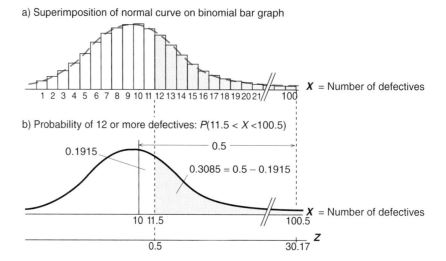

To continue to solve this problem, we use a normal curve with $\mu = 100(0.1) = 10$ screws and $\sigma = \sqrt{100(0.1)(0.9)} = \sqrt{9} = 3$ screws. Converting 11.5 and 100.5 to Z values, we obtain $(11.5 - 10)/3 = 0.5$ and $(100.5 - 10)/3 = 30.17$. Figure 6-37 shows the areas 0.1915 and 0.5 that we obtain for these Z values from the table ($Z = 30.17$ is so many standard deviations from the mean that we consider the area to be 0.5, the total area on one side of the curve). Differencing the areas results in a normal approximation of 0.3085 for the binomial probability of 12 or more defective screws, as shown in Figure 6-37. Thus, there is about a 31% chance of finding 12 or more defective screws in the sample of 100 if the defective rate is 10%. Mr. Abernathy's rule to decide whether or not to return a shipment will cause about one-third of the good shipments to be returned, a rather large return rate when nothing is abnormal. In addition, the cost of returning and obtaining a new shipment makes the decision rule costly. A higher cutoff point for returning shipments, say 14 or 15, would lower the probability of the mistake.

Note also that we could find

$$1 - P(X < 12) = 1 - P(X = 0 \text{ or } X = 1 \text{ or } \cdots \text{ or } X = 11),$$

which is approximately equal to $1 - P(-0.5 < X < 11.5)$ when using the normal curve. In either case, the normal approximation for the binomial probability is much simpler than using the binomial formula over and over again. In many instances, the normal approximation may suffice, and it will certainly be of use later in the estimation process.

6.7.3 Conditions for Good Normal Approximations

There are several rules for determining if a normal approximation for a binomial probability will be close to the exact value. The rules vary with the desired closeness. Two characteristics of the binomial random variable affect this closeness, p and n. The following discussion demonstrates the effects of p and n on the approximation and conditions for improving the approximation.

Effect of p for Given n The closer p is to 1/2 for a given n, the better the approximation will be. Remember in Example 6-10 that the approximation of the probability of two heads in three tosses is 0.3749 and the actual value is 0.375 for a difference of 0.0001. In this case p is exactly 0.5. If we leave n at three but change p to 0.4, then $P(2) = 0.288$ exactly. The normal random variable approximates this probability as 0.3002, or a difference of 0.0122.

Section 6.7 Normal Approximation of a Binomial Probability Distribution

Figure 6-38 demonstrates that, as *p* moves away from 0.5 (*n* is constant at three), the symmetry of the resulting binomial graph dissipates into skewness, which, appropriately, will be harder to approximate with the symmetric normal curve.

Effect of *n* for Given *p* Figure 6-39 demonstrates the impact of *n*, the number of trials, on the shape of the bar graph for the binomial distribution for a given *p*, in this case *p* = 0.05.

Question: Based on the figure, do you think the approximation will generally improve as *n* increases or decreases for a given *p* value?

Answer: The approximation will generally improve as *n* increases for a given *p*, though *p* is distant from 0.5, as it is in Figure 6-39 where *p* = 0.05. Even a skewed binomial distribution becomes more symmetric as *n* increases.

A General Rule A general rule that provides reasonable estimates incorporates the demonstrated effects of *n* and *p*.

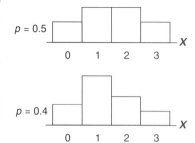

Figure 6-38
Effect of changing *p* with constant *n* (3) on the shape of a binomial probability distribution

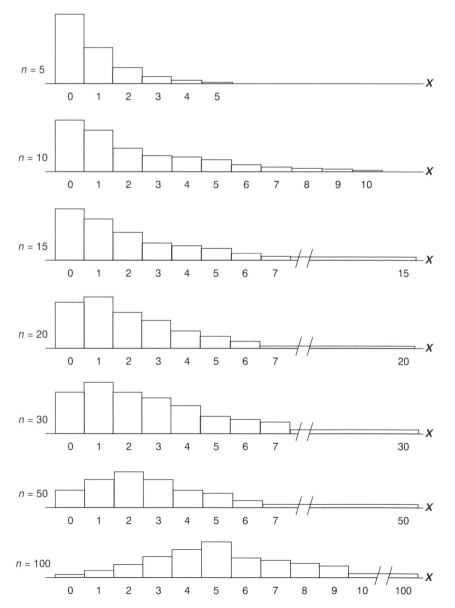

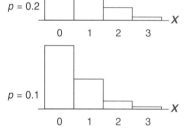

Figure 6-39
Effect of changing *n* with constant *p* (0.05) on the shape of a binomial probability distribution

> **General Rule for Good Approximation of Binomial Probability with Normal Probability**
> Both np and $n(1 - p)$ should be at least 5.

In Example 6-9 with the screws, n is 100 and p is 0.1, so $np = 10$ and $n(1 - p) = 90$, so the approximation should be fairly close. In Example 6-11 with the golfers, $n = 40$ and $p = 0.05$, so $np = 2$ and $n(1 - p) = 38$. Because 2 is smaller than 5, we should not expect a close normal approximation: The result is off by about 0.04, although 0.04 is not too bad in some cases. Remember, this is a general rule. In Example 6-10 with the coin tosses, we found very close approximations when $p = 0.5$ with a small n.

■ **Example 6-11 (continued):** If $p = 0.05$, a reasonable approximation requires what value of n?

■ **Solution:** In this case $n(1 - p) = n(0.95)$ will be larger than $np = n(0.05)$. The conditions for a reasonable approximation require the last expression, $0.05n$, to be at least 5. Thus, $0.05n \geq 5$. Upon solving, we obtain $n \geq 100$.

Section 6.7 Problems

101. If 40% of parents have problems communicating with their children, use the normal distribution to approximate the probability that at most 50 of 100 parents will say that they have such problems.

102. If Mexican food is considered fashionable by 67% of the population, what is the probability that more than half of your invitations to fashionable dinners, received independently, will include Mexican cuisine if you receive 60 invitations? 100? 150?

103. If 75% of employers feel that speaking skills are necessary for the professional success of an employee, what is the probability that more than 170 of 250 randomly selected employers will rate these skills as important?

104. If 20% of Nevada citizens are natives and familiar with the geography, what is the probability that between 20 and 35 (inclusive) of 150 randomly selected Nevadans are natives and will correctly identify the Nevada landmark described to them?

105. If a child only listens to 90% of parental instructions, what is the probability that the child will obey only 40 of the next 50 instructions?

106. If 5% of schoolchildren are gifted, what is the likelihood that, at most, 30 students in a random sample of 800 will be gifted?

Summary and Review

The number of possible normal random variables is infinite, because each variation in the value of the mean, μ, or the standard deviation, σ, generates a different normal random variable. However, any one of them can be transformed to a standard normal random variable using

$$Z = \frac{X - \mu}{\sigma},$$

where X is the (nonstandard) normal random variable. This Z value measures the distance of an X value from the mean in standard deviation units. Consequently, we can determine an X value from its Z value by using $X = \mu + Z\sigma$.

The transformation formulas enable us to use the standard normal table to determine probabilities or X values for any normally distributed set of values.

We can approximate a binomial probability distribution with a normal distribution. Actually, we approximate the area in a bar of the graph of the binomial probability distribution with the area under the normal curve above the interval that corresponds to the base of the bar. Determining the base of the bar necessitates making a continuity correction, that is, adding and subtracting 0.5 to each binomial outcome, to determine the base. To work with the normal curve, we use the shortcut formulas for the mean and standard deviation of a binomial random variable: $\mu = np$ and $\sigma = \sqrt{np(1 - p)}$. This approximation substitutes for numerous calculations with the binomial formula when several outcomes are involved in one problem, such as the probability of 300 or more successes out of 1,000 trials. The approximation improves as p approaches 0.5 and as n gets larger. A general rule for a reasonable approximation is that $n(1 - p)$ and np be at least 5.

Multiple-Choice Problems

107. Use the conversion formula to locate $X = 560$ on the Z axis if X is normally distributed with mean 800 and standard deviation 120.
  a. 560.   d. −2.
  b. −240. e. 2.
  c. 240.

108. Suppose the Z value for $X = 100$ is 1.65. Another set of values, Y, is normally distributed with the same mean as X but has a smaller σ. If $Y = 100$ is converted to a Z, what will the Z value be?
  a. Larger than 1.65, because 100 must now be more standard deviations from the mean.
  b. Smaller than 1.65, because a smaller σ means Z values are tightly clustered.
  c. Equal to 1.65, because X and Y have the same mean, and 100 must be equidistant from the same mean.
  d. Smaller than 1.65, because a smaller standard deviation positions the values closer to the mean.
  e. Negative, because $Y = 100$ must be below the mean.

109. X is normally distributed with $\mu = 4$ and $\sigma = 0.2$. Find $P(4.1 < X < 4.35)$.
  a. 0.5.   d. 0.1915.
  b. 0.2684. e. 0.6514.
  c. 0.4599.

110. X is normally distributed with $\mu = 25$ and $\sigma = 5$. Find x so that $P(X < x) = 0.6$.
  a. 26.25. d. 0.2257.
  b. 23.75. e. 0.1554.
  c. 0.25.

Suppose daily water usage in a certain community is normally distributed with $\mu = 100{,}000$ gallons and $\sigma = 5{,}000$ gallons. Use this information to answer Questions 111–112.

111. If the water utility should be able to handle at least 98% of patrons' desired usage levels, for what daily capacity must it plan and build?
  a. 2.05 gallons.   d. 39,750 gallons.
  b. −2.05 gallons.   e. 100,250 gallons.
  c. 110,250 gallons.

112. What proportion of the time will the community need more than 108,000 gallons?
  a. 1.6.   d. 0.0548.
  b. −1.6.   e. 0.4452.
  c. 0.9452.

Sheila's bowling scores follow a normal pattern with a mean of 200 points and a standard deviation of 10 points. Use this information to answer Questions 113–116.

113. What is the probability that you will need to bowl 220 or higher to defeat her?
  a. 0.4772. d. 2.
  b. 0.1915. e. 0.0228.
  c. 0.9772.

114. Determine the 60th percentile of Sheila's bowling scores.
  a. 202.5 points. d. 2.5 points.
  b. 197.5 points. e. −2.5 points.
  c. 0.25 points.

115. The middle 90% of her scores are between what two approximate scores?
  a. ±1.96 points.   d. 183.6 and 216.4 points.
  b. ±1.64 points.   e. 198.36 and 201.64 points.
  c. ±16.4 points.

116. What percent of her scores are within 1.85 standard deviations of the mean score, 200?
  a. 46.78%. d. 95.00%.
  b. 37.00%. e. 95.44%.
  c. 93.56%.

117. Use the normal distribution to approximate the probability of getting 60 successes in 200 trials of a binomial experiment where the probability of success on one trial is 0.3.
  a. 0.5.   d. 0.2.
  b. 0.0319. e. 0.4681.
  c. 0.0638.

118. Use the normal distribution to approximate the probability of obtaining between 30 and 50 (including 30 and 50) defective items in a set of 150 randomly selected items from a large shipment if the defective rate is 0.25.
  a. 0.9274. d. 0.4909.
  b. 0.9131. e. 0.4929.
  c. 0.0726.

119. When we use the normal curve to approximate a binomial probability, we are actually
  a. Assured of not erring by more than 0.001.
  b. Approximating the area of a rectangle from a bar graph.
  c. Substituting one continuous distribution for another.
  d. Approximating the tail area of a bell-shaped curve.
  e. Estimating the revised probability calculation.

120. The continuity correction factor is necessary because
  a. The approximation must be corrected to equal the exact probability.
  b. A continuous probability distribution is always based on a graduated scale.

c. The normal probability factor is usually larger than the binomial factor.
d. The mean and standard deviation of a normal distribution differ from these values for a binomial distribution.
e. A discrete distribution is being approximated by a continuous distribution, which has no probability for a single point.

121. What interval corresponds to the base of the bar for $X = 49$ in a binomial distribution?
    a. 49.
    b. 48 to 50.
    c. 48.5 to 49.5.
    d. 48.05 to 49.95.
    e. $P(X = 49)$.

122. Use the normal distribution to approximate the probability of there being 60 or fewer male children born out of the next 130 if the probability of the baby being male is 0.5.
    a. 0.7152.
    b. 0.2152.
    c. 0.2148.
    d. 0.1894.
    e. 0.3106.

123. The normal approximation of a binomial will be better if
    a. p is close to 1.
    b. n is close to 50.
    c. n is close to 0.5.
    d. p is close to 0.5.
    e. p gets infinitely large.

In 1990, 15.0% of face lifts were performed on men. Use this information to answer Questions 124–126. (Source: American Society of Plastic and Reconstructive Surgeons and the American Academy of Cosmetic Surgery, as reported in Rodney Ho, "Men Try to Put a New Face on Careers," *Wall Street Journal*, August 28, 1991, p. B1.)

124. What is the probability that 20 or more out of 100 randomly selected cosmetic surgeons will say their last face lift was performed on a man?
    a. About 1.26.
    b. About 0.3962.
    c. About 0.1038.
    d. About 0.8962.
    e. About 0.0808.

125. What is the probability that between 10 and 22 (inclusive) face lift patients among 125 randomly selected 1990 face lift patients of a certain reconstructive surgeon are likely to be male?
    a. About 0.096.
    b. About 0.4857.
    c. About 0.8162.
    d. About 0.7767.
    e. About 0.1634.

126. What is the probability that more than 175 face lift patients in a random sample of 200 face lift patients from 1990 will be women?
    a. About 0.1379.
    b. About 0.1611.
    c. About 0.1867.
    d. About 0.3621.
    e. About 0.7242.

Word and Thought Problems

127. What distance does a Z value measure? In what units?
128. Suppose point spreads (team's score-rival's score) for a particular college team are normally distributed with mean 0 and standard deviation 5. What is the probability the spread in the next game will be
    a. Greater than 10.
    b. Greater than 3.
    c. Between −8 and 5.
    d. Positive.
    e. Negative.
    f. Equal to zero.
129. Daily sales in a department store are normally distributed with mean $1,000 and standard deviation $25. Find the probability that tomorrow's sales will be
    a. Greater than $1,037.50.
    b. Less than $980.
    c. Between $940 and $960.
    d. Between $940 and $1,000.
    e. Between $940 and $1,025.
130. Explain why an outlier is a point with a Z value of 2.5 or larger.
131. A set of grades is normally distributed with a mean of 75 and a standard deviation of 8.
    a. What is the probability a student makes a 90 or more?
    b. What is the probability a student makes 60 or less?
    c. What percent of the students make between 70 and 80?
    d. Suppose the professor decides the top 10% of the grades should be As. What is the lowest score one could receive and still obtain an A?
    e. If the next 20% should receive Bs, what is the lowest score one could receive and still obtain a B?
132. Turnaround time at a computer center is normally distributed with mean 20 minutes and standard deviation 5 minutes.
    a. What proportion of the computer runs are returned in a half hour or more?
    b. What proportion of the runs are returned between five and 10 minutes after submission?
    c. What is the slowest time of the fastest 33% of the turnaround times?
    d. What is the fastest time of the slowest 33% of the turnaround times?
    e. What percentiles did you find in Parts c and d?
    f. What is the 60th percentile of the turnaround times?
133. Suppose a piece of machinery does not function an average of two hours per month with a standard deviation of 0.1 hour. Assume nonfunctioning time is normally distributed.

Word and Thought Problems

  a. What is the probability the machine will be down between 1.82 and 2.2 hours in a given month?
  b. What is the probability the machine will be down at least 2.25 hours?
134. The number of sick days employees take in a certain firm is normally distributed with a mean of 4.5 days and a standard deviation of 1.5 days.
  a. What proportion of the employees take at least seven days?
  b. What proportion of the employees take less than five days?
  c. The 20% who take the most sick leave use at least how many days?
  d. If the 10% who take the least amount of leave and the 10% who take the most leave are not considered, what are the lower and upper limits on the amount of leave the middle 80% use?
135. Scores on a national exam are considered to be normally distributed with a mean of 278 and a standard deviation of 35.
  a. A student who scores 300 is in what percentile?
  b. What proportion score 400 or more?
  c. What is the 30th percentile?
  d. The middle 95% of the scores are between which two values?
  e. What proportion of the scores are within one standard deviation of the mean?
  f. 95% of the scores are within how many standard deviations of the mean?
136. Many companies rely on surveys to determine information about consumers. Assume you collect information from a random sample of 1,000 consumers and determine the likelihood that more than 500 of them would fit in the categories mentioned below. (Source: John Koten, "You Aren't Paranoid If You Feel Someone Eyes You Constantly," *The Wall Street Journal*, March 29, 1985, pp. 1, 22.)
  a. Conformists if 38% of all consumers are conformists.
  b. Have "problem" dandruff if 25% of all consumers have problem dandruff.
  c. Pump their own gasoline if 70% of all consumers pump their own gasoline.
  d. Put water on their toothbrush before the paste if 47% of all consumers do so.
137. Estimate the standard deviation using the range.
  a. What is the area under the normal curve between Z values +3 and −3?
  b. Can you suggest a way of estimating the standard deviation for a set of data when you know its range, based on your answer to Part a?
  c. Suppose you base your estimate on the area under the standard normal curve between +2 and −2. How would you estimate the standard deviation now?
  d. Which way is likely to be more accurate? Does your answer depend on whether the data set is a sample or a population? Does it depend on the distribution of the data?
138. If the number of days from introduction of a bill in the Senate until a vote on the bill is normally distributed with a mean of 62.12 days and a standard deviation of 1 day, what proportion of the bills take between 62 and 63 days to come to a vote?
139. If 50% of the American public watches television between 6 and 8 hours per day, what is the probability that between 825 and 875 (including 825 and 875) of 1,700 randomly selected people will say they watch this much television per day?
140. a. Suppose that 48% of the nation does not support a presidential decision, and the other 52% does. If the president's assistants only have time to review 4,500 randomly selected letters received about the decision, what is the probability that they will find that more than half support the president?
  b. However, only 3% of the nation is disappointed enough to write letters condemning the decision, and twice as many (6%) are supportive enough to write, approving the decision. The other 91% do not feel strongly enough to express an opinion. Now, what is the probability that the president's assistants will find that more than 50% of the randomly selected letters support the president?
141. Suppose IQ scores on an intelligence test are normally distributed with a mean of 100 and a standard deviation of 15. Find:
  a. $P(96 < IQ < 118)$.
  b. $P(118 < IQ < 125)$.
  c. $P(100 < IQ < 110)$.
  d. The endpoints of the interval containing the middle 80% of the IQ scores.
  e. The 80th percentile of IQ scores.
  f. The percentile rank corresponding to a score of 115.
  g. The proportion of the scores within two standard deviations of the mean.
  h. The distances from the mean, in standard deviations for the endpoints of the interval that contains the middle 98% of the scores.
142. a. Find the exact probability of tossing two heads in 10 flips of a coin, using the binomial formula.
  b. Check your answer to Part a, using the binomial table.
  c. Approximate this probability using the normal distribution.
143. a. A supplier has the part you order in stock 90% of the time. What is the probability the supplier will stock at least 75 of the next 100 parts you order?
  b. What is the probability the supplier will stock at least 90? 95?
  c. What is the probability the supplier will have between 85 and 95 of the 100 parts (include 85 and 95)?
144. The author of a new etiquette book bases the rules on a random survey of 5,000 people. If more than half the

people in the sample behave similarly in a certain situation, then that behavior will be described as "respectable." Find the probability that a behavior will be called respectable by this rule, when only 48% of the general public would behave in this manner. If the true percentage is 45%, what is the probability? 52%?

145. Suppose results of cholesterol tests for the general population are normally distributed with a mean of 187 and a standard deviation of 40.
  a. What percentage of the population have cholesterol levels greater than 240?
  b. What percentage have levels lower than 150?
  c. What is the 58th percentile of the values?
  d. The middle 99% are between what two values?
  e. If only high-risk persons (say, with values above 240) are considered, would the mean and standard deviation of values likely increase or decrease from the corresponding values for the general population? Explain.
  f. The middle 80% of the values are within how many standard deviations of the mean?
  g. What proportion of the values are within 2.02 standard deviations of the mean?

146. If a data set is unimodal, will it resemble a normal distribution? Will your answer change if the data set is symmetric also?

147. a. If 75% of the customers who make airline reservations actually use the tickets and an airline makes 320 reservations for a flight that accommodates 255 passengers, find the probability there will be more passengers than accommodations.
  b. The airline can change plane sizes or remodel to accommodate more passengers. How many accommodations should there be if the airline continues to make 320 reservations but wants the probability of more passengers than accommodations to be 0.01?
  c. Suppose a company official reasons that 75% of the reservations will result in passengers, so for a plane with accommodations for 300 the company should sell 400 [0.75(400) = 300] tickets. Demonstrate to this official the problems that will result from such a policy.
  d. If the proportion of reservations that result in actual passengers increases, should the airline make more or fewer reservations for a given level of accommodations?
  e. Demonstrate your answer to Part d by returning to Part a and finding the probability of more passengers than accommodations if 90% of the reservations result in passengers, rather than 75%. Tell how this result corroborates your conclusion in Part d.

148. a. Let $n = 100$ and $p = 0.2$. Approximate the probability of more than 25 successes, $P(X > 25)$.
  b. If the same problem is stated with number of failures, Y, rather than successes, then you will get the same answer for the probability of more than 25 successes as you do for _____ failures out of 100 attempts.
  c. What are p and n for the revised problem in Part b?
  d. Demonstrate that $P(X > 25) = P(Y < 75)$. Calculate the answers and show your findings on a diagram.
  e. Demonstrate graphically that $P(X \leq 25)$ is not necessarily equal to $P(X < 25)$. Explain the differences.

149. If 80% of the people who say they will attend a party show up, what is the probability that 50 or fewer people will show up when 70 accept the invitation?

150. A company wishes to establish a warranty time for full replacement of its product. Information in its customer data base suggests the average time before replacement of the product is 2.79 years with a standard deviation of 0.72 year. If the times are normally distributed and the company only wants to replace 1% of the products it sells, what should the length of the warranty be?

151. People's preferences in 1989 for ice cream flavors varied. Vanilla, with 30% of the population, outdistanced chocolate, selected as the second favorite by only 9%, producing a 21 percentage point difference. Other preferences include butter pecan with 6%, strawberry with 5%, and chocolate chip with 4%. (Source: "Favorite Flavors," *The Sylva Herald*, May 31, 1990, p. 11A.)
  a. Find the probability that the first person chosen from two randomly selected people will prefer chocolate and the second will favor vanilla.
  b. Find the probability that three or fewer people in a randomly selected sample of 10 people will prefer strawberry.
  c. Find the probability that 135 or fewer people will prefer vanilla in a random sample of 500 people.

152. The following data consists of the high school dropout rates in 1988 for the 50 states and the District of Columbia.

```
25.1  34.5  38.9  22.8  34.1  25.3  15.1  28.3  41.8  42.0  39.0
30.9  24.6  24.4  23.7  14.2  19.8  31.0  38.6  25.6  25.9  30.0
26.4   9.1  33.1  26.0  12.7  14.6  24.2  25.9  22.6  28.1  37.7
33.3  11.7  20.4  28.3  27.0  21.6  30.2  35.4  20.4  30.7  34.7
20.6  21.3  28.4  22.9  22.7  15.1  11.7
```

Source: U.S. Department of Education, as reported in *The Chronicle of Higher Education: Almanac*, August 28, 1991, p. 4.

  a. Make a frequency distribution of the data. Aim for seven classes using the suggestions from Chapter 2, then construct the corresponding histogram. Describe the distribution of the data. Do you think we can approximate the distribution with the normal distribution, based on your histogram?
  b. Find the mean and standard deviation of this data set assuming it is a sample of state rates over many years. Use the normal distribution to determine the boundaries of the interval containing the middle 95% of the values.

Review Problems

153. The time students are absent from classes in a school year is normally distributed with a mean of 4.2 days and a standard deviation of 0.9 days.
  a. Should a teacher suspect a problem if a student misses 12 or more days during the year? What about the student who never misses?
  b. If the school system institutes a program to encourage attendance that decreases the mean number of student absences to 2.7 days with no change in normality or standard deviation, what is the probability of a student missing 4.2 or more days in a year?

154. A plane is following a due-west route as closely as possible. Suppose the degrees off course of the plane are standard normally distributed (south is negative and north is positive).
  a. What percentage of the time is the plane off course by more than 0.57° to the north?
  b. What percentage of the time is the plane off course by no more than 1.87° to the south?
  c. The middle 99% of the times when the pilot checks the course, the plane is off course by at most how many degrees in either direction?

155. Temperatures are normally distributed with mean 85° and standard deviation of 10°. What is the temperature on the coolest day in the hottest 6.3% of the days?

156. People who have had diabetes more than 17 years have a 95% chance of having retinopathy. If a diabetic person has retinopathy, there is a 28 to 30% chance he or she will need laser eye surgery. A diabetic with more than a 17-year history of the disease has scheduled an examination with the ophthalmologist. What is the probability this person will have retinopathy and need laser eye surgery? (Source: Dr. Richard Beauchemin during meeting of Jackson County Diabetic Support Group, Sylva, North Carolina, September 10, 1991.)

157. a. If the number of burglaries attempted by a certain burglar in a metropolitan area during the year is normally distributed with a mean of 600 and a standard deviation of 185, find the probability of no more than 400 attempts during a year.
  b. A special burglary-prevention device is guaranteed to impede or prevent 75% of all burglary attempts in a year. When attempts pass the 75% mark, the device breaks down. If the device is used in this area, how many foiled burglaries would there be before the device breaks down and burglars will succeed?

158. The accuracy of an archeological procedure for dating artifacts from 2000 B.C. to 1000 A.D. is assumed to be normally distributed with a mean error of 0 years and a standard deviation of 250 years.
  a. Find the probability an estimated date is off by no more than 100 years in either direction.
  b. Find the time span of the middle 98% of all errors in dates.
  c. Find the probability a date is off by more than 500 years.
  d. Find the proportion of errors within 1.56 standard deviations of the mean.
  e. The middle 85.3% of the errors are within how many standard deviations of the mean?

159. Approximate the probability of throwing 325 or more heads in 600 flips of a fair coin.

160. Suppose there are 30 windows at a large, fast-food establishment. 80% of the restaurant's customers order a single chicken sandwich. Only 26 chicken sandwiches can be ready at one time. What is the probability at least one customer will have to wait for a chicken sandwich if there is a customer ordering at each window?

161. A firm randomly selects 50 items from a production line every day and checks them for defects. If six or more are defective, the process is checked for malfunctions. When working correctly, the process generates a lot with 4% defectives.
  a. Find the chances of conducting a malfunction check of the process when the process is actually working correctly.
  b. Rework the problem assuming the usual defect rate is 10%.
  c. Assume a 10% defect rate and a cutoff value of 12. Rework the problem, and pick the more reasonable cutoff value between this answer and the one in Part b.

162. Assume the annual population growth rate in a suburb is standard normally distributed, where the growth rate is the ratio of net additions over old population. Net additions equals births and immigration minus deaths and emigration. Find
  a. The interval containing the middle 60% of the values.
  b. The proportion of time the growth rate exceeds 1.25.
  c. The proportion of time the growth rate is between −2.36 and 0.39.
  d. The 65th percentile of growth rates.

163. The number of runaway children per month in a state is normally distributed with a mean of 840 and a standard deviation of 122.7.
  a. Find the probability of no more than 500 runaways next month.
  b. Find the probability of between 500 and 1,000 runaways next month.
  c. Find the smallest number of runaways in the months with the largest 10% of numbers of runaways.
  d. The middle 99% of the number of runaways are within how many standard deviations of the mean?
  e. What proportion of the runaway values are within 0.74 standard deviations of the mean?

164. Following are probability distributions for the random variables men's and women's shoe sizes, M and F, respectively.

F = Female Shoe Size	$P(F = f_i)$	M = Male Shoe Size	$P(M = m_i)$
4.5	0.007	6.5	0.046
5.0	0.021	7.0	0.076
5.5	0.047	7.5	0.097
6.0	0.065	8.0	0.118
6.5	0.114	8.5	0.130
7.0	0.139	9.0	0.132
7.5	0.151	9.5	0.119
8.0	0.136	10.0	0.103
8.5	0.117	10.5	0.073
9.0	0.096	11.0	0.060
9.5	0.038	11.5	0.006
10.0	0.040	12.0	0.023
10.5	0.011		
11.0	0.014		
11.5	0.003		
12.0	0.003		

Source: W. A. Rossi and R. Tennant, *Professional Shoe Fitting*, National Shoe Retailers Association, New York, 1984, as reported in E. Giles and P. H. Vallandigham, "Height Estimation from Foot and Shoeprint Length," *Journal of Forensic Sciences*, Vol. 36, No. 4, July 1991, p. 1144.

   a. Find the mean and standard deviation for M and F. Does it appear that the probability distributions might be approximated by a normal random variable with the mean and standard deviations just calculated?
   b. Compare P(male wears size 8 or smaller) and P(male wears size 10 or larger) from the given probability distributions and the result assuming the sizes are normally distributed with the means and standard deviations from Part a. Compare P(female wears size 6 or smaller) and P(female wears size 9 or larger) from both sources.
   c. Repeat the comparisons from Part b for the proportion within one standard deviation of the mean, two standard deviations, and three standard deviations (rounding internal endpoints to the nearest size or half-size).
   d. If you only want the probability approximated to the nearest tenth, do your comparisons in Parts b and c appear similar? How many places past the decimal can you go before they do not seem to work?
   e. Based on your answers to Parts b–d, do you feel like a normal approximation works well here?

165. If team winning percentage records are normally distributed with a mean of 0.5 and a standard deviation of 0.15,
   a. What is the probability a randomly chosen team will have a record between 0.75 and 0.8?
   b. A team with a record of 0.67 will be in what percentile?
   c. Find the 48th percentile.
   d. What proportion of the teams have records within 0.35 standard deviations of the mean?
   e. The middle 95% of the records are within how many standard deviations of the mean?

166. The length of service of legislators is normally distributed with a mean of 8.8 years and standard deviation of 2.8 years.
   a. What proportion serve more than 12 years?
   b. What values encompass the middle 95% of years served?
   c. The middle 95% of the values are within how many standard deviations of the mean, 8.8?
   d. What proportion of the values are within three standard deviations of the mean?

167. A retailer has determined that after Christmas 2% of the pre-Christmas December customers will return for refunds and 5% for exchanges. Other customers are satisfied. The retailer uses this information to decide whether to retain a special clerk for refunds in January. If the retailer believes there is a 75% chance or more that 550 or more customers will return for either reason, then the clerk is retained.
   a. If there are 10,000 customers in December, what is the probability 550 or more will return for refunds?
   b. What is the probability 550 or more will return for exchanges?
   c. What is the probability 550 or more will return?
   d. Use the information in Parts a–c to make the decision about hiring the clerk.

168. An old outdoorsman owns a canoe-rental agency on a scenic waterway noted for sightings of a rare bird. All other outfitters in the area claim that 60% of their customers see the bird on a canoe trip. The outdoorsman claims that at least 90 of his last 100 customers saw the bird. Is this claim believable, given the information from all of the other outfitters? State your assumptions for arriving at an answer.

169. Water temperatures in a water heater are normally distributed with a mean of 112° Fahrenheit and a standard deviation of 9°.
   a. Find the proportion of time the temperature is more than 100°.
   b. Find the proportion of time the temperature is between 115 and 125 degrees.
   c. Find the endpoints of the middle 95% of the temperature readings.
   d. The middle 92% of the temperatures are within how many standard deviations of the mean?
   e. What proportion of the temperatures are within 1.05 standard deviations of the mean?

170. a. Which of the two types of random variables, discrete or continuous, is the binomial? Which is the normal? What is the difference between the two types? Relate this difference to the need for a continuity correction factor when approximating a binomial probability with a normal random variable.
   b. What two pieces of information do we need to know about a normal random variable before we can solve

problems? How do we determine this information when we approximate a binomial probability with a normal random variable?

c. What role does the standard normal play in this approximation procedure?

171. A teacher believes that questions on a test should be constructed so that at least 75% of the students who have studied should answer them correctly. What is the exact probability that no more than five of 10 students will answer a question correctly, even when the question satisfies the teacher's criterion of 75% exactly?

172. To estimate the population count of an endangered species of fish in a natural harbor, a group of professional fishers randomly select sites to patrol the harbor at a specified time. Each will lower their net at the specified time, then count the number of the endangered species they catch before returning the fish to the sea. They intend to total the counts from the different sites to estimate the total number of endangered fish of this species in the harbor. Is the set of site totals a sample or population? Is the set of possible site count values a random variable? If so, is it discrete or continuous? Is this process descriptive or inferential statistics?

173. If 75% of college students consider a high income as one of their objectives in life, what is the probability that a survey of 700 randomly selected students will find that fewer than 550 students rate this objective as important to them?

174. If a time-out discipline procedure tends to work on 75% of all four-year-olds, what is the probability that a counselor will have success using this method with at most 120 of the 180 four-year-olds being counseled?

175. If 80% of all crimes are committed by recidivists (repeat offenders), what is the likelihood that a police investigator can determine the culprit of at least 100 of 120 offenses by using fingerprints and national fingerprint files?

176. The amount of time per day spent studying a given subject by the individual student is normally distributed, with a mean of two hours and a standard deviation of 0.25 hours.

a. Find the proportion of time the student studies between 1.5 and 3 hours.

b. Find the endpoints of the interval containing the middle 98% of study times.

c. Find the proportion of time the study time is within 1.5 standard deviations of the mean.

177. Kiln temperatures are normally distributed. When the thermostat is set at 1,200 degrees, the actual thermometer readings have a mean of 1,225 and a standard deviation of 15. If the thermostat is set at 1,200, what is the likelihood the kiln will get too hot (greater than 1,250 degrees)?

Chapter 7

Sampling Distributions

In this chapter we synthesize the earlier topics in the book. We introduce how to estimate or infer population values from sample information. We treat $\overline{X}$ as a random variable and explore its behavior, including its probability distribution, called its sampling distribution, its mean or expected value, and its standard deviation, called its standard error.

Objectives

Unit I develops and demonstrates the notion of a sampling distribution. These distributions describe the behavior of certain sample values, like the sample mean, $\overline{X}$, when we randomly select sample items from a population. Upon completing the unit, you should be able to

1. Understand what information is provided in an entry of a random number table.
2. Select observations for a random sample and recognize a nonrandom procedure.
3. State three reasons for using samples to obtain information about the population, rather than using the population itself.
4. Construct a sampling distribution.
5. Use a sampling distribution to find probabilities for values of a statistic.
6. Understand the meaning of random sample, representative sample, stratified random sample, parameter, statistic, sampling distribution, destructive sampling, and infinite population.

In Unit II we summarize all possible $\overline{X}$ values for a given sample size from a particular population with a mean and standard deviation. Then, we use the central limit theorem to describe the general probability distribution for large samples. Upon completing the unit, you should be able to

1. State the relationship between the population mean μ and the mean of the $\overline{X}$ values.
2. State the relationship between the population standard deviation, σ, and the standard deviation of the $\overline{X}$ values, $\sigma_{\overline{x}}$.
3. State the central limit theorem and explain its importance for making statistical inferences.
4. Solve probability problems about values of $\overline{X}$, using the central limit theorem.
5. Describe the sampling distribution of $\overline{X}$ when the population is normal.
6. State how sampling from a finite population affects the sampling distribution of $\overline{X}$, and describe why no correction is necessary in many cases.
7. Understand the meaning of standard error of a statistic, finite population, and finite population correction factor.
8. Recognize $\mu_{\overline{x}}$ and $\sigma_{\overline{x}}$.

Unit I: Introduction to Sampling Distributions

Preparation-Check Questions

1. We have a sample consisting of prices of homes sold in a certain town in June 1993. Which of the following populations does *not* contain our sample?
   a. All houses sold in this town.
   b. June house sales in this town.
   c. Houses sold in 1993.
   d. Houses sold in 1993 in this town.
   e. Houses sold in the first two weeks of June.
2. A representation or listing of all possible values of a random variable and the probabilities associated with each value would be a(n)
   a. Expected value.
   b. Binomial situation.
   c. Random variable.
   d. Frequency distribution.
   e. Probability distribution.
3. Which of the following symbols represents the sample mean?
   a. μ.
   b. $\overline{X}$.
   c. $E(X)$.
   d. σ.
   e. p
4. There are 10 marbles in an urn, three red, three gray, and four speckled. If one reaches in the urn and selects a marble, the probability it is speckled is 4/10. This probability is correct, because
   a. The single draw is an independent outcome.
   b. The single draw is a dependent outcome.
   c. The 10 marbles are equally likely to be drawn.
   d. The 10 marbles are *not* mutually exclusive.
   e. The single draw has been revised using conditional probability.

Use the sample (12, 18, 12, 22) to answer Questions 5–6.

5. The sample median is
   a. 15.   c. 12.   e. 22.
   b. 16.   d. 18.
6. The sample standard deviation is
   a. 5.0.   d. 4.0.
   b. 24.0. e. 4.2.
   c. 4.9.

Answers: 1. e; 2. e; 3. b; 4. c; 5. a; 6. c.

Introduction

Waste Treatment Government environmental regulations require a factory to mix its liquid waste with water until there are no more than 300 waste particles per cubic centimeter (ppcc) before releasing the blend into a river. The company must randomly collect a liter of the blend twice a day and record the mean of the two waste particle counts. Supervisor Rosie does not suspect that there is a variation in the treatment process currently. However, the equipment erratically spurts the waste particles into the system, so that one-third of the blend is actually 500 ppcc, one-third is 100 ppcc, and one-third is the limit, 300 ppcc, as shown in Figure 7-1. What are the chances that the mean of the two test counts will be far enough above 300 ppcc to alert Rosie to the problem?

A statistician studied this and other contingencies when the government agency formulated the regulations. Basing his research on blends with an equal distribution of 100 ppcc, 300 ppcc, and 500 ppcc, the statistician constructed the following probability distribution for the mean of the two counts, $\overline{X}$. ($\overline{X}$ symbolizes the sample mean, because the two counts are a sample of all the possible counts that could occur during the day.)

$\overline{x}$ (ppcc)	$P(\overline{x})$
100	1/9
200	2/9
300	3/9
400	2/9
500	1/9
	1

Figure 7-1
Population and samples of size 2 for the waste treatment situation

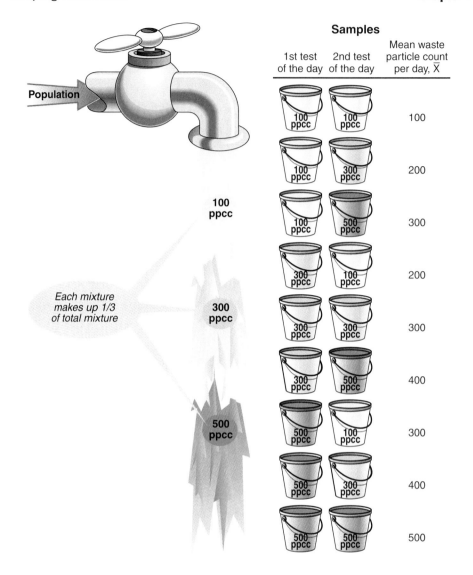

The mean count will exceed 300 ppcc (be 400 or 500) on about one-third of the days. This does not guarantee that Rosie will discover the treatment-processing problem, but the chances are good and certainly not remote. If presented with this table Rosie might very well become curious about its validity. She might also wonder what she could do to improve the chances of obtaining a sample mean that would alert her to processing problems.

This unit shows how the statistician derived this probability distribution. Constructing a few of these yourself should help you understand the behavior of a sample value, such as $\overline{X}$, and the information contained in its probability distribution, such as the one shown on the table. Armed with such information, you will be better prepared to interpret or to suspect sample results and the population inferences that follow.

7.1 The Sampling Procedure

Many survey questionnaires are carefully or scientifically administered, and still their results may indicate very different circumstances, desires, or characteristics of people in general. Every election year the media and the candidates barrage the public with all

Section 7.1   The Sampling Procedure

sorts of polls. Who's in the lead and the amount of the lead can vary quite a bit from poll to poll, even when objective experts conduct the polls. Differences may exist, because political leanings can be volatile during an election and polls survey people at different times. However, even when the experts do the surveying at the same time, and even when they do their utmost to ask objective questions in an objective way to avoid ambiguous interpretations, different results occur because of the method each uses to collect the sample data.

7.1.1 Purposes of Sampling

As we pull our gear together from previous chapters and journey into a different area of statistics, recall that in *inferential statistics* we generalize or estimate information about a population by using information in a sample. Population characteristics or values, such as μ, are called *parameters*, and sample values, such as $\overline{X}$, are called *statistics*. Thus, the inferential process consists of using statistics to estimate parameters.

Question: Which of the following is a parameter and which is a statistic: σ, S?

Answer: σ is the population standard deviation, so it is a parameter. The sample standard deviation, S, is the statistic. You may already suspect that we will use S to estimate σ.

Making a statistical inference is simple and straightforward, but, in practice, an inference can produce results with a wide range of reliability. If we know how to measure the reliability, we can change aspects of the estimation process, such as the sample size, to improve the reliability of the estimates. In some cases, we will be willing to sacrifice some reliability in exchange for saving time or money in generating the estimates.

In some of these cases, the measurement process destroys the sampled items, such as ruining a car bumper when we measure the impact of a crash at 5 miles per hour. Such cases constitute *destructive sampling*. In situations like these, it is imperative that we use sample information—not the population itself—to answer a question about the population; otherwise, we would destroy the population. Avoiding destructive sampling, saving time, and saving money are the common reasons for employing samples to answer questions about the population.

7.1.2 Random Sampling

When we collect a sample for an inference, our objective is to obtain a *representative* sample, one that mirrors the population in all of its important characteristics except size. If females tend to be more altruistic than males, then a sample that is overpopulated with males is likely to misrepresent the level of altruism in the population.

■ **Example 7-1:** Suppose left-handed people are more likely to buy small foreign cars than right-handed people. If a sample contains only right-handed people, how will this lack of representation of left-handed people affect an estimate of the percentage of the U.S. population likely to buy a small foreign car? (Source: John Koten, "You Aren't Paranoid If You Feel Someone Eyes You Constantly," *Wall Street Journal*, March 29, 1985, p. 1.)

■ **Solution:** The estimated percentage is likely to be too small. If left-handed people were in the sample, they would be more likely to purchase such a car, which raises the overall estimate from the sample.

One way to ensure representation of all important population elements is to select items from each important segment of the population. The proportion of each segment in the sample would reflect the proportion of each segment in the population. Such a sample is called a *stratified sample.* For instance, if 5% of the population is left-handed, then 5% of the sample would be left-handed people as well. Finally, we want to choose the elements of each segment in the sample in a way that results in a subsample that is *representative* of that segment. For instance, we do not want only left-handed females in the sample, if gender and dexterity are important characteristics to represent in the sample.

Random selection is the basis for most samples used in objective real-world analysis and for the samples used in this book. We choose items without regard to any characteristics of the item itself, of items chosen previously, or of those to be chosen. More formally, we choose a *random sample* in a way that allows every sample of n observations an equal chance of selection. Consequently, every element in the population will have an equal chance of being included in the sample, and the sample is likely to be representative. We also use random sampling procedures when we collect observations in more complex sampling schemes, such as the segments represented in a *stratified random sample.* (See Problem 47 at the end of Unit I for more information and details about stratified samples.) The term *sample* will imply a random sample in the remainder of this book.

Question: We need a sample of invoices quickly, so we grab the first five in the file. Is this a random sample?

Answer: It is not unless we file invoices randomly. If we organize the file in a manner that would distort findings, such as putting large orders at the front of the file, then the first five are not a random sample. Samples that include smaller orders at the back of the file do not have an equal chance of being selected. Smaller orders do not have an equal chance of being included in the sample.

Question: We take every twentieth name from a town's phone book until we obtain the desired sample size. Is this a random sample of people in the town?

Answer: No. Some people do not own phones; some are unlisted; others have multiple listings. Some listings are for single individuals; others are for large families. So the chances of being included vary among the people in the town. If the population is telephone owners or people with access to a telephone, the chances of achieving a random sample improve.

Frequently, to avoid patterns we use a random number table like the one in Appendix I or a computerized random number generator to pick objects for a sample. Figure 7-2 illustrates the lack of an intentional pattern to the random digits listed in the random number table. We match the random numbers that we select with names or objects on a list of population names or objects. The resulting subset is the sample.

To choose random values from the table that will correspond to population items, we must know N, the population size. N is the largest usable random value we can match with a population value.

■ **Example 7-2:** An administrator wants to select a random sample of eight students from a campus of 53,655 students. What is the largest usable random value? How many digits must we assemble to form each of the eight random values that correspond to one of the 53,655 students?

■ **Solution:** $N = 53,655$ is the maximum value. Because this value is composed of five digits, we must assemble five digits each time to correspond to students in the population.

To obtain a figure with five digits, we begin anywhere in the random number table and select five digits. They may be in a row (read left or right, skipping or not skipping digits), in a column (read up or down), on a diagonal, or some other selection pattern, as long as we select five digits.

Section 7.1 The Sampling Procedure

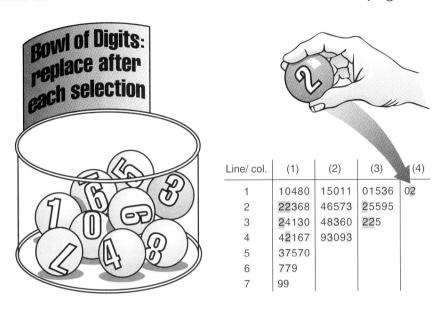

Figure 7-2
No pattern to digits' positions in a random number table

■ **Example 7-2 (continued):** Suppose we obtain 50000, 05000, 00005, 50005, 55000, 42836, 01012, and 53656 for the sample of eight students ($n = 8$) from the 53,655 students. Which of these random numbers are useful values?

■ **Solution:** All of them are usable except 53,656 and 55,000, which are larger than $N = 53,655$. We will need two more usable values from the random number table for n to be eight.

After we assemble n usable random values, we go to those numbered people or objects to collect our information. We might go to a master list of objects, such as the students, and pair the random value, such as 00005, with the fifth student name on the master list. Figure 7-3 shows the correspondence between the usable values and names on the list of 53,655 students. We would contact each highlighted student for information or a sample value. On a production line, we would check the fifth item on the conveyor belt during some time period for the desired information.

Boxes 7-1 through 7-4 discuss various issues related to accuracy of inferences, including some influences where inaccuracies creep in in spite of random selection of subjects.

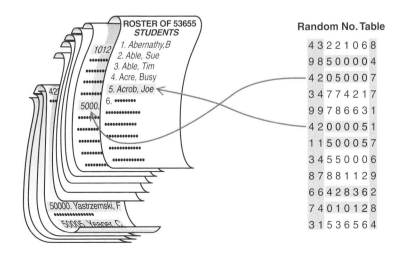

Figure 7-3
Procedure for obtaining a random sample using a random number table

Box 7-1 Time to Quit

There are times when the costs of collecting or improving the quality of data appear to outweigh the benefits of the information. Such was the case for a government agency that investigates diseases and produces information regarding contagious diseases and potential epidemics. The Centers for Disease Control (CDC) in Atlanta discontinued five years of efforts to conduct a national survey to more precisely assess the number of AIDS-infected Americans, the National Household Seroprevalence Study. Before making this decision, the CDC directed a pilot survey in Dallas, Texas, and Pittsburgh, Pennsylvania. A pilot survey is a test conducted on a small sample to ascertain problems with the survey design, including the questionnaire, responses, participation, and interviewers. The CDC uncovered a nonresponse problem, that is, persons selected for a study chose not to respond or could not be located to obtain their responses. The CDC considered this problem serious enough to harm and bias the conclusions beyond a useful limit. As a result, the CDC canceled further efforts and expenditures.

Sometimes the information sought in a survey is very sensitive and personal. A survey may ask people about their watching X-rated movies, about their sexual behavior, and whether they have tested positive for AIDS. In the CDC study, respondents supplied blood samples as well. In order to obtain cooperation and accurate answers, researchers had to take special care to protect the privacy of participants. To ensure confidentiality surveys did not record respondents' names, and results from all surveys were summarized, rather than reporting responses that might identify individuals. In spite of these efforts, researchers still found it difficult to get people to participate. This became one of those situations, in which researchers had to abandon further interviewing or efforts to improve the data-collection process rather than obtain results whose quality is unacceptable or results that are not worth the costs.

Source: Marsha F. Goldsmith, "Centers for Disease Control's Survey Succumbs," Journal of the American Medical Association, Vol. 265, No. 7, February 20, 1991, p. 838.

Section 7.1 Problems

7. Suppose you randomly select a starting point in the student telephone directory and continue selecting every tenth student until you have a random sample of 50 students. You plan to ask each one about their satisfaction with the current class registration procedure that is based on alphabetical order. How might this sample selection method affect the general findings about satisfaction with registration?

8. Are responses to newspaper advertisements for experimental subjects likely to produce a random sample of subjects? Explain.

9. Start in row 36 of column 7 of the random number table in Appendix I, and continue down the column, starting with the leftmost digit each time, until you obtain a list of five usable random values that you can associate with one of 482 elements in a population.

10. Start in column 1 of row 14 (Appendix I), and continue across the row, starting with the leftmost digit in each column, and then continue this pattern with the following rows, until you obtain 10 different usable random values that you can associate with one of the 51 values for Pupil Expenditures in the State SAT Data Base in Appendix C. What are the Pupil Expenditure values in the sample that results?

11. Use the random number table (Appendix I), and start with the rightmost digit in row 30 to obtain a usable value for College Dormitory Population from the North Carolina Census Data Base (Appendix E) and column 14 (the 8 in 69,618), and read leftward. Continuing on an upward diagonal to the left from this point, the next starting position would be the 3 in 34,693 of row 29 and column 13 then the 4 in 64,584 in row 28 and column 12, until you obtain two usable values. Compose the sample of dormitory values.

7.2 Sampling Distributions

Consider two different statistical experiments we can conduct with a population. We can select one item from the population and determine if it satisfies some event. In the waste treatment example, we randomly select one liter of the mixture and determine its waste particle content, X. Because the three levels are equally likely to occur, we can summarize the possible outcomes with the following probability distribution of X, where the X values flow from the spigot in Figure 7-1 (p. 290).

Box 7-2 Leading Questions and Elicited Responses

Random sampling increases the likelihood of obtaining a good representation of a population. However, the accuracy of a sample's reflection of the population may be spoiled by using a *leading question* on a survey. To gain a true picture of the opinions or other parameters of a population the survey must ask objective questions; that is, questions that do not bias the answers of the respondents.

A question can be asked (knowingly or unknowingly) in a way that elicits a certain response. Often such a question adds information that lets the respondent know the interviewer's desired answer. For example, suppose a financial institution that concentrates on tax-exempt municipal bonds develops a survey to collect information on public support for maintaining the tax-exempt status of these bonds. A question on their survey begins by defining tax-exempt municipal bonds. It continues that these bonds are used to finance public benefits such as education, highways, and parks, before asking for respondents' opinions regarding the tax-exempt status. It does not mention any costs to the public for continuing the tax-exempt status of these bonds. This would be considered a leading question. A better question would inquire if the respondent is familiar with tax-exempt municipal bonds. If not, then a straightforward definition could be stated, such as: "these are bonds issued by municipal governments that are free of federal income tax for bondholders."

It is tempting for special interest groups to educate people about their own opinions and to solicit favorable responses. Sometimes, busy readers and legislators may not be aware that a survey may be biased. They may not read carefully or ascertain the context of responses, absorbing only the summarized results, based on subjective and nonscientific methodology. Once a senator, accompanied by a television news team and other media, questioned constituents waiting in a long line at a service station in response to rising prices and dwindling supplies of gasoline. Some people had been waiting for hours. He solicited opinions on rationing legislation he supported in Congress and concluded that many favored his program (Senator Jacob Javitts, June 1979, on a New York television station). We all enjoy finding people who agree with us, but to convince others that these people represent a sizable proportion of some population, we must gather evidence more objectively or scientifically.

One tactic that improves the likelihood of procuring objective results is to let a third party, not directly interested in the issue or a particular result, design the survey and collect the data. Some national polling companies build and maintain reputations for statistical accuracy. They try to avoid controversy from poorly constructed questions or interviewer-interviewee relationships. Other polling companies do not enjoy such favorable reputations. A senator noted in a committee hearing that there is always a pollster who can find the result you want by changing the question (Senator Orrin Hatch, Senate Judiciary Committee Hearings on Sexual Harassment Charges of Judge Clarence Thomas, October 13, 1991). There is no perfect question or perfect survey. As informed consumers, we need to be wary and scrutinize the methodology of pollsters, especially when important decisions depend on survey results. As poll designers, we need to construct surveys and collect information cautiously.

Responses to leading questions result in dubious inferences, conclusions about the population based on sample information. The sample may overrepresent people who are in favor of the interviewer's opinion. Respondents may opt for no response or respond favorably to avoid disappointing the interviewer. As a result, we cannot be sure what population is represented by those who do respond or if the responses represent legitimate expressions of opinion. If the questions or interview do not state or hint at a preferred response, we will make more accurate inferences about the population being sampled. Measuring the accuracy of the inference then becomes a question of statistical methodology.

Try constructing some survey questions. Compose an objective question, then a leading question, to obtain information on the following events. Check your objective question for biases.

a. Respondent's opinion of a particular political candidate.
b. Respondent's opinion of participating in a poll or survey.
c. Respondent's likelihood of donating time for a charitable purpose.
d. Student's evaluation of the educational process in a particular class.
e. Teenager's feeling about the military draft.
f. Doctor seeking information about patient's symptoms before a diagnosis.

x	$P(x)$
100	1/3
300	1/3
500	1/3

The other statistical experiment is to compute the mean of a random sample of n items from the population. We label this random variable, $\overline{X}$, because the possible values are all sample means. A probability distribution of $\overline{X}$ concisely summarizes the $\overline{X}$

Box 7-3 Acceptable Survey Methods and Findings

The General Accounting Office (GAO) of the U.S. Government faulted the 1987–1988 Nationwide Food Consumption study for several reasons: an insufficient response rate (the ratio of the number of respondents to the total number of attempted surveys) of 34%, insufficient training of interviewers, inadequate spacing of interviews to control for seasonal variation, a complex questionnaire that was difficult and time-consuming to complete, and inadequate compensation to motivate respondents' cooperation ($2 for a survey that required up to three hours). Each of these deficiencies in the survey is one that statisticians attempt to avoid or overcome.

How to construct a survey is an important statistical topic. Shorter surveys usually elicit more responses. However, longer, well-designed ones can provide a richness of detail and information lacking in shorter surveys.

The response rate raises questions about the population and about the inferences we make about it. Everyone would agree that a 100% response rate for randomly selected respondents is the ideal. But no one expects perfection. Different analysts offer different subjectively chosen "adequate" response rates. When some people do not respond, we cannot be certain of the composition of the population. An important segment may not participate, such as workers who do not have three hours to spare for a food study. Consequently, there will be some uncertainty about the inferences we draw from the sample of those who do respond.

In spite of the large sample in the Nationwide Food Consumption survey (over 3,000 respondents), the GAO faulted the response rate. The results of the survey are crucial for large-scale decision making; thus, greater accuracy is required. Survey users include: the Environmental Protection Agency (EPA) for regulating pesticide residue levels; the U.S. Department of Agriculture (USDA) for regulating availability of products for food stamp and school lunch programs; and the Food and Drug Administration (FDA) for regulating "average serving size" on food labels. With such important decisions riding on the information acquired from the survey in addition to the study's implementation problems noted above, the GAO concluded that the results were faulty. The USDA agreed to review the survey and issue warnings about the limitations of the data when publishing the results.

Source: "Bungled Survey Could Affect Wide Range of Agencies," Knight-Ridder article in Asheville Citizen-Times, September 11, 1991, pp. 1A,3A.

---

values that result from the different random samples of a given size available from this population (potentially an infinite number of such samples). This probability distribution of a statistic is called its *sampling distribution*. The sampling distribution represents all possible values of the statistic, $\overline{X}$ for instance, and the probability that each value occurs. For the waste treatment example, the sampling distribution of $\overline{X}$ is

$\overline{x}$	$P(\overline{x})$
100	1/9
200	2/9
300	3/9
400	2/9
500	1/9
	1

Figure 7-1 also shows the experiment that generates the $\overline{X}$ values that appear in the last column of the figure and again in the $\overline{x}$ column of the sampling distribution just shown. Understanding the nature of these two experiments will clarify relationships between the two that are important for inferential statistics.

7.2.1 Constructing a Sampling Distribution

The nature of a sampling distribution depends on (1) the *statistic* of interest; (2) the *population distribution* from which the sample originates; and (3) the particular *sample size*, n, being investigated. Changing any one of these three factors has important effects on the sampling distribution.

Box 7-4 Are Chemists Girl Crazy?

Chemists, and scientists generally, are quick to chastise the public for entertaining notions for which there's little more than anecdotal evidence, so surely there's plenty of data showing that chemists specialize in daughters.

You hadn't heard this one? For decades, members of the profession, when not exchanging or debating the latest research findings, have marveled over an apparent trend that they believe to be unique to their community: Chemists, some like to say, produce more girl babies than boy babies.

A girl thing. "I believe it," hazards University of California, Berkeley, polymer chemist Bruce Novak, a father of two daughters who first heard the idea when he was in graduate school. "Many people think this is true," concurs Brent Iverson, a bioorganic chemist at the University of Texas in Austin and the father of three girls. "When I was an undergraduate I worked for a guy who had four daughters, and my wife worked for someone (a chemist) with five daughters." Electrochemist Nate Lewis of the California Institute of Technology was so sure it was true that when his wife, also a chemist, became pregnant with their first child a year ago, "I was expecting a girl." It was a boy.

That won't change the mental chemistry of the true believers, even though some are considerably less global in their claims. For example, several theoretical chemists told *Science* that the tendency to have daughters shows up only among theoretical chemists. Then there were the NMR spectroscopists and x-ray crystallographers who claimed similar honors for their own specialties. "My version of the myth is that organic chemists have more girls," said Robert Bergman, father of two boys and an organic chemist at Berkeley. At least he used the word "myth," but before you biologists and physicists sneer out loud at this seemingly gullible group of reasoning beings, think about fighter pilots.

A similar story had been circulating about the pilots when Bertis Little, assistant professor of obstetrics and gynecology at the University of Texas Health Science Center in Dallas, and his colleagues took off on a study of the intrepid breed. In the July 1987 *Aviation, Space and Environmental Medicine,* Little et al. announced that, in a small survey of 62 pilots and astronauts exposed to high G forces, they found that about 60% of their offspring were girls. Because the bias didn't show up among 220 bomber pilots, transport pilots, and other personnel whose aeronautic routines involve far less fast and sharp maneuvering, Little and his colleagues suggested that exposure to high G forces might influence the survival of the sex-determining chromosomes.

But even the highest fliers in the chemistry community experience few G's in their daily lives—and no similar study has been done for chemists and their offspring. So what's the evidence? Hoping to remove any stain from the reputations of the earnest chemists who were willing to go public with their beliefs, *Science* asked the U.S. Census Bureau if the agency might be able to extract telling numbers from its databases. Maybe, was the response, but the search fee would be hefty, and the wait would be at least a year. How about the American Chemical Society? Unfortunately, it doesn't keep track of the gender of its members' offspring. Princeton University's Office of Population Research? No dice.

Science's Own Sex Survey. So *Science* boldly ventured where no group has gone before: A totally nonscientific questionnaire was sent to about 250 chemists at eight small and large chemistry departments, yielding 140 usable responses. Thanks to the tireless efforts of Gayla Bradfield, a secretary in the chemistry department of the Indiana University, Bloomington, who was well aware of the importance of the survey, all 45 members of that department complied. That rules out statistical biases that might arise through self-selection if, say, chemists with more daughters were especially apt to respond. Of the Indiana chemists, 34 had children in a collective ratio of 53 girls to 41 boys, or a 56% to 44% split, respectively. But when these numbers are pooled with those from the other hundred or so returned questionnaires, the apparent preponderance of female offspring virtually disappears—out of 326 children, 158 or about 51.5% were reported to be girls.

Too shaky a sample for you? *Science* plunged on by combing the 1985 edition of *Who's Who in Frontiers of Science and Technology* for information. Of the 151 chemists found in a random search amounting to 10% of the volume, 29 listings had no information about children and 6 proved unusable because the children's names (for example, Leslie or Jody) didn't reveal gender. Of the remaining 116 listings, the gender tally came to a near dead heat—144 boys to 147 girls.

And there's more: A random sampling of 10% of the nearly 1,400 listings of chemists in the more general and up-to-date *Who's Who* database indicated a slight and insignificant preponderance (51.5%) of boys.

So pending better data, the verdict looks to be, in chemistry anyway, "myth." Who's to blame for the tall tale? John D. Roberts, a long-time physical organic chemist at Caltech, affectionately points to Robert B. Woodward, a giant in the history of organic chemistry who won a Nobel Prize for his work in 1965 and died in 1979. Woodward had three daughters in a row. That "caused the legend to grow, and he fostered it," notes Roberts (father of three boys and a girl). Evidently, sample sizes of three are perfect for starting a myth. Oh, and you'll want to know that Woodward's fourth and last child was a boy.

Source: Ivan Amato, "Are Chemists Girl Crazy?" Science, Vol. 257, July 10, 1992, pp. 158–159, with permission.

In the examples in this unit, we will select samples from *infinite populations,* populations that are so large that each value is repeated a very large or infinite number of times. In the waste treatment situation, there are infinite points in the blend where we could observe any one of the three levels of waste particles. In a battery-production process, the number of batteries produced with a lifetime of 25 hours is so large that reducing the number by one does not appreciably change the proportion of all batteries that last 25 hours. There are occasions when we collect samples from a population that is limited or finite; for instance, we might want a sample of 50 students from a school of 200 students. We will relax the assumption of infinite populations in Unit II, but for now it simplifies the development of the sampling distribution and related concepts.

Example 7-3: Suppose there are three colors of paint available for painting a room: red (R), yellow (Y), and blue (B) and the painter can use as many cans of each as she needs. She will select two cans of paint at random and combine them to form the color for the room. We want to construct the sampling distribution of the room color, C, that results from this process. (Because C is not a number we might call this sample information a qualitative statistic. A physicist, engineer, or designer might instead quantify the resulting colors using light frequencies from the spectrum.)

Solution: To construct this distribution, we must know every possible value of C or the color resulting from combining two randomly selected cans. The first step is to determine all possible random samples of size two. If the symbols R, Y, and B are each placed on many, say 1000 each, pieces of paper and two are randomly selected, possible results include YR, RY, and RR among others. The order of the letters reflects the order of selection, so RY and YR are different outcomes, though they both are elements of the event, C = orange.

Question: What are the nine possible combinations or samples of size two that can occur?
Answer: All possible samples of size two and the ensuing color, C, are

RR = red RY = orange RB = purple
YR = orange YY = yellow YB = green
BR = purple BY = green BB = blue.

The possible values of the statistic (in this case the color of the room), C, are red, orange, yellow, green, blue, and purple. The other part of a probability distribution is the probability associated with each value of the statistic C.

Question: Because random selection of sample items ensures that the outcomes of both selections are independent, what is the probability the resulting sample will be BY, blue first and yellow second?
Answer: P(blue first and yellow second) = P(blue)P(yellow) = (1/3)(1/3) = 1/9.

The same value, one-ninth, is the probability that *any* particular combination out of the nine possibilities will result. *This is always the case with a random sampling process; each of the possible samples is equally likely to occur.* The painter is just as likely to select RR as RY or YB. We can use this information to determine the probabilities for each value of C.

Question: What is the probability the color of the room will be orange?
Answer: $P(C = \text{orange}) = 2/9$. Because there are nine equiprobable samples and two of these samples produce orange, RY or YR, we know from the equiprobable outcome method in Chapter 4 that the probability C will be orange is 2/9. Alternatively,

Section 7.2 Sampling Distributions 299

$P(C = \text{orange}) = P(RY \text{ or } YR) = P(RY) + P(YR)$ (because RY and YR are mutually exclusive, the color result is the same for RY and YR, but the selection order is not): $= 1/9 + 1/9 = 2/9$.

Question: Complete the probabilities for the sampling distribution of C.

Answer:

c	$P(c)$
Red	1/9
Orange	2/9
Yellow	1/9
Green	2/9
Blue	1/9
Purple	2/9
	1

Figure 7-4 illustrates the generation of this sampling distribution.

■ **Example 7-4:** Use the three waste-particle levels (100, 300, and 500 ppcc) as the population to construct the sampling distribution of the mean waste particle count, $\overline{X}$, for samples of size 2 ($n = 2$).

■ **Solution:** There are three values in the population (100, 300, and 500). Because we are investigating the results for samples with two observations, there are nine possible random samples, as shown in Figure 7-1 (p. 290).

(100, 100) (100, 300) (100, 500)
(300, 100) (300, 300) (300, 500)
(500, 100) (500, 300) (500, 500).

Figure 7-4
The sampling distribution of C

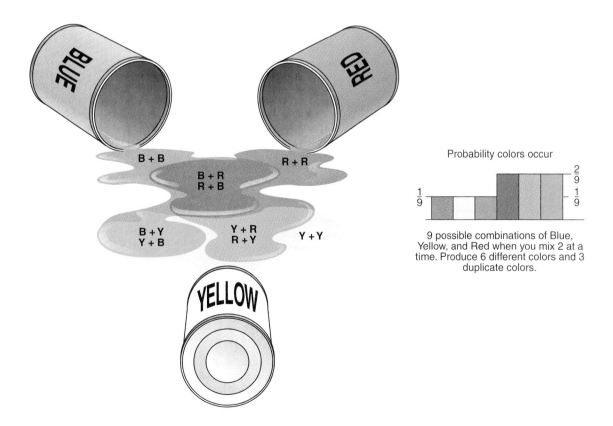

Probability colors occur

9 possible combinations of Blue, Yellow, and Red when you mix 2 at a time. Produce 6 different colors and 3 duplicate colors.

Next, we calculate the sample mean, $\overline{X}$, for each sample. For instance, $\overline{X}$ for the sample (100, 300) is (100 + 300)/2 = 200. The result for each sample appears below.

Sample	$\overline{x}$
(100, 100)	100
(100, 300)	200
(100, 500)	300
(300, 100)	200
(300, 300)	300
(300, 500)	400
(500, 100)	300
(500, 300)	400
(500, 500)	500

From these calculations, we see that the possible $\overline{X}$ values are 100, 200, 300, 400, or 500. Using the equiprobable outcomes method and the fact that each sample is equally likely to occur, we obtain the following sampling distribution of $\overline{X}$. For instance, to find $P(\overline{X} = 400)$, we see that two of the nine (mutually exclusive equiprobable) samples have a mean of 400. These are (300, 500) and (500, 300).

$\overline{x}$	$P(\overline{x})$
100	1/9
200	2/9
300	3/9
400	2/9
500	1/9
	1

From these results we can see that $\overline{X}$ values close to 300 are most likely and that $\overline{X}$ values furthest from 300 are least likely.

■ **Example 7-5: Effect of Changing the Sample Size, *n*:** Suppose we randomly test the waste mixture three times during the day rather than twice. The population is still (100, 300, 500), and $\overline{X}$ is the statistic. Adding another test gives us 27 possible random samples. The following table lists the samples and their corresponding means. Complete the sampling distribution of $\overline{X}$.

Sample	$\overline{x}$ (ppcc)	Sample	$\overline{x}$ (ppcc)
100, 100, 100	100.0	300, 300, 500	366.7
100, 100, 300	166.7	300, 500, 100	300.0
100, 100, 500	233.3	300, 500, 300	366.7
100, 300, 100	166.7	300, 500, 500	433.3
100, 300, 300	233.3	500, 100, 100	233.3
100, 300, 500	300.0	500, 100, 300	300.0
100, 500, 100	233.3	500, 100, 500	366.7
100, 500, 300	300.0	500, 300, 100	300.0
100, 500, 500	366.7	500, 300, 300	366.7
300, 100, 100	166.7	500, 300, 500	433.3
300, 100, 300	233.3	500, 500, 100	366.7
300, 100, 500	300.0	500, 500, 300	433.3
300, 300, 100	233.3	500, 500, 500	500.0
300, 300, 300	300.0		

Section 7.2 Sampling Distributions 301

■ **Solution:**

$\bar{x}$	$P(\bar{x})$
100.0	1/27
166.7	3/27
233.3	6/27
300.0	7/27
366.7	6/27
433.3	3/27
500.0	1/27
	1

Changing the sample size, n, but not the population or statistic, usually changes the sampling distribution. Notice again that values near 300 are most likely and those furthest from 300, least likely. Notice also that estimates of μ based on $\bar{X}$ can be quite different, depending on the actual sample that we select, but the sampling distribution tells us which values of $\bar{X}$ are likely to occur and which are not likely to occur.

■ **Example 7-6: Effect of Changing the Statistic—The Sample Variance:** Suppose that the statistic in the waste treatment example for samples of size 2 ($n = 2$) is the sample variance, S^2. Recall that

$$S^2 = \frac{\sum_{1}^{n}(X_i - \bar{X})^2}{n - 1}.$$

For sample (100, 300), $\bar{X} = (100 + 300)/200 = 200$, so

$$S^2 = \frac{(100 - 200)^2 + (300 - 200)^2}{2 - 1} = \frac{20{,}000}{1} = 20{,}000.$$

The result would be the same for the sample (300, 100).

Question: What is S^2 for the samples (100, 100), (300, 300), and (500, 500)?
  Answer: Because there is no variation in each sample's values, S^2 is 0 for each sample.

Find S^2 for the remaining samples: (100, 500), (500, 100), (300, 500), and (500, 300). Then, construct the sampling distribution of S^2. (Remember the $\bar{X}$ in the S^2 formula can vary from sample to sample.)

■ **Solution:**

Sample	s^2
100, 100	0
100, 300	20,000
100, 500	80,000
300, 100	20,000
300, 300	0
300, 500	20,000
500, 100	80,000
500, 300	20,000
500, 500	0

Using the equiprobable outcomes method with these results, we obtain the following sampling distribution for S^2.

s^2	$P(s^2)$
0	3/9
20,000	4/9
80,000	2/9
	1

Constructing these examples of different sampling distributions provides added opportunities to understand the experiment that generates a sampling distribution. The result is a representation of all possible values of a statistic that might occur from a sampling procedure and the probability that each value occurs. When presented with a sampling distribution, we do not have to proceed through the arduous task of developing the distribution itself, but we need to understand the background work represented in the final product. We will use this understanding to determine the probability of potential values of a statistic in the next section. We also know that changing the population, the statistic, or the sample size, n, can change the resulting sampling distribution.

7.2.2 Solving Probability Problems with Sampling Distributions

Remember that a sampling distribution is a special probability distribution. We can solve probability problems using the sampling distribution in the same way that we use any other probability distribution.

■ **Example 7-4 (continued):** Use the following sampling distributions for the waste treatment situation to find the probability that

a. The mean waste particle level for a particular day will equal or exceed 300 ppcc.
b. The mean waste particle level will be within 75 ppcc of 300 ppcc, that is, $P(225 \leq \overline{X} \leq 375)$.
c. In which situation, $n = 2$ or $n = 3$, are the $\overline{X}$ values more tightly clustered about 300?

	$n = 2$		$n = 3$
$\overline{x}$	$P(\overline{x})$	$\overline{x}$	$P(\overline{x})$
100	1/9	100.0	1/27
200	2/9	166.7	3/27
300	3/9	233.3	6/27
400	2/9	300.0	7/27
500	1/9	366.7	6/27
	1	433.3	3/27
		500.0	1/27
			1

■ **Solution: a.** When $n = 2$, $P(\overline{X} \geq 300) = P(\overline{X} = 300 \text{ or } 400 \text{ or } 500) = 3/9 + 2/9 + 1/9 = 6/9$, because the outcomes are mutually exclusive. When $n = 3$, $P(\overline{X} \geq 300) = P(\overline{X} = 300 \text{ or } 366.7 \text{ or } 433.3 \text{ or } 500) = 7/27 + 6/27 + 3/27 + 1/27 = 17/27$.
b. When $n = 2$, $P(225 \leq \overline{X} \leq 375) = P(\overline{X} = 300) = 3/9$. When $n = 3$, $P(225 \leq \overline{X} \leq 375) = P(\overline{X} = 233.3 \text{ or } 300 \text{ or } 366.7) = 6/27 + 7/27 + 6/27 = 19/27$.
c. The $\overline{X}$ values are more tightly clustered about 300 when $n = 3$, as shown by the answers to Part b. More than two-thirds of the $\overline{X}$ values are with 75 units of 300 when $n = 3$, and only one-third when $n = 2$.

It is easy to see from these problems that sampling distributions condense a lot of background work and information into a very useful format for considering the potential outcomes of sampling and inferential processes. In the next unit we will employ a simpler approach to obtain sampling distributions of $\overline{X}$ for specific situations.

Multiple-Choice Problems

Section 7.2 Problems

12. A computer company president must discuss the impact of electrical surges (sudden impulses of large quantities of electricity in the current) on the firm's products (two different desktop computer models) with a government agency. Basing his estimates on past experience with the devices, a technician expects that surges of 100 and 200 volts, respectively, would destroy essential elements of models A and B. The president and technician decide to select two computers randomly from the firm's inventory and increase the electrical input until each computer breaks in order to obtain a cheap and speedy estimate of the mean surge size that ruins the computers.
   a. Construct the probability distribution for the sample mean, $\overline{X}$.
   b. Construct the sampling distribution for samples of three rather than two computers.
   c. Which values of $\overline{X}$ are most likely to occur?
   d. The technician randomly selects two computers from the population and determines their mean destructive surge voltage. Find the probability the mean will be greater than or equal to 150 $[P(\overline{X} \geq 150)]$. Then, solve the same problem assuming the technician randomly selects three computers rather than two.

13. Suppose we want to know the minimum waste-particle level each day for the factory in Example 7-4 (p. 299). The statistic or value we look for in the sample is the smallest value in the sample, *MIN*. The minimum value for (100, 100) is 100. For (100, 300), *MIN* = 100. For (300, 500), *MIN* = 300.
   a. Find the sampling distribution of *MIN* for samples of size 2.
   b. Use the list in Example 7-5 (p. 300) to construct the sampling distribution of *MIN* for three test samples of the waste mixture.
   c. Compare these results with the sampling distribution for $\overline{X}$ from the same population and sample sizes. Describe the general shapes of the sampling distributions, or sketch a curve to depict the shapes.
   d. Find the likelihood that a day's sample of two test results will yield a value of 500 for *MIN*. Then, find the probability the value of *MIN* will be larger than 200. Repeat these problems for a sample of three test results.

14. The tensile strength of rolls of wire from a particular supplier vary. About *two-thirds* of the rolls can withstand 50 pounds per square inch (psi) when stressed, and the remainder can withstand 150 psi. If you select two rolls at random, what is the probability the mean tensile strength will be 50 psi? Form the possible samples of size 2 for the two wire types, and find their means. Then, use probability rules to find the probability for each mean and form the sampling distribution for samples of size 2. Finally, use the sampling distribution to find the probability the mean will be at most 75.

Summary and Review

A sampling distribution is a probability distribution of a statistic. The sampling distribution depends on the statistic, the sample size (*n*), and on the population distribution. We focus on the statistic $\overline{X}$, the sample mean, in order to learn about its behavior in a random sampling procedure, especially which values of $\overline{X}$ are likely to occur and which are not. It is necessary to understand the behavior of $\overline{X}$ when we employ the sample mean, $\overline{X}$, to estimate the population mean, μ.

So far, constructing a sampling distribution requires finding all possible values of the statistic and the probability associated with each of these values. The information represented by such distributions allows us to solve probability problems as we would with any other probability distribution. The only difference is that now the random variable is the value of a statistic calculated from a random sample.

We assume the sample is chosen from an infinite population in order to simplify the relationships between the population and the distribution of the possible values of the statistic. In the next unit, we will extend the sampling distribution concept to more complex situations, such as larger finite populations, larger samples, or continuous populations, where finding *all* of the individual $\overline{X}$ values and probabilities is impractical.

Multiple-Choice Problems

15. A sampling distribution is a
   a. Frequency distribution of a parameter.
   b. Frequency distribution of expected values.
   c. Probability distribution of a statistic.
   d. Probability distribution of standard deviations.
   e. Normal distribution of $\overline{X}$ values.

16. Which of the following changes will usually *not* cause the sampling distribution to change?
   a. The population values increase by 10%.
   b. The statistic changes from $\overline{X}$ to the median.
   c. The population distribution being studied changes.
   d. The sample size changes.
   e. Any of the preceding changes will usually change the sampling distribution.

17. A random sample
  a. Is distributed similarly to the population.
  b. Is the easiest sample to collect.
  c. Is not a destructive sampling.
  d. Gives each population element an equal chance of being included.
  e. Requires each population element to be represented when the selection process is complete.

Use the following information to answer Questions 18–22.

The population of responses on a survey question regarding the condition of downtown streets is (1, 2, 4, 5), where 1 corresponds to superior, 3 to average, and 5 to substandard. No respondents replied average or 3. If we assume that there is a very large number of respondents for each of the remaining values, the population of responses produces the following 16 possible random samples of size 2:

```
(1, 1)   (1, 2)   (1, 4)   (1, 5)
(2, 1)   (2, 2)   (2, 4)   (2, 5)
(4, 1)   (4, 2)   (4, 4)   (4, 5)
(5, 1)   (5, 2)   (5, 4)   (5, 5)
```

18. The possible values of $\overline{X}$ are
  a. 1, 2, 4, 5.
  b. 1.5, 2.5, 3.5, 4.5.
  c. 1, 2, 3, 4, 5.
  d. 2, 3, 4.
  e. 1, 1.5, 2, 2.5, 3, 3.5, 4, 4.5, 5.

19. Complete the sampling distribution of $\overline{X}$ for $\overline{X} = 4$, $\overline{X} = 4.5$, and $\overline{X} = 5$, respectively.

$\overline{x}$	$P(\overline{x})$
1	
1.5	
2	
2.5	
3	
3.5	
4	___
4.5	___
5	___

  a. 1/16, 2/16, 1/16.
  b. 1/16, 2/16, 3/16.
  c. 1/16, 2/16, 4/16.
  d. 1/16, 2/16, 2/16.
  e. 2/16, 1/16, 2/16.

20. The mean of the population (1, 2, 4, 5) is 3. What are the chances of obtaining a random sample of two values from the population and obtaining a sample mean equal to 3?
  a. 0.0.
  b. 0.25.
  c. 0.20.
  d. 0.125.
  e. 0.50.

21. Let *MAX* be the largest value in a sample. Which of the following is the correct term to describe *MAX*?
  a. Statistic.
  b. Parameter.
  c. Expected value.
  d. Standard deviation.
  e. Median.

22. Remember that *MAX* is the largest value in a sample. Find $P(MAX < 4)$ for the street response data.
  a. 1/16.
  b. 12/16.
  c. 4/16.
  d. 1.
  e. 9/16.

23. If we select a random sample from a population that contains different values, then
  a. Any statistic will provide a reliable estimate of μ.
  b. Any statistic will provide an exact estimate of μ.
  c. μ is a good estimate of $\overline{X}$.
  d. An $\overline{X}$ not equal to μ can occur.
  e. Statistics have a normal sampling distribution.

24. If election polls are done objectively with random sampling, different results are possible, because
  a. The polls occur on different days.
  b. The phrasing of questions differs, which elicits a seemingly different response from the same individual on the same topic.
  c. Sample sizes differ.
  d. Samples from the same population can and usually do differ.
  e. All of the above.

25. The sampling distribution of $\overline{X}$ contains information useful for
  a. Selecting sample observations randomly.
  b. Selecting sample observations that result in a representative sample.
  c. Determining the probability distribution of the mean of the population being studied.
  d. Finding the probability that the sample mean occurs within an interval of values.
  e. Ensuring that average persons are included in a sample.

Use the following information to answer Questions 26–27.

A university administrator wants a random sample of 60 students taken from a student body of 6,235 students. He uses a random number table to select students from the registrar's listing of current students.

26. What is the largest usable value in the random number table that he can associate with a student?
  a. 60.
  b. 6,235.
  c. 6,175.
  d. 4.
  e. 6,295.

27. How many digits form a single candidate for a usable value to associate with a student?
  a. 6,235 digits.
  b. 60 digits.
  c. 6 digits.
  d. 4 digits.
  e. Between 1 and 4 digits.

28. If you drop a pencil on a random number table and the point lands on a digit, find the probability it will land on a 7, 8, or 9.
  a. 0.1.
  b. 0.9.
  c. 0.3.
  d. 0.5.
  e. 0.001.

Word and Thought Problems

Use the following sampling distribution of $\overline{X}$ in inches to answer questions 29–30.

$\overline{x}$	$P(\overline{x})$
1	0.05
2	0.08
3	0.17
4	0.40
5	0.17
6	0.08
7	0.05

29. $P(\overline{X}$ is odd$) =$
   a. 0.44.   d. 0.50.
   b. 0.56.   e. 0.25.
   c. 0.57.

30. Find $P(\overline{X}$ is within 2 inches of the outcome 4 inches). Include the boundaries of this interval.
   a. 0.90.   d. 0.80.
   b. 0.45.   e. 0.16.
   c. 0.40.

31. Frequently, a store will advertise or display a basket of goods with their prices and total value for the basket along with the same information for the basket of goods taken from a competitor's store. The implied inference is that prices are typically lower at the store rather than at its competitor's store. Which of the following problems most likely makes this inference inaccurate?
   a. The mean is often not calculated.
   b. The median should be shown rather than the mean or mode.
   c. The comparison is based on the same basket of goods at each store.
   d. The comparison is based on a sample rather than the population.
   e. The sample of goods in the basket is probably not random.

To prepare for accidental spills a Hazardous Waste Spill Control committee considered the following information from a sample of 155 trucks randomly selected from trucks with hazardous materials placards at the Asheville, North Carolina, area weigh stations. Use this information to answer Questions 32–34.

Material	Percent of Total Cargoes
Flammable	40
Radioactive	22
Corrosive	16
Explosives	6
Other (such as nonflammable gas or poisons)	16

Source: North Carolina Division of Motor Vehicles and Land-of-Sky Regional Council, as reported in Ed Brackett, "WNC Spill Threat May Be Reduced," *Asheville Citizen-Times*, March 17, 1991, p. 11A.

32. The article reports that a local official agreed with the results regarding a large number of gasoline trucks, based on accident and incident experience in the area over the years. If we agree and assume that these figures accurately reflect the population, find the probability a random sample of three trucks from the population will all carry explosives.
   a. 0.0002.   c. 0.8306.   e. 0.1798.
   b. 0.1800.   d. 0.0600.

33. Again assume the figures accurately reflect the population, and find the probability a random sample of three trucks from the population will consist of one truck with radioactive cargo and two carrying corrosive material.
   a. 0.0106.   c. 0.3596.   e. 0.5504.
   b. 0.0169.   d. 0.0056.

34. If the information in the table reflects the distribution of hazardous cargoes carried by trucks involved in accidents, as the official asserts, which of the following inferences based on this information is correct?
   a. The probability a traffic accident involves a truck with flammable material is 0.40.
   b. If a truck is involved in an accident, the probability it carries corrosives is 0.16.
   c. Prohibiting radioactive cargoes in the area will lower the accident rate for accidents involving hazardous materials by 22%.
   d. Explosive cargoes account for 6% of all cargoes trucked in the Asheville area.
   e. 60% of accidents involving flammable materials occur between automobiles, not trucks.

Word and Thought Problems

35. What is a sampling distribution?

36. Suppose we select a random sample of two waste-particle test results from the population of test results (100 ppcc, 300 ppcc, 500 ppcc). Find $P(\overline{X} < 400)$ and $P(S^2 < 20{,}000)$. Use the sampling distributions developed in Examples 7-4 and 7-6 (pp. 300 and 301).

37. Find the sampling distribution of the sample median of the waste-treatment test results in Example 7-4 (p. 299) when $n = 2$. Repeat the exercise for $n = 3$. The text presents all possible random samples for $n = 3$ in Example 7-5 (p. 300).

38. Name three ways to change the sampling distribution.

39. Consider the population and samples in the information for Multiple-Choice Problems 18–22.
   a. What is μ? How many of the possible $\overline{X}$ values are the same as μ? What is the probability you will obtain an $\overline{X}$ within one unit of μ (include the boundaries of this interval)?

b. What is the maximum value in the population? If *MAX* is the maximum value in a sample, how many of the possible sample *MAX* values are the same as the true maximum population value? Find the probability *MAX* will be within one unit of the true maximum population value.

40. Use the random number table to select 10 North Carolina counties for a random sample of the number of people in nursing homes from the North Carolina Census Data Base (Appendix E). Calculate the sample mean, standard deviation, median, and mode(s). Compare your results to the actual values for all 100 counties: $\mu = 470.1$ persons, $\sigma = 514.8$ persons, median = 319.5 persons, mode = 0. Speculate on the result of a similar comparison of actual values with a random sample from 50 counties rather than 10 counties.

41. Distinguish between the distribution of a population and a sampling distribution.

42. What are three reasons for using a sample to obtain information about a population, rather than using the population itself? Relate these reasons to NASA testing a space shuttle part before an actual flight of the shuttle.

43. The number of songs on four different tapes are 10, 14, 10, and 8.
   a. Describe a method of choosing tapes so that you will obtain a random sample of three tapes.
   b. Suppose you replace a tape in the population after it is chosen for a sample before you select the next tape. That is, the same tape can appear in the same sample more than once. Construct the sampling distribution of the sample mean for samples of two tapes by using this assumption. Is this replacement sampling like sampling from a finite or infinite population? Explain.
   c. Repeat Part b without replacing a tape after it is chosen, but start the formation of each new sample with the complete population. Describe the differences in the sampling distributions that result from Parts b and c.

44. Explain why each of the following involves or does not involve destructive sampling.
   a. Determining the proportion of chicken eggs that have been fertilized by removing the yolk from a random sample of eggs.
   b. Testing an experimental teaching technique on a randomly selected class.
   c. Allowing some, but not all, financial institutions to use special checking accounts.
   d. Building dams on some streams to determine the effect on plant life.
   e. Using Pavlov's experiment on dogs, in which bells are rung when the dogs eat, over a period of time, to measure salivation when bells ring in the absence of food.
   f. Congressional subsidies for construction at randomly selected sites.
   g. Sampling of shoppers by soft drink manufacturers to determine product preference.
   h. Asking voters whom they intend to vote for in an upcoming election.

45. A random sample of 2,000 objects with a mean of 980, modes of 800 and 950, a median of 960, and a standard deviation of 500, is selected from a normally distributed population with a mean of 1,350 and a standard deviation of 380. List the values in this statement that are statistics. List the values that are parameters. What are the median and mode(s) of the population?

46. Let the random variable X be the single digit that occupies a randomly selected position in the random number table. Describe the probability distribution of X. Find $E(X)$ and the standard deviation of X.

47. Stratified random sampling is a two-step sampling procedure used to improve the likelihood of the sample being representative of the population. First, we identify mutually exclusive segments (subpopulations) of the population, which are relevant to the parameter under investigation. We call these subgroups *strata*. We use the respective size of each stratum, N_i, to find each stratum's proportion of N, the total population size. This proportion is N_i/N. Then, we select a random sample of each stratum so that n_i/n, the proportion of the total sample size, n, in the ith stratum, is identical to the ith stratum's proportion of the total population, N_i/N. For example, suppose a psychologist conducts a study of job stress in an industry composed of 20,000 workers, 5,000 of whom are female (25%). If the psychologist uses a random sample of 400 workers, stratified by sex, then 25% of the 400, or 100 observations, would be randomly selected females. The other 300, or 75%, would be randomly selected males.
   a. Relate this process to weighting observations in determining the weighted mean.
   b. When would the number of observations from each stratum be equal? Would a simple random sample (unstratified) produce the same result?
   c. If a political scientist needs a random sample of 50 members of Congress to study overall sentiment regarding a certain issue, obvious strata are the two houses of Congress, which consists of 100 senators and 435 representatives. Find the number of observations to include in the two subsamples.
   d. A political scientist wants a random sample of 800 state legislators stratified by state to study the overall sentiment regarding a constitutional amendment. How many legislators should be selected from each state? Suggest a way of selecting the subsamples that ensures randomness.

48. An alumni director selects a random sample of 8,000 alumni of an institution and mails each person a survey to determine the effect of the education received at the institution on the professional life of the alumni. About 500 return completed surveys. What is the sample size, n, when the available data are analyzed? Is the sample of 500 random? Would your answer change if *all* living alumni, rather than a random sample, were mailed the survey?

Word and Thought Problems

49. Think of a sampling situation in which interviewers drop people who are not at home to answer their telephone from a random sample. This action may bias the conclusions drawn from the sample information. Discuss potential biases that may result if interviewers allow whoever answers the telephone to respond to the survey when the person being called is not at home to respond.

50. Discuss the randomness and representativeness of survey results generated by respondents telephoning numbers with area code 900 numbers to register their response. What about newspaper surveys, where readers fill out a survey in the paper and mail it in? Should data on *copies* of the survey form in the newspaper be discarded? Explain.

51. The incidence of diabetes among Americans is 5%, and the incidence of related foot problems among diabetic patients is 25%. Use this information to determine (a) the number of diabetics and (b) the number of diabetics with foot problems likely to appear in a random sample of 400 Americans. (Source: G. P. Kosak, C. S. Hoar, J. L. Rowbotham, et al., *Management of Diabetic Foot Problems*, Philadelphia: W. B. Saunders, 1984, in Jonathan J. Scarlet, and Mark R. Blais, "Statistics on the Diabetic Foot," *Journal of the American Podiatric Medical Association*, Vol. 79, No. 6, June 1989, p. 306.)

52. Use the population in Problem 43 to find the sampling distribution of the sample range for samples containing two observations drawn with replacement. Repeat this exercise when there is no replacement.

53. Let W be the number of times the value 10 appears in the sample for the samples of size 2 for the population in Problem 43. Find the sampling distribution of W.

54. Interviewers sometimes conduct telephone polls with respondents selected by a computer generating random telephone numbers not necessarily listed in the telephone book. *If the unit to be interviewed is the household, not a specific individual in the household*, why is the procedure not truly random?

Unit II: Describing the Statistic, $\overline{X}$, as a Random Variable

Preparation-Check Questions

55. The sampling distribution of $\overline{X}$
  a. Is a random sample of $\overline{X}$ values.
  b. Is a probability distribution of $\overline{X}$.
  c. Shows how $\overline{X}$ changes as n changes.
  d. Shows the relationship between $\overline{X}$ and σ.
  e. Is a frequency distribution of μ.

56. If (10, 10, 25) is a population, find its standard deviation, σ.
  a. 15.
  b. 75.
  c. 8.7.
  d. 7.1.
  e. 50.

57. If the data set in Question 56 were a sample rather than a population, how would the standard deviation calculation (S now, instead of σ) differ?
  a. The square root would have been omitted.
  b. The median would have been used rather than the mean, $\overline{X}$.
  c. The calculation would have used fewer of the observations.
  d. The denominator would have been one less than before.
  e. There would have been no difference.

58. If X is normally distributed with mean 200 and $\sigma = 50$, find $P(X > 275)$.
  a. 0.0668.   d. 0.8664.
  b. 0.4332.   e. 0.5668.
  c. 0.9332.

59. What proportion of the X values in Question 58 are within 1.96 standard deviations of the mean?
  a. 0.975.   d. 0.05.
  b. 0.025.   e. 0.95.
  c. 0.475.

60. To find the standard deviation of a random variable, X, we calculate $\sigma = \sqrt{E(X - \mu)^2}$ or $\sqrt{E(X^2) - \mu^2}$, which
  a. Is labeled σ, because it represents the standard deviation for every outcome for every trial of the experiment.
  b. Represents the average deviation of all values from the mean of the random variable.
  c. Means find $\sqrt{\sum_1^k (x - \mu)^2 P(x)}$ or $\sqrt{\sum_1^k x^2 P(x) - \mu^2}$.
  d. Cannot be larger than the largest deviation.
  e. All of the above.

61. Use the following sampling distribution of $\overline{X}$ to determine the proportion of $\overline{X}$ values within 37.5 units of 162.5 (include the boundaries of this interval).

$\overline{x}$	$P(\overline{x})$	$\overline{x}$	$P(\overline{x})$
100	0.10	175	0.25
125	0.15	200	0.15
150	0.25	225	0.10

  a. 0.25.   c. 0.80.   e. 0.625.
  b. 0.50.   d. 0.20.

Answers: 55. b; 56. d; 57. d; 58. a; 59. e; 60. e; 61. c.

Introduction

Quality Control A firm knows that in order to satisfy quality standards the mean contents of containers of its product should be 12 ounces and the standard deviation of the contents should be 0.15 ounce. Employees conduct continual checks of the process by collecting random samples of 20 containers each hour and finding the mean content of the sampled containers. They need to know which values of the sample means are likely to occur and which are unlikely to occur when the process is working according to the standards. Such information would allow them to avoid shutting down the process for unnecessary repairs when the sample mean strays from the 12-ounce standard. However, constructing the sampling distribution of $\overline{X}$ in this instance, as we did for the discrete values in the last unit, is impossible. In this case, values are continuous, there are infinitely many measurable values for the weight of the contents. Thus, there are infinitely many possible random samples. Somehow, we must be able to represent the possible $\overline{X}$ values and their probabilities in order to conduct the quality tests and know when the results indicate a problem with the production process.

Water Pollution A government agency collects water samples from 40 different spots in a stream to test for the mean number of pollutants per cubic centimeter. Unlike the waste treatment situation in Unit I where there were only three possible waste treatment levels in the population, now the population of stream pollutant levels is infinite. Hence, it is impossible to assemble all possible random samples of size 40. The agency knows the typical pollutant level in safe water, the standard deviation of pollutant levels in safe water, and the point when the water is no longer safe. They need to know the sampling distribution of sample means in order to avoid two mistakes: calling water unsafe when the sample mean is high but the water really meets safety standards, and calling water safe when the sample mean is low but the water does not meet the safety standards.

In Unit I, a multitude of possible values for a statistic emerge when we randomly select a sample from a population. As with other random variables we are interested in the description of the possible values of the statistic; specifically, its probability distribution, its mean, and its standard deviation.

The probability distribution, called the sampling distribution of the statistic, consists of values of a statistic calculated from all the possible random samples of a given size, n. In most sampling situations we need to choose a large sample from a large population. It is impractical, if not impossible, to find all the possible random samples, when the population size, N, and the sample size, n, are large, and then to calculate the statistic for each one. Can we still comment on the likelihood that the estimate is close to the population mean, μ?

7.3 The Mean of $\overline{X}$

The random variable $\overline{X}$ has a mean, just as any random variable has a mean. This mean is the expected value or mean outcome from repeating the experiment (select a random sample of n items and compute its mean, $\overline{X}$) an infinite number of times. The mean of the sampling distribution of $\overline{X}$ would be $\mu_{\overline{x}} = E(\overline{X})$, a mean of sample means. To find this mean, we calculate $\mu_{\overline{x}} = E(\overline{X}) = \Sigma_1^k \overline{x} P(\overline{x})$, where k is the number of possible values of $\overline{x}$.

■ **Example 7-4 (continued):** The table below shows the sampling distribution of $\overline{X}$, the mean waste particle count (ppcc) for samples of two test counts ($n = 2$) from the population (100, 300, 500). Calculate $\mu_{\overline{x}}$.

$\bar{x}$	$P(\bar{x})$
100	1/9
200	2/9
300	3/9
400	2/9
500	1/9
	1

■ **Solution:** $\mu_{\bar{x}} = 100(1/9) + 200(2/9) + 300(3/9) + 400(2/9) + 500(1/9)$
$= 300$ ppcc.

Thus, the average of the $\bar{X}$ values is 300 ppcc, when Supervisor Rosie calculates the mean of a large number of random samples of two test counts. The mean of the population (100, 300, 500) is $\mu = (100 + 300 + 500)/3 = 300$ ppcc, the same as $\mu_{\bar{x}}$, the mean of the $\bar{X}$ values. The fact that $\mu_{\bar{x}}$ and μ are equal is a relationship that will be important to us for tracking the behavior of $\bar{X}$.

> **Fact:** The population mean and the mean of all possible $\bar{X}$ values for samples of a given size, n, from the population are identical.
>
> $$\mu_{\bar{x}} = \mu.$$

This mathematical statement says that the *average of the $\bar{X}$ values obtained from repeating the sampling procedure an infinite number of times is the population mean, μ, the value we wish to estimate.*

Question: Does $\bar{X} = \mu$ for every sample?
Answer: Usually not. In Example 7-4 $\mu = 300$, but $\bar{X}$ can be 100, 200, 400, or 500 in addition to 300, so $\mu_{\bar{x}} = \mu$ does *not* say that a single sample mean must equal the population mean.

7.4 The Standard Error of the Mean

In addition to its mean, a random variable also has a standard deviation. When the random variable is a statistic, we call this standard deviation a *standard error of the statistic*. For the sample mean we call it the *standard error of the mean* and represent it with $\sigma_{\bar{x}}$.

As we know from Chapter 5, the standard deviation for a discrete random variable is

$$\sigma = \sqrt{E[(X - \mu)^2]} = \sqrt{\sum_{1}^{k}(x - \mu)^2 P(x)},$$

or, using the shortcut form, the standard deviation is

$$\sigma = \sqrt{E(X^2) - \mu^2} = \sqrt{\sum_{1}^{k} x^2 P(x) - \mu^2},$$

where there are k values of X. For the statistic $\bar{X}$, we substitute $\bar{X}$ for X and $\bar{x}$ for x to make the relationship

$$\sigma_{\bar{x}} = \sqrt{E[(\bar{X} - \mu_{\bar{x}})^2]},$$

or in the shortcut form

$$= \sqrt{E(\bar{X}^2) - \mu_{\bar{x}}^2}.$$

Example 7-4 (continued): Find $\sigma_{\bar{x}}$ for the waste treatment problem. Remember that $\mu_{\bar{x}} = 300$ ppcc.

Solution:

$\bar{x}$	$P(\bar{x})$	$(\bar{x} - 300)$	$(\bar{x} - 300)^2$	$(\bar{x} - 300)^2 P(\bar{x})$
100	1/9	−200	40,000	40,000/9
200	2/9	−100	10,000	20,000/9
300	3/9	000	0	0
400	2/9	100	10,000	20,000/9
500	1/9	200	40,000	40,000/9
				120,000/9 = 13333.3333

$$\sigma_{\bar{x}} = \sqrt{13333.3333} = 115.5 \text{ ppcc.}$$

The average deviation of the $\bar{X}$ values about the mean is 115.5 ppcc. The standard deviation of the population (100, 300, 500) is

$$\sigma = \sqrt{\frac{\sum_{1}^{N}(X - \mu)^2}{N}} = \sqrt{\frac{(100 - 300)^2 + (300 - 300)^2 + (500 - 300)^2}{3}}$$

$$= 163.3 \text{ ppcc.}$$

Notice that $\sigma \neq \sigma_{\bar{x}}$. Figure 7-5 illustrates that *the population standard deviation, σ, measures the dispersion of individual population values about μ, whereas $\sigma_{\bar{x}}$ measures the dispersion of $\bar{X}$ values about μ.* Because $\bar{X}$ combines several individual population values into one *central* value, the dispersion of $\bar{X}$ values is less than the dispersion of the population values, so $\sigma_{\bar{x}} \leq \sigma$. Note that 100, 300, and 500 each occur one-third of the time in the population. $\bar{X}$ values of 100, 300, and 500 occur one-ninth, three-ninths, and one-ninth of the time, respectively. They concentrate more in the neighborhood of 300.

Question: Suppose, instead, that $\bar{X}$ values were more dispersed about the population mean than the population values, X. Which estimate would tend to be closer to the population mean, μ: a single value randomly selected from the population, or the mean of a randomly selected sample of values from the population?

Figure 7-5
Comparison of the population standard deviation, σ, and the standard error of the mean, σ_x, shows that $\bar{X}$ values are less dispersed about the population mean, μ, than population values, X

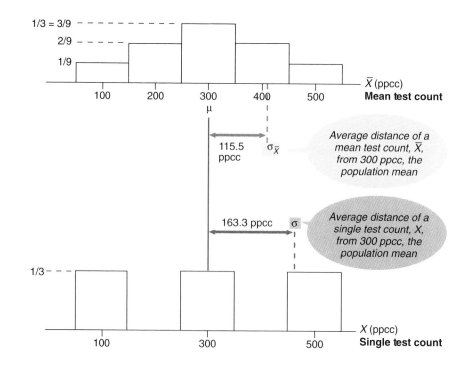

Answer: A single value would tend to be closer, if $\overline{X}$ were more dispersed. This would be counterintuitive, because it would mean that we would get a better estimate from a single piece of information about the population, than we get from a larger set of information. Hence, it is not surprising that $\sigma_{\overline{x}}$ and σ are different, and, indeed, that $\sigma_{\overline{x}}$ is the smaller of the two.

The following box shows the actual relationship between $\sigma_{\overline{x}}$ and σ.

> **Fact:** The relationship between the population standard deviation and the standard error of the mean is
> $$\sigma_{\overline{x}} = \frac{\sigma}{\sqrt{n}}.$$

From Example 7-4 with the waste-particle counts we know $\sigma = 163.3$ ppcc, $\sigma_{\overline{x}} = 115.5$ ppcc, and $n = 2$. If we calculate $\sigma_{\overline{x}}$ from the relationship in the box, we obtain $\sigma_{\overline{x}} = \sigma/\sqrt{n} = 163.3/\sqrt{2} = 115.5$ ppcc, the same result we obtain from the longer expected-value calculation. Thus, if we have a magnitude for σ, we use this formula to find $\sigma_{\overline{x}}$ and measure the dispersion of all possible $\overline{X}$ values about μ.

Question: As the sample size, n, increases, do you expect the resulting $\overline{X}$ values to be closer to or farther from μ?

Answer: Intuitively, *as n gets larger, the sample contains more information from the population. Consequently, the sample contents approach the population. It follows that $\overline{X}$ values should be closer to μ.* Mathematically, as n increases, $\sigma_{\overline{x}} = \sigma/\sqrt{n}$ will decrease (the denominator is larger, so the fraction is smaller). A smaller standard error means that the sample mean, $\overline{X}$, values cluster more tightly about the population mean, μ.

■ **Example 7-4 (continued):** If $n = 3$ instead of 2, find $\sigma_{\overline{x}}$. Remember that $\sigma = 163.3$ ppcc.

■ **Solution:** $\sigma_{\overline{x}} = \sigma/\sqrt{n} = 163.3/\sqrt{3} = 94.3$ ppcc, which is smaller than $\sigma_{\overline{x}} = 115.5$ ppcc when $n = 2$.

At this point we can summarize or condense all possible $\overline{X}$ values from a random sampling situation to $\mu_{\overline{x}} = \mu$ and $\sigma_{\overline{x}} = \sigma/\sqrt{n}$.

Sections 7.3 and 7.4 Problems

Fill in the blank cells of each row.

Problem	μ	σ	n	$\mu_{\overline{x}}$	$\sigma_{\overline{x}}$
62.	239	422	16		
63.	89.6	53.9	32		
64.		789.2	2500	1004.3	
65.			100	34.42	8.11
66.	9293	25			2.5

7.5 Sampling Distribution of $\overline{X}$: The Central Limit Theorem

It would be extremely difficult, and often impossible, to develop a sampling distribution of $\overline{X}$ for a very large population, by determining the mean of each possible random sample. For example, imagine all the possible samples of size 100 from a population with a million elements. The *central limit theorem* circumvents this problem, which is one reason it is one of the most important theorems in statistics. The theorem is stated in the following box. (We assume a finite standard deviation for the population, which will be the case in almost all circumstances.)

> **Central Limit Theorem**
> As the sample size, n, increases, the sampling distribution of $\overline{X}$ approaches a normal distribution, no matter how the population being sampled is distributed.

Usually *when $n > 30$, we consider the sampling distribution of $\overline{X}$ to be approximately normal*. Thirty is not a magical or theoretical value. Rather, it is a rule of thumb for obtaining a good approximation, as we demonstrate shortly.

This theorem says that *if n is large, we can use a normal distribution to approximate the sampling distribution of $\overline{X}$, regardless of how the population is distributed*. There is no need to know if the population is skewed, symmetric, normal, or even bimodal.

Figure 7-6 illustrates this concept by using three different populations, including in Part a of the figure the (100, 300, 500) population of waste particle counts from Example 7-4 (p. 299). The first row of the figure displays sketches of the different populations rather than a histogram for each of the three population values to indicate that the result holds for any population with a similar distribution. The rows that follow display sketches of the sampling distribution of $\overline{X}$ for $n = 2, 3, 4, 5, 6, 7$, and 30, respectively. (Remember these are infinite populations, so there are plenty of the values 100, 300, and 500 available for large samples.) Note how rapidly the distributions, even when bimodal (Part b) and skewed (Part c), approach a normal distribution.

Figure 7-7 summarizes and illustrates the sampling distribution of $\overline{X}$ along with the mean of the $\overline{X}$ values, μ, and the standard deviation of these values, $\sigma_{\overline{x}}$. Ordinarily, when we convert X to Z values, we use

$$Z = \frac{X - \mu}{\sigma},$$

using the mean and standard deviation of the random variable X. For the sampling distribution of $\overline{X}$, the $\overline{X}$ values correspond to X, $\mu_{\overline{x}}$ corresponds to and equals μ, and $\sigma_{\overline{x}}$ corresponds to σ. So, the transformation formula becomes

$$Z = \frac{\overline{X} - \mu}{\sigma_{\overline{x}}},$$

which we can write as

$$Z = (\overline{X} - \mu)/\sigma_{\overline{x}}.$$

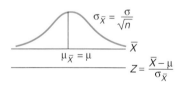

Figure 7-7
Sampling distribution of $\overline{X}$ for large n

■ **Example 7-7:** A financial analyst selects a random sample of 64 bank balances from an account with a mean, μ, of $100 and a standard deviation, σ, of $40. Find the probability that the sample's mean balance, $\overline{X}$, is between $90 and $110, that is, $P(90 \leq \overline{X} \leq 110)$.

Section 7.5 Sampling Distribution of $\overline{X}$: The Central Limit Theorem

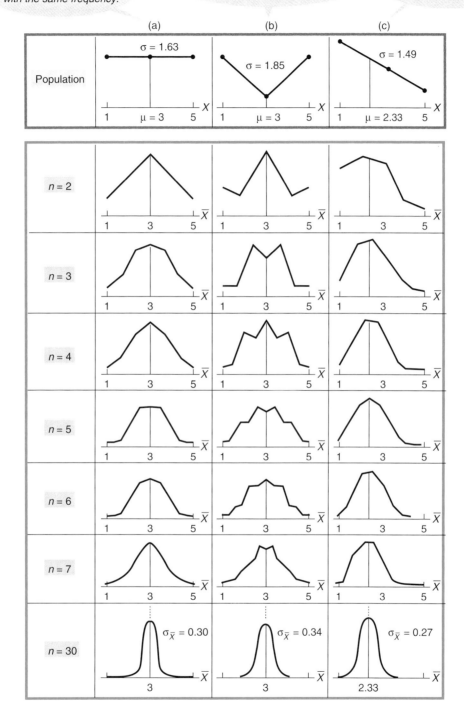

Figure 7-6
The effects of increasing n on the sampling distribution of $\overline{X}$

■ **Solution:** Because $n = 64$ is greater than 30, we assume that the possible $\overline{X}$ values are normally distributed. The mean for the curve is $\mu_{\overline{x}} = \mu = \$100$. The standard deviation for the curve is $\sigma_{\overline{x}} = \sigma/\sqrt{n} = 40/\sqrt{64} = \$5$. Converting $90 and $110 to Z values by using $Z = (\overline{X} - \mu)/\sigma_x$ (see Figure 7-8), we obtain $P(90 \leq \overline{X} \leq 110) = P(-2 \leq Z \leq 2) = 2(0.4772) = 0.9544$. Thus, about 95.44% of the time, the *mean* bank balance will be between $90 and $110.

Figure 7-8
Solution diagram for $P(90 \leq \overline{X} \leq 110)$

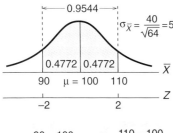

Figure 7-9
Solution diagram for $P(\overline{X} > 500)$ and $P(X > 500)$

a) Solution diagram for $P(\overline{X} > 500)$

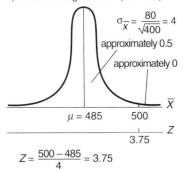

b) Solution diagram for $P(X > 500)$

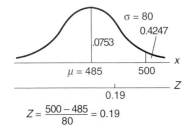

Question: Can we describe the population distribution here?

Answer: No, the problem statement provides no information about the population shape, only its mean and standard deviation. It may be skewed, bimodal, or irregular in some other way.

■ **Example 7-8:** An educational researcher selects a random sample of 400 students' scores from the population of scores on a national exam. The population mean is 485 points, and its standard deviation is 80 points. Find $P(\overline{X} > 500)$.

■ **Solution:** Because 400 observations form a large sample, we assume that the sample means are normally distributed with mean 485 and standard error, $\sigma_{\overline{x}} = \sigma/\sqrt{n} = 80/\sqrt{400} = 4$ as shown in Part a of Figure 7-9. We convert the score 500 to $Z = 3.75$. Then $P(\overline{X} > 500)$ becomes $P(Z > 3.75)$. Because $Z = 3.75$ exceeds the Z values given in the table, $P(0 < Z < 3.75)$ is approximately 0.5. Using this assumption, $P(\overline{X} > 500)$ is approximately 0.

Notice the different answer we would get to the last problem if (1) the scores are normally distributed and (2) we select one student's score, X, at random. The problem becomes $P(X > 500)$, the proportion of the population of scores that exceed 500. This is the same type of problem that we solved in Unit II of Chapter 6. Part b of Figure 7-9 shows that the standard deviation for the curve is now 80, but the mean does not change. The answer is 0.4247. If we compare this result with the result for $\overline{X}$, we find that we are more likely to obtain a single value larger than 500 from the population than a sample mean larger than 500. This is true, because $\overline{X}$ values are less dispersed than the single observations from the population ($\sigma_{\overline{x}} = 4$ while $\sigma = 80$).

■ **Example 7-9:** What percentage of $\overline{X}$ values from large samples are within 1.96 standard errors of the population mean, μ?

■ **Solution:** The middle 95% of any set of normal outcomes are with 1.96 standard deviations of the mean. In this case, the outcomes are $\overline{X}$, and the standard deviation is the standard error of the mean, $\sigma_{\overline{x}}$ (see Figure 7-10).

■ **Example 7-10:** The middle 98% of $\overline{X}$ values from large samples are within how many standard errors of the population mean, μ?

■ **Solution:** The interval that encloses $\overline{X}$ values from large samples within 2.33 standard errors of the population mean spans the middle 98% of $\overline{X}$ values (see Figure 7-11).

Question: An author frets that she uses the word "incredible" too often in her manuscripts. She decides to estimate the mean number of times the word appears per manuscript by using a sample mean. She believes the actual distribution of the number

Figure 7-10
95% of $\overline{X}$ values are within 1.96 standard errors of the population mean, μ

Figure 7-11
The middle 98% of $\overline{X}$ values are within 2.33 standard errors of the population mean, μ

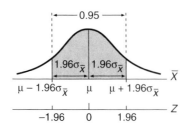

of times the word appears per manuscript is skewed left; that is, a few manuscripts use the word sparingly, but most of the manuscripts overuse the word. To use a normal distribution to describe the distribution of possible sample means, what must she do?

Answer: She must do two things: Select manuscripts randomly, and obtain more than 30 manuscripts for the sample, so the sampling distribution will be approximately normal.

Section 7.5 Problems

67. For which of the following sample sizes can we assume that the sampling distribution of $\overline{X}$ is approximately normal if we have no information about the population distribution? Use the rule of thumb.
   a. 36. **b.** 10. **c.** 1,000. **d.** 100. **e.** 25.
   f. 1. **g.** 4. **h.** 50. **i.** 16.

68. Find $P(350 < \overline{X} < 450)$, if we select a sample of 40 observations from a population with a mean of 325 and a standard deviation of 658.

69. Find $P(22.3 \leq \overline{X} < 45.6)$, if we select 88 observations from a population with a mean of 50 and a standard deviation of 102.4.

70. What percentage of $\overline{X}$ values from large samples are within 1.16 standard errors of the population mean, μ? Within 0.75 standard errors?

71. The middle 90% of $\overline{X}$ values from large samples are within how many standard errors of the population mean, μ? What about the middle 75%?

7.6 Sampling Distribution of $\overline{X}$: Normal Population

The central limit theorem provides information about the sampling distribution of $\overline{X}$ when the population is distributed in any manner, as long as n is large. *If we possess additional information that the population is normal, then the sampling distribution of $\overline{X}$ will be normal no matter how many elements are in the sample.*

Question: If the author in the previous question knows that the number of times she uses "incredible" per manuscript is normally distributed rather than skewed, how many manuscripts must she select for her sample to use a normal distribution for the sampling distribution of $\overline{X}$?

Answer: Any number of manuscripts will work now. The $\overline{X}$ values will be normally distributed.

Section 7.6 Problems

72. What percentage of $\overline{X}$ values from samples of 16 observations will fall between 950 and 1,050 if the population is normal with mean 1,000 and standard deviation 200?

73. Find $P(\overline{X} < 0.2)$ if $\overline{X}$ is the mean of a sample of five observations from a standard normal population.

74. Find $P(-0.1 \leq \overline{X} \leq 0.1)$ if $\overline{X}$ is the mean of a sample of 10 observations from a *standard normal* population.

75. Find $P(-1 \leq \overline{X} \leq 1)$ if $\overline{X}$ is the mean of a sample of four observations from a *standard normal* population.

76. What percentage of $\overline{X}$ values are within one standard error of the population mean, μ, if the sample is small but selected from a normal population?

77. Find the lower boundary (smallest value in the interval) of the largest 40% of $\overline{X}$ values from samples of 20 observations taken from a normal population with mean 111 and standard deviation 22.

7.7 Sampling Distribution of $\overline{X}$: Finite Populations

The conclusions so far assume that once we select items for the sample from infinite populations, repetitions of the same value can occur in the sample, because there is a large (or infinite) number of observations of each value available for selection. If a population is *finite*, the number of observations of each value is limited.

Question: Consider an observation from a population with 100 observations ($N = 100$) and one from an infinite population. Which one contains a larger chunk of the total information in the population?

Answer: Each observation from the finite population represents 1/100 of its population, whereas each observation from the infinite population represents an infinitesimal part of its population.

Intuitively, a sample of n observations from a finite population is more informative about the population than a sample of n observations from an infinite population. Consequently, the $\overline{X}$ values from the finite population will cluster more tightly about the population mean, μ, which diminishes the standard error of the mean, $\sigma_{\bar{x}}$, as shown in Figure 7-12. The formula for the standard error of the mean for samples of size n from a finite population with N observations is

$$\sigma_{\bar{x}} = \frac{\sigma}{\sqrt{n}} \sqrt{\frac{N-n}{N-1}}.$$

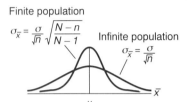

Figure 7-12
Effects of a finite population on $\sigma_{\bar{x}}$

■ **Example 7-4 (continued):** Consider again the population of waste treatment counts (100, 300, 500). If n equals 2 and the population is finite (it contains only these three values), construct the sampling distribution of $\overline{X}$ and calculate the standard error of the mean (using the expected-value formula and the new formula we just introduced).

■ **Solution:** Refer to Figure 7-1 (p. 290). Three samples are lost that were possible with an infinite population, the ones where the same value is repeated. The three samples lie along the diagonal of the following samples.

(100, 100)	(100, 300)	(100, 500)
(300, 100)	(300, 300)	(300, 500)
(500, 100)	(500, 300)	(500, 500)

The following table shows the remaining six random samples and the sampling distribution of $\overline{X}$.

Random Sample	$\bar{x}$	$P(\bar{x})$
(100, 300) and (300, 100)	200	2/6 = 1/3
(100, 500) and (500, 100)	300	1/3
(300, 500) and (500, 300)	400	1/3

First, we find $\sigma_{\bar{x}}$ for this sampling distribution by using $\sigma_{\bar{x}} = \sqrt{E(\overline{X} - \mu)^2}$. Recall that $\mu_{\bar{x}} = \mu = 300$ ppcc.

$$\sigma_{\bar{x}} = \sqrt{(200 - 300)^2 \left(\frac{1}{3}\right) + (300 - 300)^2 \left(\frac{1}{3}\right) + (400 - 300)^2 \left(\frac{1}{3}\right)}$$
$$= 81.65 \text{ ppcc}.$$

Now, we find $\sigma_{\bar{x}}$ by using $\sigma = 163.30$ ppcc and the formula

$$\sigma_{\bar{x}} = \frac{\sigma}{\sqrt{n}} \sqrt{\frac{N-n}{N-1}}$$
$$= \frac{163.3}{\sqrt{2}} \sqrt{\frac{3-2}{3-1}} = 81.65 \text{ ppcc}.$$

Notice that 81.65 ppcc, the corrected standard error, is smaller than 115.47 ppcc, the uncorrected standard error that we calculated earlier, just as we anticipated.

We modify $\sigma/\sqrt{n}$ with the factor $\sqrt{(N-n)/(N-1)}$ for finite populations. We call this factor $\sqrt{(N-n)/(N-1)}$ the *finite population correction factor*. The expression $(N-n)/(N-1)$ is the approximate proportion of the population elements that are not

Problems

in the sample. If a sample contains approximately 10% of the population, then approximately 90% remains unselected, and the finite population correction factor is $\sqrt{0.9} = 0.95$.

If the number of elements in the sample, n, is only a small portion of the population elements (5% or less is often used), then the proportion of observations left in the population is close to 1. The correction factor $\sqrt{(N-n)/(N-1)}$ that results is approximately 1, so its effect on the standard error is small or inconsequential in most practical situations. Consequently, if $n \leq 0.05N$ (*that is, less than 5% of the population is in the sample*), we use $\sigma_{\bar{x}} = \sigma/\sqrt{n}$, *even when sampling from a finite population*. We assume that samples contain 5% or less of the population for the remainder of the book to avoid calculations that have negligible effect on magnitudes.

This section says that the formula for $\sigma_{\bar{x}}$ is simple when the sample size is small relative to the size of the population. We can use a sample with more than 30 observations and still not violate the 5% requirement ($n \leq 0.05N$) if the population is very large. Newspapers or other media report findings from samples of 5,000 Americans, but 5,000 is a very small proportion of the entire U.S. population. Often we do not want or need such large samples to obtain adequate or very good information about the population mean, μ.

Section 7.7 Problems

78. Complete each row in the following table. Then use the results to answer Questions i–iv.
  i. The finite population correction factor is really the size of the standard error, $\sigma_{\bar{x}}$, corrected for a finite population relative to the uncorrected standard error from a finite population. What does it mean when this value is close to 1?
  ii. In which rows do the samples contain less than 5% of the population? What is the value of the correction factor in each case?
  iii. What are the correction factors in the remaining rows?
  iv. Does it appear that adjustments for the standard error for samples that contain less than 5% of the population are indeed small?

79. Use the new sampling distribution for the finite population of waste treatment test counts (that is, without replacement) for Example 7-4 (p. 316) to find the probability that the mean of the sample is within 100 ppcc of the population mean, 300. The answer to this problem when we assume the population is infinite is 7/9. Should the finite answer exceed 7/9? Explain.

80. If an organization has 1,000 employees and the management wants a random sample of 100 employees, which formula should we use for the standard error of the mean?

81. Suppose we are sampling items from a manufacturing process. Which formula should we use for $\sigma_{\bar{x}}$?

82. Which of the following sample sizes would permit our use of the uncorrected standard error if the finite population consists of 250 observations?
  a. 25.   e. 31.
  b. 50.   f. 12.
  c. 10.   g. 5.
  d. 30.

ID	Population Size N	Sample Size n	Proportion of Population in the Sample n/N	Finite Population Correction Factor	σ	$\sigma_{\bar{x}}$ Infinite Population	$\sigma_{\bar{x}}$ Finite Population
a	Infinite	160,000	Approximately 0	1	200	0.5	0.5
b	Infinite	16			200	50	50
c	100	36	0.36	0.80	120	20	16
d	100	4	0.04		120	60	59.1
e	5,000	2,500			150		
f	5,000	225			150		

Summary and Review

According to the central limit theorem, the sampling distribution of $\overline{X}$ approaches a normal distribution as n increases. A rule of thumb states that it is approximately normal if $n > 30$. This normality does not depend on the population distribution, which makes this theorem very basic for statistical inferences and perhaps the most important theorem in statistics.

To determine normal probabilities, we must know the mean and standard deviation. The mean of $\overline{X}$ is $\mu_{\overline{x}} = \mu$, the population mean. The standard deviation of $\overline{X}$, which measures the "average" distance of an $\overline{X}$ from μ, is called the standard error of the mean. The formula for this standard error is $\sigma_{\overline{x}} = \sigma/\sqrt{n}$.

In addition, if we know that the population is normal, then the sampling distribution is normal, no matter how many observations are in the sample.

If we sample from a finite population, each randomly chosen item brings a larger proportion of the population information to the sample. More information in a sample of size n increases the probability that the estimate will be close to μ. A smaller standard error reflects this increased chance of more accurate estimates,

$$\sigma_{\overline{x}} = \frac{\sigma}{\sqrt{n}} \sqrt{\frac{N-n}{N-1}}.$$

However, when the sample contains a relatively small part of the population (less than 5%), the finite population correction factor has very little impact on the magnitude of the standard error, $\sigma_{\overline{x}}$.

Using the central limit theorem to establish the normal distribution of $\overline{X}$ values, along with the information about the mean and standard error of these values, we can solve probability problems about the value of $\overline{X}$ from a single random sample, such as $P(\overline{X} > \overline{x})$. We also know the proportion of $\overline{X}$ values that will fall within a specific number of standard errors of the population mean, μ, from repeated sampling. These concepts and theorem about the behavior of $\overline{X}$ are essential in our forthcoming estimations when μ is unknown and $\overline{X}$ is its estimator.

Multiple-Choice Problems

83. The central limit theorem is important, because
   a. We can describe the sampling distribution of $\overline{X}$ if the sample is sufficiently large without knowing the population distribution.
   b. It describes how to approximate the standard error of the mean.
   c. It is the basis for describing the binomial random variable and situations where it is applicable.
   d. It describes the population distribution, mean, and standard deviation.
   e. It simplifies the expected-value formula.

84. Suppose a population consists of 1,000 ones, 200 threes, and 1,000 fives, as pictured in the top part of Figure 7-13. If you select a random sample of 50 elements from the population, which of the diagrams in the lower part of the figure best depicts the sampling distribution of $\overline{X}$?
   a. A.   b. B.   c. C.   d. D.   e. E.

85. Which of the following expressions is equal to $\mu_{\overline{x}}$?
   a. $\overline{X}$.
   b. $(\sigma/\sqrt{n})\sqrt{(N-n)/(N-1)}$.
   c. $\sigma/\sqrt{n}$.
   d. σ.
   e. μ.

A biologist selects a random sample of 100 sites to locate and count the strands in spider webs. The sample mean is 839 strands, the population mean is 909 strands, and the population standard deviation is 64 strands. Use this information to answer Questions 86–87.

86. What is $E(\overline{X})$?
   a. 100 strands.   d. 64 strands.
   b. 839 strands.   e. 6.4 strands.
   c. 909 strands.

87. What is the standard error of the mean in this situation?
   a. 100 strands.   d. 64 strands.
   b. 839 strands.   e. 6.4 strands.
   c. 909 strands.

88. We know that $\mu_{\overline{x}} = \mu$. What does this statement say?
   a. The sample mean equals the population mean.
   b. The mean of all possible $\overline{X}$ values is the population mean.
   c. If the sampling process is repeated a large number of times, the sample mean will equal the population mean.
   d. If a sample is large enough, the sample mean and population mean are equal.
   e. If the sample contains a very small percentage of the population elements, the sample and population have the same mean.

89. Which of the following will cause $\overline{X}$ values to be more dispersed about μ (that is, cause $\sigma_{\overline{x}}$ to increase)?
   a. Increasing the sample size.
   b. Sampling from an infinite population instead of a finite population.
   c. Reducing the population standard deviation.
   d. Taking a different sample.
   e. Sampling from any nonnormal population.

Multiple-Choice Problems

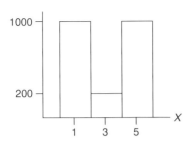

Sampling distribution of $\bar{x}$ choices

a) 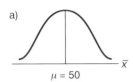 $\mu = 50$

b) 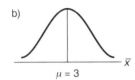 $\mu = 3$

c) 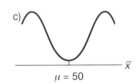 $\mu = 50$

d) 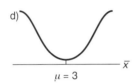 $\mu = 3$

e) 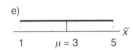 $\mu = 3$

Figure 7-13
Diagrams for Problem 84

A random sample of 121 elements is selected from a population with a mean of 250 and a standard deviation of 110. Use this information to answer Questions 90–92.

90. Find $P(\bar{X} > 269)$.
  a. 1.0.   d. 0.0287.
  b. 0.4713.   e. 0.9426.
  c. 0.9713.

91. What percentage of the $\bar{X}$ values are within 20 units (that is, two standard errors) of the population mean, μ?
  a. 18%.   d. 4.56%.
  b. 95.44%.   e. 14.28%.
  c. 97.22%.

92. The middle 70% of the $\bar{X}$ values are within how many standard errors of the population mean, μ?
  a. 1.04 standard errors.   d. 0.39 standard errors.
  b. 0.52 standard errors.   e. 0.70 standard errors.
  c. 0.84 standard errors.

93. In which of the following cases is the central limit theorem unnecessary?
  a. When nothing is known about the distribution of the population.
  b. When the population consists of a few different values, but there are an infinite number of each of them.
  c. When the population is normal.
  d. When sampling is done from a finite population.
  e. When the population mean is unknown.

94. The correction factor for the standard error when sampling from a finite population is $\sqrt{(N - n)/(N - 1)}$. What does the expression under the radical represent?
  a. The conversion of a normal random variable to a standard normal random variable.
  b. A sample estimate, because $N - 1$ is in the denominator.
  c. The proportion of the population included in the sample.
  d. The approximate proportion of the population not included in the sample.
  e. The correction for using the normal to approximate the binomial.

95. For which population distribution does the sampling distribution of $\bar{X}$ most rapidly approach a normal distribution as n increases?
  a. Skewed.   d. Unimodal.
  b. Bimodal.   e. Symmetric, unimodal.
  c. Symmetric.   f. Normal.

96. The reason we will not use the correction factor for the standard error of the mean when the sample contains less than 5% of the population is because
  a. The factor is approximately 1.
  b. Efforts to satisfy the 5% condition result in samples too small to apply the central limit theorem.
  c. 5% of a large population, such as all U.S. citizens, creates a cumbersome sample.
  d. The factor is negative in the 5% region, producing a nonsensical value for the standard error.
  e. The formulas and procedure become overly complex, and the accuracy declines.

97. Together, the central limit theorem and the fact that the mean of the $\bar{X}$ values equals the population mean, μ, imply that
  a. An $\bar{X}$ calculated from a large sample will be equal to μ.
  b. $\bar{X}$ values from large samples are more tightly clustered about μ than any other value.
  c. The correction factor for the standard error is unnecessary.
  d. Small sample estimates are inaccurate.
  e. The sample must contain more than 30 items to avoid a 5% risk of inaccuracy.

98. We expect more accurate inferences when we increase the sample size, n, because
  a. The standard error of the mean, $\sigma_{\bar{x}}$, increases.
  b. The central limit theorem indicates that the normal distribution will become more dispersed.
  c. The sample contains more information about the population.

d. The area under the tails of the normal curve increases.
e. There is less need for applying the correction factor to the standard error of the mean, $\sigma_{\bar{x}}$.

99. Because national scores on an advanced placement test in history are bimodal, the head of the history department suspects that the mean of a random sample of 75 of these scores is unlikely to reflect the true mean score. Which of the following statements correctly assesses this situation?
   a. The department head is correct, because 75 is likely to be less than 5% of the set of national scores.
   b. While the sample mean will not indicate the bimodal nature of the population, it is likely to be close to the mean of all the scores according to the central limit theorem.
   c. Because of the bimodal nature of the population, the sample median is more likely to reflect the mean score for all of the students.
   d. Because of the bimodal nature of the population, the sample mode or modes are more likely to reflect the mean score for all of the students.
   e. The bimodal population distribution results in a bimodal sampling distribution of $\bar{X}$, but this does not decrease the likelihood of a sample containing elements near both modes and a mean in the middle.

100. The standard error of the mean
   a. Equals the ratio of the population standard deviation to the square root of the sample size.
   b. Decreases when the sample size increases.
   c. Measures the dispersion of $\bar{X}$ values about the population mean, μ.
   d. Reflects the added likelihood of an accurate inference by decreasing when samples contain more population information.
   e. All of the choices are correct.

IQ scores are normally distributed with a mean score of 100 and a standard deviation of 15 points. Use this information to answer Questions 101–102.

101. How many observations must a sample of scores contain before we invoke a normal distribution for the sampling distribution of $\bar{X}$?
   a. More than 30.
   b. At least 30.
   c. More than 5% of the population.
   d. Less than 5% of the population.
   e. 1.

102. What is the standard error of the mean for a sample with only one observation?
   a. 15. c. 1.5. e. 0.
   b. 1.   d. 100.

Word and Thought Problems

103. Define μ, $\bar{X}$, and $\mu_{\bar{x}}$.
104. Define σ, S, and $\sigma_{\bar{x}}$.
105. Find the standard error of the mean, $\sigma_{\bar{x}}$, for each of the diagrams in Figure 7-6 (p. 313). Describe the effect on $\sigma_{\bar{x}}$ of increasing the sample size, n.
106. What is the central limit theorem? Why is it important to statistics?
107. Consider the effect of increasing the number of elements in a random sample.
   a. What happens to the sampling distribution of $\bar{X}$?
   b. What happens to $\mu_{\bar{x}}$?
   c. What happens to $\sigma_{\bar{x}}$?
   d. If we select a random sample and calculate $\bar{X}$, is $\bar{X}$ more likely to be close to μ when n is small or large? Explain.
108. Consider increasing the number of elements in a random sample from a finite population.
   a. Would any of your answers to Parts a–d of Problem 107 change?
   b. If the sample contains all of the population elements, what is the sample mean, $\bar{X}$? What is the value of $\sigma_{\bar{x}}$? Interpret this value.
   c. How many samples can contain every element in the population? Is your answer reasonable and in accord with your answers to Part b?
109. A random sample of 200 elements is selected from a population with $\mu = 1{,}500$ and $\sigma = 110$.
   a. Find $P(\bar{X} \geq 1508)$.
   b. Describe the probability distribution of the population if you can.
   c. Describe the sampling distribution of $\bar{X}$.
   d. What proportion of the $\bar{X}$ values are within 1.96 standard errors of the population mean, μ?
110. A random sample of 560 elements is selected from a population with a mean of 2,000 and a standard deviation of 800.
   a. What is the probability the sample mean is within 100 units of the population mean?
   b. The middle 80% of the $\bar{X}$ values are within how many standard errors of μ?
   c. The middle 80% of the $\bar{X}$ values are between what two values of $\bar{X}$?
111. Suppose $\bar{X}$ is a random variable that represents the outcome of choosing one value from a population that is normally distributed with a mean of 100 and standard deviation of 50.
   a. Find $P(X > 110)$. Note: This problem is about a population value, not a sample mean.
   b. If $\bar{X}$ is a random variable representing the mean of 625 randomly chosen X values, find $P(\bar{X} > 110)$.
   c. Give an intuitive explanation for the different answers to Parts a and b.

Word and Thought Problems

112. The sample median is another possible estimator of the population mean. For large samples from populations that are approximately normal, the sampling distribution of sample medians is approximately normal, with mean μ. The standard error of the median is approximately $1.25\sigma_{\bar{x}}$. Use this information to select the better estimator of μ, the sample mean or the sample median. Justify your choice.

113. Frequently, polls require television viewers to call in responses via a 900 telephone number. Why is the central limit theorem not applicable to the results of such polls?

114. What does $\sigma_{\bar{x}}$ tell us about the random variable $\bar{X}$?

115. Find the mean and standard error of the mean for $\bar{X}$ in Problem 43 Parts b and c of Unit I (p. 306). Describe the differences in answers, if any, and give an intuitive explanation for differences or similarities.

116. Find the standard error of the median for $n = 3$ in Problem 37 in Unit I (p. 305) using the expected-value formula.

117. Consider a sample of size 1.
    a. Compare the sampling distribution of $\bar{X}$ with the population distribution.
    b. How are the mean of $\bar{X}$ and the population mean related?
    c. How are the standard error of the mean and the population standard deviation related in this case?

118. Sometimes the discussion of sample size in this unit seems contradictory. Larger values of n simplify the sampling distribution of $\bar{X}$. Smaller values of n relative to N simplify the standard error formula. Why are the desirable values of n not contradictory, that is, how can we satisfy both situations simultaneously?

119. The director of an organization selects a random sample of 100 times from the number of hours donated by each of 500 volunteers. The standard deviation of all 500 times is 50 hours and the mean, μ, is 10 hours. This sample contains more than 5% of the population.
    a. Find the proportion of $\bar{X}$ values within 2.5 standard errors of the actual mean time of all volunteers.
    b. Translate the 2.5 standard errors into actual units of time, then express the range of an interval from 2.5 standard errors below the true mean to 2.5 standard errors above the true mean, in hours. Do this first by using the finite population correction factor to determine the standard error of the mean. Then assume an infinite population and do not incorporate the correction factor for the standard error.
    c. Based on your two answers to Part b, do $\bar{X}$ values appear to be more tightly clustered about μ if you use or do not use the finite population correction factor? Interpret your answer by substituting the idea of "proportion of population information included in the sample" for corrected and uncorrected. Is the result what you expected? Explain.
    d. If n is quadrupled to 400, what is the range in hours of the interval of $\bar{X}$ values within 2.5 standard errors of μ? Again, use the finite population correction factor, then rework the interval, assuming an infinite population.
    e. Based on your answers to Parts b and d, where n has quadrupled, did the range decrease more for sample means with or without correcting the standard error? Can you give an intuitive explanation for this result?

120. Use your random sample of 10 values from the North Carolina Nursing Home data selected for Problem 40 in Unit I (p. 306). Estimate the standard error of the mean with your results by using $S/\sqrt{n}$. Compare your estimated value with the actual value, $\sigma/\sqrt{n} = 514.8/\sqrt{10} = 162.8$. Is your sample mean close to the mean of all 100 values? Use the actual values to find the probability of obtaining a sample mean larger than the mean of your sample of 10 values (assume the population is normal).

121. Consider the computer output of descriptive statistics in Figure 3-37 (p. 138). Estimate $\sigma_{\bar{x}}$ by using $S/\sqrt{n}$ for each salesperson. Compare your answers to the standard errors (labeled SEMEAN) in the Minitab output.

122. The mean score on a standardized economics test of 902 students in states that do not mandate classes in economics education is 21.50 points (maximum possible is 46 points), with a standard deviation of 8.37 points. Suppose a government education specialist selects a random sample of 45 of these students for analysis.
    a. Find the probability that the mean of this sample is within 2.8 standard errors of 902.
    b. The middle 95% of $\bar{X}$ values from samples of 40 observations are within how many standard errors of the population mean, μ?
    c. Why are we able to approximate these answers with a normal curve?
    d. The mean score for 634 students in states that mandate courses in economics education is 21.40 points with a standard deviation of 8.77 points. Which sample is most likely to have a smaller sample mean, a random sample of 31 of these students or a random sample of 45 students from the non-mandate states? Demonstrate your answer.
    e. Consider the situation described in Part d. The mean of the mandated population is smaller than the mean of the non-mandated population, so we expect the mean of a sample from the mandated population to be smaller than the mean of a sample from the non-mandated population. Which of the following actions would decrease the chances of sample mean magnitudes reversing order: increasing n, decreasing n, or neither affects the overlap. Explain and demonstrate your answer. (Source: Sherrie L.W. Rhine, "The Effect of State Mandates on Student Performance," *American Economic Review*, Vol. 79, No. 2, May 1989, p. 233.)

Review Problems

123. What is the central limit theorem? How do you think we would determine sampling distributions without it?
124. Describe an instance in which destructive sampling is the only practical method for learning about the population.
125. An infinite population consists of the values 20 and 40.
    a. What proportion of sample means will be 20 for $n = 2$?
    b. What proportion of sample means will be 20 for $n = 3$?
    c. What proportion of S^2 values will be 0 for $n = 2$?
    d. What proportion of S^2 values will be 0 for $n = 3$?
126. Define sampling distribution, sample mean, and standard error. Then, describe the information we learn from each that is useful for making inferences.
127. Explain intuitively and technically the effect of increasing the sample size, n, on the accuracy of statistical estimates.
128. Suppose the average term served by national leaders around the world in the last century was 2.34 years with a standard deviation of 3.95 years. Find the boundaries of the interval of the middle 90% of sample means for samples of 68 leaders' terms. Express the length of this entire interval in standard errors.
129. Someone claims that the mean grade on a test is 78, but you collect a random sample of students' grades from the class and find its mean is 80. Can you be sure that the population mean is not 78? Explain.
130. $\overline{X}$, the statistic, is also a random variable. What are the descriptive measures for this random variable that are equivalent to the frequency distribution, mean, and standard deviation covered in Chapters 2 and 3?
131. The following table shows the time between the identification of a bank as an "extremely high" failure risk and the actual failure for 437 banks that failed between January 1, 1985, and May 10, 1991.

Time to Failure After Identification of Risk (months)	Number of Banks
0–3	74
4–6	53
7–9	62
10–12	92
13–18	95
19–30	51
31–58	10

    Source: U.S. House Banking Committee, as reported in Kenneth H. Bacon, "Faster Bank Closings Will Cut FDIC Losses," *Wall Street Journal*, Sept. 23, 1991, p. A1, with permission.

    a. Estimate the mean of the population of the 437 failed banks' times from the frequency distribution by using the grouped data methodology from Chapter 3 (p. 91).
    b. If we randomly select three banks' times (with replacement), find the probability all three values will be between 10 and 12 months.
    c. If all three values are between 10 and 12 months, estimate the sample mean. Compare the result with the solution to Part a. How close is the value of the statistic to the population value?
132. If the average length of larceny trials is 3.6 hours with a standard deviation of 5.1 hours, find the probability the mean of a sample of 450 trials is between 3 and 3.5 hours.
133. Describe an instance in which measuring some characteristic about an item makes it useless, harms it, or destroys it. Would it be possible or practical to study the population in these cases? What do we call this type of sampling?
134. The 90th percentile of $\overline{X}$ values for samples of 64 households' annual dental expenditures is $172.8. The standard error is $10. What is the population mean? The population standard deviation?
135. If the average daily high temperature in a New England town is 63.6° with a standard deviation of 10.2°, find the probability that the mean of a random sample of 120 days will not exceed 66°.
136. Use the random number table to select a random sample of 3 ozone values from the Air Quality Database. What are the mean and standard deviation of the sample? Estimate the standard error of the mean.
137. To investigate the relationship between religious orientation and mental health, an educational psychologist collected data from 268 undergraduates at the University of Minnesota. The students took a battery of tests to measure psychological traits. The following table lists each test and the psychological trait it measures along with the mean score and standard deviation for the 268 students. Assume that these results hold for all American students and that you select a random sample of 50 American students as you answer the following questions.

    a. Complete the standard error column of the table.
    b. What is the probability the mean of CES-D scores for the sample of 50 students will exceed 18?
    c. Find the boundaries of the middle 95% of sample means for SGT-RW shame and for guilt scores.
    d. Can you describe the population distribution for the 268 students or for all American students for any variable listed above? Explain.
    e. Find the probability the sample mean of PSI: Conflictual (mother) scores will be less than 72 or greater than 76.
    f. Assume SWBS scores on both segments are approximately normally distributed. Find the probability a randomly selected student will score less than 45 on the EWB *and* that outcome will be followed by an independently selected random student with a RWB score above 45.
    g. Fourteen of the 268 students did not identify with any formal religious group. Find the *exact*

Test	Trait Measured (author's descriptions)	μ	σ	$\sigma_{\bar{x}}$
CES-D	Depression and emotional distress	16.8	9.4	
SGT-RW				
Shame	Shame proneness	61.4	9.5	
Guilt	Guilt proneness	77.6	12.1	
SWBS				
EWB	Sense of life purpose, direction, and satisfaction	44.0	5.5	
RWB	(1) Belief that God loves them and (2) Fulfillment and meaningfulness of relationship with God	44.6	7.7	
PSI: Functional separation	Management of practical and personal affairs without parental help			
Mother		30.9	9.4	
Father		34.8	9.8	
PSI: Attitudinal separation	(1) Self-image distinct from parents (2) Own beliefs, attitudes, and values			
Mother		26.3	11.0	
Father		28.0	13.0	
PSI: Emotional separation	Freedom from need for parental approval, closeness, togetherness, and emotional support			
Mother		40.7	12.9	
Father		42.7	13.0	
PSI: Conflictual separation	Freedom from excessive guilt, anxiety, mistrust, responsibility, inhibition, resentment, and anger in relationships with parents			
Mother		76.0	15.7	
Father		77.9	17.0	

Source: P. Scott Richards, "Religious Devoutness in College Students: Relations with Emotional Adjustment and Psychological Separation from Parents," *Journal of Counseling Psychology*, Vol. 38, No. 2, 1991, pp. 189–96.

probability that all of the 50 students in the sample have a religious preference.

  h. The educational psychologist recruited the 268 students from General Psychology classes with the offer of extra credit for participating. Is this group of students a random sample of University of Minnesota students? Explain. Might it be representative? Explain.

138. An anthropologist needs to estimate the average length of time to complete an excavation in an Arizona region in order to file a grant application for financial assistance from a cultural foundation. Her own experience and that of her colleagues suggest times are bimodal and skewed to the right, so she is concerned that the mean of a random sample of the foundation's projects in the same area may not be an accurate estimate of the true mean time for her project. Do you agree or disagree? Explain.

139. Sometimes, critics screen a movie before its release to the public. Suppose most of these critics like a movie, so the production company proceeds, expecting a box-office success. However, it flops. What was wrong with the inferential process in this case?

140. Suppose the average error of student responses, when asked to name the year that the first human walked on the moon, is 1.7 years with a standard deviation of 7.5 years. The event occurred in 1969.

  a. Describe the distribution of the population for the sampling distribution of $\bar{X}$ to be normal if it is based on samples of 12 observations?

  b. What percentage of the means of random samples of 225 student response errors fall between one and two years?

  c. Find the boundaries of the interval containing the middle 95% of the sample means for $n = 225$.

  d. What percent of the $\bar{X}$ values fall within 1.96 standard errors of the population mean, $\mu = 1.7$ years?

141. What does a Z value indicate about a point's location relative to the mean?

142. Health screenings or wellness exams aim to detect health problems early or prevent them altogether. To this end, participants describe their exercise activity and alcohol consumption. Because many people do not keep detailed records or follow a set routine, they may recollect these activities from the last few days or weeks as a sample of their overall level of activity. Suppose seven individuals recollect the last two weeks (14 days) as follows:
  a. One glass of wine each day.
  b. Ran 2 miles every day.
  c. Ran 30 minutes each Monday, Wednesday, and Friday; swam 20 minutes each Tuesday and Thursday; and did no exercise on weekends.
  d. A six-pack on Friday nights—no other alcoholic beverages.
  e. A binge on Friday (cannot remember how much); otherwise, no alcoholic beverages.
  f. A beer or two every day or so.
  g. A marathon hike one Saturday; otherwise, walk to and from school (about 1 mile total) on weekdays.

Use these recollections to answer the following questions.

  i. Which measure would best depict the typical or average daily level of activity; the mean, median, mode, or weighted mean? Answer this question for each recollection, and explain each answer.
  ii. Calculate the mean, median, and mode for each recollection, if possible.
  iii. What information would the standard deviation provide about an activity? Alcohol consumption? Is knowing the standard deviation more useful when the average is large or small? Explain.
  iv. Suggest the direction of bias from using recollections as estimators for amount of exercising. Do the same for consumption of alcohol.

Chapter 8

Statistical Inference: Estimation

This chapter extends what we have already learned about statistical inference and demonstrates how to use a sample to estimate two population parameters, the population mean, μ, and the population success rate, p. We can express such estimates either as a single value (a point estimate) or as an interval of values. The accuracy of either type of estimate rests on the sampling distribution and standard error of the statistic we use for the estimate. The confidence level associated with an interval estimate or with the margin of error of a point estimate describes the accuracy of the estimate.

Objectives

The numerical value of any inference about a population that is based on a sample will seldom be exact. In Unit I we express the uncertainty of such an inference by reporting the estimate, not as a single value, but as lying somewhere within an interval of values called a confidence interval. We will find confidence intervals for the population mean. Upon completing the unit, you should be able to

1. Construct and interpret a confidence interval for μ from a point estimate, $\overline{X}$, and state the conditions that ensure the interpretation applies.
2. Predict and demonstrate the effects of a change of the sample size or the confidence level on the length of the confidence interval.
3. Use the T table to find probabilities and values of a random variable with a T probability distribution.
4. Interpret Minitab computer output depicting confidence intervals.
5. Understand the meaning of estimator, point estimate, unbiased estimator, consistent estimator, confidence interval estimate, estimated standard error, and T statistic.

6. Recognize symbols for risk level, $100\alpha\%$; confidence level, $100(1 - \alpha)\%$; estimated standard error of the mean, $\hat{\sigma}_{\overline{x}}$; general value of a T random variable (T statistic) with $n - 1$ degrees of freedom, $T_{(n-1)}$; and a specific value of T, t.

In Unit II we examine the errors that occur naturally when we infer population values from a sample. We will pay particular attention to the difference between $\overline{X}$ and μ when we use $\overline{X}$ from a large sample to estimate μ. Upon completing the unit you should be able to

1. Determine the maximum error with confidence $100(1 - \alpha)\%$.
2. Determine the sample size necessary to attain a given level of confidence that the error does not exceed a given magnitude.
3. Determine the confidence level associated with a statement about the size of the maximum error.
4. Become familiar with the terms, sampling error, nonsampling error, and margin of error.
5. Recognize the symbol for the maximum sampling error, E.

Chapter 8 Statistical Inference: Estimation **327**

In Unit III we develop a new statistic, $\hat{p}$, the proportion of sample objects that fall in a certain category, called a success, to estimate p, the proportion of successes in the population. To perform the estimation procedures from the earlier units, we must first discover the sampling distribution, mean, and standard error of the new statistic, $\hat{p}$. Upon completing the unit, you should be able to

1. Describe the approximate sampling distribution of $\hat{p}$ when the sample size is large.
2. Find a $100(1 - \alpha)\%$ confidence interval for the population success rate, p.
3. Determine a maximum sampling error, E, sample size, n, or confidence level, $100(1 - \alpha)\%$, when $\hat{p}$ is the estimator of p.
4. Employ Minitab computer output to approximate and perform the operations mentioned in the two preceding objectives.
5. Understand the meaning of success rate and standard error of the sample success rate.
6. Distinguish the symbols for the population success rate, p; the sample success rate (estimator of p), $\hat{p}$; the standard error of $\hat{p}$, $\sigma_{\hat{p}}$, and the estimated standard error of $\hat{p}$, $\hat{\sigma}_{\hat{p}}$.

Unit I: Confidence Interval Estimates of the Population Mean, μ

Preparation-Check Questions

1. According to the central limit theorem, what is true if the sample size, n, is large (greater than 30 by the rule of thumb)?
   a. The mean of all $\overline{X}$ values is 0.
   b. The standard error of the mean and the population standard deviation are equal.
   c. The sampling distribution of $\overline{X}$ is approximately normal.
   d. Sampling must be done without replacement for accuracy.
   e. Binomial trials have an individual success rate close to 1.
2. Find z so that $P(-z < Z < z) = 0.95$.
   a. 1. c. 0.4750. e. 1.96.
   b. 0.9750. d. 1.645.
3. A random sample of 36 elements is selected from a population with a mean of 876 and a standard deviation of 54. Find $P(858 < \overline{X} < 894)$.
   a. 0.9544. c. 0.33. e. 0.6293.
   b. 0.0456. d. 0.2586.
4. Given the information in Question 3, what percentage of the $\overline{X}$ values should be within 27 units (that is, three standard errors) of the population mean?
   a. 300%. c. 100%. e. 3%.
   b. 95.44%. d. 99.74%.
5. The standard error of the mean, $\sigma_{\bar{x}}$, measures
   a. The dispersion of $\overline{X}$ values about the population mean.
   b. The dispersion of values in a sample.
   c. The dispersion of values in a population.
   d. The average distance between $\overline{X}$ values.
   e. The degree of difficulty in obtaining a sample standard deviation identical to the population standard deviation.
6. A probability distribution of a statistic is a
   a. Standard error.
   b. Sampling distribution.
   c. Frequency distribution.
   d. Confidence interval.
   e. Standard deviation.
7. If $n > 30$ or the sampling distribution of $\overline{X}$ is approximately normal, which of the following would transform an $\overline{X}$ value to a standard normal outcome (Z)?
   a. $\overline{X} \pm z\sigma_{\bar{x}}$.
   b. $\overline{X} \pm z\hat{\sigma}_{\bar{x}}$.
   c. $E\sqrt{n}/\sigma$.
   d. $(\sigma/\sqrt{n})\sqrt{(N - n)/(N - 1)}$.
   e. $(\overline{X} - \mu)/(\sigma_{\bar{x}})$.

Answers: 1. c; 2. e; 3. a; 4. d; 5. a; 6. b; 7. e.

Introduction

Heat Cost Susan is remodeling an older home and considering a new method of heating. To estimate the cost of this new method, she decides to find the mean monthly expenditure for heating similar-size houses with this method. Realizing the impossibility of finding expenditures for every such home in the area and because she must contract for the installation soon, she decides, instead, to ask the local utility company for a random sample of heating bills (without names or addresses) for similar homes that employ the new method.

Question: Suggest a statistic to estimate the mean monthly heating expenditure for all such houses.

Answer: Possibly your answer is the sample mean, $\overline{X}$. This is a good answer, as we shall see, but how do we know that the sample median or some other statistic would not provide a closer estimate?

Susan also wants to know how accurate and reliable the estimate is. Many possible $\overline{X}$ values can occur in a random sampling procedure: How close to the population mean is an estimate likely to be? Is it close enough to satisfy a decision maker who uses the estimate, such as Susan?

This unit addresses questions such as these and uses information about the pattern of $\overline{X}$ values to make inferences about the population mean, μ.

8.1 Choosing an Estimator for μ

Now we want to estimate the population mean, μ. We will use the sample mean, $\overline{X}$, from a random sample as the estimate. This procedure seems simple enough, but we can do more than estimate μ. Using the sampling distribution of $\overline{X}$, we can also describe the accuracy of the estimate or of the process that generated the estimate.

We call the statistic we use to estimate a population parameter an *estimator*. $\overline{X}$ is an estimator of μ. You may wonder whether another estimator would do a better job of estimating μ.

Question: Can you think of other possible estimators of μ?

Answer: There are other possibilities, such as the sample median, the sample mode, the largest value in the sample, the midpoint between the largest and smallest value, or some other measure you might create.

Statisticians use various criteria to decide among these estimators. One of these criteria is that the estimator should be unbiased. An estimator is an *unbiased estimator* of a parameter if the expected value of the estimator is the parameter. This criterion implies that, on average, infinite repetitions of the estimation process yield the parameter being estimated. $\overline{X}$ is an unbiased estimator of μ, because we know from Unit II of Chapter 7 that the mean of the values of $\overline{X}$ (the estimator) is μ (the parameter being estimated). Symbolically, we can write $E(\overline{X}) = \mu$.

Another important criterion is consistency. An estimator is a *consistent estimator* of a parameter if the probability the estimate is close to the parameter increases as the sample size, n, increases. We know that $\overline{X}$ is a consistent estimator of μ, because as n increases, the standard error of the mean, $\sigma_{\overline{x}} = \sigma/\sqrt{n}$, decreases. Thus, the $\overline{X}$ values cluster more tightly about μ, which means the chances of obtaining an $\overline{X}$ close to μ increase as n increases.

Again, there are other criteria that we could investigate; absence of bias and consistency are only two. Upon investigating other criteria, $\overline{X}$ turns out to be the "best" unbiased estimator of μ. Now, we will examine the estimation process and the accuracy that results from it.

8.2 Confidence Interval for Population Mean, μ: Known Population Standard Deviation, σ

In most sampling situations, there are many possible random samples and corresponding $\overline{X}$ values. Consequently, we should not expect a single $\overline{X}$ to produce an exact, error-free estimate of μ. This result is not surprising, because a sample contains information from only part of the population. The whole idea of statistical inference is to produce a good, educated guess about the population without complete information. The reliability of this "guess" increases as we follow procedures that increase the probability that the sample estimate will be close to the actual population value.

Because of the central limit theorem, we know that the sampling distribution of $\overline{X}$ approaches a normal distribution when the sample is large (usually $n > 30$). The area under the normal curve is concentrated around μ, indicating that there is a good chance an $\overline{X}$ from a random sample will lie close to μ. To reflect the uncertainty of the inferential process and, incidentally, to avoid the impression that the estimate is exact, we can provide more than a single $\overline{X}$ as an estimate.

Because the inferential process is inherently uncertain, we can report an interval of values centered about $\overline{X}$ rather than a point estimate. A *point estimate* is a single-value estimate. If the sample mean is 52 meters, then the value, 52 meters, is a point estimate. Extending the point estimate to obtain lower and upper endpoints of an interval of values generates a range of possible estimates, called a *confidence interval estimate*. For example, the 52 meters might expand to a confidence interval covering the values 42 meters to 62 meters. These endpoints are not arbitrary. Their values provide specific accuracy guides based on the sampling distribution of the statistic.

8.2.1 Large Sample from Any Population

Derivation To understand the procedure for constructing a confidence interval, we begin by considering the sampling distribution of $\overline{X}$ for a large sample shown in Figure 8-1. Let α symbolize the unshaded area in the tails of the curve, so $1 - \alpha$ symbolizes the shaded area in this figure. If $1 - \alpha$ is 0.95, then the middle 95% (which is $100(1 - \alpha)\%$) of the $\overline{X}$ values lie along the axis under the shaded region (between A and B), and $100\alpha\%$, or 5%, lie outside the middle 95%. As we discovered in Chapter 7, these $\overline{X}$ values must be within 1.96 standard deviations of the mean. In this case, the mean is μ, and the standard deviation is $\sigma_{\overline{x}}$.

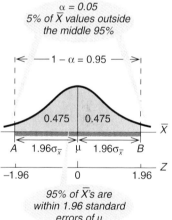

Figure 8-1
Ninety-five percent of $\overline{X}$'s are within 1.96 standard errors of μ

Question: Suppose $1 - \alpha = 0.80$. This middle 80% of the $\overline{X}$ values must lie within how many standard errors of the population mean, μ?
Answer: The middle 80% of the $\overline{X}$ values lie within 1.28 standard errors of μ.

Assuming $1 - \alpha = 0.95$ for the moment, suppose the $\overline{X}$ from a particular random sample falls in the range from A to B in Figure 8-1. If $1.96\sigma_{\overline{x}}$ is added to and subtracted from $\overline{X}$ to form an interval, this interval must contain μ. See $\overline{X}_1$ and the corresponding interval in Figure 8-2. If we follow this procedure to form an interval for any $\overline{X}$ that falls within the middle 95% of $\overline{X}$ values (between A and B), such as $\overline{X}_2$ in the figure, the interval will include μ. However, if the $\overline{X}$ falls in one of the tails of the curve, outside the middle 95% of $\overline{X}$ values, as $\overline{X}_3$ does, the interval will not include μ.

Figure 8-3 demonstrates the general case for any level of confidence $100(1 - \alpha)\%$. $100(1 - \alpha)\%$ of the $\overline{X}$ values lie within z standard errors of μ, $(P[-z < Z < z] = 1 - \alpha)$. The interval constructed using $\overline{X} \pm z\sigma_{\overline{x}}$ should contain the true population mean, μ, $100(1 - \alpha)\%$ of the time.

In summary if $100(1 - \alpha)\%$ is the level of confidence or the probability of being "correct," then we use the following boxed procedure to construct a confidence interval

Figure 8-2

The relationship between 95% confidence intervals and the sampling distribution of $\overline{X}$

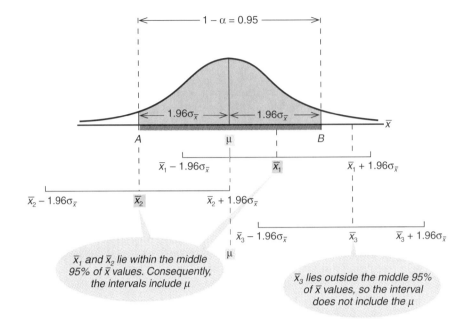

for μ for the circumstances listed in the box. Notice that we usually verbalize the level of confidence $100(1 - \alpha)\%$ as a percentage but express it as a proportion, $1 - \alpha$, when we determine probabilities or areas under a curve.

A $100(1 - \alpha)\%$ confidence interval estimate of the population mean, μ, is given by

$$\overline{X} \pm z\sigma_{\overline{x}},$$

where we find z from $P(-z < Z < z) = 1 - \alpha$ and $\sigma_{\overline{x}} = \dfrac{\sigma}{\sqrt{n}}$. This formula assumes that the following conditions hold:

1. The sample is large (usually $n > 30$).
2. σ is known.

Figure 8-3

The relationship between confidence intervals and the sampling distribution of $\overline{X}$

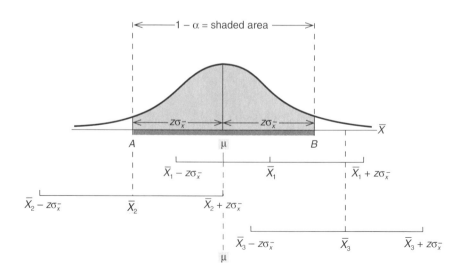

Example 8-1: A panel consisting of a communications expert, a marketing analyst, a psychologist, a sociologist, an anthropologist, and a political scientist each selected a random sample of 35 politicians' interviews on news talk shows to settle a dispute over the frequency of politicians' use of the term "middle class." They agreed that the standard deviation is approximately 54 mentions of the term per speech, because the standard deviation of all six samples was very close to 54. However, the six sample means varied substantially; 69, 96, 87, 86, 55, and 62. Demonstrate that the conditions in the procedure box hold and calculate a 99% confidence interval for the mean usage of "middle class" from each analyst's data.

Solution: The situation for each analyst satisfies the boxed assumptions. Each sample is large, because $n = 35$, and the analysts agree that σ is 54. To complete the formula we need to know the values of $\overline{X}$, z, and $\sigma_{\overline{x}}$. The $\overline{X}$ values vary with each analyst. The z and $\sigma_{\overline{x}}$ values will be the same, because each analyst needs a 99% confidence interval, while σ and n, the values used to calculate $\sigma_{\overline{x}}$, are the same.

Question: What z satisfies $P(-z < Z < z) = 0.99$?
Answer: 2.575.

The standard error is $\sigma_{\overline{x}} = 54/\sqrt{35} = 9.1277$ mentions, so the confidence intervals for the mean usage of "middle class" for each analyst are as follows. The usual format for presenting a confidence interval is to enclose the lower and upper limits of the interval in parentheses.

$69 \pm 2.575(9.1277)$, or (45.5 mentions, 92.5 mentions).
$96 \pm 2.575(9.1277)$, or (72.5 mentions, 119.5 mentions).
$87 \pm 2.575(9.1277)$, or (63.5 mentions, 110.5 mentions).
$86 \pm 2.575(9.1277)$, or (62.5 mentions, 109.5 mentions).
$55 \pm 2.575(9.1277)$, or (31.5 mentions, 78.5 mentions).
$62 \pm 2.575(9.1277)$, or (38.5 mentions, 85.5 mentions).

Question: There are several small estimates and several large estimates for the mean number of mentions of the term. However, based on the six intervals, can any analyst argue that the value 75 mentions is unlikely to be the true population mean?
Answer: No, because each interval includes the value 75.

Example 8-2: The utility company in the introductory heat cost situation provides Susan with a random sample of 100 heating bills, along with the information that previous estimates in Susan's region indicate that $\sigma = \$50$. The sample mean is $130. Find a 98% confidence interval for μ, the mean heating bill.

Solution: $P(-z < Z < z) = 0.98$ implies that $z = 2.33$. Because, $\sigma = \$50$ and $n = 100$, $\sigma_{\overline{x}} = 50/\sqrt{100} = 5$. Combining these values with $\overline{X} = 130$, we obtain $130 \pm 2.33(5) = 130 \pm 11.65 = (\$118.35, \$141.65)$ or $(\$118.4, \$141.6)$ as the 98% confidence interval estimate for the mean heating bill.

Example 8-3: A construction firm needs estimates of the mean time required for three major types of jobs in order to improve scheduling. The management randomly selects 40 records on each of the three major types of jobs. Management experience suggests that the standard deviation for driveway jobs is 2.0 hours, 4.0 hours for patio jobs, and 6.0 hours for deck jobs. Suppose the three sample means are all 12.0 hours. Find an 80% confidence interval for the mean time for each type of job.

Solution: To construct the confidence intervals, we begin with $P(-z < Z < z) = 0.8$, implying that $z = 1.28$. Using $\sigma_{\overline{x}} = \sigma/\sqrt{n}$, the standard errors are $2.0/\sqrt{40} = 0.3162$ for driveways, $4.0/\sqrt{40} = 0.6325$ for patios, and $6.0/\sqrt{40} = 0.9487$ for decks. The 80% confidence intervals for the mean time to complete each type of job are:

Driveway: 12 ± 1.28(0.3162) = (11.60 hours, 12.40 hours).
Patio: 12 ± 1.28(0.6325) = (11.19 hours, 12.81 hours).
Deck: 12 ± 1.28(0.9487) = (10.79 hours, 13.21 hours).

Question: Which sample is likely to provide the most accurate information about its population mean? Justify your answer.

Answer: The standard deviations for completion times for the different job types differ, and, hence, the standard errors differ. The driveway values are more tightly clustered about their mean than the patio or deck values. Consequently, the driveway sample is more likely to contain values close to the mean, μ, and a more accurate estimate is likely to result. Alternatively, we can compare ranges, because the confidence level is 80% for all three cases. The driveway interval covers the smallest range, which confirms that the chances of being close to μ are greatest for driveway jobs. The patio standard deviation is larger and the range of its confidence interval larger, because times for these jobs are more dispersed than driveway job times. Finally, the confidence interval for the mean time for deck jobs is largest, because these times are the most dispersed of the three data sets.

Interpretation To interpret a confidence interval estimate, we must distinguish two cases, general and specific. In the general case, before we actually collect data and calculate endpoints, we can say the probability is $1 - \alpha$ that the interval $[(\overline{X} - z\sigma_{\overline{x}}), (\overline{X} + z\sigma_{\overline{x}})]$ will contain μ. α is the *risk* that the interval does not contain the population mean.

Note that μ is a constant for the population. The magnitudes that vary in this process are the endpoints, $\overline{X} \pm z\sigma_{\overline{x}}$, not μ. The intervals $\overline{X} \pm z\sigma_{\overline{x}}$ can take on different values depending on the $\overline{X}$ that results from the sampling process. A certain proportion $(1 - \alpha)$ of these intervals contains the population mean, μ. If $1 - \alpha$ is 0.95, then 95% of the intervals contain μ, as shown in Figure 8-2.

After we produce a specific confidence interval estimate, such as ($118.4, $141.6) from Example 8-2, it is tempting, although incorrect, to say that the *probability* is 0.98 that μ is between $118.4 and $141.6. But μ is a fixed value, not a random variable. It is either in the interval or outside it. The confidence interval varies, not μ. To be correct, we must say we have used an estimation technique that will produce an interval containing μ a certain proportion of the time (0.98 in the example)—which is why we say that the "confidence" not the "probability" is 98% that the interval contains μ.

Question: Interpret the 99% confidence interval for the mean number of mentions of the term "middle class" from Example 8-1.

Answer: In each case, we are 99% confident that the interval encloses the actual mean number of mentions of the term by politicians.

Intuition To obtain a better feel for the information relayed by a confidence interval, try answering the following question.

Question: The length or width of a confidence interval is the distance from the lower to the upper endpoint. If we increase the level of confidence $100(1 - \alpha)\%$ and everything else remains fixed ($\overline{X}$, σ, n), will the length of the interval increase or decrease?

Answer: Figure 8-4 demonstrates that the length of the confidence interval will increase. We expect this, because as $100(1 - \alpha)\%$ increases, we can be more confident that the interval contains μ. A larger interval that contains more possible values for μ means increased confidence. If we guess someone's age, we have a greater chance of being correct by guessing 20 to 80 than 45 to 55. Notice that *a confidence interval does not zero in on a target as the confidence level increases.*

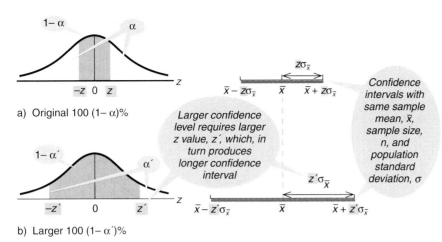

Figure 8-4
An increasing confidence level increases the length of the confidence interval

■ **Example 8-4:** Using $n = 144$, $\overline{X} = 250$, and $\sigma = 24$, we obtain the 90% confidence interval (246.7, 253.3). Find the 95% confidence interval using the same $\overline{X}$, n, and σ. Compare the lengths of the two confidence intervals.

■ **Solution:** For a 95% confidence interval, $z = 1.96$, $n = 144$, and $\sigma = 24$ gives $\sigma_{\overline{x}} = 24/\sqrt{144} = 2$, so the confidence interval is $250 \pm 1.96(2) = (246.1, 253.9)$. This interval is longer than the 90% interval, and we are more sure that the interval contains μ.

Question: If we increase the sample size while $\overline{X}$, $100(1 - \alpha)\%$, and σ remain fixed, will the confidence interval expand or contract?
   Answer: Figure 8-5 shows that the interval contracts. A larger sample size means that we have more information from the population. Increased information improves the accuracy of the estimation process. If we fix the level of confidence the smaller interval reflects the improved accuracy.

■ **Example 8-4 (continued):** With $n = 144$, $\sigma = 24$, and $\overline{X} = 250$, we obtained (246.7, 253.3) as the 90% confidence interval. Find the 90% confidence interval if n is 576, σ remains 24, and $\overline{X}$ is 250. Compare the result with the 90% interval constructed with 144 observations.

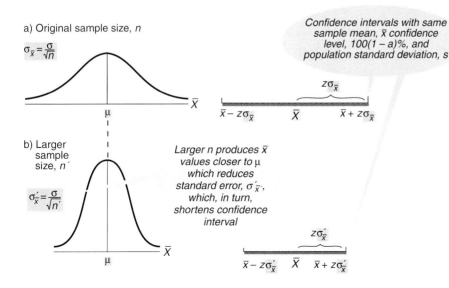

Figure 8-5
Increasing the sample size shortens the confidence interval

■ **Solution:** Increasing n causes the sample means to be less dispersed, which a smaller standard error reflects, $\sigma_{\bar{x}} = 24/\sqrt{576} = 24/24 = 1$. When n was 144, $\sigma_{\bar{x}} = 2$. Now the confidence interval is $250 \pm 1.645(1) = (248.4, 251.6)$. The added reliability of the estimate, reflected in the smaller $\sigma_{\bar{x}}$, results in a smaller confidence interval. The smaller interval reflects the increased amount of population information employed in the estimation at an unchanged level of confidence.

Section 8.2.1 Problems

8. Find a 95% confidence interval for μ if $\bar{X} = 43.6$, $n = 124$, and $\sigma = 18.3$. Interpret the result.

9. Suppose prior evidence from the construction firm in Example 8-3 indicates the population standard deviations are the following listed values. The sample mean from a sample of 40 values of project completion times for each product is also listed.

> Driveways: $\bar{X} = 7.0$ hours, $\sigma = 1.0$ hour.
> Patios: $\bar{X} = 10.0$ hours, $\sigma = 2.5$ hours.
> Decks: $\bar{X} = 12.0$ hours, $\sigma = 3.3$ hours.

Find an 80% confidence interval for the mean completion time for each job type. Interpret the result. Which mean is best estimated with the available information? Explain.

10. Jenni wants to estimate the average number of minutes of videos (not counting commercials and commentaries) she can watch per hour of viewing. Over a month, she views 34 hours of videos at random times. The mean video time per hour is 45.3 minutes with a standard deviation of 6.2 minutes. Assume the sample standard deviation accurately reflects the population standard deviation. Then find and interpret a 98% confidence interval for the mean video time per hour.

11. Michael rides his bicycle an average of 18.0 miles per day, according to a random sample of his mileage on 44 days. The standard deviation for all days is 5.0 miles. Find and interpret a 90% confidence interval for the mean daily mileage.

12. Predict the change, if any, in the length of the 95% confidence interval from Problem 8 when the sample contains only 100 objects. Give an intuitive explanation for your prediction. Verify your answer by computing the new interval.

8.2.2 Normal Population

Because the sampling distribution of $\bar{X}$ is normal when the population is normal, regardless of sample size (see Section 7.6), we can use the earlier confidence interval formula from the last section when a situation fits these conditions.

> A $100(1 - \alpha)\%$ confidence interval for the population mean, μ, when the sample is taken from a normal population with a known standard deviation σ, is
> $$\bar{X} \pm z\sigma_{\bar{x}}.$$

■ **Example 8-5:** A graph of minimum distances (in meters) required for persons with good (20:20) vision to read road signs strongly suggests that such distances are normally distributed with a standard deviation of 10.2 meters. The differences occur because of obstacles, such as letter size, color, number of signs at one spot, and clutter of advertisements or store marquees. The mean reading distance for a municipality's road and traffic signs must be 35 meters or more to qualify for a grant to improve readability. A municipal official wishes to see if the city qualifies for the grant. He measures the minimum reading distance for four randomly selected signs by a person certified to have 20:20 vision to see if the city should pursue the grant. The mean distance for the official's sample is 22.8 meters. Does this result suggest that the grant should not be pursued, assuming a 95% confidence level?

■ **Solution:** The 95% confidence interval for the mean minimum reading distance is $22.8 \pm 1.96(10.2/\sqrt{4}) = (12.8 \text{ meters}, 32.8 \text{ meters})$. The interval is entirely below the 35 meter cutoff, which suggests that the mean distance for the municipality's signs does not meet the grant requirements.

Section 8.2.2 Problems

13. If IQ scores are normally distributed with a standard deviation of 15 points, find a 98% confidence interval for the mean score by using the mean score of 112.3 for a random sample of 20 people. Interpret the result. If the population mean is 100 points, is your interval correct or incorrect? Explain.

14. A firm's records indicate that the length of a washing machine part that it manufactures is normally distributed with a mean of 2.25 feet and a standard deviation of 0.06 feet. A random sample of 50 currently produced parts has a mean of 2.17 feet. Find a 95% confidence interval estimate of the mean length. Interpret the interval. Does the interval indicate that the current manufacturing process is possibly producing a mean output equal to the historical mean? Explain.

15. Find and interpret a 90% confidence interval for the local mean reaction time for a chemistry experiment. Studies show that the reaction time is normally distributed with a standard deviation of 0.002 seconds everywhere, but the mean time varies with local atmospheric conditions. A random sample of eight mixtures in a local lab yields a mean reaction time of 10.523 seconds.

8.3 Confidence Interval for Population Mean, μ: Unknown Population Standard Deviation, σ

In practice, we usually do not know the population standard deviation, σ. Instead, we substitute the best available estimate, the sample standard deviation, S, when we calculate $\sigma_{\bar{x}}$, the standard error of the mean. (*Statisticians often use a caret or hat over any symbol for a parameter to denote an estimate of the parameter.*) Thus, $\hat{\sigma}_{\bar{x}} = S/\sqrt{n}$.

Another consideration arises when we substitute S for σ in the confidence interval formula, we add to the uncertainty of the confidence interval estimate. The normal distribution no longer adequately reflects the uncertainty of the inference. Instead, we must use the T probability distribution whose outcomes are more dispersed about the center than normally distributed outcomes.

8.3.1 The T Probability Distribution

T is a random variable whose values represent the outcomes of a statistical experiment. Figure 8-6 shows that these values fall in a certain pattern called a T probability distribution that resembles a standard normal (Z) curve. T values are symmetric about the mean 0, but the curve is flatter than the normal curve with more area in the tails. T values extend to infinitely large negative values and to infinitely large positive values and the total area under the curve is 1.

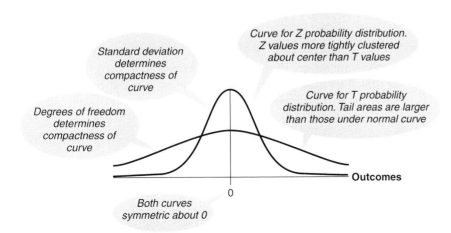

Figure 8-6

Comparison of T and normal probability distribution

Figure 8-7
An intuitive illustration of the meaning of degrees of freedom

Figure 8-8
A comparison of given areas in the T and Z tables

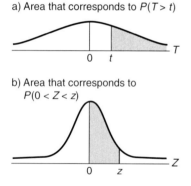

a) Area that corresponds to $P(T > t)$

b) Area that corresponds to $P(0 < Z < z)$

To find probabilities for T values, we need a value called *degrees of freedom*, just as we must know μ and σ to work with the normal curve. Although we specify the degrees of freedom for specific procedures in this book, Figure 8-7 illustrates degrees of freedom—intuitively, the number of values in a data set that are free to be any value when summary sample values are already specified. The rule for finding the degrees of freedom when we make confidence intervals for μ, is to use $n - 1$ (the sample size less 1).

Degrees of freedom determine the exact shape, the compactness, of the curve. *As the sample size increases, the degrees of freedom, $n - 1$, increases, and the T distribution approaches a standard normal distribution*, just as the sampling distribution of $\overline{X}$ approaches a normal distribution as n increases. As n increases past 30 observations, we often approximate unknown T values with values from the normal table. Statistical software packages, on the other hand, routinely employ exact T values for larger samples.

Figure 8-8a shows the probability for an interval of T values as the area under the curve above the interval. The shaded area in the figure is the $P(T > t)$. We use the T table in Appendix J to determine areas. This table provides areas in the tail of the distribution like that in Figure 8-8a. Figure 8-8b contrasts the tail area in the T table with the areas between 0 and z in the standard normal table.

The curve changes every time the sample size changes, because degrees of freedom ($n - 1$) automatically change. Rather than develop a new table for each degrees of freedom value, we condense each such table to a single row in the T table in the appendix. The table lists a few T values for each degrees of freedom value. The left-hand column of the table locates the appropriate degrees of freedom (df) row.

Question: Locate the row for 21 degrees of freedom. What T values are listed in this row?
Answer: 0.257, 0.686, 1.323, 1.721, 2.080, 2.518, 2.831, and 3.819.

Figure 8-9 displays where these T values lie along the horizontal axis of the T distribution with 21 degrees of freedom. We represent the random variable with a subscript to note its degrees of freedom, in this case $T_{(21)}$.

The table lists eight areas under the tail of the curve in the top row of the table compared to a full page of areas in the normal table. These areas are 0.40, 0.25, 0.10, 0.05, 0.025, 0.01, 0.005, and 0.0005.

■ **Example 8-6:** A T random variable with 21 degrees of freedom describes a sales forecaster's errors (the difference between the forecast and actual values). Draw the diagram for finding the probability that a randomly selected error will exceed $2.080 or $P(T_{(21)}) > 2.080)$. Then find this probability using the T table.

Section 8.3 Confidence Interval for Population Mean, μ: Unknown σ

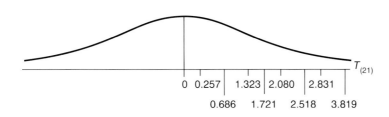

Figure 8-9
Table values of the T random variable with 21 degrees of freedom

■ **Solution:** Figure 8-10 shows the diagram and table entry for finding the probability that $T_{(21)}$ exceeds $2.080 is 0.025.

Using the symmetry of the T distribution about 0 and the fact that the total area under the curve must be 1, we can solve many types of problems with the T table, just as we do with the standard normal table. *Recall that we use the closest value when the exact value is not available in the abbreviated table.*

■ **Example 8-7:** A meteorologist's errors from forecasting daily high temperatures (forecast value minus actual value) follow a T distribution with 16 degrees of freedom.

a. Find the probability the error will be between $-1.725°$ and $1.725°$ Fahrenheit, $P(-1.725 < T_{(16)} < 1.725)$.
b. Find the boundaries of an interval containing the middle 95% of error values. $P(-t < T_{(16)} < t) = 0.95$.

■ **Solution: a.** Figure 8-11a displays three steps for finding that $P(-1.725 < T_{(16)} < 1.725) = 0.90$. Approximately 90% of the errors are between $-1.725°$ and $1.725°$ Fahrenheit.

b. Figure 8-11b illustrates three steps to locate T values on the axis that correspond to the known probabilities or areas. The areas in the table are tail areas, so we must determine a tail area beyond t from the given information before we can determine t. We combine the given information and the properties of the T curve to find that $P(-2.120 < T_{(16)} < 2.120) = 0.95$. The middle 95% of the errors are between $-2.120°$ and $2.120°$ Fahrenheit.

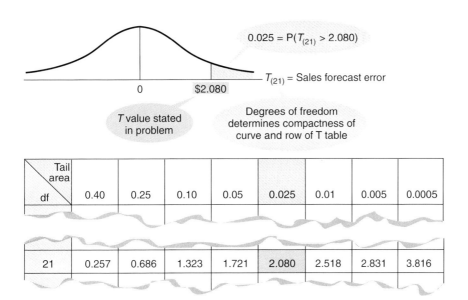

Figure 8-10
Solution diagram and T table entry for $P(T_{(21)} > 2.080)$

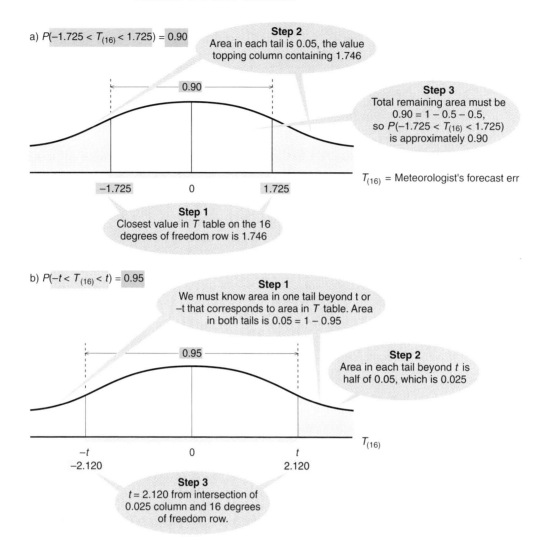

Figure 8-11
Solution diagrams for meterologist's forecast errors

The solutions to problems such as Part b of the preceding example provide the information we will need to construct a confidence interval for the population mean, μ, with a small sample.

Section 8.3.1 Problems

Use the T table to solve the following problems.

16. $P(-2.5 < T_{(11)} < 2.5)$.
17. $P(-2.947 < T_{(15)} < 2.947)$.
18. $P(-0.727 < T_{(5)} < 0.727)$.
19. $P(-1.25 < T_{(25)} < 1.25)$.
20. Find t so that $P(-t < T_{(23)} < t) = 0.98$.
21. Find t so that $P(-t < T_{(29)} < t) = 0.95$.
22. Find t so that $P(-t < T_{(8)} < t) = 0.80$.
23. Find t so that $P(-t < T_{(10)} < t) = 0.99$.

8.3.2 Normal Population

In order to proceed with a confidence interval for the population mean, μ, we must use the T distribution to incorporate the increased uncertainty introduced when we substitute the sample standard deviation, S, for the population standard deviation, σ. This

Section 8.3 Confidence Interval for Population Mean, μ: Unknown σ **339**

distribution applies when we select samples from a normal population, but it also works well when the population is symmetric with a single mode (unimodal) or when the sample is large.

If we attempt, when the population is normal, to transform $\overline{X}$ to a Z or standardized value, by subtracting μ and dividing the difference by the estimated standard error, $\hat{\sigma}_{\overline{x}}$, we produce $(\overline{X} - \mu)/\hat{\sigma}_{\overline{x}}$, a T statistic with $n - 1$ degrees of freedom, not a Z. This result only approximates the Z score, because we must estimate the standard error. Now, Figure 8-12 shows that *$100(1 - \alpha)$% of the $\overline{X}$ values will fall within t estimated standard errors ($\hat{\sigma}_{\overline{x}}$) of μ*, where $\hat{\sigma}_{\overline{x}} = S/\sqrt{n}$. Consequently, $100(1 - \alpha)$% of the intervals computed using $\overline{X} \pm t\hat{\sigma}_{\overline{x}}$ will include μ. The following box summarizes the argument.

> A $100(1 - \alpha)$% confidence interval for the population mean, μ, when the sample is taken from a normal (or a symmetric and unimodal) population with an unknown standard deviation is given by
>
> $$\overline{X} \pm t\hat{\sigma}_{\overline{x}},$$
>
> where $\hat{\sigma}_{\overline{x}} = S/\sqrt{n}$ and $P(-t < T_{(n-1)} < t) = 1 - \alpha$ [or $P(T_{(n-1)} > t) = \alpha/2$].

We locate the value t on a diagram such as Figure 8-12 by determining the area in one of the tails, $\alpha/2$, then locating the intersection of this area's column and the $n - 1$ degrees of freedom row in the T table.

■ **Example 8-8:** A team of medical researchers chose a sample of 29 people with cholesterol levels above a certain level for a medical experiment. The average age of these subjects was 53 years, and the standard deviation was 13 years. Use this information to determine a 95% confidence interval for the mean age of people with these high cholesterol

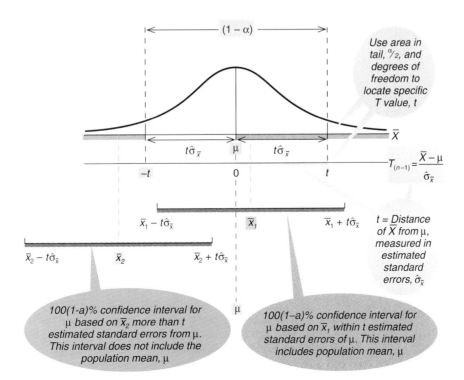

Figure 8-12
$100(1 - \alpha)$% of $\overline{X}$ values within t estimated standard errors of μ

Figure 8-13
Solution diagram for $P(-t < T_{(11)} < t) = 0.95$

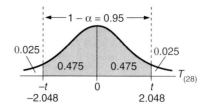

levels. Assume the ages of this population are normally distributed. (Source: Margaret M. Cobb, Howard S. Teitelbaum, and Jan L. Breslow, "Lovastatin Efficacy in Reducing Low-Density Lipoprotein Cholesterol Levels on High- vs. Low-Fat Diets," *Journal of the American Medical Association*, Vol. 265, No. 8, February 27, 1991, pp. 997–1001.)

■ **Solution:** $\overline{X} = 53$ years. $S = 13$ years, and $n = 29$, so $\hat{\sigma}_{\overline{x}} = 13/\sqrt{29} = 2.4140$ years. The intersection of the 0.025 [$\alpha/2$] column and the 28 degrees of freedom row in the T table produces $t = 2.048$ (see Figure 8-13). Thus, the 95% confidence interval is

$$53 \pm 2.048(2.4140) = 53 \pm 4.9440 \text{ or } (48.1 \text{ years}, 57.9 \text{ years}).$$

We are 95% confident that this interval contains the mean age of all people in this high-cholesterol population.

■ **Example 8-9:** Find a 98% confidence interval for the mean of maximum hurricane wind speeds. The maximum wind speeds from five randomly selected hurricanes are 105, 110, 98, 112, and 100 mph. Assume that the wind speeds are symmetric and unimodal.

■ **Solution:** First, we must calculate $\overline{X}$ and S from this sample of five observations ($n = 5$) by using the formulas from Chapter 3. A calculator is convenient for such computations.

$$\overline{X} = \frac{105 + 110 + 98 + 112 + 100}{5} = 105 \text{ mph};$$

$$S = \sqrt{\frac{\sum_{1}^{n}(X - \overline{X})^2}{n - 1}}$$

$$= \sqrt{\frac{(105 - 105)^2 + (110 - 105)^2 + (98 - 105)^2 + (112 - 105)^2 + (100 - 105)^2}{5 - 1}}$$

$$= 6.0828 \text{ mph}.$$

Using $S = 6.0828$ mph, we estimate $\sigma_{\overline{x}}$ as $\hat{\sigma}_{\overline{x}} = S/\sqrt{n} = 6.0828/\sqrt{5} = 2.7203$ mph. From the T table we obtain $t = 3.747$ (intersection of 4 degrees of freedom row and 0.01 column). Thus,

$$\overline{X} \pm t\hat{\sigma}_{\overline{x}} = 105 \pm 3.747(2.7203) = 105 \pm 10.2 = (94.8 \text{ mph}, 115.2 \text{ mph}).$$

We are 98% confident that the interval encloses the mean maximum hurricane wind speed.

Question: Why would an increase in n tend to shorten the range of the confidence interval?

Answer: Intuitively, a larger sample includes more information about the population, which produces a better chance of obtaining a good estimate of μ. If $1 - \alpha$ and S stay the same, a smaller range will reflect this intuition. Mathematically, when n changes, two things change in the confidence interval formula, $\overline{X} \pm t\hat{\sigma}_{\overline{x}}$. First, n itself changes. As n increases, $\hat{\sigma}_{\bar{x}} = S/\sqrt{n}$ decreases, shortening the confidence interval. Now, examine the 0.01 column (for $1 - \alpha = 0.98$) in the T table for the second change.

Question: What happens to the value of t as n increases ($1 - \alpha$ is constant at 0.98)?

Answer: As n increases, the degrees of freedom increase, and the t decreases, causing $t\hat{\sigma}_{\bar{x}}$ to decrease more—leading again to a smaller confidence interval.

■ **Example 8-9 (continued):** Suppose we use 10 wind speeds, rather than five, to find the 98% confidence interval for μ, but $\bar{X}$ and S remain the same values, 105 and 6.1, respectively. Find the interval and compare it with the result when $n = 5$: (94.8 mph, 115.2 mph).

■ **Solution:** $\hat{\sigma}_{\bar{x}} = 6.1/\sqrt{10} = 1.9290$ compared with 2.7280 when $n = 5$. Now $t = 2.821$ (nine degrees of freedom and 0.01 in the tail) produces:

$$105 \pm 2.821(1.9290) = 105 \pm 5.4 = (99.6 \text{ mph}, 110.4 \text{ mph}).$$

We are 98% confident that this smaller interval includes the actual mean maximum wind speed. When n was 5, we were 98% confident the interval included the true mean, but the larger range reflected the lesser amount of information provided in a smaller sample, as suggested by the preceding question and answer.

Section 8.3.2 Problems

24. An astronomer has 14 measurements of the distance of a particular star from earth. The mean is 205.6 light-years with a standard deviation of 36.2 light-years. Find and interpret a 95% confidence interval for the mean distance if the astronomer is willing to accept that the distance measures are normally distributed, based on his work with other space measurements.

25. A random sample of 19 daily newspapers produces a mean of eight local news reporters per newspaper with a standard deviation of 4.3 reporters. If the number of reporters for all dailies is normally distributed, find a 90% confidence interval for the mean number of reporters per newspaper. Interpret the result.

26. A legislator requests an estimate of the mean expenditure on drug therapy for inmates in state prisons. Her aide collects the following annual expenditures for a random sample of current prisoners: $9,000, $12,000, $8,500, $1,200, $8,500. Use this information to compute a 90% confidence interval for the mean expenditure for all prisoners. Interpret the result. What do we assume about the population when we present this interval estimate?

27. A city commission decides that the city needs an economic development director. Taking a random sample of eight counties in the same region, a clerk finds that these counties pay their development directors an average salary of $28,000 with a standard deviation of $2,500. Find a 99% confidence interval for the mean salary of such directors. Assume the distribution of all such salaries is symmetric and unimodal. Interpret the result.

28. The number of cats at Albertine's vary because strays come and go constantly. Albertine is a cat lover and supplies food for all cats that show up at feeding time. Over the past year, she purchased an average of 12 pounds of cat food each week, with a standard deviation of 2.5 pounds. Find a 98% confidence interval for the mean amount of food she purchases each week. What is the population in this problem? Is the sample randomly selected from this population?

8.3.3 Large Samples

As the sample size increases, the accuracy of confidence intervals and other inferences about the population mean, μ, depends less on the distribution of the population, even when we do not know the population standard deviation, σ. So we often proceed with the formula $\bar{X} \pm t\hat{\sigma}_{\bar{x}}$ for a confidence interval when we have a large sample (usually $n > 30$) from a population with an unknown distribution. The stated level of confidence will be a better approximation when the population distribution is closer to a normal or bell shape. Larger samples reduce the need for this condition, and smaller samples require stricter adherence to this condition.

In addition, when the sample size increases, the T distribution approaches a normal distribution.

Question: What is the limit of t in the 0.01 column of the T table as n increases?
Answer: 2.326 or 2.33 to two decimal places.

2.33 is the value we obtain from the normal table for a 98% confidence interval. Similarly, following down the column of t values in the 0.025 column (for a 95% confidence interval), we reach the limit 1.960. Thus, we see evidence that $(\overline{X} - \mu)/\hat{\sigma}_{\overline{x}}$ approaches a standard normal random variable. After n exceeds 30, the difference between the t and z values is small enough that we will use the normal table to approximate the t.

Because the T distribution approaches a Z with less dependence on the distribution of the population, we can use the following procedure for a large sample confidence interval for μ.

When the sample is large (usually $n > 30$), we often approximate the $100(1 - \alpha)\%$ confidence interval for the population mean,

$$\overline{X} \pm t\hat{\sigma}_{\overline{x}};$$

where

$$\hat{\sigma}_{\overline{x}} = \frac{S}{\sqrt{n}}, \text{ and } P(T_{(n-1)} > t) = \alpha/2$$

with

$$\overline{X} \pm z\hat{\sigma}_{\overline{x}},$$

where

$$P(-z < Z < z) = 1 - \alpha.$$

■ **Example 8-10:** A historian wishes to estimate the mean number of Confederate casualties per engagement, but many of the records have been lost or destroyed. However, a museum possesses rosters from both before and after 36 different fights, enabling the historian to determine the losses from the battles. The mean casualty count of the 36 battles is 800 casualties, and the standard deviation is 60 casualties. Use this information to construct a 95% confidence interval for the mean Confederate casualties per battle.

■ **Solution:** σ is unknown, but $S = 60$, so we estimate $\sigma_{\overline{x}}$ with $\hat{\sigma}_{\overline{x}} = 60/\sqrt{36} = 10$. Because $n = 36 > 30$, we use $\overline{X} \pm z\sigma_{\overline{x}}$, or $800 \pm 1.96(10) = (780.4$ casualties, 819.6 casualties) for the 95% confidence interval for the mean casualty count. We are 95% confident that this interval contains the true mean casualty count.

Section 8.3.3 Problems

29. During one week 72 randomly selected customers at a fried chicken restaurant consumed a mean of 1.38 pieces of chicken. The standard deviation was 0.25 piece. Find a 99% confidence interval for the mean number of pieces consumed. What populations can inferences refer to in this situation?

30. A random sample of 40 rental advertisements produces a mean number of bedrooms of 1.2 and a mean rent of $275. The respective standard deviations are 0.15 bedrooms and $37. Construct a 95% confidence interval for the mean bedroom count and a 90% confidence interval for the mean rent. Interpret the results.

31. Professor Ponder is writing a grant proposal to continue a project and must project travel expenditures. A random sample of 32 weeks of previous travel has a mean of 483 miles per week with a standard deviation of 105 miles. Find and interpret a 95% confidence interval for the mean mileage per week.

32. An instruction specialist is concerned that an elementary school textbook may require more complex thinking and reading ability than is usual for elementary students. Find an 80% confidence interval for the book's mean sentence length if a random sample of 36 sentences has a mean of 14.18 words per sentence with a standard deviation of 1.49 words. Interpret the result. If she believes the mean should not be more than 10 words per sentence, will the confidence interval estimate support her belief or alleviate her concerns?

33. A random sample of 100 professional golfers on a given day score a mean of 85 strokes, with a standard deviation of eight strokes. Find a 95% confidence interval for the mean score. The course is advertised as par (average) 72 strokes. What do your results indicate about the advertised par?

8.4 A Flowchart for Constructing Confidence Intervals for μ

The flowchart in Figure 8-14 is useful for deciding which table and formula to use when we calculate a confidence interval for the population mean, μ. The first question about σ is one we seldom ask in practice, because generally we do not know the population standard deviation, σ. Consequently, the right-hand side of the flowchart will receive more use, and the first question we ask is usually whether the sample is large or small, which often becomes whether the sample contains more than 30 observations.

If $n > 30$, we can use a z from the normal table to approximate a t from the T table in the formula $\bar{X} \pm t\hat{\sigma}_{\bar{x}}$. In this book, we will use the more familiar normal table, though the results from either table will differ very little, especially as n becomes very large.

If the sample is small ($n \leq 30$), we have less information from the population, so we must satisfy another condition about the distribution of the population to produce a reliable confidence interval. If the distribution is not (1) normal or (2) symmetric and unimodal, then there are no general formulas to use. Thus, we end at the box containing the question mark. If we use the regular confidence interval formulas in this case, the confidence level associated with the resulting interval may or may not be $100(1 - \alpha)\%$. If the population is normally distributed (or symmetric and unimodal), then the T distribution provides the t value we need to expand the point estimate into an interval estimate.

The two formulas on the right side of the flowchart are the primary situations that we encounter in this book, $\bar{X} \pm z\hat{\sigma}_{\bar{x}}$ and $\bar{X} \pm t\hat{\sigma}_{\bar{x}}$. Thus, the sample size will be an important factor in deciding on z or t. The estimated standard error in both cases is $\hat{\sigma}_{\bar{x}} = S/\sqrt{n}$.

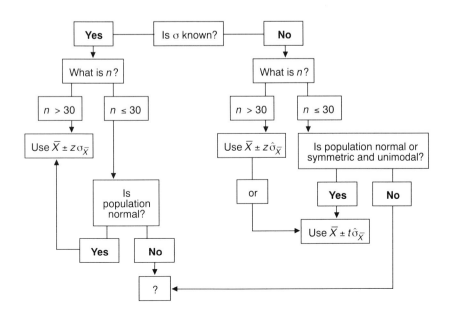

Figure 8-14
A flowchart for constructing confidence interval for μ

Figure 8-15
Minitab output for 98% confidence interval for mean of hurricane maximum wind speeds (Example 8-9)

```
             N      MEAN     STDEV    SE MEAN     98.0 PERCENT C.I.
SPEED        5    105.00      6.08       2.72    (  94.81,  115.19)
```

8.5 Confidence Intervals and the Computer

The Minitab statistical package produces confidence intervals for two situations. If σ is known, the normal probability distribution provides a z for the computation. If σ is unknown, the interval depends on the T probability distribution. It is up to the user to be aware of the effect of the population distribution and the sample size on the estimation.

■ **Example 8-9 (continued):** Figure 8-15 displays the Minitab output for a 95% confidence interval from the data set of hurricane maximum wind speeds. Our earlier result is (94.8 mph, 115.2 mph). The Minitab interval is (94.81 mph, 115.19 mph).

■ **Example 8-11:** The Division of Wildlife Management in North Carolina collects and maintains records on hunting or harvesting of big game in order to improve decision making concerned with hunting seasons and other regulations. The division collects the number of antlered bucks harvested per square mile (b/sqm) by county for all 100 counties. The population mean for this variable for all 100 counties during the 1990–1991 harvest was 1.773 b/sqm, and the standard deviation was 1.54 b/sqm. Figure 8-16 reports two different 95% confidence intervals based on two separate random samples independently selected from the 100 counties. Do these intervals contain the true population mean? (Source: "Big Game Tag Report-Number 15," *Wildlife in North Carolina*, Vol. 55, No. 7 July, 1991, center insert.)

■ **Solution:** The large sample produces an interval that contains the true mean, 1.773 b/sqm. However, the interval in Part b does not. Apparently, four relatively large data values compose the second random sample, because the mean is large and the sample standard deviation is small, 0.1241 compared to 1.395 for the larger sample. Though the risk of obtaining a sample with a mean in the tails of the curve is small (0.05 here), it does still happen.

Question: Other than σ being known and unknown, what is another reason for the difference in the two intervals in Figure 8-16?

Answer: The sample sizes differ. In Part a, $n = 32$. In Part b, $n = 4$. Consequently, the standard deviations and standard errors differ. Also, the computer forms the interval with a Z value in Part a, because the sample is large and σ is known. In Part b, the computer uses a T value.

To be 95% confident that the intervals cover the true mean, the values for all 100 counties need to be at least symmetric and unimodal. Judging by Part c of the figure, a histogram of all 100 values, it appears that this assumption is not accurate, so the level of confidence is less than 95%.

Question: Given that 32 observations comprise 32% of the 100 population values, this sample violates the assumption that the sample contain 5% or less of the population values. This is a common assumption that avoids computations that reduce the computed value of the standard error very little—a general assumption that we made in Section 7.7 for situations in this book. If we still want a 95% confidence interval and also want to incorporate the effect of the sample containing 32% of the population, do you think the range of the interval will increase or decrease? Explain.

Problems

a) Large Sample Output With σ Known

THE ASSUMED SIGMA =1.54

	N	MEAN	STDEV	SE MEAN	95.0 PERCENT C.I.
SAMPLE	32	1.773	1.395	0.272	(1.240, 2.307)

b) Small Sample Output With σ Unknown

	N	MEAN	STDEV	SE MEAN	95.0 PERCENT C.I.
SMALSAMP	4	2.9200	0.1241	0.0620	(2.7225, 3.1175)

c) Histogram of BCK/SQMI   N = 100

Midpoint	Count	
0.0	6	******
0.5	15	***************
1.0	15	***************
1.5	11	***********
2.0	7	*******
2.5	12	************
3.0	10	**********
3.5	8	********
4.0	7	*******
4.5	4	****
5.0	1	*
5.5	1	*
6.0	1	*
6.5	1	*
7.0	1	*

Figure 8-16
Minitab output for random samples from North Carolina hunting data

Answer: Although violating an assumption sounds as though we should produce a poorer result, in this case, the violation occurs because we use more not less information from the population. Intuitively, we expect a larger sample to produce a better inference, not a poorer one. So, the stated confidence interval is actually larger than it needs to be. If we incorporate the correction factor for $\sigma_{\bar{x}}$ for the sample containing a large proportion of the population, $\sigma_{\bar{x}}$ will be smaller (see Section 7.7). Consequently, $z\sigma_{\bar{x}}$ will be smaller, and the range of the confidence interval decreases. Ignoring this correction factor means that we are conservative when we say there is 95% confidence that the interval contains the true mean. The confidence level is actually larger.

Section 8.5 Problems

34. A baker randomly selects 230 doughnuts from a large batch to estimate the mean shelf life of the product, the time before its taste and texture begin to wane. Figure 8-17 shows the computer computations for a 99% confidence interval for the mean shelf life. Verify the interval results from the descriptive statistics. Interpret the interval.

35. A dream analyst asks 300 randomly selected clients to recall the number of characters in their latest dream. Figure 8-18 shows the results for a 90% confidence interval of the mean number of characters recalled. Interpret the result. Was a value from the T table or normal table used to compute this interval?

36. Figure 8-19 shows a 95% and a 98% confidence interval for the mean pollen count in a metropolitan area. Use the descriptive statistics to compute and verify the computer-generated intervals. Interpret the intervals.

Figure 8-17
Minitab output for baked goods' shelf life

```
           N      MEAN    STDEV   SE MEAN    99.0 PERCENT C.I.
SHELFLIF  230    4.3022   0.3463   0.0228    ( 4.2429,  4.3615)
```

Figure 8-18
Minitab output for recalled dream characters

```
           N      MEAN    STDEV   SE MEAN    90.0 PERCENT C.I.
CHARACT   300    3.593    2.005    0.116     ( 3.402,   3.784)
```

Figure 8-19
Minitab output for pollen counts

```
           N      MEAN    STDEV   SE MEAN    95.0 PERCENT C.I.
POLLEN     55    3.273    2.392    0.323     ( 2.626,   3.919)

           N      MEAN    STDEV   SE MEAN    98.0 PERCENT C.I.
POLLEN     55    3.273    2.392    0.323     ( 2.499,   4.046)
```

Figure 8-20
Minitab output for mortgage rates

```
           N      MEAN    STDEV   SE MEAN    95.0 PERCENT C.I.
INTRATES 2000    8.5237   0.4961   0.0111    ( 8.5019,  8.5454)

THE ASSUMED SIGMA =0.510

           N      MEAN    STDEV   SE MEAN    95.0 PERCENT C.I.
INTRATES 2000    8.5237   0.4961   0.0114    ( 8.5013,  8.5461)
```

37. Figure 8-20 shows two 95% confidence intervals for the mean of 2,000 banks' mortgage rates on a specific day. The first uses the sample information entirely, and the second assumes a standard deviation of 0.51 percentage points. The 0.51 is a value that has been stable for many weeks while the mean varied, because the surveyed banks tend to change their rates by about the same amount each week.
   a. Does the evidence in the printout concur with the assumption about the population standard deviation?
   b. Interpret each of the confidence intervals.
   c. Can we state with 95% confidence that the average rate is lower than a week before when it was 9.2%? Explain.
   d. Both intervals are 95% confidence intervals formed from 2,000 observations. The first interval uses the sample standard deviation 0.4961, and the second uses the larger specified value 0.51. Both are essentially the same value. What role do t and z play in making one interval wider than the other?

Summary and Review

Because $\overline{X}$ satisfies various statistical criteria, it is considered a "good" estimator of μ. One of these criteria is lack of bias, which means the expected value of $\overline{X}$, $E(\overline{X})$, is the population mean, μ, or the average of the $\overline{X}$ values from random samples of size n is μ. Another criterion is consistency, which means that the likelihood of obtaining an $\overline{X}$ near μ increases as the sample size increases.

We can report inferences or estimates as intervals rather than points. An interval incorporates and displays more of the uncertainty involved in the estimation process than simply reporting a single value with no information about its reliability.

When the population standard deviation, σ, is known the $100(1 - \alpha)\%$ confidence interval is $\overline{X} \pm z\sigma_{\overline{x}}$ if the sample is large (usually $n > 30$) or if the population is normal. $100(1 - \alpha)\%$ of an infinite set of $100(1 - \alpha)\%$ confidence intervals of a particular population mean constructed from this formula will include μ. The risk that an interval does not include μ is $100\alpha\%$.

There is a trade-off when we possess less information from the population to construct a confidence interval for μ. When the population standard deviation, σ, is unknown, the procedure requires additional information about the population. The population must be normally distributed (or symmetric and unimodal). In this case, we substitute the sample standard deviation, S, for σ to estimate $\sigma_{\bar{x}}$ and use the formula $\bar{X} \pm t\hat{\sigma}_{\bar{x}}$.

The T distribution is symmetric but flatter than the standard normal. As the degrees of freedom increase, it approaches a standard normal. For this reason, when the sample is large ($n > 30$), we often employ the z value to approximate the t when we construct a confidence interval. Because an unknown σ is the usual situation we confront, sample size usually determines if we use a z or t; z for confidence intervals from large samples, and t when the sample is small.

If $1 - \alpha$ changes and everything else ($\bar{X}$, σ or S, and n) remains constant, the length of the confidence interval changes in the same direction as $1 - \alpha$. Because $100(1 - \alpha)\%$ represents the level of confidence—in order to be more confident the interval contains μ—the interval should be longer or contain more values. We can also demonstrate this fact mathematically.

If n changes and everything else ($\bar{X}$, σ or S, and $1 - \alpha$) remains constant, the length of the interval changes in the opposite direction from n. The sample size n represents the amount of information collected from the population. With more information (and the same $\bar{X}$, σ or S, and $1 - \alpha$), the interval of possible values containing μ shortens. In effect, the estimator is zeroing in on μ.

Box 8-1 provides an interesting practical application of confidence intervals for the effects of secondary cigarette smoke. The statistic is different, but the interpretation of the confidence interval follows what we do for the mean.

Box 8-1 Statistics and Nonsmokers' Exposure to Smoke

Controversy continues over the connection between exposure to other people's smoke and lung cancer. Assessing the risks is difficult, because nonsmokers inhale a different physical substance with different chemical properties than the smoke inhaled by the smoker (called mainstream smoke). Environmental tobacco smoke (ETS) is a diluted combination of sidestream smoke, smoke from the burning end of the cigarette between puffs, and exhaled mainstream smoke.

Measuring mainstream smoke's qualities and makeup is easy, because it leaves the cigarette and enters the smoker directly. The characteristics of ETS are more difficult to capture, especially the exact exposure level of an experimental subject. Instead, the number of active smokers in a household usually proxies as a measure of the ETS exposure for the nonsmokers in the household. For example, a father or mother who smokes exposes a child to ETS less than if both parents smoke.

The authors of an article present the results of 30 such studies. The authors of each of the 30 studies estimated the effect of ETS with the average relative risk of lung cancer for ETS-exposed nonsmokers. Relative risk is the ratio of the incidence of lung cancer among ETS-exposed nonsmokers to the incidence among an unexposed control group. A ratio of 1 implies the same risks for both groups, less than 1 implies greater risk for the unexposed group, and larger than 1 implies greater risk for the exposed group.

In addition, the authors consider a ratio that is not much larger than 1 (generally between 1 and 3) to be *weak evidence* of increased risk. Such values may differ from 1 when nonsampling errors occur or when other variables, such as occupational exposure and diet, affect the measurement, but the study does not account for these variables. *Strong evidence*, more convincing evidence that ETS is the culprit, is a ratio greater than 5.

Confidence intervals for the relative risk incorporate the uncertainty of each estimate when we compare the estimated ratio to the value 1, 3, or 5. These comparisons suggest the merit of further investigation to reveal whether nonsampling error and other intervening factors are responsible for the result.

The following table displays 35 confidence intervals from the 30 studies (some studies produced more than one estimate). The intervals on either side of the estimates are unequal, because the statistic is a ratio with an asymmetric sampling distribution. Twenty-eight intervals include the value 1 or smaller values, which creates suspicion that ETS does not cause lung cancer. The other seven intervals that include only values greater than 1—marked with an asterisk—provide statistical evidence that the ratio is larger than 1; that is, ETS and the incidence of lung cancer are related. Because these are 95% confidence intervals, there is a 5% chance we are wrong when we conclude that the ratio must be larger than 1 from any one of these seven intervals.

Statistically speaking, these seven confidence intervals that exclude the value 1 and smaller values are the strongest indication that the true ratio is greater than 1. However, these intervals may still include values in the *weak evidence* category (1–3) discussed by the authors. For instance, the confidence interval associated with study number 14 is entirely within the 1-to-3 specification. Thus, there may be reasons other than ETS exposure that increased the risks for the experimental group.

There is no *strong evidence* of an increased risk (a confidence interval with all its values greater than 5) according to these studies. Notice that the upper limits of seven confidence intervals (marked with a plus) surpass the value 5. Further analysis of these studies reveals that small sample size is one factor that accounts for the wide interval; the largest sample among these studies investigates only 18 cases. Others investigate only two or three. Thus, what

(continued on next page)

appears to be *strong evidence* or a reason to suspect that *strong evidence* exists is based on a small amount of information. The low volume of information provided in small samples enlarges standard errors and, consequently, expands the confidence interval. Indeed, all seven of these intervals include risk-ratio values smaller than one, which allows the possibility that the risk is as great or possibly greater for the control group.

The authors conclude, "No matter how the data from all of the epidemiological studies are manipulated, recalculated, "cooked," or "massaged," the risk from exposure to spousal smoking and lung cancer remains weak.... No matter how these data are analyzed, no one has reported a strong risk relationship for exposure to spousal smoking and lung cancer."

In summary, the matter is unsettled. Current research does not demonstrate or rule out a noncontroversial, substantial effect of ETS on the incidence of lung cancer among nonsmokers.

(What conclusions would we likely draw, if we only wanted to be 90% confident; that is, risk being wrong 10% of the time?)

Studies of ETS and Lung Cancer in Nonsmokers

Study Number	Year of Study	Sex of Nonsmoker	Number of Cases	Relative Risk	95% Confidence Interval
1	1982	F	34	0.75	(0.43, 1.30)
2	1983	F	38	2.13	(1.18, 3.83)*
3	1983	F	14	2.07	(0.81, 5.26)†
		M	2	1.97	(0.38, 10.29)†
4	1984	F	13	0.79	(0.25, 2.45)
		M	5	1.00	(0.20, 5.07)†
5	1984	F	33	0.80	(0.34, 1.81)
		M	5	0.51	(0.15, 1.74)
6	1985	F	92	1.12	(0.94, 1.60)
7	1985	F	29	1.20	(0.50, 3.30)
8	1986	F	73	1.52	(1.00, 2.5)
		M	3	2.10	(0.5, 5.6)†
9	1986	F	22	1.03	(0.37, 2.71)
		M	8	1.31	(0.38, 4.59)
10	1987	F	19	1.68	(0.39, 2.97)
11	1987	F	189	1.19	(0.6, 1.4)
12	1987	F	14	1.78	(0.6, 5.4)†
13	1987	F	51	1.55	(0.87, 3.09)
14	1987	F	115	1.65	(1.16, 2.35)*
15	1987	F	33	1.20	(0.70, 2.10)
16	1988	F	34	2.16	(1.03, 4.53)*
17	1988	F	18	2.55	(0.91, 7.10)†
18	1988	F	17	—	—
19	1988	F	37	2.01	(1.12, 1.83)*
20	1988	F	90	1.10	Not Available
21	1990	F	45	0.74	(0.32, 1.68)
22	1990	F	129	0.93	(0.55, 1.57)
23	1990	M	13	1.20	(0.54, 2.68)
		F	35	0.90	(0.46, 1.76)
24	1990	F	91	2.11	(1.09, 4.08)*
25	1990	F	64	0.94	(0.62, 1.40)
26	1990	F	17	1.20	(0.40, 2.90)
27	1990	F	205	0.7	(0.6, 0.9)
28	1981	F	88	1.17	(0.85, 1.89)
					(0.77, 1.61)
29	1984	F	6	1.00	(0.59, 17.85)†
		M	4	3.25	
30	1984b	F	163	1.45	(1.04, 2.02)*
	1984a		7	2.28	(1.19, 4.22)*

Legend: * Confidence interval includes only values greater than 1;
† confidence interval includes values greater than 5.

Note: Study Number 18 did not report an estimate, but did report results that would form a 95% confidence interval including values smaller than 1.

Source: Gary L. Huber, Robert E. Brockie, and Vijay K. Mahajan, "Passive Smoking: How Great a Hazard?" *Consumers' Research*, July 1991, pp. 10–15, 33–34, with permission.

Multiple-Choice Problems

38. Which of the following is *not* true about a confidence interval for μ?
  a. It is a way to report a statistical inference that does not imply that a single estimated value is the true value.
  b. The size changes as the components (σ or S, $1 - \alpha$, and n) change values.
  c. The length changes as $\overline{X}$ changes.
  d. It is possible that the interval estimate might miss μ.
  e. It is centered on $\overline{X}$ but is expected to include μ.

39. Generally, what is the probability a sample mean will equal the population mean, $P(\overline{X} = \mu)$?
  a. $1 - \alpha$.
  b. Level of confidence.
  c. $1 - (1 - \alpha)$.
  d. 0.
  e. 1.

40. An unbiased estimator
  a. Always equals the population parameter being estimated.
  b. Is the ideal estimator but impractical to use.
  c. Impossible to calculate.
  d. On average equals the population parameter being estimated.
  e. Results from a random sample.

41. Which of the diagrams in Figure 8-21 portrays an unbiased estimator ($\hat{\Theta}$) for the parameter, Θ?
  a. A.   b. B.   c. C.   d. D.   e. E.

42. We use the normal distribution to find z in the formula $\overline{X} \pm z\hat{\sigma}_{\overline{x}}$. Which of the following conditions determines whether we can use the normal distribution or the T distribution?
  a. The sample contains more than 5% of the population observations.
  b. $n > 30$.
  c. The population is finite.
  d. We are estimating μ.
  e. We are sampling with replacement.

43. A study found that the average level of caffeine use among creative individuals (artists, writers, and musicians) over 26 years of age is 1.962 based on a scale from 0 to 3, where 0 represents none, 1 occasionally, 2 moderately, and 3 heavily. The analyst chose each type of creative individual from a different population, such as participants of the Iowa Writers Workshop who had published works or artists currently displaying their work in the Chicago area. Assume the sample of 31 artists is a random sample of all artists over 26 years of age, and find a 95% confidence interval for the mean response for caffeine usage. The sample standard deviation was 0.895. (Source: Barbara Kerr, Jeff Shaffer, Cindi Chambers, and Kirk Hallowell, "Substance Use of Creatively Talented Adults," *Journal of Creative Behavior*, Vol. 25, No. 2, 1991, pp. 145–53.)
  a. 1.962 ± 0.895.
  b. $1.962 \pm (1.96)(0.895/\sqrt{31})$.
  c. $1.962 \pm (1.96)(0.895)$.
  d. $1.962 \pm (1.645)(0.895/2)$.
  e. $1.962 \pm (1.645)(0.895)$.

44. A random sample of 49 prices of ski outfits is selected to determine the approximate mean price. The sample mean and standard deviation are $720 and $140, respectively. Find a 98% confidence interval for μ, the mean price of all ski outfits.
  a. ($680.8, $759.2).
  b. ($393.8, $1046.2).
  c. ($580, $860).
  d. ($673.4, $766.6).
  e. Information insufficient to compute.

45. A 95% confidence interval for μ is (550, 650), which means that
  a. $P(550 < \mu < 650) = 0.95$.
  b. There is a 95% chance μ falls between 550 and 650.

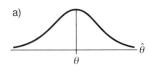
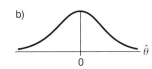
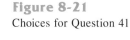

Figure 8-21
Choices for Question 41

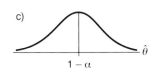
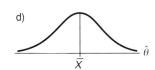

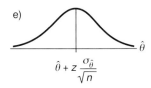

c. 95% of the time the procedure provides intervals that contain μ.
d. $P(550 < \overline{X} < 650) = 0.95$.
e. There is a 5% chance that $\overline{X}$ is not included in the interval.

46. Suppose you decrease the level of confidence without changing any value from the population or sample. Then,
a. You cannot change the level of confidence without changing anything else in the population or sample.
b. The confidence interval expands.
c. The confidence interval is less likely to include μ.
d. You will not need as many elements in your sample.
e. You can substitute S for σ.

47. As n increases,
a. The standard error of the mean decreases.
b. The likelihood that a random $\overline{X}$ is within two units of μ decreases.
c. The confidence interval expands.
d. We have obtained less information about the population.
e. The normality of the $\overline{X}$ values may disappear.

48. A random sample of nine times taken to complete a chore has a mean of 20 minutes. All times spent to complete the chore are normally distributed with a standard deviation of 12 minutes. Find a 99% confidence interval for μ.
a. (19.8 minutes, 20.2 minutes).
b. (11 minutes, 29 minutes).
c. (8 minutes, 32 minutes).
d. (9.7 minutes, 30.3 minutes).
e. Cannot be done with $n < 30$.

49. One reason the sample mean is the usual estimator for the population mean is that
a. It equals the population mean.
b. Its standard error is the same as the population standard deviation.
c. The average of all possible sample means is the population mean.
d. It is unaffected by extreme values.
e. It occurs more often than the median or mode.

50. Find $P(T_{(27)} > 2.771)$.
a. 0.995.
b. 0.495.
c. 0.005.
d. 0.505.
e. 0.490.

51. Find t such that $P(-t < T_{(11)} < t) = 0.8$.
a. 2.201.
b. 1.796.
c. 1.363.
d. 2.718.
e. 0.697.

52. If you use the T probability distribution with 16 degrees of freedom to form a confidence interval for μ, how many observations should be in the sample?
a. 16.   d. 18.
b. 17.   e. More than 30.
c. 15.

53. Which of the following is *not* a condition for using the T table to form a confidence interval for μ?
a. Normal population.
b. Symmetric and unimodal population.
c. $n > 30$.
d. S substituted for σ.

Use the following information to answer Questions 54–56.

To study the ability of displaced spiders to return to their nests, a biologist designed an experiment with female crab spiders. First, he placed the females on milkweed leaves shortly before they laid their eggs. Then, he displaced each spider in a series of three moves at specified time intervals that did not allow the spiders an opportunity to lay a thread to guide them back to the nest. The first move situated each spider at the base of the stem containing its nest leaf, the second move was identical to the first, but it occurred a day after the first. The third and final move placed the spiders on the ground nearer the nest plant than any other milkweed plant. Fourteen spiders averaged returning from the first move in 40.0 minutes with a standard deviation of 32.4 minutes. Fourteen spiders averaged a 27.6 minutes return in the second instance, with a standard deviation of 23.6 minutes. Finally, five returned from the third move in an average 75.8 minute span with a standard deviation of 42.3 minutes. (Source: Douglass H. Morse, "Homing by Crab Spiders *Misumena Vatia* (Araneae, Thomisidae) Separated from Their Nests," *The Journal of Arachnology*, Vol. 19, No. 2, pp. 111–14.)

54. Without calculating confidence intervals for the average return time for each move, predict which interval or intervals will be the longest.
a. The first and second move intervals, because they are based on the same number of observations, which is more than the third.
b. The third, because its mean is the largest.
c. The third, because standard deviation is the largest.
d. The third, because its standard deviation is the largest and its sample size is smallest, making its t value and its standard error the largest.
e. The third, because its mean and standard deviation are both larger than the others and, because its sample is smaller, which makes its t value the largest.

55. Find a 95% confidence interval for the average time for the first move. Assume all times for the first move are normally distributed.
a. (7.6 minutes, 72.4 minutes).
b. (21.30 minutes, 58.70 minutes).
c. (21.43 minutes, 58.57 minutes).
d. (23.03 minutes, 56.97 minutes).
e. (−32.4 minutes, 32.4 minutes).

56. Find an 80% confidence interval for the average time for the third move, by assuming that all times for the third move are normally distributed.
a. (46.80 minutes, 104.80 minutes).
b. (33.50 minutes, 118.10 minutes).
c. (56.88 minutes, 94.72 minutes).
d. (47.88 minutes, 103.72 minutes).
e. (24.21 minutes, 100.01 minutes).

Multiple-Choice Problems

57. A random sample of four watermelons is selected from a large shipment to determine the mean weight. The weights are 16, 10, 14, and 20 pounds. Find a 90% confidence interval for the mean weight of all watermelons in the shipment. (Assume the weights of all watermelons in the shipment are normally distributed.)
  a. (10.1 lbs., 19.9 lbs.).   **d.** (11.8 lbs., 18.2 lbs.).
  b. (10.6 lbs., 19.4 lbs.).   **e.** (11.6 lbs., 18.4 lbs.).
  c. (5.2 lbs., 24.8 lbs.).

58. A random sample of 65 observations is selected from a normal population. The sample mean and standard deviation are 291 and 83, respectively. If you want to form a 95% confidence interval for μ, which probability distribution would you use?
  a. Use the binomial distribution.
  b. Use the T distribution.
  c. Use the standard normal distribution or Z distribution.
  d. Z and T are both acceptable unless exactness is required.
  e. The conditions do not fit any of the usual situations, so the distribution is unknown.

59. A confidence interval formed with the T distribution will be wider than one based on the Z distribution.
  a. Because the use of the T distribution implies a situation in which we are working with less population information than when we use Z.
  b. Except when n is very large.
  c. Because the T curve appears standard normal except flatter with more area in the tails.
  d. When the sample is small.
  e. All of the above.

60. Figure 8-22 displays output regarding the College variable, the number of college dorm residents in a county, from the North Carolina Census Data Base (Appendix E). The printout includes a confidence interval calculated from a random sample of four observations selected from the 100 values that consists of the values; 0, 1644, 0, 591. Which of the following is a problem with the confidence interval?
  a. The estimated standard error is half of the sample standard deviation.
  b. The sample mean is incorrect.
  c. The confidence interval does not contain the population mean.
  d. The population is not normal.
  e. The population standard deviation is unknown.

	N	MEAN	MEDIAN	TRMEAN	STDEV	SEMEAN
COLLEGE	100	713	0	382	1765	176

	MIN	MAX	Q1	Q3
COLLEGE	0	8481	0	593

Figure 8-22
Minitab output for College variable from the North Carolina Census Data Base (Appendix E)

```
Histogram of COLLEGE    N = 100
Each * represents 2 obs.

Midpoint     Count
       0        72    ************************************
    1000        17    *********
    2000         2    *
    3000         1    *
    4000         3    **
    5000         0
    6000         2    *
    7000         0
    8000         3    **
```

	N	MEAN	STDEV	SE MEAN	95.0 PERCENT C.I.
SAMPLE	4	558.750	775.287	387.644	(-674.906, 1792.406)

Word and Thought Problems

61. What is a point estimate? What would you propose as estimators for the following parameters: standard deviation, proportion of successes, difference in means of two different populations? Explain your choices. What might a statistician look at in evaluating your choices?

62. What is a confidence interval? Interpret the level of confidence. Determine and interpret the risk level of a confidence interval estimate if the level of confidence is 0.8.

63. a. Find a 95% confidence interval for μ if a random sample of 75 elements (selected from a population with a standard deviation of 5) has a mean of 63.
  b. Find a 90% confidence interval, using the information in Part a.
  c. Find an 80% confidence interval, using the information in Part a.
  d. Use the results of Parts a–c to generalize about the effect of the confidence level on the size of a confidence interval. What values in the formula do not change between parts?

64. a. Find a 98% confidence interval for μ if a random sample of 600 elements (selected from a population with a standard deviation of 10) has a mean of 200.
  b. Find a 98% confidence interval using the information in Part a, but change n to 400.
  c. Find a 98% confidence interval using the information in Part a, but change n to 200.
  d. Use the results of Parts a–c to generalize about the effect of the sample size on the size of a confidence interval. What values in the formula do not change between parts?

65. A lawn mower manufacturer wants to determine how often its mower's air filter should be cleaned for maximum performance. Fifty mowers are randomly selected from the production line and subjected to "normal" mowing conditions. The air filters are constantly monitored to determine when the amount and quality of the air passing through the filter have reached certain limits. The mowing times to reach this limit are recorded. The mean time for the 50 mowers is 25 hours. The manufacturer has found a standard deviation of 3.5 hours for mowing times for similar mowers in earlier studies.
  a. Find a 95% confidence interval estimate for the mean mowing time between air filter cleanings.
  b. What is the population in this problem?

66. A paint company intends to advise its customers about drying time for a particular brand of paint. A random sample of 45 cans yields a mean drying time of 2.2 hours.
  a. What other information do you need to calculate a 95% confidence interval for the mean drying time?
  b. Suppose the missing number is 0.2. Now calculate the confidence interval.
  c. What other information would you want to know if you were a customer trying to estimate the drying time?

67. A dairy farmer wants to determine the mean number of artificial inseminations necessary to impregnate a cow. A random sample of ten cows is taken from the herd. The mean number of inseminations for these cows is 3.6 with a standard deviation of 0.55. Assume the number of necessary inseminations for all cows is a symmetric, unimodal population. Find a 90% confidence interval for the mean number of inseminations.

68. Compare confidence interval lengths for two estimates where the only difference between the two situations is the value of the population standard deviation.
  a. What happens to the size of the confidence interval as the standard deviation decreases?
  b. Give an intuitive explanation and a mathematical explanation for your answer to Part a.
  c. As a practical matter, suggest ways for decreasing the standard deviation. Apply your suggestions to Problems 65, 66, or 67.

69. What happens to the curve for a T distribution as the degrees of freedom increase? Explain how this change affects the choice of z or t for a confidence interval.

70. Use your random sample of 10 values from the North Carolina Nursing Home data selected for Problem 40 in Chapter 7, Unit I (p. 306). Estimate an 80% confidence interval for the mean of all 100 values, assuming all 100 values are normally distributed. Does your interval contain the true population mean, 470.1 persons?

71. An auditor randomly selects 20 invoices and finds taxes have been miscalculated on some. The mean is $12 underpaid with a standard deviation of $3. Find a 90% confidence interval for the mean of the mischarges. Assume mischarges are normally distributed.

72. A study to examine caffeine usage among creative individuals used a sample consisting of 4 writers, 22 artists, and 4 musicians, all 26 years old or younger. Each described his or her usage level with a value between 0 and 3, where 0 represents none, 1 occasional, 2 moderate, and 3 heavy. The mean response for the entire sample was 2.173, and the standard deviation was 0.765. Find a 95% confidence interval for the mean response. (Source: Barbara Kerr, Jeff Shaffer, Cindi Chambers, and Kirk Hallowell, "Substance Use of Creatively Talented Adults," *Journal of Creative Behavior*, Vol. 25, No. 2, 1991, pp. 145–53.)

73. An estimator of the population variance, σ^2, is the sample variance, $S^2 = \Sigma(X - \overline{X})^2/(n-1)$. However, if the denominator of this expression is n, rather than $n-1$, the expected value of the newly defined $\tilde{S}^2$ is $((n-1)/n)\sigma^2$.
  a. Consider the factor $(n-1)/n$ and explain how the new $\tilde{S}^2$ performs on average when estimating σ^2 (overestimates, underestimates, or exact estimates of σ^2).
  b. The expected value of the usual S^2 (the one with $n-1$ in the denominator) is σ^2. What term describes estimators with this property?

Word and Thought Problems

74. A random sample of 20 stolen automobiles has a mean value of $9,268 with a standard deviation of $1,651. Assuming the values of all stolen automobiles are normally distributed, find a 98% confidence interval for the mean value of a stolen car.

75. A company selects a random sample of razor blades to learn how many smooth shaves can be performed before each blade becomes too dull to use. The results are 3.5, 1, 5.25, 3, 8, 6.5, 4, 12.25, 3.75 shaves. Find a 95% confidence interval for the mean number of shaves per blade. Assume the number of shaves of all blades is normally distributed.

76. Find an 80% confidence interval for the average streak in college basketball by using the following random sample of streaks: 1, 1, 5, 2, 7, 1, 1, 1, 3. A *streak* is the number of consecutive games with identical outcomes (win, loss, or tie). Assume that the distribution of all streak values is symmetric and unimodal.

77. Preliminary sample results from 72 interviews of business firms yielded the following summary statistics regarding the time (in months) that passes between a significant change in market conditions and a subsequent price change.

Change in Market Conditions	Mean Lag	Standard Deviation
Increase in demand	3.23	2.93
Decrease in demand	3.60	3.93
Increase in costs	3.17	2.90
Decrease in costs	3.97	4.47

Source: Alan S. Blinder, "Why Are Prices Sticky? Preliminary Results from an Interview Study," *American Economic Review*, Vol. 81, No. 2, May 1991, p. 93, with permission.

   a. Use this information to construct 95% confidence intervals for the mean lags for all U.S. firms for each market condition.
   b. Using results from Part a, determine which event or events appear likely to result in a price change in less than four months.
   c. Which event's lag is most unpredictable?

78. A random sample of 1,000 visitors to a national park showed that the mean distance traveled to visit the park was 588 miles with a standard deviation of 154 miles. Find a 99% confidence interval estimate of the mean distance traveled. What is the population in this problem?

79. The commuting times for five randomly selected workers in a company are 45, 15, 22, 86, and 31 minutes. Assume commuting times for employees of this company are normally distributed. Find a 95% confidence interval for the mean commuting time.

80. To estimate the average number of young animals per herd of an endangered species, a biologist randomly selects sites in a region and counts the young members of herds sighted near the chosen locations. The mean from 25 sites is 10.6, with a standard deviation of 3.1. If the count of offspring is normally distributed, find an 80% confidence interval for the mean number of young per herd.

81. Sara's speech class is making five-minute speeches. There is one speaker ahead of Sara and 12 minutes remain in the class period. For the eight speeches delivered so far, the mean time from the start of one speech until the start of the next is 5.47 minutes with a standard deviation of 0.09 minutes. Assume these eight times are a random sample from a normal population. Construct an 80% confidence interval for the mean time. Is it likely that there will be sufficient time for Sara to deliver her entire speech during this class period?

82. A random sample of 18 of Jason's video game scores has a mean of 10,005 with a standard deviation of 4,968. If his scores are normally distributed, find a 95% confidence interval for his mean score.

83. Figure 8-23 displays a Minitab computed confidence interval based on the 100 Military values (number of county residents living in military quarters) in the North Carolina Census Data Base (Appendix E) as a sample of such values for all points in time. The lower endpoint is a negative value. This seems unreasonable. Explain why. Does a negative value signal a mistaken calculation?

84. Figure 8-24 illustrates several confidence intervals of the mean size of library staffs based on a random sample selected from staff sizes of 106 research libraries in the United States and Canada. The sample values are

      204   206   608   143   596

Each row of output is labeled A through F, and the level of confidence is removed. Study the intervals, then explain the direction of change in the confidence level as one progresses from A to F that causes the changes in the intervals. (Source: Association of Research Libraries, as reported in *The Chronicle of Higher Education: Almanac*, August 28 1991, p. 37.)

85. Professor Robinson wants to schedule her time in order to finish five projects during this week. A random sample of 14 of her projects has a mean time per project of 7.45 hours, with a standard deviation of 2.06 hours. If all project times are normally distributed, find a 99% confidence interval for the mean project time.

	N	MEAN	STDEV	SE MEAN	95.0 PERCENT C.I.
MILITARY	100	583.780	3786.214	378.621	(-167.655, 1335.215)

Figure 8-23

Minitab output of confidence interval from Military variable from the North Carolina Census Data Base (Appendix E)

Figure 8-24
Minitab output for sample from 106 research libraries in the United States and Canada

```
                                                        A
              N      MEAN      STDEV    SE MEAN    CONFIDENCE INTERVAL
LIBRARY       5    351.400    230.202   102.949    ( 65.485, 637.315)


                                                        B
              N      MEAN      STDEV    SE MEAN    CONFIDENCE INTERVAL
LIBRARY       5    351.400    230.202   102.949    ( 111.159, 591.641)


                                                        C
              N      MEAN      STDEV    SE MEAN    CONFIDENCE INTERVAL
LIBRARY       5    351.400    230.202   102.949    ( 131.862, 570.938)


                                                        D
              N      MEAN      STDEV    SE MEAN    CONFIDENCE INTERVAL
LIBRARY       5    351.400    230.202   102.949    ( 193.529, 570.271)


                                                        E
              N      MEAN      STDEV    SE MEAN    CONFIDENCE INTERVAL
LIBRARY       5    351.400    230.202   102.949    ( 228.934, 473.866)


                                                        F
              N      MEAN      STDEV    SE MEAN    CONFIDENCE INTERVAL
LIBRARY       5    351.400    230.202   102.949    ( 254.549, 448.251)
```

Unit II: Errors of Point Estimates of the Population Mean, μ

Preparation-Check Questions

86. If we calculate $\overline{X}$ values from random samples with $n > 30$, what proportion of these values will be within 1.5 standard errors of μ?
  a. 0.1336.   **d.** 0.8664.
  b. 0.1064.   **e.** 0.5668.
  c. 0.2128.

87. A random sample of 625 citizens of a large town is selected to determine the mean level of education. The sample mean is 13.75 years, and the standard deviation is 0.75 year. Find a 98% confidence interval for the mean level of education of the citizens.
  a. (13.000 years, 14.500 years).
  b. (13.680 years, 13.820 years).
  c. (12.000 years, 15.500 years).
  d. (13.740 years, 13.760 years).
  e. (12.750 years, 14.750 years).

88. If we use information from the same sample to compute several confidence intervals at different confidence levels, what happens as the confidence level increases?
  a. The standard error increases.
  b. The standard error decreases.
  c. The width of the confidence interval increases.
  d. The width of the confidence interval decreases.
  e. Nothing changes.

Answers: 86. d; 87. b; 88. c.

Introduction

Heat Cost Recall that Susan wants to obtain data on heating bills from the local utility company in order to estimate the mean household heating bill for a new heating method. Instead, she finds that the utility company already conducted an earlier study and that the mean determined from the study is $92.35. The utility company study is based on a sample, which means that there is uncertainty associated with the estimate, $92.35. What are some questions she might want to ask about the study? Here are some possibilities, based on previous chapters.

1. How was the sample selected?
2. How many observations were in the sample?
3. What is the 95% confidence interval for the mean bill?
4. What value does the study use for the population standard deviation, σ, in the confidence interval?

If the researcher conducted the study properly, all of these questions are answerable. In addition, when anyone reports a point estimate of a mean, we should expect or request the following information:

1. The method of sample selection (including a description of the population being sampled and the time of the sampling) and the number of observations in the sample.
2. The possible error; that is, the distance between the estimate and the true value being estimated.
3. The confidence or level of uncertainty associated with the estimate and its error.

Understanding such information when someone presents it to us and calculating it to present to someone else are the subjects of this chapter.

8.6 Sampling Error versus Nonsampling Error

Generally, random sampling can produce numerous $\overline{X}$ values from the same population. Consequently, we do not expect the estimate, based on incomplete information, to equal the value that incorporates the entire population. But we do employ procedures that increase the likelihood of the estimate being close.

We call this difference between an estimate and the parameter being estimated that naturally results from sampling the *sampling error*. When we estimate the population mean with the sample mean, this error is $\overline{X} - \mu$.

Question: If we estimate the population standard deviation, σ, with the sample standard deviation, S, express the sampling error using symbols.
Answer: $S - \sigma$.

A *nonsampling error* is any difference between the parameter and the estimate that is not attributable to sampling. For instance, measurement error is a source of nonsampling error. Measurement errors result from events such as an interviewee giving the wrong information (knowingly or unknowingly) to an interviewer, an interviewer incorrectly recording a response, a clerk mistranscribing responses, and an analyst mistabulating responses when summarizing them in reports. Some nonsampling errors occur because of nonresponse, such as when people do not return mail surveys or do not respond to all of the questions. In such cases, the resulting set of sample information may not be a random sample from the desired population. Sometimes questions are not phrased well enough to elicit the desired information, or they are phrased to elicit a

Box 8-2 Nonsampling Errors

The sampling errors we explore in this chapter occur because we employ only a sample of the population information to describe a population characteristic. Nonsampling errors originate from other sources of mistaken characterizations of the population. Sources of such errors and examples are listed below (the numbers in parentheses refer to the references listed at the end of this box).

1. *Elicited responses.* Box 7-2 discusses leading questions, those framed to elicit a certain response. Other questions are not likely to elicit honest responses, such as admitting personal behaviors that would embarrass the respondent, for example, one's birth control method or frequency of use. Instances where analysts specify a limited set of choices as responses may also be misleading. For example, a poll requested students to specify which clothes would be preferred during a specified year. The result showed that 90% specified a certain brand of jeans, which happened to be the only jeans on the list (2). Similar misleading results can occur on polls that ask voters about a political party's best candidate before the primaries and candidates' entry announcements take place. If questionnaires do not allow respondents to specify their choice, a person on the list may appear more popular than is the case. This can occur even when there is a blank for "other," if respondents tend to choose from listed candidates without thinking beyond the list.

2. *Nonrandom or nonrepresentative sample.* Generalizations to populations that are not well represented in the sample can misdirect readers and viewers. A diet-products company inferred that people on diets can maintain the weight loss from a study of 20 graduates of its program who also back the products in company advertisements (2). Phone surveys that request respondents to phone a 900 area code number are likely to overrepresent well-off zealots in the general population. One such survey resulted in 5,640 calls (about 72%) from offices owned by a single individual out of 7,800 calls (2). A famous incident in statistics lore is the 1936 Literary Guild survey of its subscribers, car-owners, and those with telephones in their homes to predict the outcome of the presidential election. These respondents tended to be wealthy and to favor the Republican candidate, Alf Landon, unlike the majority of the voters who elected Franklin Roosevelt.

3. *Sample responses not indicative of real-world responses.* The problem with surveying intentions that may, and often do, differ from subsequent behavior is well-known in statistics. A respondent may think well of a proposed new product but fail to purchase it when the opportunity arrives. Similar anomalies occur when favorable results from tests of advertising copy (tests of consumer recall/persuasion for a product) do not precede increases of sales of the advertised products (4). Other renowned clashes between survey and actual results include Coca-Cola taste tests for its New Coke (1) and consumer preference for pink soap in tests but not in stores (3).

4. *Subjective employment of statistical techniques and the results reported.* Any analyst faces the choice of the appropriate statistical technique from many possibilities in a given situation. In addition, there may be latitude in assumptions and specifications for any technique that will affect the results. A predisposed reporter may select favorable results rather than objective findings (perhaps all findings, potential findings, or flaws should be part of the report). These choices may dominate studies commissioned by a particular side of some debate, such as the effect of disposable and cloth diapers on the environment or the effect of breathing undisturbed asbestos in a building (2).

5. *Timing changes.* Respondents' awareness and feelings regarding an issue change over time, so the timing of the poll is important for evaluating the results. The same question may depict a very different electorate before and after a stunning primary election result.

Several of these examples paint a poor picture for trusting survey and poll evidence. Obviously, some people do misuse, even abuse, numbers (2). Whether their work is really statistics is questionable. Honest mistakes do occur. In any event, the source of the survey is a reasonable question to consider when evaluating the findings. In fact, these examples emphasize the need for more information about the study before automatically accepting poll results, even when the study reports a margin of error and its associated confidence level.

Sources: 1. Ronald Alsop, "Coke's Flip-Flop Underscores Risks of Consumer Taste Tests," *Wall Street Journal,* July 18, 1985, p. 23.
2. Cynthia Crossen, "Studies Galore Support Products and Positions, but Are They Reliable?" *Wall Street Journal,* November 14, 1991, pp. A1, A8.
3. John Koten, "You Aren't Paranoid If You Feel Someone Eyes You Constantly," *Wall Street Journal,* March 29, 1985, pp. 1, 22.
4. Joanne Lipman, "Research Tactic Misses the Big Question: Why?" *Wall Street Journal,* November 4, 1991, pp. B1, B6.

---

particular response. Good survey methods must be used in order to avoid nonsampling error. Box 8-2 provides more details and examples about nonsampling errors.

From now on in this book, we will assume that any error is a sampling error, not a nonsampling error. However, the result of any statistical inference in real life may be less accurate than described because of nonsampling errors, and we should always be alert to such problems.

8.7 Components of Sampling-Error Assessment

When we report a point estimate rather than a confidence interval, we use the sampling error to reflect the accuracy and the uncertainty of the estimate. The sampling error is also useful when we determine the number of observations to include in a sample. In discussing the sampling error (shortened to *error*) for $\overline{X}$ values, we limit the discussion to large samples ($n > 30$) and focus on three topics:

1. The maximum error, E.
2. The appropriate sample size, n.
3. The confidence that error assertions are correct, $100(1 - \alpha)\%$.

8.7.1 The Maximum Error, E

When we explore a particular population, its mean, μ, is fixed. However, each value of $\overline{X}$ creates a sampling-error value. So the behavior of the sampling error depends on the sampling distribution of $\overline{X}$. Because our discussion focuses on large samples, the normal distribution will adequately describe this sampling distribution. Figure 8-25 illustrates this distribution and shows that the sampling error is the distance along the axis between a particular $\overline{X}$ and μ.

Question: If $\overline{X}$ happens to be the value A (or B) in the figure, what is the magnitude of the sampling error in estimated standard errors ($\hat{\sigma}_{\overline{x}}$)? (Ignore the sign of the error.)
Answer: The distance from $\overline{X}$ to μ would be approximately $z\hat{\sigma}_{\overline{x}}$.

This is the extreme. About $100(1 - \alpha)\%$ of the $\overline{X}$ values are closer to μ. Then, the error will be smaller than $z\hat{\sigma}_{\overline{x}}$, as shown for $\overline{X} = C$ in Figure 8-25. Thus, $1 - \alpha$ is the probability that the sampling error, $\overline{X} - \mu$, is at most $z\hat{\sigma}_{\overline{x}}$. Notice that the sign of the error does not matter, only the distance.

We can write the last statement using the symbol E for the maximum error.

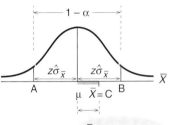

Figure 8-25
Sampling errors and sampling distribution of $\overline{X}$ for a large sample

---

The maximum possible sampling error of an estimate of μ is

$$E = \text{Maximum error} = z\hat{\sigma}_{\overline{x}}, \quad \text{where } \hat{\sigma}_{\overline{x}} = S/\sqrt{n}.$$

The confidence level attached to this statement is $100(1 - \alpha)\%$.

■ **Example 8-11:** A school board uses a random sample of 36 second-graders to determine the mean age of second-graders in the school system. They use the sample mean, 7.38 years, as their estimate of μ.

Question: Do we have enough information to comment about the potential size of the error of their estimate for the mean with 95% confidence?
Answer: No, we need S (or σ) to solve $E = z\hat{\sigma}_{\overline{x}} = z(S/\sqrt{n})$. An estimate of σ from a previous study or the sample standard deviation, S, from the school board's sample would enable us to continue.

Suppose they calculate S from the 36 observations to be 0.48 year. Now what can they say with 95% confidence about the size of the error?

■ **Solution:** $n = 36$ and $S = 0.48$, so $\hat{\sigma}_{\overline{x}} = 0.48/\sqrt{36} = 0.08$. We want to be 95% confident, so we find a value of z such that the middle 95% of the $\overline{X}$ values will be within z standard errors of μ, $P(-z < Z < z) = 0.95$. For a 95% level of confidence, $z = 1.96$. Finally,

$$E = 1.96\hat{\sigma}_{\overline{x}} = 1.96(0.08) = 0.157 \text{ year}.$$

Thus, the board can be 95% confident that the error of their estimate will be at most 0.157 year if they use 7.38 years for the estimate of the mean age. Alternatively, they could say they are 95% confident that the sample mean is within 0.157 year of the true mean. Only 5% of the time should they obtain a larger error if they use this estimation procedure: the risk of a larger error is only 5%.

Question: If we want to be more sure about the error statement, say 98% rather than 95%, what will happen to the size of the maximum error, E? Give an intuitive explanation for the answer.

Answer: The maximum possible error, E, will increase. If we want to be more confident (that is, take less risk), we claim that the error may be a larger value.

Question: Now illustrate this prediction mathematically by finding E at the 98% confidence level and comparing the results with the maximum error claim with 95% confidence.

Answer: $z = 2.33$ when the confidence level is 98%. S and n do not change, so $\hat{\sigma}_{\bar{x}}$ does not change from 0.08. Thus, $E = z\hat{\sigma}_{\bar{x}} = 2.33(0.08) = 0.186$ year. This value is larger than the 0.157 year obtained when we used a 95% confidence level.

Section 8.7.1 Problems

89. The mean number of overtime hours during the year for a random sample of 250 employees is 4.33 hours with a standard deviation of 1.92 hours. Estimate the mean number of overtime hours for all of the firm's workers, and comment on the possibility of error with 80% confidence.

90. Analysis of a random sample of 114 vials of a solution reveals that the average percentage of the solution consisting of inert ingredients is 14.58% with a standard deviation of 5.59 percentage points. What can be said with 90% confidence about the potential error, if the chemist uses 14.58% as an estimate of the mean percentage of inert ingredients in all such vials?

91. Some researchers wished to examine the relationship between the complexity of a person's consideration of vocations and the person's choice of an academic major. The tests to measure this complexity produced scores between 2 and 24. Higher scores indicated that a subject used more attributes of a vocation to evaluate different career opportunities and that a subject was able to differentiate more vocations. They divided 80 subjects into two groups. One group of 45 subjects received information about possible attributes for judging vocations. This group's mean score was 11.3 with a standard deviation of 3.8. Each of the 35 subjects in the other group determined his or her own attributes for judging. This group's mean score was 18.2 with a standard deviation of 3.7. (Source: John F. Leso and Greg J. Neimeyer, "Role of Gender and Construct Type in Vocational Complexity and Choice of Academic Major," *Journal of Counseling Psychology*, Vol. 38, No. 2, April 1991, pp. 182–88.)
   a. Assess the possible errors for estimates of the mean differentiation scores at the 90% confidence level for each of the two groups.
   b. Use the sample means and maximum errors to suggest whether or not the population means could be equal. Explain your reasoning.
   c. Predict whether the maximum error E will increase, decrease, or remain the same value if n increases for each sample, while the confidence level and sample statistics, $\overline{X}$ and S, do not change. Then, demonstrate your answer by recalculating E for each sample, but double the sample size in each case.

8.7.2 The Appropriate Sample Size, n

Now we want to know how many observations to include in a sample so that the potential error from an estimate will not exceed a specified value. Suppose the manager of a company wants the statistical division to estimate the mean customer purchase, μ, and be 99% sure that the estimate is within $10 of the true amount. How many observations do they need for the sample?

Solving $E = z(S/\sqrt{n})$ for n, we obtain

$$n = \left(\frac{zS}{E}\right)^2.$$

To find n with this formula, we need some information about a standard deviation, S (or σ), because we do not have a sample yet. There are several possible sources for a value to substitute for the standard deviation in the formula. First, previous studies may provide a good approximation of σ, depending on the similarity of the populations of the previous and present study. Another possibility is to collect a small sample and use its standard deviation in place of the unknown one. Then, after we collect the corresponding n observations, we can recalculate S as a check on the earlier value. (There are tests—beyond the scope of this book—to determine if the evidence indicates a significant difference between the two S values.) Finally, in some instances there are useful relationships between the range and the standard deviation. For example, we know almost all of the population values will be within three standard deviations of the mean if the population is normal or approximately so. In practice, this means that the range is likely to be about six standard deviations. Thus, if we know or approximate the range, we can estimate the standard deviation as one-sixth of the range.

■ **Example 8-12:** Suppose we want to estimate the mean number of credit hours taken by current students at a university. Any student can take as few as one credit hour or as many as 21 credit hours. Estimate the standard deviation of credit hours values. Assume that these values are normally distributed.

■ **Solution:** Because the values are normally distributed, the estimate of the standard deviation would be 3.33 hours = $(21 - 1)/6$.

Question: Suppose we know that some extreme population values are missing from the available data set. Perhaps we have only the middle 95% of the values. What would the estimate of the standard deviation be now?

Answer: If the population is normal or approximately normal, then the middle 95% of the observations fall within two (1.96 to be exact) standard deviations of μ. The range of the middle 95% of the values is about four standard deviations, so the estimate of σ would be one-fourth of the range.

In summary, three potential sources for an estimate of the standard deviation are

1. Values of S from previous studies or estimates.
2. S from a small preliminary sample taken from the population of interest.
3. One-sixth of the range of the population or one-fourth of the range of the middle 95% of the population values if the population is normal or approximately normal. (Problem 131 details a similar estimation procedure for σ based on Chebyshev's inequality, which we use when we doubt that the population is normal or approximately normal.)

Once we secure an estimate of the standard deviation, we can experiment with a few other similar or reasonable values to learn how they affect the required sample size, n.

■ **Example 8-13:** A biologist experimented with female crab spiders to learn about the behavior of spiders displaced from their nests. To prevent the spiders from laying guide threads in the vicinity of their nest, he put the females on nesting leaves shortly before they laid their eggs and moved them soon afterward. Three different moves presented the spiders with different challenges for recovering the nest leaf. The first two moves for each spider left them at the base of the stem with the nest leaf (the second move occurred a day after the first move), and the third left them on the ground nearer the nest plant than any other milkweed plant. Fourteen spiders averaged returning from the first move in 40.0 minutes with a standard deviation of 32.4 minutes. Fourteen spiders took an average of 27.6 minutes to return in the second instance, with a standard deviation of 23.6 minutes. Finally, five returned from the last move in an average 75.8 minute span with a standard deviation of 42.3 minutes. If the biologist treats this information as preliminary and wants eventually to be 95% confident that the estimated mean times are within 5 minutes of the true mean time, how many spiders should the sample include to satisfy this criterion for all three phases of the experiment? (Source: Douglass H. Morse, "Homing by Crab Spiders *Misumena Vatia* (Araneae, Thomisidae) Separated from Their Nests," *The Journal of Arachnology*, Vol. 19, No. 2, pp. 111–114.)

■ **Solution:** Solving for n for each move, we obtain

$$\text{First move: } n = \left(\frac{z\sigma}{E}\right)^2 = \left(\frac{1.96 \times 32.4}{5}\right)^2 = 161.31 \text{ spiders.}$$

$$\text{Second move: } n = \left(\frac{1.96 \times 23.6}{5}\right)^2 = 85.58 \text{ spiders.}$$

$$\text{Third move: } n = \left(\frac{1.96 \times 42.3}{5}\right)^2 = 274.95 \text{ spiders.}$$

Because n is the number of observations in the sample, we must round each answer to an integer value. We always round up, no matter what the fraction, in order to use enough observations to ensure the prior specifications (95% confident that the maximum error, E, is 5 minutes). Here the n values would be 162, 86, and 275 for each respective move in the experiment. However, the biologist wants to conduct the entire experiment with one sample that meets the requirements for all three cases. Obviously, 275 spiders are necessary.

■ **Example 8-14:** Back to determining the sample size in order to estimate the mean customer purchase. The statistics team wants to be 99% sure that the estimate is within $10 of the actual mean. Previous studies indicate that the standard deviation is $24.

a. Find the necessary sample size to perform this task.
b. Suppose the team restricts the limit on the error to $5 but still wants to be 99% confident. What size must the sample be?

■ **Solution: a.** $z = 2.575$, $S = 24$, and $E = 10$, so
$$n = \left(\frac{zS}{E}\right)^2 = \left(\frac{2.575 \times 24}{10}\right)^2 = 38.19 \text{ or } 39 \text{ customers.}$$
b. $z = 2.575$, $S = 24$, and $E = 5$, so
$$n = \left(\frac{zS}{E}\right)^2 = \left(\frac{2.575 \times 24}{5}\right)^2 = 152.77 \text{ or } 153 \text{ customers.}$$

When the limit on the error is $10, n needs to be 39 customers. If the limit is $5, then n needs to be 153 customers.

Question: Did the required sample size, n, increase or decrease? Why do you think n increased (decreased) when the allowable error, E, decreased?

Answer: The sample size, n, increased. When E decreases, the statistics team wants a better estimate. They require that $\overline{X}$ be much closer to the population value, μ. If $1 - \alpha$ and the standard deviation do not change, then the only avenue to improve the estimate is to obtain more information from the population, that is, increase the sample size.

Question: At a given level of confidence, $100(1 - \alpha)\%$, and given allowable error, E, if the standard deviation increases, will we need more or fewer sample observations to estimate? Give an intuitive explanation for the change in the required n.

Answer: A larger standard deviation indicates that the population values are more dispersed. Thus, the mean will be harder to estimate when the standard deviation is

Section 8.7.2 Problems

92. Determine the number of households that a historian needs from old census records to estimate the mean household size in a region within 0.25 person of the true mean with 99% confidence. An archivist published the standard deviation of household sizes for the nation during the same period as 1.32 persons, and there is no reason to suspect that variation in household size in the region differed from the national value.

93. How many observations of student art pieces should a local arts commission evaluate to estimate the mean cost of materials per piece in area art classes? The commission wants to be 90% confident that the estimate is within $0.50 of the actual mean cost. Standard deviations of such costs from similar studies in other localities are all about $4.

94. A national swim association wants to estimate the current mean time to swim 50 meters for 9- to 11-year-old competitive swimmers. They would like to be 98% confident that the estimate is no more than 1 second from the true mean. Previous years' records show that the *range* for most swimmers, excluding slow finishers, was 45 seconds. How many swimmers should they include in a sample to estimate the mean time and meet these standards?

95. A legal rights group suspects that a certain judge discriminates on the basis of sex when determining bonds. The group plans to estimate the mean fine for men and the mean fine for women with 95% confidence that the error of each estimate is at most $5,000. The group assents to the use of the national standard deviation of bonds, $25,000, for men and women combined, although they suspect that the variation is actually greater for men than for women. Use the available information to determine the number of observations of bond values they will need for each of the samples. Then, predict which sample is too large and which is too small to meet the specified conditions on the error of the estimate.

8.7.3 Confidence Level Associated with Claimed Error, $100(1 - \alpha)\%$

Often, we find estimates reported on television and in print media along with the maximum possible sampling error, which is usually called the *margin of error*. Generally, these assertions about the size of the error lack a complementary confidence level to reflect the uncertainty associated with the claim. In this section, we use the basic formula $E = z\sigma/\sqrt{n}$ to determine this confidence level.

■ **Example 8-15:** A newspaper article reports that the estimated mean number of crimes in towns of a certain size is 258,615. The report adds that the estimate is off by at most 2,000 crimes. What is missing from this report and why is the missing information important?

■ **Solution:** The level of confidence associated with the statement is missing. There is no guarantee that the estimate is within 2,000 crimes of the true mean, μ. The only way to be 100% sure is to use every observation in the population (which is impossible in cases involving destructive sampling). If we were to use the entire population, then there would be no error and no need for the 2,000 error statement.

The level of confidence that we seek is $100(1 - \alpha)\%$. Remember that $1 - \alpha$ is the proportion of the $\overline{X}$ values within z standard errors of μ or $P(-z < Z < z) = 1 - \alpha$. To determine $1 - \alpha$, we must first find z. Solving the E equation for z, we obtain $z = \dfrac{E}{\hat{\sigma}_{\overline{x}}}$. Now, substitute $S/\sqrt{n}$ for $\hat{\sigma}_{\overline{x}}$, $z = \dfrac{E\sqrt{n}}{S}$. Then, the confidence level, $100(1 - \alpha)\%$, is based on the area under the normal curve between $-z$ and $+z$, the shaded area in Figure 8-26. This is also $P(-z < Z < z)$.

The following box summarizes the two-step procedure for determining the confidence level for a correct error assertion.

Figure 8-26

Determining $1 - \alpha$ from error assertion

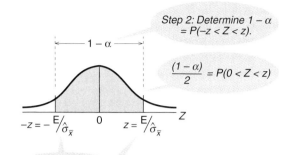

Procedure to Determine Confidence Level, $100(1 - \alpha)\%$, for Error Assertion

1. Determine z. Use the formula $z = E/\hat{\sigma}_{\bar{x}}$ or $z = (E\sqrt{n})/S$.
2. Determine the confidence level. Use the expression $1 - \alpha = P(-z < Z < z)$.

■ **Example 8-15 (continued):** If the newspaper reporter uses a sample of 100 towns to make the assertion that the estimate is off by no more than 2,000 crimes, what is the confidence level associated with the assertion? Recall that $\bar{X} = 258,615$ and $S = 8,000$.

■ **Solution:** Step 1: $z = (E\sqrt{n})/S = (2,000\sqrt{100})/8,000 = 2.5$.
Step 2: $1 - \alpha = P(-2.5 < Z < 2.5) = 2(0.4938) = 0.9876$.
So the reporter can be 98.76% confident of the claim that the error of the estimate is no more than 2,000 crimes.

■ **Example 8-16:** Speech analysts experimented with child stutterers to learn if they fixated (dwelled longer) on words during silent reading that they also later stuttered on during oral reading. The mean number of times that stutterers fixated on words during the silent reading that they also later stuttered on during oral reading was 1.58 fixations with a standard deviation of 0.45 fixation. The mean duration of fixations on these words was 656.91 msec with a standard deviation of 809.44 msec. Assume that the number of stuttered words is 1,500 words among all the subjects and that the analysts claimed that the margin of error was 0.025 fixation and 50 msec for the respective estimates. Determine the level of confidence associated with such statements. (Source: Klaas Bakker, Gene J Brutten, Peggy Janssen, and Sjoeke Van Der Meulen, "An Eyemarking Study of Anticipation and Dysfluency among Elementary School Stutterers," *Journal of Fluency Disorders*, Vol. 16, No. 1, February 1991, pp. 25–33.)

■ **Solution:** The estimated standard error for the number of fixations on each stuttered word is $S/\sqrt{n} = 0.45/\sqrt{1,500} = 0.0116$ fixation, so $z = E/\hat{\sigma}_{\bar{x}} = 0.025/0.0116 = 2.15$. $P(-2.15 < Z < 2.15) = 2(0.4842) = 0.9684$. For duration, the standard error is $809.44/\sqrt{1,500} = 20.8997$ msec, so $z = 50/20.8997 = 2.39$ and $P(-2.39 < Z < 2.39) = 2(0.4916) = 0.9832$. Thus, the respective confidence levels associated with the error statements are 96.84% and 98.32%.

Problems

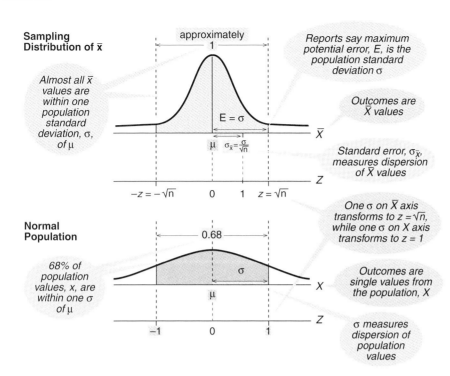

Figure 8-27
Almost all $\overline{X}$ values are within one population standard deviation of μ, and only 68% of a normal population's values are this close

Example 8-17: A report claims that if an analyst knows the population standard deviation, σ, then he or she can be assured that any large sample estimate, $\overline{X}$, will be within one population standard deviation, σ, of the mean, μ. What is the probability this statement is true?

Solution: The claim is that the maximum error is $E = \sigma$. Because $\sigma_{\overline{x}} = \sigma/\sqrt{n}$, we have

$$z = \frac{E}{\sigma_{\overline{x}}} = \frac{\sigma}{\sigma/\sqrt{n}} = \frac{\sigma\sqrt{n}}{\sigma} = \sqrt{n}.$$

If $n > 30$, then $z > \sqrt{30} = 5.48$. A z that is larger than 5.48 is larger than the largest z in the normal table provided in Appendix H, so the probability is approximately 1. That is, almost any $\overline{X}$ from a large sample will be within one population standard deviation, σ, of the mean, μ. Figure 8-27 reminds us that only 68% of the population values are within one standard deviation of the mean when the population is normal.

Section 8.7.3 Problems

96. A random sample of 100 former track stars shows that the mean retirement age is 22.34 years. The standard deviation is 2.79 years. What is the confidence level associated with a claim that this estimate is within 0.25 year of the true mean retirement age?

97. A candidate claims that the margin of error is $400 for an estimate of the mean contribution toward her reelection. The estimate is the mean of a random sample of 139 contributions. The sample standard deviation is $1,650. Find the confidence level associated with the margin of error statement.

98. A botanist measured the petal lengths of two random samples of 100 flowers randomly selected from lots in the local area. The scientist altered the genes of the flowers in the first sample but did not disturb those in the second sample. He estimated mean petal lengths for altered and unaltered flowers from the means of the two samples and claimed the error for both estimates would not exceed 0.5 millimeter.

a. What must be true if the confidence level associated with the error statement for each set of flowers is the same?

b. Suppose the standard deviation of petal lengths is 2.4 millimeters for the altered flowers and 2.0 millimeters for the unaltered flowers. Find the level of confidence associated with the claims for each error.

Summary and Review

To evaluate an inference we require certain information in addition to the point estimate; a limit for the sampling error (or margin of error) and the confidence level that reflects the uncertainty of the limit.

The sampling error is given by $\overline{X} - \mu$. E symbolizes the maximum error. From the sampling distribution of $\overline{X}$ for large samples, we find that $100(1 - \alpha)\%$ is the confidence level associated with the estimate ($\overline{X}$) lying within $E = z\hat{\sigma}_{\overline{x}}$ units of the population mean, μ. Three important findings result from this basic premise.

1. A limit, E, on the sampling error with a confidence level $100(1 - \alpha)\%$ associated with the limit. $E = z\hat{\sigma}_{\overline{x}}$.

2. The sample size, n, required to achieve a desired confidence level, $100(1 - \alpha)\%$, that $\overline{X}$ is within E units of μ. $n = (zS/E)^2$.

3. The confidence, $100(1 - \alpha)\%$, that an estimate ($\overline{X}$) is within E units of μ. $1 - \alpha = P(-z < Z < z)$, where $z = E/\hat{\sigma}_{\overline{x}}$.

Finding the required sample size requires a standard deviation before we actually select a sample. Several alternative substitute values include sample standard deviations from previous studies, a sample standard deviation from a small preliminary sample, or a fraction of the range (there are several ways to determine the fraction).

Multiple-Choice Problems

99. Suppose that a random sample of 100 observations has a mean of 200 and a standard deviation of 50. An analyst reports the results along with the assertion of a 95% chance the estimate is off by no more than 9.80. Another analyst says the 95% confidence interval estimate for μ is (190.20, 209.80). What is the difference in information content between the two reports?
    a. The first analyst gives a more exact estimate by zeroing in on a value.
    b. The second analyst is more likely to be correct, because the inference includes more than one estimate.
    c. The first analyst includes some idea of the potential error, and the confidence interval estimate minimizes the possibility of an incorrect estimate.
    d. The first analysis implies that a larger sample is used than in the confidence interval estimate.
    e. There is no real difference in information content between the two reports.

100. A forester desires information regarding the growth of a new stand of pines. Four hundred trees are randomly selected and their heights measured at the beginning and end of the summer to determine their growth. Previous studies of pines suggest a standard deviation of growths of 160 mm. The mean growth for the 400 trees is 212 mm. If 212 mm is the estimate of mean growth, what can the forester say with 80% confidence is the limit of the error?
    a. 8 mm.   d. 269 mm.
    b. 20.48 mm. e. 10.24 mm.
    c. 204.80 mm.

101. If we want to be more confident about a statement about error size but want to use the same sample and statistics, then
    a. The limit on the error must increase if we are to be more confident with the same information.
    b. The stated error size will decrease, because increased confidence means we can be more exact.
    c. The z value will necessarily decrease.
    d. The chances of our statement being incorrect will increase, because an increase in the level of confidence results in a stronger statement from the given information.
    e. We cannot change the confidence level without changing the sample or the statistic.

102. You need to estimate the mean useful life of a brand of battery to within 100 hours with 95% confidence. Previous studies indicate that the standard deviation for these battery lives is 450 hours. How many observations should your sample contain?
    a. 550 batteries.   d. 78 batteries.
    b. 100 batteries.   e. 30 batteries.
    c. 52 batteries.

103. Sam needs a standard deviation value to use in the formula to choose a sample size. He uses the standard deviation from a small preliminary sample, because he has no other information about the standard deviation. He chooses a value for n. Then he collects a random sample with n observations and recalculates S and finds that it is much larger than the small sample S. If he wants to maintain the level of confidence and error limit, what should he do?
    a. Do nothing, because both S values are estimates.
    b. Do nothing, because the new S means he needs fewer observations than he has already collected.
    c. Increase the confidence level or decrease the error limit, because of the added accuracy indicated by the larger S.
    d. Obtain more observations, because the increase in S indicates that μ will be harder to estimate at the same confidence level and error limit.
    e. Increase the sample size until the S value decreases to its former value.

Multiple-Choice Problems

Use the following information to work Problems 104–105.

An astronomy association claims that the result of a survey of 289 of its members indicates that the mean measure of the distance of a certain star from Earth is 30,000 light-years, with a standard deviation of 6,583 light-years. The association says that the estimate is within 1,000 light-years of the actual value.

104. What confidence level is associated with this statement?
   a. 0.9500.
   b. 0.9902.
   c. 0.9544.
   d. 2.575.
   e. 0.4951.

105. If the astronomy association claimed a smaller margin of error than 1,000 and used the same sample, what would happen?
   a. The confidence that the error was this small would decrease, because the association claims to be more exact with the same information.
   b. The level of confidence would decrease as the association claims to be less exact.
   c. The confidence associated with the claim would increase, because the association is zeroing in on the true value.
   d. The level of confidence would increase, because the smaller error limit does not use up the sample information and trades off with the confidence level.
   e. The error limit cannot be changed without altering the sampling procedure or statistic.

106. On a product's label the manufacturer states that the average life is 2,100 hours. What else do we need to know to assess this estimate?
   a. The sampling procedure.
   b. The maximum error.
   c. The level of confidence.
   d. The largest possible sampling error.
   e. All of the above.

Use the following information to answer Questions 107–108.

In the heat cost situation, Susan wants information on the average heating bill for houses in the region. She requests information on heating bills only for similar-size houses, say 1,600 square feet.

107. Which of the following is a likely consequence?
   a. The standard deviation of heating bills for these houses is smaller than the standard deviation for all houses in the area.
   b. The standard deviation of heating bills will be zero, because all the houses are the same size.
   c. The sample size must decrease because the population is obviously smaller than before.
   d. The sample mean is no longer normally distributed.
   e. The estimate from the new sample should provide a better estimate of the mean heating bill for all houses in the area than when sampling from the original population which contained more values.

108. Given the correct answer to Problem 107, the required sample size to be 95% confident the estimate is within $10 of the true mean for all 1,600 square foot houses is _____ the required sample from the original population.
   a. Larger than.
   b. Smaller than.
   c. The same as.

109. The sampling error
   a. Is the standard error of the statistic under investigation.
   b. Is the discrepancy between the distribution of values in a sample and the distribution of values in the population.
   c. Occurs when surveyors incorrectly record responses in interviews.
   d. Is the difference in an estimate and the parameter being estimated that arises from sampling rather than using the entire population.
   e. Is the width of a confidence interval.

110. The mean response time for a random sample of 60 survey responses is 2.3 weeks, with a standard deviation of 2 weeks. What can you say with 95% confidence about the size of the error, if you use 2.3 weeks as an estimate of the mean response time for all participants?
   a. The error is at most 0.51 week.
   b. The maximum error is 2 weeks.
   c. The error is no more than 3.92 weeks.
   d. The maximum error is 0.42 week.
   e. The error is at least 3.28 weeks.

111. The role of the central limit theorem in this chapter is to
   a. Ensure that the sampling distribution of $\bar{X}$ is approximately normal.
   b. Prevent the error size from expanding without bounds.
   c. Prevent the standard error from being infinite.
   d. Allow us to work with small samples.
   e. Eliminate nonsampling error.

Use the following information to answer Questions 112–113.

Speech analysts experimented with child stutterers to learn if they fixated or dwelled longer on words read silently on which they later stuttered. They also collected information for comparison purposes on nonstuttered words. The mean number of fixations on nonstuttered words during the silent reading was 1.49 fixations with a standard deviation of 0.52 fixation. The mean duration of fixation on these words was 521.83 msec with a standard deviation of 554.64 msec. Assume that the number of nonstuttered words is 1,500 words among all experimental subjects. (Source: Klaas Bakker, Gene J. Brutten, Peggy Janssen, and Sjoeke Van Der Meulen, "An Eyemarking Study of Anticipation and Dysfluency among Elementary School Stutterers," *Journal of Fluency Disorders*, Vol. 16, No. 1, February 1991, pp. 25–33.)

Figure 8-28
Minitab output for 95% confidence interval of library staff data with known $\sigma = 138.2$ persons

```
                THE ASSUMED SIGMA =138

             N      MEAN     STDEV    SE MEAN    95.0 PERCENT C.I.
STAFFSIZ    32     341.2     156.3      24.4     (  293.2,   389.1)
```

112. What can the analysts claim with 95% confidence about the margin of error for estimates of the mean number of fixations and the mean duration of fixations?
   a. The errors are less than or equal to 0.022 fixation and 23.486 msec.
   b. The maximum errors are 0.026 fixation and 28.069 msec.
   c. The least possible errors are 0.52 fixation and 554.64 msec.
   d. The errors are at most ±1.019 fixations and ±1,087.094 msec.
   e. The errors are more than 1.019 fixations and 1,087.084 msec.

113. The researchers want to be 95% confident the errors of the estimates are at most 0.01 fixation and 20 msec for nonstuttered words. If they use the information given above, how many words that are not likely to be stuttered should they include for the children to read, if they want to accomplish both goals to limit the errors of the estimates?
   a. 102.   d. 7,273.
   b. 2,704.   e. 10,388.
   c. 10,387.

114. Nonsampling errors result from
   a. Transcription errors.
   b. Nonrandom samples.
   c. Leading questions.
   d. Low survey response rates.
   e. All of the above.

115. The owner of a recreation park estimates the mean number of trips down a water slide in an hour by one slider to be 6.2 trips, based on a random sample of 50 sliders observed one day. The standard deviation is 2.9 slides. What is the level of confidence associated with this estimate, if it is asserted that the estimate is within one slide of the true mean?
   a. 0.9500.   d. 0.9962.
   b. 0.4927.   e. 0.8530.
   c. 0.9854.

116. Figure 8-28 shows a confidence interval estimate of the mean library staff size based on a random sample selected from 106 research libraries in the United States and Canada. If we use a point estimate of the mean staff size, what would the estimate be? Use the confidence interval in the printout to find what we can say is the maximum error with 95% confidence. (Source: Association of Research Libraries, as reported in *The Chronicle of Higher Education: Almanac,* August 28, 1991, p. 37.)
   a. 293.2 and 47.95 people.
   b. 341.2 and 47.95 people.
   c. 341.2 and 156.3 people.
   d. 389.1 and 95.9 people.
   e. 341.2 and 24.4 people.

117. How many classified ads should be included in a random sample to determine the mean number of days needed to sell an item if the estimate is to be within one day of the true mean with 80% confidence? Similar studies over the years by the newspaper indicate that the standard deviation is five days.
   a. 82 ads.   d. 97 ads.
   b. 25 ads.   e. 68 ads.
   c. 41 ads.

118. The margins of victory in professional football games are normally distributed with a mean of *PS*, the point spread, and a standard deviation of 13.86 points. A sportswriter selects a random sample of 40 margins of victory for teams favored by five points. What is the probability the sampling error produced by this sample's mean will be at most two points? (Source: Hal Stern, "On the Probability of Winning a Football Game," *American Statistician*, Vol. 45, No. 3, August 1991, pp. 179–83.)
   a. 0.91.
   b. 0.3186.
   c. 0.6372.
   d. 0.9876.
   e. 0.1114.

Word and Thought Problems

119. a. What can we say with 90% confidence about the size of the error if we use $\overline{X}$ from a sample of 69 items to estimate μ? (Previous studies indicate $\sigma = 25$.)
   b. What can we say with 99% confidence about the error in Part a? Use the results of your answer here and to Part a to generalize about the effect of the confidence level on the limit of the error. What parts of the formula do not change between the two computations?
   c. Rework Part a by using $n = 269$. Use the results of your answer here and to Part a to generalize about the effect of the sample size on the limit of the error. What parts of the formula do not change between the two computations?
   d. Rework Part a by using $\sigma = 10$. Use the results of your answer here and to Part a to generalize about the effect of the standard deviation on the limit of

Word and Thought Problems

the error. What parts of the formula do not change between the two computations? Suggest ways or examples of ways to change the standard deviation in an estimation project.

120. a. A public service commission gives you three days to estimate the mean discrepancy between utility meter measures of customers' water usage and actual usage. You plan to obtain a random sample of meters, pass 1,000 gallons of water through each meter, and record each meter's measure of the quantity of water that passes through it. Your estimate of the mean discrepancy cannot be more than 2 gallons from the actual mean discrepancy, because a concerned citizens' lobby that investigates overcharges will attend the meeting. How many meters should you study to be 95% sure that your sampling error is within the acceptable limit? (Previous studies indicate the standard deviation of meter discrepancies is 20 gallons when quantities of 1,000 gallons are used.)

b. Suppose you want to take more risks, say to be 80% sure, because there is so little time to gather information. How large should the sample be to satisfy this new constraint? Does n increase or decrease? Why do you think n increased (decreased)?

121. A zoologist asserts that, based on a sample of 225 adults of a certain species, the mean weight of the adult is 64 ounces (plus or minus 0.5 ounce). The zoologist uses a standard deviation of 3 ounces. What is the confidence level associated with this assertion? What is the risk that the assertion is wrong?

122. An appliance manufacturing firm estimates its product's average life as 10,000 hours. The firm asserts that this estimate is off by no more than 50 hours. This assertion is based on a sample of 400 items selected from a population with a standard deviation of 500 hours. What level of confidence is associated with this assertion? What level of risk?

123. A professor suggests that you estimate the mean course grade he has given in the past, using A = 4.0, B = 3.0, C = 2.0, and D = 1.0. Time for this process is minimal, so you want as small a sample as possible to be 95% sure that your estimate is within 0.03 point of the actual mean. The professor suggests that you use a standard deviation of 0.75 point. How many sample observations do you need?

124. What are the ways we obtain information about the standard deviation to use in the formula to find the required sample size? Suppose we select a sample of the required size based on one of these ways, compute the standard deviation of this sample, and find that the latest standard deviation value differs dramatically from the one used to compute the required sample size. Will the sample suffice, or must we expand it? Explain your answer.

125. A random sample consists of 67 randomly selected water specimens taken from one stream. The mean content of a particular pollutant in the sample is 4 parts per thousand (ppt), with a standard deviation of 1 ppt. What can be said with 98% confidence about the maximum possible error if this information is used to estimate the mean amount of pollutant in the stream?

126. Reuse your random sample of 10 values from the North Carolina Nursing Home data in Appendix E selected for Problem 40 in Chapter 7, Unit I (p. 306), and Problem 70 of this chapter. Assume that all 100 values are approximately normally distributed. The population standard deviation is 517.4 people. Find the level of confidence that your sample mean is no more than 400 people away from the actual mean, 470.1 people. Do you need the T distribution to work this problem? Explain your answer.

127. a. How many subsidized families should be included in a random sample to estimate the mean amount of rent subsidy per family in a region? The estimate should be within $25 of the true amount with 90% confidence. Nationally, the standard deviation of subsidies is about $100.

b. What is the population in this study?

c. After selecting the sample and computing E by using the sample standard deviation rather than the national value, the result is $E = \$20$. Do the sample and the results satisfy the original estimation error goals? Is any further work necessary? If so, explain what needs to be done.

d. Suppose instead that $E = \$35$ if we use the sample standard deviation rather than the national value. Do the sample and results satisfy the original estimation error goals? Is any further work necessary? If so, explain what needs to be done.

128. A random sample of 200 farmers in one state yields a mean soybean planting of 820 acres per farm with a standard deviation of 122 acres. What can be said about the error with 99% confidence, if this information is used to estimate the mean acreage in soybeans for all of the state's farmers?

129. How many patients should be included in a sample in order to be 95% confident that the estimated mean time spent in a waiting room is within five minutes of the true value? A small random sample of these times has a standard deviation of 17 minutes.

130. An appliance manufacturing firm estimates its product's average life as 50,000 hours. It asserts that this estimate is off by no more than 100 hours. This assertion is based on a random sample of 1,600 items, selected from a population with a standard deviation of 4,000 hours. What level of confidence is associated with this assertion? What level of risk?

131. We can approximate the likely relationship between the range of a population and its standard deviation by using Chebyshev's inequality (from Chapter 3). The relationship helps provide an estimate of σ to use in determining an appropriate sample size for inferences. We can state it as P(the distance between a population value, X, and the population mean, μ, is less than k standard deviations) $\geq 1 - 1/(k^2)$ or $P(X$ is less than k σs from $\mu) \geq 1 - 1/(k^2)$, where $k > 1$. This statement allows us to set a lower bound on the probability that a value is within a certain distance, measured in standard deviations, from the mean. If $k = 2$, then the chance of obtaining an X less than two σs from μ is at least $3/4 = 1 - 1/(2)^2$. Another way to say this is that at least 75% of the observations are within two σs of μ. Thus, if we know that an interval of values covers approximately the middle 75% of the population range, then range/4 is the estimate of σ. *Notice that the shape of the population distribution is not important in this calculation.*

   a. What proportion of the population values are within three standard deviations of the population mean?

   b. Using your results from Part a, estimate σ if the middle 90% of the population values are between 90 and 100.

132. The mean jump in a bid at an auction is estimated to be $20, based on the mean of a random sample of 80 bids recorded on a tape made at the auction. The standard deviation of these jumps is $14. What can we say with 98% confidence about the size of the error in this estimate?

133. A random sample of 100 defunct rock bands shows that the mean lifetime of a band in the sample is 1.37 years before they break up. The standard deviation is 1.98 years for these bands. What is the confidence level associated with a claim that this estimate is within 0.5 year of the true mean lifetime?

134. A teenager is given a curfew. In a random sample of 30 nights, the teenager is 10 minutes early 6 times, on time 15 times, and 10 minutes late 9 times. Use this information to estimate the mean arrival time (relative to curfew). Then, describe the accuracy of your estimate with 99% confidence.

135. A student claims that all her professors give tests on the same day. A random sample of 36 test days during the year reveals that the mean is 0.72 tests per day, with a standard deviation of 0.6 tests per day. What confidence do we have that this estimate is off by no more than 0.3? If the student takes four courses at a time, is it likely that all of her professors will give tests on the same day? What is the population in this problem?

136. How many nights' expenses should Todd include in a random sample of expenditures on video games to determine with 95% confidence the mean amount he spends on video games to within $0.50? The least he has ever spent is $1, and the most is $13.

137. You are to determine the necessary sample size to estimate μ by using $\overline{X}$. The sample will be drawn from a population with a standard deviation of 600. You want to be 95% sure about the results. What else must be specified before you can work the problem?

138. Several medical researchers experimented with patients with high cholesterol levels to explore the synergistic effects of low-fat diets and a cholesterol drug for reducing cholesterol levels. The following chart lists some results. Assume that cholesterol levels in each population are normally distributed.

Descriptors of Experimental Groups	Total Cholesterol Level (mmol/L)		
	Mean	Standard Deviation	n
High-fat diet/placebo	7.99	1.76	19
High-fat diet/drug	6.13	1.45	19
Low-fat diet/placebo	6.63	1.86	16
Low-fat diet/drug	5.09	1.47	16

Source: Margaret M. Cobb, Howard S. Teitelbaum, and Jan L. Breslow, "Lovastatin Efficacy in Reducing Low-Density Lipoprotein Cholesterol Levels on High- vs. Low-Fat Diets," *Journal of the American Medical Association*, Vol. 265, No. 8, February 27, 1991, pp. 997–1001.

   a. Use this information to determine the maximum possible sampling error with 95% confidence if the researchers use the sample means as estimates for the populations depicted above.

   b. The mean total cholesterol level for all participants was 8.00 mmol/L at the beginning of the experiment. Does there appear to be evidence of synergistic effects? Do your error calculations *suggest* that some of the means may be indistinguishable?

Unit III: Estimating *p*: The Population Proportion of Successes

Preparation-Check Questions

139. Thirty percent of the time, a computer programmer finds a mistake in the first try of a new program. This percentage has not changed with years of experience. Use the normal curve to approximate the probability that 100 or more of the next 300 programs will reveal a mistake on the first try. (Round σ to two places past the decimal.)

   a. 0.3333.   d. 0.0918.
   b. 0.3000.   e. 0.1151.
   c. 0.3849.

(continued on next page)

140. If X is normally distributed with a mean of 219 and a standard deviation of 20, what proportion of the outcomes are within 2.5 standard deviations of μ?
  a. 0.9876.   d. 0.0956.
  b. 0.4938.   e. 1.
  c. 0.1034.

141. A random sample of 64 observations has a mean of 36 and a standard deviation of 16. Find a 98% confidence interval for μ.
  a. (20, 52).   d. (32.08, 39.92).
  b. (−1.28, 73.28). e. (4.64, 66.36).
  c. (31.34, 40.66).

142. What is the probability distribution of a statistic called?
  a. T statistic.
  b. Sampling distribution.
  c. Expected value.
  d. Frequency distribution.
  e. Approximate distribution.

143. The sample mean is the estimate of the mean cost of braces for teeth. The sample contains 100 randomly selected patients' costs and has a mean and standard deviation of $2,500 and $800, respectively. What is the limit of the error of this estimate with 95% confidence?
  a. $800.   d. $1,568.
  b. $2,400.   e. $80.
  c. $156.80.

Answers: 139. e; 140. a; 141. c; 142. b; 143. c.

Introduction

Sidewalk Ordinance A local official proposed an ordinance that would require landowner assessments in order to pay for future sidewalk repair costs. A local group asks people to sign a petition against the ordinance. During a television interview, the group leader claims that most people disagree with the ordinance, because 450 of the 800 people contacted so far signed the petition.

Question: Why might this information overstate the group's support?

Answer: Because the petitioners do not describe their sampling procedure, it may not be signed by randomly contacted local citizens. In fact, the signees may consist of friends of the group and other people likely to sign, making the results misrepresent the local citizenry.

To estimate the actual support for the ordinance, the official decides to collect a random sample of opinions from names on voter registration lists. He will ask voters if they support, do not support, or are indifferent to the proposed ordinance.

Information from both the official's poll and the citizen's poll estimate the *proportion of voters who support the ordinance*. The subject of this unit is this new parameter, a population proportion (p). In this unit, we will suggest a sample size for the official's estimate adequate to satisfy certain conditions for the sampling error. We will also present the estimate as a point and as an interval along with its accuracy.

8.8 An Estimator of p

So far, we have estimated values of the population mean, μ. Now, we will focus on the proportion (or percentage) of successes in the population, p, where a success is an item that fits in a certain category, such as being in favor of the sidewalk ordinance. We call this proportion the *population success rate*. Common success rates that we estimate include manufacturing defect rates and public preferences on political and social issues.

The procedure for estimating a population success rate, p, is similar to estimating μ. Again, we begin by selecting a random sample.

Suppose that a random sample of 2,000 items contains 100 defectives. Then the proportion of defectives in the sample is 100/2,000 = 0.05, or 5%. Here, a success means being defective. (Remember *success* designates the event of interest, which is not always a positive event.) The proportion of successes in the sample, the *sample success rate*, $\hat{p}$ is the estimator of the population success rate, p.

$$\hat{p} = \frac{X}{n} = \frac{100}{2,000} = 0.05, \text{ or } 5\%,$$

where X is the number of successes in the sample and n is the sample size.

Question: A random sample of 1,000 voters shows that 560 support the incumbent candidate. What is the estimate of the candidate's support among all voters?

Answer: $\hat{p} = \dfrac{X}{n} = \dfrac{560}{1,000} = 0.56.$

So we estimate that 56% of the voters favor the incumbent.

The use of $\hat{p}$ to estimate p is similar to the use of $\overline{X}$ to estimate μ. Now, we must determine the possible values of $\hat{p}$ and the probability that each occurs when we compute $\hat{p}$ from a randomly selected sample. That is, we must find the sampling distribution of $\hat{p}$.

8.9 Sampling Distribution of the Sample Success Rate

The random variable $\hat{p} = X/n$ is similar to a binomial random variable. p is the proportion of successes in the population. If we randomly pull one item from this population, the probability we obtain a success is p. For example, if we randomly select an item from a production line with a 10% defect rate, the probability that the item will be defective is 0.1. Of course, in practical situations, we do not know that the defect rate is 10%. We try to estimate the rate.

The number of successes in a sample, X, will be a binomial random variable when we randomly select items for the sample from a large population. Each selection is a trial with two possible outcomes, success or failure. When we select items from a large population, changes in the population proportion of successes, p, are negligible between selections. Random sampling assures us that outcomes are independent, that is, that the outcome of one trial does not affect the outcome of any other. Hence, a count of the number of successes in a random sample from a large population satisfies the conditions for a binomial random variable (see Unit II of Chapter 5 to review these conditions).

However, we want more than a count of the number of successes in the sample, X. We want the proportion of successes in the sample, $\hat{p}$, which is the fraction, X/n. But because the sample size, n, is fixed when we count successes, the probability for a value of $\hat{p} = X/n$ will be the same as the probability for the X value. Hence, the sampling distribution of $\hat{p}$ is the binomial distribution with n trials and with a probability of success on each trial of p. The next example will demonstrate the similarities and differences of these two statistics.

■ **Example 8-18:** The random variable X represents the number of heads in three tosses of a fair coin. The three tosses form a random sample of three coin tosses ($n = 3$), where the population success rate, p, is 0.5. Using the binomial formula or binomial table, we can find the binomial probability distribution for X with $n = 3$ and $p = 0.5$.

x	$P(x)$
0	0.125
1	0.375
2	0.375
3	0.125

Transform each X value to an X/n value, and determine the probability for each of these new values.

Solution: Another way to say "two heads out of three tosses" is "two-thirds of the three tosses are heads." Notice that $2/3 = X/n = \hat{p}$. Because n is fixed for any given experiment or sample, in this case $n = 3$, we simply divide each outcome by this n value, producing the first column of the following chart. Also, because n is a constant and because we are simply rewording our description of the experimental outcome, the value of $P(x/n)$ is the same as for $P(x)$. Thus, the probability distribution of $\hat{p} = X/n$ uses the probability values that correspond to $P(x)$.

Proportion of Heads in a Sample of Three Tosses $x/n = \hat{p}$	Number of Heads in a Sample of Three Tosses x	$P(x) = P(x/n) = P(\hat{p})$
$0 = 0/3$	0	0.125
$1/3$	1	0.375
$2/3$	2	0.375
$1 = 3/3$	3	0.125

Recall from Chapter 6 that we can use the normal random variable to approximate a binomial random variable. Because $P(x/n) = P(x)$, the normal random variable approximates the new statistic, $\hat{p}$ as well. This approximation is particularly useful when we report estimates of p, because we will need the probability for many different $\hat{p}$ values when we form a single confidence interval or margin of error.

Figure 8-29 replicates the conversion of x to x/n that we illustrated in the chart for the coin tosses. The second axis represents X values (from the top axis) divided by n, which are the $\hat{p}$ values. We can use the diagram to approximate binomial probabilities and $\hat{p}$ probabilities with normal probabilities.

Question: To find normal probabilities, what two characteristics of the random variable must we know?

Answer: We must know the mean and standard deviation of the values of the random variable.

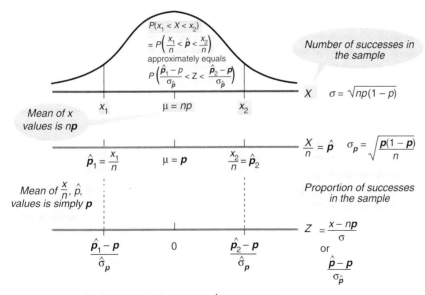

Figure 8-29

Normal approximation and relation between random variables X and $X/n = \hat{p}$

Figure 8-30
Approximate sampling distribution of $\hat{p}$

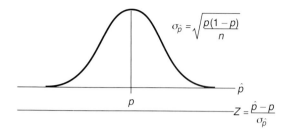

The shortcut formula for the mean of a binomial random variable, X, is $\mu = np$. Figure 8-29 shows that this mean is a value on the X axis. If we transform the mean, np, to a $\hat{p}$ value on the X/n axis, we obtain $\mu/n = np/n = p$. Hence, the mean of the $\hat{p}$ values is p.

Similarly, $\sigma = \sqrt{np(1-p)}$ is the standard deviation of the X values. To express this value in the same units as $\hat{p}$, we must divide it by n.

$$\frac{\sqrt{np(1-p)}}{n} = \sqrt{\frac{np(1-p)}{n^2}} = \sqrt{\frac{p(1-p)}{n}}.$$

This value is called the *standard error of the sample proportion*, *standard error of the sample success rate*, or *standard error of $\hat{p}$*. Figure 8-30 extracts the necessary information from Figure 8-29 to display the approximate sampling distribution of $\hat{p}$ that we will use to evaluate these estimates of p.

> When the sample size, n, is large, the sampling distribution of $\hat{p}$ is approximately normal with a mean of p and a standard error of $\sigma_{\hat{p}} = \sqrt{\dfrac{p(1-p)}{n}}$.

Box 8-3 shows an alternative derivation of this sampling distribution that is based on the work we have already done with $\overline{X}$.

■ **Example 8-18 (continued):** $\hat{p}$ is the proportion of heads in three tosses of a coin, so $p = 0.5$ and $n = 3$. What is the standard error of $\hat{p}$, $\sigma_{\hat{p}}$, for this random variable?

■ **Solution:** $\sigma_{\hat{p}} = \sqrt{p(1-p)/n} = \sqrt{0.5(1-0.5)/3} = \sqrt{0.25/3} = 0.2887$.

■ **Example 8-19:** Find the probability that between 0.1 and 0.2 of the items in a sample of 100 items from a shipment will be defective if the defect rate of the shipment is 0.2.

■ **Solution:** The defect rate is $p = 0.2$ and $n = 100$, so

$$\sigma_{\hat{p}} = \sqrt{\frac{p(1-p)}{n}} = \sqrt{\frac{0.2(0.8)}{100}} = 0.0400.$$

$P(0.1 < \hat{p} < 0.2) = P(-2.5 < Z < 0) = 0.4938$ (see Figure 8-31).

Question: The middle 95% of sample success rates are within about how many standard errors of p when the sample is large?

Answer: The middle 95% of $\hat{p}$ values (or the values of any other normal random variable) are within 1.96 standard errors of p.

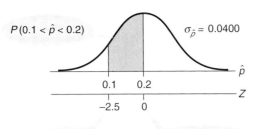

Figure 8-31
Solution diagram for $P(0.1 < \hat{p} < 0.2)$

Question: As n increases, will the chances of obtaining a $\hat{p}$ close to p increase, decrease, or remain the same?

Answer: The probability that $\hat{p}$ is close to p increases as the sample size increases. Intuitively, a larger sample incorporates more population information in the estimate, which improves the chances of obtaining a $\hat{p}$ in the vicinity of p. Mathematically, as n increases, $\sigma_{\hat{p}} = \sqrt{p(1-p)/n}$ will decrease, and the $\hat{p}$ values become more tightly clustered about p.

Remember that the normal approximation of binomial probabilities improves as p approaches 0.5 and n becomes large. When p is small, n must be large to obtain a reasonable approximation. The term *large* changes as p changes, but usually we require more than the 30 observations that were used when estimating μ. Using the condition np and $n(1-p)$ must both exceed 5 to obtain a good approximation (from Section 6.7), we find that we need a sample size of 100 when $p = 0.05$. More is better. When the sample is small, a special chart is available for determining confidence intervals, but we use large samples and the normal approximation in this book.

Box 8-3 The Central Limit Theorem and the Sampling Distribution of $\hat{p}$

We estimate the population proportion of successes with the sample success rate, the ratio of the number of successes in the sample to the number of observations in the sample. Suppose we assign each observation, X, a value, 0 if it is a failure, and 1 if it is a success. Then $\overline{X} = \Sigma X/n$ will be the sample success rate, $\hat{p}$, because ΣX is the number of ones in the sample. Likewise, the population mean is $\mu = \Sigma X/N = p$, the population success rate. We do the same thing when we talk of a baseball player's batting "average," which is nothing more than the proportion of times a batter gets on base after hitting a fair ball (the ratio of the number of hits to the number of times at bat—excluding times at bat when the player gains a base without hitting the ball). Thus, we can describe the behavior of $\hat{p}$, by using the information we already know about the behavior of $\overline{X}$.

First, we apply the central limit theorem, so that large samples compel the sampling distribution of $\overline{X}$ or $\hat{p}$ to approach a normal distribution. Large implies $n > 30$ and this arbitrary cutoff should still provide adequate approximations as long as p is between 0.16 and 0.84 (according to our rule for good approximations, which states that np and $n(1-p)$ should be greater than 5). We need larger samples for p values outside this range. We assume very large samples when we estimate p to ensure a good approximation for $\hat{p}$ values with the normal distribution.

To use the normal curve we need a mean and standard deviation, in this case a mean and standard error of $\hat{p}$. We know the mean of the $\overline{X}$ values is the population mean, μ. Hence, the mean of the $\hat{p}$ values is p. Likewise, the standard error of the sample success rate, $\sigma_{\hat{p}}$, would be $\sigma/\sqrt{n}$ or, after some algebra, $\sqrt{\dfrac{p(1-p)}{n}}$. If we must estimate the population standard deviation in order to estimate the standard error, we obtain $\hat{\sigma}_{\hat{p}} = \sqrt{\dfrac{p(1-\hat{p})}{n-1}}$. Again, because we use large samples, so that n and $n-1$ are about the same value, we approximate $\sigma_{\hat{p}}$ well with $\sqrt{\dfrac{\hat{p}(1-\hat{p})}{n}}$. There is usually little distortion from substituting n for $n-1$, even when n is barely larger than 30.

Thus, using this alternative approach, defining values of outcomes so that $\overline{X}$ is the same as $\hat{p}$, we can develop the same sampling distribution of $\hat{p}$ shown in Figure 8-30.

Section 8.9 Problems

Assume that samples are sufficiently large to approximate probabilities well with the normal distribution, and solve the following problems.

144. The middle 80% of $\hat{p}$ values are within how many standard errors of p?

145. The middle 90% of $\hat{p}$ values are within how many standard errors of p?

146. If a random sample of 800 items is selected from a production line with a 6% defect rate, what is the probability that the sample's defect rate will lie between 4 and 8%?

147. If we select a random sample of 2,000 citizens from a district where 75% of the electorate favors a candidate, what is the probability a popularity poll will yield an approval rating of 77% or more?

8.10 Confidence Interval for p

Because the sampling distribution of $\hat{p}$ is approximately normal for large samples, constructing a confidence interval for p duplicates constructing one for μ.

$100(1 - \alpha)\%$ of $\overline{X}$ values are within z standard errors ($\sigma_{\overline{x}}$) of μ. Similarly, $100(1 - \alpha)\%$ of $\hat{p}$ values are within z standard errors ($\sigma_{\hat{p}}$) of p. Thus, $100(1 - \alpha)\%$ of the intervals, $\hat{p} \pm z\sigma_{\hat{p}}$ should contain the true value, p. Refer to Figure 8-32.

Question: If we substitute for $\sigma_{\hat{p}}$ in $\hat{p} \pm z\sigma_{\hat{p}}$, we obtain $\hat{p} \pm z\sqrt{\dfrac{p(1-p)}{n}}$. What part of this formula presents a problem when we try to substitute sample information?

Answer: The formula requires p, which is unknown. In fact, we are trying to estimate p.

To resolve this problem, we substitute the point estimate, $\hat{p}$, in place of p, just as we substitute S for σ in earlier formulas. The result is $\hat{\sigma}_{\hat{p}} = \sqrt{\dfrac{\hat{p}(1-\hat{p})}{n}}$. The resulting confidence interval procedure is as follows:

Figure 8-32
Comparison of sampling distributions for $\overline{X}$ and $\hat{p}$

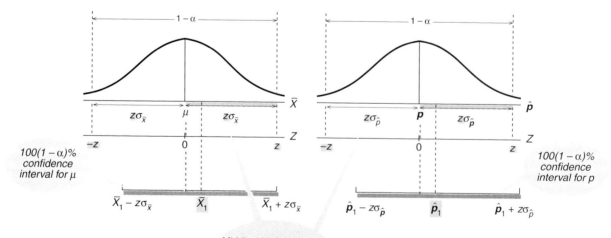

Middle $100(1 - \alpha)\%$ of normally distributed values are within Z standard errors of the mean of the values

Section 8.10 Confidence Interval for p **375**

> To form an approximate $100(1 - \alpha)\%$ confidence interval for the population's proportion of successes, p, when n is large, use
>
> $$\hat{p} \pm z\hat{\sigma}_{\hat{p}},$$
>
> where
>
> $$\hat{p} = X/n, \quad P(-z < Z < z) = 1 - \alpha, \quad \text{and} \quad \hat{\sigma}_{\hat{p}} = \sqrt{\hat{p}(1 - \hat{p})/n}.$$
>
> A guide for the sample size necessary for a good approximation is that $n\hat{p}$ and $n(1 - \hat{p})$ should both exceed 5.

Example 8-20: A random sample of 400 items contains 40 defectives. Find a 95% confidence interval for the manufacturer's defect rate.

Solution: $n = 400$, and $X = 40$, so $\hat{p} = 40/400 = 0.1$ and

$$\hat{\sigma}_{\hat{p}} = \sqrt{\frac{\hat{p}(1 - \hat{p})}{n}} = \sqrt{\frac{0.1(1 - 0.1)}{400}} = 0.015.$$

$z = 1.96$, so the confidence interval for p is $0.1 \pm 1.96(0.015) = 0.1 \pm 0.029 = (0.071, 0.129)$. We are 95% confident that this interval that covers rates between 7.1% and 12.9% contains the true defect rate.

Example 8-21: A research firm surveyed 500 randomly selected women for a cosmetics firm to learn about attitudes and behavior related to appearance. The following results were reported

a. 64% evaluate people by their looks.
b. 83% think people overemphasize attractiveness.
c. 83% use soap daily.
d. 51% use moisturizers daily.
e. 49% use lipstick or lip gloss daily.

Find and interpret a 98% confidence interval for the proportion of successes for each estimate. Can we be 98% confident that over 50% of the population are successes for each definition of success in Parts a–e? (Source: "How You Look Is Who You Are, Survey Suggests," *Asheville Citizen-Times*, October 8, 1989, p. 19A.)

Solution: $z = 2.33$ for a 98% confidence interval. The standard errors and confidence intervals vary based on the sample success rate, $\hat{p}$.

a. $\hat{\sigma}_{\hat{p}} = \sqrt{0.64(0.36)/500} = 0.0215$, and the confidence interval is $0.64 \pm 2.33(0.0215) = 0.64 \pm 0.050 = (0.590, 0.690)$.
b.–c. $\hat{\sigma}_{\hat{p}} = \sqrt{0.83(0.17)/500} = 0.0168$, and the confidence intervals are $0.83 \pm 2.33(0.0168) = 0.83 \pm 0.039 = (0.791, 0.869)$.
d.–e. $\hat{\sigma}_{\hat{p}} = \sqrt{0.51(0.49)/500} = 0.0224$, and $2.33(0.0224) = 0.052$. In Part d the confidence interval is $0.51 \pm 0.052 = (0.458, 0.562)$, and in Part e it is $0.49 \pm 0.052 = (0.438, 0.542)$.

To interpret these confidence intervals, we say we are 98% confident that each interval contains the true proportion of successes. (We define success differently for each interval.) For Parts a–c we are 98% confident that the true value is greater than 0.5, because the 98% confidence interval includes only values that are greater than 0.5.

Section 8.10 Problems

148. A random sample of 430 items from a production line contains 51 defectives. Use this information to find and interpret a 99% confidence interval for the defect rate.

149. A poll of 500 randomly selected students finds that 40% of them have the same opinion on an issue. Find a 90% confidence interval for the proportion of all students with the same opinion, and interpret the result.

150. Twenty-five percent of a random sample of 100 voters favor the incumbent candidate. Find a 95% confidence interval for the actual proportion of the voters who favor this candidate. Interpret the result.

151. A medical researcher found that 33% of a random sample of 100 vaccinated patients became ill. Find an 80% confidence interval for the proportion of all vaccinated patients who become ill from the vaccination. Interpret the result.

152. Find a 98% confidence interval for the proportion of packages that a delivery service delivers to the wrong location if 358 packages in a random sample of 1,200 packages are delivered erroneously. Interpret the result.

8.11 Evaluating Point Estimates of p

Instead of making a confidence interval estimate of p, we can use $\hat{p}$ as a point estimate along with information about the confidence level associated with the maximum sampling error.

Question: What is the sampling error when $\hat{p}$ is used to estimate p?
Answer: Error $= \hat{p} - p$.

8.11.1 The Maximum Error, E

To establish a limit on the sampling error with confidence $100(1 - \alpha)\%$ for estimates of p, we duplicate the earlier argument we used to develop the limit for estimates of μ. Figure 8-33 shows that when the sample is large, $\hat{p}$ is approximately normal and, hence, $100(1 - \alpha)\%$ of possible sampling errors are at most $z\sigma_{\hat{p}}$. With symbols, we say $E = z\sigma_{\hat{p}}$. Again, we must use an estimate of the standard error, $\hat{\sigma}_{\hat{p}}$, because usually we do not know p in the $\sigma_{\hat{p}}$ formula, so $E = z\hat{\sigma}_{\hat{p}} = z\sqrt{\hat{p}(1 - \hat{p})/n}$. This result is similar to the formula $E = z\hat{\sigma}_{\bar{x}}$ for the maximum sampling error when $\bar{X}$ estimates μ.

> To establish a limit on the sampling error with $100(1 - \alpha)\%$ confidence when $\hat{p}$ from a large sample is the estimate of p, use $E = z\hat{\sigma}_{\hat{p}}$,
>
> where $\hat{\sigma}_{\hat{p}} = \sqrt{\hat{p}(1 - \hat{p})/n}$ and $P(-z < Z < z) = 1 - \alpha$.

Figure 8-33
Sampling errors and sampling distribution of $\hat{p}$ (large samples)

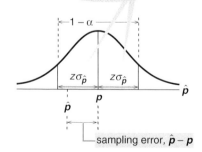

Problems

■ **Example 8-22:** Suppose that the IRS audits a firm. The IRS randomly samples 150 travel vouchers and disallows 5% of them. To save the time and expense of looking at all the travel vouchers, the IRS applies the 5% figure to all the firm's travel vouchers.

a. At the 95% confidence level, what is the maximum error that the IRS might be making (the error can be for or against the firm)?
b. Suppose the IRS inspects 100 more vouchers (250 in all) and finds that 4% should be disallowed. What happens to the error limits at the 95% level?

■ **Solution: a.** $\hat{p} = 0.05$, and $n = 150$, so $\hat{\sigma}_{\hat{p}} = \sqrt{0.05(0.95)/150} = 0.0178$. For 95% confidence $z = 1.96$, so $E = 1.96(0.0178) = 0.035$. There is a 95% chance that the IRS has missed by 3.5 or fewer percentage points.
b. $\hat{p} = 0.04$ and $n = 250$, so $\hat{\sigma}_{\hat{p}} = \sqrt{0.04(0.96)/250} = 0.0124$. Again, $z = 1.96$, so $E = 1.96(0.0124) = 0.024$. The error limit decreases by about one percentage point.

We will employ the expression $E = z\hat{\sigma}_{\hat{p}}$ to answer two other types of questions, sample sizes and levels of confidence, just as we did in Unit II for estimates of the mean.

Section 8.11.1 Problems

153. Twenty percent of 1,500 American adult respondents to a survey favor prohibiting sales of all alcoholic beverages to alleviate alcohol abuse. What can we say with 99% confidence about the size of the error of this estimate? Interpret the result. (Source: "New Prohibition?" *Asheville Citizen-Times*, February 8, 1991, p. 5A.)

154. A poll of 1,226 adults revealed that 49% believe that the devil may sometimes possess earthlings. Find a 95% confidence level limit on the sampling error of this estimate. Interpret the result. (Source: "Demons Begone," *Asheville Citizen-Times*, April 5, 1991, p. 1A.)

155. During a recession a poll of 56 senior loan officers of large domestic banks found that about
  a. 14% tightened residential mortgage standards.
  b. 17% were less willing to make general-purpose consumer loans.
  c. 25% tightened commercial real estate loans.
  d. 15 experienced less demand for commercial and industrial loans.

  Use an 80% confidence level to find the maximum sampling error of each of these estimates. Interpret the result. (Source: "New Survey Says Fewer Banks Tightening Lending Standards," *Asheville Citizen-Times*, April 4, 1991, p. 8B.)

156. One year when Christmas and the New Year fell on Wednesdays, 58% of a sample of 545 employers were giving employees three or more paid holidays. Find a 90% confidence level limit on the sampling error of this estimate. Interpret the result. (Source: "Not So Generous," *Wall Street Journal*, December 10, 1991, p. A1.)

8.11.2 The Appropriate Sample Size, n

We can solve the formula $E = z\hat{\sigma}_{\hat{p}} = z\sqrt{\hat{p}(1-\hat{p})/n}$ for n, the necessary sample size to establish the limit E on the sampling error with $100(1-\alpha)\%$ confidence. We obtain

$$n = \hat{p}(1-\hat{p})\left(\frac{z}{E}\right)^2$$

Because we determine sample size prior to sampling, no $\hat{p}$ is available for the formula. We can use reliable information about likely values of p, say from other studies. We call this likely value of p, $\tilde{p}$. When no information is available, a conservative solution is to find the largest possible sample size, n, for the given allowable error, E, and confidence level, $100(1-\alpha)\%$.

Question: Try some values for $\hat{p}$ to see if you can determine the largest possible value of the expression $\hat{p}(1-\hat{p})$.

Answer: The largest possible value of $\hat{p}(1 - \hat{p})$ is $0.25 = 0.5(1 - 0.5)$. If we let $\hat{p}$ be any value other than 0.5, then $\hat{p}(1 - \hat{p})$ will be smaller.

Thus, if we have no clue about the value of p, we use $n = 0.25(z/E)^2$. Sometimes, we do have information that limits the value of p, call this limit $\tilde{p}$, such as a legal limit to the percentage of students that can be from out of state. If we use $\tilde{p}$, then we will need a smaller sample. The expression $\tilde{p}(1 - \tilde{p})(z/E)^2$ will be smaller than $0.25(z/E)^2$. Just remember to substitute an approximate value of p when the information is available.

To approximate the required sample size, n, in order to be $100(1 - \alpha)\%$ confident that the maximum sampling error is E, use $p(1 - p)\left(\dfrac{z}{E}\right)^2$, where $P(-z < Z < z) = 1 - \alpha$ and $p = 0.5$ (except when reliable information suggests another likely value for p).

■ **Example 8-23:** A student organization wants to estimate student interest in weekly concerts.

a. How many students should be in a random sample to be 80% confident that the error is at most 0.06?

b. Suppose that about 40% of the students regularly attend monthly concerts, according to the number of IDs shown at the performances. Assuming that a smaller percentage would regularly attend weekly shows, find the sample size needed to be 80% confident that the error is at most 0.06.

■ **Solution: a.** $E = 0.06$, and $z = 1.28$. We have no information about the likely value of p, so we use $p = 0.5$ and $n = 0.25(z/E)^2 = 0.25(1.28/0.06)^2 = 113.8$. Remember to round up to be 80% confident, so $n = 114$ students.

b. $E = 0.06$, $z = 1.28$. The experience with student IDs suggests that 0.4 is an upper limit on p, so $n = p(1 - p)(z/E)^2 = 0.4(0.6)(1.28/0.06)^2 = 109.2$, or 110 students. The use of 0.4 for p reduced the required sample size from 114 to 110.

Section 8.11.2 Problems

157. A newspaper reports that
   a. 40% of workers claim they would work harder for more pay.
   b. 30% of workers claim they would work harder for more recognition.
   c. 18.5% of take-home pay goes straight to paying off installment loans.
   d. 5% of people have more than one job.
   e. 12.5% of people work Sundays.
   What sample size is required to reestimate each percentage with 90% confidence that the error will be no more than 2.5 percentage points? Suppose you did not know about the newspaper report and find the required sample sizes. (Source: "Numbers Paint Another Picture of Union's State," *Asheville Citizen-Times*, February 1, 1990, p. 2A.)

158. The article mentioned in Problem 157 also lists the following percentages:

   a. 20% of people are "significantly obese."
   b. 26.8% of Americans walk for exercise.
   c. 22% of college men chew tobacco or use snuff.
   d. 29% of medical bills are paid by patients themselves.
   Determine the required sample size to reestimate each percentage with 95% confidence that the error will be at most one percentage point. Then recompute the required n, allowing the maximum error to increase to two percentage points. What happens to the required n, when the allowable error increases?

159. Researchers claim that 51% of workers lived in large metropolitan areas in 1986 and that the percentage seems to be rising. To reestimate the percentage for this year, with 95% confidence that the error is at most three percentage points, how many observations will we need? (Source: "Decision Makers Cluster in Metropolitan Areas," *Wall Street Journal*, July 19, 1991, p. B1.)

Box 8-4 Reporting Technical Aspects of Polls

Many *popular media* reports of polls supply a margin of error but omit the level of confidence or risk associated with the combined estimate and margin of error statement. One exception is the *Wall Street Journal,* which usually reports this information and frequently much more to inform readers completely to improve their understanding and evaluation of the statistics. The quote below describes the sampling procedure for a survey concerning voter attitudes.

How Poll Was Conducted

The *Wall Street Journal/*NBC News poll was based on nationwide telephone interviews of 1,500 registered voters conducted Friday through Tuesday by the polling organizations of Peter Hart and Robert Teeter.

The sample was drawn from 315 randomly selected geographic points in the continental U.S. Each region was represented in proportion to its population. Households were selected by a method that gave all telephone numbers, listed and unlisted, an equal chance of being included. One registered voter, 18 years or older, was selected from each household by a procedure to provide the correct number of male and female respondents. The results of the survey were minimally weighted by age and occupation to ensure that the poll accurately reflects registered voters nationwide.

Chances are 19 of 20 that if all registered voters with telephones had been surveyed using the same questionnaire, the findings would differ from these poll results by no more than 2.6 percentage points in either direction. The margin of error for subgroups would be larger.

Source: Rich Jaroslovsky, "Voters Voice Dismay About Nation's Course a Year before Election," Wall Street Journal, November 1, 1991, P. A6, with permission.

8.11.3 Confidence Associated with Claimed Error, $100(1 - \alpha)\%$

Question: Frequently poll results are reported on television, on radio, or in newspapers and magazines. Less frequently, the possible error in the proportion is mentioned. What other information is needed and seldom provided for a complete report of the estimate and error?

Answer: The level of confidence that the error is within the mentioned limits. Box 8-4 shows a counterexample. *The Wall Street Journal* provided many details about the sampling procedure and accuracy.

Computing the uncertainty of the error of a point estimate, $\hat{p}$ requires the same two steps as with point estimates, $\overline{X}$, of μ. We must determine z and then $100(1 - \alpha)\%$ from z. Solving $E = z\hat{\sigma}_{\hat{p}}$ for z produces $z = E/\hat{\sigma}_{\hat{p}}$. Then, $1 - \alpha$ is the area under the normal curve between $+z$ and $-z$. The following box summarizes these steps.

To calculate the $100(1 - \alpha)\%$ confidence level associated with the error, E, when using $\hat{p}$ from a large sample to estimate p,

1. Solve for z using $z = E/\hat{\sigma}_{\hat{p}}$, where $\hat{\sigma}_{\hat{p}} = \sqrt{\hat{p}(1 - \hat{p})/n}$.
2. Solve for $1 - \alpha$ by using $1 - \alpha = P(-z < Z < z)$.

■ **Example 8-24:** A politician hires a company to conduct a random survey of voters in the district. The company finds that 54% of 100 sampled voters support the candidate. It claims that the error is at most three percentage points, so the worst-case scenario seems to be $54\% - 3\% = 51\%$ of the voters support the candidate.

a. How confident should the candidate be of winning the election?
b. If the candidate requests results at the 95% confidence level, what is the limit on the error?

■ **Solution: a.** $n = 100$ and $\hat{p} = 0.54$, so $\hat{\sigma}_{\hat{p}} = \sqrt{0.54(0.46)/100} = 0.0498$.
$E = 0.03$, so $z = 0.03/0.0498 = 0.60$.

Figure 8-34
Solution diagram to determine
$1 - \alpha = P(-0.6 < Z < 0.6)$

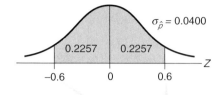

From Figure 8-34, we can see that $1 - \alpha = P(-0.6 < Z < 0.6) = 2(0.2257) = 0.4514$. Thus, the candidate can only be about 45% confident that the lead is enough for victory.

b. $z = 1.96$, and $\hat{\sigma}_{\hat{p}}$ remains 0.0498, so $E = z\hat{\sigma}_{\hat{p}} = 1.96(0.0498) = 0.0976$, or 9.76 percentage points. This result may concern the candidate more.

Section 8.11.3 Problems

160. A poll of 819 Southerners shows that 81% enjoy southern cooking without reservation. The report of this finding includes a claim that the margin of error is 3.5 percentage points. What level of confidence is associated with this statement? Interpret the result. (Source: "Southerners Love Bad-for-You Food," *Asheville Citizen-Times*, July 29, 1992, p. 3C.)

161. Fifty-six percent of a random sample believe their childrens' diets are sufficiently healthy. The sample consists of 853 adults from a national population estimated to contain 71.1 million adults in households with at least one child under age 18 in August of 1991. The report states that the margin of error is 3.4 percentage points. Find the confidence level associated with this error. Interpret the result. (Source: "Are Kids Eating Better than Nutritionists Think?" *Asheville Citizen-Times*, August 25, 1991, p. 4C.)

162. A national poll of grocery shoppers asked respondents to pick the personality whose weekly groceries they thought they would prefer. The personalities were Roseanne Arnold, Joan Collins, Bill Cosby, Jane Fonda, Madonna, and Sylvester Stallone. Thirty-three percent of the respondents chose Jane Fonda. Find the level of confidence associated with a margin of error of 2.5 percentage points if there were 1,000 respondents. Interpret the result. (Source: "Most Shoppers Prefer Fonda's Groceries over Madonna's," *Asheville Citizen-Times*, July 29, 1992, p. 2C.)

163. The poll in Problem 162 also found that 29% of the shoppers craved chocolate when asked to choose their biggest craving from the list: chocolate, pretzels and chips, ice cream, pizza, fried foods, or cookies. Find the level of confidence associated with a margin of error of 2.5 percentage points if there were 1,000 respondents. Interpret the result.

164. Recompute the confidence level in Problems 162 and 163 by using a margin of error of 3.5 percentage points rather than 2.5. Generalize the effect of increasing the margin of error on the level of confidence. Is this generalization intuitively correct? What do you hold constant when you recompute the level of confidence?

Summary and Review

We use the sample proportion of successes, also called the sample success rate, $\hat{p} = X/n$, to estimate the proportion of successes in the population, p. The sampling distribution of $\hat{p}$ is a modification of the binomial, because the random variable is the proportion of successes out of n trials, rather than the actual number of successes. The n trials are the items in the sample. For large samples, $\hat{p}$ is approximately normal with mean p and standard error $\sigma_{\hat{p}} = \sqrt{\dfrac{p(1-p)}{n}}$. This random variable is the basis for inferences about p.

We estimate $\sigma_{\hat{p}}$ with $\hat{\sigma}_{\hat{p}} = \sqrt{\hat{p}(1-\hat{p})/n}$ in the following formulas to present large-sample estimates of p.

1. A $100(1-\alpha)\%$ confidence interval for p. $\hat{p} \pm z\hat{\sigma}_{\hat{p}}$.
2. A limit, E, on the sampling error with a confidence level $100(1-\alpha)\%$ associated with the limit. $E = z\hat{\sigma}_{\hat{p}}$
3. The sample size, n, required to achieve a desired confidence level, $100(1-\alpha)\%$, that $\hat{p}$ is within E percentage points of p. $p(1-p)(z/E)^2$, where $p = 0.5$, unless reliable information suggests another likely value for p.
4. The confidence, $100(1-\alpha)\%$, that an estimate $\hat{p}$ is within E percentage points of p. $1 - \alpha = P(-z < Z < z)$, where $z = E/\hat{\sigma}_{\hat{p}}$.

Similar formulas for estimating the population mean can be found in Figure 8.14 (confidence interval, p. 343) and in the Summary and Review section of Unit II (E, n, $100(1-\alpha)\%$, p. 364).

Multiple-Choice Problems

165. This unit of the chapter concerns the estimation of which population parameter?
  a. The number of observations.
  b. The mean.
  c. The standard deviation.
  d. The proportion of successes.
  e. The level of confidence.

166. Which of the following describes the *approximate* sampling distribution of $\hat{p}$ that we employ in inferences about p?
  a. Binomial with $\mu = np$ and $\sigma = \sqrt{np(1-p)}$.
  b. Approximate normal with $\mu = np$ and $\sigma = \sqrt{np(1-p)}$.
  c. Standard normal.
  d. Normal with $\mu = p$ and standard error $\sqrt{p(1-p)/n}$.
  e. Normal with mean μ and standard error $\sigma/\sqrt{n}$.

167. Law-enforcement officials randomly stop 800 drivers and test them for intoxication. Suppose that 10% of the drivers on the road are, in fact, legally intoxicated. Which of the diagrams in Figure 8-35 would we use to find $P(\hat{p} < 0.12)$?
  a. Diagram a.   d. Diagram d.
  b. Diagram b.   e. Diagram e.
  c. Diagram c.

168. In a random sample of 647 vacationers, 58% indicated that their favorite pastime is fishing. Find a 98% confidence interval for the proportion of vacationers who favor fishing.
  a. (0.535, 0.625).
  b. (0.561, 0.599).
  c. (0.542, 0.618).
  d. (0.480, 0.680).
  e. Insufficient information to do the calculation.

169. Suppose we selected the sample in Problem 168 from vacationers who emphasize outdoor recreation rather than just vacationers. Suppose also that the sample still contains 647 observations, but now 88% opt for fishing rather than 58%. The resulting 98% confidence interval is (0.850, 0.910) compared with the longer interval (0.535, 0.625) in Problem 168. Why is the new interval shorter?
  a. We increased n in the denominator.
  b. We decreased the dispersion in the population by obtaining a more similar group of vacationers, which should increase the accuracy of our estimate.
  c. We estimated a different statistic, $\overline{X}$, not the sample success rate.
  d. The normal curve is a better approximation with $p = 0.88$ than with $p = 0.58$.
  e. There is no explanation; it is just a mathematical quirk.

170. Twenty-six percent of 1,200 randomly selected shoppers say that customer service is poor. What can we say with 80% confidence about the error, if we use the sample results to estimate the dissatisfied proportion of all shoppers?
  a. The error is at most about 0.016.
  b. The error is 0.016.
  c. The maximum error is 0.025.
  d. The error is in the neighborhood of 0.025.
  e. The error is at least 20%.

171. How many students must we include in a sample to be 87% confident that the sample proportion of out-of-state students is no more than 0.08 from the entire student body proportion?
  a. 151 students.   d. 41 students.
  b. 27 students.   e. 200 students.
  c. 90 students.

172. If we decrease the limit on the error E in Problem 171, what happens to the required sample size?
  a. It increases.   c. It remains the same.
  b. It decreases.   d. E cannot be changed.

173. A poll claims that 75% (±four percentage points) of the electorate intend to vote after finding that 300 of 400 randomly selected registered voters intend to vote. What is the confidence level associated with this statement?
  a. 0.7500.   c. 0.5468.   e. 0.4678.
  b. 0.0400.   d. 0.9356.

174. In a March 1990 *Wall Street Journal*/NBC News poll, 22% of the respondents expressed "a lot" of concern over the reunification of the two Germanies. The sample consisted of 1,003 registered voters. The report

Figure 8-35
Choices for Problem 167

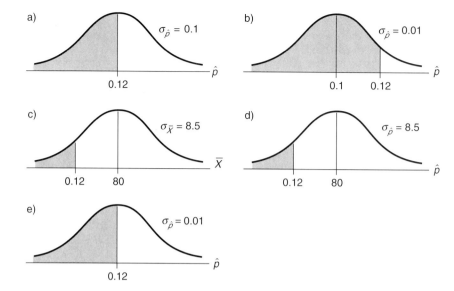

states that the error is no more than 3.2 percentage points. What level of confidence do we associate with this error magnitude for the results of the reunification question? (Source: David Shribman, "Americans, while Voicing Some Deep Concerns, Widely Favor German Reunification, Poll Shows," *Wall Street Journal*, March 16, 1990, p. A16.)
a. 95%.
b. 49.29%.
c. 2.45%.
d. 1.31%.
e. 98.58%.

175. According to the Federal Aviation Administration (FAA), about 400 of 47,000 licensed passenger plane pilots have been convicted of alcohol-related driving violations. If we consider this a random sample from such pilots over the history of commercial passenger flights and calculate a 99% confidence interval for the proportion convicted, the result is (0.0074, 0.0096). What can we interpret this result to mean? (Source: "FAA: Nearly 400 Pilots Caught Driving Drunk," *Asheville Citizen-Times*, September 24, 1991, p. 3A.)
a. There is a 99% chance that a pilot has been convicted.
b. There is about a 1% risk that more than 1% of these pilots have been convicted.
c. There is almost no chance of flying with a sober pilot.
d. There is 99% confidence that the FAA has drawn an incorrect conclusion.
e. There is 99% confidence that the true proportion is outside this interval.

Use the following information to answer Questions 176–178.

We allow respondents to indicate their opinion on a scale of 1 to 5, before estimating the proportion who respond a certain value, such as 4.

176. How many observations should we include to be 95% confident that our estimate is off by no more than 10 percentage points?
a. 97.
b. 384.
c. 385.
d. 62.
e. 61.
f. 96.

177. If we average the responses for all individuals in the sample, how many observations should we include to be 95% confident that our estimate is within 0.25 scale units of the true opinion rating? Recent studies indicate that the standard deviation of responses to this question on other surveys is 1.05.
a. 100.
b. 47.
c. 68.
d. 5.
e. 48.

178. Reconsider Question 177. Suppose there was no information from recent studies, and determine an upper limit on σ if the population is not normal. Use this value to recalculate the required sample. The respective answers to these questions are
a. 4, 984.
b. 4, 32.
c. 2, 20.
d. 2, 246.
e. 1, 15.

179. Find a 98% confidence interval for the proportion of volleyball teams that win over half of their games if 72 of 133 randomly selected teams have accomplished this feat.
a. (0.50, 0.58).
b. (61, 205).
c. (0.4407, 0.6420).
d. (0.4000, 0.6000).
e. (0.4567, 0.6260).

180. Which of the following will *increase* the necessary sample size in order to be $100(1 - \alpha)\%$ confident that the sampling error is at most E when estimating p?
a. Legal constraints that p must be between 0.75 and 0.80.
b. Legal constraints that p must be between 0.05 and 0.08.

c. Increasing the tolerable error beyond E.
d. Increasing the confidence level associated with the maximum error specification above $100(1 - \alpha)\%$.
e. Decrease the z value in the formula by increasing the risk associated with the tolerable error.

181. A psychologist finds that 42% of a random sample of 85 patients in a rehabilitation center respond within two days to recreation therapy. What is the maximum possible sampling error associated with this estimate? The psychologist wants to be 90% confident about the statement.
   a. 3.5 percentage points.   d. 21.0 percentage points.
   b. 10.5 percentage points. e. 8.8 percentage points.
   c. 17.6 percentage points.

182. An automobile firm determined that 400 out of 850 randomly selected individuals would be willing to purchase a yellow car. Find an 80% confidence interval for the proportion of people willing to buy yellow cars.
   a. (0.4619, 0.4793).   d. (−1.28, 1.28).
   b. (0.4370, 0.5041).   e. (0.4487, 0.4925).
   c. (0.4535, 0.4877).

183. A certain firm claims that its remedy cures 90% (±0.5%) of all maladies, based on a random sample of over 10,000 patients. What is the confidence level for this statement?
   a. 0.0030.
   b. 0.4525.
   c. 0.9500.
   d. 0.9050.
   e. 0.8726.

184. A community wants to estimate the current local illiteracy rate within 10 percentage points, with 95% confidence. The illiteracy rate in the last census (six years ago) was 15%, but volunteers have conducted adult education programs since that year, so they expect the rate has decreased. How many randomly selected citizens should they include in a sample to estimate the current illiteracy rate?
   a. 385.   d. 48.
   b. 96.   e. 49.
   c. 97.

Word and Thought Problems

185. Suppose 50 items out of 1,000 randomly sampled items from a production line are found to be defective. Find a 95% confidence interval for the defect rate in this production process. Interpret the result.

186. Draw and label a diagram showing the approximate sampling distribution of $\hat{p}$. Include the formula for the standard error.

187. a. Suppose 55 out of 100 randomly selected students prefer that athletic scholarships be equally distributed among all sports. Find a 95% confidence interval for the proportion of all students in favor of this idea.
   b. Is there a clear indication that a majority of the students favor the idea?
   c. At what level of confidence could you say that a majority are in favor of the idea? What would be the risk that your statement is incorrect?

188. Suppose 400 out of 600 randomly selected persons prefer container A over container B. Find a 95% confidence interval for the proportion of the populace preferring container A. Interpret the result.

189. a. Suppose 20 out of 175 randomly selected meters give erroneous readings. Find a 95% confidence interval for the proportion of defective meters. Interpret the result.
   b. If we use 0.1143 as a point estimate of the defect rate, what can we say with 95% confidence about the size of the sampling error? With 90% confidence?

190. In Problems 185, 187, and 188, find what you can say with 95% confidence about the size of the error if you estimate a value rather than an interval for the population success rate with the sample success rate.

191. As a political analyst, you wish to be 99% sure that your estimate of the proportion of the vote the former senator will receive in the next election is within 0.03 of the true value. How many voters do you need for your random sample?

192. A shopping mall director wishes to obtain an estimate of the percentage of Saturday morning shoppers who come from an adjoining county in order to better target her marketing program. How many license plates should she randomly check to be 80% sure she is within five percentage points of the true percentage?

193. a. How many bowlers must be randomly interviewed to ascertain all bowlers' interest in a new bowling alley in a town? The prospective investor wishes to be 90% sure the estimate is within 0.02 of the true proportion or level of interest.
   b. Just by listening to bowlers in the present alley complain about the existing facilities, the investor finds that about 60% seem dissatisfied. Does this information change your answer to Part a? Why or why not?

194. From a random sample of 2,000 alumni, a university official asserts that 10% of the alumni locate within the state after graduation. She claims the estimate is within 0.01 of the true percentage. What is the confidence level associated with her assertion? Interpret your answer.

195. A political activist claims 59% of the electorate will oppose a local referendum. He allows a 2% error and bases his claim on the results of a random survey of 563 voters. What is the confidence level associated with the claim? Interpret the answer.

196. a. Using a random number table or random number generator on a computer, collect a random sample of 100 digits. Find a 95% confidence interval for the proportion of fours in the table or on the generator. Interpret the interval.

b. What proportion of the entire population of digits should be fours in a set of random digits? Does your confidence interval contain this proportion?

c. If a group of 2,000 analysts calculate 95% confidence intervals from samples drawn from the same set of random digits, how many do you expect to contain the true proportion of fours?

197. a. A dairy farmer wants to estimate the probability that an artificially inseminated cow from the herd will become pregnant. A publication on the process suggests that 60% of inseminated cows are impregnated. How many cows should the farmer select for a random sample of the herd to estimate the probability of impregnation within four percentage points with 95% confidence?

b. If the farmer ignores the publication information, how would the answer to Part a change? Demonstrate the answer by solving for the appropriate sample size for the new circumstances.

198. a. A charitable organization mailed 2,000 requests for funds for a special project to randomly selected individuals from a mailing list. Twelve hundred donors responded. Construct a 90% confidence interval for the donor rate for this project from the entire mailing list. Interpret the result.

b. What is the confidence level for a claim that the sample donor rate is within two percentage points of the true rate? Interpret this answer.

c. Fifty-six of the 1,200 donations arrived in the first two weeks after the requests were mailed. The mean of these donations is $25, with a standard deviation of $5. Assume this is a random sample. Use this information to make a statement about the size of the error when estimating the mean donation with 90% confidence.

d. How many observations should the charity include in a random sample to estimate the mean response time within one week, with 95% confidence? For previous years, the standard deviation of response times is 4.62 weeks.

199. a. Thirty-five percent of a random sample of 400 voters favor candidate X. Find a 95% confidence interval for the proportion of the entire electorate who favor candidate X. Interpret this interval.

b. Find a 95% confidence interval for the proportion of the electorate who do not favor candidate X. Compare this result with that of Part a. Again, provide an interpretation of the interval.

200. Discuss the effect of a larger n on the width of a confidence interval, the size of E, and the level of confidence, $100(1 - \alpha)\%$, associated with an error statement.

201. When studying pregnant women from 21 different states during the years 1987 and 1988, analysts produced the following 95% confidence intervals for prevalence of alcohol consumption (number who consume/number in the sample) in the past month: 0.22 ± 0.04 in 1987 and 0.20 ± 0.03 in 1988.

a. The sample proportion of successes decreases between the two years. Given the confidence intervals, can we be sure the prevalence rate is different for the two years? Explain.

b. What is the estimated standard error of the sample proportion, $\hat{\sigma}_{\hat{p}}$, in each year?

c. Among pregnant women who currently smoke, alcohol consumption prevalence confidence intervals (95%) are 0.42 ± 0.10 in 1987 and 0.37 ± 0.10 in 1988. Again, can we be sure the prevalence of alcohol consumption is different in these subpopulations? Explain.

d. What is $\hat{\sigma}_{\hat{p}}$ for the situation described in Part c? (Source: Mary Serdula, David F. Williamson, Juliette S. Kendrick, Robert F. Anda, and Tim Byers, "Trends in Alcohol Consumption by Pregnant Women," *Journal of the American Medical Association*, Vol. 265, No. 7, February 20, 1991, pp. 876–79.)

202. Thirty-eight percent of a random sample of 1,500 college students prefer a live band over a DJ. What can we say with 98% confidence about the error if we use this information to estimate the percentage of all students who prefer a live band over a DJ?

203. A congressional representative ponders how to vote on a major piece of controversial legislation that appears to contain an equal number of good and bad points. To learn about constituents' feelings, the representative mails a survey asking a random sample of 250 district voters to express their feelings about the legislation (strong in favor, strongly opposed, or no sentiment). The legislator plans to vote the constituents' sentiments expressed by more than half the respondents.

a. What is the likelihood of more than 50% of the 200 respondents strongly favoring the bill, if, in fact, 40% of the constituents strongly favor the bill, 40% strongly oppose the bill, and 20% are indifferent.

b. What is the likelihood of more than 50% of the 103 respondents with strong feelings being in favor of the bill, given the breakdown of constituents' sentiments shown in Part a?

c. Describe some caveats you might warn the representative about before accepting the survey results as proof of constituents' viewpoints.

d. The responses are equally divided, so the representative decides to sample a small segment of swing voters from the last election. Sixty-five percent of the sampled segment *oppose* the bill. What is the probability more than 50% of the sample of 100 voters will *favor* the bill?

204. Figure 8-36 displays Minitab output for a variable called SAMPLE. The tally in the center of the figure

Review Problems

```
              N        MEAN      MEDIAN     TRMEAN     STDEV     SEMEAN
SAMPLE       153      0.7059     1.0000     0.7299    0.4571    0.0370

             MIN       MAX         Q1         Q3
SAMPLE      0.0000   1.0000      0.0000     1.0000
```

Figure 8-36

Minitab output of confidence interval of population proportion of successes

```
SAMPLE    COUNT
  0         45
  1        108
  N=       153
```

```
              N        MEAN       STDEV    SE MEAN       95.0 PERCENT C.I.
SAMPLE       153      0.7059     0.4571    0.0370      ( 0.6328,  0.7789).
```

indicates there are 153 SAMPLE values: 45 zeros and 108 ones, where a 1 indicates a success, such as a defective, and a zero indicates a failure.
  a. Use the tallies to construct a 95% confidence interval for the proportion of successes in the population from which the values originated.
  b. Compare your $\hat{p}$ and $\hat{\sigma}_{\hat{p}}$ from Part a with the mean and standard error of the mean in the descriptive statistics section of the figure. Then compare your 95% confidence interval with that provided in the figure. Do your comparisons corroborate the proposition in Box 8-3 (p. 373) that proportions are really means?
205. Park rangers from Lucky Lake classify fifteen of 258 netted fish as junk fish. Calculate an 80% confidence interval for the proportion of junk fish in the lake. Interpret the result.
206. A newspaper claims the result of its poll of 500 randomly selected voters has a margin of error of 3.5 percentage points. According to the poll, 48% of the sampled voters favored the proposed state constitutional amendment. What is the level of confidence associated with this error statement? Interpret your answer.
207. A random sample of 480 citizens of a community contains 42 victims of crime. What can we say with 90% confidence about the size of the error if we use the sample results to estimate the rate of victimization in the entire community.
208. Five hundred and sixty-eight of 2,012 randomly selected shoppers say they make no effort to find a bargain price for an item for which they are shopping. What can be said about the error with 95% confidence if the sample result is used to estimate the proportion of noncomparison shoppers in the entire population?
209. Twenty-six percent of a random sample of 1,300 students subscribe to a sports-oriented publication. Find a 90% confidence interval for the subscription rate for such publications among all students. Interpret the result.
210. How many children should be selected for a random sample to estimate the proportion of all children who want to be astronauts in order to be within 0.1 of the true proportion with 99% confidence?
211. A commission wants to estimate support for various issues in a given locality. How many citizens should be included in the sample to be 80% sure the estimates are within 5 percentage points of the true percentage who support each issue?
212. Twenty-two percent of 4,020 randomly selected individuals hold the opinion that the vice-presidential nominee of a party should be the second highest vote getter in the primaries. Find a 90% confidence interval for the true percentage of all individuals who hold this opinion. Interpret the result.

Review Problems

213. Magazines, newspapers, and television usually present estimates or survey results as point estimates rather than confidence intervals. What impression or information does a confidence interval give to the observer that is missing when the media reports only a point estimate?
214. a. Consider inferences about μ made with $\overline{X}$. Enter the appropriate word "increase" or "decrease" in each open cell of the following box to describe the impact of an increase of the row variable on the computed column variable, assuming none of the other variables involved in the calculation changes. For

example, an increase in σ, when nothing else changes, causes the ensuing confidence interval's width to increase, because a more dispersed population increases the difficulty of obtaining good estimates. Mathematically, the standard error, $\sigma_{\bar{x}} = \sigma/\sqrt{n}$, increases, so the interval covered by $\bar{X} \pm z\sigma_{\bar{x}}$ must increase. Ignore cells containing an "X." The last column is the confidence level associated with a specific error statement.

An Increase of	Confidence Interval Width	E	n	$100(1 - \alpha)\%$
The sample size, n			X	
Tolerable error, E	X	X		
Confidence level, $100(1 - \alpha)\%$				X
Risk, $100\alpha\%$				
Population standard deviation, σ	Increases			

b. Give an intuitive and mathematical explanation for each response in Part a.
c. How would your responses in Part a change if we considered decreases in the row variables rather than increases?
d. If we were estimating p rather than μ, would any of your responses in Part a change? Explain.

215. Over the last 50 years there have been an average of 12 tropical storms per year in a certain oceanic region. The standard deviation is one storm. Use this information to form a 90% confidence interval for the mean number of tropical storms in a typical year. What assumption do you violate when forming the confidence interval from this sample? Does your estimate mean there will be 12 tropical storms this year?

216. A local television station decides to show back-to-back reruns of "M*A*S*H," basing its decision on a random survey of 200 local residents. The station claims 40% preferred "M*A*S*H" over a list of other possible series reruns, with an error of 2.5 percentage points. What is the confidence level for this statement? Interpret the result.

217. Streaking Stanley is slightly senile and can recall only four of his marathon times: 2.8 hours, 3.05 hours, 3.1 hours, and 3.5 hours. Estimate his mean marathon time as a 95% confidence interval, based on this information. What do you assume to interpret this 95% interval?

218. A random sample of 100 overdue accounts has a mean of $26.78 and a standard deviation of $2.13. There is 98% confidence that the sample mean is within how many dollars of the population mean?

219. Wally wants to organize service club members into a tutoring service. He wants to know the effect students expect on their grades from tutoring. Ultimately, he decides to survey students to learn what they expect from four hours of tutoring at $6 per hour. Responses fall along a scale stretching from -100 to $+100$ percentage points, where 0 is to maintain the grade on the previous test. For example, -5 means to keep the student's grade from dropping more than 5 points from the last test score and $+5$ means to improve that grade by 5 points.

a. What population should Wally use?
b. He resorts to selecting students from all over campus. A convenience sample of his friends suggests the population standard deviation of expectations may be in the neighborhood of 25 points. Find the number of students to contact to estimate the mean expectation within 8 points with 95% confidence.
c. Suggest a way to select students to contact for the sample.
d. The following list shows 38 interview responses. Use this information to estimate the mean expected grade effect, and comment on the error possibilities with 95% confidence. Does the estimate meet the intended specifications about error size?

-58	-22	1	-22	26	19
-46	22	-22	-22	-18	10
15	4	-42	-43	30	-13
-69	-6	52	30	-64	13
-24	-52	27	-68	-6	-79
29	10	-38	14	20	-16
32	-23				

e. Nine of the 38 respondents indicated they were interested in purchasing four hours of tutoring at $6 per hour in at least one of their courses. Estimate the proportion of all students willing to purchase four hours and comment on the error possibilities with 95% confidence.

220. What happens to the curve for a T distribution as the degrees of freedom increase? Explain how this change affects the choice of z or t for a confidence interval.

221. According to a poll, Americans average one hour and 22 minutes consuming meals per day: 19 minutes for breakfast, 28 for lunch, and 35 for dinner. (Source: "What Do Polls Say about Consumers?" *Asheville Citizen-Times*, August 25, 1991, p. 4C.)

a. Suppose the sample consists of 426 Americans, and the sample standard deviations for the four measures are 15 minutes, 5 minutes, 7 minutes, and 12 minutes, respectively. Find the maximum error associated with each estimate with 80% confidence.
b. Compare the size of the maximum errors of any one or all of the meals with the size of the maximum error for the total time spent at meals. Which is largest? Explain intuitively what you have observed.

222. a. Select a random sample of five single digits from the random number table in Appendix I. Find an 80% confidence interval for the mean digit in the table using this sample.

Review Problems

```
              N      MEAN    MEDIAN    TRMEAN     STDEV    SEMEAN
CASES        10    16.549    16.935    16.686     1.558     0.493

            MIN       MAX        Q1        Q3
CASES    13.750    18.250    15.090    18.153
```

Figure 8-37
Minitab output for U.S. champagne and sparkling wine consumption

```
              N      MEAN     STDEV   SE MEAN       80.0 PERCENT C.I.
CASES        10    16.549     1.558     0.493    ( 15.868,   17.230)
```

b. What is the probability distribution of the population? Verify that you do or do not satisfy the usual assumptions for a confidence interval for the population mean in Part a.

c. What is the population mean? Does your confidence interval contain the true mean?

223. IQ scores are normally distributed with mean 100 and standard deviation 15.

  a. If a confidence interval to estimate the mean of all IQ values does not contain 100, such as (104, 112), does this imply that the confidence interval procedure was not followed correctly? Explain your answer.

  b. Decreasing the sample size, n, increases the range of a $100(1-\alpha)\%$ confidence interval. If we continue to return to the population and select new samples each with fewer observations, will the corresponding confidence intervals eventually include 100? Explain.

224. As part of an experiment to learn about the effect of sight in the seasonal reproductive cycle of American tree sparrows, analysts weighed the ovaries of seeing and blind females after controlled exposure to light. The analysts reported the weight of each group of birds as the sample mean ± standard error of the mean, $\overline{X} \pm \hat{\sigma}_{\overline{x}}$. For the sightless group the result is 6.11 ± 1.55 mg and for the seeing group 7.82 ± 1.18. There are 12 birds in each group. (Source: Fred E. Wilson, "Neither Retinal nor Pineal Photoreceptors Mediate Photoperiodic Control of Seasonal Reproduction in American Tree Sparrows (*Spizella arborea*)," *The Journal of Experimental Zoology*, Vol 259, No. 1, July 1991, pp. 117–27.)

  a. What level of confidence do we associate with these confidence intervals?

  b. Is it possible that the mean ovarian weight for all female American Tree Sparrows is identical, even though the sample means differ? Explain. How do the two confidence intervals confirm your answer?

225. Figure 8-37 provides Minitab output for the following set of ten U.S. champagne and sparkling-wine consumption values (expressed in millions of liter cases). We want to estimate the mean consumption over a longer period. (Source: Estimated from a chart from Jobson's Wine Marketing Handbook, as reported in Kathleen Deveny, "Desperate Vintners to Glum Consumers: Cheer Up and Buy Some Holiday Bubbly," *Wall Street Journal*, December 12, 1991, pp. B1, B5.)

13.75	15.12	17.12	18.25	18.25
18.12	17.25	16.75	15.88	15.00

  a. Verify the computer results, using the values.

  b. Someone claims that the mean is 20 million cases. Does this confidence interval provide evidence to support or refute this claim? Explain your answer.

226. Figure 8-38 shows computer output for the population of highest mean SO_2 concentrations and for a random sample of 35 values from the population. The data are taken from the Air Quality Data Base (Appendix B). The sample output contains omitted values and purposefully misprinted values, but the upper endpoint of the confidence interval, the standard deviation, and n are correct.

  a. Complete and correct the output.

  b. Use the output to determine if we violate any assumptions to make the level of this confidence interval 80%.

227. Determining "mean global temperature" to consider the greenhouse effects is a problematic statistical as well as scientific endeavor. Scientists collect temperature readings in many different locations, but certainly not every point on the globe. Some locations are closely situated and others are more distant, for instance, where population is dense versus not dense. Also locations have changed as well as the number of locations over the last century. (Source: Richard F. Gunst, "The Scientific Controversy over Global Warming," *Stats*, No. 6, Fall 1991, pp. 3–7.)

  a. Discuss how location can cause outliers in the data set.

  b. Discuss how location might distort a mean of the readings.

  c. Suggest an alternative measure.

228. A random sample of Shane's sleep time includes 6, 9, and 8 hours. Assuming that all of his sleep times are

Figure 8-38
Minitab output concerning maximum SO_2 concentration

a) The population

	N	MEAN	MEDIAN	TRMEAN	STDEV	SEMEAN
SO2	63	23.37	29.00	21.71	24.57	3.10

	MIN	MAX	Q1	Q3
SO2	0.00	93.00	0.01	43.00

```
Histogram of SO2    N = 63

Midpoint    Count
    0         30    ******************************
   10          0
   20          0
   30          6    ******
   40         15    ***************
   50          7    *******
   60          2    **
   70          1    *
   80          1    *
   90          1    *
```

b) The sample

	N	MEAN	MEDIAN	TRMEAN	STDEV	SEMEAN
SO2SAMP	35	100.35	29.00	20.65	23.59	_____

	MIN	MAX	Q1	Q3
SO2SAMP	0.00	76.00	0.01	43.00

	N	MEAN	STDEV	SE MEAN	80.0 PERCENT C.I.
SO2SAMP	35	100.35	23.59	_____	(_____ , 27.56)

normally distributed, use this information to construct an 80% confidence interval for Shane's mean sleep time. Interpret the result.

229. A medical researcher believes he has developed a new test for a disease that will more accurately detect the presence of the disease than previous tests. To obtain information concerning its improvement over the old test, he obtains permission from 500 patients about the world with the disease to have the test administered. The old test correctly predicted the presence of the disease 85% of the time, so it would have predicted about 425 of these 500 cases. The researcher will continue developing and promoting the new test if it correctly predicts 450 or more of the 500.

   a. Does the procedure seem reasonable? Explain.
   b. What are the chances of 450 or more correct predictions when the true prediction rate is the same as the old test, 85%?
   c. If there are 450 correct predictions with the new test, calculate a 90% confidence interval for the predictive accuracy rate. Interpret the result.
   d. If the researchers want to estimate the predictive accuracy rate of the new test within 2 percentage points with 95% confidence, how many patients should be administered the test?
   e. If the researchers claim the accuracy rate is 90%, based on 450 correct predictions, and the margin of error is at most 2% points, what is the confidence level associated with this assertion?

230. The number of observations in a sample is very influential on the outcome of many statistical problems about μ.

   a. Discuss the effect of a larger n versus a smaller n on the standard error of the mean and the implication for using $\overline{X}$ to estimate μ.
   b. Give an intuitive explanation for this implication.

Review Problems

   c. Given your conclusions, what effect should an increase in *n* have on the length of a $100(1 - \alpha)\%$ confidence interval for μ?
   d. How can we use sample size to control the confidence of making at most a given error when using the sample mean to estimate μ?
   e. What effect does the cost of sampling have on the choice of *n*?

231. How many students should be included in a random sample to estimate within two percentage points, with 95% confidence, the proportion of all students who plan to purchase a yearbook? In the past, 60% have purchased the book.

232. Marcia's mean hourly count of pages read and studied in a random sample of 48 hours of study is 42.8 pages with a standard deviation of 5.5 pages. Find a 98% confidence interval for the mean hourly page count. Interpret the result.

233. There are many similarities between confidence intervals for the population mean, μ, with large and small samples. Whether the population standard deviation, σ, is known or unknown, we require additional information about the population to calculate a confidence interval for μ when $n \leq 30$. What else must we know? What difference will a known or unknown standard deviation make on the computations?

234. National intelligence headquarters has just obtained new information on the maximum speed attainable by new enemy aircraft. An agent slipped into a classified computer section momentarily and randomly scanned five flight reports. He transmitted the speeds to national headquarters. There is little likelihood of obtaining more observations soon. Can this group confidently piece this small set of numbers into an estimate of the enemy's new mean fleet speed? Explain.

235. Instead of sampling all high school students to determine the proportion planning on college, a researcher samples high school seniors. Should the error associated with the new estimate be larger or smaller than an estimate based on all high school students? Explain intuitively and mathematically.

Chapter 9

Hypothesis Testing

Hypothesis testing is another form of statistical inference. We estimate a population parameter, such as its mean μ, from a sample, just as we did for point and confidence interval estimation techniques. In this case, our objective is to determine which of two values or groups of values, suggested as the actual population parameter, is supported by a sample from the population. This chapter presents methods for testing hypotheses for various situations.

Objectives

Unit I introduces the fundamental concepts of hypothesis testing, then formally develops and demonstrates them for a large-sample test of a hypothesis that the population mean, μ, is a particular value, μ_0. The alternative hypothesis claims that μ is not equal to μ_0 but can be larger or smaller than μ_0. This type of test is termed a two-tailed test of the population mean. Upon completing the unit, you should be able to

1. State the basic logic behind hypothesis testing.
2. Do a large-sample, two-tailed test for μ.
3. Interpret the result of the test and state whether the actual test statistic value that occurs is likely or unlikely to occur if the population mean, μ, were actually μ_0.
4. Define null and alternative hypotheses, Type I and II errors, significance level, significant difference, decision rule, critical value, rejection and acceptance regions, and test statistic.
5. Distinguish the critical value, c; the significance level, α; the probability of a Type II error, β; the null hypothesis, H_0; and the alternative hypothesis, H_1.

In Unit II we investigate several new situations that extend the basic ideas of hypothesis testing introduced in Unit I. Upon completing the unit, you should be able to

1. Test whether a parameter deviates in one direction (larger or smaller) from the value specified in H_0 (a one-tailed test).
2. Perform large- and small-sample tests for the population mean, μ, when the population standard deviation is not known and a large-sample test for the population success rate, p.
3. Perform these tests by using given measurement units or by transforming the values to Z or T values.
4. Set up H_0 and H_1 for a problem.

Unit III develops and demonstrates the p-value, a method of reporting hypothesis test results that summarizes most details of the test in a single value. Upon completing the unit, you should be able to

1. Calculate a p-value for a test for μ and p.
2. State intuitively what a p-value measures.
3. Use a p-value to make a decision for a hypothesis test.
4. Define p-value.

Unit I: Hypothesis Testing Basics

Preparation-Check Questions

1. If random samples of 625 elements each are selected from a population with mean 500 and standard deviation 100, what percentage of the possible sample means will be smaller than 490 or larger than 510?
   a. 92.64. d. 49.38.
   b. 1.24.   e. 98.76.
   c. 0.62.

2. A random sample of 625 elements is selected from a population with mean 500 and standard deviation 100 (the same situation as in Problem 1). Find the value of c such that $P(\overline{X} > c) = 0.05$.
   a. 664.0. d. 696.0.
   b. 506.6. e. 493.4.
   c. 507.8.

3. $P(A|B)$ means the probability that
   a. A and B occur.
   b. A or B occurs.
   c. A occurs given that B occurs.
   d. B occurs given that A occurs.
   e. A divided by B equals a specified value.

Answers: 1. b; 2. b; 3. c.

Introduction

Cost Estimate An employee gives you, the project director, an estimate for a job by combining the average costs of labor and material. He then claims that the average cost for such a job is $7,320. Because this employee's estimates of average costs have been wrong and cost you money in the past, you are suspicious of the validity of the estimate. Before you confront and contradict him, you decide to collect evidence formally that shows you are right and he is wrong.

What kind of evidence do you need? At what point will you have enough evidence to convince the employee that the real job cost will differ from his estimate? Will simply offering a counterestimate, a mean from a random sample of costs of similar jobs, provide sufficient contradictory evidence? Perhaps you would rather not challenge the employee's estimate, unless you can convince yourself that your evidence is overwhelmingly at odds with his estimate. After all, your data form a sample of similar jobs, not a population.

9.1 The Basic Logic behind Hypothesis Testing

One way we can present convincing evidence that the cost estimate of the job, $7,320, is wrong, would be to perform a hypothesis test. A *hypothesis test* is an analysis of sample data in order to choose between two competing statements about the population: "the average job cost will be $7,320," "the average job cost will not be $7,320." We call each of the competing statements a *hypothesis* and hypothesis testing provides us with an acceptable scientific procedure for decision making. It is also based on a simple idea we use in everyday reasoning.

Suppose your roommate tells you her class grade is a B. She randomly pulls some, but not all, of her tests and papers from her desk and shows the following grades: (B, B, B). The evidence does support her claim.

Question: Suppose the grades were (C, C, D) or (A, A, A). Do these grades support her claim?

Answer: No. Neither set of grades averages close to a B.

Question: Suppose the grades are (A, B, A) or (B, B, C). Do these grades support her claim?

Answer: These grade sets make it harder to refute your friend's claim of a B average, especially the latter set.

In this example the competing hypotheses are (1) the average grade is a B and (2) the average grade is not a B. If your roommate's claim is correct, her sample of grades should lead you to believe that a B is possible from combining the individual grades. There is no question that (B, B, B) supports your friend's claim. If you give your friend the benefit of the doubt, then you will not say she is wrong when you find tests grades such as (A, B, A) or (B, B, C). Remember, you only have a sample of her grades, not all of them, and it may not be a random sample. However, you will probably reject her statement when you see grades like (C, C, D) or (A, A, A).

The simple idea behind hypothesis testing follows the same logic. Initially, we assume one hypothesis is true or give it the benefit of the doubt. This "assumed" hypothesis includes a definitive statement about an unknown value, such as "your roommate's average grade is a B" ($\mu = B$). It may also include a less specific statement, such as "your roommate's average grade is a B or better" ($\mu \geq B$). The alternative hypothesis contradicts the "assumed" hypothesis; "your roommate's average grade is not a B" ($\mu \neq B$), or "your roommate's average grade is lower than a B" ($\mu < B$). Once we formulate the "assumed" and alternative hypotheses, we examine the evidence (sample data) to see which hypothesis it supports. Sometimes the evidence will clearly support one of the hypotheses. Questionable evidence that does not clearly support either hypothesis causes us to favor the "assumed" hypothesis, because it gets the benefit of the doubt. However, *evidence that would be unlikely to occur, if the "assumed" hypothesis were true, but it occurs anyway, causes us to reject the "assumed" hypothesis and choose the alternative hypothesis.* This is the basic logic behind the technical hypothesis testing procedure.

Question: A salesperson says the set of four new tires sold to you last week are top-of-the-line. However, three flats have occurred without a road incident since the purchase. Does this evidence cause you to believe the tires are not top-of-the-line?

Answer: Most people would say the evidence contradicts the salesperson's claim and you should return the tires.

After any of the flats you could decide that the salesperson's claim that the tires were top-of-the-line was false. Of course, you risk being wrong, because your four tires are only a sample of all tires of this brand and style. An alternative explanation for the result is your hapless purchase of a set of tires that includes one or more faulty tires even though these tires are generally top-of-the-line.

Question: When do you experience the greatest risk of wrongly accusing the salesperson of a false claim; if you make the accusation after the first flat, the second, third, or fourth? Can you ever be certain that the salesperson's claim is false? Explain.

Answer: The greatest risk of a false accusation occurs after a single flat, because you have only one bad tire out of four. The risk decreases after each subsequent flat, because the proportion of bad tires in your sample increases. That is, the amount of evidence that contradicts the salesperson's claim increases. The risk never completely disappears, unless you check every tire this firm sells. Only then would you know for certain the quality of every tire.

Box 9-1 provides another everyday example of a hypothesis test. It is not technical, but the logic and common sense of the method are evident.

Box 9-1 Maps and the Basic Idea of Hypothesis Testing

Highway travelers and tourists depend on road maps to reach their destinations. When we purchase or accept a road map, we display some confidence in its cartographer and expect it to provide accurate and current information about alternative routes. Suppose you take a trip, or maybe you are a traveling salesperson with distant routes to cover. If your first trip according to an expensive, detailed map leaves you floundering in Timbuktu rather than arriving at your destination, you may blame the error on the remote nature of the trip or some map-reading mistake on your part. If you get lost on a second journey, you may retract your faith in the map's accuracy. The cost in money and time of not reaching your destination promptly may be sufficient at this point to convince you to dispense with the map altogether.

More specifically, suppose a man enjoys ferrying across rivers in remote areas rather than crossing bridges. He possesses an atlas showing the location of ferries that prompt him to travel with his family out of their way in search of a river ride. After several disappointing side journeys without locating the ferries, his confidence in the atlas wanes. In addition, his family's confidence in his ability to locate a geographic objective dissipates, if not disappears. Finally he relents and buys a new, up-to-date atlas.

Section 9.1 Problems

4. A person claims to be able to predict the outcome of most tournament basketball games. A competing hypothesis would be that the person is unable to do this. If the person predicts the winners of a slate of 100 games, would you believe this person's claim if he correctly guesses all 100 games? none of the games? 25 of the games? 50 of the games? 75 of the games? If you give this person the benefit of the doubt, will any of your answers change? If you do not give him the benefit of the doubt, will any of your answers change? When do we face the greatest risk of mistakenly accusing the person of a false claim?

5. Some people claim there was a conspiracy to assassinate President Kennedy in 1963 that involved more than Lee Harvey Oswald. The competing hypothesis is that Oswald acted alone. Oswald was killed before resolving the issue, and there is no conclusive evidence to prove either statement. Discuss the importance of the benefit of the doubt when people reject one hypothesis or the other.

6. Compare the following decisions with the simple idea behind hypothesis testing:
   a. A decision to trust a friend to keep secrets.
   b. A decision to purchase a refrigerator from a firm that advertises "customer satisfaction guaranteed."
   c. A decision to purchase your lunch from a fast food restaurant that advertises "our hamburgers are hot off the grill."

7. Make a general statement to summarize the relationship between the amount of contradictory evidence we require before rejecting a claim and the risk of being wrong when we reject the claim.

9.2 Basic Terminology of Hypothesis Testing

We call the "assumed" hypothesis, the one that we give the benefit of the doubt, the *null hypothesis* and symbolize it with H_0. We denote the other hypothesis, or *alternative hypothesis*, with H_1. These hypotheses will be statements about population parameters, such as the population mean, variance, and success rate:

$$H_0: \mu = 500 \qquad H_0: \sigma^2 \leq 25 \qquad H_0: p \geq 0.6$$
$$H_1: \mu \neq 500 \qquad H_1: \sigma^2 > 25 \qquad H_1: p < 0.6$$

Observe that H_0 always includes the equal sign. The reason for this will become evident later.

Note also that H_0 and H_1 are contradictory. Either H_0 or H_1 is true, and you must examine the evidence (in a sample) to decide which hypothesis the evidence supports. Traditionally, we state the two options for this decision as "reject H_0" and "cannot reject H_0."

You may be tempted to change the phrase "cannot reject H_0" to "accept H_0." This phrase is sometimes used, although it can be problematic. One reason is that H_0 is given the benefit of the doubt, and evidence that does not solidly support either hypothesis causes us to decide in favor of the null hypothesis, H_0. With such weak evidence for H_0, we prefer to avoid the risk of incorrectly accepting it. Ideally, we collect more data and further examine the validity of the two hypotheses instead. This is especially true when H_0 is a statement we do not believe; in fact, it is usually a proposition that we wish to disprove.

Often, the sample's evidence is ambiguous, it does not clearly support either hypothesis. Hence, we need a *decision rule* to indicate what evidence will be sufficient to reject the null hypothesis, H_0. We extract the evidence from the sample in the form of a *test statistic,* the statistic that we employ to indicate or estimate the population parameter being tested. *If the test statistic is sufficiently different from the population value proposed in H_0, we will reject H_0.* If H_0 is the roommate's claim that her grades average to a B, the decision rule might be to *reject H_0 if the sample of her grades cannot average a B, $\overline{X} \ne B$* (ordinarily we use values not letters or words in a formal test). Thus, a sample like (C, C, D) or (A, A, A) cannot average to a B, so we would reject H_0. Some evidence, such as (B, B, B) clearly supports H_0. More often evidence will not clearly support or refute H_0. It is conceivable, but not obvious, that samples such as (A, B, A) or (B, B, C) could average to a B. Because H_0 receives the benefit of the doubt, we decide not to reject H_0 when faced with such ambiguous evidence.

The decision rule forms regions, the *acceptance region,* the sample evidence that causes us not to reject H_0, and the *rejection region(s),* the sample evidence that causes us to reject H_0. If we position the average grades along a number line, we can picture the acceptance and rejection regions as shown in Figure 9-1. Notice that the acceptance region consists of potential test statistic values that support H_0 (sample means that are or may average a B), whereas the rejection region(s) consist of potential test statistic values that are sufficiently different from the population value proposed in H_0 to convince us to reject H_0 (sample means that cannot average a B).

Question: Suppose the salesperson's claim that the tires you purchase are top-of-the-line is H_0. Your sample consists of the four tires in the set you purchased. State a decision rule that requires a count of flat tires that is also the strongest evidence you can collect from your set of tires that the claim is false. What is the test statistic?

Answer: The most convincing evidence that the salesperson's claim is false would be four flat tires out of four. Thus, the decision rule would be to reject H_0 if four out of the four tires go flat. The test statistic is a count of the number of flats in the sample of four tires or, stated a bit differently, the proportion of tires in the sample that go flat.

Because either statement, H_0 or H_1, can be right or wrong, four situations can occur in the process of deciding between the two hypotheses. These situations are represented in the following diagram.

	True Statement	
Sample Evidence Supports a Decision to	H_0	H_1
Not reject H_0	1	3
Reject H_0	2	4

Box 1 represents the situation where H_0 is a true statement, and the evidence (the test statistic's value) supports a decision not to reject H_0.

Question: Which of the four boxes represent wrong decisions?
Answer: Boxes 2 and 3.

Box 2 represents the situation where, unknown to you, H_0 is true, but the test statistic leads you to reject H_0. For instance, if your roommate's grades do average to a B

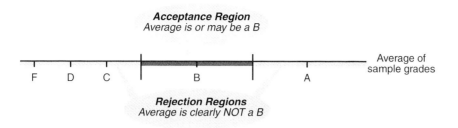

Figure 9-1

The acceptance and rejection regions reflect the decision rule

(H_0), but you observe a sample such as (C, C, B), the sample may lead you to conclude that the average is not a B (reject H_0). We call this mistake, rejecting a true null hypothesis, a *Type I error*.

	True Statement	
Sample Evidence Supports a Decision to	H_0	H_1
Not reject H_0		
Reject H_0	Type I error	

The other wrong decision is a *Type II error*. It occurs when we do not reject H_0, but, in fact, H_1 is true. You would make this error if the test statistic leads you to conclude that your roommate's grades average a B (not reject H_0), when, in fact, they do not (H_1).

	True Statement	
Sample Evidence Supports a Decision to	H_0	H_1
Not reject H_0		Type II error
Reject H_0		

H_0 and H_1 are statements about the population, but the decision rule dictates whether or not we reject H_0 based on sample evidence only. Consequently, we risk making a Type I or Type II error. The risk, or probability, of a Type I error is labeled α, which stands for the *level of significance of the test*.

$$\alpha = P(\text{Type I error}) = P(\text{reject } H_0 \text{ when } H_0 \text{ is true}).$$

If we use conditional probability notation, we can write

$$\alpha = P(\text{reject } H_0 \mid H_0 \text{ is true}).$$

If H_0 is true, then α is the probability that the decision maker will obtain sample data that will lead to a decision to reject H_0 and, thus, commit a Type I error. Figure 9-2 illustrates a significance level of 0.20 for the roommate's grade example. There is a 20% chance of obtaining a sample that, according to the decision rule, causes us to reject the roommate's claim of a B average, when, in fact, the average is a B.

If we substitute "obtain a test statistic that is sufficiently distant from the proposed value in H_0" for "reject H_0" in the definition of the significance level, we can understand the meaning of significantly different. *Whenever we reject H_0 at the α significance level, the test statistic is sufficiently different from the proposed value in H_0 for us to decide that the proposed value is false. There is a risk level of α that this decision is wrong when H_0 is true.* In short, when we reject H_0, we conclude that the population parameter is *significantly different* from the proposed value in H_0.

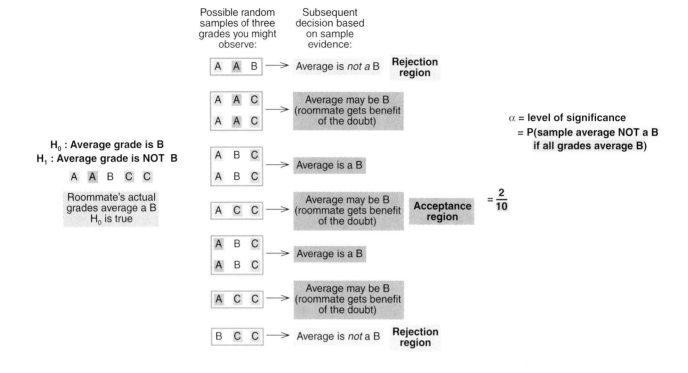

Figure 9-2
The significance level of a test of your roommate's claim that her average grade is a B

How much must the test statistic and proposed value differ in order for us to reject H_0? We answer this question when we choose the level of significance, α, for the test. We select a significance level based on how serious the consequences of making an error would be—how much risk of a mistake the decision maker can afford. If α is very small, then we want a small risk that we make a Type I error (rejecting H_0 when H_0 is true), so we need to make the acceptance region around the proposed value large and the rejection regions small. See Figure 9-3a. If we are more willing to risk a Type I error, then we can expand the range of values that causes us to reject H_0, leaving fewer values in the acceptance region. See Figure 9-3b. In practice, we frequently use levels or 0.10, 0.05, or 0.01.

Although we emphasize Type I errors, there is also a risk of making a Type II error whenever we cannot reject H_0. If H_1 is true, β is the probability that the test statistic leads the decision maker not to reject H_0, resulting in a Type II error.

$$\beta = P(\text{Type II error}) = P(\text{cannot reject } H_0 \text{ when } H_1 \text{ is true}).$$

$1 - \beta$ is the proportion of decisions that will be correct when H_1 is true. $1 - \alpha$ is the proportion of decisions that will be correct when H_0 is true.

Sample Evidence Supports a Decision to	True Statement	
	H_0	H_1
Not reject H_0	$1 - \alpha$	β
Reject H_0	α	$1 - \beta$

Question: Does it make sense to test at a level of significance greater than 0.5?
Answer: No. If you do, you would make a Type I error more than 50% of the time when H_0 is true. Tossing a coin to decide whether H_0 is true would produce a Type I error 50% of the time without the trouble of collecting a sample and making calculations.

a) A small α to avoid rejecting a true H_0

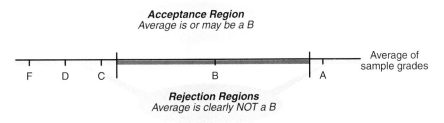

b) A large α to allow for more risk (less concern) for rejecting a true H_0

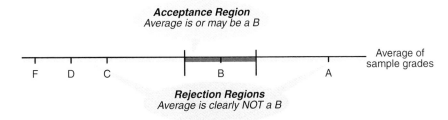

Figure 9-3
The effect of the choice of the level of significance, α, on the size of the acceptance and rejection regions

Question: Suppose you set up the following hypotheses about a defendant in a trial:

H_0: Defendant is innocent.

H_1: Defendant is guilty.

Think of what a Type I error and a Type II error would be in this circumstance. Does the U.S. judicial system attempt to minimize a Type I or Type II error?

Answer: A Type I error is to decide a defendant is guilty who is actually innocent. The accused person is presumed innocent and given the benefit of the doubt. The prosecutor must "prove beyond a reasonable doubt" that H_0 is false (that is, prove that the defendant is guilty). A Type II error results when the jury decides a defendant is innocent who is actually guilty. This is an error that the judicial system attempts to avoid also, but the court system is constructed to minimize a Type I error. If we reverse the statements of H_0 and H_1, the system would minimize a Type II error.

Now with the logic of hypothesis testing and the basic terminology in mind, we will formalize the procedure. We want to produce decisions methodically that others can reproduce given the same information. Boxes 9-2 and 9-3 describe practical situations where hypothesis testing logic applies.

Section 9.2 Problems

8. Define the following terms: null hypothesis, alternative hypothesis, Type I error, Type II error, decision rule, level of significance, acceptance region, rejection region, and test statistic.
9. Which hypothesis gets the benefit of the doubt when we perform a hypothesis test?
10. Identify the symbols α, β, H_0, and H_1.
11. Recall the person from Problem 4 who claims to be able to predict the outcome of most tournament basketball games.
    a. If this claim is H_0, what is H_1?
    b. Suppose the person forecasts the outcome of a random sample of 100 tournament games. Write a decision rule that you would use, based on the 100 predictions, to decide if the person's claim is true. What is the acceptance region? What is the rejection region? What is the test statistic?
    c. State in words what a Type I and a Type II error would be in these circumstances. Be specific.
12. A denture cleaner manufacturer states that the average annual wear on dentures from daily use of its cleaner is 0.002 cubic millimeters. To test this claim, we randomly

Box 9-2 Using Hypothesis Testing to Make a Practical Decision

Concerned about disciplinary violations on school buses, the St. Louis Board of Education instituted an experimental program during the 1991 spring term. A bus was equipped with a video system to entertain students during the ride and thus curb violations. Officials kept records for the bus with and the buses without the video system. These were the experimental and control groups, respectively. (The control group is the basis for comparison in an experiment, because it does not experience the change or "treatment" the experimental group does.)

Although comparisons between the experimental and control groups are a major focus of the study (statistical comparison of such groups is the subject of Chapter 10), a test for a difference from the average violation rate established before the experiment began could be useful as well, assuming methods and behavior for reporting incidents did not change between the two periods. Such a comparison is useful for reporting interim results before the experiment is completed—for example, to a board meeting or public relations committee—especially when a decision must be made whether to continue to finance a venture.

The average number of daily violations of a less serious nature, such as throwing objects, shoving, head or arms outside windows, or tampering with the emergency door, was 14.7 violations per day over the previous three school years, as reported by drivers and principals. Because officials expect the video system to reduce violations, the hypotheses are $H_0: \mu \geq 14.7$ and $H_1: \mu < 14.7$. The test could be performed using a random sample of daily violation reports from the spring term.

If officials compare results obtained for the entire spring term with the previous three-year-period average, the result would be a population comparison between known values. If they consider days during the spring term to be a random sample of the present and future days, then we would test for a significant difference.

The actual experiment was more complex than described here. There were other variables that could affect the results, and officials tried to avoid any special or systematic influence from these variables—for example, they assigned different drivers to the video bus on different days to avoid differential effects of bus-driver personalities and behavior. Can you think of other variables you would try to account for if you were conducting the experiment?

Do you think publicity about the experiment and the board's objectives affected the outcome? How would you expect students to behave on the video bus if they understood the purpose of the experiment? What about students on the buses without videos?

Delivering the message about the discipline problems on the bus may have been more important to the board than strict control of the experiment, especially if publicity could not be controlled. Multiple regression analysis (discussed in Chapter 11) is another statistical technique that helps avoid problems when we cannot control other variables that can affect results.

Source: Information provided by Harry Acker, Jr., Audiovisual Services of the St. Louis Public School System, May 30, 1991.

---

select 200 denture wearers as subjects and find the space their dentures displace in a beaker of water. The subjects agree to use the denture cleaner daily for a year, after which we measure the space displacement once again. The result is a random sample of 200 wear values.
   a. Suggest and justify a test statistic for this test.
   b. If H_0 is the claim that the mean wear is 0.002 cubic millimeters, a counterclaim would be that the mean wear is not 0.002 cubic millimeters. Express H_0 and H_1 with symbols.
   c. State in words what a Type I and Type II error would be in these circumstances.
   d. Suggest and explain a decision rule for this test.
13. What happens to the size of the acceptance and rejection regions when the significance level, α, decreases? Does decreasing α mean we want it to be easier or more difficult to reject H_0? Think of a reason why we would not decrease α to a very small value, even 0.

9.3 The Hypothesis Testing Procedure

Now we are ready to work through a test. Basically, we must determine if the distance between a test statistic and the population value proposed in the null hypothesis, H_0, is sufficient to reject H_0. The result will determine if the population parameter is significantly different from the proposed value.

Box 9-3   Errors and Speed Traps

Typically, we hear that traffic patrols provide drivers some leeway above the stated speed limit before they issue a speeding citation. To use police resources wisely, they must weigh the danger of different over-the-limit speeds against the cost of detecting and citing corresponding speeders. By not apprehending every driver who speeds by only a small amount, they are free to pursue other beneficial tasks for a community.

Another explanation for the leeway is that police want to avoid falsely accusing an innocent person of speeding. Suppose there is a question about the accuracy of their measuring instrument. They can only detect a vehicle's speed within some limit with a certain level of confidence, such as being 95% sure the radar reads within 3 mph of the actual speed. If the radar says the vehicle is traveling at 66 mph in a 65-mph zone, the officers cannot be sure the vehicle is speeding. It may actually be moving at 64 or 65 mph.

If the traffic patrollers give drivers the benefit of the doubt and require ample evidence of speeding before stopping a driver, they avoid a Type I error. Because they are usually interested in speeds above the limit, the test is upper-tailed. If we use the numbers in the preceding paragraph and a decision rule that says to stop a driver when the radar indicates a speed 3 mph or more over the posted limit, such as 68 mph or more in a 65 mph zone, the probability of a Type I error occurring is 0.025. If they were on the lookout for vehicles traveling too slowly, a potential menace as well, the test would be lower-tailed and would require the driver to be going 3 mph or more below the posted minimum limit according to the radar instrument.

A Type II error in this example occurs when speeders are not cited. We do not know a specific value for the probability that this occurs, β, but we do know that its likelihood decreases as the speed above the posted limit increases. It is easier to detect a speeder going 20 mph over the limit than one going 5 mph over the limit. The chances of not detecting the faster driver should be lower than the chances of not detecting the slower speeder.

Another known aspect of the probability of a Type II error is the tradeoff between the probabilities of Type I and Type II errors when these probabilities are the only values that change in a situation. By giving drivers more than the 3-mph leeway in this example—say they give 5 mph—the chances of a Type I error (citing a nonspeeder) decrease. Consequently, some speeders are not going to be cited—a Type II error—and this error will occur more frequently.

The general form for the hypotheses in this unit are

$$H_0: \mu = \mu_0$$
$$H_1: \mu \neq \mu_0,$$

where we specify a value for μ_0. Although the basic idea and procedural steps are the same for most hypothesis tests, some details of testing situations will vary, such as the population characteristic being tested or the sample size.

■ **Example 9-1: Cost Estimate:** Recall that you, a project director, doubt an employee's estimate of $7,320 for a job cost. To avoid contradicting the employee's estimate without substantial evidence that it is incorrect, you decide to perform a hypothesis test. From a list of actual costs of jobs similar to the job your employee estimated, you randomly select 49 to test the estimate. The mean and standard deviation of the sample are $\overline{X} = \$7{,}500$ and $S = \$700$. Obviously, your sample mean, $7,500, is different from the employee's estimate, $7,320. Is the $180 difference between the two values sufficient to uphold a conclusion that the mean estimate for all such jobs would be significantly different from $7,320?

■ **Solution:** A step-by-step solution that incorporates the terminology and concepts from earlier sections follows.

Step 1   State the hypotheses.

$$H_0: \mu = \$7{,}320 \text{ (the employee's estimate)}$$
$$H_1: \mu \neq \$7{,}320.$$

Notice the equal sign in H_0. H_1 says $\mu \neq \$7{,}320$, because if H_0 is incorrect, the population mean could be larger or smaller than \$7,320. We have no reason to expect that it can only be larger than \$7,320 or that it can only be smaller than \$7,320.

Here we can also appreciate the importance of placing the equal sign in our statement of the null hypothesis, H_0. If H_0 were $\mu \neq \$7{,}320$, there would not be a single proposed value for the population parameter. Consequently, there would not be a single distance between the test statistic value and the proposed population value to check for significance. There would be no specific rejection and acceptance regions, they would vary with the decision maker's choice of a value that is not \$7,320. To avoid this ambiguous situation, we always place the equal sign in H_0 and use the value specified in H_0 to conduct the test. *Whenever the statement to be tested proposes a single value for the population parameter, such as the mean cost of these jobs is \$7,320, H_0 will be that the parameter equals this value and H_1 will be that the parameter does not equal this value.*

Step 2 Choose the significance level α.

Let α be 0.01. The significance level is the risk of incorrectly rejecting H_0, in this case, of falsely contradicting our employee. Because we are very concerned to avoid such a mistake, we use $\alpha = 0.01$, a small risk level. (We are more willing to risk accepting a wrong estimate than to risk undermining our employee.)

Step 3 Determine the appropriate test statistic.

We will use the sample mean, $\overline{X}$, to estimate the population mean, μ.

Step 4 Determine and sketch the sampling distribution of the test statistic.

We must describe the sampling distribution of $\overline{X}$ in order to determine values of $\overline{X}$ that are likely to occur when H_0 is true ($\mu = \$7{,}320$) and those that are unlikely to occur unless H_1 is true ($\mu \neq \$7{,}320$).

The sampling distribution of $\overline{X}$ is approximately normal for a large sample according to the central limit theorem. Because the sample contains 49 values, $n = 49$, we can assume that the $\overline{X}$ values follow a normal distribution. The mean for the $\overline{X}$ values is $\mu = \$7{,}320$, because we assume that H_0 is true. The estimated standard error is $\hat{\sigma}_{\overline{x}} = S/\sqrt{n} = 700/\sqrt{49} = \$100$. Notice that we use the sample standard deviation, S, to form this estimate of $\sigma_{\overline{x}}$, because we do not know the population standard deviation, σ. This is the situation we encounter most of the time, because if we raise a question about the value of the population mean, there is little chance that we will know the population standard deviation. Figure 9-4 shows a diagram that summarizes this information.

Step 5 Determine the general form of the decision rule and the corresponding rejection region(s).

The decision rule dictates which values of the test statistic, $\overline{X}$, are sufficiently different from the proposed μ_0 value from H_0, so that we reject H_0.

Question: If $\overline{X}$ is close to \$7,320, would you reject H_0?

Answer: Because $\overline{X}$ is an indication of μ, if $\overline{X}$ is close to \$7,320, it would indicate H_0 ($\mu = \$7{,}320$) may be true and the employee's estimate cannot be rejected with the available data. If $\overline{X}$ is sufficiently different from \$7,320, H_0 would be rejected.

Because the true mean costs of all such jobs, μ, may be larger or smaller than \$7,320, the rejection region consists of two parts. Figure 9-5 shows the rejection regions of small $\overline{X}$ values that are sufficiently different from \$7,320 and the large $\overline{X}$ values that are sufficiently different from \$7,320. We call the boundaries of these regions, values c_1 and c_2, *critical values*.

Question: Describe the values of $\overline{X}$ in the rejection region. Use the symbols $\overline{X}$, c_1, and c_2.

Answer: We reject H_0 if $\overline{X} \leq c_1$ or $\overline{X} \geq c_2$.

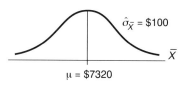

Figure 9-4
The sampling distribution of $\overline{X}$ for the job cost test

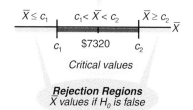

Figure 9-5
The decision rule and corresponding acceptance and rejection regions for the job cost test

Section 9.3   The Hypothesis Testing Procedure   401

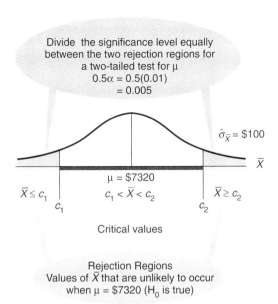

Figure 9-6
The position of the rejection and acceptance regions in the sampling distribution diagram

Step 6 Determine the critical value(s) numerically.

The rejection regions are in the tails of the normal sampling distribution of $\overline{X}$, because they must include $\overline{X}$ values that are unlikely to occur when H_0 is true ($\mu = \$7{,}320$). Figure 9-6 combines the sampling distribution of Figure 9-4 with the rejection and acceptance regions from Figure 9-5. Because the rejection region falls in both tails of the curve whenever H_1 contains $\neq$, we call such a test a *two-tailed test*.

The area in the two tails of the sampling distribution curve (above the rejection regions) is the numerical probability that we reject H_0 when H_0 is true. This probability is, by definition, the significance level, $\alpha = 0.01$. By convention, *we divide the significance level equally between the two regions for a two-tailed test for μ*. In this case 0.005, or half of 0.01, is allocated to each tail area as shown in Figure 9-6.

The next step to determine the values of c_1 and c_2 is to find the T or Z values that correspond to c_1 and c_2 (we will approximate T with Z for large samples), so that we can transform back to dollar values. Figure 9-7 shows that the area between the mean and either critical value is 0.4950. We find that the Z values that correspond to these areas are ± 2.575 from the Z table (Appendix H). We solve for c_1 and c_2 with the transformation formula $Z = (\overline{X} - \mu)/\hat{\sigma}_{\overline{x}}$.

Figure 9-7
The numerical critical values

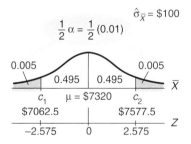

$$\frac{c_1 - 7{,}320}{100} = -2.575 \qquad \frac{c_2 - 7{,}320}{100} = 2.575$$

$$c_1 - 7{,}320 = -257.5 \qquad c_2 - 7{,}320 = 257.5$$

$$c_1 = 7{,}062.5 \qquad c_2 = 7{,}577.5.$$

Now we know c_1 and c_2, so we have determined the acceptance and rejection regions as shown in Figure 9-7. If $\overline{X} \leq \$7062.5$ or $\overline{X} \geq \$7577.5$, we will reject H_0. Otherwise, we cannot reject H_0.

Question: If H_0: $\mu = \$7{,}320$ is true, what is the probability we will obtain a sample mean, that is, $\$7{,}062.5$ or smaller, or a sample mean that is $\$7{,}577.5$ or larger?

Answer: The probability is $\alpha = 0.01$.

Step 7 Make a decision whether or not to reject H_0 by comparing the actual values of the test statistic with the critical values.

Figure 9-8
Locating the test statistic, $\overline{X}$, to make a decision

$\overline{X} = \$7500$ in the Acceptance Region (between $7062.5 and $7577.5)

The mean of the sample of 49 job costs, $\overline{X}$ is $7,500, which is between $7,062.5 and $7,577.5, so we are unable to reject H_0. See Figure 9-8. We cannot reject H_0 because our evidence is not sufficient to contradict the employee's estimate, the mean job cost is not significantly different from $7,320 at this significance level. Many different $\overline{X}$ values can result from the sampling process when $\mu = \$7,320$. However, the result $\overline{X} = \$7,500$ is not so unusual or unlikely when $\mu = \$7,320$, so we are not ready to dispense with H_0 in favor of the alternative.

Question: In not rejecting H_0, what type of error might we be making, Type I or Type II?

Answer: Type II. When we do not reject H_0, H_1 might be true—which describes a Type II error.

Question: If the sample mean from the 49 job cost values were $6,934.6 instead of $7,500, what would the decision be? Explain.

Answer: We should reject H_0, because $\overline{X} = \$6,934.6$ is smaller than $c_1 = \$7,062.5$. $\overline{X} = \$6,934.6$ would fall under the left tail of the graph in Figure 9-8.

A sample mean of $6,934.6 is unlikely to occur if the actual mean is $7,320, so we would reject H_0. We would contradict the employee's estimate of $7,320 while risking a 1% chance of this being the incorrect action. Essentially, we would say that the curve is misplaced along the $\overline{X}$ axis. The sample evidence would indicate that the population mean was not equal to $7,320, so the curve that describes the true situation should be drawn with a different μ, for example, $\mu = \$6,900$ (see Figure 9-9).

The following box summarizes the steps to perform a hypothesis test.

Hypothesis Testing Procedure

1. State the hypotheses, H_0 and H_1, about the population parameters. The equal sign belongs in H_0.
2. Choose the level of significance, α.
3. Determine the appropriate test statistic.
4. Find the appropriate standard error, degrees of freedom (see Unit II), or other information needed for determining the sampling distribution of the test statistic *assuming H_0 is true*. Draw a diagram of this sampling distribution.
5. Determine the location of the rejection region or regions from the decision rule that corresponds to the statements of H_0 and H_1. Position the critical values on the diagram.
6. Determine the actual numerical value(s) for the critical value(s) to use to make the decision. These may be measured in the original units of the situation or in transformed values, such as Z values (the next unit discusses transformations).
7. Decide whether or not to reject H_0 by comparing the test statistic actually obtained from the sample (or its transformation) with the critical value(s) from Step 6.

Figure 9-9
Rejecting H_0 means that the true sampling distribution has a different μ, specifically one that makes $\overline{X} = \$6,934.6$ a more likely observation

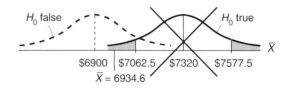

Section 9.3 — The Hypothesis Testing Procedure

■ **Example 9-2:** The water pressure that a delicate underwater seismic sensor can withstand needs to average 325 pound per square inch (psi) in order for an experiment to run smoothly. If the device could withstand significantly more pressure, the geologists and engineers would employ a cheaper model. If it could not withstand this pressure, they would need a better model. The mean of a random sample of 100 of the devices' resistance limits is 321.6 psi with a standard deviation of 16.5 psi. Does this evidence indicate the device may be inadequate ($\mu \neq 325$) at a 0.05 significance level?

■ **Solution:** The test will be two-tailed, to detect significant variation in either direction from the necessary standard (325 psi).

$$H_0: \mu = 325$$
$$H_1: \mu \neq 325.$$

Refer to Figure 9-10a for the diagram that results when we assume that H_0 is true ($\mu = 325$ psi). It summarizes the sampling distribution and given information. Remember to use $\hat{\sigma}_{\bar{x}} = S/\sqrt{n} = 16.5 \text{ psi}/\sqrt{100} = 1.65$ psi to describe the dispersion of the $\bar{X}$ values, not the sample standard deviation, $S = 16.5$ psi. The critical values correspond to the z values ± 1.96 (splitting $\alpha = 0.05$ between the two tails, leaves an area of 0.475 between the mean and each critical value, an area that corresponds to a Z value of 1.96). Through the transformation formula $Z = (c - \mu)/\hat{\sigma}_{\bar{x}}$, we obtain:

$$\frac{c_1 - 325}{1.65} = -1.96 \qquad \frac{c_2 - 325}{1.65} = 1.96$$
$$c_1 = 321.77 \text{ psi} \qquad c_2 = 328.23 \text{ psi}.$$

Because the observed sample mean, $\bar{X}$, is 321.6 psi < 321.77 psi = c_1, we reject H_0. The evidence indicates that the device is inadequate to withstand the required average pressure. The mean pressure it can withstand is significantly different from 325 psi.

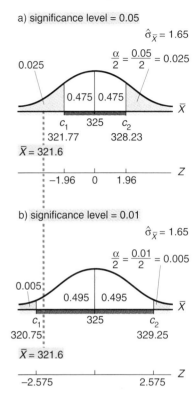

Figure 9-10
Diagrams for the underwater seismic sensor example

Question: Are you likely to obtain a sample mean of 321.6 psi if the population mean is 325 psi?
Answer: No. 321.6 psi is in the shaded region, so the probability of getting an $\bar{X}$ that is 321.6 or smaller must be less than α, the total shaded area of 0.05. In fact, the upper limit on this probability is even smaller. It is less than 0.025, because the shaded area in one tail is only 0.025.

It is important to choose the significance level near the start of the process. Otherwise, the level can be changed and manipulated to achieve any decision the decision maker desires. For instance, if the significance level in the preceding example were 0.01 rather than 0.05, the critical values would become 320.75 and 329.25. In this case, we would fail to reject H_0, because the sample mean 321.6 would be between the two critical values. See Figure 9-10b.

Question: When you use a smaller value for α, what type of error, I or II, becomes more likely?
Answer: A Type II error. The acceptance region expands with a smaller α (compare the acceptance regions in Figure 9-10 Parts a and b), making it more likely that we will not reject H_0, even if H_0 is wrong. This increases the probability of a Type II error.

Section 9.3 Problems

Determine if you can reject H_0 in each of the following cases:

14. H_0: $\mu = 25$
    H_1: $\mu \neq 25$
    $\alpha = 0.03$
    $n = 215$
    $S = 35.3$
    a. $\overline{X} = 20.5$.
    b. $\overline{X} = 23.7$.
    c. $\overline{X} = 39.9$.

15. H_0: $\mu = 444$
    H_1: $\mu \neq 444$
    $n = 100$
    $S = 50$
    $\overline{X} = 450$
    a. $\alpha = 0.05$.
    b. $\alpha = 0.10$.
    c. $\alpha = 0.20$.

16. H_0: $\mu = 0$
    H_1: $\mu \neq 0$
    $\overline{X} = -1.71$
    $S = 12.00$
    $\alpha = 0.1$
    a. $n = 169$.
    b. $n = 100$.
    c. $n = 49$.

17. H_0: $\mu = 123$
    H_1: $\mu \neq 123$
    $\alpha = 0.01$
    $n = 100$
    $\overline{X} = 139$
    a. $S = 50$.
    b. $S = 100$.
    c. $S = 200$.

18. An arachnoidologist claims that the female spider of a certain species produces an average of 500 eggs under any environmental conditions. You wish to test this claim for the same species, but in a habitat contaminated by air pollutants. You believe that pollution may alter reproductive behavior, causing spiders to produce either a lower or higher average number of eggs than they do under normal conditions. A random sample of 400 spiders' egg counts has a mean of 490 eggs and a standard deviation of 60 eggs. Use this information to test the arachnoidologist's claim at the 0.05 significance level. Interpret your decision.

19. A school system claims that the average student score on a standardized test is 7. Possible scores range from 0 to 9. Use the following random sample of 50 scores to test this claim at the 0.05 level. Interpret your decision.

9	9	7	4	5	2	5	4	8	8
4	8	9	4	8	7	9	5	6	3
9	7	6	2	6	2	2	5	6	7
7	4	8	9	6	8	4	5	9	3
4	8	4	1	7	4	9	6	9	8

9.4 The Computer and Hypothesis Testing for Population Means

Testing for the value of a population mean is a common procedure, so most statistical packages routinely process sample data and provide information for conducting such a test. Sometimes the information is part of a detailed set of descriptive statistics of the sample, so the computer saves you tedious computations. In other cases, the program allows the user to specify the value being tested, μ_0, and then conducts the test specified by the user.

Figure 9-11 shows examples of Minitab output for testing for population means. The circled items that we have already studied include the following (we will explore the other circled items in later units).

1. Sample mean, $\overline{X}$.
2. Sample standard deviation, S.
3. Estimated standard error of the mean, $\hat{\sigma}_{\overline{x}} = S/\sqrt{n}$.
4. μ_0 specified by Minitab user.
5. Direction of test specified by Minitab user (N.E. is "not equal," $\neq$; G.T. is "greater than," >; and L.T. is "less than," <).

■ **Example 9-3:** The output in Figure 9-11 displays statistics calculated from a random sample of 200 differences between assessed values and corresponding market values of houses, based on a random sample of 200 homes that sold within six weeks of

Figure 9-11
Minitab output for test of difference of assessed and market values

```
                    4      5    4
        TEST OF MU = 0 VS MU N.E. 0

              1       2       3        6        7
         N   MEAN   STDEV  SE MEAN     T     P VALUE
DIF    200   -896   7096     502    -1.79    0.076
```

being assessed by a local tax assessor ($n = 200$ is shown below the "N" in the output). A plus value means the tax assessor's estimate was higher than the selling price, and a minus, that it was lower.

Question: If we want to determine if there is any tendency to over- or underassess, what are the hypotheses?

Answer: If there is a tendency in either direction, the mean of the differences should be negative or positive. If there is no bias, the mean difference would be zero. Thus, the hypotheses are H_0: $\mu = 0$ and H_1: $\mu \neq 0$.

Use the computer-generated mean and estimated standard error to perform this test at the 0.05 level.

■ **Solution:** The computer generates the necessary information from the 200 observations in the sample. We know that $\overline{X} = -\$896$ and $\hat{\sigma}_{\overline{x}} = \$502$. To obtain the critical values we solve $(c - \mu)/\hat{\sigma}_{\overline{x}} = \pm z$ for c_1 and c_2.

$$\frac{c_1 - 0}{502} = -1.96 \quad \text{and} \quad \frac{c_2 - 0}{502} = 1.96; \quad c_1 = -\$983.9 \quad c_2 = \$983.9.$$

Because $\overline{X} = -\$896$ is between the two critical values, we cannot reject H_0. That is, the sample does not provide sufficient evidence to demonstrate a tendency to over- or underassess.

■ **Example 9-4:** An engineering firm designed a new thermostat to maintain the temperature in a building at an average of 68°. To test the success of this design feature, the firm collected a random sample of 45 temperature readings from the building. Use the computer output from these values, shown in Figure 9-12, to perform the test at the 0.1 level.

■ **Solution:** The hypotheses are H_0: $\mu = 68$ and H_1: $\mu \neq 68$ as shown on the output. Using the standard error provided by the output, 0.483, we proceed to solve $(c - \mu)/\hat{\sigma}_{\overline{x}} = \pm z$ for c_1 and c_2.

$$\frac{c_1 - 68}{0.483} = -1.645 \quad \text{and} \quad \frac{c_2 - 68}{0.483} = 1.645; \quad c_1 = 67.2055 \quad c_2 = 68.7945.$$

Because $\overline{X} = 69.178$ degrees is larger than 68.7945 degrees, we reject H_0 and decide that the thermostat is not meeting the design specification at this significance level.

Section 9.4 Problems

20. Figure 9-13 shows Minitab output for a test to determine if the average level of particulate matter in the 63 metropolitan statistical areas (MSA) in the Air Quality Data Base (Appendix B) is different from the National Ambient Air Quality Standard (NAAQS) of 50 $\mu g/m^3$. Because there is no data for some MSAs, assume this

```
TEST OF MU = 68.000 VS MU N.E. 68.000

              N      MEAN    STDEV   SE MEAN      T    P VALUE
TEMP         45    69.178    3.242     0.483   2.44     0.019
```

Figure 9-12
Minitab output for test of thermostat design

```
TEST OF MU = 50.000 VS MU N.E. 50.000

              N      MEAN    STDEV   SE MEAN      T    P VALUE
PM10         63    41.873   13.891     1.750  -4.64    0.0000
```

Figure 9-13
Minitab output for PM10 test

Figure 9-14
Minitab output for ozone test

```
TEST OF MU = 0.12000 VS MU N.E. 0.12000

              N      MEAN     STDEV    SE MEAN       T    P VALUE
OZONE        63    0.13524   0.04231   0.00533    2.86    0.0058
```

Figure 9-15
Minitab output for test of mean effective yield from money market mutual funds

```
TEST OF MU = 3.5000 VS MU N.E. 3.5000

              N      MEAN     STDEV    SE MEAN       T    P VALUE
MMMF         68    3.5278    0.3946    0.0478     0.58    0.56
```

sample is a random sample of all MSA values, and perform the test at the 0.2 level. Interpret your decision.

21. Use the Minitab output in Figure 9-14 to determine if the mean ozone level for the 63 MSAs supports a conclusion that the average concentration is significantly different from the NAAQS of 0.12 parts per million (ppm). Assume the 63 MSAs are a random sample of all MSAs, and test at the 0.02 level. Relate the decision to the specifics of this problem. (Source: "National Air Quality and Emissions Trends Report, 1989," Executive Summary and Chapter 4 Excerpts, U.S. Environmental Protection Agency, Office of Air and Radiation, Office of Air Quality Planning and Standards, Technical Support Division, Research Triangle Park, NC, Table 4–3.)

22. Figure 9-15 displays the Minitab output for a test to determine if the mean effective annual yield (in percent) of tax-exempt money market mutual funds is 3.5%. The yields of 68 randomly selected funds produced the statistics that the figure shows. Test this hypothesis at the 0.01 significance level, and interpret the decision. (Source: *Wall Street Journal,* Thursday, July 11, 1991, p. c21.)

Summary and Review

Statistical hypothesis testing is the process of choosing which of two competing hypotheses, H_0 and H_1, is supported by sample evidence. These hypotheses are statements about a population parameter. We use the evidence (the value of a test statistic) to decide which hypothesis is valid. Because we derive test statistic values from a sample, we must consider the possibility of an erroneous decision. The probabilities of these errors are defined as follows.

$\alpha = P(\text{Type I error}) = P(\text{reject } H_0 \text{ when } H_0 \text{ is true})$
$\beta = P(\text{Type II error}) = P(\text{cannot reject } H_0 \text{ when } H_1 \text{ is true})$.

The Type I error is the primary focus of much of hypothesis testing. The level of significance of the test, α, is preset by the decision maker to control for the possibility of making a Type I error. The level of significance determines the critical value(s), boundaries of the acceptance and rejection regions. The acceptance region includes values of the test statistic ($\overline{X}$) that are compatible with the null hypothesis, H_0, being true. The values of the test statistic that are unlikely to occur if H_0 is true compose the rejection region. There is a high probability of obtaining a sample mean, $\overline{X}$, in the neighborhood of the population mean, μ. For the test, H_0 specifies a specific value, μ_0, for μ, and the neighborhood around μ_0 is the acceptance region. When we obtain an $\overline{X}$ outside this neighborhood, we reject H_0.

Figure 9-16 summarizes the information for the two-tailed test.

$$H_0: \mu = \mu_0 \quad \text{and} \quad H_1: \mu \neq \mu_0.$$

In the remainder of this chapter and in future chapters, we will test hypotheses in different situations: (1) different conditions of tests for μ; (2) tests for different parameters; and (3) different forms of hypotheses.

Multiple-Choice Problems

23. A Type I error occurs when you
    a. Cannot reject H_0 when H_0 is true.
    b. Cannot reject H_0 when H_1 is true.
    c. Reject H_0 when H_0 is true.
    d. Reject H_0 when H_1 is true.
    e. None of the above.

24. The level of significance of a test is
    a. The probability of a Type II error.
    b. α.
    c. $1 - \alpha$.
    d. β.
    e. $1 - \beta$.

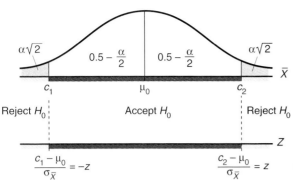

Figure 9-16
Summary diagram for a two-tailed test from a large sample

25. The rejection region corresponds to
   a. Test statistic values that are unlikely to occur if H_0 is true.
   b. Test statistic values that are compatible with a true null hypothesis.
   c. Test statistic values identical to the value or values listed in H_0.
   d. Values of $\overline{X}$ that equal μ_0.
   e. Population values with a low probability of occurring.
26. If the hypotheses are H_0: $\mu = 53$ and H_1: $\mu \neq 53$, which of the regions labeled in the following diagram correspond to the rejection region?

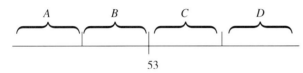

   a. A.
   b. B.
   c. C.
   d. D.
   e. Both B and C.
   f. Both A and D.
   g. A, B, C, and D (everything but the value 53).
27. Suppose we perform a two-tailed test of $\mu = 200$ at the 0.10 significance level. Which of the diagrams in Figure 9-17 is appropriate for this test? The shaded area corresponds to α.
   a. Diagram A.   c. Diagram C.   e. Diagram E.
   b. Diagram B.   d. Diagram D.
28. Find the critical values for the following situation.
   H_0: $\mu = 128$ and H_1: $\mu \neq 128$; $n = 100$, $S = 85$, $\alpha = 0.05$, $\overline{X} = 123$.
   a. 114.1 and 141.9.   d. 83.3 and 116.7.
   b. 111.3 and 144.7.   e. 86.1 and 113.9.
   c. -38.6 and 294.6.
29. If we reject H_0, we expose ourselves to making
   a. A Type I error.
   b. A Type II error.
   c. A correct decision when H_0 is true.
30. We determine that the critical values for a test of μ are 125 and 150. Which of the following $\overline{X}$ values would cause us to fail to reject H_0?

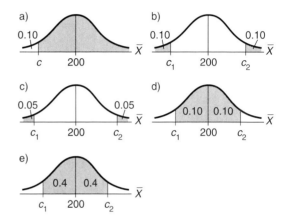

Figure 9-17
Problem 27 choices

   a. 161.   d. 143.
   b. 120.   e. 152.
   c. 118.

31. A manufacturer needs a part to meet detailed specifications. He will not return shipments of the part as long as the mean volume is 2,200 cubic centimeters. To save time and expense, the manufacturer will randomly select 100 items of the shipment and use the sample mean to estimate the shipment mean. The manufacturer knows that sample data will sometimes lead to a rejection of a good batch but considers this mistake of returning an acceptable shipment tolerable if it occurs for no more than 5% of the shipments. Find a range of acceptable sample means that will accomplish this goal. (Based on past experience, according to the firm, the standard deviation of volumes is about 150 cubic centimeters.)
   a. 2,175.3 cc and 2,224.7 cc.
   b. $-1,856.0$ cc and 2,444.0 cc.
   c. 2,050.0 cc and 2,350.0 cc.
   d. 2,190.0 cc and 2,210.0 cc.
   e. 2,170.6 cc and 2,229.4 cc.

Figure 9-18
Minitab output for military mail delivery test

```
TEST OF MU = 10.000 VS MU N.E. 10.000
              N      MEAN    STDEV   SE MEAN     T    P VALUE
TIME         52    11.577    5.031     0.698   2.26    0.028
```

32. If we test a hypothesis about the population standard deviation, σ, which of the following would be the test statistic?
    a. $\bar{X}$.   d. S.
    b. μ.   e. $\hat{p}$.
    c. p.

33. Which of the following segments must correspond to the rejection region for H_0: $\mu = 250$ and H_1: $\mu \neq 250$?

    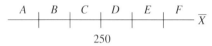

    a. Segments C and D.   e. Segment A.
    b. Segments B and E.   f. Segment C.
    c. Segments A and F.   g. Segment F.
    d. Segments B, C, D, and E.

34. When doing a hypothesis test, we give _____ the benefit of the doubt. Consequently, if the test statistic we obtain is _____ to occur when H_0 is true, we reject H_0. Choose the answer that correctly fills in the blanks.
    a. H_0, likely.   c. H_1, likely.
    b. H_0, unlikely.   d. H_1, unlikely.

35. A group of 52 soldiers in a war zone compared their latest mail for delivery time based on postmark and arrival date. The values are described in the computer output shown in Figure 9-18. Use this information to test an officer's claim that the average delivery time is 10 days. Test at the 0.05 level.
    a. Reject, $11.577 > 10$.
    b. Reject, $11.577 > 11.368$.
    c. Cannot reject, $8.632 < 10 < 11.368$.
    d. Cannot reject, $0.149 < 10 < 19.861$.
    e. Cannot reject, $0.149 < 11.577 < 19.861$.

36. Jill decides something is wrong with a pair of dice after 50 tosses each result in a 2 or a 3. This is like a hypothesis test, because
    a. The probability of obtaining a 2 or a 3 is 0.05, the level of significance.
    b. The two hypotheses are that the result is a 2 or that the result is a 3, either of which can get the benefit of the doubt.
    c. Jill's decision that something is wrong is the null hypothesis, the 50 tosses is the value of the test statistic, and the probability of not obtaining a 2 or a 3 is the significance level.
    d. An extraordinary number of tosses of twos or threes from a pair of dice indicates that Jill is not tossing an ordinary pair of dice.
    e. Evidence that is unlikely to occur in the ordinary course of events causes us to believe the assumption that events are ordinary.

Use the following information to solve Problems 37–46. The table below lists hemodialysis session lengths for 597 patients in an experiment to study the effect of length of sessions on patient mortality. Use this data to test the hypothesis that the mean of the session lengths for all hemodialysis patients is 3 hours and 30 minutes (210 minutes) at the 0.10 significance level. The session times are averages of several sessions for each patient expressed to the nearest quarter hour.

Length of Session (mean time in minutes)	Number of Subjects
105	1
120	2
135	0
150	3
165	10
180	73
195	31
210	101
225	78
240	264
255	7
270	14
285	6
300	6
315	1

Source: Philip J. Held, Nathan W. Levin, Randall R. Bovbjerg, Mark V. Pauly, and Louis H. Diamond, "Mortality and Duration of Hemodialysis Treatment," *Journal of the American Medical Association*, Vol. 265, No. 7, February 20, 1991, pp. 871–75.

37. What is H_1 for this test?
    a. $\alpha \neq 0.05$.   c. $\mu = 210$.   e. $\sigma \neq 210$.
    b. $\alpha = 0.10$.   d. $\mu \neq 210$.

38. What is the value of the test statistic for this test?
    a. $\bar{X} = 223.0$ minutes.   d. $\mu = 210$ minutes.
    b. $\mu = 223.0$ minutes.   e. $\bar{X} = 209.8$ minutes.
    c. $\bar{X} = 210$ minutes.

39. Why can we use the normal distribution for the sampling distribution of the test statistic, $\bar{X}$?
    a. The proportion of successful hemodialyses is a binomial random variable that is approximately normal, especially when the proportion is close to 0.5.
    b. The 597 values in the sample compose a large sample, and the values in a large sample must be normally distributed.
    c. Because each of the 597 values is a mean time for one of the patient's sessions, the distribution of means for different patients is about normal for large samples of patients.
    d. We know the population standard deviation, so there is no need to question the usual normality assumption.
    e. The sample is large, and the central limit theorem says that the sampling distribution of $\bar{X}$ is approximately normal for large samples.

Word and Thought Problems

40. What is the estimated standard error for the sampling distribution?
  a. 210 minutes.   **d.** 67.1 minutes.
  b. 745.3 minutes. **e.** 1.1 minutes.
  c. 27.3 minutes.

41. What is the area between the mean and a critical value? We use this area to find a Z value to correspond to a critical value.
  a. 0.475.   **d.** 0.45.
  b. 0.495.   **e.** 0.49.
  c. 0.4.

42. What are the Z values that correspond to the critical values?
  a. ±1.96.   **d.** ±2.33.
  b. ±2.575. **e.** ±1.28.
  c. ±1.645.

43. When we transform the Z values back to critical values, what is the value that we use for ϕ in the expression $(c - \phi)/\hat{\sigma}_{\bar{x}}$?
  a. 210.   **d.** 597.
  b. 209.8. **e.** 0.1.
  c. 223.

44. What are the critical values for this test?
  a. Any value that is not 210 minutes.
  b. 182.7 and 237.3 minutes.
  c. 165.1 and 254.9 minutes.
  d. 208.9 and 211.1 minutes.
  e. 208.2 and 211.8 minutes.

45. The decision is
  a. We cannot reject H_0, because $\overline{X} = 223.0$ is between 208.2 and 211.8.
  b. We reject H_0, because $\overline{X} = 223.0$ is between 208.2 and 211.8.
  c. We cannot reject H_0, because $\overline{X} = 223.0 \neq 210$.
  d. We reject H_0, because $\overline{X} = 223.0 \neq 210$.
  e. We reject H_0, because $\overline{X} = 223.0 > 211.8$.

46. Interpret the decision for this specific situation.
  a. The mean session length for all hemodialysis patients is significantly different from 210 minutes.
  b. The mean session length for all hemodialysis patients is not significantly different from 210 minutes.
  c. The mean session length for all hemodialysis patients is indistinguishable from 210 minutes.
  d. The evidence supports the hypothesis that the mean session length for all hemodialysis patients is 210 minutes.
  e. Given the benefit of the doubt, we cannot deny that the mean session length is 210 minutes.

Word and Thought Problems

47. Madras Brake Company has developed a new set of brakes. They want to test these brakes to see if they meet a certain safety standard: a mean stopping distance of less than 150 feet. If this standard is not met, the company will decide the brakes are unsafe (H_0) and will scrap them. Otherwise, the company will decide the brakes are safe (H_1), and will manufacture and sell them.
  a. State the Type I and Type II errors in terms of "brakes are safe" or "brakes are not safe."
  b. Describe how each error can cost the company. How can each error cost the general public?

48. a. Suppose H_0 is true. If α is 0.05, what is happening the other 95% of the time?
  b. Suppose H_1 is true. If β is 0.25, what is happening the other 75% of the time?
  c. Does $\alpha + \beta$ represent anything? If so, what?

49. A psychic claims to be able to predict earthquakes along any fault.
  a. If she predicts an earthquake will occur during a given three-day period along the San Andreas Fault, but it does not occur, have we "proved" that she is not a good earthquake predictor? Explain.
  b. Suppose you examine a random sample of 100 of her earthquake predictions. If only 5 earthquakes occurred as she predicted, do you have sufficient evidence to convince most people that she is a good earthquake predictor? If she correctly predicted 90 out of 100, would your answer change? Explain.
  c. Suppose she claims 60% accuracy, that is, she will be correct 60% of the time. Would correctly predicting five out of the 100 earthquakes contradict her claim for most people? If she were to correctly predict 90 out of 100, would your answer change? Explain.

50. A friend suggests to you that history is an interesting subject, so you sign up for a history course. Subjectively, you rate each lecture as interesting or not interesting. Discuss how you would reach a conclusion about the overall level of interest of the class before the end of the term based on your subjective evaluation of the lectures presented thus far. If you give your friend's opinion the benefit of the doubt while examining your own subjective evaluations, would your decision process change? If so, how? Relate this nonobjective decision process to the logic of hypothesis testing presented in this chapter.

51. A local sports editor claims to be good at picking winners of horse races. Given a slate of 100 upcoming races, he projects the winner of each race. Think what possible results would convince most people that the editor is "good" or "poor" at selecting winners. Then discuss what results you think would be acceptable before you could write a convincing letter for publication in the paper condemning the editor's claim. "Convincing" here means that you must persuade the public beyond a reasonable doubt that he is not a good picker. Relate your methodology to the hypothesis testing logic presented in this chapter.

52. Sportswriters and sportscasters have indicated that the university has a winning team this season.
  a. What win-loss percentage do you think makes the analysis correct?

b. Is it possible to have a "winning" team that fails to meet your criteria over a short nonrandom span of games? If you and the sports media judge the team incorrectly based on a short span of games, what type of error would occur? Explain how your answer depends on whether the media people get the benefit of the doubt.

53. If you were to test H_0: $\mu = 200$ and H_1: $\mu \neq 200$, show on a number line where the critical values would be located in relation to 200 (just position the critical values, you need not solve for a specific value). Where is the rejection region? The acceptance region? Repeat this problem for H_0: $\mu \leq 200$ and H_1: $\mu > 200$ (HINT: there will be only one critical value). Repeat this problem for H_0: $\mu \geq 200$ and H_1: $\mu < 200$.

54. Use $\alpha = 0.08$ and $\beta = 0.4$, to find the following:
  a. P(reject H_0 given that H_0 is true).
  b. P(reject H_0 given that H_1 is true).
  c. P(cannot reject H_0 given that H_0 is true).
  d. P(cannot reject H_0 given that H_1 is true).

55. If you decide to reject H_0, what kind of error might you make? What do we call the probability of this error? Reanswer these questions, but assume that you fail to reject H_0.

56. If you are testing H_0: $\mu = 1{,}000$ against H_1: $\mu \neq 1{,}000$ and obtain critical values of 980.2 and 1,019.8, would you fail to reject H_0, reject it, or be unable to make a decision in the following cases?
  a. $\overline{X} = 1{,}005$.
  b. $\overline{X} = 1{,}000$.
  c. $\overline{X} = 950$.

57. a. Suppose a contractor claims that her estimates are exact, on average, implying that her errors will average to zero. A random sample of differences between final and estimated costs for 38 of the contractor's customers has a mean of $-\$253.25$ (overestimates) and a standard deviation of $\$75.78$. Use this information to test the contractor's claim at the 0.01 level. Interpret the decision.
  b. If the sample standard deviation, S, were larger, we might fail to reject the contractor's claim. Determine the value of S that will produce a new critical value equal to $\overline{X} = -\$253.25$. Although a larger S may cause us to be unable to reject her claim of being correct on average, what does a larger S tell us about her ability to estimate?

58. Suppose that automobile industry analysts predict that when gasoline prices increase drastically, new and used car sales will change. Before a recent drastic gasoline price increase, the mean car sales for a local dealer were 10.2 new cars and 15.6 used cars. A random sample of sales on 36 days, collected after the gasoline price increase, produces a mean of 10.5 new cars and 18 used cars. The sample standard deviations for car sales are 3 and 6, respectively. Test each type of car to detect a significant change in sales (up or down)? Use $\alpha = 0.20$. Interpret the decisions.

59. A certain newspaper's editorial policy changes after a new editor takes charge. The mean daily sales prior to the change were 25,000 newspapers. For the first 50 days after the change, the mean daily sales value is 24,382 papers, and the standard deviation is 3,549 papers. Is there sufficient evidence to support the claim that mean daily sales changed after the editorial change? Use $\alpha = 0.05$ to detect a significant change in either direction. Interpret the decision. What assumption(s) did you make to work the problem?

60. Mr. Fernandez is supposed to be an expert on historic art. If 10 fakes are planted in a show that Mr. Fernandez attends, discuss what kinds of behavior or data would convince you that he is *not* an expert, if you give him the benefit of the doubt.

61. Test at the 0.1 level a claim that the average number of field goals in a professional football game is two, if a random sample of 60 games has a mean of 2.95 field goals with a standard deviation of 0.56. Interpret the decision.

62. Can you reject at the 0.02 level a theater manager's claim that the mean run for movies in the theater is 12 days, using the following random sample of 43 movies? Interpret the decision.

15	14	27	11	7	7	14	6	9
4	6	2	9	3	14	14	13	13
3	7	7	16	15	8	7	12	7
7	11	10	5	21	3	5	10	
18	8	8	7	14	19	7	7	

63. A random sample of 50 state banks has a mean of 112 employees per bank with a standard deviation of 45. Test at the 0.01 level the hypothesis that the mean employee count is 100. Interpret the decision.

64. a. A secretary claims that the department makes an average of 800 copies per day. Test this claim at the 0.2 level if a random sample of 100 days has a mean of 760 copies with a standard deviation of 350 copies. Interpret the decision.
  b. The standard deviation in the problem is relatively large, because there are days when the copier almost never stops and other days when departmental copy requirements slacken or when repairs result in large intervals of downtime. What effect does a large population standard deviation have compared to a smaller one when you are estimating the mean number of copies? Does a large population standard deviation make rejecting H_0 more difficult or easier? Does a large population standard deviation alter the significance level?

65. A rock group claims that its carefully choreographed show lasts for 2.5 hours on average. Test this claim at the 0.05 level if a random sample of 36 shows has a mean time of 2.75 hours with a standard deviation of 0.12 hour. Interpret your decision.

66. A professor of photojournalism believes that the average student will produce 22 usable, interesting photographs from a roll of film with 36 shots. Thirty-two randomly selected students from different photojournalism programs in the state were assigned to cover the state

```
TEST OF MU = 22.000 VS MU N.E. 22.000

              N      MEAN     STDEV    SE MEAN      T     P VALUE
PHOTOS       32     21.094    5.761     1.018     -0.89    0.38
```

Figure 9-19
Minitab output for photojournalism test

```
TEST OF MU = 10.000 VS MU N.E. 10.000

              N      MEAN     STDEV    SE MEAN      T     P VALUE
PIECES       49      9.551    0.937     0.134     -3.35    0.0016
```

Figure 9-20
Minitab output for candy piece count test

legislature on a certain day. Each student was given a roll of film with 36 shots. The resulting photographs were judged for interest and usability by all of the photojournalism professors in the state. The number of usable photographs was recorded for each student. Figure 9-19 displays a computer analysis of these data. Use this information to test at the 0.2 level for an average of 22 good photographs. Interpret your decision.

67. A librarian says that an average of 20 patrons per day ask how to use the electronic card catalog. A random sample of 80 days has a mean of 25 requests with a standard deviation of two requests. Can the librarian's statement be refuted at the 0.2 level?

68. Test at the 0.01 level to determine if a tax assessor's estimate of $62,000 for the mean price of a three bedroom house is correct, if a random sample of 45 such houses has a mean price of $62,500 with a standard deviation of $2,000. Interpret your decision.

69. A college catalog suggests using $700 as an estimate of average food expense during the term. Comment on this statement if a random sample of 50 students from last term spent a mean of $625.57 with a standard deviation of $80.92. Allow a 5% chance of making a Type I error. Interpret your decision.

70. A mechanic claims that an average tune-up requires 0.75 hour. Test this claim at the 0.05 level if a random sample of 64 tune-ups shows a mean time of 1.1 hour with a standard deviation of 0.3 hour. Interpret your decision.

71. An association of balloonists claims the average cost of attending balloon-pilot school is $200 per hour. A random sample of 49 such schools shows a mean cost of $192.26 per hour with a standard deviation of $45.50. Test the association's claim at the 0.10 level. Interpret your decision.

72. The following table lists times (in minutes) that pairs of randomly selected three-year-old children spent sharing a single available colorful toy without conflict. They were isolated in a room with other old and drab-colored articles. Test at the 0.1 level the psychologist's claim that children would play quietly for a mean time of eight minutes.

13	9	5	1
5	10	4	1
8	12	8	8
11	6	5	9
10	6	9	10
11	6	13	9
4	5	15	7
8	8	4	7
7	4	1	
9	8	5	

73. a. Small packs of chocolate candies were distributed in a statistics class to test whether the mean number of pieces in a pack is 10. The results are shown in Figure 9-20. Use this information to perform the test at the 0.01 level. Interpret your decision.

b. To acquire the above sample, the professor bought several large bags that contain the small packs at a local grocery store. Knowing this, do you consider the information used in this test to be a random sample? Explain your answer.

Unit II: Hypothesis Testing Variations

Preparation-Check Questions

74. Find $P(T_{(11)} < 2.201)$.
  a. 0.01. **d.** 0.975.
  b. 0.05. **e.** 0.95.
  c. 0.025.

75. A random sample of 16 items is selected from a normal population. The sample mean and standard deviation are 463 and 40, respectively. Find a 95% confidence interval for the population mean.

(continued on next page)

a. 463 ± 21.3.
b. 463 ± 21.2.
c. 463 ± 19.6.
d. 463 ± 78.4.
e. 463 ± 85.2.

76. The rejection region for a hypothesis test corresponds to
   a. Values that are unlikely to occur if H_0 is true.
   b. Values close to the value specified in H_0.
   c. Values that show up frequently when H_0 is true.
   d. Values close to $\overline{X}$.
   e. Values close to μ.

77. Use the following information to test the hypotheses $H_0: \mu = 1{,}562$, and $H_1: \mu \neq 1{,}562$.
   $n = 625.$ $\overline{X} = 1{,}550.$ $S = 300.$ $\alpha = 0.05.$
   What is the decision?
   a. Reject H_0, because $1{,}550 \neq 1{,}562$.
   b. Cannot reject H_0, because $1{,}538.5 < 1{,}550 < 1{,}585.5$.
   c. Cannot reject H_0, because $1{,}262 < 1{,}550 < 1{,}862$.
   d. Cannot reject H_0, because $974 < 1{,}550 < 2{,}150$.
   e. Cannot reject H_0, because $1{,}526.5 < 1{,}562 < 1{,}573.5$.

78. If more than 20% of a random sample of 400 balloons from a large shipment are white, the shipment is returned. Find the probability a shipment will be returned when the proportion of white balloons in the shipment is actually 0.15.
   a. 0.9938.   d. 0.0124.
   b. 0.4938.   e. 0.0026.
   c. 0.9876.

Answers: 74. d; 75. a; 76. a; 77. b; 78. e.

Introduction

Absentee Rate   A principal claims that the absentee rate at his school is 10%, but members of a concerned parents' organization suspect the rate is much higher. They must collect sufficient evidence to convince the principal that their suspicions are correct. This process requires hypothesis testing, but in this situation they are testing for a population proportion. Notice also that they only suspect the rate to be higher than 10%, not lower. In the examples covered in Unit I, the true value could be in either direction, below or above the specified value.

This unit covers these two topics, testing for proportions and one-tailed tests, as well as other extensions of the hypothesis testing procedure.

9.5 One-Tailed Tests

Recall that the basic strategy for a hypothesis test is to assume that the null hypothesis, H_0, is true and give this hypothesis the benefit of the doubt when examining sample evidence. We assume H_0 is true when performing the test, in spite of the fact that H_1 may be what we think is true. Next, we collect a random sample. If the test statistic from the sample is a value that is not likely to occur when H_0 is true, we reject H_0. Because we give H_0 the benefit of the doubt, we cannot reject H_0 when the value of the test statistic is close to μ_0 (the value specified in H_0) or when the value is ambiguously compatible with either H_0 or H_1 being true. Thus, we only reject H_0 when the sample evidence leads us to believe *beyond a reasonable doubt* that H_0 must be false. The level of significance, α, determines the boundaries of the acceptance and rejection regions, called critical values. These boundaries determine when the value of a test statistic, such as $\overline{X}$, is sufficiently different from the value specified in H_0 to decide H_0 is wrong. When we reject H_0, we say the population parameter is significantly different from the value specified in the null hypothesis, μ_0. The value of α that we specify is also the level of risk or the probability that we will make the mistake of rejecting H_0 when H_0 is true.

In the preceding unit, we studied hypotheses of the form:

$$H_0: \mu = \mu_0$$
$$H_1: \mu \neq \mu_0.$$

H_1 says that if μ is not μ_0, then it could be a value either smaller or larger than μ_0.

Now, consider situations where μ varies only in a single direction in H_1. These hypotheses are written as follows.

$$H_0: \mu \leq \mu_0 \qquad H_0: \mu \geq \mu_0$$
$$\text{or}$$
$$H_1: \mu > \mu_0 \qquad H_1: \mu < \mu_0$$

Notice again that *the equal sign is always in H_0. H_1 is the statement we try to demonstrate*. This rule for H_1 applies, unless we wish to demonstrate that the population parameter is a single value as we did in Unit I. In this case, we must make H_0 what we wish to demonstrate in order for the equal sign to be in H_0. We will discuss the reasons for this choice of H_0 and H_1 and the implications later in this unit.

■ **Example 9-5:** An exercise physiologist wants to demonstrate that the average person walks more than 800 miles per year. State the null and alternative hypotheses, then generally locate the rejection and acceptance regions (use c for the critical value, because we do not know a specific value yet).

■ **Solution:** Because we make H_1 what the physiologist wants to demonstrate, that the mean is more than 800, the hypotheses would be

$$H_0: \mu \leq 800$$
$$H_1: \mu > 800.$$

Again, the test statistic used as an indicator of μ is $\overline{X}$, the sample mean. To find the general location of the acceptance and rejection regions we must first answer the following question.

Question: Which values of $\overline{X}$ would make us believe $H_1: \mu > 800$ and reject H_0, those larger than 800, smaller than 800, or an $\overline{X} = 800$?

Answer: Large values of $\overline{X}$ are compatible with $H_1: \mu > 800$. An $\overline{X}$ smaller than or equal to 800 certainly would not make us want to reject $H_0: \mu \leq 800$. Some $\overline{X}$ values larger than 800, but close to 800, will not cause us to reject H_0, because we give H_0 the benefit of the doubt. Remember, if the population mean, μ, is 800, sample means will still vary, although most of them will lie in the neighborhood of 800. See Figure 9-21. Only when $\overline{X}$ is convincingly larger than 800—so much larger that such a value is unlikely when $\mu = 800$—do we reject H_0. Because only larger values cause us to reject H_0, we will need only one critical value to separate the unlikely larger $\overline{X}$ values from all other $\overline{X}$ values. The significance level, α, will determine the exact position of c and, hence, the exact magnitude of a value that is "unlikely."

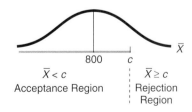

Figure 9-21
Decision rule for upper-tailed test for the annual walking average example

■ **Example 9-5 (continued):** The physiologist conducts a study of the walking habits of a random sample of 180 people in order to perform the test at the 0.05 significance level. The sample mean and standard deviation are 821.6 and 130.1 miles per year, respectively. Follow the steps from the last unit to complete this test. We use $\mu = 800$ in the diagrams when we assume H_0 is true. The μ values that are less than 800 will be accounted for shortly.

■ **Solution:** We have already stated the hypotheses and determined the test statistic in the preceding solution. The significance level is 0.05. Because the sample is large

Figure 9-22

α for the upper-tailed test

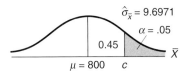

Figure 9-23

Test statistic in rejection region

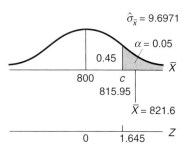

($n = 180$), the sampling distribution of $\overline{X}$ is approximately normal. We always assume that the population value specified in H_0 ($\mu = 800$ miles in this case) is the true value when sketching the diagram or determining critical values. The standard error is $\hat{\sigma}_{\overline{x}} = S/\sqrt{n} = 130.1/\sqrt{180} = 9.6971$ miles.

Question: Sketch a diagram for this test using the decision rule to reject H_0 if $\overline{X} \geq c$. Shade the area corresponding to the significance level above the rejection region.

Answer: Figure 9-22 shows this diagram. Notice that all of the area for the significance level 0.05 is in one tail of the curve. This area is the probability we reject H_0 when H_0 is true, and we only reject H_0 when $\overline{X}$ is large for this set of hypotheses. Hence, we call a test with a single rejection region a *one-tailed test*. We also call it an *upper-tailed test* when the region consists of values larger than the value proposed in H_0 and a *lower-tailed test* when the values are smaller.

Next, we must determine an actual value for the critical value, c. The Z value that corresponds to an area of 0.05 in the tail (0.45 between the mean and the critical value, c) is 1.645. We solve the following expression for c

$$Z = \frac{\overline{X} - \mu}{\hat{\sigma}_{\overline{x}}} = \frac{c - 800}{9.6971} = 1.645,$$

so $c = 815.95$ miles. If $\overline{X} \geq 815.95$, we reject H_0 at the 0.05 significance level. $\overline{X}$ values 815.95 or larger should occur only 5% of the time, if the population mean, μ, is 800 (see Figure 9-23). We must reject H_0 at the 0.05 significance level, because, as Figure 9-23 shows, $\overline{X} = 821.6$ miles > 815.95 miles. The data support the hypothesis that people walk more than 800 miles per years on average.

Now that we have completed the test, consider what happens when another μ value specified in H_0, a population mean less than 800, is true, rather than $\mu = 800$.

Question: Pick a μ value smaller than 800, and draw the curve for the sampling distribution of $\overline{X}$. Position 800 and 815.95 on the $\overline{X}$ axis.

Answer: See Figure 9-24a.

Question: If we continue to use 815.95 as the critical value, is the area in the rejection region of the new curve larger or smaller than for the curve in Figure 9-23, where $\mu = 800$?

Answer: Part a of Figure 9-24 shows that the area must be smaller, because 815.95 is further from the mean of the new curve. If we superimpose the two curves, you can see that the new area is smaller, as shown in Part b of the figure.

Figure 9-24

A decision to reject H_0 works for all values specified in H_0

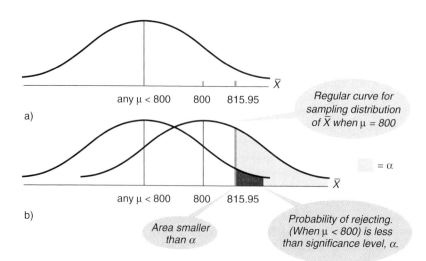

Question: If the sample mean, $\overline{X}$, is 821.6, would we reject the proposition that the population mean, μ, is *less than* 800?

Answer: Yes. $\overline{X} = 821.6$ is even further from the μ values that are smaller than 800 than it is from $\mu = 800$.

Thus, *if we reject H_0 for $\mu = 800$, we will reject it for any $\mu < 800$*—which is the reason we use $\mu = 800$ to find the critical value. Similarly, when the null hypothesis takes the form $H_0: \mu \geq \mu_0$ and we reject H_0 for $\mu = \mu_0$, we will reject it for all μ values larger than μ_0 as well.

■ **Example 9-6:** An official claims there is an average of 100 or more parking spaces on weekdays between 8 A.M. and noon on campus. Your frustrated search for parking spaces at these times leads you to believe the average is quite a bit smaller than 100. Attempt to contradict the official's claim with a hypothesis test performed at the 0.01 significance level. You and a squadron of friends randomly select 100 times between 8 A.M. and noon (on various weekdays) to count empty spaces. The result is a sample with a mean of 94.3 empty spaces and a standard deviation of 20.1 empty spaces.

■ **Solution:** Because you want to contradict the official's claim, make H_1 the contradictory claim.

$$H_0: \mu \geq 100$$
$$H_1: \mu < 100.$$

The test statistic is still $\overline{X}$, and the significance level, α, is 0.01.

Question: Where is the critical value relative to $\mu = 100$ on the $\overline{X}$ axis?

Answer: Now $\overline{X}$ values that are significantly less than 100 support H_1 and cause us to reject H_0. Again, $\mu = 100$ is given the benefit of the doubt, so some $\overline{X}$ values that are slightly smaller than 100 will not be sufficiently different to warrant our rejecting H_0. The following diagram shows the positions of c and $\mu = 100$ on the $\overline{X}$ axis. Because there is a single rejection region that consists of values much smaller than 100, this is a lower-tailed test.

```
                |              |
────────────────┼──────────────┼────── X̄
                c             100
         Reject H₀     |  Cannot reject H₀
       ←── X̄ ≤ c ──→   |   ←── X̄ > c ──→
```

The sampling distribution is approximately normal with mean 100 if H_0 is true, and $\hat{\sigma}_{\overline{x}} = S/\sqrt{n} = 20.1/\sqrt{100} = 2.01$. The critical value, c, on the $\overline{X}$ axis is the same as -2.33 on the Z axis when $\alpha = 0.01$ for a lower-tailed test.

$$\frac{c - \mu}{\hat{\sigma}_{\overline{x}}} = \frac{c - 100}{2.01} = -2.33,$$

and $c = 95.32$ empty spaces (see Figure 9-25). We reject H_0 at the 0.01 significance level, because your sample mean, $\overline{X}, = 94.3 < 95.32 = c$. The evidence does not support the official's claim.

Question: Suppose the sample mean is 105 empty spaces. What is the decision?

Answer: We cannot reject H_0, because $105 > 95.32$. In fact $105 > 100$, so the sample information readily supports the official's claim.

Whenever the test statistic satisfies H_0, we automatically know that we will be unable to reject H_0, because we give H_0 the benefit of the doubt and require strong evidence before we consider rejecting it. If the test statistic agrees with H_0, the sample evidence does not even contradict H_0. This rule assumes that the significance level is never greater than 0.5, an unrealistically high risk level.

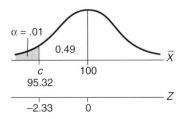

Figure 9-25

Critical value for a lower-tailed test for the parking space example

Section 9.5 Problems

79. Suppose we want to be more certain that we do not falsely contradict the official's claim about parking spaces from Example 9-6, so we reduce α to 0.002. Can we still reject the official's claim?

80. If you test H_0: $\mu \geq 1{,}000$ against H_1: $\mu < 1{,}000$ and obtain a critical value of 980.2, would you reject H_0, not reject it, or be unable to make a decision in the following cases?
    a. $\overline{X} = 1{,}005$; b. $\overline{X} = 1{,}000$; c. $\overline{X} = 950$.

81. Can you reject H_0 in the following cases?
    a. H_0: $\mu \geq 250$
    H_1: $\mu < 250$
    $\alpha = 0.03$
    $n = 36$
    $S = 60$
    i. $\overline{X} = 245$.
    ii. $\overline{X} = 240$.
    iii. $\overline{X} = 235$.

    b. H_0: $\mu \leq 333$
    H_1: $\mu > 333$
    $\alpha = 0.05$
    $S = 84$
    $\overline{X} = 350$
    i. $n = 36$.
    ii. $n = 49$.
    iii. $n = 144$.

    c. H_0: $\mu \leq 666$
    H_1: $\mu > 666$
    $\alpha = 0.05$
    $n = 100$
    $\overline{X} = 680$
    i. $S = 50$.
    ii. $S = 100$.
    iii. $S = 200$.

82. Can you reject at the 0.05 level the null hypothesis, a claim that the average full-time student carries fewer than 15 class hours if a random sample of 100 full-time students carry a mean of 14.2 hours with a standard deviation of 0.6 hour? Interpret your decision.

83. A company experimented with longer breaks to be sure that employees' average productivity would increase from this change before they instituted the change for all employees. The mean output with the regular break time was 120 units per hour, and this is the null hypothesis. A random sample of 58 employees with a longer break yielded a mean output of 124.6 units per hour with a standard deviation of 5.9 units. Does this information justify the conclusion that longer breaks result in greater mean output at the 0.1 level?

9.6 Shortcuts and Testing on the Z Axis

If we proceed directly from the hypotheses and other given information to the corresponding diagrams in Figure 9-26, we can easily decide whether or not to reject H_0. We can quickly find critical values from the significance level α and its corresponding Z value from the following list.

Significance Level (α)	Common z Values for Hypothesis Testing	
	Two-Tailed Test	One-Tailed Test
0.2	1.28	0.84
0.1	1.645	1.28
0.05	1.96	1.645
0.01	2.575	2.33
0.005	2.81	2.575
0.001	3.29	3.09

Figure 9-27 shows the diagram we used to test H_0: $\mu \leq 800$ and H_1: $\mu > 800$ at the 0.05 significance level in the preceding section. Notice that the critical value, $c = 815.95$ miles, is the same as 1.645 in Z units. As long as we know the mean and standard deviation for the normal random variable, we can easily locate any specific value on the other axis by using the transformation formula $Z = (\overline{X} - \mu)/\hat{\sigma}_{\overline{x}}$. This is true for the value of the test statistic, $\overline{X} = 821.6$, as well. It becomes $(821.6 - 800)/9.6971 = 2.23$, as Figure 9-27 shows. Because 2.23 exceeds 1.645, the critical value

Figure 9-26
Diagrams corresponding to one-tailed and two-tailed tests of hypotheses about the population mean, μ

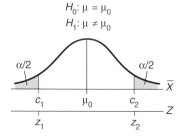
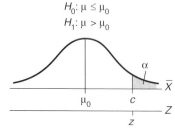
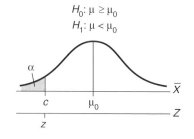

Problems

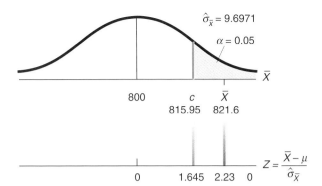

Figure 9-27
One-to-one correspondence between expressions for the test statistic and critical value on X axis and Z axis

expressed in Z terms, we automatically know that the test statistic will exceed the critical value on the $\overline{X}$ axis as well. We can decide whether or not to reject H_0 without transforming c from Z units to the original, or $\overline{X}$, units.

So far we have determined the critical value, c, on the $\overline{X}$ axis and compared the test statistic, $\overline{X}$, with this value to make the decision. This process usually makes hypothesis testing more understandable. However, as we have just seen, we will reach the same conclusion more directly if we convert the test statistic value, $\overline{X}$, to a Z and compare it with the critical value expressed in Z terms, because of the one-to-one correspondence between values on the $\overline{X}$ and Z axes.

■ **Example 9-7:** A pharmaceutical firm believes it has developed a new drug that is more effective in delaying the relapse and onset of headache pain than currently marketed products. Products presently on the market delay relapse an average of two hours. A random sample of 400 headache patients suffered a relapse an average of 2.10 hours after taking the new product. The sample standard deviation is 0.52 hour. Test H_0: $\mu \leq 2$ versus H_1: $\mu > 2$ at the 0.1 level on the Z axis to determine if the evidence substantiates the firm's belief.

■ **Solution:** The sampling distribution is approximately normal, because $n = 400$. The mean is 2 if H_0 is true and the estimated standard error is $\hat{\sigma}_{\overline{x}} = S/\sqrt{n} = 0.52/\sqrt{400} = 0.026$. Because this is an upper-tailed test at the $\alpha = 0.1$ level, we find that the critical value on the Z axis is 1.28. Next we must convert $\overline{X}$ to a Z value:

$$Z = \frac{(\overline{X} - \mu)}{\sigma_{\overline{x}}} = \frac{(2.10 - 2)}{0.026} = 3.846.$$

We reject H_0, because $3.846 > 1.28$, which means that $\overline{X} = 2.10$ must fall within the rejection region (see Figure 9-28).

Question: Interpret this decision.
Answer: At the 0.10 significance level, the evidence supports the company's claim that the product delays relapse for more than two hours.

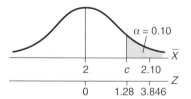

Figure 9-28
Diagram for the pharmaceutical headache remedy example

Section 9.6 Problems

84. Suppose that the test statistic for an upper-tailed test is $Z = -2.5$. Compare this value with the critical values expressed in Z values in the table in this section to determine the significance levels that would result in rejecting H_0.

85. Perform the test in Problem 83 on the Z axis to confirm your previous work.

86. Perform the test in Problem 82 on the Z axis to confirm your previous work.

87. A group of zoologists are conducting a test to demonstrate that the mean migration distance covered by a given bird species is less than 950 miles. They tag a random sample of 289 birds and record the number lodging at different nature preserves for the winter. The result is a sample mean distance of 900 miles and a standard deviation of 340 miles. Make a decision about H_0: $\mu \geq 950$ at the 0.05 level. Interpret the decision.

88. A business wants to demonstrate that its mean delivery time is 45 days. A random sample of 800 deliveries occurs in a mean time of 49 days with a standard deviation of 12 days. Is this compatible with the firm's claim, if we perform a two-tailed test at the 0.01 significance level? Interpret the decision.

9.7 Small-Sample Test for μ

All the examples to this point have employed large samples (usually $n > 30$) to test for the population mean. Suppose now that we need to use a sample with 30 or fewer observations, perhaps to avoid the cost of obtaining expensive pieces of information or to minimize destructive sampling. We can handle this situation easily by compensating for the loss of information in the form of less sample data with more information about the distribution of the population. More specifically, the procedures discussed in this section apply to samples drawn from normal populations, although the results tend to be relatively accurate if the population is symmetric and unimodal.

If the population distribution is neither normal nor symmetric and unimodal, the sampling distribution we are about to employ and the decisions we make are invalid, at least at the significance level that we state. When the population distribution is unknown, we use nonparametric testing procedures. Chapter 13 discusses this topic.

Question: Can you guess which probability distribution to use for the sampling distribution of each of the random variables, $(\overline{X} - \mu)/\sigma_{\bar{x}}$ and $(\overline{X} - \mu)/\hat{\sigma}_{\bar{x}}$, when the sample is small and the population is normal or symmetrical and unimodal?

Answer: If you consult Figure 8-14 (p. 343), you will find that you should use the standard normal probability distribution (Z random variable), when you know the population standard deviation, σ, and, hence, the standard error of the mean, $\sigma_{\bar{x}}$. Use the T probability distribution with $n - 1$ degrees of freedom when you must substitute the sample standard deviation, S, for an unknown σ to estimate $\sigma_{\bar{x}}$ using $S/\sqrt{n} = \hat{\sigma}_{\bar{x}}$. Estimating this standard error is the typical situation.

■ **Example 9-8:** A sociologist conducted a hypothesis test to determine if the average population count of isolated rural communities in a region is 888 people. A random sample of 17 such communities yields a mean count of 912 people with a standard deviation of 42 people. Assuming such counts are normally distributed, test H_0: $\mu = 888$ and H_1: $\mu \neq 888$ at the 0.05 significance level.

■ **Solution:** We need $\hat{\sigma}_{\bar{x}} = S/\sqrt{n} = 42/\sqrt{17} = 10.1865$ to transform $\overline{X}$

$$\frac{\overline{X} - \mu}{\hat{\sigma}_{\bar{x}}} = \frac{912 - 888}{10.1865} = 2.4.$$

The critical value is 2.120 (from the T table in Appendix J, splitting $\alpha = 0.05$ for each tail, and using 16 degrees of freedom). The decision is to reject H_0, because $2.4 > 2.120$ (see Figure 9-29).

Question: Interpret this decision.
Answer: The information contradicts the hypothesis that the mean population of these communities is 888.

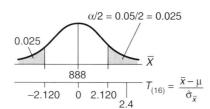

Figure 9-29
Small sample test for μ on T axis for the rural population mean example

Section 9.7 Problems

89. Perform a hypothesis test for H_0: $\mu \geq 470.2$ and H_1: $\mu < 470.2$ at the 0.1 significance level. A random sample of 22 observations from the normal population yields a mean of 455.8 and a standard deviation of 82.1.

90. A pharmaceutical firm wishes to demonstrate that its new product for curing a rare disease will produce a faster cure than will others on the market. The average cure time for existing products is 500 days. Because the disease is rare and the treatment risks not completely known, researchers use a random sample of only 10 patients for a pilot study. The new product has a mean cure time of 450 days with a standard deviation of 75 days for the 10 subjects. If cure times are normally distributed, which seems reasonable to the medical researchers, test H_0: $\mu \geq 500$ against H_1: $\mu < 500$ at the 0.05 level. Interpret the decision.

91. Jenni is allowed an average of 10 minutes or less on the telephone per call. A random sample of 25 calls shows a mean call time of 12 minutes with a standard deviation of 1 minute. Does this evidence justify the conclusion that she is exceeding her limit if we want to make the decision with a 1% chance of falsely accusing her (H_1: $\mu > 10$ and $\alpha = 0.01$)? Assume a normal population.

92. A fringe benefits expert wishes to demonstrate that the average level of fringe benefits is less than 30% of employees' gross salary (H_1: $\mu < 30$). A random sample of nine employees yields the following values of benefit packages (expressed as a percentage of each employee's gross salary): 29, 35, 15, 10, 32, 30, 18, 38, 10. Does this information justify the expert's opinion at the 0.01 significance level? The expert's previous research on this topic makes it seem reasonable to assume that benefit package values are normally distributed.

9.8 Test for Population Proportion, p

Another parameter of recurrent interest is the population proportion of successes, p. Now we will link the information we have learned about estimating this parameter with the basic testing procedure to evaluate statements about this parameter.

Question: What test statistic do we use to estimate the population success rate, p?

Answer: The sample proportion of successes, $\hat{p}$ (the number of successes in the sample divided by the number of observations in the sample), is the test statistic. *Recall also that we round four places past the decimal when we compute population and sample success rates.*

Recall also from Chapter 8 that $\hat{p}$ is a binomial random variable. For a large sample, the sampling distribution of $\hat{p}$ is approximately normal with mean p and standard error $\sigma_{\hat{p}} = \sqrt{[p(1-p)]/n}$. In this situation *large sample* does not mean more than 30 observations, as you will see in the following summary of the new circumstances.

1. Population and sample elements fall into one of two outcome categories, success or failure.
2. Sample observations are independently and randomly selected, so that values are not affected by the value of any other item in the sample.
3. The probability of success, p, stays the same for each element in the sample (the effect on p when we withdraw an element from the population is negligible).
4. The sample is appropriately large (np and $n[1-p]$ must both be greater than 5) to ensure a good normal approximation for binomial probabilities associated with values of the test statistic.

■ **Example 9-9:** A principal claims the absentee rate is 10%, but the parent group suspects a higher rate. Give the principal the benefit of the doubt, and state the hypotheses. A random sample of 400 students on a roll call contains 60 absentees. Can the parent group contradict the principal at the 0.05 level?

■ **Solution:** Because we give the principal the benefit of the doubt and determine whether the evidence is sufficient to contradict the principal's statement, the hypotheses are $H_0: p \leq 0.10$ and $H_1: p > 0.10$. The sampling distribution of $\hat{p}$ is approximately normal, because $400(0.10) = 40$ and $400(0.90) = 360$ are both greater than 5. If we assume H_0 is true ($p = 0.10$), the mean of the possible $\hat{p}$ values is 0.10, and the standard error is $\sigma_{\hat{p}} = \sqrt{[0.1(0.9)]/400} = 0.015$. *Notice that we substitute the value of p from H_0 in the $\sigma_{\hat{p}}$ formula.* Unlike confidence intervals for p, we do not substitute $\hat{p}$ for p, because p is given when we assume that H_0 is true. The sample success rate is $\hat{p} = \dfrac{60}{400} = 0.15$, which we transform to a Z value

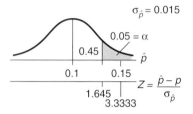

Figure 9-30
Diagram for absentee rate example

$$\frac{\hat{p} - p}{\sigma_{\hat{p}}} = \frac{0.15 - 0.1}{0.015} = 3.3333.$$

Because $3.3333 > 1.645$ (the critical value from the Z table that corresponds to an area of 0.05 in the tail), we reject H_0 (see Figure 9-30). The evidence contradicts the principal's claim.

Section 9.8 Problems

93. Test $H_0: p = 0.92$ and $H_1: p \neq 0.92$ at the 0.05 level if a random sample of 4,000 observations includes 3,600 successes.

94. Check the fairness of a coin by testing for the proportion of heads tossed with the coin. We have $H_0: p = 0.5$ and $H_1: p \neq 0.5$. A random sample of 1,225 coin tosses includes 600 heads. Should you reject H_0 at the 0.02 level? Interpret your decision.

95. If a random sample of 1,000 Bowling Green residents contains 535 persons who prefer coffee to tea, is this sufficient evidence to conclude that more than half the people in Bowling Green prefer coffee at the 0.1 level? $H_1: p > 0.5$.

96. If 53% of the 647 floaters who made a trip down the river did so in June, is this sufficient evidence to contradict an agency claim that 55% of the annual floats occur in June? Do a two-tailed test at the 0.01 level. Interpret the decision.

9.9 Setting Up Hypotheses

It is often more difficult to determine what statements to use for H_0 and H_1 before you conduct the test than to do the actual calculations and make the decision. The following sections describe a strategy for setting up the hypotheses, while reinforcing the basic notion of hypothesis testing.

9.9.1 Determine the Parameter to Be Tested

Check the problem or situation for key words, such as mean, average, rate, percent, or proportion, that indicate the parameter of interest. A question about the *mean* or *average* will usually contain one or both of these words. The words *rate*, *percent*, or *proportion* signal tests for the population success rate, p. This is not always the case, because you can test for a mean success rate or mean proportion, as we did in Problem 92.

In a real-world situation, if we cannot form the appropriate test statistic from the available sample information, then we may collect new data or reformulate the old data to fit the test statistic that corresponds to the desired parameter. Alternatively, we may reformulate the questions or hypotheses to use a parameter that corresponds to the available sample information.

9.9.2 Determine the Value Hypothesized for the Parameter

What value for the parameter is under question or dispute? Be careful at this point to pick the hypothesized value of the parameter, not the sample value. For instance, a test to demonstrate that the mean is greater than 100 may be decided by a sample mean that is 125. The hypotheses should be statements about 100, not 125.

9.9.3 Determine the Appropriate Equality or Inequality Symbol

This is probably the most difficult step, until you become accustomed to formulating hypotheses. The basic rule is to *make H_1 the hypothesis that the decision maker thinks is true or the hypothesis that he or she wants to demonstrate*.

The rationale for this rule originates in the methodology of hypothesis testing. Rejecting H_0 is the preferred or stronger decision. Because H_0 receives the benefit of the doubt and requires strong evidence to the contrary before we refute it ("beyond a reasonable doubt"), weak and ambiguous evidence causes us to decide in favor of H_0 (fail to reject H_0). Hence, we prefer to reject H_0—decide in favor of what we think is true, after subjecting our belief to a strenuous test.

In situations where there are competing interests at stake, it is important to establish which individual or group is the decision maker. Disputing groups may consider opposite statements to be the truth. If a firm believes that its product's average shelf life is two or more weeks and wants to demonstrate this claim, it would let H_0 be $\mu \leq 2$, because it believes that $\mu > 2$ (H_1). A competitor might conjecture that the product does not last so long. To substantiate this claim, the competitor would let H_0 be $\mu \geq 2$, giving the first firm's claim the benefit of the doubt, and then seek evidence to substantiate that $\mu < 2$. A consumer advocate or third party, who only believes that 2 is wrong but does not know if the value is too large or too small, would perform a two-tailed test.

Note that no matter what the direction for the test, the equal sign is always placed in H_0, so the curves for the sampling distribution of the test statistic can be anchored to a specific value. Try to draw a diagram for a test if H_0 is $\mu < 100$. There is no unique solution, because there are many values smaller than 100.

There is a consequence of placing the equal sign in H_0 when we want to demonstrate that a parameter is a specific value, such as $\mu = \mu_0$. This statement must be the null hypothesis for a two-tailed test. Not rejecting H_0 means we have found no evidence strong enough to contradict our statement. This decision is less conclusive, because we decide in favor of H_0 when the evidence is weak or ambiguous.

Sometimes we form hypotheses from claims. If the decision maker wants to obtain statistical support for a claim, then the claim becomes the alternative hypothesis, H_1. If the decision maker wants to disprove the claim, the claim becomes H_0. If the problem simply says to test a claim, we will assume that we are to disprove the claim (we make it H_0).

Sometimes we use the old or accepted value of a parameter for H_0, and H_1 includes the new value. These choices allow us to test for a change in the value. This method represents a conservative attitude toward change that requires abundant evidence before we adopt new values.

Sometimes we state H_0 and H_1 so that a Type I error is the most costly error, because we control the chances that it occurs. In this case, we would consider alternative positions for a particular statement and then employ the specification that makes the costliest error a Type I error.

Although the strategies for determining H_0 and H_1 mentioned in the few preceding paragraphs (attitude toward a claim, change in a value, and avoiding a costly error) are useful at different times, we will primarily rely on making H_1 what the decision maker believes is correct. This strategy is the basis for the following chart to help you decide whether a statement is H_0 or H_1:

If the question is
Whether there is sufficient evidence *or*
Whether the evidence is strong enough to   **then, the statement is**

conclude	maintain		
contend	corroborate		
believe	establish		
decide	support	that a statement is true,	H_1
determine	validate		
infer	verify		
assert	confirm		
affirm	substantiate		

reject	disprove		
refute	contradict		
deny	discredit	a statement,	H_0
dispute	discard		
abandon	invalidate		

The following box summarizes a general strategy for formulating H_0 and H_1.

1. Determine the parameter being tested.
2. Determine the value hypothesized for the parameter.
3. Determine the appropriate equality or inequality symbol to make H_1 what the decision maker wishes to demonstrate is true.

Box 9-4 provides an interesting application of stating the hypotheses.

Section 9.9 Problems

97. Consider a test of a candidate's claim that she received better than 50% of the vote in each precinct on average.
  a. What is the parameter being tested? You have figures for the percentage of the vote she received in each of 250 different precincts, among 5,000 total precincts.
  b. If you know the percentage that she received in each precinct, why are you unable to calculate the overall success rate for a single, combined sample consisting of all voters in the 250 precincts?
  c. You find that the mean percentage from the sample of 250 precincts is 48%. What is the value of μ that belongs to H_0 and H_1?
  d. State the hypotheses if the candidate performs the test.
  e. An opponent might conjecture that she did not do that well. State the hypotheses he would use to substantiate his counterclaim.
  f. A third party only believes that 50% is wrong but does not know if the value is too large or too small. State the hypotheses this person would use.

98. A college catalog states that the average entrance exam score is above 75.
  a. State the two hypotheses, if the college is required to substantiate this claim.
  b. A rival college wishes to dispute the claim. What are the hypotheses now?
  c. A student does not believe anything he reads in the catalog. How would he state the hypotheses?

99. Suppose a certain student is required to spend an average of at least two hours per day in a study hall. A monitor randomly chooses days and records the time the student spends in the study hall. The monitor wants to avoid accusing the student of spending less time than is required, if the student is actually meeting the requirement. He is willing to take a 5% risk of making

Box 9-4 Hypothesis Testing to Verify a Theoretical Prediction

The basis of much of economic theory is the assumption that rational people consider alternatives and act in a way that optimizes their objectives. They act in their own self-interest. Some would call this greed and disassociate themselves from what they consider selfish rather than rational behavior. Economists, however, are sometimes credited with such analytic behavior, at least more than the average person. The reason for this behavior is unknown, but some people suggest that economists are born thinking this way or that they learn this behavior from study and practice, especially in financial decision making. Supposedly, they weigh the costs and benefits of alternatives, even calculating cost-benefit ratios, and proceed with the best alternative.

To obtain more information on this phenomenon, John R. Carter and Michael D. Irons, professor and former student, respectively, at College of the Holy Cross, conducted an experiment, called an "ultimatum game." Pairs of students divided $10 among themselves or received nothing if they did not agree on a split. One member, the Proposer, made a one-time offer to the other student, the Responder, for a specific amount as the Responder's share of the $10. The Responder could agree or decline this single offer. Before random assignment to a pair, all participants were surveyed to learn the minimum amount they would accept as a Responder and the amount they would propose if selected to be a Proposer. Offers could be in $0.50 increments, so any Responder would be better off to agree to as little as $0.50 rather than not agree and go away empty-handed from the ultimatum encounter. In fact, economic theory suggests that a rational Responder will take any offer of $0.50 or more, and a rational Proposer, expecting rationality from the Responder, will propose $0.50 for the Responder and retain $9.50. Both players will have more with this settlement than if they do not agree.

Consequently, we would expect people who behave in an economic, rational, self-interested mode, to record the $0.50 and $9.50 values on the survey, while other people would record other values. Then, if group members are economically rational, the mean responses would be $0.50 and $9.50. We might state this more generally by saying that a population of economically rational people, on average, will respond $0.50 and $9.50.

Consider the $0.50 mean response for a moment. If we want to disprove that the group is rational on average, this statement would be the null hypothesis, and we would attempt to show that the mean response for the group is significantly different from $0.50. In fact, because $0.50 is the smallest positive amount that a participant can obtain, we expect that the mean response will be greater than $0.50 if it is not $0.50, so the test will be upper-tailed. If we want to demonstrate that the group is rational, $\mu = 0.50$, we would still perform an upper-tailed test for 0.50, because the equal sign goes with H_0. But failing to reject H_0 will not be as strong a validation of the group's rational economic behavior as would be the case if we reject H_0 when the rational behavior statement is H_1. By similar reasoning, we would have H_0: $\mu = \$9.50$ and H_1: $\mu < \$9.50$ if we test for the proposed amount.

The sample results are listed in the following table. In all cases, we reject H_0. The sample results show that both groups did not act in strict accord with economically rational behavior, although the magnitudes of the sample means suggest that economists behaved this way more than noneconomists.

	Sample Results			
	Economists ($n = 43$)		Noneconomists ($n = 49$)	
	$\overline{X}$	S	$\overline{X}$	S
Minimum Acceptable Amount of Respondents	1.70	1.70	2.44	1.67
Proposer Amount Kept	6.15	1.37	5.44	0.87

Carter and Irons used a different statistical technique, regression analysis, that employs more information than presented here to understand their sample results, but this formulation of the test is useful for understanding how hypothesis techniques are employed. In Chapter 10, we could use the two sample means presented here and perform a test to determine if the population mean responses are different for economists and noneconomists, regardless of whether either mean is $0.50 or not. Such a test would provide more information about a comparison of economists to other people than is done here.

Source: John R. Carter and Michael D. Irons, "Are Economists Different, and If So, Why?" Journal of Economic Perspectives, Vol. 5, No. 2, Spring 1991, pp. 171–77.

this mistake, because he will only have a random sample of the student's times, not complete documentation.
   a. What are the hypotheses?
   b. What is the significance level?

100. A historian wants to know if a random sample of courthouse entries on assessed values of land a century earlier is sufficient evidence to deny a claim that the average value per acre 100 years ago was less than $15. Is this claim H_0 or H_1?

Figure 9-11 (repeated)
Minitab output for test of difference of assessed and market values

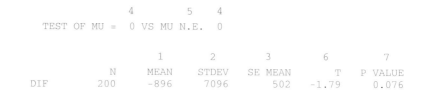

```
                        4        5        4
         TEST OF MU =   0  VS  MU N.E.    0

                   1        2        3        6         7
              N   MEAN    STDEV   SE MEAN    T     P VALUE
   DIF      200   -896    7096      502    -1.79    0.076
```

9.10 More about Computers and Hypothesis Testing for Population Means

Section 9.4 gives you the basic information for using computer output to test for a population mean. Now, we use another piece of information from the printout, the *T*-ratio, to save ourselves some computations. The information we can identify so far in Figure 9-11 (repeated here) includes the

1. Sample mean, $\overline{X}$.
2. Sample standard deviation, S.
3. Estimated standard error of the mean, $\hat{\sigma}_{\overline{x}} = S/\sqrt{n}$.
4. μ_0 specified by the Minitab user.
5. Direction of test specified by the Minitab user (N.E. is $\neq$, G.T. is $>$, and L.T. is $<$).
6. *T*-ratio = $(\overline{X} - \mu_0)/\hat{\sigma}_{\overline{x}}$ (Minitab user specifies μ_0. Many statistical packages do not allow the user to specify a mean other than zero, so the printed *T*-ratio is $(\overline{X} - 0)/\hat{\sigma}_{\overline{x}} = \overline{X}/\hat{\sigma}_{\overline{x}}$.)

The *T-ratio* or *T-statistic*, the number of estimated standard errors by which the sample mean, $\overline{X}$, differs from the hypothesized mean, is the test statistic that we compare with critical values on the *T* or *Z* axis to test hypotheses. When the sample is large, we often determine critical values from the standard normal table. However, the test statistic is generally called a *T*-ratio, because the denominator is the *estimated* standard error of the mean, $\hat{\sigma}_{\overline{x}}$.

■ **Example 9-3 (continued):** The output displayed in Figure 9-11 presents statistics calculated from a random sample of 200 differences between assessed values and corresponding market values of houses, based on a random sample of 200 homes that sold within six weeks of being assessed by a local tax assessor. If we perform a test to determine if there is any discrepancy on average (plus or minus) in the assessment process, we have $H_0: \mu = 0$ and $H_1: \mu \neq 0$.

Question: Use the mean and estimated standard error to form the *T*-ratio for the test to see if it matches the *T*-ratio in the output.
  Answer: $(\overline{X} - 0)/\hat{\sigma}_{\overline{x}} = -896/502 = -1.78$. This value is not exactly the same as the *T*-value in the output, -1.79, because of rounding.

Use -1.79 to perform the test at the 0.05 level.

■ **Solution:** The critical z values are ± 1.96 for a two-tailed test for the population mean with a large sample at the 0.05 level. Because -1.79 is now within the acceptance region ($-1.96 < -1.79 < 1.96$), we cannot reject H_0, the same decision we made when working on the $\overline{X}$ axis in Section 9.4.

■ **Example 9-10:** An airline company claims that its flights depart at most 10 minutes late, on average. However, a random sample of 14 flights by this airline during one week has a mean difference in actual and expected departure time of

Problems

```
TEST OF MU = 10.000 VS MU N.E. 10.000

                N       MEAN      STDEV    SE MEAN          T    P VALUE
LATETIME       14     15.571     11.467      3.065       1.82     0.046
```

Figure 9-31
Minitab output for airline departure test

15.571 minutes as shown on the computer output in Figure 9-31. Is this sample evidence sufficient to refute the company's claim at the 0.05 level?

■ **Solution:** Yes. The hypotheses are H_0: $\mu \leq 10$ and H_1: $\mu > 10$, because we are trying to refute the company's claim. Assuming all of the company's LATETIME values are normally distributed, the critical value for an upper-tailed test from the T table (Appendix J) with 13 degrees of freedom is 1.771. The T-ratio from the computer printout is 1.82, which exceeds 1.771, so we reject H_0.

Question: Interpret this decision.
Answer: The data contradict the company's claim and indicate that the departures are more than 10 minutes late on average.

Question: Calculate the T-ratio for a test of H_0: $\mu \geq 12.50$, using the printout in Figure 9-31. Then make a decision at the 0.01 level.
Answer: The T-ratio is $(15.571 - 12.50)/3.065 = 1.002$. The critical value from the T table with 13 degrees of freedom is 2.650, so we cannot reject H_0, $1.002 < 2.650$.

Section 9.10 Problems

101. Figure 9-32 shows Minitab output for a test to determine if the average level of particulate matter in the 63 MSAs in the Air Quality Data Base (Appendix B) is below the National Ambient Air Quality Standard of 50 μg/m^3. Because there are no data for some MSAs, assume this sample is a random sample of all MSA values, and perform the test at the 0.2 level. Interpret your decision. Compare your decision with the one made in Problem 20.

102. Use the Minitab output in Figure 9-33 to determine if the mean for the 63 MSAs supports a conclusion that the average concentration is significantly higher than the NAAQS of 0.12 ppm. Assume the 63 MSAs are a random sample of all MSAs, and test at the 0.02 level. Interpret the decision, and relate it to the decision in Problem 21. What is the smallest level of significance, α, that would cause us to reject H_0?

103. Figure 9-34 displays the Minitab output for a test to determine if the mean effective annual yield (in percent) of tax-exempt money market mutual funds is greater than 3.5%. The yields of 68 randomly selected funds produced the statistics that the figure shows. Perform this test at the 0.01 significance level, and interpret the decision. What is the smallest level of significance, α, that would cause us to reject H_0? (Source: *Wall Street Journal*, Thursday, July 11, 1991, p. c21.)

```
TEST OF MU = 50.000 VS MU L.T. 50.000

                N       MEAN      STDEV    SE MEAN          T    P VALUE
PM10           63     41.873     13.891      1.750      -4.64     0.0000
```

Figure 9-32
Minitab output for one-tailed test for PM10 standard

```
TEST OF MU = 0.12000 VS MU G.T. 0.12000

                N       MEAN      STDEV    SE MEAN          T    P VALUE
OZONE          63    0.13524    0.04231    0.00533       2.86     0.0029
```

Figure 9-33
Minitab output for one-tailed test for ozone standard

```
TEST OF MU = 3.5000 VS MU G.T. 3.5000

                N       MEAN      STDEV    SE MEAN          T    P VALUE
MMMF           68     3.5278     0.3946     0.0478       0.58       0.28
```

Figure 9-34
Minitab output for one-tailed test for money market mutual fund average annual yield

Summary and Review

A test is one-tailed or two-tailed according to the hypotheses as shown in Figure 9-26. We use a **one-tailed test** to check a statement when it may err in a single direction, too small or too large. μ_0 is the mean, μ, in the one-tailed diagrams, because if we reject H_0 when $\mu = \mu_0$, we will reject it as well for the other values specified in H_0.

Often we transform the test statistic from the original units of measure to a Z- or T-statistic. Either way the decision will be the same, because there is a one-to-one correspondence between the original and the transformed axes.

We perform tests for different parameters and for different assumptions for a given parameter. For small-sample tests for population means and large-sample tests for population success rates, the basic idea remains the same. Reject H_0 when the value of the test statistic you obtain is unlikely to occur if H_0 is true. The test statistic and its sampling distribution vary depending on the parameter tested and the known information, such as the population distribution or standard deviation, for a given situation.

To set up a hypothesis test, the decision maker must determine (1) the parameter being tested; (2) the value hypothesized for the parameter; and (3) the appropriate directional symbols ($\neq$, $<$, or $>$) for H_1. The primary rule is to make H_1 the hypothesis that the decision maker thinks is true or wishes to demonstrate as true. The equal sign always goes in H_0.

Multiple-Choice Problems

104. To reject H_0 in a one-tailed test means that
   a. There is not much difference between the test statistic and the population parameter specified in H_0.
   b. The statistic varies significantly from the parameter in a single, predetermined direction.
   c. We expose ourselves to making a Type II error.
   d. The statistic and parameter are significantly identical.
   e. The specified value (μ_0 or p_0) has been rejected, but we are not sure whether to reject for the other values specified with the inequality in H_0.

Use this information to work Problems 105–108:

H_0: $\mu \leq 500$ and H_1: $\mu > 500$.
$\overline{X} = 520$, $n = 100$, $\sigma = 150$, $\alpha = 0.05$.

105. The probability we reject H_0 for this set of hypotheses when H_0 is true is
   a. β.
   b. α.
   c. 0.95.
   d. 0.05.
   e. Level of significance.
   f. Responses a, b, and e.
   g. Responses b, d, and e.

106. The form for the decision rule is
   a. If $\overline{X} \leq c$, reject H_0.
   b. If $\overline{X} \geq c$, reject H_0.
   c. If $\overline{X} \leq 500$, reject H_0.
   d. If $\overline{X} \geq 500$, reject H_0.
   e. If $\overline{X} \leq 520$, reject H_0.
   f. If $\overline{X} \geq 520$, reject H_0.

107. The critical value on the $\overline{X}$ axis is
   a. 1.64.
   b. -1.64.
   c. 524.7.
   d. 475.3.
   e. 746.
   f. 254.

108. The decision is
   a. Reject H_0, because $520 < 524.7$.
   b. Cannot reject H_0, because $520 < 524.7$.
   c. Reject H_0, because $520 > 524.67$.
   d. Cannot reject H_0, because $520 > 524.7$.

109. Which of the diagrams in Figure 9-35 is the correct graph for the following hypothesis test?

   H_0: $\mu \geq 256$ and H_1: $\mu < 256$.
   $\sigma = 50$, $\alpha = 0.01$, $n = 100$, $\overline{X} = 243$.

   a. Diagram A.   d. Diagram D.
   b. Diagram B.   e. Diagram E.
   c. Diagram C.   f. Diagram F.

110. The critical value for a test is 888. The hypotheses are H_0: $\mu \geq 1,000$, H_1: $\mu < 1,000$. The sample mean and standard deviation are 850 and 600, respectively. What would the decision be in this case?
   a. Reject H_0.
   b. Cannot reject H_0.
   c. Cannot make a decision.

111. A computer center operator wants to test if the mean amount of time a user spends on the console is more than 15 minutes. He randomly selects 20 users and

Figure 9-35
Choices for Question 109

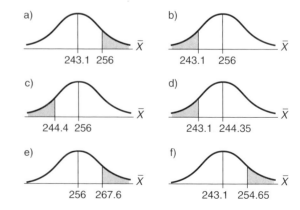

notes their times. The mean and standard deviation are 20.28 minutes and 5.36 minutes, respectively. Assume a normal population and a 0.01 level of significance to test the following hypotheses.

$$H_0: \mu \leq 15 \qquad H_1: \mu > 15.$$

(The following numbers are transformed to units that indicate the distance from the mean in standard deviations.)
  a. Reject H_0, because $20.28 > 15$.
  b. Cannot reject H_0, because $1.198 < 2.528$.
  c. Reject H_0, because $4.405 > 2.528$.
  d. Cannot reject H_0, because $1.198 < 2.539$.
  e. Reject H_0, because $4.405 > 2.539$.

112. When do we use the T distribution?
  a. Small-sample tests for p when we must estimate p.
  b. Large-sample tests for p when n approaches N.
  c. Small-sample tests for μ when we must estimate $\hat{\sigma}_x$.
  d. Large-sample tests for μ when $\hat{\sigma}_x$ is known.
  e. When the population distribution is unknown.

113. A political candidate claims that a slight majority (51%) of the electorate favors her candidacy. Test this claim at the 0.05 level against the alternative that less than a majority favor the candidacy. A random sample of 2,500 voters contains 1,250 who favor the candidacy.
  a. Cannot reject the claim, because $z = -1 > -1.645$.
  b. Cannot reject the claim, because $\hat{p} = 0.50 < 0.51$.
  c. Cannot reject the claim, because $z = -1 > -1.96$.
  d. Reject the claim, because $\hat{p} = 0.50 < 0.51$.
  e. Reject the claim, because $\hat{p} = 0.50 > 0.49$.

114. Suppose you develop a new system for producing an item that you believe is more efficient than the old system. μ_0 is the mean output from the old system. Choose the appropriate H_1 if you make H_1 what you think is true.
  a. $\mu \leq \mu_0$.   d. $\mu < \mu_0$.
  b. $\mu \geq \mu_0$.   e. $\mu = \mu_0$.
  c. $\mu > \mu_0$.   f. $\mu \neq \mu_0$.

115. Suppose you now test your new system (from Problem 114) according to the defect rate. p_0 is the defect rate of the old system. If you believe your system produces fewer defective items, choose the H_1 that reflects what you think is true.
  a. $p \leq p_0$.   d. $p < p_0$.
  b. $p \geq p_0$.   e. $p = p_0$.
  c. $p > p_0$.   f. $p \neq p_0$.

116. A government agency investigating ethical business behavior believes that businesspeople misrepresent the quality of their product for one of two motivations. Some overstate their defect rates in an effort to be truthful and not disappoint customers, and others understate the rate to mislead customers whom they expect will never check. Suppose your company claims a 4% defect rate for its product. Which H_1 reflects what the government agency thinks is true?
  a. $p \geq 0.04$.   d. $p > 0.04$.
  b. $p \leq 0.04$.   e. $p = 0.04$.
  c. $p < 0.04$.   f. $p \neq 0.04$.

117. 20% of an instant drink mix needs to be a certain ingredient when the mixture is purchased, in order for it to combine correctly with water and taste good when the customer drinks it. Test at the 0.05 level to see if the contents are correctly combined, when 200 containers of the mixture are checked. The mean percentage of the critical ingredients in the sample containers is 23.2%, with a standard deviation of 3.8 percentage points.
  a. Reject the shipment, because $23.2 > 20$.
  b. Reject the shipment, because $11.91 > 1.96$.
  c. Cannot reject the shipment, because $-1.96 < 0.84 < 1.96$.
  d. Cannot reject the shipment, because $-1.96 < 0.232 < 1.96$.
  e. Reject the shipment, because $3.2 > 1.96$.

118. A chemist has developed an experimental catalyst that she thinks speeds up the average starting time for a reaction that clears petroleum spills. The average time for the best alternative on the market is 40.1 minutes. To demonstrate her conjecture, she has replicated experiments that measure the time (in minutes) for the catalyst to start a reaction that is sufficient to dissolve a petroleum distillate at 12 parts per cubic centimeter of sea water. The computer output for her tests is shown in Figure 9-36. Can she use these results to demonstrate her conjecture at the 0.1 level?
  a. Reject, $-5.51 < -1.96$.
  b. Cannot reject, $-5.51 < -1.96$.
  c. Cannot reject, $36.1 \neq 40.1$.
  d. Reject, $36.1 < 40.1$.
  e. Reject, $-5.51 < -1.28$.

119. Suppose hospital procedures require nurses to spend an average of 15 or more minutes per patient bath. If the manager wants to avoid making the mistake of accusing a nurse of not following the procedure when the nurse is complying, what is H_0?
  a. $\mu > 15$.   c. $\mu \neq 15$.   e. $\mu \leq 15$.
  b. $\mu < 15$.   d. $\mu \geq 15$.   f. $\mu = 15$.

120. A manufacturer tells you that his product exceeds certain standards according to hypothesis tests that he conducted. Evaluate the believability of this statement, assuming that he cannot reject H_0, which states that the product meets or exceeds the standard.

```
TEST OF MU = 40.100 VS MU L.T. 40.100

              N        MEAN      STDEV    SE MEAN        T     P VALUE
TIME        128       36.102     8.208      0.725     -5.51     0.0000
```

Figure 9-36

Minitab output for experimental catalyst test

a. The evidence is convincing, because we could not reject H_0.
b. The product may meet but not exceed the standard, because H_0 includes the equal sign.
c. The evidence is weak, because the manufacturer receives the benefit of the doubt, so dubious sample results reflect favorably on the product.
d. The evidence is sufficient to convince even the most doubtful, because of the stricter standards imposed on the null hypothesis.
e. The evidence is strong no matter which hypothesis states that the produce exceeds the standard.

121. Seventy-five percent of a sample of 402 companies said employees' interpersonal skills (talking and listening with customers and other workers) is a key problem. A year earlier only 62% said communication was a problem. Perform a test at the 0.005 level to demonstrate that the percentage identifying communication as a key problem is significantly higher this year. (Source: "Communication Skills Lacking, Businesses Say," *Asheville Citizen-Times*, September 21, 1992, p. 5B.)
a. Reject H_0, because 75% is so much larger than 62%.
b. Reject H_0, because $5.370 > 2.575$.
c. Cannot reject H_0, because a test cannot compare different values from different years.
d. Cannot reject H_0, because $5.370 > 2.575$.
e. Cannot reject H_0, because 0.13 is smaller than 1.96.

Use the following information to work Problems 122–125.

In order to avoid stress in a formal setting where students are reciting the alphabet, a kindergarten teacher randomly selects times to quiz a student having difficulty learning the alphabet. She believes the student can correctly order more than the first 18 letters, which is the average of the other students in the class. She keeps a record of the number of correctly ordered letters recited by the student. These values are:

18   16   22   19   15   22   20   16   18   19.

Use this information to test at the 0.1 level to determine whether the teacher is correct and the student can correctly order more than 18 letters on an average try. Assume that the results of all such recitations would be normally distributed.

122. H_1 is
a. $\mu > 18$.   d. $p \geq 0.18$.
b. $p > 0.18$.   e. $\mu \neq 18$.
c. $\mu \geq 18$.   f. $p \neq 0.18$.

123. The decision is
a. To reject H_0, because $18.5 > 18$.
b. To reject H_0, because $0.7 < 1.383$.
c. Not to reject H_0, because $0.7 < 1.383$.
d. Not to reject H_0, because $0.7 < 1.28$.
e. Not to reject H_0, because $1.2 < 1.28$.
f. Not to reject H_0, because $0.7 < 1.645$.

124. The decision not to reject H_0 means that the
a. Teacher is definitely wrong about the student.
b. Teacher is definitely right about the student.
c. Data are sufficient to conclude that the teacher's belief is correct.
d. Data are not sufficient to support the teacher's belief.

125. What is the minimum significance level that would allow us to reject H_0 with the data?
a. 0.05.
b. A value smaller than 0.05.
c. 0.25.
d. A value larger than 0.25.
e. A value smaller than 0.25.

Word and Thought Problems

126. A company claims that its cereal contains an average of at least 1,000 chocolate chips in every box. A consumer group finds that a random sample of 75 boxes has a mean of 950 chips per box, with a standard deviation of 40 chips. Can the consumer group reject the company's claim at the 0.01 level?

127. Handmade hamburger patties at Sally's Grill are supposed to weigh 1/4 pound before cooking. Sally regularly checks her employees' patties by selecting 32 random patties and finding the mean weight. Patty's 32 patties had a mean weight of 0.275 pound with a standard deviation of 0.057 pound. Should Sally rebuke Patty, if Sally does not like to be wrong more than 5% of the time when she rebukes her employees?

128. Suppose the standard for safe brakes is that the mean stopping distance be less than 150 feet at 55 mph. The company tests 100 sets of brakes on similar automobiles at 55 mph. If the sample of 100 brakes has a mean stopping distance of 138 feet with a standard deviation of 15 feet, can the company decide the brakes are safe at the 0.05 level?

129. Discuss the effect on the range of test statistic values in the rejection region, when the
a. Significance level increases and nothing else changes.
b. Sample size increases and nothing else changes.
c. Population standard deviation increases and nothing else changes.

130. A random sample of 15 checks paid to a restaurant yields a mean value of $9.50 with a standard deviation of 70 cents. Use this information to demonstrate that the mean check amount is less than $10. Use $\alpha = 0.05$. Assume a normal population. Interpret the decision.

131. The average break time in Busy Factory is supposed to be 15 minutes or less. Management is concerned that employees are taking longer breaks. They time a random sample of 49 breaks. The mean is 20 minutes

Word and Thought Problems

with a standard deviation of 10 minutes. Is the conclusion that breaks are overlong justified at the 0.02 level?

132. A professor claims the mean score on a test is 75. Your grade and those of three of your friends are 75, 60, 65, and 72. Can you reject the professor's claim at the 0.05 level? Does this sample appear to be random? What assumptions must you make about the distribution of population values to work this problem?

133. An economic forecaster sometimes over- and sometimes underforecasts a company's sales. These forecasts produce positive and negative errors, respectively. Use the following sample of forecast errors to determine if the errors cancel over the long run (have a mean of zero): $-1, 3, 0, -1, 1$. Assume a normal population. Use $\alpha = 0.1$.

134. A parts supplier claims its defect rate is at most 10%. A random sample of 185 items yields 22 defectives. Is this sufficient information to deny the company's claim at the 0.04 level?

135. A manufacturer claims that her product exceeds certain standards according to a hypothesis test that she conducted. Assume the alternative hypothesis reflects this claim. Evaluate the believability of this statement in light of the fact that no significance level is reported.

136. a. A company is willing to accept shipments only if 10% or fewer items are defective. An employee randomly selects 100 items from each shipment and test each for quality. If 12% or more of the items in the sample are defective, the company returns the shipment. Given that an entire shipment is actually only 10% defective, what is the probability that the company will mistakenly return it?
   b. Suppose that 9% of the shipment is actually defective, and find the probability the company returns it, using the 12% sample defect rate for 100 items. Repeat this calculation for 5% actual defectives. What happens to the probability as the defect rate decreases?
   c. State the H_0 and H_1 behind Part a for the company checking the shipment. By using 10% (0.10) as p_0, the actual defect rate specified in H_0, will the probability of rejecting H_0 when $p = p_0$ be larger than, smaller than, or the same as the probability of rejecting H_0 for any other value of p included in H_0 ($p \leq 0.10$)? (Assume the critical value does not change.) Hint: use your answer to Part b.

137. Suppose the telephone company claims that it has sufficient capacity for you to obtain a long-distance line at least 99% of the time. If you have been unable to obtain a line five of the past 100 times you tried, can you reject the telephone company's claim at the 0.1 level? (Assume randomness.) Describe a potential problem that could result from the size of the sample.

138. Use the output in Figure 3-39 (p. 142) to do a lower-tailed, upper-tailed, and two-tailed test of the hypothesis that the mean coupon rate is 50 cents at a 0.05 level.

139. A hypnotist claims to have a 97% success rate for curing smoking. A random sample of 180 of the hypnotist's patients attempting to quit smoking contains 170 currently not smoking. Is there sufficient information to reject the hypnotist's claims at the 0.05 level; that is, to decide that his success rate is smaller than claimed?

140. Use your sample of 10 county nursing home census counts from Problem 40 of Chapter 7 (p. 306) and do a two-tailed test of $\mu = 470.1$ people at the 0.05 level. If 100 students in a class do this problem, how many would you expect to reject the true null hypothesis?

141. Can you reject a claim that the average age of members of Congress is at least 50, if a random sample of 25 members has a mean age of 48.2, with a standard deviation of 3.1 years? Assume all members' ages are normally distributed; test at the 0.01 level.

142. A firm advertises in a student newspaper that if a student reads and studies its instructional package, he or she can expect to score better than 600 on a national test.
   a. Assume you work for the firm and must justify this claim to a national education group or consumer group. Set up H_0 and H_1 if the mean is the parameter to be used. Justify your choice of hypotheses.
   b. Would you set up the test differently if you worked for a national consumer group? Explain your answer, and justify your choice of hypotheses.

143. An eye doctor tests for edema (swelling) by moving different lenses in front of the eye and determining if the patient can distinguish differences between the images seen through the different lenses. (The lenses are sequenced so that the image should become more distorted as the lenses change.) If the eye is healthy, it can detect a difference quickly (say, after two lens changes). If not, the images will appear similar over a wide range of lenses. If H_0 is a healthy eye, write a decision rule corresponding to the eye doctor's diagnosis process. How does this process give H_0 the benefit of the doubt?

144. Can you reject an English professor's claim that at least three-fourths of the time, the spelling "judgment" is used rather than "judgement" if in a random sample of 100 occurrences of the word it is spelled without the "e" 68 times? Test at the 0.01 level.

145. a. The director of a local morning television talk show with several hosts requires that these people average at most 0.5 "unprofessional moments" per show. Unprofessional moments include verbal mistakes, overlong pauses, or use of "uh." Is the director justified in reprimanding a person, if, in a random sample of 36 shows, this person has a mean of 0.75 such moments per show with a standard deviation of 0.1? The director is willing to risk correcting a person's overall performance 1% of the time when it is actually satisfactory in order to avoid accounting for all unprofessional moments in every show.

b. Compare the situations of two hosts, based on random samples of 36 shows for each host. One consistently had bad moments, so the mean and standard deviation are 0.75 and 0.1, respectively. Another had runs of either perfect days or perfectly awful days. The mean for this host is also 0.75, but the standard deviation is 1. Using the same risk level, should the second host be reprimanded?

c. Compare your answer to Parts a and b. Then discuss the advisability of using the director's decision process for improving or maintaining the quality of the television program. What other parameter would you suggest investigating?

146. A chemist regularly orders bulk amounts of a certain chemical that is supposed to contain at most 3% impurities on average. Two hundred different one-gram samples are randomly selected from each order, and the average proportion of impurities is calculated. What is the smallest value of a sample's mean proportion of impurities that should be allowed to decide a shipment is unacceptable, if acceptable shipments can be judged unacceptable only 5% of the time? It is assumed that the standard deviation of proportion of impurities per gram is one percentage point.

147. Can you reject at the 0.01 level the hypothesis that at least 50% of the customers at a local eatery order soft drinks, if 45% of a random sample of 250 customers order soft drinks?

148. Test at the 0.1 level the hypothesis that the mean time to get from one class to the next is at most 10 minutes if the mean of a random sample of 16 times is 11.75 minutes with a standard deviation of four minutes. Assume all times are normally distributed. Interpret this decision.

149. Can you reject at the 0.01 level the hypothesis that at most 40% of the union voters voted for the Republican presidential candidate, if a random sample of 200 union voters on their way from the voting place contains 90 who voted Republican?

150. What information do sampling distributions provide about the estimator or test statistic that is necessary for evaluating errors when making statistical inferences?

151. A surgeon is being sued for medical malpractice by a group of gall bladder patients. One of the plaintiffs' claims is that this surgeon's patients require longer than the national average of 5 days to recover. The plaintiff's lawyer had randomly selected 15 of the surgeon's previous gall bladder patients and determined their recovery times from records and interviews with these patients. Their recovery time averaged 6.1 days. To refute this claim of a longer recovery time, an expert statistical witness examines the plaintiffs' data, the values listed as follows. Use this information to demonstrate at the 0.05 level that the plaintiff's value is not convincing evidence that the surgeon's patients' mean recovery time is greater than 5 days if the surgeon is considered innocent until proven guilty and in spite of the sample mean, 6.1 days, being greater than 5. Assume recovery times are normally distributed.

4 10 6 5 5 5 6 4 5 5 5 5 15 7.

152. a. A high school vocational counselor casually noticed that students professing an interest in sports management on a counseling form tend to perform better than the average student on the state math exam. The average score for all students is 68 points. Use a 0.01 significance level and the information in the computer printout in Figure 9-37, based on a random sample of state students interested in sports management, to determine if the counselor can confirm this observation.

b. The standard deviation for all students' scores is 18.15. Does the sample standard deviation, S, recorded in the printout seem reasonable, in light of the fact that the sample is a subset of all students? Explain.

c. The true standard deviation for the population of all students interested in sports management is unknown. The two values we know, 18.15 and 12.315, are only clues. Given this fact, which of the two standard deviation values should the counselor use in order to avoid claiming that the sports management mean is higher if it is not? Explain and then demonstrate your answer with the available information.

153. A nutritionist, who assumes that all food contains more sodium than is healthy until demonstrated otherwise, is testing a new chicken casserole recipe for sodium content. You are the statistical lab assistant. Use the computer printout in Figure 9-38 of sodium content in milligrams for recipes that have passed the tastiness

Figure 9-37
Minitab output for math score test

```
TEST OF MU = 68.000 VS MU G.T. 68.000

             N      MEAN     STDEV    SE MEAN      T      P VALUE
MATHSCOR    88    72.443    12.315     1.313     3.38    0.0005
```

Figure 9-38
Minitab output for sodium content test

```
TEST OF MU = 2.0000 VS MU L.T. 2.0000

            N      MEAN     STDEV    SE MEAN      T       P VALUE
CONTENT    21    1.5276    0.1901    0.0415    -11.39    0.0000
```

Word and Thought Problems

test, assuming these recipes are a random sample of all possible tasty recipes. Convince the nutritionist that these tasty casseroles contain less than 2 milligrams of sodium, on average. Test at the 0.01 level, and assume that sodium content for all tasty recipes is normally distributed.

154. The aircraft involved in the 10 major crashes of domestic jet flights of established U.S. air carriers during the period 1975–1979 had the following load factors (percentage of available seats filled by passengers).

86.6 95.3 81.5 100.0 100.0 50.7 99.3 89.4 44.3 100.0

The U.S. Department of Transportation figures reveal that the mean load factor for domestic jet flights of established carriers during this period was approximately 59.4%. (Source: Arnold Barnett and Todd Curtis, "An Unfortunate Pattern Observed in U.S. Domestic Jet Accidents," *Flight Safety Digest,* October 1991, pp. 1–8.)

a. Is the sample of 10 crashes random?
b. What is the probability of obtaining a random sample of 10 flights with an average load factor of 84.71% or greater if the mean is 59.4% and load factors are normally distributed?
c. Does the result in part b indicate that more heavily loaded jets are more likely to be involved in major crashes at the 0.05 significance level?

Unit III: *P*-Values (Optional Unit)

Preparation-Check Questions

155. A random sample of 400 items is selected from a population where 10% of the items are defective. Find the probability that at least 12% of the sample items are defective.
   a. 0.4082.   d. 0.1056.
   b. 0.8944.   e. 0.0918.
   c. 0.9082.

156. A random sample of 16 items has a mean of 633 and a standard deviation of 4. The sample is collected from a normal population. Use this information to test H_0: $\mu = 630$ against H_1: $\mu \neq 630$ at the 0.10 level.
   a. Reject H_0, because $3 > 1.746$.
   b. Reject H_0, because $3 > 1.753$.
   c. Cannot reject H_0, because $-1.746 < 0.75 < 1.746$.
   d. Cannot reject H_0, because $-1.753 < 0.75 < 1.753$.
   e. Reject H_0, because $3 > 1.341$.

Answers: 155. e; 156. b.

Introduction

Day-Care Situation   A personnel department is researching a new program for company subsidized day-care payments for workers with young children. The department claims that the actual mean annual child-care expenditure per worker is greater than the budgeted figure of $2,500, based on a test at the 0.05 level. Upper-level managers do not want to pay more than an average subsidy of $2,500 per worker, unless the actual mean day-care cost *is* significantly higher. They have decided to increase the subsidy level only if the sample mean used in the department's test has at most a 1% chance of occurring when the actual mean is $2,500 or less. Thus, they want the test performed at the 0.01 level.

The department's rejection of a mean expenditure of $2,500 or less at the 0.05 level tells management nothing about what the decision would be if they tested at the 0.01 level. Is the sample mean close to the critical value or in the extreme end of the tail of the curve?

This unit discusses a different method of reporting test results, the *p*-value. *P*-values contain more information about sample results and allow people other than the original analyst to reject or not reject H_0 at any level of significance. Many statistical software packages routinely report *p*-values. We can use these reported values to make decisions without long calculations and statistical tables.

9.11 Test Information from P-Values

The level of significance of a test, α, represents the risk of making a Type I error. Different people prefer different levels of risk, and the same hypothesis that is not rejected at one level of significance can be rejected at another. The significance level, α, can simply be increased until the test statistic falls in the rejection region.

■ **Example 9-11:** Suppose that a random sample of 100 workers with children in day care shows a mean day-care cost of $2,600 and a standard deviation of $500. Verify the department's claim that the mean exceeds $2,500 at the 0.05 level with this information.

■ **Solution:** Because the department wants sufficient evidence that the mean is greater than 2,500 before it changes the budgeted subsidy, the hypotheses are $H_0: \mu \leq 2,500$ and $H_1: \mu > 2,500$. The sampling distribution is approximately normal (because $n = 100$), with a mean, μ, of 2,500 if H_0 is true and estimated standard error

$$\hat{\sigma}_{\bar{x}} = \frac{S}{\sqrt{n}} = \frac{500}{\sqrt{100}} = 50.$$

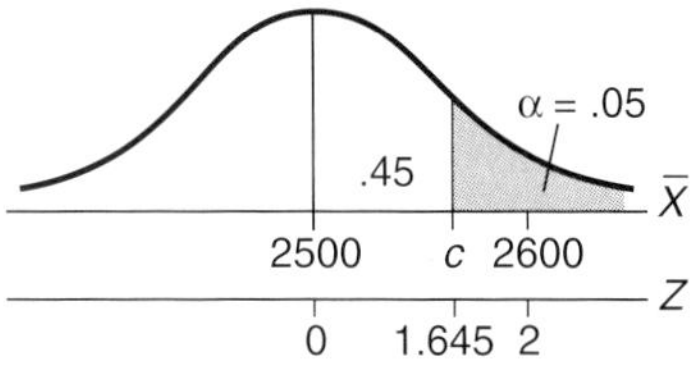

Transforming $\bar{X} = 2,600$ into a Z value, we obtain

$$\frac{\bar{X} - \mu}{\hat{\sigma}_{\bar{x}}} = \frac{2,600 - 2,500}{50} = 2.$$

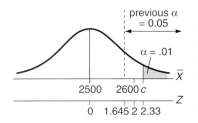

Figure 9-39
Diagram for the day-care subsidy test at the 0.05 level

The critical value from the standard normal table for an upper-tailed test at the 0.05 level is 1.645. Thus, we reject H_0, because $2 > 1.645$. See Figure 9-39. The department's claim is supported by the evidence.

Figure 9-40
Diagram for the day-care subsidy test at the 0.01 level

Question: Now perform the test at the 0.01 level.
Answer: The only difference is the critical value, now 2.33, which does not lead us to reject H_0, because $2 < 2.33$ (see Figure 9-40).

Thus, the sample mean of $2,600 causes us not to reject H_0 at the 0.01 level and to reject H_0 at the 0.05 level. Clearly, it is important to know the significance level in understanding or using test results. Otherwise, we do not know if the test statistic is close to or far away from the critical value. We might ask if H_0 is "strongly" or "weakly" rejected (or not rejected, as the case may be).

The *p*-value reports test results in a way that reflects the "strength" of the evidence. In fact, the decision is left to the person reviewing the results.

9.11.1 P-Value Definition and Interpretation

A *p-value* is the probability of obtaining the same value of the test statistic that we do obtain, or values that are even less likely, if the value proposed in H_0 is true. We say that these improbable values are more *extreme* than the actual test statistic value. They offer stronger evidence in favor of H_1 (of rejecting H_0).

■ **Example 9-11 (continued):** The hypotheses for the day-care situation are $H_0: \mu \leq 2,500$ and $H_1: \mu > 2,500$. The department's actual test statistic is the sample mean, $\bar{X} = 2,600$, and the population mean proposed in H_0 is $\mu_0 = 2,500$. The 2,600 is 100 units beyond the 2,500 specified in H_0. Locate the more extreme values of the test statistic, and find the *p*-value.

■ **Solution:** $\bar{X}$ values greater than 2,600 are stronger evidence that H_1 ($\mu > 2,500$) is true. These are the more extreme $\bar{X}$ values. See Figure 9-41a. The blue area in

Section 9.11   Test Information from *P*-Values   ■   **433**

Figure 9-41
Extreme values and corresponding *p*-values when H_1 is $\mu > 2{,}500$ for the day-care subsidy situation

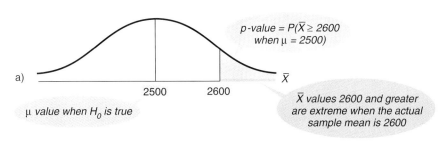

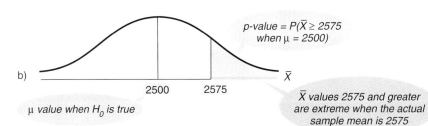

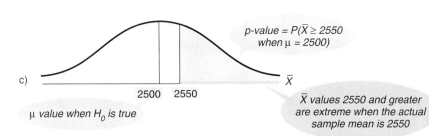

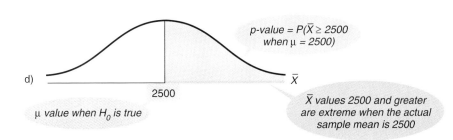

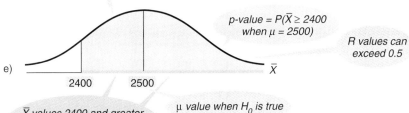

Figure 9-41a is the probability of obtaining 2,600 or one of these extreme values as a sample mean when H_0 is true. Thus, the blue area is the *p*-value.

If $\overline{X}_0$ is the observed value of $\overline{X}$, the value of the test statistic actually obtained, and μ_0 is the parameter specified in H_0, then for an upper-tailed test

p-value = P(obtain a sample mean, $\overline{X}$, as extreme or more extreme than the one actually obtained, $\overline{X}_0$, when H_0 is true)
= $P(\overline{X} \geq \overline{X}_0$, when H_0 is true)
= $P(\overline{X} \geq \overline{X}_0$, when $\mu = \mu_0)$.

■ **Example 9-11 (continued):** Draw the appropriate diagram, and shade the area corresponding to the *p*-value for the test of H_0: $\mu \leq 2{,}500$ if the sample mean you obtain to perform this test, $\overline{X}_0$, is 2,575. Repeat the sketch assuming $\overline{X}_0$ is 2,550, 2,500, then 2,400.

■ **Solution:** See Figure 9-41 b–e.

The direction of extremeness is determined by H_1. Because larger values of the test statistic offer stronger support for an upper-tailed alternative hypothesis, H_1: $\mu > \mu_0$, the *p*-value corresponds to larger test statistic values. Thus, the direction of extreme values lies to the right of $\overline{X}_0$. For a lower-tailed alternative hypothesis, H_1: $\mu < \mu_0$, extreme values lie to the left of $\overline{X}_0$. We will discuss the two-tailed situation later.

Question: Use H_0: $\mu \geq 1{,}414$ and H_1: $\mu < 1{,}414$. Draw the appropriate diagram, and shade the area corresponding to the *p*-value if the sample mean you obtain to perform this test, $\overline{X}_0$, is 1,387.
Answer: See Figure 9-42.

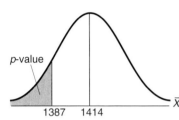

Figure 9-42
P-value for a lower-tailed test

We can measure the credibility of the null hypothesis, H_0, with a *p*-value.

Question: Consider the following potential $\overline{X}_0$ values that might be obtained when testing H_0: $\mu \leq 2{,}500$ and H_1: $\mu > 2{,}500$. The values are 2,400, 2,500, 2,550, 2,575, and 2,600. Are these values ordered from the value that most supports H_1 to the value that most supports H_0 or vice versa?
Answer: The order represents decreasing support for H_0, because smaller $\overline{X}$ values support the null hypothesis that $\mu \leq 2{,}500$, while larger $\overline{X}$ values support the alternative hypothesis that $\mu > 2{,}500$.

Upon reexamining Figure 9-41, we find that the *p*-value associated with each of these values increases as the support for H_0 increases (as $\overline{X}_0$ becomes smaller in the day-care test going down the page in Figure 9.41). This conclusion implies that the *p-value measures the credibility of H_0.* The larger the *p*-value (closer to 1), the more the test statistic supports H_0. The smaller the *p*-value (closer to 0), the more the test statistics supports H_1.

Question: If the *p*-value is 0.8, which is supported, H_0 or H_1?
Answer: H_0. Because the *p*-value is large, H_0 has a lot of credibility.

Actually, the *p*-value is the *maximum* probability of obtaining the same or a more extreme value of the test statistic if H_0 is true. If we use another value that makes H_0 true, such as $\mu = 2{,}300$ when H_0 is $\mu \leq 2{,}500$, to find a *p*-value, the probability of the same or a more extreme value (the shaded area in the diagram) would be smaller. We use the μ_0 value specified in H_0 in computations, because the maximum probability favors H_0 in the same way that we give H_0 the benefit of the doubt before rejecting it. Thus, the *p*-value measures the maximum credibility of H_0.

Section 9.11.1 Problems

157. Are the extreme values to the left or right of $\overline{X}_0$ when the alternative hypothesis is $\mu > 100$? Do these values offer more support to H_0 or H_1? Explain.

158. Are the extreme values to the left or right of $\hat{p}_0$ when the alternative hypothesis is $p < 0.4$?

159. Draw and shade the p-value for a test of H_0: $\mu \geq 343$ and H_1: $\mu < 343$ if the actual sample mean, $\overline{X}$, is 300. Repeat the exercise for actual $\overline{X}$ values of 330, 343, and 350. Label the extreme values in each case. Which $\overline{X}$ value most strongly supports H_0? Will it produce the largest or smallest p-value of the four?

160. Draw and shade the p-value for a test of H_0: $p \leq 0.85$ and H_1: $p > 0.85$ if the actual sample success rate, $\hat{p}$, is 0.90. Repeat the exercise for actual sample success rates of 0.87, 0.85, and 0.80. Label the extreme values in each case. Which $\hat{p}$ value most strongly supports H_1? Will it produce the largest or smallest p-value of the four values?

9.11.2 P-Value Calculation

Calculating p-values is the reverse of finding critical values. When we find critical values, we establish rejection regions of unlikely outcomes from a given probability, α. To find a p-value, we find the probability of unlikely or extreme outcomes from a specific outcome value, $\overline{X}_0$. In both cases, we solve the problem with a diagram that assumes that H_0 is true.

■ **Example 9-11 (continued):** Find a numerical result for the p-value for the day-care situation, if the mean of the sample of 100 employees' day-care costs is $2,600 and the standard deviation is $500. Recall the hypotheses, H_0: $\mu \leq 2,500$, H_1: $\mu > 2,500$.

■ **Solution:** The problem is to find the probability of obtaining a sample mean, $\overline{X}$, that is 2,600 or more extreme (in this case, larger) than 2,600, when $\mu = 2,500$ (see Figure 9-41a). This solution is a straightforward normal probability problem with $\mu = 2,500$ and $\hat{\sigma}_{\bar{x}} = S/\sqrt{n} = 500/\sqrt{100} = 50$. We convert 2,600 to $Z = (2,600 - 2,500)/50 = 2$ and find $P(Z > 2) = 0.0228$, the p-value.

Continuing in this manner, we can calculate the p-value for the remaining $\overline{X}_0$ values we have been considering: 2,400, 2,500, 2,550, and 2,675, as well as two more extreme values, 2,625 and 2,650. The corresponding p-values are listed in the following table.

$\overline{X}_0$	p-Value	$\overline{X}_0$	p-Value
2,400	0.9772	2,600	0.0228
2,500	0.5000	2,625	0.0062
2,550	0.1587	2,650	0.0013
2,575	0.0668		

Section 9.11.2 Problems

161. Find the p-value for H_0: $\mu \geq 23.7$ versus H_1: $\mu < 23.7$ if the mean of 165 items is 20.8 and the standard deviation is 34.1. Find the p-value when $\overline{X}$ is 23. When $\overline{X}$ is 24. When $\overline{X}$ is 25.

162. The Internal Revenue Service (IRS) suggests that the average tip in Italian restaurants is 15.8%, 15% in Chinese restaurants, and 14.8% in Mexican restaurants. Suppose a tax lawyer collects three random samples of 100 tips, one sample from each type of restaurant, and finds that in all three cases the mean tip is 15% and the standard deviation is 1%. What is the p-value for each type of restaurant if the lawyer seeks to dispute the IRS findings and show that the percentages are lower? (Source: Data from Associated Press article; "Lasagna Attracts Tippers," *Asheville Citizen*, August 28, 1989, p. 2.)

163. The director of a youth soccer league wants every player to play at least one quarter in each game and 75% to play at least two quarters in each game. Before he corrects the coaches unnecessarily, he collects a random sample of 250 players one Saturday to find what percentage played at least two quarters. Find the p-value for the director if 175 played at least two quarters. Interpret the result. Does the result appear to indicate that the result is unlikely if coaches follow the director's wishes? Explain.

9.11.3 Decisions with the *P*-Value

Now the question is how to use a *p*-value to make a decision. We know that the *p*-value measures the credibility of H_0, but how much credibility is necessary to fail to reject H_0? How much credibility must we lack before we reject H_0?

Figure 9-43 answers this question by using the day-care information to test at the 0.05 level. We calculated the critical value on the $\overline{X}$ axis earlier to be \$2,582. $\overline{X}$ values that equal or exceed the critical value, such as 2,600, produce *p*-values that equal or do not exceed the significance level, α. We reject H_0 in these cases. $\overline{X}$ values that are less than 2,582 result in our not rejecting H_0 and in *p*-values that are larger than α. In summary, *when the p-value, the measure of the credibility of H_0, sinks to α or below, H_0 is no longer acceptable. It is rejected*. We can also determine from this argument that the *p*-value is the minimum α that will cause us to reject H_0. The following rule emerges.

> **Decision Rule with *P*-Values**
> When *p*-value $> \alpha$, we cannot reject H_0. When *p*-value $\leq \alpha$, reject H_0.

Question: The *p*-value of a test is 0.08. Can we reject H_0 at the 0.05 level?
Answer: H_0 cannot be rejected, because the *p*-value $= 0.08 > 0.05 = \alpha$.

■ **Example 9-12:** A city council is concerned about low attendance at publicly funded youth dances. The program director says the average attendance is more than 550 at each dance. The council is skeptical. To test this claim, the council randomly selects five dances and finds the mean attendance is 529.26 people with a standard deviation of 16.71 people. Assuming attendance at all dances is normally distributed, find and use the *p*-value to perform this test at the 0.05 level.

■ **Solution:** Giving the benefit of the doubt to the program director, and because the council believes attendance is low, the hypotheses are H_0: $\mu \geq 550$ and H_1: $\mu < 550$. We must use the T distribution with $n - 1 = 4$ degrees of freedom for the sampling

Figure 9-43
The rejection region corresponds to *p*-values that are $\alpha = 0.05$ or smaller

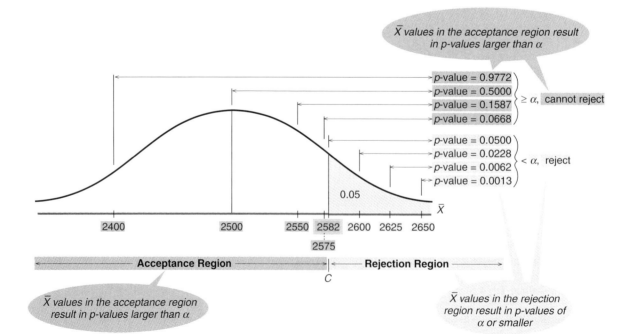

distribution of $\overline{X}$, because the sample is small ($n = 5$) and the population is normal. The estimated standard error is $S/\sqrt{n} = 16.71/\sqrt{5} = 7.4729$. Because H_1 is $\mu < 550$, small values of $\overline{X}$ support H_1, making the direction of extremeness to the left. In particular, we focus on $\overline{X}$ values smaller than 529.26. The p-value is the probability of obtaining an $\overline{X}$ smaller than 529.26, when H_0 is true ($\mu \geq 550$). We convert $\overline{X} = 529.26$ to $T = (\overline{X} - \mu)/\hat{\sigma}_{\overline{x}} = (529.26 - 550)/7.4729 = -2.7753$. The probability of a T value with four degrees of freedom being smaller than -2.7753 is approximately 0.025 (see Figure 9-44). We reject H_0, because the p-value $= 0.025 < 0.05 = \alpha$. The data dispute the program director's claim.

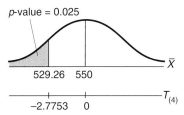

Figure 9-44
Diagram for the p-value for the dance attendance example

Section 9.11.3 Problems

164. Use the p-values from Problems 161–163 to make decisions at the 0.1 significance level. Interpret the decisions.

165. For which of the following p-values would we reject H_0 at the 0.05 significance level: 0.09, 0.001, 0.01, 0.1, 0.005, 0.05?

166. Compute and use the p-value to determine if you can reject H_0 in the following cases? Compare the decisions with those you made in Problem 81 of Unit II.

   a. $H_0: \mu \geq 250$ **b.** $H_0: \mu \leq 333$ **c.** $H_0: \mu \leq 666$
      $H_1: \mu < 250$ $H_1: \mu > 333$ $H_1: \mu > 666$
      $\alpha = 0.03$ $\alpha = 0.05$ $\alpha = 0.05$
      $n = 36$ $S = 84$ $n = 100$
      $S = 60$ $\overline{X} = 350$ $\overline{X} = 680$

   i. $\overline{X} = 245$.   **i.** $n = 36$.   **i.** $S = 50$.
   ii. $\overline{X} = 240$.   **ii.** $n = 49$.   **ii.** $S = 100$.
   iii. $\overline{X} = 235$.   **iii.** $n = 144$.   **iii.** $S = 200$.

167. A convenience store manager relocated the snack food shelves during remodeling. The manager feels the relocation may have been a mistake, because she perceives that a smaller percentage of customers are purchasing snack foods. Previously, 60% did so. To check her feeling in a manner that avoids relocating the shelves again unnecessarily, find the p-value. Then make and interpret a decision at the 0.01 significance level. During a certain week, 520 of 900 randomly selected customers purchased snack foods.

9.12 Two-Tailed P-Values

When we do a two-tailed test, the extremeness is in two directions, because values of the test statistic that are very small or very large—compared to the value specified in H_0—support H_1. For a test of the mean, we ask what is the probability of obtaining a sample mean, $\overline{X}$, as far or farther from the proposed mean, μ_0, as the actual $\overline{X}_0$. We do not specify the direction, just the distance.

■ **Example 9-13:** A manufacturing process usually produces an average output of 5,000 units per hour. When market conditions are tight, the manager desires an output very close to 5,000, not more and not less, to avoid losses from excess inventory or from empty-handed customers. Employees routinely check the process with a sample of 100 hours of output. One sample has a mean of 4,800 and a standard deviation of 1,200. Use this information to calculate the associated p-value.

■ **Solution:** The manager is interested in deviation from 5,000 in either direction, so we have a two-tailed test for $\mu = 5,000$. $\overline{X}_0 = 4,800$ is 200 units from the 5,000 specified in H_0. Thus, values at least 200 units from 5,000 are at least as extreme as the sample result. Such extreme values can be greater than 5,200 as well as less than 4,800, as shown in Figure 9-45. The mean is 5,000 if H_0 is true and the estimated standard error is $\hat{\sigma}_{\overline{x}} = 1,200/\sqrt{100} = 120$. We convert the distance from the mean, 200, to $(4,800 - 5,000)/120 = -200/120 = -1.67$. The probability of a Z value falling at or below -1.67 or one falling at or above 1.67 is 0.095, as shown in Figure 9-45.

For symmetric curves, we can simply find the probability for extreme values in one tail and double the result. In the last example the p-value is twice the probability that $\overline{X}$ is 200 or more units from 5,000 in a single direction.

Figure 9-45
Two-tailed p-value for the average output example: The sum of the shaded areas

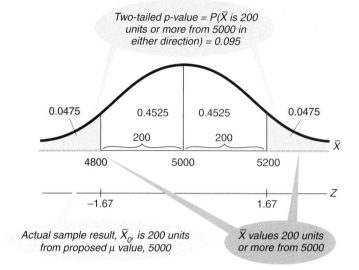

Question: If $\alpha = 0.10$, can the manager from Example 9-13 reject H_0?
Answer: Yes. P-value = $0.095 < 0.10 = \alpha$, so reject H_0. The manufacturing process does not average the dictated 5,000 units of output per hour.

Section 9.12 Problems

Determine if you can reject H_0 in each of the following cases by computing and using the p-value. Compare your decisions with those in Problems 14–17 in this chapter.

168. H_0: $\mu = 25$
 H_1: $\mu \neq 25$
 $\alpha = 0.03$
 $n = 215$
 $S = 35.3$
 a. $\bar{X} = 20.5$.
 b. $\bar{X} = 23.7$.
 c. $\bar{X} = 39.9$.

169. H_0: $\mu = 444$
 H_1: $\mu \neq 444$
 $n = 100$
 $S = 50$
 $\bar{X} = 450$
 a. $\alpha = 0.05$.
 b. $\alpha = 0.10$.
 c. $\alpha = 0.20$.

170. H_0: $\mu = 0$
 H_1: $\mu \neq 0$
 $\bar{X} = -1.71$
 $S = 12.00$
 $\alpha = 0.1$
 a. $n = 169$.
 b. $n = 100$.
 c. $n = 49$.

171. H_0: $\mu = 123$
 H_1: $\mu \neq 123$
 $\alpha = 0.01$
 $n = 100$
 $\bar{X} = 139$
 a. $S = 50$.
 b. $S = 100$.
 c. $S = 200$.

172. A farmer believes that the optimum time to harvest a large field of a certain vegetable (to avoid numerous pickings and loss from under- and overripeness) is when 80% of the crop is ripe. Although he will harvest immediately if more than 80% are ripe, he is interested in the optimum time, because if more than 80% are ripe he believes he will have harvested less than the potential. The farmer finds that 74 of 100 randomly picked vegetables from the field are ripe.
 a. Find the two-tailed p-value for determining if the entire field is at the optimum harvest stage.
 b. The farmer is willing to use this decision process to determine whether the field is optimal. He understands that he risks making the mistake of deciding that the field is not optimal when it actually is. He is willing to make this error 10% of the time. Should the field be harvested or not based on this sample?

9.13 P-Values for the Population Proportion of Successes (Small Samples)

Obtaining a certain number or proportion of successes in a random sample involves a binomial random variable. Remember that this situation requires two possible outcomes on each of n independent observations of a population, with a constant probability

Section 9.13 P-Values for the Population Proportion of Successes (Small Samples)

of success, p, for each observation (the p here is not the same p in the term p-value). When the true population proportion, p, is not close to 0.5 and n is small, the normal approximation is not likely to produce a good estimate of the binomial probability. Specifically, if np or $n(1 - p)$ is smaller than 5, the approximation will be poor. In such cases, we can easily perform a more accurate test with the binomial formula or binomial table. We simply find the p-value.

9.13.1 One-Tailed Test

If X_0 is the number of successes in a random sample of n objects and p_0 is the proposed value of p in the hypotheses, then the p-value will be the probability that the number of successes in the sample, X, is as extreme or more extreme than X_0. If H_1 is $p < 0.4$, and a random sample of nine objects contains one success, then the p-value will be the probability that X is 1 or less when $p = 0.4$.

$$p\text{-value} = P(X \leq 1, \text{ given that } p = 0.4).$$

The normal approximation would probably not work well here, because $9(0.4) = 3.6$ is smaller than 5. From the binomial table in Appendix G, we find that this p-value is $P(X = 0) + P(X = 1) = 0.0101 + 0.0605 = 0.0706$. If the significance level were 0.05, we would not be able to reject H_0, because $0.0706 > 0.05$.

■ **Example 9-14:** After a prominent official's home was vandalized, a citizen's group became concerned that the police paid too much attention to a relatively minor crime at the expense of a declining arrest rate for armed robbery. The arrest rate for armed robberies was 0.75 prior to the vandalism according to previous police news releases. The citizen's group randomly sampled the local newspapers to learn of reported armed robberies and found from the police records that 68% of the 19 located armed robberies resulted in arrests. Find the p-value, and decide at the 0.05 level whether the evidence is sufficient to indicate a drop in armed robbery arrest rates.

■ **Solution:** Because the group believes the arrest rate dropped, H_1 is $p < 0.75$ and the p-value will be the probability of a sample success rate of 0.68 the actual sample value or less if $p = 0.75$. Because the binomial table uses the number of successes, X, rather than the proportion of successes, $\hat{p} = X/n$, we must determine the X value from our sample first. Sixty-eight percent of 19 is 12.92, so the committee apparently rounded $\hat{p} = 13/19$ to 68%. The p-value is $P(X \leq 13, \text{ given that } p = 0.75) = 0.3322$, according to the binomial table. Because this p-value is greater than 0.05, we cannot reject the null hypothesis. There is insufficient evidence to claim that the arrest rate has dropped.

9.13.2 Two-Tailed Test

For a two-tailed p-value, we must consider both tails of the distribution to determine extreme test statistic values. Unless $p = 0.5$, the distribution is asymmetric, so the probabilities in the two tails are unequal. However, it is still convenient, and it works to find the one-tailed p-value as before and double it.

We can do so, because in order to make a decision, we only care about values of the test statistic and the rejection region on one side of the curve, once we know which side the test statistic occupies. Then the sample result is either far enough in this tail to reject or it is not.

To find the direction of extremeness for the single tail, we must first ascertain the mean or middle of the distribution or the expected sample outcome when H_0 is true. We do this by multiplying the p value from H_0, p_0, by n. If the sample X is smaller than np_0, smaller values are extreme. If the sample X value is larger than np_0, then larger values are extreme.

Question: What if the sample X equals np_0?

Answer: We would automatically not reject, because the test statistic agrees with the null hypothesis, which gets the benefit of the doubt.

■ **Example 9-15:** A high school biology class with nine students wants to test whether 30% of their specimens to be dissected will not conform exactly with the diagrams in their books as their manual suggests. The sample of nine specimens contains five with deformities. Use this information to find the two-tailed p-value, and test at the 0.1 level.

■ **Solution:** H_0 is $p = 0.3$. If H_0 is true, the expected number of successes is $np_0 = 9(0.3) = 2.7$. The actual number of successes is 5, which is larger than 2.7; thus, larger values are extreme. Thus, the two-tailed p-value is twice the probability of obtaining five or more deformities in the 9 sampled animals. From the binomial table, we find

$$\begin{aligned}\text{two-tailed } p\text{-value} &= 2[P(X \geq 5 \text{ given that } p = 0.3)] \\ &= 2[P(X = 5 \text{ or } X = 6 \text{ or } X = 7 \text{ or } X = 8 \text{ or } X = 9 \text{ when } p = 0.3)] \\ &= 2(0.0988) = 0.1976.\end{aligned}$$

This value is larger than $\alpha = 0.1$, so we cannot reject H_0.

Section 9.13 Problems

173. If H_0 is $p \geq 0.35$ and a random sample of 12 objects contains no successes, what is the p-value? What is the decision at the 0.005 significance level?

174. If H_0 is $p = 0.80$ and a random sample of 17 objects contains 16 successes, what is the p-value and the decision at the 0.05 level?

175. Use the p-value to perform a hypothesis test to demonstrate (at the 0.05 level) that at least 50% of letters from the front arrived over a year later during the U.S. Civil War, if a sample of 10 diaries after the war reveal that nine families received such letters and one did not. Interpret the decision.

176. Bear sightings in neighborhoods signal a lack of food in the wild. After several such sightings, rangers randomly select 12 streams to check for sufficient fish stocks and 12 berry patches to check for sufficient fruit. Use this information to perform two hypothesis tests in an attempt to demonstrate that neither stock is ample at the 0.05 level if rangers consider 60% or more of sites sufficiently stocked with food to be ample. Six streams were sufficient. Two patches were sufficient. Interpret each decision.

177. A political commentator claims that the surprise entry of a third-party candidate will entice 15% of the electorate to vote for the candidate. Can we reject this claim at the 0.05 level if one of 17 randomly selected people on the street say they will vote for this candidate.

9.14 Computers and *P*-Values

Computer outputs that include p-values allow us to evaluate the results easily, without consulting probability tables for critical values. Some packages automatically deliver p-values (especially two-tailed p-values) for a test of zero as the proposed mean. Minitab offers the user the opportunity to specify different values as well as directions for one-tailed tests.

The final circled value indicated in Figure 9-11 (p. 404) is number 7, the p-value. The circled 7 in Figure 9-46 also indicates the location of the p-value. The text above the numbers alerts the user to the nature of the p-value, one- or two-tailed and the direction if it is one-tailed.

Figure 9-46
Minitab output for the test of forecast errors

```
TEST OF MU = 0.000 VS MU N.E. 0.000
                                                        ⑦
         N     MEAN    STDEV    SE MEAN     T    P VALUE
ERROR    5     0.400   1.673    0.748      0.53   0.62
```

Section 9.14   Computers and *P*-Values   ■ **441**

Box 9-5   Using a Two-Tailed *P*-Value to Do a One-Tailed Test

Several software packages do not offer the flexibility of a user-specified μ_0 for hypothesis testing or a user-specified direction for the test. Some do automatically provide two-tailed *p*-values for common tests, such as output for testing for a mean of zero. In such cases, it is tempting to half the two-tailed *p*-value and compare the result with the significance level, α, for a quick one-tailed test. But this procedure will not always work. An additional preliminary check is required.

The first step is to determine whether the test statistic agrees with H_1. If H_1 is $\mu < 0$, then a negative value of $\overline{X}$ agrees with H_1. In Figure 9-11 (p. 404) this is true, because $-896 < 0$. If H_1 is $\mu > 0$, this is not true. In this case, the test statistic agrees with H_0, which means that we fail to reject H_0, because H_0 receives the benefit of the doubt (provided α is less than 0.5). Actually, there is no need to proceed here, because we have made a decision.

If the test statistic agrees with H_1, the second step is to half the two-tailed *p*-value. This half is the one-tailed *p*-value. If the statistic does not agree with H_1, then the one-tailed *p*-value is 1 minus this half, or $1 - 0.5$(two-tailed *p*-value). The one-tailed *p*-value for the data in Figure 9-11, if we know that $\overline{X}$ agrees with H_1, is $0.5(0.076) = 0.038$. This value can then be compared with α to make a decision for the one-tailed test.

The $1 - 0.5$(two-tailed *p*-value) solution is useful only if you want to know the *p*-value. Otherwise, the decision in this case is made in the first step. However, it does demonstrate why using simply half of the two-tailed *p*-value can lead to erroneous decisions. If the test statistic agrees with H_0 in a one-tailed test, the *p*-value will be larger than 0.5. The figures on the right and computer output on the next page demonstrate these ideas.

The steps for converting a two-tailed *p*-value to a one-tailed *p*-value are as follows:

Procedure to Convert a Two-Tailed *P*-Value to a One-Tailed *P*-Value:

1. Determine if the test statistic agrees with H_1.
2. Use 0.5(two-tailed *p*-value) if the statistic agrees with H_1, and use $1 - 0.5$(two-tailed *p*-value) if it does not.

Determining a one-tailed *p*-value from a two-tailed *p*-value for an upper-tailed test

a) Two-tail *p*-value

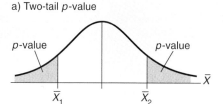

b) One-tail *p*-value when H_1 is satisfied

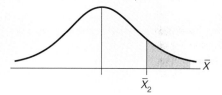

c) One-tail *p*-value when H_1 is not satisfied

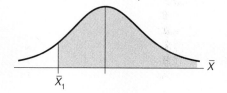

(continued on next page)

■ **Example 9-16:** Figure 9-46 displays output for testing for a mean error of zero for an economic forecaster. These forecasts produce positive and negative errors, respectively, and result in the following sample of forecast errors to determine if the errors cancel over the long run (have a mean of zero): $-1, 3, 0, -1, 1$. Use the information in the printout to perform a two-tailed test for a mean error of zero, at the 0.1 level. Assume a normal population.

■ **Solution:** We cannot reject H_0, because $0.62 > 0.1$. The evidence does not enable us to refute that the mean error is zero.

Often the desired test is one-tailed, rather than two-tailed, like the *p*-value automatically provided in some statistical software packages. In such cases, it is convenient to know how to convert a two-tailed *p*-value into a one-tailed *p*-value. When doing this, it is also important to avoid the mistake of simply taking half of the two-tailed *p*-value to be the one-tailed *p*-value. Box 9-5 discusses the two steps required to accomplish this test.

Box 9-5 Using a Two-Tailed *P*-Value to Do a One-Tailed Test (continued)

Minitab output for comparing one-tailed and two-tailed *p*-values when using the same data set

```
(A) TEST FOR DIFFERENCE IN ASSESSED AND MARKET VALUES
-------------------------------------------------------------------------------
(1) TWO-TAIL TEST
    TEST OF MU =   0 VS MU N.E.   0

                N        MEAN       STDEV     SE MEAN           T      P VALUE
    DIF       200        -896        7096         502       -1.79        0.076

(2) LOWER-TAIL TEST
    TEST OF MU =   0 VS MU L.T.   0

                N        MEAN       STDEV     SE MEAN           T      P VALUE
    DIF       200        -896        7096         502       -1.79        0.038

(3) UPPER-TAIL TEST
    TEST OF MU =   0 VS MU G.T.   0

                N        MEAN       STDEV     SE MEAN           T      P VALUE
    DIF       200        -896        7096         502       -1.79         0.96


(B) TEST FOR FORECAST ERRORS
-------------------------------------------------------------------------------
(1) TWO-TAIL TEST
    TEST OF MU = 0.000 VS MU N.E. 0.000

                N        MEAN       STDEV     SE MEAN           T      P VALUE
    ERROR       5       0.400       1.673       0.748        0.53         0.62

(2) LOWER-TAIL TEST
    TEST OF MU = 0.000 VS MU L.T. 0.000

                N        MEAN       STDEV     SE MEAN           T      P VALUE
    ERROR       5       0.400       1.673       0.748        0.53         0.69

(3) UPPER-TAIL TEST
    TEST OF MU = 0.000 VS MU G.T. 0.000

                N        MEAN       STDEV     SE MEAN           T      P VALUE
    ERROR       5       0.400       1.673       0.748        0.53         0.31
```

Section 9.14 Problems

178. The output in Part A of Box 9-5 (shown above) displays statistics calculated from a random sample of 200 differences between assessed values and corresponding market values of houses, based on a random sample of 200 homes that sold within six weeks of being assessed by a local tax assessor. Use the two-tailed *p*-value to test whether there is any discrepancy on average in the assessment process at the 0.05 level.

179. Refer to Figure 9-31 (p. 425), the statistical output for an upper-tailed test of an airline company's claim that its flights depart at most 10 minutes late, on average. Use this output to determine if this sample evidence is sufficient to refute the company's claim at the 0.05 level.

180. Is there sufficient evidence in the sample, as displayed in the *p*-value in Figure 9-34 (p. 425), to reject the hypothesis that the average annual yield is at most 3.5%? Test at the 0.01 level.

Summary and Review

A *p*-value is a concise way to report the sample results for a hypothesis test. It measures the credibility of H_0. The larger the *p*-value, the more credible H_0 is. Specifically, a *p*-value is the maximum probability that a test statistic is at least as extreme as the actual test statistic value when H_0 is true. The direction of extremeness is determined by H_1, so a *p*-value greater than 0.5 is associated with a test statistic falling on the side of the curve that agrees with H_0 (for a one-tailed test).

When H_0 is rejected or not rejected at a certain level, we do not know whether the decision is close or obvious, "weak" or "strong." A *p*-value incorporates the sample results in such a way that each decision maker can make such judgments at any significance level and not be restricted to the original tester's decision or risk level. The decision not to reject or to reject at any given level of significance is based on the following rule: Reject H_0 if the *p*-value is less than or equal to the level of significance, α. Do not reject H_0, if the *p*-value is larger than the level of significance.

The *p*-value is especially useful, because some computer outputs report *p*-values, which means we no longer need to search for critical values in statistical tables.

To construct a two-tailed *p*-value when the curve is symmetric we double the extreme area in one tail. This works as well for a small-sample test of population proportions, another situation where *p*-values facilitate the calculation and decision process. The regular population proportion procedure breaks down in this case, because the sampling distribution is usually not approximately normal for small samples.

Multiple-Choice Problems

181. The *p*-value
  a. Equals P(less extreme sample results given that H_1 is true).
  b. Equals $1 - \alpha$.
  c. Measures the credibility of H_0.
  d. Is the maximum α that will result in rejecting H_0 for the sample results.
  e. Is the same as β.

182. A random sample of 24 elements is selected from a normal population to test H_0: $\mu \leq 303$ and H_1: $\mu > 303$. The sample mean and standard deviation are 328 and 49, respectively. Find the *p*-value.
  a. 0.40.   d. 0.01.
  b. 0.99.   e. 0.25.
  c. 0.05.

183. A Minitab output for testing a mean reports results for testing H_0: $\mu \geq \mu_0$ versus H_1: $\mu < \mu_0$. The *p*-value is 0.25. For which of the following possible significance levels will you *not* reject H_0?
  a. 0.60.   d. 0.50.
  b. 0.40.   e. 0.75.
  c. 0.10.

184. From the diagrams in Figure 9-47, choose the correct one for finding the *p*-value when we have 370 successes out of 1,000 observations and

$$H_0: p = 0.35.$$
$$H_1: p \neq 0.35.$$

Note: Shaded areas represent *p*-values.

185. The *p*-value is 0.99 for testing H_0: $p = 0.8$ and H_1: $p \neq 0.8$, when a random sample of 4,000 items contains 3,260 successes, which means that
  a. H_0 has very little credibility.

Figure 9-47
Choices for Question 184

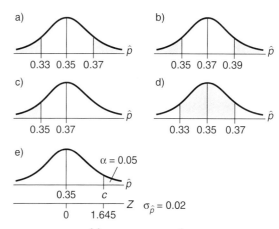

blue area = *p*-value

  b. The chance of obtaining a sample proportion at least as extreme as 0.8150, when $p = 0.8$, is 0.99.
  c. $P(\hat{p} \geq 3{,}260/4{,}000$, given that $p = 0.8) = 0.99$.
  d. $P(\hat{p} \leq 3{,}260/4{,}000$, given that $p = 0.8) = 0.99$.
  e. If $\alpha = 0.01$, then we should reject H_0.

186. A random sample of 400 ages of employees in a large international corporation is used to refute the claim of a government age-discrimination council that the average worker age for the corporation is less than 25 years. The sample mean and standard deviation are 25.57 years and 4.13 years, respectively. Find the *p*-value.
  a. 0.0029.   c. 0.4971.   e. 0.9942.
  b. 0.9971.   d. 0.0058.

Figure 9-48

Minitab output for the differences in actual and optimal harvest time

```
TEST OF MU = 0.000 VS MU N.E. 0.000

            N      MEAN    STDEV    SE MEAN       T    P VALUE
DIF       261    -1.088   11.322      0.701   -1.55      0.12
```

187. Which of the following (if any) is not an advantage of using *p*-values to report test results?
  a. Any level of test can now be performed easily.
  b. The "strength" or "weakness" of test decisions is readily apparent.
  c. The population distribution is unimportant.
  d. The credibility of a sample value smaller than μ_0 in $H_0: \mu \leq \mu_0$ now has an upper limit.
  e. All of the above are advantages.

Use the information in Figure 9-48 to answer Questions 188–192.

Assume that the data values are 261 differences between the actual times when a crop is harvested and the optimum harvesting time. It is expected that these differences would average to zero for all farmers.

188. What is the value of the test statistic, $\overline{X}$, for testing for a mean of 0?
  a. 11.32.   d. 261.
  b. 0.70.   e. 0.12.
  c. −1.088.

189. What is the positive critical value for a two-tailed test at the 0.05 level and the corresponding decision?
  a. 1.645, cannot reject.   d. 1.96, reject.
  b. 1.645, reject.   e. 1.28, cannot reject.
  c. 1.96, cannot reject.

190. What is the two-tailed *p*-value and the decision at the 0.01 level?
  a. 0.38, cannot reject.   d. 0.62, cannot reject.
  b. 0.24, cannot reject.   e. 0.76, cannot reject.
  c. 0.12, cannot reject.

Use the procedure from Box 9-5 to work Problems 191–193. Problems 191 and 192 still refer to Figure 9-48.

191. For an upper-tailed test with the same sample mean, the *p*-value is
  a. 0.94.   d. 0.06.
  b. 0.24.   e. 0.88.
  c. 0.12.

192. For a lower-tailed test with the same sample mean, the *p*-value is
  a. 0.94.   d. 0.06.
  b. 0.24.   e. 0.88.
  c. 0.12.

193. The *p*-value is 0.16 for a two-tailed test of $\mu = 25$ as calculated by a computer. However, you want to do a one-tailed test at the 0.10 level. What is your decision?
  a. Cannot reject H_0, because 0.08 > 0.10.
  b. Reject H_0, because 0.08 > 0.10.
  c. Cannot reject H_0, because 0.16 > 0.10.
  d. Reject H_0, because 0.16 > 0.10.
  e. There is not enough information to make the decision here.

194. A company uses hypothesis testing procedures for workers to check if a product meets a standard. Employees find the mean of a random sample of items from a production process to make the decision. What is one advantage of the traditional method, where management determines the critical values for the employees to use for comparison with the sample results, over the management determining a significance level for comparison with the sample results for the *p*-value method.
  a. Critical values are more accurate although time-consuming to locate compared to *p*-values.
  b. *P*-values are more computation intensive, because the employees must compute the sample mean.
  c. Population and sample dispersion measures do not affect critical values, but they delay decisions based on *p*-values.
  d. Employees merely compute the sample mean, a familiar computation, and compare it with the critical values, whereas *p*-value computations require probability calculations.
  e. Critical values are theoretical values from statistical tables with no need for revision, whereas significance levels are subjective and, hence, subject to constant scrutiny and fluctuation.

Use the following information to answer Problems 195–201.

A researcher studied 970 characters from 30 daytime public broadcasting television programs, 30 prime time network programs, and 30 Saturday morning network programs (a total of 90 television programs directed toward families and children). A newspaper summary says that the researcher found that few television characters work and the distribution of the job types among the working characters is different from the actual distribution. Specifically, the study found that

1. 58% of the characters' roles either do not involve them working or place them in fictional jobs, such as a ghostbuster.
2. 81% of the Saturday morning roles fit one of these descriptions.
3. 75% of the characters who do work are employed in white-collar positions (the actual percentage of U.S. workers is 56% according to the U.S. Bureau of Labor Statistics). (Source: "Study: TV Children Have Poor Role Models," *Asheville Citizen-Times*, July 21, 1992, p. 2A.)

Word and Thought Problems

195. An upper-tailed *p*-value for the white-collar results under list item 3 where the null hypothesis reflects the government figure is
  a. 0.5. **c.** 0.19. **e.** 0.56.
  b. Approximately zero. **d.** 0.75.

196. The result in Problem 195 means that
  a. The government figure does not reflect real life.
  b. We are unlikely to observe the distribution of television character work roles in real life.
  c. The researchers' sample of television characters is a nonrandom sample of television characters.
  d. The test statistic and value proposed in H_0 are practically the same value.
  e. The credibility of H_0 is extremely high.

197. If we assume the television sample is a random sample of real-life characters and test if the government figure is correct at the 0.05 significance level, the decision would be to
  a. Reject H_0, because $0 < 0.05$.
  b. Not reject H_0, because $0 < 0.05$.
  c. Reject H_0, because $0.75 \neq 0.56$.
  d. Not reject H_0, because $0.75 \neq 0.56$.
  e. Cannot reject H_0, because $0 < 0.025$.

198. Use the 81% value from list item 2, and assume the sample contains 16 observations. Find the *p*-value to refute a claim that the true percentage of Saturday morning characters fitting these roles is 75% or less.
  a. 0.3518. **d.** 0.6075.
  b. 0.4050. **e.** 0.4019.
  c. 0.7500.

199. Would the evidence in Problem 198 be sufficient evidence to refute a claim that the true percentage of Saturday morning characters fitting these roles is 75% or less if we test at the 0.05 level?
  a. Yes, because $0.4050 > 0.05$.
  b. Yes, because $0.4050 > 0.025$.
  c. Yes, because $0.4050 > 0.10$.
  d. No, because $0.4050 > 0.025$.
  e. No, because $0.4050 > 0.05$.
  f. No, because $0.4050 > 0.10$.

200. If 58% of 12 random selected television characters involve them working or place them in fictional jobs, such as a ghostbuster, what would the two-tailed *p*-value be for a test of $H_0: p = 0.95$?
  a. 0.0004. **d.** 0.0098.
  b. 0.0002. **e.** 0.9998.
  c. 0.0001.

201. If the significance level is 0.05 for the test in Problem 200, what is the decision?
  a. Reject H_0, because $0.0004 < 0.025$.
  b. Cannot reject H_0, because $0.0004 < 0.025$.
  c. Reject H_0, because $0.0004 < 0.05$.
  d. Cannot reject H_0, because $0.0004 < 0.05$.
  e. Reject H_0, because $0.0004 < 0.10$.
  f. Cannot reject H_0, because $0.0004 < 0.10$.

Word and Thought Problems

202. Consider the hypotheses $H_0: p \leq 0.03$, $H_1: p > 0.03$. The *p*-value for this problem is 0.04. For which of the following significance levels will H_0 be rejected: 0.002, 0.004, 0.008, 0.02, 0.04, 0.08, 0.2, 0.4, 0.8?

203. Consider the hypotheses $H_0: \mu = 80$, $H_1: \mu \neq 80$. The *p*-value for this problem is 0.04. For which of the following significance levels will H_0 be rejected: 0.002, 0.004, 0.008, 0.02, 0.04, 0.08, 0.2, 0.4, 0.8?

204. Is it possible for the *p*-value to exceed 1? 0.5? Explain your answers.

205. The two-tailed *p*-value for a test of $H_0: \mu = 0$ is 0.98. Does this evidence support H_0? If you reject H_0 for this problem, what do you decide about μ?

206. 483 of 622 randomly sampled job recruiters rate communications skills as the most important factor in forming their initial impression of job candidates. Find the *p*-value for testing the hypothesis that 80% or more of all recruiters share this belief, and make a decision if $\alpha = 0.1$. Interpret the decision.

207. A new piece of equipment is advertised as being more productive than older models that had a mean output of 3,000 units per hour. A random sample of 139 hours' output from the new machine has a mean of 3,100 units, with a standard deviation of 482 units. Verify the company's claim at the 0.05 level by constructing and using the *p*-value.

208. In a random sample of 500 people in their twenties, the number who enjoy watching wrestling on television is 100. Television executives claim that at least 25% of this age group enjoy watching wrestling on television.
  a. Calculate the *p*-value to test this claim.
  b. Test the claim at the 0.05 level. Interpret the decision.
  c. What is the minimum value of α that will cause rejection of the claim?

209. A random sample of 10 used cars on a large lot has a mean mileage of 55,961, with a standard deviation of 5,800 miles. The dealer claims that the average mileage of all their cars is at most 50,000. Calculate the *p*-value to test this claim, then perform the test at the 0.025 significance level. Assume the mileages for all cars on the lot are normally distributed. Interpret the decision.

210. A random sample of 100 elevator-shoe purchases has a mean of 1.05 inches of elevation with a standard deviation of 0.2 inch.
  a. Use this information to construct the two-tailed *p*-value to test the hypothesis that the mean elevator-shoe purchase has 1 inch of elevation.
  b. Make a decision at the 0.01 level. Interpret the decision.

211. The *p*-value used by an opposition group for testing a politician's claim that a slight majority (51%) of citizens favor a tax increase is 0.044.

Figure 9-49
Minitab output for the fund-raising problem

```
TEST OF MU =   0.0 VS MU N.E.   0.0

              N      MEAN    STDEV   SE MEAN      T    P VALUE
PROFIT       33     -24.3     66.9      11.6    -2.09    0.045
```

a. If the sample proportion is 0.38, what does the 0.044 represent?
b. What is the decision at the 0.05 level?

212. The saturation rate for VCRs (percentage of customers who own one) in a random sample of households is 45%. This results in a p-value of 0.09 for testing that the rate is at most 40%. Test at the 0.05 level and interpret the decision.

213. What are some advantages of p-values over traditional methods of reporting hypothesis test results?

214. A random sample of viewers of adult television comedies reveals that 25% of the viewers are between the ages of 6 and 19. This results in a two-tailed p-value of 0.18 when testing that the percentage is actually 30%. Use this information to test the 30% value at the 0.2 level. Interpret the decision.

215. A firm is concerned about the size of loads delivered by its hauling contractor and doubts the contractor's claim that the average load is more than 800 pounds. The firm randomly selects 20 loads and finds the mean weight is 893.6 pounds with a standard deviation of 200 pounds. Assuming the weights of all loads are normally distributed, find the p-value. Give the benefit of the doubt to the contractor. If the significance level is 0.05, use the p-value to make a decision. Interpret the decision.

216. Use the p-value in Figure 9-36 (p. 427) to determine if the new catalyst works faster than the old ones. Test at the 0.1 level. Compare this decision and the work involved to do the test with a critical value. See multiple-choice Problem 118 in Unit II (p. 427).

217. Find the p-value, and test a claim that the percentage of a lawyer's clients who are successful in court is at least 90% if a random sample of 10 clients contains seven successes. Use $\alpha = 0.05$. Interpret the decision.

218. Determine if more than 45% of residents favor land-use planning for the local planning board by finding the p-value when five of 10 randomly selected residents favor the planning. Use $\alpha = 0.01$. Interpret the decision.

219. Test to determine if 15% of a doctor's patients are diabetic as contended by a local medical organization for all its doctors' patients if a random sample of eight of the doctor's patients contains four who are diabetic. Use $\alpha = 0.1$. Interpret the decision.

220. Repeat Problems 217 through 219 using the normal approximation. Compare the p-values and decisions, and comment on the use of the normal approximation rather than the exact binomial values in these two problems.

Use the procedure detailed in Box 9-5 (pp. 441–442) for Problems 221–223.

221. A computer output from a statistical routine includes the two-tailed p-value for a test of H_0: $\mu = 0$, where $\overline{X}_0$ is calculated from the data you input into the computer. Call this p-value, pv. However, you want to change your hypothesis and do an upper-tailed test. What additional information do you need to ascertain if the p-value for the upper-tailed test is $0.5pv$ or $1 - 0.5pv$?

222. The average kilowatt-hour (KWH) electric usage for July for a random sample of 120 utility customers is 1,652 KWH, with a standard deviation of 376 KWH. Find the two-tailed p-value for testing that the mean usage is 1,550 KWH. Then, perform two-tailed, lower-tailed, and upper-tailed tests at the 0.05 level.

223. A club is concerned about its fund-raising events, yet the officers claim that, overall, the amount of money raised is more than the cost of conducting the fund-raisers. A random sample of 33 events' profits (revenues from the events minus costs) is fed into the computer, with the output displayed in Figure 9-49. Remember that profits can be positive or negative. Use the results to test the officers' claim for the dissident club members at the 0.05 level.

Review Problems

Note: An asterisk precedes p-value problems.

224. The mean number of cavities over the last five years is 3.6 for a random poll of users of Brand X toothpaste. Can you automatically assume that the manufacturer, who claims the average is 2, is wrong? Explain.

225. A certain newspaper's editorial policy shifts when a new editor takes charge. The mean advertising revenues prior to the shift were $6,000. For a random sample of 50 days after the shift, advertising revenues average $6,500 with a standard deviation of $1,486. Have average revenues changed since the shift? Use $\alpha = 0.05$ to detect a significant change in either direction. What assumption(s) did you make to work the problem?

226. a. A fast-food chain models its dining areas to accommodate the size of the average local family. 0.5% of the inhabitants within a 5-mile radius of the restaurants are randomly sampled to ascertain the number of people in the immediate family. Then the mean family size, $\overline{X}$, is calculated. If $\overline{X} \leq 3.25$, model A is used for the dining area. $3.25 < \overline{X} < 3.75$

Review Problems

results in model B, and $\overline{X} \geq 3.75$ results in model C. Report the probability that this decision procedure will result in models A or C being selected for a site when the true mean family size is 3.5 in the following localities.

Site	Population within 10-Mile Radius (N)	$n = 0.005 N$	S (people)
1	8,526	43	2.2
2	55,023	226	3.8
3	106,002	531	5.3

b. Does it seem reasonable that σ is bounded by a value (has an upper limit) such as 10? Explain.

c. Judging from your answer to Parts a and b (and given that mean family size is 3.5), what restaurant model is likely to be used in densely populated areas?

d. If sampling is expensive, so the food chain is considering using smaller samples to estimate mean family size, what happens to the probability of choosing models A or C when a smaller sample is used?

e. How small must the sample size be for the probability to be 0.2 of choosing A or C when $\mu = 3.5$ and $\sigma = 2$?

227. If you reject H_0, what type of error might you be making? Is it possible to make Type I and Type II errors simultaneously with a single decision?

228. a. A publication reports that an average corporation will hire 5% more college graduates in the current year than during the year before. Can you refute this claim at the 0.05 level, using a random survey of 500 corporations with a mean expected graduate hiring increase of 4.25% with a standard deviation of 6.75%?

*b. Calculate the p-value for this test, and use it to demonstrate the decision.

229. A coach's goal for her camp participants in a weight lifting program requires an average increase in weights lifted of 4 pounds per week for the group. Past experience suggests the standard deviation of weight increases is 0.15 pounds. Determine critical values that the coach can use for checking progress by randomly sampling 36 participants each week. The coach does not want to falsely accuse all camp participants of not lifting according to the plan more than 1% of the time.

230. A speech pathologist believes that remedial treatment for an impediment requires an average of less than 14 sessions. To confirm this belief, she collects a random sample of 10 clients' records of treatment for this impediment. If she uses these values (12, 16, 10, 7, 11, 10, 9, 20, 12, 12) to test her belief at the 0.1 level, what will her decision be?

231. A supplier of new parts claims that its defect rate is 5%. The first 5 shipments of 100 parts each (500 total parts) contain a total of 75 defectives. Is this sufficient information to deny the company's claim at the 0.05 level? Is this sample random? Why is randomness important?

232. Jason is supposed to practice the violin at least 30 minutes every day. A random sample of 20 days yields a mean practice time of 25 minutes with a standard deviation of two minutes. Does this evidence justify the conclusion that Jason is not practicing the required amount of time? (Use $\alpha = 0.1$.) Assume that the overall distribution of daily practice times is normal.

233. a. Suppose that schoolteachers who quit teaching tend to take less stressful jobs. An educational association survey of 446 teachers who quit last May showed that 161 left education altogether. The most popular new career for these 161 teachers was selling computers. Test at the 0.2 level the association's hypothesis that 40% of all teachers who quit last May left education entirely. Interpret the decision. What is the population for this problem?

*b. Calculate the p-value for this problem, and use it to demonstrate the decision.

234. A hypnotist claims to have at least a 98% success rate for helping clients lose weight. A random sample of 250 of the hypnotist's overweight patients contains 240 satisfied with their weight loss. Is this sufficient information to reject the hypnotist's claim at the 0.05 level; that is, decide that the success rate is smaller than claimed?

235. Can you reject a runner's claim that her average time to run 100 meters is at most 12 seconds if a random sample of 10 of her 100-meter sprint times has a mean of 12.5 seconds, with a standard deviation of 1 second? Test at the 0.05 level. What assumptions are necessary to ensure a 0.05 level?

236. Test at the 0.02 level the hypothesis that 55% of the student body is female if 100 students in a random sample of 200 students are female. Interpret the decision.

237. Test if the average cover charge for clubs in an area is $5 if a random sample of 12 nightspots has a mean cover charge of $6 with a standard deviation of $0.75. Test at the 0.02 level, and assume all cover charges are normally distributed. Interpret the decision.

***238.** The two-tailed p-value for a test of H_0: $\mu = 49.10$ is 0.98. Does this evidence support H_0? If you reject H_0 for this problem, what do you decide about μ?

***239.** A local welfare department claims that more of the county population should be classified as impoverished, according to an upper-tailed test at the 0.05 level. You work for a government watchdog group that is wary of starting extra job-training programs in an area, because resources are limited and the objective is to assist only the most impoverished areas. You want to avoid sending funds to an area unless there is a greater need in that area than in other areas. The common poverty rate is 10% or less. In other words, you will make funds available if a program-seeking locality's poverty rate is so large that it has only a 1% chance of occurring when the true local rate is 10% or less (H_0). Thus, you want the rate tested at the 0.01 level, rather than at 0.05. The sample rate for the county mentioned previously is 12%, based on a random sample of 650 individuals. Find the upper-tailed *p*-value. Use it to verify the local department's claim at the 0.05 level. Then make a decision at the 0.01 level for the watchdog agency.

*240. Gold prospectors in the West average finding at most 2 ounces of gold per month according to an association of prospectors. A random sample of 25 prospectors results in a *p*-value for testing this claim of 0.075.
  a. Can you reject the claim at the 0.05 level?
  b. What do you assume about monthly gold amounts that validates a *p*-value of 0.075?

*241. Fifty-five percent of a random sample of preschool children's mothers have a full-time job. These results produce a two-tailed *p*-value of 0.03 for testing that the true percentage is 50%. Use this information to perform the test at the 0.01 level.

*242. Find the *p*-value, and attempt to refute a claim at the 0.1 level that a laboratory snake will strike when provoked at least 60% of the time if five snakes in a random sample of 20 laboratory animals struck when provoked.

243. A pizza restaurant wants to advertise that the special luncheon pizza, which varies with the day of the week, will be ready in five minutes or less. Before making such a claim public, the management wants to verify it. Twenty weekdays are set aside for employees to concentrate on speed while maintaining quality of the daily special. The manager records time between ordering and serving each special. The result is 1,215 values with an average time of 4.3 minutes and a standard deviation of 0.4 minutes.
  a. Is there sufficient evidence for the management to make the claim with a 5% risk of making an incorrect claim?
  b. The manager is also considering not charging a customer for any special that does not meet the five-minute limit. He assumes the 4.3 and 0.4 are fairly accurate, because the data set is so large. If serving times are normally distributed, what proportion of special pizza customers will not receive their order within the specified time? If the restaurant serves about 50 specials per day, how many of these are likely to be late?
  c. The proportion in Part b is 0.0401. The sample proportion of late pizzas from the 1,215 recorded times is 0.0485. Use this value to check the accuracy of the 0.0401 and be sure that the proportion is not larger than 0.0401. Perform this test at the 0.05 level. Does this test also provide evidence that the population of order time values is normally distributed?
  d. Is this sample of 1,215 times random? Explain. Does your answer affect your conclusions in Parts a, b, and c?

Chapter 10

Test for the Difference between Two Population Means

Determining the significance of the difference in a characteristic between two otherwise identical groups or situations is a basic procedure in science, business, and in daily decision making. In this chapter we will develop and use a test to determine if a difference in the means of samples from two such populations provides adequate evidence that the two population means differ as well.

Objectives

This unit demonstrates the basics of the test for the difference between two population means. Upon completing the unit you should be able to

1. State the hypotheses and interpret conclusions of this test in a form useful for decision making.
2. Describe the sampling distribution, the mean, and the standard error of the statistic, $\overline{X}_A - \overline{X}_B$, the difference in two sample means.
3. Perform a test for the difference between two population means using critical values and p-values for samples that are large, independent, and random.
4. Define independent random sample and standard error of the difference between means.
5. Recognize σ_{dif}, the standard error of the test statistic for the difference between two sample means, and $\hat{\sigma}_{dif}$, the estimated standard error of the difference.

Unit II extends the basics to different situations that require some modification of the procedure we used in the first unit. Upon completing this unit you should be able to

1. Describe the sampling distribution, mean, and standard error of the test statistic for the difference between two sample means when the samples are small, independent, random, and selected from normal populations with equal variances.
2. Use critical values and p-values to perform and interpret the decision for a test for a difference in means in the situation listed in item 1.
3. Describe and perform the procedure for testing for a difference in two population means when observations from the two samples are paired and then interpret the decision.
4. Use computer output to perform and interpret tests for differences in means for each situation covered in Units I and II, as well as the small, independent sample situation when the population variances are unequal.
5. Recognize S_P^2, a pooled estimate of the population variance; D, the difference in values for an item or person in a paired-difference test; $\overline{D}$, the mean of the n paired differences in a sample; S_D, the standard deviation of a sample of paired differences; $\sigma_{\overline{D}}$, the standard error of the mean of paired differences; $\hat{\sigma}_{\overline{D}}$, an estimate of the standard error of the mean of paired differences; μ_D, the mean of a population of paired differences; and σ_D, the standard deviation of a population of paired differences.

Unit I: Basics of the Test

Preparation-Check Questions

1. When we estimate the population standard deviation, we use
   a. S.
   b. S^2.
   c. The standard error of the mean.
   d. The range.
   e. The significance level.
2. To estimate the spread of all possible sample means, we use
   a. The sample standard deviation.
   b. S.
   c. $S/\sqrt{n}$.
   d. The population standard deviation.
   e. The sampling error.
3. To obtain the variance from the standard deviation, we
   a. Take the square root of the standard deviation.
   b. Square the standard deviation.
   c. Average the deviations.
   d. Divide the standard deviation by the square root of the sample size, n.
   e. Divide the sum of squares by $n - 1$ rather than N.
4. How do we standardize an outcome or value of a random variable?
   a. Subtract the mean.
   b. Square all values.
   c. Subtract 0 and divide by 1.
   d. Subtract the mean and divide the result by the standard deviation.
   e. Multiply by the standard deviation and add the result to the mean.
5. If the difference between a test statistic and a hypothesized population value is large, compared to the expected difference when H_0 is true, then we
   a. Reject H_0.
   b. Cannot reject H_0.
   c. Decide both are true.
   d. Decide both are false.
   e. Cannot decide.
6. When testing hypotheses with a p-value, H_0 is rejected if
   a. $\alpha = p$-value.
   b. $\alpha \neq p$-value.
   c. $\alpha < p$-value.
   d. $\alpha \leq p$-value.
   e. $\alpha \geq p$-value.

Answers: 1. a; 2. c; 3. b; 4. d; 5. a; 6. e.

Introduction

Productivity Differences between Pools The advertising and promotion department of a large corporation has two secretarial pools. Both perform the identical function of producing technical reports, diagrams, and memoranda, but they are located on different floors of the corporate offices. The supervisor randomly assigns work to any available secretary. Before investing a large sum of money on new software to assist the secretaries in their work, the top manager decides to experiment and determine if the proposed software actually increases productivity. The manager will outfit one pool with the new software and compare that pool's productivity with the productivity of the other pool.

Because the pools engage in similar activities, the manager assumes there is no reason to expect that differences in output will arise from their performing different tasks. To check this assumption, before initiating the experiment, the manager calculates the mean output in pages for each pool, using random samples of 50 days of output from records of each pool. The means are 1,534 and 1,323 pages per day. Can the manager simply say that the productivity rates differ for the two pools because these two values differ?

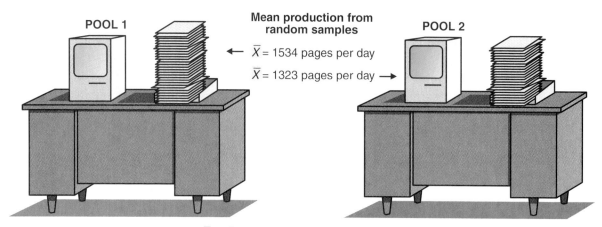

Figure 10-1
Is a difference in sample means significant?

■ **Solution:** No. Remember sampling error. There is a chance that the difference between the two values occurs because the manager investigates a sample, rather than the entire population of output for each pool. Other possible sources of error include a different number of secretaries in each pool on a given day (sick and vacation days), different workloads on the chosen days (mostly diagrams versus mostly prose or heavy versus light workloads), and different days of the week (if, say, Mondays, are busier than Fridays). See Figure 10-1.

After checking the records for these days, the manager eliminates the nonsampling factors mentioned above as sources of the sample differences. If these factors could not be eliminated, random sampling would reduce the chances of their producing a

> **Box 10-1** China: Mass Medical Laboratory
>
> China's longtime lack of major contact with the outside world and with the consequences of industrialization makes its people, especially the ones in rural areas, ideal subjects for studying the connection of health with dietary habits and life-style. Historically and currently, they depend very little on meat for food, eating instead a high-fiber diet. The medical profession contends that both of these foods dramatically affect people's health, especially with regards to heart disease and cancer.
>
> Therefore, Chinese history affords medical researchers a unique opportunity to study long-term effects of a low-fat, high-fiber diet on the incidence of these diseases and other health problems. Research in the United States or Great Britain, on the contrary, must involve subjects with a history of the opposite consumption pattern, high fat and low fiber. These circumstances make it difficult to assess the role of a diet that goes against habits formed over a lifetime, against the grain (so to speak).
>
> Other Chinese assets for such studies include factors that serve as controls for the experiment while the people undergo the treatment, the current changes in diet and life-style. Their culture promotes homogeneity in a given region with sometimes large degrees of heterogeneity between regions. The people tend to be nonmobile (more than 90% live and die in the same province). There is not much diversity of food choice beyond local products and local water. These factors generate very little intraregion variation but sometimes dramatic interregion variation.
>
> In addition, the large Chinese population has cooperated when researchers collect blood and urine samples and when they complete a health questionaire. This cooperation facilitates the collection of large data sets, 8,000 people in 69 counties in 25 provinces. Later stages of the study include Taiwan, a middle step between China and the West regarding life-style and dietary habits, allowing comparison of various levels of the treatment.
>
> Not surprisingly, medical experts from China, Great Britain, and the United States are taking advantage of these large-scale possiblities to study the relationship of diet, living habits, and health. The benefits accrue to those who may require life-style changes to avoid or inhibit health problems as well as to the Chinese whose life-styles are changing with the advent of industrialization. Results so far corroborate many maintained hypotheses, such as increased meat consumption increases the incidence of cardiovascular disease and lowered plasma cholesterol levels decrease deaths from coronary disease. The lack of evidence also jeopardizes other ideas; for instance, that an association exists between low plasma cholesterol levels and cancer or that deleterious digestive effects result from very high-fiber diets.
>
> *Source:* Anne Simon Moffat, "China: A Living Lab for Epidemiology," *Science, Vol. 248, May 1990, pp. 553–55.*

systematic effect. The remaining explanations for a difference in the pools' mean outputs are productivity differences or sampling error. In order to credit the difference to productivity, the manager must determine if the difference in pool means is a significant difference (the conclusion we reach from a greater difference between sample means than we ordinarily expect from the sampling error that results when we use samples to infer information about a population).

The manager can also determine if a significant difference in mean outputs exists after the experimental pool uses the new software, as shown in Figure 10-1. If a significant difference exists before the experiment, then the test for the difference in two population means can determine if the difference has increased or decreased significantly after the experiment.

Box 10-1 describes a large-scale experiment. This scale avoids researchers observing a difference in sample means that can be explained by some cause other than the one they suspect.

10.1 The Hypotheses

Rather than test for the value of one population mean, we now compare two population means to determine if they are significantly different. We do not specify the exact values of the means—only the difference between them.

Question: How would you state the hypothesis "the means do not differ" using μ_1 for the mean of the first population and μ_2 for the mean of the second population?

Answer: If there is no difference, the means are the same, $\mu_1 = \mu_2$.

When we speak of the "difference" between two values, we usually mean the result of subtracting one from the other. So we also write the above expression as $\mu_1 - \mu_2 = 0$. We will continue using this form.

Usually, we employ this test to determine if there is a difference and, sometimes, which mean is larger. The null and alternative hypotheses can be one of three forms:

Two-tailed	Upper-tailed	Lower-tailed
$H_0: \mu_1 - \mu_2 = 0$	$H_0: \mu_1 - \mu_2 \leq 0$	$H_0: \mu_1 - \mu_2 \geq 0$
$H_1: \mu_1 - \mu_2 \neq 0$	$H_1: \mu_1 - \mu_2 > 0$	$H_1: \mu_1 - \mu_2 < 0.$

Of course, we can reverse the order of the populations in each set of hypotheses to be $\mu_2 - \mu_1$. In either case, we must be sure to keep subscripts in the same order as we proceed through any solution.

Question: If you cannot reject H_0, how would you interpret the decision for each set of hypotheses to someone unfamiliar with hypothesis testing techniques?

Answer: If we cannot reject H_0 for a two-tailed test, we are unable to demonstrate a difference between the means of the two populations. For a lower-tailed test, H_0: $\mu_1 - \mu_2 \geq 0$, the data did not show that the mean of population 2 is significantly larger than that of population 1. For an upper-tailed test, H_0: $\mu_1 - \mu_2 \leq 0$, the data do not support a conclusion that the mean of population 1 is greater than the mean of population 2.

■ **Example 10-1:** Suppose the secretarial pool from the productivity experiment without the new software is population 1 and the pool with the software is population 2. If the manager believes the use of the software will improve productivity, what should H_1 be?

■ **Solution:** $\mu_1 - \mu_2 < 0$ or, equivalently, $\mu_2 - \mu_1 > 0$.

Section 10.1 Problems

7. Show with symbols the hypothesis when a decision maker believes that (μ_1 and μ_2 are both positive)
   a. The mean of population 1 is smaller than the mean of population 2.
   b. The mean of population 1 is more than 100 units larger than the mean of population 2.
   c. The mean of population 1 is at most 25 units larger than the mean of population 2.
8. In experiments we usually have two samples, a control group and an experimental group. The experimental group undergoes a treatment or is somehow differentiated from the control group. Otherwise, they are alike, or attributes are randomly distributed throughout both samples so as not to produce a systematic effect that interferes with the effect of the treatment. Then we measure some characteristic of both groups to learn if they differ with respect to this characteristic. Use μ_C and μ_E, and express the null and alternative hypotheses when we test for
   a. A positive experimental effect from the treatment.
   b. A negative experimental effect from the treatment.
   c. Any experimental effect from the treatment.

10.2 Test for Large Independent Random Samples

When we want to test for a difference in population means, situations will offer various kinds and amounts of information, such as sample size and selection method, that will affect technical aspects of the test. In this unit, we explore one of the simpler testing situations.

10.2.1 The Test Statistic

The hypotheses contain statements about populations, but we make the decision using sample information. When doing tests for a single population mean, we use the sample mean, $\overline{X}$, as the test statistic to indicate the value of the population mean.

Question: Suggest a test statistic for $\mu_1 - \mu_2$.

Answer: We use $\overline{X}$ to estimate μ, so it is only natural to use the difference in two sample means to estimate the difference in two corresponding population means. Thus, the test statistic or estimator for $\mu_1 - \mu_2$ is $\overline{X}_1 - \overline{X}_2$ (where $\overline{X}_1$ is the mean of a sample from the first population and $\overline{X}_2$ is the mean of a sample from the second population).

Before we can make a decision, we must understand the behavior of this new test statistic, or, more technically, its sampling distribution.

10.2.2 The Sampling Distribution of $\overline{X}_1 - \overline{X}_2$

The test statistic, $\overline{X}_1 - \overline{X}_2$ combines two sample means, statistics whose individual behavior we already know. The behavior of $\overline{X}_1 - \overline{X}_2$ is similar, but we need one additional assumption, the samples must be independent.

Two samples are *independent* when knowledge of the observations in one sample provides no information about observations in the other sample. Usually, when an item selected for one sample does not influence the selection of items for the other sample, the samples are independent. Generally, if we select the first sample randomly and then select the second randomly, our samples will be independent.

From several statistical theorems, we develop the following:

Sampling Distribution of $\overline{X}_1 - \overline{X}_2$ for Large, Independent Samples

As the sizes of independent samples, n_1 observations from population 1 and n_2 observations from population 2, increase, the sampling distribution of $\overline{X}_1 - \overline{X}_2$ approaches a normal population. Again, the population distributions do not matter.

We will continue to use our rule of thumb for a large sample. When n_1 and n_2 are both greater than 30, they constitute large samples.

To work with a normal random variable, we need to know its mean and standard deviation. The mean of the difference in sample means, $\overline{X}_1 - \overline{X}_2$, is the mean of all possible $(\overline{X}_1 - \overline{X}_2)$ values, μ_{dif}. This mean is $\mu_1 - \mu_2$.

Mean of the Difference in Two Independent Sample Means

$$\mu_{\text{dif}} = \mu_1 - \mu_2.$$

The standard deviation is called the standard error of the difference in means (shortened to the *standard error of the difference*), σ_{dif}. To derive the standard error, σ_{dif}, we first square $\sigma_{\overline{x}}$ to obtain the variance of the random variable $\overline{X}$. For $\overline{X}_1$ we obtain

$$\text{Variance of } \overline{X}_1 = (\sigma_{\overline{x}})^2 = \left(\frac{\sigma_1}{\sqrt{n_1}}\right)^2 = \frac{\sigma_1^2}{n_1}.$$

Only the subscript changes when we find the variance for the second mean. Next, we sum the variances for the two means to obtain the variance of $(\overline{X}_1 - \overline{X}_2)$.

$$\text{Variance of } (\overline{X}_1 - \overline{X}_2) = \text{Variance of } \overline{X}_1 + \text{Variance of } \overline{X}_2$$

$$(\sigma_{dif})^2 = (\sigma_{\bar{x}_1})^2 + (\sigma_{\bar{x}_2})^2 = \frac{\sigma_1^2}{n_1} + \frac{\sigma_2^2}{n_2}.$$

This expression holds as long as the samples are independent.

Finally, to obtain the standard error of the difference, we take the square root of the variance.

Standard Error of the Difference in Two Independent Sample Means

$$\sigma_{dif} = \sqrt{\frac{\sigma_1^2}{n_1} + \frac{\sigma_2^2}{n_2}}.$$

Question: Usually we will not know σ_1 or σ_2, so how can we estimate σ_{dif}?

Answer: We substitute sample variances for unknown population variances, S_1^2 for σ_1^2 and S_2^2 for σ_2^2, to obtain:

Estimated Standard Error of the Difference in Two Independent Sample Means

$$\hat{\sigma}_{dif} = \sqrt{\frac{S_1^2}{n_1} + \frac{S_2^2}{n_2}}.$$

Generally large samples ensure that the normal distribution will still approximate probabilities for standardized values of $\overline{X}_1 - \overline{X}_2$ when we substitute S^2 for σ^2 in σ_{dif}.

In summary, if we select independent random samples from each of two populations with means μ_1 and μ_2, respectively, and variances σ_1^2 and σ_2^2, respectively, where n_1 and n_2 are both greater than 30, then the sampling distribution of $\overline{X}_1 - \overline{X}_2$ is approximately normal with mean $\mu_1 - \mu_2$ and an estimated standard error of $\hat{\sigma}_{dif} = \sqrt{(S_1^2/n_1) + (S_2^2/n_2)}$ as shown in Figure 10-2.

Section 10.2.1 and 10.2.2 Problems

9. A personnel committee selects a random sample of 50 secretaries for an experiment that requires them to attend a workshop on productivity. They test the secretaries before and after the workshop, and then compare the means of the scores from the different times. Are these samples independent?

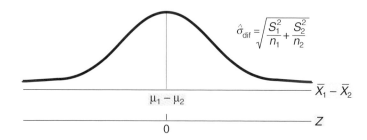

Figure 10-2
Sampling distribution of $\overline{X}_1 - \overline{X}_2$ for large samples and unknown σ_{dif}

10. To compare average incomes in two different cities, a company contacts a random sample of 150 citizens of each city to offer them free products in return for answering a questionnaire that includes income questions.
   a. Are the samples independent? Explain.
   b. The distributions of the income values of citizens in each city are probably skewed to the right, as are most sets of incomes values. What effect does this have on the use of the normal approximation for $\overline{X}_1 - \overline{X}_2$? Explain.
   c. The results of the questionnaire for the first city are a sample mean of 15 thousand dollars and a standard deviation of 4.5 thousand dollars. (We express the values without the zeros to make the calculations easier.) Similar sample results for the second city are a mean of 15.75 thousand dollars and a standard deviation of 8 thousand dollars. What is the value of the test statistic, $\overline{X}_1 - \overline{X}_2$, and the estimated standard error of the difference, $\hat{\sigma}_{\text{dif}}$?
   d. Actually, there is no difference in the mean income in the two cities. What is the probability of observing the -0.75 thousand-dollar difference or a more negative difference?

11. A psychologist selects a random sample of 62 high school students and measures the mean time it takes them to complete a personality test. Independently, she also selects a random sample of 58 college students and measures their mean time to complete the task. Suppose the actual mean time for *all* high school students is 34.2 minutes and for *all* college students, 31.6 minutes. The standard deviations are 2.2 and 3.8 minutes, respectively. Find the probability that the difference between the sample means will exceed 5 minutes: $P(\overline{X}_1 - \overline{X}_2 > 5)$, where high school is population 1 and college is population 2.

12. It may seems strange that we use $\overline{X}_1 - \overline{X}_2$ as the test statistic, but we add the variance of $\overline{X}_1$ to the variance of $\overline{X}_2$ to find the variance of $\overline{X}_1 - \overline{X}_2$. Demonstrate the effect on the range with two populations that each consists of the values (0, 1, 2). Form a new population of differences in values of the two populations. Then compare the range of the original populations with the range of the combined population of differences. What are the implications for the variance of the population of differences, when its range increases?

10.2.3 The Decision

Now that we know the test statistic and its sampling distribution, we can test a hypothesis. We follow the same general procedure as before. Start by assuming H_0 is true, and sketch a diagram accordingly. Determine the critical value on the $\overline{X}_1 - \overline{X}_2$ axis or on the Z axis. Then compare the test statistic with the critical value to make the decision. Alternatively, we can determine the *p*-value and compare it to α to make the decision.

Example 10-1 (continued): To determine if new software will increase productivity in the secretarial pools, the manager independently selects a random sample of 36 output values from each pool. The sample from the pool without the software has a mean of 1,783 pages and a standard deviation of 312 pages. The mean and standard deviation of the second sample are 2,010 and 547 pages, respectively. Is this sufficient evidence to decide at the 0.05 significance level that the new software increases productivity? Assume that the productivities were identical prior to the experiment.

Solution: Following the steps for hypothesis tests from Chapter 9 (p. 402), we find:

Step 1: Because we want to know if there is sufficient evidence to decide that the software increases productivity, H_1 is $\mu_1 - \mu_2 < 0$, where the pool with the software is population 2. Hence, we have $H_0: \mu_1 - \mu_2 \geq 0$.
Step 2: The problem statement establishes the significance level at 0.05.
Step 3: We know that the test statistic for a test of $\mu_1 - \mu_2$ is $\overline{X}_1 - \overline{X}_2$. In this case, $\overline{X}_1 - \overline{X}_2 = 1,783 - 2,010 = -227$ pages.
Step 4: Because both samples contain more than 30 observations and they are independently selected, we treat the test statistic as a normal random variable. Assuming H_0 is true produces a mean of 0 pages. The estimated standard error, $\hat{\sigma}_{\text{dif}} = \sqrt{(312)^2/36 + (547)^2/36} = 104.9541$ pages. Figure 10-3 displays the sampling distribution for $\overline{X}_1 - \overline{X}_2$.

Figure 10-3

Diagram for office productivity experiment

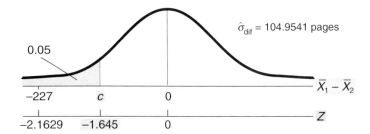

Step 5: When $\overline{X}_2$, the mean output of the software pool, is larger than the mean of the other pool, $\overline{X}_1$, the test statistic, $\overline{X}_1 - \overline{X}_2$ will be negative. More negative values of $\overline{X}_1 - \overline{X}_2$ support the alternative hypothesis that $\mu_1 - \mu_2 < 0$. Hence, the rejection region is under the left tail of the normal curve, as shown in Figure 10-3.

Step 6: The critical value from the Z table (Appendix H) with an area of 0.05 in the tail and an area of 0.45 between the mean and critical value is -1.645.

Step 7: When we convert the -227-page difference in means from Step 3 to a Z value, we obtain $(-227 - 0)/104.9541 = -2.1629$, which is smaller than the critical value, -1.645, so we reject H_0. See Figure 10-3 again.

Question: Interpret this decision for the manager.

Answer: The rejection of H_0 means that we decide that μ_1 is significantly smaller than μ_2, which means that the mean productivity is higher for the pool with the new software.

■ **Example 10-2:** An alumnus from school A randomly selects a sample of 42 games played by the school's volleyball team over several years, while an alumna selects independently 50 games played by her school, B. Then they both determine the mean number of spikes per game, 8.7 and 6.8 spikes for A and B, respectively. The alumnus from school A infers from this sample information that team A has the larger mean number of spikes per game. The sample standard deviations are 2.6 and 4.1 spikes for A and B, respectively.

a. Demonstrate that this difference in means is sufficient to corroborate the alumnus's belief with a 10% chance of making a Type I error.

b. Find the two-tailed p-value for these sample results, and use the answer to conduct a two-tailed test at the 0.10 significance level.

■ **Solution: a.** H_0 is $\mu_A - \mu_B \leq 0$ and H_1 is $\mu_A - \mu_B > 0$, because we want to demonstrate that μ_A is greater than μ_B, as the alumnus believes. The samples are both large and independently selected, so $\overline{X}_1 - \overline{X}_2$ is a normal random variable. If H_0 is true, the mean is 0. The estimated standard error, $\hat{\sigma}_{dif} = \sqrt{((2.6)^2/42) + ((4.1)^2/50)} = 0.7051$ spikes. The critical Z value is 1.28 for an upper-tailed test at the 0.1 level. $\overline{X}_A - \overline{X}_B = 8.7 - 6.8 = 1.9$ spikes, which becomes $Z = (1.9 - 0)/0.7051 = 2.69$. The decision is to reject H_0, because $2.69 > 1.28$. Thus, the evidence supports H_1, the alumnus's belief. See Figure 10-4.

b. The two-tailed p-value is the probability that $\overline{X}_A - \overline{X}_B$ is at least 1.9 spikes from 0 given that $\mu_A - \mu_B = 0$. Symbolically, this is

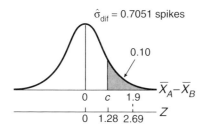

Figure 10-4
Diagram for volleyball teams

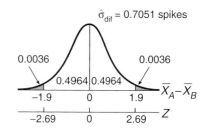

Figure 10-5
Two-tailed P-value for volleyball team example

$P(\overline{X}_A - \overline{X}_B$ is at least 1.9 units from 0 when $\mu_A - \mu_B = 0)$. Using the Z value for 1.9 that we calculated in Part a, 2.69, we can restate the expression as

$P(Z$ is more than 2.69 units from 0 when $\mu_A - \mu_B = 0)$
$= P(Z \leq -2.69) + P(Z \geq 2.69) = 2P(Z \geq 2.69)$
$= 2(0.0036) = 0.0072$.

(See Figure 10-5.) We reject H_0, because the p-value $= 0.0072 < 0.10 = \alpha$. The data demonstrate a significant difference in the mean number of spikes for the two schools.

Box 10-2 shows that we can use a test for the difference in means to explore different gender reactions to family social situations.

Section 10.2.3 Problems

Throughout the chapter, problems marked with an asterisk involve p-values and are optional.

13. Test the following hypotheses at the 0.05 significance level.
$H_0: \mu_1 - \mu_2 = 0$, $H_1: \mu_1 - \mu_2 \neq 0$.
$\overline{X}_1 = 450$ $S_1 = 25$ $n_1 = 46$.
$\overline{X}_2 = 475$ $S_2 = 35$ $n_2 = 63$.

14. Test the following hypotheses at the 0.01 significance level.
$H_0: \mu_1 - \mu_2 \leq 0$, $H_1: \mu_1 - \mu_2 > 0$.
$\overline{X}_1 = 691.3$ $S_1 = 2.5$ $n_1 = 39$.
$\overline{X}_2 = 679.7$ $S_2 = 6.1$ $n_2 = 95$.

15. The manager from the productivity experiment in Example 10-1 collected random samples of 50 observations from each secretarial pool to determine if there was a difference between the mean productivities of the pools before experimenting with the new software. The sample mean for the pool that did not get the software during the experiment was 1,534 pages and for the software pool it was 1,323 pages. The respective sample standard deviations were 45 pages and 63 pages.
  a. Use this information to test for a difference at a 0.01 level of significance. Interpret the decision.
  b. After the experiment the manager found that the software pool's mean output was significantly larger than the pool without software. Based on the means before and after the experiment, does the new software appear to increase productivity?

16. Among the advantages of using video surgical techniques are reduced patient costs. Dr. Maurice E. Arregui and colleagues at St. Vincent Hospital and Health Care Center, Indianapolis, found that the average hospital charges for 484 video gallbladder surgical patients was $3,772, and the average for 788 traditional gallbladder surgery patients was $6,736. We want to know if there is sufficient evidence to conclude that the average cost of a traditional gallbladder surgery is at least $2,800 more than the average cost of video surgery at the 0.05 level. Assume the standard deviations of both measures are $2,000. (Source: "Gentler Surgical Method Cuts Damage to Chest," *Wall Street Journal*, June 27, 1991, p. B1.)
  a. Perform the test using critical values. Interpret the decision.
  *b. Perform the test using p-values.

Box 10-2 Gender Differences among Adolescents from Divorced Homes

To learn more about factors that influence children's adjustment to divorce, a study focuses on adolescents between the ages of 10 and 18. The study also considers gender differences in adjustments.

One aspect of the study concerns children "being caught between parents." This concept includes various facets of the parent-parent and parent-child relationships, such as the child being an intermediary messenger or informer between parents as well as hesitating to express emotions or information to one parent about the other parent. The researchers measure this variable by combining a series of responses to questions related to this concept during a telephone interview. The possible values range between 0 (low) and 12 (high). Among the 522 respondents, the mean is 5.1 with a standard deviation of 3.0 for females and 4.4 with a standard deviation of 2.7 for males. The resulting T statistic for the difference in means, -2.69, has 520 degrees of freedom (p-value $= 0.007$). Thus, there is a significant gender difference in the mean value of the "caught" variable at the 0.05 significance level.

The adjustment outcome, depression or anxiety, refers to mental and physical symptoms related to these states, such as head and stomach aches or sensation of loneliness, overtiredness, and desire to run away. To measure this variable, the researchers sum responses to 10 items asking the adolescents how often they experienced the symptoms in the previous month. Choices vary between 0 for never and 3 for three or more times. The mean result for females is 16.8 and 13.7 for males. The T statistic is -5.12 (p-value ≤ 0.0001), so by this measure there is a significant gender difference in depression and anxiety for adolescents from divorced homes.

Another adjustment outcome is deviance, behaviors that can be considered negative or maladjustments to varying degrees, including smoking cigarettes, consuming drugs and alcohol, damaging school property, and getting in trouble with the police. Allowable responses for the 15 behaviors were 1 (never in last 12 months) to 4 (often in the last 12 months). The resulting means are 21.2 and 22.4 for females and males, respectively. The T statistic is 2.37 (p-value ≤ 0.05), so again there is a significant gender difference at the 0.05 level.

In the cases of these three variables, there are significant differences in male and female adolescents' adjustments to divorced parents. The implication is that the averages are larger for females for being caught and suffering from depression or anxiety, whereas males have a larger mean deviance. The results for two other variables, measuring "closeness to father" and "closeness to mother," also yield significant gender differences. For the father the results are 31.8 for females and 34.3 for males (T statistic $= 3.49$, p-value ≤ 0.001). Corresponding results for the mother are 35.4 and 36.9, respectively (T statistic $= 2.31$, p-value ≤ 0.05).

Based on these results, the researchers control for gender in later analyses of the relation between family and children characteristics and adjustment of children to the divorce, especially regarding the mediator or being caught between the two parents.

Source: Christy M. Buchanan, Eleanor E. Maccoby, and Sanford M. Dornbusch, "Caught between Parents: Adolescents' Experience in Divorced Homes," Child Development, Vol. 62, No. 5, 1991, pp. 1008–1029.

Summary and Review

We use a test for the difference between two population means when we compare two aspects of a variable, such as the difference in mean miles per gallon for two types of automobiles. This is a test for the difference between two parameters, rather than a test for a single value of either mean. Two-tailed, upper-tailed, and lower-tailed tests are possible. It is important to maintain the order of the population subscripts throughout the testing procedure in order to avoid erroneous decisions.

The appropriate test statistic is $\overline{X}_1 - \overline{X}_2$. This unit describes the procedure to use for large, independently selected random samples. Random samples are independent when the items selected for one sample provide no information about the other sample's items. In this case, the sampling distribution of $\overline{X}_1 - \overline{X}_2$ is approximately normal. The mean of the test statistic is $\mu_1 - \mu_2$, and the standard error is $\sigma_{\text{dif}} = \sqrt{(\sigma_1^2/n_1) + (\sigma_2^2/n_2)}$, which is frequently estimated with $\hat{\sigma}_{\text{dif}} = \sqrt{(S_1^2/n_1) + (S_2^2/n_2)}$. Knowing this information about the test statistic allows us to conduct hypothesis tests, because we can describe the values of $\overline{X}_1 - \overline{X}_2$, differences in sample means, that are likely and not likely to occur when H_0 is true. We conduct these tests using critical or p-values following the usual procedures from Chapter 9 for the new test statistic and its sampling distribution.

Multiple-Choice Problems

17. If the mean outputs from two production techniques are μ_{new} and μ_{old}, what would H_1 be if we believed that the new technique would result in higher output?
    a. $\mu_{new} - \mu_{old} = -100$.
    b. $\mu_{new} - \mu_{old} = 0$.
    c. $\mu_{new} - \mu_{old} < 0$.
    d. $\mu_{new} - \mu_{old} \neq 0$.
    e. $\mu_{new} - \mu_{old} > 0$.

18. If the means from two samples from different populations are 100 and 200, can we say that the population means are different?
    a. Yes, because the difference is more than 1.96.
    b. Yes, because $100 \neq 200$.
    c. No, because the difference must be larger than 196.
    d. No, because $100 \neq 200$.
    e. No, because there may be sampling error.

19. If H_1 is $\mu_1 - \mu_2 \neq 0$, what critical value would you obtain from the Z table if the significance level is 0.01?
    a. 1.96.
    b. 2.33.
    c. 1.645.
    d. 2.575.
    e. 1.28.

Use the following information to answer Questions 20–26.

Highway engineers select independent random samples of 40 and 150 observations, respectively, from highway downtimes, periods of time that stretches of interstate highways are partially or wholly barricaded because of construction. The first sample comes from last year's values (population 1) and the second from five years earlier (population 2). The sample means are 53 and 47 days, respectively, with corresponding sample standard deviations of 10 and 15 days.

20. If highway engineers believe that downtimes have decreased over time, H_1 will be
    a. $\mu_1 - \mu_2 < 0$.
    b. $\mu_1 - \mu_2 = 0$.
    c. $\mu_1 - \mu_2 > 0$.
    d. $\mu_1 - \mu_2 \neq 0$.
    e. $\mu_1 - \mu_2 = 6$.

21. $\hat{\sigma}_{dif}$ equals
    a. -5 days.
    b. 5 days.
    c. 125 days.
    d. 2 days.
    e. 11 days.

22. What is the critical value on the Z axis at the 0.05 significance level?
    a. 1.645.
    b. -1.645.
    c. 1.96.
    d. -1.96.
    e. 6.

23. The estimate of the difference in means is
    a. -6 days.
    b. 6 days.
    c. 3 days.
    d. -3 days.
    e. -110 days.

24. The decision for the test is
    a. Cannot reject H_0, because $3 > -1.645$.
    b. Reject H_0, because $3 > 0$.
    c. Cannot reject H_0, because $-3 < -1.645$.
    d. Reject H_0, $-3 < -1.645$.
    e. Reject H_0, because $-3 < 0$.

25. Highway engineers interpret the decision to mean that periods during which highways have been barricaded have
    a. Decreased over the five-year period.
    b. Increased over the five-year period.
    c. Not changed over the five-year period.
    d. Not decreased over the five-year period.
    e. Not increased over the five-year period.

26. If the engineers obtain a random sample of 100 stretches of highway and find the downtimes for the same 100 stretches in both years,
    a. Nothing changes in the procedure, the decision stands.
    b. The samples are no longer independent, so the procedure breaks down.
    c. The standard error is twice the standard error for a single mean.
    d. The standard error is reduced, so the engineers are more likely to reject.
    e. The standard error is reduced, so the engineers are not likely to be able to reject.

27. Which of the following assumptions is unnecessary to perform the tests conducted in this unit?
    a. Randomly selected samples.
    b. Large samples.
    c. Independent samples.
    d. H_0 is true.
    e. Normal populations.

28. To indicate a significant difference in population means,
    a. The observed difference in sample means must be a value that seldom occurs when H_0 is true.
    b. The value specified in H_0 must not be zero.
    c. The difference in sample means must fall within the middle $(1 - \alpha)\%$ of differences that occur when H_0 is true.
    d. The observed difference in sample means must be zero.
    e. The observed difference in sample means must be the same as the value specified in H_0.

Use the following information to answer Questions 29–36.

The following values comprise random samples of earthquake intensities in the central region of the United States and in the region comprising California and Western Nevada (call this the western region) during the period 1971–1980. The samples are independent. The values represent the arabic equivalent of roman numerals used in the modified Mercalli intensity scale (1 is weakest and 12 is strongest). (Source: Jerry L. Coffman, Carl A. von Hake, and Carl W. Stover, eds., *Earthquake History of the United States*, Publication 41-1, rev. ed., repr. 1982 with Supplement (1971–1980), U.S. Department of Commerce, National Oceanic and Atmospheric Administration, and U.S. Department of the Interior, Geological Survey, Boulder, Col., 1982, pp. 11a, 35a–36a.)

CENTRAL

5	5	6	6	5	5	6	5	5	5
6	5	5	5	6	5	7	6	5	6
5	6	5	6	5	5	5	5	5	5
5	5	5	5						

WESTERN

5	6	7	6	5	6	6	8	6	6
11	6	6	5	5	6	7	5	6	6
5	6	6	6	6	9	6	6	6	6
6	6	6	6	6	6	6	6	6	6
6	7	6	6	6	6				

29. If we believe that the mean intensity in the western region is more than one point greater than the central region's mean, what is H_1?
   a. $\mu_W - \mu_C = 1$.
   b. $\mu_W - \mu_C \geq 1$.
   c. $\mu_W - \mu_C > 1$.
   d. $\mu_W - \mu_C < 1$.
   e. $\mu_W - \mu_C \neq 1$.

30. What is the point estimate of the difference in the mean intensities?
   a. 1.0000.
   b. 0.8286.
   c. 0.0000.
   d. −0.8286.
   e. −1.0000.

31. What is $\hat{\sigma}_{dif}$?
   a. 0.0306.
   b. 0.1749.
   c. 1.3069.
   d. 1.1432.
   e. 0.1174.

32. What is the value of the test statistic on the Z axis?
   a. 4.74.
   b. −4.74.
   c. −0.98.
   d. 0.98.
   e. 0.8286.

33. The decision at the 0.1 significance level is
   a. Reject H_0, because $-0.98 < 1.28$.
   b. Reject H_0, because $0.8286 \neq 0$.
   c. Reject H_0, because $0.8286 > 0$.
   d. Cannot reject H_0, because $-0.98 < 1$.
   e. Cannot reject H_0, because the test statistic agrees with H_0.

34. A decision that we cannot reject H_0 means that the mean intensity in the western region
   a. Is greater than the mean in the central region.
   b. Is the same as the mean in the central region.
   c. Is less than the mean in the central region.
   d. Does not exceed the mean in the central region by more than one point.
   e. Exceeds the mean in the central region by more than one point.
   f. Differs from the mean in the central region by one point.

*__35.__ The p-value for this problem is
   a. 0.1635.
   b. 0.6730.
   c. 0.8365.
   d. 0.3365.
   e. 0.3270.

*__36.__ We cannot reject H_0 at the 0.1 significance level with the p-value from Problem 35 because
   a. $0.8365 > 0.1$.
   b. $0.8365 > 0.05$.
   c. $0.4182 > 0.01$.
   d. $0.4182 > 0.05$.
   e. $1 - 0.8340 > 0.1$.

37. Samples are independent when
   a. Their sample means differ.
   b. We record two measurements on a given set of items or people to form the two samples.
   c. One records successes, and the other records quantities that are measurable.
   d. The choice of items for one provides no information about the values chosen for the other.
   e. Every item in a single population of measures does not have an equal chance of being included.

38. If we do not maintain the order of subscripts throughout the test procedure,
   a. The standard error can decrease, resulting in a larger test statistic, making us more likely to reject a true H_0.
   b. The test statistic will have the wrong sign, which could cause the wrong decision for a one-tailed test.
   c. The test statistic will have the wrong sign, which could cause the wrong decision for a two-tailed test.
   d. The test statistic will have the wrong sign, which could cause the wrong decision for any test.
   e. Nothing really happens, but steps are more difficult to follow.

Word and Thought Problems

39. Define μ, $\overline{X}$, $\mu_1 - \mu_2$, and $\overline{X}_1 - \overline{X}_2$.

40. Define σ, S, $\sigma_1 - \sigma_2$, σ_{dif}, and $\hat{\sigma}_{dif}$.

41. Test at the 0.05 level to determine if the mean number of defectives is greater for a night shift. Random samples of sizes 41 and 81 are independently selected from day and night shifts, respectively, $\overline{X}_{day} = 4$, $S_{day} = 1$, $\overline{X}_{night} = 5$, and $S_{night} = 0.25$. Interpret your decision.

42. A manufacturer is testing a new device to extend the life of a product. A random sample of 100 units without the device has a mean life of 752 hours with a standard deviation of 76 hours. Similar values for a random sample of 125 units with the device are 770 hours and 98 hours, respectively. Can we say there is a significant improvement in product life at the 0.01 level? Interpret your decision.

43. An economist investigates salaries before and after unionization of an industry to determine the union's effect on wages. He selects a random sample of 50 workers before unionization and another random sample of 50 workers afterward in order to test at the 0.05 level to determine if mean salaries increase after unionization. The sample mean and standard deviation from the earlier period are $5 and $0.50, respectively. Values for the later period are $6 and $0.10, respectively. Interpret the decision.

Word and Thought Problems

44. To test for the impact of the three-point basket rule on a university basketball team's scoring, a physical educator selected independent random samples of scores from 50 games before and 50 after the rule change. The before sample mean and standard deviation are 64.6 and 10.85 points, respectively. Values for the after sample are 73.2 and 15.76 points, respectively. Test for a significant difference in mean scoring, at the 0.1 level. Interpret your decision.

45. If the means of two populations are different, we will have the most difficulty rejecting H_0: $\mu_1 - \mu_2 = 0$ when the population standard deviations fit which of the following situations: both are small, both are large, they differ (one small, one large), or the magnitudes of the population standard deviations are immaterial for making the decision. Explain your answer.

46. Suppose someone tells you that people between the ages of 30 and 39 are more charitable than those between 40 and 49, based on comparing mean times spent as volunteers during the year from samples of each age group. How would you explain to this person that just comparing sample means is not convincing evidence?

47. Test to demonstrate that fertilizer improves the mean output of a gardener's tomato plants if one group of plants is fertilized and another is not. A random sample of 32 days for the fertilized group has a mean output of 10.2 pounds of tomatoes, with a standard deviation of 1 pound. A random sample of 36 days from the unfertilized group has a mean output of 8.5 pounds, with a standard deviation of 1.5 pounds.
    a. Test at the 0.02 level and interpret your decision.
    *b. What is the p-value for this test? Demonstrate that the decision does not change, if made with a p-value rather than a critical value.

48. Small differences in sample means indicate little, if any, difference in population means. Explain how small differences in sample means do not guarantee identical population means.

49. Brand X dishwashing liquid claims that users can wash more dishes with it than with Brand Z. Likewise, Z claims that it lasts longer than X. A random sample of 100 bottles of X washes a mean of 1,583 dishes with a standard deviation of 206 dishes. A random sample of 100 bottles of Z washes 1,710 dishes on average with a standard deviation of 52 dishes.
    a. Assume that you manufacture brand X. Set up the hypotheses to demonstrate your product's superiority at the 0.05 level, and determine if the evidence supports your product's claim.
    b. Repeat Part A, assuming that you manufacture brand Z.
    c. Assume you work for a consumer group. Test to determine if there is any difference in the products at the 0.05 level.
    d. Interpret and summarize the findings in these three parts.

50. Verify a claim that the means of two classes' performances on a test are not significantly different at the 0.01 level. A random sample of 36 students from the first class has a mean score of 76.3, with a standard deviation of 10.2 points. The other class sample consists of 40 students with a mean score of 70.9 and a standard deviation of 15.7 points. Interpret your decision.

51. A certain basketball player has scored a mean of 22 points per game for all games in which he has participated. In a random sample of 35 games in which he did not participate, the mean total points scored by his team is 101 with a standard deviation of 15 points. In a random sample of 40 games in which he did participate, the mean total points scored by the team is 118 with a standard deviation of 23 points. Based on this evidence, can you demonstrate that the team at least partially makes up for his absence in terms of points? That is, is the difference in means with and without the player less than the mean number of points that the player scores, 22, at the 0.05 significance level? Interpret the decision. Assume independence.

*52. The two-tailed p-value for a test of H_0: $\mu_1 - \mu_2 = 0$ is 0.90. Does this evidence support H_0? If you reject H_0 for this problem, what do you decide about $\mu_1 - \mu_2$?

53. a. Test at the 0.01 significance level to determine if there is a difference in mean attendance at movie theaters in different regions of the country.

    $\overline{X}_E = 15.22$ $\overline{X}_W = 13.56$
    $S_E = 1.4$ $S_W = 2.0$
    $n_E = 32$ $n_W = 49$.

    (Values are in millions of patrons. E = East and W = West.)
    *b. Find the p-value for this problem. Then demonstrate and interpret your decision.

54. The mean salary for doctors is $143,000 with a standard deviation of $20,000, based on a random sample of 42 doctors. The mean salary is $105,000 with a standard deviation of $30,000, in a random sample of 31 lawyers. Test to see if there is a difference in the mean salaries of the two groups at the 0.05 level. Interpret your results.

55. Some states mandate economics education in high school. To study the effects of this requirement on the amount of learning that occurs, researchers collected information on students in states with and without the requirement. Part of the study involved scores on a standardized economics test both before and after the economics instruction. The researchers calculated a variable called GAP, which measures the relative amount of learning that occurred during the course. The potential difference in a student's before and after score is the difference between the maximum possible score on the test and the student's before score. The GAP is the ratio of the difference in a student's before and after scores (after − before) to the potential difference. The mean GAP measure for 634 students from mandate states was 0.05 with a standard deviation of 0.28. Similar measures for 902 students from nonmandate states were −0.02 and 0.29. (Source: Sherrie L.W.

Rhine, "The Effect of State Mandates on Student Performance," *American Economic Review*, Vol. 79, No. 2, May 1989.
   a. Interpret the mean of −0.02 for nonmandate students.
   b. Does the evidence support the hypothesis that mandate state students learn more economics? Test at the 0.05 level. Explain your results.

56. A fungus started destroying dogwood trees in the North Carolina mountains in 1987. To determine if proper tree maintenance improves the trees' vigor and prevents or retards damage and loss of dogwoods, the U.S. Forest Service is conducting experiments with wild, native dogwoods and cultivated varieties. The amount of water and fertilizer varies among the trees. Suppose 56 trees that die receive on average 3.01 gallons of water per week and 88 that survive average 2.72 gallons. Is this sufficient evidence to claim that trees that average receiving more water are more susceptible to damage or death from the fungus? Test at the 0.01 level. Assume the standard deviations are 0.85 gallons for both sets of trees. Interpret your decision. (Source: "Dogwood Threat Studied," *Asheville Citizen-Times*, May 28, 1991, p. 2B.)

57. A talent agent selected a random sample of 40 days from the last six months and recorded each day the number of requests for a folk musician and the number for classical musicians. The results were means of 15 and 20 requests for the folk and classical, respectively. The corresponding standard deviations were 15 and 9 requests. The agent wants support to explain to potential clients that he can place more classical musicians on average.
   a. Can he perform a test for difference in means using procedures from this unit? Explain your answer.
   b. Assume there are no problems and perform the test at the 0.01 level using critical values. Interpret your decision.
   *c. Repeat the test from Part b using p-values.

58. The following data sets are independent, random samples of distances students travel to reach their schools. Use this information to test for a difference in mean distances at the 0.05 level. Interpret your decision.

SCHOOL 1

6	10	15	4	10	12	11	8	14
6	10	12	3	9	19	0	11	10
15	5	13	2	17	10	12	1	6
9	17	13	6	10	1	10	5	18
3	3	8	7	2	12	13	15	5
6	16	7	16	3				

SCHOOL 2

13	9	18	14	9	15	14	18	13
21	10	7	8	7	18	8	11	8
12	10	10	5	15	11	6	10	5
18	11	14	10	13	3	9	13	4
4	6	12	16	13				

Unit II: Variations of the Test

Preparation-Check Questions

Use the following information to answer Questions 59–62.

$n_1 = 240$ $\overline{X}_1 = 8{,}370$ $S_1 = 549$
$n_2 = 173$ $\overline{X}_2 = 6{,}509$ $S_2 = 912$.

59. If you test $\mu_1 \leq 0$ at the 0.05 significance level, what will the test statistic be on the Z axis?
   a. 240.   c. 15.25.   e. 1,861.
   b. 8,370.   d. 236.19.

60. The decision for the test from Question 59 will be
   a. Reject H_0, because 236.19 > 1.96.
   b. Cannot reject H_0, because 236.19 > 1.96.
   c. Reject H_0, because 236.19 > 1.645.
   d. Cannot reject H_0, because 236.19 > 1.645.
   e. Reject H_0 because 8,370 > 0.

61. If you test $\mu_1 - \mu_2 \leq 0$, the estimated standard error, $\hat{\sigma}_{\text{dif}}$, will be
   a. 363.   c. 77.8692.   e. 1,461.
   b. 181.   d. 730.5.

62. The value of the test statistic on the Z axis for testing $\mu_1 - \mu_2 \leq 0$ will be
   a. 7,439.5.   c. 23.90.   e. 2.45.
   b. 1,861.   d. 95.54.

63. When making inferences about the population mean with a small sample ($n \leq 30$),
   a. We require more information about the population in order to gauge the uncertainty of the estimate.
   b. Inferences are less reliable, because we are working with less population information.
   c. We often require the T probability distribution to reflect the increase in uncertainty of inferences.
   d. The standard error will tend to be large to reflect the increased dispersion of sample means about the population mean.
   e. All of the above choices are correct answers.

64. Samples are independent if
   a. The items selected for one sample are not related to items selected for the other.
   b. They are subsets of the same population.
   c. They are subsets of different populations.
   d. They cannot share common items.
   e. We exclude certain population items from one or both samples.

Answers: 59. d; 60. c; 61. c; 62. c; 63. e; 64. a.

Introduction

Viewership Experiment The producer of a local television station's news magazine program grew concerned over the uninteresting themes and boring execution of segments related to local news, culture, and events. Finally, he required all cast and crew members responsible for the program to register for a local history course and a folklore course offered by a nearby college and then to integrate their new experience and related research in future programs. To demonstrate the personal and station benefits of the enhanced stories, the producer reports the following findings. Viewership of the program averaged 265,000 before the change and 410,000 afterward, according to a polling firm. This consulting firm based their findings on earlier work with 12 randomly selected programs and on seven randomly selected programs since the completion of the courses. On a personal level, the producer suggests the cast and crew are more satisfied with their jobs and more confident of their professionalism and future career potential. He cites averages of the following variables for each of the 15 cast and crew members for before and after: job offers, extra time voluntarily spent on each program, and amount of weekly fan mail.

Question: There are plenty of before versus after data here, but what problems do we encounter if we try to use the procedures from Unit I to test for differences in means for each of these variables for the 15 crew members or a difference in mean viewership before and after the course?

Answer: First, the samples are all small. Viewership samples include 12 and 7 observations, and the crew and cast samples include 15 members in each of the six samples (before and after for three different variables). In none of the cases can we employ the large sample ($n > 30$) approximation of the sampling distribution. In addition, the cast and crew samples are dependent (not independent), because the same 15 people comprise each of the six samples of individual enrichment variables.

We could use the Unit I procedure, but we could not be sure the significance level is really the level we state when we announce the decision, because the normal probability distribution does not approximate the behavior of $\overline{X}_1 - \overline{X}_2$ very well for small, dependent samples from unknown population distributions. In this unit we will study the procedures to address the two problems of small samples and dependent samples.

10.3 Test for Small Independent Samples

When the samples are small, the test statistic is still $\overline{X}_1 - \overline{X}_2$. As you might imagine, the smaller amount of information about the population in the samples will increase the uncertainty of the inferential process, $\overline{X}$ values are more dispersed. Consequently, to proceed we will need more information about the population and the T distribution rather than the standard normal. Less population information in the form of smaller sample sizes, n, tends to increase σ_{dif} and diminish the degrees of freedom for the T distribution, both of which expand the acceptance region, making a larger difference necessary to establish a significant difference. *The situation we consider in this section employs small, independent random samples from normal populations.*

Figure 10-6 shows the sampling distribution when we known σ_1^2 and σ_2^2. When we do not know these variances (a common situation), but we do know or have reason to believe the population variances are equal, that is $\sigma^2 = \sigma_1^2 = \sigma_2^2$, then the standard error of the difference becomes

$$\sigma_{\text{dif}} = \sqrt{\frac{\sigma_1^2}{n_1} + \frac{\sigma_2^2}{n_2}} = \sqrt{\frac{\sigma^2}{n_1} + \frac{\sigma^2}{n_2}} = \sqrt{\sigma^2\left(\frac{1}{n_1} + \frac{1}{n_2}\right)}.$$

Figure 10-6
Sampling distribution of $\overline{X}_1 - \overline{X}_2$ for small samples, known population variances, and normal populations

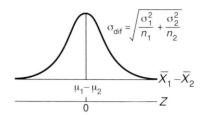

Each of the two samples produces an estimate (S_1^2 and S_2^2) of the common variance, σ^2. To make the best use of all the available information when we estimate σ^2, we pool or combine the two estimates. The operation is really a weighted averaging of the two S^2 values to produce an estimate somewhere between the two.

Question: Samples can differ in size, n. How might sample size affect the weight we attach to each S^2 in the pooling process?

Answer: The larger sample presumably contains more information about the population and, hence, the population variance, σ^2. Thus, we will assign weights to each S^2 that reflect sample size, with larger weight assigned to the larger sample's S^2.

To find the pooled estimator of σ^2 from the weighted mean of the two estimates, we use

$$S_P^2 = \frac{(n_1 - 1) S_1^2 + (n_2 - 1) S_2^2}{(n_1 - 1) + (n_2 - 1)},$$

which becomes

The Pooled Estimator of the Population Variance, σ^2, when $\sigma_1^2 = \sigma_2^2$

$$S_P^2 = \frac{(n_1 - 1) S_1^2 + (n_2 - 1) S_2^2}{n_1 + n_2 - 2}.$$

■ **Example 10-3:** The standard deviations for the samples collected before and after the news program changes in the viewership experiment (p. 465) are 72 thousand viewers and 125 thousand viewers, respectively. (We express the values without the zeros to make the calculations easier.) Recall that the before sample contains 12 values and the after sample contains seven. Assuming the population variance does not change during the two periods, calculate the pooled estimate of the variance, S_P^2. *Remember to use variances in the formula, not standard deviations.*

■ **Solution:**

$$S_P^2 = \frac{(n_A - 1) S_A^2 + (n_B - 1) S_B^2}{n_A + n_B - 2} = \frac{(7 - 1)(125)^2 + (12 - 1)(75)^2}{(7 + 12 - 2)} = 9154.4118.$$

S_P^2 is an estimate of the common population variance, σ^2. We need to substitute it for σ^2 in order to estimate σ_{dif}.

Estimated Standard Error of the Difference when $\sigma_1^2 = \sigma_2^2$

$$\hat{\sigma}_{dif} = \sqrt{S_P^2 \left(\frac{1}{n_1} + \frac{1}{n_2} \right)}.$$

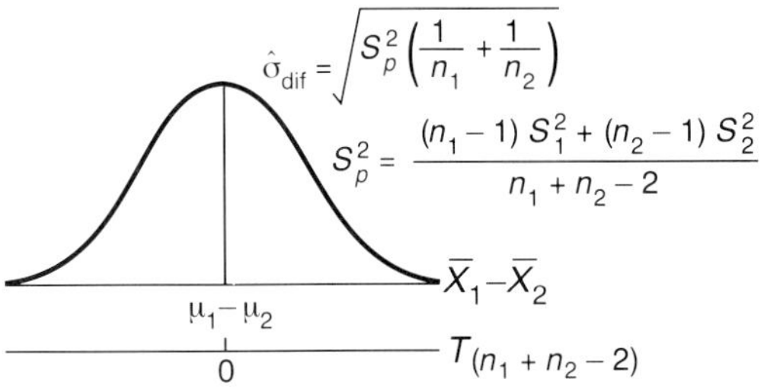

Section 10.3   Test for Small Independent Samples

Figure 10-7
Sampling distribution for $\overline{X}_1 - \overline{X}_2$ for small, independent samples, unknown but equal population variances, and normal populations

$$\hat{\sigma}_{dif} = \sqrt{S_p^2\left(\frac{1}{n_1} + \frac{1}{n_2}\right)}$$

$$S_p^2 = \frac{(n_1 - 1)S_1^2 + (n_2 - 1)S_2^2}{n_1 + n_2 - 2}$$

■ **Example 10-3 (continued):** Continue this problem by finding $\hat{\sigma}_{dif}$. Recall that $n_B = 12$, $n_A = 7$, and $S_P^2 = 9{,}154.4118$.

■ **Solution:** $\hat{\sigma}_{dif} = \sqrt{9154.4118\left(\frac{1}{7} + \frac{1}{12}\right)} = 45.5043$ thousand viewers

Question: Describe the two measures that S_P^2 and $\hat{\sigma}_{dif}$ estimate.

Answer: S_P^2 is an estimate of the common population variance, σ^2. This value gauges the dispersion of the original values in each population. $\hat{\sigma}_{dif}$ estimates the dispersion of the $(\overline{X}_1 - \overline{X}_2)$ values.

When the population variances are equal, the value for degrees of freedom is $n_1 + n_2 - 2$. We think of using $n_1 + n_2$ observations to estimate two parameters, $\overline{X}_1$ and $\overline{X}_2$, so we lose two degrees of freedom in the process. Figure 10-7 summarizes the new situation for testing for a difference in population means when the circumstances include:

1. Small, independent, random samples.
2. Normal populations (or symmetric and unimodal population).
3. Common population variances, $\sigma^2 = \sigma_1^2 = \sigma_2^2$.

The only differences in the new diagram and testing procedure from the large sample procedure are (1) determining the pooled estimator of the common population variance, S_P^2, then (2) substituting it in the formula to estimate the standard error, $\hat{\sigma}_{dif} = \sqrt{S_P^2[(1/n_1) + (1/n_2)]}$, and (3) using the T table (Appendix J) rather than the standard normal table (Appendix H).

■ **Example 10-3 (continued):** Demonstrate that viewership significantly increased after the program changes, using $\hat{\sigma}_{dif} = 45.5043$ thousand viewers that we just calculated along with the previous information: $\overline{X}_B = 265$ thousand viewers, $\overline{X}_A = 410$ thousand viewers, $n_B = 12$ and $n_A = 7$. Past studies indicate that viewership is approximately normally distributed. Test at the 0.01 level.

■ **Solution:** To demonstrate an increase in viewership the hypotheses are H_0: $\mu_A - \mu_B \leq 0$ and H_1: $\mu_A - \mu_B > 0$. The difference in sample mean is $\overline{X}_A - \overline{X}_B = 410 - 265 = 145$ thousand viewers. When we standardize this value we obtain $(145 - 0)/45.5043 = 3.2$. For $\alpha = 0.01$, the critical value is $T_{(7+12-2)} = T_{(17)} = 2.567$. Because $3.2 > 2.567$, we reject H_0, which confirms the producer's claim.

■ **Example 10-4:** A service center wants to warn customers with service contracts that attempting their own repairs often results in more costly repairs. The center selects random samples of repair costs for a certain problem from jobs where the customer tried first and those where the center did all of the work. Fifteen values from repairs involving customer fixes have a mean cost of $27.28 with a standard deviation of $11.45. The other sample of 12 observations has a mean of $18.02 with a standard deviation of $6.35. Does the evidence support the service center's belief at the 0.05

level? The service center believes repair costs for both types are normally distributed and dispersed identically, but the means are different.

■ **Solution:** Because the service center believes that the mean of jobs involving customer repair, μ_C, is larger than the mean of jobs with no customer repair, μ_S, the hypotheses are H_0: $\mu_C - \mu_S \leq 0$ and H_1: $\mu_C - \mu_S > 0$. *Identically dispersed populations imply the population variances are equal,* so

$$S_P^2 = \frac{(15-1)(11.45)^2 + (12-1)(6.35)^2}{15 + 12 - 2} = \$91.1593.$$

Consequently, $\hat{\sigma}_{dif} = \sqrt{91.1593(1/15 + 1/12)} = \$3.6978$. The difference in sample means is $\overline{X}_C - \overline{X}_S = 27.28 - 18.02 = \$9.26$, which converts to $(9.26 - 0)/3.6978 = 2.50$. The decision is to reject H_0, because $2.50 > 1.708$, the critical value from the T table with 25 degrees of freedom and an area of 0.05 in the tail. Thus, the mean cost of jobs that follow customer repair attempts is significantly larger than the mean cost of jobs with no customer repair involved.

Question: What are the populations in this repair problem? Do we have all of the populations we need to determine the cheapest repair method? Explain.

Answer: While we have information about two methods, service center fixes from scratch and service center fixes of customer repairs, we lack information on costs for successful customer repairs. If we did have information for this type of repair, we would need another statistical technique, analysis of variance, to determine if the three population means differ. Chapter 12 covers this technique.

If we do not know that the population variances are equal, then we must consider the information available in the sample standard deviations. There is a hypothesis test that we could perform to test for a difference in population variances, which is beyond the scope of this book. However, we can form a rough rule of thumb based on this test for small independent samples from normal populations that says that the *population variances are not significantly different if the larger sample variance is less than twice as large as the smaller sample variance.*

When we cannot reasonably assume that the population variances are equal, then we calculate $\hat{\sigma}_{dif}$ as we do for large samples:

$$\hat{\sigma}_{dif} = \sqrt{\frac{S_1^2}{n_1} + \frac{S_2^2}{n_2}}.$$

The procedure differs from the large-sample case, because we use a T distribution to reflect the decrease in information. So far, this appears very simple and follows similar concessions we make for estimating and testing for a single μ with small samples. However, the one complication is to determine the degrees of freedom value. Because this calculation can be quite tedious and not very intuitive, we will delay tests for this situation until the section on using the computers, Section 10.5. If you do not have access to a computer, there is a general rule of thumb that produces a conservative (makes it harder to reject H_0) critical value or p-value. This rule says to use the smaller of $n_1 - 1$ or $n_2 - 1$ for the degrees of freedom.

Section 10.3 Problems

65. Test for a difference in the means of two normal populations whose variances are equal with the following information. Use a 0.05 significance level.
$\overline{X}_1 = 450$ $S_1 = 25$ $n_1 = 26$
$\overline{X}_2 = 475$ $S_2 = 35$ $n_2 = 3$.

66. Populations 1 and 2 are normal and dispersed identically. Try to demonstrate that the mean of population 1 is larger than the mean of population 2 with the following information. Use a 0.1 significance level.
$\overline{X}_1 = 36.9$ $S_1 = 3.1$ $n_1 = 18$
$\overline{X}_2 = 35.1$ $S_2 = 4.8$ $n_2 = 10$.

67. One part of a larger study of bulimia in high school girls involved body image dissatisfaction. The researchers used samples of 15 girls identified as bulimic through testing and interviews and 15 who were not. A score on an exercise indicated each girl's dissatisfaction with her body. Two measures resulted, composites of dissatisfaction with distinct body parts. The mean scores for BATH, dissatisfaction with buttocks, abdomen, thighs, and hips, were 4.37 and 3.26 for bulimic and nonbulimic, respectively. The corresponding standard deviations were 0.67 and 0.74. The means for ROB, the measure for the other body parts, were 3.06 and 2.20, respectively, with corresponding standard deviations of 0.65 and 0.50. (Source: C. M. Dacey, W. M. Nelson, III, V. F. Clark, and K. G. Aikman, "Bulimia and Body Image Dissatisfaction in Adolescence," *Child Psychiatry and Human Development*, Vol. 21, No. 3, Spring 1991, pp. 179–84.)

a. Use this information to determine if bulimic high school girls are more dissatisfied with their bodies. Test at the 0.05 level. Assume the populations are normally distributed and population standard deviations are equal. Interpret the decision.

*b. The *p*-values for the two tests we just completed are 0.0001 and 0.0002, respectively. What must α be in order to reject H_0?

68. Random samples of the interest rates on ten different days from before and ten different days after a major shift in Federal Reserve policy reveal an average rate of 10.1% with a standard deviation of 2.4 percentage points before the change and an average rate of 8.2% with a standard deviation of 1.8 percentage points after the change. Previous research indicates that interest rates tend to be normally distributed. Use this information to determine if the average interest rate is indeed lower after the policy shift. Test at the 0.01 significance level.

10.4 Paired Differences (Dependent Samples)

When samples are not independent, then one sample's values contain information about the other sample's values. When we select items in one sample to match or associate one-for-one with observations in the other sample, we can perform a paired-difference test. For example, we might collect information from the same people before and after a training exercise. These observations are paired and useful for determining if the means of the before and after values differ.

This dependency complicates the calculation of $\hat{\sigma}_{dif}$. We avoid this complication by forming a single sample that consists of difference values, D, of the paired values from the two samples. If B and A represent the first person's performance before and after training, respectively, then $D = B - A$ (or $D = A - B$). We treat each D as we would an X in an ordinary sample and test for a mean of the differences, μ_D, following the procedures from Chapter 9 for testing hypotheses about the mean of a single population.

■ **Example 10-5:** Eight randomly selected students complete a grammar exercise before receiving an explanation of proper grammar for the situations on the quiz. Afterward, they retake a quiz on the same situations. The before and after quiz scores appear below. Determine the paired differences, D, for this sample. Then calculate the mean and standard deviation of the sample of D values, $\overline{D}$ and S_D.

Student		1	2	3	4	5	6	7	8
Scores	Before	25	32	16	17	16	20	25	24
	After	30	35	30	17	22	19	29	29

■ **Solution:** The D column is the difference in the before and after values, $D = B - A$. The mean of the D values is the sum of the D values divided by 8, the number of students, or D values, in the sample, $\overline{D} = -36/8 = -4.5$. We use the formula for the sample standard deviation from Chapter 3 or a calculator to find the standard deviation of the eight values, $S_D = \sqrt{146.00/(8-1)} = 4.5670$, or 4.6. The following table shows the calculation for the D values as well as intermediate steps for $\overline{D}$ and S_D.

Figure 10-8
Sampling distribution of $\overline{D}$

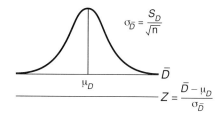

Student	Scores Before	Scores After	D	$D - \overline{D}$	$(D - \overline{D})^2$
1	25	30	−5	−0.5	0.25
2	32	35	−3	1.5	2.25
3	16	30	−14	−9.5	90.25
4	17	17	0	4.5	20.25
5	16	22	−6	−1.5	2.25
6	20	19	1	5.5	30.25
7	25	29	−4	0.5	0.25
8	24	29	−5	−0.5	0.25
			−36		146.00

We will use the sample of eight D values to perform tests for the mean of the population of D values, μ_D, the mean for all students who engage in the before and after grammar training experiment. *Remember, the direction of the test for D must be consistent with the order of calculating D.* If $D = B - A$ (where B = before and A = after), then $D < 0$ is the same as $B - A < 0$. If the grammar analyst believes the mean quiz score increases after the explanation, then the hypotheses are H_0: $\mu_D \geq 0$ and H_1: $\mu_D < 0$. If we reverse the order to $A - B$, H_0 would be $\mu_D \leq 0$ and H_1 would be $\mu_D > 0$ for the grammar analyst.

As we continue, think of each D as an X in an ordinary sample. The test statistic, $\overline{D}$, is an estimate of μ_D, just as $\overline{X}$ is an estimate of μ. The sample standard deviation, S_D, estimates σ_D, the standard deviation of the population of D values for all student subjects in such an experiment. Then to estimate $\sigma_{\overline{D}}$, the standard error of the $\overline{D}$ values, we use $S_D/\sqrt{n}$, where n is the number of pairs. Figure 10-8 displays the large sample situation for a paired-difference test.

■ **Example 10-5 (continued):** Use the sample of D values to determine if mean quiz scores do increase after instruction. The hypotheses are H_0: $\mu_D \geq 0$ and H_1: $\mu_D < 0$, where D represents individual D values, before minus after score. Use $\overline{D} = -4.5$ and $S_D = 4.6$. Test at the 0.05 level.

■ **Solution:** First, notice that we are testing for the population mean with a small sample and unknown standard deviation. We need a normal (or symmetric, unimodal) population of D values to use the T probability distribution. *The D values will be normal if the before and after values are each normally distributed.* The degrees of freedom will be $n - 1$, which is $7 = 8 - 1$ in this test. $\hat{\sigma}_{\overline{D}} = S_D/\sqrt{n} = 4.6/\sqrt{8} = 1.6263$, so $\overline{D}$ becomes $(-4.5 - 0)/1.6263 = -2.7669$ or -2.77 on the T axis. The critical value is $T_{(7)} = -1.895$. We reject H_0, because $-2.77 < -1.895$. Thus, the data support the grammar analyst's belief that the mean score increases after instruction.

Section 10.4 Paired Differences (Dependent Samples)

■ **Example 10-6:** A veterinary research center randomly selects five breeds of dogs for an experiment to determine if a special dietary supplement affects canine red blood cell count. Next, the researchers randomly select two dogs of each breed, one receives the supplement and the other does not. The change in red blood cell count between the beginning of the experiment and 72 hours later is recorded for each dog. Explain why this situation is a paired-difference problem. Then use the following data to test at the 0.01 level for any difference in the red blood cell count due to the dietary supplement.

	Supplement	
Pedigree	With	Without
1	100	70
2	−20	30
3	220	0
4	360	250
5	50	0

■ **Solution:** If red blood cell count may vary with breed (some breeds may have a high count whether they receive the supplement or not), then the samples are dependent. The researchers design the experiment so that, once they select a breed of dog, there will be a dog of that breed in each sample. Thus, the samples contain paired observations.

The hypotheses are H_0: $\mu_D = 0$ and H_1: $\mu_D \neq 0$. The data and calculations follow. Using $D = X_{with} - X_{without}$, $\overline{D} = 360/5 = 72$. $S_D = \sqrt{40,480/(5-1)} = 100.5982$, so $\hat{\sigma}_{\overline{D}} = 100.5982/\sqrt{5} = 44.9889$.

	Supplement				
Pedigree	With	Without	D	$D - \overline{D}$	$(D - \overline{D})^2$
1	100	70	30	−42	1,764
2	−20	30	−50	−122	14,884
3	220	0	220	148	21,904
4	360	250	110	38	1,444
5	50	0	50	−22	484
			360		40,480

The critical values are ±4.604 from the T distribution with 4 degrees of freedom, assuming the populations are normal or approximately so. Converting $\overline{D} = 72$ to a T value produces $(72 - 0)/44.9889 = 1.600$, which is between the critical values, so we cannot reject H_0. That is, the data do not support the hypothesis that the supplement affects the red blood cell count.

■ **Example 10-3 (continued):** The television news producer cites the averages for each of the following three variables for the 15 cast and crew members for before and after their training: job offers, extra time voluntarily spent on each program, and amount of weekly fan mail. He contends that the cast and crew are better off on average for all three measures, because the means of the differences for each employee are positive, as postulated, and the upper-tailed p-values are 0.043, 0.01, and 0.038, respectively. Do the data support the producer's claims at the 0.05 level? Assume normal populations. Explain.

■ **Solution:** The p-values are all less than $0.05 = \alpha$, which indicates support for his claims, so the producer can reject the null hypothesis in all three cases.

Section 10.4 Problems

69. Use the following data to test $H_0: \mu_D \leq 0$ versus $H_1: \mu_D > 0$ at the 0.05 significance level. Assume the populations are normal.

Experimental Subject	Before Measure	After Measure
1	25	26
2	49	43
3	74	69
4	15	11
5	48	41
6	78	77

70. Use the following data to test for a difference due to the treatment at the 0.01 significance level. Assume the populations are normal.

Experimental Subject	Before Treatment	After Treatment
1	2.85	2.60
2	7.49	8.43
3	7.40	6.99
4	1.51	3.11
5	4.85	4.13

71. Many chemical reactions occur more rapidly when the chemicals are heated. Unlike conventional methods, microwave dielectric heating causes some solids and liquids to transform electromagnetic energy into heat, thus opening the possibility of faster reactions. To investigate this possibility, chemists compared results of conventional and microwave heating on the reaction that causes the change of pinacol into pinacolone on a phyllosilicate solid support. The experiment includes results for six different phyllosilicate solid supports, heated conventionally for 15 hours at 100°C or using microwaves at 450 watts of power for 15 minutes. The times were chosen to realize optimum yields. Use the results that follow to test at the 0.1 level for a significantly higher percentage of conversion with microwave heating. (Source: B. P. Barnsley, L. Reilly, J. Jones, and J. Eshman, "First Australian Symposium on Microwave Power Applications," Wollongong, 1989, p.49, and E. Gutierrez, A. Loup, G. Bram, and E. Ruiz-Hitzky, *Tetrahedron Letters*, Vol. 30, 1989, p. 945, both reported in D. Michael P. Mingos and David R. Baghurst, "Applications of Microwave Dielectric Heating Effects to Synthetic Problems in Chemistry," *Chemical Society Reviews*, Vol. 20, No. 1, 1991, pp. 1–47.)

Phyllosilicate Solid Support	Pinacol changed to Pinacolone (%)	
	Conventional Method	Microwave Technique
Na^+	5	38
Ca^{2+}	2	23
Cu^{2+}	30	94
La^{3+}	80	94
Cr^{3+}	99	98
Al^{3+}	98	99

72. The manager of the secretarial pools questions the test results that indicate that the new software increases productivity, because factors other than the software may have influenced the productivity measures. For instance, the computer pool may have had a higher mean before and after the experiment, or different outputs may reflect different workloads. The manager wants to compare the mean gain in productivity per worker regardless of pool. Each worker in a random sample drawn from secretaries in each pool would have a gain value equal to the post-experiment output minus the pre-experiment output (both will receive the new software). Use the following results for 12 secretaries to determine if the software increases mean output. Test at the 0.01 level, and assume the populations are normal.

a. Perform the test using critical values.

*__b.__ Calculate an approximate *p*-value, and use it to perform the test.

Secretary	Output Week before Receiving Software	Output after Software
1	54	88
2	125	89
3	56	91
4	72	112
5	48	63
6	91	90
7	55	55
8	111	98
9	97	120
10	90	131
11	75	95
12	84	104

10.5 The Computer and Tests for Differences in Means

We can use the computer to make the cumbersome calculations for tests for difference between means, such as the standard errors or the degrees of freedom when the population variances differ. The emphasis shifts to our understanding and interpreting what the computer accomplishes for us.

10.5.1 Independent Samples

Comparing means is a common statistical procedure, so Minitab has a routine that performs this task for independent samples. The Minitab user specifies which of the two assumptions regarding population variances, equal or unequal, to use. Figure 10-9 shows Minitab output from both assumptions for the situation in Example 10-7. The reader would not know which assumption applies without being told or without noticing the pooled estimator printed at the bottom of the computer output. The boxed numbers in Figure 10-9 refer to the following:

1. Number of observation in each sample, n.
2. Sample means, $\overline{X}$.
3. Sample standard deviations, S.
4. Estimated standard error of the mean for each population, $\hat{\sigma}_{\overline{x}}$.
5. Value of the test statistic when testing for a difference of 0.
6. Type of test:

   NE = population means are not equal (two-tailed test),
   LT = first listed population's mean is less than the mean of the second listed population (lower-tailed test),
   GT = first listed population's mean is greater than the mean of the second listed population (upper-tailed test).

7. P-value for testing for difference of 0 (direction of test indicated in item 6).
8. Degrees of freedom for test statistic.
9. Pooled estimator of population standard deviation, S_P (notice this is not the variance, S_P^2), for test when $\sigma_1^2 = \sigma_2^2$.
10. Confidence interval for $\mu_1 - \mu_2$.

■ **Example 10-7:** An automobile dealer believes that drivers who lease cars drive them more the first year than do drivers who buy a car. To check his hypothesis, he collects independent random samples from both groups when they return their cars for annual servicing. He enters the result in the computer, and Figure 10-9 results. Use the Minitab output to find the following:

```
a) Equal Population Variances

    TWOSAMPLE T FOR BUY VS LEASE    3      4
          1   N    2   MEAN      STDEV    SE MEAN
    BUY       44     15167       4385      661
    LEASE     32     17438       4507      797
                                             10
    95 PCT CI FOR MU BUY - MU LEASE: (-4325, -217)

                           6     5        7       8
    TTEST MU BUY = MU LEASE (VS NE): T= -2.20 P=0.0307 DF= 74
                     9
    POOLED STDEV =       4437


b) Unequal Population Variances

    TWOSAMPLE T FOR BUY VS LEASE    3      4
          1   N    2   MEAN      STDEV    SE MEAN
    BUY       44     15167       4385      661
    LEASE     32     17438       4507      797
                                             10
    95 PCT CI FOR MU BUY - MU LEASE: (-4339, -202)

                           6     5        7       8
    TTEST MU BUY = MU LEASE (VS NE): T= -2.19 P=0.0318 DF= 65
```

Figure 10-9

Minitab output for comparing population means for mileage of leased versus purchased cars

a. Whether the samples are large or small.
b. The point estimate of the difference in means, $\mu_L - \mu_B$ (where the subscript L represents lease and B represents buy).
c. $\hat{\sigma}_{dif}$.
d. The standardized value of the test statistic.
e. The dealer's decision for his hypothesis at the 0.05 significance level. Interpret the decision.
f. The p-value and the decision for a two-tailed test at the 0.01 level. Interpret the decision.

■ Solution: a. The sample of owned car mileages consists of 44 observations and that of leased car mileages consists of 32 observations. Both contain more than 30 observations, which we consider large. Thus, we will conduct the test for large, independent samples from Unit I.
b. $\overline{X}_L - \overline{X}_B = 17{,}438 - 15{,}167 = 2{,}271$ miles.
c. $\hat{\sigma}_{dif} = \sqrt{(\sigma_{\bar{x}_L})^2 + (\sigma_{\bar{x}_B})^2} = \sqrt{(797)^2 + (661)^2} = 1{,}035.4371$ miles. Notice that *computations for the standard error of the difference are the same for large, independent samples and for small, independent samples when the population variances are unequal.*
d. The standardized value is $2{,}271/1{,}035.4371 = 2.19$. This value corresponds to the circled 5 in Figure 10-9b (the signs differ, because Minitab used $\mu_B - \mu_L$, but we use $\mu_B - \mu_L$).
e. The hypotheses are H_0: $\mu_L - \mu_B \leq 0$ and H_1: $\mu_L - \mu_B > 0$. The critical value (from the Z table because both samples are large) is 1.645 at the 0.05 level. The decision is to reject H_0, because $2.19 > 1.645$. The evidence indicates that the mean first-year mileage on leased cars is, indeed, greater than that on purchased cars.
f. The two-tailed p-value is $0.0318 > 0.01$, so we cannot reject H_0. The evidence does not support any difference in population means at the 0.01 significance level. The decisions in e and f differ, because e is a one-tailed test at the 0.05 significance level and f is a two-tailed test at the 0.01 level. It is more difficult to reject in the second case.

Figure 10-10

Minitab output for repair costs example

```
TWOSAMPLE T FOR C VS S
      N       MEAN      STDEV    SE MEAN
C    15       27.3       11.5       2.96
S    12      18.02       6.35       1.83

95 PCT CI FOR MU C - MU S: (1.640, 16.88)

TTEST MU C = MU S (VS NE): T=2.50 P=0.019 DF=25

POOLED STDEV =            9.581
```

```
TWOSAMPLE T FOR C VS S
      N       MEAN      STDEV    SE MEAN
C    15       27.3       11.5       2.96
S    12      18.02       6.35       1.83

95 PCT CI FOR MU C - MU S: (2.042, 16.47)

TTEST MU C = MU S (VS NE): T=2.66 P=0.014 DF=22
```

Box 10-3 Hog Socialization

Credible agricultural research on hogs depends on controlling factors that can influence experimental results. The existence of a relation between social conditions and eating could destroy many studies of weight gain, an important segment of hog research. If pigs eat and gain weight differently when penned as a group from when penned separately, then experiments must control for this factor. Thus, the living conditions for all experimental pigs need to be the same or the pigs should experience randomly assigned living conditions. Analysts usually describe their efforts to control other factors that may affect the experimental outcomes when they report and interpret the outcomes. In fact, if all pigs experience the same condition, the direct population being studied is more narrow than all pigs or than pigs living with a different number of other pigs. Group living conditions are the most likely occurrence on an actual farm. However, penning pigs separately facilitates experimentation, because analysts can easily measure each pig's consumption, a difficult task when they pen the pigs as a group.

To test the importance of this grouping factor, pigs were randomly assigned to samples to experience a series of different living conditions (alone or in groups of four in one of three trials), while controlling sample characteristics for weaning, sex, age, and parentage. A comparison of mean body weights and the associated p-value for tests for the three trials are shown here ($\overline{X}$ is the mean body weight, the G subscript denotes grouped pigs, and the A subscript stands for pigs penned alone).

Trial	$\overline{X}_G - \overline{X}_A$	p-Value
1	−1.02	0.1730
2	−2.00	0.0185
3	−0.77	0.0975

From these mixed findings, the analyst concludes that there is no evidence that penning a pig separately is detrimental to its eating habits or weight gain compared to grouped penning. Thus, analysts can easily measure food consumption of individually penned pigs without unfavorably affecting weight gain measures and then apply the results to realistic farm conditions.

Source: Maria E. Jamtgaard, "Porking Out!" Stats, No. 5, Spring 1991, pp. 14–15. Copyright 1991 by the American Psychological Association. Reprinted by permission.

■ **Example 10-4 (continued):** Figure 10-10 displays the computer output of the service center versus consumer repair costs. Compare the test statistic and degrees of freedom values in the output with the computations in Section 10.3 (p. 467). Although the output is for a two-tailed test, in which case would $H_0: \mu_C - \mu_S \leq 0$ receive the most support, when the population variances are equal or unequal?

■ **Solution:** The computations agree with the T and degrees of freedom output. A negative or small positive test statistic value supports H_0. Smaller values offer more support, so the 2.50 from the equal variance assumption, smaller than the 2.66 from assuming unequal variances, is the most supportive case for H_0. Alternatively, larger p-values reflect a more credible H_0. The T values are both positive (agree with H_1), so we can compare the two-tailed p-values as well as using half of the two-tailed p-values. In either case, the result is larger when we assume the variances are equal, so this is the most supportive case for H_0.

The difference in pigs' eating habits, based on how many pigs share the same dinner table or trough, is the subject of Box 10-3. This box offers another important application of the testing procedure.

10.5.2 Paired Differences

Figure 10-11 shows output for testing for a mean difference with the paired-difference data of Example 10-5 from Section 10.4 regarding the before and after grammar scores.

```
TEST OF MU = 0.00 VS MU N.E. 0.00
         N       MEAN      STDEV     SE MEAN        T      P VALUE
DIF      8      -4.50       4.57        1.61    -2.79       0.027
```

Figure 10-11

Minitab output for paired-difference test for before and after scores

We must use computation commands to have the program compute individual differences, D, then perform a T test on the differences as if they were from a single sample. Hence, the output is familiar from Chapter 9.

Question: If the D values were after scores minus before scores, instead, which information in the computer output would change?

Answer: All of the D values would change sign but not magnitude, so $\overline{D}$ would be 4.50 rather than -4.50, and the T statistic would be 2.79, rather than -2.79. Now we simply perform the test on the opposite end of the curve, but the conclusion would stand. The other statistics remain the same, because they depend on magnitudes and distances, not directions or signs.

Question: What is the minimum level of significance that results in rejecting H_0 for a two-tailed test?

Answer: If $\alpha \leq 0.027 = p$-value, then reject H_0.

Question: Using the p-value shown in Figure 10-11, what is the result of a two-tailed test at the 0.03 level?

Answer: Reject, $0.027 < 0.03$.

Section 10.5 Problems

73. Figure 10-12 displays a Minitab output for a test for a difference in the mean price change of stocks classified as volatile (their prices changed by at least 3%) versus the mean change of the nonvolatile stocks. Test at the 0.05 significance level. Assume the populations are normal. (Source: Data taken from Associated Press Data reported in *USA Today*, October 9, 1992, pp. 4B–5B.)
    a. What is the value of $\overline{X}_{\text{Volatile}} - \overline{X}_{\text{Nonvolatile}}$?
    b. Assume the population variances are equal and perform a two-tailed test with critical values. Interpret the decision. Do the variances meet the rough rule-of-thumb requirement?
    c. Assume the population variances are unequal and perform the test. Does your decision change? If so, interpret the new decision.
    *d. Repeat Parts b and c using the p-values. The T values of the test statistic are about the same in both cases (0.34 versus 0.30), yet the p-value for the unequal variances case is larger. Why?

74. Figure 10-13 shows the Minitab computations for a test for difference in the mean expenditure on repairs for cars under an extended warranty versus repairs for cars where owners are responsible for all of the expenses. Use this information to demonstrate that the mean expenditure is

Figure 10-12
Minitab output to test for difference in mean stock price change

```
TWOSAMPLE T FOR VOLATILE VS NOTVOL
             N      MEAN      STDEV     SE MEAN
VOLATILE     8      0.297     0.984     0.35
NOTVOL      11      0.182     0.498     0.15

95 PCT CI FOR MU VOLATILE - MU NOTVOL: (-0.61, 0.84)

TTEST MU VOLATILE = MU NOTVOL (VS NE): T= 0.34  P=0.74  DF= 17

POOLED STDEV =            0.738




TWOSAMPLE T FOR VOLATILE VS NOTVOL
             N      MEAN      STDEV     SE MEAN
VOLATILE     8      0.297     0.984     0.35
NOTVOL      11      0.182     0.498     0.15

95 PCT CI FOR MU VOLATILE - MU NOTVOL: (-0.74, 0.97)

TTEST MU VOLATILE = MU NOTVOL (VS NE): T= 0.30  P=0.77  DF= 9
```

```
TWOSAMPLE T FOR SELFPAY VS WARRANTY
              N       MEAN      STDEV    SE MEAN
SELFPAY     100        739        397       40
WARRANTY    100       1776        656       66

95 PCT CI FOR MU SELFPAY - MU WARRANTY: (-1189, -886)

TTEST MU SELFPAY = MU WARRANTY (VS LT): T= -13.53  P=0.0000  DF= 162


TWOSAMPLE T FOR SELFPAY VS WARRANTY
              N       MEAN      STDEV    SE MEAN
SELFPAY     100        739        397       40
WARRANTY    100       1776        656       66

95 PCT CI FOR MU SELFPAY - MU WARRANTY: (-1189, -887)

TTEST MU SELFPAY = MU WARRANTY (VS LT): T= -13.53  P=0.0000  DF= 198

POOLED STDEV =             542
```

Figure 10-13

Minitab computations for test of difference in mean repair expenditure for older vehicles

larger for vehicles under warranty at the 0.01 significance level.

75. Figure 10-14 displays Minitab output for an analysis of the mean number of Democrats versus the mean number of Republicans in Western North Carolina counties. Part a shows the actual data as well as descriptive statistics for the two populations. Minitab used random samples with replacement (Democrats = DMRNDSMP and Republicans = RPRNDSMP) from the actual data in the test computations in Part b. Part c shows the result of a paired-difference test for a random sample of five counties values for Democratic and Republican registrants. (Source: Data from counties' Boards of Elections and U.S. Census, as reported in "Figures In, Number Up," *Asheville Citizen-Times*, October 15, 1992, p. 9A.)

   a. State the alternative hypotheses for these tests. Study the actual data to determine if H_0 or H_1 is true. Do you think assuming normal populations is reasonable here?

   b. Perform the test at the 0.05 significance level using all three sets of computations. Interpret the decisions.

   c. Can you explain the discrepancy in the decisions with the sampling method (independent versus paired) after studying the actual data?

76. Figure 10-15 displays the Minitab computations to determine if the average high temperature varied from one day to the next. Part a shows an analysis based on samples of high temperatures from randomly selected cities about the nation on one day or the next. Part b shows an analysis based on a comparison of the temperatures on the two days for 15 randomly selected cities. Use this information to conduct tests for a difference at the 0.05 significance level. (Source: Data from Weather Services Corporation, reported in "Four-Day Highlights," *USA Today*, October 9, 1992, p. 14A.)

```
a. Actual Data and Descriptive Statistics for Populations of Registered
   Democrats and Republicans

   ROW   DEMOCRAT   REPUBLIC       ROW   DEMOCRAT   REPUBLIC
    1      1820       7380          10     8372       6562
    2     64746      36469          11    12993       5480
    3     23094      15304          12     1738       8188
    4      7690       5697          13     5506       4409
    5      2617       2426          14    19651       8217
    6     21186       7731          15     9148       6751
    7     18113      23568          16    12067      11982
    8      2856       2578          17     6663       4715
    9     10744       5402

                   N      MEAN    MEDIAN    TRMEAN    STDEV    SEMEAN
   DEMOCRAT       17     13471      9148     10835    14880     3609
   REPUBLIC       17      9580      6751      8264     8611     2088

                 MIN      MAX        Q1        Q3
   DEMOCRAT     1738    64746      4181     18882
   REPUBLIC     2426    36469      5059     10100
```

Figure 10-14

Minitab computations for test of difference in mean number of Republicans and Democrats in Western North Carolina Counties

(continued on next page)

Figure 10-14 (continued)
Minitab computations for test of difference in mean number of Republicans and Democrats in Western North Carolina Counties

b. Computations for Random Samples from Each Party's Registrants

```
TWOSAMPLE T FOR DMRNDSMP VS RPRNDSMP
           N      MEAN     STDEV    SE MEAN
DMRNDSMP   5      8145     8678     3881
RPRNDSMP   5     13611    13474     6026

95 PCT CI FOR MU DMRNDSMP - MU RPRNDSMP: (-23009, 12077)

TTEST MU DMRNDSMP = MU RPRNDSMP (VS GT): T= -0.76 P=0.76 DF= 6


TWOSAMPLE T FOR DMRNDSMP VS RPRNDSMP
           N      MEAN     STDEV    SE MEAN
DMRNDSMP   5      8145     8678     3881
RPRNDSMP   5     13611    13474     6026

95 PCT CI FOR MU DMRNDSMP - MU RPRNDSMP: (-21999, 11067)

TTEST MU DMRNDSMP = MU RPRNDSMP (VS GT): T= -0.76 P=0.77 DF= 8

POOLED STDEV =     11333
```

c. Computations for Paired-Difference Test for Random Sample of Five Counties

```
TEST OF MU =    0.000 VS MU G.T.    0.000

            N      MEAN      STDEV    SE MEAN      T    P VALUE
DEM-REP     5    5273.400   4490.379  2008.159   2.63    0.029
```

Figure 10-15
Minitab computations for high-temperature analysis

a. Independent Random Samples of High Temperatures for Two Days

```
TWOSAMPLE T FOR YESTERDY VS TODAY
           N      MEAN     STDEV    SE MEAN
YESTERDY  24     71.79     8.79      1.8
TODAY     36     68.7     10.5       1.8

95 PCT CI FOR MU YESTERDY - MU TODAY: (-2.1, 8.3)
TTEST MU YESTERDY = MU TODAY (VS NE): T= 1.18 P=0.24 DF= 58
POOLED STDEV =     9.88


TWOSAMPLE T FOR YESTERDY VS TODAY
           N      MEAN     STDEV    SE MEAN
YESTERDY  24     71.79     8.79      1.8
TODAY     36     68.7     10.5       1.8

95 PCT CI FOR MU YESTERDY - MU TODAY: (-2.0, 8.1)
TTEST MU YESTERDY = MU TODAY (VS NE): T= 1.22 P=0.23 DF= 54
```

b. Paired Differences of Two Days' High Temperatures for Random Sample of Cities

```
TEST OF MU = 0.000 VS MU N.E. 0.000

            N      MEAN     STDEV    SE MEAN      T    P VALUE
HIGHPAIR   15    -4.133     7.736    1.997     -2.07    0.057
```

Summary and Review

This unit presents the proper procedures for small, independent, random samples and for dependent samples where values in one sample are paired with those of the other sample. Most differences between the small and large independent sample tests center on the standard error of the difference, σ_{dif}, and the appropriate degrees of freedom to use with the T distribution. In the case of pairs, once we form a sample of differences from the pairs, we conduct paired-difference tests just like single sample tests in Chapter 9.

The flowchart in Figure 10-16 shows the different aspects of the tests described in this chapter. The flowchart assumes that population variances are unknown, a common occurrence, and small sample information assumes normal (or symmetric and unimodal) populations. When the population variances are known, we can use the normal distribution for large independent samples or for small independent samples from normal populations. We would simply substitute the known variances for the sample variances in the formula for the standard error of the difference, $\sigma_{dif} = \sqrt{(\sigma_1^2/n_1) + (\sigma_2^2/n_2)}$.

Multiple-Choice Problems

77. Independent, random samples for a test for differences in the mean times that highways are partially or wholly barricaded because of construction for last year versus five years ago contain 12 and 8 observations. There is evidence that the standard deviation of these times did not change between the two sampled years, although the sample standard deviations are 10 and 15 days, respectively. The corresponding sample means are 53 and

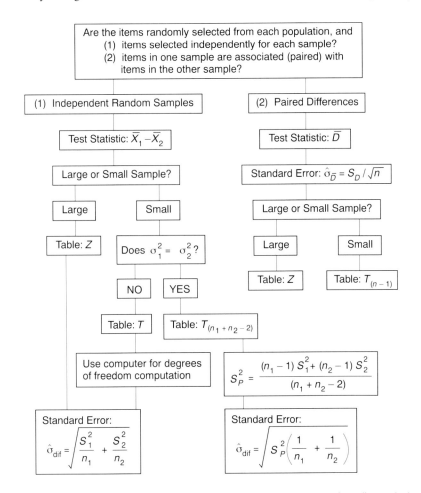

Figure 10-16
Flowchart for testing for difference between population means

Note: Flowchart assumes populations' variances are unknown and small samples' values originate in normal or approximately normal populations.

47 days. Which of the following changes should you make in the testing method used for a large sample test in order to test the small-sample data correctly?
  a. Use $T_{(18)}$, rather than Z.
  b. Use $T_{(11)}$, rather than Z.
  c. Use 148.6111 in $\hat{\sigma}_{dif}$.
  d. Values from the two samples must be paired before we proceed.
  e. $\hat{\sigma}_{dif}$ is 148.6111.
  f. $\hat{\sigma}_{dif}$ is 6.0381.
  g. a and c are correct.
  h. b and f are correct.

78. If both samples for the situation in Problem 77 were to contain the same highway stretches, what changes should you make in the testing method?
  a. Use $T_{(18)}$, rather than Z.
  b. Use $T_{(11)}$, rather than Z.
  c. Use 148.6, rather than 100 and 225, in $\hat{\sigma}_{dif}$.
  d. Values from the two samples must be paired before we proceed.
  e. $\hat{\sigma}_{dif}$ is 148.6111.
  f. $\hat{\sigma}_{dif}$ is 6.0381.
  g. a and c are correct.
  h. b and f are correct.

79. An engineer plans to use the mean outputs of 100 machines before and after lubrication to test for a difference caused by lubrication. Which of the following causes a problem for using $\overline{X}_A - \overline{X}_B$ as the test statistic?
  a. The samples are too large.
  b. The samples are too small.
  c. The population standard deviations are unknown.
  d. There is no real difference in the means.
  e. The samples are not independent.

Figure 10-17 shows a computer output for comparing the means of outstanding consumer credit as a percentage of personal income based on independent, random samples from two months. Use this information to answer Questions 80 and 81.

80. What is the point estimate of $\mu_1 - \mu_2$?
  a. 1.101.
  b. −1.101.
  c. 12.776.
  d. 11.675.
  e. 0.464.

81. Is the difference between the means significant at the 0.2 level?
  a. No, because 1.1101 is very small.
  b. Yes, because 1.1101 > 0.
  c. No, because −1.1101 < 0.
  d. No, because 6.08 > 1.28.
  e. Yes, because −6.08 < −1.28.

82. The weights employed in the pooled estimator of a common population variance
  a. Are inversely related to sample sizes.
  b. Reflect the amount of population information included in each separate estimate of the population variance.
  c. Actually decrease the overall sample size and degrees of freedom used in the test.
  d. Tend to increase the variance estimate to reflect the increased dispersion from combining two independent random variables.
  e. Give heavier weight to smaller samples.

Use the following information to answer Questions 83–92.

To examine seed production of the Atamasco Lily, botanists studied a random sample from 200 plants in Georgia. Some flowers were covered with bridal veil to keep pollinators out, so the botanists could control the method of pollination, self- or cross-pollination. Twenty self-pollinated flowers produced a mean of 19.4 seeds with a standard deviation of 9.8 seeds. Twenty-three cross-pollinated flowers produced a mean of 24.7 seeds with a standard deviation of 14.0 seeds. (Source: Steven B. Broyles and Robert Wyatt, "The Breeding System of *Zephyranthes atamasco* (Amaryllidaceae)," *Bulletin of the Torrey Botanical Club*, Vol. 118, No. 2, 1991, pp. 137–40.

83. Estimate the difference in mean seed production from the two pollination methods. Use cross-pollination as the first population.
  a. 5.3.   d. 4.2.
  b. 24.7. e. −4.2.
  c. 19.4.

84. What is H_1 for a test to determine if the type of pollination makes a difference (S = self-pollinated, C = cross-pollinated)?
  a. $\mu_C - \mu_S = 0$.
  b. $\mu_C - \mu_S \neq 0$.
  c. $\mu_C - \mu_S \geq 0$.
  d. $\mu_C - \mu_S \leq 0$.
  e. $\mu_C - \mu_S < 0$.
  f. $\mu_C - \mu_S > 0$.

Figure 10-17
Minitab output comparing mean credit-income ratio for two months

```
TWOSAMPLE T FOR MONTH1 VS MONTH2
           N     MEAN    STDEV   SE MEAN
MONTH1    31    11.675   0.906    0.16
MONTH2    31    12.776   0.442    0.079

95 PCT CI FOR MU MONTH1 - MU MONTH2: (-1.47, -0.736)

TTEST MU MONTH1 = MU MONTH2 (VS NE):  T=-6.08  P=0.0000  df= 44
```

Multiple-Choice Problems

85. If the variances are unequal and $\hat{\sigma}_{dif}$ is 3.6502, what is the standardized value of the test statistic?
   a. 4.2.
   b. 5.3.
   c. 1.4169.
   d. 1.4520.
   e. 5.1105.

86. Because the degrees of freedom value exceeds 30, we will approximate the critical T value with a value from the Z table. What is the decision for a test to determine if the type of pollination makes a difference at the 0.01 level?
   a. Reject H_0, because $5.3 \neq 0$.
   b. Reject H_0, because $1.45 \neq 0$.
   c. Cannot reject H_0, because 1.45 is between ± 2.33.
   d. Reject H_0, because 1.45 is between ± 2.575.
   e. Cannot reject H_0, because 1.45 is between ± 2.575.

87. If the population variances are equal, estimate the value of this variance.
   a. 149.6771.   d. 12.2343.
   b. 3.6502.   e. 3.7405.
   c. 13.3237.

88. If the population variances are equal, find $\hat{\sigma}_{dif}$.
   a. 149.6771.   d. 12.2343.
   b. 3.6502.   e. 3.7405.
   c. 13.3237.

89. If the population variances are equal, what is the standardized value of the test statistic?
   a. 4.2.   d. 1.4520.
   b. 5.3.   e. 5.1105.
   c. 1.4169.

90. If the population variances are equal, what is the value for degrees of freedom?
   a. 20.   d. 43.
   b. 23.   e. 41.
   c. 39.

91. We are unable to reject H_0 for a two-tailed test at the 0.05 level, whether the population variances are equal or not. Interpret the decision not to reject H_0.
   a. Cross-pollination increases the mean seed production.
   b. Self-pollination increases the mean seed production.
   c. Type of pollination affects mean seed production, but we cannot be sure which is better.
   d. Type of pollination does not affect mean seed production.

*92. The botanists performed both cross- and self-pollination on flowers on the same plant when possible to allow pairwise comparisons. The result from 12 pairs is a standardized test statistic of 0.89 and a two-tailed p-value of 0.3925. Does this information change the previous decisions that we could not reject H_0 at the 0.05 level?
   a. No, because 0.89 is between ± 1.96.
   b. No, because 0.89 is between ± 2.228.
   c. No, because 0.89 is between ± 2.201.
   d. Yes, because p-value $= 0.3925 > 0.05$.
   e. Yes, because half of the p-value $0.1963 > 0.05$.

Use the following information to answer Questions 93–98.

Figure 10-18 displays output for testing for a paired difference between the mean years in a state legislature for state representatives in the 1993 session and the 1990 session. The differences are paired observations from 100 legislative districts. Years in the legislature equals 1993 minus first year of election for each of the representatives of the 100 districts in the 1993 session and 1990 minus first year of election in the 1990 session.

93. What is $\overline{D}$?
   a. 9.66.   d. 0.04.
   b. 2.01.   e. 50.
   c. 2.08.

94. What is S_D?
   a. 0.966.   d. 93.40.
   b. 9.665.   e. 0.097.
   c. 4.84.

95. What is $\hat{\sigma}_{\overline{D}}$?
   a. 0.966.   c. 4.84.   e. 0.097.
   b. 9.665.   d. 93.40.

*96. What is the decision for a two-tailed test at the 0.05 level? (Use the p-value.)
   a. Reject H_0, because $0.04 < 0.05$.
   b. Cannot reject H_0, because $2.01 > 0$.
   c. Reject H_0, because $0.02 < 0.05$.
   d. Cannot reject H_0, because $0.08 > 0.05$.
   e. Cannot reject H_0, because $0.04 < 0.05$.

97. What is the decision for a two-tailed test at the 0.01 significance level? (Use critical values.)
   a. Reject H_0, because $2.08 > 2.33$.
   b. Cannot reject H_0, because $2.08 < 2.575$.
   c. Reject H_0, because $2.01 < 2.33$.
   d. Cannot reject H_0, because $2.01 \neq 0$.
   e. Cannot reject H_0, because $2.01 < 2.575$.

98. Interpret a decision to reject H_0.
   a. The average time in the legislature in 1990 is at least as long as that in 1993.
   b. The average time in the legislature in 1990 is not different from that in 1993.
   c. The average time in the legislature in 1990 is shorter than that in 1993.
   d. The average time in the legislature in 1990 is longer than that of 1993.
   e. The average time in the legislature in 1990 is different from that in 1993.

```
TEST OF MU = 0.000 VS MU N.E. 0.000

              N       MEAN     STDEV    SE MEAN       T    P VALUE
EXPDIF      100      2.010     9.665      0.996    2.08     0.040
```

Figure 10-18
Minitab output for test of difference in legislative experience

Word and Thought Problems

99. Define μ_D and $\overline{D}$.

100. Define σ, S, S_p^2, $\sigma_1 - \sigma_2$, σ_{dif}, $\hat{\sigma}_{\text{dif}}$, $\sigma_{\overline{D}}$, and $\hat{\sigma}_{\overline{D}}$.

101. Tests for differences in population means aim to determine the effect of some variable on the values in the two samples. For instance, we might test to determine if there is a gender difference in mean weekly grocery purchases. How can we avoid other variables, such as family size, affecting the results and making the decisions ambiguous? Explain.

102. A group of developers randomly select five cars to find the effect of a new type of fuel filter under actual driving conditions. They record the number of gallons that pass through the filter before the filter totally clogs on each car, using the old and new filter types. They believe the times are normally distributed for each filter type. Use the following data to test the developers' hypothesis that the new filter is better at the 0.05 level. Interpret your decision. How would your work differ if you were testing for the producer of the old filter?

Car	Filter Old	Filter New
1	2,100	2,110
2	2,000	2,210
3	2,700	3,105
4	2,900	3,900
5	2,150	3,130

103. Frequently, pundits claim that presidential administrations work hard to improve economic conditions or at least economic statistics during election years. Can you reject the hypothesis that the unemployment rate is lowest in election years if the mean of a random sample of 11 election-year rates is 4.73% with a standard deviation of 0.51 percentage points and similar values for a random sample of seven nonelection years are 5.31% and 0.72 percentage points, respectively? Test at the 0.1 level. Assume the population is normal with equal variances. Interpret your decisions.

104. A local nightclub occasionally provides nonalcoholic nights when it provides live entertainment and allows people sixteen or older to enter. Test at the 0.01 level for a difference in mean gross receipts on these nights versus regular nights. The mean for a random sample of ten of the special nights is $1,802, and the standard deviation is $252, while a random sample of eight regular nights has a mean and standard deviation of $2,100 and $339, respectively. Assume that the populations are normally distributed with equal variances. Interpret your results.

105. Test at the 0.02 level to determine if the average bill for office visits to private doctors for six randomly selected sets of symptoms is significantly different from the fees at a fast-service medical clinic where the next available doctor treats the next patient. Use the following data. Assume the fee populations are normally distributed. Interpret your decision.

Symptom Set	Doctor Fee ($)	Clinic Fee ($)
1	22.00	19.00
2	27.50	26.50
3	22.00	22.00
4	48.75	36.25
5	28.00	29.50
6	18.00	21.00

106. Test at the 0.01 level to determine if the prices in shoe store A are higher than those in shoe store B. Independent, random samples of 19 pairs of shoes each from A and B have means of $28.63 and $25.38, respectively. The corresponding standard deviations are $5.89 and $8.12. Assume populations are normal with equal variances. Interpret your decision.

107. X and Y are normally distributed random variables with means of 25 and 50, respectively. Their variances are 42 and 58, respectively. They are also independent. Consider the random variable W, which is the difference between X and Y; that is, $W = X - Y$.
 a. What is the mean of W?
 b. What is the standard deviation of W?

108. A physicist is conducting the same experiment in two different environments to determine if air pressure affects the results. The results for the two environments for 10 randomly selected trials of the experiment follow. A different technician conducts each pair of trials. Theory suggests that environment 2 should produce higher results. Test this hypothesis at the 0.01 level. Assume results in each environment are normally distributed. Interpret your decision.

Trial	Environment 1	Environment 2
1	10	10
2	12	13
3	9	8
4	12	16
5	12	12
6	10	9
7	11	15
8	12	15
9	10	14
10	9	10

109. The following compilations of historical data for two infantry cantons illustrates the difficulty of recruiting tall, native soldiers for the Prussian army. Finding tall soldiers was a major priority for Frederick William I, which, some historians hypothesize, led to increased recruitment of foreigners, because of the lack of tall native recruits. Cantonist refers to male citizens of a canton (including soldiers). The compilations classify

Word and Thought Problems

the manpower of two infantry cantons, where H = Hacke's canton in 1783 and D = Diericke's in 1805. The "Cantonists (%)" and "Native Soldiers (%)" columns are relative frequency distributions. Assuming there were 5,000 Cantonists, we find there would be 545 native soldiers in Hacke's group and 378 in Diericke's. Assume the data sets of Cantonists and native soldiers are independent, random samples of men in the respective cantons and soldiers from the cantons during this historical period to answer the following questions. Perform all tests at the 0.05 significance level. (Source: B. v. Bagensky, *Regimentsbuch des Grenadier-Regiments König Friedrich Wilhelm IV. (1. pommerschen) Nr. 2, von 1679–1891* [Berlin, 1892], pp. 52–53, 56–57, and R. Kopka von Lossow, *Geschichte des Grenadier-Regiments König Frederich I. (4. ostpreussischen) Nr. 5* (Berlin, 1889–1901, 2:139, in Willerd R. Fann, "Foreigners in the Prussian Army, 1713–1756: Some Statistical and Interpretive Problems," *Central European History*, Vol. 23, No. 1, March 1990, pp. 76–84.

Height (inches)	% of Cantonists		% of Height Class Who Were Soldiers		% of Native Soldiers	
	H	**D**	**H**	**D**	**H**	**D**
Over 70	1.3	0.6	100.0	95.5	11.7	7.7
70	1.8	1.1	99.3	93.2	16.7	14.9
69	3.6	3.0	96.5	82.6	31.5	34.3
68	4.3	4.5	88.4	62.3	35.1	34.6
67	6.5	4.7	8.3	14.4	5.0	8.5
Under 67	82.5	86.1	0.0	0.0	0.0	0.0
n	5,000	5,000			545	378

a. Approximate the mean and variance of heights for each of the four samples (columns) of data under "Cantonists (%)" and "Native Soldiers (%)" *using expected-value formulas*. Assume that the heights of those over 70 inches are about 72 inches and that the heights of those under 67 are about 63.
b. Determine if the mean heights of males are different in the two time periods associated with Hacke and Diericke.
c. Determine if the mean height of native soldiers in Hacke's canton is larger than that of Diericke's.
d. Determine if the average percentage taken from each height class for soldiers is different. Use the data in "% of Height Class Who Were Soldiers" columns.

110. A 20-member commission meets frequently. Test at the 0.1 level to determine if attendance increases if meetings are not held on Fridays. Independent, random samples of five meetings held on Fridays and 25 meetings held on other days have mean attendance of 6.7 and 7.3 members, respectively. The population variances are believed to be unequal because Friday attendance is more erratic, so the data are fed into the computer and produce Figure 10-19. Assume that the populations of attendance values are normally distributed. Interpret your decision.

111. Can you reject at the 0.1 level a claim that losing gubernatorial candidates have lost fewer previous elections than winning gubernatorial candidates? A random sample of 20 winners has a mean of 1.3 previous losses with a standard deviation of 0.6. A random sample of 25 losers has a mean of 0.83 previous losses with a standard deviation of 0.22. Work this problem assuming normal populations with equal variances. Does the assumption about population variances seem reasonable? Explain. Interpret the decisions.

112. School boards compare random samples of students' records from two schools to determine if there is a difference between the mean scores for each school on a standardized exam. They feed the data into a computer, and the output in Figure 10-20 results. Use this information to perform two-tailed, lower-tailed, and upper-tailed tests for $\mu_A - \mu_B$ at the 0.03 level. Interpret your decisions.

113. Data for wins by six different conferences in the 1985 and 1986 NCAA Basketball Tournaments, taken from the NCAA Probation Database in Appendix D, are used for a paired-difference test to determine if the mean number of wins is different for conferences in the two years. Use the computer output in Figure 10-21 to perform two-tailed, lower-tailed, and upper-tailed tests for the differences between means at the 0.05 level if each D value is the 1986 value minus the 1985 value. What must you assume about the populations? Interpret your decisions.

114. A random sample of 100 citizens of a state spent a mean of $5 on lottery tickets during the month of June with a standard deviation of $8. The same random sample of citizens spent a mean of $9.25 in July with a standard deviation of $13.80. If the difference between the amounts each person spent in the two months is determined, the standard deviation of all the differences is $11. Is there sufficient evidence to support a

```
TWOSAMPLE T FOR FRIDAY VS NOFRIDAY
              N      MEAN     STDEV    SE MEAN
FRIDAY        5      6.7      1.8      0.80
NOFRIDAY     25      7.3      0.8      0.16

95 PCT CI FOR MU FRIDAY - MU NOFRIDAY: (-2.88, 1.68)

TTEST MU FRIDAY = MU NOFRIDAY (VS LT): T=-0.73  P=0.2526  df= 4
```

Figure 10-19

Minitab output for test of difference in committee attendance

Figure 10-20
Minitab output for test for difference in school scores

```
Note: Identical results for both assumptions about the population variances

TWOSAMPLE T FOR A VS B
      N      MEAN     STDEV    SE MEAN
A    20      84.3      11.6      2.60
B    20      76.2      11.5      2.58

95 PCT CI FOR MU A - MU B: (0.6430, 15.47)

TTEST MU A = MU B (VS NE): T=2.20 P=0.034 DF=38

POOLED STDEV =    11.6
```

Figure 10-21
Minitab computations for test for difference in conference wins

```
TEST OF MU = 0.00 VS MU N.E. 0.00

           N     MEAN     STDEV    SE MEAN      T      P VALUE
DIF8685    6    -0.50     7.01      2.86      -0.17     0.87
```

hypothesis that the mean expenditure varies by month at the 0.2 level? Interpret your decision.

115. Botanists studying the effect of cross- and self-pollination of the Atamasco Lily found that the mean percentage of cross-pollinated seeds that germinated 32 weeks after planting was 77.7% with a standard deviation of 22.9 percentage points in a sample of 23 flowers. Similar information for a sample of 20 self-pollinated flowers was 62.3% and 36.1 percentage points, for the mean and standard deviation, respectively. Assume the population variances are equal, and test the hypothesis that the mean germination rate is higher for cross-pollinated flowers at the 0.05 level. Interpret your decision. (Source: Steven B. Broyles and Robert Wyatt, "The Breeding System of *Zephyranthes atamasco* (Amaryllidaceae), *Bulletin of the Torrey Botanical Club*, Vol. 118, No. 2, 1991, pp. 137–40.)

116. A city is vying to take over the title of "Windiest City." A random sample of 10 days from the contender has a mean maximum wind speed of 28 miles per hour (mph) with a standard deviation of 8 mph. A random sample of 15 days from the city currently holding the title has a mean of 26 mph with a standard deviation of 9 mph. Is there sufficient evidence to agree that the contender is windier than the title holder at the 0.005 level? Assume the populations are normally distributed and dispersed equally. Interpret your decision.

117. Expenditures on ice cream in September and January for a random sample of eight students are presented below. Does this evidence contradict a claim that ice cream purchases by students are nonseasonal (mean purchases are the same for every season) at the 0.05 level? Interpret your decision.

Ice Cream Expenditures

Student	September($)	January($)
1	8.13	5.06
2	0.89	1.84
3	3.22	3.01
4	4.24	0.00
5	2.59	0.89
6	2.98	0.00
7	3.68	0.89
8	5.25	4.25

118. A random sample of 20 apartment dwellers has a mean of 1.2 children per household with a standard deviation of 2.2. A random sample of 10 homeowners has a mean of 1.8 children per household with a standard deviation of 1.7. Is there sufficient evidence to support a belief that the average number of children per household in houses is larger than that for apartment dwellers at the 0.05 level? Assume the populations are normally distributed with equal variances. Interpret your decision.

Review Problems

119. Two firms in the same industry are being considered as merger candidates. The mean profits over the last 40 quarters are being compared in the merger-decision process to determine if there is a difference. The sample mean profit for firm 1 is $253,000, and for firm 2 is $297,000. Management believes that this information indicates that firm 2 is more profitable. Can you demonstrate that the difference between sample means is great enough to corroborate management's thinking with only a 10% chance of making a Type I error? The

```
TWOSAMPLE T FOR SMALL VS LARGE
          N      MEAN    STDEV   SE MEAN
SMALL    104     1746     4707      462
LARGE     84     1295     1565      171

95 PCT CI FOR MU SMALL - MU LARGE: (-524, 1424)

TTEST MU SMALL = MU LARGE (VS NE): T=0.91 P=0.36 DF=130
```

Figure 10-22

Minitab computations for test for difference in mean sales

standard deviation of the paired differences is $99,322.714. Interpret your decision.

120. Test to see if the average length of vacation varies during recessions if a random sample of 210 motel customers during a recession stays an average of 3.75 days with a standard deviation of 4.21 days. During a nonrecessionary period an independent random sample of 182 customers has a mean stay of 5.83 days with a standard deviation of 9.66 days. Test at the 0.01 significance level. Interpret your decision.

121. An analysis of the effects of divorce and separation on children's behavior differed from most analyses of this effect that only study kids after the disintegration of the family. The new study analyzed data for 1,700 American families interviewed when the children were between 7 and 11 years old and then four years later. The researchers also studied 17,000 British families with children 7 years old and then when the children were eleven. To measure the effect, they used reading and math test scores as well as parent and teacher ratings of the children's behavioral and emotional problems. Thus, they could compare results for kids whose parents divorced or separated during the four-year interim and those who did not. The researchers found that many problems existed prior to the termination of the family. The researchers speculate that family problems before, but related, to the eventual rift could cause or exacerbate problems before the actual termination of the family. Thus, the breakup may exacerbate preexisting problems. They emphasize conflict rather than divorce as the culprit. Describe the estimator or estimators, the statistical test, and the hypotheses that the researchers might use in this study. (Source: "Children's Troubles May Precede Divorces," *Wall Street Journal*, June 7, 1991, p. B1.)

122. To compare the values of different types of recreations to society, recreation managers select random samples of 15 visitors each from visitors at a beach park and at a mountain park. The managers determine the distance each visitor traveled to reach the recreation site. They assume that longer distances imply more value for the particular recreation site. The mean distances are 420.2 and 391.6 miles for the beach and mountain, respectively, with corresponding standard deviations of 755.6 and 823.1 miles. Because the possible distances people travel to these sites are large, the managers believe the differences in standard deviations is probably not significant. A statistical test confirms their suspicion. Use this information, and try to demonstrate a preference for the beach at the 0.05 significance level. Previous research indicates recreation travel distances are normally distributed. Interpret your decision.

123. Figure 10-22 displays information for comparing mean sales values for firms that spend large or small proportions of their sales values on advertising. "Small" is less than 1%. Use the information in the figure to do lower-tailed, upper-tailed, and two-tailed tests for $\mu_L - \mu_S$ at a 0.05 level. Interpret each decision.

124. Test to demonstrate a difference greater than $1,000 in the mean annual per capita tax receipts during administration 1 over administration 2, if a random sample of 500 taxpayers' payments during administration 1 is $2,823 with a standard deviation of $479 and a random sample of 350 taxpayers' payments during administration 2 is $1,547 with a standard deviation of $612. Assume normal populations with equal variances, and test at the 0.05 significance level. Interpret the decision.

125. When studying family and children situations related to children adjusting to divorce, researchers constructed measures of parental relationships. The following list records the *large*-sample means for males and females and corresponding T statistics for testing for a gender difference. Test for a difference in each case at the 0.05 significance level. Interpret your decision. (Source: Christy M. Buchanan, Eleanor E. Maccoby, and Sanford M. Dornbusch, "Caught between Parents: Adolescents' Experience in Divorced Homes," *Child Development*, Vol. 62, No. 5, 1991, pp. 1008–29.)

Discord: $\bar{X}_M = 4.1$ $\bar{X}_F = 4.1$ T-statistic $= -0.30$.
Cooperative communication: $\bar{X}_M = 4.7$ $\bar{X}_F = 4.3$ T-statistic $= 1.65$.
Hostility: $\bar{X}_M = 5.5$ $\bar{X}_F = 5.6$ T-statistic $= -0.73$.
(Rounding causes the T-statistic for Discord to not be zero.)

*126. A broker claims that the daily Dow-Jones Index's closing average (that is, its value when the market closes or the end of the trading day) is a good indicator of behavior of closing prices of most stocks for that day. On a given day, determine the direction (up or down) and magnitude of change of the Dow-Jones closing average. Call this magnitude *closing*.

a. Use a random number generator or table to select 20 stocks from *The Wall Street Journal* listings. Determine the average magnitude of the change in closing prices of these stocks (using − for down and + for up) and their standard deviation. Use these values to calculate a *p*-value for H_0: $\mu = closing$, then make a decision at the 0.05 level. Interpret your decision.

b. On the same day, randomly select 50 stocks, and determine the direction of change. If the Dow-Jones is a good indicator, then we expect that 50% or more of the stocks will follow its direction of change. Use the 50 observations to find the *p*-value for this problem. Make a decision at the 0.05 level. Interpret your decision.

c. Suppose 36 stocks are randomly selected one day and 42, the next. The mean and standard deviation of the first sample are 45 and 5, respectively. For the second sample, these values are 53 and 14, respectively. If the difference in *closings* from the first day to the second is +12, use the sample information to test the broker's claim (H_0: $\mu_2 - \mu_1 = 12$) at the 0.05 level by finding the *p*-value. Interpret your decision.

127. A dairy farmer wants to determine the mean number of artificial inseminations necessary to impregnate a cow. A random sample of 10 cows is taken from the herd. The mean number of inseminations for these cows is 3.6 with a standard deviation of 0.55. Assume the number of necessary inseminations for all cows is a symmetric, unimodal population. Find a 90% confidence interval for the mean number of inseminations.

128. A greeting card company believes that they sell at least 50,000 more Christmas cards than Valentine cards on average during a single season. Test this claim at the 0.1 level using the data from the independently and randomly selected seasons below. Interpret your decision. What assumptions did you make to work the problem?

$\overline{X}_C = 1,296$ thousand cards $\overline{X}_V = 750$ thousand cards
$S_C = 482$ thousand cards $S_V = 656$ thousand cards
$n_C = 20$ seasons $n_V = 18$ seasons.

129. Criminal justice researchers collected independent random samples of seven days of robbery data before and seven days after media publicity surrounding video film displaying apparent police brutality. They want to compare the mean number of robberies committed during the two periods. They speculate that the media blitz has disturbed the situation so that the means have changed, although they are not sure of the direction of change. Use the information listed below to test for a change in the means at the 0.01 level. Assume that the number of robberies is normally distributed during both time periods. Interpret your decision. What assumptions did you make to work the problem?

$\overline{X}_B = 2,856$ robberies $\overline{X}_A = 2,697$ robberies
$S_B = 1,282$ robberies $S_A = 1,392$ robberies.

130. A librarian wants to determine if the mean number of books that students check out varies between English and Psychology majors. He randomly selects a freshman, sophomore, junior, senior, first-year graduate student, and a second-year graduate student from each of the majors. By checking the records he determines the number of books each checks out during the most recent term. Use the following results and test for a difference in the mean number of books checked out at the 0.01 significance level. Assume the number of books checked out is normally distributed. Interpret your decision.

Student	Number of Books Checked out	
	English Major	Psychology Major
Freshman	8	3
Sophomore	10	6
Junior	12	13
Senior	12	18
Graduate 1	26	31
Graduate 2	27	27

131. The creation of a Prussian national army, rather than one with many foreign mercenaries, interests historians. Using historical information and data, they debate whether Frederick William I started the creation of a Prussian national army followed by a reversal of this trend by Frederick the Great after 1740. A problem arises, because prior to 1740 records did not distinguish foreigners from natives, so historians estimate the foreign contingent from recruitment patterns and a few accessible statistics. For example, 21% of the enlisted men on the May 1715 muster roll of the *Leibkompagnie* of the Dohna regiment were foreigners. The author of this study states, "One company, of course, does not necessarily typify the whole army." Discuss this statement and the role of statistics for inferring historical events. (Source: Willerd R. Fann, "Foreigners in the Prussian Army, 1713–1756: Some Statistical and Interpretive Problems," *Central European History*, Vol. 23, No. 1, March 1990, pp. 76–84.)

132. Can you support a claim that the average time to work a problem in a math text increases as you proceed through the book, at the 0.05 level, if the mean time to work a random sample of 10 problems from the first half of the book is 6.62 minutes with a standard deviation of 3.16 minutes? A random sample of six problems from the last half averages 8.25 minutes with a standard deviation of 2.88 minutes. Assume times are normally distributed and that the variances do not differ. Interpret your result.

133. A college will only accept students who score above 75 on a 100-point national exam. A high school claims that after adopting an innovative teaching approach, its students' average score increased by more than eight points and now exceeds the college's requirement. To verify these statements, the principal collects a random sample of 100 students' scores before and 100 from

after the change. The results are (A = after and B = before):

$\bar{X}_A = 75.8$ points $\bar{X}_B = 67.1$ points
$S_A = 12.2$ points $S_B = 14.5$ points.

The testing agency indicates that scores tend to be normally distributed in almost any subgroup of students taking the test, but the dispersion of scores can be quite variable among different groups.

a. Use this information and a 0.05 significance level to perform and interpret the statistical work for the principal to demonstrate the claim that the average is now greater than 75 points.
b. Use this information and a 0.05 significance level to perform and interpret the statistical work for the principal to demonstrate the claim about the change in the average.
c. The principal located 45 students who took the test before and after the change. Pairing their scores results in a mean difference of 8.9 points (after − before) and a standard deviation of these differences of 2.9 points. Use this information and perform a test to verify the principal's claim about the change in averages. Use $\alpha = 0.05$ and interpret your decision. What are some potential problems with this comparison that are not likely to surface when the principal uses independent random samples?
d. A principal from an alternative private high school nearby disputes this principal's claims. Does any of the data support this new rival's claim? Explain your answer.

134. The mean time between haircuts is 2.8 weeks with a standard deviation of 0.3 week for a random sample of 158 men. Can we reject the hypothesis that the mean interval between haircuts is at most 2.75 weeks using these data? Use $\alpha = 0.1$. Interpret your decision.

135. In an effort to improve quality and international competition, some American businesses are reducing the number of their suppliers to fewer firms that produce better parts and inputs. Make any necessary assumptions and use the following information to form a 95% confidence interval estimate of the mean difference in the number of suppliers (previous − current). Then list any assumptions you make.

Company	Number of Suppliers	
	Previous	Current
Xerox	5,000	500
Motorola	10,000	3,000
Digital Equipment	9,000	3,000
General Motors	10,000	5,500
Ford Motor	1,800	1,000
Texas Instruments	22,000	14,000
Rainbird	520	380
Allied-Signal Aerospace	7,500	6,000

Source: Reprinted by permission of the *Wall Street Journal*, © 1991 Dow Jones & Company, Inc. All Rights Reserved Worldwide.

136. Test to determine if installing a virus detection program reduces the monthly average number of downtimes needed to find and repair the damage from a computer virus that goes undetected. Assume downtimes are normally distributed and their variances did not change after the program installation. Test at the 0.01 level, using the data from the following independently selected random samples.

$\bar{X}_B = 15.3$ $\bar{X}_A = 8.3$
$S_B = 22.6$ $S_A = 10.0$
$n_B = 36$ $n_A = 44$.

*__137.__ A random sample of 12 new health customer's medical expenditures is collected for the year prior to their becoming insured and the year after. The values follow. Does this information confirm the insurance company's hypothesis at the 0.05 level that mean expenditures are higher when the customers are insured than when they are uninsured? Construct the p-value to make the determination. Assume medical expenditures are normally distributed. Interpret your decision.

Customer	Before Expenditure ($)	After Expenditure ($)
1	800	3,677
2	79	125
3	1,500	1,244
4	1,781	2,085
5	5,433	19,883
6	18,500	25,866
7	654	342
8	8,499	866
9	3,755	6,344
10	1,488	7,495
11	250	188
12	166	157

138. Figure 10-23 shows computer output of tests for a difference in population means. In each case, the populations are normally distributed, the mean of the first population is 100, the mean of the second population is 200, and the population standard deviations (variances) are equal. However, the population standard deviations decrease as you proceed through the figure.

a. Study the results, then write a statement that generalizes the effect of decreasing the population standard deviations on $\hat{\sigma}_{dif}$ and on our ability to detect the true differences in the population means with this test.
b. What proportion of the cases have pooled standard deviations (not variances) closer to the true population standard deviation, σ, than S_1 or S_2?

Figure 10-23
Minitab output for tests for difference in population means

a) $\mu_1 = 100$ $\quad\quad \mu_2 = 200 \quad\quad \sigma_1 = \sigma_2 = 5000$

```
TWOSAMPLE T FOR C1 VS C2
         N       MEAN      STDEV    SE MEAN
C1      20        -46       4682       1047
C2      20       -116       6302       1409

95 PCT CI FOR MU C1 - MU C2: (-3484, 3625)
TTEST MU C1 = MU C2 (VS NE): T= 0.04 P=0.97 DF= 38
POOLED STDEV =           5551
```

b) $\mu_1 = 100$ $\quad\quad \mu_2 = 200 \quad\quad \sigma_1 = \sigma_2 = 2500$

```
TWOSAMPLE T FOR C3 VS C4
         N       MEAN      STDEV    SE MEAN
C3      20       -724       2052        459
C4      20        271       3104        694

95 PCT CI FOR MU C3 - MU C4: (-2680, 689)
TTEST MU C3 = MU C4 (VS NE): T= -1.20 P=0.24 DF= 38
POOLED STDEV =           2631
```

c) $\mu_1 = 100$ $\quad\quad \mu_2 = 200 \quad\quad \sigma_1 = \sigma_2 = 1250$

```
TWOSAMPLE T FOR C5 VS C6
         N       MEAN      STDEV    SE MEAN
C5      20         65        924        207
C6      20       -130       1408        315

95 PCT CI FOR MU C5 - MU C6: (-567, 958)
TTEST MU C5 = MU C6 (VS NE): T= 0.52 P=0.61 DF= 38
POOLED STDEV =           1191
```

d) $\mu_1 = 100$ $\quad\quad \mu_2 = 200 \quad\quad \sigma_1 = \sigma_2 = 625$

```
TWOSAMPLE T FOR C7 VS C8
         N       MEAN      STDEV    SE MEAN
C7      20        145        740        166
C8      20         86        628        141

95 PCT CI FOR MU C7 - MU C8: (-381, 499)
TTEST MU C7 = MU C8 (VS NE): T= 0.27 P=0.79 DF= 38
POOLED STDEV =            687
```

e) $\mu_1 = 100$ $\quad\quad \mu_2 = 200 \quad\quad \sigma_1 = \sigma_2 = 312.5$

```
TWOSAMPLE T FOR C9 VS C10
         N       MEAN      STDEV    SE MEAN
C9      20        193        250         56
C10     20        187        321         72

95 PCT CI FOR MU C9 - MU C10: (-178, 191)
TTEST MU C9 = MU C10 (VS NE): T= 0.07 P=0.95 DF= 38
POOLED STDEV =            288
```

f) $\mu_1 = 100$ $\quad\quad \mu_2 = 200 \quad\quad \sigma_1 = \sigma_2 = 156.25$

```
TWOSAMPLE T FOR C11 VS C12
         N       MEAN      STDEV    SE MEAN
C11     20         72        160         36
C12     20        184        161         36

95 PCT CI FOR MU C11 - MU C12: (-214, -9)
TTEST MU C11 = MU C12 (VS NE): T= -2.20 P=0.034 DF= 38
POOLED STDEV =            160
```

Review Problems

g) $\mu_1 = 100 \qquad \mu_2 = 200 \qquad \sigma_1 = \sigma_2 = 78.125$

```
TWOSAMPLE T FOR C13 VS C14
        N      MEAN     STDEV    SE MEAN
C13    20      87.7      70.2       16
C14    20     189.5      62.0       14

95 PCT CI FOR MU C13 - MU C14: (-144, -59)
TTEST MU C13 = MU C14 (VS NE): T= -4.86 P=0.0000 DF= 38
POOLED STDEV =        66.2
```

h) $\mu_1 = 100 \qquad \mu_2 = 200 \qquad \sigma_1 = \sigma_2 = 39.0625$

```
TWOSAMPLE T FOR C15 VS C16
        N      MEAN     STDEV    SE MEAN
C15    20      89.4      42.6      9.5
C16    20     216.8      27.4      6.1

95 PCT CI FOR MU C15 - MU C16: (-150.3, -104.5)
TTEST MU C15 = MU C16 (VS NE): T= -11.24 P=0.0000 DF= 38
POOLED STDEV =        35.8
```

i) $\mu_1 = 100 \qquad \mu_2 = 200 \qquad \sigma_1 = \sigma_2 = 19.53125$

```
TWOSAMPLE T FOR C17 VS C18
        N      MEAN     STDEV    SE MEAN
C17    20     102.0      18.3      4.1
C18    20     197.7      17.2      3.9

95 PCT CI FOR MU C17 - MU C18: (-107.0, -84.3)
TTEST MU C17 = MU C18 (VS NE): T= -17.03 P=0.0000 DF= 38
POOLED STDEV =        17.8
```

j) $\mu_1 = 100 \qquad \mu_2 = 200 \qquad \sigma_1 = \sigma_2 = 9.765625$

```
TWOSAMPLE T FOR C19 VS C20
        N      MEAN     STDEV    SE MEAN
C19    20    103.71      7.95      1.8
C20    20    199.01      9.26      2.1

95 PCT CI FOR MU C19 - MU C20: (-100.8, -89.8)
TTEST MU C19 = MU C20 (VS NE): T= -34.92 P=0.0000 DF= 38
POOLED STDEV =        8.63
```

Figure 10-23 (continued)
Minitab output for tests for difference in population means

Chapter **11**

Correlation and Regression Analysis

Several statistical techniques explore relationships among variables to explain and predict phenomena that we observe in the world around us. The correlation procedure provides information about the association of two variables and whether they both change together in a systematic and predictable manner. The regression procedure proceeds beyond correlation to incorporate, establish, measure, and predict the effect of one variable on the other. These techniques prove useful in many arenas, including situations where analysts study experimental data as well as nonexperimental investigations where analysts cannot isolate subjects and control changes but still need to probe the interrelationships among variables.

Objectives

In Unit I we discuss the correlation procedure to measure the strength of the association between two variables. The correlation coefficient that results from the computations does not incorporate any causal relationship between the two variables, that is, we do not know if changes in one variable cause the other to change, only that the two variables tend to change in a regular, predictable manner. This unit also introduces information about collecting and picturing data sets before we proceed with the correlation and regression procedures. Upon completing the unit, you should be able to

1. Describe the organization of values in a data set to be used for correlation and regression analysis.
2. Plot a scatter diagram.
3. Calculate a correlation coefficient.
4. Interpret the coefficient.
5. Test for a significant correlation between two variables.
6. Understand the concepts of time series, cross section, upward sloping, positively sloped, downward sloping, negatively sloped, correlation coefficient, and scatter diagram.
7. Recognize the population correlation coefficient, ρ, and the sample correlation coefficient, r.

In Unit II we introduce the regression procedure. This technique assumes that a change in one variable provides information about the value of the other variable. We estimate the magnitude of the change that results. The linear function, $Y = a + bX$ approximates and describes the connection between the behavior of the two variables. Upon completing the unit, you should be able to

1. State the criterion used to establish a single estimated line for a set of data points.
2. State the conditions assumed about the model and the errors to obtain dependable estimates and test results.
3. Estimate the line and interpret the results, using words related to the phenomenon under investigation.
4. Use the estimated line to forecast Y when X is given.
5. Distinguish the true and the estimated line, as well as the true and estimated error terms, using symbols.
6. Interpret the information in a computer printout of an estimated regression line and use the information to identify unusual values in the data set that may have an unusually large effect on the estimates.
7. Define model, function, slope, intercept, independent and dependent variables, residual, error term, standard error of the estimate, and leverage point.

Section 11.1 Data Sets **491**

8. Recognize Y, an actual or observed Y value; $\overline{Y}_x$, the mean of the Y values that correspond to a specific value of X; $\hat{Y}$, a point on the estimated regression line, which is an estimate of $\overline{Y}_x$; e, the error or vertical distance of a Y value from the regression line; $\hat{e}$, the residual, which is an estimate of e; a, the intercept or constant term in the true regression line; $\hat{a}$, an estimate of a; b, the coefficient of the independent variable in the true regression line; $\hat{b}$, an estimate of b; σ_e, the standard error of the estimate; and $\hat{\sigma}_e$, an estimate of σ_e.

In Unit III we explore and expand the three uses for regression (1) to evaluate if Y changes predictably when X changes, (2) to measure the effect of changes in X on Y, and (3) to forecast Y (or the X value that produces a given Y value). Upon completing the unit, you should be able to

1. Interpret an R^2 value.
2. Describe the sampling distribution of the estimated regression coefficient, $\hat{b}$, and the T-ratio, $\hat{b}/\hat{\sigma}_{\hat{b}}$, then use this information to evaluate the model by performing a hypothesis test on b.
3. Forecast Y and X values when the other is given.
4. Calculate a prediction interval for Y.
5. State four precautions to take when forecasting values with regression.
6. Define standard error of the coefficient, T-ratio, interpolation, extrapolation, and structural change in a model.
7. Recognize R^2, the coefficient of determination; $\sigma_{\hat{b}}$, the standard error of the coefficient; and $\hat{\sigma}_{\hat{b}}$, the estimated standard error of the coefficient.

Unit I: Correlation Analysis

Preparation-Check Questions

1. A deviation is defined as $X - \overline{X}$. If the deviation is -10, this value
   a. Means that the observation is 10 units smaller than the mean.
   b. Means that the observation is 10 units larger than the mean.
   c. Is incorrect, because deviations must be positive.
   d. Is the average distance from the mean.
   e. Implies that there are negative values in the sample.

2. Find the standard deviation for the following sample: 5, 8, 10, 1.
   a. 3.39.
   b. 11.5.
   c. 3.92.
   d. 15.3.
   e. 0.

Answers: 1. a; 2. c.

Introduction

Skills Test and Productivity Experiment For several years a job-training center has been using a skills test as part of its application process. The center wishes to determine if test scores really provide useful information about the quality of future employees. That is, do higher test scores correspond to or correlate with more productive workers? If the test contributes no useful information, the center will dispense with it and save the expense of administering it.

The center keeps records of test scores and of productivity measures from a mock production process used at the completion of the training program. Finding a measure of association between two variables, such as these, is the subject of this unit.

11.1 Data Sets

11.1.1 Assembling Data Sets: Time Series or Cross Section

The first step in learning more about the joint behavior of two variables with correlation and regression procedures is to collect an appropriate data set. Rather than a single set of values for a variable, we need a data set with information for the two related variables, generally labeled X and Y. For each X value in the data set we need the corresponding Y value. We do not display the Y values randomly, rather they are aligned with their corresponding X value. The following table lists the test score and productivity measures for 10 randomly selected trainees from last year's enrollees.

Trainee	Test Score	Productivity (units per day)
1	85	32
2	70	28
3	95	41
4	60	40
5	70	26
6	80	30
7	50	22
8	75	30
9	85	35
10	65	29

When the data set consists of observations about several entities, such as several people, several business firms, several counties, or several countries, during a single time period, we call it a *cross section*. For example, the ten trainees' scores and productivities from last year is a cross section of trainees. If, instead, the data set consists of records of a single unit, such as a single person, single business firm, single county, or single country, over several time periods, we call it a *time series*. Examples include a single trainee who repeats the test and productivity sequence over several years of a training program or quarterly values of a business firm's profit over a period that spans several quarters.

Question: Which type of data set is the set of total U.S. population values from the census count conducted every 10 years?

Answer: The census figures are a times series; the single unit is the United States, and the counts occur during many different time periods.

Question: Is the data set consisting of January 1992 crime rates by county a time series or cross section?

Answer: It is a cross section, because the units are different counties from a certain time period, January 1992.

11.1.2 Displaying Data Sets: Scatter Diagrams

When we search for a relationship between two variables, a standard graph of the available data pairs (X, Y), called a *scatter diagram,* frequently helps us discover a pattern, such as when X values increase, Y values increase also, or when X values increase, Y values tend to decrease. To obtain a scatter diagram for the variables X and Y, we locate each X value on its axis and the corresponding Y value on the Y axis. We plot a single point on the diagram at the intersection of each (X, Y) pair in the data set. This diagram provides a visual impression of the relationship, if any, and the form of that relationship.

■ **Example 11-1:** Using the test score and productivity data from Section 11.1.1 (shown at the top of this page) plot a scatter diagram.

■ **Solution:** Each individual's X value (test score) corresponds to a Y value (productivity). This pair is (85, 32) for trainee 1. To plot the data pair for trainee 1, we locate the test score value of 85 on the X axis and the productivity value of 32 on the Y axis. A point is plotted where these values intersect on the diagram. Figure 11-1 displays this point (boxed) as well as the other nine points.

Section 11.1 Data Sets ■ 493

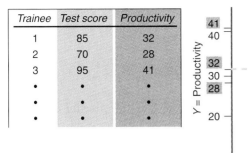
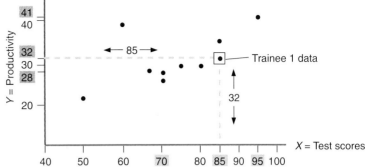

Question: Does there appear to be any association between test scores and productivity? Explain your answer.

Answer: The scatter diagram gives the impression that trainees who score higher on the test tend to be more productive. This is not true for every case, but generally it seems to be true.

Figure 11-1
Scatter diagram for trainee data

When we describe a relationship between variables, we look for the tendency of one variable to systematically change in a specific direction when the other variable changes in a specific direction. If both variables tend to change in the same direction—when one increases, the other increases, or when one decreases, the other decreases—we say that the relationship is *upward sloping* or *positively sloped*. In this case, large values of both variables tend to occur together or, equivalently, small values tend to occur together. This is true for the score and productivity data. When values of the variables tend to change in opposite directions, one decreases when the other increases, we say the relationship is *downward sloping* or *negatively sloped*. In this case, large values of one variable accompany small values of the other variable.

■ **Example 11-2:** Plot a scatter diagram for the data on gymnasts' body weights and corresponding scores in competitions. Describe any apparent pattern or relationship you find between the two variables.

Gymnast	Weights	Scores
1	100	1,030
2	110	980
3	110	1,000
4	90	1,050
5	100	1,040
6	90	1,000

■ **Solution:** Figure 11-2 displays the scatter diagram for the six data points. Although one of the points does not fit the pattern, the points' locations generally indicate a downward sloping relationship: As weight increases, the score decreases.

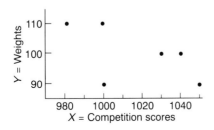

Figure 11-2
Scatter diagram for gymnast data

The scatter diagram is a first and easy step to learning more about a relationship between two variables. It quickly displays the distribution of the data points, which may signal the nature of the relationship, upward or downward sloping, or potential problems with the available data for discovering more about the relationship. For instance, the aberrant point of the sixth gymnast or "outlier" may relay important information about the relationship, such as a lightweight gymnast who, further research reveals, trains less than the other athletes. This extra information adds to our understanding of the complexity of the weight-score relationship.

Section 11.1 Problems

3. Plot a scatter diagram for the following variables, a stock's opening and closing prices, for a sample of eight consecutive days. Describe the relationship that appears on the diagram. Can you identify an outlier? Does the closing or opening price make it an outlier? Is this data set a cross section or a time series?

	Prices				Prices	
Day	Opening	Closing		Day	Opening	Closing
1	97.625	100.500		5	98.625	98.000
2	99.750	96.125		6	100.000	99.250
3	99.250	98.750		7	100.500	98.000
4	98.250	100.000		8	102.125	99.500

4. Plot a scatter diagram for the variables, the percentage of students who take the SAT and the SAT score, from the values listed below that are taken from the State SAT Score Data Base in Appendix C. First, use states where 52% or more of the high school seniors take the test. Repeat the exercise for the states where 13% or less take the test.
   a. Are these data sets time series or cross sections? Explain your answer.
   b. For each diagram, describe any general relationship between the two variables that you observe.
   c. Are there any outliers? If so, which state or states is an outlier? For each outlier determine if the value for the percentage of students taking the test or the SAT score variable or both cause the point to be an outlier.

52% or More Take SAT Test

Row	State	SAT	% Who Take
1	Connecticut	901	74
2	DC	850	68
3	Delaware	903	58
4	Georgia	844	57
5	Hawaii	885	52
6	Indiana	867	54
7	Maine	886	60
8	Maryland	908	59
9	Massachusetts	900	72
10	N. Hampshire	928	67
11	N. Jersey	891	69
12	N. York	882	70
13	N. Carolina	841	55
14	Pennsylvania	883	64
15	R. Island	883	62
16	S. Carolina	834	54
17	Vermont	897	62
18	Virginia	895	58

13% or Less Take SAT Test

Row	State	SAT	% Who Take
1	Alabama	984	8
2	Arkansas	981	6
3	Iowa	1,088	5
4	Louisiana	993	9
5	Mississippi	996	4
6	N. Dakota	1,069	6
7	Oklahoma	1,001	9
8	S. Dakota	1,061	5
9	Utah	1,031	5
10	Wyoming	977	13
11	Wisconsin	1,019	11
12	Tennessee	1,008	12
13	N. Mexico	1,007	12
14	Kansas	1,040	10
15	Kentucky	994	10
16	Michigan	968	12
17	Missouri	995	12
18	Nebraska	1,030	10

5. Plot a scatter diagram for the following research data on reaction time (in seconds) for a chemist's experiment and the room temperature (in degrees Fahrenheit) when he conducted the experiment at 10 randomly selected times during a day. Describe the apparent relationship between the two variables. Is the data set a time series or cross section?

Row	Temperature	Reaction Time
1	75	9.54
2	74	9.22
3	73	9.14
4	71	8.83
5	77	9.73
6	73	9.53
7	70	9.50
8	74	8.83
9	72	8.79
10	72	9.37

11.2 The Correlation Coefficient

The visual impression of association provided by a scatter diagram is only a start to describing the relationship between two variables. For some scatter diagrams, a relation may not appear, or the relation may appear ambiguous (see Figure 11-3). To avoid different qualitative interpretations, we develop a measure of association that produces a single quantitative value to describe any given data set.

11.2.1 A Measure of the Direction of Association

To develop the measure of association we begin by dividing the plane of the diagram into four quadrants, four sections formed when we intersect two lines at right angles. We locate the dividing lines at the mean of each variable, $\overline{X}$ and $\overline{Y}$.

■ **Example 11-1 (continued):** Refer to the trainee data (p. 492) to establish the four quadrants based on $\overline{X}$ and $\overline{Y}$. Which quadrants contain most of the data points? Relate the last answer to the upward-sloping relationship (as scores increase, productivity tends to increase, or as scores decrease, productivity tends to decrease) suggested by the scatter diagram.

■ **Solution:** The mean of the test scores is $\overline{X} = 735/10 = 73.5$ points, and the mean productivity is $\overline{Y} = 313/10 = 31.3$ units per day. With dividing lines at these values, we can label the data points in the scatter diagram according to the quadrant where they lie, as shown in Figure 11-4. Seven of the 10 points lie in Quadrants I and III in accord with an upward-sloping relationship. Quadrant I encloses pairs with large X and Y values, and Quadrant III contains pairs with small X and Y values. *A value is large or small relative to its mean.*

Question: If the relationship is downward sloping (as X increases, Y decreases, and vice versa), which quadrants should contain most of the data points? Refer to the gymnast data in Figure 11-2.

Figure 11-3
Scatter diagrams with no apparent visual association

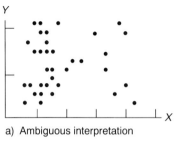

a) Ambiguous interpretation

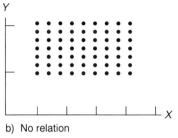

b) No relation

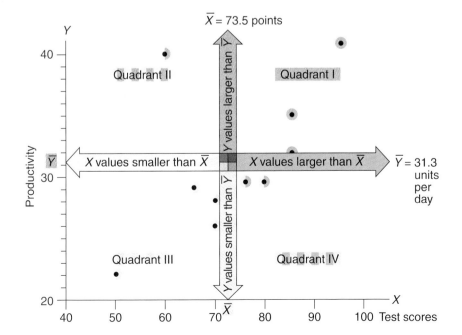

Figure 11-4
Scatter diagram with quadrants separated at sample means

Figure 11-5

Scatter diagram with quadrants for gymnast data

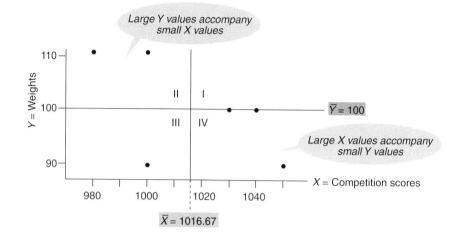

Answer: From the diagram it appears that Quadrants II and IV should include most of the pairs where one variable has a large value and the other a small value. Figure 11-5 verifies that this is the case for the gymnast data.

> Thus, most (not necessarily all) of the points are in Quadrants I and III when the relationship is upward sloping, and in II and IV when it is downward sloping.

Now, we will investigate the characteristics of points in each quadrant; more specifically, we will search for a tendency for same-size values to occur together or opposite-size values to occur together. Again we consider a value to be large when it is larger than the mean and small when it is smaller than the mean. It follows naturally to consider the deviations ($X - \overline{X}$ and $Y - \overline{Y}$) of each point in the different quadrants.

Question: Take the point for trainee 9 from Example 11-1 (85 test points, 35 units per day). What are the deviations for this data point? Recall the mean test score, $\overline{X}$, is 73.5 and the mean productivity, $\overline{Y}$, is 31.3.

Answer: $X - \overline{X} = 85 - 73.5 = 11.5$, $Y - \overline{Y} = 35 - 31.3 = 3.7$. See Figure 11-6.

For any point in Quadrant I, X is to the right of the $\overline{X}$ line, so $X - \overline{X} > 0$, Y is above the $\overline{Y}$ line, so $Y - \overline{Y} > 0$. By similar reasoning $X - \overline{X}$ and $Y - \overline{Y}$ are both negative for a point (data pair) in Quadrant III.

Figure 11-6

Deviations for trainee 9 in Quadrant I

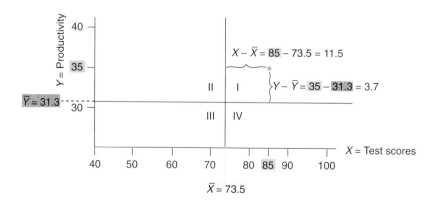

Section 11.2 The Correlation Coefficient ■ **497**

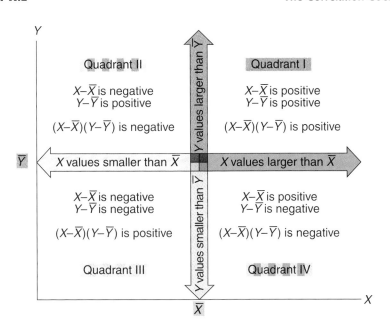

Figure 11-7
Signs of the deviations and the products of deviations associated with points in each quadrant

Question: Describe the signs of deviations in Quadrants II and IV.
Answer: In Quadrant II, X will be to the left of $\overline{X}$, so the deviation is negative. However, Y is above the $\overline{Y}$ line, so the Y deviation will be positive. Similar reasoning leads to opposite signs for the deviations in Quadrant IV.

Figure 11-7 summarizes the results of the analysis to this point. It also shows the sign that results when we combine the signs of the two deviations by multiplying the two deviations associated with any point together. Using the deviations for the ninth trainee that we calculated earlier, 11.5 and 3.7, we find $(X - \overline{X})(Y - \overline{Y}) = (11.5)(3.7) = 42.55$.

Question: If most of the points lie in Quadrants II and IV, would you expect the mean of the products of their derivations for all points in the data set to be positive or negative?
Answer: The product is negative for each point in Quadrants II and IV. Because most of the points are in these quadrants, the negative products will usually more than offset the positive products from the fewer points in Quadrants I and III. It follows that the sum (and the mean) will usually be negative.

A few extreme points in Quadrant I or III are more difficult to offset and could cause a positive sum, even though most of the points are in Quadrants II and IV. However, if most points are in Quadrants II and IV, the mean of the products will likely be negative. If most are in Quadrants I and III, we expect a positive value for the mean of the products.

Thus, the sign of the mean of the products of the deviations is an indication of an upward- or downward-sloping relationship. In addition, we would like for the magnitude of the mean product to add information about the relationship between the two variables. For the upward-sloping relationship of the trainee test scores and productivities in Example 11-1, the positive mean product is 39.95. For the downward-sloping relationship of the gymnasts in Example 11-2, the negative mean product is −116.67.

Question: Using the absolute values of the mean products, 39.95 and 116.67, which of the relationships do you think is stronger? *Stronger* means that as one variable increases, the other is more likely to change in the expected direction (according to the sign of the mean product).

Answer: The size of these numbers does not indicate strength of the relationship. The magnitude will vary with units of measure, though the relationship never changes. If the training center measures productivity as the dollar value of output rather than units, the mean product will change proportionately, though the sign will remain positive. In fact, the mean product could become larger in absolute value than the mean product for the gymnastic data. The sign does not vary with units of measure.

In summary, the mean product indicates the sloping direction of the relationship but not the strength. The correlation coefficient will do both.

11.2.2 A Measure of Strength and of Direction of Association: Calculation

One way to standardize values so that units do not matter is to use the familiar formula $Z = $ (outcome $-$ mean)/standard deviation. In this case, we can transform each X value to $Z_X = (X - \mu_X)/\sigma_X$ and each Y value to $Z_Y = (Y - \mu_Y)/\sigma_Y$. Recall that the sign of a Z value indicates whether the value is above or below the mean, so each standardized value retains the sign of the original deviation. In addition, the sign of the mean of the products of the transformed values will still indicate if the relationship is upward or downward sloping. We call this mean of the standardized values the population *correlation coefficient*, ρ (pronounced "rho").

Now, because Z values are unitless, the magnitude of ρ indicates a strong or weak linear relationship. We will be able to compare different correlation coefficients in order to determine which of several linear relationships is strongest.

With some algebra, this mean of standardized values becomes

$$\rho = \frac{N(\Sigma XY) - (\Sigma X)(\Sigma Y)}{\sqrt{[N(\Sigma X^2) - (\Sigma X)^2][N(\Sigma Y^2) - (\Sigma Y)^2]}}.$$

Normally, we only have a sample, not a population, of data points. The calculation procedure does not change, but the symbol for the sample correlation coefficient is r, and N, the population size, becomes n, the sample size.

Sample Correlation Coefficient

$$r = \frac{n(\Sigma XY) - (\Sigma X)(\Sigma Y)}{\sqrt{[n(\Sigma X^2) - (\Sigma X)^2][n(\Sigma Y^2) - (\Sigma Y)^2]}}$$

To find this value, we need to compute the sum of all of the X values in the sample, ΣX, and the sum of the squared X values, ΣX^2. We need the same two sums for the Y values, ΣY and ΣY^2. Finally, we need the sum of the products of the original X and Y values ΣXY. Then we substitute the results in the formula to obtain the correlation coefficient.

■ **Example 11-1 (continued):** Find the correlation coefficient for the trainee data (p. 492).

■ **Solution:** Let X be the trainee's test score and Y be the trainee's productivity. The following table displays the calculations necessary to obtain the correlation coefficient,

$$r = \frac{n(\Sigma XY) - (\Sigma X)(\Sigma Y)}{\sqrt{[n(\Sigma X^2) - (\Sigma X)^2][n(\Sigma Y^2) - (\Sigma Y)^2]}}$$

$$= \frac{10(23{,}405) - (735)(313)}{\sqrt{10(55{,}625) - (735)^2}\sqrt{10(10{,}115) - (313)^2}} = 0.5595, \text{ or } 0.6.$$

Section 11.2 The Correlation Coefficient

Trainee	X	Y	X^2	Y^2	XY
1	85	32	7,225	1,024	2,720
2	70	28	4,900	784	1,960
3	95	41	9,025	1,681	3,895
4	60	40	3,600	1,600	2,400
5	70	26	4,900	676	1,820
6	80	30	6,400	900	2,400
7	50	22	2,500	484	1,100
8	75	30	5,625	900	2,250
9	85	35	7,225	1,225	2,975
10	65	29	4,225	841	1,885
	735	313	55,625	10,115	23,405

■ **Example 11-2 (continued):** Find the correlation coefficient, r, for the gymnast data (p. 493).

■ **Solution:** Let Y be the weight and X be the score. The following table reveals the necessary calculations to obtain r,

$$r = \frac{n(\Sigma XY) - (\Sigma X)(\Sigma Y)}{\sqrt{[n(\Sigma X^2) - (\Sigma X)^2][n(\Sigma Y^2) - (\Sigma Y)^2]}}$$

$$= \frac{6(609,300) - (6100)(600)}{\sqrt{6(6,205,400) - (6,100)^2} \sqrt{6(60,400) - (600)^2}} = -0.5728, \text{ or } -0.6.$$

Gymnast	Y	X	Y^2	X^2	XY
1	100	1,030	10,000	1,060,900	103,000
2	110	980	12,100	960,400	107,800
3	110	1,000	12,100	1,000,000	110,000
4	90	1,050	8,100	1,102,500	94,500
5	100	1,040	10,000	1,081,600	104,000
6	90	1,000	8,100	1,000,000	90,000
	600	6,100	60,400	6,205,400	609,300

11.2.3 Information Relayed by Correlation Coefficient

The correlation coefficient is positive for the trainee data and negative for the gymnast data, as we expected after seeing the scatter diagram (Figures 11-4 and 11-5). Although their magnitudes are similar, it is not a simple task to gauge the strength of the relationship unless an r value is close to -1, 0, or $+1$. When all the data points lie on a straight, upward-sloping line, $r = 1$. If all the points lie on a downward-sloping line, $r = -1$. Figure 11-8 shows that as the points scatter and move away from a straight line, the r value moves closer to 0.

As the points scatter, we are likely to find some points in all four quadrants. Positive Z products, $Z_X Z_Y$, from Quadrants I and III will offset negative products from Quadrants II and IV. Eventually r will approach and equal 0, as the spread continues and no noticeable relationship appears, as shown in the middle scatter diagram of Figure 11-8. The positive and negative products tend to balance and offset one another completely.

When there is no straight line (linear) relationship between X and Y, r will be close to or equal to 0. However, an r close to 0 cannot be interpreted as "X and Y are unrelated." The relation may be nonlinear.

■ **Example 11-3:** Plot the following points and find r: (0, 6), (1, 3), (3, 1), (7, 0), (11, 1), (13, 3), and (14, 6).

Figure 11-8
Diagrams associated with different correlation coefficients

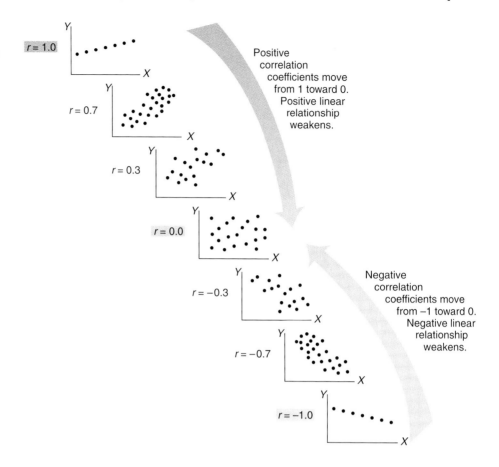

Solution: See Figure 11-9 and the following table for details. Notice how the points are balanced among the quadrants on either side of the $\overline{X} = 7$ axis, causing the products to cancel, yet a curved relationship between X and Y clearly exists:

$$r = \frac{7(140) - (49)(20)}{\sqrt{7(545) - (49)^2} \sqrt{7(92) - (20)^2}} = 0.$$

ID	X	Y	X^2	Y^2	XY
1	0	6	0	36	0
2	1	3	1	9	3
3	3	1	9	1	3
4	7	0	49	0	0
5	11	1	121	1	11
6	13	3	169	9	39
7	14	6	196	36	84
	49	20	545	92	140

Figure 11-9
Plot of a nonlinear relationship

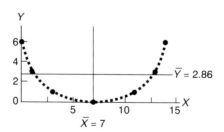

In summary, r will always be a value in the interval from -1 to $+1$. The closer r is to these endpoints, the stronger the linear relationship is. The closer the value is to 0, the weaker the linear relationship is. However, an r value close to 0 does not rule out a nonlinear relationship. A positive coefficient indicates a positive, upward-sloping relationship, whereas a negative coefficient indicates the opposite relationship.

Because the computations grow very tedious as the number of observations increase, we usually resort to computers or even sophisticated calculators. We enter the data and the electronic device reports the correlation coefficient.

Section 11.2 Problems

6. Compute the correlation coefficient from Problem 3 for the stock's opening and closing prices, for a sample of eight consecutive days. Interpret the result. Does your interpretation fit your description of the relationship from Problem 3? What effect does the outlier have on the computation?

7. Compute the correlation coefficients for the two data sets in Problem 4 for the two variables, percentage of students who take the SAT and SAT total. First, use states where 52% or more of the high school seniors in the state take the test. Repeat the exercise for the states where 13% or less take the test. Do the correlation coefficients fit your descriptions from Problem 4?

8. Compute the correlation coefficient for the reaction time and temperature data in Problem 5. Does the coefficient fit your description from Problem 5?

9. Customers can order any number of special candy bars for $2 per bar. In addition, there is a flat fee of $5 for shipping and handling that does not change with the number of bars ordered. Shellie computes the amounts needed to order one, two, three, and four bars, $7, $9, $11, and $13, respectively. Plot the corresponding pairs and find r: (1, 7), (2, 9), (3, 11), and (4, 13).

11.3 Testing to Establish Correlation between Two Variables

It is difficult to present an accurate picture of the strength of the linear relationship unless the correlation coefficient is close to -1, 0, or 1. In cases where coefficients fall in the intervals between these values, we can use hypothesis testing to evaluate the sample results and establish that the variables are correlated.

The null hypothesis for the test for correlation states that there is no correlation between the variables, $H_0: \rho = 0$, and the alternative hypothesis negates the null and states that the variables are correlated.

The particular form of H_1 depends on the direction to be tested. The decision maker may suspect a negative, a positive, or either direction of correlation. The corresponding alternative hypotheses are $H_1: \rho < 0$, $H_1: \rho > 0$, and $H_1: \rho \neq 0$.

The test statistic is based on the familiar T distribution, when $\rho = 0$ and when the X and Y populations meet certain conditions (we usually assume that they do—unless we suspect that X or Y is not normally distributed).

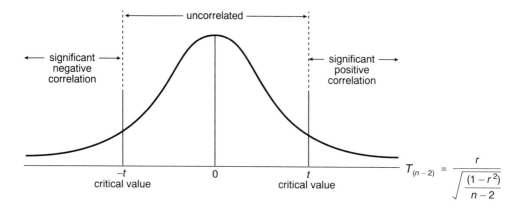

Figure 11-10

Rejection regions' locations for different correlation tests

> Test statistic
>
> $$\frac{r}{\sqrt{\frac{1-r^2}{n-2}}},$$
>
> where r is the estimate of ρ from the sample. The sampling distribution follows a T distribution with $n-2$ degrees of freedom.

The rejection region follows the usual placement according to the direction of the alternative hypothesis as shown in Figure 11-10.

■ **Example 11-1 (continued):** Test to determine if test score and productivity are significantly positively correlated at the 0.05 level, using the sample correlation coefficient, $r = 0.6$, computed from the 10 trainees' observations (p. 492). Interpret the decision.

■ **Solution:** The hypotheses are H_0: $\rho \leq 0$ and H_1: $\rho > 0$ for a test for positive correlation. The test statistic is

$$\frac{0.6}{\sqrt{\frac{1-(0.6)^2}{10-2}}} = 2.12.$$

The critical value from the T distribution with eight degrees of freedom and an area of 0.05 in the tail is 1.860. Because the test statistic, 2.12, exceeds this critical value, we reject H_0 and conclude that test scores and productivity are positively correlated (see Figure 11-11).

Notice that we conclude that the variables are correlated. We still do not know the strength of the linear relationship, but the 0.6 estimate is sufficiently large to establish the correlation. Notice also that we do not establish a cause-and-effect relationship between the two variables. Correlation only measures association, not causation. Box 11-1 explores the issue of causation, specifically superstition, with correlation.

■ **Example 11-2 (continued):** Use the sample correlation coefficient, -0.6, from the six gymnasts' body weights and competition scores (p. 493) to test for correlation between the two variables at the 0.05 level. Interpret the decision.

Section 11.3 Testing to Establish Correlation between Two Variables

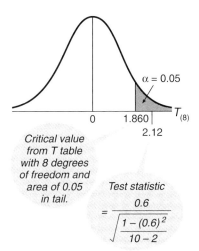

Figure 11-11

Diagram for test for positive correlation between trainees' test scores and productivity

■ **Solution:** Because the test seeks to determine any correlation, positive or negative, it is two-tailed, $H_0: \rho = 0$, $H_1: \rho \neq 0$. The critical values from the T distribution with four degrees of freedom are ± 2.776. The test statistic is

$$\frac{-0.6}{\sqrt{\dfrac{1-(-0.6)^2}{6-2}}} = -1.5,$$

which is between the critical values, so we cannot reject H_0. The data do not support a correlation between weights and scores.

Although the correlation coefficients from the last two examples are the same absolute magnitude (0.6 and -0.6), we decide that one indicates a correlation between the two variables and the other does not. This occurs for two reasons. The second example

---

Box 11-1 Superstition, Correlation, Causation, and Life Lines

Sometimes superstitions develop when people notice that some event seems to regularly follow another: A baseball player may notice that he hits well whenever he wears a certain pair of socks. Generally, there is no reasonable causative connection between such events, although belief in the connection may cause behavior that realizes the superstition: The baseball player may perform well just because wearing the socks gives him confidence.

Similarly, correlation does not imply causation. It is entirely possible to substitute values in the correlation formula and calculate a correlation coefficient for any two variables, whether or not they are related. A result close to -1 or $+1$ does not necessarily mean the variables have a cause and effect relationship, only that they are associated. A correlation coefficient close to -1 or $+1$ suggests that when one variable changes, the other changes in a predictable manner, but it does not verify a causal link between the two, or, if there is such a link, which variable changes first.

To determine if data support the contention that longer life lines in peoples' palms coincide with longer life expectancy, several researchers examined the life lines of 51 cadavers. They used the correlation coefficient to analyze the data. The correlation of left life line length with age at death was 0.056. For right hands, the correlation coefficient was 0.021. Both values are small and insignificant, leading the authors to conclude: "This study is of the greatest importance because it is one of the few instances in which soothsaying has been, in sooth, objectively tested. We happily conclude that palmistry may be used to predict life expectancy but, when it is so used, it is blessedly free of scientific worthiness or usefulness to life insurers."

Source: M. E. Wilson and L. E. Mather, "Life Expectancy," Journal of the American Medical Association, Vol. 229, No. 11, September 1974, pp. 1421–22.

Section 11.3 Problems

10. Test for a negative correlation between X and Y if the sample correlation coefficient, r, is -0.5 from a sample of 25 observations. Test at the 0.05 level. Interpret the decision.

11. Use the sample correlation coefficient, $r = -0.2451$, to test at the 0.1 level for a significant correlation between the opening and closing prices of the stocks from Problems 3 and 6. Interpret the decision.

12. Use the sample correlation coefficients from Problem 7 for the percentage of students taking the test and SAT score to perform two-tailed tests for each group. $r = 0.4$ for the 18 high-percentage states, and -0.5 for the 18 low-percentage states. Test at the 0.01 level. Interpret the decision.

13. Use the sample correlation coefficient, 0.4, from Problem 8, to test for a significant positive correlation between temperature and reaction speed at the 0.1 level. Interpret the decision.

14. While studying the social development of young physically abused children, researchers advanced several measures of these children's behavior, including observed behavior, their peers' reactions, and their teachers' perceptions. They collected this information for 14 abused children and a control group of 14 unabused children. These variables include:

- H/A = teacher's rating of child's hostile-aggressive behavior.
- A/W = teacher's rating of child's anxious-withdrawn behavior.
- H/D = teacher's rating of child's hyperactive-distractible behavior.
- RAT = peer rating of desirability of subject as playmate.
- S/INI = number of times subject initiated nonnegative interaction with peers.
- P/INI = number of times peers initiated nonnegative interactions with subject.
- P/RECP = proportion of subject initiations to which peers respond nonnegatively (negative response would mean ignoring or rejecting the initiation).
- NEGB = number of incidents of negative behavior evidenced by negative comments toward peers, by rough play, and by aggression.

Another aspect of the research on abused children concerns the effect of age on their behavior, that is, does their behavior improve with age. The following table contains values of sample correlation coefficients for each of the behavioral measures and age. Superscripts signify a correlation coefficient that differs significantly from zero, a is significant at the 0.01 level and b at the 0.05 level.

a. Interpret a negative correlation coefficient for age and negative behavior, NEGB.

b. Compare the estimates for the sample of abused children and the control sample to determine if the results indicate opposite behaviors.

c. Compare the pairs of coefficients for the abused and unabused samples for each variable. What does it mean when one is significant and the other is not?

Variables	Correlation Coefficient with Child's Age	
	Abused	Unabused
H/A	-0.101	-0.477^b
A/W	-0.489^b	0.149
H/D	-0.142	-0.655^a
RAT	-0.330	-0.195
S/INI	0.732^a	0.615^a
P/INI	0.163	0.265
P/RECP	0.079	0.493^b
NEGB	-0.469^b	-0.461^b

Source: Mary E. Haskett and Janet A. Kistner, "Social Interactions and Peer Perceptions of Young Physically Abused Children," *Child Development*, Vol. 62, No. 5, 1991, pp. 979–90. Copyright © 1991 by the Society for Research in Child Development.

Summary and Review

Correlation and regression analysis require a data set of matched observations for at least two different variables. When the observations cover a single entity over various time periods, the data set is a time series. A cross section consists of observations on several entities at one point in time.

A graph of corresponding values of two variables is called a scatter diagram. This device is frequently the first step in discovering the association of two variables by obtaining a visual impression of the simultaneous behavior of the variables.

The formula for the sample correlation coefficient corresponds to the calculation of the population correlation coefficient.

Sample Correlation Coefficient

$$r = \frac{n(\Sigma XY) - (\Sigma X)(\Sigma Y)}{\sqrt{[n(\Sigma X^2) - (\Sigma X)^2][n(\Sigma Y^2) - (\Sigma Y)^2]}}.$$

This coefficient measures the direction and strength of the linear relationship between two variables. It is always a value between -1 and $+1$. Negative values indicate a downward-sloping relationship (when one variable increases, the other decreases). Positive values indicate an upward-sloping relationship (when one variable changes, the other changes in the same direction). An r value close to 1 or -1 accompanies data points that fall close to a straight line when graphed in a scatter diagram. An r value close to 0 means there is no linear relationship. However, there may be a nonlinear relationship.

We use the correlation coefficient to determine if two variables are associated. If there is an association, one variable may be easier to measure or more accessible than the variable actually desired for a study. For example, changes in U.S. unemployment rates might be used to describe local unemployment changes if the two measures are highly correlated and the locality cannot afford the time or money to conduct its own employment survey. If one variable is known or usually available before the other, it can be used to predict the behavior of the other.

A hypothesis test that employs the test statistic $r/\sqrt{(1 - r^2)/(n - 2)}$ is useful for establishing that there is or is not correlation between two variables. However, the test settles neither the question of the strength of the linear relationship nor the direction of causation, if any.

Multiple-Choice Problems

15. We use scatter diagrams to
    a. Find the standard deviation of the data set.
    b. Graph the value of r as points approach the origin.
    c. Find the critical value for hypothesis tests about the correlation coefficient.
    d. Obtain a visual impression of the relationship between two variables.
    e. Understand confidence intervals of the correlation coefficient.

16. To use a data set for correlation or regression analysis, there must be
    a. A cross section of time periods.
    b. A pair of values, one for each variable, corresponding to each single entity or time period covered in the sample.
    c. A random set of X values and a random set of Y values, where the number of X and Y values is the same.
    d. At least 30 observations on each of the variables.
    e. Values matched according to the relationship described by a straight line.

17. A data set that contains values for several entities during a single time period is called a
    a. Cross section.
    b. Random sample.
    c. Representative sample.
    d. Time series.
    e. Unitary analysis set.

18. The correlation coefficient, r, is concerned with the product of deviations. Why do we complicate the formula beyond a simple mean of the products of the deviations?
    a. Because the products will balance and offset one another.
    b. Because the mean of these products alone will not indicate the direction or slope of the relationship.
    c. Because the mean of these products alone will not indicate the strength of the relationship.
    d. Because the products will always sum to 0.
    e. To avoid having to use absolute values.

19. In which of the scatter diagrams in Figure 11-12 will the correlation coefficient be closest to 1?
    a. Diagram a.
    b. Diagram b.
    c. Diagram c.
    d. Diagram d.
    e. Diagram e.

20. The sign of r
    a. Is negative when the relationship is weak.
    b. Indicates the direction of change of one variable when the other variable changes.
    c. Is determined by the standard deviations of X and Y.
    d. Is negative when decreases in X accompany decreases in Y.
    e. Is positive when Y values are larger than the mean.

21. If $r = 0$, then
    a. Y decreases when X decreases.
    b. The relationship between X and Y, if there is one, is not linear.
    c. There is no relationship between X and Y.
    d. Changes in Y explain changes in X.
    e. All the (X, Y) pairs fall on a straight line.

Figure 11-12
Choices for Question 19

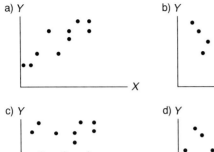
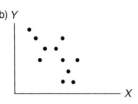
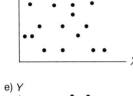
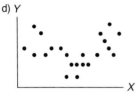
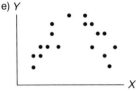

22. For variables to be correlated,
   a. Changes in one variable must cause the other variable to change values.
   b. As one variable increases, the other tends to increase as well.
   c. The correlation coefficient must be 1.
   d. All the (X, Y) pairs in a scatter diagram must fall on a straight line.
   e. As either variable increases, the other tends to change in a predictable direction.

23. Estimate the population correlation coefficient, ρ, with the sample correlation coefficient, r, from the following set. Y is number of sentences to be corrected for grammar and X is the number of mistakes in the exercise.

X	1	5	6	5
Y	10	8	9	7

   a. -7.2.   d. 0.64.
   b. 7.2.   e. -0.64.
   c. 629.

24. If data from different companies produce a correlation between the price of a part and the defect rate of the part of $+0.05$, then
   a. 5% of the time we will make a Type I error if we use the correlation coefficient.
   b. 95% of the time the value is within one standard error of the true value.
   c. The sampling error in the estimation process is 0.05.
   d. The defect rate is $0.05 \times$ price.
   e. The price is not a good indicator of quality.

Figure 11-13 displays Minitab computer output for the scatter diagram and correlation of the purchase of natural gas and electricity in 50 randomly selected municipalities. Use this information to answer Questions 25–27.

25. Based on the scatter diagram, what would you guess the sign of the correlation coefficient will be?
   a. Positive.
   b. Negative.
   c. No sign, because completely random.

26. What is the correlation coefficient?
   a. 1.   d. 0.189.
   b. 0.0001.   e. 99.
   c. 0.811.

27. The value of the correlation coefficient indicates that
   a. Municipalities that are heavy users of one form of energy are heavy users of the other.
   b. Users tend to substitute one form of energy for the other.
   c. The relationship between the two is nonlinear.
   d. There is no relationship between the two variables.
   e. The scatter diagram corroborates the independence of the variables.

28. If a correlation coefficient is -0.95, the minus sign indicates that
   a. X and Y change in the same direction.
   b. X and Y change in opposite directions.
   c. X and Y do not change.
   d. X and Y can only decrease.
   e. Either X or Y or both can only take on negative values.

29. The magnitude of the correlation coefficient -0.95 indicates that
   a. There is no linear association between X and Y.
   b. There is a very weak linear association between X and Y.
   c. There is a very strong linear association between X and Y.
   d. There is a perfect linear association between X and Y.
   e. There is a nonlinear association between X and Y.

Use the following information to work problems 30–33.

Researchers collected information on 28 children, some abused and some not, to learn more about these children's behavior, including observed behavior, their peers' reactions, and their teachers' perceptions. The measures collected include:

H/A = teacher's rating of child's hostile-aggressive behavior.
A/W = teacher's rating of child's anxious-withdrawn behavior.
H/D = teacher's rating of child's hyperactive-distractible behavior.
RAT = peer rating of desirability of subject as playmate.
S/INI = number of times subject initiated nonnegative interaction with peers.
P/INI = number of times peers initiated nonnegative interactions with subject.

Multiple-Choice Problems

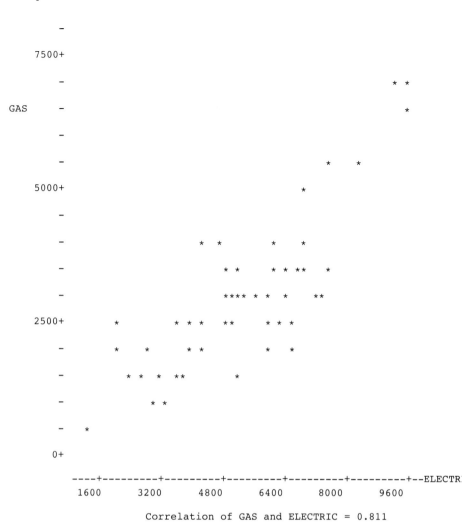

Figure 11-13
Minitab output of scatter diagram and correlation coefficient of natural gas versus electrical energy usage

Correlation of GAS and ELECTRIC = 0.811

$P/RECP$ = proportion of subject initiations that peers respond nonnegatively (negative response would be ignoring or rejecting the initiation).

$NEGB$ = number of incidents of negative behavior evidenced by negative comments toward peers, by rough play, and by aggression.

The following matrix displays the estimated correlation coefficients for these variables for all 28 children (the coefficient provides information about the variable at the head of the column and the variable labeled to the left of the row).

30. Interpret the -0.504 coefficient between negative behaviors ($NEGB$) and the number of subject-initiated nonnegative interactions (S/INI).
  a. There is a 50% chance the child who behaves poorly alone will also behave poorly with other children.
  b. There is a weak upward-sloping linear relationship between the number of negative behaviors and the number of subject initiated nonnegative interactions.
  c. Each time the child behaves negatively, the number of positive interactions decreases by 0.504 on average.

	A/W	H/D	RAT	S/INI	P/INI	P/RECP	NEGB
H/A	0.275	0.785[a]	−0.322[b]	−0.345[b]	−0.238	−0.155	0.361[b]
A/W		0.221	−0.075	−0.301	0.176	−0.097	0.222
H/D			−0.214	−0.361[b]	−0.192	−0.050	0.477[a]
RAT				−0.080	0.294	0.093	−0.323[b]
S/INI					0.260	0.476[a]	−0.504[a]
P/INI						0.204	−0.299
P/RECP							−0.450[a]

Source: Mary E. Haskett and Janet A. Kistner, "Social Interactions and Peer Perceptions of Young Physically Abused Children," *Child Development*, Vol. 62, No. 5, 1991, pp. 979–90.

d. Subjects who initiate nonnegative interactions are also likely to engage in negative behaviors.
e. Those who initiate nonnegative interactions with their peers are involved in fewer negative behavioral incidents.

31. Superscripts signify a correlation coefficient that differs significantly from zero, *a* is significant at the 0.01 level and *b* at the 0.05 level. Study the *significant coefficients' signs,* and identify any exceptions from reasonable expectations. (Hint: H/A, A/W, H/D, and NEGB are negative variables, that is, more of these reflects poorly on the child. The other variables reflect well on the child when their values increase.)
   a. There is one when positive variables are paired with other positive variables.
   b. There are none when positive variables are paired with negative variables.
   c. There are none when negative variables are paired with other negative variables.
   d. There are none among the significant coefficients.

32. Which variables have no significant coefficients?
   a. NEGB.
   b. The variables whose larger values reflect well on a child.
   c. The variables whose larger values reflect negatively on a child.
   d. A/W and P/INI.
   e. All of the variables have at least one significant coefficient.

33. A variable with no significant coefficient implies that the variable
   a. Potentially offers information about the children that is not included in any of the other variables.
   b. Is linearly related to at least one of the other experimental variables.
   c. Reflects the same behavior of the children as the other variables in the experiment.
   d. Has a greater degree of confidence associated with it than with risky, significant variables.
   e. Has some extreme values in the data set.

Use the following data on reported and confirmed cases of polio in five different years in the Americas to work Problems 34–38. Note that during this period an extensive campaign was under way to eliminate the disease by vaccination. In addition, the Pan American Health Organization offered U.S. $100 to anyone reporting a case with symptoms that proved to be polio.

Reported	Confirmed
1,100	800
1,550	1,050
1,650	650
1,986	329
2,103	128
2,360	10

Source: Pan American Health Organization, as reported in Teri Randall, "Rest of World Ready to Follow This Hemisphere's Approach to Eliminating Polio in Near Future," *Journal of American Medical Association,* Vol. 265, No. 7, 1991, pp. 839–40.

34. Ordinarily we would expect reported and confirmed cases to move in the same direction. What might cause us to expect movement in opposite directions in this case?
   a. The values are totals for two continents rather than individual data.
   b. The data are a time series rather than a cross section.
   c. Rewards encourage reporting, whereas vaccination lowers actual occurrence.
   d. People avoid the embarrassment and risk of job loss from reporting their affliction.
   e. Nothing in the available information suggests behavior that differs from the ordinary.

35. A scatter diagram of the data would lead us to expect the correlation coefficient to be
   a. Negative.
   b. Positive.
   c. About 0.
   d. Incalculable.
   e. Unable to predict from scatter diagram, because it appears random.

36. Estimate the correlation between the two variables.
   a. -0.8.
   b. 0.8.
   c. -0.9.
   d. 0.9.
   e. -1.0.

37. What is the value of the test statistic to check for correlation?
   a. -0.9.
   b. 0.9.
   c. -1.11.
   d. -4.1.
   e. -5.4.

38. What is the decision for a two-tailed test for correlation at the 0.05 level?
   a. Reject, $-4.1 < -2.776$, the variables are linearly dependent.
   b. Reject, $-4.1 < -2.776$, the variables are linearly independent.
   c. Reject, $-4.1 < -2.132$, the variables are linearly dependent.
   d. Reject, $-4.1 < -2.132$, the variables are linearly independent.
   e. Reject, $-4.1 < -1.96$, the variables are linearly dependent.
   f. Reject, $-4.1 < -1.96$, the variables are linearly independent.

Word and Thought Problems

39. If $r = -1$ for ACT scores and college performance, are ACT scores useful for predicting college performance? Explain.

40. Accountants must balance an entity's assets with its liabilities and net worth or there is a mistake in the accounts. This basic identity is ASSETS = LIABILITIES + NET WORTH. If a firm's net worth has not changed during any month for the last two years, what would the correlation coefficient be for its assets and liabilities over this span of time?

41. An r of -0.10 is calculated for the variables X = number of complaints satisfactorily resolved in a customer service department and Y = age of the department attendant (or mean age if more than one attendant is on duty).
   a. What can you say about the association between the two variables, based on the r of -0.10?
   b. What if the r value were $+0.10$?
   c. What if the r value were 0.95?

42. a. Plot the scatter diagram for (1, 1), (1, 2), (2, 1), (2, 2).
   b. What kind of relationship appears? What value of r would you expect for this relationship?
   c. Calculate r to verify your answer.

43. A company's products are sold in homes rather than stores—a homeowner often gathers an audience of friends and neighbors. A salesperson has collected the data shown below for the last six in-home group sales. Plot the scatter diagram. Use these data to calculate the correlation coefficient, and interpret the results for the salesperson. Test for correlation at the 0.01 level. Interpret the decision.

Size of Audience	Mean Sale per Person in Group ($)
5	8.50
12	7.00
15	7.50
8	8.00
6	8.00
10	7.50

44. Use the following data to answer Parts a through d.

Time of Day	Time	H = Hours Past 8 AM	E = Total Enrollment (in thousands)	AE = Average Enrollment per Class
Day	8:00	0.00	8	30
	9:15	1.25	12	30
	10:30	2.50	14	30
	11:45	3.75	13	30
	1:00	5.00	8	30
	2:15	6.25	3	30
	3:30	7.50	1	30
Night	5:00	9.00	3	30
	5:30	9.50	4	30
	6:00	10.00	2	30
	6:30	10.50	1	30

   a. Plot a scatter diagram for H versus E and one for H versus AE.
   b. Based on Part a, predict what the correlation coefficient for H and E would be (positive versus negative; close to -1, 0, or 1). Calculate the coefficient to verify your prediction.
   c. What is the correlation for H and AE? If administrators try to determine the relationship between enrollment and time of day, what can you say based on the available information?
   d. The relationship between H and E might change if we distinguish day and night classes. Predict and then find the correlation for day classes. Repeat this process for night classes.

45. Predict the correlation coefficient for the scatter diagrams where the points lie on the indicated lines in Figure 11-14.

46. Calculate the correlation coefficient for age and amount of television watched per day based on the following data. Interpret the result. Then test for correlation at the 0.1 significance level. Interpret the decision.

Age	Hours
4	3.0
6	1.5
14	2.0
16	0.5
8	1.5
19	3.5
8	2.5
10	1.0
4	0.0
17	0.5
16	2.0
19	0.0
12	3.0
10	1.0
8	1.5
14	4.0
9	0.5
16	1.0
7	2.0
8	1.0

Figure 11-14

Scatter diagrams for Problem 45

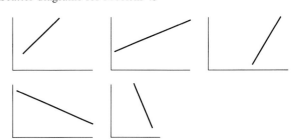

47. Suppose the correlation coefficient for income and political contribution is -0.7. Interpret this result. Is it reasonable or what you expected? Why or why not? If not, what would be a reasonable value? Explain.

48. Calculate the correlation coefficient for position on the ballot (expressed as position divided by total number of candidates, that is, proportion of candidates listed above the subject candidate) and percentage of the vote based on the following data for a random sample of candidates.

Position	Percent of Vote	Position	Percent of Vote
0.40	62	0.12	40
0.11	45	0.74	33
0.50	40	0.20	52
0.33	58	0.50	48
0.16	45	0.20	51

49. Predict the correlation coefficient for population of an area and per capita income (total income/population) for the area. What information would be helpful for the prediction?

50. Would you expect the correlation coefficient for price and quality (longevity) of an item to be positive or negative? Large or small? Explain.

51. In a study of grade inflation, researchers estimated the correlation coefficient between the average grade and the corresponding standard deviation of classes in different disciplines to be -0.886, a statistically significant result. The average correlation between students' performances in the first and second classes among departments designated as low graders is 0.6147 and among high-grader departments it is 0.3681. The values are significant for the low graders, but not always for the high graders. Discuss how these three values evidence grade inflation. (Source: Richard Sabot and John Wakeman-Linn, "Grade Inflation and Course Choice," *Journal of Economic Perspectives*, Vol. 5, No. 1, 1991, pp. 159–70.)

52. Choose 10 pairs of single random digits (X, Y) from the random number table in Appendix I.
  a. If H_0 is $\rho = 0$ and H_1 is $\rho \neq 0$, which hypothesis do you think is true? Why?
  b. Perform the test using your data pairs at the 0.01 level.

53. Calculate the correlation coefficient for hospital size (assets, A) and patient-nurse ratio (PN) based on the following data. Interpret the result. Test for correlation at the 0.05 level. Interpret the decision.

A	PN	A	PN
40	5.0	55	4.1
42	3.2	61	5.2
29	6.0	68	4.6
68	3.5	52	4.4
71	3.8	47	4.6
50	5.5		

54. Forecasters search the behavior of the government's Composite Index of Leading Indicators for signals of future changes in economic activity. The following table displays lagged observations of the Composite Index (L) and the Federal Reserve Index of Industrial Production (P) over a one-year period. $L(i)$ represents the value of L measured i months earlier than the month when P is measured. Thus, $L(3)$ for June is the value of the Index for the preceding March. Calculate r for P and L for each of the columns of L, and comment on the usefulness of the index for forecasting changes in economic activity during this period.

Month	P	L(1)	L(3)	L(6)
Jan.	107.5	145.3	144.4	144.1
Feb.	108.5	145.4	144.6	144.8
Mar.	108.9	144.1	145.3	145.0
Apr.	108.8	145.4	145.4	144.4
May	109.4	145.2	144.1	144.6
Jun.	110.1	146.0	145.4	145.3
Jul.	110.4	146.2	145.2	145.4
Aug.	110.5	146.2	146.0	144.1
Sep.	110.6	144.5	146.2	145.4
Oct.	109.9	143.2	146.2	145.2
Nov.	108.3	141.5	144.5	146.0
Dec.	107.2	139.9	143.2	146.2

Source: Various issues of *Federal Reserve Bulletin*, Board of Governors of the Federal Reserve System, Washington, D.C., and *Survey of Current Business*, U.S. Department of Commerce, Bureau of Economic Analysis, Washington, D.C. Discrepancies, usually created by revisions from later, more accurate data, were slight, so values from later issues are listed.

55. After lagging the Composite Index of Leading Indicators for three months, the computer generates the scatter diagram and correlation coefficient for the Composite Index and the Federal Reserve Index of Industrial Production. Figure 11-15 shows the output for 30 months of data. Interpret the correlation coefficient. Does this value indicate that the behavior of the Composite Index is useful for predicting the production index three months later during this time span? How do these results compare with the results in Problem 54? (A 2 in the figure means 2 data points coincide.)

56. a. Plot the scatter diagram and calculate the correlation coefficient based on the following data on $X =$ percentage of a state's high school graduates who took the SAT test and $Y =$ state's average SAT score. Interpret the result.

X	Y	X	Y
6	964	8	1,033
31	923	5	1,045
11	981	18	917
4	999	56	925
36	899	64	869
16	983	8	997
69	896	59	896
42	897	47	827
39	889	3	1,068
51	823	16	958
47	857	5	1,001

Word and Thought Problems

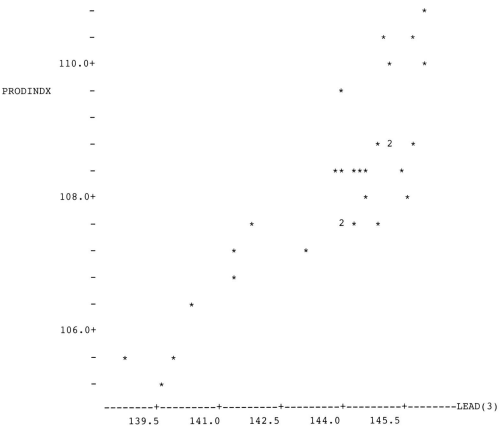

Figure 11-15
Minitab output for composite index of leading indicators and index of industrial production data

X	Y	X	Y
7	995	40	908
14	977	50	885
48	860	59	877
3	1,088	48	790
5	1,045	2	1,075
6	985	9	999
5	975	32	868
46	890	4	1,022
50	889	54	904
65	888	52	888
10	973	19	982
7	1,028	7	968
3	988	10	1,011
10	975	5	1,017

Source: *National Report on College-Bound Seniors*, College Entrance Examination Board, Educational Testing Service, Princeton, N.J., 1982 as reported in Brian Powell and Lala Carr Steelman, "Variations in State SAT Performance: Meaningful or Misleading?" *Harvard Educational Review*, Vol. 54, No. 4, November 1984, pp. 389–412.

b. Test for a negative correlation at the 0.05 level. Interpret the decision.
c. What do these results imply about ranking states' educational attainment by average performance on the SAT?

57. To investigate the relationship between religious orientation and mental health, an education psychologist collected data from 268 undergraduates at the University of Minnesota. The students took a battery of tests to measure psychological traits. The following table lists each test and the psychological trait it measures along with its correlation coefficient with the 268 students' scores on a test to evaluate intrinsic and extrinsic aspects of their religious orientation. An intrinsically religious person accepts and lives by the tenets of religious beliefs for his or her own satisfaction. Extrinsically religious people follow religious tenets to satisfy others or to satisfy other needs of their own that they believe can be achieved through religious behavior. A single asterisk signals statistical significance at the 0.01 level, two asterisks at the 0.005 level, and three asterisks at the 0.001 level. Summarize the findings for each orientation suggested by the significant correlation findings.

Test	Trait Measured (author's descriptions)	Correlation Coefficients	
		Intrinsic	Extrinsic
CES-D	Depression and emotional distress	0.07	0.07
SGT-RW			
Shame	Shame proneness	−0.01	0.17**
Guilt	Guilt proneness	0.30***	−0.04
SWBS			
EWB	Sense of life purpose, direction, and satisfaction	0.13	−0.02
RWB	1. Belief that God loves them and 2. Fulfillment and meaningfulness of relationship with God	0.77***	−0.06
PSI: Functional separation	Management of practical and personal affairs without parental help		
Mother		−0.21***	−0.10
Father		−0.16*	−0.14*
PSI: Attitudinal separation	1. Self-image distinct from parents 2. Own beliefs, attitudes, and values		
Mother		−0.25***	−0.10
Father		−0.29***	0.09
PSI: Emotional separation	Freedom from need for parental approval, closeness, togetherness, and emotional support		
Mother		−0.18***	−0.11
Father		−0.20***	−0.12
PSI: Conflictual separation	Freedom from excessive guilt, anxiety, mistrust, responsibility, inhibition, resentment, and anger in relationships with parents		
Mother		−0.08	−0.16**
Father		0.01	−0.16**

Source: P. Scott Richards, "Religious Devoutness in College Students: Relations with Emotional Adjustment and Psychological Separation from Parents," *Journal of Counseling Psychology*, Vol. 38, No. 2, 1991, pp. 189–96.

58. Use the following information on the number of complaints per 100,000 passengers flown from January to September and the corresponding on-time performance in July (percentage of flights that arrive within 15 minutes of schedule) to compute the correlation coefficient. Interpret the result. Test for correlation at the 0.01 significance level. Interpret the decision.

Number of Complaints	On-Time Performance
7.76	53
4.50	66
3.76	76
2.77	59
2.59	61
1.46	65
0.57	63
0.51	86

Source: Department of Transportation and individual airlines, as reported in Teri Agins and William M. Carley, "Delays, Bumpings and Lost Bags: Travelers Gripe about Air Travel," *Wall Street Journal*, November 10, 1986, p. 21.

Unit II: Simple Regression Models: Introduction

Preparation-Check Questions

59. A scatter diagram
  a. Is used for a shortcut calculation of the standard deviation.
  b. Is used for a shortcut calculation of the correlation coefficient.
  c. Is a picture of a frequency distribution.
  d. Is a picture of a probability distribution.
  e. Depicts the observed values of two variables.

60. The correlation coefficient
  a. Measures the strength and direction of a linear relationship.
  b. Is a product of two different mean relationships.
  c. Is used to change a variance to a standard deviation.
  d. Solves the dilemma of two variables changing in opposite directions simultaneously.
  e. Measures the time span between changes in the earlier and later variable.

61. The notation $E(X)$ represents
  a. The multiplication of E by the sample mean.
  b. The average value of X when the phenomenon is observed a large number of times.
  c. The standard deviation of X.
  d. A test statistic for the difference between two population means.
  e. A sample mean for the sampling distribution of $\bar{X}$.

62. Which of the curves in Figure 11-16 has the smallest standard deviation?
  a. A.   d. D.
  b. B.   e. E.
  c. C.

Answers: 59. e; 60. a; 61. b; 62. a.

Figure 11-16
Choices for Question 62

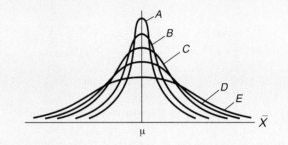

Introduction

Advertising-Sales Relationship The owner of an appliance store recognizes that advertising and other promotions are important in generating sales. She thinks her decision about the level of advertising expenditure could be improved if she knew more about the relationship between sales and advertising.

She records monthly sales and advertising data for the last 14 months in the following table and plots Figure 11-17. The lines in Figure 11-17a show that the two variables behave similarly over the time period, which strengthens the owner's feeling about a regular relationship between them. The scatter diagram in Figure 11-17b also indicates that a relationship may exist.

Period	Month	Sales (thousands of dollars)	AD (hundreds of dollars)	Period	Month	Sales (thousands of dollars)	AD (hundreds of dollars)
1	Jan.	12	1	8	Aug.	19	3
2	Feb.	10	1	9	Sep.	26	3
3	Mar.	18	3	10	Oct.	23	2
4	Apr.	22	1	11	Nov.	27	4
5	May	18	2	12	Dec.	32	6
6	Jun.	19	2	13	Jan.	15	1
7	Jul.	17	2	14	Feb.	11	1

Figure 11-17

Graphical presentation of sales and advertising data for appliance store example

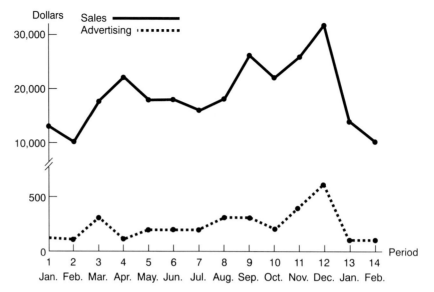

a) Sales and advertising dollars plotted over time

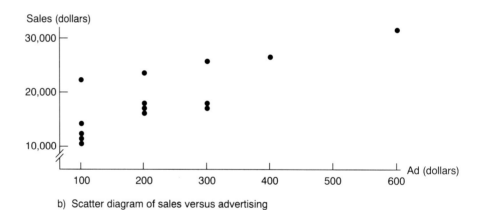

b) Scatter diagram of sales versus advertising

Question: Describe the relationship that appears to exist between advertising and sales.

Answer: There is an upward-sloping relationship (changes in advertising accompany sales changes in the same direction).

The correlation coefficient is 0.83, which also indicates a strong upward-sloping relationship. The owner would like to approximate the relationship with a straight line, such as the one drawn in Figure 11-18.

Figure 11-18

Line that approximates relationship between sales and advertising

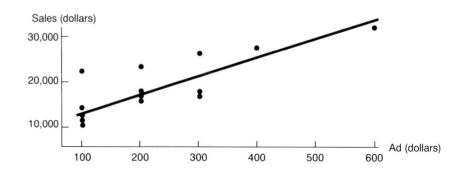

The owner has several questions about the line. Why do some points not fall on the line? Can the line be used for forecasting sales if advertising expenditures are, say, $500? How will sales change if advertising expenditure changes?

The subjects of this unit about linear regression analysis include estimating linear relationships and exploring questions such as the ones we just raised.

11.4 Linear Models

11.4.1 Functions, Predictability, and Causality

Social scientists, businesspeople, and more traditional scientists all search for explanations of the behavior of different variables. The real world is complex. Interactions among a multitude of variables are difficult to detail, so we use a simplified version of a situation that contains the major relationships. We call this simplification of a real-world phenomenon a *model*.

In the introductory example the manager considers the effect of advertising on sales. Including other effects, such as prices and seasonality, would produce a more realistic, yet more complex, model, that would complicate the estimation procedures. This unit introduces a useful method for using simplified models of the relationship between two variables called regression analysis. Section 11.13 extends this discussion to include more variables.

Unlike the correlation coefficient, which measures association but neither assumes nor implies a cause-and-effect relationship, implicit in the regression models we develop in this unit is the idea that we can predict the value of a variable Y from changes in another variable, X. Which variable initiates the process, X, and which is a consequence, Y, are part of the assumptions of the model. We can say that correlation accomplishes the less specific task of establishing whether there is a linear relationship between X and Y (when either one changes, the other changes systematically) and the strength of that relationship. Regression goes further by specifying the exact linear relationship and thereby measuring both the magnitude and the direction of any effect.

Often, we proceed beyond the assumption of predictability when we build models to explain as well as predict behavior among variables. We incorporate the hypothesis that changes in X *cause* Y to change. It is possible to produce a model that predicts well, in spite of the fact that a change in X does not directly *cause* Y to change. For instance, we can measure the impact on a student's grades when the student watches less television, but the variable study time may, in fact, determine both. Hence, it is important to remember that regression results do not *prove* that changes in X cause Y to change, so we should be attentive when we use the word *cause*. In summary, a causal theory may imply a certain specification of the model, which variable is X and which is Y, and we can employ regression analysis to estimate, substantiate, and explore what we know about these models. However, regression analysis can never prove causality. In addition, the future usefulness of regression estimates for prediction may well depend on the real connection between the two variables, not just predictability over the sample data.

Question: In the sales-advertising relationship, changes in which variable, sales or advertising, lead to changes in the other variable?

Answer: The manager, who uses advertising, assumes that changes in advertising will cause sales to change. More specifically, increases in advertising spur increases in sales. However, some firms base their advertising budget on the amount of revenue generated by sales of the product. In this case, changes in sales generate changes in advertising expenditures. A single regression model, such as the ones introduced in this book, concentrates on one direction of the effect. More complicated models incorporate both directions simultaneously, but the computation and analysis of such models is beyond the scope of this book. As we continue, we will question the effect of changes in advertising spending on sales.

It is important to remember which variable's change leads and which variable's value follows from the change. Some notation and terminology help us keep the relevant direction of prediction in mind.

We express the regression model, including its notion of predictability and causality, mathematically as $Y = f(X)$; Y is a *function* of X, when the value of Y *depends* on the value of X. If we specify a value for X, then there is a rule or method for associating a unique Y value with the given X value. $Y = f(X)$ assumes that changes in X will systematically lead to changes in Y. The *independent variable* is X, because its value is given or determined outside (independently of) the model. The *dependent variable* is Y, because the model and the given value of X determine its value; Y depends on the value of X.

Question: Which variable, sales or advertising, in the sales-advertising relationship is the independent variable and which is the dependent?
Answer: Because changes in advertising lead to sales changes, the independent variable is advertising, and the dependent variable is sales. Sales = f(advertising).

There are three reasons for performing regression analysis:

1. To determine if variables are related (is sales a function of advertising?)
2. To measure the relationship (how does each dollar of advertising affect sales?)
3. To predict a specific value of a variable (what would sales be if advertising were $100 or how much should we spend on advertising to achieve a specified level of sales?)

Because we estimate a straight line (linear function) to answer these questions, a review of linear characteristics, especially as they relate to a real-world phenomenon, is in order.

11.4.2 Linear Functions

The rule for associating a unique Y with an X value can take many forms. Often, the rules take on a mathematical form, such as $Y = f(X) = 5 + 3X$. Suppose $X = 2$. Then $Y = 5 + 3(2) = 11$.

We call equations of the form $Y = a + bX$ (where a and b are specific constant values, while X and Y are variables) *linear equations* or *linear functions*. The graph of this equation is a straight line.

When graphing, we label the horizontal axis with units of the independent variable, while the vertical axis corresponds to values of the dependent variable. The value of Y when $X = 0$ is the Y *intercept*, because the line crosses the Y axis at this point or value.

In the general linear equation, $Y = a + bX$, the value of a is the Y intercept.

Question: What is the intercept in $Y = 100 - 50X$?
Answer: The intercept is 100, because Y is 100 when X is 0, $100 = 100 - 50(0)$.

Section 11.4 — Linear Models

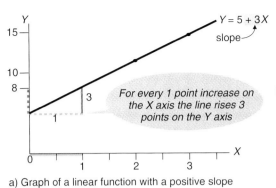
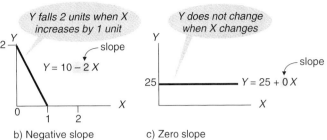

a) Graph of a linear function with a positive slope
b) Negative slope
c) Zero slope

Figure 11-19
Effect of slope (coefficient of X) on the graph of a linear function

In the general linear equation, $Y = a + bX$, the coefficient of X, b, is the *slope*.

The value of the slope describes how Y changes (direction and magnitude) when X increases by one unit. We can illustrate this concept with the linear function $Y = 5 + 3X$. The coefficient of X here is 3, so the slope of the line is 3, or the value of Y rises three points for every one unit increase (horizontally) in the value of X. *In general Y changes by b units (rises when b is positive and falls when b is negative), whenever X increases by one unit along the horizontal axis.*

Using the two points, $(2, 11)$ and $(3, 14)$, we can graph the function, as shown in Figure 11-19a. The upward or positive slope is apparent.

Question: What is the slope of $Y = 10 - 2X$? Interpret the answer. Sketch a graph of the function.

Answer: b = slope = -2. This value is negative, which means Y decreases by two units when X increases by one unit. A downward-sloping relationship, as depicted in Figure 11-19b, results.

Question: What is the slope of $Y = 25$? Interpret the answer.

Answer: The function is $Y = 25 + 0(X)$, so b = slope = 0. The Y value does not change when X changes. Thus, changes in X do not affect Y. We could say that the model is misspecified (see Figure 11-19c).

Question: Suppose monthly utility bills measured in dollars (B) are a function of the size of the home (S), measured in square feet. Interpret the value, 0.08, if the relationship is $B = 50 + 0.08\,S$.

Answer: The bill increases by 8 cents per month for each additional square foot of space in the home.

Regression analysis requires that we specify the variables X and Y, such as X = advertising and Y = sales and then estimate the values of a and b in order to estimate the regression line, $Y = a + bX$. We use data for X and Y, matched observations on the two variables for either n time periods or n entities, to produce $\hat{a}$ and $\hat{b}$, estimates of a and b, respectively. For the sales-advertising relationship, we have 14 observations on sales, each matched with the advertising value that we hypothesize affected the observed sales value. The data set is a time series, because the 14 observations on the variables cover 14 time periods or months. The estimated regression line is $\hat{Y} = \hat{a} + \hat{b}X$. Thus, $\hat{b}$ is an estimate of the effect of a unit increase in X on Y. In the sales-advertising example, $\hat{b}$ estimates the effect of a unit increase in advertising ($100 expenditure for advertising) on the sales of the firm (sales expressed in thousand-dollar

Section 11.4 Problems

63. Consider $G = a + bH$, where G is the grade on a test and H represents hours spent preparing for the test.
    a. Which is the dependent variable and which is the independent?
    b. Interpret b in the context of the grade-study time model and tell what sign (+ or −) you think it would have.
    c. If the estimated model is $\hat{G} = 55 + 6H$, what is $\hat{b}$? Interpret this specific value. Is it a reasonable value?
    d. A model is a simplification of a real-world phenomenon. What complexities are omitted in the model $G = a + bH$?

64. Interpret the coefficient in a regression model estimated in order to explain the amount of money bet on different horses in a race, $\hat{A} = 12{,}000 + 1{,}500W$, where A is the amount bet on a horse (in thousands of dollars) and W is the number of races won by the horse. Speculate on additional variables that influence A that are omitted from the model.

65. Consider the simple regression model $W = f(I)$, where W is the proportion of objects at an art show that are weavings and I is the average income of visitors to the show according to surveys randomly distributed on exiting the exhibit.
    a. What is the independent variable in this model? the dependent variable?
    b. Write the general form for a linear model here.
    c. Interpret the coefficient in this linear model. What sign would you anticipate for this model? Why?

66. Suppose that the following regression model is constructed to explain average mileage per gallon of gasoline (M): $\hat{M} = 28.3 - 3.5CL$, where CL is the size of each automobile's engine (in cubic liters). There are 40 automobiles in the sample.
    a. What is the independent variable? What is the dependent variable?
    b. Is the data set a time series or a cross section?
    c. Interpret the -3.5.
    d. Name some other variables that have been omitted to simplify the phenomenon of gasoline mileage in this model.

11.5 Criterion for the Estimated Regression Line

Returning now to the sales-advertising relationship and Figure 11-18 (p. 514), we will fit a straight line of the form $Y = a + bX$ to the data (where Y = sales and X = advertising expenditures). When there is only one independent variable in the equation or model, we call this estimation and analysis procedure *simple linear regression*. Because we do not know the values of a and b that position the line on the diagram, we must use the available data to obtain estimates instead, $\hat{a}$ and $\hat{b}$.

To obtain a unique line that "best fits" the data, we must determine a criterion that defines "best." An estimated line that minimizes the distance between the given data points and this estimated line seems desirable. The estimated regression line, $\hat{Y} = \hat{a} + \hat{b}X$, will consist of points $(X, \hat{Y})$, where we derive $\hat{Y}$ from substituting X in the estimated equation. The *residual*, $\hat{e}$, is the vertical distance between a point on the estimated line, $(X, \hat{Y})$, and a point (X, Y) from the data set. As Figure 11-20 shows, for any specific X value, we take the corresponding Y value from the data set and the $\hat{Y}$ from the estimated line to obtain the residual

$$\hat{e} = Y - \hat{Y}.$$

A logical and intuitive candidate for the regression, or estimated, line, is one that makes these residuals as small as possible, so that their sum is the smallest possible value. However, problems develop when we apply this proposal. Suppose we choose the horizontal line where Y is always equal to the mean of the Y values in the data set, $\hat{Y} = \overline{Y}$, to fit the points (see Figure 11-19c).

Question: What are $\hat{a}$ and $\hat{b}$ for this line?

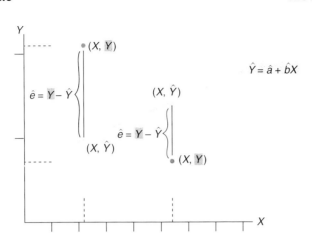

Figure 11-20
Residuals about a regression line

Answer: If the line $\hat{Y} = \hat{a} + \hat{b}X$ reduces to $\hat{Y} = \overline{Y}$, then $\hat{a} = \overline{Y}$ and $\hat{b} = 0$ or $\hat{Y} = \overline{Y} + 0(X)$.

This answer implies that changes in X do not affect Y. The sum of the residuals in this case, $\Sigma\,\hat{e}$, becomes $\Sigma\,\hat{e} = \Sigma\,(Y - \overline{Y})$. Remember from the discussion of standard deviations in Chapter 3 that the sum of deviations about a mean for a data set will always be zero. $\Sigma\,(Y - \overline{Y}) = 0$. Thus, the sum of the residuals about this line is zero, a small value; however, because $b = 0$, the line $Y = \overline{Y}$ indicates that changes in X do not affect Y. Such a model contradicts the hypothesis that Y depends on the value of X, $Y = f(X)$.

The solution criterion is similar to the solution we use to formulate the standard deviation to measure the dispersion of data about the mean. Because the deviations always sum to zero, we square the deviations. Now we square the residuals and search for *the line that will minimize the sum of the squared residuals*, $\Sigma\,\hat{e}^2$.

11.6 The Estimators, $\hat{a}$ and $\hat{b}$

We use calculus methods (omitted here) to find formulas for the estimates of a and b that form the least-squares line, the line that minimizes the sum of squared residuals. These formulas are

$$\hat{b} = \frac{n\Sigma XY - (\Sigma X)(\Sigma Y)}{n(\Sigma X^2) - (\Sigma X)^2}, \qquad \hat{a} = \frac{\Sigma Y - \hat{b}\Sigma X}{n} = \overline{Y} - \hat{b}\overline{X}$$

where n is the number of observations or data points ($n = 14$ in the sales-advertising example).

■ **Example 11-4 (Sales-Advertising Relationship Continued):**
Use the sales-advertising data presented earlier (p. 513), and calculate $\hat{a}$ and $\hat{b}$.

■ **Solution:** From the calculations shown below, we obtain

$$\hat{b} = \frac{n\Sigma XY - (\Sigma X)(\Sigma Y)}{n(\Sigma X^2) - (\Sigma X)^2} = \frac{14(713) - 32(269)}{14(100) - (32)^2} = \frac{9982 - 8608}{376} = 3.7$$

$$\hat{a} = \frac{\Sigma Y - \hat{b}\Sigma X}{n} = \frac{269 - 3.7(32)}{14} = \frac{150.6}{14} = 10.8.$$

Figure 11-21
The estimated sales = f(advertising) model

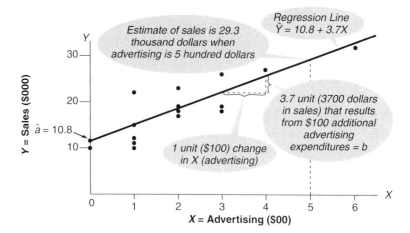

Period	Sales (Y)	AD (X)	XY	X^2
1	12	1	12	1
2	10	1	10	1
3	18	3	54	9
4	22	1	22	1
5	18	2	36	4
6	19	2	38	4
7	17	2	34	4
8	19	3	57	9
9	26	3	78	9
10	23	2	46	4
11	27	4	108	16
12	32	6	192	36
13	15	1	15	1
14	11	1	11	1
	$\Sigma Y = 269$	$\Sigma X = 32$	$\Sigma XY = 713$	$\Sigma X^2 = 100$

So the regression equation is $\hat{Y} = 10.8 + 3.7X$, as shown in Figure 11-21.

Question: Interpret the 3.7.
Answer: A single-unit increase in advertising will generate a 3.7-unit increase in sales. Because we measure sales in units of $1,000, and an advertising unit is $100, an additional $100 spent on advertising will generate $3,700 in additional sales (see Figure 11-21).

Question: If advertising is $500, estimate sales.
Answer: Estimated sales = $\hat{Y} = 10.8 + 3.7(5) = 29.3$, or $29,300. Figure 11-21 illustrates this point.

We will assume that the equation is linear when we use this procedure, although we can use the method for some nonlinear equations that we can mathematically transform to look like a linear model, a procedure that is beyond the scope of this book. We can also use the linear function to approximate segments of a curve that are relatively linear. For instance, Figure 11-22 depicts a relationship between sales and advertising that bends toward the horizontal at higher levels of advertising expenditures, because the effect of additional advertising expenditures on sales is not as great after large expenditures (which means high levels of audience exposure already). However, if we are only interested in the range of advertising expenditures for which we have data, advertising between $100 and $600, the linear assumption appears to be appropriate. Often, we can

Problems

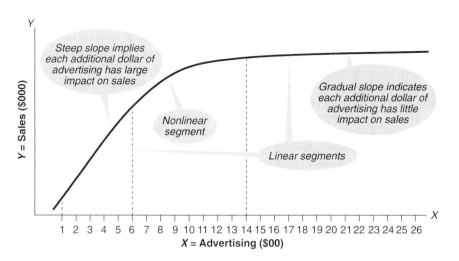

Figure 11-22
Segments of a curve can be approximately linear

examine the scatter diagram to determine if a straight line is likely to fit the data well. In Unit III, we discuss other procedures that we use to evaluate the fit.

In summary, we compute $\hat{a}$ and $\hat{b}$ using the formulas introduced here and sample data for the dependent and independent variables. The resulting estimates form a line that minimizes the sum of the squared residuals between the data points and the estimated line. No other values or combination of values for $\hat{a}$ and $\hat{b}$ will produce a smaller sum of squared residuals.

Section 11.6 Problems

67. a. Use the following data from Example 11-2 to estimate the regression line for a model where a gymnast's score depends on the gymnast's physical condition as measured by his or her weight.

Gymnast	Weights	Scores
1	100	1,030
2	110	980
3	110	1,000
4	90	1,050
5	100	1,040
6	90	1,000

  b. Interpret the regression coefficient.
  c. Forecast the score of a 105-pound gymnast.
  d. Study the scatter diagram for this data set in Figure 11-5 (p. 496). Does it appear that a straight line will provide a good fit? Explain your answer.

68. a. Estimate a regression line to determine the effect of room temperature (in degrees Fahrenheit) on the reaction time (in seconds) for the chemist from Problem 5.

Row	Temperature	Reaction Time
1	75	9.54
2	74	9.22
3	73	9.14
4	71	8.83
5	77	9.73
6	73	9.53
7	70	9.50
8	74	8.83
9	72	8.79
10	72	9.37

  b. Interpret the coefficient.
  c. Forecast the reaction time when the chemist conducts the experiment in a 72.5° room.

69. a. Use the following data on number of credit cards issued, C, expressed in millions and the collectable debt expressed as a percentage of credit card amounts due, D, to determine the effect of issuing cards on the level of collectable debt.

ID	Year	Cards	Debt
1	1981	116	98.06
2	1982	117	98.32
3	1983	124	98.62
4	1984	142	98.60
5	1985	160	97.67
6	1986	181	96.68
7	1987	206	97.40
8	1988	224	96.96
9	1989	242	96.84

Source: Credit card companies, Federal Reserve Board, and Bankcard Holders of America, as reported in "Credit Cards in the U.S.," *Asheville Citizen-Times*, January 5, 1992, p. 4B.

b. Interpret the estimated coefficient.
c. Plot a scatter diagram to demonstrate the estimated relationship. Comment on the appropriateness of the linear assumption.

70. To investigate whether forecasters produce unbiased predictions, an analyst estimated the following model $A_t = f(P_t) = a + bP_t$, where A_t is the actual value for period t and P_t is the predicted value for period t. He estimated the model for three different economic variables (inflation rate, GNP growth, unemployment rate) for five different forecasters.

a. If the forecasters produce unbiased predictions, that is, the predictions are correct on average, what are the ideal values for a and b?
b. Which of the results below suggest unbiased forecasts?

Forecaster	Inflation Rate		Real GNP Growth		Unemployment Rate	
	a	b	a	b	a	b
Economists-consensus	4.72	0.43	0.35	0.81	−0.40	1.05
Econometric models						
Consensus	5.04	0.38	0.38	0.79	−0.06	1.00
Chase Econometrics	5.42	0.32	0.60	0.77	1.47	0.75
Data Resources	4.64	0.44	0.22	0.84	−0.54	1.07
Wharton Associates	4.91	0.39	−0.05	0.92	−0.04	1.00

Source: Thomas B. Fomby, "A Comparison of Judgmental and Econometric Forecasts of the Economy: The Business Week Survey," *Economic Review*, September 1982, pp. 1–10.

11.7 The General Form and Assumptions of a Regression Model

This section describes the general form of the simple regression model and assumptions that are appropriate for obtaining good estimates from the $\hat{a}$ and $\hat{b}$ formulas introduced in the preceding section.

11.7.1 The Line of Means

We *fit* a straight line to the data points to approximate the behavior of Y when X changes, but if the relationship between X and Y is linear, you may wonder why some of the points do not fall on the line. In fact, numerous values of Y may correspond to a single value of X in a data set. In the sales-advertising example, a given advertising expenditure may generate different sales levels at different times (see periods 1, 2, 3, 13, and 14 in the data set on p. 513). Multiple points above an X value are common. Which points will fall on the line?

To answer this question, first, let the mean of all the Y values that correspond to a single X value be $\overline{Y}_X$. For instance, if advertising is $100 per month, sales can be many values, because for a given amount of advertising, people do not always buy the same amount of the product. They may happen to be in a store for some other purpose and decide to purchase the advertised product while they are there. Other times, they may postpone purchases because of unexpected interruptions of their shopping plans. Suppose that the mean of the sales values is $1,500, when advertising is $100. Similarly, imagine finding a particular $\overline{Y}_X$ value that corresponds to each of the other X values.

Question: Consider that there are numerous sales values that correspond to an advertising expenditure of $500. Would you expect the average sales values at this level of advertising would be greater than the mean sales when advertising is $100?

Answer: When advertising is $500 rather than $100, we expect that the increased level of advertising informs more people about the product or persuades more people to purchase it. Consequently, the mean sales level would be greater.

If the sales-advertising relationship is linear, a line that connects all the mean sales values will be a straight line. This line is the true regression line. For the general case, we write $\overline{Y}_X = a + bX$. Figure 11-23 shows this *line of means*.

There are several reasons why Y values do not always fall on the true regression line.

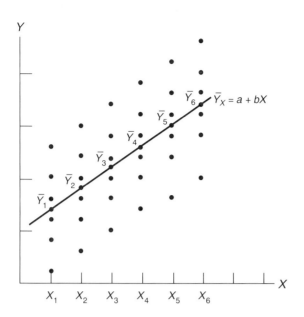

Figure 11-23
Regression line or line of means

1. *Omitted variables:* The model is simple, because it has only one independent variable. Other independent or explanatory variables might reduce the errors in the model. For example, sales depend on variables other than advertising, such as prices, competitors' prices, and competitors' advertising.
2. *Errors in measurement:* The numbers in the data set are subject to error. We can make a mistake when we transcribe or data code for the computer. The values may not measure the exact variable or concept we have in mind. For instance, advertising expenditure aggregates different types of advertising. Minutes of television commercials might explain sales levels better than newspaper advertising, but both may be lumped together in expenditure data.
3. *Response differences in subjects:* Behavior exhibited at a given level of the independent variable is not always the same. At a certain level of advertising, a group of people are exposed to the presentation. Depending on who happens to be exposed, different levels of response follow. This effect, added to the variety of responses of different people to the same stimulus (the advertisement), will result in multiple sales levels for the same level of advertising.
4. *Random effects:* Unexpected events, such as war, drought, and disaster, interrupt normal responses, causing points to deviate from the regression line. A tornado passing through town on the day of an advertisement will disrupt normal buying behavior.

For these reasons, there are usually numerous Y values that correspond to any given X value. The mean of the Y values for a given X is $\overline{Y}_X$. When the $\overline{Y}_X$ values form a straight line, $\overline{Y}_X = a + bX$, $\hat{Y}$ will be an estimate of $\overline{Y}_X$ that we derive from the estimated line, $\hat{Y} = \hat{a} + \hat{b}X$.

11.7.2 The Error Term and Its Properties

For a given X value there is a distribution of possible Y values. Consider the individual Y value shown in Figure 11-24. Remember that the residual is the difference between Y and $\hat{Y}$, $\hat{e} = Y - \hat{Y}$. Similarly, the *error term* is the difference between Y and $\overline{Y}_X$, the true mean of the Y values corresponding to X, $e = Y - \overline{Y}_X$. For every Y there is a corresponding e. Most of the assumptions about the model involve these error terms.

Figure 11-24

True and estimated models

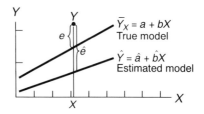

These assumptions avoid complex and unusual situations, so the formulas for $\hat{a}$ and $\hat{b}$ are relatively simple. Basically, we expect to construct a model so that the error terms will be totally random. That is, they are devoid of any information about the relationship between X and Y. If a situation violates these assumptions, the formulas for $\hat{a}$ and $\hat{b}$ would not necessarily provide the best estimates of the actual a and b. If the e values behave in a predictable or systematic pattern that drastically violates the assumptions, we may need to incorporate this information in $\hat{a}$ and $\hat{b}$ in order to improve the estimates, although less extreme violations often have very little effect on the regression results. In addition, the assumptions described below are common in regression studies. They are usually the starting point, even when more complicated models become necessary to obtain more reliable estimates of a and b.

The Probability Distribution of the Error Term, e If we separated the e values that correspond to a particular X value, these e values follow a normal distribution. Figure 11-25 shows that this assumption must hold for each X value. This assumption is the basis for tests that we will use to evaluate the estimated model in Unit III. When this is not true, we cannot apply our conclusions from the test to the model.

The Mean of the Error Term, μ_e Just as the Y values that correspond to a given X have a mean, $\overline{Y}_X$, the e values that result from each of these Y values have a mean, μ_e. Because the sum of the deviations in a data set is zero, it is appropriate that the sum of the error terms, which are really deviations, be zero. Thus, the mean is zero.

Question: What does it mean if a single e is zero?
  Answer: There is no error. The observed Y value is the mean value $\overline{Y}_X$. A nonzero value implies that the observed behavior differs from the average.

Thus, a zero mean for the e is intuitively appealing. We might say that the errors tend to cancel or that the typical e value is zero.

The Standard Deviation of the Error Term, σ_e The standard deviation measures the spread or dispersion of values about their mean. For a given X, the standard

Figure 11-25

Normally distributed error terms, e, for each X value

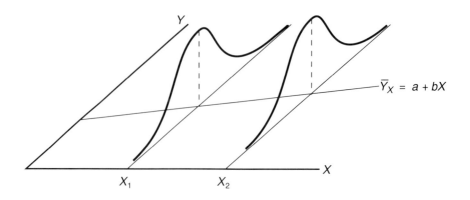

deviation of the corresponding *e* values describes deviations from the regression line. The formula would be $\sigma_e = \sqrt{[\Sigma(e - \mu_e)^2]/N} = \sqrt{\Sigma e^2/N}$, because $\mu_e = 0$ (the last assumption). We call this value the *standard error of the estimate*.

We assume that this standard deviation is the same constant value for each *X*. Notice that the points corresponding to each *X* in Figures 11-23 and 11-25 appear to have the same spread about their means.

We can better understand the importance of this assumption if we consider a situation that violates the prescribed condition. For example, let *Y* be the size of a country's international trade balance, and let *X* be the size of the country's economy as measured by its gross domestic product (GDP). We would expect countries involved in large amounts of business to generate larger trade surpluses and deficits than those generated by small countries. The result is a difference in the standard deviation of *e* values, depending on the size of the economy, *X*. Figure 11-26a displays distributions of two *e* variables where the standard deviations differ.

Question: If the data set contains *Y* values that correspond to points X_1 and X_2, which points are more valuable for estimating the regression line, ones associated with X_1 or ones associated with X_2? Use the distribution for the corresponding e_1 and e_2 values superimposed on the true regression line in Figure 11-26b.

Answer: *Y* values corresponding to X_1 are more valuable and should receive greater weight in the calculations, because e_1 is more likely to be close to zero, which is just another way of saying *Y* values associated with X_1 are more likely to be close to their mean, $\overline{Y}_X$, a point on the true regression line.

When the standard deviations are constant, no observation is more valuable than any other for locating points on the regression line. Consequently, the formulas for $\hat{a}$ and $\hat{b}$ weight all points in the data set equally.

Independent Error Terms If we know a value of *e* that corresponds to a particular *X* value, say e_1 that corresponds to X_1, then e_1 does not relay any information about the other values of *e* that correspond to the same or to any other *X* value. The error terms are independent.

Again to understand the importance of this assumption, consider the situations shown in Figure 11-27a that violate this assumption. The *e* values are predictable from one observation to the next. In the first diagram, errors change sign from one observation to the next. In the second and third diagrams, the errors tend to retain the same sign from one observation to the next. We should incorporate such predictable behavior in the estimates; otherwise, it can distort the estimates, as shown in Figure 11-27b. However, we expect the error terms in the models that we consider to be random and, hence, independent, so that there is no useful information in the error terms about any other error term or about the relationship between *X* and *Y*.

Figure 11-26
Standard error of estimate not constant

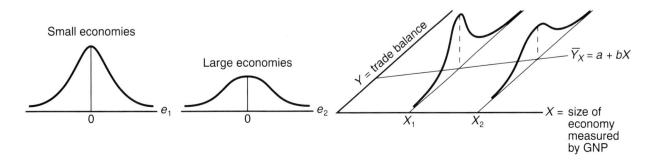

Figure 11-27

Dependent error terms and distortion of the estimated regression line

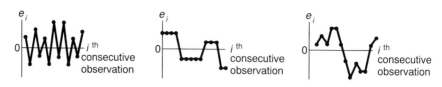

a) Dependent error relationships

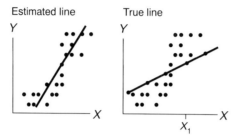

b) Comparisons of estimated and true regression line when error terms are dependent

Economic and business situations often have a problem with independence of error terms, because economic behavior tends to change slowly. In a time series of economic variables it is possible for an observation, such as spending on buildings and equipment, to be above a long-run relationship. If the independent variable, say the GDP, and the building and equipment expenditure do not change by much, we may well find a positive error in the following time period. Thus, the value of one error term is useful for predicting another error term.

The following box summarizes the discussion of the error term behavior that we assume occurs in the problems in this book.

Assumptions about the Error Term, e, in the General Regression Model

1. The error terms associated with each X value are normally distributed.
2. The mean of all of the e values above any X value is zero.
3. The standard deviation of the set of e values associated with each X is the same number.
4. The e values associated with one X value are independent of the e values associated with any other X value.

Again, if we attempt to estimate a regression line for a situation that does not satisfy any one of these four conditions, the quality of the estimates may diminish, depending on the severity of the violations. We could improve their quality by incorporating any systematic behavior of e that affects the $X - Y$ relationship into the estimates of a and b. In some cases, unreasonable estimates, such as a wrong coefficient sign, may result from such violations rather than from poor theories and unreasonable expectations.

Section 11.7 Problems

71. If the values of error terms are random, will knowing the value of X help us predict an e value? Give an intuitive description of the behavior of nonrandom error terms and what this behavior means for the model.

72. A presidential candidate's performance in the primaries is poorer in states in which he makes many public appearances prior to the vote than in states where he seldom appears. However, he has had a tendency to do

Figure 11-28
Candidate's primary performance model

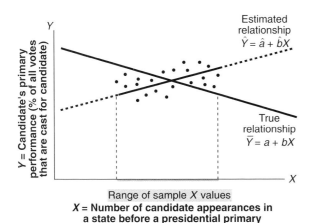

X = Number of candidate appearances in a state before a presidential primary

Figure 11-29
Scatter diagrams not suitable for the model

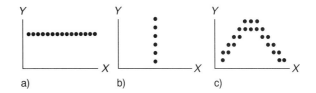

well or poorly in spurts; that is, a win in one primary spurs activities that produce more wins or good showings, or a poor showing spawns a series of other poor performances. Figure 11-28 pictures the true relation between the percentage of primary vote for the candidate and the number of preprimary appearances in the area along with a sample of observations on percentage of votes and number of appearances that might be collected for a regression analysis. Discuss the implications of using the estimated model shown in Figure 11-28 for forecasts of future primary performances of the candidate. What assumption is violated?

73. A person exhibits erratic spending behavior from one pay period to the next. During one period, he overspends; during the next period, he overreacts and cuts his spending drastically followed by overspending, then underspending, and so on. Generally, his income and spending are increasing over time. We try to explain this person's monthly spending with a regression model, spending = f(income).

  a. A particular point on the regression line might be ($100, $125). Use the notion of the line of means to interpret this point.
  b. Which assumption about the error terms appears to be violated?

74. Consider a model in which a person's expenditure for food is a dependent variable and the person's age is the independent variable. Discuss the possibility that such a model violates the assumption of a constant standard error of the estimate for each age level from youth through elderly ages.

75. a. Explain, intuitively, the problem that would result from trying to estimate the model $Y = f(X)$, if the X values in the sample are identical.
  b. What problems would result from trying to estimate the model $\overline{Y}_X = a + bX$ for the data sets in the scatter diagrams shown in Figure 11-29?
  c. Is the data set below useful for regression estimates? Explain.

X	Y
25	43
82	43
69	43
10	43
12	43

11.8 Regression and the Computer

Figure 11-30 displays basic regression output from Minitab. The computer output is based on the sales-advertising data from Example 11-4, so we can compare the computer-generated values with the earlier hand-calculated measures. The circled numbers correspond to the following items (we have not discussed some of these yet):

1. The estimation regression line, $\hat{Y} = \hat{a} + \hat{b}X$.
2. The estimated constant term or Y-intercept, $\hat{a}$.
3. The estimated regression coefficient, $\hat{b}$.
4. The estimated standard error of the coefficient, $\hat{\sigma}_{\hat{b}}$.
5. The T-ratio for a test for a zero regression coefficient, $\hat{b}/\hat{\sigma}_{\hat{b}}$.
6. The two-tailed p-value for a test for a zero regression coefficient.
7. The estimated standard error of the estimate, $\hat{\sigma}_e$.
8. The coefficient of determination, R^2.

Figure 11-30
Minitab output for regression model, sales = f(advertising)

```
   The regression equation is
①  SALES = 10.9 + 3.65  AD

   Predictor      Coef         Stdev       t-ratio           p
   Constant   ② 10.862         1.905        5.70        0.000
   AD         ③ 3.6543      ④ 0.7127     ⑤ 5.13        0.000  ⑥

⑦  s = 3.694     ⑧ R-sq = 68.7%       R-sq(adj) = 66.0%

   Analysis of Variance

   SOURCE         DF            SS           MS           F           p
   Regression      1        358.64       358.64       26.29       0.000
   Error          12  ⑨    163.72        13.64
   Total          13        522.36

⑩  Unusual Observations
   ⑪        ⑫         ⑬            ⑭                    ⑮           ⑯
   Obs.     AD       SALES          Fit      Stdev.Fit   Residual    St. Resid
     4     1.00     22.000       14.516         1.347       7.484       2.18R ⑰
    12     6.00     32.000       32.787         2.825      -0.787      -0.33 X ⑱

   R denotes an obs. with a large st. resid.
   X denotes an obs. whose X value gives it large influence.
```

9. The degrees of freedom for a T-test of the regression coefficient.
10. Unusual observations section.
11. Order of unusual observation in the data set.
12. Value of independent variable for unusual observation, X.
13. Actual value of dependent variable for unusual observation, Y in data set that corresponds to X shown in item number 12.
14. Estimated value of dependent variable for unusual observation, $\hat{Y} = \hat{a} + \hat{b}X$, where X is the value from item number 12.
15. Residual for unusual observation, $\hat{e} = Y - \hat{Y}$, where Y and $\hat{Y}$ are taken from item numbers 13 and 14, respectively.
16. Standardized residual of unusual observation.
17. Potential outliers, observations with large standardized residual designated with an "R" beside the standardized residual that pulls estimated line in its direction in order to minimize sum of squared residuals (standardization procedure is beyond the scope of this book).
18. Potential leverage points, observations with large influence on position of estimated regression line, because the X value is very different from other X values in the data set, designated with an "X" beside the standardized residual (distance calculation is beyond the scope of this book).

First of all, notice that the computer-generated regression line is very close to our own result, $\hat{Y} = 10.8 + 3.7X$. The small rounding difference in the estimated Y intercept, 3.8 versus 3.9, and in other measures on the printout often occur with regression estimates. We will continue to use 3.8 in the following discussion. The standard error of the estimate, $\hat{\sigma}_e$, is \$3.694, the "average" error or "average" distance of a Y value in the data set from the corresponding $\hat{Y}$ on the regression line.

Next, we need to notice the unusual observations section (we discuss the other items in Unit III). This additional information at the bottom of the printout assists us when we determine if some data points unduly affect the estimates; this situation occurs when outliers pull the estimated line in their direction in order to minimize the sum of squared residuals. Locate the fourth and twelfth observations, (1, 22) and (6, 32), in the sales-advertising scatter diagram in Figure 11-31. The residual associated with the fourth observation is quite large. This is comparable to being 2.18 standard deviations

Section 11.8   Regression and the Computer   ■ **529**

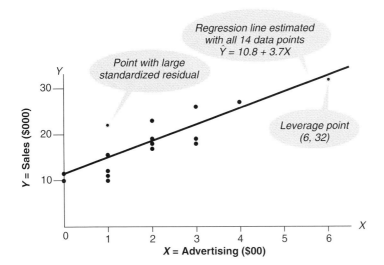

Figure 11-31
Unusual points in the sales-advertising data set

from the mean, very close to the 2.5 that we defined as an indication of an outlier in Chapter 3. Minitab denotes any observation with a standardized residual greater than 2 with an "R." We call the twelfth observation a *leverage point,* because the X value is very different, perhaps an outlier, when compared to the mass of X values in the data set. In this case, 6 is more than 2.5 standard deviations above the mean advertising value of $2.3, although Minitab uses a more extensive definition of distance and influence.

Question: Part a of Figure 11-32 is an abbreviated presentation of the computer regression estimate when we omit the fourth observation from the data set, while Figure 11-33 contrasts the results on a scatter diagram. Describe the effect of the fourth observation on the estimates.

Answer: When we compare the line estimated without observation 4 to the original line we estimated with all 14 data values, we find that including observation 4 causes the regression line to flatten. The Y intercept increases from 9.3 to our original 10.8, and the slope decreases from 4.1 to our original 3.7. The left part of the line rotates upward toward the fourth observation in order to reduce the residual associated with this point and, thus, reduce the overall sum of squared residuals for all of the points.

Question: Study the abbreviated output in Part b of Figure 11-32 where the twelfth observation, a leverage point, is omitted from the data set. Describe the effect of the twelfth observation on the estimates.

Answer: Again the original line is flattened to accommodate this point, but the difference is very slight, practically imperceptible on a scatter diagram (so we do not show it). The intercept increases from 10.4 to 10.8, and the slope decreases from 3.9 to 3.7. The right side of the line rotates downward toward the leverage point.

This leverage point demonstrates that the unusual X value does not necessarily produce a Y value that is out of line with the mass of data points as Figure 11-31 shows. The twelfth observation is generally in line with the other observations, so little adjustment in the estimate occurs to accommodate this point.

Question: Part c of Figure 11-32 and Figure 11-34 show the effect of omitting both of the unusual observations. Describe the effect of the combination of both points on the estimates.

Answer: Together the unusual observations compound the flattening effect to accommodate both points. The intercept increases from 8.2 to 10.4, while the slope decreases from 4.7 to 3.7.

Figure 11-32
Minitab output to explore the effect of unusual observations

a) Estimated Regression Line when Observation 4, (1,22), is omitted

```
The regression equation is
SALES-4 = 9.30 + 4.07 AD-4

Predictor         Coef       Stdev     t-ratio         p
Constant         9.301       1.654       5.62      0.000
AD-4            4.0675      0.5994       6.79      0.000

Unusual Observations
Obs.    AD-4     SALES-4      Fit   Stdev.Fit   Residual   St.Resid
 11     6.00      32.000    33.706      2.322     -1.706     -0.90 X

X denotes an obs. whose X value gives it large influence.
```

b) Estimated Regression Line when Observation 12, (6,32), is omitted

```
The regression equation is
SALES-12 = 10.4 + 3.92 AD-12

Predictor         Coef       Stdev     t-ratio         p
Constant        10.397       2.460       4.23      0.001
AD-12            3.917       1.109       3.53      0.005

Unusual Observations
Obs.   AD-12    SALES-12      Fit   Stdev.Fit   Residual   St.Resid
  4    1.00       22.00     14.31       1.54       7.69      2.18R

R denotes an obs. with a large st. resid.
```

c) Estimated Regression Line when Observations 4 and 12 are omitted

```
The regression equation is
S-USUAL = 8.17 + 4.68 A-USUAL

Predictor         Coef       Stdev     t-ratio         p
Constant         8.168       2.102       3.89      0.003
A-USUAL         4.6794      0.9174       5.10      0.000
```

Figure 11-33
Regression estimate with and without fourth observation

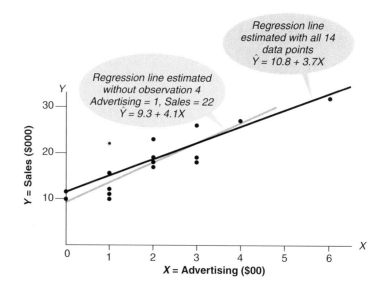

Problems

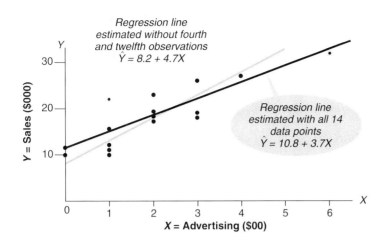

Figure 11-34
Regression estimate with and without the two unusual observations

Whether we should drop the unusual points or some combination of them is an unanswered question in statistics. Remember, it is always possible to randomly select an extreme observation for any sample, so we cannot necessarily dismiss an unusual point as lacking information about the model. Indeed, in an extreme case the mass of points may be outliers, whereas the one "unusual" point lies closest to the true regression line. In addition, we have shown how the regression procedure adjusts the estimated regression line to accommodate points. In the process, the residuals of outliers may decrease while the residuals of regular values increase, which means that the procedure may shield outliers from being identified as such. Consequently, we must be careful before we dismiss values in the data set. For sure, unusual observations deserve attention, so we can learn if there is some extraneous or systematic factor, which the model does not take into account, that affects this particular observation of Y. We may learn more about the phenomenon we are modeling when we study the outlier or unusual observation.

Computer calculations are essential in regression analysis, because very few problems are small enough to merit hand computations. This is especially true when the sample size increases or we estimate more complicated models, such as models with more than one independent variable. The speedy calculations allow us to concentrate on the interpretation and evaluation of the results. The speed also allows us to conduct experiments, such as we have done in this section, to determine the effect of unusual observations on the regression estimates. Finally, the experience with hand calculations and other information covered in this unit allows us to envision the work that the computer accomplishes for us. They also provide a conceptual and technical basis for interpretation and evaluation of the results.

Section 11.8 Problems

76. Figure 11-35 shows the result of Minitab's regression computations for the gymnast weight and score data from Example 11-2 and Problem 67. Use this information to answer the following questions.
  a. Which is the dependent and which is the independent variable in the model?
  b. What is the standard error of the estimate?
  c. What is the residual for the sixth gymnast, whose weight is 90 and whose score is 1,000?
  d. Although the scatter diagram made the sixth gymnast's data appear to be an outlier, why is it not shown in the output as an unusual value?

77. Figure 11-36 shows an abbreviated computer output for the estimated model of the number of vocabulary words that are above ninth-grade level that persons of different ages used in papers, under the assumption that greater experience with and exposure to words will generate a wider selection and choice of words to express thoughts

Figure 11-35
Minitab output for gymnast weight and score data

```
The regression equation is
SCORE = 1192 - 1.75 WEIGHT

Predictor        Coef        Stdev       t-ratio         p
Constant        1191.7       125.6          9.49      0.001
WEIGHT          -1.750       1.252         -1.40      0.235

s = 25.04       R-sq = 32.8%       R-sq(adj) = 16.0%

Analysis of Variance
SOURCE          DF          SS           MS          F          p
Regression       1        1225.0       1225.0       1.95      0.235
Error            4        2508.3        627.1
Total            5        3733.3
```

Figure 11-36
Minitab output for vocabulary model

```
The regression equation is
WORDS = 31.4 + 0.015 AGE

Predictor        Coef        Stdev       t-ratio         p
Constant        31.355       6.427         4.88      0.000
AGE             0.0147       0.1794        0.08      0.935

s = 10.59       R-sq = 0.0%        R-sq(adj) = 0.0%
```

Figure 11-37
Minitab output for reversed model of sales and advertising

```
The regression equation is
AD = - 1.32 + 0.188 SALES

Predictor        Coef        Stdev       t-ratio         p
Constant       -1.3244      0.7388        -1.79      0.098
SALES          0.18788      0.03665        5.13      0.000

s = 0.8375      R-sq = 68.7%       R-sq(adj) = 66.0%

Analysis of Variance

SOURCE          DF          SS           MS          F          p
Regression       1        18.440       18.440      26.29      0.000
Error           12         8.418        0.701
Total           13        26.857

Unusual Observations
Obs.       SALES         AD          Fit      Stdev.Fit     Residual     St.Resid
 4          22.0        1.000       2.809       0.246        -1.809       -2.26R

R denotes an obs. with a large st. resid.
```

and communicate. Use this information to answer the following questions.
  a. Which is the independent and which is the dependent variable in this model?
  b. What is the typical effect on advanced vocabulary usage as a person advances another year in age?
  c. Do there appear to be any outliers in the model? Explain.
78. Figure 11-37 shows the Minitab regression and scatter diagram output that result when we use the sales-advertising data from Example 11-4 to estimate a model with advertising as the dependent variable and sales as the independent variable.
  a. Interpret the coefficient.
  b. If we solve the original estimate model, $\hat{Y} = 10.8 + 3.7X$ for X, we obtain $\hat{X} = -2.9 + 0.3Y$. Meanwhile, the estimated equation to explain the variation in X is $\hat{X} = -1.32 + 0.188Y$. What accounts for the discrepancy?
  c. Why is the standard error of the estimate, $\hat{\sigma}_e$, for this model different from the standard error of the original model?

Summary and Review

A model or linear relationship between two variables of the form $Y = f(X) = a + bX$ is posited, which assumes that we can employ changes in X to predict changes in Y. Regression analysis allows us (1) to investigate real-world support for this hypothesis, (2) to measure the relationship (estimate a and b), and (3) to forecast Y.

Formal assumptions include the following:

$$Y = a + bX + e,$$

where Y is an actual Y value and e is the vertical distance between Y and $\overline{Y}_X$. The equation,

$$\overline{Y}_X = a + bX,$$

where $\overline{Y}_X$ is the mean of the Y values corresponding to a given X value, is the true regression line. The equation

$$\hat{Y} = \hat{a} + \hat{b}X,$$

where $\hat{Y}$ is an estimate of $\overline{Y}_X$, is the estimated regression line.

We collect data on corresponding values of X and Y. As we proceed, the values of X as well as Y must vary in the data set. Otherwise, we cannot estimate changes in Y that result from changes in X.

The formulas for $\hat{a}$ and $\hat{b}$ are:

$$\hat{b} = \frac{n\Sigma XY - (\Sigma X)(\Sigma Y)}{n(\Sigma X^2) - (\Sigma X)^2} \qquad \hat{a} = \frac{\Sigma Y - \hat{b}\Sigma X}{n}$$

These formulas produce a line that minimizes the sum of the squared residuals. If, in addition, the e values satisfy certain assumptions (listed in the following box), then the estimators, $\hat{a}$ and $\hat{b}$, satisfy other properties of good estimators.

> **Assumptions about the Error Term, e, in the General Regression Model**
> 1. The error terms associated with each X value are normally distributed.
> 2. The mean of all of the e values above any X value is zero.
> 3. The standard deviation of the set of e values associated with each X is the same number.
> 4. The e values associated with one X value are independent of the e values associated with any other X value.

We interpret the value $\hat{b}$ as the estimated change in Y when X increases by one unit. If we substitute a value of X in the estimated regression equation, $\hat{Y}$, a forecast value of Y or $\overline{Y}_X$, results.

Outliers and leverage points deserve study to ascertain the source of the unusual values, if possible. Such information may yield rewards in terms of understanding the phenomenon being modeled.

Multiple-Choice Problems

Use the following regression results to answer Questions 79–81:

$$\hat{D} = 2 + 0.5S,$$

where D = delivery time = time between ordering and receiving a catalog purchase (weeks) and S = size of order (pounds).

79. If we use $-2 + 0.5S$ instead of $+2$ in the formula, then
    a. The sign of D will be reversed for all S values.
    b. The sum of the squared residuals will increase.
    c. We must reverse the sign of 0.5 also.
    d. The mean delivery time will be zero.
    e. The slope will change.
80. Interpret the coefficient, 0.5.
    a. It is the reciprocal of 2—thus, the maximum waiting time.
    b. It is the mean delivery time when shipments are zero.
    c. It is the change in S, shipment size, when delivery time increases by one week.
    d. If the shipment size increases by 1 pound, then delivery time will increase by 1/2 week.
    e. If shipments equal 1 pound, then the maximum delivery time is 1/2 week.
81. Forecast delivery time, if you order a shipment of 10 pounds.
    a. 5 weeks.   c. 0.5 week.   e. 7 weeks.
    b. 2 weeks.   d. 2.5 weeks.
82. Use the following data to compute $\hat{b}$, where $Y = f(X)$. Y is the number of machine breakdowns and X is the number of preventative maintenance checks on the machine.

Y	X
4	0
3	1
2	0
1	10

    a. -0.2.   c. 1.   e. -3.6.
    b. 0.2.   d. 3.6.

Figure 11-38
Diagram for Question 85

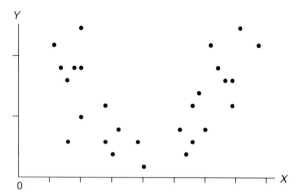

83. Interpret the estimated coefficient from Problem 82.
    a. A 1% increase in preventative maintenance reduces breakdowns by 20%.
    b. 20% of unmaintained machines tend to break down.
    c. The correlation between checks and breakdowns is −0.2, a relatively weak association.
    d. Each additional breakdown induces management to reduce preventative maintenance by an average of 0.2 checks.
    e. Each additional preventative maintenance check reduces a machine's average breakdowns by 0.2.
84. Which of the following statements is *not* true about the following data?

Y	X
26	5
26	0
26	1
26	10
26	0

    a. The assumption that Y changes when X changes is reinforced by the data.
    b. The regression equation is $\hat{Y} = 26$.
    c. Changes in X are not useful for predicting changes in Y.
    d. $\hat{a} = 26$, and $\hat{b} = 0$.
    e. Forecasts of Y are constant.
85. A set of data has been plotted in Figure 11-38. Which of the following assumptions appears to have been violated?
    a. Y is a linear function of X.
    b. The mean of the e values is zero.
    c. $Y = f(X)$.
    d. The standard deviations of the e values are all the same value.
    e. X values are given.
86. Regression analysis of $Y = f(X)$ is useful for
    a. Forecasting X values.
    b. Analyzing real-world support for the effect of changes in X on Y.
    c. Forecasting Y values when X values are known or estimated.
    d. Measuring the effect of changes in X on Y.

Figure 11-39
Choices for Question 87

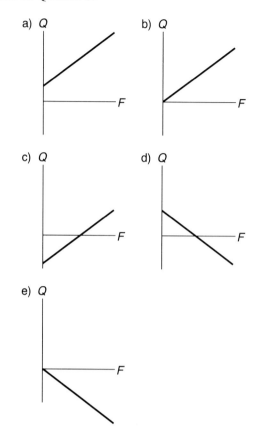

    e. Finding a line that minimizes the sum of the squared distances of each point from the line.
    f. All of the choices represent uses of regression analysis.
87. If output (Q) is a linear function of fertilizer (F), which of the diagrams in Figure 11-39 is most reasonable?
    a. A.   d. D.
    b. B.   e. E.
    c. C.
88. The standard error of the estimate
    a. Measures the spread of the X values about zero.
    b. Is small when the data points are clustered about the regression line.
    c. Is assumed to be zero.
    d. Varies in the model, according to the value of X being examined.
    e. Measures the association between X and the error term.
89. What is the difference between e and $\hat{e}$?
    a. e is the intercept, and $\hat{e}$ is the slope.
    b. e is always larger than $\hat{e}$, because the denominator of e is smaller.
    c. e is the difference between a Y value and the true regression line; $\hat{e}$ is the difference between a Y value and the estimated regression line.
    d. There is no difference; they are both the same value.
    e. There is no difference; they are both equal to zero.

Word and Thought Problems

90. A model
    a. Is a complete description of a real-world relationship.
    b. Is a physical representation of a mathematical idea.
    c. Must contain more than two variables to do any good.
    d. Is a simplification of a real-world phenomenon.
    e. Can only be constructed if the error terms are zero.
91. The difference between correlation and regression
    a. Vanishes as the sample size increases.
    b. Is that regression assumes that we can predict and measure the value of Y when X changes.
    c. Is that correlation is nonlinear and regression is linear.
    d. Is a constant correction factor.
    e. Is indistinguishable; they both measure association.

Figure 11-40 shows computer output for a scatter diagram and regression analysis of a family buffet diner's data. The waiter randomly selected and then recorded a guest's tea consumption in glasses at each event as well as the cups of sugar added to each gallon of tea at that meal. Figuring that tea consumption is a function of the sweetness of the tea, the waiter seeks to verify this conjecture, measure the effect of additional sugar, and forecast consumption for different levels of sweetness. Use this information and the printout to answer Questions 92–94.

92. Based on the scatter diagram
    a. There is no obvious extreme point.
    b. The points are randomly distributed.
    c. People shy away from excessively sweet tea.
    d. The relationship is nonlinear.
    e. Sweetness and tea consumption are directly related.
93. The estimated coefficient of SUGAR implies that
    a. The company can save about $0.64 per customer by reducing the amount of sugar in the tea by one cup.
    b. Each new customer who enters the diner allows the manager to save money by reducing the amount of sugar per glass by 0.64 cups.
    c. Additional customers allow the manager to raise the sugar by 0.64 cups per gallon.
    d. An additional cup of sugar per gallon of tea raises average consumption by about 0.64 glass.
    e. A 1% increase in the amount of sugar per gallon increases average consumption by 0.64%.
94. If there is a cup of sugar per gallon, what will the typical consumption be approximately?
    a. 1.25 glasses.   d. 0.535 glasses.
    b. 0.605 glasses.   e. 1.50 glasses.
    c. 0.644 glasses.

95. A student's study times fluctuate from day to day; if the student spends most of one evening studying, he is likely to spend very little time studying the next evening. Generally, his scores and subsequent study time have decreased as the term progressed. What problem is likely to result if we estimate the following linear model of his study time?

$$\text{Study time} = f(\text{scores}).$$

   a. Mean of error terms is not zero.
   b. The slope will be more negative than usual, because of the balancing effect of low study time.
   c. Error terms are not independent from one observation to the next.
   d. The error term contains the same information as study time.
   e. The standard deviation of the error terms varies with different study times.

Use the following information to answer Questions 96–98: $\hat{S} = 4.5 - 0.5T$, where S = hours of restful sleep during the night and T = room temperature minus 78°. ($T = 6$ when the room temperature is 84°. $T = -4$ when the temperature is 74°.)

96. Interpret the coefficient estimate, -0.5.
    a. For each additional increase in temperature above 78°, the amount of restful sleep decreases by half an hour.
    b. For each additional increase in temperature, the amount of restful sleep decreases by half an hour.
    c. For each additional hour of sleep, the temperature decreases by half a degree, as long as temperatures are above 78°.
    d. For each additional hour of sleep, the temperature decreases by half a degree.
    e. The maximum amount of restful sleep is 4.5 hours, and it decreases by half an hour if the room becomes cooler.
97. Forecast the amount of restful sleep in a room where the temperature is 82°.
    a. $-36.5°$.   d. 0.5 hour.
    b. 86.5°.   e. 8.5 hours.
    c. 2.5 hours.
98. Suggest a temperature for achieving eight hours of restful sleep.
    a. 7°.   d. 85°.
    b. $-7°$. e. 78°.
    c. 71°.

Word and Thought Problems

99. What is the difference among the following symbols: Y, $\overline{Y}_X$, and $\hat{Y}$? e and $\hat{e}$? b and $\hat{b}$? σ_e and $\hat{\sigma}_e$?
100. What criterion does regression analysis employ to find one line to fit the points in a data set?
101. Use the following set of data to answer Parts a–e.

Y Sales	X Advertising
2	1
4	0
6	1
8	5

Figure 11-40
Minitab output for sweet tea data

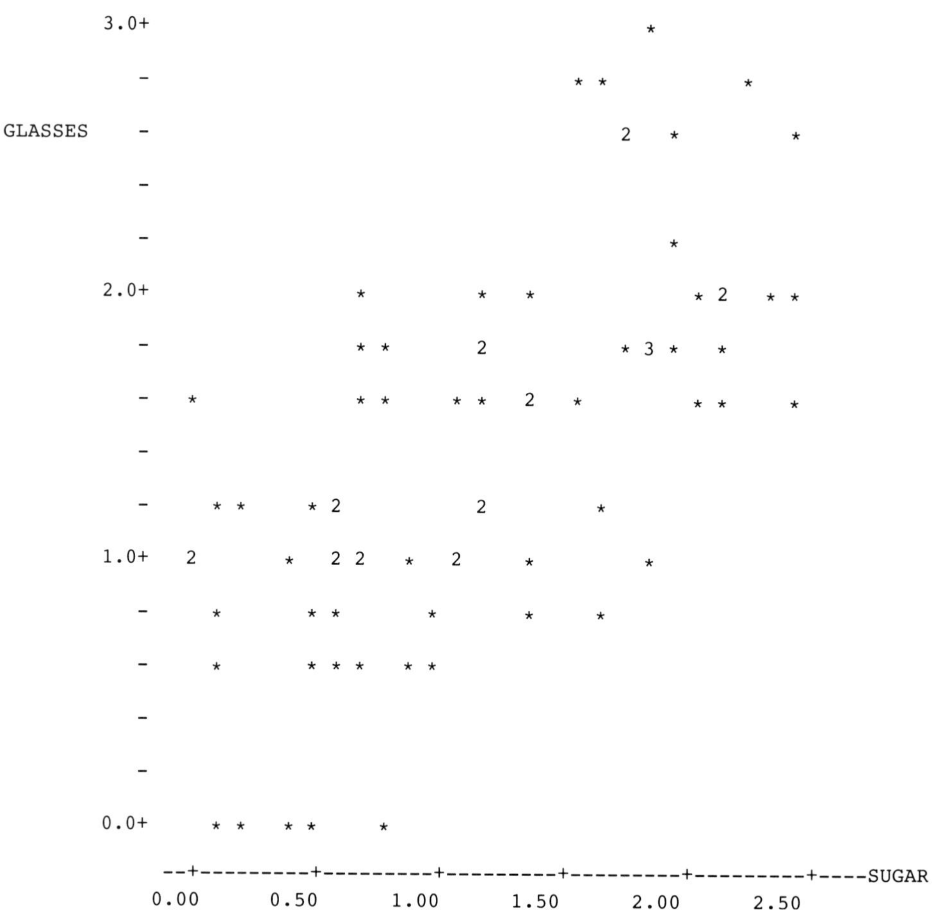

```
          3.0+                                              *
             -                                    *  *              *
  GLASSES    -                                         2   *      *
             -
             -                                              *
          2.0+                        *         *  *              * 2   * *
             -                        * *          2         * 3 *   *
             -           *            * *      * *  2   *         * *     *
             -
             -       * *      * 2              2            *
          1.0+    2        *    2 2       *    2      *          *
             -          *       * *          *       *    *
             -          *       * * *     * *
             -
             -
          0.0+       * *    * *       *
              --+---------+---------+---------+---------+---------+----SUGAR
              0.00      0.50      1.00      1.50      2.00      2.50
```

The regression equation is
GLASSES = 0.605 + 0.644 SUGAR

Predictor	Coef	Stdev	t-ratio	p
Constant	0.6050	0.1197	5.06	0.000
SUGAR	0.64384	0.08586	7.50	0.000

s = 0.5354   R-sq = 43.5%   R-sq(adj) = 42.7%

Analysis of Variance

SOURCE	DF	SS	MS	F	p
Regression	1	16.120	16.120	56.23	0.000
Error	73	20.927	0.287		
Total	74	37.047			

Unusual Observations

Obs.	SUGAR	GLASSES	Fit	Stdev.Fit	Residual	St.Resid
4	1.60	2.7500	1.6352	0.0710	1.1148	2.10R
29	1.90	3.0000	1.8283	0.0866	1.1717	2.22R
47	0.80	0.0000	1.1201	0.0704	-1.1201	-2.11R

R denotes an obs. with a large st. resid.

Word and Thought Problems

a. Plot a scatter diagram of the points.
b. Calculate $\hat{a}$ and $\hat{b}$.
c. Interpret $\hat{b}$. Does this value seem reasonable?
d. What is the estimated regression line?
e. Find $\hat{Y}$ when X is 3.

102. What role does the notion of predictability and causality play in regression analysis?

103. One method for forecasting the trend (long-run pattern) of a variable Y is to estimate $Y = f(t)$, where t is the time period. Suppose you have records of your height on each of your birthdays, as shown below. Estimate $\hat{H} = \hat{a} + \hat{b}t$ and then forecast your height on your 50th birthday. Does this value seem reasonable? Is a linear function appropriate for a lifetime? Is it appropriate for the years in the sample?

t(years)	H(inches)
1	25
2	28
3	31
4	32
5	33
6	34
7	35
8	36
9	37
10	40
11	43
12	48
13	54
14	59
15	64
16	67
17	69
18	70
19	70

104. Using information on the world record for the 1-mile run available in 1937, students in an elementary statistics class estimated the function $R = 4.33907564 - 0.00644959\, T$, where R is the world record expressed in minutes and T is the number of years after 1895 when the record was established, $T = $ year established $- 1895$.

a. Predict the year when the 4-minute mile barrier would be broken. What is the difference between the prediction and 1954, the year Roger Bannister broke the record?
b. Using the following information, update the estimated line, and predict the year the record will be 3:40 minutes (3.67 minutes).
c. Discuss the reasonableness of using a linear model.

ID	Time	Year	ID	Time	Year
1	4.93333	1864	23	4.10333	1942
2	4.60833	1865	24	4.07667	1942
3	4.48333	1868	25	4.04333	1943
4	4.48000	1868	26	4.02667	1944
5	4.43333	1874	27	4.02333	1945
6	4.40833	1875	28	3.99000	1954
7	4.38667	1880	29	3.96667	1954
8	4.35667	1882	30	3.95333	1957
9	4.32333	1882	31	3.90833	1958
10	4.30667	1884	32	3.90667	1962
11	4.30333	1894	33	3.90167	1964
12	4.28333	1895	34	3.89333	1965
13	4.26000	1895	35	3.85500	1966
14	4.25667	1911	36	3.85167	1967
15	4.24333	1913	37	3.83333	1975
16	4.21000	1915	38	3.82333	1975
17	4.17333	1923	39	3.81667	1979
18	4.15333	1931	40	3.81333	1980
19	4.12667	1933	41	3.80883	1981
20	4.11333	1934	42	3.80667	1981
21	4.10667	1937	43	3.78883	1981
22	4.10333	1942	44	3.77200	1985

Sources: Cletus O. Oakley and Justine C. Baker, "Least Squares and the 3:40-Minute Mile," *Mathematics Teacher*, Vol. 70, No. 4, April 1977, pp. 322–24, and *The World Almanac and Book of Facts*, 1992 edition (Scripps Howard, N.Y.: Pharos Books, 1992), p. 915.

105. Suppose you obtain the following regression estimates $\hat{OT} = 2 + 0.01S$, where $OT = $ overtime hours per employee and $S = $ sales over \$1,000 ($S = 0$ if sales are less than \$1,000). For instance, if sales are \$1,525, then $S = 1{,}525 - 1{,}000 = 525$. If sales = \$700, then $S = 0$.

a. Interpret the coefficient 0.01. Is the value reasonable?
b. Forecast overtime hours per employee if sales are 0, then 500, 1,000, and 1,500.

106. Use the following data to estimate the effect of rent on the number of rentals in North Carolina, then estimate the model for Buncombe County. Interpret the coefficients and compare the results. Check the scatter diagram for problems.

Rent(\$)	NC Rentals	Buncombe Rentals
125	288,186	7,233
375	360,735	9,936
625	52,038	1,088
875	4,888	156
1,125	3,869	86

Source: Estimated from a frequency distribution from U.S. Census Bureau data as reported in Paul Johnson, "When the Rent Comes Due," *Asheville Citizen-Times*, September 15, 1991, pp. 1A, 10A.

107. Suppose we estimate total national traffic fatalities, F, to be a linear function of the national speed limit, L. The resulting equation is $\hat{F} = 108(L)$.

a. Graph the relationship between F and L.
b. Interpret the model.
c. Suppose the speed limit is 65 miles per hour. What would you expect the fatality total to be?

108. Hypothesize a model to explain changes for each of the following variables. You may suggest more than one independent variable, but then try to pick out the most important one that you would use to do a simple regression.

Figure 11-41
Comparisons of estimated and true regression line when error terms are dependent

a) True line

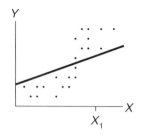

b) Estimated line

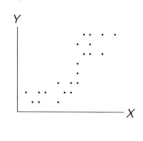

107. Suppose we estimate total national traffic fatalities, F, to be a linear function of the national speed limit, L. The resulting equation is $\hat{F} = 108(L)$.
   a. Graph the relationship between F and L.
   b. Interpret the model.
   c. Suppose the speed limit is 65 miles per hour. What would you expect the fatality total to be?

108. Hypothesize a model to explain changes for each of the following variables. You may suggest more than one independent variable, but then try to pick out the most important one that you would use to do a simple regression.
   a. Orders for an electronic product.
   b. Daily output at a factory.
   c. Amount of advertising done on a local television station.
   d. Amount of advertising done in a local newspaper.
   e. Amount you spend on groceries.
   f. Amount you spend on recreation.
   g. Amount of weight lost each week for person on diet.
   h. Batting average of professional baseball players.
   i. Number of classes students take in a term.
   j. Time you arise in the morning.
   k. Amount bet on each horse in a race.

109. C is the insect count in June, measured by the number of bugs per cubic meter at a randomly selected site. S is the severity of the previous winter, measured by the average low winter temperature at the corresponding site over the winter months. C is assumed to be a function of S. An estimate of the function is $\hat{C} = 182 + 77S$.
   a. Interpret the 77. Is the sign reasonable?
   b. Forecast the insect count if the average winter temperature for the previous winter is zero.
   c. What would have to occur so that there are no bugs? Is this a reasonable result?
   d. Sketch a nonlinear function that might better represent the relationship between winter temperature and bug population. Describe the behavior depicted in your sketch.

110. Regression analysis attempts to explain variations in the dependent variable with variations in the independent variable. Consider the model $DR = f(GDP)$, where DR = congressional Democrat-Republican ratio, and GDP = gross domestic product (measure of prosperity). A random sample of 10 different congresses is proposed for the analysis. Explain intuitively the effect of
   a. Ten identical DR values for the 10 congresses.
   b. Ten identical GDP values for the 10 corresponding congresses.
   c. A nonlinear function (sketch your idea of the functional form and describe its implication for the relation between DR and the GDP).
   d. Error terms with a nonzero average.
   e. Large variation in DR when the GDP is high, and less variation when the GDP is small.

111. The regression line is described as a line of means. Explain this concept, and relate it to the example in Problem 110.

112. Suppose error terms are dependent, as shown in Figure 11-41a, where the solid line is the true relationship and the dots mark the observed (X, Y) pairs. Error terms tend to be in either negative or positive clusters. The true line is removed in Figure 11-41b. Sketch the estimated regression line for the data points in Part b. Then sketch an identical line in the first diagram. Show the true and estimated Y values corresponding to X_1. Discuss potential forecasting problems for a regression model when this error pattern exists.

113. What is the standard error of the estimate? Would you prefer that the value be large or small in a regression context? Explain.

114. If litigation time in a civil suit is a function of the damages being sought,
   a. Which is the dependent variable? Independent?
   b. What would you expect the sign of the coefficient to be? Explain.
   c. What would you expect the intercept to be? Explain.

115. A psychologist believes that the time in hours, T, for an animal to learn a specific task is a linear function of the amount of food given as a reward, R. An estimated regression model is

$$\hat{T} = 26.2 - 4.3R.$$

Sketch the function. Erase any segments of the line that do not appear to be reasonable. Explain the remaining segments and why the omitted segments are unreasonable in the context of this model.

116. In this problem we use the percentage of operating cost supplied by state park visitors' revenue, S, as a measure of subsidies by states for their parks to determine if V, the number of day visitors (expressed in thousands), is responsive to such subsidies.
   a. What sign do you expect for the coefficient of S? Explain.
   b. Use the following data and a computer to estimate the regression equation. Interpret the coefficient.

Word and Thought Problems

Does the estimate agree with your expectations? If not, try to determine the source of the disagreement.

c. Suggest other independent variables that belong in this model to explain variation in the number of visitors.

ID	V (thousands of day visitors)	S (% of revenues)	ID	V (thousands of day visitors)	S (% of revenues)
1	2,064	34.4	26	7,632	20.4
2	4,139	92.4	27	6,967	62.7
3	628	89.8	28	10,363	49.0
4	12,064	34.9	29	13,105	43.1
5	6,233	28.6	30	24,165	60.0
6	5,991	3.2	31	23,720	49.9
7	38,325	24.7	32	6,047	80.1
8	9,845	23.2	33	3,970	45.4
9	36,879	15.8	34	5,717	56.0
10	69,569	27.7	35	800	23.0
11	8,629	77.1	36	14,005	35.8
12	34,372	8.9	37	18,683	37.9
13	20,007	80.9	38	5,126	9.6
14	10,368	64.2	39	1,874	37.9
15	6,419	42.3	40	1,381	3.6
16	9,060	40.9	41	7,243	112.2
17	12,684	15.0	42	3,213	60.4
18	841	32.1	43	1,633	25.6
19	5,263	57.8	44	3,705	26.2
20	7,464	59.5	45	2,717	22.7
21	3,337	47.7	46	39,276	24.6
22	2,216	74.4	47	36,009	32.0
23	7,509	33.5	48	66,060	28.6
24	3,448	26.5	49	5,240	7.3
25	9,129	73.5	50	18,598	22.3

Source: *1989 Annual Information Exchange*, National Association of State Park Directors, as reported in U.S. Bureau of the Census, *Statistical Abstract of the United States: 1990,* 110th ed. (Washington, D.C.: Bureau of the Census), 1990, p. 225.

117. A geographer figures that the census count of a town is a linear function of the distance from the town center to the nearest navigable waterway. Use the following data to answer the parts to this problem.

  a. Use a computer to estimate and interpret the coefficient of distance.

  b. Forecast the census for a town that is 20 miles from the nearest navigable waterway.

ID	Distance	Population	ID	Distance	Population
1	85	13,524	12	78	32,276
2	11	74,800	13	73	14,366
3	11	89,505	14	22	79,926
4	55	41,033	15	22	44,975
5	66	28,825	16	91	10,474
6	110	9,542	17	22	71,773
7	50	43,527	18	8	68,335
8	14	59,318	19	91	13,180
9	89	2,183	20	95	3,834
10	24	56,503	21	88	22,522
11	34	43,889	22	79	8,089
23	61	34,802	41	115	9,168
24	70	6,501	42	106	26,974
25	22	74,304	43	109	4,144
26	57	16,468	44	11	67,852
27	35	55,744	45	50	21,381
28	93	12,198	46	72	22,678
29	109	15,784	47	87	8,489
30	13	67,900	48	105	9,210
31	90	3,233	49	76	23,481
32	25	55,248	50	123	20,149
33	98	18,169	51	108	12,375
34	28	54,599	52	23	45,949
35	24	77,431	53	122	19,738
36	28	57,396	54	85	13,002
37	97	3,176	55	81	13,719
38	8	56,986	56	19	52,523
39	35	53,002	57	101	15,875
40	66	2,916	58	41	32,419

118. A psychologist hypothesizes that the number of hours per day an individual spends on a hobby is a linear function of the age of the individual.

  a. Discuss the reasonableness of the choice of the independent variable, the linear assumption, and the sign of the coefficient you would expect.

  b. Use the printout in Figure 11-42 to add to your discussion of linearity, and interpret the estimated coefficient.

119. a. Use the following data and the computer to estimate a linear relationship between the concession revenues and attendance at an athletic event.

  b. Then use the information to determine if the current policy of charging an entrance fee of $5 with an average attendance of 3,000 should be discarded in favor of free entry to increase overall revenue. The athletic director estimates that average attendance would triple with free entry.

  c. Does the intercept term seem reasonable in this situation? Explain.

ID	Attendance	Concessions
1	2,326	6,165
2	2,895	8,143
3	2,500	7,258
4	3,521	8,111
5	2,888	7,017
6	3,558	8,445
7	2,117	4,967
8	2,601	7,433
9	3,697	6,962
10	3,020	6,136
11	3,311	6,282
12	2,948	7,368
13	2,475	5,278
14	3,617	9,351
15	3,598	9,079
16	2,762	6,408

Figure 11-42

Minitab output for hobby data

```
           -                                              *
HOBBYHRS   -                                  *    *   * **
           -                              *  *2  *4      **     ** *
           -                           2 2      *2 * ****
    4.00+                                3 22*2****    *
           -                          *  242*  2          *
           -                      *  *  32* 2    *
           -                        **       3
           -                    3  * 2 2           *
    3.50+                      *22**2 *
           -                  ** 22*  *
           -                   25  3*2
           -                 *   3 3* *
           -                 * 2 2*2
    3.00+                 *  * 2
           -              *
             --+---------+---------+---------+---------+---------+----AGE
               15        30        45        60        75        90
```

The regression equation is
HOBBYHRS = 2.77 + 0.0222 AGE

Predictor	Coef	Stdev	t-ratio	p
Constant	2.76703	0.04276	64.71	0.000
AGE	0.0221752	0.0009489	23.37	0.000

s = 0.1748   R-sq = 80.1%   R-sq(adj) = 79.9%

Analysis of Variance

SOURCE	DF	SS	MS	F	p
Regression	1	16.685	16.685	546.10	0.000
Error	136	4.155	0.031		
Total	137	20.841			

Unusual Observations

Obs.	AGE	HOBBYHRS	Fit	Stdev.Fit	Residual	St.Resid
7	84.0	4.2238	4.6297	0.0423	-0.4059	-2.39RX
11	83.0	4.2065	4.6076	0.0414	-0.4011	-2.36RX
13	88.0	4.2476	4.7184	0.0459	-0.4708	-2.79RX
31	78.0	4.3570	4.4967	0.0370	-0.1397	-0.82 X
100	45.0	4.2440	3.7649	0.0151	0.4791	2.75R

R denotes an obs. with a large st. resid.

X denotes an obs. whose X value gives it large influence.

120. Forensic researchers estimated the following four models to forecast the height of a criminal when they knew his/her foot length. Observations from 6,682 men and 1,330 women were used.
  a. Interpret the coefficients.
  b. Explain the difference in magnitudes of the intercepts and lack of difference in the coefficients that result from estimating Height = f(foot length) with the same data set for men. Is the explanation the same for women?
  c. Forecast the height of a female criminal if she leaves a footprint that is 24 centimeters long. Do the same for a male that leaves a print 27 centimeters long. (Source: Eugene Giles and Paul H. Vallandigham, "Height Estimation from Foot and Shoeprint Length," *Journal of Forensic Sciences,* Vol. 36, No. 4, July 1991, pp. 1138–39.)

Male: centimeters: Height = 82.206 + 3.447 (foot length)
inches: Height = 32.364 + 3.447 (foot length)
Female: centimeters: Height = 75.065 + 3.614 (foot length)
inches: Height = 29.553 + 3.614 (foot length)

Unit III: Simple Regression Models: Evaluating and Forecasting

Preparation-Check Questions

121. A regression line
  a. Minimizes the sum of the squared residuals.
  b. Has a minimum value for the dependent variable (that is, a U-shaped relationship).
  c. Must pass through the origin on the axes.
  d. Expresses the effect of the dependent variable on the independent variable.
  e. Passes through all the points in the data set.

122. If $Y = 20 + 0X$ is the relationship between X and Y then
  a. X is 20 when Y is 0.
  b. The line passes through the origin.
  c. X has no effect on Y.
  d. The relationship is nonlinear.
  e. Y increases by 20 when X increases by one unit.

123. If $X = 10$ is substituted in the regression equation $Y = 150 - 2X$, the result is $Y = 130$. 130 represents
  a. The value of Y when $X = 0$.
  b. An estimate of the change in Y if X increases by two units.
  c. An estimate of the change in Y if X decreases by two units.
  d. An estimate of the mean of the Y values corresponding to $X = 10$.
  e. The Y value in the data set corresponding to $X = 10$.

124. When we reject H_0 in a hypothesis test,
  a. The population value is too large.
  b. The test statistic value from the sample is unlikely to occur if H_0 is true.
  c. We make a Type II error.
  d. The p-value is very large.
  e. The sampling distribution is a modified binomial.

Answers: 121. a; 122. c; 123. d; 124. b.

Introduction

Sales-Advertising Relationship The store owner of Example 11-4 estimated the relationship between advertising, X, and sales, Y, to be $\hat{Y} = 10.8 + 3.7X$. She knows this line will minimize the sum of the squared residuals in the data set but wonders about the validity of the model. She knows that a line estimated from a sample, when the line, in fact, originates from many unobserved data points, is subject to sampling error. In addition, formulas and computers can produce estimates that may look authentic when there is no true relationship, because they depend only on the values that a user enters. Does the owner possess enough data points to feel comfortable using the estimated line? Does the regression equation support the hypothesized model?

Figure 11-43

Regression results of sales-advertising model

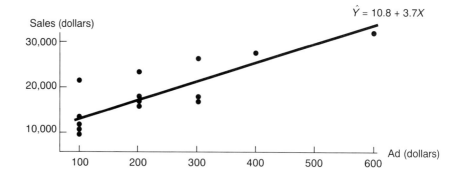

In addition, she needs the model to forecast sales. If she substitutes an advertising value in the estimated equation, a sales forecast results. But it is difficult to forecast the future, so the manager will be satisfied to predict a value close to the value that eventually occurs. What are the chances this model will provide a forecast that is close? How can a user of the model obtain better forecasts?

These questions lead to the topics of this unit. We will study how to evaluate regression results and how to use these results to predict future values of dependent variables.

11.9 Model Evaluation

Using the formulas for $\hat{a}$ and $\hat{b}$ from Unit II, we obtain an expression for the straight line that will minimize the sum of the squared residuals about the line. Figure 11-43 repeats the regression line we get using the sales-advertising data from Example 11-4.

The line appears to fit the data rather well. Is there a measure of the degree of the line's fit? If so, we could compare fits when different data sets yield similar visual impressions. Is the fit sufficient to persuade us to adopt this model? The answers to these questions will help us evaluate the results. This unit presents two evaluation methods to provide information about the answers: the coefficient of determination, R^2 (pronounced "R squared"), and a hypothesis test for the value of the regression coefficient, b.

11.9.1 R^2: The Coefficient of Determination

When we use regression analysis, we actually examine changes or variation in the dependent variable, Y. We hypothesize that we can explain or predict changes in Y with changes in the independent variable, X. *R^2 measures the proportion or percentage of variation in Y that is explained by the regression of Y on X.* Its value tells us how much of the change in Y that we observe in the sample can be accounted for by changes in X, the independent variable in the model.

A graphical presentation often helps users visualize the mathematical form of R^2. Consider the actual data point, Y, in Figure 11-44.

The distance $Y - \overline{Y}$ represents the variation in Y to be explained by the model. Actually, we square this distance in the formula to obtain $(Y - \overline{Y})^2$, but the picture simplifies and corresponds to the more complicated computations about to be discussed.

Question: If we sum these squared values for each observation in the data set, we obtain an expression similar to one we studied earlier in descriptive statistics, $\Sigma(Y - \overline{Y})^2$. This sum is the numerator of which descriptive measure?

Answer: The sum, $\Sigma(Y - \overline{Y})^2$, is the value in the numerator of the variance or standard deviation formula. This is the total variation of the variable.

Figure 11-44

Illustration of R^2 based on a single point, Y

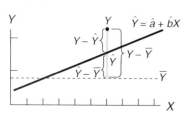

Section 11.9 — Model Evaluation

To obtain the standard deviation, we divide this sum, the total variation, by N or $n - 1$ before taking the square root to obtain an average measure of variation. In the R^2 formula, rather than employ an average, we use the sum that represents total variation:

$$\Sigma(Y - \overline{Y})^2 = \text{total variation of the dependent variable to be explained.}$$

Question: What is the name and symbol we use for $Y - \hat{Y}$? See Figure 11-44 for help.
Answer: The residual is $\hat{e} = Y - \hat{Y}$.

In a sense, this distance or residual is the portion of the variation that is not explained by the regression model. Again, the value is squared and summed for all observations:

$$\Sigma(Y - \hat{Y})^2 = \Sigma \hat{e}^2 = \text{unexplained variation.}$$

Question: What is $\Sigma(\hat{Y} - \overline{Y})^2$? Again refer to Figure 11-44.
Answer:

$$\Sigma(\hat{Y} - \overline{Y})^2 \text{ is the explained variation.}$$

We might say that by using the estimated model and accounting for the specific value of X, we move from $\overline{Y}$ to $\hat{Y}$. Generally, we expect the Y we observe to be closer to $\hat{Y}$ than $\overline{Y}$. Otherwise, we expect a Y value close to $\overline{Y}$, the mean of all the Y values in the data set, when we ignore the effect on Y of X values and variation in the X values.

Now we combine the explained and total variations to produce R^2. Recall that R^2 is the proportion or percentage of variation in Y, the dependent variable, that is explained by the model:

$$R^2 = \frac{\text{Explained variation}}{\text{Total variation}} = \frac{\Sigma(\hat{Y} - \overline{Y})^2}{\Sigma(Y - \overline{Y})^2}.$$

More intuitively, R^2 is the proportion or percentage of the changes in Y that we observe in the sample and explain with changes in X through the model.

The following box details the steps to calculate this statistic, although in most cases, we would use the computer. However, a few computations of R^2 may help you understand what different values of R^2 imply.

Steps for Calculating R^2 by Hand
1. Calculate $\overline{Y}$, the mean of the dependent variable sample values.
2. Find the total variation in the dependent variable, $\Sigma(Y - \overline{Y})^2$.
3. Calculate $\hat{Y} = \hat{a} + \hat{b}X$ for each X value in the sample.
4. Find the explained variation in the dependent variable, $\Sigma(\hat{Y} - \overline{Y})^2$.
5. Form the ratio of explained (Step 4) to total variation (Step 2).
6. To express the proportion as a percentage, multiply the result of Step 5 by 100.

■ **Example 11-4 (continued):** Determine R^2 for the estimated sales-advertising model, $\hat{Y} = 10.8 + 3.7X$ (p. 513 displays the original data set).

Solution: The following chart displays the steps and calculations. Consider the first row of the table. Because $\bar{Y} = 269/14 = 19.2$, the mean sales value, we find $Y - \bar{Y} = 12 - 19.2 = -7.2$ (Column 3). $(Y - \bar{Y})^2 = (-7.2)^2 = 51.84$ (Column 4). $\hat{Y} = 10.8 + 3.7X = 10.8 + 3.7(1) = 14.5$ (Column 5). $(\hat{Y} - \bar{Y}) = (14.5 - 19.2) = -4.7$ (Column 6), and $(\hat{Y} - \bar{Y})^2 = 22.09$ (Column 7).

(1) Sales (Y)	(2) Advertising (X)	(3) $Y - \bar{Y}$	(4) $(Y - \bar{Y})^2$	(5) $\hat{Y} = 10.8 + 3.7X$	(6) $\hat{Y} - \bar{Y}$	(7) $(\hat{Y} - \bar{Y})^2$
12	1	−7.2	51.84	14.5	−4.7	22.09
10	1	−9.2	84.64	14.5	−4.7	22.09
18	3	−1.2	1.44	21.9	2.7	7.29
22	1	2.8	7.84	14.5	−4.7	22.09
18	2	−1.2	1.44	18.2	−1.0	1.00
19	2	−0.2	0.04	18.2	−1.0	1.00
17	2	−2.2	4.84	18.2	−1.0	1.00
19	3	−0.2	0.04	21.9	2.7	7.29
26	3	6.8	46.24	21.9	2.7	7.29
23	2	3.8	14.44	18.2	−1.0	1.00
27	4	7.8	60.84	25.6	6.4	40.96
32	6	12.8	163.84	33.0	13.8	190.44
15	1	−4.2	17.64	14.5	−4.7	22.09
11	1	−8.2	67.24	14.5	−4.7	22.09
269			$\Sigma(Y - \bar{Y})^2 = 522.36$ (total variation)			$\Sigma(\hat{Y} - \bar{Y})^2 = 367.72$ (explained variation)

From these calculations, we can find R^2 = explained variation/total variation = 367.72/522.36 = 0.70. Accordingly, the model explains 70% of the total variation in the dependent variable, sales.

Question: What are the minimum and maximum values that R^2 can assume?
Answer: The least amount of variation that can be explained is 0,

$$\frac{\text{Explained variation}}{\text{Total variation}} = \frac{0}{\text{Total variation}} = 0 = \text{Minimum possible } R^2.$$

The maximum occurs when explained variation equals total variation

$$\frac{\text{Explained variation}}{\text{Total variation}} = \frac{\text{Total variation}}{\text{Total variation}} = 1.$$

Thus, $0 \leq R^2 \leq 1$, or if expressed in percentages, $0 \leq R^2 \leq 100\%$. The larger the R^2, the better the line fits the data points, that is the smaller the sum of squared residuals.

What is a desirable value for R^2? This question is not easily answered. Knowledge of and experience with the phenomenon being modeled contribute to our judgment about the R^2. In Example 11-4, where sales = f(advertising), there are several other variables that are important for explaining total sales variations, such as price, competitors' prices, and customers' incomes. Thus, we should not expect advertising to explain all or even 90% of the variation. The 70% result does not seem unreasonable under the circumstances.

Question: If a model is $\hat{Y} = 10{,}000 + 2{,}000X$, where Y = amount bet on a horse in a race and X = horse's fastest time for the race distance, does a low R^2 seem reasonable?
Answer: We would expect a low R^2 for this model, because the explained variation in the numerator should not be very large: People's bets on horses depend on many things besides race times, including birthdays, looks, colors, jockeys, and post position.

We should eye a high R^2 in this case warily. Instead we could weight other evaluation measures, such as the hypothesis test in the coming section, more heavily as well as consider the reasonableness of the estimates (are the sign and magnitude of the coefficient believable?).

Section 11.9.1 Problems

125. The R^2 value for the credit card model from Problem 69 is 0.711. Here $\hat{D} = 99.9321 - 0.01339C$, where D is the collectible debt and C is the number of credit cards issued. Interpret this R^2 value.

126. The R^2 value for a model to explain students' test scores with the number of classes that the student missed is 24.0%. Comment on the magnitude of this measure in the context of this model.

127. Calculate and interpret the R^2 for the gymnast model from Problem 67, $\hat{S} = 1,191.67 - 1.75W$, where $S =$ score in the competition and $W =$ gymnast's weight. The data are listed in the following table. Is the R^2 value reasonable for this simple model?

Gymnast	Weights	Scores
1	100	1,030
2	110	980
3	110	1,000
4	90	1,050
5	100	1,040
6	90	1,000

128. Calculate and interpret the R^2 for the chemist's experiment reaction time model from Problem 68, $\hat{Y} = 4.33 + 0.07X$, where Y is the reaction time and X is room temperature when the chemist conducts the experiment.

Row	Temperature	Reaction Time
1	75	9.54
2	74	9.22
3	73	9.14
4	71	8.83
5	77	9.73
6	73	9.53
7	70	9.50
8	74	8.83
9	72	8.79
10	72	9.37

11.9.2 Hypothesis Tests for the Regression Coefficient, b

Another way to evaluate the regression results is to test hypotheses about b, the coefficient of X in $\overline{Y}_X = a + bX$. The estimate of the coefficient, $\hat{b}$, is the logical test statistic. When the error terms satisfy the assumptions listed in the last unit, $\hat{b}$ will satisfy many of the statistical properties that make it a good estimator of b. The next step is to establish the sampling distribution of $\hat{b}$, so we will know what values of $\hat{b}$ are likely and which are unlikely to occur when H_0 is true. Then we must determine the hypotheses to test, so that a decision will relate to the evaluation of the model. We begin with the sampling distribution of $\hat{b}$.

The Sampling Distribution of the Test Statistic, $\hat{b}$ The estimate of b, $\hat{b}$, depends on the random sample that we happen to select from the population. There are many possible samples and, hence, many possible $\hat{b}$ values. The sampling distribution of $\hat{b}$ is a representation of all possible values of $\hat{b}$ and the probability associated with each of these values.

From statistical theorems we know that if the error terms satisfy the list of assumptions outlined in Unit II (p. 526), then the $\hat{b}$ values will be normally distributed with a mean b and a standard deviation, called the *standard error of the coefficient*. Figure 11-45a depicts this situation.

Figure 11-45
Comparison of sampling distribution of $\hat{b}$ when $\sigma_{\hat{b}}$ is known and unknown

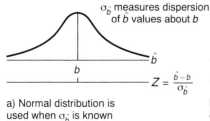

a) Normal distribution is used when $\sigma_{\hat{b}}$ is known

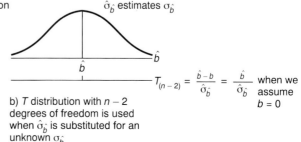

b) T distribution with $n - 2$ degrees of freedom is used when $\hat{\sigma}_{\hat{b}}$ is substituted for an unknown $\sigma_{\hat{b}}$

The calculation of this standard error can be quite involved and tedious, so we will leave its computation to the computer. However, it behaves much like other standard errors we have encountered. It measures the dispersion of $\hat{b}$ values about their mean, the true regression coefficient, b. As the sample size increases, the standard error shrinks, because the $\hat{b}$ values cluster more tightly about b. Generally we do not know the value of $\sigma_{\hat{b}}$, so we must use an estimate, $\hat{\sigma}_{\hat{b}}$.

A consequence of estimating the standard error is that the sampling distribution is no longer exactly normal. The standardization procedure, $\dfrac{\text{Outcome} - \text{Mean}}{\text{Standard Deviation}} = \dfrac{\hat{b} - b}{\hat{\sigma}_{\hat{b}}}$, produces a T statistic rather than the standard normal Z. Shortly, we will see that the b value from H_0 that we use for the test is zero, so the test statistic generally will be

$$\frac{\hat{b} - 0}{\hat{\sigma}_{\hat{b}}} = \frac{\hat{b}}{\hat{\sigma}_{\hat{b}}}.$$

This statistic is often called the *T-ratio*. We lose two degrees of freedom because we estimate two parameters, $\hat{a}$ and $\hat{b}$, with the n sample elements, so *the T-ratio = $\hat{b}/\hat{\sigma}_{\hat{b}}$ from a simple regression model follows a T distribution with $n - 2$ degrees of freedom* as shown in Part b of Figure 11-45.

Testing for a Coefficient of Zero   As we stated before, the usual null hypothesis includes $b = 0$. The pairs of possible hypotheses are:

$H_0: b = 0$ $H_0: b \leq 0$ $H_0: b \geq 0$
$H_1: b \neq 0$ $H_1: b > 0$ $H_1: b < 0$.

Question: What does the statement $b = 0$ suggest about the model?

Answer: $Y = a + 0(X)$. X does not influence Y, so the model has been misspecified by including an incorrect or unimportant independent variable.

Thus, *if we do not reject H_0, the sample data do not support the model $\overline{Y}_X = a + bX$.* However, the conclusion not to reject H_0 is not this simple. The data may not support the model for one of the following reasons: The sample is nonrepresentative, the true model is nonlinear, one or more assumptions about e have been violated, the data does not measure a variable exactly as the theory suggests (such as using national figures on unemployment as a substitute for local unemployment values), or perhaps the sequence of X affecting Y requires a lagged relationship (such as market prices this year affect farmers' decisions about crops next year). At any rate, *if we do not reject H_0, we need to investigate further before abandoning the model.*

The alternative, H_1, will be one- or two-tailed depending on what the decision maker or analyst thinks is true. Often, we can clearly discern the expected direction of change in Y when X changes, and a one-tailed test in the indicated direction is appropriate. At times, there are reasons for a change in both directions, and we do not know the dominant direction. This case requires a two-tailed test. *If we reject H_0, then the data support our hypothesis about the relationship between X and Y; the data support our model.* Rejection, which carries negative connotations in everyday activities, is generally a positive decision when we evaluate a regression coefficient.

Question: In Example 11-4 we hypothesize that sales is a function of advertising, $S = f(A)$. What should H_1 be?

Answer: $H_1: b > 0$, because we expect sales to increase when advertising increases.

Question: Suppose that sales of beer $= f$(the state of the economy) and that we measure the state of the economy with GDP. What is H_1?

Answer: Beer could be like many other products. As income (GDP) increases, consumers buy more. If this is our hypothesis, then H_1 is $b > 0$. On the other hand, we

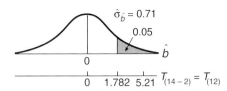

Figure 11-46
Hypothesis testing diagram for sales-advertising model

could argue that when income increases, people switch to more expensive products, such as hard liquor or expensive wines, and that beer sales will decrease. If this is what we expect to observe, then H_1 is $b < 0$. If there is no reason to believe that either effect dominates, we would use H_1: $b \neq 0$.

■ **Example 11-4 (continued):** Test for a significant positive effect of advertising on sales at the 0.05 level. $\hat{b} = 3.7$, $\hat{\sigma}_{\hat{b}} = 0.71$, and $n = 14$. Interpret the decision.

■ **Solution:** To test for a significant positive effect, the hypotheses are H_0: $b \leq 0$ and H_1: $b > 0$. The assumption that H_0 is true ($b \leq 0$) results in the diagram in Figure 11-46. The critical value at the 0.05 level is 1.782, according to the T table in Appendix J with $14 - 2 = 12$ degrees of freedom and a tail area of 0.05. We must convert the sample result, $\hat{b}$, to a T:

$$T = \frac{\hat{b}}{\hat{\sigma}_{\hat{b}}} = \frac{3.7}{0.71} = 5.21.$$

Because $5.21 > 1.782$, we reject H_0 and conclude that advertising does have a positive effect on sales ($b > 0$).

In summary, H_0 is $b = 0$, $b \leq 0$, or $b \geq 0$. Thus, when we assume that H_0 is true, we give the benefit of the doubt to the idea that we have inaccurately chosen the independent variable to explain or predict changes in the dependent variable. *We make H_1 what the decision maker or analyst thinks is true or what theory suggests is true.* H_1 is what we consider to be the correct specification of the model—how changes in X influence the value of Y. So, if we reject H_0 as we did in the preceding example, the data corroborate the theory or model. If we cannot reject H_0, the data do not support the model as specified. But we should not necessarily discard the model immediately. Further investigation may reveal an unusual sample or one of the other problems mentioned earlier.

Remember, rejecting H_0 does not sound good because "rejecting" has a negative connotation. However, because we make H_1 what we think is true, rejecting H_0 means the data support what we believe is true. Thus, we prefer to reject H_0.

Section 11.9.2 Problems

129. Suppose that the density of air pollutants is a function of the humidity level of the atmosphere. If the hypothesized explanation is that higher humidity retards the dispersion of pollutants and causes more contamination, what is H_1?

130. In Problems 69 and 125 collectible debt expressed as a percentage of credit card amounts due, D, is a function of the number of credit cards issued, C. What should H_1 be? Explain.

131. Set up the hypotheses if $Y = f(X)$, where Y is the grade on a test and X represents hours of study. Explain.

132. Test at the 0.01 level for a negative coefficient of C, credit cards issued, in the function $D = f(C)$, where $D =$ collectible debt. The estimated coefficient is -0.01339, the estimated standard error is 0.00322, and the sample consists of nine observations. Interpret the decision in terms related to the model evaluation.

133. $Y =$ number of strike days, and $X =$ median age of employees.
   a. State the implication of the three different possible sets of hypotheses about the regression coefficients.
   b. Computer results of a regression of the strike example are $\hat{Y} = 10 - 2X$. Do a two-tailed test at the 0.05 level if $n = 28$ and $\hat{\sigma}_{\hat{b}} = 1$. Interpret the decision.

11.10 Forecasting the Dependent Variable with a Simple Regression Model

Analysts frequently forecast by using estimated regression equations. To obtain a point estimate of the dependent variable, Y_f, we simply substitute a value of the independent variable, X_f, in the estimated equation and solve for $\hat{Y}_f$. (The f subscripts denote forecast values.)

■ **Example 11-4 (continued):** If advertising level is $A_f = 6$, forecast sales in $\hat{S} = 10.8 + 3.7A$.

■ **Solution:** $\hat{S}_f = 10.8 + 3.7(6) = 33$.

Such a forecast is simple. We forecast this way in the previous unit. However, in this unit we need to understand and report the uncertainty associated with regression forecasts. To reflect this uncertainty we can expand the point estimate into an interval forecast, much as we did a confidence interval estimate. Whether the forecast is a point or interval, we need to be alert to the special problems identified with regression forecasts. This section covers these topics.

11.10.1 A Prediction Interval for Y

We can form a confidence interval about the prediction (usually called a *prediction interval*) by employing the following rule.

> **Rule:** For a $100(1 - \alpha)\%$ prediction interval for the Y_f that corresponds to a given X_f, use
>
> $$\hat{Y}_f \pm t \sqrt{\hat{\sigma}_e^2 \left[1 + \frac{1}{n} + \frac{(X_f - \overline{X})^2}{\Sigma (X - \overline{X})^2} \right]},$$
>
> where $\hat{Y}_f = \hat{a} + \hat{b}X_f$ and t is from the T table (Appendix J) with $(n - 2)$ degrees of freedom and an area of $(1 - \alpha)/2$ in the tail.

■ **Example 11-4 (continued):** Use $\hat{S} = 10.8 + 3.7A$, $\Sigma (A - \overline{A})^2 = 26.86$, $\hat{\sigma}_e = 3.69$, $\overline{A} = 2.3$, and $n = 14$ to form a 95% prediction interval for S when advertising expenditures are $6.

■ **Solution:** From the problem in the introduction to this section, we know that $\hat{Y}_f = 33$ when $A_f = \$6$. Because $n = 14$, there are $12 = 14 - 2$ degrees of freedom. With 12 degrees of freedom and an area of 0.025 in each tail, the value from the T table is 2.179. Noting that $Y = S$ and $X = A$, we have

$$\hat{Y}_f \pm t \sqrt{\hat{\sigma}_e^2 \left[1 + \frac{1}{n} + \frac{(X_f - \overline{X})^2}{\Sigma (X - \overline{X})^2} \right]} = 33 \pm 2.179 \sqrt{(3.69)^2 \left[1 + \frac{1}{14} + \frac{(6 - 2.3)^2}{26.86} \right]},$$

or ($22.89, $43.11).

Remember to interpret the prediction interval in the same manner that we interpret confidence intervals from Chapter 8. Loosely, our confidence that the interval contains the actual value of the dependent variable is $100(1 - \alpha)$. More precisely, $100(1 - \alpha)\%$ of the time, this technique produces an interval that contains the actual value.

Section 11.10.1 Problems

134. Construct a 95% prediction interval for sales when advertising is $7, using the information for Example 11-14 presented in this section.

135. Form a 90% prediction interval for collectible debt (%) when the number of credit cards issued is 250 (million). Recall that $\hat{D} = 99.932114 - 0.013\,C$. The following table lists the original data set. $\hat{\sigma}_e = 0.4349$.

ID	Year	Cards	Debt
1	1981	116	98.06
2	1982	117	98.32
3	1983	124	98.62
4	1984	142	98.60
5	1985	160	97.67
6	1986	181	96.68
7	1987	206	97.40
8	1988	224	96.96
9	1989	242	96.84

136. Find an 80% prediction interval for a gymnast's score, if the gymnast weighs 100 pounds. Use the regression equation, $\hat{S} = 1{,}191.67 - 1.75W$, where S is the score in the competition and W is the gymnast's weight. The standard error of the estimate is 25.04 points. The data are listed in the following table.

Gymnast	Weights	Scores
1	100	1,030
2	110	980
3	110	1,000
4	90	1,050
5	100	1,040
6	90	1,000

11.10.2 Precautions for Regression Forecasts

Forecasting with a regression equation is simple, but we must employ it cautiously. Four reasons for employing caution follow: extrapolation, structural changes, violated assumptions, and quality of the forecast value of X_f.

Extrapolation When the value of X used for the forecast, X_f, falls within the range of the X values in the sample that we use to estimate the regression equation, we are *interpolating* when we forecast. If X_f lies outside the range of the sample X values, we are *extrapolating*. In the sales-advertising example, the advertising values are between 1 and 6. If we forecast with $X_f = 7$, we are extrapolating. *Extrapolated predictions are less certain than interpolated predictions, because they require the extra assumption that the estimated model describes the X and Y relationship outside the sample range for which we constructed it.* We usually have no data points or information to certify that the model is appropriate outside the sample range, so we take some added risks when we extrapolate.

■ **Example 11-5:** $\hat{Y} = 20 + 2X$, where Y is the yield per acre and X is the amount of fertilizer per acre. The data set looks like the one in Figure 11-47a. Relate the problem of extrapolation to this estimation situation.

■ **Solution:** The implication of the linear equation is that more fertilizer will increase the yield per acre and that there is no limit to the process. However, farmers know there is a limit, because too much fertilizer will burn the crop. The model is better for values of X from 1 to 15. Once we progress beyond 15 with fertilizer values, we might find that the data set appears as shown in Figure 11-47b. The linear model would cause us to overestimate crops in the higher range of X values. The risk when extrapolating is evident.

Question: Consider the term $(X_f - \overline{X})^2$ in the prediction interval formula (p. 547). What value for X_f will make the width of the interval smallest?

Answer: When $X_f = \overline{X}$, the central or typical X value in the sample, then $X_f - \overline{X} = 0$. (See also Problem 136.) As the X_f value moves farther away from $\overline{X}$, the $(X_f - \overline{X})^2$ term increases, causing the width of the interval to increase.

Figure 11-47
Two scatter diagrams for yield = f(fertilizer)

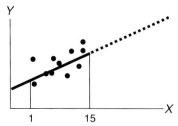

a) Extrapolating linear model beyond values in the sample

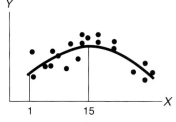

b) Extrapolation results in over-forecasts; forecast values larger than actual values

Structural Changes The functional relationship between variables can change over time. The effect on Y of a change in X can increase, decrease, disappear, or assume another form, such as $Y = f(X^2)$. If such a change occurs, there has been a *structural change* in the relationship between X and Y. If the change is of consequence, predictions made with the old invalid model will tend to be inaccurate.

Consider a regression forecasting model for attendance at professional football games that is estimated using 1982 data. During that season, the players went on strike. When the players settled the strike and returned to play, many disenchanted fans were slow to come back or did not return to the stadiums at all to watch the games. Consequently, the model for early 1982 would probably underestimate current attendance. Other structural changes would be the formation of another football league, additional teams added to the current league, or increased interest in another sport. These would affect attendance at the old professional league's games. Finally, ticket prices will have changed. If price is the independent variable, forecasters may incur extrapolation error.

Question: Why might a regression model for forecasting Russian immigration, estimated prior to 1992, produce forecasting errors?

Answer: The political system changed drastically with the disintegration of the Soviet Union, which allowed radical changes in immigration patterns.

Violated Assumptions If a situation does not satisfy the assumptions that make $\hat{a}$ and $\hat{b}$ good estimators, then poor forecasts may result. For example, if the standard deviations of the e values are different, then an estimation scheme that weights observations according to their reliability may improve forecasts.

Quality of Forecast Value of X_f In many cases, we do not know the value of the independent variable, X_f, so we must forecast this value too. This adds to the forecast error. In the sales-advertising example, management usually controls advertising expenditure, so the advertising value used for a forecast does not add to the uncertainty. Suppose instead that sales depend on GDP: Sales $= f(\text{GDP})$. GDP is not controllable and must be forecast before we can forecast sales. An accurate forecast of GDP should produce a better forecast of sales. Similarly, forecasting collectible debt as a function of the number of credit cards issued for a single company involves less uncertainty than a model for the entire credit card industry, because any one firm in the industry can control its own issues but must forecast the number its competitors will issue.

Sometimes to accommodate the forecasting function, analysts use lagged values of the independent variable to estimate the model. In other words, we say that Y in any time period depends on the X value from the previous period or earlier. In many cases, this action correctly models behavior if we know past values rather than current values when we behave or make decisions. In addition, we usually know the value of a lagged independent variable when we forecast, which reduces the inaccuracy that results when we forecast the independent variable.

Question: Suppose voter turnout, V, is a function of the unemployment rate, U. How might we model the relationship or accommodate forecasting by using available data to estimate the equation?

Answer: To avoid forecasting concurrent future values of the unemployment rate and voter turnout, we can estimate $V_t = f(U_{t-1})$, where t refers here to a time period. We might lag more than one period if data on unemployment are slow to be publicized or if we believe people make up their minds about voting at an earlier time period before the election.

Box 11-2 Estimating Historical Behavior with Regression

Usually we forecast unknown future values. In a reversal of the usual pattern, two researchers decided to use regression analysis to predict past behavior. More specifically, they estimated alcohol consumption during Prohibition, the period in American history from about 1920 to 1933 when the Eighteenth Amendment to the Constitution outlawed making, transporting, and selling alcohol (except for medical and religious reasons). Some hypotheses suggest that consumption increased dramatically during Prohibition; others suggest that it decreased.

Technically, the researchers estimated four different models with the general form, result, R, depends on alcohol consumption, C, $R = f(C)$, where R could be one of four different results of alcohol consumption. (They also controlled for changes in other influences in the social and legal setting of the relationship.) The four different consumption results are (1) incidence of cirrhosis, (2) alcoholism deaths, (3) drunkenness arrests, and (4) alcoholic psychosis. Each was the dependent variable in one of the four estimated models.

The researchers wanted to estimate consumption, C, for the Prohibition years. Obviously the data for C are not available to estimate any of the models during this period, because alcohol consumption was illegal. Instead, they used data available from other time periods and estimated the four models over various periods between 1900 and 1950, excluding 1920–1935 (omitting the non-Prohibition years 1934–1935 allows for different behavior during the first years following Prohibition before normal manufacturing and marketing processes returned). Finally, they substituted data on R that were available during the Prohibition era and solved the estimated equations for C for each of the years. (Data for alcoholic psychosis and for drunkenness arrests were not available for the entire period 1900–1950.)

The following figures, which display the findings, suggest that consumption fell dramatically at first, rose steadily, and, finally, leveled to 60 to 70% of the pre-Prohibition level. Although consumption dropped compared to pre-Prohibition levels (the consumption rate was declining before 1920), the estimates for the final years of Prohibition are similar to post-Prohibition levels until about 1940, after which consumption increased beyond the pre-Prohibition levels. The authors conclude that, given the price increases that accompanied the legal restrictions (some evidence suggest prices increased by 300%), the drop in consumption due to criminalization and subsequent changes in social attitudes appears to have been modest.

Source: Jeffrey A. Miron and Jeffrey Zwiebel, "Alcohol Consumption during Prohibition," American Economic Review, Vol. 81, No. 2, 1991, pp. 242–47.

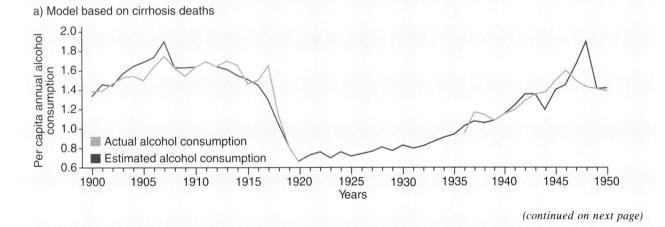

(continued on next page)

The preceding observations do not nullify regression as a forecasting tool, but they should instill some caution in the user. Any forecast of the future is likely to involve error. By checking for the preceding sources of error, we may improve the forecast. In spite of these problems, the regression equation may be the best available forecasting technique, and it is frequently used in such cases. The user simply needs to be aware of the extra uncertainty generated by these error sources.

Box 11-2 offers an interesting twist to forecasting with regression models.

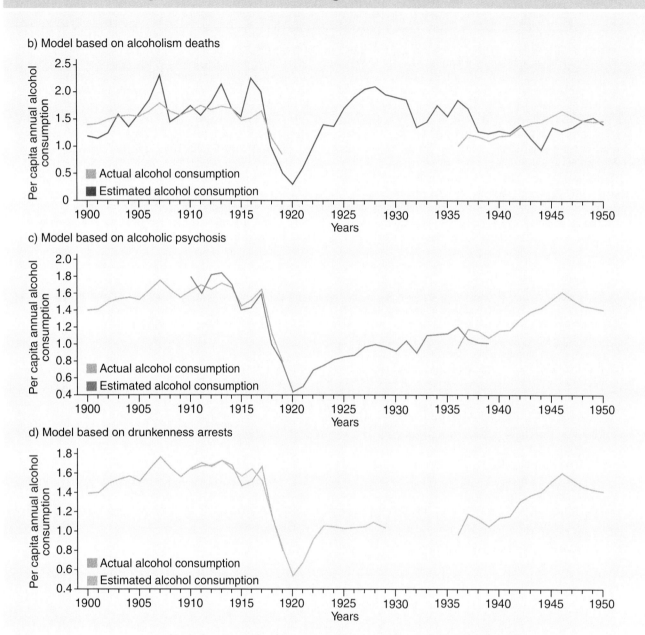

Alcohol consumption is measured in pure gallons of alcohol per capita

11.11 General Evaluation of the Model

The optimum model appears to be one with a high R^2 (or an R^2 appropriate to the complexity of the model) that produces T-ratios large enough to reject the hypothesis that the regression coefficient is zero. Frequently, models fail to rate highly by both measures. Sometimes the standards differ, a model that forecasts well may not bear the same scrutiny as one that explains the relationship among variables. We may emphasize the R^2 for the forecasting model and the hypothesis test for the explanatory model.

We use the tools presented here to start the analysis and evaluation. There are others that check for violations of assumptions, still others that examine the data and residuals for outliers. In the final analysis, the user must synthesize the information from these tools with his or her own experience, information, logic, creativity, and skill as a detective, as well as those of others, as part of the final judgment. Even when we assess a model as a poor one, we often learn more about the phenomenon than we knew when we began modeling. Ideally, we produce a usable model that accomplishes our objective better than previous models or methods.

One aspect of evaluation that we studied earlier, but do not list as an evaluation tool, is interpretation of the coefficients. This process will often alert us to a problem, such as a coefficient that surpasses the bounds of reason, based on the information we possess about the modeled phenomenon. A large T-ratio for a coefficient with the wrong sign can disconcert an analyst more than finding the coefficient insignificantly different from zero. Such signals must enter the overall evaluation of the quality of the model.

One final precaution for evaluating a model with the three tools (coefficient interpretations, R^2, and T-tests) needs emphasis. An unreasonable coefficient, a low or unexpected R^2, or a failure to reject H_0 for a T-test signals a problem. The usual interpretation of these poor evaluation results is that the dependent variable is not predictable from changes in the independent variable. Alternative sources of the problems may be the wrong mathematical form (a nonlinear function may produce better results), violations of the regression assumptions, or a problematic distribution of data, such as an outlier, a structural change during the time period spanned by the data set, or some constraint on the values that a variable may assume (capacity often constrains the attendance at events).

Evaluation requires probing each of these problem sources as well as the evaluation tools before we can judge the quality of a model, especially if the evaluation tools indicate a problem. Furthermore, models may appear to be good when, in fact, these problem sources have worked to the advantage of the model builder's hypotheses. Further investigations in these situations may inform the researcher that a model that appears to fit well according to the evaluation tools may be deceiving. The flowchart in Figure 11-48 is helpful for organizing an evaluation of a model.

11.12 Regression Analysis and the Computer

Refer again to Figure 11-30 (p. 528), the computer output for the sales-advertising data from Example 11-4, to locate on the printout the new information presented in this unit. The new measures and their circled numbers correspond to the following key:

4. The estimated standard error of the coefficient, $\hat{\sigma}_{\hat{b}}$.
5. The T-ratio for a test for a zero regression coefficient, $\hat{b}/\hat{\sigma}_{\hat{b}}$.
6. The two-tailed p-value for a test for a zero regression coefficient.
8. The coefficient of determination, R^2.
9. The degrees of freedom for a T-test of the regression coefficient.

We can use the information in Figure 11-30 to verify that the T-ratio is 5.13 and that R^2 is 67.8%. Rounding causes both to differ from the values calculated by hand earlier, 5.21 and 70%, respectively. The values are close enough to evaluate the model and make decisions in most practical circumstances.

The p-value is a two-tailed p-value for testing if $b = 0$. Once again we can convert it to a one-tailed p-value corresponding to the direction of the desired test. The following two steps are necessary to accomplish this.

Figure 11-48 Flowchart for analyzing simple regression estimates

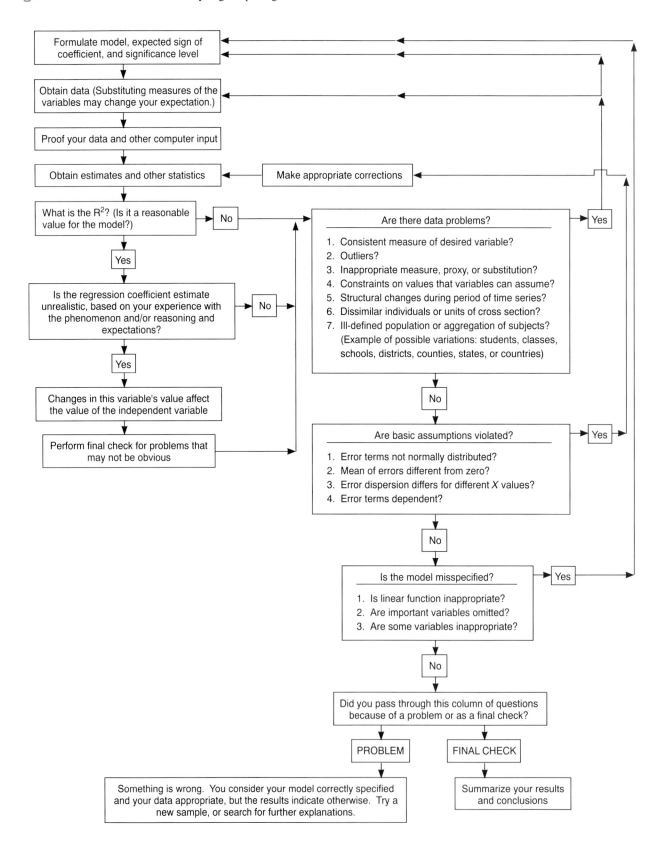

Section 11.12   Regression Analysis and the Computer

```
The regression equation is
TIME = 29.8 - 0.0897 CHEMICAL

Predictor         Coef         Stdev       t-ratio          p
Constant       29.7726        0.3307         90.03      0.000
CHEMICAL      -0.08968       0.05408         -1.66      0.101

s = 1.324         R-sq = 3.6%        R-sq(adj) = 2.3%

Analysis of Variance

SOURCE          DF             SS           MS           F          p
Regression       1          4.820        4.820        2.75      0.101
Error           74        129.709        1.753
Total           75        134.529

Unusual Observations
Obs.CHEMICAL       TIME        Fit  Stdev.Fit   Residual    St.Resid
 53     5.0      26.254     29.324      0.154     -3.070      -2.33R

R denotes an obs. with a large st. resid.
```

Figure 11-49
Minitab output for chemist data

Procedure to Convert a Two-Tailed p-Value to a One-Tailed p-Value
1. Determine if the sign of $\hat{b}$ agrees with H_1.
2. Use $0.5 \times$ (two-tailed p-value) if the sign of $\hat{b}$ agrees with H_1, or, if it does not, use $1 - 0.5 \times$ (two-tailed p-value).

Question: Interpret the two-tailed p-value of 0.0003 from Figure 11-30.
   Answer: If H_0 is true ($b = 0$), there is a 0.03% chance of obtaining a $\hat{b}$ at least as extreme (farther from 0 in either direction) than the $\hat{b} = 3.7$ obtained from the given sample. This value is very small, so H_0 receives little support from the data; consequently, we will reject H_0 when testing at the common significance levels, 0.1, 0.05, 0.01, or 0.001.

Question: How large must α be in the preceding question for us to reject H_0?
   Answer: As long as the significance level, α, is greater than or equal to 0.0003, we will reject H_0, because the p-value will be less than or equal to α.

Question: Do an upper-tailed test for the effect of advertising on sales at the 0.01 level using this p-value, and interpret the result. $\hat{b} = 3.7$
   Answer: $\hat{b} = 3.7$ agrees with H_1: $b > 0$, so the one-tailed p-value is $0.5(0.0003) = 0.00015$, which is smaller than $\alpha = 0.01$. Reject H_0. Advertising has a significant positive effect on sales. This decision verifies our earlier conclusion when we used a critical value rather than a p-value.

■ **Example 11-6:** A chemist used regression analysis to estimate the probable effect of introducing specified quantities of a chemical inhibitor on the time for a chemical reaction to occur in a solution. Use the output in Figure 11-49 to answer the following questions.

a. Because the chemist hypothesizes that the inhibitor retards the reaction, perform a lower-tailed test at the 0.05 level using a critical value: H_0: $b \geq 0$ and H_1: $b < 0$. Interpret the decision.
b. Perform the test in Part a using the p-value.
c. Interpret the R^2 value.

Figure 11-50
Minitab output for infant mortality data

```
The regression equation is
INFMORT = 86.3 - 0.00650 PCGNP

112 cases used 19 cases contain missing values

Predictor         Coef        Stdev      t-ratio         p
Constant        86.303        4.393        19.64     0.000
PCGNP       -0.0065028    0.0008146        -7.98     0.000

s = 37.76         R-sq = 36.7%      R-sq(adj) = 36.1%

Analysis of Variance

SOURCE         DF          SS          MS          F          p
Regression      1       90853       90853      63.73      0.000
Error         110      156818        1426
Total         111      247671

Unusual Observations
Obs.    PCGNP    INFMORT       Fit  Stdev.Fit   Residual   St.Resid
  2       150     171.00     85.33       4.32      85.67      2.28R
  3       350     171.00     84.03       4.23      86.97      2.32R
 85     19270      33.00    -39.01      13.61      72.01      2.04RX
100     14480      20.00     -7.86       9.90      27.86      0.76 X
111     16690      10.00    -22.23      11.59      32.23      0.90 X
124     13680       8.00     -2.65       9.29      10.65      0.29 X
128     16370       7.00    -20.15      11.35      27.15      0.75 X
129     14370       7.00     -7.14       9.81      14.14      0.39 X
```

R denotes an obs. with a large st. resid.
X denotes an obs. whose X value gives it large influence.

■ **Solution: a.** The critical value would ordinarily come from the T distribution, in this case, with 74 degrees of freedom. Because this distribution is approximately normal, we use the critical value from the Z table (Appendix H), -1.645. The T-ratio from the printout, -1.66, is smaller than -1.645, which means we should reject. However, because these values are so close, we need to use a more accurate critical value. Books of more detailed statistical tables or computer statistical packages are a good source. Using Minitab's distribution, we find that the actual critical value is -1.6657, so we cannot reject H_0. The data do not support the chemist's hypothesis that the chemical retards the reaction. The closeness of the test statistic and critical value suggests that the chemist obtain a larger sample for a more decisive result at the given significance level.

b. To do the test with the p-value, we first note that the coefficient is negative, which agrees with H_1. Half of the two-tailed p-value, 0.101, is 0.0505, which is greater than 0.05, so we cannot reject H_0. Notice that using the p-value nullifies the need to find the accurate critical value.

c. The R^2 value is 3.6%, so 3.6% of the variation in the reaction times is explained by variation in the amount of the inhibitor added to the solution.

Section 11.12 Problems

137. Figure 11-50 displays the estimated regression equation for the model INFMORT = f(PCGNP), where INFMORT represents deaths per 1,000 live births for infants under one year of age and PCGNP is the GNP per capita. Use this information to answer the following questions. (Source: United Nations Children's Fund,

Problems

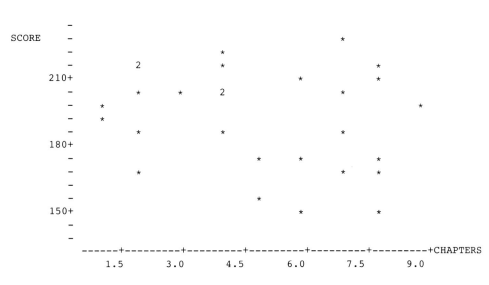

Figure 11-51
Minitab output for score material model

```
The regression equation is
SCORE = 202-2.01 CHAPTERS

Predictor        Coef        Stdev      t-ratio         p
Constant      201.637        9.759        20.66     0.000
CHAPTERS       -2.006        1.755        -1.14     0.263

s = 22.61        R-sq = 4.8%      R-sq(adj) = 1.1%

Analysis of Variance

SOURCE         DF          SS          MS         F         p
Regression      1       668.0       668.0      1.31     0.263
Error          26     13290.7       511.2
Total          27     13958.7
```

The State of the World's Children, 1988 (New York: Oxford University Press, 1988), pp. 64–69.)

a. What percentage of the variation in INFMORT is explained by the model?

b. What sign would you predict for the coefficient of PCGNP? Explain. State the hypotheses to be tested accordingly.

c. Which value allows us to decide that PCGNP is important for explaining the behavior of INFMORT?

d. Is the coefficient significantly negative at the 0.05 level? Explain.

e. Overall the model seems to perform well according to the R^2 and hypothesis test. However, the computer output indicates there may be a problem. What is it?

138. Dr. Walker, a history professor, wonders if the amount of material he makes each student responsible for on a test, M, affects the student's score, S. Specifically, he believes scores are lower when the test covers more material. He assembles data on test scores for 28 former students (with 250 possible points on each test) and the corresponding number of chapters covered on each test (from 1 to 9). He uses the data for a regression analysis to determine if M is an important variable for explaining changes in S and, if so, the magnitude of the effect.

a. Write the professor's model in functional form and the corresponding hypotheses for a test of the coefficient of M.

b. Study Figure 11-51, which displays a scatter diagram of the professor's data. Does it appear that there is a relationship between the two variables?

c. Use the result of the regression analysis in the figure to evaluate the model with the R^2 and the T-test at the 0.05 significance level. Interpret the result.

139. Figure 11-52 shows estimated models to explain state differences in college entrance test scores, part a for states where the larger number of students take the ACT test (1989) and part b where the majority of the students take the SAT (1991). ACT scores range between 1 and 36, and SAT scores fall between 400 and 1,600. The independent variable is per capita personal income (1990), which measures the opportunities for

Figure 11-52

Minitab output on standard test scores

a) ACT Scores

```
The regression equation is
ACTSCORE = 14.9 + 0.000229 ACTPCPI

Predictor        Coef        Stdev      t-ratio        p
Constant       14.870        1.321        11.25    0.000
ACTPCPI    0.00022928   0.00007964         2.88    0.008

s = 0.8921        R-sq = 24.2%      R-sq(adj) = 21.3%

Analysis of Variance

SOURCE       DF         SS         MS        F        p
Regression    1     6.5969     6.5969     8.29    0.008
Error        26    20.6917     0.7958
Total        27    27.2886

Unusual Observations
Obs.  ACTPCPI  ACTSCORE       Fit   Stdev.Fit   Residual   St.Resid
  2     21761    17.900    19.859       0.455     -1.959    -2.55RX
 14     12735    15.900    17.790       0.341     -1.890    -2.29R

R denotes an obs. with a large st. resid.
X denotes an obs. whose X value gives it large influence.
```

b) SAT Scores

```
The regression equation is
SATSCORE = 830 + 0.00264 SATPCPI

Predictor       Coef       Stdev     t-ratio        p
Constant      830.24       34.92       23.78    0.000
SATPCPI     0.002643    0.001762        1.50    0.149

s = 23.66         R-sq = 9.7%       R-sq(adj) = 5.4%

Analysis of Variance

SOURCE       DF         SS         MS        F        p
Regression    1     1259.1     1259.1     2.25    0.149
Error        21    11752.8      559.7
Total        22    13011.8

Unusual Observations
Obs.  SATPCPI  SATSCORE       Fit   Stdev.Fit   Residual   St.Resid
  4     24181    840.00    894.14        9.43     -54.14     -2.50R
 16     17156    922.00    875.58        6.57      46.42      2.04R

R denotes an obs. with a large st. resid.
```

exposure to more breadth of education and information. We expect it to have a positive influence on scores, no matter which test is taken. (Source: U.S. Department of Education; College Board; U.S. Department of Commerce; and Census Bureau, as reported in *The Chronicle of Higher Education: Almanac,* August 28, 1991, pp. 4–7, 39–104.)

a. Which model seems better according to the usual evaluation tools? Explain.
b. Notice that the coefficients and constant term for the ACT model are all smaller than their complements in the SAT model. Why is this reasonable?
c. Forecast the average score for each type of test for a state where per capita personal income is $15,000.

Section 11.13   Multiple Regression Models   **559**

11.13 Multiple Regression Models

■ **Example 11-7:** A teacher believed that the amount of material that a student must study for a test was a major influence on the student's test score, or $S = f(M)$, where S is score and M is material covered on the test. He used the regression procedure to estimate and analyze the model, $\hat{S} = 202 - 2.01M$, but he found it performed poorly according to the low R^2 and insignificant coefficient.

Next he decided that this model specification was much too simple to explain the differences among scores. He considered the fact that the 28 scores reflected not only differences in the testing situations, such as room temperature (T) and degree of detail required in answers (D), but also differences in student characteristics, such as motivation (MO), previous history courses taken (H), and learning ability (A). He formulated a more complex model using these variables:

$$S = f(M, T, D, MO, H, A).$$

The simple model does not account for these differences. In fact, it is probably impossible to set up an experiment that controls for all of these variables within a short time span.

Multiple regression analysis is the estimation and evaluation of this more complex model, one that includes more than one independent variable. The model takes the form $Y = f(X_1, X_2, X_3, \ldots, X_k)$, where each X_i is one of k different independent explanatory variables. More specifically, the linear form is

$$\overline{Y} = a_0 + a_1 X_1 + a_2 X_2 + a_3 X_3 + \cdots + a_k X_k,$$

where a_i is the coefficient of X_i. In the teacher's score model there are six ($k = 6$) independent variables with the linear form

$$\overline{S} = a_0 + a_M M + a_T T + a_D D + a_{MO} MO + a_H H + a_A A.$$

Once we move beyond a simple model and include more than one independent variable, the computation of the coefficients (and constant term, a_0) becomes quite tedious. This is increasingly true as the number of observations and independent variables increase. Without computers regression would not be a viable analytical or statistical tool in most instances. To obtain these estimates, we must add a column of values for each independent variable to the data sets we used for simple regression. The values in each row of the new columns must still correspond to the value of the dependent variable in the same row.

■ **Example 11-7 (continued):** As the teacher prepared to estimate the new model, $S = f(M, T, D, MO, H, A)$, he discovered that data from previous terms and former students were no longer available, such as room temperature. Other variables were difficult or impossible to measure, such as degree of detail required in test answers and learning ability. Instead, he decided that the best proxy for unmeasurable student characteristics was each student's score on a test of prerequisites for the course, P, which was administered at the beginning of each term. He hoped that this score would control for different student characteristics or remove their effect from the score material relationship. Two rows from the data set are shown below for the first two of 28 students in the sample. The computer computations produce the estimated model $\hat{S} = 60.89 - 1.54M + 2.00P$. Predict the score on a 250-point five-chapter test for a student with a 75 on the pretest.

Student	Score	Chapters Covered	Pretest Score
1	171	5	59
2	167	7	58
...	...	...	...

■ **Solution:** If we substitute $M = 5$ and $P = 75$ in the estimated equation, we obtain $60.89 - 1.54(5) + 2.00(75) = 203.19$ as the predicted score.

Recall that we previously mentioned three objectives for performing a regression: to determine if an independent variable significantly affects the dependent variable; if so, to determine the magnitude of the effect; and, finally, to forecast future values of the dependent variable. These objectives remain valid for multiple regression, but now there is the additional objective of *controlling for the effect of other relevant variables when determining the effect of a specific variable, say X_j, on Y*. This is similar to a controlled experiment, except we do not need to work in a laboratory. Sometimes we do not have the ability to physically set and maintain specific circumstances or to preset and control the value of some variables. Actually, including additional independent variables to control for other effects improves the chances of obtaining more accurate estimates for each coefficient and, consequently, of achieving the mentioned objectives.

Controlling for other factors makes regression analysis an important tool for social scientists and businesspeople, who must often analyze data that were not collected in a controlled environment (e.g., a laboratory). It is also a useful procedure for traditional scientists and researchers in other disciplines who summarize and analyze data that vary from uncontrolled factors and randomness inside or outside a laboratory. Multiple regression allows users to separate and control for different relevant effects, which, otherwise, might obscure a significant result.

Thus, the coefficient $\hat{a}_j$ that corresponds to the jth independent variable, X_j, is interpreted as *the effect on Y of a unit increase in X_j, when the values of all the other X variables are held constant*.

■ **Example 11-7 (continued):** Recall the estimated score model, where $\hat{S} = 60.89 - 1.54M + 2.00P$. Interpret the coefficients of M, materials covered, and of P, pretest score.

■ **Solution:** When the pretest score does not change and the number of chapters covered on a test increases by one, a student's expected score decreases by about 1.54 points. Similarly, when the number of chapters covered on a test does not change and pretest score increases by one point, then a student's expected score increases by two points.

Recall that in the original simple regression, $S = f(M)$, with no variable to control for pretest score, the effect of the number of chapters covered, M, appeared to be insignificant according to the T-test. The multiple regression model allows the teacher to compensate for some scores that are unusually high when the number of chapters covered is large or a low score when the test covers only a few chapters, a behavior that does not fit the teacher's hypothesis. These events can occur, because some students have the background and characteristics (higher pretest scores) needed to perform better than poorer students, even when the amount of material covered is large. Likewise, some poorer students will have a lower score, in spite of the smaller amount of material covered. These effects are obscured in the simple regression model.

We use R^2 and hypothesis tests for the regression coefficient to evaluate the regression results along with other evaluation tools suited to the more complex model. Evaluation, problem solving, and further analysis are beyond the scope of this book, but this section on multiple regression is a general introduction to a powerful tool for analyzing data and real-world phenomena in many disciplines.

> **Box 11-3** Speed Limits and Traffic Fatalities
>
> A reduction in traffic fatalities followed the lowering of the speed limits to 55 miles per hour. The limit was lowered to conserve gasoline and other petroleum products in response to the crisis over limited supplies of these products. After time passed and motorists adjusted to more scarce supplies, discussions of the merit of lower speed limits centered on the effect of the lower limits on traffic fatalities.
>
> To examine this effect, researchers estimated the multiple regression model:
>
> FATAL = −2.395 − 0.084 AGE
> − 0.572 CHANGE + 0.878 BEER
> + 0.102 SPEED
> − 0.750 (New England) − 0.301 (Mideast)
> + 0.893 (Southwest) + 0.497 (Southeast)
> − 0.638 (Great Lakes) − 0.239 (Plains)
> + 0.666 (Rocky Mountains) − 0.127 (Far West),
>
> where
>
> FATAL = fatality rate.
> AGE = minimum age for beer purchases.
> CHANGE = 1 for states raising minimum beer drinking age in previous year.
> BEER = average beer consumption per year among people 18 and over.
> SPEED = average speed on rural interstate highways.
>
> A regional name is 1 if the state is in the named region.
>
> Variables other than the average speed at which people drive (presumably directly related to speed limits) affect the fatalities. Thus, the researchers included other variables, such as beer consumption and legal age for beer purchases. They also controlled for regional differences.
>
> After other effects are controlled, the coefficient of SPEED is significantly positive. This coefficient suggests that the fatality rate will increase by 0.102 points when the average speed on rural interstate highways increases by 1 mph. Using this estimate, the researchers predicted that increasing the average speed from 55 to 60 mph would increase the traffic fatality rate by 0.51 points. Assuming that Americans travel about 15,503 million vehicle-miles per year, about 7,907 more Americans would die in traffic accidents each year.
>
> *Source: Paul M. Sommers, "Drinking Age and the 55 MPH Speed Limit," Atlantic Economic Journal, Vol. 13, No. 1, March 1985, pp. 43–48.*

In summary, multiple regression analysis estimates the separate effect of each independent variable on the dependent variable without any movement in the other independent variables in the model. The additional complexity of the model captures more of the influences on the dependent variable than does a simple model. The model explains more of the variation in the dependent variable and reduces the subsequent error terms. Generally, forecasts improve when based on the enriched model.

Box 11-3 describes a multiple regression model about traffic fatalities.

Summary and Review

After checking the sign and magnitude of the coefficient for reasonableness, two additional measures for evaluating regression results are R^2, the coefficient of determination, and the regression coefficient's T-ratio (or its alternative form, the p-value). R^2 measures the proportion or percentage of variation in the dependent variable that the model explains. We use the T-ratio or p-value to decide on hypotheses about the coefficient of X. The null hypothesis includes $b = 0$ (changes in X are not useful for predicting Y). The alternative hypothesis agrees with the theory about the direction that Y will change when X increases.

Computer regression programs frequently output two-tailed p-values for testing these hypotheses. To use these values for one-tailed tests, $\hat{b}$ must agree with the alternative hypothesis. Then, if half the p-value is less than or equal to the level of significance, we reject H_0.

Regression models are common forecasting tools. We obtain a point estimate, Y_f, when we substitute an X value in the estimated equation. We can also construct a prediction interval for Y_f. By checking for the following circumstances, we avoid poor forecasts when using regression models or become aware of potential sources of increased uncertainty: (1) extrapolation, (2) structural changes, (3) violated assumptions, and (4) quality of forecast of the independent variable.

This unit also introduces the notion of regression to over models with more than one independent variable: $Y = a_0 + a_1 X_1 + a_2 X_2 + a_3 X_3 + \cdots + a_k X_k$. The interpretation of coefficients in a linear model is the same as before, except that it now includes the idea of *ceteris paribus* or "everything else held constant." "Everything else" comprises all of the other independent variables included in the model. The coefficient is an estimate of the change in the dependent variable when the corresponding independent variable increases by one unit and the values of all other independent variables remain unchanged. Such an estimate reflects the use of multiple regression to emulate a controlled experiment for studies in which laboratory conditions are unavailable or impossible to implement. Multiple regression also makes the estimation of more complex models possible.

When the evaluation tools (coefficient interpretations, R^2, and coefficient test) indicate a poor or misspecified model, sources other than the theory that generated the original model may be the culprit. Various problems result from violations of the assumptions for reliable estimates of the regression model. The distribution of the data points in the sample may deter accurate estimation. The mathematical form of the model may not be linear. Any of these sources of estimation problems deserve attention before we discard a strongly held position regarding the phenomenon. Integrating the evaluation tool results and the areas just mentioned as well as personal experience and information regarding the phenomenon improves the chances of discovering problems and of producing an acceptable model. The estimation and evaluation procedures constitute regression analysis.

The new formulas in this chapter are:

1. Sample correlation coefficient:

$$r = \frac{n(\Sigma XY) - (\Sigma X)(\Sigma Y)}{\sqrt{[n(\Sigma X^2) - (\Sigma X)^2][n(\Sigma Y^2) - (\Sigma Y)^2]}}.$$

2. Test statistic for correlation: $\dfrac{r}{\sqrt{\dfrac{1-r^2}{(n-2)}}}$.

3. Regression slope coefficient:

$$\hat{b} = \frac{n(\Sigma XY) - (\Sigma X)(\Sigma Y)}{n(\Sigma X^2) - (\Sigma X)^2}.$$

4. Regression intercept or constant term:

$$\hat{a} = \frac{\Sigma Y - \hat{b}\Sigma X}{n}.$$

5. R^2 coefficient of determination: $R^2 = \dfrac{\Sigma(\hat{Y} - \overline{Y})^2}{\Sigma(Y - \overline{Y})^2}$

6. T-ratio: $\dfrac{\hat{b} - b}{\hat{\sigma}_{\hat{b}}} = \dfrac{\hat{b}}{\hat{\sigma}_{\hat{b}}}$.

7. Prediction interval:

$$Y_f \pm t \sqrt{\sigma_e^2 \left[1 + \frac{1}{n} + \frac{(X_b - \overline{X})^2}{\Sigma(X - \overline{X})^2}\right]}.$$

Multiple-Choice Problems

Throughout the chapter, problems with an asterisk denote *p*-value problems and are optional.

140. R^2 measures
 a. The amount of variation left to explain.
 b. The proportion of variation in the dependent variable explained by the model.
 c. How closely the errors fit a normal distribution.
 d. The validity of the assumptions that must be true to obtain $\hat{a}$ and $\hat{b}$.
 e. How close *b* is to zero.

141. What would R^2 be for the scatter diagram in Figure 11-53?
 a. 0.0.   c. 0.5.   e. 1.
 b. 0.25. d. 0.75.

142. Suppose we hypothesize that the number of customers entering a locally owned store in Cullowhee, North Carolina, today is a function of the temperature in Great Falls, Montana, on the same day. R^2 turns out to be 0.95. Evaluate the results.
 a. The model agrees very well with reality.
 b. The results are coincidental, because the model is nonsensical.

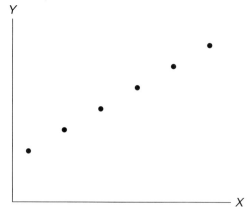

Figure 11-53
Diagram for Question 141

 c. R^2 is low, compared to what would be expected from this model.
 d. The direction of the functional relationship should be reversed to improve the results.
 e. The correlation coefficient is 0.05.

Multiple-Choice Problems

143. Do a lower-tailed test for the coefficient of X at the 0.05 level if the regression results are
$$\hat{Y} = 26 - 18X \quad \hat{\sigma}_{\hat{b}} = 12 \quad n = 20.$$
  a. Reject H_0, because $\hat{b}$ has the wrong sign.
  b. Reject H_0, because $1.5 > 0$.
  c. Cannot reject H_0, because $-1.734 < -1.5$.
  d. Cannot reject H_0, because $1.5 < 1.860$.
  e. Reject H_0, because $18 \neq 0$.

144. When we cannot reject H_0: $b = 0$, what does the decision imply about the model?
  a. A forecast with the model will produce no errors.
  b. It is a good model, because we know the value of b.
  c. Changes in X do not affect Y.
  d. A two-tailed test should be performed next.
  e. The line passes through the origin.

145. If you believe X has a positive effect on Y, which of the following is appropriate?
  a. H_1: $b > 0$. d. H_0: $b > 0$.
  b. H_1: $b < 0$. e. H_0: $b < 0$.
  c. H_1: $b \neq 0$. f. H_0: $b \neq 0$.

*146. Suppose $\hat{Y} = 25 - 10X$ and the two-tailed p-value is 0.08. The decision for a two-tailed test of the coefficient of X at the 0.1 level would be
  a. Reject H_0, because $0.1 > 0.08$.
  b. Reject H_0, because $0.05 < 0.08$.
  c. Cannot reject H_0, because $0.1 > 0.08$.
  d. Cannot reject H_0, because $0.1 > 0.04$.
  e. Cannot reject H_0, because the sign is wrong.

*147. Using the information in Problem 146, perform a lower-tailed test.
  a. Reject H_0, because $0.1 > 0.08$ and the sign of $\hat{b}$ is correct.
  b. Cannot reject H_0, because $0.1 > 0.08$ and the sign of $\hat{b}$ is correct.
  c. Reject H_0, because $0.1 > 0.04$ and the sign of $\hat{b}$ is correct.
  d. Cannot reject H_0, because $0.1 > 0.04$ and the sign of $\hat{b}$ is correct.
  e. Cannot reject H_0, because the sign of $\hat{b}$ is wrong.

148. $\hat{\sigma}_{\hat{b}}$ is a measure of
  a. The spread of the error terms about b.
  b. The spread of the error terms about 0.
  c. The change in Y when X increases.
  d. The spread of possible $\hat{b}$ values around the regression line.
  e. The spread of possible $\hat{b}$ values about b.

149. A point estimate of $\hat{Y}_f$ if $\hat{Y} = 100 - 6X$ and $X_f = 10$ would be
  a. 100.   d. 94.
  b. 60.   e. 40.
  c. -60.

150. A characteristic of 90% prediction intervals for Y_f is
  a. Approximately 90 out of every 100 prediction intervals calculated will contain the true Y_f.
  b. The endpoints do not depend on the value of $\hat{b}$.
  c. The width of the intervals decreases as the X value deviates from the range of sample X values.
  d. The width of the intervals increases as the Y value deviates from the range of sample Y values.
  e. These intervals are wider than 95% prediction intervals calculated from the same value.

151. Which of the following is likely to cause a problem when forecasting with a regression equation, where $Y = f(X)$, $Y =$ tourist dollars generated at a ski lodge, and $X =$ price of a pass? The sample came from the winter of 1992.
  a. A forecast for July 1993.
  b. Dependent error terms.
  c. A pass price of zero.
  d. Using the pass price from 1991.
  e. All of the above.

Use the following information to work Problems 152–159:

A sociologist studying rural communities hypothesizes that people who live in more sparsely populated areas are more interested in the welfare and progress of the community. To verify this conjecture she selects a random sample of 25 individuals from many sizes of communities—larger metropolitan areas to very small farm communities—who have lived in the same place for over 10 years. Each subject answers questions on a test instrument designed to measure civic interest on a scale of 0 to 13. In addition she tallies the number of times over the past month that each subject engaged in civic activities in categories, such as political, volunteer, charitable, recycling, educational, and cultural. The values of this variable range from 3 to 59. Figure 11-54 presents computer outputs for both variables.

152. The regression equation for activities estimates that
  a. An increase in civic activity signals problems and elicits migration from the community at the rate of 0.00006764 people per activity.
  b. As the size of a community increases by one inhabitant, the number of individual civic activities decreases by 0.00006764.
  c. Activity varies by 15.22, whenever new people move into town.
  d. As people move into town and increase the pool of potential activists the subsequent increase in activities is 0.00006764.
  e. The rate of change of civic activity is different in different-size communities.

153. The respective R^2 values for the activity and interest models are,
  a. 15.22% and 3.77%.
  b. 5327.8 and 326.99.
  c. 0.00006764 and 0.00001418.
  d. 26.2% and 20.2%.
  e. 0.00002369 and 0.00000587.

154. The respective T-ratios are
  a. -2.86 and -2.42.
  b. 9.52 and 6.72.
  c. -0.00006764 and -0.00001418.
  d. 0.00002369 and 0.00000587.
  e. 2 for both models.

Figure 11-54
Minitab output for civic behavior

a) Dependent Variable is Number of Civic Activities

```
The regression equation is
ACTIVITY = 33.6 - 0.000068 POP

Predictor        Coef         Stdev       t-ratio         p
Constant       33.551         3.524          9.52     0.000
POP        -0.00006764    0.00002369         -2.86     0.009

s = 15.22       R-sq = 26.2%      R-sq(adj) = 23.0%

Analysis of Variance

SOURCE        DF         SS          MS        F         p
Regression     1     1888.4      1888.4     8.15     0.009
Error         23     5327.8       231.6
Total         24     7216.2

Unusual Observations
Obs.    POP    ACTIVITY     Fit    Stdev.Fit    Residual    St.Resid
 12  388531      4.00      7.27       8.03        -3.27       -0.25 X
 21  459904      5.00      2.44       9.61         2.56        0.22 X

X denotes an obs. whose X value gives it large influence.
```

b) Dependent Variable is Score on Interest Instrument

```
The regression equation is
INTEREST = 5.86 -0.000014 POP

Predictor        Coef         Stdev       t-ratio         p
Constant       5.8633        0.8731         6.72     0.000
POP        -0.00001418   0.00000587         -2.42     0.024

s = 3.771       R-sq = 20.2%      R-sq(adj) = 16.8%

Analysis of Variance

SOURCE        DF         SS          MS        F         p
Regression     1      83.01       83.01     5.84     0.024
Error         23     326.99       14.22
Total         24     410.00

Unusual Observations
Obs.    POP    INTEREST     Fit    Stdev.Fit    Residual    St.Resid
 12  388531     1.000      0.353      1.989        0.647        0.20 X
 21  459904     2.000     -0.659      2.382        2.659        0.91 X

X denotes an obs. whose X value gives it large influence.
```

155. The critical value for a lower-tailed test for the coefficients for both models at the 0.01 level is
    a. -2.500.
    b. -2.492.
    c. -2.485.
    d. -1.714.
    e. Not the same value for both models.
156. The decisions for the tests at the 0.01 level for activities and interests, respectively, are
    a. Reject for both models.
    b. Cannot reject for both models.
    c. Reject for activities and cannot reject for interest.
    d. Reject for interests and cannot reject for activities.
    e. Inconclusive for both models.
*157. Do two-tailed tests at the 0.01 level, using the p-values.
    a. Reject for both models.
    b. Cannot reject for both models.
    c. Reject for activities and cannot reject for interest.
    d. Reject for interests and cannot reject for activities.
    e. Inconclusive for both models.
158. If all we know is the R^2 values for each model and we want to choose the best of the two models, we would

Multiple-Choice Problems

a. Choose the activity model, because it has the larger R^2.
b. Choose the interest model, because it has the smaller R^2.
c. Choose the activity model, because it has the R^2 closest to 50%.
d. Cannot say, because the R^2 values are similar and indistinguishable statistically, that is, not significantly different.
e. Cannot say, because the models attempt to explain variation in different variables, each of which measures different behavior and is dispersed differently.

159. Forecasting uncertainty increases with these models
   a. Because official census values for population are only available every 10 years.
   b. If we want to forecast for an average-size community.
   c. If we want to forecast for the average-size community associated with the average number of activities or the average level of interest in the sample.
   d. Because the R^2 values and T-ratios are small.
   e. If a plot of the error terms indicates that they are independent.

160. The regression equation is
   a. A decision to reject or not reject a statement about a population value.
   b. An estimate of a single population value.
   c. A prediction of the future outcome of an experiment.
   d. A determination of the relative merits of the outcomes of an experiment.
   e. A description or estimate of a functional relationship among two or more variables.

161. A major advantage of multiple regression over simple regression is that
   a. The relationship among the variables is linear.
   b. Fewer assumptions are required to obtain reliable estimates.
   c. Less information is required to produce more accurate forecasts.
   d. Fewer estimation problems arise, so the evaluation procedure is less complicated, though the model itself is more complex.
   e. More variables' systematic effects on the dependent variable are controlled or taken into account.

162. A scatter diagram of students' test scores versus their number of absences appears to consist of random data points, with little likelihood of a significant linear relationship. Which of the following changes could result in finding a significant effect of absences on total points in a regression model, despite the unfavorable picture in the scatter diagram?
   a. Eliminate linear dependence between scores and absences.
   b. Make scores the dependent and one of the independent variables in a new function.
   c. Make absences the dependent variable and total points the independent variable.
   d. Make scores the independent and the dependent variable for the model.
   e. Form another model to control for other variables that cause the dispersion of total point values.

163. When we cannot reject H_0 for a test for the independent variable's coefficient,
   a. The decision implies that the regression analysis vindicates the theoretical aspects of the model.
   b. We will also be unable to reject H_0 for the R^2 test, because they both test the same hypotheses.
   c. The independent variable influences the dependent variable.
   d. The sign of the coefficient is the opposite of the anticipated result, indicating the data need to be reentered in the formulas with the opposite sign.
   e. We need to investigate the data set and check for violated assumptions of the regression model to be sure the problem is not technical rather than a misspecified model.

164. The results of a regression model to explain the amount of traffic on any stretch of two-lane U.S. highway (T) is $\hat{T} = 518.2 + 10.6S - 15.6A + 103H + 127M$, where

   S = speed limit on the stretch of highway.
   A = speed limit on alternate nonfederal highway route.
   H = time of day when traffic count occurs (in hours past 6 A.M; no observances occur past 6 P.M.).
   M = population of nearest town.

   Interpret the -15.6.
   a. A decrease in the speed limit on the alternate route will lower the speed limit on the U.S. highway route.
   b. The speed limit on the alternate route must be between 15 and 16 mph lower than the U.S. highway limit.
   c. If the alternate route speed limit decreases by 1 mph, then traffic will decrease on the U.S. highway by 15.6 cars per day, so long as none of the other variables changes values.
   d. If the alternate route speed limit increases by 1 mph, traffic on the U.S. highway will drop by 15.6 cars per day, if all other variables remain the same.
   e. If the amount of traffic increases by one more car per day, the speed limit on the U.S. highway should be dropped by 15 to 16 mph, assuming that nothing else changes.

165. Multiple regression emulates a laboratory experiment for situations where
   a. The assumptions for the error terms are unsatisfied.
   b. A straight line does not adequately represent the behavior of individuals.
   c. The environmental factors pose a hazard to the scientist and other laboratory personnel.
   d. Behavior cannot be restricted to an environment where a single factor changes and all other factors are controlled.
   e. Outliers are likely to result and produce unreasonable behavior patterns that ordinary techniques cannot model adequately.

Figure 11-55
Minitab output of infant mortality data

```
The regression equation is
INFMORT = 203 - 0.00255 PCGNP - 0.905 CALORIE% - 0.510 BDAYATTN
         - 0.46 GOVTHEAL

41 cases used 90 cases contain missing values

Predictor        Coef        Stdev      t-ratio         p
Constant       202.83        34.72         5.84     0.000
PCGNP        -0.002550     0.001304       -1.95     0.058
CALORIE%       -0.9051       0.3536       -2.56     0.015
BDAYATTN       -0.5100       0.1963       -2.60     0.014
GOVTHEAL        -0.460        1.241       -0.37     0.713

s = 28.76          R-sq = 65.0%        R-sq(adj) = 61.1%

Analysis of Variance

SOURCE         DF          SS           MS         F          p
Regression      4       55351        13838     16.73      0.000
Error          36       29778          827
Total          40       85128

SOURCE         DF      SEQ SS
PCGNP           1       35019
CALORIE%        1       13773
BDAYATTN        1        6445
GOVTHEAL        1         114

Unusual Observations
Obs.    PCGNP    INFMORT        Fit   Stdev.Fit    Residual    St.Resid
   3      350     171.00     108.80        8.38       62.20       2.26R
   4      170     153.00      82.69        6.21       70.31       2.50R
  17      470     124.00      60.49        8.47       63.51       2.31R
 101     1300      18.00      34.93       21.05      -16.93      -0.86 X

R denotes an obs. with a large st. resid.
X denotes an obs. whose X value gives it large influence.
```

166. Figure 11-55 displays the estimated regression equation for the model

INFMORT = 203 − 0.00255 PCGNP
− 0.905 CALORIE% − 0.510 BDAYATTN
− 0.46 GOVTHEAL ,

where

INFMORT = deaths per 1,000 live births for infants under one year of age.
PCGNP = GNP per capita.
CALORIE% = daily per capita calorie supply as percentage of requirements.
BDAYATTN = percentage of births attended by trained health personnel.
GOVTHEAL = percentage of central government expenditures devoted to health.

(Source: United Nations Children's Fund, *The State of the World's Children, 1988* (New York: Oxford University Press, 1988), pp. 64–69, 74–75.)

The coefficient signs
a. Appear correct, because increases in each independent variable create better conditions for sustaining the life of a child.
b. Indicate dependence among the error terms, because they are all negative.
c. Show that the variables shorten the lives of the children.
d. Exhibit a downward effect of mortality rates on all of the independent variables, but more so on nutritional intake.
e. The proportion of explanation of the dependent variable decreases as the independent variables increase.

Word and Thought Problems

167. What are the possible values of R^2? Which values are better and which are worse? Describe a situation where a low R^2 may be appropriate and a situation where a large R^2 may signal an unreasonable result.

168. The estimated regression line from the following data is $\hat{Y} = 3.46 + 0.88\,X$, where Y = sales and X = advertising.

Y Sales	X Advertising
2	1
4	0
6	1
8	5

    a. Test H_0: $b \le 0$, H_1: $b > 0$ at the 0.05 significance level, if $\hat{\sigma}_{\hat{b}} = 0.54$. Interpret the decision.
    b. Evaluate the model, using the results in Part a and $R^2 = 0.57$.
    c. Find a 95% prediction interval for Y_f if $X_f = 10$ and $\hat{\sigma}_e = 2.07$.

169. Suppose you obtain the following results from estimating a linear function:

    $$\hat{Y} = 7{,}000 + 1{,}000X,$$
    $$(500)$$

    where Y = attendance at a team's ith home game and X = team's winning percentage at game time. (The standard error of the coefficient is frequently reported in parentheses below the coefficient—in this case, it is 500.) $R^2 = 0.67$, and $n = 25$.
    a. Interpret the 1,000.
    b. Interpret R^2.
    c. Test the coefficient at the 0.05 level, using H_1: $b > 0$. Interpret the decision.
    *d. Perform a two-tailed test at the 0.05 level if the two-tailed p-value is 0.057.
    *e. Use the two-tailed p-value from Part d to perform the upper-tailed test at the 0.05 level. Compare the decision with the result in Part c.

170. What are some possible problems with using a regression equation for forecasting?

171. Political science researchers argue that innovation occurs in laws covering new legal areas, so that we can expect the ease with which citizens use and understand new laws on an issue to increase over time. The researchers dub this ease-of-use characteristic "facilitation." Consequently, successive laws should encompass the experience of preceding laws and increase facilitation. More specifically, we can predict the facilitativeness of a state's original law by the date when it first adopted a new law concerning the new legal issue: Original facilitation level = f(Original adoption date). However, because states change laws and incorporate their own and others' experiences, we should not expect to predict the current facilitation status of a state's law on the issue with the date at which it originally adopted a law on the issue: Current facilitation level = f(Original adoption date).
    a. Test the propositions using computer regression estimates and the following data, evaluations of the facilitativeness of states' original living will laws, for 38 states and D.C. The researchers base the facilitation scores on a content analysis of the laws, where larger scores accompany more facilitative laws. Interpret the results of the decisions, and relate them to the hypotheses of the researchers. The standard errors of the coefficients are 0.1719 and 0.1956 for the respective models. (Source: Henry R. Glick and Scott P. Hays, "Innovation and Reinvention in State Policymaking: Theory and the Evolution of Living Will Laws," *Journal of Politics*, Vol. 53, No. 3, 1991, pp. 835–49. Latescor values were estimated from scatter diagram.)

    Origscor = facilitation score on original law.
    Latescor = later facilitation score with amended laws.
    Adoptyr = date the original law passed.

ID	State	Adoptyr	Origscor	Latescor
1	CA	76	8	8
2	AR	77	11	25
3	ID	77	6	14
4	NV	77	9	10
5	NC	77	10	10
6	NM	77	12	15
7	OR	77	11	12
8	TX	77	7	15
9	KS	79	14	14
10	WA	79	12	12
11	AL	81	13	13
12	DC	81	16	16
13	DE	82	11	12
14	VT	82	12	12
15	IL	83	10	14
16	VA	83	18	18
17	FL	84	15	15
18	GA	84	7	8
19	LA	84	18	18
20	MS	84	10	10
21	WV	84	13	13
22	WI	84	10	11
23	WY	84	12	12
24	AZ	85	13	13
25	CO	85	11	11
26	CT	85	6	6
27	IN	85	13	13
28	IA	85	13	13
29	ME	85	20	20

ID	State	Adoptyr	Origscor	Latescor
30	MD	85	16	16
31	MO	85	12	12
32	MT	85	20	20
33	NH	85	9	9
34	OK	85	8	8
35	TN	85	13	13
36	UT	85	15	15
37	AK	86	20	20
38	HI	86	13	13
39	SC	86	9	7

ID	Study Time	Grade	ID	Study Time	Grade
1	8	68	15	23	100
2	1	88	16	15	68
3	18	64	17	11	98
4	11	44	18	19	92
5	4	54	19	18	75
6	18	88	20	9	96
7	25	98	21	5	55
8	9	63	22	3	71
9	10	56	23	16	82
10	11	97	24	4	79
11	16	61	25	8	77
12	21	75	26	16	92
13	13	61	27	23	76
14	14	88	28	11	60

  b. Is the data set used in each regression a time series or cross section? Explain.
  c. Sketch scatter diagrams of the data, and relate the result to the decisions made in Part a.

172. The following function is hypothesized to be linear: $Y = f(X)$, where Y is the number of sick days taken by employee i, and X is the annual income of employee i (measured in thousands).
  a. What sign would you expect for the coefficient of X? Why?
  b. Suppose the estimated relationship is $\hat{Y} = 50 - 2X$ (with a standard error of the coefficient of 1.5) using data from 15 randomly selected employees. Perform a lower-tailed test at the 0.01 level, and interpret your result.
  c. What is a point estimate of sick days taken by an employee making $20,000 per year?
  ***d.** The two-tailed p-value for the hypothesis in Part b is 0.206. Use this information to corroborate the decision in Part b.

173. State legislative researchers estimated the following equation (the standard error of the coefficient is in parentheses):

$$\hat{Y} = 150 + 10X.$$
(4)

$R^2 = 0.44$, $n = 50$, Y = number of new businesses locating in state i, and X = measure of business taxes in state i.
  a. Interpret the R^2 value.
  b. Interpret the coefficient of X. Does it seem to be reasonable? Why or why not?
  c. Test H_1: $b \neq 0$ at the 0.1 level. Interpret the decision.
  ***d.** The two-tailed p-value is 0.016. Use this information to corroborate the results in Part c.
  e. Suppose we suspect that raising taxes causes firms to avoid locating in the state. If we perform the test at the 0.1 level now, what would the decision be? Interpret the decision.

174. a. If $Y = f(X)$, where Y is the grade of the ith student on a statistics test and X is the number of hours student i studied for the test, what do you expect the sign of the coefficient to be? Explain.
  b. Estimate the model using the following data set. Interpret the coefficient.
  c. Perform an upper-tailed test at the 0.1 level. If you did not use a computer for your computations, the standard error of the coefficient is 0.4576. Interpret the decision.
  ***d.** If the two-tailed p-value is 0.084, perform the test in Part c.
  e. Interpret the R^2 of 11.0%.

***175.** What two things must we check before making a decision for a hypothesis test, when using a two-tailed p-value to do a one-tailed test?

176. A personnel department believes the number of applicants for a position is a function of the unemployment rate. The department collects data for the last 25 open positions and obtains the following results:

$$\hat{Y} = 252 - 12.5X.$$
(5)

  a. What do you think H_1 should be? Why?
  b. Interpret the coefficient -12.5. How does its sign compare with your anticipated sign in Part a?
  c. Do a two-tailed test at the 0.05 level.
  ***d.** The two-tailed p-value is 0.02. Use this number to do a lower-tailed test for b at the 0.05 level.

177. Financial analysts collected data to investigate the hypothesis that a domestic steel company's stock price is a function of the most recent steel import figures. The results are $\hat{Y} = 120.875 - 0.125X$, $n = 29$.
$\phantom{The results are \hat{Y} = 120.875 - 0.12}$(0.073)
  a. What do you think H_1 should be? Why?
  b. Interpret -0.125. How does its sign compare with your anticipated sign in Part a?
  c. Do a two-tailed test at the 0.05 level. Interpret the decision.
  ***d.** The two-tailed p-value is 0.1. Use this number to do a lower-tailed test for b at the 0.05 level. Interpret the decision.

178. Suppose we use data from a random sample of counties to estimate $\hat{Y} = 200 + 50X$, where Y is the number of swimmers patronizing the local swimming pool per day and X represents the hours of operation. $\overline{X} = 20$, $\Sigma (X - \overline{X})^2 = 100$, and $n = 20$. The standard error of

Word and Thought Problems

the estimate about the line is 25. Find a 95% prediction interval for the number of daily users if the county parks and recreation department considers opening 10 hours per day.

179. *Rule of Thumb* Rather than spending time looking for critical values from a table, there is an easy way to check for significance when performing hypothesis tests for regression coefficients. The rule of thumb is to reject H_0 if the absolute value of the T-ratio is greater than 2, especially for larger samples. Use your tables to justify this rule of thumb. Should we ignore the sign of the coefficient when making the decision? Explain.

180. Describe the sampling distribution of $\hat{b}$ in $\hat{Y} = \hat{a} + \hat{b}X$. Include all relevant assumptions. What is the T-ratio? Describe the sampling distribution of the T-ratio, including all relevant assumptions.

181. A music educator persuaded a random sample of introductory students in the same age group to keep records on practice time. The teacher computed the average daily practice time to the nearest 10-minute interval and the progress of the student, measured in number of lesson books completed during the time span of the experiment, as shown in the following

ID	Practice	Progress
1	10	3.8
2	0	1.1
3	10	2.0
4	40	5.0
5	30	6.3
6	10	2.7
7	20	5.9
8	40	3.7
9	10	0.9
10	20	3.5
11	10	1.1
12	40	4.0
13	30	1.9
14	30	6.7
15	0	0.8
16	0	0.7
17	10	0.5
18	20	2.7
19	0	0.8

a. If the teacher models the relation between progress and practice as Progress = f(Practice), what is the expected sign of the coefficient? Explain.
b. Estimate the regression line. Interpret the coefficient.
c. Test for a positive coefficient at the 0.01 level. Interpret the result. If you do not estimate the equation with a computer, the standard error of the coefficient is 0.02448.
*d. The two-tailed p-value is 0.001. Use this value to corroborate the decision from Part c.
e. Find the predicted value and residual for the first point in the data set.

182. Roy believes that the number of cars he sees while jogging is a linear function of the time of day he jogs, always between 7 A.M. and 2:30 P.M. (measured in hours past midnight). He keeps a record for 21 days of the number of cars, C, and time, T. The result of a linear regression on these data is

$$\hat{C} = 1{,}000 - 50t. \quad (24)$$

a. Interpret the -50.
b. Do a lower-tailed test at the 0.05 level. Interpret the decision.
c. Interpret an R^2 of 0.82.
d. What is your overall evaluation of this model?
*e. The two-tailed p-value for this model is 0.05. Do an upper-tailed test using the p-value.
f. When would you be extrapolating if you were using this model? Discuss the implications of extrapolation for this model.

183. A regression line is the line that "best" fits the sample data points. Why then do we bother with the additional measures and procedures for evaluating the results if we already have the "best" fit?

184. $\hat{V} = 48 - 2.1C, \quad n = 12, \quad R^2 = 0.08,$
$\quad\quad\quad (3.0)$

where

V = percentage of registered voters who participate in an election.
C = difference in polls between two leading candidates, one week before election (percentage of voters in poll planning to vote for the leader minus percentage of the next highest candidate)

a. Interpret the -2.1. Is the result reasonable?
b. Perform a lower-tailed test for b at the 0.01 level. Interpret your decision.
c. Use R^2 and your answer to Part b to evaluate the model.
*d. How can the two-tailed p-value be 0.5, if $\hat{b}$ is not zero?

185. A banker wishes to measure the effect of poor economic conditions on consumer loan repayment behavior. Data on the percentage of consumer installment credit that is overdue and the simultaneous unemployment rate yield the regression output displayed in Figure 11-56.
a. What is the dependent variable?
b. How many data pairs are used in the calculations?
c. What is the regression equation?
d. Interpret the coefficient. Is this result reasonable?
e. Interpret the R^2 value.
f. Do an upper-tailed test using the T-ratio at the 0.1 level.
*g. What is the upper-tailed p-value and decision at the 0.1 level?

Figure 11-56
Minitab regression output for data on overdue payments

```
The regression equation is
OVERDUE = 4.85 - 0.358 UNEMPLOY

Predictor       Coef       Stdev      t-ratio       p
Constant       4.8480     0.2613       18.56     0.000
UNEMPLOY      -0.35818    0.03994      -8.97     0.000

s = 0.4677        R-sq = 76.3%       R-sq(adj) = 75.3%

Analysis of Variance

SOURCE         DF         SS          MS          F          p
Regression      1       17.589      17.589      80.42      0.000
Error          25        5.468       0.219
Total          26       23.057

Unusual Observations
Obs. UNEMPLOY    OVERDUE       Fit   Stdev.Fit   Residual   St.Resid
  5    13.6      1.0646     -0.0232    0.3112     1.0878      3.12RX
 17     4.8      4.0568      3.1288    0.1047     0.9280      2.04R

R denotes an obs. with a large st. resid.
X denotes an obs. whose X value gives it large influence.
```

h. A data pair in the sample is (UNEMPLOY = 13.6, OVERDUE = 1.0646). What is the predicted Y value for UNEMPLOY = 13.6? Use this result to locate the endpoints of a 95% prediction interval, if the mean of the sample unemployment values is 6.1407% and the sum of the squared deviations of these values is 1,155.2.

186. Figure 11-57 shows computer output for two models of time between parole and recidivists' relapses into criminal activity over a two-year period estimated by a criminal justice major for a project. The student also collected data on the criminal's time in prison before parole and the percentage of the sentence served to test if harsher punishments reduce recidivist activity. Times are measured in months.
   a. How many criminals' records does the student use in the project?
   b. What is H_1 for each model according to the student's objective?
   c. What are the estimated regression equations? Interpret each coefficient.
   d. Perform the tests for the student at the 0.05 level using critical values. Interpret the decisions.
   *e. Repeat Part d using the p-values.
   f. Interpret the R^2 values for each model. Are R^2 comparisons valid for choosing the better fit between the two models here? Explain.

g. If criminals average spending 20% of their time in prison, predict the average time before committing another crime among those who are recidivists. Stretch this estimate into an 80% prediction interval if the average percentage of the sentence spent in prison is 20.037 and the sum of the squared deviations for this variable is 26,918.

187. A regression model to explain variation in income with IQ scores performs poorly, fails the T-test, and has a low R^2. Suggest a way or ways to improve the results.

188. Data on annual per capita beer sales (in gallons) produce the following estimate:

$$\hat{S} = 32.8 - 2.65\text{GDP} + 3\text{PHARD} - 4.5\text{WORK},$$

where GDP is the gross national product, a measure of the output of the economy, expressed in trillions of dollars, PHARD is the average price of a gallon of hard liquor (dollars), and WORK is the average number of hours worked per week during the year being observed. Does beer consumption appear to be counter- or pro-cyclical, that is, does it emulate the same or opposite behavior of economic activity? Do people appear to substitute beer for hard liquor or just drink more of both when times are difficult?

Review Problems

189. If the tagged price of an item is X dollars and the sales tax is 5%, we can figure the total purchase price, T, using $T = 1.05X$. If we collect a random sample of purchases at a retail store, then record the tagged and total purchase prices, what will the correlation coefficient be for these two variables?

Review Problems

(a) Independent Variable is Months Served in Prison

```
The regression equation is
RELAPSE = 11.5 + 0.0114 MONTHS

Predictor        Coef       Stdev      t-ratio        p
Constant      11.4913      0.6328        18.16    0.000
MONTHS       0.011433     0.005008         2.28    0.026

s = 2.261         R-sq = 7.4%        R-sq(adj) = 6.0%

Analysis of Variance
SOURCE         DF           SS          MS         F        p
Regression      1       26.656      26.656      5.21    0.026
Error          65      332.418       5.114
Total          66      359.075

Unusual Observations
Obs.   MONTHS    RELAPSE       Fit   Stdev.Fit   Residual   St.Resid
  1      244     14.000     14.281      0.709     -0.281    -0.13 X
  3       30     18.000     11.834      0.502      6.166     2.80R
  6       54     17.000     12.109      0.407      4.891     2.20R
 13      241     13.000     14.247      0.695     -1.247    -0.58 X
 52       53     17.000     12.097      0.411      4.903     2.20R
 54      255     13.000     14.407      0.760     -1.407    -0.66 X

R denotes an obs. with a large st. resid.
X denotes an obs. whose X value gives it large influence.
```

(b) Independent Variable is Percent of Sentence Served

```
The regression equation is
RELAPSE = - 68.0 + 4.03 PERCENT

Predictor        Coef       Stdev      t-ratio        p
Constant      -68.038       5.083       -13.39    0.000
PERCENT        4.0339      0.2536        15.91    0.000

s = 1.063         R-sq = 79.6%       R-sq(adj) = 79.2%

Analysis of Variance
SOURCE         DF           SS          MS         F        p
Regression      1       285.69      285.69    253.06    0.000
Error          65        73.38        1.13
Total          66       359.07

Unusual Observations
Obs.   PERCENT   RELAPSE       Fit   Stdev.Fit   Residual   St.Resid
  3      21.5    18.000     18.691      0.393     -0.691    -0.70 X
  6      21.2    17.000     17.481      0.322     -0.481    -0.48 X
 16      20.3    16.000     13.851      0.146      2.149     2.04R
 47      18.7     8.000      7.396      0.363      0.604     0.60 X
 52      21.3    17.000     17.885      0.346     -0.885    -0.88 X
 55      18.7     8.000      7.396      0.363      0.604     0.60 X

R denotes an obs. with a large st. resid.
X denotes an obs. whose X value gives it large influence.
```

Figure 11-57
Minitab output for data on recidivists

190. Preliminary analyses in a study to learn more about the effects of working mothers and child-care methods on the health and development of children in Guatemala employed correlation analysis. Anthropometric measures include different nutritional conditions transformed according to international standards:
   1. H/A = height for age (long-term indicator of nutritional development).
   2. W/A = weight for age (long-term indicator that may be influenced by short-term fluctuations, such as illness or period of low food intake).
   3. W/H = weight for height (indicates disproportion in the body).

   Some of the findings are listed below. One asterisk indicates significantly different from zero when $\alpha = 0.05$. Two asterisks indicate the same decision when $\alpha = 0.01$.

Mother's Work Variable	H/A	W/A	W/H
Working and Nonworking Mothers ($n = 239$)			
Age in months	0.05	0.13*	0.08
Education	0.34**	0.26**	0.04
House quality	0.25**	0.17**	−0.03
Income per capita	0.21**	0.21**	0.11
Birth order	−0.23**	−0.18**	−0.07
Working Mothers only ($n = 112$)			
Hours of work per year	0.01	0.07	0.12
Mother's income per month	0.26**	0.25**	0.13
Mother's income as percentage of family income	0.17*	0.20*	0.14

Source: Patrice L. Engle, "Maternal Work and Child-Care Strategies in Peri-Urban Guatemala: Nutritional Effects," *Child Development*, Vol. 62, No. 5, 1991, pp. 954–65. Copyright © 1991 by the Society for Research in Child Development.

   a. Interpret the relationship indicated by the sign of the correlation coefficient for each significant estimate.
   b. We say that the closer the coefficient is to ±1, the stronger the linear relationship is, whereas the closer to 0, the less indication of a linear relationship. Study the estimates and report on the relationship between sample size and relative size of significant coefficients. In other words, suppose we establish a critical value so that correlation coefficients whose absolute value exceeds this critical value indicate a linear relationship. As n increases, does the critical value increase or decrease? What evidence in the correlation matrix supports your conclusion?

191. a. Compute the correlation coefficient for saturation and purchases of electronic products, using the following data. Interpret the estimate.
    b. Test for positive correlation at the 0.05 level. Interpret the decision.
    c. Is the data set an example of a time series or cross section? Explain.
    d. Do you think the results would be different if you had a time series for any one of the products? Explain.
    e. Do you think the results would be different if you used a different characterization of sales, such as per capita sales or the annual growth rate of sales? Explain. The growth rates (in percents) corresponding to the list of products are −4.1, 3.7, −12.0, −8.5, 10, 32.4, 29.6, 32.5. Use this information to explore your conjectures.

Product	Saturation	Sales (in thousands)
Color television sets	97	20,808
Videocassette recorders	74	10,119
Telephone answering machines	43	11,000
Cordless telephones	36	9,150
Home computers	29	5,500
Compact-disk players	28	9,155
Video camcorders	15	2,962
Projection television sets	7	351

Source: Reprinted by permission of the *Wall Street Journal*, © 1991 Dow Jones & Company, Inc. All Rights Reserved Worldwide.

192. In a pass-fail course, the percentage of the class who pass is X and the percentage who fail is Y. Together X and Y comprise 100% of the class participants, $X + Y = 100$. What is the correlation coefficient for X and Y? Will your answer change if some participants withdraw from the class before evaluation, that is, Z is the percentage who withdraw and $X + Y + Z = 100$?

193. Correlation and regression concentrate on directions of change of two variables. What is the major difference in the assumptions that affects the interpretation of each procedure's coefficient?

194. A researcher used regression analysis to learn more about categorizing two forms of adult scorpions. The adult male chela (pincer) size (CSI) is a linear function of anterior body length (ABL) confirming that the large form of adult males have larger chelae. The regression result is $CSI = -37.1 + 12.9\ ABL$. Do these results agree with the hypothesis? (Source: T. G. Benton, The Life History of *Euscorpius Flavicaudis (Scorpiones, Chactidae)*, *The Journal of Arachnology*, Vol. 19, No. 2, 1991, pp. 105–10.)

195. Use the following data on the number of credit card holders, H, expressed in millions and the collectible debt expressed as a percentage of credit card amounts due, D, to determine the effect of variation in the number of card holders on the level of collectible debt. Not every card issued goes to a new card holder, many people have more than one credit card. Interpret the estimated coefficient, and plot a scatter diagram to demonstrate the estimated relationship.

ID	Year	Holders	Debt
1	1981	62	98.06
2	1982	64	98.32
3	1983	69	98.62
4	1984	71	98.60
5	1985	72	97.67
6	1986	76	96.68
7	1987	78	97.40
8	1988	80	96.96
9	1989	83	96.84

Source: Credit card companies, Federal Reserve Board, Bankcard Holders of America, as reported in "Credit Cards in the U.S.," *Asheville Citizen-Times*, January 5, 1992, p. 4B.

196. To determine if stutterers take longer to think of word associations, researchers recorded response times to 36 word associations for 15 stutterers and 15 nonstutterers for a total of 540 responses for each group. The average response time and standard deviation for stutterers were 1.97 and 1.07 seconds, respectively. Similar values for nonstutterers are 1.75 and 0.71 seconds. Use this information to test for a difference in means between stutterers and nonstutterers at the 0.01 level. Interpret the decisions. (Source: Kathryn M. Crowe and Robert M. Kroll, "Response Latency and Response Class for Stutterers and Nonstutterers as Measured by a Word-Association Task," *Journal of Fluency Disorders*, Vol. 16, 1991, pp. 35–54.)

197. A regression equation to explain the defect rate as a function of the number of hours a shift has been at work is estimated to be

$$\hat{Y} = 0.0001 - 0.003X.$$

  a. Interpret -0.003.
  b. Forecast the defect rate for a shift that has been at work for 4.75 hours.
  c. Discuss the reasonableness of the estimated equation.

198. List the possible objectives of a regression analysis. What criteria does the regression line satisfy? What assumptions must the error terms satisfy for the procedure to produce reliable estimates and tests?

199. Suppose we want to estimate a model of the effect of the number of classes a student has missed prior to a quiz on the student's score for a 12-point quiz.
  a. Find $\hat{a}$ and $\hat{b}$ for the following data, assuming $Q = f(A)$. Plot a scatter diagram for the data to verify the sign of $\hat{b}$.

Q, Quiz Score	A, Absences
1	4
0	5
10	1
5	2
4	3

  b. Predict the quiz score of a student who misses one class.
  c. Why are $\overline{Q}$ values not all equal for different values of A?
  d. Interpret the R^2 of 0.93 for this model.
  e. We usually expect scores to drop as the number of absences increases. Do the sample results support this hypothesis at the 0.01 level?
  f. Find a 90% prediction interval for Q_f when $A_f = 2$, $\Sigma(A - \overline{A})^2 = 10$, $\hat{\sigma}_e = 1.21$, and $\overline{A} = 3$. Discuss the effect of the source of the A value used in the forecast on the accuracy of the forecast.

200. Why might a model for forecasting energy usage, which was estimated prior to the 1973 energy crisis, produce forecasting errors?

201. What is the standard error of the coefficient? Would you prefer its value to be large or small in a regression context? Explain.

202. List four precautions for forecasting with regression analysis.

203. The years, T, any coach spends in any single job are assumed to be a direct linear function of his or her winning percentage, P. The result of a regression, using data from 19 coaches, is

$$\hat{T} = 1.0 + 0.28P$$
$$(0.21)$$

  a. Interpret the 0.28.
  b. Use the standard error in parentheses below the coefficient and test the coefficient at the 0.01 level. Interpret your decision.
  ***c.** Do a two-tailed test at the 0.05 level, using the two-tailed p-value, 0.10.

204. What role does multiple regression play in the role of behavioral and social sciences that is similar to the work of natural scientists? Is regression analysis useful for natural scientists as well? Think of an example where they might employ the procedure.

205. A government official hypothesizes that the number of new businesses that are formed is a consequence of the level of the GNP, which reflects overall demand in an economy, and the level of interest rates in an economy. The official collects GNP (in billions of dollars) and new business formation data for 40 quarters. Figure 11-58 pictures the computer output of the regression analysis. Interpret the results.

206. Suppose a regression analysis is performed on the number of dates, D, a person has in a week with the number of showers, S, and the number of teeth brushings, T, during the corresponding week as the independent variables. The result, based on a random sample of 15 people, is $\hat{D} = 0.500 + 0.950\, S + 0.223\, T$. Interpret the 0.950. Is it reasonable for the coefficient not to be an integer? To be positive?

207. A criminologist hypothesizes that crime rates, C (measured as a percentage or incidence of a certain crime among the population), in metropolitan areas are

Figure 11-58

Minitab output for new business formation data

```
The regression equation is
NEWBUS = 137233 + 18.1 GNP - 8120 INTRATE

Predictor        Coef       Stdev      t-ratio       p
Constant       137233       38050        3.61      0.001
GNP            18.087        9.215       1.96      0.057
INTRATE         -8120        4915       -1.65      0.107

s = 32461        R-sq = 10.0%        R-sq(adj) = 5.1%

Analysis of Variance

SOURCE          DF           SS             MS           F        p
Regression       2      4309183488      2154591744      2.04    0.144
Error           37     38986416128      1053686912
Total           39     43295600640

SOURCE          DF         SEQ SS
GNP              1       1433913600
INTRATE          1       2875270144
```

a function of arrest rates, A (measured by the percentage of suspects that are arrested), in the area and of the amount of police presence around the city, P (measured by total hours worked by police personnel), provided by the local government. The analyst selects a random sample of 100 metropolitan areas and collects data for a regression. The results are
$\hat{C} = 36 - 0.17A - 0.11P$.

a. Suggest a policy to the governing body of a city for reducing crime based on these results. Do you think these results are useful for small towns? Explain.
b. Do you think this model is sufficiently complex to explain variation in crime rates? Explain.

208. Why might an analyst use a multiple regression model rather than a simple regression model?

209. Some forecasters use R^2 as a major tool for evaluating different regression models for a variable to be forecast. Why might R^2 measures be better evaluation tools than individual T-tests in these circumstances? Under what forecasting circumstances might T-tests be better than R^2 measures?

210. If the sign of a coefficient is opposite of what you expect, what will the decision be when you do a T-test? Explain. Why is it incorrect to look at the sign of an estimated coefficient to decide which direction to test, for example, to do a lower-tailed test, because the coefficient is negative?

211. To assess the effect of working mothers' earnings characteristics and socioeconomic status variables on the health and development of children in Guatemala, a researcher estimated the following models using multiple regression analysis with $n = 107$ children. The health and development measures are anthropometric measures transformed according to international standards:

i. H/A = height for age (long-term indicator of nutritional development).

ii. W/A = weight for age (long-term indicator that may be influenced by short-term fluctuations, such as illness or period of low food intake).

iii. W/H = weight for height (indicates disproportion in the body).

A single asterisk indicates values significantly different from 0 at the 0.05 significance level and two asterisks indicate significance at the 0.01 level.

Mother's Work Variable	H/A	W/A	W/H
Child's age in months	0.010	0.036**	0.031**
Education (years passed)	0.074*	0.046	−0.001
House quality	0.065	0.045	0.001
Income per capita	0.003	0.004	0.003
Birth order of child	−0.036	0.017	0.042
Child's gender (1 = m, 2 = f)	−0.277	−0.360	−0.255
Marital status (1 = union, 0 = nonunion)	0.919*	1.060**	0.666*
Hours of work per year	−0.000	−0.000	0.000
Mother's income per month	−0.000	−0.002	−0.001
Mother's income (% of family income)	0.019**	0.021**	0.010*
R^2	0.27	0.29	0.18

Source: Patrice L. Engle, "Maternal Work and Child-Care Strategies in Peri-Urban Guatemala: Nutritional Effects," *Child Development*, Vol. 62, No. 5, 1991, pp. 954–65. Copyright © 1991 by the Society for Research in Child Development.

a. Interpret the significant coefficients.
b. The objective of the analysis is to determine the effects of a mother's income earning characteristics on the development of the child. Why are the variables describing the child included in the model?

Review Problems

212. Use the values of the following variables from the State Health Data Base (Appendix K) and the computer to make the necessary computations, in order to answer the following questions; S for percentage of a state's adult population that has less than three different 20-minute leisure-time physical activities per week and O for the percentage of a state's adult population that is overweight.

  a. Find and interpret the correlation coefficient between the two variables.

  b. If we hypothesize that O is a linear function of S, which is the dependent, and which is the independent variable? What do you expect the sign of the coefficient to be?

  c. Estimate and interpret the regression coefficient.

  d. Test your hypothesis about the coefficient's sign from Part b. Use a 0.05 significance level.

Chapter 12

Analysis of Variance

In this chapter we learn to test for differences in means among more than two populations, as in an experiment with more than two differences among the participants. When we compare multiple pairs of means (using testing techniques from Chapter 10), the significance levels are no longer fixed or predictable. So, rather than compare two means at a time, we must turn to the important statistical technique called analysis of variance (ANOVA), the subject of this chapter.

Objectives

We use one-way analysis of variance (ANOVA) to test for differences in multiple (more than two) population means, when no other variable may interfere and cause the experimental outcome to change. Upon completing the chapter, you should be able to

1. State the hypotheses and conclusions of an ANOVA test formally, in terms of means and effect sizes, then interpret the decision in everyday terms related to a given problem or situation.
2. State the purpose of ANOVA in controlled experimental situations and relate the test statistic to this purpose.
3. State the assumptions we employ when we conduct an ANOVA procedure.
4. Determine critical values from an F table (Appendix L).
5. Perform a one-way ANOVA for a completely randomized design.
6. State the purpose of two-way ANOVA.
7. Understand the concepts of controlled experiment, experimental design, completely randomized design, factor, experimental factor, and treatment.
8. Recognize the following symbols: Y, an outcome for a subject in an experiment; $\overline{Y}$, the mean outcome for a sample of subjects treated identically in an experiment; $\overline{\overline{Y}}$, the grand mean of outcomes for all experimental subjects; μ, the mean outcome for a population of subjects treated identically in an experiment; σ^2, the variance of the population of all potential experimental subjects; n, the number of subjects in a sample of experimental subjects; SS, the sum of the squared observations for each participant in the experiment; SST, the *total* variation of experimental outcomes about the grand mean; SSW, the variation of experimental outcomes *within* samples of subjects treated identically; SSB, the variation *between* sample means and the grand mean; MSW, the overall mean dispersion of experimental outcomes about their respective sample means (also called the mean square error); and MSB, the mean dispersion of the different sample means about the grand mean.

Unit I: One-Way Analysis of Variance

Preparation-Check Questions

1. What operation do we perform on the variance to obtain the standard deviation?
   a. Square.
   b. Square root.
   c. Divide by n.
   d. Divide by $n - 1$.
   e. Multiply by 2.

2. What do the variance and standard deviation measure about values in a data set?
   a. The spread or dispersion.
   b. The center.
   c. The distance from the smallest to the largest value.
   d. Skewness.
   e. Most important value.

3. Generally, which of the following circumstances decreases the probability that an estimate will be close to the desired population parameter?
   a. A large sample.
   b. A large population standard deviation.
   c. A large significance level.
   d. A large confidence level.
   e. A large population mean.

4. Assuming H_0 is true for a particular test results in a probability distribution for a test statistic, L, that has the following unusual graph. L is an estimator of a population parameter, λ. Which region on the L axis corresponds to the rejection region for a test of H_0: $\lambda \leq 0$?

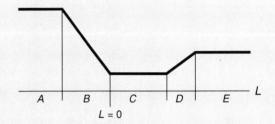

a. A.   b. B.   c. C.   d. D.   e. E.

Answers: 1. b; 2. a; 3. b; 4. e.

Introduction

Training Method Experiment Suppose that the National Aeronautics and Space Administration (NASA) wants to learn how different training methods affect the reaction speed of astronauts as they perform a certain task. Three consultants have each proposed a different training method. NASA divides 24 potential astronauts into three samples and trains each sample by one of the three methods. Afterward, each astronaut performs the task during a mock space flight, while analysts record each reaction speed. The following table lists the results (in seconds) along with the mean reaction time for each sample. The samples are different sizes for two reasons: NASA assigns fewer astronauts to more expensive methods, and some of the participants dropped out of the program. Is it possible that all three techniques train the astronauts equally well, or does this evidence indicate a difference in training methods?

	Training Method		
	1	2	3
	0.50	1.54	0.70
	0.72	1.28	1.46
	1.02	1.64	1.38
	0.48	1.20	0.78
	0.86	1.74	1.24
	1.10	1.34	1.14
	0.68		1.04
	0.92		0.88
	0.84		
	0.82		
Means	0.79	1.46	1.08

The means of the three samples are certainly different, but they are only sample means. Many different sample means can originate from the same population, which has a single mean. We have a decision to make. Are the sample means so distinct from one another that they indicate significant differences in the mean reaction speeds produced by each type of training? Or, are the sample means so close that they indicate that the mean reaction speeds are essentially equal for the three training methods? In other words, does the sample evidence substantiate a training method effect on the reaction speed?

Statisticians and experimenters employ *analysis of variance (ANOVA)* to analyze this type of data and answer such questions.

12.1 Setting Up the Problem

To analyze the association of several different values of a variable (or of several different sets of circumstances) with the outcome of a controlled experiment, we can use the technique analysis of variance (ANOVA). Specifically, we compare the means of more than two different populations.

12.1.1 The Problem with Paired Tests (Optional)

At first, you may wonder why we do not apply the techniques that determine differences between two means, discussed in Chapter 10, to differences among three or more means. For instance, using the astronaut case with three means, we would need three tests, one for each pair of training method means: $\mu_1 = \mu_2$, $\mu_2 = \mu_3$, and $\mu_1 = \mu_3$. Problems develop when we do this. The probability of a Type I error becomes ambiguous when we perform multiple tests using the same data. If $\alpha = 0.05$ for each paired test, an upper limit for the probability of at least one rejection, even though the three means are equal, would be $1 - P(\text{no rejections}) = 1 - (1 - \alpha)^3 = 1 - (1 - 0.05)^3 = 1 - (0.95)^3 = 1 - 0.8574 = 0.1426$. Thus, individual paired tests can magnify the chances of deciding the three means are unequal when they are equal beyond the individual significance levels. In this example, the result, 0.1426, is almost three times the chance we would make a Type I error if we used ANOVA (0.05), which tests for equality of all three means simultaneously.

12.1.2 Controlled Experiments

When the outcomes of an experiment or some other activity vary from trial to trial, we usually pursue an explanation. Often, by casually observing the situation, we can explain the various outcomes we observe. In other cases, more formal scientific and logical principles help us discover the cause.

For instance, NASA had completed many successful launches of the space shuttles before the January 1986 disaster. One of the first explanations offered for the disaster was that the outside temperature had been lower than for previous launches. When we can reduce the set of all possible factors that affect the outcome to only one, we assume it is the immediate cause of the variation in outcomes.

An experiment is a more formal procedure conducted in order to collect evidence to explain or to substantiate hypotheses about causal processes. Ideally we conduct it so that only two things change: the variable that we suspect will cause the outcome to vary, and the outcome itself. Such an ideal experiment is called a *controlled experiment*. A *factor* is a variable that can potentially change the outcome. We call the factor that we investigate and suspect to be the cause of the outcome variation the *experimental factor*. We control the variation in the amount or level of the experimental factor

Section 12.1 Setting Up the Problem

during an experiment. A *treatment* is a certain set of circumstances (usually expressed as a level of the experimental factor) that we allow to occur when we observe the outcome of the experiment.

For example, NASA scientists and engineers simulate space flight conditions in order to learn how shuttles, satellites, other space vehicles, and people might behave when a single change in a single variable occurs during space flight. The scientists adjust the appropriate environmental factors, such as weightlessness, so that there will be no differences between the test environment and conditions during the actual space flight that would alter their findings about the effect of the experimental factor. The experimental factor in the Training Method Experiment is the method of training the astronauts, and the treatments are the three different training methods. During the mock space flight, the scientists control all environmental factors, in order to eliminate any differences from actual space flight that might influence the reaction of the astronauts.

However, chance or random occurrences may alter the levels of factors, making *completely* controlled experiments impossible. Chance is especially troublesome for social scientists who study human behavior. Mechanical and electronic apparatuses as well as physical principles tend to behave predictably under repeated laboratory trials, but people's reactions, even in a laboratory, are not as simple. The nature of humans (as well as of animals and of plants) dictates that subjects' behaviors will vary, even when they experience the same treatment. For example, astronauts within one treatment sample, receiving the same type of training, may still react at different speeds. We consider such behavioral variation that is not systematically linked to a causal factor to be generated randomly, or by chance. The complex array of factors that may contribute to the outcome of an experiment is a troublesome one for physical scientists as well.

Consider NASA experiments that simulate actual space flight. They are expensive and complicated attempts to control as many variables as possible. NASA requires such experiments because of the potential danger to equipment and, more importantly, to lives during an actual flight. However, an individual's reaction during these tests may be different from the reaction in actual flight, because of other uncontrolled factors. For instance, the reactions in space may vary because the crew members are different, they are less rested, the task occurs during another important maneuver, or because the astronaut knows that the situation is real and not a test. Controlling for every possible factor is impossible, so we deliberately control only the most important known factors.

Another problem that prevents complete control of the variables in any experiment is the routine use of data that are collected outside a laboratory, such as a firm's sales. If we investigate the effect of various promotional strategies, such as contests and giveaways, on a firm's sales data, we cannot directly control other factors that affect customers' buying decisions. We might witness a drop in sales in spite of a special promotion, because it occurred in the midst of a recession. We can arrange experiments, so that the variation we assign to chance includes the variation of uncontrolled factors, those of secondary importance, and those we do not recognize, because they are submerged in nonlaboratory data. This arrangement bypasses any systematic effects on the results that might arise from the uncontrolled factors.

We try to avoid two major mistakes when we set up a controlled experiment. The first is to attribute the difference in outcomes to a factor that is not responsible for it. The second is to fail to recognize the importance of a true causal factor. A well-designed experiment avoids these mistakes. Subsequently, we may use the ANOVA technique to analyze the resulting data.

Question: NASA must divide the random sample of 24 potential astronauts from the Training Method Experiment among three treatment samples, each trained using a different method. Suggest a way to assign astronauts to the three samples.

Answer: We usually form samples by random assignment in this type of situation. In addition, to ensure that samples are representative of the population, we often create samples each with equal proportions of other characteristics that might affect the outcome, such as time in the astronaut program or previous flight experiences. Such complexities fall in the realm of more complicated ANOVAs, which we discuss at the end of this chapter in Section 12.6. However, simple random assignment is one way to avoid complications from factors that may be difficult to control directly.

When explicit control of nonexperimental factors is difficult or impossible, random assignment to samples ensures that all factors have an equal chance of occurring in each sample. This method implicitly controls nonexperimental factors and increases the likelihood that we can attribute differences in outcomes to different treatment levels of the experimental factor.

Experimental design is the process of selecting participants, of assigning them to treatment samples, and of obtaining prescribed information from them. We say that the design is *completely randomized* when we randomly assign subjects to different treatment samples or randomly select subjects from different treatment populations for each treatment sample. Complete randomization is an implicit way to control for nonexperimental factors that affect the outcome. Limiting the population being studied is an explicit control method. For instance, NASA might only assign former pilots to the three training groups, if reaction times were thought to vary between former pilots and nonformer pilots as well as among training methods. We use *one-way ANOVA* to analyze the resulting data to determine whether or not different treatments of a *single* experimental factor are a significant source of the variation in outcomes. *This analysis is the same whether we randomly select subjects from different treatment populations or randomly assign subjects to different treatment samples to represent different treatment populations.*

Question: Name some other factors, not previously discussed, that might affect the outcome of the astronaut experiment if NASA does not explicitly control for them in the experimental design.

Answer: Possible factors include manual dexterity, sex, height, other physical attributes, and mental alertness.

Section 12.1.2 Problems

5. An educational researcher investigated the variation in average time spent on homework outside of school by elementary school, middle school, and high school students.
  a. What is the experimental factor? What are the treatments?
  b. What are other factors that might affect average after-school study time?
  c. How can the researcher control for these other factors and for chance behaviors when collecting study time data?

6. A rescue unit wants to increase members' reaction speeds at the scene of an accident. To begin their investigation, they record individuals' reactions to a simulated accident in a laboratory.
  a. Name some controllable factors that affect the reaction speed of the subjects. Name some factors that are difficult or impossible to control.
  b. How does the researcher conduct a completely randomized design for the experiment? What is the purpose of such a design?

7. Suppose that the first astronaut training sample from the Training Method Experiment consists of only left-handed people, while the other two training samples contain only right-handed people. Engineers designed the instrument panel so that right-handed people can more easily perform the task. Given this scenario, test results show that method 1 produces much slower reaction speeds. Based on this experimental design, can we attribute the experimental results to the training method factor? Explain, then suggest two ways to improve the design.

Section 12.1   Setting Up the Problem

12.1.3 The Hypotheses

Because NASA randomly divides the astronauts in the Training Method Experiment among three training samples, they create three populations. Each population consists of all of the *potential* astronauts trained using a single method. Each treatment sample is a random sample from one of these populations. The basis for this experiment is the hypothesis that mean reaction times will differ among the three populations or treatments, training methods in this case.

The null hypothesis of ANOVA is that the means of all populations are equal. For the Training Method Experiment H_0 would be:

$$H_0: \mu_1 = \mu_2 = \mu_3,$$

where the subscripts denote a particular population or training method. This statement actually implies that all of the observations originate from the same population with the mean, μ, rather than from three different populations with different means. The alternative hypothesis is that not all of the means are equal:

H_1: at least one of the population means, μ_i, differs from the others.

Question: The sample mean reaction times in the astronaut example are all different, 0.79, 1.46, and 1.08 seconds. If training method actually produces no difference in reaction time (that is, H_0 is true and all reaction times originate from one population, which is the population of trained personnel), how do we account for the different sample means?

Answer: It is not unusual for the many possible samples from the same population to have different means, because each sample may contain different values. ANOVA addresses the question of whether there are sufficient differences among the sample means to decide that there is a statistically significant difference in population or treatment means. We label the decision that there is a difference H_1.

Another interpretation of these hypotheses depends on experimental terminology. H_0 says that the experimental factor training method has no effect on the reaction time. It follows that all population means should be identical. H_1 states that the experimental factor is a source of the variation of the outcomes. It is important to remember that a well-conceived theory accounts for causality here, not the decision from the ANOVA hypothesis test. The sample and the decision provide evidence to refute or substantiate theories, but they do not *prove* causality.

Question: Suppose NASA includes a fourth sample of astronauts who receive no task training in the experiment. Interpret a decision to not reject H_0. Compare this interpretation to the interpretation of not rejecting H_0 when the experiment includes only the three original samples. Compare the interpretations of rejecting H_0 in both the three- and four-treatment cases.

Answer: If we do not reject H_0 when there are four treatments, then we conclude that training does not affect the mean time to perform the task (all three training method means are equivalent to the result of receiving no training). Not rejecting H_0 in the three-treatment case implies that all three methods train the astronauts equally well (we implicitly assume that some training will improve reaction speed). The question is whether or not one method has more effect than another. Rejecting H_0 for the three-treatment case indicates that there is a difference among the effects of the three methods. In the four-treatment case, rejecting H_0 implies that there is at least one effective difference among the three training methods and the no-training approach.

Section 12.1.3 Problems

8. State and interpret the ANOVA hypotheses for the average study time experiment in Problem 5.

9. There are four types of volunteers in the rescue squad from Problem 6, doctors, nurses, emergency medical technicians, and firemen. State and interpret the ANOVA hypotheses for the average reaction time experiment.

10. Study the situation in Problem 7 for the Training Method Experiment. If the experimental factor is training method, would H_0 and H_1 differ from those proposed in this section? Would we expect to reject H_0 or not in the situation from Problem 7?

12.1.4 Variations That Affect the Sample Values

The answer to the question of whether or not means differ depends on the dispersion of the sample means as well as the dispersion of the individual observations within the samples.

Question: The sample mean reaction times in the astronaut example are 0.79, 1.46, and 1.08 seconds. Plot these values on a number line of reaction times. Describe the apparent differences in mean reaction speeds from the training methods, based on the sample means.

Answer: Method 1 appears to produce faster reaction times than method 3, which, in turn, appears to produce faster reactions than method 2, as shown on the following number line.

Figure 12-1a shows hypothetical positions on a number line of experimental outcomes, Y, where treatments do affect the outcome (H_1 is true). In this case $\mu_1 < \mu_2 < \mu_3$. The rank and magnitudes of the means may vary. For instance, in some situations μ_2 may be larger than μ_3, and the distances between the means may not be equal. Part b of the figure represents a case in which H_1 is true but the differences among the effects of the treatments are smaller than in Part a. Part c demonstrates the situation

Figure 12-1
Population means when H_0 and H_1 are true

a) H_1 is true and the means are widely dispersed.

b) H_1 is true, but the means are less dispersed.

c) H_0 is true. The means are identical, which indicates a single population.

Section 12.1 — Setting Up the Problem

that when H_0 is true, the various treatment means are equivalent, the experimental factor does not affect the outcome.

Figure 12-1 depicts population means. However, an experiment generates sample means and actual data points, such as those in the astronaut example, both of which are spread about these population means. We do not know the population means, but we try to discover if they are different using the ANOVA technique and the sample information.

Question: The positions of all of the astronaut data are shown above the number line in Figure 12-2. Describe the apparent differences in mean reaction speeds from the training methods, based on the sample means and individual values in the graph.

Answer: The reaction speeds from method 2 exceed all of the speeds from method 1, a likely indication of a true difference between methods 1 and 2. The differences are less obvious with method 3, especially between 1 and 3. The sample means are closer and the values from method 3 are very dispersed, so that they overlap with values in the other samples.

Question: Which of the situations shown in Figure 12-1 (Part a or b) depicts a difference among treatment effects that we are most likely to detect? Assume that Y values are normally distributed about their respective population means.

Answer: There is more distance between the μ values in Part a than in b, so the differences among these treatment means should be easiest for us to detect. This is true, because normally distributed values tend to cluster about the mean. See Figure 12-3.

Question: Reexamine Figure 12-3, and determine if smaller standard deviations increase or decrease the likelihood of detecting a difference in means.

Answer: Smaller standard deviations mean tighter clusters of Y values around their respective μ values. This fact increases the likelihood that a single sample's values will cluster about this population mean. There is less chance that values in samples from different populations will overlap. Hence, the likelihood of detecting the difference in population means increases as the standard deviations decrease. Compare Figure 12-4 with Figure 12-3. Notice also that $\overline{Y}$ values are likely to lie close together when H_0 is true (Part c of both figures), which increases the likelihood of not rejecting H_0.

Figures 12-1, 12-3, and 12-4 help us understand and anticipate the decisions from performing ANOVA with actual experimental data. ANOVA searches for differences in treatment means. After studying these figures and the related discussion, you may suspect that distances *between* the sample means, $\overline{Y}$ values, and distances between data points, Y, *within* each sample will be important. Distances between means show up in Part a or b of any one of these figures. Distances within populations differ between Figure 12-3 and Figure 12-4 that shows more compact curves. We will compare summary estimates of these two distances, between sample means and within samples, to determine whether or not to reject H_0. We summarize these distances with a dispersion measure, the variance.

Figure 12-2
Samples of astronaut reaction speeds

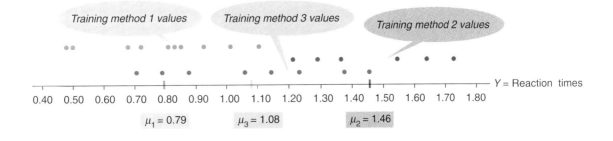

Figure 12-3
Population distributions when H_0 and H_1 are true

a) H_1 is true and the means are widely dispersed.

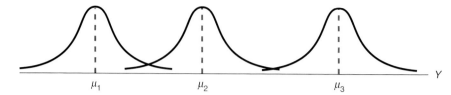

b) H_1 is true and the means are less dispersed.

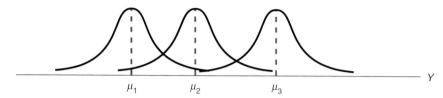

c) H_0 is true. The means are identical, which indicates a single population.

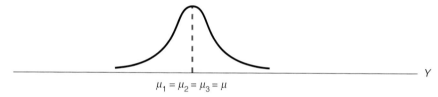

Figure 12-4
Population distributions with smaller standard deviations than in Figure 12-3

a) H_1 is true and the means are widely dispersed.

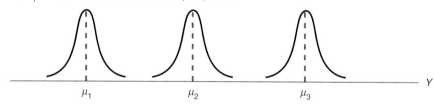

b) H_1 is true and the means are less dispersed.

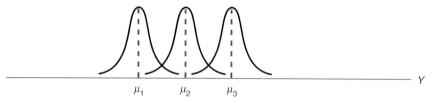

c) H_0 is true. The means are identical, which indicates a single population.

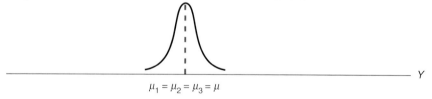

12.1.5 Measuring the Variation and Its Components

We begin by pooling the observations from all of the data sets to form one sample with n observations. From this set of values we measure the total variation of the n experimental outcomes. To calculate this measure we need to know the mean of all n experimental outcomes, the *grand mean*, $\overline{\overline{Y}}$.

Example 12-1: Figure 12-5 shows the following data as a single sample formed from the 24 astronaut experiment values. Calculate $\overline{\overline{Y}}$ and the variance, S^2, of this composite sample of Y values:

Figure 12-5
Experimental samples combined and treated as a sample from a single population

Training Method		
1	2	3
0.50	1.54	0.70
0.72	1.28	1.46
1.02	1.64	1.38
0.48	1.20	0.78
0.86	1.74	1.24
1.10	1.34	1.14
0.68		1.04
0.92		0.88
0.84		
0.82		

Solution: Using the calculations shown below, we find $\overline{\overline{Y}} = 25.30/24 = 1.0542$ seconds. We use the shortcut formula for S^2 from Chapter 3, so that

$$S^2 = \Sigma(Y - \overline{\overline{Y}})^2/(n-1) = \frac{n\Sigma Y^2 - (\Sigma Y)^2}{n(n-1)} = \frac{24(29.4660) - (25.30)^2}{24(24-1)} = 0.1215.$$

Training Method	Y	Y^2
Sample 1	0.50	0.2500
	0.72	0.5184
	1.02	1.0404
	0.48	0.2304
	0.86	0.7396
	1.10	1.2100
	0.68	0.4624
	0.92	0.8464
	0.84	0.7056
	0.82	0.6724
Sample 2	1.54	2.3716
	1.28	1.6384
	1.64	2.6896
	1.20	1.4400
	1.74	3.0276
	1.34	1.7956
Sample 3	0.70	0.4900
	1.46	2.1316
	1.38	1.9044
	0.78	0.6084
	1.24	1.5376
	1.14	1.2996
	1.04	1.0816
	0.88	0.7744
Total	25.30	29.4660

The shortcut formula facilitates ANOVA calculations that become tedious otherwise. Either formula, definitional or shortcut, produces the average squared deviation, which, directly or indirectly, requires that we sum the squared deviations for each value in the sample, $\Sigma(Y - \overline{\overline{Y}})^2$, which we label SST for total sum of squares

> **SST** = Total sum of squares
>
> = sum of the squared deviations of the experimental outcomes about the grand mean, $\overline{\overline{Y}}$
>
> = $\Sigma(Y - \overline{\overline{Y}})^2$ Definitional formula
>
> = $SS - n(\overline{\overline{Y}})^2$ Shortcut formula,
>
> where SS = sum of squares of all experimental outcomes = ΣY^2.

SST is a measure of the *total* variation of the experimental outcomes about the grand mean, $\overline{\overline{Y}}$, rather than the *average* variation that the standard deviation and variance measure. For instance, the total variation *SST* for the Training Method Experiment is $29.4660 - 24(1.0542)^2 = 2.7939$ (from the shortcut formula), but the variance that we calculated earlier, S^2, is $2.7939/(24 - 1) = 0.1215$, the average variation.

The measures we calculated, $\overline{\overline{Y}} = 1.0542$, $S^2 = 0.1215$, and SST = 2.7939, are estimates of the mean of all of the Y values, μ, their variance, σ^2, and their total variation, respectively. When H_1 is true, this variation incorporates two possible sources of dispersion: (1) chance variation, or the spread of the Y values about their respective treatment means, μ_i, as shown for each population in Figures 12-3 and 12-4, and (2) variation due to the effect of the treatments, as shown in the distinct treatment population means, μ_1, μ_2, and μ_3 in Parts a and b of both figures.

Question: It is possible that all of the experimental data in the astronaut experiment originates from a single population, in spite of the different treatments. This means that there is no treatment effect (H_0 is true). Given this assumption, describe the two components of SST, the total variation of the Y values.

Answer: The variation in Y is solely chance variation, because there is no treatment effect to cause variation. The outcome Y would likely be the same, regardless of the sample to which NASA assigns the participant. Figures 12-3c and 12-4c depict this situation.

> **Analysis of variance** breaks down the variation of experimental data into the two components, variation due to effects of different treatments and random variation within the treatment sample.

To explore these relationships, we express the deviation of an individual observation symbolically. Consider $(Y - \overline{\overline{Y}})$, the deviation of the outcome value Y from the grand mean $\overline{\overline{Y}}$. If we add $\overline{Y}$, the mean of the treatment sample that contains the outcome Y, to the deviation and then subtract it also, the result is equation (1):

$$(Y - \overline{\overline{Y}}) = (Y - \overline{Y}) + (\overline{Y} - \overline{\overline{Y}}) \qquad (1)$$

Deviation of Y value from $\overline{\overline{Y}}$ Deviation within treatment sample   Deviation between treatment sample mean, $\overline{Y}$, and $\overline{\overline{Y}}$

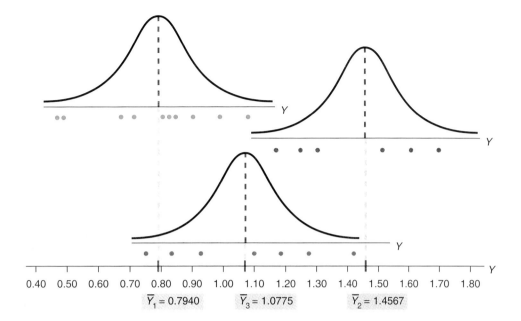

Figure 12-6
Astronaut reaction speeds separated as treatment samples from different normal populations all with the same standard deviation

$(Y - \overline{Y})$ represents the deviation of a value in one treatment sample from $\overline{Y}$, the mean of that sample. This deviation is analogous to the deviation of an individual observation from the mean of one of the experimental samples in Figure 12-6. (Note that we now include more places past the decimal for the $\overline{Y}$ values, because these are intermediate steps in the ANOVA calculations.)

$(\overline{Y} - \overline{\overline{Y}})$ symbolizes the deviation of the sample mean of one treatment sample from the grand mean. This deviation is analogous to measuring the spread of μ_1, μ_2, and μ_3 in Figure 12-6 by treating each μ as an observation in the data set consisting of all possible treatment means.

For instance, the deviation of the first value in the first astronaut-training sample from the grand mean is

$$(Y - \overline{\overline{Y}}) = 0.50 - 1.0542 = -0.5542.$$

This observation's deviation *within* the first treatment sample is

$$(Y - \overline{Y}) = 0.50 - 0.7940 = -0.2940.$$

Finally, the deviation *between* the first sample's mean and the grand mean is

$$(\overline{Y} - \overline{\overline{Y}}) = 0.7940 - 1.0542 = -0.2602.$$

Thus, the deviation from the grand mean is the sum of the deviation *within* the sample and the deviation *between* the sample mean and the grand mean as equation (1) suggests: $-0.5542 = -0.2940 + (-0.2602)$.

If we square both sides of (1) and sum the results over all observations, we find that the *total* sum of squares (SST) is the sum of the squared deviations *within* each treatment sample (SSW) plus the sum of the squared deviations *between* each treatment sample mean and the grand mean (SSB) with each squared *between* deviation weighted by the number of observations in the treatment sample. This relationship parallels Equation (1), except that we use the sum of the squared terms rather than the sum of the individual terms (the algebra is beyond the scope of this book). The following box and Figure 12-7 summarize this important relationship and shortcut formulas for calculating each sum of square value for a set of experimental outcomes.

Figure 12-7

Decomposition of SST, the total variation of experimental outcomes

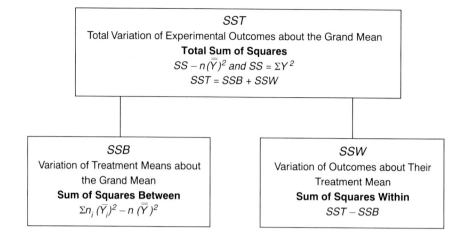

Formula for Decomposition of Variation of Experimental Outcomes
$$SST = SSB + SSW,$$
where **SST** = Total sum of squares = $SS - n(\bar{\bar{Y}})^2$ and $SS = \Sigma Y^2$,
SSB = Sum of squares between
= Weighted sum of squared deviations of each treatment sample mean from the grand mean
= $\Sigma n_i(\bar{Y}_i - \bar{\bar{Y}})^2$ Definitional formula
= $\Sigma n_i(\bar{Y}_i)^2 - n(\bar{\bar{Y}})^2$ Shortcut formula (subscripts refer to ith treatment sample)
SSW = Sum of squares within
= Sum of squared deviations of each outcome from its sample mean
= Sum of $(Y - \bar{Y})^2$ for all outcomes
= $SST - SSB$ Shortcut formula

■ **Example 12-1 (continued):** Calculate the sum of squares *within*, SSW, and the sum of squares *between*, SSB, for the astronaut data. Recall that $\bar{\bar{Y}} = 1.0542$, $\bar{Y}_1 = 0.7940$, $\bar{Y}_2 = 1.4567$, $\bar{Y}_3 = 1.0775$, SST = 2.7939.

■ **Solution:** $SSB = \Sigma n_i(\bar{Y}_i)^2 - n(\bar{\bar{Y}})^2$
$= [10(0.7940)^2 + 6(1.4567)^2 + 8(1.0775)^2] - 24(1.0542)^2$
$= 1.6522.$
$SSW = SST - SSB = 2.7939 - 1.6521 = 1.1418.$

These three types of sums of squares figure prominently when we decide whether or not to reject H_0 for the analysis of variance.

Question: If H_0 is true and all of the treatment population means are equal, which of the two sum-of-squares values, SSW or SSB, should be smaller? Explain.

Answer: SSB should be smaller, because there is no treatment effect when the population means are equal (H_0 is true). SSW should approach SST, because both of these values reflect only chance variation when H_0 is true. SST and SSW are not likely to be exactly equal, because we use samples rather than populations, so SSB will most likely not be zero.

Section 12.1.5 Problems

11. Begin a search for evidence of grade inflation (different mean grades) in several disciplines by finding SST, SSB, and SSW for the following random samples of students' grades from different courses in the five listed disciplines.

Music	Psychology	Political Science	Economics	Math
97	81	76	72	78
75	79	73	62	72
80	81	73	65	79
88	86	82	43	73
83	86	67	78	75
76	87	76	63	69
84	81	71		
		76		

12. Start an ANOVA on the following data to verify that the average tenure in teenagers' jobs depends on the type of job. Find SST, SSB, and SSW.

Grocery	Fast Food	Retail	Theater
47	21	19	12
36	16	35	24
43	60	17	6
32	27	12	1
58	9	10	19
28	48	60	15

13. One-way ANOVA will be used to compare three methods of completing a task on a production line. Find SST, SSB, and SSW for the following random sample of times.

Method 1	Method 2	Method 3
5.3	7.5	7.0
4.0	4.8	5.6
6.9	4.4	5.0
3.7	4.8	3.5
4.9	4.8	4.3
5.0	6.5	4.7
4.5	4.0	2.4
6.1	2.1	3.6

12.2 The Test Statistic

In order to make a decision using ANOVA, we use a test statistic that incorporates information about the two factors: spread of the individual data points about their treatment means; and the spread of the treatment means about the grand mean. As we mentioned earlier, the smaller the spread *within* the treatment samples (compare Figures 12-3 and 12-4) and the larger the spread *between* the different treatment means (compare Parts a and b of Figures 12-3 and 12-4), the easier it is to identify a significant difference in treatment means.

12.2.1 SSW and MSW

To begin the discussion of the test statistic for ANOVA, we consider SSW, which measures the total variation from within the treatment samples. It is the sum of the squared deviations of each observation from its treatment mean.

Question: Does this measure contain chance variation, treatment variation, or both?
Answer: Because all participants in a given treatment sample get the same treatment, the different outcomes within the sample cannot reflect treatment effects. Thus, chance must generate the sum of squares within each sample. (Recall that random assignment of individuals to the different samples reduces the risk of any unsuspected variable causing a systematic variation in the sample outcomes.)

Identical treatment means (the H_0 in ANOVA) implies that the individual observations all originate from the same population. (See Part c of Figures 12-3 and 12-4.) This implies, in turn, that the variance of a single treatment sample, S^2, estimates the common population variance, σ^2. Notice that the three curves in Figure 12-3a all appear to have the same variance. This is also true in Figures 12-4a and 12-6. Summing the internal variations from each sample parallels the pooling process of Chapter 10, where we obtained S_p^2 by taking a weighted mean of two sample variances. *Consequently, SSW is a pooled estimator of the total chance variation.*

If we divide SSW by n, the result is an estimate of the average squared deviation due to chance, or σ^2. As is true for sample estimates of population variances, dividing by sample size, n, produces underestimates on average (a biased estimator). However, dividing by $n - 1$ produces a good estimate on average (an unbiased estimator). In particular, using the squared deviations from a single treatment sample to estimate σ^2 requires dividing their sum by $n_i - 1$. To compensate for bias when pooling treatment sample estimates, we divide the sum of the squared deviations from all treatments by the sum of the denominators from each of the treatment samples, $\Sigma(n_i - 1) = n - J$, where n is the total number of experimental observations and J is the number of treatment samples in the experiment. We can relate subtracting J to losing a degree of freedom for each treatment mean that we estimate. In Chapter 10, we lost two degrees of freedom when we estimated the difference in two population means and the denominator of S_p^2 was $n_1 + n_2 - 2$ or $n - 2$.

To summarize, *$SSW/(n - J)$ estimates the average variation in the experimental outcomes due to chance.* We also call it the *mean square within* (or sometimes, *mean square error*), or *MSW*. Such variation occurs even when the treatments do not affect outcomes (H_0 is true); that is, when all outcomes originate from the same population.

For the astronaut training experiment SSW = 1.1418, and n represents the 24 observations that result from combining the $J = 3$ treatment samples, so MSW = SSW/$(n - J) = 1.1418/(24 - 3) = 0.0544$. This is an estimate of the chance variation that we would find when we observe a set of astronaut reaction times.

12.2.2 SSB and MSB

Next we consider SSB, the total variation between the treatment sample means. SSB is also the sum of all of the squared deviations of treatment means from the grand mean. We weight deviations based on more information (larger samples) more heavily in this calculation. $\overline{Y} - \overline{\overline{Y}}$ *is an estimate of the effect of one particular treatment on an outcome, Y.* Thus, SSB measures the portion of the total squared deviation of Y values from the grand mean that we attribute to treatment effects.

As with SSW and MSW, the mean squared deviation due to treatment is SSB divided by $J - 1$, the number of treatment means, $\overline{Y}$, minus 1. We call the result, SSB/$(J - 1)$, *MSB*, the *mean square between samples*. MSB is an estimator of the dispersion of $\overline{Y}$ values about μ, the common population mean.

Question: What do we call a standard deviation of $\overline{Y}$ values?

Answer: We call it the standard error of the mean, $\sigma_{\overline{y}}$. MSB is an estimator of $\sigma_{\overline{y}}^2$, the variance of $\overline{Y}$ values or the average squared distance of $\overline{Y}$ values from μ, rather than the standard deviation or standard error.

Question: If H_0 is true (all observations originate from the same population), will all $\overline{Y}$ values be identical? Why or why not?

Answer: Even when we randomly select samples from the same population, the $\overline{Y}$ values from different samples will vary, because of the chance selection of any population element for a particular sample. The sampling distribution of $\overline{Y}$ describes the possible $\overline{Y}$ values and their likelihood of occurrence.

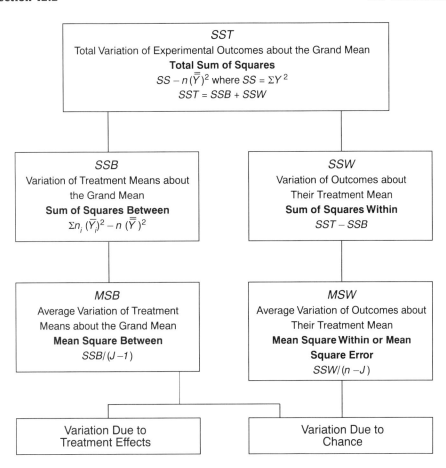

Figure 12-8
Estimates of different sources of variation in experimental outcomes

Thus, *the treatment means, $\overline{Y}$ values, can vary for two reasons, the effects of chance and the effects of different treatment exposures. MSB includes both.* (We might think of the weighting factor, sample size, in SSB as inflating the deviation to reflect both variations.) *In cases where there is no treatment effect (H_0 for ANOVA is true), MSB estimates the variation in the experimental outcomes due to chance alone, σ^2.* Figure 12-8 summarizes the estimation roles of MSW and MSB as well as the types of variation each includes.

For the astronaut data, SSB = 1.6522 and $J = 3$ for the three training methods in this experiment. So MSB = SSB/($J - 1$) = 1.6522/(3 - 1) = 0.8261. This is an estimate of the variation due to treatment effects of the different training methods that also includes variation due to chance selection of observations for different samples.

12.2.3 The Ratio of Mean Squared Deviations

Consider the ratio, MSB/MSW. We can express the information this ratio includes or estimates as:

$$\frac{\text{MSB}}{\text{MSW}} = \frac{\text{Average variation in } Y \text{ due to treatment and chance}}{\text{Average variation in } Y \text{ due to chance}}.$$

The equality does not strictly hold, because these are estimates based on samples, not perfect information. When H_0 is true (there are no treatment effects), the two mean squares are both estimators of the chance variation, σ^2. Therefore, we expect that the ratio will approach the value 1, as shown in the following box.

If H_0 is true,

$$\frac{MSB}{MSW} = \frac{\text{Average variation in } Y \text{ due to chance}}{\text{Average variation in } Y \text{ due to chance}} \text{ is approximately } \frac{\sigma^2 \text{ estimate}}{\sigma^2 \text{ estimate}} = 1.$$

Question: If H_1 is true (treatment means are different), should the MSB/MSW ratio be large or small?

Answer: If H_1 is true, varying the experimental factor increases the variation among sample averages. Thus, MSB contains treatment variation as well as chance variation, so MSB/MSW is likely to be large.

The MSB/MSW ratio for the astronaut data is MSB/MSW = 0.8261/0.0544 = 15.1857, which is larger than 1. The decision whether or not to reject H_0 becomes a question of whether the MSB/MSW ratio is so large that we no longer accept sampling error as a believable explanation for the large ratio. At that point, we believe that treatment effects cause the large ratio, and, subsequently, we reject H_0. Rejection occurs if the probability of obtaining the large ratio is small when H_0 is true. To determine the critical value, we must know the sampling distribution of the test statistic (in this case, the ratio of mean squared deviations).

When H_0 is true, the sampling distribution of MSB/MSW is a probability distribution that we have not used before, an F distribution. A discussion of the F distribution follows, but first we summarize the assumptions that we must make to ensure that the F distribution is appropriate.

ANOVA Assumptions
1. Sample observations are independent; that is, the outcome associated with one subject provides no information about the outcome associated with any other subject in the experiment. We satisfy this assumption when we randomly select subjects and assign them to different treatment samples or when we randomly select them from one of the treatment populations.
2. Populations are normally distributed, with means $\mu_1, \mu_2, \ldots, \mu_J$ and a common variance, σ^2.

Section 12.2 Problems

14. Continue the search for evidence of grade inflation (different mean grades) in five different disciplines from the 34 observations in Problem 11 by computing the value of the test statistic from the following information: SSB = 1,643.6401 and SSW = 1,319.0952. Does the value of this statistic offer any hope that H_1 is true? Explain.

15. Continue the ANOVA on the 24 observations of teenagers' job tenures in four types of jobs from Problem 12. Find the value of the test statistic, and explain whether the test statistic indicates that H_1 may be true. SSB = 2,396.4583 and SSW = 4,726.5000.

16. Find the value of the test statistic for the one-way ANOVA based on the 24 task-completion times on the production line for three methods in Problem 13. SSB = 1.1908 and SSW = 40.0875. Does this test statistic offer any hope that H_1 may be true? Explain.

12.3 The F Distribution

A random variable with an F probability distribution is continuous, so we need a table and the graph of an F statistic to determine probabilities and outcomes. F values are skewed to the right and always positive, as shown in Figure 12-9. Two values determine the exact shape of the curve: the degrees of freedom associated with the numerator, and the degrees of freedom associated with the denominator. Fraction terminology, numerator and denominator, is useful, because an F statistic is based on a ratio.

There are infinite possible degrees of freedom combinations, so it is impossible to produce an F table that covers every case. The F table provided in this book (see Appendix L) contains entries for several degrees of freedom and for six areas, or probabilities, in the right tail of the graph. There are six tables, one for each of six frequently used significance levels, or tail areas, 0.10, 0.05, 0.025, 0.01, 0.005, and 0.001.

Consider the 0.05 table. Suppose a random variable, F, has 8 degrees of freedom associated with its numerator and 24 with its denominator, expressed as $F_{(8, 24)}$. Locate the 8 by reading across the top of the table, and the 24 by reading down the leftmost column. Find the intersection of the column under 8 and the row across from 24. The value at the intersection is 2.36. Figure 12-9a pictures this value for F. We write this information as $P(F_{(8, 24)} \geq 2.36) = 0.05$.

Question: Find a particular outcome or F value, f, such that $P(F_{(24, 8)} > f) = 0.05$.
Answer: $f = 3.12$, as shown in Figure 12-9b. We find the intersection of the "24" column and the "8" row in the F table in Appendix L with 0.05 in the tail.

These questions illustrate the importance of correctly specifying the degrees of freedom. If we accidently swap the degrees associated with the numerator and denominator, the mistake will usually result in using an incorrect entry from the table.

Question: Find f, so that $P(F_{(8, 24)} \geq f) = 0.01$.
Answer: To answer this question, we must use the 0.01 table, where we find f is 3.36. See Figure 12-9c, which also illustrates the previous answer of 2.36 (for the 0.05 area) to contrast the two results.

If the F table does not include a specific degrees of freedom value required in a problem, then use the closest value that is smaller than the desired value. This value gives the benefit of the doubt to H_0 by expanding the acceptance region, because smaller degrees of freedom yield larger critical values. If we use a larger than desired degrees-of-freedom value instead, the critical value will be too small, which would increase the chance of rejecting H_0 when it is true (increase α).

Figure 12-9
F diagrams

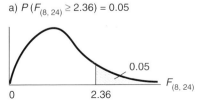

a) $P(F_{(8, 24)} \geq 2.36) = 0.05$

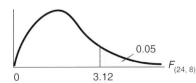

b) Diagram for $P(F_{(24, 8)} > 3.12) = 0.05$

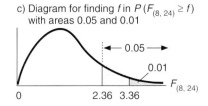

c) Diagram for finding f in $P(F_{(8, 24)} \geq f)$ with areas 0.05 and 0.01

Section 12.3 Problems

17. The outcomes of an experiment follow an F distribution with 10 degrees of freedom for the numerator and 12 degrees of freedom for the denominator. Find:
  a. $P(F_{(10,12)} > 5.09)$.
  b. $P(F_{(10,12)} > 7.29)$.
  c. f, so that $P(F_{(10,12)} > f) = 0.05$.
  d. f, so that $P(F_{(10,12)} > f) = 0.005$.

18. An F statistic has 24 degrees of freedom associated with the numerator and 7 with the denominator. Find:

  a. The proportion of F values that exceed 7.65.
  b. The proportion of F values that exceed 2.58.
  c. f, so that the proportion of F values larger than f is 0.05.
  d. f, so that the proportion of F values larger than f is 0.01.

12.4 The Decision for ANOVA

Now we can determine when the value of the F-ratio, MSB/MSW, from ANOVA is larger than what we would expect to observe, if there were no treatment effects. When H_0 is true, this statistic has an F distribution with $J - 1$ degrees of freedom associated with the numerator and $n - J$ degrees of freedom associated with the denominator, where J is the number of treatments or samples and n is the total number of experimental observations. The level of significance, α, is the probability that the F-ratio is unexpectedly large when H_0 is true. ANOVA always involves an upper-tailed test, because only large values of the F-ratio cause us to reject H_0. See Figure 12-10.

■ **Example 12-1 (continued):** For the astronaut example, the F-ratio is 15.1857, based on 24 observations from three different samples. Locate the critical value for testing at the 0.05 level. What is the rejection region? Make and interpret a decision.

■ **Solution:** The degrees of freedom associated with the numerator is $J - 1 = 3 - 1 = 2$. The degrees of freedom associated with the denominator is $n - J = 24 - 3 = 21$. So we locate the critical value, $F_{(2, 21)} = 3.47$ in the 0.05 F table in Appendix L. The rejection region consists of F values larger than 3.47. We reject H_0, because $15.1857 > 3.47$, as Figure 12-11 shows. Rejecting H_0 means that the sample evidence supports the conclusion that training method does affect astronaut reaction speed. Section 12.5 discusses a way to determine the difference in the methods.

■ **Example 12-2:** The F-ratio from an ANOVA of the effect of different types of computers on the performance speed for a certain complex calculation is 2.62. What is

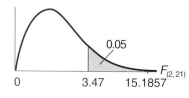

Figure 12-11
Diagram for ANOVA test of astronaut example

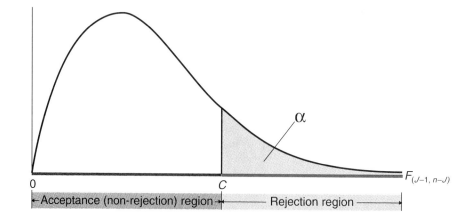

Figure 12-10
Diagram for ANOVA test, the sampling distribution of the F-ratio = MSB/MSW

Section 12.4   The Decision for ANOVA   595

the decision if the experiment involves six types of computers and a total of 35 computers? Test at the 0.025 level, and interpret the decision.

■ **Solution:** H_0 would be that the mean time is the same for all six computer types or $\mu_1 = \mu_2 = \mu_3 = \mu_4 = \mu_5 = \mu_6$. H_1 is that at least one of these means is different from the others. $n = 35$ and $J = 6$, so the critical value is $F_{(5, 29)} = 3.04$ according to the F table with a 0.025 tail area. We cannot reject H_0, because the test statistic, 2.62, is smaller than 3.04. We are unable to detect a difference in average performance speed among types of computers at the 0.025 significance level.

This extended discussion may make the ANOVA procedure seem rather long before we finally make a decision. However, the following box summarizes the procedure and shows that the steps are straightforward.

The ANOVA Procedure
1. Calculate the grand mean, $\overline{\overline{Y}}$, for all n observations.
2. Sum the square of each observation to find SS.
3. Calculate the total sum of squares, $\text{SST} = \text{SS} - n(\overline{\overline{Y}})^2$.
4. Calculate the sum of squares between, $\text{SSB} = \Sigma n_i(\overline{Y}_i)^2 - n(\overline{\overline{Y}})^2$.
5. Calculate the mean sum of squares between, $\text{MSB} = \text{SSB}/(J - 1)$.
6. Calculate the sum of squares within, $\text{SSW} = \text{SST} - \text{SSB}$.
7. Calculate the mean sum of squares within, $\text{MSW} = \text{SSW}/(n - J)$.
8. Form the test statistic, MSB/MSW.
9. Find the critical value in the F table (Appendix L) with a tail area of α (the significance level), using $J - 1$ degrees of freedom associated with the numerator and $n - J$ degrees of freedom associated with the denominator.
10. If the test statistic (F-ratio) exceeds the critical value ($F_{(J-1, n-J)}$), reject H_0. If not, we cannot reject H_0.

Symbols

n = total number of observations from the experiment.
J = number of treatment samples.
n_i = number of observations in the ith treatment sample.

■ **Example 12-3:** Nielsen television network ratings drive many television network and advertiser decisions and, consequently, affect the choices available to television viewers. Often, ratings battles center on the top shows, but we might also study the typical or average rating of shows on each network to compare the different networks. The following table lists ratings for three random samples of prime time programs from one week, one sample from each of the three networks, ABC, CBS, and NBC. Figure 12-12 diagrams the data. The sample means differ, but do the different ratings posted by Nielsen indicate that the networks (not individual programs) rate differently on average? Test at the 0.025 level, and interpret the decision. Comment on the relationship between Figure 12-12 and the decision.

Network	Randomly Selected Ratings
ABC	10.5, 8.4, 6.5, 10.4, 15.8
CBS	8.0, 7.6, 10.2, 12.5, 4.9, 12.8, 9.5
NBC	12.5, 10.5, 12.6, 15.9, 6.4, 8.1, 18.9, 16.0

Source: *USA Today*, June 8, 1988, p. 3D.

Figure 12-12
Nielsen network-rating data

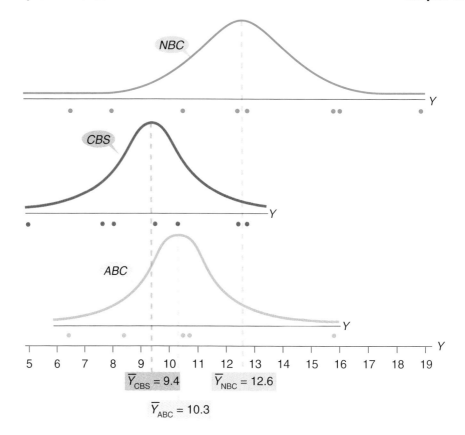

■ **Solution:** H_0 is that the average network ratings are identical, $\mu_{ABC} = \mu_{CBS} = \mu_{NBC}$. H_1 is that at least one of these means differs from the others. The following table shows the detailed calculations for the following steps. The boxed values are the sum of each sample's values that we use to determine each sample's mean, $\overline{Y}_i$.

Treatment Sample = Network Sample	Y = Rating	Y^2	$\overline{Y}_i$	$\overline{Y}_i^2$	$n_i \overline{Y}_i^2$
ABC	10.5	110.25			
	8.4	70.56			
	6.5	42.25			
	10.4	108.16			
	15.8	249.64			
	51.6		10.3200	106.5024	532.5120
CBS	8	64.00			
	7.6	57.76			
	10.2	104.04			
	12.5	156.25			
	4.9	24.01			
	12.8	163.84			
	9.5	90.25			
	65.5		9.3571	87.5561	612.8929

(continued on next page)

Section 12.4 — The Decision for ANOVA

Treatment Sample = Network Sample	Y = Rating	Y^2	$\overline{Y}_i$	$\overline{Y}_i^2$	$n_i \overline{Y}_i^2$
NBC	12.5	156.25			
	10.5	110.25			
	12.6	158.76			
	15.9	252.81			
	6.4	40.96			
	8.1	65.61			
	18.9	357.21			
	16	256.00			
	100.9		12.6125	159.0752	1272.6013
Totals	218.0	2638.86			2418.0061

$\overline{\overline{Y}} = 10.9$

Step 1: Grand mean is the mean of all 20 sample values, $\overline{\overline{Y}} = 218.0/20 = 10.9$.
Step 2: SS is the sum of the 20 squared values $= \Sigma Y^2 = 2{,}638.86$.
Step 3: SST $=$ SS $- n(\overline{\overline{Y}})^2 = 2{,}638.86 - 20(10.9)^2 = 262.66$.
Step 4: SSB $= \Sigma n_i(\overline{Y}_i)^2 - n(\overline{\overline{Y}})^2 = 2{,}418.0061 - 20(10.9)^2 = 41.8061$. (Rounding causes the sum to be 2418.0061 rather than 2418.0062.)
Step 5: MSB $=$ SSB$/(J - 1) = 41.8061/(3 - 1) = 20.9031$.
Step 6: SSW $=$ SST $-$ SSB $= 262.66 - 41.8061 = 220.8539$.
Step 7: MSW $=$ SSW$/(n - J) = 220.8539/(20 - 3) = 12.9914$.
Step 8: Test statistic $=$ F-ratio $=$ MSB/MSW $= 20.9031/12.9914 = 1.6090$.
Step 9: Critical value $= F_{(2, 17)} = 4.62$ from the 0.025 part of the F table.
Step 10: We cannot reject H_0, because $1.609 < 4.62$.

The data do not support the existence of a difference in mean rating among the networks. An examination of Figure 12-12 reveals that ABC and CBS averages are very close and that their sample values are similarly dispersed. The NBC mean looks larger, but its sample values are very dispersed, making it difficult to find a significant difference.

Question: When one program in a certain time slot receives a high rating, it increases the likelihood that the ratings for the other two shows in the same time slot will be lower. Does this mean that the ratings observations in the example are not independent?

Answer: No, because programs were randomly selected for the experiment without matching their time slots, so the effect of a program's being in the same time slot as another is part of the chance variation.

Box 12-1 provides an application of ANOVA in the study of sociological phenomena.

Box 12-1 Child Nutritional Development and Working Mothers

Much of the research concerning the effect of working mothers on children focuses on such nonphysical aspects as school accomplishments, psychological development, and social development, including attachments. Researchers assume that changing work roles within families will not hinder or affect physical development in countries where a lack of necessities is not a primary concern. On the other hand, poverty dominates many other countries, where the effect of changing social conditions on physical development elicits more immediate concern.

To add to the growing evidence on this topic, a researcher collected information regarding working mothers and the nutritional development of their young children in Guatemala. Using one-way ANOVA, she analyzed the impact of

(continued on next page)

Box 12-1 (continued)

several variables on development of children between 8 and 35 months of age, measured by three anthropometric variables (transformed against international standards):

H/E = Height for the child's age (a long-term indicator of nutritional development).
W/A = Weight for the child's age (a long-term indicator that may be influenced by short-term fluctuations, such as illness or period of low food intake).
W/H = Weight for the child's height (indicates disproportion in the body).

Several experimental factors in the analyses included:

WT = Work type (formal or informal). Some analyses included a no work category as well. Formal implies legal or institutional business arrangements, such as owning a business or working in a store or factory, or training.
WA = Work amount (full-time, part-time, or occasional) Some analyses also used a no work category.
CS = Child-care strategies (take child to work, leave child in another person's care [adult or preteen], leave child in another person's home, work at home with child, some analyses also used a no work category).

The following table lists several of the significant ANOVA results. Missing comparisons are not significant.

Outcome	Treatment	Test Statistic	Degree of Freedom	Maximum P-Value
H/A	WT	3.85	2,236	0.02
W/A	WT	4.18	2,236	0.01
W/H	CS	3.10	4,232	0.02

Source: Patrice L. Engle, "Maternal Work and Child-Care Strategies in Peri-Urban Guatemala: Nutritional Effects," Child Development, Vol. 62, No. 5, 1991, pp. 954–65. Copyright © 1991 by the Society for Research in Child Development.

Section 12.4 Problems

19. Does the F-ratio of 9.0338 from the 34 observations in Problem 11 (p. 589) provide sufficient evidence for us to conclude that there is grade inflation (different mean grades) in five different disciplines at the 0.05 significance level? Assume normally distributed grades with equal variances.
20. Continue the ANOVA on the 24 observations of teenagers' job tenures in four types of jobs from Problem 12 (p. 589). The value of the test statistic is 3.3802. Test at the 0.10 level, and interpret the decision. What assumptions do we need to make?
21. Complete the one-way ANOVA for the average completion times on the production line for three methods from Problem 13 (p. 589). The 24 times in the samples produce a test statistic value of 0.3119. Test at the 0.01 level. What assumptions do you make? Interpret the conclusion.
22. Test at the 0.005 level for a difference in mean voter turnout in different regions of the state using the turnouts from randomly selected precincts in three regions. Assume turnouts are normally distributed with equal variances. Interpret the decision.

East	Central	West	East	Central	West
34	32	41	35	41	19
61	49	31	47	34	34
53	35	27	47	37	22
57	31	26	48	52	32
54	42	50	59	48	
49	38	37	45	44	
63	37	29	43	40	
44	36	36	51		
43	33	36	54		
39	49	28	41		
55	41	39	36		

12.5 One-Way ANOVA on the Computer

When there are several treatments or a large data set, a computer not only handles ANOVA calculations quickly, it also provides more information than we would ordinarily compute in such situations.

Figure 12-13 displays Minitab output for the astronaut data. The circled numbers correspond to the following information:

1. Degrees of freedom associated with the numerator of the test statistic, the F-ratio.
2. SSB.
3. MSB.

Section 12.5 — One-Way ANOVA on the Computer

Figure 12-13 Minitab output for astronaut training method experiment

```
ANALYSIS OF VARIANCE
SOURCE      DF        SS         MS         F          p
FACTOR  ①  2    ②  1.6533   ③ 0.8266   ⑦ 15.20   ⑧ 0.000
ERROR   ④ 21    ⑤  1.1423   ⑥ 0.0544
TOTAL      23    ⑨  2.7956

                                        ⑫ INDIVIDUAL 95 PCT CI'S FOR MEAN
                        ⑩                  BASED ON POOLED STDEV
LEVEL     N      MEAN        STDEV      ---------+---------+---------+---------
METHOD1  10     0.7940       0.2031     (----*-----)
METHOD2   6     1.4567       0.2152                           (------*-----)
METHOD3   8     1.0775       0.2776                 (-----*-----)
                                        ---------+---------+---------+---------
⑪ POOLED STDEV = 0.2332                          0.90      1.20      1.50
```

4. Degrees of freedom associated with the denominator of the F-ratio.
5. SSW.
6. MSW.
7. F-ratio $=$ MSB/MSW.
8. P-value for ANOVA test.
9. SST.
10. Treatment means.
11. Pooled standard deviation for combined treatment samples $= \sqrt{\text{MSW}}$.
12. Confidence intervals for treatment means.

The values in the figure approximate the results of the calculations we performed earlier in Section 12.2. Differences may occur because of rounding.

Rejecting H_0 in experiments implies that at least one treatment effect is different from the others. This can be useful information, but questions about the magnitude of the differences naturally follow.

Question: How would you estimate the difference between the means of two different treatments?

Answer: The most logical estimator of the difference is $(\overline{Y}_1 - \overline{Y}_2)$, the same estimator we used in Chapter 10. A larger difference in these means indicates the possibility of a greater difference between treatment effects.

■ **Example 12-1 (continued):** Estimate and interpret the paired-differences between the means of the astronaut data. Interpret the estimates and then rank the treatment effects.

■ **Solution:** $(\overline{Y}_1 - \overline{Y}_2) = 0.794 - 1.4567 = -0.6627$ seconds.
$(\overline{Y}_1 - \overline{Y}_3) = 0.794 - 1.0775 = -0.2835$ seconds.
$(\overline{Y}_2 - \overline{Y}_3) = 1.4567 - 1.0775 = 0.3792$ seconds.

Method 2 produces an average response speed that is 0.6627 second slower than that produced by method 1, whereas the method 3 average is only 0.2835 second slower. The implication is that method 1 produces faster reaction times than does method 3, which, in turn, produces faster times than does method 2.

ANOVA does not reveal the magnitude of the effect of each different treatment. It is possible that only one treatment mean is different. It is also possible that the means are all different. Point estimates of the differences can be expressed as $\overline{Y}_k - \overline{Y}_q$, where k and q are two different treatment samples, but this does not tell us the significance of each difference. Frequently, some of the sample treatment differences will be significant, and some result solely from sampling error. To answer such questions better, we must know which treatment means are significantly different. Advanced statistical texts often describe techniques for statistical answers to such questions.

The Minitab output displays confidence intervals for the individual treatment means using the pooled estimator of $\sigma^2 = \sqrt{MSW}$ from ANOVA, instead of individual sample variances, to estimate the standard error of the mean, $\hat{\sigma}_{\bar{y}}$. This information provides some guidance about the relative size of the effects and differences, but it is not the technique we use to determine differences or ranks at a given significance level.

■ **Example 12-4:** Figure 12-14 shows the Minitab output of an ANOVA for data from an experiment in which the 50 outcomes are heights of biscuits and the treatments are five different brands of biscuit mixes. Use this output to answer the following questions.

a. What is the value of the test statistic?
b. What is the decision at the 0.005 significance level?
c. How can we interpret the decision?
d. Estimate the differences between brands, and rank the brands by ascending sample mean height. Which pair of means are most likely to be significantly different?

■ **Solution:** a. The test statistic is 8.89.

b. The hypotheses are H_0: $\mu_1 = \mu_2 = \mu_3 = \mu_4 = \mu_5$ and H_1: at least one of the five brand means differs from the others. We reject H_0, because $8.89 > 4.37$, the critical value from the F distribution with 4 and 40 degrees of freedom. Forty is used, because the actual value, 45, is not included in the table, but the correct critical value would be even smaller than 4.37. Alternatively, the p-value is 0.0001, which is less than 0.005, so we reject.

c. The choice of brand of biscuit mix will affect the height of the biscuit, or, the mean heights for all brands are not the same.

d. If we arrange the sample means in ascending order, we obtain $\bar{Y}_4 = 0.725$, $\bar{Y}_5 = 0.826$, $\bar{Y}_3 = 0.896$, $\bar{Y}_1 = 1.126$, and $\bar{Y}_2 = 1.232$. The treatment effect estimates would be:

$$\bar{Y}_1 - \bar{Y}_2 = 1.126 - 1.232 = -0.106.$$
$$\bar{Y}_1 - \bar{Y}_3 = 1.126 - 0.896 = 0.230.$$
$$\bar{Y}_1 - \bar{Y}_4 = 1.126 - 0.725 = 0.401.$$
$$\bar{Y}_1 - \bar{Y}_5 = 1.126 - 0.826 = 0.300.$$
$$\bar{Y}_2 - \bar{Y}_3 = 1.232 - 0.896 = 0.336.$$
$$\bar{Y}_2 - \bar{Y}_4 = 1.232 - 0.725 = \mathbf{0.507}.$$
$$\bar{Y}_2 - \bar{Y}_5 = 1.232 - 0.826 = 0.406.$$
$$\bar{Y}_3 - \bar{Y}_4 = 0.896 - 0.725 = 0.171.$$
$$\bar{Y}_3 - \bar{Y}_5 = 0.896 - 0.826 = 0.070.$$
$$\bar{Y}_4 - \bar{Y}_5 = 0.725 - 0.826 = -0.101.$$

Figure 12-14
Minitab output for biscuit brand experiment

```
ANALYSIS OF VARIANCE
SOURCE       DF        SS          MS         F        p
FACTOR        4      1.7881      0.4470      8.89    0.000
ERROR        45      2.2629      0.0503
TOTAL        49      4.0511

                                 INDIVIDUAL 95 PCT CI'S FOR MEAN
                                 BASED ON POOLED STDEV
LEVEL      N      MEAN     STDEV -------+---------+---------+---------
BRAND1    10     1.1260    0.2663                      (-----*-----)
BRAND2    10     1.2320    0.3860                           (----*-----)
BRAND3    10     0.8960    0.0450           (-----*-----)
BRAND4    10     0.7250    0.1514  (-----*-----)
BRAND5    10     0.8260    0.0809       (-----*-----)
                                 -------+---------+---------+---------
POOLED STDEV =   0.2242                 0.75      1.00      1.25
```

The means most likely to be significant are the two with the greatest distance between them. In this situation, these two means result from brands 2 and 4. We establish this finding from the first and last means on the ordered list of means or, equivalently, from the largest absolute estimated difference listed in bold print on page 600. (ignore sign of the distance).

Box 12-2 displays an ANOVA and a more detailed analysis to determine which means are higher and which are lower.

Section 12.5 Problems

23. A behavioral psychologist, who specializes in helping smokers quit, believes the smoker's job type will affect the time it takes to stop smoking among those who do stop. Use the computer output in Figure 12-15 for an ANOVA of stopping times (in weeks) to test for these factors' effects at the 0.01 level. The jobs types are sedentary, active, or unemployed. Assume normal populations with equal variances. Interpret the decision.

24. Cut flowers sold in bouquets appear fresh for various lengths of time. A florist places 15 randomly selected bouquets in one of four locations. One location receives direct sunlight, another receives indirect sunlight, the third only artificial light, and the last receives none. Use this information and the computer printout in Figure 12-16 to conduct a one-way ANOVA at the 0.01 significance level. Assume the populations involved are normally distributed with identical variances. Interpret the decision.

25. An automobile service center employs three mechanics. Determine if there is a significant difference in mean completion time for repair jobs among the mechanics, using the computer output in Figure 12-17 based on randomly collected completion times for the three mechanics. Make a decision at the 0.05 level, and interpret it. Assume populations are normal and variances are equal.

Figure 12-15
Minitab output for one-way ANOVA for smoker data

```
ANALYSIS OF VARIANCE
SOURCE        DF        SS         MS        F         p
FACTOR         2      34.1       17.0     1.13     0.337
ERROR         27     406.1       15.0
TOTAL         29     440.2

                              INDIVIDUAL 95 PCT CI'S FOR MEAN
                              BASED ON POOLED STDEV
LEVEL         N       MEAN    STDEV -+---------+---------+---------+-----
SEDNTARY     10     10.400    4.195      (---------*---------)
ACTIVE       10      9.800    3.048 (---------*---------)
UNEMPLOY     10     12.300    4.270              (---------*---------)
                                    -+---------+---------+---------+-----
POOLED STDEV =       3.878          7.5       10.0      12.5      15.0
```

Figure 12-16
Minitab output for one-way ANOVA of flower data

```
ANALYSIS OF VARIANCE
SOURCE        DF        SS         MS        F         p
FACTOR         3     105.92      35.31     8.03     0.000
ERROR         56     246.27       4.40
TOTAL         59     352.18

                              INDIVIDUAL 95 PCT CI'S FOR MEAN
                              BASED ON POOLED STDEV
LEVEL         N       MEAN    STDEV ----+---------+---------+---------+--
DIRECT       15      5.600    2.772                   (------*------)
INDIRECT     15      4.000    2.478             (------*------)
ARTIFICL     15      2.733    1.624       (------*------)
NONE         15      2.133    1.060   (------*------)
                                    ----+---------+---------+---------+--
POOLED STDEV =       2.097          1.6       3.2       4.8       6.4
```

Box 12-2 Parenting

The effect of parenting style on children is an old but continuing question. Anecdotal evidence abounds, as well as philosophical and theoretical hypotheses. To determine if there are sufficient data to indicate that products of different styles do indeed diverge, researchers subject different styles to statistical tests. In the study mentioned here, researchers identified four parenting styles and four possible outcomes or objectives of parenting. They analyzed the effect of these styles on the different outcomes with an ANOVA that directly controls and tests for multiple experimental or influential factors.

Combinations of the levels of two different variables defined the four parenting styles. Acceptance/involvement (A/I) meant the "extent to which the adolescent perceives his or her parents as loving, responsive, and involved." Strictness/Supervision (S/S) signified "parental monitoring and supervision." The four styles that resulted were:

1. AV = authoritative—high levels of A/I and S/S.
2. AN = authoritarian—low A/I and high S/S.
3. I = indulgent—high A/I and low S/S.
4. N = neglectful—low A/I and S/S.

The researchers used the responses of about 10,000 adolescent participants to questionnaires to pare the group to 4,100 participants. Then they classified these teens' parent or parents into one of the four parenting style categories.

The four sets of outcome variables are psychosocial development, academic competence, internalized stress, and problem behavior. Each includes several dimensions that are listed in the table below.

After controlling for gender, race, and extent of parental education, the analysts determined that mean outcomes vary significantly among parenting styles. They proceeded to perform one-way analysis on each dimension of the four outcomes to determine the effects of parenting style. The following table records the results, all of which are significant at the 0.001 level.

Outcome	Dimension	F
Psychosocial development	Self-reliance	6.71
	Work orientation	24.65
	Social competence	10.89
Academic competence	Grade point average	14.21
	Academic competence	13.58
	School orientation	38.08
Internalized stress	Somatic symptoms	6.43
	Psychological symptoms	16.02
Problem behavior	Drug and alcohol use	25.94
	School misconduct	20.59
	Delinquency	9.16

Further analysis to rank the parenting styles produces the rankings in the figure. Higher rows represent higher or better ranks, so if a style is on one row and not on lower rows it is significantly better than all styles listed below it. For instance, under self-reliance AV is significantly higher than AN and N, but not I, whereas I is not significantly different from any of the other styles. AN and N are not significantly different from each other or from I but are significantly lower than AV. Internalized stress and problem behavior dimensions are reverse ordered, so that the lower means are better (higher row on the diagram).

Overall, authoritative parenting produces better outcomes than neglectful parenting on average. In every diagram authoritative is on a row above and not including neglectful. In fact, authoritative is on the top row and neglectful on the bottom row of every diagram. Authoritarian and indulgent parenting produce mixed results and usually fall between the two extreme forms of parenting. Generally, authoritarianism produces fewer problem behaviors, whereas indulgence engenders higher psychosocial development.

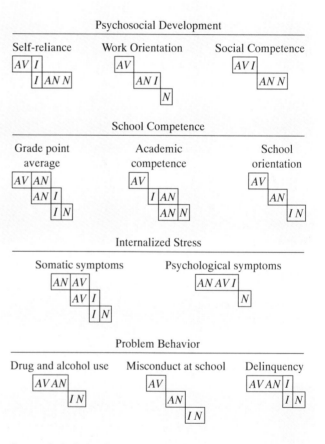

Source: Susie D. Lamborn, Nina S. Mounts, Laurence Steinberg, and Sanford M. Dornbusch, "Patterns of Competence and Adjustment among Adolescents from Authoritative, Authoritarian, Indulgent, and Neglectful Families," Child Development, Vol. 62, No. 5, 1991, pp. 1049–65.

```
ANALYSIS OF VARIANCE
SOURCE      DF      SS        MS       F       p
FACTOR      2       0.048     0.024    0.03    0.974
ERROR       15      13.717    0.914
TOTAL       17      13.764
                                       INDIVIDUAL 95 PCT CI'S FOR MEAN
                                       BASED ON POOLED STDEV
LEVEL       N       MEAN      STDEV   ----------+---------+---------+------
MECH1       6       1.4500    1.1807         (---------------*---------------)
MECH2       6       1.4333    0.6408         (---------------*---------------)
MECH3       6       1.3333    0.9688  (---------------*---------------)
                                      ----------+---------+---------+------
POOLED STDEV =      0.9563                    1.00       1.50      2.00
```

Figure 12-17

Minitab output for one-way ANOVA of mechanics' repair times

12.6 Two-Way Analysis of Variance

■ **Example 12-5:** Medical researchers experiment to determine if four brands of headache remedies provide different levels of relief for patients with chronic headaches. Relief is the time before onset of another headache. They randomly assign twelve subjects to each of the four factor levels. However, they suspect that the use of a blood pressure medication, drug B, affects relief levels from any of the four brands. What can they do to control the effects of drug B in the experiment?

■ **Solution:** Random assignment of subjects to different treatment samples would control for this effect. Another possibility is to use only subjects who take the same amount of drug B. For example, all participants could take drug B, or no participants could take drug B. A third possibility would be to select samples with the same distribution of drug B usage levels among the subjects as in the population of patients with chronic headaches.

Randomization, the first solution, does not ensure that the researchers will control the effects of drug B on the outcome. It only gives every drug B usage level an equal chance of inclusion in each sample. Thus, we reduce the likelihood that drug B use will systematically affect the results of the experiment or the interpretation of the results.

The second solution, to use only patients who take drug B, produces conclusions about relief from the headache remedies that apply only to drug B users. Thus, controlling for a second factor by limiting the population also limits the researchers' ability to generalize the results.

The third solution controls for systematic effects of drug B, but the results do not provide information about the actual effects of using the drug. Substantiating the effects of different usage levels of drug B is useful information for medical treatment.

A final solution permits explicit control of the effects of drug B. The researchers can use a second experimental factor. To do so, they randomly select subjects for one of 12 samples from different combinations of brand and drug B usage levels. These drug B levels are: (1) no intake of drug B in the last 30 days; (2) no intake within the last seven days but some intake within last 30 days, (3) current intake of drug B or within last seven days. The relief times expressed in hours are listed in the following table.

Drug B Usage	Brands			
	1	2	3	4
None	8.68	6.23	4.92	7.43
	8.14	6.73	5.21	6.76
	11.57	4.31	6.34	4.63
	8.98	5.80	5.00	5.87

(continued)

Drug B Usage	Brands			
	1	2	3	4
Last 30 Days	8.62	8.25	6.35	7.38
	3.35	7.88	8.41	5.58
	4.95	5.65	6.42	7.37
	5.16	6.47	4.70	7.28
Currently	7.89	3.78	7.72	5.46
	5.62	1.27	5.16	6.44
	7.08	0.83	2.60	6.02
	6.91	5.85	5.82	8.64

A one-way ANOVA is not the best analysis of these experimental results, because it ignores available information about each observation, specifically, drug B usage of each participant. Instead we use two-way ANOVA to analyze experimental data with information about two experimental factors.

These medical researchers wish to study the effects of more than one factor on outcomes. In other cases, it is difficult or impossible to control all but one factor. While it may seem that we violate the concept of the ideal experiment—varying only one factor affecting the outcome—by allowing more than one factor to change, there are benefits to incorporating the effects of more than one experimental factor.

In a one-factor experiment, we presume that other factors are randomly distributed in the treatment samples. Thus, any variation in outcome caused by these factors becomes part of the chance, or within-sample, variation and, ideally, is identical within each sample.

When we deliberately choose some subjects from each treatment sample to experience different levels of the second factor, the experiment can account for more of the variation in outcomes, which reduces the chance variation. This technique is similar to the reduction in variation we can achieve when we consider subgroups of a population. For instance, the variation of all dairy product prices is larger than the variation within dairy subgroups, such as milk prices alone or cheese prices alone.

In the context of ANOVA, each treatment sample represents a population whose observations vary because of chance. Deliberately placing subjects into subsamples of the treatment sample reduces the variation that we attribute to chance in the one-way analysis. Variation in the second factor *explains* or accounts for some of the outcome variation that the one-factor experiment attributes to chance.

When we reduce the chance or unexplained variation, we attribute the remaining variation to different treatments. More treatment variation produces a larger value for the test statistic. The reduction in the chance variation better reveals the importance of the experimental factors. Figure 12-18 shows that we can detect a difference between means more easily when there is less variation within the treatments (chance variation).

Parts a and b of Figure 12-18 represent one- and two-factor experimental situations, respectively. Alternatively, we could perform one-way ANOVA on the situation represented by the top row of Part b, alone, or on the situation shown in the bottom row, alone. For instance, we could perform a one-way ANOVA for all four headache remedies but only use subjects who have never taken drug B. Each of these one-way ANOVAs would provide evidence about the effect of different remedies on a limited population (those who experience a given level of the second factor). These ANOVAs control for levels of the second factor by ignoring other levels of the second factor. Consequently, we must constrain conclusions to the more narrow population as mentioned earlier in this section.

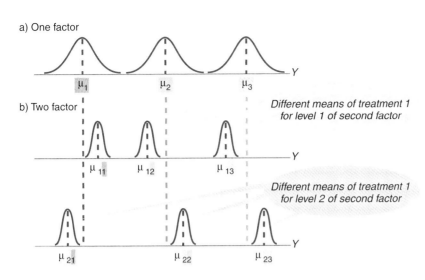

Figure 12-18
Treatment populations for one factor versus two factors

Two-factor or two-way ANOVA, on the other hand, employs all of the information on both factors to determine their impacts on or association with the outcome. Notice the smaller variation of the curves in Part b. This occurs because they represent subgroups of the treatments of the first factor; hence, the data contain less chance variation. Thus, two-way ANOVA differentiates the effects of the two experimental factors more clearly. Box 12-3 shows how forensic science researchers employ ANOVA to develop techniques for estimating the age of deceased persons.

Box 12-3 Deadly Statistics

Determining age at death, especially of an adult, is a difficult task for forensic anthropologists. Conventional estimation techniques consider changes in the skeletal structure, but loss or destruction of pertinent skeletal remains hampers or even prevents the use of these techniques. For instance, natural decay and predator activity may eliminate relevant pieces of the skeleton. The complications multiply when estimation techniques depend on personal characteristics, such as gender and race.

Scientists develop these estimation techniques using statistical analyses. A recently developed technique suggests that changes in texture and organization or degenerative transformations on the earlike form of the ilium (hip bone) signify certain age groups. Two researchers tested a new technique with a sample taken from a collection of more than 1,600 skeletons kept by the Smithsonian Institution, dubbed the Terry Collection after the initial preservationist. One question they sought to answer was the effect of gender and race on the accuracy of age identification. These facts as well as the death age were known for all skeletons selected for the sample.

After estimating the death age with the new technique, they subjected the estimation errors to two-way ANOVA.

Initially, the analysis indicated that average error magnitudes depend on race and not on gender. However, further investigation with other statistical techniques and closer examination of the makeup of the skeleton sample led them to believe that age was the complicating factor rather than race. Blacks dominated the younger death age skeletons and whites the older ones in the original sample.

To correct this problem, the researchers truncated the sample to balance race and gender characteristics over an age range. This time the ANOVA yielded no support for gender or race dependencies. Supplementary analysis supported the dependence of accuracy on the age of the person at death. The technique overestimates death age for younger subjects and underestimates the death age for older ones. They developed a correction method to adapt the technique for all ages, but, overall, found the estimated age ranges were too large to be useful for forensic science purposes.

Source: Katherine A. Murray and Tracy Murray, "A Test of the Auricular Surface Aging Technique," Journal of Forensic Sciences, Vol. 36, No. 4., July 1991, pp. 1162–69.

Summary and Review

ANOVA is a procedure for testing for differences among more than two population means. In a controlled experimental situation, the decision denotes whether or not different treatments of an experimental factor affect the outcome. Often behavioral scientists find conducting a controlled experiment difficult, if not impossible, because individuality complicates stock responses to treatments and because data often originate in nonlaboratory settings. Such experiments must be carefully designed to randomize the placement of subjects in treatment samples, in order to implicitly control other factors.

We use the one-way ANOVA procedure for experiments with only one experimental factor and J treatment levels for the factor. The hypotheses are

$H_0: \mu_1 = \mu_2 = \cdots = \mu_J$
H_1: at least one mean is different from the others.

The test statistic is the ratio between the weighted mean of the squared deviations among the treatment sample means and the mean of the squared deviations within the treatment samples. The numerator is a measure of chance variation as well as variation among treatment means, unless H_0 is true, in which case, it measures only chance variation. The denominator is a measure of chance variation among outcomes. The test is always an upper-tailed test, because only large values of the test statistic indicate that treatment effects have increased the variation in the numerator of the ratio.

The sampling distribution of this ratio is an F distribution, with $J - 1$ degrees of freedom associated with the numerator and $n - J$ degrees of freedom associated with the denominator.

Rejecting H_0 indicates that at least one treatment mean is different from the others. We can use the means of the treatment samples to estimate difference effects and to rank the treatments by their effects on outcomes; however, the statistical significance of such comparisons is an advanced topic.

Two-way ANOVA extends one-way ANOVA to allow for the effects of a second factor. Not only do we test for the importance of a second factor in a two-factor experiment; the importance of the first factor is usually more distinct, because we explicitly control more factors in the experiment. We no longer need to trust randomization to equally distribute the effects of the second factor across the treatment samples of the first factor. The total sum of squares to be explained, and the sum of squares due to the first factor remain the same with the same data set. However, the sum of squares within usually diminishes. Thus, the chance variation included in the mean sum of squares from the first factor diminishes, producing a larger MSB/MSW. When we use two-way ANOVA, we improve the chances of learning if the second factor affects outcomes, especially in cases in which its effect is masked by the chance variation in a one-way ANOVA.

New Equations from Chapter 12:

$\text{SS} = \Sigma Y^2$.
$\text{SST} = \text{SS} - n\bar{\bar{Y}}^2$.
$\text{SSB} = \Sigma n_i \bar{Y}_i^2 - n\bar{\bar{Y}}^2$.
$\text{MSB} = \text{SSB}/(J - 1)$.
$\text{SSW} = \text{SST} - \text{SSB}$.
$\text{MSW} = \text{SSW}/(n - J)$.
$F\text{-ratio} = \text{MSB}/\text{MSW}$.

Multiple-Choice Problems

Throughout the chapter, problems marked with an asterisk are p-value problems and are optional.

26. The term experimental design refers to
    a. A prototype product that is subjected to tests for marketability and safety.
    b. A plan for obtaining information in a manner that ensures that systematic variation, if it exists, results from known factors.
    c. A special diagram for conducting hypothesis tests on experimental data.
    d. A data collection method that requires a laboratory setting for accuracy of information and interpretation of results.
    e. An explanation of test results when observations are dependent within and among samples.

27. When several factors may affect an outcome, what can be done to avoid a systematic effect from variables other than the experimental factor?
    a. Use similar subjects that do not differ in levels of the uncontrolled factors.
    b. Limit treatments to two levels of the experimental factor.
    c. Assign subjects randomly to different treatment samples.
    d. Use a two-way ANOVA to incorporate and control explicitly the effect of a second factor.
    e. Choices a, c, and d are all correct responses.

Multiple-Choice Problems

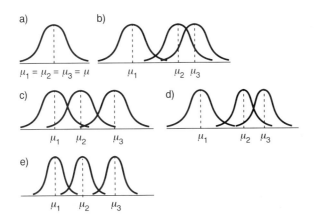

Figure 12-19
Choice of diagrams that portray a true ANOVA H_0

28. Which of the diagrams in Figure 12-19 best depicts the situation when H_0 in ANOVA is true?
    a. Diagram a.   d. Diagram d.
    b. Diagram b.   e. Diagram e.
    c. Diagram c.
29. If we reject H_0 when performing ANOVA, the correct interpretation is that
    a. There is no good estimator of the variation within the experimental data.
    b. The experimental factor is not important in explaining variation in outcomes.
    c. Mean outcome varies with treatments.
    d. The sample observations are either dependent or nonrandom.
    e. All treatment means are identical.
30. An automobile body shop owner tried three different paint application methods over the last 12 months. He used one from October through February, another from March through May, and the final one from June through September. He randomly selects 30 records from the period, consisting of 10 complete paint jobs on midsize cars using each method, to learn if the total time to complete the job (including drying time) differs with the painting method. Is there a problem in this experimental design that would prevent the use of one-way ANOVA?
    a. Application method and time-of-year (temperature) effects may be confounded.
    b. Jobs are selected, not assigned randomly to treatment samples.
    c. Job times are independent, so they are of no value in determining average outcomes.
    d. Drying time and labor time are dependent.
    e. There are too few observations to obtain reliable variation estimates.

Use the following information to answer Questions 31–38.

Twenty-four cardiac patients with the same ailment and the same mean systolic blood pressure over the past year are randomly assigned to four treatment samples. The members of each sample sit in one of four types of chair in the examination room before each person's blood pressure is measured by the same nurse. The resulting systolic values are listed in the table. Assume all populations are normally distributed with equal variances.

Treatments			
Rocking Chair	Straight Chair	Recliner	Stool
122	144	144	128
135	152	130	135
130	127	118	144
140	139	136	149
151	155	175	161
133	131	129	135

31. What is the experimental factor?
    a. Cardiac condition.   d. Systolic blood pressure.
    b. Age.   e. Chair type.
    c. Blood pressure.
32. Find $\overline{\overline{Y}}$.
    a. 3,343.   d. 465,652.
    b. 469,589.   e. 139.2917.
    c. 11,175,649.
33. Find SS.
    a. 3,343.   d. 465,652.
    b. 469,589.   e. 139.2917.
    c. 11,175,649.
34. Find SST.
    a. 3,936.9583. d. 465,826.
    b. 469,589.   e. 27,547.56.
    c. 2,794,953.
35. Find SSB and SSW.
    a. 465,826 and 465,652.   d. 27,547.56 and 30,276.
    b. 173.4583 and 3,763.5000. e. 3,936.9583 and 0.
    c. 57.8194 and 188.1750.
36. What is the value of the test statistic for this problem?
    a. 1.00.   d. 1.10.
    b. 0.0460. e. 21.7182.
    c. 0.3073.
37. What is the decision if the test is conducted at the 0.01 level?
    a. Cannot reject, because $1 < 4.94$.
    b. Reject, because $1 < 2.87$.
    c. Reject, because $0.3073 < 2.87$.
    d. Cannot reject, because $0.3073 < 4.94$.
    e. Cannot reject, because $0.05 > 0.01$.

Figure 12-20
Minitab output for one-way ANOVA of hair-color duration

```
ANALYSIS OF VARIANCE ON DURATION
SOURCE      DF       SS        MS       F        p
COLOR        3     82.74     27.58    6.11    0.002
ERROR       41    185.03      4.51
TOTAL       44    267.78

                                   INDIVIDUAL 95 PCT CI'S FOR MEAN
                                   BASED ON POOLED STDEV
LEVEL    N      MEAN    STDEV    --+---------+---------+---------+----
  1     10     2.700    1.059    (--------*-------)
  2     10     5.800    2.150                         (-------*--------)
  3     10     5.800    2.530                         (-------*--------)
  4     15     6.133    2.326                           (------*------)
                                  --+---------+---------+---------+----
POOLED STDEV =         2.124       1.6       3.2       4.8       6.4
```

38. Interpret the decision.
    a. The mean blood pressure is lower when the patient sits in a rocking chair.
    b. The mean blood pressure varies by treatment sample.
    c. Most, but not all, of the means are equal.
    d. There is a lot of variation in blood pressure readings that this experiment did not detect.
    e. Apparently, the type of chair does not significantly affect blood pressure readings.

Use the following information to answer questions 39–43.

Figure 12-20 shows a computer output from performing ANOVA on 45 observations of the duration of a hair-color rinse, measured from time of one application to the next application. The manufacturer conducts the test to confirm that rinse color affects duration. Assume that times are normally distributed with equal variances.

39. The decision at the 0.01 significance level is
    a. Cannot reject H_0, because $6.11 > 4.31$.
    b. Cannot reject H_0, because $6.11 < 26.41$.
    c. Reject H_0, because $6.11 > 4.31$.
    d. Reject H_0, because $27.58 > 4.51$.
    e. Reject H_0, because $6.11 < 26.41$.
*40. Use the *p*-value to perform the ANOVA at the 0.05 level. The decision is
    a. Reject H_0, because $6.11 > 2.84$.
    b. Reject H_0, because $0.002 < 0.05$.
    c. Cannot reject H_0, because $6.11 > 2.84$.
    d. Cannot reject H_0, because $0.002 < 0.05$.
    e. Cannot reject H_0, because $0.001 < 0.05$.
41. Interpret this decision.
    a. Average durations vary depending on the rinse color.
    b. Duration of application does not affect rinse color.
    c. Rinse color means vary with duration of application.
    d. The mean durations are all the same.
    e. Colors 1 and 2 are different from colors 3 and 4.
42. Based on the treatment means, we can hypothesize that
    a. $\mu_1 = \mu_2 = \mu_3 = \mu_4$.
    b. $\mu_1 < \mu_2 < \mu_3 < \mu_4$.
    c. $\mu_1 < \mu_2 = \mu_3 = \mu_4$.
    d. $\mu_2 < \mu_1 < \mu_4 = \mu_3$.
    e. $\mu_1 < \mu_2 = \mu_3 < \mu_4$.

43. We can interpret the confidence intervals in the output to indicate that
    a. The means are indistinguishable.
    b. An application of color 1 does not last as long as applications of the other colors.
    c. The effects of colors 2 and 3 on duration are indistinguishable, but color 1 results in a significantly shorter duration, and color 4 results in a significantly longer duration.
    d. The duration is different for each color.
    e. Application of color 1 lasts longest.
44. A newspaper article suggests that some types of magazines, especially exercise and nutrition, fare better at maintaining advertising during recession. In order to employ ANOVA to test this hypothesis among five magazine categories, (1) news, (2) health, (3) craft, (4) home repair, and (5) women's services, which of the following must be true? (Source: Patrick M. Reilly, "Rodale Finds Clean Living Gives a Healthy Tone to Ad Levels," *Wall Street Journal*, August 12, 1991, pp. B1, B5.)
    a. We must select five random samples each with the same number of observations to avoid one category having undue influence on the results.
    b. We need random samples of any four of the five categories to ensure the degree of freedom is one less than the number of populations.
    c. Random samples of monthly percentage changes in advertising for each of five randomly selected magazines, one in each category, will bypass any problem of dependent samples.
    d. If any magazine selected for a sample has a large readership, then the rest of that sample's selections must be from smaller, more specialized categories to ensure that the standard deviations are identical.
    e. We must restrict small samples to only one representative from a large publisher if we want to generalize the results to all publishers.
45. Two-way ANOVA
    a. Explicitly controls for variation in a second experimental factor.
    b. Reduces the chance variation.

c. Tends to increase the test statistic for the effects of the first experimental factor.
d. Tests for a difference in mean outcomes for two different factors.
e. All of the above are true for two-way ANOVA.

46. A factor in ANOVA is
   a. An experimental outcome that we constrain in some statistical way.
   b. A variable whose variation we associate with variation in average outcome.
   c. A rule for the experimental design.
   d. A variable that is held constant during the experiment in a well-controlled experiment.
   e. A randomly assigned participant.

47. ANOVA calculations for an experiment with six treatments and 81 participants include the following: SST = 500, and SSB = 200. What is the value of the test statistic?
   a. 300.
   b. 2.5.
   c. 0.4.
   d. 10.
   e. 5.99.

48. What is the decision at the 0.05 level, based on the information in Question 47?
   a. Reject H_0, because 10 > 2.37.
   b. Reject H_0, because 10 > 3.74.
   c. Reject H_0, because 10 > 2.25.
   d. Reject H_0, because 10 > 4.43.
   e. Cannot reject H_0, because 10 > 3.74.

49. A decision not to reject the null hypothesis in ANOVA means that
   a. The experimental factor is responsible for the outcome changes.
   b. A factor has not been randomized in the selection or assignment process.
   c. Changes in outcome have not been explained in this experiment.
   d. Treatment means are all different.
   e. At least one treatment mean is different from the others.

*50. If the p-value for a one-way ANOVA is 0.08, what is the decision at the 0.05 significance level?
   a. Cannot reject H_0, because 0.08 > 0.05.
   b. Reject H_0, because 0.04 < 0.05.
   c. Reject H_0, because 0.04 > 0.025.
   d. Cannot reject H_0, because 0.08 > 0.025.
   e. Reject H_0, because 0.04 < 0.10.

Word and Thought Problems

51. If the null hypothesis for ANOVA is true, what does MSB measure? If H_0 is false, what does MSB measure? Repeat these questions substituting MSW for MSB.

52. How would you conduct a controlled experiment, using a completely randomized design to learn if average student performance varies with the duration of a class period? Clearly state your expected outcomes, factors, the experimental factor, and treatments, as well as how you would measure values.

53. We pool variations within different treatment samples in ANOVA. Which ANOVA assumption makes this pooling reasonable? What do SSW and MSW measure about the populations? Relate pooling, its justifying assumption, and the value estimated using the "within" variation.

54. Another ANOVA assumption is that populations are normally distributed about their means. Suppose we assumed bimodal probability distributions instead. How would this affect our ability to discern differences in means? Use sketches to demonstrate your answer.

55. Suppose you frequently purchase gallons of milk, always from one of three places, a convenience store and two supermarkets. To learn if the average cost of the milk varies with the store where you shop, because milk prices fluctuate often, you decide to record prices and perform an ANOVA. Although there is no pattern to where you shop, to ensure that purchase locations are random for the experiment, you choose the place for each purchase by tossing a die. A 1 or 2 represents the convenience store, a 3 or 4 represents supermarket A, and a 5 or 6 represents supermarket B. You record the following prices over a 10-week period. Use this information to perform an ANOVA at the 0.05 level. Assume prices are normally distributed with equal variances. Interpret the decision.

Convenience	Supermarket A	Supermarket B
1.89	1.79	1.99
2.29	1.55	1.69
1.99	2.19	1.79
2.19	2.09	2.10
2.29	1.79	2.15
	1.89	2.03
	2.01	2.09
		1.79

56. Suppose medical technicians check three different durations of sustained stationary bicycle riding (10, 20, and 30 minutes) at 15 mph for effects on heart rate. The total test sample includes 15 randomly selected subjects. The technicians check heart rates at the end of each ride. The results are

Duration (minutes)		
10	20	30
80	111	125
108	101	83
76	72	99
86	88	105
92		96
		108

a. Suggest other factors that we should consider in order to narrow the population and prevent extraneous factors from influencing the conclusion.
b. Perform an ANOVA test on these data at the 0.025 significance level. Assume normal populations with equal variances.
c. Interpret your conclusion.

57. John ordered eight backpacking items from four different distributors over the last year, two from each distributor. He recorded the delivery times shown here.

Distributors

1	2	3	4
2.0	1.0	1.0	0.5
2.5	3.0	2.5	1.5

a. Perform an ANOVA at the 0.05 level. Assume normal populations with equal variances.
b. Interpret your decision.
c. Comment on the design of this experiment.

58. To determine if the average duration of commercial breaks differs among several television stations, Linda spends one evening randomly choosing stations from among the five that she normally watches. She watches the station chosen using the sampling procedure until the next commercial break, which she times. She then randomly chooses another station and continues the process. The results are recorded below.

Television Station

1	2	3	4	5
0.9	2.1	5.5	1.3	4.0
1.1	1.4	3.6	1.5	3.0
	1.8		2.0	2.0

a. Perform a one-way ANOVA at the 0.01 level, and interpret your decision. Assume normal populations with equal variances.
b. Suggest other factors to explain the time variation.

59. a. A data set contains only one observation for each treatment. Can we perform ANOVA? Give an intuitive and a mathematical explanation for your answer.
b. One treatment in a set of 10 treatments contains only one observation, but the other treatments all have at least two observations. Can you perform ANOVA? Explain.

60. A home economist measures the fat content of cooked chicken drumsticks coated with one of four different batters. Do a one-way ANOVA on the results listed below at the 0.05 level. Assume normal populations with equal variances. Interpret the decision.

Recipes

1	2	3	4
22	19	27	36
31	21	28	22
22	21	28	23

(continued)

Recipes

1	2	3	4
17	19	35	31
46	27	20	29
38	32	35	24
31	20	39	28
26	32	21	25
30	32	20	30
21	35	20	32

61. Fourteen friends wonder if actual and estimated waiting times for tables vary among six different restaurants. They randomly assign themselves to one of the six restaurants, where each measures the difference between actual waiting time and waiting time in seconds estimated for the customer upon entry into the restaurant. Do a one-way ANOVA using a significance level of 0.05. Assume normal populations with equal variances. Interpret your decision.

Restaurant

1	2	3	4	5	6
61	1	99	24	73	70
18	90	57	12	91	91
		91			
		33			

62. Take three samples of five single digits from the random number table in Appendix I.
a. If we use these values to perform a one-way ANOVA, which is true, H_0 or H_1?
b. Perform the test at the 0.05 level, and interpret your decision.
c. Do we meet all of the assumptions for performing ANOVA? Explain.

63. An educator uses a random sample of children to test if the number of lessons a child must take before she or he can swim across a pool is related to the frequency of class meetings. The data are in the table.

Frequency of Lessons

Every Weekday	Every M–W–F	Every T–Th	Every W
23	18	25	5
19	2	11	8
25	16	10	20
19	19	8	20

a. Test at the 0.025 level for equality of means. Assume normal populations with equal variances.
b. Interpret your decision.
c. Suggest other factors that might influence the outcome.

64. Sometimes viewers and commentators question the objectivity of Olympic games judges when assessing competitors from another country. Suppose the following scores in a certain event are awarded by four judges for athletes from a country other than their own.

Judges' Scores

1	2	3	4
3.3	6.2	4.8	3.9
5.4	7.1	7.0	4.1
5.2	6.8	5.2	5.9
3.0	6.9	4.8	5.9
3.5	6.4	4.2	5.9
4.4	5.6	6.0	5.7

a. Assume normal populations with equal variances and perform an ANOVA at the 0.05 level.
b. Interpret your decision. Do the data support the contention of biased judgments?
c. What other factors could have affected the recorded values? Has the sampling procedure used here controlled for these factors?

65. Schools across a state wonder about the performance of their students, relative to other schools' students. The media frequently reports and compares student performance means across schools. Suppose that three schools in a county report their tenth-graders' results, along with the variance (calculated with the sample formula) and the number of tenth-grade participants. We can find SS by manipulating the shortcut formula for S^2, $SS = \Sigma(n_i - 1)S_i^2 + \Sigma n_i(\overline{Y}_i)^2$. Also, $\overline{\overline{Y}} = (\Sigma n_i \overline{Y}_i)/n$. Use these formulas and the following data to perform ANOVA on the results reported for each school. Test at the 0.05 level. Assume normal populations with equal variances. Interpret the decision.

School (i)

	1	2	3
$\overline{Y}_i$	72.6	75.0	63.9
S_i^2	2.9	4.2	8.6
n_i	102	87	125

66. Figure 12-21 depicts the results of a one-way ANOVA for NCAA Basketball Tournament wins per team over the period 1982–1986 by different athletic conferences. Interpret the results using a 0.05 significance level. Hypothesize a ranking of the conferences by average wins per team. The data are taken from the NCAA Tournament Data Base in Appendix M.

67. To learn if students' performances vary with her advance description of an exam, a professor uses different descriptors in different classes and no descriptors in one class before students take a common exam simultaneously.
a. Use the data in the table and computer output in Figure 12-22 to check at the 0.05 level for the effect of the descriptions on student performance. Assume normal populations and equal variances.
b. Interpret the results. Make recommendations for the professor's pretest disclosures.

Professor's Descriptor

Easy	Difficult	None
90	92	90
85	76	85
87	98	95
85	65	75
70	78	76
72	70	70
78	55	60
65	50	62
58	40	68

68. Use the ANOVA output in Figure 12-23 to test for differences in research library holdings (in millions of volumes) among four types of schools: state supported, religious, independent, and mixed (government and independent). Test at the 0.1 level. (Source: Association

```
ANALYSIS OF VARIANCE ON WINPRTM
SOURCE     DF        SS          MS         F        p
CNF        10        22.312      2.231      6.40     0.000
ERROR      44        15.340      0.349
TOTAL      54        37.652

                                    INDIVIDUAL 95 PCT CI'S FOR MEAN
                                    BASED ON POOLED STDEV
LEVEL      N         MEAN        STDEV  ---+---------+---------+---------+---
1          5         2.1833      0.4308                       (----*----)
2          5         1.8400      0.8119                      (----*-----)
3          5         0.7800      0.4253       (-----*----)
4          5         0.7033      0.3924       (----*----)
5          5         1.8667      0.5055                       (-----*----)
6          5         0.2500      0.2500 (----*-----)
7          5         0.3000      0.4472  (----*----)
8          5         1.2133      0.6100              (----*----)
9          5         0.6833      0.2075       (----*----)
10         5         1.3667      1.0165               (-----*----)
11         5         0.5667      0.8300      (-----*----)
                                        ---+---------+---------+---------+---
POOLED STDEV =       0.5905              0.0        1.0       2.0       3.0
```

Figure 12-21

Minitab output for one-way ANOVA of percentages of wins per team

Figure 12-22

Minitab output for one-way ANOVA for exam descriptor experiment

```
ANALYSIS OF VARIANCE
SOURCE      DF        SS         MS       F       p
FACTOR       2       285        142    0.67   0.520
ERROR       24      5080        212
TOTAL       26      5365
                                      INDIVIDUAL 95 PCT CI'S FOR MEAN
                                      BASED ON POOLED STDEV
LEVEL        N      MEAN      STDEV  ------+---------+---------+---------+
C11          9     76.67      11.05              (------------*------------)
C12          9     69.33      19.09   (------------*------------)
C13          9     75.67      12.20             (------------*------------)
                                      ------+---------+---------+---------+
POOLED STDEV =    14.55               64.0      72.0      80.0      88.0
```

Figure 12-23

ANOVA output for research library holdings

```
ANALYSIS OF VARIANCE
SOURCE      DF        SS         MS       F       p
FACTOR       3      8.70       2.90    0.77   0.516
ERROR       41    154.12       3.76
TOTAL       44    162.82
                                      INDIVIDUAL 95 PCT CI'S FOR MEAN
                                      BASED ON POOLED STDEV
LEVEL        N      MEAN      STDEV  --+---------+---------+---------+----
STATE       25     2.690      1.505            (--*---)
RELIGION     3     1.913      0.135  (------------*-----------)
INDEPEND    14     3.437      2.685                 (----*----)
MIXED        3     3.365      1.732         (----------*----------)
                                      --+---------+---------+---------+----
POOLED STDEV =    1.939               0.0       2.0       4.0       6.0
```

of Research Libraries, as reported in "Holdings of University Research Libraries in U.S. and Canada, 1989–90," *The Chronicle of Higher Education Almanac*, Vol. 38, No. 1, August 1991, p. 37.)

69. A random sample of 15 stutterers completed a test requiring them to press a buzzer when they associated a word with another word that was read to them. Researchers measured their response times to 36 such words using tape recordings and electronic analysis. Researchers omitted observations if 20 seconds elapsed without a response, so there are 523 observations in the total sample. Twelve of the words were verbs, twelve were nouns, and twelve were adjectives. (Source: Kathryn M. Crowe and Robert M. Kroll, "Response Latency and Response Class for Stutterers and Nonstutterers as Measured by a Word-Association Task," *Journal of Fluency Disorders*, Vol. 16, 1991, pp. 35–54.)

a. Use the ANOVA output in Figure 12-24 to determine if mean response time varies with word class. Assume normal populations with equal variances. Test at the 0.05 level. Level 1 refers to verbs, level 2 to nouns, and level 3 to adjectives. Interpret the conclusion.

b. Regardless of the decision in Part a, have we demonstrated that stutterers' behavior differs from nonstutterers'? If so, explain how. If not, explain how you think we might determine whether there is such a difference.

Figure 12-24

ANOVA output for stutterers' associations with different word classes

```
ANALYSIS OF VARIANCE ON TIMES
SOURCE      DF        SS         MS       F       p
CLASS        2      4.61       2.31    1.95   0.143
ERROR      520    613.94       1.18
TOTAL      522    618.56
                                      INDIVIDUAL 95 PCT CI'S FOR MEAN
                                      BASED ON POOLED STDEV
LEVEL        N      MEAN      STDEV  --------+---------+---------+--------
1          172     2.049      1.072              (----------*---------)
2          175     1.855      0.921   (----------*---------)
3          176     2.059      1.242               (---------*----------)
                                      --------+---------+---------+--------
POOLED STDEV =    1.087               1.80      1.95      2.10
```

Review Problems

70. Design an experiment to determine if majoring in some disciplines prepares students better than others for taking a standardized test for entering graduate study. Describe the populations involved, different factors involved, the experimental factor, sample selection, controls, data to be collected, and statistical technique to use for analysis. What assumptions are necessary to properly perform the technique? Do you think your experiment will satisfy these assumptions? Explain.

Review Problems

71. How does the test for a difference in means in this chapter differ from the one in *Chapter 10*? What assumptions are different?

72. Why must we control factors in an experiment before collecting data? Suggest ways to control for other factors while allowing the experimental factor to vary.

73. While exploring the relationship between working mothers' situations and physical development of children, a researcher used one-way ANOVA to compare differences among several other variables that she expected might influence the relationship. She did not want the samples to differ in values of these variables, because different means might result from these differences rather than work type. She categorized work types as formal, informal, or nonworker. Informal workers did not regularly engage in conventional work places, jobs or trades but rather were self-employed at home, often irregularly. One-way ANOVA of different variables, where the treatments are job types, yielded the results listed in the table. Make and interpret decisions at the 0.01 significance level for each ANOVA. Is there a significant difference for any of these variables among the samples composed of different work types?

Outcome	Test Statistic	df	Maximum p-value
Mother's education	22.98	2,237	0.0001
Per capita family income	15.25	2,237	0.0100
House Quality	9.19	2,237	0.0100
Children's Birthdate	4.05	2,237	0.0500

Source: Patrice L. Engle, "Maternal Work and Child-Care Strategies in Peri-Urban Guatemala: Nutritional Effects," *Child Development*, Vol. 62, No. 5, 1991, pp. 954–65.

74. A study involved two samples of 14 observations, one containing abused children and one with nonabused children, with the objective of learning about development differences between the two types of children.
   a. First, compare the traits of each sample to see if there is evidence of a significant difference at the 0.1 level other than abuse between the two samples using the information in the table. Assume normal populations with equal variances.

Sample	Variable	Mean	Standard Deviation
Abused	Age (months)	53.7	8.64
Nonabused	Age (months)	54.6	10.52
Abused	IQ	88.7	6.32
Nonabused	IQ	90.9	7.38
Abused	Mean monthly family income	767	733
Nonabused	Mean monthly family income	1,228	871

   b. Test for a difference in means between abused and nonabused children for the following variables at the 0.1 level:

RAT = peer rating of desirability of subject as playmate.
H/A = teacher's rating of child's hostile or aggressive behavior.
RGH = rough play.
AGG = aggressive behavior involving property or territory.

Assume normal populations with equal variances.

Sample	Variable	Mean	Variance
Abused	RAT	−0.542	0.637
Nonabused	RAT	0.246	0.735
Abused	H/A	9.357	24.471
Nonabused	H/A	5.431	11.108
Abused	RGH	0.0075	0.0002
Nonabused	RGH	0.0104	0.0004
Abused	AGG	0.193	0.0095
Nonabused	AGG	0.085	0.0053

Source: Mary E. Haskett and Janet A Kistner, "Social Interactions and Peer Perceptions of Young Physically Abused Children," *Child Development*, Vol. 62, No. 5, 1991, pp. 979–90. Copyright © 1991 by the Society for Research in Child Development.

75. Use the following information to test for a difference in the mean amount of time theater directors spend researching productions of five different plays. Test at the 0.05 level, assuming normal populations with equal variances. Interpret the decision.

	Play 1	Play 2	Play 3	Play 4	Play 5
1	18	27	42	6	8
2	18	26	13	6	9
3	18	24	60	5	20
4	16	24	17	4	12
5	18	29	38	6	
6	14	25		5	
7	18	22		7	
8		30		6	
9		23		3	
10		32		3	
11				5	
12				6	

76. An advertising agency selects three random samples each with 10 potential customers of a product and exposes each sample to a different advertisement for the product. The sample members then answer questions about the product. Assume populations are normally distributed with equal variances, and perform the following test at the 0.10 level. Determine if the mean information content or message varies among the advertisements, using the participant scores below.

Sample 1	Sample 2	Sample 3
92	87	76
75	98	83
90	68	86
90	76	86
82	80	86
100	81	85
88	89	82
81	89	82
87	89	83
95	88	89

77. A random sample of 35 movies nominated for best picture awards since 1928 had a mean run in a local theater of 4.3 days, with a standard deviation of 2.1 days. Find an 80% confidence interval for the mean run of nominated movies. Does the sample contain less than 5% of the population? (Assume there have been five nominees per year since 1929.) A friend generally needs 10 days to schedule a free night for a movie. Suppose she learns about a showing on the first day of the run. Is it likely (can she be 80% confident) that on average the theater will still be running the movie, if she schedules a free night in 10 days? Explain.

78. Test at the 0.1 level to determine if the average winning time for the mile race in an athletic conference has changed over the last five years, based on the following random samples of winning times (in minutes). Assume winning times are normally distributed with equal variances.

Year 1	Year 2	Year 3	Year 4	Year 5
4.33	3.99	4.22	4.09	4.33
4.80	4.25	4.58	4.58	4.85
4.07	5.01	4.23	4.37	5.22
4.50	4.18	4.84	4.88	3.97
4.55	4.15	4.92	5.12	4.65

79. To learn if different levels of abstraction for the cues in a word-association test account for the differences in the corresponding sample means of response times, researchers performed an ANOVA. A random sample of 15 stutterers' responses to 36 different words in three categories of abstraction (low = 1, medium = 2, high = 3) resulted in 523 observations and in the computer output in Figure 12-25. Use this output to make and interpret a decision at the 0.1 significance level. Assume response times are normally distributed with equal variances. Do the average times rank in the same order you would have expected? Explain. (Source: Kathryn M. Crowe, and Robert M. Kroll, "Response Latency and Response Class for Stutterers and Nonstutterers as Measured by a Word-Association Task," *Journal of Fluency Disorders*, Vol. 16, 1991, pp. 35–54.)

80. To learn more about depression in nursing home patients, medical researchers collected individual data, including a psychiatric diagnosis regarding depression, on 454 admissions to several nursing homes near Baltimore. Patients fell into one of three categories, depressive disorder, depressive symptoms (less several cases), and

Figure 12-25
ANOVA output for stutterers' associations with different levels of abstraction

```
ANALYSIS OF VARIANCE ON TIMES
SOURCE      DF        SS        MS        F        p
ABSTRCT      2      9.69      4.84     4.14    0.016
ERROR      520    608.87      1.17
TOTAL      522    618.56

                                    INDIVIDUAL 95 PCT CI'S FOR MEAN
                                    BASED ON POOLED STDEV
LEVEL        N       MEAN    STDEV  -------+---------+---------+---------
  1        173      1.831    0.839  (--------*-------)
  2        174      1.966    0.946          (-------*-------)
  3        176      2.162    1.379                  (-------*-------)
                                    -------+---------+---------+---------
POOLED STDEV =      1.082             1.80      2.00      2.20
```

no depression. The observations in each category numbered 57, 82, and 315, respectively. The researchers checked whether the means of two other patient characteristics varied with state of depression. The first was the score on the Mini-Mental State Examination (MMSE), a test of the patient's awareness and judgment. The other was a measure of activities of daily living (ADL), performing routine health and hygiene functions. The mean scores and F statistics are shown on the table. Make and interpret decisions at the 0.05 significance level.

Depression Diagnosis

Variable	Depression Disorder	Depressive Symptoms	No Depression	F
MMSE	18.0	20.0	14.7	11.10
ADL	14.1	12.6	15.0	3.70

Source: Barry W. Rovner, Pearl S. German, Larry J. Brant, Rebecca Clark, Lynda Burton, and Marshal F. Folstein, "Depression and Mortality in Nursing Homes," *Journal of the American Medical Association*, Vol. 265, No. 8, February 27, 1991, pp. 993–96.

81. Test at the 0.05 level to determine if average acceleration measurements (in feet/sec^2) differ among four sets of physics lab partners. Each set performed the experiment and measured the acceleration four times with the following results. Assume such measurements are normally distributed with equal variances. Interpret the decision.

Group 1	Group 2	Group 3	Group 4
45.3	43.3	42.6	45.0
42.1	43.8	48.8	45.0
48.6	44.2	45.0	45.8
47.7	44.1	45.8	43.6

82. To study the effectiveness of different treatments for low back pain, several researchers at the University of North Carolina at Chapel Hill, plan to collect the following data:

   1. Information from a telephone survey of 2,500 randomly selected North Carolinians to learn if they experienced back pain in the last year and what was done, if anything, to alleviate the pain.
   2. Patients' continuing experience with back pain, treatments, and speed of relief from these treatments, for 1,500 patients from 25 orthopedists, 50 internists and family practitioners, 25 physicians in health maintenance organizations, and 50 chiropractors.

   The study excludes pregnant women, patients with several illnesses, and automobile accident victims. (Source: "UNC Researchers Launch Study to Improve Back Pain Treatments," *The Sylva Herald and Ruralite*, January 9, 1992, p. 6A.)
   a. Can we employ one- or two-way ANOVA to analyze the data in (1)? (2)? Explain.
   b. What are possible experimental factors that can be analyzed with an ANOVA for the data?
   c. What is the role of excluding some types of individuals, such as pregnant women, from the study?

83. Check if average beach cabin prices differ at different locations along the coast based on the following random samples. Assume prices are normally distributed with equal variances, and test at the 0.1 level. Interpret the decision. Do you think the only factor affecting prices is location? Explain.

N. Carolina	S. Carolina	Maine	Maryland	Georgia	Florida
845.90	753.59	772.49	851.21	1,306.78	980.56
605.29	691.12	607.15	784.69	714.90	805.19
516.30	734.13	897.82	935.73	857.71	1,142.90
741.00	613.02	656.99	743.54	855.03	431.23
701.29	935.68	701.58	999.19	792.44	1,015.33

84. A restaurant owner believes the average Tuesday night sales between 10 and 11 P.M. over the last five years exceeds $110. He randomly selected ten Tuesdays from those years' records. The mean and standard deviation of the sample figures are $125 and $7.75, respectively. Determine if these data confirm the owner's hypothesis. Assume a normal population of 10 to 11 P.M. Tuesday sales, and test at the 0.1 level.

85. The United Nations Population Division five-year estimates of nations' annual mortality rates of children under five years of age per 1,000 live births are used to classify different countries in one of four categories: very high (over 170), high (95–170), middle (26–94), and low (25 and under). Use the following random samples of different variables from each category to determine if the average values of the variable varies in the different categories. Assume normal populations with equal variances, and test at the 0.01 significance level. Interpret your decision. The variables are:
   a. Per capita GNP.
   b. Daily per capita calorie supply as percentage of requirements.

Per Capita GNP

Very High	High	Medium	Low
470	1,080	2,100	940
280	470	680	7,420
150	1,110	1,090	7,010
310	1,250	790	4,290
230	770	1,640	6,230
550	6,730	190	16,370
470	7,170	2,080	14,370
300	530	2,150	1,430
160	2,550	1,650	6,020
160		820	6,520
290			

Daily per Capita Calorie Supply as Percentage of Requirements

Very High	High	Medium	Low
109	97	101	130
93	119	101	130
92	109	111	119
85	118	98	124
105	108	126	140
124	127	134	114
78	118	110	118
88	94	103	126
85	100	128	114
103	89	111	131
85	132	117	114

Source: United Nations Children's Fund, *The State of the World's Children, 1988* (New York: Oxford University Press, 1988), pp. 72–73.

86. A study of differences among three types of child-care arrangements, family day care in a neighborhood woman's home, community day care centers, and kibbutzim, produced the following results. Use this ANOVA information to make decisions at the 0.01 significance level, then interpret each decision.
    a. This first set of statistics concerns mothers' education levels and several attributes of caregivers.

Variable	F	df
Mother's education	1.46	2, 73
Caregiver's education	3.94	2, 49
Caregiver's age	1.01	2, 47
Caregiver's experience	7.98	2, 49
Caregiver's training	7.30	2, 41

   b. The second set of statistics explore attributes of the educational environment established in each child-care arrangement.

Variable	F	df
Quality of physical environment	16.82	2, 48
Directed educational activity	3.63	2, 45
Emotional tone (relaxed, tense)	12.14	2, 46
Caregiver's positive interaction	4.60	2, 50
Caregiver's restrictions	4.49	2, 50

   c. Finally, measures of children's behavior, presumably a reaction to the different arrangements or to different qualities of the environments provided, yield the following statistics.

Variable	F	df
Tendency to participate in		
Positive learning experiences	7.68	2, 82
Social behavior experiences	5.78	2, 82

Source: Miriam K. Rosenthal, "Daily Experiences of Toddlers in Three Child Care Settings in Israel," *Child and Youth Care Forum*, Vol. 20, No. 1, February 1991, pp. 37–58.

87. Think of the earliest forms of primitive people. Imagine, then describe the variables that they might have first estimated.

88. What aspect of experimental design is the subject of the following discussion of using statistics to examine the growth of the British Labor Party?

    > One way forward might be to compare a party's inflated (or deflated) municipal vote some years into a parliament, with its vote at a similar point in the previous parliament, rather than with its vote at the last general election (which would have been fought under very different conditions). The difference between the two municipal polls would often indicate the extent to which a party's position had changed, all other things being equal. Used this way, municipal results could help to identify the extent and sources of a party's support, and the condition of its growth or decline. (Source: Duncan Tanner, "Elections, Statistics, and the Rise of the Labour Party, 1906–1931," *The Historical Journal*, Vol. 34, No. 4, 1991, pp. 893–908.)

89. **Relation between ANOVA and Test for Difference between Means (*From Chapter 10*).** The following samples meet the assumptions required for one-way ANOVA. Use them to answer the following questions.

A	B
10	40
50	60
30	30
50	30

   a. What is the null hypothesis for ANOVA in this situation? For a test of difference between two means?
   b. Perform one-way ANOVA at the 0.05 level. Interpret your decision.
   c. The ANOVA assumptions are the same as the assumptions for testing for a difference between means in which of the following situations: unequal variances, equal variances, or paired-differences? Do a two-tailed test for equality of means at the 0.05 level. Interpret your decision.
   d. Compare your test statistics, critical values, and decisions from Parts b and c. Show that $F_{(1, n_1+n_2-2)} = T^2_{(n_1+n_2-2)}$. Compare the first column (one degree of freedom associated with numerator) of the F table for a tail area of 0.05 with the 0.025-area column in the T table (Appendix J).

90. Test at the 0.01 level to determine if the mean evaluations of grammar, originality, research, and communication for papers in a freshman English competency qualifier performance vary. Assume normal populations with identical variances. The data set consists of randomly selected scores from sets of scores for each criteria, and computer output, Figure 12-26, follows. Interpret your decision.

Review Problems

```
ANALYSIS OF VARIANCE
SOURCE      DF         SS         MS         F         p
FACTOR       3     1871.8      623.9      9.47     0.000
ERROR      116     7643.5       65.9
TOTAL      119     9515.3

                                  INDIVIDUAL 95 PCT CI'S FOR MEAN
                                  BASED ON POOLED STDEV
LEVEL          N       MEAN    STDEV  ---+---------+---------+---------+---
GRAMMAR       30     78.500    9.930                       (------*-------)
ORIGINAL      30     69.933    6.448  (-------*------)
RESEARCH      30     70.500    6.796    (------*-------)
COMMUICA      30     77.667    8.786                     (------*-------)
                                       ---+---------+---------+---------+---
POOLED STDEV =       8.117            68.0      72.0      76.0      80.0
```

Figure 12-26
Minitab output for one-way ANOVA for English evaluation data

Chapter 13

Chi-Square and Nonparametric Tests

While this book concentrates on procedures for analyzing sets of numbers, this chapter discusses numerical procedures suitable for analyzing information that is not measurable; the data do not express an amount. Rather, we know about a quality, category, or rank. Often, when we cannot meet the assumptions of familiar test procedures that we have already studied, such as the test for a mean or a difference in two or more means, we convert measurements to categories or ranks and pursue these new techniques.

Objectives

In Unit I we discuss the chi-square test for independence between two variables. We seek to establish a dependent relationship between two variables. Upon completing the unit, you should be able to

1. State the advantages and disadvantages of nonparametric tests over parametric tests.
2. Perform a chi-square test for independence.
3. State when this type of test is appropriate.
4. Determine critical values from a chi-square table.
5. State the hypotheses and conclusions of a chi-square test in a usable form.
6. Understand the concepts of parametric test, nonparametric test, observed frequency, and expected frequency.
7. Distinguish $\chi^2_{(R-1)(C-1)}$, the symbol for a chi-square random variable with $(R-1)(C-1)$ degrees of freedom; f_o and f_e, an observed and expected frequency, respectively, in a table for a chi-square test.

Unit II covers three more nonparametric tests that substitute for parametric tests when the situation does not meet the conditions necessary for using a parametric test. Upon completing the unit, you should be able to

1. Do the following for each of the Sign, Wilcoxon-Mann-Whitney rank sum, and Kruskal-Wallis tests:
   a. State the parametric test for which each substitutes.
   b. State the hypotheses associated with each test.
   c. Calculate the test statistic and describe its sampling distribution.
   d. Make and interpret a decision for each test.
2. Recognize M, the hypothesized median; SR, a sum of ranks associated with values in a sample; n_i, the number of observations in sample i when a test involves more than one sample; n_T, the total number of observations in a combined sample; μ_{SR}, the mean of the possible sample SR values; σ_{SR}, the standard error of the possible sample SR values; and H, the Kruskal-Wallis test statistic.

Unit I: The Chi-Square Test for Independence of Two Variables

Preparation-Check Questions

1. If outcomes A and B are independent, then
   a. $P(A|B) = P(B|A)$.
   b. Knowing the value of B helps us determine the value of A and vice versa.
   c. If B happens, A cannot happen.
   d. There is no advantage in knowing B when we are primarily interested in A.
   e. $P(A \text{ or } B) = P(A) + P(B)$.

2. If $P(A) = 0.6$, $P(B) = 0.2$, and the two outcomes are independent, what is $P(A \text{ and } B)$?
   a. 0.8.
   b. 0.4.
   c. 0.32.
   d. 0.88.
   e. 0.12.

3. When performing hypothesis tests, we use a statistic calculated from sample information. We reject the null hypothesis when this statistic's value
   a. Is improbable, assuming the null is true.
   b. Is probable, assuming the null is true.
   c. Is improbable, assuming the alternative is true.
   d. Is probable, assuming the alternative is true.
   e. Is the value hypothesized in the null.

4. If the probability of a defective is 0.02, and we examine 100 items, how many defectives would we expect to find?
   a. 0.02.
   b. 0.
   c. 2.
   d. 100.
   e. 98.

5. Not rejecting H_0 when H_1 is true
   a. Is a Type II error.
   b. Is a Type I error.
   c. Is a correct decision.
   d. Has a probability symbolized by α.
   e. Indicates a need for the decision maker to alter the significance level.

Answers: 1. d; 2. e; 3. a; 4. c; 5. a.

Introduction

Loan Applicant Evaluation To determine if the social status of borrowers affects financial institutions' attitudes toward them, a sociologist collected data on past loan recipients. The researcher measured variables for the analysis from anonymous data from a financial institution, rather than a survey. For example, the applicant's job type substituted for social status. Measuring the financial institution's attitude was difficult, but the researcher believed data on the loan repayment behavior of individuals would substitute well for attitudes, because institutions favor those who repay on time. The table on the next page displays the number of recipients from each employment category and their particular repayment behavior. The sociologist analyzed this information to learn if the two variables, employment status and repayment behavior, are related. There are 1,279 individuals in the sample.

Question: Would the researcher record a number for each individual's employment status and repayment behavior?

	Repayment Behavior				
Employment	Due Date or Before	During Grace Period	Late	Default	Total
Unemployed	10	25	36	54	125
Part-time	86	80	34	36	236
Minimum wage	23	55	43	23	144
Blue collar	96	23	29	15	163
White collar	165	98	33	61	357
Self-employed	87	88	26	53	254
Total	467	369	201	242	1,279

Answer: There are no numbers to record for individuals. Rather these variables, employment status and repayment behavior, describe qualities of the individual, not quantities. A person is unemployed or is not unemployed. The information available about each individual only supplies categories to describe the repayment behavior of each not a quantity such as the number of weeks early or late.

Recall from Chapter 1 that we call *quantitative variables* those variables whose values represent amounts or measures of the variable, rather than a category or quality. On the other hand, *qualitative variables* are those for which we cannot measure amounts. Instead, we classify each observation into some category that describes a quality (gender, for instance) or ranks the observation relative to the other observations in a data set. Ranks are values, but the intervals between two consecutive ranks are not consistent, which makes comparisons of amounts or distances between the two values meaningless.

We could assign a different number to different categories to identify an observation that possesses a particular quality of a variable. Although such a numerical assignment provides information about the quality of a certain item, interpretation and measurement problems can occur when we employ these quantities in further quantitative analyses.

Question: Suggest values for the classes of repayment behavior.

Answer: The possibilities are infinite, but a problem arises because of the interval between categories. For example, suppose we assign a value of 1 to paying on or before due date, 2 to grace period, 3 to late, and 4 to default. Because of the way people think about numbers, this numbering system implies that an increase of one unit from early payment (1) to grace period (2) is comparable to a shift from late (3) to default (4). Clearly the shift from 3 to 4 represents a more drastic difference in circumstances than a shift from 1 to 2.

This unit describes a technique that employs qualitative data to determine if two qualitative variables are related. *Incidentally, we can use this technique for quantitative variables when a situation satisfies the assumptions necessary for a valid test.*

13.1 Nonparametric and Parametric Tests

The tests described in previous chapters proceed from assumptions about the population where the sample originates to questions or hypotheses about the parameters of a population, such as its mean, μ. Such tests are called *parametric tests*. Parametric tests for μ, differences in population means, and ANOVA all proceed from normal populations.

Section 13.2   Setting Up the Problem   621

Because the tests we will develop in this and the next unit do not make specific assumptions about the distribution or parameters of the population, they are called *nonparametric tests*. The questions asked or the hypotheses posed, H_0 and H_1, resemble those for parametric tests. For instance, in the next unit we will study a test about the median that substitutes for a test for the mean when analysts dispute or question the normality of the population under study. Earlier chapters of this book describe correlation and regression coefficients. We use both to measure relationships between two or more variables. Alternatively, we can use the test developed in this unit to learn if one variable contains any information useful in predicting the value of another variable. These tests are useful in two general situations where parametric tests would be invalid: (1) when a situation violates the assumptions of parametric tests, or (2) when a measurement problem exists.

While the nonparametric tests can handle both qualitative and quantitative data, there are drawbacks when we substitute a nonparametric test for a parametric test in a situation that satisfies the assumptions of the parametric test. First, because nonparametric tests do not use actual magnitudes from each observation, information is lost.

Question: Consider the following two columns of data. One is actual heights, and the second is ranks of the height values. Which column tells us more about the heights in the group?

Height (inches)	Heights (ranking)
57	5
68	4
49	6
74	1
70	2
69	3

Answer: The first column contains more information. Notice we could derive the second column from the first, but not the reverse.

If we lose, or cannot use, available information, our risk of making an error when we choose between H_0 and H_1 increases.

Question: Because we choose the significance level, α, and thus control for the chances of making a Type I error (rejection of a true null hypothesis) when we perform a test, which risk increases?

Answer: The risk or probability of a Type II error (not rejecting a false null hypothesis) increases.

This increased risk of a Type II error is the second drawback to using a nonparametric test in a situation where a parametric test would be valid.

13.2 Setting Up the Problem

The *chi-square test of independence* is a test that checks for a relationship between two variables. It is suitable for qualitative variables. As we noted earlier, we cannot measure differences among the categories of a qualitative variable. Thus, this test does not produce a measure of the effect of a change in one variable on the other, it does not demonstrate a cause and effect. It simply establishes that two variables are dependent or independent.

13.2.1 The Hypotheses

> **Hypotheses of the Chi-Square Test for Independence**
> H_0: The two variables are independent.
> H_1: The two variables are dependent.

Question: What does it mean to say that employment status and repayment behavior are independent? dependent?

Answer: Two variables are independent if the value of one provides no information about the value of the other. In this case, knowledge of a person's employment status would not help us determine his or her repayment behavior. If these variables were dependent, then employment status and financial behavior are related.

Another way of stating these hypotheses derives from the formal definition of independence of events A and B, A and B are independent if $P(A|B) = P(A)$ and $P(B|A) = P(B)$. For instance, if the variables in the example are independent, then

$$P(\text{default}) = P(\text{default} | \text{unemployed}).$$
$$= P(\text{default} | \text{part-time}).$$
$$= P(\text{default} | \text{blue collar}).$$
$$= P(\text{default} | \text{white collar}).$$
$$= P(\text{default} | \text{self-employed}).$$

We could write similar statements for the other repayment behaviors or for the employment status conditional on the repayment behavior. Together, these statements mean that a person's employment status does not affect the probability that the person repays the loan and that repayment behavior indicates nothing about an individual's employment status. The test result tells us something about whether the variables are associated, but it does not measure how strongly or in what direction they are related, nor does it imply that one variable *causes* the other variable to attain a particular value.

Question: Suppose we do a chi-square test for the two variables, response to different types of commercials and sex of respondent. What does a decision to reject H_0 say about the relation of the two?

Answer: The variables are dependent: Females respond differently from males to different types of commercials.

13.2.2 Expectations When H_0 Is True

After we set up the hypotheses, the next step is to obtain data for the two variables. We select a random sample and assemble the data into a *contingency table,* an arrangement that displays the distribution of sample values relative to two different variables that may be dependent, or *contingent,* on each other. The table for the loan application situation (p. 620) is a contingency table. Each cell (intersection of a column and row) in the table contains a tally of items that fall in a particular class (Due Date or Before, During Grace Period, Late, or Default versus Unemployed, Part-time, Minimum wage, Blue collar, White collar, or Self-employed) of each of the two variables (Employment and Repayment Behavior) simultaneously. We call this tally the *observed frequency.* In the loan applicant table there are 36 individuals in the cell that corresponds to being employed part-time and defaulting on a loan.

To make the presentation easier, we let

E = employment status.
B = repayment behavior.
f_o = observed frequency of a cell.

Section 13.2 Setting Up the Problem

If we randomly select an applicant from the loan applicant sample, we can use the equiprobable outcomes rule from Chapter 4 to obtain the probability that the person is self-employed, $P(E = \text{self-employed}) = 254/1{,}279 = 0.1986$. Similarly, the probability the person is in default is $P(B = \text{default}) = 242/1{,}279 = 0.1892$.

Question: If we assume that H_0 is true, then E and B are independent outcomes. Assume that H_0 is true and find $P(E = \text{self-employed and } B = \text{default})$.

Answer: $P(E = \text{self-employed and } B = \text{default}) = P(E = \text{self-employed}) \times P(B = \text{default}) = 0.1986 \times 0.1892 = 0.0376$ (see Rule 2" from Chapter 4, p. 178).

If the variables are independent (H_0 is true) and there are 1,279 individuals in the sample, approximately 0.0376 (3.76%) of the 1,279 individuals, or 48.1 individuals, should lie in this cell.

Example 13-1: Assume that E and B are independent (H_0 is true), and find the expected number of individuals who would fall in the $E = $ blue collar and $B = $ pay by due date cell in the loan applicant contingency table.

Solution: Let n be the number of observations in the sample that are arranged in the contingency table. In this case, $n = 1{,}279$. Then,

$$\text{Expected number of individuals} = P(E = \text{blue collar}) \times P(B = \text{due date}) \times n$$
$$= \frac{163}{1{,}279} \times \frac{467}{1{,}279} \times 1{,}279.$$

Note that we can cancel values in the preceding expression:

$$= \frac{163}{1{,}279} \times \frac{467}{\cancel{1{,}279}} \times \cancel{1{,}279} = \frac{163 \times 467}{1{,}279} = 59.5$$

If we describe the information represented by the values in the last expression in the solution, we obtain

$$\text{Expected number of individuals} = \frac{\left[\begin{array}{c}\text{Number of blue}\\\text{collar in sample}\end{array}\right] \times \left[\begin{array}{c}\text{Number who pay by}\\\text{due date in sample}\end{array}\right]}{\text{Total number of individuals in sample}}$$

$$= \frac{\left[\begin{array}{c}\text{Row total}\\\text{for blue collar}\end{array}\right] \times \left[\begin{array}{c}\text{Column total}\\\text{for due date}\end{array}\right]}{\text{Total number of individuals in sample}}.$$

In like manner, we could have obtained the expected number of individuals for the self-employed and default cell

$$\frac{\left[\begin{array}{c}\text{Row total for}\\\text{self-employed}\end{array}\right] \times \left[\begin{array}{c}\text{Column total}\\\text{for default}\end{array}\right]}{\text{Total number of individuals in sample}} = \frac{254 \times 242}{1{,}279} = 48.1.$$

We can continue to find the expected number of individuals for each cell in the same manner. We call each result the *expected frequency,* symbolized by f_e, because we expect this number of observations to fall in the cell if H_0 is true.

To Find the Expected Frequency for a Cell, f_e
1. Find the total number of observations in the row containing the cell.
2. Find the total number of observations in the column containing the cell.
3. Multiply the values from Steps 2 and 3, then divide this product by the total number of sample observations, n.

Example 13-1 (continued): Find the expected frequencies for each cell in the table of applicant data.

Solution: The following table shades the expected frequency for each cell below the observed frequency. Note the difference between the observed and expected frequencies in each cell.

Employment	Repayment Behavior				Total
	Due Date or Before	During Grace Period	Late	Default	
Unemployed	10	25	36	54	125
	45.6411	36.0633	19.6443	23.6513	
Part-time	86	80	34	36	236
	86.1704	68.0876	37.0884	44.6536	
Minimum wage	23	55	43	23	144
	52.5786	41.5450	22.6302	27.2463	
Blue collar	96	23	29	15	163
	59.5160	47.0266	25.6161	30.8413	
White collar	165	98	33	61	357
	130.3511	102.9969	56.1040	67.5481	
Self-employed	87	88	26	53	254
	92.7428	73.2807	39.9171	48.0594	
Total	467	369	201	242	1,279

For a valid result, the chi-square test for independence requires that f_e be at least 5 for each cell. Sometimes we allow a situation to violate this condition to a small degree. For instance, some analysts proceed with the test if no more than 20% of the cells violate the requirement. But we obtain more reliable results if we do not violate this requirement. Frequently, we solve the problem by combining classes, as long as the new groupings are reasonable. For instance, if several expected frequencies for "During Grace Period" and "Late" were smaller than 5, we could combine these categories into "After Due Date" to increase the expected frequencies.

Section 13.2 Problems

6. Medical researchers collected the following data on gallstone disease to determine the prevalence of this disease among elderly people. Fill in the expected frequency for each cell of the following table. Do all of the expected frequencies exceed 5? If not, suggest reasonable combinations of classes to overcome this difficulty if possible. (Source: Jack Ratner, Andre Lisbona, Marvin Rosenbloom, Max Palayew, Sharon Szabolcsi, and Tessie Tupaz, "The Prevalence of Gallstone Disease in Very Old Institutionalized Persons," *Journal of the American Medical Association*, Vol. 265, No. 7, February 20, 1991, pp. 902–03.)

Gallstone Disease Diagnosis	Age			Total
	50–79	80–89	90 or more	
Positive	17	31	24	72
Negative	16	23	6	45
Total	33	54	30	117

Section 13.3 The Test

7. A climatologist believes that changes in mean temperature of a locality are useful for predicting local population changes. Specifically, she contends that a change in the mean low temperature for the present month compared with the same month last year affects the birth rate nine months from now. To test this hypothesis, she randomly selects 174 international sites from those with available temperature and birth information. She records whether the change in the mean low temperature from one year earlier was down 10 to 30°F, down 0.1 to 9.9°, up 0.0 to 9.9°, or up 10 to 30° along with data that indicate the change in the number of births 9 months later this year over the number of births in the same month of the preceding year. If the current month is January, we would find the change in mean low temperature from the preceding January and the change in births from last October to the coming October. Fill in the expected frequencies for the following contingency table. Do all of the expected frequencies exceed 5? If not, suggest reasonable combinations of classes to overcome this difficulty if possible.

Birth Change	Mean Low Temperature Change (degrees Fahrenheit)				Total
	Down 10–30	Down 0.1–9.9	Up 0.0–9.9	Up 10–30	
Increase	5	12	10	36	63
Decrease	10	10	43	48	111
Total	15	22	53	84	174

8. In an examination of factors affecting the use of research by Congress, staffers identified some useful studies and rated the usefulness of these studies (0 = don't remember, 1 = not at all, ..., and 5 = used extensively). The General Accounting Office (GAO) was responsible for many of the studies. The following contingency table is based on the responses of 100 staffers. We assume that a study whose usage cannot be recalled must have been of little consequence, so we can group the 0 through 2 responses into a single class labeled "of little or no use" in order to avoid expected frequencies below 5. Fill in the expected frequencies. Do all of the expected frequencies exceed 5? If not, suggest reasonable combinations of classes to overcome this difficulty, if possible. (Source: John F. Boyer and Laura I. Langbein, "Factors Influencing the Use of Health Evaluation Research in Congress," *Evaluation Review*, Vol. 15. No. 5, 1991, pp. 507–32.)

Study's Source	Congressional Usage (0–5 Response)				Total
	0–2	3	4	5	
GAO	8	30	18	11	67
Non-GAO	8	9	10	6	33
Total	16	39	28	17	100

9. Forensic scientists collected data to study the presence of benzoylecgonine (BE), a primary by-product of cocaine metabolism, externally on an individual's hair rather than use an internal detection method. Fill in the expected frequencies of the following contingency table so that we can use it to determine if the identification of BE depends on hair texture. The table is based on the researchers' data on 48 prisoners. Two chemical analyses of each sample of the prisoner's hair detected the presence of BE above a cutoff value (1.25 nanograms per sample) for 22 prisoners. Do all of the expected frequencies exceed 5? If not, suggest reasonable combinations of classes to overcome this difficulty if possible. (Source: Scott A. Reuschel and Frederick P. Smith, "Benzoylecgonine (Cocaine Metabolite) Detection in Hair Samples of Jail Detainees Using Radioimmunoassay (RIA) and Gas Chromatography/Mass Spectrometry (GC/MS)," *Journal of Forensic Sciences*, Vol. 36, No. 4, July 1991, pp. 1179–85.)

Presence of BE	Hair Texture			Total
	Very Curly	Curly	Straight	
Yes	5	1	16	22
No	3	6	17	26
Total	8	7	33	48

13.3 The Test

This section presents the test statistic, the sample value that we will use to make the decision to reject or not reject H_0, and the sampling distribution of this statistic. Armed with this information, we can make a decision about the independence of the two variables.

13.3.1 The Chi-Square Test Statistic

There are discrepancies between the observed and expected frequencies in the table of f_o and f_e values for the loan applicant example. Remember, we calculate the f_e values after assuming H_0 is true. The differences in the table can arise from sampling error or from H_0 being false. If the differences are generally small, we attribute them to sampling error; that is, we assume the differences would disappear if we had the population instead of a sample. Large discrepancies are unlikely if the variables are truly independent (H_0 true).

To decide if the discrepancies are large or small overall, we need a statistic that takes the preceding information into account. That statistic is the chi-square statistic, symbolized with χ^2.

Chi-Square Test Statistic

$$\chi^2 = \sum_{\substack{\text{all} \\ \text{cells}}} \frac{(f_o - f_e)^2}{f_e}$$

We find the weighted squared difference between the expected and observed frequencies for each cell and then sum the cell results. The weights are inversely proportional to the size of the expected frequency, that is, a difference of 10 counts more if the expected frequency is 20 than it does if the expected frequency is 100. When there is a large discrepancy between f_o and f_e, the statistic tends to be large, indicating that H_0 is false.

The value for the first cell in the loan applicant data (p. 622) is

$$\frac{(10 - 45.64)^2}{45.64} = 27.83.$$

If we perform this operation for each cell of the loan applicant contingency table, we obtain $\chi^2 = 199.3$. The value that each cell contributes to the test statistic is shaded in the contingency table on page 627 along with the observed and expected frequencies in the same positions as before. Their sum, the value of the test statistic, $\chi^2 = 199.3$, is listed in the last line of the table.

To know if this value, 199.3, is unusually large, we need to know the sampling distribution of the test statistic. Then we can determine which values of χ^2 are likely to occur when H_0 is true and which are not. Then we reject H_0 if the value of the test statistic that we obtain from our sample is unlikely to occur when the null hypothesis is true. This test statistic's sampling distribution is approximately a chi-square probability distribution. We must learn about this new probability distribution before we continue.

Section 13.3.1 Problems

10. Find the value of the chi-square statistic for the contingency table for the gallstone data in Problem 6 (p. 624).

11. Find the value of the chi-square statistic for the contingency table for the temperature-birth data in Problem 7 (p. 625).

12. Find the value of the chi-square statistic for the contingency table for the congressional research usage data in Problem 8 (p. 625).

Employment	Repayment Behavior				Total
	Due Date or Before	During Grace Period	Late	Default	
Unemployed	10	25	36	54	125
	45.6411	36.0633	19.6443	23.6513	
	27.8321	3.3940	13.6177	38.9427	
Part-time	86	80	34	36	236
	86.1704	68.0876	37.0884	44.6536	
	0.0003	2.0842	0.2572	1.6770	
Minimum wage	23	55	43	23	144
	52.5786	41.5450	22.6302	27.2463	
	16.6397	4.3576	18.3352	0.6618	
Blue collar	96	23	29	15	163
	59.5160	47.0266	25.6161	30.8413	
	22.3651	12.2755	0.4470	8.1367	
White collar	165	98	33	61	357
	130.3511	102.9969	56.1040	67.5481	
	9.2101	0.2424	9.5144	0.6348	
Self-employed	87	88	26	53	254
	92.7428	73.2807	39.9171	48.0594	
	0.3556	2.9566	4.8522	0.5079	
Total	467	369	201	242	1279
$\chi^2 = 199.2978$					

13.3.2 The Chi-Square Distribution

The chi-square random variable is continuous, and only positive values are possible. Figure 13-1 displays its general shape, which is skewed to the right, although the exact shape changes as the degrees of freedom change. To locate outcomes and areas on the diagram, we use a chi-square table (see Appendix N). A portion of the table is shown.

Degrees of Freedom	Probability χ^2 Exceeds Critical Values Listed				
	0.1	0.05	0.025	0.01	0.005
1	2.71	3.84	5.02	6.63	7.88
2	4.61	5.99	7.38	9.21	10.6
3	6.25	7.81	9.35	11.3	12.8
.	.	.	.	.	.
.	.	.	.	.	.
.	.	.	.	.	.
11	17.3	19.7	21.9	24.7	26.8
12	18.5	21.0	23.3	26.2	28.3
13	19.8	22.4	24.7	27.7	29.8
14	21.1	23.7	26.1	29.1	31.3
15	22.3	25.0	27.5	30.6	32.8

The first step in using the chi-square table is to determine the degrees of freedom. For this test, *the statistic has $(R - 1) \times (C - 1)$ degrees of freedom, where R is the number of classes of the variable listed as rows in the left column of the contingency table and C is the number of classes of the variable listed as columns of the table.* There are

Figure 13-1
The general shape of the chi-square probability distribution

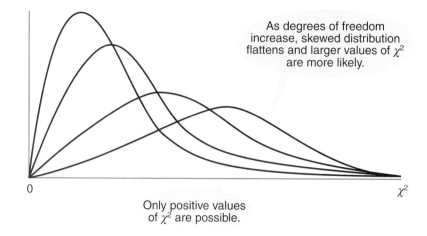

Figure 13-2
Diagrams that summarize specific information from the chi-square table

a) Diagram for $P(\chi^2_{(15)} > 22.3)$

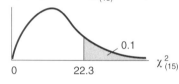

b) Diagram for $P(\chi^2_{(13)} > 24.7)$

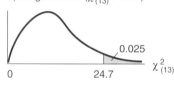

c) Diagram for $P(\chi^2_{(2)} > 10.6) = 0.005$

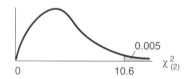

$(R - 1)(C - 1)$ degrees of freedom, because in any table with R rows and C columns we can freely specify cell values in $R - 1$ rows and $C - 1$ columns. The last row and column must contain appropriate values so that row and column totals sum to the given values. Thus, we can freely choose values for a table with dimensions $(R - 1)$ by $(C - 1)$ or $(R - 1)(C - 1)$ cells. For the loan applicant case, there are $24 = 6(4)$ cells in the table, but only $15 = (6 - 1)(4 - 1)$ degrees of freedom. So we locate 15 degrees of freedom in the degrees-of-freedom (first) column of the table.

Figure 13-2 shows the probabilities, or areas, and outcomes that the chi-square table provides. We locate outcome values across each degrees-of-freedom row. At the top of each column is the area in the tail of the curve. The smallest outcome (chi-square value) under this tail area is listed in the intersection of the area column and the degrees-of-freedom row. Find the value 22.3 in the 15-degrees-of-freedom row. At the top of this column is the value 0.1. Thus, the area under the curve to the right of 22.3 is 0.1. We can write this statement as $P(\chi^2_{(15)} > 22.3) = 0.1$, where the subscript of χ denotes the degrees of freedom. (See Figure 13-2a.)

Question: Find $P(\chi^2_{(13)} > 24.7)$.
Answer: We find 24.7 in the 13 degrees-of-freedom row. At the top of the column that contains 24.7 is the value 0.025, so $P(\chi^2_{(13)} > 24.7) = 0.025$. See Figure 13-2b.

Question: Find c so that $P(\chi^2_{(2)} > c) = 0.005$.
Answer: $c = 10.6$ (intersection of 2 degrees-of-freedom row and 0.005 column). See Figure 13-2c.

With the information in this table we can detect when test statistic values are unusually large when the variables are independent.

Section 13.3.2 Problems

13. If a contingency table has five classes of one variable and seven of the other, how many degrees of freedom will the test statistic have?

14. Find the value c, such that the probability that the value of a chi-square random with 22 degrees of freedom will exceed c is 0.05.

15. Find the value c, such that the probability that the value of a chi-square random with 18 degrees of freedom will exceed c is 0.01.

16. Find c, such that $P(\chi^2_{(8)} > c) = 0.025$.

13.3.3 Making the Decision

Now we are ready to make a decision. Remember, the test statistic reflects the total amount of discrepancy in the sample between what we expect to observe if the variables are independent and what we actually observe. A large test statistic indicates that the variables may be dependent rather than independent. The critical value marks the smallest value in the rejection region, chi-square values that are unusually large for two variables that are independent (that is, when H_0 is true). So we reject H_0 if the test statistic exceeds the critical value.

■ **Example 13-1 (continued):** The sociologist from the loan applicant situation is willing to risk a Type I error 5% of the time when she decides if job type and repayment behavior are independent. Find the critical value. Recall that there are six categories of employment and four for repayment behavior. The chi-square test statistic is 199.3. Make and interpret a decision.

■ **Solution:** The degrees of freedom are $(R-1)(C-1)$ or $(6-1)(4-1) = 15$ for this situation. From the chi-square table, we find the critical value 25.0 for a significance level of 0.05 (the area in the tail is 0.05), as displayed in Figure 13-3.

Only 5% of the time should we obtain a test statistic value larger than 25.0 if these variables are truly independent. A value that falls in the rejection region indicates that there is sufficient evidence to attribute the discrepancies between f_o and f_e values to H_0 being false. That is, we decide the variables are, indeed, related or dependent.

Because the test statistic value we derived from the loan applicant contingency table is 199.3, which is larger than 25.0 (as shown in Figure 13-3), we reject H_0. The data support H_1, the hypothesis that job type and repayment behavior are related. The sociologist may state this decision in terms that the general public can understand as "a bank's loan officer could properly use the employment status of an applicant in making loan decisions in order to reduce the default rate."

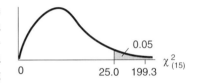

Figure 13-3
Diagram for chi-square test for loan applicant table

Question: If the chi-square test statistic is 5.622 from a 5 × 5 table, can we reject H_0 at the 0.01 level?

Answer: No. The critical value is 32.0 with a 0.01 significance level and $(5-1)(5-1) = 16$ degrees of freedom. Because 5.622 is smaller than 32.0, we cannot reject H_0.

The following box outlines the steps for performing the chi-square test for independence.

Procedure for Chi-Square Test for Independence of Two Variables

1. Obtain a random sample of n items from the population of interest.
2. Form a contingency table consisting of $R \times C$ cells that classify and tally sample items to report f_o, the observed frequency of items that exhibit the same qualities or values of both variables simultaneously. R and C symbolize the number of rows and columns, respectively, or the number of categories of the two variables.
3. Find the expected frequency for each cell using the total observations in the row and column that intersect at the specific cell

$$f_e = \frac{\text{(Row total)}\,\text{(Column total)}}{n}.$$

(continued on next page)

Procedure for Chi-Square Test for Independence of Two Variables (continued)

4. Find the value of the test statistic

$$\chi^2 = \sum_{\substack{\text{all} \\ \text{cells}}} \frac{(f_o - f_e)^2}{f_e}$$

5. Find the critical value from the chi-square table that has $(R - 1)(C - 1)$ degrees of freedom and an area equal to the significance level in the tail.
6. Reject H_0 if the test statistic is larger than the critical value, and conclude that the two variables are dependent.

We can use p-values, as well, to make decisions. The p-value in this case indicates the probability of obtaining a test statistic larger than the value we actually obtain from our contingency table. The p-value for the 199.3 is approximately 0, which is, of course, smaller than 0.05, so we would reject H_0.

Example 13-2: Owners of a tourist attraction ask guests to rate their experience as poor, average, or good. The owners note the date of the visit on the rating sheet. They randomly select 1,000 evaluations and classify the data according to rating and time of visit (weekend or weekday). If time of visit affects the rating, the owners intend to adopt special programs for the poorly rated days. Use the following data to conduct a chi-square test of independence to determine if the sample evidence indicates that the variables are related. Use a 0.025 significance level, and interpret the decision for the owners.

	Rating			
Time	Poor	Average	Good	Total
Weekday	98	256	188	542
Weekend	43	202	213	458
Total	141	458	401	1,000

Solution: Step 1: The owners have already chosen a random sample of 1,000 evaluations.
Step 2: The contingency table lists the observed frequencies.
Step 3: The expected frequencies for each cell are:

Cell at			
Row	Column	Calculation	f_e
1	1	542(141)/1,000	76.4220
1	2	542(458)/1,000	248.2360
1	3	542(401)/1,000	217.3420
2	1	458(141)/1,000	64.5780
2	2	458(458)/1,000	209.7640
2	3	458(401)/1,000	183.6580

Step 4: The following contingency table lists both the observed and expected frequencies as well as each cell's contribution to the test statistic, $(f_o - f_e)^2/f_e$ in the same positions as before. The

chi-square test statistic in the last line of the table, 22.5, is the sum of the cell contributions.

Time	Rating			Total
	Poor	Average	Good	
Weekday	98	256	188	542
	76.4220	248.2360	217.3420	
	6.0926	0.2428	3.9613	
Weekend	43	202	213	458
	64.5780	209.7640	183.6580	
	7.2100	0.2874	4.6878	
Total	141	458	401	1,000
$\chi^2 = 22.4819$				

Step 5: The critical value for a significance level, α, of 0.025 with $(R - 1)(C - 1) = (2 - 1)(3 - 1) = 2$ degrees of freedom is 7.38.

Step 6: Because $22.5 > 7.38$, we reject H_0, which means that we should advise the owners that time of visits and rating are related. Their plan for special programs to improve the rating during the week may be in order.

Note: When each variable has only two classes, the resulting table produces a test statistic with only one degree of freedom, $(2 - 1)(2 - 1) = 1$. In such a case, we must correct the test statistic in order to improve the chi-square approximation or use a test of difference in two proportions. This book does not cover these modifications.

Section 13.3.3 Problems

17. Recall the medical researchers from Problem 6 that collected data on gallstone disease to determine if there is a relationship between elderly people's ages and having gallstone disease. The contingency table consists of three classes of age and two for having or not having the disease. The chi-square test statistic is 6.1 from the table. Make and interpret a decision at the 0.05 significance level.

18. The climatologist from Problem 7 believes that changes in mean temperature of a locality are useful for predicting local population changes. From a contingency table with four classes for temperature changes and two for a change in birth rates, she obtains the chi-square test statistic value, 11.8. Make and interpret a decision at the 0.005 significance level.

19. Problem 8 detailed an examination of factors affecting the use of GAO research and non-GAO research by Congress, based on staffers' rating of different research studies among four categories (0 through 2 = of little or no use, ..., 5 = used extensively). The chi-square test statistic value is 4.0. Use this information to make and interpret a decision at the 0.01 significance level.

20. In a study of the personality characteristics of first-time and repeat offenders arrested for driving while intoxicated (DWI), researchers classified participants according to personality types based on responses to a personality style questionnaire and according to their DWI record, as shown in the following contingency table. Verify their finding that "the distribution of participants by the four personality types for first-time offenders was significantly different from the distribution of the repeat offenders." Use a 0.01 significance level. (Source: James R. Reynolds, Joseph T. Kunch, and Corrine S. Cope, "Personality Differences of First-Time and Repeat Offenders Arrested for Driving while Intoxicated," *Journal of Counseling Psychology*, Vol. 38, No. 3, 1991, pp. 289–95.)

	Personality Type				
	Stability-oriented		Change-oriented		
Record	Extravert	Introvert	Extravert	Introvert	Total
First-time	68	34	49	21	172
Repeat	34	23	4	3	64
Total	102	57	53	24	236

Figure 13-4

Minitab output for loan applicant situation

```
          Expected counts are printed below observed counts
           EARLY    ON TIME     LATE   DEFAULT    Total
  1  1     10         25         36        54      125
     2  45.64      36.06      19.64     23.65

  2        86         80         34        36      236
         86.17      68.09      37.09     44.65

  3        23         55         43        23      144
         52.58      41.54      22.63     27.25

  4        96         23         29        15      163
         59.52      47.03      25.62     30.84

  5       165         98         33        61      357
        130.35     103.00      56.10     67.55

  6        87         88         26        53      254
         92.74      73.28      39.92     48.06

  Total   467        369        201       242     1279
                                                3
  ChiSq =  27.832  +  3.394  + 13.618  + 38.943  +
            0.000  +  2.084  +  0.257  +  1.677  +
           16.640  +  4.358  + 18.335  +  0.662  +
           22.365  + 12.276  +  0.447  +  8.137  +
            9.210  +  0.242  +  9.514  +  0.635  +      4
       5    0.356  +  2.957  +  4.852  +  0.508  = 199.298
  df = 15
```

13.4 The Chi-Square Test and the Computer

As the number of categories of the two variables increases, manual computation of the chi-square statistic becomes more cumbersome. In such cases, we often resort to computer calculations. Figure 13-4 shows Minitab output for the loan applicant situation. The circled values correspond to the following:

1. Observed cell frequency, f_o for the first cell.
2. Expected cell frequency, f_e for the first cell.
3. Cell contribution to test statistic, $(f_o - f_e)^2/f_e$ for the first cell.
4. Test statistic for table, $\chi^2 = \sum_{\text{all cells}} [(f_o - f_e)^2/f_e]$.
5. Degrees of freedom for the table.

Question: Can you judge the contribution of a cell to the test statistic simply by observing its f_o or f_e? Demonstrate your answer with the output in Figure 13-4.

Answer: f_o and f_e are not always guides for gauging the cell contribution. For instance, f_o ranges from 10 to 165 and f_e from 19.6 to 130.4, yet the smallest contribution is 0.00 from the part-time (level 2) and early combination, which has 86 and 86.2 as the observed and expected frequencies, respectively. Neither of these is an extreme. The largest cell contribution, from the unemployed (level 1) and default combination, similarly does not have an extreme f_o. However, its f_e is one of the smaller values that, in turn, produces a large weight for the squared discrepancy from this particular cell.

Section 13.4 Problems

21. Verify your test statistic from Problems 8 and 12 with the computer output of the analysis in Figure 13-5. Level 1 is GAO studies, and level 2 is non-GAO studies.

22. Verify your test statistic and decision in Problem 20 with the computer output of the analysis in Figure 13-6. Level 1 is first-time DWI offenders, and level 2 is repeat offenders.

Figure 13-5
Minitab output for congressional research usage contingency table

```
Expected counts are printed below observed counts

        USAGE0-2   USAGE3    USAGE4    USAGE5    Total
  1        8         30        18        11       67
            10.72     26.13     18.76     11.39

  2        8          9        10         6       33
             5.28     12.87      9.24      5.61

Total     16         39        28        17      100

ChiSq =   0.690  +  0.573  +  0.031  +  0.013  +
          1.401  +  1.164  +  0.063  +  0.027  =  3.962
df = 3
```

Figure 13-6
Minitab output for DWI contingency table

```
Expected counts are printed below observed counts

         S-O/E     S-O/I     C-O/E     C-O/I     Total
  1        68        34        49        21       172
            74.34     41.54     38.63     17.49

  2        34        23         4         3        64
            27.66     15.46     14.37      6.51

Total    102        57        53        24       236

ChiSq =   0.541  +  1.369  +  2.786  +  0.704  +
          1.453  +  3.680  +  7.486  +  1.891  = 19.909
df = 3
```

Figure 13-7
Minitab output for chemists and girl babies contingency table

```
Expected counts are printed below observed counts

         SCIENCE   WHOSWHO   DATABASE   Total
  1       158       147        72        377
           162.35    144.92     69.72

  2       168       144        68        380
           163.65    146.08     70.28

Total    326       291       140        757

ChiSq =   0.117 +  0.030  +  0.074  +
          0.116 +  0.030  +  0.074  =  0.440
df = 2
```

23. Folklore among chemists suggests that chemists tend to have more girl babies. Use the Minitab output in Figure 13-7 from three different surveys, a nonscientific *Science* magazine survey, a random search of chemists from the 1985 edition of *Who's Who in Frontiers of Science and Technology*, and *Who's Who* updated data base, to determine if there is a significant relationship between survey, the source of the findings, and rate of female children among chemists (level 1 is female children, and level 2 is male). Test at the 0.05 level. Interpret the decision. (Source: Ivan Amato, "Are Chemists Girl Crazy?" *Science*, Vol. 257, July 10, 1992, pp. 158–59.)

24. Suppose a random sample of 2,985 females in Europe yields the contingency table and statistics shown in Figure 13-8. The sampled females fell into one of three work categories and lived in one of five countries. Use this information to determine if the distribution of female work patterns is different in the five countries represented in the samples. Test at the 0.01 significance level, and interpret the decision. Each level represents a different country.

Figure 13-8
Minitab output for female work distribution contingency table

```
         Expected counts are printed below observed counts

          WHIT-COL  BLUE-COL  HOMEMAKR    Total
     1        127       254       264      645
            95.72    233.37    315.91

     2         84       177       254      515
            76.43    186.33    252.24

     3         91       210       304      605
            89.79    218.89    296.32

     4         66       245       225      536
            79.55    193.93    262.52

     5         75       194       415      684
           101.51    247.48    335.01

 Total       443      1080      1462     2985

 ChiSq =  10.219 +   1.824 +   8.530 +
           0.750 +   0.467 +   0.012 +
           0.016 +   0.361 +   0.199 +
           2.307 +  13.449 +   5.363 +
           6.924 +  11.556 +  19.099  =  81.077
 df = 8
```

Summary and Review

Parametric hypothesis tests require assumptions about population distributions or parameters. Nonparametric tests provide alternative tests for situations that do not satisfy the assumptions for parametric tests or that involve qualitative variables. Nonparametric hypotheses are usually not the same as those for parametric tests, but they are similar. We use them to obtain similar information or a conclusion that implies a parametric hypothesis. When their assumptions hold, statisticians prefer parametric tests, because nonparametric tests employ less detailed information and, consequently, increase the risk of making a Type II error.

We use the chi-square test for independence to determine if two variables are dependent. The test is especially useful for qualitative variables, but we can use it with quantitative variables as well when we group values of similar magnitudes.

The hypotheses of the test are

H_0: The variables are independent.
H_1: The variables are dependent.

If H_0 is true, there is no relation between the variables. H_1 implies that there is a relation, so that if we know the value of one of the variables, we are better able to determine or predict the value of the other variable.

The test statistic is $\chi^2 = \sum_{\text{all cells}} [(f_o - f_e)^2/f_e]$. Its sampling distribution is a chi-square probability distribution with $(R-1)(C-1)$ degrees of freedom, where R and C are the number of categories of each variable in the contingency table. To obtain a reliable result from the test, the expected frequency, f_e, should be greater than 5 in each cell. If this is not true, we try to combine groups into meaningful classes to satisfy this condition. Also, a corrected test statistic is necessary for a table with only two categories for both of the two variables exhibited in the contingency table.

Multiple-Choice Problems

25. A qualitative variable is one
   a. Whose value indicates a category or quality, rather than an amount of something.
   b. That can only assume positive integer values.
   c. Whose magnitude measures the exact proportion of some quality compared to another level of the same quality.
   d. That is a better measure of a quantity when comparing two measures.
   e. Whose magnitude has no meaning.

26. The chi-square test is used to determine
   a. If the means of two variables differ.
   b. If two variables are dependent.

Multiple-Choice Problems

c. If two qualitative variables have the same number of categories.
d. The best way to quantify two different qualitative variables.
e. If the squared value of a variable is significantly different from zero.

27. What is H_0 in the chi-square test?
    a. $\mu_1 = \mu_2$.
    b. Two variables are independent.
    c. The parametric assumptions are satisfied.
    d. μ = the same value as before the circumstances changed.
    e. $b = 0$.

28. If we reject H_0 for the chi-square test, we say that the two variables are dependent. Suppose the two variables are brand and sales of an appliance and we reject H_0. Interpret this result.
    a. The standard deviations of sales and brands are approximately the same, because the test indicates identical sums of squares.
    b. The increasing sales causes the retailer to offer more expensive brands for sale.
    c. Sales and brands appear to be related, but direction of causation is not indicated by the chi-square test.
    d. The more brands that are offered, the more sales will increase.
    e. The two variables are not related, so knowledge of the brand of the appliance is not helpful for predicting sales of the appliance.

29. Which of the following pairs of variables requires the chi-square test to check for a relationship, because it includes a qualitative variable?
    a. Wage rate and years of education.
    b. Income and profession.
    c. Attendance and scores at a sporting event.
    d. Age and retirement savings.
    e. Number of parties during a school term and grades.

Use the following information to answer Questions 30–35.

An arts and crafts retailer is considering locating a new store in a different town. Before doing so, an employee gathers information to determine if altering the setting of this type of store influences profits. A random sample of stores of this type yields the information in the following table. "Small" means profits are less than $20,000, and "large" means profits are at least $20,000.

Setting	Profit				
	Loss	Break Even	Small	Large	Total
Downtown	1	0	24	31	56
Mall	21	0	34	26	81
Farmhouse	8	0	29	18	55
Total	30	0	87	75	192

30. The first step to perform a chi-square test for this situation is to
    a. Calculate the test statistic.
    b. Incorporate the $20,000 into the values in the table.
    c. Find exact values for the profits.
    d. Make the correction for small degrees of freedom.
    e. Omit the break-even column, since the f_e requirement is not met.

Use the corrected version of the preceding table (omit break-even column) to answer Questions 31–35.

31. What is f_e for the first cell?
    a. 8.75.   d. 322,560.
    b. 0.11.   e. 30.
    c. 1,680.

32. The degrees of freedom for the test are
    a. 3.
    b. 9.
    c. 4.
    d. 6.
    e. 192.

33. What is the critical value for a test at the 0.05 significance level?
    a. 7.81.   d. 9.49.
    b. 12.6.   e. 11.1.
    c. 1.96.

34. The value of the test statistic is 18.72. What is the decision with regard to H_0 at the 0.05 significance level?
    a. Cannot reject, because $0.5(18.72) < 9.49$.
    b. Cannot reject, because $18.72 > 9.49$.
    c. Reject, because $0.5(18.72) < 9.49$.
    d. Reject, because $18.72 > 9.49$.
    e. Not enough information to make a decision.

35. Interpret the results of the test in Question 34.
    a. The probability of making a large profit does not depend on the setting of the store.
    b. The setting of the store does not depend on the profit level of the store.
    c. Knowledge of the setting of the store helps us determine the profit position of the store.
    d. A farmhouse store's profits differ from a mall store's profits, but the location is not the determining factor.
    e. To increase the profits by one unit, the store must move 18.72 blocks closer to population centers.

36. A cell in a contingency table for type of specialty shop and age group of most patrons has a row total of 23, a column total of 23, and there are 3,673 shops in the sample. Which of the following statements is true for this case?
    a. Because the row and column totals are equal, the variables must be independent.
    b. Because $n > 30$, the test is valid.
    c. The observed frequency is 0.14.
    d. The expected frequency is smaller than 5, so the table should be rearranged if possible or the test not done if a sizable portion of the cells have small expected frequencies.
    e. The numbers must be incorrect, since $R \times C$ can never be greater than n.

37. An automobile association conducted a survey of 183 new car buyers to determine the type of car purchased and the location chosen for most maintenance services. The row total for a particular cell is 56, the column total is

Figure 13-9

Minitab output of contingency table for number of children versus activity

```
Expected counts are printed below observed counts

              CHILD0    CHILD1    CHILD2    CHILD3    CHILD4UP    Total
     1          19         9        15        13         15         71
              18.36      6.37     14.61     15.55      16.11

     2          19         4        15        17         16         71
              18.36      6.37     14.61     15.55      16.11

     3          60        21        48        53         55        237
              61.28     21.26     48.78     51.90      53.78

   Total        98        34        78        83         86        379

   ChiSq =   0.022  +   1.086  +   0.010  +   0.418  +   0.077  +
             0.022  +   0.881  +   0.010  +   0.135  +   0.001  +
             0.027  +   0.003  +   0.012  +   0.023  +   0.028  = 2.757

   df = 8
```

18, and the cell's observed frequency is 10. Find the value that this cell contributes to the test statistic.
  a. 5.51.   c. 0.66.   e. 3.66.
  b. 2.02.   d. 0.82.

38. Nonparametric tests often replace parametric tests when
  a. The test is for two variables.
  b. The data is qualitative.
  c. The *p*-value is missing.
  d. The population appears to be normally distributed
  e. The significance level cannot be controlled.

39. A problem with using a nonparametric test when conditions suit a parametric test is
  a. The chances of making a Type II error increase.
  b. We must assume that sample values behave a certain way that is not indicated in the actual data values.
  c. The significance level is unknown.
  d. Nonparametric tests require more detailed assumptions about the population.
  e. Not all situations suitable for parametric tests are suitable for nonparametric tests.

40. Which of the following statements about the test statistic for the chi-square test of independence is true?
  a. Large values of the statistic are likely to result in not rejecting H_0.
  b. Large values of the statistic reflect large discrepancies between the observed and expected frequencies.
  c. Large values of the statistic result from large expected frequencies.
  d. Large values of the statistic result from large observed frequencies.
  e. Large values of the statistic result from having a small table, that is, fewer categories of the two characteristics under investigation.
  f. Large values of the statistic result from having a small sample.

41. If the chi-square test statistic is 7.75, in which of the following table sizes would we decide that the two characteristics are dependent at the 0.05 level?
  a. 20 classes of one variable and 2 of the other.
  b. 10 classes of one variable and 3 of the other.
  c. 7 classes of one variable and 5 of the other.
  d. 5 classes of both variables.
  e. 3 classes of one variable and 2 of the other.

42. If we cannot reject H_0 when doing a chi-square test for a possible relationship between A = achievement (high or low) and S = self-image (high, average, or low), which of the following statements is indicated by the decision?
  a. The two characteristics have an opposing relationship, that is a high A is more likely to occur with a low B.
  b. The two characteristics are directly related, that is, a high A is more likely to occur with a high B.
  c. Neither variable offers information about the likely values of the other variables.
  d. $P(A|B) = P(B|A)$.
  e. A is not likely to occur if B occurs.

Use the following information to answer Questions 43–44.

We can use statistics on the number of children in a family and the family's activity during a park visit to examine the relationship between the two variables. Figure 13-9 shows computer output of the results for a chi-square test of independence.

43. What is the critical value and decision at the 0.05 level?
  a. $2.757 < 15.5$, cannot reject H_0.
  b. $2.757 < 25.0$, cannot reject H_0.
  c. $379 > 15.5$, reject H_0.
  d. $379 > 25.0$, reject H_0.
  e. $0.9487 < 15.5$, cannot reject H_0.

44. This decision means that
  a. Family size is useful information for directing families to certain areas of the park.
  b. The likelihood of picnicking increases as family size increases.
  c. The chances a family will ride horses are not affected by the size of the family.
  d. The chances a family will hike increase if the family picnics.
  e. Large families use fewer of the park facilities.

Word and Thought Problems

45. What is the difference between a parametric and a nonparametric test? State the advantages and disadvantages of using nonparametric rather than parametric tests.

46. Find f_e for the empty cell below.

	A_1	A_2	Total
B_1	83	42	125
B_2	16		107
B_3	25	25	50
Total	124	158	282

47. If we decide not to reject the null hypothesis when doing a chi-square test for gender and major discipline of students, what does our decision imply about gender and major? Interpret the decision using words other than "dependent" or "independent."

48. At a large antique show, demonstrators randomly selected 478 pieces of furniture to be stripped of the old finish, using one of five brands of stripping fluid. Judges rated the pieces as good, average, or poor, based on before-and-after photographs. Determine if brand of stripping fluid and quality of finish removal are independent, by performing a chi-square test at the 0.05 level. Interpret the decision for the antique show attendees.

			Brand		
Rating	A	B	C	D	E
Good	33	52	35	19	16
Average	69	54	19	18	25
Poor	23	30	25	48	12

49. Determine if athletic conference and winning percentage for games against opponents outside the conference are related, using the following data for 100 teams. Test at the 0.01 significance level. Interpret the decision.

Outside Conference Winning Percentage	A	B	C	D	E	F	G	H	I	J
Greater than 50%	9	6	8	1	0	9	7	4	2	4
50% or less	1	5	2	9	10	0	3	6	8	6

50. Determine if the specific doctor and type of treatment for a certain ailment are related, using the following data on 509 patients with this ailment. Test at the 0.01 significance level. Interpret the decision.

Treatments	Dr. Wade	Dr. Wisley	Dr. Morgan
1	16	82	63
2	87	26	71
3	65	29	70

51. Test at the 0.1 significance level for prejudice against pedigree by judges in dog shows, based on the following data for 1,008 dogs. Interpret the decision.

Previous Champs	Blue	Red	White	None
Father	63	59	89	48
Mother	95	10	92	29
Both	15	87	9	16
Neither	32	60	79	50
Other	17	57	10	91

52. Supervisors of a large factory are interested in the apparent tendency for night-shift workers to respond to the rhythm of background music. They randomly select work periods and pipe one of three music speed categories (slow, medium, and fast) into the work area. The following table below records the number of times output is above the old average value for each shift in each speed category over several months. Is there sufficient evidence to indicate that music speed and output of workers on each shift are related at the 0.05 level? Interpret the results.

	Music Speed		
Shift	Slow	Medium	Fast
Day	20	20	25
Evening	21	24	25
Night	18	21	24

53. A store owner believes types of customer purchases are related to the days of the week. Knowledge of such a relationship will help in ordering and inventorying perishables. The owner randomly selects one of each of the seven days of the week from a month and lists the number of customers who purchased each type of product in the following table, according to cash-register records. Is there sufficient evidence to indicate that types of purchases depend on the day of the week at the 0.1 level? Interpret the results.

Day	Exotic Produce	Everyday Produce	Fancy Meat	Everyday Meat	Non-perishable
Sun.	50	80	80	60	200
Mon.	0	70	50	60	150
Tues.	10	40	80	60	100
Wed.	40	120	60	70	220
Thur.	40	150	40	70	280
Fri.	80	120	100	70	250
Sat.	50	100	100	60	150

54. A convention-center bar collected evidence on dependence between type of convention and type of drink consumed. At recent conventions, the number of orders for each drink category and type of convention were recorded. Determine from the following table whether there is a relationship at the 0.05 level? Interpret the results.

Figure 13-10

Minitab output for repair visits and customer size contingency table

```
Expected counts are printed below observed counts

              0 VISITS   1 VISIT   2 VISITS   3 VISITS   4 VISITS    Total
      1          40         7          5          6         10        68
                25.35     11.75       9.27       6.18      15.45

      2          22        11         15          6         22        76
                28.33     13.13      10.36       6.91      17.27

      3          20        20         10          8         18        76
                28.33     13.13      10.36       6.91      17.27

   Total         82        38         30         20         50       220

   ChiSq =    8.473  +  1.917  +  1.969  +  0.005  +  1.925  +
              1.413  +  0.345  +  2.074  +  0.120  +  1.294  +
              2.448  +  3.598  +  0.013  +  0.172  +  0.031  = 25.797

   df = 8
```

	Drink Type		
Convention	Beer	Wine	Hard Liquor
Farmers	15	50	10
Beauticians	20	18	22
Theologians	18	50	75
Coaches	2	25	50
Mayors	25	25	25

55. Are gender of respondent and evaluation of a product related, based on the information in the following table? (Use a 0.01 significance level.) Interpret the results.

	Evaluation		
Gender	Good	Average	Poor
Male	50	200	75
Female	100	100	75

56. A large restaurant chain randomly selects a sample of 101 individuals to determine if the source of vegetables affects customer reaction to the meal. Each person randomly receives a meal including frozen, canned, or fresh vegetables. Afterwards, the customers rate the meal as good, average, or poor. Use the following results to test for independence of the variables at the 0.05 level. Interpret the results.

	Vegetable Source		
Rating	Frozen	Canned	Fresh
Good	12	10	21
Average	18	6	15
Poor	8	9	5

57. To determine if time of planting bulbs is related to bloom activity in the spring, Carrie plants 200 bulbs in random spots at four different times, as shown in the following table. Given the results listed, check for a relation at the 0.05 significance level. Interpret the decision.

Planting Time	Result in Spring	
	Bloom	No Bloom
Summer	24	16
Fall	47	12
Winter	21	20
Spring	42	18

58. Numerous recent repair visits concern a computer firm. Many of the repairs are simple for those acquainted with computer hardware in general, such as customers with more computer equipment. The firm obtains information on the number of visits for easy repairs and the size of business. They measure size approximately with the customer's expenditure on computer equipment. If there is a relationship between number of visits for simple repairs and size of customer, the computer firm may discontinue marketing to certain firms or rewrite its service agreement. Figure 13-10 shows the computer output for a chi-square test of independence.
a. What is the value of the test statistic?
b. What is the decision at the 0.01 significance level?
c. Interpret the decision.

59. To compare opinions about ethics between marketing managers of small and large firms, researchers mailed a survey to firms listed in *Fortune 500* and a random sample of small firms. Conduct chi-square tests at the 0.05 significance level for the pairs of variables and information displayed in Figure 13-11, where level 1 is small and level 2 is large. Interpret the decisions. (Source: Judith A. Wiles, Charles R. Wiles, and Peter J. Gordon, "Ethical Attitudes of Marketing Managers: A Comparison of Small Businesses and Fortune 500 Companies," *Business Insights*, Vol. 5, No. 2, 1989, pp. 34–38).

60. To check for dependencies among different variables for subjects in a study, the authors reported the following results for the following variables (*df* denotes degrees of freedom):

Word and Thought Problems

a) Education Levels versus Firm Size (HIGHSCH = high school, COL/TECH = college/technical, COLGRAD = college/graduate, GRADSCH = graduate school)

Expected counts are printed below observed counts

```
              HIGHSCH  COL/TECH  COLGRAD  GRADSCH   Total
         1        25       116      258      261     660
               23.67    106.06   254.18   276.10

         2         2         5       32       54      93
                3.33     14.94    35.82    38.90

     Total        27       121      290      315     753

    ChiSq =    0.075 +   0.932 +  0.057 +  0.825 +
               0.534 +   6.617 +  0.407 +  5.857 = 15.306
    df = 3
    1 cells with expected counts less than 5.0
```

Figure 13-11
Minitab output for contingency table for ethics data

b) Ethical Ratings for Accepting a $50 Gift from Supplier at Christmas (DEFNO = definitely no, SOMENO = somewhat no, NEITHER = neither yes nor no, YES = yes)

Expected counts are printed below observed counts

```
              DEFNO   SOMENO   NEITHER     YES    Total
         1      259      166       179      60      664
             261.39   173.68    170.17   58.77

         2       39       32        15       7       93
              36.61    24.32     23.83    8.23

     Total     298      198       194      67      757

    ChiSq =   0.022 +  0.339 +   0.459 +  0.026 +
              0.156 +  2.422 +   3.274 +  0.184 = 6.881
    df = 3
```

c) Ethical Ratings for Giving a Prospective Customer a $50 Gift (NOJUDGE = no judgment)

Expected counts are printed below observed counts

```
               WRONG  NOJUDGE   RIGHT    Total
         1       471      146      40      657
              474.55   142.98   39.47

         2        70       17       5       92
               66.45    20.02    5.53

     Total      541      163      45      749

    ChiSq =   0.027 +  0.064 +  0.007 +
              0.190 +  0.456 +  0.050 = 0.793
    df = 2
```

1. Type of work (formal versus informal, where formal implies legal or institutional business arrangements, such as owning a business or working in a store or factory, or training, while informal lacks these tendencies; some analyses also use no work).
2. Amount of work (full-time, part-time, occasional; some analyses also use no work).
3. Marital status (in a union or not).
4. Child-care strategies (take child to work, leave child in another person's care [adult or preteen], leave child in another person's home, work at home with child, some analyses use no work).

Variables Compared	df	χ test statistic	maximum p-value
Type & amount of work	2	24.67	0.000
Type of work & marital status	2	12.16	0.001
Amount of work & marital status	2	37.16	0.000
Child-care strategy & type of work	3	13.89	0.003

Source: Patrice L. Engle, "Maternal Work and Child-Care Strategies in Peri-Urban Guatemala: Nutritional Effects," *Child Development*, Vol. 62, No. 5, 1991, pp. 954–65. Copyright © 1991 by the Society for Research in Child Development.

a. Make and interpret a decision at the 0.05 level for each comparison.
b. Verify the chi-square test statistic for amount versus type of job, using the following table.

Type of Work	Amount of Work		
	Full-Time	Part-Time	Occasional
Formal	18	22	4
Informal	8	25	35

61. Using the data in the following table, test to determine if deaths from the drug TMF versus overdoses of other drugs in the Pittsburgh area vary with year. Assume that data are a random sample of drug deaths for recent years. Test at the 0.1 level and interpret the decision. Can we be sure about our decision at the 0.1 level? Explain your answer.

Drug	1986	1987	1988
TMF	0	0	15
Non-TMF	18	34	41

Source: Jonathan Hibbs, Joshua Perper, and Charles L Winek, "An Outbreak of Designer Drug-Related Deaths in Pennsylvania," *Journal of the American Medical Association*, Vol. 265, No. 8, February 1991, pp. 1011–13.

62. Test to see if stance on an issue depends on political affiliation using the following results of a random sample of 2,123 registered voters. Test at the 0.025 level and interpret the decision.

Political Affiliation	Stance		
	Pro	Con	No opinion
Independent	123	82	69
Democrat	450	401	211
Republican	397	188	202

63. Researchers investigated the existence of a relationship between levels of depression and mortality of nursing home patients. Samples consisted of admissions to eight nursing homes in the Baltimore area that fell into one of three states or categories of depression, depression disorder (DD), depressive symptoms (DS), or no depression (ND). Determine if mortality rates and state of depression are related using the following information. Test at the 0.05 level, and interpret your decision.

	Depression Diagnosis		
	DO	DS	ND
Death	27	20	94
Survive	30	62	221

Source: Barry W. Rovner, Pearl S. German, Larry J. Brant, Rebecca Clark, Lynda Burton, and Marshal F. Folstein, "Depression and Mortality in Nursing Homes," *Journal of the American Medical Association*, Vol. 265, No. 8, February 27, 1991, pp. 993–96.

64. Determine if the number of women and men directors or officers of *Fortune 500* firms is independent of the industry at the 0.01 significance level using the following information. Interpret the decision.

Industry	Females	Males
Aerospace	15	485
Apparel	16	173
Chemicals	31	1,238
Computers	27	682
Electronics	30	998
Food	38	1,220
Industrial and farm equipment	16	701
Metals	14	536
Pharmaceuticals	29	473
Publishing and printing	31	479
Soaps and cosmetics	25	269
Textiles	8	230
Transportation equipment	1	117

Source: Mary Ann Von Glinow, University of Southern California, in "Few Women at the Top," *Wall Street Journal*, December 30, 1991, p. B1.

Introduction

Unit II: More Nonparametric Tests

Preparation-Check Questions

65. Nonparametric tests are most useful when
  a. Only a small sample is available.
  b. The hypothesis is about the standard deviation.
  c. The data are quantitative.
  d. The situation violates the assumptions for parametric tests.
  e. Estimating parameters from normal populations.

66. In which of the diagrams is the median (M) of the data points (labeled x) appropriately placed?
  a. $-x-x\text{———}xx\underset{M}{\text{—}\mid\text{—}}x\text{———}$
  b. $-x-x\underset{M}{\text{—}\mid\text{—}}x-x-x-x-x-$
  c. $\underset{M}{\text{—}\mid\text{—}}x-x-x-x-x-x-$
  d. $-x-x-x-x\text{———}\underset{M}{\text{—}\mid\text{—}}$
  e. $-x-x-x\underset{M}{\text{—}\mid\text{—}}x-x-x-$

67. Which of the following assumptions must we satisfy when performing tests for means with small samples?
  a. Success or failure on each trial.
  b. Normal populations.
  c. Sample size must be at least 30.
  d. The standard deviation is smaller than the mean.
  e. The standard deviation is a known quantity.

68. Without using a continuity correction, find the approximate probability that 55% or more of 500 tosses of a fair coin will result in heads. (Round σ two places past the decimal.)
  a. 250.   e. 0.0125.
  b. 0.1118. f. 0.4875.
  c. 0.5.   g. 0.9875.
  d. 0.55.   h. 2.24.

69. When you are working with a normal distribution, what two things must you know in order to find probabilities?
  a. Degrees of freedom for numerator and denominator.
  b. μ and σ.
  c. p and $\hat{p}$.
  d. If the chi-square or F distribution can be substituted.
  e. S and n.
  f. R and C.

Answers: 65. d; 66. e; 67. b; 68. e; 69. b.

Introduction

Construction Cost Situation In an earlier chapter, an employee believes the average cost of a particular construction job is $7,320. Suppose that when attempting to collect data to test this claim, we find that contractors and their previous customers are reluctant to present cost figures or cannot remember them. However, 30 of those contacted will relate whether the cost was larger or smaller than $7,320. Can we test the claim with this information alone?

Productivity Difference Situation In another example a corporate manager forms hypotheses about the difference in mean output from two secretarial pools. Another manager who uses secretaries from both pools wants to test this hypothesis quickly. He does not know exact output values for each secretary, but he can mentally rank them, based on his prior experience. In addition, his samples of output figures are small and he doubts that outputs are normally distributed in each pool. Instead, his experience suggests outputs in each pool are bimodal; that is, there are two sets of clerks in each pool, a very productive group and a group that performs only adequately. This situation violates the assumptions for performing the T-test for the difference in means. In either case—with rankings or with a bimodal population—is there a valid method for testing with the available information?

Training Method Situation In the astronaut example, NASA compares astronauts' reaction time when they perform a certain task after they experience different training methods. One of the assumptions of the ANOVA technique is that the times for each training method are normally distributed with equal variances. Suppose that normality in one or more of these sets of times is questionable. Can we no longer use the data for determining differences in mean performance times?

This unit provides alternative tests for cases similar to these three. The new procedures substitute for situations that violate parametric test assumptions or situations when samples contain less informative data, such as ranks rather than values that express an amount.

13.5 The Sign Test

This test is based on the median. It substitutes for a parametric test for the population mean. Actually, it is a test for a proportion, where the definition of the median dictates the specific proportion.

13.5.1 Alternative Test for Population Central Tendency

Parametric tests for the population mean require either a large sample ($n > 30$) or a normal population when the sample is small. They also require quantitative data for calculating the sample mean. The sign test is an alternative when the situation does not satisfy these conditions.

Question: Name two other alternative measures of central tendency that we studied in Chapter 3.
 Answer: The median and the mode are the two alternatives.

Question: When are the mean and median identical?
 Answer: The values are identical when the data are perfectly symmetric, although they can also be equal for some nonsymmetric data sets.

The two measures are not usually the same value, because most data sets are not perfectly symmetric. However, if the data are not skewed, we usually expect the two values to be fairly close. So a test to determine the median may adequately answer questions about the population center or mean in cases where parametric test results could be misleading.

While the parametric version tests for μ, the nonparametric alternative tests for the population median, M. *We specify a value for the median, M_0, and test the hypotheses $H_0: M = M_0$, or $M \geq M_0$, or $M \leq M_0$. The corresponding alternative hypotheses are $M \neq M_0$, $M < M_0$, or $M > M_0$.*

13.5.2 The Test Statistic

The sign test checks the population median, M; however, the test statistic is not the sample median, $\hat{M}$. The test statistic is based on the expected consequence of the population median designated in H_0, specifically, the proportion of observations above (or below) the median, p. If the sample result adheres to this expectation, we do not reject the stipulated value; if not, we reject the value in favor of an alternative value of the median, or center of the data set.

Question: The median is the middle value of the data set, after arranging the values in ascending or descending order. What percent of the values are on each side of the median?
 Answer: Approximately, half or 50% of the observations will be on each side of the median.

Section 13.5 The Sign Test 643

We base the sign or median test on this equal division of the nonmedian values in the data set. *If p is the proportion of values on either side of the population median, then p must be about 0.5.* So we translate the hypotheses about M into familiar hypotheses about p. The transformed hypotheses are

$$H_0: p = 0.5 \qquad H_0: p \geq 0.5 \qquad H_0: p \leq 0.5$$
$$H_1: p \neq 0.5 \qquad H_1: p < 0.5 \qquad H_1: p > 0.5$$

Question: What is the test statistic for a test of p?
Answer: $\hat{p}$, the sample proportion of successes, estimates the population success rate.

Question: Is the proportion of values that lie above or the proportion that lie below the median in a sample likely to be exactly 0.5? Explain.
Answer: The population median splits nonmedian population values approximately equally. Because a sample contains only a portion of the population, the values in the sample will probably not divide exactly, just as the sample mean is generally not equal to the population mean. However, if we randomly select the sample values from the population, then we can expect that approximately half of the values will be above and half below the population median.

Because $\hat{p}$ is the proportion of successes in the sample, we must define "success" to conduct the test. Specifically, we must count the number of sample values above (or below) the median to form $\hat{p}$, the test statistic. Using "above the median" as a success will make rejection regions for one-tailed tests easier to visualize, as we shall see. *So when we assume that H_0 is true (M_0 is the population median), $\hat{p}$ will be the proportion of non-M_0 sample values that fall above M_0. Notice that sample values equal to M_0 do not count in the numerator or denominator of this proportion.*

Question: Sketch a number line, then place the hypothesized value of the median, M_0, and eight sample values on the number line, so that $\hat{p}$ the proportion of sample values greater than M_0 will be 0.5. Repeat the diagram for $\hat{p} = 0.25$, and for $\hat{p} = 0.75$.
Answer: The following diagrams depict only a few of many possibilities. Notice that in the second column of number lines, $\hat{p}$ is the proportion of sample values that are not equal to M_0; whereas, in the first column, none of the sample values is equal to M_0. In the first case, $\hat{p} = 4/8 = 0.5$, and in the second case, $\hat{p} = 3/6 = 0.5$. We exclude any value that equals the hypothesized median, M_0, when we form $\hat{p}$.

Question: Examine the number line diagrams to determine if a $\hat{p}$ smaller than 0.5, say 0.25, indicates that the actual median, M, is larger than, smaller than, or equal to M_0. Explain.

Answer: A small $\hat{p}$ occurs when a larger proportion of the sample values fall below M_0 than fall above it. The median would have to be smaller than M_0 to divide the sample points equally. Thus, the sample information indicates that the population median is smaller than M_0, $M < M_0$. Similarly, a large $\hat{p}$ indicates $M > M_0$. To summarize: *both large and small values of $\hat{p}$ indicate that the hypothesized value, M_0, is incorrect.*

■ **Example 13-3 (Construction Cost Situation—page 639):** An employee estimates that the mean cost of a certain job is $7,320. Suppose that 17 of 30 contractors estimate their cost would be more than $7,320. What is $\hat{p}$, and what does its value indicate about the actual median? No respondent says the cost would be exactly $7,320.

■ **Solution:** $\hat{p} = 17/30 = 0.5667 > 0.5$, which suggests that the $7,320 figure is low.

Notice that we only need to know whether a sample observation is above or below M_0. If we give values above M_0 a plus sign and those below M_0 a minus sign, then the test statistic depends on the number of positive signs; hence, the name of the test, sign test.

Section 13.5.2 Problems

70. We hypothesize that the median grade on a test is 78. In a sample of 110 students, most will only confide their position relative to 78, not their actual score. Twenty-five scored more than 78, 10 scored 78, and 75 less than 78. What is $\hat{p}$? What does the value of $\hat{p}$ indicate about the true median?

71. To demonstrate that the typical health and sanitation rating for local restaurants is above A^-, a consumer group selects a random sample of 150 restaurants and finds that 102 have A or A^+ ratings, 27 have B^+ or lower ratings, and the rest have an A^- rating. What is $\hat{p}$, and what does this value indicate about the hypothesized A^- value?

72. The president stated in a speech that the typical economic growth forecast for the domestic economy is 2.5%. The opposition party believes the typical forecast is above 2.5%. To check the opposition party's belief, a reporter selects and contacts a random sample of 90 economists. Fourteen agreed with the president, 40 thought that 2.5% was too low, and the remainder considered the value too high. What would the test statistic be in this situation? If the president is wrong, does the sample evidence indicate that his value is too small or too large?

13.5.3 The Sampling Distribution

The sign test is a special case of a test for p. The hypothesized value in H_0 and H_1 is always 0.5, and we interpret the decision in terms of the position or value of the median, not the proportion. Otherwise, the methodology is the same as before. The sampling distribution of $\hat{p}$, the test statistic, is approximately normal for large samples, especially when $p = 0.5$. (When the sample is small or we require an exact answer, we can use the binomial table in Appendix G.)

Question: What is the mean of the $\hat{p}$ values when H_0 is true?

Answer: *The mean of the possible $\hat{p}$ values is p, 0.5,* the value specified in H_0 for all sign tests.

Question: What is the standard error of the $\hat{p}$ values when H_0 is true?

Answer: $\sigma_{\hat{p}} = \sqrt{\dfrac{p(1-p)}{n}} = \sqrt{0.5(1-0.5)/n} = 0.5/\sqrt{n}.$

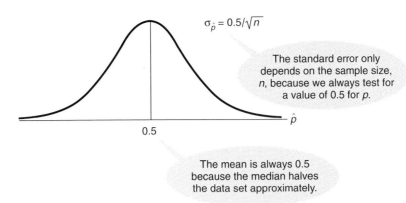

Figure 13-12
The sampling distribution of $\hat{p}$ for the sign test

Figure 13-12 displays the sampling distribution of $\hat{p}$ that we use for sign tests.

■ **Example 13-3 (continued):** There are 30 responses about the costs to complete a construction job. What are μ and $\sigma_{\hat{p}}$?

■ **Solution:** $\mu = 0.5$ and $\sigma_{\hat{p}} = 0.5/\sqrt{30} = 0.0913$.

13.5.4 The Decision and Its Interpretation

As discussed in Section 13.5.2, the specified value of M_0 appears to be close, if not equal, to the true median when $\hat{p}$ is close to 0.5. A $\hat{p}$ that is smaller than 0.5 makes us suspicious that the true median may be smaller than M_0. When $\hat{p}$ is larger than 0.5, we suspect that the true median may be larger than M_0. These conclusions lead directly to the rejection regions for the various forms of hypotheses.

Question: If H_1 is $M < M_0$, sketch the rejection region on a normal curve.

Answer: We suspect that $M < M_0$ when $\hat{p}$ is small (less than 0.5), so the rejection region is in the left tail of the normal curve. Figure 13-13 displays the different forms of the hypotheses and corresponding rejection regions for the sign test. Notice that we can still locate the rejection region by determining which values of the test statistic make H_1 appear to be true.

■ **Example 13-3 (continued):** Use the $\hat{p} = 17/30 = 0.5667$ from the contractor estimates relative to \$7,320 to test for a median cost greater than \$7,320 at the 0.05 level. Interpret the decision.

■ **Solution:** $\hat{p} = 0.5667$ is indeed greater than 0.5. The question is whether it is far enough from 0.5 to reject H_0 and decide that p is greater than 0.5, which, in turn,

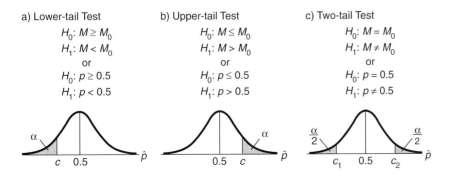

Figure 13-13
Rejection regions for the sign test

Figure 13-14
Decision in contractor example

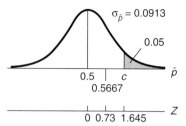

implies that the median, M, is larger than M_0. Recall that $p = 0.5$ and $\sigma_{\hat{p}} = 0.0913$. When we convert 0.5667 to a Z, we obtain $(0.5667 - 0.5)/0.0913 = 0.73$. Consequently, we cannot reject H_0, because $0.73 < 1.645$, the critical value from the Z table (Appendix H) with an area 0.45 between the mean and critical value. Thus, the data are not sufficient evidence to deny that the median or average cost is \$7,320 (see Figure 13-14).

The following box outlines the sign test procedure.

Sign Test Procedure
1. Determine n, the number of values in a random sample not equal to M_0, the hypothesized median value from H_0.
2. Determine $\hat{p}$, the proportion of the n observations that are larger than M_0.
3. Use a normal distribution, with a mean of $p = 0.5$ and a standard error, $\sigma_{\hat{p}}$, of $0.5/\sqrt{n}$ to form the rejection region according to the direction indicated in H_1.
4. Determine the position of $\hat{p}$ relative to the critical value.
5. If H_0 is rejected, conclude that the population median is not M_0. The relative magnitude of M compared to M_0 is determined by the direction indicated in H_1.

■ **Example 13-4:** Fifty-seven teenagers rank a new rock group as above average, below average, or average. The survey results include 33 above averages, seven averages, and 17 below averages. Does this evidence suggest that this group is destined to become popular (above average)? Use $\alpha = 0.05$. Interpret the decision.

■ **Solution:** $\hat{p} = 33/50 = 0.66$ and $\sigma_{\hat{p}} = 0.5/\sqrt{50} = 0.0707$. We reject H_0, because when we transform 0.66 to a Z, we obtain $(0.66 - 0.5)/0.0707 = 2.26$, which exceeds 1.645, the critical value. The median rating is above average according to this decision. Based on this popularity poll, the group's future looks bright.

Section 13.5.4 Problems

73. In Problem 70 we hypothesized that the median grade on a test is 78. From a sample of 110 students, we found that 25% of the 100 values were larger than 78 (10 values were 78). Use this information to complete the test at the 0.01 level. Interpret the conclusion.
74. To demonstrate that the typical health and sanitation rating for local restaurants is above A^- in Problem 71, a consumer group selects a random sample of restaurants and finds that 79.07% of the 129 ratings that are not A^- are above A^-. Use this information to complete the test at the 0.05 significance level. Interpret the decision.
75. In Problem 72 a reporter collected a random sample of economists' projections for the economy to verify the opposition party's claim that the typical forecast was higher than the value claimed by the president, 2.5%. Of the 76 economists who did not agree exactly with the president 52.63% thought that 2.5% was too low. Use this information to complete this test at the 0.01 significance level. Interpret the conclusion.

13.6 Wilcoxon-Mann-Whitney Rank Sum Test

This test replaces a parametric test for the difference in population means. We will use the ranks of observations rather than actual measurements as we proceed.

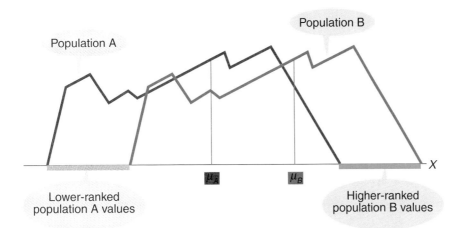

Figure 13-15
Identical nonnormal populations in different positions on the number line

13.6.1 An Alternative Test for a Difference between Two Population Means with Independent Samples

A parametric test for a difference between two population means requires information about the populations and sampling procedures. We must assume that we have independent random samples from each population and that we have either large samples or normal populations. Sometimes we assume equal variances. Quantitative data are a necessity. The nonparametric alternative, the Wilcoxon-Mann-Whitney rank sum test, assumes that the underlying values in each of the populations are continuous, not discrete, and that the population distributions are the same ($\sigma_1^2 = \sigma_2^2$), although they may be located in different positions on the number line. These different positions translate into different means when we meet all of the other assumptions. Thus, H_0 and H_1 will refer to the familiar differences in means from Chapter 10, $\mu_1 - \mu_2$, and the value we test for is a difference of zero, which really states that the means are equal. In addition, the test requires that we use independent random samples from the two populations.

This test does not require normal populations or actual data values, only their ranks. Relaxing the normality assumptions makes this test especially useful for small-sample situations. Using ranks makes testing possible in large- and small-sample situations where we do not have actual data values, such as the manager who ranks the productivity of different clerks without knowing an actual measure of the productivities.

Figure 13-15 displays two identical nonnormal populations in different positions, hence with different means.

Question: If we combine values from both populations into a single data set and arrange them in ascending order, which population will have larger ranks (the ranks start with 1 for the smallest value)?

Answer: Population B consists of larger values, so it follows that its ranks will be larger.

This comparison of the ranks from a combination of values in the two independent samples, each randomly selected from one of the identical populations, is the basis for the Wilcoxon-Mann-Whitney test. Using this information, we can decide whether the population means differ, and, if so, which mean appears to be larger.

Box 13-1 provides an interesting scientific application of the Wilcoxon-Mann-Whitney test and a variation of the chi-square test from Unit I.

Box 13-1 Chi-Square Test for Goodness of Fit

Successfully applying parametric tests often requires that we deal with normal populations. Although the population is normal, the distribution of a random sample of values from a normal population may not be normal. The difference in the population and sample distributions is similar to sampling error, because if we enlarge the sample to include the entire population, the discrepancy between the distributions would disappear.

We can use a chi-square statistic to compare observed and expected frequencies in a sample to test the hypothesis that a population is normal, just as we do for the test for independence. In this case, the statistic compares observed frequencies with the expected frequencies in different intervals of outcomes for a normally distributed random variable. Large values of the test statistic indicate relatively large discrepancies between the observed and expected frequencies, leading us to reject the normal population hypothesis. We call this procedure the goodness-of-fit test.

When investigating the length of mature male scorpions' chelae (claws or pincers on the second set of appendages), a frequency distribution caused the researcher to suspect a bimodal distribution. The chi-square goodness-of-fit test supported the conclusion. In addition, the chi-square goodness-of-fit test confirmed a nonnormal distribution of body sizes, specifically anterior body lengths. Consequently, the researcher investigated the humps in the distribution or modes and found two forms of mature male scorpions that merited separate investigation, small mature males and large mature males. Subsequent Wilcoxon-Mann-Whitney tests substantiate differences in body weights and body sizes between the two forms.

The chi-square goodness-of-fit test revealed no significant deviation from normality for adult females' chela and prosoma (anterior) body measurements. However, the researcher reasoned that two forms might exist for females as well but be less distinguishable than the male forms. Wilcoxon-Mann-Whitney results revealed that females categorized as large produced significantly heavier broods, although the number of offspring in the broods was not significantly different.

Source: T. G. Benton, The Life History of Euscorpius Flavicaudis (Scorpiones, Chactidae), The Journal of Arachnology, Vol. 19, No. 2, 1991, pp. 105–10.

13.6.2 The Test Statistic

As usual with nonparametric tests, the actual magnitudes of sample values do not figure directly in the test statistic computation. As its name implies, the Wilcoxon-Mann-Whitney rank sum test uses the ranks of the sample values to form the test statistic.

The first step in the process is to combine both samples and arrange the combined $(n_1 + n_2)$ values in ascending order. Then we assign ranks to the values, starting with 1 for the smallest value and continuing to $n_1 + n_2$ for the largest value. When a value repeats in the sample, we find the mean rank for the positions occupied by these observations and assign this mean rank to all of the repeat values. For instance, if the value 56.6 occurs four times as the fourth through seventh smallest values in the combined sample, we would assign the rank 5.5 to each of the four values. (The test developed in this section assumes there are no ties, but it provides useful results if there are only a few ties. The correction for ties is small or negligible when samples are large.)

■ **Example 13-5:** The following data correspond to the number of cigarettes smoked per day by 22 randomly selected smokers who died at the same age, 12 died of cancer, and 10 died without ever having cancer. There is reason to believe that the relevant populations are skewed rather than normally distributed. Combine and order the original values. Underline the cancer group values.

Sample	Cigarettes Smoked per Day											
Cancer	63	42	48	41	21	1	42	53	32	19	65	72
Noncancer	27	0	0	38	21	17	43	10	3	12		

■ **Solution:** The data are ordered below and cancer values underlined. Brackets beneath identical values indicate values tied for a rank. We assign mean ranks to the ties in the last row of rankings. For example, the first and second values in the data set are the same, 0, so we average their ranks, 1 and 2, to obtain 1.5. We assign this rank to both values.

Actual Values	0 0 1 3 10 12 17 19 21 21 27 32 38 41 42 42 43 48 53 63 65 72
Matched Rankings	1 2 3 4 5 6 7 8 9 10 11 12 13 14 15 16 17 18 19 20 21 22
Final Rankings	1.5 1.5 3 4 5 6 7 8 9.5 9.5 11 12 13 14 15.5 15.5 17 18 19 20 21 22

The next step is to choose one of the samples and find the sum of the ranks (SR), that correspond to its values. This is the test statistic for the Wilcoxon-Mann-Whitney test.

■ **Example 13-5 (continued):** Find the rank sum, SR, for each sample in the smoker data. In an actual test, we would do this for only one of the samples.

■ **Solution:** If we use the cancer sample the result is

$$SR = 3 + 8 + 9.5 + 12 + 14 + 15.5 + 15.5 + 18 + 19 + 20 + 21 + 22 = 177.5.$$

The second sample, noncancer, produces

$$SR = 1.5 + 1.5 + 4 + 5 + 6 + 7 + 9.5 + 11 + 13 + 17 = 75.5.$$

To understand more about the values of SR that concur with H_0 and those that do not, we begin by considering the mean rank of the values in the combined sample. Let n_T be the number of observations in the combined sample, $n_1 + n_2$. Then the mean rank must be the midpoint of the set of consecutive ranks in order to balance the distances of points from the mean on both sides. Thus, the mean rank is $(n_T + 1)/2$. (This is the same formula we used to locate the position of the median in Chapter 3.)

Question: What is the mean rank of the combined sample of the 22 smokers?

Answer: The mean rank is $(22 + 1)/2 = 11.5$. Study the line labeled "matched rankings" above to see that this number does balance the distances of points from the mean on both sides.

Suppose we assign this mean rank, 11.5, to each value in one of the samples, say the noncancer sample with 10 victims. Then to sum these ranks for the noncancer sample we would add 11.5 ten times or we could just use $10(11.5) = 115$. This is the sum of ranks that we would expect for this sample if the noncancer smokers' values are evenly or randomly distributed throughout the combined sample. Similarly, the sum for the cancer smokers would be $12(11.5)$, or 138. In summary, *if a sample consists of n_i values that are evenly or randomly distributed throughout the combined sample, then the sum of the mean ranks for this sample is $n_i(n_T + 1)/2$.*

Question: If the noncancer smokers tended to smoke fewer cigarettes per day than cancer smokers, would the sum of the actual ranks for the noncancer smokers likely be smaller, larger, or equal to 115? Explain.

Answer: If noncancer smokers smoked fewer cigarettes per day, their ranks in the combined sample will tend to be small, rather than randomly distributed throughout the combined sample. Consequently, the sum of their ranks, SR, will be smaller than 115, the sum of the mean ranks (the sum we would expect if the noncancer values were randomly distributed in the combined sample). If the noncancer smokers smoked more cigarettes per day than the cancer smokers, their ranks would tend to be larger and the sum of their ranks larger than 115.

If the population means are identical (H_0 is true), we expect the values from each sample to be distributed randomly throughout the combined sample. Consequently, we expect that the sum of the actual ranks of each sample will be close to the sum of the mean ranks for that sample, $n_i(n_T + 1)/2$. A rank sum, SR, that departs drastically from this value signals a sample whose values tend to be larger or smaller than those in the other sample. Such a finding would lead us to believe that the sources of the samples, their populations, are indeed in different positions and, hence, that the population means are different.

If we arrange values in ascending order, then when a sample's SR exceeds the sum of its mean ranks, we expect that this sample's population mean is larger than the other sample's population mean. Similarly, an SR smaller than the sum of the mean ranks signals a smaller population mean.

To determine when the sum of the actual ranks, SR, is drastically (significantly) different from the sum of the mean ranks, we must know which values of the test statistic, SR, are likely to occur when H_0 is true and which are not. The sampling distribution of SR contains this information. The next section details this distribution.

Section 13.6.2 Problems

76. A group of doctors ranked a random sample of 25 potential heart recipients from least to most urgent according to the patients' medical files. The underlined ranks below belong to patients who underwent bypass surgery during the last year.

1 2 3 4 <u>5</u> <u>6</u> 7 8 9 <u>10</u> 11 12 <u>13</u> 14 15
16 17 18 19 <u>20</u> <u>21</u> 22 <u>23</u> 24 <u>25</u>

  a. Calculate SR for the bypass group and the sum of the mean ranks.
  b. Calculate SR for the nonbypass group and the sum of the mean ranks.
  c. Based on the statistics, which population appears to have a larger typical or average sense of urgency according to the doctors? Explain.

77. Use the following information from a random sample of 30 current year automobile models' ranks by a consumer group from least to most maintenance problems to find SR and the sum of the mean ranks for small cars (underlined ranks) and large cars. Which type of car appears to average more maintenance problems? Explain.

1 2 3 4 <u>5</u> <u>6</u> 7 8 9 10 11 12 <u>13</u> 14 15 16 17
18 19 20 21 <u>22</u> <u>23</u> 24 25 <u>26</u> <u>27</u> 28 29 30

78. Housing prices are often skewed to the right in any geographic area. Use the following price data (expressed in thousands of dollars) from two areas to compute an SR and sum of mean ranks for each area. In which area does the average price appear to be higher, based on the SR and sum of mean rank values? Explain.

Area 1	Area 2
103	92
75	90
125	67
88	95
65	98
72	101
72	89
75	95
110	85
81	105
80	89
92	
86	

79. School administrators rank a random sample of students' grades in a required class, A, B, C, D, or F, then separate the grades into two groups according to whether the teaching method was cooperative learning, which is a student team approach, or some other method.

Grade in Required Course	Number of Students Who Made the Listed Grade in a	
	Cooperative Learning Class	Class Employing Another Method
F	0	1
D	5	10
C	5	5
B	0	10
A	3	8

  a. Use this information to find SR and the sum of the mean ranks for the group who experienced the cooperative approach.
  b. Find SR and the sum of the mean ranks for the students who experienced other teaching methods.
  c. After observing the data and the computed values from Parts a and b, do you think the samples are from the same population? Explain.

13.6.3 The Sampling Distribution of SR

When H_0 is true (the population means are equal), *SR will be approximately normally distributed if the samples each contain at least 10 observations* ($n_1 \geq 10$ and $n_2 \geq 10$). To work with any normal distribution, we need a mean and standard deviation.

We have already calculated the mean of the SR values for a single sample when we assume its values are randomly distributed throughout the combined sample. We can show that this mean would also be the mean of all the possible rank sums for samples of the same size. That is, the sum of the mean ranks that we calculated for the n_1 observations in the first sample, $n_1(n_T + 1)/2$, is also the mean of all the potential SR values, μ_{SR}, for samples of size n_1 in the combined sample of n_T observations. We obtain similar results for the second sample, so we conclude the following.

$$\mu_{SR} = n_i(n_T + 1)/2 .$$

μ_{SR} represents the rank sum that we would expect to occur when H_0 is true and the population means are identical. If H_0 is true, then the placement of cancer sample values in the combined smoker sample is random and unlikely to follow a distinct pattern, such as falling in the first or the last n_1 positions (n_1 is the number of smokers with cancer in the combined sample).

Notice that μ_{SR} depends on the size of the sample selected for the test statistic. It will be different for each sample unless both samples contain the same number of observations.

Question: Give an intuitive explanation for the cancer mean SR exceeding the noncancer mean SR, even when the population means are identical.

Answer: When the populations are identical, we expect the observations from both samples to be randomly distributed throughout the combined sample. Both samples should have about the same distribution of ranks and, hence the same average of these ranks. However, there are more values in the cancer sample ($n_1 > n_2$), so there will be more ranks to sum, which would produce a larger sum even when each sample's values are evenly distributed in the combined sample.

The standard error of the *SR* values is the same for both samples. Omitting its development, we simply state that

$$\sigma_{SR} = \sqrt{\frac{n_1 n_2 (n_T + 1)}{12}} .$$

■ **Example 13-5 (continued):** Find σ_{SR} for the smoker problem.

■ **Solution:** $\sigma_{SR} = \sqrt{\dfrac{12(10)(22 + 1)}{12}} = 15.1658$

Figure 13-16 summarizes the sampling distribution of SR. Again, this distribution holds when the samples are independent random samples from populations with the same variances, and each sample contains at least 10 observations with no (or few) ties.

13.6.4 The Decision and Its Interpretation

If the population means are identical (H_0 is true), then SR will usually be close to its expected value or mean, μ_{SR}. If SR is not close to μ_{SR}, then H_0 appears to be false and we may be able to reject it. Rejecting H_0 means that we decide that the two population means differ somehow.

To associate a sample's SR result with the hypotheses, we need to recall that an SR value larger than μ_{SR} indicates that the mean of this sample's population is the larger

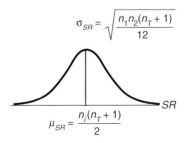

Figure 13-16
Sampling distribution of SR (n_1 and n_2 at least 10)

Figure 13-17
Hypotheses and corresponding rejection regions for Wilcoxon-Mann-Whitney test

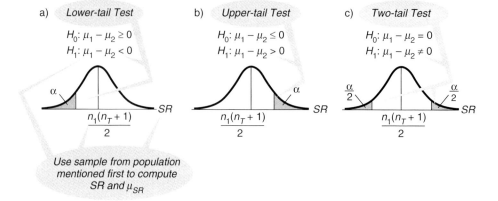

mean. Similarly, an SR value smaller than μ_{SR} indicates that the sample's population mean is the smaller mean. We make use of these relationships so that when we choose a sample to compute SR, the decision and the interpretation will be direct. If we state the hypotheses with $\mu_1 - \mu_2$, then we use the first sample to compute SR and μ_{SR}, because the position of SR relative to μ_{SR} will tell us how μ_1 fares relative to μ_2. So if H_1 states that $\mu_1 - \mu_2 < 0$, we should look for an SR smaller than μ_{SR} to support this hypothesis. Figure 13-17 shows the hypotheses and the corresponding rejection regions. As these examples show, *the rejection region lies under a tail that corresponds to the direction indicated in H_1 if we use the SR that corresponds to the population mentioned first in the hypotheses.*

■ **Example 13-5 (continued):** Use the mean and standard deviation for the cancer sample rank sum, $\mu_{SR} = 138$ and $\sigma_{SR} = 15.1658$, along with SR = 177.5, to demonstrate that the average number of cigarettes smoked is greater for the cancer smoker. Test at the 0.05 level, and interpret the decision.

■ **Solution:** Because we want to demonstrate that the average is greater for the cancer population, we make H_1 state that $\mu_1 - \mu_2 > 0$, where the subscripts 1 and 2 refer to the cancer and noncancer samples, respectively. The rejection region is under the upper tail of the normal curve. 177.5 becomes $z = (177.5 - 138)/15.1658 = 2.60$ and $2.60 > 1.645$, the critical value from the Z table with an area of 0.45 between the mean and critical value. Thus, we reject H_0 (see Figure 13-18). The evidence indicates that the mean number of cigarettes smoked is larger for the cancer patients than it is for the noncancer patients.

The following box summarizes the necessary steps to perform the Wilcoxon-Mann-Whitney rank sum test.

Figure 13-18
Decision for Wilcoxon-Mann-Whitney test for smoker data

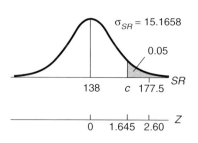

Wilcoxon-Mann-Whitney Rank Sum Test Procedure

1. Select two independent, random samples, each of which contains at least 10 observations.
2. Pool both samples' values into a single sample.
3. Arrange the pooled values in ascending order.
4. Assign ranks to values, beginning with 1 for the lowest. If there are identical values in the combined sample, assign the mean of the ranks that would be assigned to each of the identical values had they not been tied.
5. Pick the sample that corresponds to the population mentioned first in H_0 and H_1 to use for the test.
6. Find SR = rank sum for sample chosen in Step 5.
7. Use a normal distribution with the mean, $\mu_{SR} = n_i(n_T + 1)/2$, and standard deviation, $\sigma_{SR} = \sqrt{n_1 n_2 (n_T + 1)/12}$, to form the rejection region, where n_i = number of observations in the chosen sample.
8. Determine whether SR is in the rejection region.
9. If H_0 is rejected, conclude that the population means differ as stated in H_1.

■ **Example 13-6 (Productivity Difference Situation—page 641):** A manager compares two secretarial pools' productivities to determine if there is a difference in mean output. Suppose that the only available data are the manager's ranking of individuals' productivity levels in the combined pool of secretaries, as listed in the following data. Pool 1 ranks are underlined. Use this information to determine if $\mu_2 - \mu_1 < 0$ at the 0.01 level. Interpret the decision.

1 <u>2 3</u> 4.5 <u>4.5</u> 6 <u>7 8 9</u> 10 12 <u>12 12</u> 14 <u>15</u> 16 17 18 19 <u>20 21</u> 22 <u>23</u> 24 <u>25 26</u> 27 <u>28</u> 29
30 <u>31 32 33 34</u> 35 <u>36</u> 37 <u>38</u> 39 40 41 42 <u>43</u> 44 <u>45</u> 46 47 48 49 50

■ **Solution:** We use the second sample to compute SR, because the hypotheses name μ_2 first, $H_1: \mu_2 - \mu_1 < 0$.

$$SR = 1 + 4.5 + 4.5 + 6 + 10 + 12 + 14 + 16 + 17 + 18 + 19 + 22 + 24 + 27$$
$$+ 29 + 30 + 35 + 37 + 39 + 40 + 41 + 42 + 44 + 50$$
$$= 582.$$

$n_2 = 24$ and $n_T = 50$, so $\mu_{SR} = 24(50 + 1)/2 = 612$, and

$$\sigma_{SR} = \sqrt{\frac{24(26)(50+1)}{12}} = 51.4976.$$

We do not reject H_0, because $(582 - 612)/51.4976 = -0.58 > -2.33$. There is no evidence of a significant difference between the mean productivities of the two pools.

Box 13-2 uses a substitute test for the Wilcoxon-Mann-Whitney test when samples are not independent.

Section 13.6.4 Problems

80. We can state H_1 for Example 13-6 with the secretarial pools as $\mu_1 - \mu_2 > 0$, that is, the first pool is more productive on average. Determine if the decision is the same for this statement of the hypotheses at the 0.01 level.

81. In Problem 76 a group of doctors ranked a random sample of 25 potential heart recipients from least to most urgent according to the patients' medical files, where 12 of the patients underwent bypass surgery during the last year and 13 did not. The SR and μ_{SR} values for both samples are shown in the following table.

Sample	SR	μ_{SR}
Bypass	161	156
No bypass	164	169

Box 13-2 Gender and Children's Musical Preferences

There are many variables that affect an individual's sensation and response to music: characteristics of listeners, the environment where the listening occurs, and the music itself. Music and psychology researchers study many of these dimensions to better understand people's reactions to music. This study focuses on the effect of vocal vibrato, pulsing or vibrating sounds as opposed to a single unwavering sound, and the performer's gender on the response of fourth- through sixth-graders.

These young subjects listened to 24 matched recordings (equal numbers of high and low vibrato selections across several musical styles and across both sexes of performers). Then they rated each recording according to their perception of the sex of the performer ("man" to "woman"), the strength of the performance ("weak" to "strong"), amount of vibrato ("smooth voice" to "wavy voice"), and their preference for the music ("like" to "dislike"), using a 1 to 7 scale for each of the four dimensions. If most subjects tend to rank dimensions in the correct direction, then rankings may be skewed. In addition, the qualitative nature of the values means that nonparametric procedures are appropriate.

To learn about the effect of listener's gender on reactions, the researchers analyzed the ratings on each of the dimensions for the male and female subjects separately. Differences in ratings are dependent, because the same subject rates both categories of each variable, such as male and female performances of low and high vibrato. The Wilcoxon-Mann-Whitney test assumes independent samples in the same manner that the parametric tests did. However, there is an alternative nonparametric test for matched pairs of dependent observations, just as there is an alternative parametric test for matched pairs. In this case, we call the test the Wilcoxon matched-pairs signed-ranks test.

Using this procedure, the analysis indicates that listeners of both sexes:

1. Perceive the performance of male singers as stronger, though females rate males lower than males rate males.
2. Prefer low vibrato.
3. Favor male performers.

The conclusions are somewhat limited by the students' apparent lack of ability to discern levels of vibrato as well as trained musicians do. Chi-square tests (not the test for independence in this chapter) indicate a disparity between the subjects' and trained musicians' rankings for one-fourth of the recordings. The subjects identified the gender of the singer more accurately than vibrato levels. In only one case do the test results indicate a disparity between subjects' perception of the performer's sex and reality.

Source: Albert LeBlanc and Carolyn Sherrill, "Effect of Vocal Vibrato and Performer's Sex on Children's Music Preference," Journal of Research in Music Education, Vol. 34, No. 4, 1986, pp. 222–37.

Use this information to test for any difference in average sense of urgency at the 0.05 significance level. Interpret the decision.

82. In Problem 77 we used information from a random sample of 30 current year automobile models' ranks by a consumer group from least to most maintenance problems to find SR and the sum of the mean ranks for small cars and large cars. Use the results reported below to test for more maintenance problems on average for large cars ($\mu_L - \mu_S > 0$) at the 0.05 significance level. Interpret the decision.

Sample	SR	μ_{SR}
Small	184	201.5
Large	281	263.5

83. Because housing prices are often skewed to the right in a particular area, we can use the Wilcoxon-Mann-Whitney test to analyze the prices from areas 1 and 2 in Problem 78. Use the following statistics to determine if we can reject the hypothesis that the average price in area 1 is smaller than that of area 2 at the 0.1 significance level. Interpret the decision.

Sample	SR	μ_{SR}
Area 1	138.5	162.5
Area 2	161.5	137.5

84. In Problem 79 school administrators ranked a random sample of students' grades in a required course and then separated the grades into two groups according to whether the teaching method was cooperative learning or not. Is the average grade significantly lower in the cooperative style class? Test at the 0.05 significance level and interpret the decision.

Sample	SR	μ_{SR}
Cooperative	278.5	312.0
Not cooperative	849.5	816.0

85. Psychologists use responses to the Rorschach Inkblot test to assess different aspects of an individual's personality. One researcher employed such evaluations of 70 subjects from a children's residential treatment center. Thirty-five of these children were later hospitalized again for psychological problems and 35 were not. The researcher classified each subject's thought organization as mature or immature. The researcher measured a subject's object relations ability (separating autonomous figures) in three different ways: median score, low score, and high score. The purpose of the research was to learn if these scores were good predictors of later psychiatric rehospitalization. The researcher hypothesized that higher scores for object relations and for thought organization would predict rehospitalization as well. One reason for using nonparametrics to analyze this data set is the uncertain meaning of the scores. They rank or classify levels more than they quantify the amount or strength of the classifications, especially relative to one another (would a person with a score of four relate objects twice as often as a subject with a score of two?). Use the following rank sums for the rehospitalized group to test the predictability for each measure of object relations and then for thought organization at the 0.05 significance level. Interpret the decisions.

Variable	SR
Thought organization	1,389
Object relations	
Median	1,364
Low	1,454
High	1,430

Source: Steven B. Tuber, "Children's Rorschach Scores as Predictors of Later Adjustment," *Journal of Consulting and Clinical Psychology*, Vol. 51, No. 3, 1983, pp. 379–85.

13.7 The Kruskal-Wallis Test

In much the same way that one-way ANOVA extends a test for differences between two population means, the Kruskal-Wallis test extends the Wilcoxon-Mann-Whitney rank sum test.

13.7.1 An Alternative Test for One-Way ANOVA

The conditions required for one-way ANOVA include normal populations with identical standard deviations. When a situation does not meet either or both of the normality conditions or when the data is ordinal (ranked), the Kruskal-Wallis test substitutes for a parametric one-way ANOVA.

The ANOVA null hypothesis asserts that the means for all treatments are equal, where there are at least three treatments. The alternative is that the means differ. Because we extend the assumptions for the Wilcoxon-Mann-Whitney test, continuous populations with the same shape (same standard deviations), to all the treatment populations, the Kruskal-Wallis hypotheses are stated the same way as the ANOVA hypotheses. This test also requires independent random samples from each treatment population.

H_0: $\mu_1 = \mu_2 = \cdots = \mu_J$

H_1: at least one of the means is different

13.7.2 The Test Statistic

The analogy with the Wilcoxon-Mann-Whitney rank sum test continues as we find the value of the test statistic. Suppose there are three treatments or samples. We pool the sample values into a single set of values and rank them in ascending order. Ties among values are resolved by assigning to each set of tied values the mean of their corresponding rank values. Then we calculate the sum of ranks for each of the three samples, SR_1, SR_2, and SR_3. The following discussion assumes that there are no, or very few, identical values among the samples in order to avoid an adjustment to the test statistic for numerous ties.

The formula we use for the actual test statistic, H, is a shortcut that derives from theory and algebra. It incorporates all the SR values, and it has a known sampling distribution. In the next section we will discuss how values of H result from identical populations and from different populations.

> **Kruskal-Wallis Test Statistic**
>
> $$H = \frac{12}{n_T(n_T + 1)} \left[\sum_{i=1}^{k} \frac{(SR_i)^2}{n_i} \right] - 3(n_T + 1),$$
>
> where the i subscript refers to a specified sample in the analysis and n_T is the total number of observations in the combined sample, $n_1 + n_2 + n_3 + \cdots + n_k$ for k samples.

■ **Example 13-7 (Training Method Situation—page 642):** The following table of experimental reaction times of astronauts groups the times according to the training method each experienced (there are three methods). Determine the value of the test statistic, H.

	Method	
1	2	3
0.50	1.54	0.70
0.72	1.28	1.46
1.02	1.64	1.38
0.48	1.20	0.78
0.86	1.74	1.24
1.10	1.34	1.14
0.68		1.04
0.92		0.88
0.84		
0.82		

■ **Solution:** We pool and rearrange samples in ascending order.

Value	Method Sample	Rank	Value	Method Sample	Rank
0.48	1	1	1.04	3	13
0.50	1	2	1.10	1	14
0.68	1	3	1.14	3	15
0.70	3	4	1.20	2	16
0.72	1	5	1.24	3	17
0.78	3	6	1.28	2	18
0.82	1	7	1.34	2	19
0.84	1	8	1.38	3	20
0.86	1	9	1.46	3	21
0.88	3	10	1.54	2	22
0.92	1	11	1.64	2	23
1.02	1	12	1.74	2	24

From this arrangement, we obtain the following values:

$$SR_1 = 1 + 2 + 3 + 5 + 7 + 8 + 9 + 11 + 12 + 14 = 72.$$
$$SR_2 = 16 + 18 + 19 + 22 + 23 + 24 = 122.$$
$$SR_3 = 4 + 6 + 10 + 13 + 15 + 17 + 20 + 21 = 106.$$

Using $n_1 = 10$, $n_2 = 6$, and $n_3 = 8$, so that $n_T = 10 + 6 + 8 = 24$, we obtain

$$H = \frac{12}{24(24 + 1)} \left[\frac{(72)^2}{10} + \frac{(122)^2}{6} + \frac{(106)^2}{8} \right] - 3(24 + 1) = 13.0713.$$

Section 13.7.2 Problems

86. An educational researcher selects three random samples of times for students to complete final exams (in minutes), one sample for each of three periods during finals week: first day, middle day, and last day. Compute the Kruskal-Wallis test statistic, H, for the data.

First Day	Middle Day	Last Day
25	126	73
150	42	49
93	89	121
89	76	132
67	27	89
45	75	95
135	98	97
115	51	56
60	56	82

87. A physicist selects random samples of the power (measured in horsepower) of experimental engines developed by students from five different majors. Use the following information to compute the Kruskal-Wallis test statistic, H.

Physics	Engineering	Chemistry	Math	Industrial Technology
1.7	0.6	1.9	1.9	2.3
1.1	1.8	1.6	2.0	0.8
3.0	2.1	1.0	1.4	0.9
2.2	2.0	2.5	1.3	1.9
0.5	1.9	1.5	1.2	0.7

88. A manager randomly selects 30 employees and ranks them on perceived productivity in teamwork situations. Then the employer sorts the employees according to the objective of their team assignment to determine if average productivity differs with type of objective. Compute the Kruskal-Wallis test statistic, H, for the following ranks.

Personnel Matters	Customer Satisfaction	Strategic Planning
1	4	3
2	5	9
6	8	10
7	13	11
14	16	12
15	19	18
17	20	21
22	23	25
27	24	26
	28	29
		30

13.7.3 The Sampling Distribution, the Decision, and Its Interpretation

To proceed we must find which values of the test statistic, H, are likely to occur when H_0 is true, and which values are more likely when H_1 is true.

Question: If the population means are identical (H_0 is true), what can we say about the mean rank for each sample?

Answer: When the population means are identical, the ranks we assign to each sample are random and will differ only because of chance. Thus, the mean ranks should be similar; ideally, they would be equal.

Example 13-8: The SR values are equal in the following example. The means of the SR values are also equal, because the sample sizes are identical.
a. Find H.
b. Exchange the 5 in sample 2 and the 6 in sample 1. Then recompute H.
c. Further exchange the 12 in sample 2 and the 13 in sample 1. Recompute H.
d. Finally, assume the least six values are in the first sample, the next six in the second sample, and the largest six in sample 3. Then compute H.

	Sample Ranks		
	Sample 1	Sample 2	Sample 3
	13	2	16
	10	7	15
	1	17	4
	9	14	11
	18	5	3
	6	12	8
SR	57	57	57
Mean rank	9.5	9.5	9.5

■ **Solution:** **a.** When the population means are identical and the random distribution conforms to the ideal situation of equal mean ranks for each sample, H is 0.

$$H = \frac{12}{(18)(19)} \left[\frac{(57)^2}{6} + \frac{(57)^2}{6} + \frac{(57)^2}{6} \right] - 3(19) = 0.$$

b. After the exchange in which sample 1 gains a smaller rank value of 5, its rank sum decreases from 57 to $SR_1 = 56$, while SR_2 increases to 58 from the new rank value of 6. H increases from 0 to 0.01.

$$H = \frac{12}{(18)(19)} \left[\frac{(56)^2}{6} + \frac{(58)^2}{6} + \frac{(57)^2}{6} \right] - 3(19) = 0.0117.$$

c. After the second exchange, SR_1 further decreases to 55, and SR_2 increases to 59. H increases further, from 0.01 to 0.05, as the distribution of ranks among the samples departs further from the ideal situation for identical populations.

$$H = \frac{12}{(18)(19)} \left[\frac{(55)^2}{6} + \frac{(59)^2}{6} + \frac{(57)^2}{6} \right] - 3(19) = 0.0468.$$

d. This final dispersion of the ranks is the most extreme case. The smallest values are all in one sample, the middle are all in another, and the largest are all in the third sample. H increases by a sizable amount when this occurs. $SR_1 = 1 + 2 + 3 + 4 + 5 + 6 = 21$, $SR_2 = 7 + 8 + 9 + 10 + 11 + 12 = 57$, and $SR_3 = 13 + 14 + 15 + 16 + 17 + 18 = 93$.

$$H = \frac{12}{(18)(19)} \left[\frac{(21)^2}{6} + \frac{(57)^2}{6} + \frac{(93)^2}{6} \right] - 3(19) = 15.1579.$$

The results of Parts a, b, c, and d in this solution demonstrate that H increases as the rank sums become more dispersed. When the mean rankings are similar, H is close to zero. Thus, when H_0 is true, we expect H to be close to zero. H would equal zero if it were not for sampling error. But when H_1 is true and the populations are not identical, the mean rankings are likely to be dispersed and, hence, H will be large. H_0 is rejected when the dispersion of ranks—and consequent large H value—indicates a difference in population means. How large H must be before we reject H_0 is determined by α and the sampling distribution of the test statistic, H.

> When H_0 is true and there are at least five observations in each independent, random sample, the sampling distribution of H is approximately a chi-square distribution with $J - 1$ degrees of freedom, where J is the number of samples or treatments in the experiment.

Section 13.7   The Kruskal-Wallis Test

■ **Example 13-8 (continued):** Find the rejection region and make a decision for the astronaut data, assuming the test is to be done at the 0.05 level. Recall that there are three treatment samples and that $H = 13.0713$. Interpret the decision.

■ **Solution:** Reject H_0, because $H = 13.07 > 5.99$, the value from the chi-square table (Appendix N) with an area of 0.05 in the tail and $3 - 1 = 2$ degrees of freedom. This decision indicates that mean reaction time differs according to training method.

The following box summarizes the steps for performing the Kruskal-Wallis test.

Kruskal-Wallis Test Procedure
1. Collect an independent random sample from each of the populations, or randomly assign individuals to the treatments, with at least five observations in each sample.
2. Combine the different sample values into a single sample, arranged in ascending order.
3. Assign ranks to each observation, beginning with 1 for the smallest value. Assign the mean rank of a set of tied values to each value in the set.
4. Calculate the rank sum, SR, for each sample.
5. Calculate

$$H = \frac{12}{n_T(n_T + 1)} \left[\sum \frac{(SR_i)^2}{n_i} \right] - 3(n_T + 1),$$

where the i subscripts refer to the different samples.
6. Use the chi-square distribution with $J - 1$ degrees of freedom to establish the critical value for the rejection region in the right tail. J is the number of treatments or samples in the experiment.
7. Reject H_0 if the H statistic is larger than the critical value from the table.
8. Rejecting H_0 means that the population means differ.

■ **Example 13-9:** A group of educators ranked 25 activities for their school system to accomplish over the next five-year period. The following table groups these ranks according to expected cost: very small, moderate, costly, and very expensive. Each activity's numerical rank is based on its overall assessed priority. Use this information to determine if the average ranking varies with the cost of the activity. Use a 0.025 significance level. Interpret the decision.

Cost of Activity			
Very Small	Moderate	Costly	Very Expensive
5	22	17	13
14	21	3	18
1	16	9	11
10	12	2	8
4	25	23	6
7		15	24
20		19	

■ **Solution:** Step 1:   There are at least five observations in each sample, and the activities chosen in each sample are treated as randomly selected activities from a population of activities that fit the sample's cost description.

Steps 2–3: The school system presented the 25 activities as a single sample, and the educators ranked this single sample of activities. So we already know the ranks when all 25 are combined into one sample.

Step 4: $SR_1 = 61$, $SR_2 = 96$, $SR_3 = 88$, and $SR_4 = 80$.

Step 5:

$$H = \frac{12}{(25)(25+1)} \left[\frac{(61)^2}{7} + \frac{(96)^2}{5} + \frac{(88)^2}{7} + \frac{(80)^2}{6} \right] - 3(25+1) = 5.9580.$$

Step 6: The critical value from the chi-square table with 3 degrees of freedom ($J = 4$, because there are four samples) and a tail area of 0.025 is 9.35.

Step 7: We cannot reject H_0, because $H = 5.9580 < 9.35$.

Step 8: The data do not demonstrate that the average priority varies with cost.

Section 13.7.3 Problems

89. In Problem 86 an educational researcher selects three random samples of times for students to complete final exams (in minutes), one sample for each of three periods during finals week: first day, middle day, and last day. The Kruskal-Wallis test statistic is $H = 1.2725$. Test for a difference in mean time to complete the test for the three different periods at the 0.05 significance level. Interpret the decision.

90. In Problem 87 a physicist selects random samples of the power (measured in horsepower) of experimental engines developed by students from five different majors. From the data she finds that $H = 0.9969$. Use this information to test for a significant difference in average horsepower among the students majoring in different disciplines at the 0.01 level.

91. In Problem 88 a manager randomly selects 30 employees and ranks them on perceived productivity in teamwork situations. Then the employer sorts the employees according to the objective of their team assignment (personnel matters, customer satisfaction, or strategic planning) in order to demonstrate that the type of objective affects productivity. He computes the Kruskal-Wallis test statistic, $H = 1.8446$, for the data. Can he demonstrate a significant difference in average productivity depending on team objective at the 0.1 level?

92. By separating one survey's top 25 ranked business schools according to their geographic region and listing their acceptance rates, we obtain the following data set. Use this information to perform a Kruskal-Wallis test at the 0.01 level to determine if average acceptance rates vary among regions. Interpret the decision.

Business School Acceptance Rates (%)

Upper East Coast & New England	West Coast & South	Upper Midwest (includes Western Pennsylvania)
15.0	10.2	22.6
19.4	18.2	30.0
21.3	23.0	32.4
13.4	15.4	31.2
37.1	24.7	29.0
14.9	20.7	26.8
35.0	28.1	33.0
35.9	31.9	
	31.7	
	28.1	

Source: College Research Group, as reported in "Top 25 Business Schools," *U.S. News & World Report*, April 29, 1991, p. 68.

13.8 The Computer and Nonparametric Tests

This section presents examples of Minitab output for the three tests from this unit, Sign, Wilcoxon-Mann-Whitney rank sum, and Kruskal-Wallis.

13.8.1 Computer Output for the Sign Test

Figure 13-19 is the Minitab printout for a sign test to determine if the median salary for top hitters or pitchers in baseball is $1,250,000. Each of the two data sets is a set of salaries. The circled values correspond to the following:

Section 13.8   The Computer and Nonparametric Tests

```
                            1              2
       SIGN TEST OF MEDIAN = 1250000 VERSUS N.E. 1250000
                  3     4      5      6       7        8
                  N   BELOW  EQUAL  ABOVE  P-VALUE   MEDIAN
       HITTERS    44    18     1     25    0.3604   1375000
       PITCHERS   45    30     0     15    0.0357    925000
```

Figure 13-19

Minitab output for sign test of baseball salaries

1. M_0, the specified median value being tested.
2. The alternative hypothesis, H_1, specified by the user, where
   "N.E." represents "not equal,"
   "G.T." represents "greater than,"
   "L.T." represents "less than."
3. The number of observations in the sample, n.
4. The number of observations smaller than M_0.
5. The number of observations equal to M_0.
6. The number of observations larger than M_0.
7. The p-value for the user specified direction for the test.
8. The actual median for the n sample observations.

If the sample contains 50 or fewer observations, the p-value is exact; that is, the computer prints the exact binomial probability. For more than 50 observations, the computer prints a normal approximation.

■ **Example 13-10:** Assuming each set of salaries is a random sample of salaries of top hitters or pitchers, use the results in Figure 13-19 to conduct the sign test at the 0.05 level. Then corroborate your decision with the p-value. (Source: Lawrence Hadley, and Elizabeth Gustafson, "Major League Baseball Salaries: The Impacts of Arbitration and Free Agency," *Journal of Sport Management*, Vol. 5, 1991, pp. 111–27.)

■ **Solution:** The test is two-tailed in the printout, because of the "N.E." listed in the top line. $\hat{p} = 25/43 = 0.5814$ for hitters (we use 43 because one of the hitter's salaries is \$1,250,000) and $15/45 = 0.3333$, for pitchers. The standard deviation for hitters is $0.5/\sqrt{43} = 0.0762$, and the transformed test statistic is 1.0675, which is between ± 1.96. Consequently, we cannot reject H_0. The data do not support the hypothesis that the median hitter's salary is different from \$1,250,000. For pitchers, the standard deviation is $0.5/\sqrt{45} = 0.0745$, and the transformed test statistic is -2.2361, which is smaller than -1.96. Therefore, we reject H_0 and decide that the median pitcher's salary does not equal \$1,250,000. The p-value for hitters exceeds 0.05, and the p-value for pitchers does not, results that substantiate the earlier decisions.

13.8.2 Computer Output for the Wilcoxon-Mann-Whitney Rank Sum Test

Figure 13-20 displays the printout for the Wilcoxon-Mann-Whitney rank sum test of the cigarette data from Example 13-5. The circled numbers correspond to the following information:

1. Sample labels.
2. Sample size.
3. Sample median.
4. SR used for test: corresponding to first sample listed on printout.
5. The alternative hypothesis, H_1, specified by the user, where

   "N.E." represents "not equal,"
   "G.T." represents "greater than," and
   "L.T." represents "less than."

Figure 13-20
Minitab output for Wilcoxon-Mann-Whitney test of smoking data

```
          Mann-Whitney Confidence Interval and Test
         ①                ②                    ③
          CANCER    N =  12     MEDIAN  =       42.00
          NONCANCR  N =  10     MEDIAN  =       14.50
          POINT ESTIMATE FOR ETA1-ETA2 IS       25.50
          95.6 PCT C.I. FOR ETA1-ETA2 IS (5.00,42.00)
        ④ W = 177.5                                                  ⑥
        ⑤ TEST OF ETA1 = ETA2  VS.  ETA1 N.E. ETA2 IS SIGNIFICANT AT 0.0101
        ⑦ THE TEST IS SIGNIFICANT AT 0.0101 (ADJUSTED FOR TIES)
```

Figure 13-21
Minitab output for Wilcoxon-Mann-Whitney test of secretarial pools

```
          Mann-Whitney Confidence Interval and Test

          POOL2        N =  24        Median  =         23.00
          POOL1        N =  26        Median  =         27.00
          Point estimate for ETA1-ETA2 is                -3.00
          95.1 pct c.i. for ETA1-ETA2 is (-11.00, 7.00)
          W = 582.0
          Test of ETA1 = ETA2 vs. ETA1 L.T. ETA2 is significant at 0.2834
          The test is significant at 0.2834 (adjusted for ties)

          Cannot reject at alpha = 0.05
```

6. *P*-value for test for alternative hypothesis stated in number 5.

7. *P*-value for test for alternative hypothesis stated in number 5 after adjusting for ties.

■ **Example 13-5 (continued):** Use Figure 13-20 to perform a Wilcoxon-Mann-Whitney rank sum test for a difference in average number of cigarettes smoked per day at the 0.02 significance level.

■ **Solution:** Using $\mu_{SR} = 12(22 + 1)/2 = 138$ and $\sigma_{SR} = \sqrt{[12(10)(22 + 1)]/12} = 15.1658$, the test statistic, 177.5, converts to $Z = (177.5 - 138)/15.1658 = 2.60$. The test statistic exceeds 2.33, the critical value for a two-tailed test at the 0.02 significance level, so we reject H_0 and conclude that there is a significant difference in average number of cigarettes smoked at the 0.02 significance level.

Question: When you perform a one-tailed Wilcoxon-Mann-Whitney rank sum test for the difference between means, $\mu_{cancer} > \mu_{noncancer}$ (or $\mu_1 > \mu_2$), which population should you state first in H_1: $\mu_i - \mu_j$ if you are using Minitab? State the complete H_1.

Answer: For Minitab, use the variable printed first in the output for *i* and the second for *j*. In this case, we would use H_1: $\mu_{cancer} - \mu_{noncancer} > 0$.

■ **Example 13-6 (continued):** Figure 13-21 shows the Minitab output for testing for a difference in productivity of the secretarial pools. State H_1 for the presentation in the printout. If the test had been in the opposite tail, "G.T.," what would the *p*-value be? Would the decision change in this case?

■ **Solution:** The order of the means for H_1 will be $\mu_2 - \mu_1$ (ETA1 still refers to the first variable listed in the printout, in this case pool 2). For a test in the opposite tail, we want H_1: $\mu_2 - \mu_1 > 0$. The *p*-value would be $0.7166 = 1 - 0.2834$, and we would continue to be unable to reject H_0.

13.8.3 Computer Output for the Kruskal-Wallis Test

Figure 13-22 displays the output for the Kruskal-Wallis test for the astronaut reaction time and training method data from Example 13-7. The circled numbers in the figure indicate the following information:

```
        1          2         3         4          5
      LEVEL      NOBS     MEDIAN   AVE. RANK   Z VALUE
        1         10      0.8300      7.2       -3.10
        2          6      1.4400     20.3        3.13
        3          8      1.0900     13.3        0.37
      OVERALL    24                  12.5
                  6                    7
             8         9        10
          H = 13.07  d.f. = 2  p = 0.002
```

Figure 13-22
Minitab output for Kruskal-Wallis test of astronaut data

1. Treatment sample labels (the analyst must supply the meaning of the level values).
2. Sample sizes.
3. Sample medians.
4. Sample mean ranks.
5. Standardized distance of sample mean ranks from mean rank for all observations.
6. Total number of observations.
7. Mean of ranks for all observations.
8. H.
9. Degrees of freedom for Kruskal-Wallis test.
10. P-value.

■ **Example 13-7:** Confirm the earlier decision to reject H_0 at the 0.05 level for the Kruskal-Wallis test using the p-value in Figure 13-22.

■ **Solution:** We reject H_0, because $0.002 < 0.05$. Average reaction time varies with training method.

■ **Example 13-11:** To enlighten the argument over the quality differences in cantaloupes from different farming regions in Virginia, county agricultural agents from five regions each randomly selected eight cantaloupe growers in their region. They requested the farmers identify their best specimen, which the agents promptly purchased. Expert tasters ranked the 40 cantaloupes for epicurean quality. Perform a Kruskal-Wallis test at the 0.01 level with these results, using the computer output in Figure 13-23. Interpret the decision.

■ **Solution:** The critical value for four degrees of freedom is 13.3. $H = 15.25$ exceeds this value, so we reject H_0. Also the p-value is 0.004, which is less than 0.01, which signals rejection. Thus, the typical quality varies with region.

Section 13.8 Problems

93. Suppose Figure 13-24 shows Minitab output for analyzing pulse rates using the three tests in this unit.
    a. Use the sign test and the printout in Part a of the figure to determine if the median pulse rate is greater than 100 at the 0.05 significance level after a 10-minute exercise session. Interpret the decision.
    b. Use the Wilcoxon-Mann-Whitney rank sum test and the output in Part b of the figure to determine if average pulse rate is different at the 0.05 significance level for pushups versus pullups. Interpret the decision.
    c. Use the Kruskal-Wallis test and the printout in Part c of the figure to test for a difference in average pulse rate at the 0.05 level for different lengths of exercise time. Interpret the decision.

94. Suppose Figure 13-25 shows Minitab output for analyzing military enlistment periods using the three tests in this unit.
    a. Use the sign test and the printout in Part a of the figure to determine, at the 0.1 significance level, if the median Marine enlistment period is less than nine years. Interpret the decision.

Figure 13-23
Minitab output for Kruskal-Wallis test of cantaloupe data

```
         LEVEL    NOBS    MEDIAN    AVE. RANK    Z VALUE
           1        8      34.50       33.1        3.42
           2        8      25.50       23.8        0.88
           3        8      14.50       15.3       -1.42
           4        8      16.00       17.0       -0.95
           5        8      12.50       13.4       -1.93
         OVERALL   40                  20.5

         H = 15.25   d.f. = 4   p = 0.004
```

Figure 13-24
Minitab output for analysis of pulse rate data

a) SIGN TEST OF MEDIAN = 100.0 VERSUS G.T. 100.0

```
                N    BELOW   EQUAL   ABOVE    P-VALUE    MEDIAN
    10MINUTE   45     20       0      25      0.2757     102.0
```

b) Mann-Whitney Confidence Interval and Test

```
    PUSHUPS    N = 24    Median =    102.50
    PULLUPS    N = 21    Median =    101.00
    Point estimate for ETA1-ETA2 is    0.00
    95.1 pct c.i. for ETA1-ETA2 is (-6.00, 6.00)
    W = 549.0
    Test of ETA1 = ETA2 vs. ETA1 n.e. ETA2 is significant at 0.9546
    The test is significant at 0.9546 (adjusted for ties)

    Cannot reject at alpha = 0.05
```

c)
```
         LEVEL    NOBS    MEDIAN    AVE. RANK    Z VALUE
           1       45     102.0       42.1        -7.19
           2       45     118.0      115.7         3.74
           3       45     112.0      101.6         1.66
           4       45     113.0      102.6         1.79
         OVERALL  180                 90.5

         H = 53.73   d.f. = 3   p = 0.000
         H = 53.79   d.f. = 3   p = 0.000 (adj. for ties)
```

Figure 13-25
Minitab output for analysis of military enlistment periods

a) SIGN TEST OF MEDIAN = 9.000 VERSUS L.T. 9.000

```
                N    BELOW   EQUAL   ABOVE    P-VALUE    MEDIAN
    MARINES   100     46      23      31      0.0553     9.000
```

b) Mann-Whitney Confidence Interval and Test

```
    MARINES    N = 100    Median =    9.0000
    ARMY       N = 100    Median =    4.5933
    Point estimate for ETA1-ETA2 is    4.0864
    95.0 pct c.i. for ETA1-ETA2 is (3.6983, 4.4126)
    W = 15019.0
    Test of ETA1 = ETA2 vs. ETA1 n.e. ETA2 is significant at 0.0000
    The test is significant at 0.0000 (adjusted for ties)
```

c)
```
         LEVEL    NOBS    MEDIAN    AVE. RANK    Z VALUE
           1      100      9.000      339.6        13.89
           2      100      4.593      102.3        -9.81
           3      100      5.301      161.4        -3.90
           4      100      5.684      198.7        -0.18
         OVERALL  400                 200.5

         H = 228.35   d.f. = 3   p = 0.000
         H = 228.50   d.f. = 3   p = 0.000 (adj. for ties)
```

b. Use the Wilcoxon-Mann-Whitney rank sum test and the output in Part b of the figure to determine if average military tenure of enlistees in the Marines and Army differ at the 0.1 significance level. Interpret the decision.

c. Use the Kruskal-Wallis test and the printout in Part c of the figure to test at the 0.05 level for a difference in average enlistment period for four different branches of the military. Interpret the decision.

Summary and Review

This unit presents three nonparametric tests, sign, Wilcoxon-Mann-Whitney rank sum, and Kruskal-Wallis. These tests substitute for parametric tests discussed earlier in this book, when situations do not satisfy the conditions and assumptions necessary for a valid result.

The sign test is a nonparametric test for the median of a population. It is based on the fact that approximately half of the nonmedian sample values should be above the median, and half should be below it. If more than half the sample values exceed a hypothesized value for the median, M_0, then the median may be larger than M_0. If less than half are above, the median may be smaller than hypothesized.

The Wilcoxon-Mann-Whitney rank sum test is a nonparametric substitute for the test for a difference between two population means using independent random samples. We rank the pooled sample values and use these ranks to determine whether the two original populations have the same mean. The test statistic is the sum of one sample's ranks. For convenience we usually choose the sample from the population that we state first in H_1 (choose population 1 if H_1 is stated as $\mu_1 - \mu_2$). The rank sum statistic is normally distributed if both samples contain at least 10 observations. A test statistic that is larger or smaller than anticipated when H_0 is true indicates that the two population means differ. More specifically, a rank sum that is larger than expected leads us to believe that the mean of the population represented by the statistic is larger than the other population mean. A rank sum that is smaller than expected reverses the relative magnitudes of the population means.

The Kruskal-Wallis test parallels a one-way ANOVA, but we employ ranks rather than actual magnitudes. Much as ANOVA extends the parametric test for a difference between two population means, the Kruskal-Wallis test extends the nonparametric Wilcoxon-Mann-Whitney test for a difference between two population means. We rank the pooled sample values. Each sample's rank sum, SR, contributes to H, the test statistic. When H_0 is true, the mean SR for each sample should be similar (ideally speaking, identical), which produces an H of 0 or close to 0. When H_1 is true, H becomes large as the rank sums differ.

The following table summarizes details of the four test procedures from this chapter. If the samples are small, there are tables of more exact critical values for the tests covered in this unit, but this book does not cover these situations.

Characteristic	Chi-Square Test for Independence	Sign Test	Wilcoxon-Mann-Whitney Rank Sum Test	Kruskal-Wallis Test
Parametric Equivalent	Correlation, Regression, One-way ANOVA	Test for population mean	Tests for difference between population means	One-way ANOVA
Nonparametric Hypothesis	H_0: Variables A and B are independent H_1: Variables A and B are dependent	H_1: $M \neq M_0$ H_1: $M < M_0$ H_1: $M > M_0$, where M is population median	H_1: $\mu_1 - \mu_2 = 0$ H_1: $\mu_1 - \mu_2 > 0$ H_1: $\mu_1 - \mu_2 < 0$	H_0: $\mu_1 = \mu_2 = \cdots = \mu_J$ H_1: at least one of the means is different
Assumptions and Requirements	f_e at least 5 for most or all cells	Large random sample	1. Independent random samples from continuous populations 2. Samples both contain at least 10 observations	1. Independent random samples from continuous populations 2. Samples all contain at least 5 observations

(continued on next page)

Characteristic	Chi-Square Test for Independence	Sign Test	Wilcoxon-Mann-Whitney Rank Sum Test	Kruskal-Wallis Test
Assumptions and Requirements *(continued)*			3. No, or few, identical values tied for rank in the combined sample 4. Population variances are equal	3. No, or few, identical values tied for rank in the combined sample 4. Population variances are equal.
Test Statistic	$\sum_{\text{all cells}} \dfrac{(f_o - f_e)^2}{f_e}$	$\hat{p}$ = proportion of non-M_0 sample values that are greater than M_0	SR = rank sum of sample mentioned first in parametric H_1	$H = \dfrac{12}{n_T(n_T + 1)} \left[\sum \dfrac{(SR_i)^2}{n_i} \right] - 3(n_T + 1)$
Sampling Distribution	Chi-square with $(R - 1)(C - 1)$ degrees of freedom, where R and C are the number of categories of the two variables	Approximate normal with mean 0.5 and $\sigma_{\hat{p}} = 0.5/\sqrt{n}$	Approximately normal with mean $\mu_{SR} = n_i(n_T + 1)/2$ and $\sigma_{SR} = \sqrt{\dfrac{n_1(n_2)(n_T + 1)}{12}}$ where $n = 1$ or 2	Chi-square with $J - 1$ degrees of freedom, where J is the number of treatments or samples in the experiment

Multiple-Choice Problems

95. The sign test described in this unit, substitutes for which parametric test
   a. Paired-difference between two population means.
   b. Difference between two population means using independent samples.
   c. The mean of a population.
   d. One-way analysis of variance.
   e. Two-way analysis of variance.

96. The Wilcoxon-Mann-Whitney rank sum test is an alternative test for
   a. Paired-difference between two population means.
   b. Difference between two population means using independent samples.
   c. The mean of a population.
   d. One-way analysis of variance.
   e. Two-way analysis of variance.

97. The Kruskal-Wallis test is an alternative test for
   a. Paired-difference between two population means.
   b. Difference between two population means using independent samples.
   c. The mean of a population.
   d. One-way analysis of variance.
   e. Two-way analysis of variance.

98. Which of the nonparametric tests presented in this unit uses a test statistic that is normal or approximately normal?
   a. Sign.
   b. Wilcoxon-Mann-Whitney rank sum.
   c. Kruskal-Wallis.
   d. Sign and Wilcoxon-Mann-Whitney rank sum.
   e. Sign and Kruskal-Wallis.
   f. Kruskal-Wallis and Wilcoxon-Mann-Whitney rank sum.
   g. All of the test statistics in this unit are normally distributed.
   h. None of the test statistics in this unit are normally distributed.

99. The basic idea of the sign test is that
   a. The number of observations in a data set on each side of the hypothesized median should be the same.
   b. The average rank of two samples from the same population should be the same.
   c. The average rank of three or more samples from the same population should be the same.
   d. The expected frequency is the product of sample size and the probability of the data falling into one of the categories.
   e. The median is normally distributed about μ but has a larger standard deviation than the mean.

100. The basic idea of the Wilcoxon-Mann-Whitney rank sum test is that
   a. The number of observations in a data set on each side of the hypothesized median should be the same.
   b. The average rank of two samples from the same population should be the same.
   c. The average rank of three or more samples from the same population should be the same.
   d. The expected frequency is the product of sample size and the probability of the data falling into one of the categories.
   e. The median is normally distributed about μ but has a larger standard deviation than the mean.

Multiple-Choice Problems

101. The basic idea of the Kruskal-Wallis test is that
  a. The number of observations in a data set on each side of the hypothesized median should be the same.
  b. The average rank of two samples from the same population should be the same.
  c. The average rank of three or more samples from the same population should be the same.
  d. The expected frequency is the product of sample size and the probability of the data falling into one of the categories.
  e. The median is normally distributed about μ but has a larger standard deviation than the mean.

102. Which of the three tests presented in this unit is based on a binomial random variable?
  a. Sign.
  b. Wilcoxon-Mann-Whitney rank sum.
  c. Kruskal-Wallis.
  d. Chi-square.
  e. None of the above.

103. If M_0, the hypothesized median, results in the following diagram from the sample data, the true median probably

$$—xxx\underset{M_0}{\vert}—x-x-x-x-x-x-x-x—$$

  a. Equals M_0.
  b. Is smaller than M_0.
  c. Is larger than M_0.
  d. Is nonexistent.
  e. Cannot be determined based on this information.

104. Which test statistic presented in this unit is neither normal nor approximately normal?
  a. $\hat{p}$.
  b. SR.
  c. H.
  d. All of them are normal or approximately normal.
  e. None of them are normal or approximately normal.

105. In the Wilcoxon-Mann-Whitney rank sum test, suppose that the samples are nonrandom and dependent. The result is that
  a. $\hat{p}$ is no longer binomial.
  b. Rank values are not governed solely by chance.
  c. p is no longer 0.5.
  d. The medians of both populations are identical.
  e. SR will have an F rather than a chi-square distribution.

106. How many sample observations do we require to properly employ the normal approximation in a Wilcoxon-Mann-Whitney rank sum test?
  a. The number of observations that makes both pn and $(1 - p)n$ greater than 5.
  b. More than 30.
  c. At least 5 in each sample.
  d. More than the degrees of freedom needed for the test.
  e. At least 10 in each sample.

Use this information to answer questions 107–111.
The numbers of leadership changes in four countries during five randomly selected decades are represented by the following data set.

	Countries			
	1	2	3	4
	2	7	15	5
	0	10	1	11
	12	2	4	9
	3	14	1	13
	10	8	6	6

107. Which test do we use to detect a difference in the average number of leadership changes in the four countries?
  a. Sign.
  b. Wilcoxon-Mann-Whitney rank sum.
  c. Kruskal-Wallis.
  d. Any of the above.
  e. None of the above.

108. What rank values would we assign to the "1" observations in the combined sample?
  a. 18.5.   f. 18.
  b. 1.   g. 1.5.
  c. 19.   h. 2 and 3.
  d. 1 and 2. i. 3.
  e. 2.5.   j. 2.

109. The value of the H statistic is
  a. 52.5.   d. 210.
  b. 2.5.   e. 10.
  c. 2.44.

110. The decision at the 0.05 level is
  a. Cannot reject, because $2.44 > 1.96$.
  b. Cannot reject, because $2.44 < 9.49$.
  c. Cannot reject, because $2.44 < 7.81$.
  d. Reject, because $2.44 > 1.96$.
  e. Reject, because $2.44 > 1.645$.

111. The decision in Problem 110, when interpreted means that
  a. The average number of leadership changes is the same in all four countries for the decades studied.
  b. The average number of leadership changes is different in each of the four countries for the decades studied.
  c. The distribution of leadership changes is different for two of the four countries for the decades studied.
  d. The distribution of leadership changes is different for each of the countries for the decades studied.
  e. The average number of leadership changes in all four countries is affected by the decade chosen.

112. What is the value of the H statistic when the means of the rank sums, $\overline{SR}_i$, are the same value for all treatments?
  a. 0.   d. n_T/number of treatments.
  b. 1.   e. Larger than 1.
  c. n_T.

113. In a Wilcoxon-Mann-Whitney rank sum test, which of the following is always the same for both samples?
  a. SR.   c. σ_{SR}.   e. X_i.
  b. μ_{SR}.   d. n.

A psychologist randomly selects 20 former patients, 10 females and 10 males, and determines each patient's length of time in therapy (number of visits). Use the following data to answer Questions 114–120.

Number of Visits

Female	Male
48	34
27	61
4	25
8	2
50	6
79	15
15	5
21	17
10	39
9	44

114. Which test is used to determine if there is a difference between average number of visits for male and female patients?
  a. Sign.
  b. Wilcoxon-Mann-Whitney rank sum.
  c. Kruskal-Wallis.
  d. Any of the above.
  e. None of the above.

115. If the psychologist believes that $\mu_F - \mu_M > 0$, what test statistic should be used for this test?
  a. SR_M.   f. $\overline{X}_M$.
  b. SR_F.   g. $\overline{X}_F - \overline{X}_M$.
  c. $\overline{SR}_M$.   h. $SR_F - SR_M$.
  d. $\overline{SR}_F$.   i. $\overline{SR}_F - \overline{SR}_M$.
  e. $\overline{X}_F$.

116. What is the mean of the test statistic?
  a. 105.   d. 420.
  b. 0.   e. 1.
  c. 210.

117. What is the standard error?
  a. 100.   d. 10.
  b. 175.   e. 13.23.
  c. 1.

118. What is SR_F?
  a. 210.   c. 10.75.   e. 105.
  b. 102.5.   d. 107.5.

119. What is the decision at the 0.05 level?
  a. Cannot reject, because $2.5 < 2.575$.
  b. Cannot reject, because $0.19 < 1.96$.
  c. Cannot reject, because $0.19 < 1.645$.
  d. Reject, because $107.5 > 105$.
  e. Reject, because $2.5 > 1.645$.

120. The data indicate that
  a. The average number of visits is the same for male and female patients.
  b. The average number of visits for females is greater than that for males.
  c. The average number of visits for males is greater than that for females.
  d. The population distributions are situated in different locations on the number line.
  e. The population distributions are different for each sex.

Forty-two of 108 randomly selected persons say that they brush their teeth more than twice a day, and eight persons say exactly twice a day. Use this information in Questions 121–124 to test if the median number of times people brush is less than twice a day.

121. What is the value of the test statistic?
  a. 0.5.   d. 0.61.
  b. 0.39.   e. 0.42.
  c. 0.58.

122. What are the mean and the standard deviation, respectively, of the test statistic?
  a. 0.5 and 10.   d. 50 and 10.
  b. 0.5 and 0.05.   e. 0.42 and 0.053.
  c. 54 and 10.

123. What is the decision at the 0.05 significance level?
  a. Reject, because $-1.66 < -1.645$.
  b. Reject, because $-1.6 > -1.96$.
  c. Reject, because $0.42 < 0.5$.
  d. Cannot reject, because $-1.6 > -1.645$.
  e. Cannot reject, because $0.58 > 0.5$.

124. The interpretation of this decision is that the median number of times that people brush their teeth per day is
  a. Two.
  b. Less than two.
  c. Greater than two.
  d. Less than or equal to two.
  e. Greater than or equal to two.

Figure 13-26 presents the Wilcoxon-Mann-Whitney rank sum test computations for data on enrollments in adult education programs. Use the results to answer

Figure 13-26
Minitab output for Wilcoxon-Mann-Whitney test of adult education enrollment in the South and West

```
Mann-Whitney Confidence Interval and Test

SOUTH        N = 17       Median =          26.3
WEST         N = 13       Median =          12.9
Point estimate for ETA1-ETA2 is              13.2
95.1 pct c.i. for ETA1-ETA2 is (2.9, 37.9)
W = 319.0
Test of ETA1 = ETA2 vs. ETA1 n.e. ETA2 is significant at 0.0213
```

Multiple-Choice Problems

```
LEVEL     NOBS    MEDIAN   AVE. RANK   Z VALUE
  1         9     760.0      30.4       1.11
  2        12     646.0      22.2      -0.90
  3        16     644.5      20.2      -1.78
  4        13     833.0      31.7       1.79
OVERALL    50                25.5

H = 6.15   d.f. = 3   p = 0.105
H = 6.15   d.f. = 3   p = 0.105 (adj. for ties)
```

Figure 13-27
Minitab output for Kruskal-Wallis test of regional tax data

```
SIGN TEST OF MEDIAN = 64.00 VERSUS G.T. 64.00

        N    BELOW   EQUAL   ABOVE   P-VALUE   MEDIAN
AGE    56     20       2      34     0.0384    66.00
```

Figure 13-28
Minitab output for sign test of ages of senior citizen club

Questions 125–129 in order to determine if enrollments in adult education programs differ between the Southern and Western regions of the United States (including the District of Columbia). (Source: *1986 State and Metropolitan Area Data Book* [Washington, D.C.: Government Printing Office, 1986], p. 532.)

125. Which area appears to have the larger enrollment?
  a. South.
  b. West.
  c. They have equal enrollments.
  d. The information on the printout is insufficient to answer the question.

126. If a two-tailed test is performed at the 0.05 level, the decision is
  a. Reject, because $2.32 > 1.96$.
  b. Reject, because $2.32 > 1.645$.
  c. Reject, because $3.19 > 1.96$.
  d. Cannot reject, because $1.46 < 1.96$.
  e. Cannot reject, because $1.46 < 1.645$.

127. If H_1 were $\mu_S < \mu_W$, and SR_W rather than SR_S were given in the printout, state the form of $\mu_i - \mu_j$ that we use to perform the test.
  a. $\mu_S - \mu_W < 0$.
  b. $\mu_S - \mu_W > 0$.
  c. $\mu_W - \mu_S < 0$.
  d. $\mu_W - \mu_S > 0$.
  e. $\mu_S - \mu_W \neq 0$.

128. The decision at the 0.05 level for the hypothesis as stated to answer Problem 127 is
  a. Reject, because $-2.32 < -1.645$.
  b. Reject, because $-2.32 < -1.96$.
  c. Reject, because $146 < 201.50$.
  d. Cannot reject, because $-2.32 < 1.96$.
  e. Cannot reject, because SR is below the mean and the test is upper-tailed.

129. Which of the following is the interpretation of the decision in Problem 128?
  a. Average enrollment in the South is larger than average enrollment in the West.
  b. Average enrollment in the West is larger than average enrollment in the South.
  c. Average enrollments in the two regions are not significantly different.
  d. Average enrollments in the two regions are significantly different.
  e. Average enrollment in the South is at least as large as average enrollment in the West.

Figure 13-27 shows the Kruskal-Wallis test output for data on taxes per capita for different regions of the United States (not including the District of Columbia). Levels 1 through 4 correspond to Northeast, Midwest, South, and West, respectively. Use this information to answer Questions 130–132. (Source: *1986 State and Metropolitan Area Data Book* [Washington, D.C.: Government Printing Office, 1986], p. 567.)

130. A ranking of the four areas from highest to lowest taxes, based on the sample ranks, is
  a. Northeast > Midwest > South > West.
  b. Northeast > West > Midwest > South.
  c. West > South > Northeast > Midwest.
  d. Northeast > South > Midwest > West.
  e. West > Northeast > Midwest > South.

131. The result of the Kruskal-Wallis test at the 0.05 level is
  a. Cannot reject, because $6.15 < 67.50$.
  b. Cannot reject, because $6.15 < 7.81$.
  c. Cannot reject, because $6.15 < 9.49$.
  d. Reject, because $6.15 > 1.96$.
  e. Reject, because $6.15 > 1.645$.

132. This decision can be interpreted as
  a. Per capita tax payments are different in different regions.
  b. Per capita tax payments are the same in all four regions.
  c. Per capita tax payments are the same in the West and Northeast but have a different value in the South and Midwest.
  d. Average per capita tax payments are different in different regions.
  e. Average per capita tax payments are the same in all four regions.

Use the printout in Figure 13-28 to perform a sign test to show that the average age is greater than 64 for a senior citizen club where the minimum age is 60 and there is no maximum. Questions 133–135 refer to this situation.

133. A reason to perform a sign test rather than a parametric test for the mean is because
 a. Age is a qualitative variable.
 b. The data set is small.
 c. The population of members' ages is skewed to the right.
 d. Members' ages are not independent, because of the membership conditions.
 e. There will be too many ties, because many people will be the same age.

134. The decision for the sign test at the 0.05 level is
 a. Reject, because more than half of the 56 observations are above 64.
 b. Reject, because $1.90 > 1.645$.
 c. Reject, because $0.6296 > 0.5$.
 d. Cannot reject, because $-1.96 < 1.90 < 1.96$.
 e. Cannot reject, because the median, 66, is within 5 of 64.

135. The interpretation of the decision is
 a. The median age is 64.
 b. The median age is normally distributed.
 c. The median age exceeds 64.
 d. The median age does not exceed 64.
 e. The ages are normally distributed.

Word and Thought Problems

136. What is the test statistic and sampling distribution for each of the three nonparametric tests presented in this unit?

137. Compare the lower-tailed sign test with tossing a coin n times and finding the probability of getting a head on x percent or less of the tosses.

138. On a survey, 220 respondents checked income categories above $25,000 for present income, and 310 checked income categories below $25,000. Use this information to test at the 0.01 level for a median income of $25,000 in the population. Interpret the decision.

139. The lengths of prison sentences for larceny are skewed in a certain jurisdiction. Does the following sample of sentences lengths (in years) corroborate an analyst's belief that the median sentence is shorter than two years? Test the hypothesis at the 0.01 significance level. Interpret the decision.

1 0.5 0.5 2 2.5 3 0.5 0.25 1 1 1 1 3 5 0.5
1 0.5 2 0.25 0.75 0.75 2.5 5 7 10 5 3 0.5 1
1 2 2 2 0.5 3 0.5 0.25

140. a. Use the following sample information to determine if the median birth month is later than July. Test at the 0.01 level, and interpret the decision. Assume that 1,000 birth certificates were investigated to produce the monthly percentages.

Month	Births (%)
Jan.	7.9
Feb.	7.4
Mar.	8.4
Apr.	7.9
May	8.3
Jun.	8.4
Jul.	8.8
Aug.	9.0
Sep.	8.8
Oct.	8.5
Nov.	8.1
Dec.	8.3

b. Use the following information from the same sample to determine if the median birth day is Wednesday. Test at the 0.01 level, and interpret the decision.

Day	Births (%)
Sun.	11.6
Mon.	14.6
Tue.	15.5
Wed.	15.3
Thu.	15.3
Fri.	15.4
Sat.	12.3

c. Use the following information from the sample to determine if the median term of a pregnancy is less than 39 weeks. Test at the 0.01 level, and interpret the decision.

Term (weeks)	Births (%)
27	0.7
28	0.3
29	0.3
30	0.3
31	0.3
32	1.3
33	1.3
34	1.3
35	1.3
36	3.7
37	13.4
38	13.4
39	13.4
40	22.1
41	15.1
42	11.8

Source: Center for Health Statistics, as estimated from graphics by S. Durant, Knight-Ridder Tribune, as reported in "Births and Newborns: A Profile," *Asheville Citizen-Times*, Jan. 20, 1992, p. 1C.

Word and Thought Problems

```
Histogram of RETURNS    N = 230
Each * represents 5 obs.

Midpoint    Count
       0       28   ******
     100      136   ****************************
     200       52   ***********
     300        9   **
     400        3   *
     500        1   *
     600        0
     700        0
     800        1   *



SIGN TEST OF MEDIAN = 110.0 VERSUS L.T. 110.0

              N    BELOW   EQUAL   ABOVE   P-VALUE   MEDIAN
RETURNS     230      124       3     103    0.0922    99.50
```

Figure 13-29
Minitab output for sign test for income tax return data

141. Eighty percent of a survey's 500 respondents say that they watch television at least one hour per day. The other 20% say they watch less than one hour per day.
  a. Can you use this information to test $H_0: M \leq 1$ against the alternative, $H_1: M > 1$? Explain.
  b. Suppose 80% said that they watched television more than 1 hour per day. Perform the test at the 0.005 level. Interpret the decision.

142. a. Consider the computer output in Figure 13-29 which describes a random sample of 230 income tax refunds. Is there a problem for conducting a parametric test to determine if the mean return is less than $110? Explain.
  b. Conduct a sign test for the value using the output at the 0.1 level. Interpret the decision.

143. Use the following sample of coat prices to test for $\mu = \$100$, then test for $M = \$100$, at a 0.05 significance level. Interpret the decisions.

```
 90   90  100  100  100  100  110  110  110  110
110  110  120  120  120  120  140  140  500  500
```

Can you explain the difference between the two decisions? Why might we use the sign test rather than the parametric test for the mean?

144. The following random samples of race times for two runners are independent. Assume that race times are skewed. Use the Wilcoxon-Mann-Whitney rank sum method to test at the 0.05 level for a difference in the average times of the two runners. Interpret the decision. Times are in seconds.

Runner 1 10.2 12.1 15.8 11.0 10.5 13.6 14.5 12.8 13.6 14.0
Runner 2 12.7 12.6 13.4 10.3 14.4 14.2 12.0 12.6 14.8 12.2 11.2 11.6

145. Use the following evaluations of the facilitativeness of living will laws, the ease with which people can make and execute living wills, in two independent random samples of 10 states. The researchers base the evaluations on a content analysis of the laws, and larger scores accompany more facilitative laws. The first sample contains scores of the original law passed by the selected states, while the second sample contains scores based on amended living will laws. Use this information to test at the 0.10 level for an increase in the level of facilitation after amendments. Interpret your decision. (Source: Henry R. Glick and Scott P. Hays, "Innovation and Reinvention in State Policymaking: Theory and the Evolution of Living Will Laws," *Journal of Politics*, Vol. 53, No. 3, 1991, pp. 835–49.)

Original Law Scores 13 13   6   8 12 18 10 12 15   9
Amended Law Scores   12 25 14 20 12   8 10   8 12   9

146. Figure 13-30 shows the computer output for a Wilcoxon-Mann-Whitney rank sum test to show the average enrollment in adult education programs in the Northeast is less than the Midwest enrollment. Use this information to perform the test at the 0.01 significance level. Interpret the decision. Do the histograms support the decision to use the nonparametric rather than a parametric test? Explain. (Source: *1986 State and Metropolitan Area Data Book* [Washington, D.C.: Government Printing Office, 1986], p. 532.)

147. The number of visits to a general practitioner in a given year for a randomly selected group of 20- to 29-year-olds and for a group of 30- to 39-year-olds are listed in the following table. Use this data set and try to show at the 0.01 level that the average number of visits for the older age group is larger. The populations are skewed to the right. Interpret the decision.

Age	Number of Annual Visits
20–29	2 1 0 10 5 0 1 8 25 12 7 4 6 18
30–39	0 4 13 8 3 9 15 12 2 16

Figure 13-30
Minitab output for Wilcoxon-Mann-Whitney test of adult education data

```
Histogram of NRTHEAST    N = 9

Midpoint    Count
   0          2    **
  10          2    **
  20          2    **
  30          1    *
  40          1    *
  50          0
  60          0
  70          0
  80          0
  90          1    *


Histogram of MIDWEST     N = 12

Midpoint    Count
   0          2    **
  10          3    ***
  20          2    **
  30          2    **
  40          1    *
  50          1    *
  60          0
  70          0
  80          1    *


              Mann-Whitney Confidence Interval and Test

NRTHEAST    N =   9      Median =         20.40
MIDWEST     N =  12      Median =         18.55
Point estimate for ETA1-ETA2 is          -2.80
95.7 pct c.i. for ETA1-ETA2 is (-20.62, 16.32)
W = 95.0
Test of ETA1 = ETA2 vs. ETA1 l.t. ETA2 is significant at 0.4018

Cannot reject at alpha = 0.05
```

148. The top 20 teams in an annual college football team ranking are represented by the following data. The underscored rank values indicate that the team finished in the top 20 the previous year. Test at the 0.005 level to determine if the average ranking of the teams that made both this year's and last year's top 20 list is better (closer to 1 or first place) than the average ranking of the teams that only appeared on this year's list. Interpret the decision.

1 2 $\underline{3}$ 4 5 6 7 8 $\underline{9}$ 10 $\underline{11}$ 12 $\underline{13}$ $\underline{14}$ 15 16 17 $\underline{18}$ 19 20

149. How do you determine which SR to use in the test statistic when performing a Wilcoxon-Mann-Whitney rank sum test?

150. Suppose we separate the 500 largest U.S. business firms ranks (in ascending order by sales) into eight industries. The sum of the ranks for the eight industries are 10,804; 13,598; 21,121; 16,934; 15,550; 14,387; 17,854; and 15,002. The corresponding sample sizes are 50, 70, 120, 60, 50, 80, 30, and 40. Use this information to test at the 0.05 level for any difference in average business size among the eight industries. Interpret the decision.

151. The top 25 box-office successes among movies released in a certain year are ranked in ascending order of gross receipts as follows:

1 2 $\underline{3}$ $\underline{4}$ 5 $\underline{6}$ $\underline{7}$ $\underline{8}$ $\underline{9}$ 10 $\underline{11}$ $\underline{12}$ 13 $\underline{14}$ 15 16 17 18 19 $\underline{20}$ $\underline{21}$ 22 23 24 25

The underscored values indicate that the movie is a comedy.

a. Test at the 0.1 level to determine if comedies earn more (have a higher rank value) on average than other types of movies. Use the Wilcoxon-Mann-Whitney rank sum test, and interpret the decision.

b. Repeat the test using the Kruskal-Wallis test statistic. Does the decision change? Explain why these results are reasonable.

```
LEVEL    NOBS    MEDIAN    AVE. RANK    Z VALUE
  1        9      4978       22.3        -0.82
  2       12      4193       19.5        -1.72
  3       17      4505       24.0        -0.67
  4       13      6078       37.1         3.11
OVERALL   51                 26.0

H = 10.33    d.f. = 3    p = 0.016
H = 10.33    d.f. = 3    p = 0.016 (adj. for ties)
```

Figure 13-31

Minitab output for Kruskal-Wallis test of crime rate data

152. A travel agency has collected data on the number of times each of six sites has been included on itineraries each year for the last five years. Using the results listed in the table, test at the 0.01 level to determine if any sites are more popular on average than others throughout the five-year period. Interpret the decision.

		Sites			
1	2	3	4	5	6
10	6	11	15	3	42
17	12	28	22	16	38
5	18	13	25	33	30
27	24	20	19	7	10
29	31	12	14	3	5

153. A medical doctor reviews his patient files for the last six years to determine the number of visits for each of four different childhood diseases. Use the data to test at the 0.05 level for a difference in average number of visits for each of the diseases. Interpret the decision.

	Visits for		
Chicken Pox	**Measles**	**Mumps**	**Tonsillitis**
125	98	57	143
190	153	218	155
210	356	226	187
200	142	63	196
315	130	257	193
298	161	267	201

154. A chemistry professor records the percentages of students in each of six different labs who correctly complete five randomly selected experiments. Use this information to test at the 0.05 level for a difference in average completion rates in the six labs. Interpret the decision.

		Labs			
1	2	3	4	5	6
90	92	87	85	94	97
80	81	76	63	65	72
70	69	86	84	59	61
60	40	45	62	64	58
88	44	66	75	50	55

155. Crime rates in different regions of the United States are compared in the computer output in Figure 13-31. Perform a Kruskal-Wallis test at the 0.1 level using this information. Interpret the decision. (Source: *1986 State and Metropolitan Area Data Book* [Washington, D.C.: Government Printing Office, 1986], p. 541.)

156. To learn if a drug affects rats' loss of fear or avoidance behavior, researchers administered different doses of the drug to three different samples and a placebo to a fourth sample, the control, during the extinction phase of the experiment. During this phase the animals learn that the danger is reduced or no longer exists. After the extinction phase and a further time lag, the researchers record the length of time before the rats venture from a safe position back into the situation where they first encountered the fear-producing experience. We call these retention times. Use the following Kruskal-Wallis H statistics to analyze the variance at the 0.05 significance level. Interpret each decision. All samples contained 14 rats. (Source: Sonia del Cerro and José Borrell, "β-Endorphin Impairs Forced Extinction of an Inhibitory Avoidance Response in Rats," *Life Sciences*, Vol. 41, No. 5, 1987, pp. 579–84.)

  a. H for analyzing the retention times among the four samples that experience the extinction phase is 9.62. In this case, the researchers inject the doses 10 minutes prior to the extinction phase.

  b. H when the drug administration occurs immediately after the extinction phase is 8.5 for the four groups that experience the extinction phase.

  c. In the final experiment, the researchers administered a different drug or drug combination after the extinction phase. One sample received the new drug, one the old drug, and one a combination of the two drugs. The control group received a saline solution. H is 7.84 for this situation.

Review Problems

Asterisked questions denote *p*-value problems, which are optional.

157. The three nonparametric tests in Unit II substitute for which parametric tests? Which conditions and assumptions of each of these parametric tests do test situations often violate? List the assumptions of the nonparametric tests. State a null and its alternative hypotheses for each test.

158. List the assumptions and conditions necessary to perform the Chi-Square test for independence. State the null and alternative hypotheses for this test.

159. A politician claims that the average criminal spends less than one-fourth of his or her actual sentence in prison. A random sample of proportion of sentence spent in prison by a group of former prisoners consists of the following values. Each value is expressed as a percentage.

5   6   9   10   10   10   15   17   17   17   20   21   23
25   25   25   25   27   29   33   33   33   42   48   51   53
68   77   82

 a. Do you think that the data for the percentages of sentence served for *all* former prisoners are normally distributed? Explain.
 b. Use the sign test at the 0.05 level to try to disprove the politician's claim. Interpret the decision.

160. a. To study the existence and incidence of "tonic immobility" (animals feigning death to avoid attack, TI), a researcher observed cockfights (without artificial or natural spurs on the roosters). A match usually consists of several pittings, bouts lasting from the referee's start until one of two events occurs, entangled spurs or 10-second lapse without an attack. The researcher hypothesizes that longer matches tire roosters and increase the likelihood of a rooster using tonic immobility. Use the following results to test for independence of match length and occurrences of TI at the 0.05 level. When comparing bout length (less or more than 11.2 pittings) in 86 bouts with the *number of bouts* where TI did and did not occur, the chi-square test statistic is 31.1 and the *p*-value is 0.001. When comparing bout length with the *number of pittings* involving instances of TI, there were 1,528 pittings, which produced a chi-square test statistic of 36.8 and a *p*-value of 0.001. Do the findings in both tests, one with a sample of 86 bouts, the other with a sample of 1,528 pittings, support or contradict one another?
 b. To ascertain survival rates of rooster participants in cock fights, a researcher obtained the following information from several owners. Does it appear that the hypothesis that outcome of the fight provides information about survivorship is true? Explain. Using this information and an adjustment for instances with only one degree of freedom, the test statistic is 31.7. Verify the hypothesis at the 0.05 level. (Source: Harold A. Herzog, "Immobility in Intraspecific Encounters: Cockfights and the Evolution of 'Animal Hypnosis,'" *The Psychological Record*, Vol. 28, 1978, pp. 543–548.)

Outcome	Survived	Perished
Winner	34	2
Loser	6	21

161. One hundred congressional staffers rated the usefulness of studies potentially worthwhile to Congress for enacting health care legislation (0 = don't remember, 1 = not at all, . . . , and 5 = used extensively). The GAO was responsible for many of the studies. Determine if the degree of usefulness to the staffers differed between GAO and non-GAO studies by performing a chi-square test on the following responses. Test at the 0.05 level, and interpret the decision.

Source of Study	Congressional Staff Usage						Total
	0	1	2	3	4	5	
GAO	0	0	5	9	24	29	67
Non-GAO	0	0	3	7	14	9	33
	0	0	8	16	38	38	100

Source: John F. Boyer and Laura I. Langbein, "Factors Influencing the Use of Health Evaluation Research in Congress," *Evaluation Review*, Vol. 15, No. 5, 1991, pp. 507–32.

162. A random sample of prime interest rates in presidential election years, during congressional election (non-presidential election) years, and in non-election years are listed in the following table. Assume that interest rates tend to be skewed to the right, a few large interest rates lie beyond the mass of rates, and use the nonparametric alternative test for a difference in average interest rate among the three categories of years. Test at the 0.005 level. Interpret the decision.

Prime Interest Rates (%)		
Presidential Election Years	Non-Presidential Election Years	Non-Election Years
2.1	7.7	3.5
4.7	3.0	2.0
5.0	2.9	5.1
11.0	2.5	9.6
8.2	9.1	9.4
	10.3	10.8
		7.9
		6.5
		5.8

163. A patient's responses to the Rorschach Inkblot test include information about several behavioral variables, one of which is object relatedness, which concerns the

Review Problems

patient's separation and connection of perceived figures in the inkblot. To determine if object relatedness might be affected by overall responses on the test for the subjects of the experiment, a researcher subtracted the number of object relations responses from the total number of responses to the test for each of 70 patients, 35 who were rehospitalized and 35 who were not. The Wilcoxon-Mann-Whitney rank sum test statistic is 1,235 when comparing the non-object-relatedness values for the two groups. Test for a difference in the populations at the 0.05 significance level. Interpret the results. (Source: Steven B. Tuber, "Children's Rorschach Scores as Predictors of Later Adjustment," *Journal of Consulting and Clinical Psychology*, Vol. 51, No. 3, 1983, pp. 379–85.)

*164. Two possible defense mechanism of lizards in the presence of a predator are tail autotonomy, which is the loss of their tail, and tonic immobility, an immobile state resembling death. The ejected tail subsequently thrashes about, distracting the predator while the lizard escapes. A tonic immobile state averts attack by not providing any location signals. To verify the opposition of these tactics, a researcher measured the length of tonic immobility in lizards with and without autotomized tails. The *p*-value for a two-tailed Wilcoxon-Mann-Whitney rank sum test is 0.02. Make and interpret the decision for this test at the 0.05 significance level. (Source: Harold A. Herzog and Hugh Drummond, "Tail Autotomy Inhibits Tonic Immobility in Geckos," *Copeia*, No. 3, 1984, pp. 763–64.)

165. Use the following information on number of children per family from a random sample of 351 families to test a claim that the median number of children is two. Use a 0.05 significance level. Interpret the decision.

Number of Children	Frequency
0	92
1	75
2	121
3	42
4	15
More than 4	6
	351

166. A consumer group has ranked 23 computers for all-around quality. The ranking is listed with underscored rankings indicating that the computer comes with a color monitor. Use the data to test at the 0.05 level to determine if the average all-around quality is the same for computers that include color monitors and those that do not. Interpret the decision.

1 2 3 4 5 6 7 8 9 <u>10</u> 11 12 13 14 <u>15</u> 16 17 <u>18</u> <u>19</u> 20 21 22 <u>23</u>

167. To ascertain a shade of red that satisfies most of a school's students and alumni, a committee chooses a random sample of 362 participants from these groups to select their favorite shade from nine swatches. The committee uses these results to test if the shade they have chosen is indeed the median choice for all students and alumni. Test their decision at the 0.1 level, if 189 select a lighter shade, 150 a darker shade, and 23 the committee's shade.

168. Using one survey's top 25 ranked business schools as a sample, we can categorize schools according to their geographic region and then record their ranks. We obtain the following data set. Use this information to perform a Kruskal-Wallis test at the 0.01 level. Interpret the decision.

Business School Rankings

Upper East Coast & New England	West Coast & South	Upper Midwest (includes Western Pennsylvania)
1	2	4
3	7	6
5	9	10
8	14	13
11	15	18
12	16	20
19	17	22.5
25	21	
	22.5	
	24	

Source: College Research Group, as reported in "Top 25 Business Schools," *U.S. News & World Report*, April 29, 1991, p. 68.

169. Using the smoking data described in Example 13-5 (p. 648), perform a *parametric* test at the 0.05 level to determine if $\mu_{cancer} - \mu_{noncancer} > 0$. Assume normal populations with identical standard deviations. Interpret the decision. Compare your decision with the decision based on the nonparametric test discussed in Unit II. Explain the difference between the decisions, if there is one.

170. Figure 13-32 shows the computer output for a chi-square test for independence between the congressional staffers' ratings of staff-identified important studies and researcher-identified important studies. The studies concerned health care evaluation, which could affect legislation. Test for independence at the 0.01 level for each of the following characteristics, using the indicated part of the figure. Higher ratings indicate a more positive rating for the quality being investigated. Level 1 refers to staff-identified studies, and level 2 refers to researcher-identified studies. The computer analysis eliminated "Don't Know" responses. Make any adjustments and rework the problem when the situation does not satisfy the assumptions for the chi-square test. (Source: John F. Boyer and Laura I. Langbein, "Factors Influencing the Use of Health Evaluation Research in Congress," *Evaluation Review*, Vol. 15. No. 5, 1991, pp. 507–32.)

   a. Figure 13-32a, methodological credibility or quality of studies.

Figure 13-32
Minitab output for chi-square tests of health studies' impact

Expected counts are printed below observed counts

a) Methodological Credibility

	RATE=1	RATE=2	RATE=3	RATE=4	RATE=5	Total
1	2	2	19	44	33	100
	1.49	3.73	22.39	42.54	29.85	
2	0	3	11	13	7	34
	0.51	1.27	7.61	14.46	10.15	
Total	2	5	30	57	40	134

ChiSq = 0.173 + 0.803 + 0.513 + 0.050 + 0.332 +
        0.507 + 2.363 + 1.508 + 0.148 + 0.977 = 7.375
df = 4
* WARNING * 1 cells with expected counts less than 1.0
          * Chisquare approximation probably invalid
4 cells with expected counts less than 5.0

b) Researcher's Reputation

	RATE=2	RATE=3	RATE=4	RATE=5	Total
1	4	15	30	51	100
	3.76	17.29	27.07	51.88	
2	1	8	6	18	33
	1.24	5.71	8.93	17.12	
Total	5	23	36	69	133

ChiSq = 0.015 + 0.304 + 0.318 + 0.015 +
        0.047 + 0.922 + 0.963 + 0.045 = 2.628
df = 3
2 cells with expected counts less than 5.0

c) Level of Communication

	RATE=1	RATE=2	RATE=3	RATE=4	RATE=5	Total
1	11	15	36	1	4	67
	14.73	16.20	32.40	0.74	2.95	
2	9	7	8	0	0	24
	5.27	5.80	11.60	0.26	1.05	
Total	20	22	44	1	4	91

ChiSq = 0.942 + 0.089 + 0.401 + 0.094 + 0.378 +
        2.631 + 0.247 + 1.120 + 0.264 + 1.055 = 7.221

df = 4
* WARNING * 2 cells with expected counts less than 1.0
          * Chisquare approximation probably invalid
4 cells with expected counts less than 5.0

b. Figure 13-32b, researcher's reputation.
c. Figure 13-32c, level of communication between researcher and users during the research process.

171. The Minitab Output in Figure 13-33 results from three random samples each with five months of hotel occupancy rates for the Asheville, North Carolina, area. Use the output to perform a Kruskal-Wallis test at the 0.1 level. Interpret the decision. (Source: Asheville Convention and Visitors Bureau, as reported in *Asheville Citizen-Times*, February 16, 1992, p. B1.)

172. A random sample of a forecaster's errors (differences between true and forecasted value) includes the following values:

```
LEVEL     NOBS    MEDIAN    AVE. RANK    Z VALUE
1989       5      58.00        8.2         0.12
1990       5      70.00        9.6         0.98
1991       5      47.00        6.2        -1.10
OVERALL   15                   8.0

H = 1.46    d.f. = 2    p = 0.482
H = 1.47    d.f. = 2    p = 0.481  (adj. for ties)
```

Figure 13-33

Minitab output for Kruskal-Wallis test for occupancy rates

```
-25   53   75  -110   23   -5    0   13  -26   82   33
-53  -12   -9    18    3    3   79  -92  -20  -22
```

Use these values to determine if the median forecast error is different from 0. Test at the 0.05 level. Interpret the decision.

173. A researcher grouped 27 counseling student supervisees in their first counseling sessions according to their stage of ego development, measured by responses on an ego-assessment instrument (test). Three stages of development emerged from this process: self-aware, conscientious, and individualistic. To ascertain the impact of ego development on supervisees' thoughts about a counseling session, supervisees immediately reviewed a videotaped session of themselves with a client and recalled aloud their thoughts and feelings while counseling during the session. The researcher separated thoughts according to the subject or idea to develop scoring units, which were subsequently categorized into a subdimension of one of six dimensions: time, place, focus, locus, orientation, and mode. The basic hypothesis was that supervisees with higher ego development would perform better in the counseling session. More specifically, the hypotheses tested stated that supervisees at higher ego levels would report the following for each of the six dimensions:

1. More thoughts about the past and present that recognize and incorporate patterns.
2. More thoughts that connect relevant out-of-session knowledge and experience, such as an experience in another session related to the current session.
3. More thoughts that focus on the client and client-counselor interaction, rather than their own or their supervisor's personality or behavior.
4. More thoughts that incorporate theoretical understanding of internal dynamics, rather than external events.
5. More thoughts that indicate professional judgment rather than personal judgment.
6. More thoughts that indicate purposeful and objective analysis of the situation for planning as well as more positive thoughts concerning themselves and their clients.

The following table lists the resulting Kruskal-Wallis test statistics. In addition, the table includes two combined categories of thoughts, negative-client-focused and negative-counselor-focused. Use this information to test the hypotheses at the 0.05 significance level.

Dimension	Sub-dimension	H	P-value
Time	Past	1.31	0.52
	Present	2.50	0.29
Place	Out-of-session	1.04	0.60
Focus	Counselor	1.75	0.42
	Client-counselor	0.78	0.68
Locus	Internal	3.31	0.19
Orientation	Professional	0.48	0.48
Mode	Neutral	5.02	0.08
	Planning	1.34	0.51
	Negative	8.14	0.02
Combined	Negative-client-focused	11.58	0.003
	Negative-counselor-focused	5.20	0.07

Source: L. DiAnne Borders, "Developmental Cognitions of First Practicum Supervisees," *Journal of Counseling Psychology*, Vol. 36, No. 2, 1989, pp. 163–69.

174. Figure 13-34 shows the Wilcoxon-Mann-Whitney rank sum results for a test for difference in average incomes between residents in two counties. Use the information to make a decision at the 0.05 level. Interpret the decision. What potential problem with income distributions could make a nonparametric test more valid?

175. Test at the 0.01 level to determine if intentions to use and actual use of a new skating rink are related, using the data in the table. A marketing research firm obtained the intentions data in personal interviews before the rink opened. Rink owners encouraged customers to write their names along with the date on their ticket stubs and deposit them in a box for daily prize drawings for the first six months the rink operated. The researchers matched customers' names on stubs with interview information to obtain the following table of values. Interpret the decision.

Actual Average Monthly Use	Intended Average Monthly Use					
	0	1	2	3	4	5 or more
0	28	31	15	11	5	6
1	22	53	36	20	12	4
2	19	64	67	43	12	15
3	8	9	56	23	22	9
4	14	92	71	43	51	26
5 or more	3	16	62	33	43	21

Figure 13-34
Minitab output for Wilcoxon-Mann-Whitney test for income data

```
Mann-Whitney Confidence Interval and Test

C1                  N = 37        Median =        37062
C2                  N = 53        Median =        36185
Point estimate for ETA1-ETA2 is              2877
95.1 pct c.i. for ETA1-ETA2 is (-123, 4877)
W = 1896.0
Test of ETA1 = ETA2 vs. ETA1 n.e. ETA2 is significant at 0.0821
The test is significant at 0.0818 (adjusted for ties)

Cannot reject at alpha = 0.05
```

176. Using a random sample of 50 single digits from the random number table in Appendix I, test at the 0.1 level to determine if the median of the sample is 4.5. Is H_0 true?

Appendix A Rounding Rules

The first rounding rule concerns what to do with excess digits at the end of a number. To round 495.6783 to the nearest hundredth, we must consider whether the number is closer to 495.67 or 495.68. Because 495.6783 is more than halfway between the two values, we round 495.6783 to 495.68. If it had been less than halfway, we would use the lower or truncated value, so 495.6744 becomes 495.67. If the number to be rounded had been exactly halfway between the two candidate values, we would round so that the last remaining digit is even, 495.6750 becomes 495.68, while 495.6850 would also be rounded to 495.68. This rule may seem confusing at first, but it attempts to offset any long-term bias from repeated calculations from rounded values, by rounding up approximately half of the time and down half of the time. Note that the IRS always rounds up if the value is halfway between the two possible rounded values.

Example A-1: Round the following values.

1. 12,355.5 to the nearest hundred.
2. 83.32 to the nearest tenth.
3. 99.9995 to the nearest hundredth.
4. 14.5 to the nearest whole number.
5. 0.5 to the nearest whole number.

Solution:
1. 12,355.5 becomes 12,400, because 12,355.5 is more than halfway between 12,300 and 12,400.
2. 83.32 becomes 83.3, because 83.32 is less than halfway between 83.3 and 83.4.
3. 99.9995 becomes 100.00, because 99.9995 is more than halfway between 99.99 and 100.00.
4. 14.5 becomes 14, because 14.5 is exactly halfway between 14 and 15 and the last digit of 14 is even.
5. 0.5 becomes 0, because 0.5 is exactly halfway between 0 and 1 and 0 is an even digit.

The second rule is to round the *final* answer, not intermediate steps. We avoid rounding until the final answer for two reasons. Calculators easily carry several places past the decimal through a long series of calculations; however, the solution we report should not appear to be measured so accurately when actually it is not. For instance, we would seldom measure a length to the accuracy indicated by a result of 39.2334115 inches. This rule also avoids divergence in final answers from earlier rounding in intermediate steps. For example, suppose we want the solution to $18.68 + [15.23(8.996)^2]$ rounded to the nearest tenth. If we start by rounding all of the values to the nearest tenth and round after each step, the solution would be 1,249.9. If we do not round before starting, but round after each step, we obtain 1,250.8. Finally, following the second rule produces 1,251.2.

A note is necessary here. Intermediate steps of a solution show rounded figures for ease of presentation, but the actual computations continue with unrounded values. The final answers result from unrounded intermediate steps.

The final rule is to round the final answer one more place than the number of places associated with the values in the problem. If the given values in a situation are 500, 480, and 1,052, the answer would be rounded one place past the decimal. An exception to this rule often occurs when a final answer has no excess digits. Thus, a final answer found by summing the three values might be expressed as 2,032 rather than 2,032.0.

Sometimes problem values are not all reported with the same degree of accuracy. To avoid misrepresenting the degree of accuracy in our solution, the rounding position will depend on the *least detailed, approximated* number given in the problem. The *least-detailed number* is the one with the fewest digits past the decimal (or the one with the most zeros at the end if a number has no places past the decimal). If a problem's values are 582 with 60.8 or 582 with 60.8382, the least detailed value is 582, and the final answer would be rounded one place past the decimal in both cases. If the 582 was 500, the final answer would be rounded to the nearest tens place. We will assume that all numbers are *rounded* or otherwise *approximated,* unless there is a reason to know that the value is exact, such as a count of items. Because we can append zeros at the end of integer values or other values that are known to be exact in order to hold as many positions as needed, we ignore such values and concentrate only on approximated values when determining the least detailed value. For example, if five people out of 20 answer yes on a survey question, then the proportion who answer yes is 5/20 or 0.25, an exact proportion. If the same problem involves a calculation with a value such as 17.8826, then we can think of 0.25 as 0.2500, because it is exact, and we round the answer five places past the decimal.

Question: We are told or know that an average score is exactly 75.6 and the only other value in the problem is $8.02. How many places past the decimal will the solution carry?

Answer: We will round the solution three places past the decimal, because the least detailed value is 8.02 (75.6 is exact), and 8.02 carries two places past the decimal.

The general rounding rules are listed in the following box.

Rounding Rules

1. Consider the two numbers that end at the desired position that are potentially closest to the actual number to be rounded.
   a. If the actual number is more than halfway between the two values, round up.
   b. If the actual number is less than halfway between the two values, truncate the excess digits.
   c. If the actual number is exactly halfway between the two values, round so that the last remaining digit is even.
2. Round the final answer, not intermediate steps.
3. Round to one more place than the number of places associated with the problem values. If the number of places varies among the problem values, then use the number of places associated with the least detailed approximated problem value.

Appendix B   Air Quality Data Base

PM10 = annual mean concentration of particulate matter with aerodynamic diameters smaller than 10 micrometers (in $\mu g/m^{3*}$; applicable NAAQS** is 50 $\mu g/m^3$).
SO_2 A.M. = annual mean concentration of sulfur dioxide (in ppm†; applicable NAAQS** standard is 0.03 ppm).
SO_2 24-hr = second highest 24-hour average concentration of sulfur dioxide (in ppm†; applicable NAAQS** is 0.14 ppm).
CO = second highest nonoverlapping 8-hour average concentration of carbon monoxide (in ppm†; applicable NAAQS** is 9 ppm).
NO_2 = annual mean concentration of nitrogen dioxide (in ppm†; applicable NAAQS** is 0.053 ppm).
O_3 = second highest daily maximum 1-hour average concentration of ozone (in ppm†; applicable NAAQS** is 0.12 ppm).
Pb = maximum quarterly average concentration of lead (in $\mu g/m^{3*}$; applicable NAAQS** is 1.5 $\mu g/m^3$).

Metropolitan Statistical Area	1987 Population	PM10	SO_2 A.M.	SO_2 24-hr	CO	NO_2	O_3	Pb
1 Anaheim-Santa Ana, CA	2,219,000	47	0.004	0.013	11	0.047	0.24	0.08
2 Atlanta, GA	2,657,000	34	0.008	0.046	8	0.029	0.12	0.04
3 Bakersfield, CA	505,000	79	0.007	0.022	9	0.033	0.16	0.06
4 Baltimore, MD	2,303,000	44	0.013	0.044	9	0.035	0.13	0.11
5 Baton Rouge, LA	538,000	28	0.007	0.056	4	0.019	0.16	0.09
6 Beaver County, PA	191,000	34	0.023	0.128	4	0.020	0.10	0.27
7 Bergen-Passaic, NJ	1,294,000	40	0.012	0.045	8	0.035	0.12	0.05
8 Boston, MA	2,842,000	36	0.014	0.058	6	0.032	0.12	0.08
9 Bridgeport-Milford, CT	444,000	36	0.014	0.051	5	0.026	0.18	0.07
10 Buffalo, NY	958,000	29	0.014	0.068	5	0.024	0.11	0.04
11 Charleston, WV	261,000	35	0.015	0.076	3	0.021	0.10	0.04
12 Chicago, IL	6,199,000	48	0.011	0.037	7	0.034	0.12	0.16
13 Cincinnati, OH-KY-IN	1,438,000	45	0.016	0.084	5	0.030	0.12	0.11
14 Cleveland, OH	1,851,000	52	0.018	0.061	8	0.034	0.12	0.17
15 Dallas, TX	2,456,000	36	0.005	0.018	5	0.021	0.13	1.76
16 Denver, CO	1,645,000	35	0.007	0.033	11	0.040	0.11	0.02
17 Detroit, MI	4,362,000	52	0.015	0.058	8	0.026	0.14	0.09
18 El Paso, TX	573,000	69	0.015	0.059	13	0.022	0.14	0.42
19 Fort Worth-Arlington, TX	1,269,000	25	0.001	0.007	6	0.013	0.13	0.04
20 Fresno, CA	597,000	76	0.004	0.013	12	0.032	0.15	0.07
21 Hartford, CT	748,000	30	0.011	0.046	9	0.020	0.14	0.09
22 Houston, TX	3,228,000	30	0.009	0.055	8	0.028	0.23	0.05
23 Huntington-Ashland, WV-KY-OH	323,000	42	0.017	0.085	6	0.013	0.12	0.06
24 Indianapolis, IN	1,229,000	43	0.014	0.056	5	0.023	0.12	1.13
25 Jacksonville, FL	878,000	37	0.006	0.060	7	0.015	0.11	0.05
26 Jersey City, NJ	547,000	37	0.016	0.052	7	0.031	0.12	0.08
27 Johnstown, PA	252,000	33	0.016	0.089	4	0.019	0.10	0.31
28 Kansas City, MO-KS	1,546,000	47	0.006	0.020	7	0.015	0.11	0.11
29 Lancaster, PA	404,000	29	0.007	0.037	4	0.018	0.10	0.05
30 Los Angeles-Long Beach, CA	8,505,000	64	0.005	0.021	18	0.057	0.33	0.14
31 Manchester, NH	146,000	24	0.009	0.050	7	0.022	0.10	0.03
32 Memphis, TN-AR-MS	972,000	33	0.009	0.033	10	0.026	0.12	0.19
33 Middlesex-Somerset-Hunterdon, NJ	966,000	34	0.010	0.037	5	0.024	0.13	0.38
34 Milwaukee, WI	1,389,000	40	0.007	0.032	6	0.029	0.15	0.07
35 Minneapolis-St. Paul, MN-WI	2,336,000	33	0.010	0.072	11	0.009	0.10	1.04
36 New Haven-Meriden, CT	519,000	44	0.016	0.088	6	0.028	0.15	0.08
37 New Orleans, LA	1,321,000	39	0.004	0.019	7	0.022	0.11	0.09
38 New York, NY	8,529,000	66	0.021	0.081	12	0.049	0.13	0.12
39 Newark, NJ	1,891,000	38	0.013	0.056	9	0.038	0.13	0.41
40 Norfolk-Virginia Beach-Newport News, VA	1,346,000	29	0.008	0.039	7	0.020	0.10	0.12
41 Oakland, CA	1,968,000	37	0.003	0.017	6	0.025	0.13	0.21
42 Oklahoma City, OK	975,000	27	0.006	0.016	9	0.015	0.11	0.07
43 Orlando, FL	935,000	32	0.002	0.006	5	0.013	0.11	0.02
44 Oxnard-Ventura, CA	628,000	40	0.001	0.007	4	0.027	0.17	0.04
45 Philadelphia, PA-NJ	4,866,000	46	0.015	0.065	12	0.040	0.16	0.41
46 Pittsburgh, PA	2,105,000	43	0.024	0.106	8	0.028	0.13	0.10
47 Providence, RI	643,000	39	0.015	0.047	8	0.024	0.13	0.21

Air Quality Data Base
Appendix B

Metropolitan Statistical Area	1987 Population	PM10	SO$_2$ A.M.	SO$_2$ 24-hr	CO	NO$_2$	O$_3$	Pb
48 Riverside-San Bernardino, CA	2,119,000	93	0.004	0.023	8	0.045	0.28	0.08
49 Sacramento, CA	1,336,000	43	0.006	0.020	13	0.025	0.14	0.08
50 St. Louis, MO-IL	2,458,000	76	0.018	0.102	8	0.026	0.13	2.29
51 Salt Lake City-Ogden, UT	1,055,000	56	0.019	0.103	8	0.034	0.15	0.15
52 San Diego, CA	2,286,000	44	0.006	0.018	10	0.032	0.19	0.07
53 San Francisco, CA	1,590,000	36	0.003	0.015	8	0.026	0.09	0.15
54 Santa Barbara-Santa Maria-Lompoc, CA	341,000	34	0.002	0.010	7	0.027	0.16	0.05
55 Springfield, MA	517,000	31	0.013	0.055	8	0.029	0.13	0.06
56 Steubenville-Weirton, OH-WV	149,000	45	0.035	0.127	13	0.023	0.11	0.00
57 Tampa-St. Petersburg-Clearwater, FL	1,965,000	32	0.011	0.039	6	0.022	0.10	0.13
58 Tucson, AZ	619,000	52	0.002	0.008	7	0.023	0.10	0.06
59 Tulsa, OK	733,000	36	0.007	0.038	7	0.020	0.12	0.20
60 Vallejo-Fairfield-Napa, CA	404,000	32	0.002	0.010	10	0.019	0.11	0.10
61 Washington, DC-MD-VA	3,646,000	43	0.014	0.046	9	0.031	0.13	0.10
62 Wheeling, WV-OH	173,000	34	0.026	0.076	5	0.019	0.11	0.07
63 Wilmington, DE-NJ-MD	559,000	35	0.018	0.049	5	0.034	0.13	0.16

Source: Table 4-3, "1989 Metropolitan Statistical Area (MSA) Air Quality Factbook: Peak Statistics for Selected Pollutants by MSA," *National Air Quality and Emissions Trends Report, 1989: Executive Summary and Chapter 4-Excerpts,* U.S. Environmental Protection Agency, Office of Air and Radiation, Office of Air Quality Planning and Standards, Technical Support Division, Research Triangle Park, North Carolina, February 1991, pp. 4–16; 4–25.

*μg/m^3 = microgram per cubic meter.
**NAAQS = National Ambient Air Quality Standard.
†ppm = parts per million.

Appendix C State SAT Data Base

Variable Definitions

The following data are for each of the 50 states and the District of Columbia. These data were extracted from reports published by the North Carolina Department of Public Instruction and from the *Statistical Abstracts of the United States*. For each state, you have the following information. Professor Robert Pittman of the School of Education of Western Carolina University compiled this data set and allowed its reprint.

Variable	Description
State	State or District of Columbia
Percentage taking	Percentage of the high school seniors in the state taking the SAT
Teacher salaries	Average teacher's salary in the state in 1988
Pupil expenditure	Average per pupil expenditure in the state in 1988
Per capita income	Per capita income in the state in 1988
College educated	Percentage of the state's adult population with a bachelor's degree in 1980
Children poverty	Percentage of the children in the state under the age of 18 who are classified as living in poverty in 1979
SAT score	Average SAT score obtained by the seniors in the state in 1990
Nonmetropolitan population	Level of the state's population classified as living in nonmetropolitan areas in 1988 1. Predominantly urban population 2. Intermediate urban/rural 3. Predominantly rural population
Unemployment	Level of unemployment in the state in 1985 1. Low unemployment 2. Medium unemployment 3. High unemployment

State SAT Data Base

State	Percentage Taking	Teacher Salaries	Pupil Expenditure	Per Capita Income	College Educated	Children Poverty	SAT Score	Nonmetropolitan Population	Unemployment
AL	8	23,320	2,752	10,068	12.2	23.6	984	2	3
AK	42	40,424	7,038	15,724	21.1	12.1	914	3	3
AZ	25	27,388	3,265	12,428	17.4	16.5	942	2	2
AR	6	20,340	2,410	9,998	10.8	23.4	981	3	3
CA	45	33,159	3,994	15,138	19.6	15.2	903	1	2
CO	28	28,651	4,359	13,158	23.0	11.5	969	1	1
CT	74	33,487	6,141	17,784	20.7	11.4	901	1	1
DC	68	34,705	5,643	16,708	27.5	27.0	850	1	3
DE	58	29,575	4,994	13,572	17.5	15.6	903	2	1
FL	44	25,198	4,389	13,314	14.9	18.5	884	1	2
GA	57	26,177	2,939	11,918	14.6	21.0	844	2	2
HI	52	28,785	3,894	13,461	20.3	13.0	885	2	1
ID	17	22,242	2,814	10,404	15.8	14.3	968	3	2
IL	16	29,663	4,217	14,085	16.2	14.9	994	1	3
IN	54	27,386	3,616	11,856	12.5	11.9	867	2	2
IA	5	24,867	3,846	12,213	13.9	11.5	1,088	3	3
KS	10	24,647	4,262	12,757	17.0	11.4	1,040	3	1
KY	10	24,274	3,355	10,354	11.1	21.6	994	3	3
LA	9	21,209	3,211	9,933	13.9	23.4	993	2	3
ME	60	23,425	4,276	12,092	19.2	15.8	886	3	1
MD	59	30,933	4,871	15,128	20.4	12.5	908	1	1
MA	72	30,019	5,396	15,972	20.0	13.1	900	1	1
MI	12	32,926	4,122	13,065	14.3	13.3	968	1	3
MN	14	29,900	4,513	13,436	17.4	10.2	1,019	2	2
MS	4	20,669	2,760	8,868	12.3	30.4	996	3	3
MO	12	24,703	3,566	12,659	13.9	14.6	995	2	2
MT	20	23,798	4,061	10,788	17.5	13.8	987	3	2
NE	10	23,246	3,641	12,479	15.5	12.1	1,030	3	1
NV	24	27,600	3,829	13,896	14.4	10.0	921	1	3
NH	67	24,019	3,990	15,363	18.2	9.4	928	2	1

State SAT Data Base

State	Percentage Taking	Teacher Salaries	Pupil Expenditure	Per Capita Income	College Educated	Children Poverty	SAT Score	Nonmetropolitan Population	Unemployment
NJ	69	30,720	6,910	17,280	18.3	14.1	891	1	1
NM	12	24,351	3,880	10,113	17.6	22.1	1,007	3	3
NY	70	34,500	6,864	14,660	17.9	19.0	882	1	2
NC	55	24,900	3,911	11,239	13.2	18.3	841	3	1
ND	6	21,660	3,353	11,504	14.8	14.3	1,069	3	1
OH	22	27,606	4,019	12,453	13.7	13.2	949	1	3
OK	9	22,006	3,051	10,089	15.1	15.7	1,001	2	2
OR	49	28,060	4,574	11,960	17.9	12.0	923	2	3
PA	64	29,174	5,063	13,085	13.6	13.9	883	1	3
RI	62	32,858	5,456	13,522	15.4	13.6	883	1	1
SC	54	24,241	3,075	10,238	13.4	21.0	834	2	2
SD	5	19,750	3,159	11,377	14.0	20.0	1,061	3	1
TN	12	23,785	3,189	11,209	12.6	20.6	1,008	2	3
TX	42	25,655	3,462	11,886	16.9	18.7	874	1	2
UT	5	22,621	2,658	9,727	19.9	10.7	1,031	2	1
VT	62	23,397	4,949	12,153	19.0	13.9	897	3	1
VA	58	27,436	4,145	13,979	19.1	14.9	895	2	1
WA	44	28,116	4,083	13,773	19.0	11.5	923	1	3
WV	15	21,736	3,895	9,587	10.4	18.4	933	3	3
WI	11	28,998	4,991	12,715	14.8	10.4	1,019	2	2
WY	13	27,260	6,885	10,454	17.2	7.7	977	3	2

Source: Reports published by the North Carolina Department of Public Instruction and from the U.S. Bureau of the Census, *Statistical Abstract of the United States: 1982–3, 1984, 1989, and 1990* (Washington, D.C.: Bureau of the Census).

Appendix D   NCAA Probation Data Base

This data set consists of observations of 10 variables recorded for 85 different college and university teams. To be included, teams must have played Division I football in 1953 or 1954. Most variables concern characteristics or activities over the years 1953 through 1983. Professor Brian Goff of the Economics Department of Western Kentucky University has allowed the use of this data set. The data were compiled from various sources, including U.S. House of Representatives, Subcommittee Hearings; *NCAA Enforcement Program* (1978, pp. 1480–1520); *Washington Post* (various issues 1978–1983); *College Football U.S.A. 1869–1971* (1972, pp. 502–8); *World Almanac* (1960, 1970, 1980, various pages); *American Universities and Colleges* (1982, various pages).

Variables	Description
Team	School represented by team
WINPCT	Number of wins from 1953 to 1983 divided by number of wins and losses over the same period (Southern Mississippi's value is the mean of the other WINPCT values).
Top 20	Number of times team is ranked in top 20 at end of a season (1953–1983).
Probation	Code for team on probation anytime during 1953–1983 (1 = yes, 0 = no).
Enrollment	School enrollment using available information closest to 1968–1969 academic year
Origin	Year school was established
Independent	Code for team that is independent—not affiliated with a conference (1 = independent) 0 = conference)
Religious	Code for team that is affiliated with a religious organization (1 = affiliated, 0 = unaffiliated)
Private	Code for team that is affiliated with a private, rather than public, institution (1 = private, 0 = public)
National TV	Number of national TV appearances by team over 1953–1983
Regional TV	Number of regional TV appearances by team over 1953–1983

Row	Team	WINPCT	Top 20	Probation	Enrollment	Origin	Independent	Religious	Private	National TV	Regional TV
1	Air Force	0.403	4	0	3,300	1954	1	1	1	4	19
2	Alabama	0.746	23	0	17,177	1831	0	0	0	34	16
3	Arizona	0.525	2	1	24,100	1885	0	0	0	1	16
4	Arizona State	0.745	11	1	22,131	1885	0	0	0	2	15
5	Arkansas	0.693	16	0	10,900	1871	0	0	0	19	28
6	Army	0.504	7	0	3,700	1802	1	1	1	32	15
7	Auburn	0.677	18	1	13,000	1856	0	0	0	10	15
8	Baylor	0.464	6	0	6,258	1845	0	1	1	2	17
9	Boston	0.598	1	0	10,000	1863	1	1	1	4	13
10	Brigham Young	0.511	5	0	25,100	1875	0	1	1	2	20
11	California	0.416	2	1	34,799	1868	0	0	0	7	24
12	Cincinnati	0.484	0	0	29,410	1819	0	0	0	0	2
13	Clemson	0.589	8	1	6,471	1889	0	0	0	1	17
14	Colorado State	0.389	0	0	16,300	1870	0	0	0	0	8
15	Colorado	0.548	8	1	30,877	1876	0	0	0	4	16
16	Duke	0.510	6	0	8,000	1838	0	1	1	6	15
17	East Carolina	0.600	1	0	9,797	1907	0	0	0	0	5
18	Florida	0.612	9	0	19,672	1853	0	0	0	9	20
19	Florida State	0.566	6	0	17,500	1857	1	0	0	4	18
20	Georgia	0.636	13	0	21,500	1785	0	0	0	13	23
21	Georgia Tech.	0.560	10	0	8,000	1885	0	0	0	15	14
22	Houston	0.601	10	1	25,000	1927	0	0	0	6	12
23	Illinois	0.470	5	1	30,407	1867	0	0	0	7	29
24	Indiana	0.343	2	0	50,800	1820	0	0	0	0	14
25	Iowa	0.451	7	0	20,000	1847	0	0	0	12	18
26	Iowa State	0.430	1	0	19,800	1858	0	0	0	0	9
27	Kansas	0.422	3	1	15,691	1866	0	0	0	1	15
28	Kansas State	0.309	0	1	13,000	1863	0	0	0	1	7
29	Kentucky	0.430	3	1	28,300	1865	0	0	0	1	10
30	Louisiana State	0.648	15	0	28,348	1860	0	0	0	10	22
31	Louisville	0.514	1	0	9,700	1798	0	0	0	0	2
32	Maryland	0.538	8	0	49,200	1856	0	0	0	5	17
33	Memphis State	0.558	1	0	18,000	1912	0	0	0	0	2
34	Miami	0.539	8	1	17,500	1925	1	0	1	13	15
35	Michigan	0.694	18	0	38,770	1817	0	0	0	28	28
36	Michigan State	0.576	11	1	45,200	1855	0	0	0	17	25
37	Minnesota	0.493	6	1	51,850	1851	0	0	0	9	23

Row	Team	WINPCT	Top 20	Probation	Enrollment	Origin	Independent	Religious	Private	National TV	Regional TV
38	Mississippi	0.664	14	0	6,012	1844	0	0	0	6	17
39	Mississippi St.	0.440	6	1	9,200	1878	0	0	0	1	7
40	Missouri	0.598	10	0	44,000	1839	0	0	0	10	26
41	Navy	0.537	6	0	4,100	1845	1	1	1	32	16
42	Nebraska	0.688	17	0	19,500	1869	0	0	0	23	20
43	New Mexico	0.479	0	0	14,830	1889	0	0	0	0	12
44	New Mexico St.	0.438	1	0	10,000	1888	0	0	0	0	3
45	North Carolina	0.533	7	0	15,610	1795	0	0	0	3	29
46	N. Carolina St.	0.493	6	0	10,392	1862	0	0	0	1	18
47	Northwestern	0.314	1	0	17,259	1851	0	0	1	5	16
48	Notre Dame	0.701	20	0	7,200	1842	1	1	1	42	26
49	Ohio State	0.781	26	0	45,080	1870	0	0	0	28	28
50	Oklahoma	0.782	25	1	22,511	1890	0	0	0	30	26
51	Oklahoma State	0.468	2	1	17,000	1890	0	0	0	1	8
52	Oregon	0.447	1	1	14,000	1876	0	0	0	3	17
53	Oregon State	0.455	6	0	15,500	1868	0	0	0	2	13
54	Penn. State	0.758	20	0	49,327	1855	1	0	0	22	26
55	Pittsburgh	0.541	12	0	25,880	1787	1	0	0	21	18
56	Purdue	0.580	10	0	38,500	1869	0	0	0	9	28
57	Rice	0.359	5	0	3,027	1891	0	0	1	2	11
58	Rutgers	0.587	3	0	32,100	1864	1	0	0	0	2
59	S. Carolina	0.486	1	1	15,500	1801	0	0	0	1	11
60	S. California	0.717	25	1	26,754	1880	0	0	1	36	21
61	S. Methodist	0.479	7	1	9,700	1911	0	1	1	6	24
62	S. Mississippi	0.533	0	1	10,075	1910	1	0	0	0	4
63	Stanford	0.505	7	0	11,250	1891	0	0	1	13	26
64	Syracuse	0.562	10	0	23,400	1870	1	0	1	5	19
65	Temple	0.475	0	0	31,300	1884	0	0	0	0	1
66	Tennessee	0.642	12	0	38,143	1794	0	0	0	8	19
67	Texas	0.746	21	0	35,678	1883	0	0	0	37	26
68	Texas-El Paso	0.361	1	0	9,131	1913	0	0	0	0	0
69	Texas A&M	0.468	7	1	13,700	1876	0	0	0	18	21
70	Texas Christian	0.367	4	0	8,700	1873	0	1	1	2	9
71	Texas Tech.	0.516	4	0	19,034	1923	0	0	0	4	21
72	Tulane	0.368	2	0	8,256	1834	0	0	1	0	7
73	Tulsa	0.524	1	1	7,220	1894	0	0	0	1	11
74	UCLA	0.662	16	0	27,500	1919	0	0	0	33	23
75	Utah State	0.572	1	0	9,000	1888	0	0	0	0	7
76	Utah	0.490	1	0	24,000	1850	0	0	0	1	8
77	Vanderbilt	0.359	0	0	5,700	1873	0	0	1	1	5
78	Virginia	0.306	0	0	9,600	1819	0	0	0	0	4
79	Virginia Tech.	0.580	2	0	8,227	1872	0	0	0	0	10
80	Wake Forest	0.306	0	0	3,300	1834	0	1	1	0	7
81	Washington	0.568	11	0	32,000	1861	0	0	0	8	30
82	Washington St.	0.393	1	0	13,400	1892	0	0	0	1	18
83	West Virginia	0.581	8	0	16,400	1867	0	0	0	0	12
84	Wisconsin	0.466	6	0	61,300	1849	0	0	0	4	22
85	Wyoming	0.582	3	1	6,883	1886	0	0	0	0	15

Appendix E North Carolina Census Data Base

Column	Variable	Column	Variable
Units = persons*			
1	ID number	8	Mental hospital population = MENT
2	County name	9	Juvenile institution population = JUV
3	Total population = TOTAL	10	College dormitory population = COL
4	Total male population = TOTMALE	11	Military quarters population = MIL
5	Total female population = TOTFEM	12	Homeless emergency shelter population = SHEL
6	Correctional institutional population = COR	13	Homeless visible in street locales = STRT
7	Nursing home population = NURS		

(1) ID	(2) County	(3) Total	(4) TOTMALE	(5) TOTFEM	(6) COR	(7) NURS	(8) MENT	(9) JUV	(10) COL	(11) MIL	(12) SHEL	(13) STRT
1	Alamance	108,213	51,315	56,898	199	829	0	5	1,644	0	9	0
2	Alexander	27,544	13,712	13,832	96	188	0	0	0	0	0	0
3	Alleghany	9,590	4,625	4,965	11	179	0	0	0	0	0	0
4	Anson	23,474	10,995	12,479	146	198	0	0	0	0	0	0
5	Ashe	22,209	10,748	11,461	9	211	0	0	0	0	0	0
6	Avery	14,867	7,441	7,426	108	125	0	85	594	0	0	0
7	Beaufort	42,283	19,953	22,330	6	428	0	0	0	0	47	0
8	Bertie	20,388	9,408	10,980	31	22	0	0	0	0	0	0
9	Bladen	28,663	13,481	15,182	118	391	0	0	0	0	0	0
10	Brunswick	50,985	24,934	26,051	28	276	0	0	0	7	22	0
11	Buncombe	174,821	82,978	91,843	649	1,674	87	184	1,420	0	175	25
12	Burke	75,744	37,049	38,695	580	624	0	28	0	0	0	0
13	Cabarrus	98,935	47,841	51,094	267	706	0	237	266	0	22	5
14	Caldwell	70,709	34,665	36,044	236	496	0	0	0	0	15	0
15	Camden	5,904	2,960	2,944	0	0	0	0	0	0	0	0
16	Carteret	52,556	25,968	26,588	238	432	0	0	34	81	13	0
17	Caswell	20,693	10,096	10,597	369	255	0	0	0	0	0	0
18	Catawba	118,412	57,536	60,876	294	780	35	25	789	0	33	0
19	Chatham	38,759	18,807	19,952	40	326	0	0	0	0	0	0
20	Cherokee	20,170	9,688	10,482	24	172	0	0	0	0	0	0
21	Chowan	13,506	6,275	7,231	18	215	0	0	0	0	0	0
22	Clay	7,155	3,485	3,670	4	0	0	0	0	0	0	0

North Carolina Census Data Base

(1) ID	(2) County	(3) Total	(4) TOTMALE	(5) TOTFEM	(6) COR	(7) NURS	(8) MENT	(9) JUV	(10) COL	(11) MIL	(12) SHEL	(13) STRT
23	Cleveland	84,714	40,623	44,091	213	702	0	0	748	0	17	0
24	Columbus	49,587	23,379	26,208	180	381	0	63	0	0	0	0
25	Craven	81,613	40,610	41,003	98	488	0	9	0	2,718	0	2
26	Cumberland	274,566	142,001	132,565	295	918	105	96	1,287	18,090	102	29
27	Currituck	13,736	6,920	6,816	133	84	0	0	0	0	0	0
28	Dare	22,746	11,306	11,440	26	114	0	0	0	5	4	0
29	Davidson	126,677	62,097	64,580	286	854	0	119	0	0	18	0
30	Davie	27,859	13,670	14,189	109	171	0	0	0	0	6	1
31	Duplin	39,995	19,182	20,813	184	361	0	0	0	0	0	0
32	Durham	181,835	85,765	96,070	733	1,280	0	52	6,157	0	125	1
33	Edgecombe	56,558	25,845	30,713	253	54	0	0	0	0	20	0
34	Forsyth	265,878	125,379	140,499	615	2,704	0	88	4,476	1	263	27
35	Franklin	36,414	17,483	18,931	146	368	0	0	606	0	0	0
36	Gaston	175,093	84,130	90,963	325	1,091	0	0	535	0	79	17
37	Gates	9,305	4,597	4,708	102	0	0	0	0	0	0	0
38	Graham	7,196	3,573	3,623	6	12	0	0	0	0	0	0
39	Granville	38,345	18,988	19,357	1,017	370	570	212	0	0	0	0
40	Greene	15,384	7,657	7,727	616	88	0	0	0	0	0	0
41	Guilford	347,420	164,204	183,216	618	2,258	0	20	8,481	0	247	2
42	Halifax	55,516	26,352	29,164	914	422	0	0	0	0	38	0
43	Harnett	67,822	32,928	34,894	648	587	0	0	1,190	0	21	3
44	Haywood	46,942	22,382	24,560	124	575	0	52	0	0	0	2
45	Henderson	69,285	33,116	36,169	124	750	0	78	0	0	39	0
46	Hertford	22,523	10,444	12,079	17	256	0	0	583	0	0	3
47	Hoke	22,856	11,448	11,408	1,108	136	0	0	0	0	0	0
48	Hyde	5,411	2,650	2,761	19	0	0	0	0	9	0	0
49	Iredell	92,931	44,927	48,004	201	593	18	0	0	0	18	0
50	Jackson	26,846	12,947	13,899	11	269	0	0	2,696	0	0	0
51	Johnston	81,306	39,256	42,050	274	610	3	0	0	0	11	0
52	Jones	9,414	4,429	4,985	2	0	0	0	0	0	0	0
53	Lee	41,374	19,948	21,426	239	408	0	0	0	0	6	0
54	Lenoir	57,274	26,722	30,552	29	427	0	157	0	0	9	0
55	Lincoln	50,319	24,748	25,571	215	302	0	0	0	0	0	0
56	Macon	23,499	11,272	12,227	4	219	10	0	0	0	0	0
57	Madison	16,953	8,316	8,637	7	97	0	0	698	0	0	0
58	Martin	25,078	11,777	13,301	80	172	0	0	0	0	4	0
59	McDowell	35,681	17,389	18,292	114	302	0	0	0	0	0	0
60	Mecklenbrg	511,433	245,660	265,773	256	3,049	0	194	6,252	0	538	108
61	Mitchell	14,433	6,930	7,503	6	124	0	0	0	0	0	0
62	Montgomery	23,346	11,763	11,583	707	198	0	72	0	0	0	0
63	Moore	59,013	28,156	30,857	164	632	0	174	0	0	0	0
64	Nash	76,677	36,382	40,295	204	549	0	32	290	0	16	0
65	New Hanovr	120,284	57,071	63,213	344	908	8	4	1,768	0	92	8
66	Northamptn	20,798	10,059	10,739	562	180	0	0	0	0	0	0
67	Onslow	149,838	89,617	60,221	434	509	5	0	0	33,175	25	0
68	Orange	93,851	44,471	49,380	220	527	23	0	8,321	0	30	5
69	Pamlico	11,372	5,457	5,915	13	94	0	0	0	13	0	0
70	Pasquota	31,298	14,694	16,604	53	315	0	0	946	4	59	4
71	Pender	28,855	14,010	14,845	123	205	20	0	0	0	0	0
72	Perquimans	10,447	5,000	5,447	0	144	0	0	0	0	0	0
73	Person	30,180	14,432	15,748	95	193	0	0	0	0	12	1
74	Pitt	107,924	51,312	56,612	129	667	15	10	4,292	24	72	7
75	Polk	14,416	6,757	7,659	11	203	0	0	0	0	0	0
76	Randolph	106,546	52,052	54,494	337	711	0	0	0	0	20	0
77	Richmond	44,518	21,260	23,258	449	365	0	0	0	0	42	1
78	Robeson	105,179	49,714	55,465	424	721	24	43	639	1	7	2
79	Rockingham	86,064	40,970	45,094	108	662	11	0	0	0	4	0
80	Rowan	110,605	53,692	56,913	796	1,271	132	35	1,044	0	24	0
81	Rutherford	56,918	27,371	29,547	209	512	0	0	0	0	0	0
82	Sampson	47,297	22,473	24,824	120	284	0	0	0	0	0	0
83	Scotland	33,754	15,764	17,990	129	264	0	0	659	0	0	0
84	Stanly	51,765	24,940	26,825	141	324	0	5	484	0	0	0
85	Stokes	37,223	18,278	18,945	117	190	0	0	0	0	0	0
86	Surry	61,704	29,562	32,142	50	595	4	0	0	0	0	0
87	Swain	11,268	5,614	5,654	35	163	0	17	0	0	7	0
88	Transylvan	25,520	12,472	13,048	12	166	0	0	591	0	0	0
89	Tyrell	3,856	1,842	2,014	4	0	0	0	0	0	0	0

Appendix E

North Carolina Census Data Base

(1) ID	(2) County	(3) Total	(4) TOTMALE	(5) TOTFEM	(6) COR	(7) NURS	(8) MENT	(9) JUV	(10) COL	(11) MIL	(12) SHEL	(13) STRT
90	Union	84,211	41,286	42,925	153	302	0	0	1,046	0	30	2
91	Vance	38,892	18,354	20,538	127	390	0	0	0	0	0	0
92	Wake	423,380	207,214	216,166	3,581	1,823	803	17	7,678	0	215	2
93	Warren	17,265	8,272	8,993	69	263	0	0	0	0	0	0
94	Washington	13,997	6,648	7,349	114	110	0	0	0	0	0	0
95	Watauga	36,952	18,063	18,889	76	203	0	0	4,150	0	20	0
96	Wayne	104,666	52,073	52,593	544	777	627	6	258	4,250	28	0
97	Wilkes	59,393	29,052	30,341	142	444	0	0	0	0	3	2
98	Wilson	66,061	30,871	35,190	39	623	0	0	644	0	30	0
99	Yadkin	30,488	14,799	15,689	114	310	0	0	0	0	0	0
100	Yancey	15,419	7,463	7,956	90	94	0	0	0	0	0	0

*The data do not include Indian reservations and tribal designated areas.

Source: "Summary Tape File 1A: Eight Page County Profiles," North Carolina State Library and North Carolina State Data Center, from Census Tape STF1A, Washington, D.C.: Bureau of the Census, 1991.

Data for these categories (no. 14–56) are available on the data disk.

Appendix F

Counting: Combinations and Permutations

At times the number of outcomes in a probability computation may become large and difficult to tract. Several counting rules simplify the process and eliminate the need for protracted lists.

Suppose a local television station wishes to choose among five old program series, S1, S2, S3, S4, and S5, to fill three nonprime time slots, 4:30, 6:30, and 11:30 P.M.

Question: All five programs are available for each slot (the programmer is willing to fill all three slots with the same series). How many evening lineups, sequences of three series, are possible in this circumstance? Notice that the order of the shows on the list is important. The set S2 S5 S3 implies that the lineup is program S2 in the 4:30 slot, S5 at 6:30, and S3 at 11:30, whereas S5 S3 S2 means S5 shows at 4:30, S3 at 6:30, and S2 at 11:30, an entirely different lineup for the evening.

Answer: Figure F-1 shows the five alternatives for each slot, because programs can repeat during the evening. Any one of the five shows can follow each choice from the five possible series for the first slot, so there are 25 possible lineups for the first two slots. Again any of the five shows can follow each of these two-show lineups, so there are 125 possible programming sequences. Figure F-1 shows all of these possible lineups. The boxes below summarize the situation.

Number of Alternatives			Number of Sequences
Slots			Lineups
1	2	3	
5	5	5	125

Question: Suppose only two of the five shows, S1 and S4, are suitable for kids after school, the first slot. Three, S1, S2, and S4, are family oriented and appeal to the 6:30 audience as well. Any one of the five can be shown in the late evening. How many lineups are possible now?

Answer: Figure F-2 shows the possible sequences. There are two alternatives for the first slot, each followed by three alternatives for the second slot for a total of six lineups for the two shows. Finally, there are five alternatives to follow each of these six lineups, for a total of 30 lineups for the three shows for the evening. The following diagram summarizes the counting.

Number of Alternatives			Number of Sequences
Slots			Lineups
1	2	3	
2	3	5	30

Notice what happens if we multiply the numbers in the alternative boxes on the left of the diagram. The product is the number in the sequence box, the total number of possible sequences. In fact, we could place an equal sign between the three boxes on the left and the lineup box on the right. These multiplications demonstrate the first counting rule.

Counting Rule 1

If there are n_i alternative selections for the ith position in a sequence of m choices, then the total number of distinct sequences is $n_1 \times n_2 \times n_3 \times \cdots \times n_m$.

"Distinct" means that we differentiate positions in the sequences and we consider the positions important. In the example, just knowing that the manager selects shows 1, 4, and 5, does not tell us which show occupies which slot, so this is not a distinct sequence.

Appendix F — Counting: Combinations and Permutations

Figure F-1 Number of arrangements of five programs in three slots when programs can repeat during the day

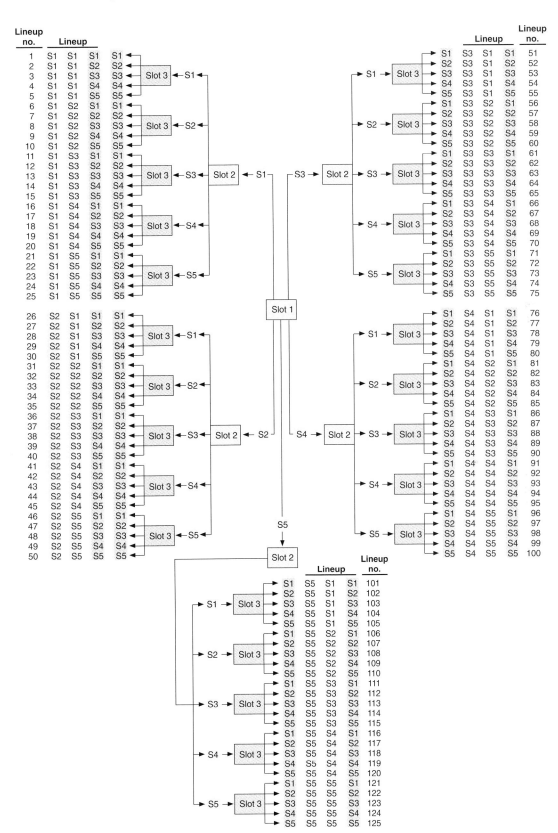

Figure F-2
Number of arrangements for three time slots when two shows are suitable for the first spot, three for the second, and five for the third

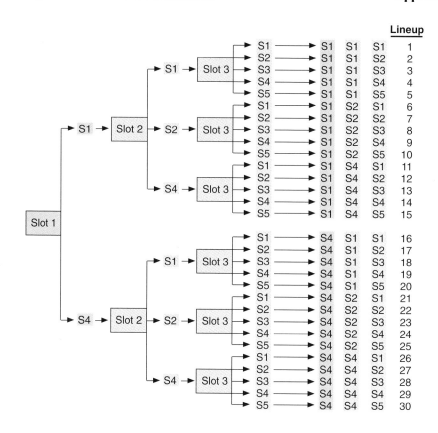

Question: If there are five slots to fill, all five programs are suitable for any of the slots, but no program can be shown more than once during the day, how many different arrangements are possible?

Answer: Using the blocks again, we find there are 120 arrangements. After choosing one of the five for the first slot, only four choices remain for the second slot, then three for the third slot, two for the fourth, and finally only one for the last slot.

Alternative Slots						Sequences
1	2	3	4	5		Lineups
5	4	3	2	1	=	120

A shorthand way to write $5 \times 4 \times 3 \times 2 \times 1$ is 5! (five factorial). $n!$ means find the product of n and each positive integer smaller than n. $8! = 8 \times 7 \times 6 \times 5 \times 4 \times 3 \times 2 \times 1$. 0! is defined to be 1, an important point to remember for some situations, such as the binomial probabilities in Chapter 5.

Question: If there were three slots and only three shows available for each slot with no repeats, how many lineups would be possible?

Answer:

Number of Alternatives				Number of Sequences
Slots				
1	2	3		Lineups
3	2	1	=	6

Again, we could write this answer as 3! We generalize from the last two situations to the second counting rule.

Appendix F — Counting: Combinations and Permutations

> **Counting Rule 2**
> When we arrange n alternatives into a sequence with n positions without repeating the same alternative, there are a total of $n!$ distinct possible sequences.

Question: If there are three slots to fill, all five programs are suitable viewing for any of the slots, but no program can be shown more than once during the day, how many different arrangements are possible?

Answer:

Number of Alternatives Slots				Number of Sequences Lineups
1	2	3		
5	4	3	=	60

Figure F-3 displays these 60 alternatives.

Question: If there were 10 shows with four slots to fill (without repeating the same show), how many lineups would be possible?

Answer:

Alternative Slots				Sequences Lineups
1	2	3	4	
10	9	8	7	= 5,040

Imagine listing all of these arrangements!

Notice in the television programming example that, when we filled five slots with five non-repeating shows, there were $5! = 120$ possible lineups. When we fill three slots from five shows, as we did in the previous question, we stop short of completing the 5! procedure. In fact, we stop before filling the last two of the five slots. If we fill these two slots the total possible lineups increases by 2! or by a factor of 2. Thus, 1/2 of the lineups are no longer possible when we fill only three slots. Using the box diagrams we can picture what is happening with the factorial and mathematical procedures.

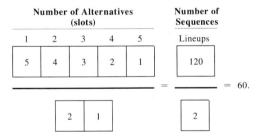

We can see that the answer is really $\dfrac{5!}{(5-3)!}$. When we arrange n alternatives into a sequence with x positions without repeating the same alternative, the total number of distinct sequences is called the *permutations* of x taken from n objects. Duplicate objects in the data set are not the same as allowing an object to repeat. If there are two nines in the set of alternatives, then nine can appear in a sequence twice. But the same nine cannot repeat. There could not be a sequence with three nines in this situation.

> **Counting Rule 3: The Number of Permutations of x Objects Taken from n Objects**
> When we arrange n alternatives into a sequence with x positions without repeating the same alternative, a total of $n!/(n-x)!$ distinct ordered sequences are possible.

Figure F-3
Permutations of five programs taking three at a time

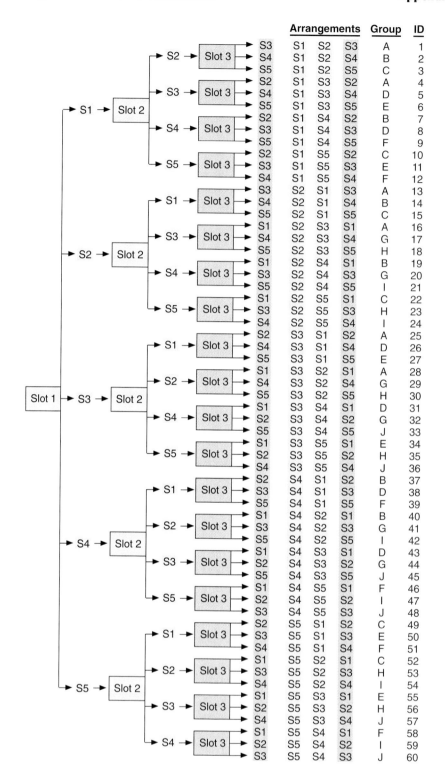

In the last question with 10 shows available to fill four slots, the number of permutations would be

$$\frac{10!}{(10-4)!} = \frac{10!}{6!} = \frac{10 \times 9 \times 8 \times 7 \times 6 \times 5 \times 4 \times 3 \times 2 \times 1}{6 \times 5 \times 4 \times 3 \times 2 \times 1}$$

Notice that the 6, 5, 4, 3, 2, and 1 in the numerator and denominator cancel, leaving our former answer, $10 \times 9 \times 8 \times 7 = 5{,}040$ arrangements.

The order of selection determined the time slot a program filled in each case so far. Suppose instead that we are only interested in the number of possible sets of three shows without regard to the specific time slot. Looking back at Figure F-3, the diagram that displays the 60 ordered arrangements, we can locate duplicate arrangements, that is, sets of the same three shows, just in different orders. We call these groups *combinations*. Different letters designate the 10 different possible combinations.

For any set of three programs, such as S1, S2, and S4 (labeled with the letter "B"), there are $3! = 6$ ordered arrangements, as shown in the diagram, so there will be six times as many ordered arrangements as unordered arrangements. Thus, one-sixth of the 60 permutations in the figure gives us the 10 combinations. Using factorial notation, this is

$$\left(\frac{1}{3!}\right)\left(\frac{5!}{(5-3)!}\right) = \frac{5 \times 4 \times 3 \times 2 \times 1}{(3 \times 2 \times 1)(2 \times 1)} = 10$$

which leads us to the final rule.

> **Counting Rule 4: The Number of Combinations of x Objects Taken from n Objects**
> When we arrange n alternatives into x positions without regard to order and without repeating the same alternative, there are a total of $\dfrac{n!}{x!(n-x)!}$ possible sets of X alternatives called combinations.

■ **Example F-1:** Use Rule 4 to find the number of combinations of 10 shows when the programmer selects shows for four time slots.

■ **Solution:** There are 10 alternatives (n) and four positions (x), so the number of combinations is

$$\frac{10!}{4!(10-4)!} = \frac{10 \times 9 \times 8 \times 7 \times 6 \times 5 \times 4 \times 3 \times 2 \times 1}{(4 \times 3 \times 2 \times 1)(6 \times 5 \times 4 \times 3 \times 2 \times 1)} = 210.$$

Again, a great number of values in the numerator and denominator cancel.

■ **Example F-2:** If the television programmer randomly selects from the five shows to fill the three slots (without repeating the same show), then each of the 10 combinations is equally likely to be chosen. What is the probability that S1 fills one of the three slots?

■ **Solution:** We only care about the selection of S1 and not its particular slot. The 10 combinations of three programs are the 10 different (mutually exclusive) possible outcomes (unordered sets of three shows). A success in this case is a set that includes S1. We can use a backdoor approach to find the number of sets that include S1 by first finding the number that do not include S1. The number of combinations of three shows that we can select from the four shows S2, S3, S4, and S5 is

$$\frac{4!}{(3!)(4-3)!} = \frac{4 \times 3 \times 2 \times 1}{(3 \times 2 \times 1)(1)} = 4.$$

These four combinations are included in the 10 possible combinations that result when we select from all five shows, but they are the ones that do not include S1. So six of the 10 combinations must include S1, and the probability of choosing a set that includes S1 will be 6/10. Check this with Figure F-3.

We employ these rules in different situations involving uncertainty. Specifically, we use the combinations formula in Chapter 5 as part of the binomial probability formula.

Problems

1. How many different arrangements of a 10-question multiple-choice test are possible?
2. **a.** If a teacher composes a test with three questions from a set of 20 questions, how many possible tests are possible? Assume that the order of the questions on the test is unimportant.
   b. If there are five questions out of the 20 that you are unprepared to answer, how many three-question tests can the teacher compose so that you are prepared for all three questions?
   c. If you have no reason to suspect that some of the 20 questions are more likely to appear on the test than others, what is the probability of your finding that the test consists of questions for which you are prepared?
3. A physician will randomly choose 20 subjects from 25 applicants for a medical experiment. How many sets of 20 subjects are possible, if order of selection is unimportant? If the order of selection determines the dosage level the subject receives, how many different arrangements are possible? Which answer is the number of combinations, and which is the number of permutations?
4. A police sergeant needs a lineup of six people for a witness to identify a criminal.
   a. If there are six detainees in the local jail, how many possible arrangements are possible for the lineup if order is important? If order is not important?
   b. If there are nine detainees in the local jail, how many possible arrangements are possible if order is unimportant? If order is important?
   c. The criminal is among the nine detainees. What is the likelihood the lineup will include the criminal if the sergeant chooses six detainees randomly among the nine?
5. A storeowner has 12 spaces for different candies. There are five types of peanut and chocolate candies, but the owner wants this type in only the three leftmost spaces. There are 10 types of hard candy, which the owner wishes to occupy the remaining nine spaces. No one type can occupy more than one space. How many different arrangements of the candy are possible?
6. How many arrangements of candidates' names on a ballot are possible for a slate of four candidates? How many arrangements include candidate A first?
7. Fifteen friends each contribute $1 into a kitty and place their business card in a fish bowl. Then one of the friends randomly selects three cards. The first name drawn receives $2, the second $3, and the third $10. How many arrangements of winners are possible? How many sets of three winners without regard to amount are possible? What is the probability John, a participant, wins something?
8. A sheriff in the nineteenth century must select a posse of four people to pursue a bank robber. There are six deputies. How many posses can the sheriff form? If the order of selection determines the direction that the deputy searches, how many arrangements are possible?
9. A professor must select four students from a class of 10 to make presentations during the next class period. How many sets of four students are possible? If the order of selection determines the order of presentation, how many presentation sequences are possible?
10. How many arrangements according to chair (first chair, second chair, and so on, with lower-ordered chairs preferred) of seven violinists in an orchestra are possible? If there are nine violinists qualified for the seven chairs, how many sets of seven violinists are possible without regard to chair? If the order of selection determines the particular chair, how many arrangements are possible?
11. How many ways can a cinema manager allocate six films among six auditoriums? If 10 films are available, how many sets of six films can the manager show without regard to the particular auditorium?
12. Six male and four female applicants qualify for three new positions—all with the same job description—in a firm. If three are randomly chosen for the positions, what is the probability they will all be male?

Solutions to Problems (Odd)

1. $10! = 3,628,800$ tests.
3. Order unimportant: $25!/(20!)(5!) = 53,130$ sets of subjects = number of combinations.
   Order important: $25!/5! = 1.2926 \times 10^{26}$ sets of subjects = number of permutations.

5. We assume the order, not just being chosen for display, is important.
Number of arrangements of peanuts and chocolate = 5!/2! = 60 arrangements.
Number of arrangements of hard candy = 10!/1! = 3,682,800 arrangements.
Total number of arrangements = 60(3,682,800) = 217,728,000 arrangements.

7. Number of arrangements of winners = 15!/12! = 2730 arrangements.
Number of sets of three winners without regard to amounts = 15!/(3!)(12!) = 455 sets.
Number of arrangements where John does not win = 14!/(3!)(11!) = 364 sets.
Number of arrangements where John does win something = 455 − 364 = 91 sets.
P(John wins something) = 91/455 = 0.2

9. Number of sets of students = 10!/(4!)(6!) = 210 arrangements.
Number of sets with selection order determining presentation order = 10!/6! = 5040 arrangements.

11. Number of allocations of 6 films = 6! = 720 allocations.
Number of allocations of 10 films to 6 cinemas without regard to particular cinema chosen = 10!/(6!)(4!) = 210 allocations.

Appendix G Binomial Distribution Tables

1. $(0.01 \leq p \leq 0.10$ or $0.90 \leq p \leq 0.99)$ on pages G1–G3
2. $(0.10 \leq p \leq 0.90)$ on pages G3–G6

Any X values not shown or $\cdots$ correspond to probabilities of 0.0000.

n	X	X	→0.01 ←0.99	0.02 0.98	0.03 0.97	0.04 0.96	0.05 0.95	0.06 0.94	0.07 0.93	0.08 0.92	0.09 0.91	0.10 0.90
2	0	2	0.9801	0.9604	0.9409	0.9216	0.9025	0.8836	0.8649	0.8464	0.8281	0.8100
	1	1	0.0198	0.0392	0.0582	0.0768	0.0950	0.1128	0.1302	0.1472	0.1638	0.1800
	2	0	0.0001	0.0004	0.0009	0.0016	0.0025	0.0036	0.0049	0.0064	0.0081	0.0100
3	0	3	0.9703	0.9412	0.9127	0.8847	0.8574	0.8306	0.8044	0.7787	0.7536	0.7290
	1	2	0.0294	0.0576	0.0847	0.1106	0.1354	0.1590	0.1816	0.2031	0.2236	0.2430
	2	1	0.0003	0.0012	0.0026	0.0046	0.0071	0.0102	0.0137	0.0177	0.0221	0.0270
	3	0	$\cdots$	$\cdots$	$\cdots$	0.0001	0.0001	0.0002	0.0003	0.0005	0.0007	0.0010
4	0	4	0.9606	0.9224	0.8853	0.8493	0.8145	0.7807	0.7481	0.7164	0.6857	0.6561
	1	3	0.0388	0.0753	0.1095	0.1416	0.1715	0.1993	0.2252	0.2492	0.2713	0.2916
	2	2	0.0006	0.0023	0.0051	0.0088	0.0135	0.0191	0.0254	0.0325	0.0402	0.0486
	3	1	$\cdots$	$\cdots$	0.0001	0.0002	0.0005	0.0008	0.0013	0.0019	0.0027	0.0036
	4	0	$\cdots$	$\cdots$	$\cdots$	$\cdots$	$\cdots$	$\cdots$	$\cdots$	$\cdots$	0.0001	0.0001
5	0	5	0.9510	0.9039	0.8587	0.8154	0.7738	0.7339	0.6957	0.6591	0.6240	0.5905
	1	4	0.0480	0.0922	0.1328	0.1699	0.2036	0.2342	0.2618	0.2866	0.3086	0.3280
	2	3	0.0010	0.0038	0.0082	0.0142	0.0214	0.0299	0.0394	0.0498	0.0610	0.0729
	3	2	$\cdots$	0.0001	0.0003	0.0006	0.0011	0.0019	0.0030	0.0043	0.0060	0.0081
	4	1	$\cdots$	$\cdots$	$\cdots$	$\cdots$	$\cdots$	0.0001	0.0001	0.0002	0.0003	0.0004
6	0	6	0.9415	0.8858	0.8330	0.7828	0.7351	0.6899	0.6470	0.6064	0.5679	0.5314
	1	5	0.0571	0.1085	0.1546	0.1957	0.2321	0.2642	0.2922	0.3164	0.3370	0.3543
	2	4	0.0014	0.0055	0.0120	0.0204	0.0305	0.0422	0.0550	0.0688	0.0833	0.0984
	3	3	$\cdots$	0.0002	0.0005	0.0011	0.0021	0.0036	0.0055	0.0080	0.0110	0.0146
	4	2	$\cdots$	$\cdots$	$\cdots$	$\cdots$	0.0001	0.0002	0.0003	0.0005	0.0008	0.0012
	5	1	$\cdots$	$\cdots$	$\cdots$	$\cdots$	$\cdots$	$\cdots$	$\cdots$	$\cdots$	$\cdots$	0.0001
7	0	7	0.9321	0.8681	0.8080	0.7514	0.6983	0.6485	0.6017	0.5578	0.5168	0.4783
	1	6	0.0659	0.1240	0.1749	0.2192	0.2573	0.2897	0.3170	0.3396	0.3578	0.3720
	2	5	0.0020	0.0076	0.0162	0.0274	0.0406	0.0555	0.0716	0.0886	0.1061	0.1240
	3	4	$\cdots$	0.0003	0.0008	0.0019	0.0036	0.0059	0.0090	0.0128	0.0175	0.0230
	4	3	$\cdots$	$\cdots$	$\cdots$	0.0001	0.0002	0.0004	0.0007	0.0011	0.0017	0.0026
	5	2	$\cdots$	$\cdots$	$\cdots$	$\cdots$	$\cdots$	$\cdots$	$\cdots$	0.0001	0.0001	0.0002
8	0	8	0.9227	0.8508	0.7837	0.7214	0.6634	0.6096	0.5596	0.5132	0.4703	0.4305
	1	7	0.0746	0.1389	0.1939	0.2405	0.2793	0.3113	0.3370	0.3570	0.3721	0.3826
	2	6	0.0026	0.0099	0.0210	0.0351	0.0515	0.0695	0.0888	0.1087	0.1288	0.1488
	3	5	0.0001	0.0004	0.0013	0.0029	0.0054	0.0089	0.0134	0.0189	0.0255	0.0331
	4	4	$\cdots$	$\cdots$	0.0001	0.0002	0.0004	0.0007	0.0013	0.0021	0.0031	0.0046
	5	3	$\cdots$	$\cdots$	$\cdots$	$\cdots$	$\cdots$	$\cdots$	0.0001	0.0001	0.0002	0.0004
9	0	9	0.9135	0.8337	0.7602	0.6925	0.6302	0.5730	0.5204	0.4722	0.4279	0.3874
	1	8	0.0830	0.1531	0.2116	0.2597	0.2985	0.3292	0.3525	0.3695	0.3809	0.3874
	2	7	0.0034	0.0125	0.0262	0.0433	0.0629	0.0840	0.1061	0.1285	0.1507	0.1722
	3	6	0.0001	0.0006	0.0019	0.0042	0.0077	0.0125	0.0186	0.0261	0.0348	0.0446
	4	5	$\cdots$	$\cdots$	0.0001	0.0003	0.0006	0.0012	0.0021	0.0034	0.0052	0.0074
	5	4	$\cdots$	$\cdots$	$\cdots$	$\cdots$	$\cdots$	0.0001	0.0002	0.0003	0.0005	0.0008
	6	3	$\cdots$	$\cdots$	$\cdots$	$\cdots$	$\cdots$	$\cdots$	$\cdots$	$\cdots$	$\cdots$	0.0001
10	0	10	0.9044	0.8171	0.7374	0.6648	0.5987	0.5386	0.4840	0.4344	0.3894	0.3487
	1	9	0.0914	0.1667	0.2281	0.2770	0.3151	0.3438	0.3643	0.3777	0.3851	0.3874
	2	8	0.0042	0.0153	0.0317	0.0519	0.0746	0.0988	0.1234	0.1478	0.1714	0.1937
	3	7	0.0001	0.0008	0.0026	0.0058	0.0105	0.0168	0.0248	0.0343	0.0452	0.0574
	4	6	$\cdots$	$\cdots$	0.0001	0.0004	0.0010	0.0019	0.0033	0.0052	0.0078	0.0112
	5	5	$\cdots$	$\cdots$	$\cdots$	$\cdots$	0.0001	0.0001	0.0003	0.0005	0.0009	0.0015
	6	4	$\cdots$	$\cdots$	$\cdots$	$\cdots$	$\cdots$	$\cdots$	$\cdots$	$\cdots$	0.0001	0.0001

Appendix G

Binomial Distribution Tables

G2

n	X	X	0.01 / 0.99	0.02 / 0.98	0.03 / 0.97	0.04 / 0.96	0.05 / 0.95	0.06 / 0.94	0.07 / 0.93	0.08 / 0.92	0.09 / 0.91	0.10 / 0.90
11	0	11	0.8953	0.8007	0.7153	0.6382	0.5688	0.5063	0.4501	0.3996	0.3544	0.3138
	1	10	0.0995	0.1798	0.2434	0.2925	0.3293	0.3555	0.3727	0.3823	0.3855	0.3835
	2	9	0.0050	0.0183	0.0376	0.0609	0.0867	0.1135	0.1403	0.1662	0.1906	0.2131
	3	8	0.0002	0.0011	0.0035	0.0076	0.0137	0.0217	0.0317	0.0434	0.0566	0.0710
	4	7	...	...	0.0002	0.0006	0.0014	0.0028	0.0048	0.0075	0.0112	0.0158
	5	6	...	...	...	...	0.0001	0.0002	0.0005	0.0009	0.0015	0.0025
	6	5	...	...	...	...	...	...	...	0.0001	0.0002	0.0003
12	0	12	0.8864	0.7847	0.6938	0.6127	0.5404	0.4759	0.4186	0.3677	0.3225	0.2824
	1	11	0.1074	0.1922	0.2575	0.3064	0.3413	0.3645	0.3781	0.3837	0.3827	0.3766
	2	10	0.0060	0.0216	0.0438	0.0702	0.0988	0.1280	0.1565	0.1835	0.2082	0.2301
	3	9	0.0002	0.0015	0.0045	0.0098	0.0173	0.0272	0.0393	0.0532	0.0686	0.0852
	4	8	...	0.0001	0.0003	0.0009	0.0021	0.0039	0.0067	0.0104	0.0153	0.0213
	5	7	...	...	...	0.0001	0.0002	0.0004	0.0008	0.0014	0.0024	0.0038
	6	6	...	...	...	...	...	...	0.0001	0.0001	0.0003	0.0005
13	0	13	0.8775	0.7690	0.6730	0.5882	0.5133	0.4474	0.3893	0.3383	0.2935	0.2542
	1	12	0.1152	0.2040	0.2706	0.3186	0.3512	0.3712	0.3809	0.3824	0.3773	0.3672
	2	11	0.0070	0.0250	0.0502	0.0797	0.1109	0.1422	0.1720	0.1995	0.2239	0.2448
	3	10	0.0003	0.0019	0.0057	0.0122	0.0214	0.0333	0.0475	0.0636	0.0812	0.0997
	4	9	...	0.0001	0.0004	0.0013	0.0028	0.0053	0.0089	0.0138	0.0201	0.0277
	5	8	...	...	...	0.0001	0.0003	0.0006	0.0012	0.0022	0.0036	0.0055
	6	7	...	...	...	...	...	0.0001	0.0001	0.0003	0.0005	0.0008
	7	6	...	...	...	...	...	...	...	...	...	0.0001
14	0	14	0.8687	0.7536	0.6528	0.5647	0.4877	0.4205	0.3620	0.3112	0.2670	0.2288
	1	13	0.1229	0.2153	0.2827	0.3294	0.3593	0.3758	0.3815	0.3788	0.3698	0.3559
	2	12	0.0081	0.0286	0.0568	0.0892	0.1229	0.1559	0.1867	0.2141	0.2377	0.2570
	3	11	0.0003	0.0023	0.0070	0.0149	0.0259	0.0398	0.0562	0.0745	0.0940	0.1142
	4	10	...	0.0001	0.0006	0.0017	0.0037	0.0070	0.0116	0.0178	0.0256	0.0349
	5	9	...	...	...	0.0001	0.0004	0.0009	0.0018	0.0031	0.0051	0.0078
	6	8	...	...	...	...	...	0.0001	0.0002	0.0004	0.0008	0.0013
	7	7	...	...	...	...	...	...	...	...	0.0001	0.0002
15	0	15	0.8601	0.7386	0.6333	0.5421	0.4633	0.3953	0.3367	0.2863	0.2430	0.2059
	1	14	0.1303	0.2261	0.2938	0.3388	0.3658	0.3785	0.3801	0.3734	0.3605	0.3432
	2	13	0.0092	0.0323	0.0636	0.0988	0.1348	0.1691	0.2003	0.2273	0.2496	0.2669
	3	12	0.0004	0.0029	0.0085	0.0178	0.0307	0.0468	0.0653	0.0857	0.1070	0.1285
	4	11	...	0.0002	0.0008	0.0022	0.0049	0.0090	0.0148	0.0223	0.0317	0.0428
	5	10	...	...	0.0001	0.0002	0.0006	0.0013	0.0024	0.0043	0.0069	0.0105
	6	9	...	...	...	...	...	0.0001	0.0003	0.0006	0.0011	0.0019
	7	8	...	...	...	...	...	...	...	0.0001	0.0001	0.0003
16	0	16	0.8515	0.7238	0.6143	0.5204	0.4401	0.3716	0.3131	0.2634	0.2211	0.1853
	1	15	0.1376	0.2363	0.3040	0.3469	0.3706	0.3795	0.3771	0.3665	0.3499	0.3294
	2	14	0.0104	0.0362	0.0705	0.1084	0.1463	0.1817	0.2129	0.2390	0.2596	0.2745
	3	13	0.0005	0.0034	0.0102	0.0211	0.0359	0.0541	0.0748	0.0970	0.1198	0.1423
	4	12	...	0.0002	0.0010	0.0029	0.0061	0.0112	0.0183	0.0274	0.0385	0.0514
	5	11	...	...	0.0001	0.0003	0.0008	0.0017	0.0033	0.0057	0.0091	0.0137
	6	10	...	...	...	...	0.0001	0.0002	0.0005	0.0009	0.0017	0.0028
	7	9	...	...	...	...	...	...	...	0.0001	0.0002	0.0004
	8	8	...	...	...	...	...	...	...	...	...	0.0001
17	0	17	0.8429	0.7093	0.5958	0.4996	0.4181	0.3493	0.2912	0.2423	0.2012	0.1668
	1	16	0.1447	0.2461	0.3133	0.3539	0.3741	0.3790	0.3726	0.3582	0.3383	0.3150
	2	15	0.0117	0.0402	0.0775	0.1180	0.1575	0.1935	0.2244	0.2492	0.2677	0.2800
	3	14	0.0006	0.0041	0.0120	0.0246	0.0415	0.0618	0.0844	0.1083	0.1324	0.1556
	4	13	...	0.0003	0.0013	0.0036	0.0076	0.0138	0.0222	0.0330	0.0458	0.0605
	5	12	...	...	0.0001	0.0004	0.0010	0.0023	0.0044	0.0075	0.0118	0.0175
	6	11	...	...	...	...	0.0001	0.0003	0.0007	0.0013	0.0023	0.0039
	7	10	...	...	...	...	...	...	0.0001	0.0002	0.0004	0.0007
	8	9	...	...	...	...	...	...	...	...	...	0.0001
18	0	18	0.8345	0.6951	0.5780	0.4796	0.3972	0.3283	0.2708	0.2229	0.1831	0.1501
	1	17	0.1517	0.2554	0.3217	0.3597	0.3763	0.3772	0.3669	0.3489	0.3260	0.3002
	2	16	0.0130	0.0443	0.0846	0.1274	0.1683	0.2047	0.2348	0.2579	0.2741	0.2835
	3	15	0.0007	0.0048	0.0140	0.0283	0.0473	0.0697	0.0942	0.1196	0.1446	0.1680
	4	14	...	0.0004	0.0016	0.0044	0.0093	0.0167	0.0266	0.0390	0.0536	0.0700
	5	13	...	...	0.0001	0.0005	0.0014	0.0030	0.0056	0.0095	0.0148	0.0218

(entries for n = 18 continued on next page)

n	X	X	0.01 / 0.99	0.02 / 0.98	0.03 / 0.97	0.04 / 0.96	0.05 / 0.95	0.06 / 0.94	0.07 / 0.93	0.08 / 0.92	0.09 / 0.91	0.10 / 0.90
18	6	12	...	...	...	...	0.0002	0.0004	0.0009	0.0018	0.0032	0.0052
	7	11	...	...	...	...	...	...	0.0001	0.0003	0.0005	0.0010
	8	10	...	...	...	...	...	...	...	...	0.0001	0.0002
19	0	19	0.8262	0.6812	0.5606	0.4604	0.3774	0.3086	0.2519	0.2051	0.1666	0.1351
	1	18	0.1586	0.2642	0.3294	0.3645	0.3774	0.3743	0.3602	0.3389	0.3131	0.2852
	2	17	0.0144	0.0485	0.0917	0.1367	0.1787	0.2150	0.2440	0.2652	0.2787	0.2852
	3	16	0.0008	0.0056	0.0161	0.0323	0.0533	0.0778	0.1041	0.1307	0.1562	0.1796
	4	15	...	0.0005	0.0020	0.0054	0.0112	0.0199	0.0313	0.0455	0.0618	0.0798
	5	14	...	...	0.0002	0.0007	0.0018	0.0038	0.0071	0.0119	0.0183	0.0266
	6	13	...	...	...	0.0001	0.0002	0.0006	0.0012	0.0024	0.0042	0.0069
	7	12	...	...	...	...	...	0.0001	0.0002	0.0004	0.0008	0.0014
	8	11	...	...	...	...	...	...	...	0.0001	0.0001	0.0002
20	0	20	0.8179	0.6676	0.5438	0.4420	0.3585	0.2901	0.2342	0.1887	0.1516	0.1216
	1	19	0.1652	0.2725	0.3364	0.3683	0.3774	0.3703	0.3526	0.3282	0.3000	0.2702
	2	18	0.0159	0.0528	0.0988	0.1458	0.1887	0.2246	0.2521	0.2711	0.2818	0.2852
	3	17	0.0010	0.0065	0.0183	0.0364	0.0596	0.0860	0.1139	0.1414	0.1672	0.1901
	4	16	...	0.0006	0.0024	0.0065	0.0133	0.0233	0.0364	0.0523	0.0703	0.0898
	5	15	...	...	0.0002	0.0009	0.0022	0.0048	0.0088	0.0145	0.0222	0.0319
	6	14	...	...	...	0.0001	0.0003	0.0008	0.0017	0.0032	0.0055	0.0089
	7	13	...	...	...	...	...	0.0001	0.0002	0.0005	0.0011	0.0020
	8	12	...	...	...	...	...	...	...	0.0001	0.0002	0.0004
	9	11	...	...	...	...	...	...	...	...	...	0.0001

2. ($0.10 \leq p \leq 0.90$)

Any X values not shown or $\cdots$ correspond to probabilities of 0.0000.

n	X	X	0.10 / 0.90	0.15 / 0.85	0.20 / 0.80	0.25 / 0.75	0.30 / 0.70	0.35 / 0.65	0.40 / 0.60	0.45 / 0.55	0.50 / 0.50
2	0	2	0.8100	0.7225	0.6400	0.5625	0.4900	0.4225	0.3600	0.3025	0.2500
	1	1	0.1800	0.2550	0.3200	0.3750	0.4200	0.4550	0.4800	0.4950	0.5000
	2	0	0.0100	0.0225	0.0400	0.0625	0.0900	0.1225	0.1600	0.2025	0.2500
3	0	3	0.7290	0.6141	0.5120	0.4219	0.3430	0.2746	0.2160	0.1664	0.1250
	1	2	0.2430	0.3251	0.3840	0.4219	0.4410	0.4436	0.4320	0.4084	0.3750
	2	1	0.0270	0.0574	0.0960	0.1406	0.1890	0.2389	0.2880	0.3341	0.3750
	3	0	0.0010	0.0034	0.0080	0.0156	0.0270	0.0429	0.0640	0.0911	0.1250
4	0	4	0.6561	0.5220	0.4096	0.3164	0.2401	0.1785	0.1296	0.0915	0.0625
	1	3	0.2916	0.3685	0.4096	0.4219	0.4116	0.3845	0.3456	0.2995	0.2500
	2	2	0.0486	0.0975	0.1536	0.2109	0.2646	0.3105	0.3456	0.3675	0.3750
	3	1	0.0036	0.0115	0.0256	0.0469	0.0756	0.1115	0.1536	0.2005	0.2500
	4	0	0.0001	0.0005	0.0016	0.0039	0.0081	0.0150	0.0256	0.0410	0.0625
5	0	5	0.5905	0.4437	0.3277	0.2373	0.1681	0.1160	0.0778	0.0503	0.0313
	1	4	0.3280	0.3915	0.4096	0.3955	0.3601	0.3124	0.2592	0.2059	0.1563
	2	3	0.0729	0.1382	0.2048	0.2637	0.3087	0.3364	0.3456	0.3369	0.3125
	3	2	0.0081	0.0244	0.0512	0.0879	0.1323	0.1811	0.2304	0.2757	0.3125
	4	1	0.0004	0.0022	0.0064	0.0146	0.0284	0.0488	0.0768	0.1128	0.1563
	5	0	...	0.0001	0.0003	0.0010	0.0024	0.0053	0.0102	0.0185	0.0313
6	0	6	0.5314	0.3771	0.2621	0.1780	0.1176	0.0754	0.0467	0.0277	0.0156
	1	5	0.3543	0.3993	0.3932	0.3560	0.3025	0.2437	0.1866	0.1359	0.0938
	2	4	0.0984	0.1762	0.2458	0.2966	0.3241	0.3280	0.3110	0.2780	0.2344
	3	3	0.0146	0.0415	0.0819	0.1318	0.1852	0.2355	0.2765	0.3032	0.3125
	4	2	0.0012	0.0055	0.0154	0.0330	0.0595	0.0951	0.1382	0.1861	0.2344
	5	1	0.0001	0.0004	0.0015	0.0044	0.0102	0.0205	0.0369	0.0609	0.0938
	6	0	...	...	0.0001	0.0002	0.0007	0.0018	0.0041	0.0083	0.0156
7	0	7	0.4783	0.3206	0.2097	0.1335	0.0824	0.0490	0.0280	0.0152	0.0078
	1	6	0.3720	0.3960	0.3670	0.3115	0.2471	0.1848	0.1306	0.0872	0.0547
	2	5	0.1240	0.2097	0.2753	0.3115	0.3177	0.2985	0.2613	0.2140	0.1641
	3	4	0.0230	0.0617	0.1147	0.1730	0.2269	0.2679	0.2903	0.2918	0.2734
	4	3	0.0026	0.0109	0.0287	0.0577	0.0972	0.1442	0.1935	0.2388	0.2734
	5	2	0.0002	0.0012	0.0043	0.0115	0.0250	0.0466	0.0774	0.1172	0.1641
	6	1	...	0.0001	0.0004	0.0013	0.0036	0.0084	0.0172	0.0320	0.0547
	7	0	...	...	...	0.0001	0.0002	0.0006	0.0016	0.0037	0.0078

(entries for n = 7 continued on next page)

Appendix G

Binomial Distribution Tables

n	X	X	0.10 / 0.90	0.15 / 0.85	0.20 / 0.80	0.25 / 0.75	0.30 / 0.70	0.35 / 0.65	0.40 / 0.60	0.45 / 0.55	0.50 / 0.50
8	0	8	0.4305	0.2725	0.1678	0.1001	0.0576	0.0319	0.0168	0.0084	0.0039
	1	7	0.3826	0.3847	0.3355	0.2670	0.1977	0.1373	0.0896	0.0548	0.0313
	2	6	0.1488	0.2376	0.2936	0.3115	0.2965	0.2587	0.2090	0.1569	0.1094
	3	5	0.0331	0.0839	0.1468	0.2076	0.2541	0.2786	0.2787	0.2568	0.2188
	4	4	0.0046	0.0185	0.0459	0.0865	0.1361	0.1875	0.2322	0.2627	0.2734
	5	3	0.0004	0.0026	0.0092	0.0231	0.0467	0.0808	0.1239	0.1719	0.2188
	6	2	...	0.0002	0.0011	0.0038	0.0100	0.0217	0.0413	0.0703	0.1094
	7	1	...	...	0.0001	0.0004	0.0012	0.0033	0.0079	0.0164	0.0313
	8	0	...	...	...	...	0.0001	0.0002	0.0007	0.0017	0.0039
9	0	9	0.3874	0.2316	0.1342	0.0751	0.0404	0.0207	0.0101	0.0046	0.0020
	1	8	0.3874	0.3679	0.3020	0.2253	0.1556	0.1004	0.0605	0.0339	0.0176
	2	7	0.1722	0.2597	0.3020	0.3003	0.2668	0.2162	0.1612	0.1110	0.0703
	3	6	0.0446	0.1069	0.1762	0.2336	0.2668	0.2716	0.2508	0.2119	0.1641
	4	5	0.0074	0.0283	0.0661	0.1168	0.1715	0.2194	0.2508	0.2600	0.2461
	5	4	0.0008	0.0050	0.0165	0.0389	0.0735	0.1181	0.1672	0.2128	0.2461
	6	3	0.0001	0.0006	0.0028	0.0087	0.0210	0.0424	0.0743	0.1160	0.1641
	7	2	...	...	0.0003	0.0012	0.0039	0.0098	0.0212	0.0407	0.0703
	8	1	...	...	...	0.0001	0.0004	0.0013	0.0035	0.0083	0.0176
	9	0	...	...	...	...	...	0.0001	0.0003	0.0008	0.0020
10	0	10	0.3487	0.1969	0.1074	0.0563	0.0282	0.0135	0.0060	0.0025	0.0010
	1	9	0.3874	0.3474	0.2684	0.1877	0.1211	0.0725	0.0403	0.0207	0.0098
	2	8	0.1937	0.2759	0.3020	0.2816	0.2335	0.1757	0.1209	0.0763	0.0439
	3	7	0.0574	0.1298	0.2013	0.2503	0.2668	0.2522	0.2150	0.1665	0.1172
	4	6	0.0112	0.0401	0.0881	0.1460	0.2001	0.2377	0.2508	0.2384	0.2051
	5	5	0.0015	0.0085	0.0264	0.0584	0.1029	0.1536	0.2007	0.2340	0.2461
	6	4	0.0001	0.0012	0.0055	0.0162	0.0368	0.0689	0.1115	0.1596	0.2051
	7	3	...	0.0001	0.0008	0.0031	0.0090	0.0212	0.0425	0.0746	0.1172
	8	2	...	...	0.0001	0.0004	0.0014	0.0043	0.0106	0.0229	0.0439
	9	1	...	...	...	...	0.0001	0.0005	0.0016	0.0042	0.0098
	10	0	...	...	...	...	...	...	0.0001	0.0003	0.0010
11	0	11	0.3138	0.1673	0.0859	0.0422	0.0198	0.0088	0.0036	0.0014	0.0005
	1	10	0.3835	0.3248	0.2362	0.1549	0.0932	0.0518	0.0266	0.0125	0.0054
	2	9	0.2131	0.2866	0.2953	0.2581	0.1998	0.1395	0.0887	0.0513	0.0269
	3	8	0.0710	0.1517	0.2215	0.2581	0.2568	0.2254	0.1774	0.1259	0.0806
	4	7	0.0158	0.0536	0.1107	0.1721	0.2201	0.2428	0.2365	0.2060	0.1611
	5	6	0.0025	0.0132	0.0388	0.0803	0.1321	0.1830	0.2207	0.2360	0.2256
	6	5	0.0003	0.0023	0.0097	0.0268	0.0566	0.0985	0.1471	0.1931	0.2256
	7	4	...	0.0003	0.0017	0.0064	0.0173	0.0379	0.0701	0.1128	0.1611
	8	3	...	...	0.0002	0.0011	0.0037	0.0102	0.0234	0.0462	0.0806
	9	2	...	...	...	0.0001	0.0005	0.0018	0.0052	0.0126	0.0269
	10	1	...	...	...	...	...	0.0002	0.0007	0.0021	0.0054
	11	0	...	...	...	...	...	...	...	0.0002	0.0005
12	0	12	0.2824	0.1422	0.0687	0.0317	0.0138	0.0057	0.0022	0.0008	0.0002
	1	11	0.3766	0.3012	0.2062	0.1267	0.0712	0.0368	0.0174	0.0075	0.0029
	2	10	0.2301	0.2924	0.2835	0.2323	0.1678	0.1088	0.0639	0.0339	0.0161
	3	9	0.0852	0.1720	0.2362	0.2581	0.2397	0.1954	0.1419	0.0923	0.0537
	4	8	0.0213	0.0683	0.1329	0.1936	0.2311	0.2367	0.2128	0.1700	0.1208
	5	7	0.0038	0.0193	0.0532	0.1032	0.1585	0.2039	0.2270	0.2225	0.1934
	6	6	0.0005	0.0040	0.0155	0.0401	0.0792	0.1281	0.1766	0.2124	0.2256
	7	5	...	0.0006	0.0033	0.0115	0.0291	0.0591	0.1009	0.1489	0.1934
	8	4	...	0.0001	0.0005	0.0024	0.0078	0.0199	0.0420	0.0762	0.1208
	9	3	...	...	0.0001	0.0004	0.0015	0.0048	0.0125	0.0277	0.0537
	10	2	...	...	...	...	0.0002	0.0008	0.0025	0.0068	0.0161
	11	1	...	...	...	...	...	0.0001	0.0003	0.0010	0.0029
	12	0	...	...	...	...	...	...	...	0.0001	0.0002
13	0	13	0.2542	0.1209	0.0550	0.0238	0.0097	0.0037	0.0013	0.0004	0.0001
	1	12	0.3672	0.2774	0.1787	0.1029	0.0540	0.0259	0.0113	0.0045	0.0016
	2	11	0.2448	0.2937	0.2680	0.2059	0.1388	0.0836	0.0453	0.0220	0.0095
	3	10	0.0997	0.1900	0.2457	0.2517	0.2181	0.1651	0.1107	0.0660	0.0349
	4	9	0.0277	0.0838	0.1535	0.2097	0.2337	0.2222	0.1845	0.1350	0.0873
	5	8	0.0055	0.0266	0.0691	0.1258	0.1803	0.2154	0.2214	0.1989	0.1571
	6	7	0.0008	0.0063	0.0230	0.0559	0.1030	0.1546	0.1968	0.2169	0.2095
	7	6	0.0001	0.0011	0.0058	0.0186	0.0442	0.0833	0.1312	0.1775	0.2095

(entries for n = 13 continued on next page)

Binomial Distribution Tables

Appendix G

n	X	X	0.10 / 0.90	0.15 / 0.85	0.20 / 0.80	0.25 / 0.75	0.30 / 0.70	0.35 / 0.65	0.40 / 0.60	0.45 / 0.55	0.50 / 0.50
13	8	5	...	0.0001	0.0011	0.0047	0.0142	0.0336	0.0656	0.1089	0.1571
	9	4	...	...	0.0001	0.0009	0.0034	0.0101	0.0243	0.0495	0.0873
	10	3	...	...	...	0.0001	0.0006	0.0022	0.0065	0.0162	0.0349
	11	2	...	...	...	...	0.0001	0.0003	0.0012	0.0036	0.0095
	12	1	...	...	...	...	...	...	0.0001	0.0005	0.0016
	13	0	...	...	...	...	...	...	...	...	0.0001
14	0	14	0.2288	0.1028	0.0440	0.0178	0.0068	0.0024	0.0008	0.0002	0.0001
	1	13	0.3559	0.2539	0.1539	0.0832	0.0407	0.0181	0.0073	0.0027	0.0009
	2	12	0.2570	0.2912	0.2501	0.1802	0.1134	0.0634	0.0317	0.0141	0.0056
	3	11	0.1142	0.2056	0.2501	0.2402	0.1943	0.1366	0.0845	0.0462	0.0222
	4	10	0.0349	0.0998	0.1720	0.2202	0.2290	0.2022	0.1549	0.1040	0.0611
	5	9	0.0078	0.0352	0.0860	0.1468	0.1963	0.2178	0.2066	0.1701	0.1222
	6	8	0.0013	0.0093	0.0322	0.0734	0.1262	0.1759	0.2066	0.2088	0.1833
	7	7	0.0002	0.0019	0.0092	0.0280	0.0618	0.1082	0.1574	0.1952	0.2095
	8	6	...	0.0003	0.0020	0.0082	0.0232	0.0510	0.0918	0.1398	0.1833
	9	5	...	...	0.0003	0.0018	0.0066	0.0183	0.0408	0.0762	0.1222
	10	4	...	...	...	0.0003	0.0014	0.0049	0.0136	0.0312	0.0611
	11	3	...	...	...	...	0.0002	0.0010	0.0033	0.0093	0.0222
	12	2	...	...	...	...	...	0.0001	0.0005	0.0019	0.0056
	13	1	...	...	...	...	...	...	0.0001	0.0002	0.0009
	14	0	...	...	...	...	...	...	...	...	0.0001
15	0	15	0.2059	0.0874	0.0352	0.0134	0.0047	0.0016	0.0005	0.0001	...
	1	14	0.3432	0.2312	0.1319	0.0668	0.0305	0.0126	0.0047	0.0016	0.0005
	2	13	0.2669	0.2856	0.2309	0.1559	0.0916	0.0476	0.0219	0.0090	0.0032
	3	12	0.1285	0.2184	0.2501	0.2252	0.1700	0.1110	0.0634	0.0318	0.0139
	4	11	0.0428	0.1156	0.1876	0.2252	0.2186	0.1792	0.1268	0.0780	0.0417
	5	10	0.0105	0.0449	0.1032	0.1651	0.2061	0.2123	0.1859	0.1404	0.0916
	6	9	0.0019	0.0132	0.0430	0.0917	0.1472	0.1906	0.2066	0.1914	0.1527
	7	8	0.0003	0.0030	0.0138	0.0393	0.0811	0.1319	0.1771	0.2013	0.1964
	8	7	...	0.0005	0.0035	0.0131	0.0348	0.0710	0.1181	0.1647	0.1964
	9	6	...	0.0001	0.0007	0.0034	0.0116	0.0298	0.0612	0.1048	0.1527
	10	5	...	...	0.0001	0.0007	0.0030	0.0096	0.0245	0.0515	0.0916
	11	4	...	...	...	0.0001	0.0006	0.0024	0.0074	0.0191	0.0417
	12	3	...	...	...	...	0.0001	0.0004	0.0016	0.0052	0.0139
	13	2	...	...	...	...	...	0.0001	0.0003	0.0010	0.0032
	14	1	...	...	...	...	...	...	...	0.0001	0.0005
	15	0	...	...	...	...	...	...	...	...	...
16	0	16	0.1853	0.0743	0.0281	0.0100	0.0033	0.0010	0.0003	0.0001	...
	1	15	0.3294	0.2097	0.1126	0.0535	0.0228	0.0087	0.0030	0.0009	0.0002
	2	14	0.2745	0.2775	0.2111	0.1336	0.0732	0.0353	0.0150	0.0056	0.0018
	3	13	0.1423	0.2285	0.2463	0.2079	0.1465	0.0888	0.0468	0.0215	0.0085
	4	12	0.0514	0.1311	0.2001	0.2252	0.2040	0.1553	0.1014	0.0572	0.0278
	5	11	0.0137	0.0555	0.1201	0.1802	0.2099	0.2008	0.1623	0.1123	0.0667
	6	10	0.0028	0.0180	0.0550	0.1101	0.1649	0.1982	0.1983	0.1684	0.1222
	7	9	0.0004	0.0045	0.0197	0.0524	0.1010	0.1524	0.1889	0.1969	0.1746
	8	8	0.0001	0.0009	0.0055	0.0197	0.0487	0.0923	0.1417	0.1812	0.1964
	9	7	...	0.0001	0.0012	0.0058	0.0185	0.0442	0.0840	0.1318	0.1746
	10	6	...	...	0.0002	0.0014	0.0056	0.0167	0.0392	0.0755	0.1222
	11	5	...	...	...	0.0002	0.0013	0.0049	0.0142	0.0337	0.0667
	12	4	...	...	...	...	0.0002	0.0011	0.0040	0.0115	0.0278
	13	3	...	...	...	...	...	0.0002	0.0008	0.0029	0.0085
	14	2	...	...	...	...	...	...	0.0001	0.0005	0.0018
	15	1	...	...	...	...	...	...	...	0.0001	0.0002
	16	0	...	...	...	...	...	...	...	...	...
17	0	17	0.1668	0.0631	0.0225	0.0075	0.0023	0.0007	0.0002	...	...
	1	16	0.3150	0.1893	0.0957	0.0426	0.0169	0.0060	0.0019	0.0005	0.0001
	2	15	0.2800	0.2673	0.1914	0.1136	0.0581	0.0260	0.0102	0.0035	0.0010
	3	14	0.1556	0.2359	0.2393	0.1893	0.1245	0.0701	0.0341	0.0144	0.0052
	4	13	0.0605	0.1457	0.2093	0.2209	0.1868	0.1320	0.0796	0.0411	0.0182
	5	12	0.0175	0.0668	0.1361	0.1914	0.2081	0.1849	0.1379	0.0875	0.0472
	6	11	0.0039	0.0236	0.0680	0.1276	0.1784	0.1991	0.1839	0.1432	0.0944
	7	10	0.0007	0.0065	0.0267	0.0668	0.1201	0.1685	0.1927	0.1841	0.1484
	8	9	0.0001	0.0014	0.0084	0.0279	0.0644	0.1134	0.1606	0.1883	0.1855
	9	8	...	0.0003	0.0021	0.0093	0.0276	0.0611	0.1070	0.1540	0.1855

(entries for n = 17 continued on next page)

Appendix G

Binomial Distribution Tables

n	X	X	→0.10 →0.90	0.15 0.85	0.20 0.80	0.25 0.75	0.30 0.70	0.35 0.65	0.40 0.60	0.45 0.55	0.50 0.50
17	10	7	...	...	0.0004	0.0025	0.0095	0.0263	0.0571	0.1008	0.1484
	11	6	...	...	0.0001	0.0005	0.0026	0.0090	0.0242	0.0525	0.0944
	12	5	...	...	...	0.0001	0.0006	0.0024	0.0081	0.0215	0.0472
	13	4	...	...	...	...	0.0001	0.0005	0.0021	0.0068	0.0182
	14	3	...	...	...	...	...	0.0001	0.0004	0.0016	0.0052
	15	2	...	...	...	...	...	...	0.0001	0.0003	0.0010
	16	1	...	...	...	...	...	...	...	...	0.0001
	17	0	...	...	...	...	...	...	...	...	...
18	0	18	0.1501	0.0536	0.0180	0.0056	0.0016	0.0004	0.0001	...	...
	1	17	0.3002	0.1704	0.0811	0.0338	0.0126	0.0042	0.0012	0.0003	0.0001
	2	16	0.2835	0.2556	0.1723	0.0958	0.0458	0.0190	0.0069	0.0022	0.0006
	3	15	0.1680	0.2406	0.2297	0.1704	0.1046	0.0547	0.0246	0.0095	0.0031
	4	14	0.0700	0.1592	0.2153	0.2130	0.1681	0.1104	0.0614	0.0291	0.0117
	5	13	0.0218	0.0787	0.1507	0.1988	0.2017	0.1664	0.1146	0.0666	0.0327
	6	12	0.0052	0.0301	0.0816	0.1436	0.1873	0.1941	0.1655	0.1181	0.0708
	7	11	0.0010	0.0091	0.0350	0.0820	0.1376	0.1792	0.1892	0.1657	0.1214
	8	10	0.0002	0.0022	0.0120	0.0376	0.0811	0.1327	0.1734	0.1864	0.1669
	9	9	...	0.0004	0.0033	0.0139	0.0386	0.0794	0.1284	0.1694	0.1855
	10	8	...	0.0001	0.0008	0.0042	0.0149	0.0385	0.0771	0.1248	0.1669
	11	7	...	...	0.0001	0.0010	0.0046	0.0151	0.0374	0.0742	0.1214
	12	6	...	...	...	0.0002	0.0012	0.0047	0.0145	0.0354	0.0708
	13	5	...	...	...	...	0.0002	0.0012	0.0045	0.0134	0.0327
	14	4	...	...	...	...	...	0.0002	0.0011	0.0039	0.0117
	15	3	...	...	...	...	...	...	0.0002	0.0009	0.0031
	16	2	...	...	...	...	...	...	...	0.0001	0.0006
	17	1	...	...	...	...	...	...	...	...	0.0001
	18	0	...	...	...	...	...	...	...	...	...
19	0	19	0.1351	0.0456	0.0144	0.0042	0.0011	0.0003	0.0001	...	...
	1	18	0.2852	0.1529	0.0685	0.0268	0.0093	0.0029	0.0008	0.0002	...
	2	17	0.2852	0.2428	0.1540	0.0803	0.0358	0.0138	0.0046	0.0013	0.0003
	3	16	0.1796	0.2428	0.2182	0.1517	0.0869	0.0422	0.0175	0.0062	0.0018
	4	15	0.0798	0.1714	0.2182	0.2023	0.1491	0.0909	0.0467	0.0203	0.0074
	5	14	0.0266	0.0907	0.1636	0.2023	0.1916	0.1468	0.0933	0.0497	0.0222
	6	13	0.0069	0.0374	0.0955	0.1574	0.1916	0.1844	0.1451	0.0949	0.0518
	7	12	0.0014	0.0122	0.0443	0.0974	0.1525	0.1844	0.1797	0.1443	0.0961
	8	11	0.0002	0.0032	0.0166	0.0487	0.0981	0.1489	0.1797	0.1771	0.1442
	9	10	...	0.0007	0.0051	0.0198	0.0514	0.0980	0.1464	0.1771	0.1762
	10	9	...	0.0001	0.0013	0.0066	0.0220	0.0528	0.0976	0.1449	0.1762
	11	8	...	...	0.0003	0.0018	0.0077	0.0233	0.0532	0.0970	0.1442
	12	7	...	...	...	0.0004	0.0022	0.0083	0.0237	0.0529	0.0961
	13	6	...	...	...	0.0001	0.0005	0.0024	0.0085	0.0233	0.0518
	14	5	...	...	...	...	0.0001	0.0006	0.0024	0.0082	0.0222
	15	4	...	...	...	...	...	0.0001	0.0005	0.0022	0.0074
	16	3	...	...	...	...	...	...	0.0001	0.0005	0.0018
	17	2	...	...	...	...	...	...	...	0.0001	0.0003
	18	1	...	...	...	...	...	...	...	...	...
	19	0	...	...	...	...	...	...	...	...	...
20	0	20	0.1216	0.0388	0.0115	0.0032	0.0008	0.0002	...	...	...
	1	19	0.2702	0.1368	0.0576	0.0211	0.0068	0.0020	0.0005	0.0001	...
	2	18	0.2852	0.2293	0.1369	0.0669	0.0278	0.0100	0.0031	0.0008	0.0002
	3	17	0.1901	0.2428	0.2054	0.1339	0.0716	0.0323	0.0123	0.0040	0.0011
	4	16	0.0898	0.1821	0.2182	0.1897	0.1304	0.0738	0.0350	0.0139	0.0046
	5	15	0.0319	0.1028	0.1746	0.2023	0.1789	0.1272	0.0746	0.0365	0.0148
	6	14	0.0089	0.0454	0.1091	0.1686	0.1916	0.1712	0.1244	0.0746	0.0370
	7	13	0.0020	0.0160	0.0545	0.1124	0.1643	0.1844	0.1659	0.1221	0.0739
	8	12	0.0004	0.0046	0.0222	0.0609	0.1144	0.1614	0.1797	0.1623	0.1201
	9	11	0.0001	0.0011	0.0074	0.0271	0.0654	0.1158	0.1597	0.1771	0.1602
	10	10	...	0.0002	0.0020	0.0099	0.0308	0.0686	0.1171	0.1593	0.1762
	11	9	...	...	0.0005	0.0030	0.0120	0.0336	0.0710	0.1185	0.1602
	12	8	...	...	0.0001	0.0008	0.0039	0.0136	0.0355	0.0727	0.1201
	13	7	...	...	...	0.0002	0.0010	0.0045	0.0146	0.0366	0.0739
	14	6	...	...	...	...	0.0002	0.0012	0.0049	0.0150	0.0370
	15	5	...	...	...	...	...	0.0003	0.0013	0.0049	0.0148
	16	4	...	...	...	...	...	...	0.0003	0.0013	0.0046
	17	3	...	...	...	...	...	...	...	0.0002	0.0011
	18	2	...	...	...	...	...	...	...	...	0.0002

Appendix H — Standard Normal Distribution Table

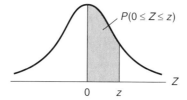

The values in the body of the table are the areas between the mean, 0, and the value of Z. If we want to find the area under the standard normal curve between $Z = 0$ and $Z = 1.96$, we find the $Z = 1.90$ row and .06 column (for $Z = 1.90 + .06 = 1.96$) and read 0.4750 at the intersection.

z	.00	.01	.02	.03	.04	.05	.06	.07	.08	.09
.00	.0000	.0040	.0080	.0120	.0160	.0199	.0239	.0279	.0319	.0359
.10	.0398	.0438	.0478	.0517	.0557	.0596	.0636	.0675	.0714	.0753
.20	.0793	.0832	.0871	.0910	.0948	.0987	.1026	.1064	.1103	.1141
.30	.1179	.1217	.1255	.1293	.1331	.1368	.1406	.1443	.1480	.1517
.40	.1554	.1591	.1628	.1664	.1700	.1736	.1772	.1808	.1844	.1879
.50	.1915	.1950	.1985	.2019	.2054	.2088	.2123	.2157	.2190	.2224
.60	.2257	.2291	.2324	.2357	.2389	.2422	.2454	.2486	.2517	.2549
.70	.2580	.2611	.2642	.2673	.2704	.2734	.2764	.2794	.2823	.2852
.80	.2881	.2910	.2939	.2967	.2995	.3023	.3051	.3078	.3106	.3133
.90	.3159	.3186	.3212	.3238	.3264	.3289	.3315	.3340	.3365	.3389
1.00	.3413	.3438	.3461	.3485	.3508	.3531	.3554	.3577	.3599	.3621
1.10	.3643	.3665	.3686	.3708	.3729	.3749	.3770	.3790	.3810	.3830
1.20	.3849	.3869	.3888	.3907	.3925	.3944	.3962	.3980	.3997	.4015
1.30	.4032	.4049	.4066	.4082	.4099	.4115	.4131	.4147	.4162	.4177
1.40	.4192	.4207	.4222	.4236	.4251	.4265	.4279	.4292	.4306	.4319
1.50	.4332	.4345	.4357	.4370	.4382	.4394	.4406	.4418	.4429	.4441
1.60	.4452	.4463	.4474	.4484	.4495	.4505	.4515	.4525	.4535	.4545
1.70	.4554	.4564	.4573	.4582	.4591	.4599	.4608	.4616	.4625	.4633
1.80	.4641	.4649	.4656	.4664	.4671	.4678	.4686	.4693	.4699	.4706
1.90	.4713	.4719	.4726	.4732	.4738	.4744	.4750	.4756	.4761	.4767
2.00	.4772	.4778	.4783	.4788	.4793	.4798	.4803	.4808	.4812	.4817
2.10	.4821	.4826	.4830	.4834	.4838	.4842	.4846	.4850	.4854	.4857
2.20	.4861	.4864	.4868	.4871	.4875	.4878	.4881	.4884	.4887	.4890
2.30	.4893	.4896	.4898	.4901	.4904	.4906	.4909	.4911	.4913	.4916
2.40	.4918	.4920	.4922	.4925	.4927	.4929	.4931	.4932	.4934	.4936
2.50	.4938	.4940	.4941	.4943	.4945	.4946	.4948	.4949	.4951	.4952
2.60	.4953	.4955	.4956	.4957	.4959	.4960	.4961	.4962	.4963	.4964
2.70	.4965	.4966	.4967	.4968	.4969	.4970	.4971	.4972	.4973	.4974
2.80	.4974	.4975	.4976	.4977	.4977	.4978	.4979	.4979	.4980	.4981
2.90	.4981	.4982	.4982	.4983	.4984	.4984	.4985	.4985	.4986	.4986
3.00	.4987	.4987	.4987	.4988	.4988	.4989	.4989	.4989	.4990	.4990

Source: From Billingsley et al., *Statistical Inference for Management and Economics*, 3d ed. Copyright © 1986 by Allyn and Bacon. Reprinted by permission.

Appendix I Random Number Table

Line/Col.	(1)	(2)	(3)	(4)	(5)	(6)	(7)	(8)	(9)	(10)	(11)	(12)	(13)	(14)
1	10480	15011	01536	02011	81647	91646	69179	14194	62590	36207	20969	99570	91291	90700
2	22368	46573	25595	85393	30995	89198	27982	53402	93965	34095	52666	19174	39615	99505
3	24130	48360	22527	97265	76393	64809	15179	24830	49340	32081	30680	19655	63348	58629
4	42167	93093	06243	61680	07856	16376	39440	53537	71341	57004	00849	74917	97758	16379
5	37570	39975	81837	16656	06121	91782	60468	81305	49684	60672	14110	06927	01263	54613
6	77921	06907	11008	42751	27756	53498	18602	70659	90655	15053	21916	81825	44394	42880
7	99562	72905	56420	69994	98872	31016	71194	18738	44013	48840	63213	21069	10634	12952
8	96301	91977	05463	07972	18876	20922	94595	56869	69014	60045	18425	84903	42508	32307
9	89579	14342	63661	10281	17453	18103	57740	84378	25331	12566	58678	44947	05585	56941
10	85475	36857	43342	53988	53060	59533	38867	62300	08158	17983	16439	11458	18593	64952
11	28918	69578	88231	33276	70997	79936	56865	05859	90106	31595	01547	85590	91610	78188
12	63553	40961	48235	03427	49626	69445	18663	72695	52180	20847	12234	90511	33703	90322
13	09429	93969	52636	92737	88974	33488	36320	17617	30015	08272	84115	27156	30613	74952
14	10365	61129	87529	85689	48237	52267	67689	93394	01511	26358	85104	20285	29975	89868
15	07119	97336	71048	08178	77233	13916	47564	81056	97735	85977	29372	74461	28551	90707
16	51085	12765	51821	51259	77452	16308	60756	92144	49442	53900	70960	63990	75601	40719
17	02368	21382	52404	60268	89368	19885	55322	44819	01188	65255	64835	44919	05944	55157
18	01011	54092	33362	94904	31273	04146	18594	29852	71585	85030	51132	01915	92747	64951
19	52162	53916	46369	58586	23216	14513	83149	98736	23495	64350	94738	17752	35156	35749
20	07056	97628	33787	09998	42698	06691	76988	13602	51851	46104	88916	19509	25625	58104
21	48663	91245	85828	14346	09172	30168	90229	04734	59193	22178	30421	61666	99904	32812
22	54164	58492	22421	74103	47070	25306	76468	26384	58151	06646	21524	15227	96909	44592
23	32639	32363	05597	24200	13363	38005	94342	28728	35806	06912	17012	64161	18296	22851
24	29334	27001	87637	87308	58731	00256	45834	15398	46557	41135	10367	07684	36188	18510
25	02488	33062	28834	07351	19731	92420	60952	61280	50001	67658	32586	86679	50720	94953
26	81525	72295	04839	96423	24878	82651	66566	14778	76797	14780	13300	87074	79666	95725
27	29676	20591	68086	26432	46901	20849	89768	81536	86645	12659	92259	57102	80428	25280
28	00742	57392	39064	66432	84673	40027	32832	61362	98947	96067	64760	64584	96096	98253
29	05366	04213	25669	26422	44407	44048	37937	63904	45766	66134	75470	66520	34693	90449
30	91921	26418	64117	94305	26766	25940	39972	22209	71500	64568	91402	42416	07844	69618
31	00582	04711	87917	77341	42206	35126	74087	99547	81817	42607	43808	76655	62028	76630
32	00725	69884	62797	56170	86324	88072	76222	36086	84637	93161	76038	65855	77919	88006
33	69011	65797	95876	55293	18988	27354	26575	08625	40801	59920	29841	80150	12777	48501
34	25976	57948	29888	88604	67917	48708	18912	82271	65424	69774	33611	54262	85963	03547
35	09763	83473	73577	12908	30883	18317	28290	35797	05998	41688	34952	37888	38917	88050
36	91567	42595	27958	30134	04024	86385	29880	99730	55536	84855	29080	09250	79656	73211
37	17955	56349	90999	49127	20044	59931	06115	20542	18059	02008	73708	83517	36103	42791
38	46503	18584	18845	49618	02304	51038	20655	58727	28168	15475	56942	53389	20562	87338
39	92157	89634	94824	78171	84610	82834	09922	25417	44137	48413	25555	21246	35509	20468
40	14577	62765	35605	81263	39667	47358	56873	56307	61607	49518	89656	20103	77490	18062
41	98427	07523	33362	64270	01638	92477	66969	98420	04880	45585	46565	04102	46880	45709
42	34914	63976	88720	82765	34476	17032	87589	40836	32427	70002	70663	88863	77775	69348
43	70060	28277	39475	46473	23219	53416	94970	25832	69975	94884	19661	72828	00102	66794
44	53976	54914	06990	67245	68350	82948	11398	42878	80287	88267	47363	46634	06541	97809
45	76072	29515	40980	07391	58745	25774	22987	80059	39911	96189	41151	14222	60697	59583
46	90725	52210	83974	29992	65831	38857	50490	83765	55657	14361	31720	57375	56228	41546
47	64364	67412	33339	31926	14883	24413	59744	92351	97473	89286	35931	04110	23726	51900
48	08962	00358	31662	25388	61642	34072	81249	35648	56891	69352	48373	45578	78547	81788
49	95012	68379	93526	70765	10593	04542	76463	54328	02349	17247	28865	14777	62730	92277
50	15664	10493	20492	38391	91132	21999	59516	81652	27195	48223	46751	22923	32261	85653

Source: William H. Beyer, *Handbook of Tables for Probability and Statistics,* 2nd ed. (Boca Raton, Fla: CRC Press, 1968), with permission.

Appendix J Student's *T* Distribution Table

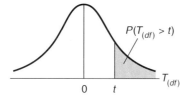

Values in body of table are values of t such that $P(T_{(df)} > t)$ = tail area at head of column. If degrees of freedom (df) is 20, then $P(T > 0.257) = 0.40$.

df* \ Tail Areas	.40	.25	.10	.05	.025	.01	.005	.0005
1	.325	1.000	3.078	6.314	12.706	31.821	63.657	636.619
2	.289	.816	1.886	2.920	4.303	6.965	9.925	31.598
3	.277	.765	1.638	2.353	3.182	4.541	5.841	12.924
4	.271	.741	1.533	2.132	2.776	3.747	4.604	8.610
5	.267	.727	1.476	2.015	2.571	3.365	4.032	6.869
6	.265	.718	1.440	1.943	2.447	3.143	3.707	5.959
7	.263	.711	1.415	1.895	2.365	2.998	3.499	5.408
8	.262	.706	1.397	1.860	2.306	2.896	3.355	5.041
9	.261	.703	1.383	1.833	2.262	2.821	3.250	4.781
10	.260	.700	1.372	1.812	2.228	2.764	3.169	4.587
11	.260	.697	1.363	1.796	2.201	2.718	3.106	4.437
12	.259	.695	1.356	1.782	2.179	2.681	3.055	4.318
13	.259	.694	1.350	1.771	2.160	2.650	3.012	4.221
14	.258	.692	1.345	1.761	2.145	2.624	2.977	4.140
15	.258	.691	1.341	1.753	2.131	2.602	2.947	4.073
16	.258	.690	1.337	1.746	2.120	2.583	2.921	4.015
17	.257	.689	1.333	1.740	2.110	2.567	2.898	3.965
18	.257	.688	1.330	1.734	2.101	2.552	2.878	3.922
19	.257	.688	1.328	1.729	2.093	2.539	2.861	3.883
20	.257	.687	1.325	1.725	2.086	2.528	2.845	3.850
21	.257	.686	1.323	1.721	2.080	2.518	2.831	3.819
22	.256	.686	1.321	1.717	2.074	2.508	2.819	3.792
23	.256	.685	1.319	1.714	2.069	2.500	2.807	3.767
24	.256	.685	1.318	1.711	2.064	2.492	2.797	3.745
25	.256	.684	1.316	1.708	2.060	2.485	2.787	3.725
26	.256	.684	1.315	1.706	2.056	2.479	2.779	3.707
27	.256	.684	1.314	1.703	2.052	2.473	2.771	3.690
28	.256	.683	1.313	1.701	2.048	2.467	2.763	3.674
29	.256	.683	1.311	1.699	2.045	2.462	2.756	3.659
30	.256	.683	1.310	1.697	2.042	2.457	2.750	3.646
40	.255	.681	1.303	1.684	2.021	2.423	2.704	3.551
60	.254	.679	1.296	1.671	2.000	2.390	2.660	3.460
120	.254	.677	1.289	1.658	1.980	2.358	2.617	3.373
∞	.253	.674	1.282	1.645	1.960	2.326	2.576	3.291

df = degrees of freedom

Source: Table is taken from Table III of Fisher & Yates; STATISTICAL TABLES FOR BIOLOGICAL, AGRICULTURAL AND MEDICAL RESEARCH Published by Longman Group UK Ltd., 1974.

Appendix K State Health Data Set

Variable Definitions

The following data are for 44 states and the District of Columbia. These data were extracted from the Behavioral Risk Factor Surveillance System (BRFSS), as reported in *Morbidity and Mortality Weekly Report—CDC Surveillance Summaries,* December 1991, and from *Statistical Abstracts of the United States,* 1991. Not all states participated in the BRFSS—hence those states were omitted. For each state, the following information is provided. Professor Robert Pittman of the School of Education of Western Carolina University compiled this data set and allowed its reprint.

Variable	Description
State	State or District of Columbia
A. No physical (% population)	Estimate of the percentage of the state's adult population who report doing no exercise, recreation, or physical activity within the past month
B. Sedentary (% population)	Estimate of the percentage of the state's adult population who have less than three different 20-minute leisure time physical activities per week
C. Smoke (% population)	Estimate of the percentage of the state's adult population who smoke regularly
D. Overweight (% population)	Estimate of the percentage of the state's adult population who are overweight
E. Drinking (% population)	Estimate of the percentage of the state's adult population who consumed the equivalent of five or more alcoholic drinks on a single occasion in the last month
F. Drinking Category	Level of consumption of alcoholic drinks 1 = among lower 1/3 of 44 states and DC in alcohol consumption 2 = among middle 1/3 of 44 states and DC in alcohol consumption 3 = among upper 1/3 of 44 states and DC in alcohol consumption
G. Drink/Drive (% population)	Estimate of the percentage of the state's adult population who report driving after having too much to drink in the past month
H. No Seat Belt (% population)	Estimate of the percentage of the state's adult population who indicate that they frequently do *not* use seat belts
I. Urban	Percentage of the state's total population who live in urban areas
J. Infant Mortality	Infant mortality rate/1,000 live births
K. Hospital Beds	Number of hospital beds available in the state per 100,000 population
L. Doctors	Number of doctors in the state per 100,000 population
M. Medical Access	Access to medical treatment—combination of the number of doctors and the number of hospital beds per 100,000 population 1 = among lower 1/3 of 44 states and DC in access (least) 2 = among middle 1/3 of 44 states and DC in access 3 = among upper 1/3 of 44 states and DC in access (most)
N. Car Accidents	Death rate in the state due to motor vehicle accidents per 100,000 population
O. Older Population	Percentage of the state's population age 65 or over
P. Low Birth Weight	Percentage of low birth-weight births in the state
Q. Income	Per capita income for the state, 1989
R. Income Category	State per capita income rating 1 = among lower 1/3 of 44 states and DC in income 2 = among middle 1/3 of 44 states and DC in income 3 = among upper 1/3 of 44 states and DC in income
S. Income/Education Category	A rating based on a combination of the per capita income and the percentage of the state's population who have graduated from college 1 = among lower 1/3 of 44 states and DC in income/education 2 = among middle 1/3 of 44 states and DC in income/education 3 = among upper 1/3 of 44 states and DC in income/education
T. Heart Disease	Death rate in the state due to heart disease per 100,000 population
U. Cancer	Death rate in the state due to malignancies per 100,000 population
V. Stroke	Death rate in the state due to cerebrovascular diseases per 100,000 population
W. Accidents	Death rate in the state due to accidents and adverse effects per 100,000 population
X. Pulmonary	Death rate in the state due to chronic pulmonary diseases per 100,000 population
Y. Flu	Death rate in the state due to pneumonia/flu per 100,000 population
Z. Diabetes	Death rate in the state due to diabetes per 100,000 population
1. Suicide	Death rate in the state due to suicide per 100,000 population
2. Cirrhosis	Death rate in the state due to chronic liver disease and cirrhosis per 100,000 population

State Health Data Set

Variable	Description
3. Atherosclerosis	Death rate in the state due to atherosclerosis per 100,000 population
4. Homicide	Death rate in the state due to homicide per 100,000 population
5. HIV	Death rate in the state due to HIV infection per 100,000 population
6. Waste Sites	Number of hazardous waste sites in the state
7. College	Percentage of the adult population in the state who have graduated from college

State Health Data Base

	A	B	C	D	E	F	G	H	I	J	K	L	M	N	O	P	Q	R	S	T	U	V	W	X	Y	Z	1	2	3	4	5	6	7
AL	34	60	22	24	10	1	2	39	67	12	454	151	2	25	13	8	11,634	1	1	324	206	71	55	35	29	19	12	10	9	12	2	12	12
AZ	21	51	21	21	14	2	3	30	79	10	291	191	1	25	13	6	13,625	2	2	252	180	44	49	42	29	16	20	13	9	9	4	11	22
CA	24	54	20	21	17	3	3	13	96	9	288	242	2	18	11	6	16,779	3	3	250	169	57	37	34	33	12	15	15	8	11	6	88	26
CO	19	45	21	16	17	3	3	17	82	10	326	202	1	16	14	6	15,050	3	3	196	135	40	34	39	30	11	18	10	11	6	5	16	27
CT	26	52	22	23	17	3	3	23	92	9	300	293	3	13	14	8	20,687	3	3	319	212	54	31	29	34	16	10	10	8	6	6	15	28
DC	52	73	19	27	6	1	2	26	100	23	775	591	3	12	13	7	19,337	3	3	356	264	65	39	24	45	22	8	35	4	50	40	0	35
DE	27	54	23	26	15	2	1	31	66	12	304	191	1	18	12	14	14,949	3	2	310	218	45	44	30	23	22	13	10	6	5	4	20	19
FL	32	55	24	24	13	2	2	20	91	11	416	203	2	24	18	7	15,049	3	2	380	253	72	44	43	27	17	17	14	10	14	12	51	20
GA	37	62	24	20	9	1	2	23	65	13	402	167	2	25	10	8	13,464	3	2	267	171	64	50	29	26	14	8	17	6	13	8	13	18
HI	32	62	21	18	19	3	4	5	76	5	253	225	1	13	11	7	15,584	3	1	167	139	39	55	18	18	13	8	7	4	3	6	7	24
ID	28	59	20	22	10	1	1	29	20	9	323	120	1	24	12	5	12,041	1	1	237	159	60	55	29	34	16	18	7	8	4	1	9	17
IL	32	60	24	21	16	2	4	29	83	11	424	210	3	15	12	8	16,060	3	3	341	202	60	38	30	33	16	11	11	9	10	4	37	21
IN	27	61	27	26	13	2	3	28	69	11	414	151	2	17	12	7	13,451	2	1	318	204	70	39	37	29	22	12	8	12	6	2	35	14
IA	34	61	22	25	13	2	2	24	44	9	516	145	3	18	15	5	13,183	2	3	359	218	78	41	43	44	16	13	7	18	2	1	21	17
KY	42	69	29	23	10	1	1	46	47	11	427	161	2	21	13	7	11,873	2	1	338	215	69	46	41	36	23	13	9	10	6	1	17	15
LA	29	59	25	24	16	2	4	24	70	11	431	184	2	20	13	9	11,348	1	1	303	191	58	45	46	23	20	13	7	9	13	6	11	17
ME	36	60	27	24	10	1	1	41	36	8	382	173	2	16	13	9	14,110	2	1	330	227	65	40	47	30	23	13	11	14	3	1	9	19
MD	30	62	22	23	8	1	1	13	93	11	289	325	3	16	11	5	17,489	3	3	276	201	51	33	30	28	17	9	10	6	11	6	10	27
MA	23	50	24	19	18	3	3	46	90	8	394	322	3	12	14	6	18,636	3	3	344	223	64	30	34	43	16	11	12	11	4	5	25	28
MI	32	59	29	26	18	3	3	21	80	11	372	180	3	18	12	7	14,757	2	2	327	194	57	35	31	30	22	12	12	13	12	2	78	17
MN	25	55	21	21	21	3	2	24	68	8	476	212	3	14	15	5	14,771	3	3	377	188	70	35	43	36	16	11	7	10	3	2	42	22
MS	39	66	24	26	11	1	2	48	30	12	509	125	2	28	12	9	10,102	1	1	353	201	76	55	31	27	19	11	8	8	12	3	2	16
MO	33	61	26	23	16	2	3	27	66	10	480	190	3	20	14	7	14,008	2	1	359	213	71	44	41	38	16	13	7	8	4	4	24	22
MT	18	52	19	20	19	3	4	28	24	9	558	150	3	23	13	6	12,390	1	2	257	192	57	47	48	31	19	18	11	11	4	1	10	21
NE	25	55	23	24	17	3	5	51	49	9	574	168	3	18	14	6	13,268	2	2	321	198	72	39	36	46	16	10	6	16	3	2	6	20
NH	20	47	22	21	16	2	2	40	56	8	323	186	1	17	11	5	17,829	3	3	278	195	56	28	34	28	16	13	11	10	3	2	16	24
NM	28	51	22	22	15	2	1	12	48	10	274	173	1	35	11	7	11,302	3	2	190	145	42	57	31	24	17	23	13	7	11	2	10	21
NY	33	63	23	20	12	1	2	20	91	11	417	307	3	13	13	8	17,293	3	1	376	216	54	32	31	39	18	7	14	9	13	22	83	23
NC	40	61	28	24	9	1	1	16	57	13	335	179	1	22	12	8	12,979	3	1	299	196	74	50	32	30	20	12	11	6	9	3	22	18
ND	27	56	20	23	17	3	4	60	40	11	694	167	3	12	14	5	11,937	2	2	309	193	65	36	30	32	17	11	9	15	2	0	2	22
OH	33	69	26	23	9	1	3	24	79	10	418	191	2	16	13	7	13,994	3	1	345	212	62	35	38	30	19	18	9	9	5	2	33	18
OK	41	66	27	23	11	1	3	26	59	9	398	145	1	20	13	7	11,591	3	3	340	197	71	44	37	36	16	13	9	12	8	2	11	17
OR	22	49	22	22	12	2	2	33	69	9	289	196	2	22	14	5	13,422	2	2	281	211	73	45	43	33	16	17	10	14	5	4	8	20
PA	27	55	24	25	18	3	3	26	85	10	435	227	3	16	15	7	14,811	2	2	401	237	67	38	36	43	22	12	10	9	6	4	95	19
RI	26	55	26	22	18	3	2	49	93	8	328	244	2	10	15	6	15,189	3	2	354	242	64	30	34	26	29	10	13	9	5	4	11	20
SC	34	70	25	25	10	1	2	14	61	12	328	156	1	28	11	9	11,665	3	1	294	177	75	56	27	22	16	10	9	5	10	3	23	17
SD	29	57	21	23	16	2	4	57	30	10	558	138	2	21	12	5	12,335	3	2	364	190	73	40	34	43	13	10	9	10	5	1	14	18
TN	39	61	27	24	6	1	1	26	68	11	480	187	3	22	13	8	12,854	3	1	328	201	76	49	36	33	15	12	8	9	10	2	14	16
TX	29	54	23	22	19	3	5	15	82	9	362	169	1	20	10	7	13,687	3	2	243	154	49	39	24	36	12	13	8	7	12	7	28	22
UT	23	49	17	19	11	1	2	14	78	8	267	177	2	18	9	6	11,425	1	1	175	101	41	35	22	24	13	14	5	4	3	2	12	24
VT	25	51	22	20	21	3	4	34	23	7	315	244	2	21	12	5	13,945	3	1	289	183	57	42	45	33	17	16	11	10	3	4	8	27
VA	26	59	23	20	16	2	4	16	73	10	334	204	1	16	11	5	15,936	3	3	269	185	59	37	27	29	13	16	9	7	8	4	20	27
WA	20	52	22	19	18	3	1	16	82	11	267	206	1	22	11	7	15,438	1	1	253	179	60	38	40	28	14	16	10	10	6	4	45	24
WV	40	68	27	25	9	1	1	46	36	9	463	164	2	25	15	6	10,901	1	1	397	232	70	48	37	37	21	11	9	11	6	1	5	11
WI	25	54	25	23	27	3	6	29	67	8	398	181	2	17	13	5	13,933	2	2	321	202	70	35	33	37	16	14	8	9	4	1	39	19

Source: Extracted from the Behavioral Risk Factor Surveillance System (BRFSS), as reported in *Morbidity and Mortality Weekly Report—CDC Surveillance Summaries*, December 1991, and from U.S. Bureau of the Census, *Statistical Abstract of the United States, 1991* 111th ed. (Washington, D.C.: Bureau of the Census, 1991). Not all states participated in the BRFSS—hence, those states were omitted.

Appendix L   F Distribution Tables

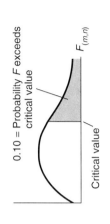

0.10 = Probability F exceeds critical value

$F_{(m,n)}$

Critical value

1. 0.10 Tail Area

n \ m	1	2	3	4	5	6	7	8	9	10	12	15	20	24	30	40	60	120	∞
1	39.86	49.50	53.59	55.83	57.24	58.20	58.91	59.44	59.86	60.19	60.71	61.22	61.74	62.00	62.26	62.53	62.79	63.06	63.33
2	8.53	9.00	9.16	9.24	9.29	9.33	9.35	9.37	9.38	9.39	9.41	9.42	9.44	9.45	9.46	9.47	9.47	9.48	9.49
3	5.54	5.46	5.39	5.34	5.31	5.28	5.27	5.25	5.24	5.23	5.22	5.20	5.18	5.18	5.17	5.16	5.15	5.14	5.13
4	4.54	4.32	4.19	4.11	4.05	4.01	3.98	3.95	3.94	3.92	3.90	3.87	3.84	3.83	3.82	3.80	3.79	3.78	3.76
5	4.06	3.78	3.62	3.52	3.45	3.40	3.37	3.34	3.32	3.30	3.27	3.24	3.21	3.19	3.17	3.16	3.14	3.12	3.10
6	3.78	3.46	3.29	3.18	3.11	3.05	3.01	2.98	2.96	2.94	2.90	2.87	2.84	2.82	2.80	2.78	2.76	2.74	2.72
7	3.59	3.26	3.07	2.96	2.88	2.83	2.78	2.75	2.72	2.70	2.67	2.63	2.59	2.58	2.56	2.54	2.51	2.49	2.47
8	3.46	3.11	2.92	2.81	2.73	2.67	2.62	2.59	2.56	2.54	2.50	2.46	2.42	2.40	2.38	2.36	2.34	2.32	2.29
9	3.36	3.01	2.81	2.69	2.61	2.55	2.51	2.47	2.44	2.42	2.38	2.34	2.30	2.28	2.25	2.23	2.21	2.18	2.16
10	3.29	2.92	2.73	2.61	2.52	2.46	2.41	2.38	2.35	2.32	2.28	2.24	2.20	2.18	2.16	2.13	2.11	2.08	2.06
11	3.23	2.86	2.66	2.54	2.45	2.39	2.34	2.30	2.27	2.25	2.21	2.17	2.12	2.10	2.08	2.05	2.03	2.00	1.97
12	3.18	2.81	2.61	2.48	2.39	2.33	2.28	2.24	2.21	2.19	2.15	2.10	2.06	2.04	2.01	1.99	1.96	1.93	1.90
13	3.14	2.76	2.56	2.43	2.35	2.28	2.23	2.20	2.16	2.14	2.10	2.05	2.01	1.98	1.96	1.93	1.90	1.88	1.85
14	3.10	2.73	2.52	2.39	2.31	2.24	2.19	2.15	2.12	2.10	2.05	2.01	1.96	1.94	1.91	1.89	1.86	1.83	1.80
15	3.07	2.70	2.49	2.36	2.27	2.21	2.16	2.12	2.09	2.06	2.02	1.97	1.92	1.90	1.87	1.85	1.82	1.79	1.76
16	3.05	2.67	2.46	2.33	2.24	2.18	2.13	2.09	2.06	2.03	1.99	1.94	1.89	1.87	1.84	1.81	1.78	1.75	1.72
17	3.03	2.64	2.44	2.31	2.22	2.15	2.10	2.06	2.03	2.00	1.96	1.91	1.86	1.84	1.81	1.78	1.75	1.72	1.69
18	3.01	2.62	2.42	2.29	2.20	2.13	2.08	2.04	2.00	1.98	1.93	1.89	1.84	1.81	1.78	1.75	1.72	1.69	1.66
19	2.99	2.61	2.40	2.27	2.18	2.11	2.06	2.02	1.98	1.96	1.91	1.86	1.81	1.79	1.76	1.73	1.70	1.67	1.63
20	2.97	2.59	2.38	2.25	2.16	2.09	2.04	2.00	1.96	1.94	1.89	1.84	1.79	1.77	1.74	1.71	1.68	1.64	1.61
21	2.96	2.57	2.36	2.23	2.14	2.08	2.02	1.98	1.95	1.92	1.87	1.83	1.78	1.75	1.72	1.69	1.66	1.62	1.59
22	2.95	2.56	2.35	2.22	2.13	2.06	2.01	1.97	1.93	1.90	1.86	1.81	1.76	1.73	1.70	1.67	1.64	1.60	1.57
23	2.94	2.55	2.34	2.21	2.11	2.05	1.99	1.95	1.92	1.89	1.84	1.80	1.74	1.72	1.69	1.66	1.62	1.59	1.55
24	2.93	2.54	2.33	2.19	2.10	2.04	1.98	1.94	1.91	1.88	1.83	1.78	1.73	1.70	1.67	1.64	1.61	1.57	1.53
25	2.92	2.53	2.32	2.18	2.09	2.02	1.97	1.93	1.89	1.87	1.82	1.77	1.72	1.69	1.66	1.63	1.59	1.56	1.52
26	2.91	2.52	2.31	2.17	2.08	2.01	1.96	1.92	1.88	1.86	1.81	1.76	1.71	1.68	1.65	1.61	1.58	1.54	1.50
27	2.90	2.51	2.30	2.17	2.07	2.00	1.95	1.91	1.87	1.85	1.80	1.75	1.70	1.67	1.64	1.60	1.57	1.53	1.49
28	2.89	2.50	2.29	2.16	2.06	2.00	1.94	1.90	1.87	1.84	1.79	1.74	1.69	1.66	1.63	1.59	1.56	1.52	1.48
29	2.89	2.50	2.28	2.15	2.06	1.99	1.93	1.89	1.86	1.83	1.78	1.73	1.68	1.65	1.62	1.58	1.55	1.51	1.47
30	2.88	2.49	2.28	2.14	2.05	1.98	1.93	1.88	1.85	1.82	1.77	1.72	1.67	1.64	1.61	1.57	1.54	1.50	1.46
40	2.84	2.44	2.23	2.09	2.00	1.93	1.87	1.83	1.79	1.76	1.71	1.66	1.61	1.57	1.54	1.51	1.47	1.42	1.38
60	2.79	2.39	2.18	2.04	1.95	1.87	1.82	1.77	1.74	1.71	1.66	1.60	1.54	1.51	1.48	1.44	1.40	1.35	1.29
120	2.75	2.35	2.13	1.99	1.90	1.82	1.77	1.72	1.68	1.65	1.60	1.55	1.48	1.45	1.41	1.37	1.32	1.26	1.19
∞	2.71	2.30	2.08	1.94	1.85	1.77	1.72	1.67	1.63	1.60	1.55	1.49	1.42	1.38	1.34	1.30	1.24	1.17	1.00

2. 0.05 Tail Area

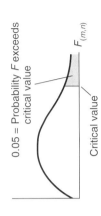

0.05 = Probability F exceeds critical value

Critical value $F_{(m,n)}$

n \ m	1	2	3	4	5	6	7	8	9	10	12	15	20	24	30	40	60	120	∞
1	161.4	199.5	215.7	224.6	230.2	234.0	236.8	238.9	240.5	241.9	243.9	245.9	248.0	249.1	250.1	251.1	252.2	253.3	254.3
2	18.51	19.00	19.16	19.25	19.30	19.33	19.35	19.37	19.38	19.40	19.41	19.43	19.45	19.45	19.46	19.47	19.48	19.49	19.50
3	10.13	9.55	9.28	9.12	9.01	8.94	8.89	8.85	8.81	8.79	8.74	8.70	8.66	8.64	8.62	8.59	8.57	8.55	8.53
4	7.71	6.94	6.59	6.39	6.26	6.16	6.09	6.04	6.00	5.96	5.91	5.86	5.80	5.77	5.75	5.72	5.69	5.66	5.63
5	6.61	5.79	5.41	5.19	5.05	4.95	4.88	4.82	4.77	4.74	4.68	4.62	4.56	4.53	4.50	4.46	4.43	4.40	4.36
6	5.99	5.14	4.76	4.53	4.39	4.28	4.21	4.15	4.10	4.06	4.00	3.94	3.87	3.84	3.81	3.77	3.74	3.70	3.67
7	5.59	4.74	4.35	4.12	3.97	3.87	3.79	3.73	3.68	3.64	3.57	3.51	3.44	3.41	3.38	3.34	3.30	3.27	3.23
8	5.32	4.46	4.07	3.84	3.69	3.58	3.50	3.44	3.39	3.35	3.28	3.22	3.15	3.12	3.08	3.04	3.01	2.97	2.93
9	5.12	4.26	3.86	3.63	3.48	3.37	3.29	3.23	3.18	3.14	3.07	3.01	2.94	2.90	2.86	2.83	2.79	2.75	2.71
10	4.96	4.10	3.71	3.48	3.33	3.22	3.14	3.07	3.02	2.98	2.91	2.85	2.77	2.74	2.70	2.66	2.62	2.58	2.54
11	4.84	3.98	3.59	3.36	3.20	3.09	3.01	2.95	2.90	2.85	2.79	2.72	2.65	2.61	2.57	2.53	2.49	2.45	2.40
12	4.75	3.89	3.49	3.26	3.11	3.00	2.91	2.85	2.80	2.75	2.69	2.62	2.54	2.51	2.47	2.43	2.38	2.34	2.30
13	4.67	3.81	3.41	3.18	3.03	2.92	2.83	2.77	2.71	2.67	2.60	2.53	2.46	2.42	2.38	2.34	2.30	2.25	2.21
14	4.60	3.74	3.34	3.11	2.96	2.85	2.76	2.70	2.65	2.60	2.53	2.46	2.39	2.35	2.31	2.27	2.22	2.18	2.13
15	4.54	3.68	3.29	3.06	2.90	2.79	2.71	2.64	2.59	2.54	2.48	2.40	2.33	2.29	2.25	2.20	2.16	2.11	2.07
16	4.49	3.63	3.24	3.01	2.85	2.74	2.66	2.59	2.54	2.49	2.42	2.35	2.28	2.24	2.19	2.15	2.11	2.06	2.01
17	4.45	3.59	3.20	2.96	2.81	2.70	2.61	2.55	2.49	2.45	2.38	2.31	2.23	2.19	2.15	2.10	2.06	2.01	1.96
18	4.41	3.55	3.16	2.93	2.77	2.66	2.58	2.51	2.46	2.41	2.34	2.27	2.19	2.15	2.11	2.06	2.02	1.97	1.92
19	4.38	3.52	3.13	2.90	2.74	2.63	2.54	2.48	2.42	2.38	2.31	2.23	2.16	2.11	2.07	2.03	1.98	1.93	1.88
20	4.35	3.49	3.10	2.87	2.71	2.60	2.51	2.45	2.39	2.35	2.28	2.20	2.12	2.08	2.04	1.99	1.95	1.90	1.84
21	4.32	3.47	3.07	2.84	2.68	2.57	2.49	2.42	2.37	2.32	2.25	2.18	2.10	2.05	2.01	1.96	1.92	1.87	1.81
22	4.30	3.44	3.05	2.82	2.66	2.55	2.46	2.40	2.34	2.30	2.23	2.15	2.07	2.03	1.98	1.94	1.89	1.84	1.78
23	4.28	3.42	3.03	2.80	2.64	2.53	2.44	2.37	2.32	2.27	2.20	2.13	2.05	2.01	1.96	1.91	1.86	1.81	1.76
24	4.26	3.40	3.01	2.78	2.62	2.51	2.42	2.36	2.30	2.25	2.18	2.11	2.03	1.98	1.94	1.89	1.84	1.79	1.73
25	4.24	3.39	2.99	2.76	2.60	2.49	2.40	2.34	2.28	2.24	2.16	2.09	2.01	1.96	1.92	1.87	1.82	1.77	1.71
26	4.23	3.37	2.98	2.74	2.59	2.47	2.39	2.32	2.27	2.22	2.15	2.07	1.99	1.95	1.90	1.85	1.80	1.75	1.69
27	4.21	3.35	2.96	2.73	2.57	2.46	2.37	2.31	2.25	2.20	2.13	2.06	1.97	1.93	1.88	1.84	1.79	1.73	1.67
28	4.20	3.34	2.95	2.71	2.56	2.45	2.36	2.29	2.24	2.19	2.12	2.04	1.96	1.91	1.87	1.82	1.77	1.71	1.65
29	4.18	3.33	2.93	2.70	2.55	2.43	2.35	2.28	2.22	2.18	2.10	2.03	1.94	1.90	1.85	1.81	1.75	1.70	1.64
30	4.17	3.32	2.92	2.69	2.53	2.42	2.33	2.27	2.21	2.16	2.09	2.01	1.93	1.89	1.84	1.79	1.74	1.68	1.62
40	4.08	3.23	2.84	2.61	2.45	2.34	2.25	2.18	2.12	2.08	2.00	1.92	1.84	1.79	1.74	1.69	1.64	1.58	1.51
60	4.00	3.15	2.76	2.53	2.37	2.25	2.17	2.10	2.04	1.99	1.92	1.84	1.75	1.70	1.65	1.59	1.53	1.47	1.39
120	3.92	3.07	2.68	2.45	2.29	2.17	2.09	2.02	1.96	1.91	1.83	1.75	1.66	1.61	1.55	1.50	1.43	1.35	1.25
∞	3.84	3.00	2.60	2.37	2.21	2.10	2.01	1.94	1.88	1.83	1.75	1.67	1.57	1.52	1.46	1.39	1.32	1.22	1.00

3. 0.025 Tail Area

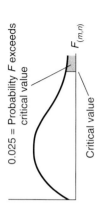

0.025 = Probability F exceeds critical value

$F_{(m,n)}$

Critical value

n \ m	1	2	3	4	5	6	7	8	9	10	12	15	20	24	30	40	60	120	∞
1	647.8	799.5	864.2	899.6	921.8	937.1	948.2	956.7	963.3	968.6	976.7	984.9	993.1	997.2	1001	1006	1010	1014	1018
2	38.51	39.00	39.17	39.25	39.30	39.33	39.36	39.37	39.39	39.40	39.41	39.43	39.45	39.46	39.46	39.47	39.48	39.49	39.50
3	17.44	16.04	15.44	15.10	14.88	14.73	14.62	14.54	14.47	14.42	14.34	14.25	14.17	14.12	14.08	14.04	13.99	13.95	13.90
4	12.22	10.65	9.98	9.60	9.36	9.20	9.07	8.98	8.90	8.84	8.75	8.66	8.56	8.51	8.46	8.41	8.36	8.31	8.26
5	10.01	8.43	7.76	7.39	7.15	6.98	6.85	6.76	6.68	6.62	6.52	6.43	6.33	6.28	6.23	6.18	6.12	6.07	6.02
6	8.81	7.26	6.60	6.23	5.99	5.82	5.70	5.60	5.52	5.46	5.37	5.27	5.17	5.12	5.07	5.01	4.96	4.90	4.85
7	8.07	6.54	5.89	5.52	5.29	5.12	4.99	4.90	4.82	4.76	4.67	4.57	4.47	4.42	4.36	4.31	4.25	4.20	4.14
8	7.57	6.06	5.42	5.05	4.82	4.65	4.53	4.43	4.36	4.30	4.20	4.10	4.00	3.95	3.89	3.84	3.78	3.73	3.67
9	7.21	5.71	5.08	4.72	4.48	4.32	4.20	4.10	4.03	3.96	3.87	3.77	3.67	3.61	3.56	3.51	3.45	3.39	3.33
10	6.94	5.46	4.83	4.47	4.24	4.07	3.95	3.85	3.78	3.72	3.62	3.52	3.42	3.37	3.31	3.26	3.20	3.14	3.08
11	6.72	5.26	4.63	4.28	4.04	3.88	3.76	3.66	3.59	3.53	3.43	3.33	3.23	3.17	3.12	3.06	3.00	2.94	2.88
12	6.55	5.10	4.47	4.12	3.89	3.73	3.61	3.51	3.44	3.37	3.28	3.18	3.07	3.02	2.96	2.91	2.85	2.79	2.72
13	6.41	4.97	4.35	4.00	3.77	3.60	3.48	3.39	3.31	3.25	3.15	3.05	2.95	2.89	2.84	2.78	2.72	2.66	2.60
14	6.30	4.86	4.24	3.89	3.66	3.50	3.38	3.29	3.21	3.15	3.05	2.95	2.84	2.79	2.73	2.67	2.61	2.55	2.49
15	6.20	4.77	4.15	3.80	3.58	3.41	3.29	3.20	3.12	3.06	2.96	2.86	2.76	2.70	2.64	2.59	2.52	2.46	2.40
16	6.12	4.69	4.08	3.73	3.50	3.34	3.22	3.12	3.05	2.99	2.89	2.79	2.68	2.63	2.57	2.51	2.45	2.38	2.32
17	6.04	4.62	4.01	3.66	3.44	3.28	3.16	3.06	2.98	2.92	2.82	2.72	2.62	2.56	2.50	2.44	2.38	2.32	2.25
18	5.98	4.56	3.95	3.61	3.38	3.22	3.10	3.01	2.93	2.87	2.77	2.67	2.56	2.50	2.44	2.38	2.32	2.26	2.19
19	5.92	4.51	3.90	3.56	3.33	3.17	3.05	2.96	2.88	2.82	2.72	2.62	2.51	2.45	2.39	2.33	2.27	2.20	2.13
20	5.87	4.46	3.86	3.51	3.29	3.13	3.01	2.91	2.84	2.77	2.68	2.57	2.46	2.41	2.35	2.29	2.22	2.16	2.09
21	5.83	4.42	3.82	3.48	3.25	3.09	2.97	2.87	2.80	2.73	2.64	2.53	2.42	2.37	2.31	2.25	2.18	2.11	2.04
22	5.79	4.38	3.78	3.44	3.22	3.05	2.93	2.84	2.76	2.70	2.60	2.50	2.39	2.33	2.27	2.21	2.14	2.08	2.00
23	5.75	4.35	3.75	3.41	3.18	3.02	2.90	2.81	2.73	2.67	2.57	2.47	2.36	2.30	2.24	2.18	2.11	2.04	1.97
24	5.72	4.32	3.72	3.38	3.15	2.99	2.87	2.78	2.70	2.64	2.54	2.44	2.33	2.27	2.21	2.15	2.08	2.01	1.94
25	5.69	4.29	3.69	3.35	3.13	2.97	2.85	2.75	2.68	2.61	2.51	2.41	2.30	2.24	2.18	2.12	2.05	1.98	1.91
26	5.66	4.27	3.67	3.33	3.10	2.94	2.82	2.73	2.65	2.59	2.49	2.39	2.28	2.22	2.16	2.09	2.03	1.95	1.88
27	5.63	4.24	3.65	3.31	3.08	2.92	2.80	2.71	2.63	2.57	2.47	2.36	2.25	2.19	2.13	2.07	2.00	1.93	1.85
28	5.61	4.22	3.63	3.29	3.06	2.90	2.78	2.69	2.61	2.55	2.45	2.34	2.23	2.17	2.11	2.05	1.98	1.91	1.83
29	5.59	4.20	3.61	3.27	3.04	2.88	2.76	2.67	2.59	2.53	2.43	2.32	2.21	2.15	2.09	2.03	1.96	1.89	1.81
30	5.57	4.18	3.59	3.25	3.03	2.87	2.75	2.65	2.57	2.51	2.41	2.31	2.20	2.14	2.07	2.01	1.94	1.87	1.79
40	5.42	4.05	3.46	3.13	2.90	2.74	2.62	2.53	2.45	2.39	2.29	2.18	2.07	2.01	1.94	1.88	1.80	1.72	1.64
60	5.29	3.93	3.34	3.01	2.79	2.63	2.51	2.41	2.33	2.27	2.17	2.06	1.94	1.88	1.82	1.74	1.67	1.58	1.48
120	5.15	3.80	3.23	2.89	2.67	2.52	2.39	2.30	2.22	2.16	2.05	1.94	1.82	1.76	1.69	1.61	1.53	1.43	1.31
∞	5.02	3.69	3.12	2.79	2.57	2.41	2.29	2.19	2.11	2.05	1.94	1.83	1.71	1.64	1.57	1.48	1.39	1.27	1.00

0.01 = Probability F exceeds critical value

$F_{(m,n)}$ Critical value

4. 0.01 Tail Area

n \ m	1	2	3	4	5	6	7	8	9	10	12	15	20	24	30	40	60	120	∞
1	4052	4999.5	5403	5625	5764	5859	5928	5982	6022	6056	6106	6157	6209	6235	6261	6287	6313	6339	6366
2	98.50	99.00	99.17	99.25	99.30	99.33	99.36	99.37	99.39	99.40	99.42	99.43	99.45	99.46	99.47	99.47	99.48	99.49	99.50
3	34.12	30.82	29.46	28.71	28.24	27.91	27.67	27.49	27.35	27.23	27.05	26.87	26.69	26.60	26.50	26.41	26.32	26.22	26.13
4	21.20	18.00	16.69	15.98	15.52	15.21	14.98	14.80	14.66	14.55	14.37	14.20	14.02	13.93	13.84	13.75	13.65	13.56	13.46
5	16.26	13.27	12.06	11.39	10.97	10.67	10.46	10.29	10.16	10.05	9.89	9.72	9.55	9.47	9.38	9.29	9.20	9.11	9.02
6	13.75	10.92	9.78	9.15	8.75	8.47	8.26	8.10	7.98	7.87	7.72	7.56	7.40	7.31	7.23	7.14	7.06	6.97	6.88
7	12.25	9.55	8.45	7.85	7.46	7.19	6.99	6.84	6.72	6.62	6.47	6.31	6.16	6.07	5.99	5.91	5.82	5.74	5.65
8	11.26	8.65	7.59	7.01	6.63	6.37	6.18	6.03	5.91	5.81	5.67	5.52	5.36	5.28	5.20	5.12	5.03	4.95	4.86
9	10.56	8.02	6.99	6.42	6.06	5.80	5.61	5.47	5.35	5.26	5.11	4.96	4.81	4.73	4.65	4.57	4.48	4.40	4.31
10	10.04	7.56	6.55	5.99	5.64	5.39	5.20	5.06	4.94	4.85	4.71	4.56	4.41	4.33	4.25	4.17	4.08	4.00	3.91
11	9.65	7.21	6.22	5.67	5.32	5.07	4.89	4.74	4.63	4.54	4.40	4.25	4.10	4.02	3.94	3.86	3.78	3.69	3.60
12	9.33	6.93	5.95	5.41	5.06	4.82	4.64	4.50	4.39	4.30	4.16	4.01	3.86	3.78	3.70	3.62	3.54	3.45	3.36
13	9.07	6.70	5.74	5.21	4.86	4.62	4.44	4.30	4.19	4.10	3.96	3.82	3.66	3.59	3.51	3.43	3.34	3.25	3.17
14	8.86	6.51	5.56	5.04	4.69	4.46	4.28	4.14	4.03	3.94	3.80	3.66	3.51	3.43	3.35	3.27	3.18	3.09	3.00
15	8.68	6.36	5.42	4.89	4.56	4.32	4.14	4.00	3.89	3.80	3.67	3.52	3.37	3.29	3.21	3.13	3.05	2.96	2.87
16	8.53	6.23	5.29	4.77	4.44	4.20	4.03	3.89	3.78	3.69	3.55	3.41	3.26	3.18	3.10	3.02	2.93	2.84	2.75
17	8.40	6.11	5.18	4.67	4.34	4.10	3.93	3.79	3.68	3.59	3.46	3.31	3.16	3.08	3.00	2.92	2.83	2.75	2.65
18	8.29	6.01	5.09	4.58	4.25	4.01	3.84	3.71	3.60	3.51	3.37	3.23	3.08	3.00	2.92	2.84	2.75	2.66	2.57
19	8.18	5.93	5.01	4.50	4.17	3.94	3.77	3.63	3.52	3.43	3.30	3.15	3.00	2.92	2.84	2.76	2.67	2.58	2.49
20	8.10	5.85	4.94	4.43	4.10	3.87	3.70	3.56	3.46	3.37	3.23	3.09	2.94	2.86	2.78	2.69	2.61	2.52	2.42
21	8.02	5.78	4.87	4.37	4.04	3.81	3.64	3.51	3.40	3.31	3.17	3.03	2.88	2.80	2.72	2.64	2.55	2.46	2.36
22	7.95	5.72	4.82	4.31	3.99	3.76	3.59	3.45	3.35	3.26	3.12	2.98	2.83	2.75	2.67	2.58	2.50	2.40	2.31
23	7.88	5.66	4.76	4.26	3.94	3.71	3.54	3.41	3.30	3.21	3.07	2.93	2.78	2.70	2.62	2.54	2.45	2.35	2.26
24	7.82	5.61	4.72	4.22	3.90	3.67	3.50	3.36	3.26	3.17	3.03	2.89	2.74	2.66	2.58	2.49	2.40	2.31	2.21
25	7.77	5.57	4.68	4.18	3.85	3.63	3.46	3.32	3.22	3.13	2.99	2.85	2.70	2.62	2.54	2.45	2.36	2.27	2.17
26	7.72	5.53	4.64	4.14	3.82	3.59	3.42	3.29	3.18	3.09	2.96	2.81	2.66	2.58	2.50	2.42	2.33	2.23	2.13
27	7.68	5.49	4.60	4.11	3.78	3.56	3.39	3.26	3.15	3.06	2.93	2.78	2.63	2.55	2.47	2.38	2.29	2.20	2.10
28	7.64	5.45	4.57	4.07	3.75	3.53	3.36	3.23	3.12	3.03	2.90	2.75	2.60	2.52	2.44	2.35	2.26	2.17	2.06
29	7.60	5.42	4.54	4.04	3.73	3.50	3.33	3.20	3.09	3.00	2.87	2.73	2.57	2.49	2.41	2.33	2.23	2.14	2.03
30	7.56	5.39	4.51	4.02	3.70	3.47	3.30	3.17	3.07	2.98	2.84	2.70	2.55	2.47	2.39	2.30	2.21	2.11	2.01
40	7.31	5.18	4.31	3.83	3.51	3.29	3.12	2.99	2.89	2.80	2.66	2.52	2.37	2.29	2.20	2.11	2.02	1.92	1.80
60	7.08	4.98	4.13	3.65	3.34	3.12	2.95	2.82	2.72	2.63	2.50	2.35	2.20	2.12	2.03	1.94	1.84	1.73	1.60
120	6.85	4.79	3.95	3.48	3.17	2.96	2.79	2.66	2.56	2.47	2.34	2.19	2.03	1.95	1.86	1.76	1.66	1.53	1.38
∞	6.63	4.61	3.78	3.32	3.02	2.80	2.64	2.51	2.41	2.32	2.18	2.04	1.88	1.79	1.70	1.59	1.47	1.32	1.00

5. 0.005 Tail Area

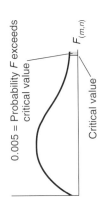

0.005 = Probability F exceeds critical value

$F_{(m,n)}$

Critical value

n \ m	1	2	3	4	5	6	7	8	9	10	12	15	20	24	30	40	60	120	∞
1	16211	20000	21615	22500	23056	23437	23715	23925	24091	24224	24426	24630	24836	24940	25044	25148	25253	25359	25465
2	198.5	199.0	199.2	199.2	199.3	199.3	199.4	199.4	199.4	199.4	199.4	199.4	199.4	199.5	199.5	199.5	199.5	199.5	199.5
3	55.55	49.80	47.47	46.19	45.39	44.84	44.43	44.13	43.88	43.69	43.39	43.08	42.78	42.62	42.47	42.31	42.15	41.99	41.83
4	31.33	26.28	24.26	23.15	22.46	21.97	21.62	21.35	21.14	20.97	20.70	20.44	20.17	20.03	19.89	19.75	19.61	19.47	19.32
5	22.78	18.31	16.53	15.56	14.94	14.51	14.20	13.96	13.77	13.62	13.38	13.15	12.90	12.78	12.66	12.53	12.40	12.27	12.14
6	18.63	14.54	12.92	12.63	11.46	11.07	10.79	10.57	10.39	10.25	10.03	9.81	9.59	9.47	9.36	9.24	9.12	9.00	8.88
7	16.24	12.40	10.88	10.05	9.52	9.16	8.89	8.68	8.51	8.38	8.18	7.97	7.75	7.65	7.53	7.42	7.31	7.19	7.08
8	14.69	11.04	9.60	8.81	8.30	7.95	7.69	7.50	7.34	7.21	7.01	6.81	6.61	6.50	6.40	6.29	6.18	6.06	5.95
9	13.61	10.11	8.72	7.96	7.47	7.13	6.88	6.69	6.54	6.42	6.23	6.03	5.83	5.73	5.62	5.52	5.41	5.30	5.19
10	12.83	9.43	8.08	7.34	6.87	6.54	6.30	6.12	5.97	5.85	5.66	5.47	5.27	5.17	5.07	4.97	4.86	4.75	4.64
11	12.23	8.91	7.60	6.88	6.42	6.10	5.86	5.68	5.54	5.42	5.24	5.05	4.86	4.76	4.65	4.55	4.44	4.34	4.23
12	11.75	8.51	7.23	6.52	6.07	5.76	5.52	5.35	5.20	5.09	4.91	4.72	4.53	4.43	4.33	4.23	4.12	4.01	3.90
13	11.37	8.19	6.93	6.23	5.79	5.48	5.25	5.08	4.94	4.82	4.64	4.46	4.27	4.17	4.07	3.97	3.87	3.76	3.65
14	11.06	7.92	6.68	6.00	5.56	5.26	5.03	4.86	4.72	4.60	4.43	4.25	4.06	3.96	3.86	3.76	3.66	3.55	3.44
15	10.80	7.70	6.48	5.80	5.37	5.07	4.85	4.67	4.54	4.42	4.25	4.07	3.88	3.79	3.69	3.58	3.48	3.37	3.26
16	10.58	7.51	6.30	5.64	5.21	4.91	4.69	4.52	4.38	4.27	4.10	3.92	3.73	3.64	3.54	3.44	3.33	3.22	3.11
17	10.38	7.35	6.16	5.50	5.07	4.78	4.56	4.39	4.25	4.14	3.97	3.79	3.61	3.51	3.41	3.31	3.21	3.10	2.98
18	10.22	7.21	6.03	5.37	4.96	4.66	4.44	4.28	4.14	4.03	3.86	3.68	3.50	3.40	3.30	3.20	3.10	2.99	2.87
19	10.07	7.09	5.92	5.27	4.85	4.56	4.34	4.18	4.04	3.93	3.76	3.59	3.40	3.31	3.21	3.11	3.00	2.89	2.78
20	9.94	6.99	5.82	5.17	4.76	4.47	4.26	4.09	3.96	3.85	3.68	3.50	3.32	3.22	3.12	3.02	2.92	2.81	2.69
21	9.83	6.89	5.73	5.09	4.68	4.39	4.18	4.01	3.88	3.77	3.60	3.43	3.24	3.15	3.05	2.95	2.84	2.73	2.61
22	9.73	6.81	5.65	5.02	4.61	4.32	4.11	3.94	3.81	3.70	3.54	3.36	3.18	3.08	2.98	2.88	2.77	2.66	2.55
23	9.63	6.73	5.58	4.95	4.54	4.26	4.05	3.88	3.75	3.64	3.47	3.30	3.12	3.02	2.92	2.82	2.71	2.60	2.48
24	9.55	6.66	5.52	4.89	4.49	4.20	3.99	3.83	3.69	3.59	3.42	3.25	3.06	2.97	2.87	2.77	2.66	2.55	2.43
25	9.48	6.60	5.46	4.84	4.43	4.15	3.94	3.78	3.64	3.54	3.37	3.20	3.01	2.92	2.82	2.72	2.61	2.50	2.38
26	9.41	6.54	5.41	4.79	4.38	4.10	3.89	3.73	3.60	3.49	3.33	3.15	2.97	2.87	2.77	2.67	2.56	2.45	2.33
27	9.34	6.49	5.36	4.74	4.34	4.06	3.85	3.69	3.56	3.45	3.28	3.11	2.93	2.83	2.73	2.63	2.52	2.41	2.25
28	9.28	6.44	5.32	4.70	4.30	4.02	3.81	3.65	3.52	3.41	3.25	3.07	2.89	2.79	2.69	2.59	2.48	2.37	2.29
29	9.23	6.40	5.28	4.66	4.26	3.98	3.77	3.61	3.48	3.38	3.21	3.04	2.86	2.76	2.66	2.56	2.45	2.33	2.24
30	9.18	6.35	5.24	4.62	4.23	3.95	3.74	3.58	3.45	3.34	3.18	3.01	2.82	2.73	2.63	2.52	2.42	2.30	2.18
40	8.83	6.07	4.98	4.37	3.99	3.71	3.51	3.35	3.22	3.12	2.95	2.78	2.60	2.50	2.40	2.30	2.18	2.06	1.93
60	8.49	5.79	4.73	4.14	3.76	3.49	3.29	3.13	3.01	2.90	2.74	2.57	2.39	2.29	2.19	2.08	1.96	1.83	1.69
120	8.18	5.54	4.50	3.92	3.55	3.28	3.09	2.93	2.81	2.71	2.54	2.37	2.19	2.09	1.98	1.87	1.75	1.61	1.43
∞	7.88	5.30	4.28	3.72	3.35	3.09	2.90	2.74	2.62	2.52	2.36	2.19	2.00	1.90	1.79	1.67	1.53	1.36	1.00

6. 0.001 Tail Area

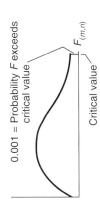

0.001 = Probability F exceeds critical value
$F_{(m,n)}$
Critical value

n \ m	1	2	3	4	5	6	7	8	9	10	12	15	20	24	30	40	60	120	∞
1	4053*	5000*	5404*	5625*	5764*	5859*	5929*	5981*	6023*	6056*	6107*	6158*	6209*	6235*	6261*	6287*	6313*	6340*	6366*
2	998.5	999.0	999.2	999.2	999.3	999.3	999.4	999.4	999.4	999.4	999.4	999.4	999.4	999.5	999.5	999.5	999.5	999.5	999.5
3	167.0	148.5	141.1	137.1	134.6	132.8	131.6	130.6	129.9	129.2	128.3	127.4	126.4	125.9	125.4	125.0	124.5	124.0	123.5
4	74.14	61.25	56.18	53.44	51.71	50.53	49.66	49.00	48.47	48.05	47.41	46.76	46.10	45.77	45.43	45.09	44.75	44.40	44.05
5	47.18	37.12	33.20	31.09	29.75	28.84	28.16	27.64	27.24	26.92	26.42	25.91	25.39	25.14	24.87	24.60	24.33	24.06	23.79
6	35.51	27.00	23.70	21.92	20.81	20.03	19.46	19.03	18.69	18.41	17.99	17.56	17.12	16.89	16.67	16.44	16.21	15.99	15.75
7	29.25	21.69	18.77	17.19	16.21	15.52	15.02	14.63	14.33	14.08	13.71	13.32	12.93	12.73	12.53	12.33	12.12	11.91	11.70
8	25.42	18.49	15.83	14.39	13.49	12.86	12.40	12.04	11.77	11.54	11.19	10.84	10.48	10.30	10.11	9.92	9.73	9.53	9.33
9	22.86	16.39	13.90	12.56	11.71	11.13	10.70	10.37	10.11	9.89	9.57	9.24	8.90	8.72	8.55	8.37	8.19	8.00	7.81
10	21.04	14.91	12.55	11.28	10.48	9.92	9.52	9.20	8.96	8.75	8.45	8.13	7.80	7.64	7.47	7.30	7.12	6.94	6.76
11	19.69	13.81	11.56	10.35	9.58	9.05	8.66	8.35	8.12	7.92	7.63	7.32	7.01	6.85	6.68	6.52	6.35	6.17	6.00
12	18.64	12.97	10.80	9.63	8.89	8.38	8.00	7.71	7.48	7.29	7.00	6.71	6.40	6.25	6.09	5.93	5.76	5.59	5.42
13	17.81	12.31	10.21	9.07	8.35	7.86	7.49	7.21	6.98	6.80	6.52	6.23	5.93	5.78	5.63	5.47	5.30	5.14	4.97
14	17.14	11.78	9.73	8.62	7.92	7.43	7.08	6.80	6.58	6.40	6.13	5.85	5.56	5.41	5.25	5.10	4.94	4.77	4.60
15	16.59	11.34	9.34	8.25	7.57	7.09	6.74	6.47	6.26	6.08	5.81	5.54	5.25	5.10	4.95	4.80	4.64	4.47	4.31
16	16.12	10.97	9.00	7.94	7.27	6.81	6.46	6.19	5.98	5.81	5.55	5.27	4.99	4.85	4.70	4.54	4.39	4.23	4.06
17	15.72	10.66	8.73	7.68	7.02	6.56	6.22	5.96	5.75	5.58	5.32	5.05	4.78	4.63	4.48	4.33	4.18	4.02	3.85
18	15.38	10.39	8.49	7.46	6.81	6.35	6.02	5.76	5.56	5.39	5.13	4.87	4.59	4.45	4.30	4.15	4.00	3.84	3.67
19	15.08	10.16	8.28	7.26	6.62	6.18	5.85	5.59	5.39	5.22	4.97	4.70	4.43	4.29	4.14	3.99	3.84	3.68	3.51
20	14.82	9.95	8.10	7.10	6.46	6.02	5.69	5.44	5.24	5.08	4.82	4.56	4.29	4.15	4.00	3.86	3.70	3.54	3.38
21	14.59	9.77	7.94	6.95	6.32	5.88	5.56	5.31	5.11	4.95	4.70	4.44	4.17	4.03	3.88	3.74	3.58	3.42	3.26
22	14.38	9.61	7.80	6.81	6.19	5.76	5.44	5.19	4.99	4.83	4.58	4.33	4.06	3.92	3.78	3.63	3.48	3.32	3.15
23	14.19	9.47	7.67	6.69	6.08	5.65	5.33	5.09	4.89	4.73	4.48	4.23	3.96	3.82	3.68	3.53	3.38	3.22	3.05
24	14.03	9.34	7.55	6.59	5.98	5.55	5.23	4.99	4.80	4.64	4.39	4.14	3.87	3.74	3.59	3.45	3.29	3.14	2.97
25	13.88	9.22	7.45	6.49	5.88	5.46	5.15	4.91	4.71	4.56	4.31	4.06	3.79	3.66	3.52	3.37	3.22	3.06	2.89
26	13.74	9.12	7.36	6.41	5.80	5.38	5.07	4.83	4.64	4.48	4.24	3.99	3.72	3.59	3.44	3.30	3.15	2.99	2.82
27	13.61	9.02	7.27	6.33	5.73	5.31	5.00	4.76	4.57	4.41	4.17	3.92	3.66	3.52	3.38	3.23	3.08	2.92	2.75
28	13.50	8.93	7.19	6.25	5.66	5.24	4.93	4.69	4.50	4.35	4.11	3.86	3.60	3.46	3.32	3.18	3.02	2.86	2.69
29	13.39	8.85	7.12	6.19	5.59	5.18	4.87	4.64	4.45	4.29	4.05	3.80	3.54	3.41	3.27	3.12	2.97	2.81	2.64
30	13.29	8.77	7.05	6.12	5.53	5.12	4.82	4.58	4.39	4.24	4.00	3.75	3.49	3.36	3.22	3.07	2.92	2.76	2.59
40	12.61	8.25	6.60	5.70	5.13	4.73	4.44	4.21	4.02	3.87	3.64	3.40	3.15	3.01	2.87	2.73	2.57	2.41	2.23
60	11.97	7.76	6.17	5.31	4.76	4.37	4.09	3.87	3.69	3.54	3.31	3.08	2.83	2.69	2.55	2.41	2.25	2.08	1.89
120	11.38	7.32	5.79	4.95	4.42	4.04	3.77	3.55	3.38	3.24	3.02	2.78	2.53	2.40	2.26	2.11	1.95	1.76	1.54
∞	10.83	6.91	5.42	4.62	4.10	3.74	3.47	3.27	3.10	2.96	2.74	2.51	2.27	2.13	1.99	1.84	1.66	1.45	1.00

Source: *Biometrika Tables for Statisticians*. Reprinted by permission of the Biometrika Trustees.
*Multiply these entries by 100.

Appendix M NCAA Tournament Data Base

This data set consists of annual (1982–1987) values of 8 variables for 11 different college athletic conferences over the indicated years related to the NCAA Basketball Tournament. Professor Brian Goff of the Economics Department of Western Kentucky University compiled this data set and has allowed its reprint. Values are taken from *NCAA Championships* and *Louisville Courier Journal*.

Variables	Description
Conference	Abbreviation for the conference, where ACC = Atlantic Coast BE = Big East B8 = Big Eight B10 = Big Ten MET = Metro MOV = Missouri Valley PAC = Pacific Coast Ten SEC = Southeastern SUN = Sunbelt SWC = Southwest WAC = Western Athletic
Year	Year corresponding to recorded values of other variables
Teams	Number of conference's teams selected in the given year to participate in the tournament
Seed	Mean seed (position) of conference's participating teams in the given year (1 to 16 possible). (Metro was assigned the value 16 in 1987, because the team did not participate that year)
Wins	Number of tournament games won by conference's participants in the given year
WINPCT	Percentage of tournament games played that were won by conference's participants in the given year (47 total tournament games in 1982 and 1983 and 63 games in other years)
WINPRTM	WINS/TEAMS, or the number of wins per participating team in the conference in the given year (period "." for Metro in 1987 signifies missing data or no value, because no teams participated from this conference that year)

Row	Conference	Year	Teams	Seed	Wins	WINPCT	WINPRTM
1	ACC	1987	6	5.5	5	0.079365	0.83333
2	ACC	1986	6	3.7	13	0.206349	2.16667
3	ACC	1985	5	3.2	12	0.190476	2.40000
4	ACC	1984	5	4.7	8	0.126984	1.60000
5	ACC	1983	4	5.6	11	0.234043	2.75000
6	ACC	1982	4	5.2	8	0.170213	2.00000
7	BE	1987	5	3.6	14	0.222222	2.80000
8	BE	1986	4	4.2	4	0.063492	1.00000
9	BE	1985	6	6.6	18	0.285714	3.00000
10	BE	1984	4	6.5	7	0.111111	1.75000
11	BE	1983	5	4.9	6	0.127660	1.20000
12	BE	1982	4	5.6	9	0.191489	2.25000
13	B8	1987	4	6.0	5	0.079365	1.25000
14	B8	1986	5	6.4	7	0.111111	1.40000
15	B8	1985	3	5.7	2	0.031746	0.66667
16	B8	1984	2	4.5	1	0.015873	0.50000
17	B8	1983	3	6.1	1	0.021277	0.33333
18	B8	1982	2	4.5	2	0.042553	1.00000
19	B10	1987	6	4.5	12	0.190476	2.00000
20	B10	1986	6	5.7	4	0.063492	0.66667
21	B10	1985	6	5.3	4	0.063492	0.66667
22	B10	1984	3	3.9	4	0.063492	1.33333
23	B10	1983	5	6.2	3	0.063830	0.60000

Appendix M NCAA Tournament Data Base

Row	Conference	Year	Teams	Seed	Wins	WINPCT	WINPRTM
24	B10	1982	4	6.8	1	0.021277	0.25000
25	MET	1987	0	16.0	0	0.000000	
26	MET	1986	3	4.0	7	0.111111	2.33333
27	MET	1985	2	5.5	2	0.031746	1.00000
28	MET	1984	2	7.1	4	0.063492	2.00000
29	MET	1983	2	3.2	4	0.085106	2.00000
30	MET	1982	2	3.2	4	0.085106	2.00000
31	MOV	1987	2	11.0	0	0.000000	0.00000
32	MOV	1986	2	8.5	1	0.015873	0.50000
33	MOV	1985	4	8.7	1	0.015873	0.25000
34	MOV	1984	2	7.8	1	0.015873	0.50000
35	MOV	1983	1	7.8	0	0.000000	0.00000
36	MOV	1982	1	3.9	0	0.000000	0.00000
37	PAC	1987	2	7.0	1	0.015873	0.50000
38	PAC	1986	2	10.5	0	0.000000	0.00000
39	PAC	1985	4	7.7	0	0.000000	0.00000
40	PAC	1984	2	7.8	2	0.031746	1.00000
41	PAC	1983	2	6.5	1	0.021277	0.50000
42	PAC	1982	2	7.1	0	0.000000	0.00000
43	SEC	1987	6	7.0	8	0.126984	1.33333
44	SEC	1986	5	6.2	11	0.174603	2.20000
45	SEC	1985	5	8.0	6	0.095238	1.20000
46	SEC	1984	4	7.1	3	0.047619	0.75000
47	SEC	1983	4	6.8	5	0.106383	1.25000
48	SEC	1982	3	8.2	2	0.042553	0.66667
49	SUN	1987	2	10.5	1	0.015873	0.50000
50	SUN	1986	4	7.5	3	0.047619	0.75000
51	SUN	1985	3	7.0	2	0.031746	0.66667
52	SUN	1984	2	9.7	1	0.015873	0.50000
53	SUN	1983	2	9.7	1	0.021277	0.50000
54	SUN	1982	2	9.1	2	0.042553	1.00000
55	SWC	1987	3	9.3	1	0.015873	0.33333
56	SWC	1986	1	13.0	0	0.000000	0.00000
57	SWC	1985	3	6.7	2	0.031746	0.66667
58	SWC	1984	3	5.6	5	0.079365	1.66667
59	SWC	1983	2	3.2	5	0.106383	2.50000
60	SWC	1982	2	6.5	4	0.085106	2.00000
61	WAC	1987	3	9.5	3	0.047619	1.00000
62	WAC	1986	2	12.0	0	0.000000	0.00000
63	WAC	1985	3	12.0	1	0.015873	0.33333
64	WAC	1984	2	7.8	1	0.015873	0.50000
65	WAC	1983	1	3.0	2	0.042553	2.00000
66	WAC	1982	1	10.4	0	0.000000	0.00000

Appendix N Chi-Square Distribution Table

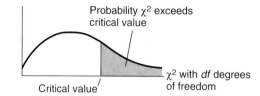

Probability χ^2 exceeds critical value values listed below

df*	0.1	0.05	0.025	0.010	0.005
1	2.71	3.84	5.02	6.63	7.88
2	4.61	5.99	7.38	9.21	10.6
3	6.25	7.81	9.35	11.3	12.8
4	7.78	9.49	11.1	13.3	14.9
5	9.24	11.1	12.8	15.1	16.7
6	10.6	12.6	14.4	16.8	18.5
7	12.0	14.1	16.0	18.5	20.3
8	13.4	15.5	17.5	20.1	22.0
9	14.7	16.9	19.0	21.7	23.6
10	16.0	18.3	20.5	23.2	25.2
11	17.3	19.7	21.9	24.7	26.8
12	18.5	21.0	23.3	26.2	28.3
13	19.8	22.4	24.7	27.7	29.8
14	21.1	23.7	26.1	29.1	31.3
15	22.3	25.0	27.5	30.6	32.8
16	23.5	26.3	28.8	32.0	34.3
17	24.8	27.6	30.2	33.4	35.7
18	26.0	28.9	31.5	34.8	37.2
19	27.2	30.1	32.9	36.2	38.6
20	28.4	31.4	34.2	37.6	40.0
21	29.6	32.7	35.5	38.9	41.4
22	30.8	33.9	36.8	40.3	42.8
23	32.0	35.2	38.1	41.6	44.2
24	33.2	36.4	39.4	43.0	45.6
25	34.4	37.7	40.6	44.3	46.9
26	35.6	38.9	41.9	45.6	48.3
27	36.7	40.1	43.2	47.0	49.6
28	37.9	41.3	44.5	48.3	51.0
29	39.1	42.6	45.7	49.6	52.3
30	40.3	43.8	47.0	50.9	53.7

df = degrees of freedom

Source: *Biometrika Tables for Statisticians*. Reprinted by permission of the Biometrika Trustees.

Answers

CHAPTER 1

Section 1.1 Problems (Odd)

 1 a. Fifty percent of adult Iowans prefer the general over the other 11 celebrities for a famous neighbor. The sample is the 810 adult Iowans, and the population is all adult Iowans.
   b. A majority of adult Iowans prefer someone among the eleven people on the list other than the Arnolds for a famous neighbor. The sample is the 810 adult Iowans, and the population is all adult Iowans.
 3 a. About 58% of the characters in television programs directed toward families and children either do not work or work in unrealistic jobs. The sample is the 970 characters, and the population is all characters in television programs directed toward families and children. A similar inference from the report is that few television characters work.
   b. About 81% of the characters in Saturday morning network television programs directed toward families and children either do not work or work in unrealistic jobs. The sample is the characters from the 30 Saturday morning programs, and the population is all characters in Saturday morning television programs directed toward families and children. The report does not mention a specific inference for Saturday morning shows, but the conclusion here backs the general conclusion that few television characters work.
   c. Seventy-five percent of the working characters from television programs directed toward families and children hold white-collar jobs. The sample is the working characters that are part of the set of 970 characters and the population is all working characters on this type of television program. The report infers that the distribution of the job types among the working characters from this type of program is different from the actual distribution. The 56% figure is an estimate that adds specific information to demonstrate the difference in television workers and real-world workers that might affect children's expectations about the world and their future.

Section 1.2 Problems (Odd)

 5 Qualitative, ordinal.   7 Qualitative, nominal.
 9 Quantitative, interval.

Section 1.3 Problems (Odd)

11 20   13 1.0   15 13.5   17 194.0   19 17,026
21 3,328.6

Multiple-Choice Problems

22 b	23 b	24 e	25 d	26 a	27 c	28 e
29 c	30 a	31 b	32 c	33 d	34 e	35 d
36 c	37 e	38 a	39 a	40 a	41 b	42 d

Word and Thought Problems (Odd)

43 The sample is the 44 hospital administrators. The population is hospital administrators from large metropolitan hospitals. The inference is that the chances of being hired as a hospital administrator improve greatly with a degree in hospital administration. The inference seems reasonable. Additional information that would be helpful includes the sampling procedure (is it random?), the definition of a large metropolitan hospital, and what degrees are counted as "related to hospital administration."

45 The sample is the one meal that has been served to you. Possible populations include all meals that you would be served or all meals that are served in the restaurant. The inference is that all meals in the population will be similar, so you will not return. We expect the accuracy of the inference to improve with more observations in the sample. Enlarging the sample will require more money, more time spent in the restaurant, and possibly more frustration (or sickness) from the poor service or food. This potential cost must be weighed against the alternative of learning that the restaurant is satisfactory.

47 The grade can be a descriptive or inferential statistic. If used to condense the details of a student's performance for the course, it is descriptive. If used to indicate how the student would perform if the class were repeated, how the student performs in all classes, or how all students in the class performed or will perform, it is an inferential statistic.

49 $\sum_{i=1}^{4} X_i =$ total repetitions of the experiment.

51 $\sum_{i=1}^{4} X_i Z_i = 17$. The total change in volume for all six repetitions is 17 cubic centimeters.

53 $(\sum_{i=1}^{4} Z_i)^2 = 25$; $\sum_{i=1}^{4} (Z_i^2) = 27$. The two expressions are not equal in this or the general case, $\sum_{i=1}^{n} (X_i)^2 \neq (\sum_{i=1}^{n} X_i)^2$.

55 a. The 14 tickets that were checked form the sample.
   b. The population can be all tickets during any time period that includes the 14 checked tickets.
   c. That a large proportion (11/14) of such tickets are not prosecuted.

57 The leg is a sample of the child's body, which is the population. A random sample might be more informative.

59 The population is the garment to be laundered or the piece of furniture to be stained. The sample is the tested area, and the inference is that the entire garment or piece of furniture will yield the result of the tested area.

61 The population is the members of the tested species that also have the disease. The inference is that what happens to the experimental animals will happen to the population. Other populations can be formed that encompass this species and one or more other species, such as humans. The inference is that the reactions of the experimental species sample will be the same or similar to reaction of all species in the population.

63 They are descriptive statistics when the only desired information is a summary of what has already occurred. They are inferential when they are guesses for an extended set of occurrences.

65 a. 57.18.   b. 13,027.8.   c. 81.6.   d. 26.69.   e. 8.79.
   f. 6.

67 a. (1) Almost no black students were frequent smokers and (2) there was a large racial difference among high school smokers.
   b. (1) Blacks in the sample of 11,000 surveyed students and (2) the 11,000 surveyed students.
   c. (1) Black U.S. high school students and (2) all U.S. high school students.

CHAPTER 2

UNIT I

Section 2.1 Problems (Odd)

1 a. 14 **b.** 3 **c.** Cannot tell **d.** Cannot tell
 e. low **f.** 0.09 and 0.13

3 States with few students taking the test scored higher than those where most students took the test. Most of the scores for the low-percentage states are in the lower part of their listed scores, while the opposite is true for the high-percentage states.

5

PM10 ($\mu g/m^3$)	Number of MSAs
20–34	21
35–49	31
50–64	5
65–79	5
80–94	1
	63

The patterns are similar.

7

SAT Scores	Number of State Scores
830–849	3
850–869	2
870–889	5
890–909	7
910–929	1
	18

Both patterns show most scores in the higher levels of 830–929.

Section 2.2 Problems (Odd)

9

Peak Ozone Level (ppm)	Proportion of MSAs
0.09–0.13	0.6984
0.14–0.18	0.2222
0.19–0.23	0.0317
0.24–0.28	0.0317
0.29–0.33	0.0159

11

SAT Scores	Proportion of State Scores
950–974	0.0556
975–999	0.3889
1,000–1,024	0.2222
1,025–1,049	0.1667
1,050–1,074	0.1111
1,075–1,099	0.0556

Section 2.3 Problems (Odd)

13

Peak Ozone Level (ppm)	Number of MSAs That Are "More than" Stated Level
0.08	63
0.13	19
0.18	5
0.23	3
0.28	1
0.33	0

15

SAT Scores	Number of States Where Scores Are "Less than" Stated Score
950	0
975	1
1,000	8
1,025	12
1,050	15
1,075	17
1,100	18

Multiple-Choice Problems

17 b **18** c **19** e **20** e **21** a **22** d **23** a
24 c **25** d **26** e **27** e **28** b **29** b **30** c
31 a **32** d **33** c **34** a **35** d **36** c **37** c

Word and Thought Problems (Odd)

39 a. Limits are 0.50 and 0.99, boundaries are 0.495 and 0.995, midpoint is 0.745.
  b. 60
  c. Cannot tell
  d.

Fossil Age (millions of years)	Relative Frequency of Fossils
0.50–0.99	0.0261
1.00–1.49	0.0196
1.50–1.99	0.0326
2.00–2.49	0.0561
2.50–2.99	0.0391
3.00–3.49	0.1252
3.50–3.99	0.1630
4.00–4.49	0.2803
4.50–4.99	0.2581
	1.0001

  e.

Fossil Age (millions of years)	Number of Fossils That Are "at Least" Stated Age	Proportion of Fossils That Are "at Least" Stated Age
0.50	767	1.0000
1.00	747	0.9739
1.50	732	0.9544
2.00	707	0.9218
2.50	664	0.8657
3.00	634	0.8266
3.50	538	0.7014
4.00	413	0.5385
4.50	198	0.2581
5.00	0	0.0000

  f. Yes, more than 0.49, 0.99, 1.49, 1.99, and so on.

41 The first frequency distribution shows that no students correctly measured the reaction and that quite a few stragglers were slow to notice the end of the reaction. The second frequency distribution makes it appear that some students may have correctly measured the reaction time. The second frequency distribution also shows that the number of stragglers remained fairly constant, with a drop only after five minutes. The skewness is no longer apparent. The first distribution has more classes, which provides more detail, resulting in a more accurate description of the distribution of the data.

Answers

43 Outlier

45 a. Classes tended to follow the appropriate procedure. One class appears to be an outlier.
  b. Use an open-ended class for the outlier, and spread the other 31 values among five newly constructed classes.

47 a and b.

Hourly Earnings ($)	Number of States	Relative Frequency of States
6.30–6.99	7	0.1373
7.00–7.69	10	0.1961
7.70–8.39	4	0.0784
8.40–9.09	15	0.2941
9.10–9.79	8	0.1569
9.80–10.49	4	0.0784
10.50–11.19	1	0.0196
11.20–11.89	2	0.0392
	51	1.0000

c. $7.69
d. $11.18, $11.23, and $11.74, which are concentrated between $6.35 and $10.07.

e.

Hourly Earnings ($)	Number of States
6.25–6.74	5
6.75–7.24	6
7.25–7.74	6
7.75–8.24	2
8.25–8.74	14
8.75–9.24	5
9.25–9.74	5
9.75–10.24	5
10.25 or more	3
	51

49

Radios per 1,000 People	Number of Nations
0–99	35
100–199	27
200–299	26
300–399	9
400–499	7
500–599	6
600–699	4
700–799	3
800–899	5
900–999	3
1,000–1,099	1
1,100 or more	2
	128

51 a and b.

Effective Annual Yield (%)	Number of Funds	Proportion of Funds
1.80–2.09	1	0.0147
2.10–2.39	0	0.0000
2.40–2.69	0	0.0000
2.70–2.99	3	0.0441
3.00–3.29	10	0.1471
3.30–3.59	25	0.3676
3.60–3.89	16	0.2353
3.90–4.19	11	0.1618
4.20–4.49	2	0.0294
	68	1.0000

c.

Effective Annual Yield (%)	Number of Funds with Effective Annual Yield of Stated Value or More
1.80	68
2.10	67
2.40	67
2.70	67
3.00	64
3.30	54
3.60	29
3.90	13
4.20	2
4.50	0

d. The outlier is 1.82.

Effective Annual Yield	Number of Funds
Less than 2.80	1
2.80–2.99	3
3.00–3.19	8
3.20–3.39	11
3.40–3.59	16
3.60–3.79	13
3.80–3.99	10
4.00–4.19	4
4.20–4.39	2
	68

53 a.

Daily per Capita Calorie Supply Requirements (%)	Number of Nations
65–74	2
75–84	7
85–94	24
95–104	22
105–114	21
115–124	16
125–134	22
135–144	8
145–154	3
	125

Daily per Capital Calorie Supply (% of requirement)	Number of Nations with Supply "Less than" Stated Percentage
65	0
75	2
85	9
95	33
105	55
115	76
125	92
135	114
145	122
155	125

b. Knowing the numbers who get less than they need and those who get more than they need would quickly illuminate the food, at least caloric, situation internationally.

55 a.

Milligrams of Sodium per Three Tablespoons of Peanut Butter	Number of Varieties
0–39	9
40–79	0
80–119	0
120–159	0
160–199	10
200–239	15
240–279	3
	37

Milligrams of Sodium per Three Tablespoons of Peanut Butter	Number of Varieties with "at Least" Stated Amount of Sodium
0	37
40	28
80	28
120	28
160	28
200	18
240	3
280	0

b.

Milligrams of Sodium per Three Tablespoons of Peanut Butter	Number of Varieties
Less than 160	9
160–174	4
175–189	3
190–204	3
205–219	3
220–234	12
235–249	1
250–264	2
	37

57 a. Outlier

b.

Number of Weeks on Best-Seller List	Number of Books
0–9	14
10–19	5
20–29	0
30–39	1
	20

Number of Weeks on Best-Seller List	Number of Books
0–4	8
5–9	6
10–14	4
15–19	1
20 or more	1
	20

Both distributions show a skewed data set: There are relatively few books that are on the best-seller list for a very long time.

UNIT II

Section 2.4 Problems (Odd)

63 See Figure A2-63; skewed to the right.

65 See Figure A2-65; skewed to the right.

Section 2.5 Problems (Odd)

67 See Figure A2-67.

69 See Figure A2-69.

Section 2.6 Problems (Odd)

71 The ogive slopes downward to the right as shown in Figure A2-71.

73 The ogive slopes upward to the right as shown in Figure A2-73. An "at least" ogive will slope downward to the right.

Figure A2-63
Frequency polygons for peak ozone levels

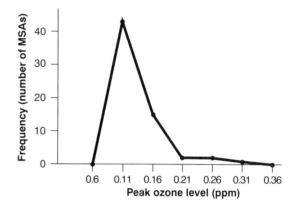

Figure A2-65
Frequency polygons for SAT scores for low-percentage states

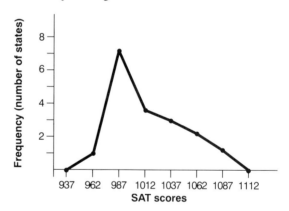

Figure A2-67
Histogram for peak ozone levels

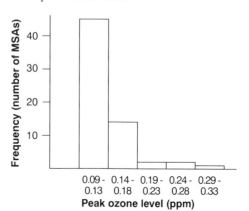

Figure A2-69
Histogram for SAT scores for low-percentage states

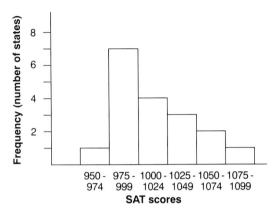

Figure A2-71
Ogive for peak ozone levels

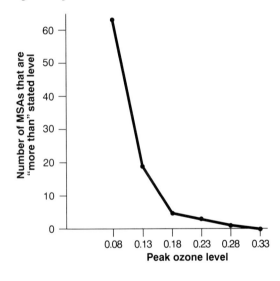

Figure A2-73
Ogive for SAT scores for low-percentage states

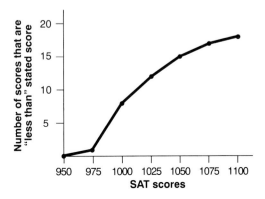

Section 2.7 Problems (Odd)

75

Peak Ozone Levels (ppm)		Number of MSAs
Stem	Leaves	Frequency
0.0	9	1
0.1	0, 0, 0, 0, 0, 0, 0, 0, 0, 1, 1, 1, 1, 1, 1, 1,	19
	1, 1, 1, 2, 2, 2, 2, 2, 2, 2, 2, 2, 2, 2, 3, 3,	24
	3, 3, 3, 3, 3, 3, 3, 3, 3, 3, 3	
	4, 4, 4, 4, 5, 5, 5, 5	8
	6, 6, 6, 6, 7	5
	8, 9	2
0.2		0
	3	1
	4	1
		0
	8	1
0.3		0
	3	1

77

SAT Scores		Number of State Scores
Stem	Leaves	Frequency
96	8	1
97	7	1
98	1, 4	2
99	3, 4, 5, 6	4
100	1, 7, 8	3
101	9	1
102		0
103	0, 1	2
104	0	1
105		0
106	1, 9	2
107		0
108	8	1

Section 2.8 Problems (Odd)

79 a. Skewed to the right. **b.** 18.
  c. It is correct, because the data values are integers to the nearest tens place.

81 a. It is approximately symmetric, with two points where most of the values occur, around 880–900 and 1,000.
  b. No.
  c. The 18 low-percentage state scores combine with the 18 high-percentage state scores to produce the two humps that occur as single humps in the individual data sets. The other 15 state scores do not upset this pattern.

Multiple-Choice Problems

82 a	**83** e	**84** c	**85** c	**86** c	**87** b	**88** c
89 d	**90** a	**91** a	**92** b	**93** b	**94** b	**95** a
96 c	**97** a	**98** b	**99** d	**100** c	**101** d	**102** c

Word and Thought Problems (Odd)

103 a. The stem-and-leaf display.
  b. The stem-and-leaf display.
  c. A frequency curve.

105 a. A frequency polygon or frequency curve will dip down along the horizontal axis, then rise again at the outlier's position at the beginning or end of the diagram.
  b. A histogram will have several consecutive bars with no height, then a bar with a positive height at the outlier's position at the beginning or end of the diagram.

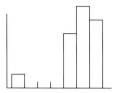

c. The ogive will have a long horizontal segment near the beginning or end of the curve. When the curve does slant, its slope may be steep.

d. There will be several consecutive stems with no leaves next to a stem with leaves at the beginning or end of the display. The middle diagram of Figure 2.15 (p. 60) shows an example.

107 a. The distribution is approximately evenly distributed, following an up-and-down pattern with a gradual trailing-off for a larger number of eye blinks.

Number of Eye Blinks

Stem	Leaves	Frequency
0	1, 2, 6, 6, 8	5
1	0, 5, 8	3
2	2, 3, 3, 4, 4, 5	6
3	0, 1, 9	3
4	0, 2	2
5	0, 3, 4, 7	4
6	0, 3, 5, 9	4
7	2, 4, 6	3
		30

b.

Eye Blinks	Percent of Subjects
0–9	16.67
10–19	10.00
20–29	20.00
30–39	10.00
40–49	6.67
50–59	13.33
60–69	13.33
70–79	10.00
	100.00

c. They provide the same impression.

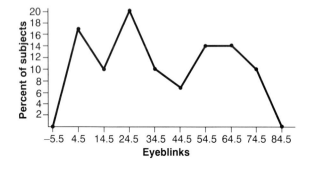

109 The outlier causes the last three points to be near horizontal.

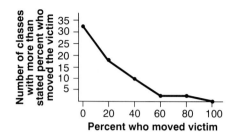

111

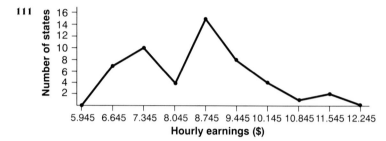

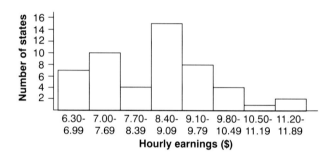

113 Yes, as a matter of fact, they did originate from the same data set, student measures of reaction time, Problem 41, Unit I (p. 43). These two diagrams demonstrate again the potential misrepresentation that can occur with a frequency distribution and its visual counterparts.

115 Yes. If the outliers are equally spaced on either side of the middle of the data set, then the data set can still be symmetric.

117 a.

Fair Attendance (millions)

Stem	Leaves	Frequency
0.	57, 58, 58, 58, 59, 59, 60, 61, 62, 63, 64, 64, 64, 65, 68, 69, 69, 72, 72, 72, 72, 79, 80, 81, 83, 85, 91, 98	28
1.	00, 00, 10, 20, 20, 20, 20, 40, 60, 70	10
2.	90	1
3.	30	1

b.

Fair Attendance (millions)	Number of Fairs
0.50–0.74	21
0.75–0.99	7
1.00–1.24	7
1.25–1.49	1
1.50–1.74	2
More than 1.74	2
	40

119 An open-ended class would allow the values bunched in the first class to be spread into several classes, providing more detail.

Answers

121 The data set is skewed to the right. Without a careful inspection of the labels, the second display dispels the impression that there is an outlier.

Review Problems (Odd)

123 a. Class limits overlap, and the class sizes are unequal.
b. The data are about evenly distributed except for the last class. We could say that the data are skewed to the right.

125 The first bar represents a class with a size that is twice as large as the other classes. In addition, the class limits are more accurate than the data.

127 a. 141 times. **b.** Cannot tell. **c.** 20–29.

d.

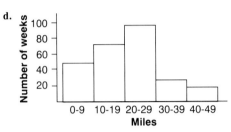

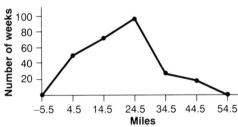

e. The data are skewed.

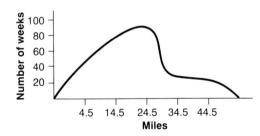

129 The total number of observations in the data set.

131 a.

Stem	Leaf	Frequency
1	6, 7, 8	3
2	0, 4, 7, 7, 8	5
3	0, 2, 3, 4, 4, 8, 9, 9	8
4	0, 1, 3, 6, 6, 7, 7, 7, 8	9
5	0, 2, 5	3

b.

Wins	Number of Teams
10–19	3
20–29	5
30–39	8
40–49	9
50–59	3
	28

c.

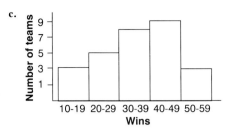

d.

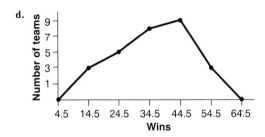

e.

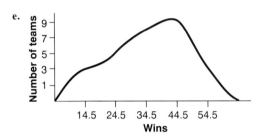

133 The sample is likely to be evenly distributed. The frequencies in the frequency distribution would be approximately equal. Similarly, the frequency polygon would be close to a straight horizontal line, and the bars in the histogram should all have about the same height. The ogive would be close to a straight line that slants up or down.

135 a.

Number of African-Americans (thousands)	Number of States
1–500	32
501–1,000	3
1,001–1,500	10
1,501–2,000	3
2,001–2,500	2
2,501–3,000	1
	51

b.

Number of Hispanics (thousands)	Number of States
1–1,300	47
1,301–2,600	2
2,601–3,900	0
3,901–5,200	1
5,201–6,500	0
6,501–7,800	1
	51

c. Both distributions are skewed to the right.

137

Movie Lengths (minutes)	Proportion of Movies
50–74	0.0789
75–99	0.4474
100–124	0.3842
125–149	0.0632
150–174	0.0211
175–199	0.0053
	1.0001

Minutes	Proportion of Movies with at Least Stated Lengths
50	1.0000
75	0.9211
100	0.4737
125	0.0895
150	0.0263
175	0.0053
200	0.0000

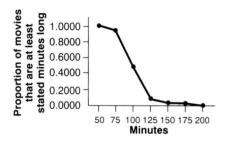

139 Both sets are slightly skewed to the right.

141 Symmetric.

CHAPTER 3

UNIT I

Section 3.1 Problems (Odd)

Problem	Mean	Median	Mode(s)
5	143.6	53	1
7	0.1154	0.0745	0.044, 0.111
9	6.68	7.8	No mode

Section 3.2 Problems (Odd)

11 a. Yes. The mean may be slightly larger than the median because the data are slightly skewed right.
  b. The low-percentage states' scores are skewed to the right. The high percentage states' scores are smaller than the low-percentage states' scores and skewed to the left. Thus, the low-percentage states' scores pull their already larger mean toward even larger values, whereas the high-percentage states' scores pull their smaller mean toward even smaller scores. As a result, the typical scores will appear closer when we compare the medians.

13 Pay raises are likely to be skewed to the right.

Section 3.3 Problems (Odd)

15 0.132   **17** 1,013.3 for low-percentage states. 882.0 for high-percentage states.

Section 3.4.1 Problems (Odd)

19 21,503.5 people is the 1st quartile, and 81,459.5 people is the 3rd quartile

21 760.5 points is the 1st quartile, and 1,269.0 points is the 3rd quartile

Section 3.4.2 Problems (Odd)

23 The 40th percentile is 30,334 people, the 45th percentile is 36,683 people, 26,846 is the 34th percentile.

25 The 5th percentile is 495.0; the 87th percentile is 1,434.5; 1,273 is the 76th percentile.

Section 3.5 Problems (Odd)

27 The data set is skewed to the right because the median and trimmed mean are smaller than the mean, and the largest value in the data set is much larger than any of the location measures.

29 The data appear to be approximately symmetric.

Multiple-Choice Problems

30 b	**31** d	**32** e	**33** a	**34** b	**35** e	**36** b
37 c	**38** b	**39** a	**40** d	**41** d	**42** b	**43** c
44 e	**45** b	**46** a	**47** b	**48** c	**49** b	**50** e

Word and Thought Problems (Odd)

51 The mean strays from the mass of the data when outliers are present.

53

Part	Mean	Median	Mode(s)
a	57.9	62	21
b	4.9	5	0, 9
c	5	5	5
d	−0.15	0.90	None

55 The mean, median, and mode are 7.82, 8.40, and 7.9, respectively. The difference in these values indicates a lack of symmetry.

57 The national median is also the district median which makes the district seem typical. If more than half of the district students score less than the national median, a problem appears to exist.

59 The weighted mean is 78.6 points, and the simple mean is 74.4 points. The weighted mean is probably preferred.

61 a. 14.6 months.
  b. The classes are not equal-size, and the limits are not multiples of 5. The limits and sizes have been chosen to coincide with age groupings in years and six-month intervals that may be useful medically.

63 $14.465.

65 The weights for observations in a simple or unweighted mean are all the value 1.

67 a. 290   **b.** 97.5

69 a. The number of points determined by the grade are the X values. The number of credit hours for the course is the corresponding W.
  b. The quality points are between 0 and 4, and credit hours are between 1 and the maximum allowable hours students can attempt.
  c. 2.89
  d. 1.91
  e. For the low or "average" achiever whose record includes Ds, the four-point system produces a higher quality-point average, because the D contributes to the quality-point average.

Answers

71 The data set is likely to be skewed to the right.

73 The mean is smaller than the smallest value in the data set, and the third quartile is the largest value in the data set.

75 If there are many small local stations and only a few large ones in metropolitan areas, then there are likely to be extreme values for employment and advertising.

UNIT II

Section 3.7 Problems (Odd)

Prob.	Range	Quartile 1st	Quartile 3rd	Interquartile Range	Sum of Deviations	Mean Absolute Deviation
81	745	1.5	296	294.5	0	153.8
83	0.421	0.045	0.114	0.069	0	0.0748
85	13.6	1.2	10.1	8.9	0	4.18

Section 3.8 Problems (Odd)

87 a. $\sigma = 6.46$ percentage points.
  b. The "average" distance between the mean market share and a specific market share is about 6.46 percentage points.
  c. Every value in the data set is between 0 and 25.82.

89 a. $\sigma = 7.7$ thousand students.
  b. The "average" deviation of the enrollment values in this data set is 7,700 students.
  c. No, because there are no enrollment figures more than 2.5 standard deviations from the mean.

Section 3.9 Problems (Odd)

91 $S = 209.3$ closings $> \sigma = 201.7$ closings.

93 $S = 0.1140$, and $S^2 = 0.0130$. The average distance of an incarceration rate from the mean is 0.1140, and the average squared distance is 0.0130.

95 $S = 5.01$ million flyers. The average distance of an airline's frequent flyer count from the mean is 5.01 million flyers.

Section 3.10 Problems (Odd)

97 The answers agree.

99 The answers agree.

Section 3.11 Problems (Odd)

101 $\overline{X} = \$15.395$, $S = \$0.3523$.

103 $\sigma = 2.8432$ million players.

Section 3.12 Problems (Odd)

105 a. The standard deviation in the figure differs, because (1) Minitab assumes that the data set is a sample in its descriptive statistical output; and (2) the solution to 103 is an approximation of the standard deviation from a frequency distribution, whereas the computer output uses the 47 individual values.
  b. The data set is skewed to the right with several extremely large outliers.

107 Yes. The first three distributions are skewed to the right, and the SAT scores appear symmetric.

Multiple-Choice Problems

108 b	**109** d	**110** e	**111** c	**112** a	**113** a
114 d	**115** c	**116** a	**117** c	**118** d	**119** d
120 c	**121** b	**122** a	**123** a	**124** b	**125** c
126 e	**127** e	**128** d	**129** c	**130** d	

Word and Thought Problems (Odd)

131

Problem	Variance	Standard Deviation
a	$\sigma^2 = 1{,}150.41$	$\sigma = 33.9$
b	$S^2 = 20.1257$	$S = 4.5$
c	$\sigma^2 = 0$	$\sigma = 0$
d	$S^2 = 50.06$	$S = 7.08$

133 All seem to be the same type of bird at the same level of maturity.

135 Rank in parentheses. Rank 1 = most dispersed. T = tie for rank.

Data Set	Range	Interquartile Range	MAD	Variance	Standard Deviation
Positive	9(3)	3.5(2T)	2.1(3)	7.3(3)	2.7(3)
Dubious	17(1)	7.5(1)	4.1(1)	23.8(1)	4.9(1)
Negative	13(2)	3.5(2T)	2.6(2)	12.5(2)	3.5(2)

137 a. 188.0 articles. The "average" distance of a value from the mean is 188.0 articles.
  b. $\overline{X} = 289.2$ articles, and $S = 432.7$ articles. The mean and the standard deviation increase.
  c. The mean increases by the value 2,000 articles and the standard deviation does not change.
  d. Range: 761 articles. The number of articles published spans 761 units. The range becomes 1,941 articles when the value 2,000 is added to the data set, increasing the range substantially. Increasing each of the original observations by 2,000 does not affect the range, 761 articles.
  Interquartile Range: The interquartile range is 112 articles. The middle 50% of the values cover 112 units. When the value 2,000 articles is added to the data set, the interquartile range increases to 151.5 articles. The final transformation increases all values by 2,000 but does not change the interquartile range, 112 articles.

139 The one with the larger standard deviation.

141 If the standard deviation of each judge's scores is small, then the judge has awarded approximately the same score to every diver. If the standard deviation is large, then it is possible that the judge may have purposely scored some divers lower to improve the position of a favored diver. Statement a is compatible with biased judges even though judges may have different standards for judgments. Statement b indicates unbiased judges because each diver scored about the same. Statement c does not necessarily indicate biased judges, because the quality of the divers' performances may indeed be quite diverse.

143 0.4 hour.

145 The approximated values from Example 3–19 are $\overline{X} = 13{,}799.5$ lawyers and $S = 16{,}368.2$ lawyers. The actual values are $\overline{X} = 12{,}582$ lawyers and $S = 16{,}349$ lawyers.

147 $\overline{X} = 9.8409$ million years, and $S_G = 4.0$ million years.

149 $\overline{X} = 5{,}050.8$ cases, and $S = 370.1$ cases. An outlier might occur when special constitutional issues of wide interest are being litigated around the country. The Z scores are -2.3,

−0.9, 0.3, 0.1, 0.1, −0.1, 0.3, 0.2, 0.6, and 1.6. No Z score exceeds 2.5 in absolute value, so we conclude there are no outliers. The −2.3 says that the observation is 2.3 standard deviations below the mean.

151 a. The outliers are 40, 30, 20, 45, 35, and 25.
  b. The interval, ±2 standard deviations from the mean, stretches from 51.5 to 94.7 points.
  c. The interval, ±1 standard deviation from the mean, encloses 68% of the values.

153 AGE ($\sigma \neq 0$), WINS (cannot be greater than STARTS), FINISH ($\sigma \neq 0$), TIME ($\sigma \neq 0$)

155 a. Very dispersed with cluster of points in the middle.
  b. Data are compact with little dispersion.
  c. Very dispersed throughout data set.
  d. Data are relatively compact without outliers.
  e. All values are identical.
  f. The middle half of the values are identical, and a few values are very different.
  g. The smallest and largest values each make up at least one-fourth of the values in the data set.

Review Problems (Odd)

157 a. All values are percentages.

Variable	Mean	Median	Mode(s)	Range	Interquartile Range	Standard Deviation
Total	34.53	34.60	34.7, 35.0	8.80	2.35	1.79
Federal	22.30	22.20	23.2	6.70	1.95	1.38
St/Loc	12.22	11.95	11.6	7.50	1.45	1.40

  b. 70th percentile is 35.15%.

159 The mean will raise the measure of the typical income. The first student appears better with the mean, whereas the second student is better with the median.

161 a. $\overline{X}$ = 12,912,200 offenses and S = 695,700 offenses.
  b.

Offenses	Z Score
12,250	−1.0
13,408	0.7
13,424	0.7
12,974	0.1
12,109	−1.2
11,882	−1.5
12,431	−0.7
13,212	0.4
13,509	0.9
13,923	1.5

There are no outliers.

163 $550.

165 a. The mean or typical value is $45.64 million. The median or middle value is $53.6 million. The mode, the value that occurs most often, is $53.6 million. The range or length of the interval spanned by the values is $72.1 million. The standard deviation or "average" distance of a value from the mean is $24.52 million.
  b. The mean decreases to $45.12 million. The median decreases to $48.05 million. The mode and range do not change. The standard deviation decreases to $22.42 million.

167 If the range is 398.22, that is also the largest possible deviation.

169 a. B, because its standard deviation is smaller.
  b. A, because its mean is lower, and the standard deviation is larger.

171 μ: 1,375.2 meals, median: 1,311.0 meals; there is no mode; and σ = 273.7 meals. The typical monthly number of meals served is 1,375.2, and the "average" distance from the mean is 273.7 meals.

173 $\overline{X}$ = 34.53%, which is the same as the actual mean. S_G = 1.83%, which is approximately the same as the actual standard deviation of 1.79%.

175 a. The data set is skewed to the right. Yes, because the mean is larger than the median.
  b. 91, 109, and 97.
  c. About 21.0.
  d. 9.0 and 30.0.

177 There are many possible sketches that could fit the following descriptions
  a. ──── I + I ──── Median in middle of box and whiskers same length.
  b. ──── I + I ── One long and one short whisker with median near short whisker.
  c. ──────── I + I ──────── Wide box and long whiskers.
  d. ─ I + I ──── Narrow box and short (perhaps no) whiskers.

CHAPTER 4

UNIT I

Section 4.2 Problems (Odd)

3 One-sixth of infinite rolls will be a 7.

5 Sample space = {"not more than $5,000," "greater than $5,000 but not more than $10,000," "greater than $10,000 but not more than $20,000," and "greater than $20,000"}. We cannot use the equiprobable outcomes method, because we cannot be sure that an equal number of respondents selected each of the four responses. We must use the relative frequency of each response.

7 12/27

Section 4.4 Problems (Odd)

9 Yes   **11** No   **13** Yes   **15** Yes   **17** No   **19** Yes

Section 4.5 Problems (Odd)

21 3/4   **23** 1/2   **25** 98/100

27 Consider the event "professor is in his/her office." P(professor is in office | current time is 15 minutes before his/her next class) > P(professor is in office at random time). P(professor is in office | today is Sunday) < P(professor is in office at random time).

29 28/101   **31** 88/377

Multiple-Choice Problems

33 b   **34** e   **35** d   **36** c   **37** e   **38** a   **39** c
40 a   **41** c   **42** b   **43** b   **44** b   **45** c   **46** a
47 d   **48** d   **49** b   **50** c   **51** a   **52** e

Answers

Word and Thought Problems (Odd)

53 Very few heads (close to 0) or very many heads (close to 10).

55 The relative frequency method is based on the idea of enumerating outcomes from repeated trials, whereas the equiprobable outcomes method requires that we ensure or assume that the outcomes are equiprobable. The probability values represent the proportion of infinite experimental trials that will result in the event occurring.

57 1/4, 4/52, 1/52

59 1/4, 0, 1/2

61 a. 1/24 **b.** 1/23 **c.** 1/25 **d.** 1/17

63 a. 1/5 **b.** 3/5 **c.** 2/5 **d.** 1/3 **e.** 1/5 **f.** 1/5

65 a. 0 **b.** 0 **c.** 1 **d.** 1

67 No. The conditional probabilities of child involvement are unequal.

69 Technically, there is no probability for a recession next month, if there is no value approached by X/n. However, the best guess might be the last calculated value of X/n or another relative frequency conditional on recent values.

71 a. 0.0079 **b.** 0.0081 **c.** 0.1235 **d.** 0.1197
e. 0.3152 **f.** 0.0051 **g.** 0.0021

UNIT II

Section 4.7.1 Problems (Odd)

77 0.44 **79** 0.03

81 $P(J) = 2,350/10,000$, $P(F|J) = 1,250/2,350$, $P(F$ and $J) = 1,250/10,000 = 1/8$

Section 4.7.2 Problems (Odd)

83 Dependent **85** Independent

87 No, because $P(R) \neq P(R/T)$ **89** 0.000512

91 $P(S$ and $F) = 0.12$

Section 4.8.1 Problems (Odd)

93 20/45 **95** 6,200/10,000 **97** 0.85

Section 4.8.2 Problems (Odd)

99 Not mutually exclusive **101** Mutually exclusive
103 Not mutually exclusive **105** 0.78 **107** 0.30

Multiple-Choice Problems

109 a **110** a **111** b **112** a **113** c **114** d
115 a **116** b **117** c **118** e **119** d **120** d
121 a **122** b **123** a **124** c **125** a **126** c
127 b **128** c **129** b **130** a **131** c

Word and Thought Problems (Odd)

133 $P(A)$ is the proportion of experimental trials that event A occurs. The Equiprobable Outcomes Rule is often useful for determining this value.
$P(A')$ is the proportion of experimental trials that A does not occur. Rule 1 applies.
$P(A|B)$ is the probability that A occurs given that B occurs. Rule 2' applies.
$P(A$ and $B)$ is the proportion of experimental trials that the compound event "A and B" occurs. Rules 2 and 2" apply.
$P(A$ or $B)$ is the proportion of experimental trials that the compound event "A or B" occurs. Rules 3 and 3' apply.

135 With replacement: Mutually Exclusive column: No, No, No
Independence column: Yes, Yes, Yes
$P(A$ and $B)$ column: 1/104, 16/2,704, 1/2,704

Without replacement: Mutually Exclusive column: No, No, No
Independence column: No, No, No
$P(A$ and $B)$ column: 26/2,652, 12/2,652, 1/2,652

137 Because drunkenness impairs driving ability, the likelihood of a traffic accident changes when drunk drivers are on the road, so the events are dependent.

139 e, f, h (all parts), and j

141 a. No **b.** Yes **c.** 0 **d.** 0.25

143 0.1667

145 0.28

147 0.9384. We assume that the events are independent.

149 0.0002

151 a. 0.882 **b.** 0.002 **c.** 0.118

153 0.7125, no, no

155 a. 0.7 **b.** 0.3 **c.** 0.92 **d.** 0.48

Review Problems (Odd)

157 It has rained on 60% of the days similar to today or current conditions.

159 a. 98/150 **b.** Word-of-mouth **c.** 135/150 **d.** 38/98
e. 38/150 **f.** 58/150 **g.** 103/150 **h.** 113/150

161 0.15

163 0.0125

165 0.56

167 1/25

169 0.4

171 0.45

173 0.782

175 0.95; 0.05

177 a. 0.5556, 0.7500, 0.8889 **b.** 0.5556, 0.3600, 0.9375
c. 0.8400, 0.8025 **d.** 0.8889, 0.9375

179 a. No, $P($California$) \neq P($California$|$AIEA$)$. **b.** No, because we expect the top ten AIEA states not to be typical with regard to AIEA population. **c.** No, because, for practical purposes, knowing a person is from Wisconsin is not much help in determining whether the person is AIEA, and vice versa.

CHAPTER 5

UNIT I

Section 5.1 Problems (Odd)

7 When the teacher decides the total number of questions and then proceeds to write the questions, the variable is nonrandom. If the teacher writes questions with some objective in mind, such as covering certain topics, and lets the writing process determine the number of questions, the variable is random. The variable is random for students, unless (1) the teacher reveals the number of questions ahead of time or (2) the students report the number of questions to someone else. The number of questions is discrete.

9 No; yes.

11 a. Obvious candidates are 30, 40, 50, 60, 70, 80, and 90 minutes.
   b. We could assign a rank value for the difficulty of producing a project, or we could assign a value equal to the estimated cost to hire a comparable performance.

Section 5.2.1 Problems (Odd)

13 0.20
15 $P(S = 30) = 0.15$
    $P(S \le 60) = 0.84$

Section 5.2.2 Problems (Odd)

17 $E(S) = 51.9$ minutes. The average taping time for a large number of taping sessions is 51.9 minutes.
19 $E(X) = 7$ dots. The average value rolled with a pair of dice over a long-run series of rolls is seven dots.
21 3.86 or 3.9 days. The average delivery time for a letter to a private household is 3.9 days.

Section 5.2.3 Problems (Odd)

23 1.7 guests. The "average" distance between a party outcome and the mean is 1.7 guests.
25 1.1 jurors. The "average" deviation from the mean is 1.1 jurors.

Multiple-Choice Problems

27 e   28 c   29 b   30 a   31 a   32 c   33 e
34 b   35 d   36 d   37 b   38 a   39 d   40 b
41 c   42 b   43 c   44 e   45 b   46 d   47 b
48 a

Word and Thought Problems (Odd)

49 A random variable is a variable whose values represent the outcomes of a statistical experiment. A probability distribution is a representation of all possible values of a random variable and the probability associated with each value. The probabilities sum to 1.
51 A sample. It is a discrete random variable. Inferential statistics.
53 $\mu = E(R) = 0.55$ inches and $\sigma = 0.59$ inches. The standard deviation is larger than the mean, so the variation is large relative to the size of the mean.
55 a. -0.16% b. 11.84%
57 $0.06. This game.
59 The expected value is $220, and the standard deviation is $292.9. No, because the standard deviation is large.
61 -0.33 weeks
63 $B; A$
65 a. 16.7 hours   b. The one described in Part b
67 $E(A) = 440$, $\sigma = 37.4$. $E(B) = 460$, $\sigma = 34.6$. They are both 10 units from 450 on average, but B's values cluster slightly more about the mean, so the results on the production line are likely to be closer to the desired level with machine B.

69 a.

x	$P(x)$
10	0.040
30	0.010
50	0.015
100	0.935

   b. 0.960   c. 0.960   d. 95.0
   e. 95.0   f. 19.6

UNIT II

Section 5.3 Problems (Odd)

79 a. 15   b. 12   c. 1   d. 1   e. 1,140
81 0.4096
83 0.1319
85 0.1272
87 $P(0) = 0.0010$, $P(1) = 0.0146$, $P(2) = 0.0879$, $P(3) = 0.2637$, $P(4) = 0.3955$, $P(5) = 0.2373$

Section 5.4 Problems (Odd)

89 0.2336   91 0.3932, 0.2621   93 0.2508   95 0.0020

Section 5.5.1 Problems (Odd)

97 Approximately 1, 15 catfish are most likely, and 0 is least likely.
99 0.4853, 1 stream is most likely (see Figure A5-99)

Figure A5-99

Solution to Problem 99; shaded area corresponds to $P(X \ge 2)$

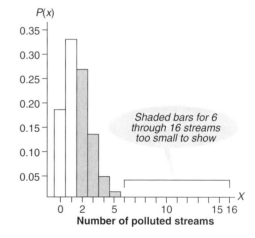

101 0.5367 (see Figure A5-101)

Sections 5.5.2 and 5.5.3 Problems (Odd)

103 a. $\mu = 36$, $\sigma = 3.175$ b. $\mu = 6$, $\sigma = 2.412$
     c. $\mu = 142.5$, $\sigma = 8.441$
105 $n = 5$, $p = 0.25$, $\mu = 1.25$ hearts, $\sigma = 0.968$ heart

Multiple-Choice Problems

107 c   108 a   109 a   110 e   111 c   112 e
113 d   114 d   115 c   116 b   117 c   118 a
119 b   120 e   121 e   122 e   123 c   124 a
125 e   126 b

Word and Thought Problems (Odd)

127 The probability of X successes in n trials, two outcomes on each trial, the outcomes are independent, and the probability of success on one trial is the same for all trials.
129 a. 0.1699   b. 0.9853   c. 0.1846

Answers

Figure A5-101
Solution to Problem 101; shaded area corresponds to $P(X \leq 14)$

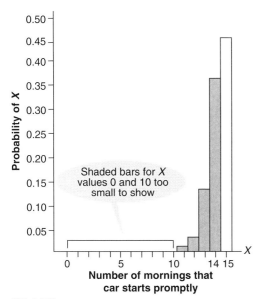

131 1,140

133 0.0002

135 0.0469

137 0.1829

139 0.0061

141 We cannot use the binomial formula, because there are more than two outcomes to each trial.

143 0.8853, 0.7164, 0.5997

145

x	$P(x)$
0	0.3277
1	0.4096
2	0.2048
3	0.0512
4	0.0064
5	0.0003

$\mu = 1$, $\sigma = 0.89$

147 0.2525

Review Problems (Odd)

149 Both use relative frequencies, but the probability distribution's relative frequencies are long run. The mean of a random variable, X, is its expected value or $\Sigma\, xP(x)$. The standard deviation of X is the square root of the variance, which is the expected value or average of the squared deviations or $\sqrt{\Sigma(x-\mu)^2 P(x)}$. The interpretations are generally the same, but the data set for the random variable is the outcomes of infinite repetitions of the experiment.

151 The expected value is negative for the player. The expected value is positive for the state.

153 a. 0.01 **b.** 0.28 **c.** 34 minutes, the average time for a student taping a session to complete the project for a large number of students
d. 16.2 minutes

155 a. Probability distribution of the random variable X **b.** It is unreasonable, because $E(X)$ is the mean of the X values and the largest possible X value is 1,100.
c. It is unreasonable, because the range of the X values is 1050, which is smaller than 2,500.

157 1.1 sightings

159 a. B **b.** A **c.** A

161 27.6 days

163 0.6770

165 3

167 0.0002

169

X = Completed Calls	$P(x)$	Y = Completed Calls between 4 and 6 P.M.	$P(y)$
0	0.0003	0	0.3732
1	0.0035	1	0.4355
2	0.0212	2	0.1693
3	0.0743	3	0.0220
4	0.1672		
5	0.2508	$\mu = 0.8$ calls	
6	0.2508		
7	0.1612	$\sigma = 0.8$ calls	
8	0.0605		
9	0.0101		

$\mu = 5.4$ calls $\sigma = 1.5$ calls

CHAPTER 6

UNIT I

Section 6.3.1 Problems (Odd)

5 0.4306 **7** 0.4977 **9** 0.8106 **11** 0.0142 **13** 0.6366
15 0.9015 **17** 0.0790
19 a. 0.4924 **b.** 0.9544 **c.** 0.1611 **d.** 0.8389

Section 6.3.2 Problems (Odd)

21 1.13 **23** -0.99 **25** 0.67 **27** 0.13
29 a. 0.39 **b.** -1.28 **c.** -2.33 and 2.33

Multiple-Choice Problems

31 c **32** c **33** e **34** a **35** b **36** e **37** a
38 b **39** d **40** e **41** d **42** a **43** b **44** c
45 e **46** b **47** c **48** b **49** d **50** a

Word and Thought Problems (Odd)

51 0.4979 **53** 0.2843 **55** 0.95 **57** 0.0082 **59** 0.0049
61 0 **63** 0.47 **65** -0.36 **67** 2.575
69 0.9544, 0.9876, 0.9940 **71 a.** 0.1587 **b.** 0.9977
c. 0.3174 **d.** -0.58 **e.** ± 1.28
73 a. 0.1977 **b.** 0.3830 **c.** 1.41
75 a. 0.9772 **b.** 0.8400 **c.** 0.6826

Unit II

Section 6.4 Problems (Odd)

77–93 Figure A6-77 shows the answers.

Figure A6-77

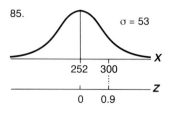

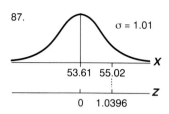

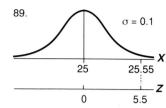

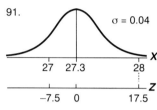

93.

σ = 5.84 yards

X = results of play (yards)

−8.1 3.22 6.14 8.1

−1.94 0 0.5 0.84

28 is 17.5 standard deviations above the mean.
27 is 7.5 standard deviations below the mean.

a) A gain of 8.1 yards is 0.84 standard deviations above the mean.
b) A loss of 8.1 yards is 1.94 standard deviations below the mean.
c) A result 0.5 standard deviation above the mean corresponds to a gain of 6.14 yards on the play.

Section 6.5 Problems (Odd)

95 a. 0.1230 **b.** 0.7088 **c.** 0.2912 **d.** 0.2765 **e.** 0.0119 **f.** 0.5386

Section 6.6 Problems (Odd)

97 a. 510 **b.** 531 **c.** 506.5 **d.** 478.5 **e.** 402, 598 **f.** 649

99 a. 7.366 **b.** 9.06 **c.** 6.607 **d.** −0.167 (loss of 0.167 yards) **e.** loss of 4.255 yards and gain of 10.695 yards **f.** −14.3

Section 6.7 Problems (Odd)

101 0.9838 **103** Approximately 0.9934
105 Approximately 0.0122

Multiple-Choice Problems

107 d **108** a **109** b **110** a **111** c **112** d
113 e **114** a **115** d **116** c **117** c **118** a
119 b **120** e **121** c **122** c **123** d **124** c
125 c **126** a

Word and Thought Problems (Odd)

127 The distance of a point from the mean of the data set in standard deviation units
129 a. 0.0668 **b.** 0.2119 **c.** 0.0466 **d.** 0.4918 **e.** 0.8331
131 a. 0.0301 **b.** 0.0301 **c.** 0.4648 **d.** 85.24 **e.** 79.16
133 a. 0.9413 **b.** 0.0062
135 a. 74th **b.** Approximately 0 **c.** 259.8 **d.** 209.4, 346.6 **e.** 0.6826 **f.** 1.96
137 a. 0.9974 **b.** Divide range by 6. **c.** Divide by 4. **d.** Given that the distribution of the data set is normally distributed the more accurate method depends on the available information and the assumptions made about the data set. If the data set is a population and the range is known, use one-sixth of the range. If the data set is a sample and the range is approximately the range of the middle 95% of the population values use one-fourth of the range.

139 Approximately 0.7850
141 a. 0.4913 **b.** 0.0676 **c.** 0.2486 **d.** 80.8, 119.2 **e.** 112.6 **f.** 84th **g.** 0.9544 **h.** ±2.33
143 a. Approximately 1 **b.** Approximately 0.5675, approximately 0.0668 **c.** Approximately 0.9328
145 a. 0.0934 **b.** 0.1788 **c.** 195 **d.** 84 **e.** The population is a smaller group of larger test values. The mean should be larger and the standard deviation smaller. **f.** 1.28 **g.** 0.9566
147. a. Approximately 0.0228 **b.** 258.05 or approximately 258 accommodations **c.** Approximately 50% of the flights will have more people show up to fly than seats. **d.** Fewer. **e.** P(more passengers than accommodations) is approximately 1. The company should not oversell so many tickets when most purchasers show up.
149 Approximately 0.0505
151 a. 0.027 **b.** 0.9989 **c.** Approximately 0.0778

Review Problems (Odd)

153 a. Yes. A student who is never absent is an unusual occurrence as well, but not of concern. **b.** 0.0475
155 100.3°
157 a. 0.1401 **b.** 724
159 Approximately 0.0228
161 a. Approximately 0.0057 (exact probability is 0.0144) **b.** Approximately 0.4052 **c.** Approximately 0.0011 (a cutoff of 12 means this type of mistake should occur less often)
163 a. 0.0028 **b.** 0.9004 **c.** 997.1 **d.** 2.575 **e.** 0.5408
165 a. 0.0247 **b.** 87th **c.** 0.4925 **d.** 0.2736 **e.** 1.96
167 a. Approximately 0 **b.** Approximately 0.0116 **c.** Approximately 1 **d.** Hire the clerk.
169 a. 0.9082 **b.** 0.2958 **c.** 94.4, 129.6 **d.** 1.75 **e.** 0.7062
171 0.0781
173 Approximately 0.9838
175 Approximately 0.2119
177 0.0475

Answers

CHAPTER 7

UNIT I

Section 7.1 Problems (Odd)

7 If the 50 selected names start with the same first letter of those allowed to register first during the most recent registration period, then the corresponding students probably easily obtained their desired class schedule and will probably express less dissatisfaction than those who register later.

9 298; 61; 206; 99; 113

11 The random values are 87 and 100, which both correspond to zeros for Swain and Yancey Counties.

Section 7.2 Problems (Odd)

13 a. $n = 2$

min	P(min)
1	5/9
3	3/9
5	1/9

b. $n = 3$

min	P(min)
1	19/27
3	7/27
5	1/27

c. These sampling distributions are skewed to the right, and the sampling distributions of $\bar{X}$ are symmetric.

d. For $n = 2$, $P(MIN = 500) = 1/9$ and $P(MIN > 2) = 4/9$. For $n = 3$, $P(MIN = 500) = 1/27$ and $P(MIN > 2) = 8/27$.

Multiple-Choice Problems

15 c **16** e **17** d **18** e **19** a **20** b **21** a
22 c **23** d **24** e **25** d **26** b **27** d **28** c
29 a **30** a **31** e **32** a **33** b **34** c

Word and Thought Problems (Odd)

35 The probability distribution of a statistic.

37 The sampling distribution for the sample median is the same as the sampling distribution for $\bar{X}$ in the $n = 2$ case.

Median	P(median)
1	1/9
2	2/9
3	3/9
4	2/9
5	1/9
	1

For the $n = 3$ case, we have

Median	P(median)
1	7/27
3	13/27
5	7/27

39 a. $\mu = 3$; 4; 10/16   **b.** 5; 7; 12/16

41 The population distribution describes the values that form the population: the values and their frequency or probability of occurrence. A sampling distribution represents all possible values of a statistic and the probability of each value, based on random samples of a given size taken from a specified population.

43 a. Use the random number table, a computerized random number generator, or, given the small size of the population, write each possible outcome on a slip of paper, and draw three slips blindly to select three observations for a random sample.

b.

$\bar{x}$	$P(\bar{x})$
8	1/16
9	4/16
10	4/16
11	2/16
12	4/16
14	1/16

Sampling with replacement is equivalent to sampling from an infinite population, because each value can be chosen indefinitely for a single sample.

c.

$\bar{x}$	$P(\bar{x})$
9	4/12
10	2/12
11	2/12
12	4/12

Sampling without replacement results in a more compact sampling distribution of $\bar{X}$, because the samples composed of repetitions of either the smallest population value, 8, or the largest population value, 14, cannot occur. These two samples would extend the interval of possible $\bar{X}$ values.

45 The statistics are $\bar{X} = 980$, modes = 800 and 950, median = 960, and $S = 500$. The parameters are $\mu = 1,350$ and $\sigma = 380$. The population median and mode are 1,350.

47 a. The weights account for the importance of each observation.
b. When each stratum forms the same proportion of the population; not necessarily.
c. 41 representatives and 9 senators.
d. 16; use a random number table with usable values that correspond to the number of legislators in each of the states.

49 If the person who answers responds, the response may not be the same as that of the person originally selected, which might bias results. Less active people may tend to stay home.

51 a. 20 diabetics   **b.** 5 diabetics with related foot problems.

53

Replacement		Nonreplacement	
w	P(w)	w	P(w)
0	4/16	0	2/12
1	8/16	1	8/12
2	4/16	2	2/12

UNIT II

Sections 7.3 and 7.4 Problems (Odd)

63 89.6, 9.53   **65** 34.42, 81.1

Section 7.5 Problems (Odd)

67 a, c, d, h   **69** 0.3391   **71** 1.645, 1.15

Section 7.6 Problems (Odd)

73 0.6736   **75** 0.9544   **77** 112.2

Section 7.7 Problems (Odd)

79 $P(200 \leq \overline{X} \leq 400) = 1$ when the population is finite. This answer is larger than 7/9, because two observations from a population of three observations should produce an $\overline{X}$ with a higher probability of being close to the mean than two observations from an infinite population.

81 $\sigma_{\overline{X}} = \sigma/\sqrt{n}$

Multiple-Choice Problems

83 a	84 b	85 e	86 c	87 e	88 b	89 b
90 d	91 b	92 a	93 c	94 d	95 f	96 a
97 b	98 c	99 b	100 e	101 e	102 a	

Word and Thought Problems (Odd)

103 μ = population mean, $\overline{X}$ = sample mean, $\mu_{\overline{X}}$ = mean of all possible $\overline{X}$ values.

105 Column a: 1.15, 0.94, 0.82, 0.73, 0.67, 0.62, 0.30
Column b: 1.31, 1.07, 0.92, 0.83, 0.76, 0.70, 0.34
Column c: 1.05, 0.86, 0.74, 0.67, 0.61, 0.56, 0.27

107 a. It approaches a normal distribution. **b.** Nothing **c.** Decreases **d.** Large, because $\sigma_{\overline{X}}$ will be smaller, meaning the $\overline{X}$ are more tightly clustered about μ.

109 a. 0.1515 **b.** Cannot **c.** Normal. **d.** 0.95.

111 a. 0.4207 **b.** Approximately 0 **c.** Sample means condense sets of values from the population into a central value, $\overline{X}$. Thus, $\overline{X}$ values cluster more tightly about μ than individual values in the population.

113 Such polls are not random.

115 Replacement: $\mu_{\overline{X}} = \mu = 10.5$, $\sigma_{\overline{X}} = 1.54$; nonreplacement: $\mu_{\overline{X}} = 10.5$, $\sigma_{\overline{X}} = 1.26$. $\mu_{\overline{X}} = \mu$ no matter the method of selection, but the standard error for nonreplacement sampling is smaller than its counterpart for replacement sampling. Nonreplacement ensures that the sample contains at least as much information from the population as does the replacement sample.

117 a. They are the same. **b.** $\mu_{\overline{X}} = \mu$. **c.** They are the same.

119 a. 0.9876 **b.** finite 38.8 to 61.2 hours; infinite 37.5 to 56.2 hours **c.** $\overline{X}$ values are more tightly clustered when sampling from finite populations. The finite population case is likely to contain a larger proportion of information from the population, because each observation is a larger proportion of the observations than observations from an infinite population. **d.** Finite 47.2 to 52.8 hours, infinite 43.8 to 56.2 hours **e.** The range is smaller for the sample from a finite population. Quadrupling the sample size from a finite population indicates that the proportion of information from the population that is in the sample is four times what it was before. Quadrupling from an infinite population adds new population information, but it does not increase the proportion of population information in the sample.

121 Your estimates should be the same as SEMEAN, except for small rounding differences.

Review Problems (Odd)

123 The theorem states that as the sample size, n, increases, the sampling distribution of $\overline{X}$ approaches a normal distribution, no matter how the population is distributed. Without this theorem or further knowledge we would be left with the construction method of Unit I. An alternative is to form the empirical sampling distribution of a statistic. A relative frequency distribution of the statistic based on many (but not infinite) samples.

125 a. 0.25 **b.** 0.125 **c.** 0.5 **d.** 0.25

127 Increasing n means that the sample is more likely to contain added population information, thus increasing the likelihood of an accurate estimate. Technically, increasing n decreases the standard error of the mean, $\sigma_{\overline{X}}$, which means that $\overline{X}$ values are less dispersed about the population mean, μ.

129 No, because it is possible to obtain a sample of the students' grades with a mean different from the mean grade for all students.

131 a. 11.6 banks **b.** 0.0093 **c.** 11 banks; 0.6 bank.

133 The life of a light bulb; possible but impractical; destructive sampling

135 0.9951 ($\sigma_{\overline{X}} = 0.9311$)

137 a. The standard errors down the column are 1.33, 1.34, 1.71, 0.78, 1.09, 1.33, 1.39, 1.56, 1.84, 1.82, 1.84, 2.22, and 2.40, respectively.
b. 0.1841, 74.25 to 80.95
c. 58.77 to 64.03 and 74.25 to 80.95
d. No, population information is not provided.
e. 0.5359
f. (0.5714)(0.4801) = 0.2743.
g. 0.0684.
h. It is probably not a random sample, because every Minnesota student is not given an equal chance of selection. To the extent that all students are equally likely to be taking the course, it may be an expedient way of obtaining a representative sample.

139 The critics are not a random sample of moviegoers.

141 The distance of the point from the mean measured in standard deviations.

CHAPTER 8

UNIT I

Section 8.2.1 Problems (Odd)

9 Driveways: (6.80 hours, 7.20 hours)
Patios: (9.49 hours, 10.51 hours)
Decks: (11.33 hours, 12.67 hours)
In each case, there is 80% confidence that the interval contains the actual mean completion time. The mean of driveway completion times is best estimated because the standard deviation is smallest.

11 (16.76 miles, 19.24 miles). We are 90% confident that the interval contains the true mean daily mileage.

Section 8.2.2 Problems (Odd)

13 (104.48 points, 120.12 points). We are 98% confident that this interval encloses the true mean score. Although the interval does not contain the population mean, 100 points, we have not calculated the confidence interval incorrectly. We just randomly selected a sample with a mean that has a low probability of occurrence.

15 (10.5218 seconds, 10.5242 seconds). We are 90% confident that the interval contains the true local mean reaction time.

Section 8.3.1 Problems (Odd)

17 0.99 **19** 0.80 **21** 2.045 **23** 3.169

Answers

Section 8.3.2 Problems (Odd)

25 (6.29 reporters, 9.71 reporters). We are 90% confident this interval contains the true mean number of reporters.

27 ($24,907.29, $31,092.71) We are 99% confident that this interval contains the actual mean salary.

Section 8.3.3 Problems (Odd)

29 (1.304 pieces, 1.456 pieces). Populations could refer to all customers during this week.

31 (446.6 miles, 519.4 miles). We are 95% confident this interval contains the actual mean mileage per week.

33 (83.4 strokes, 86.6 strokes). This evidence does not support the golf course claim.

Section 8.5 Problems (Odd)

35 We are 90% confident that the interval includes the mean count of dream characters for all of the analyst's patients. The computation employed a T value.

37 a. Yes, the values are very close
  b. We are 95% confident that each interval contains the actual mean mortgage rate for all banks.
  c. Yes, because neither interval includes 9.2%.
  d. T values are larger than the Z value for the same level of confidence and therefore produce a wider interval.

Multiple-Choice Problems

38 c	39 d	40 d	41 a	42 b	43 b	44 d
45 c	46 c	47 a	48 d	49 c	50 c	51 c
52 b	53 c	54 d	55 b	56 a	57 a	58 d
59 e	60 d

Word and Thought Problems (Odd)

61 A point estimate consists of a single value. S; sample proportion of successes; and $\overline{X}_1 - \overline{X}_2$. Lack of bias and consistency of an estimator.

63 a. (61.9, 64.1) **b.** (62.1, 63.9) **c.** (62.3, 63.7)
  d. As the level of confidence decreases, while n, $\overline{X}$, and σ do not change, the width of the confidence interval decreases.

65 a. (24.0 hours, 26.0 hours)
  b. Lawn mowers currently produced by this manufacturer

67 (3.28 inseminations, 3.92 inseminations)

69 The T distribution approaches a standard normal distribution. When n is large ($n > 30$), we can substitute z for t with little distortion.

71 ($10.8 mischarges, $13.2 mischarges)

73 a. Underestimates **b.** Unbiased estimator

75 (2.7 shaves, 7.8 shaves)

77 a. Demand increase (2.553 months, 3.907 months)
  Demand decrease (2.692 months, 4.508 months)
  Cost increase (2.500 months, 3.840 months)
  Cost decrease (2.937 months, 5.003 months)
  b. Demand increase and cost increase
  c. Cost decrease.

79 (4.8 minutes, 74.8 minutes)

81 (5.42 minutes, 5.52 minutes). Yes.

83 The variable is a count. No.

85 (5.792 hours, 9.108 hours)

UNIT II

Section 8.7.1 Problems (Odd)

89 We can be 80% confident that the error is at most 0.155 if we use 4.33 hours as the estimate.

91 a. First group, $E = 0.93$; second group, $E = 1.03$
  b. The largest possible mean score for the first population is 12.23, and the lowest possible for the second population is 17.17. Because there is no overlap, the population means probably differ.
  c. Decrease. First group, $E = 0.66$; second group, $E = 0.73$

Section 8.7.2 Problems (Odd)

93 174 pieces

95 97 men and 97 women. Men's sample is too small and women's sample is too large.

Section 8.7.3 Problems (Odd)

97 $z = 2.86$, $1 - \alpha = 0.9958$ or 99.58% confidence

Multiple-Choice Problems

99 e	100 e	101 a	102 d	103 d	104 b
105 a	106 e	107 a	108 b	109 d	110 a
111 a	112 b	113 e	114 e	115 c	116 b
117 c	118 c

Word and Thought Problems (Odd)

119 a. $E = 5.0$.
  b. $E = 7.7$. As the level of confidence increases, the maximum possible error increases if the sample size and standard deviation do not change.
  c. $E = 2.5$. As the sample size increases, the maximum possible error decreases, if the confidence level and standard deviation do not change.
  d. $E = 2.0$. As the population standard deviation decreases, the maximum possible error decreases if the sample size and confidence level do not change. We can change the standard deviation by redefining the population so that its elements are more similar.

121 $1 - \alpha = 0.9876$, risk $= 1 - (1 - \alpha) = \alpha = 0.0124$

123 2,401

125 $E = 0.3$ ppt

127 a. 44 families
  b. All subsidized families in the area
  c. Yes; no.
  d. No; yes; computation of a larger n proceeds by using S from the sample of 44 families.

129 45 patients

131 a. 0.89 **b.** 1.6667

133 0.9886

135 99.74%; no; all class days for this student during the year

137 Tolerable or maximum sampling error

UNIT III

Section 8.9 Problems (Odd)

145 1.645   **147** 0.0192

Section 8.10 Problems (Odd)

149 (0.3640, 0.4360). We are 90% confident that this interval contains the true proportion of students with the same opinion.

151 (0.2698, 0.3902). We can be 80% confident that this interval contains the actual proportion of vaccinated patients who become ill from the vaccination.

Section 8.11.1 Problems (Odd)

153 We are 99% confident that the error is no more than 2.66 percentage points, if we estimate that 20% of Americans favor prohibition to alleviate alcohol abuse.

155 We can be 80% confident the error is at most E percentage points, if we use $\hat{p}$ to estimate the p for all banks.
a. $E = 0.0594$ **b.** $E = 0.0643$ **c.** $E = 0.0741$
d. $E = 0.0757$

Section 8.11.2 Problems (Odd)

157 With information: **a.** 1,040 **b.** 910 **c.** 653 **d.** 206 **e.** 474
Without information: All cases require 1,083 observations.

159 $n = 1,067$ workers

Section 8.11.3 Problems (Odd)

161 The reporter can be 95.44% confident that the statement about the error is correct.

163 We are 91.82% confident that the error is at most 2.5 percentage points if we estimate that 29% of shoppers crave chocolate most among the listed foods.

Multiple-Choice Problems

165 d	166 d	167 b	168 a	169 b	170 a
171 c	172 a	173 d	174 e	175 b	176 a
177 c	178 a	179 c	180 d	181 e	182 e
183 d	184 e				

Word and Thought Problems (Odd)

185 (0.0365, 0.0635). We can be 95% confident that this interval contains the actual proportion of items from the production line that are defective.

187 a. (0.4525, 0.6475). **b.** No **c.** 0.6876; risk = 0.3124

189 a. (0.0671, 0.1614). We can be 95% confident that the actual proportion of defective meters is covered by this interval.
b. 95%: $E = 0.0471$ 90%: $E = 0.0396$

191 1,842

193 a. 1,692
b. $n = 1,624$ if the information is reliable and the investor uses it.

195 $z = 0.96$, $1 - \alpha = 0.6630$

197 a. 577 **b.** Larger sample required, $n = 601$

199 a. (0.3033, 0.3967). We can be 95% confident that the interval 30.33% to 39.67% contains the actual percentage of the electorate who favor candidate X.
b. (0.6033, 0.6967). We can be 95% confident that the interval 60.33% to 69.67% contains the actual percentage of the electorate who favor candidate X. The widths of the intervals are the same but centered about estimates that must sum to 1.

201 a. No, because the intervals overlap. **b.** $\hat{\sigma}_{\hat{p}} = 0.0204$ in 1987 and 0.0153 in 1988. **c.** No, because the intervals overlap. **d.** $\hat{\sigma}_{\hat{p}} = 0.0510$ for both populations

203 a. 0.0019 **b.** 0.5 **c.** Word your questions carefully, and consider the composition of the nonrespondent group.
d. Approximately zero.

205 (0.0395, 0.0768). We are 80% confident that this interval contains the true proportion of junk fish in Lucky Lake.

207 The error is at most 0.0212.

209 (0.2400, 0.2800). We are 90% confident that the interval contains the actual proportion of the students who subscribe to sports-oriented publications.

211 164

Review Problems (Odd)

213 Uncertainty of the process and estimate

215 (11.8 storms, 12.2 storms). This is not a random sample. No.

217 (2.65 hours, 3.57 hours). We assume Stanley *randomly* recalls the race times and all his race times are normally distributed.

219 a. Perhaps a list, from the registrar, of students with low grade averages or those on probation **b.** 38 students **c.** Obtain a list of 38 or more randomly selected student contacts from the registrar. **d.** The mean expectation is -10.50 points. The maximum error, E, is 10.8 points. No. **e.** $\hat{p} = 0.2368$, $E = 0.14$

221 a. Total: $E = 0.9$
Breakfast: $E = 0.3$
Lunch: $E = 0.4$
Dinner: $E = 0.7$
b. The maximum error for the total time spent at all three meals is larger than any of the three meals separately. σ decreases when looking at a single type of meal, which decreases E.

223 a. No, because only $100(1 - \alpha)\%$ of the confidence intervals should contain the true mean.
b. No, because it is possible that $\overline{X}$ will move further from 100 and S decrease in the smaller samples.

225 a. (15.868, 17.231).
b. Refute; this interval does not contain 20.

227 a. Temperatures from close locations tend to be numerous and about the same value. Meanwhile, remote areas produce fewer readings, possibly colder regions. The more extreme values, even around the equator, may produce more readings now than 100 years ago.
b. The number of readings in different locations produce outliers and differential weighting according to geographic location of the thermometer; thus, distorting a mean.
c. Possibly the median may ameliorate some of these weighting and outlier effects.

229 a. By using 450 or more correct predictions, his standards exceed the standard set by the existing test.
b. Approximately zero **c.** (0.878, 0.922)
d. Conservatively, 2,401 patients. If their test is no worse than the existing test, n is 1,225 patients. **e.** 0.8638

231 2,305 students

233 The population must be normally distributed. If σ is known, the confidence interval is based on the standard normal distribution. If σ is unknown and replaced with S, a T statistic is the basis of the interval.

235 Smaller, because segmenting the population in this situation should produce a more cohesive, less dispersed group. Mathematically, a more cohesive group implies the proportion of successes or the proportion of failures should move closer to 1 (or 0). These values reduce $\hat{\sigma}_{\hat{p}}$ and, thus, $E = z\hat{\sigma}_{\hat{p}}$.

CHAPTER 9

UNIT I

Section 9.1 Problems (Odd)

5 If there is no overwhelming factual evidence to refute either statement, the hypothesis that gets the benefit of the doubt will be the accepted hypothesis. To the extent that people give the benefit of the doubt to the hypothesis that they believe, they will tend to interpret the questionable evidence in favor of their maintained hypothesis. A more convincing strategy would be to agree to interpret questionable evidence in favor of the opposing hypothesis and search for overwhelming evidence to disprove the hypothesis.

7 As we accumulate more evidence that contradicts a claim, we encounter less risk of falsely rejecting a true claim.

Section 9.2 Problems (Odd)

9 The null hypothesis.

11 a. H_1: The person cannot correctly predict most tournament basketball games. **b.** One possible decision rule is to reject the claim if there are 25 or more incorrect predictions. The acceptance region is 0 to 24 incorrect predictions, and the rejection region is 25 to 100 incorrect predictions. The test statistic is the number of incorrect predictions. **c.** A Type I error would be to decide the person cannot correctly predict most games, when the person can do so. A Type II error would be to decide the person can correctly predict most games, when he or she cannot.

13 When the significance level decreases, the acceptance region increases in size, and the rejection region(s) decrease. A decrease in α means it will be more difficult to reject H_0. When α is very small, there is almost no chance we will reject H_0. To make $\alpha = 0$ we need all the data in the population to be sure we are not making an error.

Section 9.3 Problems (Odd)

15 Cannot reject H_0 in a, b, and c.

17 a. Reject H_0 **b.** Cannot reject H_0 **c.** Cannot reject H_0

19 Reject H_0; we decide that the sample evidence does not support the school system's claim about the average score.

Section 9.4 Problems (Odd)

21 We reject H_0. The average ozone concentration is significantly different from the standard of 0.12 ppm.

Multiple-Choice Problems (Odd)

23 c	24 b	25 a	26 f	27 c	28 b	29 a
30 d	31 e	32 d	33 c	34 b	35 b	36 d
37 d	38 a	39 e	40 e	41 d	42 c	43 a
44 e	45 e	46 a				

Word and Thought Problems (Odd)

47 a. Type I: to decide the brakes are safe when they are not; Type II: to decide the brakes are not safe when they are.
b. Type I error implies the company produces a defective product, which may result in costly law suits and fewer purchases. Type II implies the company scraps a good, possibly profitable product. From the public's point-of-view, Type I can result in accidents, while Type II means society loses a quality product.

49 a. No, because this is only a one-observation sample.
b. No, because only five correct predictions out of 100 *randomly* selected predictions is not convincing. Ninety out of 100 would likely be convincing. **c.** 5% and 90% correct are both different enough from 60% to suggest that the 60% is in error.

51 One possible solution is to expect the editor to be above p (a percentage of correct picks) or close to p. A result from the 100 races much smaller than p would be convincing evidence that the editor is a poor picker.

53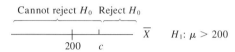

55 Type I error, α; Type II error; β.

57 a. Reject H_0. The data indicate that she tends to overestimate on average. **b.** $S = \$606.27$. A larger S means there is substantial variation in the amount of her errors.

59 Cannot reject H_0. The data do not indicate that there has been a change in average daily sales. We assumed that the 50 observations are random.

61 Reject H_0. The data indicate that the mean number of goals scored in a game is not two but probably a larger value.

63 Cannot reject H_0. The data do not substantiate a mean number of employees different from 100.

65 Reject H_0. The sample evidence substantiates a mean performance time that is significantly different from 2.5 hours.

67 Yes

69 Reject H_0. The data contradict the catalog estimate and suggest the mean food expenditure is lower than $700.

71 Cannot reject H_0. The data do not produce a value sufficiently different from $200 to decide that the $200 is wrong.

73 a. Reject H_0. The mean number of pieces is significantly different from 10. **b.** The whole set of bags may be considered a random sample, but the individual packages are not random. The small packages were probably produced about the same time, so a malfunctioning machine could have caused many of these packages' contents to be systematically less than the usual amount.

UNIT II

Section 9.5 Problems (Odd)

79 No

81 a. Cannot reject for any part of the problem.
b. i. Cannot reject H_0.
   ii. Cannot reject H_0.
   iii. Reject H_0.

c. i. Reject H_0.
  ii. Cannot reject H_0.
  iii. Cannot reject H_0.

83 Yes.

Section 9.6 Problems (Odd)

85 $Z = 5.94 > 1.28$

87 We reject H_0. The data support that the mean migration distance is less than 950 miles.

Section 9.7 Problems (Odd)

89 Cannot reject H_0

91 Yes

Section 9.8 Problems (Odd)

93 Reject H_0

95 Yes

Section 9.9 Problems (Odd)

97 a. μ **b.** We would need the total number of votes for the candidate and the total number of voters in the 250 precincts. **c.** 50 **d.** $H_0: \mu \leq 50; H_1: \mu > 50$ **e.** $H_0: \mu \geq 50; H_1: < 50$ **f.** $H_0: \mu = 50; H_1: \mu \neq 50$

99 a. $H_0: \mu \geq 2$, and $H_1: \mu < 2$ **b.** 0.05

Section 9.10 Problems (Odd)

101 Reject H_0. In this case, we have ample evidence to claim that the average concentration of particulate matter is significantly lower than the standard.

103 Cannot reject H_0. The mean yield is not significantly higher than 3.5%; $\alpha = 0.2810$ is the minimum significance level that leads us to reject H_0.

Multiple-Choice Problems

104 b	105 g	106 b	107 c	108 b	109 c
110 a	111 e	112 c	113 a	114 c	115 d
116 f	117 b	118 e	119 d	120 c	121 b
122 a	123 c	124 d	125 d		

Word and Thought Problems (Odd)

127 Yes

129 a. Expands **b.** Expands **c.** Contracts

131 Yes

133 The evidence is not sufficient to deny that the errors cancel.

135 A reasonable sample mean can cause us to reject H_0, if α is large.

137 Yes; $n(1 - p)$ does not exceed 5, so the normal approximation of the sampling distribution of $\hat{p}$ may not work well.

139 Yes

141 Reject the claim.

143 A decision rule might be that if the patient detects no difference after two lens changes, then there is edema (i.e., reject H_0 that the eye is healthy). Ideally, a perfectly healthy eye detects any lens change. The decision rule says that in practice the healthy eye may have difficulty distinguishing differences in the neighborhood of the correct setting, so edema is not diagnosed unless there is a failure to distinguish large ranges of lens changes.

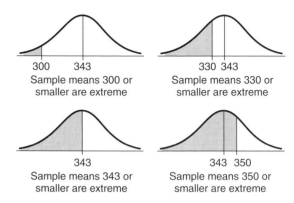

Figure A9-159
Solutions for Problem 159

145 a. There is substantial evidence to back up a reprimand by the director. **b.** Cannot reject H_0. The director would not be justified or meet his 1% criterion if he reprimanded a host in this situation. **c.** The decision process will cause the director to reprimand the host who is consistently poor but not the erratic performer, though both adversely affect the quality of the program. The problem is the large standard deviation of the erratic performer makes it more difficult to detect a significant departure from the standard, a mean of 0.5. In this case, a test of the standard deviation might be appropriate to check for consistency of behavior, even when the sample mean does not exceed the standard.

147 Cannot reject H_0. The data are not strong enough to say that less than 50% of the customers order soft drinks.

149 No

151 $1.540 < 1.761$; cannot reject H_0.

153 Reject H_0: $\mu \geq 2$, because $-11.38313 < -2.528$.

UNIT III

Section 9.11.1 Problems (Odd)

157 Right; H_1

159 See Figure A9–159; 350; largest.

Section 9.11.2 Problems (Odd)

161 0.1379; 0.3974; 0.5438; 0.6879

163 The director gives the coaches the benefit of the doubt, so H_1 is $p < 0.75$. p-value $= 0.0336$. The probability of obtaining a sample with 70% or fewer players who played less than two quarters (if 75% or more players play at least two quarters) is 0.0336. Thus, the sample result is not very likely to occur.

Section 9.11.3 Problems (Odd)

165 0.001, 0.01, 0.005, and 0.05

167 The manager makes H_1 that $p < 0.6$. The test statistic is $\hat{p} = 0.5778$. The p-value is 0.0869. The manager cannot reject the former value, 60%, because the p-value $= 0.0869 > 0.01 = \alpha$. The data do not support that a smaller percentage of customers are buying snack foods.

Answers

Section 9.12 Problems (Odd)

169 p-value $= 0.2302$.
  a. Cannot reject H_0.
  b. Cannot reject H_0.
  c. Cannot reject H_0.

171 a. p-value $= 0$; Reject H_0.
  b. p-value $= 0.1096$; Cannot reject H_0.
  c. p-value $= 0.4238$; Cannot reject H_0.

Section 9.13 Problems (Odd)

173 0.0057; we cannot reject H_0.

175 The hypotheses are H_0: $p \leq 0.5$ and H_1: $p > 0.5$. p-value $= 0.0108$, so we reject H_0. The evidence supports that more than 50% of the letters from the front arrived a year late.

177 No

Section 9.14 Problems (Odd)

179 Reject H_0.

Multiple-Choice Problems

181 c	**182** d	**183** c	**184** a	**185** b	**186** a
187 c	**188** c	**189** c	**190** c	**191** a	**192** d
193 e	**194** d	**195** b	**196** b	**197** a	**198** b
199 e	**200** a	**201** c			

Word and Thought Problems (Odd)

203 0.04, 0.08, 0.2, 0.4, and 0.8

205 Yes. If you reject H_0 here, you decide that the population mean is nonzero.

207 p-value $= 0.0071$; reject H_0; data support company claim.

209 p-value $= 0.005$; reject H_0. The evidence contradicts the car dealer's claim.

211 a. The probability of obtaining a sample proportion that is 0.38 or less if the politician is correct. **b.** Reject H_0.

213 p-values indicate the strength of the evidence used to make the decision. User can perform test at any significance level.

215 p-value $= 0.975$; cannot reject H_0. The evidence here strongly supports the contractor's claim rather than the company's belief.

217 p-value $= 0.0702$; cannot reject H_0. The data are not sufficient to contradict the 90% or better success rate of the lawyer.

219 p-value $= 0.0426$; reject H_0. The data imply that diabetic patients make up some percentage other than 15% of the doctors' patients.

221 You must know if the test statistic, $\overline{X}_0$, satisfies H_0 or H_1.

223 Reject H_0. The data contradict the officers' claim.

Review Problems (Odd)

225 Yes; we assume the first 50 days are a random sample.

227 Type I error; no

229 $c_1 = 3.936$ and $c_2 = 4.064$

231 Yes. The first five shipments are probably not random. Randomness is important, because the sampling distribution and central limit theorem are based on random samples.

233 a. Reject H_0. The data do not support the association's claim. The population is made up of the former teachers who quit last May.
  b. p-value $= 0.0930$, reject H_0.

235 No; normal population

237 Reject H_0 and decide that the mean cover charge is significantly different from $5.

239 p-value $= 0.0446$. Reject H_0 at the 0.05 significance level. Do not reject H_0 at the 0.01 significance level.

241 We cannot reject H_0.

243 a. Yes. Reject H_0 **b.** 0.0401; about two **c.** Cannot reject H_0: $p = 0.0401$ In some sense we have conducted a test of normality, because 0.0401 is what we expect to happen if the population is normal. **d.** The sample is not random. If there is a bias, say faster service during the sample period, then we may want to consider bias and its direction, which could cause a stronger or weaker rejection. Parts a, b, and c may change and the risks of incorrect decisions increase if this sample is not random.

CHAPTER 10

UNIT I

Section 10.1 Problems (Odd)

7 a. H_0: $\mu_1 \geq \mu_2$, H_1: $\mu_1 < \mu_2$ or H_0: $\mu_1 - \mu_2 \geq 0$, H_1: $\mu_1 - \mu_2 < 0$ or H_0: $\mu_2 - \mu_1 \leq 0$, H_1: $\mu_2 - \mu_1 > 0$
  b. H_0: $\mu_1 - \mu_2 \leq 100$, H_1: $\mu_1 - \mu_2 > 100$ or H_0: $\mu_2 - \mu_1 \geq -100$, H_1: $\mu_2 - \mu_1 < -100$
  c. H_0: $\mu_1 - \mu_2 \geq 25$, H_1: $\mu_1 - \mu_2 < 25$ or H_0: $\mu_2 - \mu_1 \leq -25$, H_1: $\mu_2 - \mu_1 > -25$

Section 10.2.1 and 10.2.2 Problems (Odd)

9 No

11 Approximately zero

Section 10.2.3 Problems (Odd)

13 Reject H_0.

15 a. H_0: $\mu_1 - \mu_2 = 0$ and H_1: $\mu_1 - \mu_2 \neq 0$. Reject H_0, because $19.27 > 2.575$. The mean output differed before the experiment. **b.** Yes

Multiple-Choice Problems

17 e	**18** e	**19** d	**20** a	**21** d	**22** b	**23** b
24 a	**25** d	**26** b	**27** e	**28** a	**29** c	**30** b
31 b	**32** c	**33** e	**34** d	**35** c	**36** a	**37** d
38 b						

Word and Thought Problems (Odd)

39 $\mu =$ population mean, $\overline{X} =$ sample mean,
  $\mu_1 - \mu_2 =$ difference in population means, and
  $\overline{X}_1 - \overline{X}_2 =$ difference in sample means

41 H_1: $\mu_n - \mu_d > 0$; reject H_0. The data support that the mean number of defectives from the night shift is greater than the mean number on the day shift.

43 H_1: $\mu_B - \mu_A < 0$; reject H_0. The data support the hypothesis that average wages increased after unionization.

45 Both are large; when the standard errors of sample means from each population are larger, smaller differences in the sample means are more likely to occur, which means we are more likely to be unable to reject H_0.

47 a. H_1: $\mu_F - \mu_U > 0$; reject H_0. The data support that the fertilizer improves the mean output of the tomato plants.
  b. The p-value is approximately zero. Reject H_0.

49 $\hat{\sigma}_{\text{dif}} = 21.2462$, $\overline{X}_X - \overline{X}_Z = -127$, $Z = -5.9775$.
  a. H_1: $\mu_X - \mu_Z > 0$, cannot reject H_0.
  b. H_1: $\mu_X - \mu_Z < 0$, reject H_0.
  c. H_1: $\mu_X - \mu_Z \neq 0$, reject H_0.
  d. In Part a there is no evidence to indicate that X is any better than Z. Part b verifies Z's contention. Part c indicates there is a difference, and the point estimates lead us to believe that Z is better.

51 H_1: $\mu_W - \mu_{WO} < 22$ where W = with player, WO = without; cannot reject H_0. We cannot say that the average difference between scores is smaller than the average amount the player scores, 22.

53 a. H_1: $\mu_E - \mu_W \neq 0$; reject H_0.   b. The p-value is approximately zero; reject H_0. The data indicate that there is a significant difference in mean attendance in the two regions.

55 a. The average result was for a student to score fewer points after the economics instruction than before.   b. H_1: $\mu_M - \mu_{NM} > 0$; reject H_0. The data support that students perform better in the mandate states.

57 a. No, because the observations are not independent.   b. H_1: $\mu_C - \mu_F > 0$; cannot reject H_0. The agent's data do not support his contention that he can place more classical musicians on average.   c. p-value = 0.0351, cannot reject H_0.

UNIT II

Section 10.3 Problems (Odd)

65 H_0: $\mu_1 - \mu_2 = 0$, H_1: $\mu_1 - \mu_2 \neq 0$; cannot reject H_0.

67 a. H_0: $\mu_B - \mu_{NB} \leq 0$, H_1: $\mu_B - \mu_{NB} > 0$, where B = bulimic and NB = not bulimic; BATH: Reject H_0. ROB: Reject H_0. Both tests indicate bulimic high school girls are more dissatisfied with their bodies.   b. $\alpha > 0.0001$ for BATH and $\alpha > 0.0002$ for ROB.

Section 10.4 Problems (Odd)

69 Reject H_0.

71 H_0: $\mu_D \leq 0$ and H_1: $\mu_D > 0$, using microwave outcome minus conventional outcome. Reject H_0.

Section 10.5 Problems (Odd)

73 a. 0.115   b. Cannot reject H_0. There is no significant difference in the mean change in prices of volatile and nonvolatile stock prices. No.   c. No.   d. We cannot reject H_0. The greater dispersion of the T distribution for 9 degrees of freedom means there is even more area in the tail than there would be if both tests used a T distribution with 17 degrees of freedom.

75 a. H_1 states that the mean number of Democrats per county is greater than the mean number of Republicans per county. Casual observations show that H_1 is true. The normality assumptions are violated.   b. For the independent samples tests, we are unable to reject H_0. There is not a significant difference in the number of Democrats and Republicans in each county. For the paired-difference test, we reject H_0. The mean number of Democrats is significantly greater than the mean number of Republicans on a county basis.   c. When we choose independent samples, we risk choosing a substantial number of large Republican values and of small Democrat values, because of the variation in county sizes. When we choose counties and pair their Democrat and Republican values, we retain more information on a county basis.

Multiple-Choice Problems

77 g   **78** d   **79** e   **80** b   **81** e   **82** b   **83** a
84 b   **85** d   **86** e   **87** a   **88** e   **89** c   **90** e
91 d   **92** c   **93** b   **94** b   **95** a   **96** a   **97** b
98 e

Word and Thought Problems (Odd)

99 μ_D is the mean of the population of all paired-differences. $\overline{D}$ is the mean of a sample of these differences.

101 First, we can narrow the population, so that the sample observations have a common value or characteristic that might otherwise confound the results. Second, we can pair observations so that the samples are matched with respect to the offending variable. Third, we can randomly select observations so that all the characteristics have an equal chance of showing up in each sample.

103 Reject H_0: $\mu_E - \mu_{NE} \geq 0$. The mean unemployment rate is significantly lower in election years.

105 H_1: $\mu_D \neq 0$, using doctor for the first population, and D stands for difference. Cannot reject H_0. Mean fees are not significantly different for the two health service providers.

107 a. -25 b. 10.

109 a. For X = height in H canton for "% of Cantonists",
$\overline{X} = 63.934$ inches,
$S^2 = 4.4736$ square inches.
Similar computations produce:
"% of Cantonists" D: $\overline{X} = 63.724$ and $S^2 = 3.4578$.
"% of Soldiers" H: $\overline{X} = 69.067$ and $S^2 = 1.7665$.
                   D: $\overline{X} = 68.864$ and $S^2 = 1.5095$.
  b. Yes   c. Yes   d. No

111 W = winners, NW = nonwinners, H_1: $\mu_W - \mu_{NW} > 0$. Reject H_0. The ratio of the variance of the winner population is over seven times the magnitude of the variance of the nonwinner population, so our rough rule of thumb is not satisfied. The data support that winners average losing more prior elections.

113 Two-tailed: Cannot reject H_0; Upper-tailed: Cannot reject H_0; Lower-tailed: Cannot reject H_0; normal populations; cannot demonstrate significant difference in mean number of wins between the two years.

115 Reject H_0; the mean percentage is larger for cross-pollinated seeds.

117 Reject H_0. Mean expenditure is significantly different in the two months.

Review Problems (Odd)

119 Yes; mean profits are greater for Firm 2.

121 The estimators could be mean scores on the reading and math tests, segments of those tests, or combinations. For behavior and emotional problems, the mean ratings of their problems, segments, or combinations provide sources for estimates. The test would be a paired-difference test for a difference in the means. If they test for any difference, H_1 is $\mu_D \neq 0$. If they believe problems are worse afterward, they would use H_1: $\mu_D > 0$ (using after minus before scores).

123 Lower-tailed: Cannot reject H_0; Upper-tailed: Cannot reject H_0; Two-tailed: Cannot reject H_0; we are unable to show significant difference between the means.

125 Cannot reject H_0 for any of the variables. Their means do not differ significantly.

Answers

127 (3.28 inseminations, 3.92 inseminations)

129 We cannot demonstrate a significant difference before and after; equal population variances.

131 A sample of size 1 is insufficient for drawing reliable inferences. In fact, historians collect data from diaries and other historical records, not always quantitative values, nevertheless, information that is useful for drawing inferences about historical events. The more evidence they compile, the stronger and more reliable their conclusions.

133 a. $H_1: \mu > 75$, $\hat{\sigma}_{\bar{x}} = 1.22$; cannot reject H_0. The data do not support the principal's claim. **b.** $H_1: \mu_A - \mu_B > 8$; cannot reject H_0. The data do not support the principal's claim. **c.** $H_1: \mu_D > 8$. Reject H_0. The data support the principal's claim that the average score increased by more than eight points. Potential problems include: learning from taking the test may improve scores on retake and short interim between individual's first and second scores may not be long enough to attribute difference to new teaching technique.

d. If the new principal gives the first principal the benefit of the doubt and makes the original claims the null hypotheses, then the new principal cannot reject the null in any circumstance.

135 (1,589.4 suppliers, 6,520.6 suppliers). The assumptions include normal populations and random samples.

137 No. We cannot confirm the company's hypothesis. The *p*-value is 0.10.

CHAPTER 11

UNIT I

Section 11.1 Problems (Odd)

3 Figure A11-3 shows the scatter diagram. The relationship appears to be downward sloping. The outlier is Day 8. The opening price, 102.125 is removed from the other opening prices, whereas 99.5 is closer to the customary closing prices. The data set is a time series.

Figure A11-3
Scatter diagram of opening and closing prices

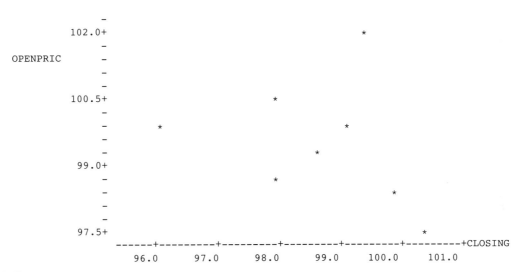

Figure A11-5
Scatter diagram for room temperatures and reaction time

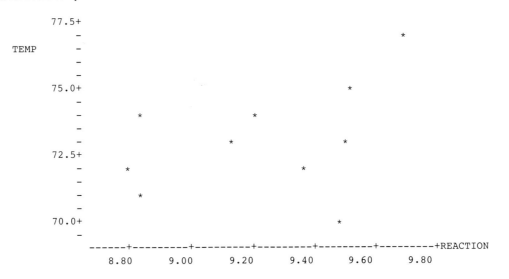

Figure A11-9
Scatter diagram for candy cost and total cost

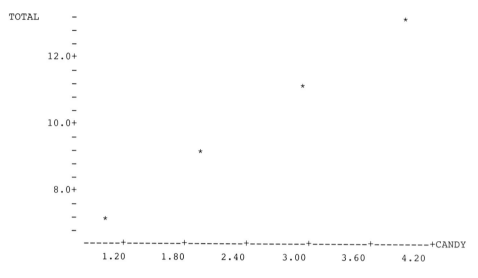

5 The scatter diagram in Figure A11-5 (on the preceding page) appears to show a relationship that is upward sloping. The data set is a cross section.

Section 11.2 Problems (Odd)

7 52% or more take SAT test: $r = 0.4$. 13% or less take SAT test: $r = -0.5$. The coefficient fits the expected relationship for each group. It is positive for the high-percentage states, and negative for the low-percentage states.

9 $r = 1.0$. See the plot in Figure A11-9.

Section 11.3 Problems (Odd)

11 We cannot reject H_0. There is no support at the 0.1 level for a correlation between the opening and closing prices of the stock.

13 We cannot reject H_0, so the correlation coefficient is not sufficiently large to support a conclusion of positive correlation between temperature and reaction time.

Multiple-Choice Problems

15 d	**16** b	**17** a	**18** c	**19** a	**20** b	**21** b
22 e	**23** e	**24** e	**25** a	**26** c	**27** a	**28** b
29 c	**30** e	**31** d	**32** d	**33** a	**34** c	**35** a
36 c	**37** d	**38** a				

Word and Thought Problems (Odd)

39 When $r = -1$, there is a perfect linear relationship between ACT scores and college performance; however, the relationship is perverse, for the two variables change in opposite directions.

41 a. There is a weak negative relationship. **b.** There is a weak positive relationship. **c.** There is a strong positive relationship.

43 Figure A11-43 shows the scatter diagram. $r = -0.8$ says there is a fairly strong negative relationship between size and mean sales. Larger parties will usually accompany smaller than average purchases per person. The test statistic is -2.7. We cannot reject H_0, which suggests that the variables are uncorrelated.

45 The first three diagrams should each have a correlation coefficient of 1; the last two diagrams should have correlation coefficients of -1.

47 Higher incomes are associated with smaller contributions and vice versa. This is an unexpected result, because wealthier

Figure A11-43
Scatter diagram of party size and purchase amount

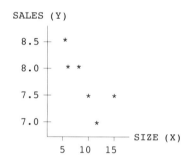

people are generally associated with being able to make larger contributions. A positive value closer to 1 seems more reasonable.

49 Whenever population changes and income does not, per capita income will tend to change in the opposite direction. Income behavior would be helpful.

51 If teachers do less to distinguish among student performances, there is less room to spread grades. Thus, as average grade increases, individual grades themselves tend to be less dispersed and the correlation between average grade and standard deviation is negative. Inflated grades provide less information about student skills, abilities, and knowledge, leaving them poorer predictors of future student performances. Students fail to move to areas where their strengths lie, opting for classes expected to raise their GPA instead. Thus, low-grading disciplines produce higher correlations, because apt students continue to score well. In high-grading disciplines, not all students continue to do well, so the correlation is smaller. Thus, grades from the low-grading disciplines tend to be better predictors of future performance in the discipline. Students discover earlier if their strength and interest lie in the low-grading disciplines.

53 $r = -0.5$. There is some evidence of a negative linear relationship between hospital size and patient-nurse ratio. We cannot reject H_0. The evidence does not support any correlation between the two variables at the 0.05 level.

Answers

55 The strong linear relationship means the leading indicators were useful indicators of future industrial output. In Problem 54, an r of 0.803 indicates that the linear relationship over these 12 periods continues over a wider time span.

57 More intrinsic orientation in students is associated with more guilt, more religious well-being, less functional separation from the mother, less attitudinal and emotional separation from both parents. More extrinsic orientation is associated with more shame, less functional separation from the father, and less conflictual separation from both parents.

UNIT II

Section 11.4 Problems (Odd)

63 a. G is dependent. H is independent.   **b.** b = change in grade if you increase study time by 1 hour; positive.
c. $\hat{b} = 6$, which means that the test score increases by approximately 6 points when the student spends an additional hour studying. This value is reasonable.   **d.** Test scores are dependent on other factors, such as number of other courses the student takes, number of related courses taken previously, and mental ability.

65 a. The independent variable is I. The dependent variable is W.
b. $W = a + bI$ **c.** The change in the weaving proportion when the average income of visitors increases by \$1. If we think wealthier people are inclined to like weavings we would anticipate a positive sign. If we think they would not want weavings, we would anticipate a negative sign.

Section 11.6 Problems (Odd)

67 a. $\hat{b} = -1.75$; $\hat{a} = 1{,}191.67$ **b.** Every additional pound that a gymnast weighs drops the predicted score by about 1.75 points.
c. 1007.92 points   **d.** Although a linear model looks appropriate, the sixth gymnast is an outlier in the data set.

69 a. $\hat{b} = -0.0133856$; $\hat{a} = 99.932114$ The estimated equation is $\hat{D} = 99.932 - 0.0134\,C$.   **b.** If credit cards issued increase by one million, collectable debt should decrease by about 0.013 percentage points.   **c.** The scatter diagram in Figure A11-69 shows that a linear model is appropriate.

Section 11.7 Problems (Odd)

71 If error terms are random, knowing X will not provide useful information for predicting the value that e will assume in any particular situation. When error terms are nonrandom, they are predictable, so they contain information about the model for the phenomenon under study.

73 a. \$125 is the mean spending level when income is \$100.
b. The error terms will be dependent.

75 a. There will be no change in X to use to explain the change in Y.
b. In Figure 11-29a, Y does not change. In Figure 11-29b, X does not change. In Figure 11-29c, the relationship is nonlinear.
c. No; there are no changes in Y to be explained.

Section 11.8 Problems (Odd)

77 a. Words is dependent, and age is independent.   **b.** Word usage increases by 0.015 words per paper.   **c.** The output indicates no unusual values.

Multiple-Choice Problems

79 b	80 d	81 e	82 a	83 e	84 a	85 a
86 f	87 a	88 b	89 c	90 d	91 b	92 e
93 d	94 a	95 c	96 a	97 c	98 c

Word and Thought Problems (Odd)

99 Y is an actual or observed Y value corresponding to a given X.
$\overline{Y}_X$ is the mean of all the Y values corresponding to a given X.
$\hat{Y}$ is an estimate of $\overline{Y}_{\hat{X}}$
e is the true error term, $(Y - \overline{Y}_X)$.
$\hat{e}$ is the residual or estimate of e, $Y - \hat{Y}$.
b is the coefficient of X in the true regression line, $\overline{Y}_X = a + bX$.
$\hat{b}$ is the estimate of b in $\hat{Y} = \hat{a} + \hat{b}X$.
σ_e is the standard error of the estimate, the standard deviation of the error terms.
$\hat{\sigma}_e$ is the estimated standard error of the estimate.

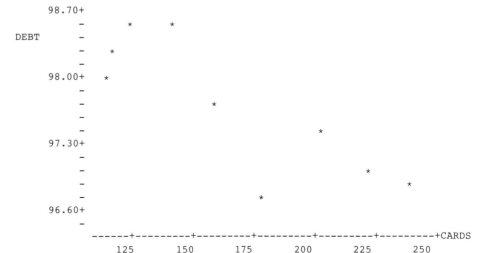

Figure A11-69
Scatter diagram of credit card data for Problem 69

Figure A11-101
Sales-advertising scatter diagram for Problem 101

101 a. Figure A11-101 shows the scatter diagram. **b.** $\hat{b} = 0.8813559$; $\hat{a} = 3.46$ **c.** An advertising increase of 1 unit increases sales by 0.88 units. If both units are dollars, it appears that $1 spent on advertising does not generate enough sales to pay for the ad. **d.** $\hat{Y} = 3.46 + 0.88X$ **e.** $\hat{Y} = 6.1$ units of sales

103 $\hat{H} = 18.7 + 2.74t$; for the 50th birthday, $\hat{H}_{50} = 155.7$ inches, which is about 13 feet. The H curve must level off at some point when growth slows, though the model may be useful for the sample period.

105 a. Each additional dollar of sales over $1,000 results in an additional 0.01 hour of overtime for each employee on average. The value is probably reasonable. **b.** $\widehat{OT} = 2$ for sales of 0, 500, and 1,000. When sales are 1,500, $\widehat{OT} = 7$.

107 a. Figure A11-107 shows the graph. **b.** Fatalities increase by 108 when the limit increases by 1. **c.** $\hat{F} = 7{,}020$ fatalities.

Figure A11-107
Graph of speed limit and traffic fatalities for Problem 107

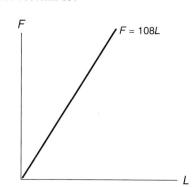

109 a. The insect count increases by 77 when average low temperature increases by 1, which is reasonable.
b. 182 bugs per cubic meter **c.** $S = -2.36°$, which means a cold winter is required to lower the insect density to zero.
d. The effect of warmer temperature decreases as the average low temperature increases and finally levels off. Figure A11-109 shows the diagram.

111 There are many possible Y values for each X value. The regression line connects the average of the Y values for each X value. There are many possible DR values that correspond to any given GDP value. If you find the mean of the DR values corresponding to each GDP value, the means form the regression line, $\overline{DR} = a + b$ GDP.

113 σ_e is the standard deviation of the error terms. We prefer a small standard error, because it is easier to estimate the true line, when points lie close to the line.

Figure A11-109
Sketch of insect count and temperatures for Problem 109

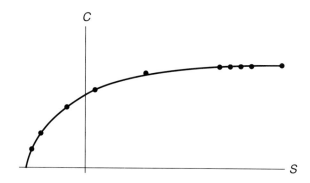

115 Figure A11-115 shows the diagram. The segment below the R axis and the segment to the left of the T axis should be erased. The remaining downward-sloping segment implies that increasing rewards reduces the time for the animal to learn the task.

Figure A11-115
Sketch of rewards and training time for Problem 115

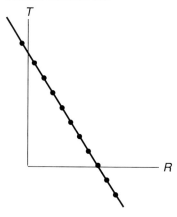

117 a. -609, so as the distance from navigable water increases by 1 mile, census count drops by 609 people on average.
b. 58,766 people.

119 a. Concessions = $1{,}945 + 1.74$ Attendance **b.** The following table illustrates the estimated revenue situation in the two alternative fee situations.

Alternative	Average Attendance	Revenue($) Entrance	Revenue($) Concessions	Revenue($) Total
Current $5 fee	3,000	15,000	7,165	22,165
No entry fee	9,000	0	17,605	17,605

Free entry would likely reduce total revenue. However, it is conceivable that they might spend more if they get in free.
c. The intercept should be zero, because there is no concession revenue if no one attends the event.

Answers

UNIT III

Section 11.9.1 Problems (Odd)

125 The model explains 71.1% of the variation in collectible debt.

127 $R^2 = 0.3281$. Variation in gymnasts' weights accounts for about 33% of the variation in competition scores. This seems to be a rather large percentage.

Section 11.9.2 Problems (Odd)

129 $H_1: b > 0$

131 $H_0: b \leq 0$ and $H_1: b > 0$; more study produces higher grades.

133 a. If you have no indication of the relation of age and willingness to strike, make $H_1: b \neq 0$. If you think older workers are less likely to strike, you would have $H_1: b < 0$. The opposite, $H_1: b > 0$, indicates a belief that older workers are more likely to strike. **b.** We cannot reject H_0. The data do not support the present model.

Section 11.10.1 Problems (Odd)

135 (95.582%, 97.588%)

Section 11.12 Problems (Odd)

137 a. 36.7% **b.** We would predict a negative coefficient to reflect the effect of an increase in prosperity on the infant mortality rate: $H_0: b \geq 0$, $H_1: b < 0$. **c.** -7.98 the T-ratio or the p-value of 0.000 **d.** Reject H_0, the coefficient is significantly negative at the 0.05 significance level. **e.** The data set may contain some outliers and leverage points.

139 a. The ACT is a better model because the R^2 and the T-ratio are larger. **b.** ACT scores are much smaller, between 1 and 36, whereas SAT scores range from 400 to 1,600. **c.** ACT = 18.335, SAT = 869.6

Multiple-Choice Problems

140 b	**141** e	**142** b	**143** c	**144** c	**145** a
146 a	**147** c	**148** e	**149** e	**150** a	**151** e
152 b	**153** d	**154** a	**155** a	**156** c	**157** c
158 e	**159** a	**160** e	**161** e	**162** e	**163** e
164 d	**165** d	**166** a			

Word and Thought Problems (Odd)

167 $0 \leq R^2 \leq 1$. Generally, values close to 0 indicate a poor model for explaining the variation in the dependent variable, whereas values close to 1 do the opposite. When a model is very simple and many important variables are missing, we should expect a low R^2. On the other hand, a large R^2 for a simple model may occur when the independent variable's movement captures the movement of the important omitted variables.

169 a. When the team's winning percentage increases by one percentage point, attendance increases by 1,000 fans on average. **b.** Variation in winning percentage explains 67% of the variation in attendance. **c.** Reject H_0. The data support the hypothesis that increased winning percentage will increase attendance. **d.** Cannot reject H_0. **e.** We reject H_0 as we did in Part c.

171 a. Hypothesis: facility = f(adoption date). $H_1: b > 0$. The regression equations are

$$\text{ORIGSCOR} = -21.0 + 0.403 \text{ ADOPTYR} \quad R^2 = 0.129.$$
$$(0.1719)$$
$$\text{LATESCOR} = 18.2 - 0.060 \text{ ADOPTYR} \quad R^2 = 0.003.$$
$$(0.1956)$$

The data support the contention that adoption year is a good predictor of the facilitativeness of original laws but do not support the predictive power of adoption year for current facilitation status. Together these decisions support the researchers' dual hypothesis. **b.** The data sets are cross sections, because they concern 39 different entities at one point in time. **c.** Figure A11-171 displays the scatter diagrams. Notice the upward-sloping appearance of the points for the original scores and the balance on either side of a near-horizontal line for the current scores.

173 a. The model explains 44% of the variation in number of businesses. **b.** The number of new businesses locating in the state increases by 10 when the state increases its business taxes by one unit. This is unreasonable, because we expect businesses to avoid high-tax areas. **c.** Reject H_0. The evidence indicates that state tax rates affect the number of businesses locating in the state. **d.** Reject H_0 again. **e.** $H_1: b < 0$, and the sign of the coefficient disagrees, so we cannot reject H_0. Consequently, the data offer no support for the inverse relationship between tax rates and the number of firms locating in the state.

175 First we must check for a coefficient sign that agrees with H_1. Then, we check if half of the two-tailed p-value is less than or equal to the significance level.

177 a. $b < 0$, because more imports should lower the price of the domestic company's stock as its business feels the impact of imports. **b.** The stock price decreases $0.125 for each unit increase in imports. This agrees with the hypothesis of Part a. **c.** We cannot reject H_0. **d.** The sign agrees with H_1 for a lower-tailed test. The p-value = 0.05, so reject H_0.

179 You cannot ignore the sign of the estimated coefficient. A positive coefficient results in not rejecting H_0 for a lower-tailed test, and a negative coefficient results in not rejecting H_0 for an upper-tailed test.

181 a. Positive because we expect that more practice induces progress.
b. Progress = 1.06 + 0.103 Practice. As average practice time increases by one minute, the progress increases by 0.103 books during the period.
c. Reject H_0. The data support the hypothesis expressed in Part a.
d. The sign of the coefficient is positive, and the p-value is 0.0005.
e. 2.09 books; 1.71 books

183 A straight line may not be the best functional form for the data. In addition, the line is based on a sample, not a population, so we must account for the uncertainty rather than declare that our work is totally accurate.

185 a. OVERDUE **b.** 27 **c.** OVERDUE = $4.85 - 0.358$ UNEMPLOY **d.** The percentage of consumer installment credit that is overdue decreases by 0.358 when the unemployment rate increases one percentage point. This seems unreasonable, because unemployed people will find paying off debts difficult. The sign of the coefficient is incorrect. **e.** The model explains 76.3% of the variation in percent overdue. **f.** The sign is wrong, so we cannot reject H_0. **g.** The sign is incorrect, so $1 - 0.5(0.000)$ is approximately 1, so again we

Figure A11-171 Scatter diagrams of facilitation score and adoption date for Problem 171

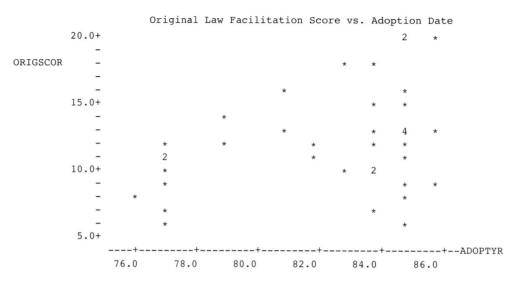

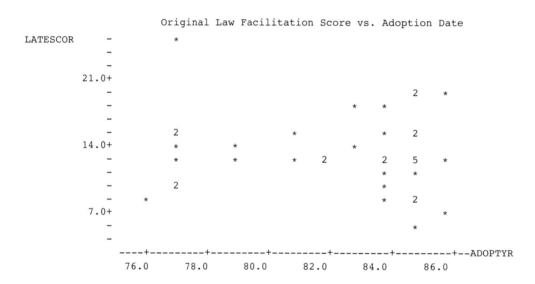

cannot reject H_0. **h.** The predicted value is -0.0232. The 95% prediction interval is $(-1.027, 0.980)$.

187 Add explanatory variables or try a nonlinear model.

Review Problems (Odd)

189 The correlation should be exactly 1.

191 a. The correlation coefficient is 0.912, which indicates a strong linear relationship between the two variables. **b.** We reject H_0 and decide that the variables are significantly correlated. **c.** The data set is a cross section, because the observations cover information about different products at a given point in time. **d.** A time series for a single product might show a negative correlation, because increased saturation of the market for the product should reduce subsequent sales as the pool of potential customers shrinks. **e.** Per capita would probably matter very little, because population remains essentially unchanged during a given time period. However, growth rate from the previous year is probably better, because there may be a larger demand for television sets than home computers. Growth in sales should decline as the market approaches saturation. The correlation coefficient in this case is -0.635.

193 Correlation measures association, whereas regression incorporates the assumption that values of Y are predictable from changes in X.

195 The regression equation is DEBT = $104 - 0.0823$ HOLDERS. As the number of holders increases by one million the collectible debt decreases by 0.0823 percentage points. Figure A11-195 shows the scatter diagram.

197 a. As the time at work during a shift increases by one hour, the defect rate decreases by 0.003 on average. **b.** -0.01415. **c.** If workers do their best work at or near the first of the shift, we would expect the defect rate to increase on average during

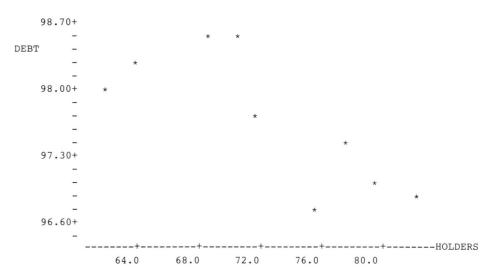

Figure A11-195 Scatter diagram of credit card data for Problem 195

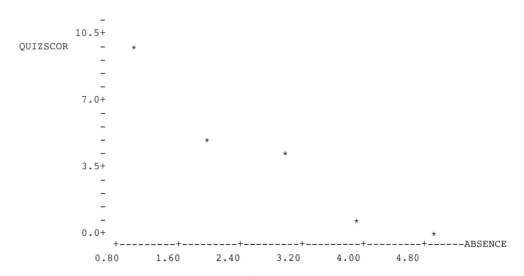

Figure A11-199 Scatter diagram of quiz scores and absences for Problem 199

the shift. Thus, the sign of the coefficient is wrong, unless workers improve throughout the shift. We obtain a negative defect rate as our forecast for 4.75 hours.

199 a. $\hat{b} = -2.4$ and $\hat{a} = 11.2$. Thus, $\hat{Q} = 11.2 - 2.4A$. The scatter diagram in Figure A11-199 corroborates the downward-sloping relationship, evidenced by the negative $\hat{b}$. **b.** 8.8 points **c.** The average of all quiz scores for a given number of absences will be larger for a lower number of absences. **d.** The model explains 93% of the variation in quiz scores. **e.** Yes **f.** The prediction interval is (3.15 points, 9.65 points). If fate dictates the student's class attendance, then the accuracy of the forecast of quiz scores is subject to the accuracy of the forecast of attendance.

201 It measures the dispersion of $\hat{b}$ values about their mean, b. Small, because the regression analysis will be more likely to estimate a $\hat{b}$ value close to the true b.

203 a. As the winning percentage increases by one percentage point, the coach's tenure increases by 0.28 year on average, if nothing else changes. **b.** We cannot reject H_0. The data do not support the assumption. **c.** Cannot reject H_0.

205 As GNP increases by $1 billion, new business formations increase by 18.087 on average, ceteris paribus. On the other hand, as interest rates rise by one percentage point, new business formations decrease by 8,120 firms on average, ceteris paribus.

207 a. Increased arrest rates and police presence decrease crime rates, so increasing the size of the police force or incentives for arresting criminals should lower the local crime rate. The use of this model by small towns is questionable, because the observations used to estimate the model came from large cities. **b.** Other important variables include likelihood of sentences once arrested, likelihood of escaping, the amount stolen in robberies and thefts, cost of forgone income from being in jail, and socioeconomic variables, so the model is probably not sufficiently complex.

209 R^2 measures are used in short-term forecasting because the behavior may be very similar to the available data points.

Long-term forecasting accuracy may be more dependent on understanding the causal properties of the phenomenon, more accurately tested with T tests for the coefficients.

211 a. *H/A* model: For each additional year of schooling for the mother, the height-for-age of a child increases by 0.074 on average, when nothing else changes. The height-for-age ratio increases by 0.919 on average when the working mother's marital status involves a union rather than a nonunion (being single) (marital status = 1 rather than 0) when all other variables remain unchanged. As the mother's income as a percentage of family income increases by one percentage point, the height-for-age ratio increases by 0.019 on average when no other variable changes. *W/A* and *W/H* models (weight-for-height ratio results are in parentheses): When the child's age increases by one month, the weight-for-age ratio increases by 0.036 (0.031) on average, when no other variable changes. The weight-for-age ratio increases by 1.060 (0.666) on average when the working mother's marital status involves a union rather than being single (marital status = 1 rather than 0) when all other variables remain unchanged. An increase of one percentage point in the percentage of family income attributable to the mother increases the weight-for-age ratio by 0.021 (0.010) on average, ceteris paribus. **b.** The child characteristics control for such things as age and gender effects on the three ratios.

CHAPTER 12

Section 12.1.2 Problems (Odd)

5 a. The experimental factor is the educational level of the students. The treatments are the three levels, elementary, middle, and high school. **b.** Other factors include the number and type of extracurricular activities, family responsibilities, number and type of subjects taken, and disposition toward studies. **c.** A completely randomized experimental design controls for these factors. The researcher can randomly select individuals from students in a particular income category, with the same course load, or with the same achievement level on a test to measure disposition.

7 Left-handed people are less likely to react as fast on a panel that is poorly designed to accommodate their dexterity. We explicitly try to control the factors that we know or believe affect the outcome with complete randomization, thus reducing the likelihood that unknown factors will influence the results. In the instrument panel case, an added approach would be to use only right-handed (or only left-handed) people in the experiment.

Section 12.1.3 Problems (Odd)

9 $H_0: \mu_D = \mu_N = \mu_{EMT} = \mu_F$; H_1: at least one of the four means differs from the others, where D = doctor, N = nurse, EMT = emergency medical technician, and F = firemen. H_0 states that the mean reaction time does not vary among the different types of volunteers, whereas H_1 implies that volunteer type does not influence average reaction speed.

Section 12.1.5 Problems (Odd)

11 SST = 2,962.7353; SSB = 1,643.6401; SSW = 1,319.0952

13 SST = 41.2783; SSB = 1.1908; SSW = 40.0875

Section 12.2 Problems (Odd)

15 MSB/MSW = 3.3802. The test statistic is greater than 1, but not by much, so there is little chance that H_1 may be true.

Section 12.3 Problems (Odd)

17 a. 0.005   **b.** 0.001   **c.** 2.75   **d.** 5.09

Section 12.4 Problems (Odd)

19 Yes

21 We cannot reject H_0. There is no evidence that the method affects the completion time on the production line at the 0.01 significance level. We assume normal populations with equal variances and independent random samples.

Section 12.5 Problems (Odd)

23 We cannot reject H_0. We cannot demonstrate a significant difference in mean quitting time among individuals with different types of jobs.

25 We cannot reject H_0. We cannot demonstrate a significant effect from different mechanics on mean repair time.

Multiple-Choice Problems

26 b	**27** e	**28** a	**29** c	**30** a	**31** e	**32** e
33 b	**34** a	**35** b	**36** c	**37** d	**38** e	**39** c
40 b	**41** a	**42** e	**43** b	**44** c	**45** e	**46** b
47 d	**48** a	**49** c	**50** a			

Word and Thought Problems (Odd)

51 If H_0 is true, MSB only reflects chance variation. If H_0 is false, MSB reflects evidence of a treatment effect along with chance variation. MSW measures chance variation in either case.

53 The assumption that all population variances are equal makes pooling reasonable. SSW measures total population chance variation, and MSW measures average chance variation or the population variance. Because the within variation from each sample measures the same variation, pooling these values uses more available information and increases the likelihood of obtaining a close estimate of σ^2.

55 We cannot reject H_0. MSB/MSW = 2.211. The mean prices at the three locations are not significantly different.

57 a. MSB/MSW = 0.62. We cannot reject H_0. **b.** Distributorship makes no difference in average delivery time. **c.** The item ordered may affect delivery time and the sample is nonrandom.

59 a. No. There is no within variation. We attribute any difference in values to treatment. This means MSB/MSW is undefined or infinitely large, and H_0 is automatically rejected. **b.** Yes. Now there is some measure of the within variation from 9 of the 10 treatment samples.

61 MSB/MSW = 2.13. We cannot reject H_0. The average discrepancy between actual and estimated waiting time does not vary among the restaurants.

63 a. MSB/MSW = 1.32. We cannot reject H_0. **b.** There does not appear to be a difference in mean number of lessons required to achieve goal of swim across pool among different frequencies of classes. **c.** Height, weight, physical ability, fear of water.

65 We reject H_0. There are significant differences in mean performance among the schools.

67 a. We cannot reject H_0. **b.** There is no significant difference in average performance for different descriptors. The professor should not be concerned.

Answers

69 a. We cannot reject H_0. There is no evidence that the word class affects the mean response time at the 0.05 significance level.   **b.** Nonstutterers are not in the sample, so we need information about their behavior to compare and determine if stutterers behave differently.

Review Problems (Odd)

71 The procedures in Chapter 10 test for a difference between *two* populations means, while the ANOVA procedure checks for a difference when there are *more than two populations* involved. In both cases we assume that the populations are normal. ANOVA assumes that samples are independent and population variances are equal. Chapter 10 covers this case as well as those where samples are paired observations on a single subject (dependent) and those where the samples are larger and population variances differ.

73 We reject H_0 for all of the variables at the 0.01 level except Children's Birthdate. With this risk in mind, we would say that the samples do not vary with regard to the Children's Birthdate, but do vary for the other three outcomes.

75 MSB/MSW = 19.6. Reject H_0. The mean research times for different plays are not the same.

77 (3.85 days, 4.75 days); no; There is an 80% chance that this interval contains the average movie run, so it is not likely the movie will be showing in 10 days when your friend is ready to go.

79 We reject H_0. The average times decrease as the level of abstraction decreases, which is reasonable. It is more difficult to think of an association for words involving more abstract concepts.

81 MSB/MSW = 0.83. We cannot reject H_0. The mean measurement does not vary among the four sets of lab partners.

83 MSB/MSW = 1.40. We cannot reject H_0. The average price does not vary with location. Other factors include size of cabin, accessories provided with cabin, and season of year.

85 a. MSB/MSW = 11.71. Reject H_0. The average per capita GNP varies with level of child mortality rate.   **b.** MSB/MSW = 10.04. Reject H_0. The average daily per capita calorie supply as a percentage of requirements varies with child mortality rate.

87 Possibilities include: amount of food needed for day, week, or winter, amount of cave space needed for family or clan, number of associates in enemy's army, and time until daylight.

89 a. $H_0: \mu_A = \mu_B$, $H_1: \mu_A \neq \mu_B$ for both tests   **b.** MSB/MSW = 0.1765. We cannot reject H_0. The means are not significantly different.   **c.** Equal variances; cannot reject H_0. This implies that the means are not significantly different.   **d.** The decisions are the same; $0.1765 = (-0.4201)^2$; $5.9874 = (2.447)^2$.

CHAPTER 13

UNIT I

Section 13.2 Problems (Odd)

7 All of the expected frequencies in the contingency table are greater than 5.

Birth Change	Mean Low Temperature Change (degrees Fahrenheit)			
	Down 10–30	Down 0.1–9.9	Up 0.0–9.9	Up 10–30
Increase	5.4310	7.9655	19.1897	30.4138
Decrease	9.5690	14.0345	33.8103	53.5862

9 Four of the expected frequencies are smaller than 5. We could form the class "not straight" from curly and very curly.

Presence of BE	Hair Texture		
	Very Curly	Curly	Straight
Yes	3.6667	3.2083	15.1250
No	4.3333	3.7917	17.8750

Section 13.3.1 Problems (Odd)

11 $\chi^2 = 11.7638$

Section 13.3.2 Problems (Odd)

13 24

15 34.8

Section 13.3.3 Problems (Odd)

17 Because 6.1 exceeds 5.99, we reject H_0 and decide that age and the presence of gallstone disease are related.

19 Because 4.0 does not exceed 11.3, we cannot reject H_0. The data fail to validate a relationship between level of congressional use and source of the research.

Section 13.4 Problems (Odd)

21 The computations are correct.

23 We cannot reject H_0 and the survey does not indicate that findings depend on the source of the study.

Multiple-Choice Problems

25 a   **26** b   **27** b   **28** c   **29** b   **30** e   **31** a   **32** c
33 d   **34** d   **35** c   **36** d   **37** e   **38** b   **39** a   **40** b
41 e   **42** c   **43** a   **44** c

Word and Thought Problems (Odd)

45 Nonparametric tests make no assumptions about the form or parameters of population distributions. They are useful when assumptions of parametric tests are not met or when qualitative variables are involved. However, the probability of type II errors increases.

47 There is no relationship between sex and major of students.

49 Reject H_0, test statistic = 41.5. Conference and winning percentage against outside opponents are related.

51 Reject H_0, test statistic = 347.3. Pedigree and prize level are related.

53 Reject H_0, test statistic = 190.8. Weekday and purchases are related.

55 Reject H_0, test statistic = 46.2. Sex and evaluation are related.

57 Reject H_0, test statistic = 10.0. Planting time and bloom result are related.

59 Reject H_0, test statistic = 15.3. Education level and firm size are dependent. Cannot reject H_0, test statistic = 6.9. Ethical judgments about accepting a $50 gift from a supplier at Christmas are independent of firm size. Cannot reject H_0, test statistic = 0.8. Ethical judgments about giving a prospective customer $50 are independent of firm size.

61 We reject H_0, test statistic = 16.2. It appears that drug type depends on the year under consideration. There is some doubt about the conclusion, because two f_e values are smaller than 5.

63 Reject H_0, test statistic = 9.0. Mortality rates and state of depression are related.

UNIT II

Section 13.5.2 Problems (Odd)

71 $\hat{p} = 0.7907$, which indicates that the actual median rating is larger than an A⁻.

Section 13.5.4 Problems (Odd)

73 The test is two-tailed, because we do not know whether the median is larger or smaller than 78, if 78 is wrong, H_0: $p = 0.5$, H_1: $p \neq 0.5$. $\sigma_{\hat{p}} = 0.05$. Test statistic = -5. Reject H_0. The data indicate that the median grade is different from 78. It is probably smaller, because $\hat{p}$ is so much smaller than 0.5.

75 The hypotheses are H_0: $M \leq M_0$ and H_1: $M > M_0$. Test statistic = 0.46. We cannot reject H_0. We are unable to demonstrate that the typical forecast is different from 2.5%.

Section 13.6.2 Problems (Odd)

77 For small cars, SR = 184, and the sum of the average ranks is 201.5. For large cars, SR = 281, and the sum of the average ranks is 263.5. Because SR for large cars exceeds its sum of average ranks, while the opposite is true for the SR for small cars, we should suspect that large cars average more maintenance problems.

79 a., b.

Grade	Number of Students Cooperative	Number of Students Other	Rank for This Grade	Rank Sum by Grade Cooperative	Rank Sum by Grade Other
F	0	1	1.0	0.0	1.0
D	5	10	9.0	45.0	90.0
C	5	5	21.5	107.5	107.5
B	0	10	31.5	0.0	315.0
A	3	8	42.0	126.0	336.0
	13	34		278.5	849.5

The sum of the average ranks for the cooperative group is 312 and 816 for the "other" group.
c. No, because SR for the cooperative group is smaller than the sum of its average ranks, while SR for the other-method group exceeds its sum of average ranks.

Section 13.6.4 Problems (Odd)

81 Using either sample, we cannot reject H_0, which means we cannot demonstrate a significant difference in average sense of urgency between the two patient groups.

83 H_0: $\mu_1 - \mu_2 \geq 0$, H_1: $\mu_1 - \mu_2 < 0$, reject H_0. The average price in area 1 is significantly smaller than the average price in area 2.

85 Higher scores for rehospitalized people mean the parametric H_1 will be $\mu_1 - \mu_2 > 0$, where the subscript 1 denotes the rehospitalized group. The computations for the subsequent Z values follow.

Variable	SR	$Z = (SR - \mu_{SR})/\sigma_{SR}$ $= (SR - 1{,}242.5)/85.1347$
Thought Organization	1,389	1.72
Object Relations		
Median	1,364	1.43
Low	1,454	2.48
High	1,430	2.20

All of these Z values, except the one for median object relations scores, are larger than 1.645, the critical value. We can associate larger average Rorschach criteria with rehospitalization as suspected with all measures, except the median score for object relations ability.

Section 13.7.2 Problems (Odd)

87 $H = 1.0320$

Section 13.7.3 Problems (Odd)

89 $H = 1.2725$; we cannot reject H_0. We are unable to demonstrate a significant difference in average test time among the three different times during the finals period.

91 $H = 1.8446$; we cannot reject H_0. The evidence is not sufficient to demonstrate a significant difference in mean productivity for teams with different objectives.

Section 13.8 Problems (Odd)

93 a. $\hat{p} = 0.5556$; converts to $z = 0.75$; we cannot reject H_0. We are unable to demonstrate that the median pulse rate is greater than 100 beats after 10 minutes of exercise at the 0.05 level.
b. SR = 549 converts to $Z = -0.0683$; we cannot reject H_0. We are unable to demonstrate a significant difference in average pulse rate after performing pullups and pushups.   **c.** $H = 53.73$. We reject H_0 and conclude that average pulse rate varies among the different lengths of the exercise period.

Multiple-Choice Problems

95 c   **96** b   **97** d   **98** d   **99** a   **100** b   **101** c
102 a   **103** c   **104** c   **105** b   **106** e   **107** c   **108** e
109 c   **110** c   **111** a   **112** a   **113** c   **114** b   **115** b
116 a   **117** e   **118** d   **119** c   **120** a   **121** e   **122** b
123 d   **124** e   **125** a   **126** a   **127** d   **128** e   **129** e
130 e   **131** b   **132** e   **133** c   **134** b   **135** c

Word and Thought Problems (Odd)

137 Both are binomial probability problems with $p = 0.5$. $\hat{p}$ is the proportion of heads in n tosses for the coin tossing problem and the proportion of samples values less than M_0 in the lower-tailed sign test. The answer to the coin tossing problem is a probability, while the probability part of the lower-tailed test is to determine an outcome or critical value for a given probability, α.

139 Cannot reject H_0, test statistic = -1.78. The data do not support the analyst's belief.

141 a. The problem is workable if none of the 500 say exactly 1 hour.   **b.** Reject H_0, test statistic = 13.4164. The data support the hypothesis that the average or median time spent watching television is greater than 1 hour.

Answers

143 μ: cannot reject H_0, test statistic = 1.8566. We assume a normal population. The data are not strong enough to reject the null that $\mu = 100$.

M: reject H_0, test statistic = 3. The data support the alternative that the median is different from 100.

The sample indicates a large dispersion of population values, so a sample mean very different from 100 is required to disprove $\mu = 100$. Most of the sample values are close to 100, so there is insufficient evidence to reject H_0. The median test does not account for the distance of values above 100. The information in the exact magnitudes and distances is not used. If the normality is questionable or unsubstantiated, the parametric test is preferable.

145 Cannot reject H_0, test statistic = 0.1512. The data do not support the contention that average facilitation increased with amended laws.

147 Cannot reject H_0, test statistic = 0.76. The data do not support the hypothesis of a larger average number of visits for the older age group.

149 Choose the sample whose mean is listed first in H_1.

151 a. Cannot reject H_0, test statistic = −0.78. The data do not support the hypothesis that comedies earn more on average.
b. Cannot reject H_0, $H = 0.60$. Both procedures test for a difference in means, but the Kruskal-Wallis can handle more samples. We expect the Kruskal-Wallis and Wilcoxon-Mann-Whitney rank sum test to agree when the Wilcoxon-Mann-Whitney test is two-tailed.

153 Cannot reject H_0, $H = 2.49$. The average number of visits does not differ for different diseases.

155 Reject H_0, $H = 10.33$. The average crime rate varies by region.

159 a. These proportions are skewed if few criminals spend their entire sentences in jail and many are paroled early.
b. Cannot reject H_0, test statistic = −0.2. We cannot discredit the politician's claim.

161 Omit the first two columns. We cannot reject H_0, test statistic = 2.655. The ratings are unrelated to the type of study.

163 Cannot reject H_0, test statistic = −0.09. There is no difference in the mean non-object-relatedness values for the two groups.

165 Reject H_0, test statistic = −6.8576. The median is different from 2. In fact, it appears to be smaller than 2.

167 Reject H_0, test statistic = −2.1182. The median shade is different from the committee selection. The median selection appears to be a lighter shade.

169 Reject H_0, test statistic = 3.09. The mean number of cigarettes smoked is greater for the cancer group. The decision is the same as the nonparametric decision.

171 Cannot reject H_0, $H = 1.46$. We cannot show that the average occupancy rate varies by year.

173 In all cases, the critical value is 5.99 from the chi-square table with two degrees of freedom, because there are three categories of ego development. In only two cases, Negative mode and Negative-client-focused, is the H statistic greater than 5.99, so that we reject. The p-values for these two cases are both smaller than 0.05.

175 Reject H_0, test statistic = 191.9621. Intended and actual use of the rink are dependent.

Review Problems (Odd)

157

Questions	Sign Test	Wilcoxon-Mann-Whitney Rank Sum Test	Kruskal-Wallis
Parametric Alternative	Population mean	Difference in population means	One-way ANOVA
Violated Assumptions	Normality quantitative	Normality quantitative	Normality quantitative
Nonparametric Assumptions	Large random sample	1. At least 10 observations per sample 2. Continuous population 3. Independent random samples 4. Equal population variances 5. No, or few, values tied for rank	1. At least 5 observations per sample 2. Continuous population 3. Independent random samples 4. Equal population variances 5. No, or few, values tied for rank
Hypotheses	H_0: $M = M_0$ H_1: $M \neq M_0$ or one-tailed versions of the test	H_0: $\mu_1 - \mu_2 = 0$ H_1: $\mu_1 - \mu_2 \neq 0$ or one-tailed versions of the test	H_0: Population means are all equal H_1: At least one population mean differs from the others

Index

A

Absentee rate, 412
Absolute value, 112
Acceptance region, 394–397
Accuracy, 31, 116, 293–297
Acker, Harry, Jr., 398
Advertising-sales relationship, 513–515, 519–521, 541–542
Age identification, 605
Agricultural research, 296, 475
AIDS, 5
Airline fatalities, 234, 237
Air Quality Data Base, B1–B2
Alcohol consumption, 551–552
Alsop, Ronald, 356
Alternative hypothesis, 393–398
Amato, Ivan, 297
American Chemical Society, 297
Analysis of variance (ANOVA). *See also* Kruskal-Wallis test
  assumptions, 592
  computers and, 598–602
  controlled experiments, 578–580
  decision for, 594–598
  F distribution, 592, 593
  hypotheses, 581–582
  measuring variation and its components, 584–588
  one-way, 580
  paired tests, 578
  problem set-up, 578–588
  test statistic, 589–592
  two-way, 603–605
  variations that affect sample values, 582–584
Analytic behavior, 423
And compound events, 171–180
ANOVA. *See* Analysis of variance
Asterisks, in box-and-whiskers diagrams, 131
Astronomical measures, 121

Averages, 75. *See also* Mean; Median; Mode; Percentiles; Quartiles
  comparing measures, 84–87
  computers and, 100–101
  unweighted measures, 76–83
  weighted measures, 88–93
Aviation fatalities, 234, 237
Aviation, Space and Environmental Medicine, 297

B

Bailey, Jeff, 156
Bakker, Klaas, 362
Bar graphs. *See* Histograms
Baxley, Beth, 274
Bayes' formula, 174–175
Bell curve, 247, 250–252
Benton, T. G., 648
Bergman, Robert, 297
Bernoulli trial, 223–224, 226
Betting odds, 162
Bias, 295
Bickman, Leonard, 57
Bimodal data sets, 82
Binomial distribution tables, G1–G6
Binomial experiments, 223–225
Binomial formula, 223–230
Binomial probability distribution, 232–235
  histogram, 234–235
  normal approximation of, 274–280
  simplification rule, 233
  table representation, 232–234
Binomial probability table, 231–232, G1–G6
Binomial random variables, 221–238, 275–276
  binomial formula, 223–230
  binomial probability table, 231–232, G1–G6

  defined, 224
  properties of, 232–238
    mean, 236–237
    probability distribution, 232–235, 274–280
    standard deviation, 237–238
Boundaries, for probability values, 159–160
Box-and-whiskers diagrams, 131–133
Boxplots, 131–133
Bradfield, Gayla, 297
Breslow, Jan L., 340
Brockie, Robert E., 348
Brutten, Gene J., 362
Buchanan, Christy M., 460

C

Carter, John R., 423
Casino gambling, and expected values, 210
Causation, 502, 503, 515–516
Census, 5
Centers for Disease Control (CDC), 294
Central limit theorem, 312–315, 329
  sampling distribution and, 373
Central tendency measures. *See* Location measures
Chebyshev's inequality, 119–120, 238, 359
Chemists, gender of offspring, 297
Children
  divorce effects, 460
  gender of, 297
  intelligence level, 38
  musical preferences by gender, 654
  nutritional development, 597–598
  parenting style effects, 601
China, medical research, 453
Chi-square distribution table, N1
Chi-square test for goodness of fit, 648

Chi-square test for independence, 619–632, 665–666
  computers and, 632
  decision for, 629–631
  distribution, 627–628
  nonparametric tests, 620–621
  parametric tests, 620–621
  problem set-up, 621–624
    hypotheses, 622
    when null hypothesis is true, 622–624
    test statistic, 625–626
Chi-square tests, 654
Class boundaries, 25–26
Class limits, 25–26, 31
Class midpoints, 25–26, 91–93
Class size, 25–26, 29, 109–110
Cobb, Margaret M., 340
Coca-Cola, 356
Coefficient of determination, 542–544
Coefficient of variation, 116, 117, 122
Coffman, Jerry L., 204
Cohen, I. Bernard, 63
Committee composition, 222–223
Complement of event A, 160–161
Completely randomized design, 580
Compound events, 170–187
  and events, 171–180
  or events, 180–185
Computer applications, 57–59
  analysis of variance, 598–602
  averages, 100–101
  chi-square test for independence, 632
  confidence intervals, 344–345
  data set distribution, 57–59
  descriptive statistics, 130–133
  dispersion measures, 130–133
  frequency distributions, 34, 57–59
  histograms, 57–59
  hypothesis testing, 404–405, 424–425
  Kruskal-Wallis test, 662–663
  location measures, 100–101
  nonparametric tests, 660–663
  population mean, tests for differences, 472–476
  p-values, 440–442
  random number generator, 292
  regression analysis, 527–531, 553–556
  standard deviation, 130–131
  Wilcoxon-Mann-Whitney rank sum test, 661–662
Conditional probabilities, 162–165
Confidence intervals, 327–348
  computers and, 344–345
  estimator for population mean, 328
  flowchart for, 343–344
  known population standard deviation, 329–334
    large sample, 329–334
    normal population, 334
  population proportion of successes, 374–375
  unknown population standard deviation, 335–343
    large samples, 341–342
    normal population, 338–341
    T probability distribution, 335–338
Confidence levels, 361–363, 379–380
Consistent estimator, 328
Constant, 14
Consumer Confidence Index, 3
Consumer preferences, 356
Consumer Price Index, 91
Contingency table, 622–624
Continuity correction factor, 277
Continuous probability distribution, 277
Continuous random variables, 200–201, 247–259
  compared with discrete random variables, 247–248
  probability distribution, 247–250
  probability distribution of normal random variable, 250–252
  standard normal random variable, 252–259
Control group, 398
Correlation analysis, 491–504. *See also* Regression analysis
  correlation coefficient, 495–501
  data sets, 491–494
  testing to establish correlation between two variables, 501–504
Correlation coefficient, 495–501
Cost estimating, 391, 399–402
Counting rules, F1–F8
Coxcombs, 63
Crispell, Diana, 177
Critical values, 400–401
Crossen, Cynthia, 356
Cross section, 491–492
Cumulative frequency distributions, 38–40
  graphing, 51–53
Cumulative relative frequency distributions, 53
Current Population Survey, 268

D

Data, 10
  accuracy, 31, 293–297
  collection, 294–295
  quality, 294
Data sets, 491–494. *See also* Dispersion measures; Frequency distributions; Location measures
  assembling, 491–492
  bimodal, 82
  comparisons, 116
  computer visualization, 57–59
  dispersion in, 109
  displaying, 492–494
  distribution of, 57–59
  of experimental outcomes, 198–201
  trimodal, 82
  unimodal, 84
Day-care payments, 431
Decision-making, 208–210, 398
Decision rule, 394, 400
  with p-values, 436–437
Degrees of freedom, 336–337, 467–468, 593
Department of Agriculture, U.S. (USDA), 296
Dependent variable, 516
  forecasting with simple regression model, 548–552
Depth values, 59
Descriptive statistics, 5, 22
  computers and, 130–133
Destructive sampling, 6, 291
Deviation, 111
Disasters, psychological effects of, 57
Discrete probability distribution, 277
Discrete random variables, 200–201
  compared with continuous random variables, 247–248
Dispersion, 108–109
Dispersion measures, 107–133. *See also* Population standard deviation
  alternative measures, 109–112
    interquartile range, 110–111
    mean absolute deviation, 111–112
    range, 109–110
  computers and, 130–133
  estimated standard deviation from grouped data, 128–129
  outliers, 108–109
  sample standard deviation, 122–124
  shortcut formulas, 125–127
Divorce, effects on adolescents, 460
Dornbusch, Sanford M., 460, 601
Downward sloping relationship, 493

E

Earthquake intensity, 204–205, 208
Economic theory, 423
Elicited responses, 295
Empirical Rule, 116–119
Engle, Patrice L., 598
Environmental Protection Agency (EPA), 296

Index

Environmental tobacco smoke (ETS), 347–348
Environment versus heredity, 38
Equality symbol, in hypothesis testing, 421–422
Equations, order of operations, 13–14
Equiprobable outcomes, 155–157
Error terms, 523–526
   independent, 525–526
   mean of, 524
   probability distribution, 524
   standard deviation of, 524–525
Estimated mean, from grouped data, 91–93
Estimated regression line, 518–519
Estimated standard deviation, from grouped data, 128–129
Estimated standard error, 339, 456, 466
Estimation
   confidence intervals of population mean, 327–348
   errors of point estimates, 354–363
   population proportion of successes, 368–380
Estimators, 328
   pooled, 466, 590
   for population mean, 328
   of population proportion of successes, 369–370
   regression analysis, 519–521
Events, 151
   complement of, 160–161
   independent, 176–178
   negation of, 160–161
Expected frequency, 623–624
Expected values, 205–210
Experimental design, 580
Experimental factor, 578–579
Experimental group, 398
Experimental outcomes, 198–201
Experiments, 150, 198–202
   analysis of variance, 578–580
   binomial, 223–225
Extrapolation, 549
Extreme values. *See* Outliers

F

Factor, 578
Failure, 223–225
Fair coin, 238
Fair game, 209
Fair Labor Standards Acts, 268
F distribution, 592, 593
F distribution tables, L1–L6
Federal Aviation Administration (FAA), 234, 237

Fighter pilots, gender of offspring, 297
File-drawer problem, 57
Finite population correction factor, 316–317
Fisheries industry, 122
Food and Drug Administration, U.S. (FDA), 4, 296
Football, probability of winning, 270
Forecasts, 89, 548–552
Forensic science, 605
Formulas, order of operations, 13–14
Foster parents, 38
Freedom, degrees of, 336–337, 467–468, 593
Frequency curves, 47–48, 51, 84–87
Frequency distributions, 24–36
   class boundaries, 25–26
   class limits, 25–26, 31
   class midpoints, 25–26
   class size, 25–26, 29, 109–110
   computers and, 34, 57–59
   constructing, 26–34
   cumulative, 38–40, 51–53
   defined, 24
   flowchart for constructing, 32
   graphic presentation, 46–63
      with computers, 34, 57–59
      frequency curves, 47–48, 51
      frequency polygons, 47–48, 51
      histograms, 49–51
      ogives, 51–53
      stem-and-leaf displays, 54–57
   open-ended classes, 34–36
   relative, 36–38, 53
   tabular presentation, 22–39
Frequency polygons, 47–48, 51, 86

G

Gambling
   expected values and, 210
   figuring odds, 162
Games of chance, 209–210
Gender
   children's musical preferences and, 654
   differences among adolescents from divorced homes, 460
   of offspring, 297
General Accounting Office (GAO), 296
Giles, E., 251
Goldsmith, Marsha F., 294
Goodness-of-fit test, 648
Gordon, Peter J., 151
Grand mean, 584–588
Groups, 91–92
Gustafson, Elizabeth, 661

H

Hadley, Lawrence, 661
Hake, Carl A. von, 204
Hatch, Orrin, 295
Health insurance, 197–198, 203–204, 208
Heating costs, 327–328, 355
Heckert, Paul, 121
Heredity versus environment, 38
Hinges, 131–133
Histograms, 49–51
   for binomial probability distribution, 234–235
      normal approximation, 274–280
   computer-generated, 57–59
   for probability distribution of continuous random variable, 247–250
   stem-and-leaf displays, 54–59
Historical behavior, estimating with regression analysis, 551–552
Hogs, agricultural research on, 475
Holtmann, A. G., 164
Hordinsky, J. R., 234
Huber, Gary L., 348
Hypothesis, 391–392, 453–454
   alternative, 393–398
   analysis of variance, 581–582
   chi-square test for independence, 622
   null, 393, 622–624
Hypothesis testing, 391–406. *See also p*-values
   computers and, 404–405, 424–425
   defined, 391–392
   equality symbol in, 421–422
   hypotheses determination, 420–422
   inequality symbol in, 421–422
   logic of, 391–393
   maps and, 393
   one-tailed tests, 412–415
   for population mean, 404–405, 424–425
      differences between two, 453–454
      small-sample test, 418–419
   and population proportion of successes, 419–420
   for practical decisions, 398
   procedure, 398–403
   for regression coefficient, 545–547
   shortcuts, 416–417
   terminology, 393–398
   to verify theoretical predication, 423
   on Z axis, 416–417

I

Idson, Todd L., 164

Inclusive *or*, 180
Independent error terms, 525–526
Independent events, 176–178
Independent samples, 455
Independent variables, 516
Inequality symbol, in hypothesis testing, 421–422
Inferential statistics, 6–9, 291
Infinite populations, 298
Insurance companies, and expected values, 210
Intelligence, heredity versus environment, 38
Intercept, 516–517
Interpolation, 549
Interquartile range, 110–111
Interval scales, 11–12
Inventory decisions, 263
IQ measures, 38, 274
Irons, Michael D., 423
Iverson, Brent, 297

J

Jamtgaard, Maria E., 475
Janssen, Peggy, 362
Jaroslovsky, Rich, 379
Javitts, Jacob, 295

K

Koten, John, 291, 356
Kruskal-Wallis test, 655–660, 665–666
 as alternative for one-way ANOVA, 655
 computer output for, 662–663
 decision and interpretation, 657–660
 sampling distribution, 657–660
 test statistic, 655–656
Kuhlman, J. J., Jr., 234

L

Lamborn, Susie D., 601
Landon, Alf, 356
Leading questions, 295, 356
Leahy, Patrick, 4
Leaves, 54–59
LeBlanc, Albert, 654
Leverage point, 529
Levine, B., 234
Lewis, Nate, 297
Life expectancy, 33–34
Life lines, 503
Lillestøl, Jostein, 122
Linear equations, 516–518

Linear functions, 516–518
Linear models, 515–518
Linear regression, 518–519
Line of means, 522–523
Lipman, Joanne, 356
Literary Guild Survey, 356
Little, Bertis, 297
Loan applicant evaluation, 619–620
Location measures, 75–101
 comparing measures, 84–87
 computers and, 100–101
 outliers, 84–85
 percentiles, 95, 97–99
 quartiles, 94–96
 unweighted central measures, 76–83
  mean, 76–78
  median, 78–81
  mode, 82–83
 weighted central measures, 88–93
  estimated mean from grouped data, 91–93
  weighted mean, 88–91
Long-run relative frequency method, 152–155
Loree, M. Ray, 274
Lower-tailed test, 414–415
Lung cancer, 347–348

M

Maccoby, Eleanor E., 460
Mahajan, Vijay K., 348
Mainstream smoke, 347–348
Maps, and hypothesis testing, 393
Margin of error, 361–363, 379
Mathematical operations, 13–17
Mathematical symbols, 14–17, 78
Mather, L. E., 503
Matthews, Thomas, 230
Maximum error, 357–358
McGuire, Richard, 4
Mean, 76–78, 84–86. *See also* Averages; Population mean; Sample mean
 binomial random variable, 236–237
 of error term, 524
 estimated from grouped data, 91–93
 grand mean, 584–588
 versus median, 79
 normal curve and, 251
 of random variables, 205–210, 236–237
 standard error of, 309–311, 335
 weighted, 88–91
Mean absolute deviation (MAD), 111–112
Mean square between samples, 590–591
Mean square error, 589–590

Mean square within, 589–590
Measurement errors, 355–356
Measurement scales, 10–12
Median, 78–81, 84–85, 95
 locating from ogive, 87
 versus mean, 79
 outliers and, 80, 100
Medical research, 453
Meulen, Sjoeke Van Der, 362
Minimum wage regulation, 268
Minitab, 57–59. *See also* Computer applications
Miron, Jeffrey A., 551
Modal party size, 82
Mode, 82–83
Model, 515
Model evaluation, 542–547
 coefficient of determination, 542–544
 hypothesis tests for regression coefficient, 545–547
Modified Mercalli Intensity Scale, 204, 208
Moffat, Anne Simon, 453
"More than" distribution, 39
Morse, Douglass H., 360
Mortality rates, 63
Mounts, Nina S., 601
Multiple regression analysis, 398, 559–561
Murray, Katherine A., 605
Murray, Tracy, 605
Musical preferences, by gender, 654
Mutual exclusivity, 150, 183–184, 227

N

National Aeronautics and Space Administration (NASA), 577–579, 581–586
National health insurance, 197–198, 203–204, 208
National Household Seroprevalence Study, 294
Nationwide Food Consumption study, 296
NCAA Probation Data Base, D1–D2
NCAA Tournament Data Base, M1–M2
Negation of event, 160–161
Negatively sloped relationship, 493
Neumann, Frank, 204
Nielsen television ratings, 595–597
Nightingale, Florence, 63
Nominal values, 10
Nonparametric tests, 418, 641–663
 chi-square test for independence, 620–621
 computers and, 660–663
 Kruskal-Wallis test, 655–660, 665–666

Index

sign test, 642–646, 665–666
Wilcoxon-Mann-Whitney rank sum test, 646–654, 665–666
Nonresponse problem, 294
Nonsampling error, 355–356
Normal approximation, 276–280
Normal curve, 247, 250–252
Normal distribution, 247, 250–252
Normal random variables, 247, 250. *See also* Continuous random variables
   normal approximation of binomial probability distribution, 274–280
   probabilities from, 267–270
   probability distribution, 250–252
   standard normal, 252–259
   transformation formula, 263–266
   X values from, 271–274
North Carolina Census Data Base, E1–E3
Norwegian fisheries, 122
Notation, 14–17, 78
   for probability, 152, 160, 162
Notes on Matters Affecting the Health, Efficiency and Hospital Administration of the British Army (Nightingale), 63
Novak, Bruce, 297
Null hypothesis, 393, 622–624
Nutritional development of children, 597–598

O

Observed frequency, 622–624
Odds, figuring, 162
Ogives, 51–53
   locating median from, 87
One-tailed tests, 412–415, 439
One-way ANOVA, 580. *See also* Analysis of variance
Open-ended classes, 34–36
Opinion polls. *See* Polls
Or compound events, 180–185
Order of operations, 13–14
Ordinal measures, 10–11
Outcomes, 150, 198–201
Outliers, 34–35
   box-and-whiskers diagrams and, 131–133
   dispersion measures and, 108–109
   interquartile range and, 110–111
   location measures and, 84–85
   median and, 80, 100
   range and, 110
   standard deviation and, 116, 120–121
   stem-and-leaf displays and, 55
Overtime regulation, 268

P

Paired tests, 578
Parameters, 291
Parametric tests, 620–621
Parenting style, effect on children, 601
Pena, Nelson, 179
Percentiles, 95, 97–99
Pesticides, exporting, 4
Pilots, gender of offspring, 297
Pilot survey, 294
Pittman, Robert, 274
Point estimates, 329
   errors of, 354–363
   of population proportion of successes, 376–380
Polar-area diagram, 63
Polls, 290–291. *See also* Surveys
   accuracy of data, 295
   nonsampling errors, 356
   technical aspects, 379
Pooled estimator, 466, 590
Population, 5, 298
Population correlation coefficient, 498–499
Population distribution, 296–302
Population mean, 309, 326
   confidence interval estimate, 327–348
      estimator for, 328
      flowchart for constructing, 343–344
      known population standard deviation, 329–334
      T probability distribution, 335–338
      unknown population standard deviation, 335–343
   errors of point estimates, 354–363
      sampling-error assessment, 357–363
      sampling error versus nonsampling error, 355–356
   formula, 78
   hypothesis testing for, 404–405, 418–419, 424–425
   test for difference between two, 450–476
      basics, 451–460
      computers and, 472–476
      hypotheses, 453–454
      large independent random samples, 454–459
      paired differences, 469–471
      small independent samples, 465–468
      variations, 464–476
Population proportion of successes, 368–380, 419–420
   confidence interval for, 374–375
   estimator of, 369–370
   hypothesis testing and, 419–420
   point estimates of, 376–380
   p-values for, 438–440
   sampling distribution of sample success rate, 370–373
Population standard deviation, 113–122, 309–311. *See also* Confidence intervals
   applications of standard deviation, 116–121
      accuracy gauge, 116
      Chebyshev's inequality, 119–120
      data set comparisons, 116
      Empirical Rule, 116–119
      outlier detection, 116, 120–121
   calculation, 113–115
   interpretation, 115–116
Population success rate, 326, 369–370
Population variance, 113–114
Positively sloped relationship, 493
Positive predictive value, 174
Practical decisions, hypothesis testing for, 398
Prediction, 89
Prediction interval, 548
Pregnancy, determining probability of, 174
Price changes, weighted measure of, 91
Probability, 149–165
   Bayes' formula, 174–175
   boundaries, 159–160
   complement of an event, 160–161
   compound events, 170–187
      and events, 171–180
      or events, 180–185
   conditional, 162–165
   continuous distribution, 277
   defined, 149–150
   determination methods, 151–158
   equiprobable outcomes, 155–157
   from normal random variables, 267–270
   negation of an event, 160–161
   notation, 152, 160, 162
   of pregnancy diagnosis, 174
   relative frequencies, 152–155
   rounding, 175
   sampling distributions and, 302–303
   of standard normal random variable, 253–255
   strategy for problems, 186–187
   subjective, 157–158
   terminology, 149–151
   visualizing with Venn diagrams, 158–159
   of winning football game, 270
   of winning scholarship, 164
Probability distributions. *See also* Sampling distributions

Probability distributions. *(continued)*
  of binomial random variable, 232–235, 274–280
  continuous, 277
  of continuous random variables, 247–250
  discrete, 277
  of error term, 524
  of normal random variables, 250–252
  of random variables, 202–205
  T probability distribution, 335–338
Probability of the event, 153–154
Probability of x successes in n trials, 224–225
Productivity, 451–453, 491, 641
p-values, 431–442
  calculation, 435
  computers and, 440–442
  converting two-tailed value to one-tailed value, 441–442, 555
  decision rule with, 436–437
  definition, 432–434
  interpretation, 432–434
  for population proportion of successes, 438–440
  test information from, 432–437
  two-tailed, 437–440

Q

Qualitative statistic, 298
Qualitative variables, 10, 620
Quality control, 308
Quantitative variables, 10–11, 620
Quartiles, 94–96

R

Randomized design, 580
Random number generator, 292
Random number table, 292–293, I1
Random sampling, 6–8, 291–293
  accuracy of data, 293–297
  stratified, 292, 306
Random variables, 197–215. *See also*
  Binomial random variables;
  Continuous random variables;
  Normal random variables;
  Sample mean
  and data set of experimental outcomes, 198–201
  defined, 198
  discrete, 200–201, 247–248
  expected value of, 205–210
  mean of, 205–210, 236–237
  probability distribution of, 202–205
  standard deviation of, 211–215, 237–238
  standard normal, 252–259
  values of, 202–215
Range, 30, 109–110
Ratio scales, 11–12
Regression analysis, 423, 513–531. *See also* Correlation analysis
  computers and, 527–531, 553–556
  estimated regression line, 518–519
  estimators, 519–521
  forecasting dependent variable, 548–552
  historical behavior estimation, 551–552
  linear models, 515–518
  model, 522–526
    error term, 523–526
    line of means, 522–523
  model evaluation, 542–547, 552–553
    coefficient of determination, 542–544
    hypothesis tests for regression coefficient, 545–547
  multiple, 398, 559–561
Regression coefficient, hypothesis testing for, 545–547
Regression forecasts, 548–552
Rejection region, 394–397, 400–401
Relative frequencies, 152–155
Relative frequency distributions, 36–38
  cumulative, 53
Representative samples, 7, 9, 291–294
Residuals, 518–519
Response rate, 296
Risk-taking, 210
Roberts, John D., 297
Roosevelt, Franklin, 356
Roper Organization, 230–232
Rounding, 14, A1–A2
  for probability, 175
  for sample sizes, 360
  Z values, 255
Rubin, Ernest, 80
Rubonis, Anthony V., 57

S

Sample mean, 289–290, 295–296, 300–302, 328
  as estimator of population mean, 328
  formula, 78
  as random variable, 307–317
    mean of, 308–309
    sampling distribution of, 312–315
    standard error of the mean, 309–311
Samples, 6, 291
  independent, 455
  representative, 7, 9, 291–294
  stratified, 292
Sample size, 296–302, 358–361
  rounding, 360
Sample space, 150
Sample standard deviation, 122–124, 335, 338–339
Sample success rate, 370–373
Sample variance, 124, 301–302
Sampling, 6, 9, 290–294, 379. *See also* Random sampling
Sampling distributions, 289–317
  central limit theorem and, 373
  construction of, 296–302
  defined, 295–296
  Kruskal-Wallis test, 657–660
  population proportion of successes, 370–373
  probability problems and, 302–303
  sampling procedure, 290–294
  statistic in, 296–302
  Wilcoxon-Mann-Whitney rank sum test, 651
Sampling errors, 452–453
  assessment, 357–363
  confidence level, 361–363
  defined, 355
  maximum error, 357–358
  versus nonsampling error, 355–356
  sample size, 358–361
  standard deviation and, 359
Scatter diagrams, 492–494
Scholarship competitions, probability of winning, 164
Science, 297
Shanken, Marvin R., 230
Sherrill, Carolyn, 654
Sidewalk ordinance, support for, 369
Sigma, 14–17
Significance level, 395–397, 400
Sign test, 642–646, 665–666
Simple linear regression, 518–519
Singer, Max, 4
Skeels, Harold M., 38
Skewness, 48, 57, 84–86
Skodak, Marie, 38
Slope, 493, 517
Smith, M. L., 234
Smoking, exposure to, 347–348
Software packages for statistics, 57–59. *See also* Computer applications
Sommers, Paul M., 561
Southern, Dan, 174
Speed limits, 561
Speed traps, 399
Spread, 108–109
Standard deviation, 109, 115–116, 122. *See also* Dispersion measures;

Index

Population standard deviation;
Sample standard deviation
applications of, 116–121
   accuracy gauge, 116
   Chebyshev's inequality, 119–120
   data set comparisons, 116
   Empirical Rule, 116–119
   outlier detection, 116, 120–121
binomial random variable, 237–238
computers and, 130–131
of error term, 524–525
estimated from grouped data, 128–129
IQ scores and, 274
normal curve and, 251
outliers, 116, 120–121
of random variables, 211–215, 237–238
sample standard deviation, 122–124, 335, 338–339
sampling error and, 359
stock prices and, 117
Standard error, estimated, 339, 456, 466
Standard error of the coefficient, 545–546
Standard error of the difference, 455–456
Standard error of the estimate, 525
Standard error of the mean, 309–311, 335
Standard error of the sample proportion, 372
Standard error of the sample success rate, 372
Standard error of the statistic, 309
Standardizing values, 120–121, 265–266
Standard normal distribution table, H1
Standard normal random variables, 252–259
   probabilities, 253–255
   Z values, 256–259
Standard score, 120–121, 274
State Health Data Set, K1–K3
State SAT Data Base, C1–C2
Statistic, 3
   qualitative, 298
   in sampling distribution, 296–302
Statistical experiment, 150, 198–202
Statistical notation. *See* Notation
Statistics, 1–5, 291
   appropriateness of, 5
   definitions, 5–9
      descriptive, 5, 22, 130–133
      inferential, 6–9, 291

misuse of, 4, 356
software for, 57–59
Steinberg, Laurence, 601
Stem-and-leaf displays, 54–59
   outliers, 55
Stern, Hal, 270
Stock prices, 117
Stover, Carl W., 204
Strata, 306
Stratified random sample, 292, 306
Stratified sample, 292
Structural change, 550
Subjective probabilities, 157–158
Subscript, 14
Success, 152, 223–225
Summation symbol, 14–17
Superstition, 503
Surveys, 290–291, 294–296, 355–356. *See also* Polls
Symbols, 14–17, 78
   for probability, 152, 160, 162
Symmetric distribution, 48

T

T distribution table, 336–337, J1
Teitelbaum, Howard S., 340
Television viewership, 465, 595–597
Terms, 15–16
Test scores, 89–90, 491
Test statistic, 394–396, 400
   analysis of variation, 589–592
   chi-square test for independence, 625–626
   Kruskal-Wallis test, 655–656
   Wilcoxon-Mann-Whitney rank sum test, 648–650
Theoretical prediction, 423
Thomas, Clarence, 295
Time series, 491–492
Tobacco smoke exposure, 347–348
Total sum of squares, 586–588
T probability distribution, 335–338, J1
Traffic fatalities, 561
Training method experiment, 577–579, 581–586, 642
T-ratio, 424–425, 546
Treatment, 579
Treible, Judy, 198
Trejo, Stephen J., 268
Trials, 150, 198–202
Trimodal data sets, 82
T-statistic, 424–425
T table, 336–337, J1
Two-factor ANOVA, 603–605. *See also* Analysis of variance

Two-tailed test, 401–403, 437–440
Two-way ANOVA, 603–605. *See also* Analysis of variance
Type I errors, 395–397, 399
Type II errors, 395–397, 399, 403

U

Ultimatum game, 423
Unbiased estimator, 328
Unimodal data sets, 84
Universe, 5
Unweighted central measures, 76–83
   mean, 76–78
   median, 78–81
   mode, 82–83
Upper-tailed test, 414
Upward sloping relationship, 493

V

Vallandigham, P. H., 251
Values, converting to proportion, 36
Variables, 10–12, 14, 198. *See also* Random variables
   dependent, 516, 548–552
   independent, 516
   qualitative, 10, 620
   quantitative, 10–11, 620
   testing for correlation, 501–504
Venn diagrams, 158–159

W

Wall Street Journal, 379
Waste treatment, 289–290
Water pollution, 308
Weighted central measures, 88–93
   estimated mean from grouped data, 91–93
   weighted mean, 88–91
Weighted mean, 88–91
Wentworth, Edward Norris, 9
Who's Who in Frontiers of Science and Technology, 297
Wilcoxon-Mann-Whitney rank sum test, 646–654, 665–666. *See also* Kruskal-Wallis test
   computer output for, 661–662
   decision and interpretation, 651–653
   sampling distribution, 651
   test statistic, 648–650
Wilcoxon matched-pairs signed-ranks test, 654

Wiles, Charles R., 151
Wiles, Judith A., 151
Will, George, 4
Wilson, M. E., 503
Winslow, Ron, 173
Wood, Harry O., 204
Woodward, Robert B., 297

Working mothers, and child nutritional development, 597–598
World Health Organization (WHO), 4

X, Y, Z

X values, 271–274

Yield per acre, 152–153
Z score, 120–121
Z values, 255–259, 274
   rounding, 255
Zwiebel, Jeffrey, 551

Standard Normal Distribution Table

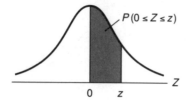

$P(0 \le Z \le z)$

The values in the body of the table are the areas between the mean, 0, and the value of Z. If we want to find the area under the standard normal curve between $Z = 0$ and $Z = 1.96$, we find the $Z = 1.90$ row and .06 column (for $Z = 1.90 + .06 = 1.96$) and read 0.4750 at the intersection.

z	.00	.01	.02	.03	.04	.05	.06	.07	.08	.09
.00	.0000	.0040	.0080	.0120	.0160	.0199	.0239	.0279	.0319	.0359
.10	.0398	.0438	.0478	.0517	.0557	.0596	.0636	.0675	.0714	.0753
.20	.0793	.0832	.0871	.0910	.0948	.0987	.1026	.1064	.1103	.1141
.30	.1179	.1217	.1255	.1293	.1331	.1368	.1406	.1443	.1480	.1517
.40	.1554	.1591	.1628	.1664	.1700	.1736	.1772	.1808	.1844	.1879
.50	.1915	.1950	.1985	.2019	.2054	.2088	.2123	.2157	.2190	.2224
.60	.2257	.2291	.2324	.2357	.2389	.2422	.2454	.2486	.2517	.2549
.70	.2580	.2611	.2642	.2673	.2704	.2734	.2764	.2794	.2823	.2852
.80	.2881	.2910	.2939	.2967	.2995	.3023	.3051	.3078	.3106	.3133
.90	.3159	.3186	.3212	.3238	.3264	.3289	.3315	.3340	.3365	.3389
1.00	.3413	.3438	.3461	.3485	.3508	.3531	.3554	.3577	.3599	.3621
1.10	.3643	.3665	.3686	.3708	.3729	.3749	.3770	.3790	.3810	.3830
1.20	.3849	.3869	.3888	.3907	.3925	.3944	.3962	.3980	.3997	.4015
1.30	.4032	.4049	.4066	.4082	.4099	.4115	.4131	.4147	.4162	.4177
1.40	.4192	.4207	.4222	.4236	.4251	.4265	.4279	.4292	.4306	.4319
1.50	.4332	.4345	.4357	.4370	.4382	.4394	.4406	.4418	.4429	.4441
1.60	.4452	.4463	.4474	.4484	.4495	.4505	.4515	.4525	.4535	.4545
1.70	.4554	.4564	.4573	.4582	.4591	.4599	.4608	.4616	.4625	.4633
1.80	.4641	.4649	.4656	.4664	.4671	.4678	.4686	.4693	.4699	.4706
1.90	.4713	.4719	.4726	.4732	.4738	.4744	.4750	.4756	.4761	.4767
2.00	.4772	.4778	.4783	.4788	.4793	.4798	.4803	.4808	.4812	.4817
2.10	.4821	.4826	.4830	.4834	.4838	.4842	.4846	.4850	.4854	.4857
2.20	.4861	.4864	.4868	.4871	.4875	.4878	.4881	.4884	.4887	.4890
2.30	.4893	.4896	.4898	.4901	.4904	.4906	.4909	.4911	.4913	.4916
2.40	.4918	.4920	.4922	.4925	.4927	.4929	.4931	.4932	.4934	.4936
2.50	.4938	.4940	.4941	.4943	.4945	.4946	.4948	.4949	.4951	.4952
2.60	.4953	.4955	.4956	.4957	.4959	.4960	.4961	.4962	.4963	.4964
2.70	.4965	.4966	.4967	.4968	.4969	.4970	.4971	.4972	.4973	.4974
2.80	.4974	.4975	.4976	.4977	.4977	.4978	.4979	.4979	.4980	.4981
2.90	.4981	.4982	.4982	.4983	.4984	.4984	.4985	.4985	.4986	.4986
3.00	.4987	.4987	.4987	.4988	.4988	.4989	.4989	.4989	.4990	.4990